Advanced Engineering Mathematics

K. A. Stroud
Formerly Principal Lecturer
Department of Mathematics, Coventry University

with additions by

Dexter J. Booth
Formerly Principal Lecturer
School of Computing and Engineering, University of Huddersfield

FIFTH EDITION

palgrave
macmillan

First edition 1986
Second edition 1990
Third edition 1996
Fourth edition 2003
This edition first published 2011 by
PALGRAVE MACMILLAN

Palgrave Macmillan in the UK is an imprint of Macmillan Publishers Limited,
registered in England, company number 785998, of 4 Crinan Street, London N1 9XW

Palgrave Macmillan in the US is a division of St Martin's Press LLC, 175 Fifth Avenue, New York, NY 10010.

Palgrave Macmillan is the global academic imprint of the above companies and has companies and representatives
throughout the world.

Palgrave® and Macmillan® are registered trademarks in the United States, the United Kingdom, Europe and other
countries

ISBN 978–0–230–27548–5   paperback

This book is printed on paper suitable for recycling and made from fully
managed and sustained forest sources. Logging, pulping and manufacturing
processes are expected to conform to the environmental regulations of the
country of origin.

A catalogue record for this book is available from the British Library.

A catalog record for this book is available from the Library of Congress.

Printed in China

Summary of contents

Contents

Programme 7   Fourier series 1   236

Programme 8   Fourier series 2   267

Programme 9   Introduction to the Fourier transform   297

Programme 10   Power series solutions of ordinary differential equations 1   334

Programme 11   Power series solutions of ordinary differential equations 2   357

Programme 12   Power series solutions of ordinary differential equations 3   378

Programme 13   Numerical solutions of ordinary differential equations   398

Programme 16   Matrix algebra   519

Programme 17   Systems of ordinary differential equations   563

Programme 21   Integral functions   735

Programme 22   Vector analysis 1   771

Preface to the first edition

The purpose of this book is essentially to provide a sound second year course in Mathematics appropriate to studies leading to B.Sc. Engineering Degrees and other qualifications of a comparable level. The emphasis throughout is on techniques and applications, supported by sufficient formal proofs to warrant the methods being employed.

The structure of the text and the techniques used follow closely those of the author's first year book, *Engineering Mathematics – Programmes and Problems*, to which this further book is a companion volume and a continuation of the highly successful learning strategies devised. As with the previous work, the text is based on a series of self-instructional programmes arising from extensive research and rigid evaluation in a variety of relevant courses and, once again, the individualised nature of the development makes the book eminently suitable both for general class use and for personal study.

Each of the course programmes guides the student through the development of a particular topic, with numerous worked examples to demonstrate the techniques and with increased responsibility passing to the student as mastery is achieved. Revision exercises are provided where appropriate and each programme terminates with a *Revision Summary* of the main points covered, a *Test Exercise* based directly on the work of the programme and a set of *Further Problems* which provides opportunity for the additional practice that is essential for ensured success. The ability to work at one's own pace throughout is of utmost importance in maintaining motivation and in achieving mastery.

In several instances, the topic of a programme is a direct extension of basic work covered in *Engineering Mathematics* and where this is so, the title page of the programme carries a brief reference to the relevant programme in the first year treatment. This clearly directs the student to worthwhile revision of the prerequisites assumed in the further development of the subject matter.

A complete set of Answers to all problems and a detailed Index are provided at the end of the book.

Grateful acknowledgement is made of the constructive suggestions and cooperation received from many quarters both in the development of the original programmes and in the final preparation of the text. Recognition must also be made of the many sources from which examples have been gleaned over the years and which contribute in no small measure to the success of the work.

Finally my sincere appreciation is due to the publishers for their patience, advice and ready cooperation in the preparation of the text for publication.

K.A. Stroud

Preface to the second edition

Since the first publication of *Further Engineering Mathematics* as core material for a typical second year engineering degree course, requests have been received from time to time for the inclusion of further topics to cover the particular requirements of individual syllabuses.

Some limit, inevitably, has to be placed on the physical size of the text, but it has been possible at least to include a programme on *Linear Optimisation* (*Linear Programming*) which was one of the subjects most frequently required.

The treatment of the additional material follows the structure of the rest of the book and the emphasis is largely on the practical use of the *simplex method* for the solution of both maximisation and minimisation problems.

The opportunity has also been taken to amend and clarify a number of minor points in the existing text and my thanks are due to those correspondents who have undertaken to write with constructive comment. Such feedback is always welcome.

K.A.S.

Preface to the third edition

With the new edition of *Further Engineering Mathematics*, the opportunity has been taken to incorporate a number of minor revisions and amendments to the previous text.

The format of the pages has been changed and the publishers have undertaken the complete resetting of the text to result in a more open presentation of the material and to facilitate the learning process still further.

Once again, my sincere thanks are due to all those correspondents who have kindly written with constructive comment concerning the book and to the publishers for their continued support, advice and cooperation throughout the preparation, production and marketing of the work.

K.A.S.

Preface to the fifth edition

It is now over 40 years since Ken Stroud first developed his approach to personalised learning with his classic text *Engineering Mathematics*, now in its sixth edition and having sold over half a million copies. Some 15 years later he followed this with *Further Engineering Mathematics* which was restyled as *Advanced Engineering Mathematics* for its fourth edition. As in all earlier editions his unique and hugely successful programmed learning style is continued in this fifth edition of *Advanced Engineering Mathematics*, and I am delighted to have been asked to contribute to it. As with the fourth edition, I have endeavoured to retain the very essence of the style, particularly the time-tested Stroud format with its close attention to technique development throughout. This methodology has, over the years, contributed significantly to the mathematical abilities of so many students all over the world.

New to this edition

To cater for continual changes in engineering mathematics the work of this edition builds upon material that was introduced in the previous edition. Notably, the new programme, *Introduction to invariant linear systems*, builds upon the new-styled *Z transforms*, now called *Difference equations and the Z transform*. It was also felt that the work on convolution was incorrectly located so it has been moved to *Laplace transforms 2* and significantly expanded to further clarify the concept. The programme on *Fourier series* has been split into two parts; the first dealing with periodic functions with period 2π and the second dealing with general periodic functions as well as half-range series. The programme, *Power series solutions of ordinary differential equations*, was very large and unwieldy and has been split into three. The first part deals with the Leibnitz–Maclaurin method with the addition of Cauchy–Euler equations, the second part deals with the Frobenius method and the third part deals with special functions. Finally, the programme on *Matrix algebra* has been split into two with the second part concentrating on systems of differential equations.

Acknowledgements

This is a further opportunity that I have had to work on the Stroud books. It is as ever a challenge and an honour to be able to work with Ken Stroud's material. Ken had an understanding of his students and their learning and thinking processes which was second to none, and this is reflected in every page of this book. As always, my thanks go to the Stroud family for their continuing support for and encouragement of new projects and ideas which are allowing Ken's hugely successful teaching methodology to be offered to a whole new range of students. I should also like to express my thanks and appreciation for the valuable feedback that has been provided by all the reviewers during the writing of this new edition and of previous editions upon which this one builds. In particular I should like to mention Professor Pete Peterson of John Tyler Community College, Virginia, USA. Engineering mathematics is not a static universe and it is always a challenge to best determine how a new edition is to be developed. Pete's encouraging comments and sympathetic treatment of the new material was greatly appreciated. Finally, I should like to thank the entire production team at Palgrave Macmillan for all their care and principally my editor Helen Bugler whose enthusiasm and professionalism I greatly value and admire.

Huddersfield   Dexter J Booth
March 2011

Hints on using the book

This book contains twenty-eight Programmes, each of which has been written in such a way as to make learning more effective and more interesting. It is almost like having a personal tutor, for you proceed at your own rate of learning and any difficulties you may have are cleared before you have the chance to practise incorrect ideas or techniques.

You will find that each Programme is divided into sections called frames. When you start a Programme, begin at Frame 1. Read each frame carefully and carry out any instructions or exercise which you are asked to do. In almost every frame, you are required to make a response of some kind, testing your understanding of the information in the frame, and you can immediately compare your answer with the correct answer given in the next frame. To obtain the greatest benefit, you are strongly advised to cover up the following frame, where necessary, until you have made your response. When a series of dots occurs, you are expected to supply the missing word, phrase, or number. At every stage, you will be guided along the right path. There is no need to hurry: read the frames carefully and follow the directions exactly. In this way, you must learn.

At the end of each Programme, you will find a **Revision summary** and a **Can you?** checklist that matches the **Learning outcomes** given at the beginning of the Programme. Read these carefully to make sure you have not missed anything. Next you will find a short **Test exercise**. This is set directly on what you have learned in the Programme: the questions are straightforward and contain no tricks. When you have completed these, return to the **Can you?** checklist as a final reminder of the contents of the Programme. To provide you with the necessary practice, a set of **Further problems** is also included. Remember that in mathematics, as in many other situations, practice makes perfect – or more nearly so.

Even if you feel you have done some of the topics before, work steadily through each Programme: it will serve as useful revision and fill in any gaps in your knowledge that you may have.

Useful background information

1 Algebraic identities

$$(a+b)^2 = a^2 + 2ab + b^2 \qquad (a+b)^3 = a^3 + 3a^2b + 3ab^2 + b^3$$

$$(a-b)^2 = a^2 - 2ab + b^2 \qquad (a-b)^3 = a^3 - 3a^2b + 3ab^2 - b^3$$

$$(a+b)^4 = a^4 + 4a^3b + 6a^2b^2 + 4ab^3 + b^4$$

$$(a-b)^4 = a^4 - 4a^3b + 6a^2b^2 - 4ab^3 + b^4$$

$$a^2 - b^2 = (a-b)(a+b)$$

$$a^3 + b^3 = (a+b)(a^2 - ab + b^2)$$

$$a^3 - b^3 = (a-b)(a^2 + ab + b^2)$$

2 Trigonometrical identities

(1) $\sin^2\theta + \cos^2\theta = 1; \quad \sec^2\theta = 1 + \tan^2\theta;$

$\cosec^2\theta = 1 + \cot^2\theta$

(2) $\sin(A+B) = \sin A \cos B + \cos A \sin B$

$\sin(A-B) = \sin A \cos B - \cos A \sin B$

$\cos(A+B) = \cos A \cos B - \sin A \sin B$

$\cos(A-B) = \cos A \cos B + \sin A \sin B$

$$\tan(A+B) = \frac{\tan A + \tan B}{1 - \tan A \tan B}$$

$$\tan(A-B) = \frac{\tan A - \tan B}{1 + \tan A \tan B}$$

(3) Let $A = B = \theta \quad \therefore \quad \sin 2\theta = 2\sin\theta\cos\theta$

$$\cos 2\theta = \cos^2\theta - \sin^2\theta$$

$$= 1 - 2\sin^2\theta = 2\cos^2\theta - 1$$

$$\tan 2\theta = \frac{2\tan\theta}{1 - \tan^2\theta}$$

(4) Let $\theta = \dfrac{\phi}{2} \qquad \therefore \quad \sin\phi = 2\sin\dfrac{\phi}{2}\cos\dfrac{\phi}{2}$

$$\cos\phi = \cos^2\frac{\phi}{2} - \sin^2\frac{\phi}{2}$$

$$= 1 - 2\sin^2\frac{\phi}{2} = 2\cos^2\frac{\phi}{2} - 1$$

$$\tan \phi = \frac{2 \tan \frac{\phi}{2}}{1 - \tan^2 \frac{\phi}{2}}$$

(5) $\sin C + \sin D = 2 \sin \dfrac{C + D}{2} \cos \dfrac{C - D}{2}$

$\sin C - \sin D = 2 \cos \dfrac{C + D}{2} \sin \dfrac{C - D}{2}$

$\cos C + \cos D = 2 \cos \dfrac{C + D}{2} \cos \dfrac{C - D}{2}$

$\cos D - \cos C = 2 \sin \dfrac{C + D}{2} \sin \dfrac{C - D}{2}$

(6) $2 \sin A \cos B = \sin(A + B) + \sin(A - B)$

$2 \cos A \sin B = \sin(A + B) - \sin(A - B)$

$2 \cos A \cos B = \cos(A + B) + \cos(A - B)$

$2 \sin A \sin B = \cos(A - B) - \cos(A + B)$

(7) Negative angles: $\sin(-\theta) = -\sin \theta$

$\cos(-\theta) = \cos \theta$

$\tan(-\theta) = -\tan \theta$

(8) Angles having the same trigonometrical ratios:

  (a) Same sine: θ and $(180° - \theta)$

  (b) Same cosine: θ and $(360° - \theta)$, i.e. $(-\theta)$

  (c) Same tangent: θ and $(180° + \theta)$

(9) $a \sin \theta + b \cos \theta = A \sin(\theta + \alpha)$

$a \sin \theta - b \cos \theta = A \sin(\theta - \alpha)$

$a \cos \theta + b \sin \theta = A \cos(\theta - \alpha)$

$a \cos \theta - b \sin \theta = A \cos(\theta + \alpha)$

where $\begin{cases} A = \sqrt{a^2 + b^2} \\ \\ \alpha = \tan^{-1} \dfrac{b}{a} \quad (0° < \alpha < 90°) \end{cases}$

3 Standard curves

(a) *Straight line*

Slope, $m = \dfrac{dy}{dx} = \dfrac{y_2 - y_1}{x_2 - x_1}$

Angle between two lines, $\tan \theta = \dfrac{m_2 - m_1}{1 + m_1 m_2}$

  For parallel lines, $m_2 = m_1$

  For perpendicular lines, $m_1 m_2 = -1$

Equation of a straight line (slope $= m$)

(1) Intercept c on real y-axis: $y = mx + c$

(2) Passing through (x_1, y_1): $y - y_1 = m(x - x_1)$

(3) Joining (x_1, y_1) and (x_2, y_2): $\dfrac{y - y_1}{y_2 - y_1} = \dfrac{x - x_1}{x_2 - x_1}$

(b) *Circle*

Centre at origin, radius r: $x^2 + y^2 = r^2$

Centre (h, k), radius r: $(x - h)^2 + (y - k)^2 = r^2$

General equation: $x^2 + y^2 + 2gx + 2fy + c = 0$

with centre $(-g, -f)$: radius $= \sqrt{g^2 + f^2 - c}$

Parametric equations: $x = r\cos\theta,\ y = r\sin\theta$

(c) *Parabola*

Vertex at origin, focus $(a, 0)$: $y^2 = 4ax$

Parametric equations: $x = at^2,\ y = 2at$

(d) *Ellipse*

Centre at origin, foci $\left(\pm\sqrt{a^2 + b^2}, 0\right)$: $\dfrac{x^2}{a^2} + \dfrac{y^2}{b^2} = 1$

where $a =$ semi-major axis, $b =$ semi-minor axis

Parametric equations: $x = a\cos\theta,\ y = b\sin\theta$

(e) *Hyperbola*

Centre at origin, foci $\left(\pm\sqrt{a^2 + b^2}, 0\right)$: $\dfrac{x^2}{a^2} - \dfrac{y^2}{b^2} = 1$

Parametric equations: $x = a\sec\theta,\ y = b\tan\theta$

Rectangular hyperbola:

Centre at origin, vertex $\pm\left(\dfrac{a}{\sqrt{2}}, \dfrac{a}{\sqrt{2}}\right)$: $xy = \dfrac{a^2}{2} = c^2$

where $c = \dfrac{a}{\sqrt{2}}$ i.e. $xy = c^2$

Parametric equations: $x = ct,\ y = c/t$

4   Laws of mathematics

(a) *Associative laws* – for addition and multiplication

$a + (b + c) = (a + b) + c$

$a(bc) = (ab)c$

(b) *Commutative laws* – for addition and multiplication

$a + b = b + a$

$ab = ba$

(c) *Distributive laws* – for multiplication and division

$a(b + c) = ab + ac$

$\dfrac{b + c}{a} = \dfrac{b}{a} + \dfrac{c}{a}$ (provided $a \neq 0$)

Numerical solutions of equations and interpolation

Learning outcomes

When you have completed this Programme you will be able to:

- Appreciate the Fundamental Theorem of Algebra
- Find the two roots of a quadratic equation and recognise that for polynomial equations with real coefficients complex roots exist in complex conjugate pairs
- Use the relationships between the coefficients and the roots of a polynomial equation to find the roots of the polynomial
- Transform a cubic equation to its reduced form
- Use Tartaglia's solution to find the roots of a cubic equation
- Find the solution of the equation $f(x) = 0$ by the method of bisection
- Solve equations involving a single real variable by iteration and use a spreadsheet for efficiency
- Solve equations using the Newton–Raphson iterative method
- Use the modified Newton–Raphson method to find the first approximation when the derivative is small
- Understand the meaning of interpolation and use simple linear and graphical interpolation
- Use the Gregory–Newton interpolation formula with forward and backward differences for equally spaced domain points
- Use the Gauss interpolation formulas using central differences for equally spaced domain points
- Use Lagrange interpolation when the domain points are not equally spaced

Introduction

1 In this Programme we shall be looking at analytic and numerical methods of solving the general equation in a single variable, $f(x) = 0$. In addition, a functional relationship can be exhibited in the form of a collection of ordered pairs rather than in the form of an algebraic expression. We shall be looking at interpolation methods of estimating values of $f(x)$ for intermediate values of x between those listed among the ordered pairs.

First we shall look at the **Fundamental Theorem of Algebra**, which deals with the factorisation of polynomials.

The Fundamental Theorem of Algebra

2 The *Fundamental Theorem of Algebra* can be stated as follows:

> **Every polynomial expression $f(x) = a_n x^n + a_{n-1} x^{n-1} + \cdots + a_1 x + a_0$ can be written as a product of n linear factors in the form**
> $$f(x) = a_n(x - r_1)(x - r_2)(\cdots)(x - r_n)$$

As an immediate consequence of this we can see that there are n values of x that satisfy the polynomial equation $f(x) = 0$, namely $x = r_1, x = r_2, \ldots, x = r_n$. We call these values the *roots* of the polynomial, but be aware that they may not all be distinct. Furthermore, the polynomial coefficients a_i and the polynomial roots r_i may be real, imaginary or complex.

For example the quadratic equation

$x^2 + 5x + 6 = 0$ can be written $(x + 2)(x + 3) = 0$ so it has the two *distinct* roots $x = -2$ and $x = -3$

$x^2 - 4x + 4 = 0$ can be written as $(x - 2)(x - 2) = 0$ so it has the two *coincident* roots $x = 2$ and $x = 2$

$x^2 + x + 1 = 0$ can be written as $(x + a)(x + b) = 0$ so it has the two roots $x = -a$ and $x = -b$

To find the numerical values of a and b we need to use the formula for finding the roots of a general quadratic equation. Can you recall what it is? If not, then refer to Frame 14 of Programme F.6 in *Engineering Mathematics, Sixth Edition*.

The solution to the quadratic equation $ax^2 + bx + c = 0$ is . . . .

The answer is in the next frame

$$x = \frac{-b \pm \sqrt{b^2 - 4ac}}{2a}$$

3

So the roots of $x^2 + x + 1 = 0$ are . . . .

Next frame

$$x = -\frac{1}{2} \pm j\frac{\sqrt{3}}{2}$$

4

Because

$$a = b = c = 1 \text{ and so } x = \frac{-b \pm \sqrt{b^2 - 4ac}}{2a} = \frac{-1 \pm \sqrt{1 - 4}}{2}$$

$$= -\frac{1}{2} \pm j\frac{\sqrt{3}}{2}$$

This quadratic equation has two distinct *complex* roots. Notice that the two roots form a *complex conjugate pair* – each is the complex conjugate of the other. **Whenever a polynomial with real coefficients a_i has a complex root it also has the complex conjugate as another root.**

So given that $x = -2 + j\sqrt{5}$ is one root of a quadratic equation with real coefficients then

the other root is . . . .

$$x = -2 - j\sqrt{5}$$

5

Because

The complex conjugate of $x = -2 + j\sqrt{5}$ is $x = -2 - j\sqrt{5}$ and complex roots of a polynomial equation with real coefficients always appear as conjugate pairs.

The quadratic equation with these two roots is . . . .

6

$$x^2 + 4x + 9 = 0$$

Because

If $x = a$ and $x = b$ are the two roots of a quadratic equation then $(x - a)(x - b) = 0$ gives the quadratic equation. That is $(x - a)(x - b) = x^2 - (a + b)x + ab = 0$.

Here, the two roots are $x = -2 + j\sqrt{5}$ and $x = -2 - j\sqrt{5}$ so that

$$\left(x - \left[-2 + j\sqrt{5}\right]\right)\left(x - \left[-2 - j\sqrt{5}\right]\right) = 0$$

That is $x^2 - x\left[-2 + j\sqrt{5} - 2 - j\sqrt{5}\right] + \left[-2 + j\sqrt{5}\right]\left[-2 - j\sqrt{5}\right] = 0$.

So $x^2 + 4x + 9 = 0$.

Notice that the coefficients are . . . .

7

Real

Relations between the coefficients and the roots of a polynomial equation

Let α, β, γ be the roots of $px^3 + qx^2 + rx + s = 0$. Then, writing the expression $px^3 + qx^2 + rx + s$ in terms of α, β, γ gives

$$px^3 + qx^2 + rx + s = p\left(x^3 + \frac{q}{p}x^2 + \frac{r}{p}x + \frac{s}{p}\right)$$

$$= . . . .$$

8

$$p(x - \alpha)(x - \beta)(x - \gamma)$$

Therefore

$$px^3 + qx^2 + rx + s = p\left(x^3 + \frac{q}{p}x^2 + \frac{r}{p}x + \frac{s}{p}\right)$$

$$= p(x - \alpha)(x - \beta)(x - \gamma)$$

$$= p(x^2 - [\alpha + \beta]x + \alpha\beta)(x - \gamma)$$

$$= p(x^3 - [\alpha + \beta]x^2 + \alpha\beta x - \gamma x^2 + [\alpha + \beta]\gamma x - \alpha\beta\gamma)$$

$$= p(x^3 - [\alpha + \beta + \gamma]x^2 + [\alpha\beta + \alpha\gamma + \beta\gamma]x - \alpha\beta\gamma)$$

Therefore, equating coefficients

(a) $\alpha + \beta + \gamma = $ . . . .

(b) $\alpha\beta + \alpha\gamma + \beta\gamma = $ . . . .

(c) $\alpha\beta\gamma = $ . . . .

9

$$\text{(a)} \quad -\frac{q}{p}; \quad \text{(b)} \quad \frac{r}{p}; \quad \text{(c)} \quad -\frac{s}{p}$$

This, of course, applies to a cubic equation. Let us extend this to a more general equation.

So on to the next frame

10

In general, if α_1, α_2, $\alpha_3 \ldots \alpha_n$ are roots of the equation

$$p_0 x^n + p_1 x^{n-1} + p_2 x^{n-2} + \ldots + p_{n-1}x + p_n \qquad = 0 \quad (p_0 \neq 0)$$

then　　sum of the roots　　　　　　　　　　　　　　$= -\dfrac{p_1}{p_0}$

　　　　　sum of products of the roots, two at a time　　$= \dfrac{p_2}{p_0}$

　　　　　sum of products of the roots, three at a time　$= -\dfrac{p_3}{p_0}$

　　　　　sum of products of the roots, n at a time　　$= (-1)^n . \dfrac{p_n}{p_0}$

So for the equation $3x^4 + 2x^3 + 5x^2 + 7x - 4 = 0$, if α, β, γ, δ are the four roots, then

(a) $\alpha + \beta + \gamma + \delta = \ldots\ldots\ldots$
(b) $\alpha\beta + \beta\gamma + \gamma\delta + \delta\alpha + \delta\beta + \gamma\alpha = \ldots\ldots\ldots$
(c) $\alpha\beta\gamma + \beta\gamma\delta + \gamma\delta\alpha + \alpha\beta\delta = \ldots\ldots\ldots$
(d) $\alpha\beta\gamma\delta = \ldots\ldots\ldots$

11

$$\text{(a)} \quad -\frac{2}{3}; \quad \text{(b)} \quad \frac{5}{3}; \quad \text{(c)} \quad -\frac{7}{3}; \quad \text{(d)} \quad -\frac{4}{3}$$

Now for a problem or two on the same topic.

Example 1

Solve the equation $x^3 - 8x^2 + 9x + 18 = 0$ given that the sum of two of the roots is 5.

Using the same approach as before, if α, β, γ are the roots, then

(a) $\alpha + \beta + \gamma = \ldots\ldots\ldots$
(b) $\alpha\beta + \beta\gamma + \gamma\alpha = \ldots\ldots\ldots$
(c) $\alpha\beta\gamma = \ldots\ldots\ldots$

12

$$(a)\ 8;\quad (b)\ 9;\quad (c)\ -18$$

So we have $\alpha + \beta + \gamma = 8$ Let $\alpha + \beta = 5$
$$\therefore 5 + \gamma = 8 \quad \therefore \gamma = 3$$

Also $\alpha\beta\gamma = -18$ $\alpha\beta(3) = -18$ $\therefore \alpha\beta = -6$
$$\alpha + \beta = 5 \quad \therefore \beta = 5 - \alpha \quad \therefore \alpha(5 - \alpha) = -6$$
$$\alpha^2 - 5\alpha - 6 = 0 \quad \therefore (\alpha - 6)(\alpha + 1) = 0 \quad \therefore \alpha = -1 \text{ or } 6$$
$$\therefore \beta = 6 \text{ or } -1$$

Roots are $x = -1, 3, 6$

13

Example 2

Solve the equation $2x^3 + 3x^2 - 11x - 6 = 0$ given that the three roots form an arithmetic sequence.

Let us represent the roots by $(a - k), a, (a + k)$

Then the sum of the roots $= 3a = \ldots\ldots\ldots\ldots$

and the product of the roots $= a(a - k)(a + k) = \ldots\ldots\ldots\ldots$

14

$$3a = -\frac{3}{2}; \quad a(a + k)(a - k) = \frac{6}{2} = 3$$

$$\therefore a = -\frac{1}{2} \qquad -\frac{1}{2}\left(\frac{1}{4} - k^2\right) = 3 \qquad \therefore k = \pm\frac{5}{2}$$

If $k = \frac{5}{2}$ $a = -\frac{1}{2}; \ a - k = -3; \ a + k = 2$

If $k = -\frac{5}{2}$ $a = -\frac{1}{2}; \ a - k = 2; \ a + k = -3$

$\therefore$ required roots are $-3, -\frac{1}{2}, 2$

Here is a similar one.

Example 3

Solve the equation $x^3 + 3x^2 - 6x - 8 = 0$ given that the three roots are in geometric sequence.

This time, let the roots be $\frac{a}{k}, a, ak$

Then $\frac{a}{k} = a + ak = \ldots\ldots\ldots\ldots$ and $\left(\frac{a}{k}\right)(a)(ak) = \ldots\ldots\ldots\ldots$

| sum of roots $= -3$;   product of roots $= 8$ | **15** |

It then follows that the roots are ...., ..., ...

| $-4, \quad 2, \quad -1$ | **16** |

The working rests on the relationships between the roots and the coefficients, i.e. if α, β, γ are the roots of the cubic equation

$$ax^3 + bx^2 + cx + d = 0$$

then   (a) $\alpha + \beta + \gamma = ...$

(b) $\alpha\beta + \beta\gamma + \gamma\alpha = ...$

(c) $\alpha\beta\gamma = ...$

| (a) $-\dfrac{b}{a}$;   (b) $\dfrac{c}{a}$;   (c) $-\dfrac{d}{a}$ | **17** |

In each of the three examples reconstruct the cubic to confirm that they are correct.

Now on to the next stage

Cubic equations

The Fundamental Theorem of Algebra tells us that every cubic expression   **18**

$$f(x) = ax^3 + bx^2 + cx + d$$

can be written as a product of three linear factors

$$f(x) = a(x - r_1)(x - r_2)(x - r_3)$$

Consequently, every cubic equation

$$f(x) = a(x - r_1)(x - r_2)(x - r_3) = 0$$

has three roots which may be distinct or coincident and which may be real or complex. However, because complex roots of a polynomial with real coefficients always appear in complex conjugate pairs we can say that every such cubic equation has

at least one ....

19

at least one real root

To find the value of this real root we can employ a formula equivalent to the formula used to find the two roots of the general quadratic. This is called Tartaglia's method but before we can proceed to look at that we must first consider how to transform the general cubic to its **reduced form**.

Next frame

20 Transforming a cubic to reduced form

In every case, an equation of the form

$$x^3 + ax^2 + bx + c = 0$$

can be converted into the reduced form $y^3 + py + q = 0$ by the substitution $x = y - \dfrac{a}{3}$.

The example will demonstrate the method.

Example 4

Express $f(x) = x^3 + 6x^2 - 4x + 5 = 0$ in reduced form.

Substitute $x = y - \dfrac{a}{3}$, i.e. $x = y - \dfrac{6}{3} = y - 2$. Put $x = y - 2$.

The equation then becomes

$$(y - 2)^3 + 6(y - 2)^2 - 4(y - 2) + 5 = 0$$
$$(y^3 - 3y^2 2 + 3y4 - 8) + 6(y^2 - 4y + 4) - 4(y - 2) + 5 = 0$$

which simplifies to ....

21

$$y^3 - 16y + 29 = 0$$

Tartaglia's solution for a real root

In the sixteenth century, Tartaglia discovered that a root of the cubic equation $x^3 + ax + b = 0$, where $a > 0$, is given by

$$x = \left\{ -\frac{b}{2} + \sqrt{\frac{a^3}{27} + \frac{b^2}{4}} \right\}^{1/3} + \left\{ -\frac{b}{2} - \sqrt{\frac{a^3}{27} + \frac{b^2}{4}} \right\}^{1/3}$$

That looks pretty formidable, but it is a good deal easier than it appears. Notice that $\dfrac{b}{2}$ and $\sqrt{\dfrac{a^3}{27} + \dfrac{b^2}{4}}$ occur twice and it is convenient to evaluate these first and then substitute the results in the main expression for x.

▶

Example 5

Find a real root of $x^3 + 2x + 5 = 0$.

Here, $a = 2$, $b = 5$ $\therefore \dfrac{b}{2} = 2.5$

$$\sqrt{\dfrac{a^3}{27} + \dfrac{b^2}{4}} = \sqrt{\dfrac{8}{27} + \dfrac{25}{4}} = \sqrt{6.5463} = 2.5586$$

Then $x = (-2.5 + 2.5586)^{1/3} + (-2.5 - 2.5586)^{1/3}$

$$= 0.3884 - 1.7166 = -1.3282 \qquad x = -1.328$$

Once we have a real root, the equation can be reduced to a quadratic and the remaining two roots determined. They are $x = 0.664 + j1.823$ and $x = 0.664 - j1.823$ (see *Engineering Mathematics, Sixth Edition*, Programme F.6).

Example 6

Determine a real root of $2x^3 + 3x - 4 = 0$.

This is first written $x^3 + 1.5x - 2 = 0$ $\therefore a = 1.5$, $b = -2$

Now you can evaluate $\dfrac{b}{2}$ and $\sqrt{\dfrac{a^3}{27} + \dfrac{b^2}{4}}$ and so determine

$$x = \ldots\ldots\ldots$$

$$\boxed{0.8796}$$

22

Because

$$\left\{ -\frac{b}{2} + \sqrt{\frac{a^3}{27} + \frac{b^2}{4}} \right\}^{1/3} = \{2.06066\}^{1/3} = 1.2725 \text{ and}$$

$$\left\{ -\frac{b}{2} - \sqrt{\frac{a^3}{27} + \frac{b^2}{4}} \right\}^{1/3} = \{-0.6066\}^{1/3} = -0.3929,$$

therefore $x = 1.2725 - 0.3929 = 0.8796$

Note: If you wish to find the real root of a cubic of the form $x^3 + ax + b = 0$ where $a < 0$ then it is best that you resort to numerical methods. Read on.

Next frame

Numerical methods

23

The methods that we have used so far to solve quadratic equations and to find the real root of a cubic equation are called *analytic methods*. These analytic methods used straightforward algebraic techniques to develop a formula for the answer. The numerical value of the answer can then be found by simple substitution of numbers for the variables in the formula. Unfortunately, general polynomial equations of order five or higher cannot by solved by analytic methods. Instead, we must resort to what are termed *numerical methods*. The simplest method of finding the solution to the equation $f(x) = 0$ is the *bisection* method.

Bisection

The bisection method of finding a solution to the equation $f(x) = 0$ consists of

Finding a value of x, say $x = a$, such that $f(a) < 0$
Finding a value of x, say $x = b$, such that $f(b) > 0$

The solution to the equation $f(x) = 0$ must then lie between a and b. Furthermore, it must lie either in the first half of the interval between a and b or in the second half.

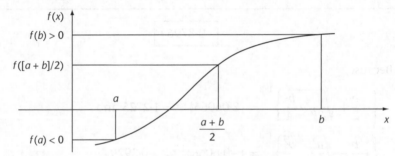

Find the value of $f([a + b]/2)$ – that is halfway between a and b.

If $f([a+b]/2) > 0$ then the solution lies in the first half and if $f([a+b]/2) < 0$ then it lies in the second half. This procedure is repeated, narrowing down the width of the interval by a half each time. An example should clarify all this.

Example 7

Find the positive value of x that satisfies the equation $x^2 - 2 = 0$.

Firstly we note that if $x = 1$ then $x^2 - 2 < 0$, and that if $x = 2$ then $x^2 - 2 > 0$, so the solution that we seek must lie between 1 and 2.

We look for the . . . .

> The mid-point between 1 and 2 which is 1·5

24

Now, when $x = 1·5$, $x^2 - 2 = 0·25 > 0$

　　　　　so the solution must lie between ....

> 1 and 1·5

25

The mid-point between 1 and 1·5 is 1·25. When $x = 1·25$, $x^2 - 2 = -0·4375 < 0$

　　　　　so the solution must lie between ....

> 1·25 and 1·5

26

The mid-point between 1·25 and 1·5 is 1·375. We now evaluate $x^2 - 2$ at this point and determine in which half interval the solution lies. This process is repeated and the following table displays the results. In each block of six numbers the first column lists the end points of the interval and the mid-point. The second column contains the respective values $f(x) = x^2 - 2$. Construct the table as follows.

(a) For each block of six numbers copy the last number in the first column into the second place of the first column of the following block. This represents the centre point of the previous interval.

(b) For each block of six numbers copy the number that represents the other end point of the new interval from the first column into the first place of the first column of the following block. Look at the signs in the second column of the first block to decide which is the appropriate number.

a	1·0000	−1·0000	1·0000	−1·0000	1·5000	0·2500	1·5000	0·2500
b	2·0000	2·0000	1·5000	−0·2500	1·2500	−0·4375	1·3750	−0·1094
$(a+b)/2$	1·5000	0·2500	1·2500	−0·4375	1·3750	−0·1094	1·4375	0·0664

a	1·3750	−0·1094	1·4375	0·0664	1·4063	−0·0225	1·4219	0·0217
b	1·4375	0·0664	1·4063	−0·0225	1·4219	0·0217	1·4141	−0·0004
$(a+b)/2$	1·4063	−0·0225	1·4219	0·0217	1·4141	−0·0004	1·4180	0·0106

a	1·4141	−0·0004	1·4141	−0·0004	1·4141	−0·0004	1·4141	−0·0004
b	1·4180	0·0106	1·4160	0·0051	1·4150	0·0023	1·4146	0·0010
$(a+b)/2$	1·4160	0·0051	1·4150	0·0023	1·4146	0·0010	1·4143	0·0003

a	1·4141	−0·0004	1·4143	0·0003	1·4142	−0·0001
b	1·4143	0·0003	1·4142	−0·0001	1·4142	0·0001
$(a+b)/2$	1·4142	−0·0001	1·4142	0·0001	1·4142	0·0000

The final result to four decimal places is $x = 1·4142$ which is the correct answer to that level of accuracy – but it has taken a lot of activity to produce it. A much faster way of solving this equation is to use an iteration formula that was first devised by Newton.

　　　　　　　　　　　　　　　　　　　　　　　Next frame

Numerical solution of equations
by iteration

27

The process of finding the numerical solution to the equation

$$f(x) = 0$$

by iteration is performed by first finding an approximate solution and then using this approximate solution to find a more accurate solution. This process is repeated until a solution is found to the required level of accuracy. For example, Newton showed that the square root of a number a can be found from the iteration equation

$$x_{i+1} = \frac{1}{2}\left(x_i + \frac{a}{x_i}\right), \quad i = 0, 1, 2, \ldots$$

where x_0 is the approximation that starts the iteration off. So, to find a succession of approximate values of $\sqrt{2}$, each of increasing accuracy, we proceed as follows. Let $x_0 = 1{\cdot}5$ – found by the first stage of the bisection method. Then

$$x_1 = \frac{1}{2}\left(x_0 + \frac{a}{x_0}\right) = 0{\cdot}5(1{\cdot}5 + 2/1{\cdot}5) = 1{\cdot}4166\ldots$$

This value is then used to find x_2.

By rounding x_1 to $1{\cdot}4167$, the value of x_2 is found to be ....

28

$$\boxed{x_2 = 1{\cdot}4142}$$

Because

$$x_2 = \frac{1}{2}\left(x_1 + \frac{a}{x_1}\right) = 0{\cdot}5(1{\cdot}4167 + 2/1{\cdot}4167) = 1{\cdot}4142\ldots$$

This has achieved the same level of accuracy as the bisection method in just two steps.

Using a spreadsheet

This simple iteration procedure is more efficiently performed using a spreadsheet. If the use of a spreadsheet is a totally new experience for you then you are referred to Programme 4 of *Engineering Mathematics, Sixth Edition* where the spreadsheet is introduced as a tool for constructing graphs of functions. If you have a limited knowledge then you will be able to follow the text from here. The spreadsheet we shall be using here is Microsoft Excel, though all commercial spreadsheets possess the equivalent functionality.

Open your spreadsheet and in cell A1 enter *n* and press **Enter**. In this first column we are going to enter the iteration numbers. In cell A2 enter the number 0 and press **Enter**. Place the cell highlight in cell A2 and highlight the block of cells A2 to A7 by holding down the mouse button and wiping the highlight down to cell A7. Click the **Edit** command on the Command bar and ▶

point at **Fill** from the drop-down menu. Select **Series** from the next drop-down menu and accept the default **Step value** of 1 by clicking OK in the Series window.

The cells A3 to A7 fill with . . . .

| the numbers 1 to 5 | **29** |

In cell B1 enter the letter x – this column is going to contain the successive x-values obtained by iteration. In cell B2 enter the value of x_0, namely 1·5.

In cell B3 enter the formula

$$= 0·5*(B2+2/B2)$$

The number that appears in cell B3 is then . . . .

| 1·416667 | **30** |

Place the cell highlight in cell B3, click the command **Edit** on the Command bar and select **Copy** from the drop-down menu. You have now copied the formula in cell B3 onto the Clipboard. Highlight the cells B4 to B7 and then click the **Edit** command again but this time select **Paste** from the drop-down menu.

The cells B4 to B7 fill with numbers to provide the display

. . . .

31

n	x
0	1·5
1	1·416667
2	1·414216
3	1·414214
4	1·414214
5	1·414214

By using the various formatting facilities provided by the spreadsheet the display can be amended to provide the following

n	x
0	1·500000000000000
1	1·416666666666670
2	1·414215686274510
3	1·414213562374690
4	1·414213562373090
5	1·414213562373090

The number of decimal places here is 15, which is far greater than is normally required but it does demonstrate how effective a spreadsheet can be. In future we shall restrict the displays to 6 decimal places.

▶

Notice that to find a value accurate to a given number of decimal places or significant figures it is sufficient to repeat the iterations until there is no change in the result from one iteration to the next.

Save your spreadsheet under some suitable name such as *Newton* because you may wish to use it again.

Now we shall look at this spreadsheet a little more closely

32 Relative addresses

Place the cell highlight in cell B3 and the formula that it contains is = 0·5*(B2+2/B2). Now place the cell highlight in cell B4 and the formula there is = 0·5*(B3+2/B3). Why the difference?

When you enter the cell address B2 in the formula in B3 the spreadsheet understands that to mean *the contents of the cell immediately above*. It is this meaning that is copied into cell B4 where the *cell immediately above* is B3. If you wish to refer to a specific cell in a formula then you must use an **absolute address**.

Place the cell highlight in cell C1 and enter the number 2. Now place the cell highlight in cell B3 and re-enter the formula

= 0·5*(B2+C1/B2)

and copy this into cells B4 to B7. The numbers in the second column have not changed but the formulas have because in cells B3 to B7 the same reference is made to cell C1. *The use of the dollar signs has indicated an absolute address.* So why would we do this?

Change the number in cell C1 to 3 to obtain the display ....

33

n	x
0	1·500000000000000
1	1·750000000000000
2	1·732142857142860
3	1·732050810014730
4	1·732050807568880
5	1·732050807568880

These are the iterated values of $\sqrt{3}$ – the square root of the contents of cell C1. We can now use the same spreadsheet to find the square root of any positive number.

Newton's iterative procedure to find the square root of a positive number is a special case of the **Newton–Raphson** procedure to find the solution of the general equation $f(x) = 0$, and we shall look at this in the next frame.

Newton–Raphson iterative method

34

Consider the graph of $y = f(x)$ as shown. Then the x-value at the point A, where the graph crosses the x-axis, gives a solution of the equation $f(x) = 0$.

If P is a point on the curve near to A, then $x = x_0$ is an approximate value of the root of $f(x) = 0$, the error of the approximation being given by AB.

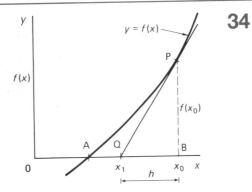

Let PQ be the tangent to the curve as P, crossing the x-axis at Q $(x_1, 0)$. Then $x = x_1$ is a better approximation to the required root.

From the diagram, $\dfrac{PB}{QB} = \left[\dfrac{dy}{dx}\right]_P$ i.e. the value of the derivative of y at the point P, $x = x_0$.

$$\therefore \frac{PB}{QB} = f'(x_0) \quad \text{and} \quad PB = f(x_0)$$

$$\therefore QB = \frac{PB}{f'(x_0)} = \frac{f(x_0)}{f'(x_0)} = h \text{ (say)}$$

$$x_1 = x_0 - h \qquad \therefore x_1 = x_0 - \frac{f(x_0)}{f'(x_0)}$$

If we begin, therefore, with an approximate value (x_0) of the root, we can determine a better approximation (x_1). Naturally, the process can be repeated to improve the result still further. Let us see this in operation.

On to the next frame

35

Example 1

The equation $x^3 - 3x - 4 = 0$ is of the form $f(x) = 0$ where $f(1) < 0$ and $f(3) > 0$ so there is a solution to the equation between 1 and 3. We shall take this to be 2, by bisection. Find a better approximation to the root.

We have $f(x) = x^3 - 3x - 4 \quad \therefore f'(x) = 3x^2 - 3$

If the first approximation is $x_0 = 2$, then

$$f(x_0) = f(2) = -2 \quad \text{and} \quad f'(x_0) = f'(2) = 9$$

A better approximation x_1 is given by

$$x_1 = x_0 - \frac{f(x_0)}{f'(x_0)} = x_0 - \frac{x_0{}^3 - 3x_0 - 4}{3x_0{}^2 - 3}$$

$$x_1 = 2 - \frac{(-2)}{9} = 2.22$$

$$\therefore x_0 = 2; \quad x_1 = 2.22$$

▶

---

rmula for $f'(x_0)$, namely $= 3*B2^2 - 3$ and copy into cells

rmula for x_1, namely $= B2 - C2/D2$ and copy into cells

The final display is ....

40

x	$f(x)$	$f'(x)$
-2		9
22222	0.30727	11.81481
06215	0.004492	11.47008
05823	1.01E-06	11.46492
05823	5.15E-14	11.46492

r in the second column is repeated then we know that hat particular level of accuracy. The required root is to 6 dp. Save the spreadsheet so that it can be used as a h problems.
ther example.

Next frame

41

$- 5x - 1 = 0$ is of the form $f(x) = 0$ where $f(1) < 0$ and olution to the equation between 1 and 2. We shall take the Newton–Raphson method to find the root to six

preadsheet as a template and make the following

ell B2 enter the number ....

42

$\boxed{1.5}$

$_0$ that is used to start the iteration

ll C2 enter the formula ....

43

$\boxed{\text{: B2^3 + 2*B2^2 - 5*B2 - 1}}$

$f(x_0) = x_0{}^3 + 2x_0{}^2 - 5x_0 - 1$. Copy the contents of o C5.

ll D2 enter the formula ....

44

$$= 3*B2\wedge2 + 4*B2 - 5$$

Because

That is the value of $f'(x_0) = 3x_0^2 + 4x_0 - 5$. Copy the contents of cell D2 into cells D3 to D5.

In cell B2 the formula remains the same as ....

45

$$= B2 - C2/D2$$

The final display is then ....

46

n	x	$f(x)$	$f'(x)$
0	1·5	−0·625	7·75
1	1·580645	0·042798	8·817898
2	1·575792	0·000159	8·752524
3	1·575773	2·21E-09	8·75228

We cannot be sure that the value 1·575773 is accurate to the sixth decimal place so we must extend the table.

Highlight cells A5 to D5, click **Edit** on the Command bar and select **Copy** from the drop-down menu.

Place the cell highlight in cell A6, click **Edit** and then **Paste**.

The seventh row of the spreadsheet then fills to produce the display

n	x	$f(x)$	$f'(x)$
0	1·5	−0·625	7·75
1	1·580645	0·042798	8·817898
2	1·575792	0·000159	8·752524
3	1·575773	2·21E-09	8·75228
4	1·575773	−8·9E-16	8·75228

And the repetition of the x-value ensures that the solution $x = 1·575773$ is indeed accurate to 6 dp.

Now do one completely on your own.

Next frame

47

Example 3

The equation $2x^3 - 7x^2 - x + 12 = 0$ has a root near to $x = 1·5$. Use the Newton–Raphson method to find the root to six decimal places.

The spreadsheet solution produces ....

n x_1 we can get a better approximation still by repeating the

$$x_1 - \frac{x_1^3 - 3x_1 - 4}{3x_1^2 - 3}$$

$f(x_1) = \ldots\ldots\ldots; \quad f'(x_1) = \ldots\ldots\ldots$

$$f(x_1) = 0·281; \quad f'(x_1) = 11·785$$

Then $x_2 = \ldots\ldots\ldots$

$$x_2 = 2·196$$

$= 2·196$

s a starter value, we can continue the process until
ree to the desired degree of accuracy.

$$x_3 = \ldots\ldots\ldots$$

$$x_3 = 2·196$$

0·002026; $\quad f'(x_2) = f'(2·196) = 11·467$

$= 2·196 - \dfrac{0·00203}{11·467} = 2·196$ (to 4 sig. fig.)

ple but effective and can be repeated again and again.
eration, usually gives a result nearer to the required root

In general $x_{n+1} = \ldots\ldots\ldots$

$$x_{n+1} = x_n - \frac{f(x_n)}{f'(x_n)}$$

sults

et and in cells A1 to D1 enter the headings n, x, $f(x)$

h the numbers 0 to 4

alue for x_0, namely 2

rmula for $f(x_0)$, namely $= B2\wedge3 - 3*B2 - 4$ and copy into

48

$$x = 1 \cdot 686141 \text{ to } 6 \text{ dp}$$

Because

Fill cells A2 to A6 with the numbers 0 to 4

In cell B2 enter the value for x_0, namely $1 \cdot 5$

In cell C2 enter the formula for $f(x_0)$, namely $= 2*B2\wedge3 - 7*B2\wedge2 - B2 + 12$ and copy into cells C3 to C6

In cell D2 enter the formula for $f'(x_0)$, namely $= 6*B2\wedge2 - 14*B2 - 1$ and copy into cells D3 to D6

In cell B3 enter the formula for x_1, namely $= B2 - C2/D2$ and copy into cells B4 to B6.

The final display is ....

49

n	x	$f(x)$	$f'(x)$
0	$1 \cdot 5$	$1 \cdot 5$	$-8 \cdot 5$
1	$1 \cdot 676471$	$0 \cdot 073275$	$-7 \cdot 60727$
2	$1 \cdot 686103$	$0 \cdot 000286$	$-7 \cdot 54778$
3	$1 \cdot 686141$	$4 \cdot 46\text{E-}09$	$-7 \cdot 54755$
4	$1 \cdot 686141$	0	$-7 \cdot 54755$

As soon as the number in the second column is repeated then we know that we have arrived at that particular level of accuracy. The required root is therefore $x = 1 \cdot 686141$ to 6 dp.

First approximations

The whole process hinges on knowing a 'starter' value as first approximation. If we are not given a hint, this information can be found by either

(a) applying the remainder theorem if the function is a polynomial
(b) drawing a sketch graph of the function.

Example 4

Find the real root of the equation $x^3 + 5x^2 - 3x - 4 = 0$ correct to six significant figures.

Application of the remainder theorem involves substituting $x = 0$, $x = \pm 1$, $x = \pm 2$, etc. until two adjacent values give a change in sign.

$$f(x) = x^3 + 5x^2 - 3x - 4$$
$$f(0) = -4; \quad f(1) = -1; \quad f(-1) = 3$$

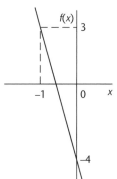

The sign changes from $f(0)$ to $f(-1)$. There is thus a root between $x = 0$ and $x = -1$.

Therefore choose $x = -0 \cdot 5$ as the first approximation and then proceed as before.

Complete the table and obtain the root

$$x = \text{............}$$

50

$$x = -0.675527$$

The final spreadsheet display is

n	x	$f(x)$	$f'(x)$
0	−0·5	−1·375	−7·25
1	−0·689655	0·11907	−8·469679
2	−0·675597	0·000582	−8·386675
3	−0·675527	1·43E-08	−8·386262
4	−0·675527	0	−8·386262

51   **Example 5**

Solve the equation $e^x + x - 2 = 0$ giving the root to 6 significant figures.

It is sometimes more convenient to obtain a first approximation to the required root from a sketch graph of the function, or by some other graphical means.

In this case, the equation can be rewritten as $e^x = 2 - x$ and we therefore sketch graphs of $y = e^x$ and $y = 2 - x$.

x	0·2	0·4	0·6	0·8	1
e^x	1·22	1·49	1·82	2·23	2·72
$2 - x$	1·8	1·6	1·4	1·2	1

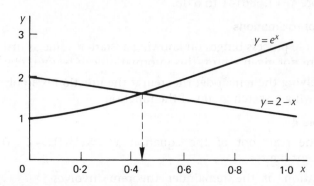

It can be seen that the two curves cross over between $x = 0.4$ and $x = 0.6$.

Approximate root $x = 0.4$

$$f(x) = e^x + x - 2 \qquad f'(x) = e^x + 1$$

$$x = \ldots\ldots\ldots$$

Finish it off

52

$$\boxed{x = 0\cdot442854}$$

The final spreadsheet display is

n	x	$f(x)$	$f'(x)$
0	0·4	−0·10818	2·491825
1	0·443412	0·001426	2·558014
2	0·442854	2·42E-07	2·557146
3	0·442854	7·11E-15	2·557146

Note: There are times when the normal application of the Newton–Raphson method fails to converge to the required root. This is particularly so when $f'(x_0)$ is very small, so before we leave this section let us consider this difficulty.

Modified Newton–Raphson method

53

If the slope of the curve at $x = x_0$ is small, the value of the second approximation $x = x_1$ may be further from the exact root at A than the first approximation.

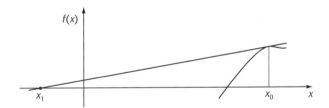

If $x = x_0$ is an approximate solution of $f(x) = 0$ and $x = x_0 - h$ is the exact solution then $f(x_0 - h) = 0$. By Taylor's series

$$f(x_0 - h) = f(x_0) - hf'(x_0) + \frac{h^2}{2!}f''(x_0) - \ldots = 0$$

(a) If we assume that h is small enough to neglect terms of the order h^2 and higher then this equation can be written as

$f(x_0 - h) \approx f(x_0) - hf'(x_0)$, that is $f(x_0) - hf'(x_0) \approx 0$ and so

$h \approx \dfrac{f(x_0)}{f'(x_0)}$ giving $x_1 = x_0 - \dfrac{f(x_0)}{f'(x_0)}$ as a better approximation

to the solution of $f(x) = 0$.

This is, of course, the relationship we have been using and which may fail when $f'(x)$ is small.

Notice: h is positive unless the sign of $f(x_0)$ is the opposite of the sign of $f'(x_0)$.

(b) If we consider the first three terms then

$$f(x_0 - h) \approx f(x_0) - hf'(x_0) + \frac{h^2}{2!}f''(x_0) \approx 0, \text{ that is}$$

$$2f(x_0) - 2hf'(x_0) + h^2f''(x_0) \approx 0$$

Since $f'(x_0)$ is small we shall assume that we can neglect it so

$$h = \pm\sqrt{\frac{-2f(x_0)}{f''(x_0)}}$$

That is $h = \sqrt{\dfrac{-2f(x_0)}{f''(x_0)}}$ unless the signs of $f(x_0)$ and $f'(x_0)$ are different when

it is $h = -\sqrt{\dfrac{-2f(x_0)}{f''(x_0)}}$. We use this result only when $f'(x_0)$ is found to be

very small. Having found x_1 from x_0 we then revert to the normal

relationship $x_{n+1} = x_n - \dfrac{f(x_0)}{f'(x_0)}$ for subsequent iterations.

Note this

54

Example 6

The equation $x^3 - 1\cdot3x^2 + 0\cdot4x - 0\cdot03 = 0$ is known to have a root near $x = 0\cdot7$. Determine the root to 6 significant figures.

We start off in the usual way.

$$f(x) = x^3 - 1\cdot3x^2 + 0\cdot4x - 0\cdot03$$
$$f'(x) = 3x^2 - 2\cdot6x + 0\cdot4$$

and complete the first line of the normal table.

n	x_n	$f(x_n)$	$f'(x_n)$	$h = \dfrac{f(x_n)}{f'(x_n)}$	$x_{n+1} = x_n - h$
0	0·7				

Complete just the first line of values.

55

We have

n	x_n	$f(x_n)$	$f'(x_n)$	$h = \dfrac{f(x_n)}{f'(x_n)}$	$x_{n+1} = x_n - h$
0	0·7	−0·044	0·05	−0·88	1·58

We notice at once that

(a) The value of x_1 is well away from the approximate value (0·7) of the root.

(b) The value of $f'(x_0)$ is small, i.e. 0·05.

To obtain x_1 we therefore make a fresh start, using the modified relationship
$x_1 = \ldots\ldots\ldots\ldots$

56

$$x_1 = x_0 \pm \sqrt{\frac{-2f(x_0)}{f''(x_0)}}$$

$$
\begin{aligned}
f(x) &= x^3 - 1{\cdot}3x^2 + 0{\cdot}4x - 0{\cdot}03 = [(x - 1{\cdot}3)x + 0{\cdot}4]x - 0{\cdot}03 \\
f'(x) &= 3x^2 - 2{\cdot}6x + 0{\cdot}4 \qquad = (3x - 2{\cdot}6)x + 0{\cdot}4 \\
f''(x) &= 6x - 2{\cdot}6
\end{aligned}
$$

n	x_0	$f(x_0)$	$f''(x_0)$	$h = \sqrt{\dfrac{-2f(x_0)}{f''(x_0)}}$	$x_1 = x_0 \pm h$
0	0·7	−0·044			

Complete the line.

57

n	x_0	$f(x_0)$	$f''(x_0)$	$h = \sqrt{\dfrac{-2f(x_0)}{f''(x_0)}}$	$x_1 = x_0 \pm h$
0	0·7	−0·044	1·6	0·2345	0·9345

Note that in the expression $x_1 = x_0 \pm h$, we chose the positive sign since at $x_0 = 0{\cdot}7$, $f(x_0)$ is negative and the slope $f'(x_0)$ is positive.

Having established that $x_1 = 0{\cdot}9345$, we now revert to the usual $x_{n+1} = x_n - \dfrac{f(x_n)}{f'(x_n)}$ for the rest of the calculation. Complete the table therefore and obtain the required root.

58

The final spreadsheet display is

n	x	$f(x)$	$f'(x)$	$f''(x)$
0	0·7	−0·044	0·05	1·6
1	0·934521	0·024625	0·590233	
2	0·892801	0·002544	0·469997	
3	0·887387	4·02E-05	0·45516	
4	0·887298	1·06E-08	0·454919	
5	0·887298	9·16E-16	0·454919	

Therefore to six decimal places the required root is $x = 0{\cdot}887298$.

Note that we only used the modified method to find x_1. After that the normal relationship is used.

▶

And now ...

To date our task has been to find a value of x that satisfies an explicit equation $f(x) = 0$. This is quite general because *any* equation in x can be written in this form. For example, the equation

$$\sin x = x - e^{3x}$$

can always be written as

$$\sin x - x + e^{3x} = 0$$

and then approached by one of the methods that we have discussed so far.

What we want to do now is to work the other way – given a value of x, to find the corresponding value of $f(x)$. If $f(x)$ is given explicitly then this is no problem, it is just a matter of substituting the value of x in the formula and working it out. However, many times a function exists but it is not given explicitly, as in the case of a set of readings compiled as a result of an experiment or practical test. We shall consider this problem in the following frames.

Next frame

Interpolation

59

When a function is defined by a well-understood expression such as

$$f(x) = 4x^3 - 3x^2 + 7$$

or

$$f(x) = 5\sin(\exp[x])$$

the values of the dependent variable $f(x)$ corresponding to given values of the independent variable x can be found by direct substitution. Sometimes, however, a function is not defined in this way but by a collection of ordered pairs of numbers.

Example 1

A function can be defined by the following set of data:

x	$f(x)$
1	4
2	14
3	40
4	88
5	164
6	274

Intermediate values, for example, $x = 2.5$, can be estimated by a process called **interpolation**.

The value of $f(2.5)$ will clearly lie between 14 and 40, the function values for $x = 2$ and $x = 3$.

Purely as an estimate, $f(2.5) = \ldots\ldots\ldots\ldots$

What do you suggest?

60

| 27 |

1 Linear interpolation

If you gave the result as 27, you no doubt agreed that $x = 2.5$ is midway between $x = 2$ and $x = 3$, and that therefore $f(2.5)$ would be midway between 14 and 40, i.e. 27. This is the simplest form of interpolation, but there is no evidence that there is a linear relationship between x and $f(x)$, and the result is therefore suspect.

Of course, we could have estimated the function value at $x = 2.5$ by other means, such as

. . . .

61

| by drawing the graph of $f(x)$ against x |

2 Graphical interpolation

We could, indeed, plot the graph of $f(x)$ against x and, from it, estimate the value of $f(x)$ at $x = 2.5$.

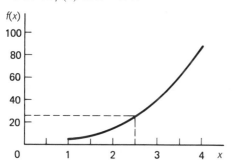

This method is also approximate and can be time consuming.

$$f(2.5) \approx 26$$

In what follows we shall look at interpolation using *finite differences*, which work well and quickly when the values of x are equally spaced. When the values of x are not equally spaced we need to resort to the more involved algebraic method called *Lagrangian interpolation* (which could also be used for equally spaced points).

Next frame

3 Gregory–Newton interpolation formula using forward finite differences

62

x	$f(x)$
$\vdots$	$\vdots$
x_0	$f(x_0)$
x_1	$f(x_1)$
$\vdots$	$\vdots$

$\Delta f_0 = f(x_1) - f(x_0)$

We assume that $x_0, x_1, \ldots$ are distinct, equally spaced apart, and $x_0 < x_1 < \ldots$

▶

For each pair of consecutive function values, $f(x_0)$ and $f(x_1)$, in the table, the *forward difference* Δf_0 is calculated by subtracting $f(x_0)$ from $f(x_1)$. This difference is written in a third column of the table, midway between the lines carrying $f(x_0)$ and $f(x_1)$.

x	$f(x)$	Δf
1	4	
		10
2	14	
		26
3	40	
$\vdots$	$\vdots$	

Complete the table for the data given in Frame 59 which then becomes . . . .

63

x	$f(x)$	Δf
1	4	
		10
2	14	
		26
3	40	
		48
4	88	
		76
5	164	
		110
6	274	

We now form a fourth column, the forward differences of the values of Δf, denoted by $\Delta^2 f$, and again written midway between the lines of Δf. These are the second forward differences of $f(x)$.

So the table then becomes . . . .

64

x	$f(x)$	Δf	$\Delta^2 f$
1	4		
		10	
2	14		16
		26	
3	40		22
		48	
4	88		28
		76	
5	164		34
		110	
6	274		

A further column can now be added in like manner, giving the third differences, denoted by $\Delta^3 f$, so that we then have . . . .

65

x	$f(x)$	Δf	$\Delta^2 f$	$\Delta^3 f$
1	4			
		10		
2	14		16	
		26		6
3	40		22	
		48		6
4	88		28	
		76		6
5	164		34	
		110		
6	274			

Notice that the table has now been completed, for the third differences are constant and all subsequent differences would be zero.

Now we shall see how to use the table. So move on

To find $f(2\cdot5)$

66

We have to find $f(2\cdot5)$. Therefore denote $\left.\begin{array}{l} x = 2 \text{ as } x_0 \\ x = 3 \text{ as } x_1 \end{array}\right\}$ $x = 2\cdot5$ as x_p

Let $h = $ the constant range between successive values of x,

i.e. $h = x_1 - x_0$

Express $(x_p - x_0)$ as a fraction of h, i.e. $p = \dfrac{x_p - x_0}{h}$, $0 < p < 1$

Therefore, in the case above, $h = 1$ and $p = \dfrac{2\cdot5 - 2\cdot0}{1} = 0\cdot5$.

All we now use from the table is the set of values underlined by the broken line drawn diagonally from $f(x_0)$.
   So we have

$$p = \ldots\ldots\ldots; \quad f_0 = \ldots\ldots\ldots; \quad \Delta f_0 = \ldots\ldots\ldots;$$
$$\Delta^2 f_0 = \ldots\ldots\ldots; \quad \Delta^3 f_0 = \ldots\ldots\ldots$$

67

$$p = 0\cdot5 \quad f_0 = 14; \quad \Delta f_0 = 26; \quad \Delta^2 f_0 = 22; \quad \Delta^3 f_0 = 6$$

Now we are ready to deal with the *Gregory–Newton forward difference interpolation formula*

$$f_p = f_0 + p\Delta f_0 + \frac{p(p-1)}{1 \times 2}\Delta^2 f_0 + \frac{p(p-1)(p-2)}{1 \times 2 \times 3}\Delta^3 f_0 + \cdots$$

This is sometimes written in operator form

$$f_p = \left\{1 + p\Delta + \frac{p(p-1)}{2!}\Delta^2 + \frac{p(p-1)(p-2)}{3!}\Delta^3 + \cdots\right\}f_0$$

which you no doubt recognise as the binomial expansion of

$$f_p = (1 + \Delta)^p \times f_0$$

Substituting the values in the above example gives

$$f(2\cdot5) = f_p = \ldots\ldots\ldots\ldots$$

68

$$24\cdot625$$

Because

$$f_p = 14 + 0\cdot5(26) + \frac{0\cdot5(-0\cdot5)}{1 \times 2}(22) + \frac{0\cdot5(-0\cdot5)(-1\cdot5)}{1 \times 2 \times 3}(6)$$

$$= 14 + 13 - 2\cdot75 + 0\cdot375$$

$$= 27\cdot375 - 2\cdot75 = 24\cdot625$$

Comparing the results of the three methods we have discussed

(a) Linear interpolation $f(2\cdot5) = 27$

(b) Graphical interpolation $f(2\cdot5) = 26$

(c) Gregory–Newton formula $f(2\cdot5) = 24\cdot625$ – the true value

Example 2

x	$f(x)$
2	14
4	88
6	274
8	620
10	1174

It is required to determine the value of $f(x)$ at $x = 5\cdot5$.
In this case

$$x_0 = \ldots\ldots\ldots\ldots \quad x_1 = \ldots\ldots\ldots\ldots$$

$$h = \ldots\ldots\ldots\ldots \quad p = \ldots\ldots\ldots\ldots$$

69

$$x_0 = 4; \quad x_1 = 6; \quad h = 2; \quad p = 0\cdot75$$

Because

$$h = x_1 - x_0 = 6 - 4 = 2$$

$$p = \frac{x_p - x_0}{h} = \frac{5\cdot5 - 4}{2} = \frac{1\cdot5}{2} = 0\cdot75$$

First compile the table of forward differences $\ldots\ldots\ldots\ldots$

x	$f(x)$	Δf	$\Delta^2 f$	$\Delta^3 f$
2	14			
		74		
4	88		112	
		186		48
6	274		160	
		346		48
8	620		208	
		554		
10	1174			

$x_0 \longrightarrow$ (at row $x=4$)
$x_1 \longrightarrow$ (at row $x=6$)

The Gregory–Newton forward difference interpolation formula is

$$f_p = (1 + \Delta)^p \times f_0$$

i.e. $f_p = \ldots\ldots\ldots$

$$f_p = \left\{ 1 + p\Delta + \frac{p(p-1)}{2!}\Delta^2 + \frac{p(p-1)(p-2)}{3!}\Delta^3 + \ldots \right\} f_0$$

$$= f_0 + p\Delta f_0 + \frac{p(p-1)}{2!}\Delta^2 f_0 + \frac{p(p-1)(p-2)}{3!}\Delta^3 f_0 + \ldots$$

So, substituting the relevant values from the table, gives

$$f(5\cdot5) = f_p = \ldots\ldots\ldots$$

$$214\cdot4$$

Because

x	$f(x)$	Δf	$\Delta^2 f$	$\Delta^3 f$
2	14			
		74		
4	88		112	
		186		48
6	274		160	
		346		48
8	620		208	
		554		
10	1174			

$x_0 \longrightarrow$ (at row $x=4$)
x_p (between rows $x=4$ and $x=6$)
$x_1 \longrightarrow$ (at row $x=6$)

$$f(5\cdot5) = f_p = 88 + 0\cdot75(186) + \frac{0\cdot75(-0\cdot25)}{1 \times 2}(160)$$
$$+ \frac{0\cdot75(-0\cdot25)(-1\cdot25)}{1 \times 2 \times 3}(48)$$
$$= 88 + 139\cdot5 - 15 + 1\cdot875 = 214\cdot375$$
$$\therefore f(5\cdot5) = 214\cdot4$$

Finally, one more.

▶

Example 3

Determine the value of $f(-1)$ from the set of function values.

x	-4	-2	0	2	4	6	8
$f(x)$	541	55	1	-53	-155	31	1225

Complete the working and then check with the next frame.

73

$$\boxed{f(-1) = 10}$$

Here is the working; method as before.

x	$f(x)$	Δf	$\Delta^2 f$	$\Delta^3 f$	$\Delta^4 f$
-4	541				
		-486			
-2	55		432		
		-54		-432	
0	1		0		384
		-54		-48	
2	-53		-48		384
		-102		336	
4	-155		288		384
		186		720	
6	31		1008		
		1194			
8	1225				

$x_0 \longrightarrow$ at -2; x_p between -2 and 0; $x_1 \longrightarrow$ at 0.

$x_0 = -2;\quad x_1 = 0;\quad x_p = -1;\qquad \therefore h = 2;\quad p = \tfrac{1}{2}$

$$f_p = f_0 + p\Delta f_0 + \frac{p(p-1)}{1\times 2}\Delta^2 f_0 + \frac{p(p-1)(p-2)}{1\times 2 \times 3}\Delta^3 f_0$$

$$+ \frac{p(p-1)(p-2)(p-3)}{1\times 2 \times 3 \times 4}\Delta^4 f_0$$

$$= 55 + \frac{1}{2}(-54) + \frac{\frac{1}{2}\left(-\frac{1}{2}\right)}{1\times 2}(0) + \frac{\frac{1}{2}\left(-\frac{1}{2}\right)\left(-\frac{3}{2}\right)}{1\times 2 \times 3}(-48)$$

$$+ \frac{\frac{1}{2}\left(-\frac{1}{2}\right)\left(-\frac{3}{2}\right)\left(-\frac{5}{2}\right)}{1\times 2 \times 3 \times 4}(384)$$

$$= 55 - 27 + 0 - 3 - 15 = 10$$

$$\therefore f_p = f(-1) = 10$$

This table of data does have its restrictions. For example, if we had wanted to find $f(2\cdot 5)$ from the table we would have run out of data because there is no $\Delta^4 f$ entry available. In such a case we can resort to a zig-zag path through the table using **central differences**.

Next frame

Central differences

The central difference operator δ is defined by its action on the expression $f(x)$ as

$$\delta f(x) = f(x + h/2) - f(x - h/2)$$

and using this operator the interpolated value of $f(x)$ near to the given value of f_0 is defined by the **Gauss forward** formula as

$$f_p = f_0 + p\delta f_{0+\frac{1}{2}} + \frac{p(p-1)}{2!}\delta^2 f_0 + \frac{(p+1)p(p-1)}{3!}\delta^3 f_{0+\frac{1}{2}}$$
$$+ \frac{(p+1)p(p-1)(p-2)}{4!}\delta^4 f_0 + \dots$$

or by the **Gauss backward** formula as

$$f_p = f_0 + p\delta f_{0-\frac{1}{2}} + \frac{(p+1)p}{2!}\delta^2 f_0 + \frac{(p+1)p(p-1)}{3!}\delta^3 f_{0-\frac{1}{2}}$$
$$+ \frac{(p+2)(p+1)p(p-1)}{4!}\delta^4 f_0 + \dots$$

There are no tabulated values at the half-interval values $x_0 + h/2$ and $x_0 - h/2$ and so these are taken to be the differences evaluated at mid-interval as given in the forward difference table. This means that the tables for the Gregory–Newton forward differences and the central differences are identical (apart, that is, from the column headings); the method of tracing through the table, however, is different. For example, to find $f(2\cdot5)$ for the example given in Frame 59

x	$f(x)$	$\delta f(x)$	$\delta^2 f(x)$	$\delta^3 f(x)$
1	4			
		10		
2	14		16	
		26		6
3	40		22	
		48		6
4	88		28	
		76		6
5	164		34	
		110		
6	274			

Here $x_0 = 2$, $f_0 = 14$, $\delta f_{0+\frac{1}{2}} = 26$, $\delta^2 f_0 = 16$, $\delta^3 f_{0+\frac{1}{2}} = 6$, $\delta^4 f_0 = 0$ and $p = 0\cdot5$. Thus

$$f_p = 14 + (0\cdot5)26 + \frac{(0\cdot5)(-0\cdot5)}{2}16 + \frac{(0\cdot5)(-0\cdot5)(-1\cdot5)}{6}6$$
$$= 14 + 13 - 2 - 0\cdot375 = 24\cdot625$$

which agrees with the value found using the Gregory–Newton forward difference formula.

▶

Try one for yourself. The given tabulated values are

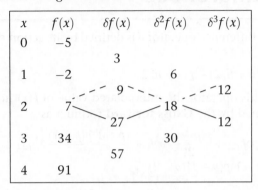

x	$f(x)$	$\delta f(x)$	$\delta^2 f(x)$	$\delta^3 f(x)$
0	-5			
		3		
1	-2		6	
		9		-12
2	7		18	
		27		12
3	34		30	
		57		
4	91			

Using the Gauss forward difference formula, the interpolated value of

$$f(2 \cdot 2) = \ldots\ldots\ldots$$

Next frame

75　　　　　　　　　　　　　　　$\boxed{10 \cdot 576}$

Because

Using $f_p = f_0 + p\delta f_{0+\frac{1}{2}} + \dfrac{p(p-1)}{2!}\delta^2 f_0 + \dfrac{p(p-1)(p+1)}{3!}\delta^3 f_{0+\frac{1}{2}} + \ldots$ and

following the solid line through the table where

$$x_0 = 2, \quad f_0 = 7, \quad \delta f_{0+\frac{1}{2}} = 27, \quad \delta^2 f_0 = 18, \quad \delta^3 f_{0+\frac{1}{2}} = 12 \text{ and } p = 0 \cdot 2,$$

then $f_p = 7 + (0 \cdot 2)27 + \dfrac{(0 \cdot 2)(-0 \cdot 8)}{2}18 + \dfrac{(0 \cdot 2)(-0 \cdot 8)(1 \cdot 2)}{6}12$

$$= 7 + 5 \cdot 4 - 1 \cdot 44 - 0 \cdot 384$$

$$= 10 \cdot 576$$

Using the Gauss backward difference formula (following the broken line)

$$f_p = f_0 + p\delta f_{0-\frac{1}{2}} + \dfrac{p(p+1)}{2!}\delta^2 f_0 + \dfrac{p(p-1)(p+1)}{3!}\delta^3 f_{0-\frac{1}{2}} + \ldots$$

where here $\delta f_{0-\frac{1}{2}} = 9$ and $\delta^3 f_{0-\frac{1}{2}} = 12$ and so

$$f_p = 7 + (0 \cdot 2)9 + \dfrac{(0 \cdot 2)(1 \cdot 2)}{2}18 + \dfrac{(0 \cdot 2)(1 \cdot 2)(-0 \cdot 8)}{6}12$$

$$= 7 + 1 \cdot 8 + 2 \cdot 16 - 0 \cdot 384 = 10 \cdot 576$$

as found with the Gauss forward difference formula.

Next frame

Gregory–Newton backward differences

We have seen that the Gregory–Newton forward difference procedure loses terms if the interpolation is for points sufficiently forward in the table. We have also seen how this difficulty can be avoided by using central differences. However, even with central differences we can run out of data before completing a full traverse of the table. In such a situation we resort to the Gregory–Newton backward difference formula

$$f_p = f_0 + p\Delta f_{-1} + \frac{p(p+1)}{2!}\Delta^2 f_{-2} + \frac{p(p+1)(p+2)}{3!}\Delta^3 f_{-3} + \cdots$$

As an example, consider the table of Frame 74.

x	$f(x)$	Δf	$\Delta^2 f$	$\Delta^3 f$
1	4			
		10		
2	14		16	
		26		6
3	40		22	
		48		6
4	88		28	
		76		6
5	164		34	
		110		
6	274			

Using this table we can calculate $f(5\cdot5)$ by tracing back through the table (see broken line) as

$$f(5\cdot5) = f_0 + (0\cdot5)\Delta f_{-1} + \frac{(0\cdot5)(1\cdot5)}{2}\Delta^2 f_{-2} + \frac{(0\cdot5)(1\cdot5)(2\cdot5)}{6}\Delta^3 f_{-3}$$

$$= 164 + (0\cdot5)76 + \frac{(0\cdot5)(1\cdot5)28}{2} + \frac{(0\cdot5)(1\cdot5)(2\cdot5)6}{6}$$

$$= 214\cdot375$$

In each of the examples that we have looked at so far the tabular display of differences eventually results in a column of zeros and this determines the number of terms in an interpolation calculation. The zeros have arisen because all the examples have been derived from polynomials. The following example deals with a tabular display of differences which does not result in a column of zeros. In this case the number of terms used in the interpolation calculation determines confidence in the accuracy of the result.

▶

Example

Use the Gregory–Newton forward difference method to find $f(0.15)$ to 4 decimal places from the following finite difference table.

x	$f(x)$	Δf	$\Delta^2 f$	$\Delta^3 f$
0	0·000000			
		0·099833		
0·1	0·099833		−0·000998	
		0·098836		−0·000988
0·2	0·198669		−0·001985	
		0·096851		−0·000968
0·3	0·295520		−0·002953	
		0·093898		−0·000938
0·4	0·389418		−0·003891	
		0·090007		
0·5	0·479426			

Here $x_0 = 0.1$, $x_1 = 0.2$, $x_p = 0.15$ and therefore $p = 0.5$, and

$$f_p = f_0 + p\Delta f_0 + \frac{p(p-1)}{2!}\Delta^2 f_0 + \frac{p(p-1)(p-2)}{3!}\Delta^3 f_0 + \dots$$

$$= 0.099833 + \frac{1}{2}(0.098836) + \left(\frac{1}{2}\right)\left(-\frac{1}{2}\right)(-0.001985)/2$$

$$+ \left(\frac{1}{2}\right)\left(-\frac{1}{2}\right)\left(-\frac{3}{2}\right)(-0.000969)/6 + \dots$$

$$= 0.099833 + 0.049418 + 0.000248 - 0.000061 + \dots$$

$$= 0.1494 \text{ to 4 dp}$$

As you can see, the calculation can continue indefinitely and termination is dictated by the number of decimal places required in the final answer.

Lagrange interpolation

78

If the straight line $p(x) = a_0 + a_1 x$ passes through the two points $(x_0, f(x_0))$ and $(x_1, f(x_1))$, where a_0 and a_1 are constants, then the equation for this line can also be written as

$$p(x) = \frac{x - x_1}{x_0 - x_1} f(x_0) + \frac{x - x_0}{x_1 - x_0} f(x_1)$$

For example, the straight line $p(x) = 3 + 2x$ passes through the two points $(1, 5)$ and $(2, 7)$. Substituting the values for the variables in the above equation demonstrates this alternative form for the equation

▶

$$p(x) = \frac{x-2}{1-2}5 + \frac{x-1}{2-1}7 = 10 - 5x + 7x - 7 = 3 + 2x$$

So, given the two data points from Frame 59, (2, 14) and (3, 40), using linear interpolation

$$f(2\cdot5) \approx p(2\cdot5) = \ldots\ldots\ldots$$

$$\boxed{27}$$

79

Because

$$p(x) = \frac{x - x_1}{x_0 - x_1}f(x_0) + \frac{x - x_0}{x_1 - x_0}f(x_1)$$

$$= \frac{x - 3}{2 - 3}14 + \frac{x - 2}{3 - 2}40 = 26x - 38$$

and so

$$f(2\cdot5) \approx p(x) = 26(2\cdot5) - 38 = 27$$

The principle of Lagrange interpolation is that a function $f(x)$ whose values are given at a collection of points is assumed to be approximately represented by a polynomial $p(x)$ that passes through each and every point. The polynomial is called the **interpolation polynomial** and it is of degree one less than the number of points given. For two data points the interpolating polynomial is taken to be a linear polynomial, as you have just seen in the last example. For three data points the interpolating polynomial is taken to be a quadratic, for four data points the interpolation polynomial is taken to be a cubic, and so on.

80

In the same manner as before it can be shown that the quadratic

$$p(x) = a_0 + a_1 x + a_2 x^2$$

that passes through the three points $(x_0, f(x_0))$, $(x_1, f(x_1))$ and $(x_2, f(x_2))$ can be written as

$$p(x) = \frac{(x - x_1)(x - x_2)}{(x_0 - x_1)(x_0 - x_2)}f(x_0) + \frac{(x - x_0)(x - x_2)}{(x_1 - x_0)(x_1 - x_2)}f(x_1)$$

$$+ \frac{(x - x_0)(x - x_1)}{(x_2 - x_0)(x_2 - x_1)}f(x_2)$$

So let's try one. Given the collection of values

x	$f(x)$
1·5	0·405
2·1	0·742
3	1·099

by Lagrangian interpolation, $f(1\cdot8) \approx \ldots\ldots\ldots$ to 2 decimal places.

81

$$\boxed{0\cdot58}$$

Because

$$p(x) = \frac{(x - x_1)(x - x_2)}{(x_0 - x_1)(x_0 - x_2)} f(x_0) + \frac{(x - x_0)(x - x_2)}{(x_1 - x_0)(x_1 - x_2)} f(x_1)$$

$$+ \frac{(x - x_0)(x - x_1)}{(x_2 - x_0)(x_2 - x_1)} f(x_2)$$

$$= \frac{(x - 2\cdot1)(x - 3)}{(1\cdot5 - 2\cdot1)(1\cdot5 - 3)} 0\cdot405 + \frac{(x - 1\cdot5)(x - 3)}{(2\cdot1 - 1\cdot5)(2\cdot1 - 3)} 0\cdot742$$

$$+ \frac{(x - 1\cdot5)(x - 2\cdot1)}{(3 - 1\cdot5)(3 - 2\cdot1)} 1\cdot099$$

$$= \frac{(x^2 - 5\cdot1x + 6\cdot3)}{0\cdot9} 0\cdot405 + \frac{(x^2 - 4\cdot5x + 4\cdot5)}{(-0\cdot54)} 0\cdot742$$

$$+ \frac{(x^2 - 3\cdot6x + 3\cdot15)}{1\cdot35} 1\cdot099$$

$$= -0\cdot11x^2 + 0\cdot958x - 0\cdot784$$

So that

$$f(1\cdot8) \approx p(1\cdot8) = 0\cdot58 \text{ to 2 decimal places.}$$

By carefully considering the interpolating polynomials for two and three data points you should be able to see a pattern. Write down what you think the interpolating polynomial should be for four data points:

....

82

$$p(x) = \frac{(x - x_1)(x - x_2)(x - x_3)}{(x_0 - x_1)(x_0 - x_2)(x_0 - x_3)} f(x_0) + \frac{(x - x_0)(x - x_2)(x - x_3)}{(x_1 - x_0)(x_1 - x_2)(x_1 - x_3)} f(x_1)$$

$$+ \frac{(x - x_0)(x - x_1)(x - x_3)}{(x_2 - x_0)(x_2 - x_1)(x_2 - x_3)} f(x_2) + \frac{(x - x_0)(x - x_1)(x - x_2)}{(x_3 - x_0)(x_3 - x_1)(x_3 - x_2)} f(x_3)$$

Use this interpolating polynomial for the data points

x	$f(x)$
1	0·368
1·2	0·301
1·3	0·273
1·5	0·223

To 2 decimal places, $f(1\cdot4) \approx$ ....

$$\boxed{0{\cdot}25}$$ **83**

Because $p(x)$

$$= \frac{(x-x_1)(x-x_2)(x-x_3)}{(x_0-x_1)(x_0-x_2)(x_0-x_3)}f(x_0) + \frac{(x-x_0)(x-x_2)(x-x_3)}{(x_1-x_0)(x_1-x_2)(x_1-x_3)}f(x_1)$$

$$+ \frac{(x-x_0)(x-x_1)(x-x_3)}{(x_2-x_0)(x_2-x_1)(x_2-x_3)}f(x_2) + \frac{(x-x_0)(x-x_1)(x-x_2)}{(x_3-x_0)(x_3-x_1)(x_3-x_2)}f(x_3)$$

$$= \frac{(x-1{\cdot}2)(x-1{\cdot}3)(x-1{\cdot}5)}{(1-1{\cdot}2)(1-1{\cdot}3)(1-1{\cdot}5)}0{\cdot}368 + \frac{(x-1)(x-1{\cdot}3)(x-1{\cdot}5)}{(1{\cdot}2-1)(1{\cdot}2-1{\cdot}3)(1{\cdot}2-1{\cdot}5)}0{\cdot}301$$

$$+ \frac{(x-1)(x-1{\cdot}2)(x-1{\cdot}5)}{(1{\cdot}3-1)(1{\cdot}3-1{\cdot}2)(1{\cdot}3-1{\cdot}5)}0{\cdot}273 + \frac{(x-1)(x-1{\cdot}2)(x-1{\cdot}3)}{(1{\cdot}5-1)(1{\cdot}5-1{\cdot}2)(1{\cdot}5-1{\cdot}3)}0{\cdot}223$$

$$= \frac{(x^3-4x^2+5{\cdot}31x-2{\cdot}34)}{(-0{\cdot}03)}0{\cdot}368 + \frac{(x^3-3{\cdot}8x^2+4{\cdot}75x-1{\cdot}95)}{0{\cdot}006}0{\cdot}301$$

$$+ \frac{(x^3-3{\cdot}7x^2+4{\cdot}5x-1{\cdot}8)}{(-0{\cdot}006)}0{\cdot}273 + \frac{(x^3-3{\cdot}5x^2+4{\cdot}06x-1{\cdot}56)}{0{\cdot}03}0{\cdot}223$$

$$= -0{\cdot}167x^3 + 0{\cdot}767x^2 - 1{\cdot}415x + 1{\cdot}183$$

So that

$$f(1{\cdot}4) \approx p(1{\cdot}4) = 0{\cdot}25 \text{ to 2 decimal places}$$

The general Lagrange interpolation polynomial for $n+1$ data points at $x_0, x_1, \ldots, x_n$ is

$$p(x) = \frac{(x-x_1)(x-x_2)(\cdots)(x-x_n)}{(x_0-x_1)(x_0-x_2)(\cdots)(x_0-x_n)}f(x_0)$$

$$+ \frac{(x-x_0)(x-x_2)(\cdots)(x-x_n)}{(x_1-x_0)(x_1-x_2)(\cdots)(x_1-x_n)}f(x_1) + \cdots$$

$$\cdots + \frac{(x-x_0)(x-x_1)(\cdots)(x-x_{n-1})}{(x_n-x_0)(x_n-x_1)(\cdots)(x_n-x_{n-1})}f(x_n)$$

This now completes the work of this Programme. What follows is a **Revision summary** and a **Can you?** checklist. Read the summary carefully and respond to the questions in the checklist. When you feel sure that you are happy with the content of this Programme, try the **Test exercise**. Take your time, there is no need to hurry. Finally, a collection of **Further problems** provides valuable additional practice.

Revision summary 1

84

1 The *Fundamental Theorem of Algebra* can be stated as follows:

Every polynomial expression $f(x) = a_n x^n + a_{n-1} x^{n-1} + \cdots + a_1 x + a_0$ can be written as a product of n linear factors in the form

$$f(x) = a_n(x - r_1)(x - r_2)(\cdots)(x - r_n)$$

2 *Relations between the coefficients and the roots of a polynomial equation*
Whenever a polynomial with *real coefficients* a_i has a complex root it also has the complex conjugate as another root.
If α, β, γ, ... are the roots of the equation

$$p_0 x^n + p_1 x^{n-1} + p_2 x^{n-2} + \cdots + p_{n-1} x + p_n = 0$$

then, provided $p_0 \neq 0$

sum of roots $= -\dfrac{p_1}{p_0}$

sum of the product of the roots, taken two at a time $= \dfrac{p_2}{p_0}$

sum of the product of the roots, taken three at a time $= -\dfrac{p_3}{p_0}$

sum of the product of the roots, taken n at a time $= (-1)^n \dfrac{p_n}{p_0}$.

3 *Cubic equations*
Reduced form
Every cubic equation of the form $x^3 + ax^2 + bx + c = 0$ can be written in reduced form $y^3 + py + q = 0$ by using the transformation $x = y - \dfrac{a}{3}$.
Tartaglia's solution
Every cubic equation with real coefficients has at least one real root that may be found analytically using Tartaglia's method. The real root of $x^3 + ax + b = 0$ when $a > 0$ is

$$x = \left\{ -\frac{b}{2} + \sqrt{\frac{a^3}{27} + \frac{b^2}{4}} \right\}^{1/3} + \left\{ -\frac{b}{2} - \sqrt{\frac{a^3}{27} + \frac{b^2}{4}} \right\}^{1/3}$$

If $a < 0$ it is best to resort to numerical methods.

4 *Numerical methods*
Bisection
The bisection method of finding a solution to the equation $f(x) = 0$ consists of

Finding a value of x such that $f(x) < 0$, say $x = a$
Finding a value of x such that $f(x) > 0$, say $x = b$.

The solution to the equation $f(x) = 0$ must then lie between a and b. Furthermore, it must lie either in the first half of the interval between a and b or in the second half.

▶

5 *Numerical solution of equations by iteration*
The process of finding the numerical solution to the equation

$$f(x) = 0$$

by iteration is performed by first finding an approximate solution and then using this approximate solution to find a more accurate solution. This process is repeated until a solution is found to the required level of accuracy.

6 *Using a spreadsheet*
Iteration procedures are more efficiently performed using a spreadsheet.

7 *Newton–Raphson iteration method*
If $x = x_0$ is an approximate solution to the equation $f(x) = 0$, a better approximation $x = x_1$ is given by

$$x_1 = x_0 - \frac{f(x_0)}{f'(x_0)}, \text{ and in general } x_{n+1} = x_n - \frac{f(x_n)}{f'(x_n)}$$

8 *Modified Newton–Raphson iteration method*
If, in the Newton–Raphson procedure $f'(x_0)$ is sufficiently small enough to cause the value of x_1 to be a worse approximation to the solution than x_0, then x_1 is obtained from the relationship

$$x_1 = x_0 \pm \sqrt{\frac{-2f(x_0)}{f''(x_0)}}$$

Subsequent iterations then use $x_{n+1} = x_n - \frac{f(x_n)}{f'(x_n)}$.

9 *Interpolation*
Linear
Graphical

10 *Gregory–Newton interpolation formulas using forward finite differences*

$$f_p = f_0 + p\Delta f_0 + \frac{p(p-1)}{2!}\Delta^2 f_0 + \frac{p(p-1)(p-2)}{3!}\Delta^3 f_0 + \cdots$$

11 *Gauss interpolation formulas using central finite differences*
Gauss forward formula

$$f_p = f_0 + p\delta f_{0+\frac{1}{2}} + \frac{p(p-1)}{2!}\delta^2 f_0 + \frac{(p+1)p(p-1)}{3!}\delta^3 f_{0+\frac{1}{2}}$$
$$+ \frac{(p+1)p(p-1)(p-2)}{4!}\delta^4 f_0 + \cdots$$

Gauss backward formula

$$f_p = f_0 + p\delta f_{0-\frac{1}{2}} + \frac{(p+1)p}{2!}\delta^2 f_0 + \frac{(p+1)p(p-1)}{3!}\delta^3 f_{0-\frac{1}{2}}$$
$$+ \frac{(p+2)(p+1)p(p-1)}{4!}\delta^4 f_0 + \cdots$$

▶

12 *Gregory–Newton interpolation formula using backward finite differences*

$$f_p = f_0 + p\Delta f_{-1} + \frac{p(p+1)}{2!}\Delta^2 f_{-2} + \frac{p(p+1)(p+2)}{3!}\Delta^3 f_{-3} + \cdots$$

13 *Lagrange interpolation*

If the straight line $p(x) = a_0 + a_1 x$ passes through the two points $(x_0, f(x_0))$ and $(x_1, f(x_1))$, where a_0 and a_1 are constants, then the interpolation polynomial (straight line) for this line can be written as

$$p(x) = \frac{x - x_1}{x_0 - x_1}f(x_0) + \frac{x - x_0}{x_1 - x_0}f(x_1)$$

The quadratic interpolating polynomial that passes through the three points $(x_0, f(x_0))$, $(x_1, f(x_1))$ and $(x_2, f(x_2))$ can be written as

$$p(x) = \frac{(x - x_1)(x - x_2)}{(x_0 - x_1)(x_0 - x_2)}f(x_0) + \frac{(x - x_0)(x - x_2)}{(x_1 - x_0)(x_1 - x_2)}f(x_1)$$
$$+ \frac{(x - x_0)(x - x_1)}{(x_2 - x_0)(x_2 - x_1)}f(x_2)$$

The cubic interpolating polynomial that passes through the four data points $(x_0, f(x_0))$, $(x_1, f(x_1))$, $(x_2, f(x_2))$ and $(x_3, f(x_3))$ can be written as

$$p(x) = \frac{(x - x_1)(x - x_2)(x - x_3)}{(x_0 - x_1)(x_0 - x_2)(x_0 - x_3)}f(x_0)$$
$$+ \frac{(x - x_0)(x - x_2)(x - x_3)}{(x_1 - x_0)(x_1 - x_2)(x_1 - x_3)}f(x_1)$$
$$+ \frac{(x - x_0)(x - x_1)(x - x_3)}{(x_2 - x_0)(x_2 - x_1)(x_2 - x_3)}f(x_2)$$
$$+ \frac{(x - x_0)(x - x_1)(x - x_2)}{(x_3 - x_0)(x_3 - x_1)(x_3 - x_2)}f(x_3)$$

The interpolating polynomial that passes through $n + 1$ data points is

$$p(x) = \frac{(x - x_1)(x - x_2)(\cdots)(x - x_n)}{(x_0 - x_1)(x_0 - x_2)(\cdots)(x_0 - x_n)}f(x_0)$$
$$+ \frac{(x - x_0)(x - x_2)(\cdots)(x - x_n)}{(x_1 - x_0)(x_1 - x_2)(\cdots)(x_1 - x_n)}f(x_1) + \cdots$$
$$\cdots + \frac{(x - x_0)(x - x_1)(\cdots)(x - x_{n-1})}{(x_n - x_0)(x_n - x_1)(\cdots)(x_n - x_{n-1})}f(x_n)$$

Can you?

Checklist 1 85

Check this list before and after you try the end of Programme test.

On a scale of 1 to 5 how confident are you that you can: Frames

- Appreciate the Fundamental Theorem of Algebra?
  Yes ☐ ☐ ☐ ☐ ☐ *No* [1] to [3]

- Find the two roots of a quadratic equation and recognise that for polynomial equations with real coefficients complex roots exist in complex conjugate pairs?
  Yes ☐ ☐ ☐ ☐ ☐ *No* [4] to [6]

- Use the relationships between the coefficients and the roots of a polynomial equation to find the roots of the polynomial?
  Yes ☐ ☐ ☐ ☐ ☐ *No* [7] to [17]

- Transform a cubic equation to reduced form?
  Yes ☐ ☐ ☐ ☐ ☐ *No* [18] to [20]

- Use Tartaglia's solution to find the real root of a cubic equation?
  Yes ☐ ☐ ☐ ☐ ☐ *No* [21] and [22]

- Find the solution of the equation $f(x) = 0$ by the method of bisection?
  Yes ☐ ☐ ☐ ☐ ☐ *No* [23] to [26]

- Solve equations involving a single real variable by iteration and use a spreadsheet for efficiency?
  Yes ☐ ☐ ☐ ☐ ☐ *No* [27] to [33]

- Solve equations using the Newton–Raphson iterative method? [34] to [52]
  Yes ☐ ☐ ☐ ☐ ☐ *No*

- Use the modified Newton–Raphson method to find the first approximation when the derivative is small?
  Yes ☐ ☐ ☐ ☐ ☐ *No* [53] to [58]

- Understand the meaning of interpolation and use simple linear and graphical interpolation?
  Yes ☐ ☐ ☐ ☐ ☐ *No* [59] to [61]

- Use the Gregory–Newton interpolation formula using forward and backward differences for equally spaced domain points? [62] to [73]
  Yes ☐ ☐ ☐ ☐ ☐ *No*

- Use the Gauss interpolation formulas using central differences for equally spaced domain points?
  Yes ☐ ☐ ☐ ☐ ☐ *No* [74] to [77]

- Use Lagrange interpolation when the domain points are not equally spaced?
  Yes ☐ ☐ ☐ ☐ ☐ *No* [78] to [83]

Test exercise 1

86

1 Given that $x = -1 + j\sqrt{3}$ is one root of a quadratic equation with real coefficients, find the other root and hence the quadratic equation.

2 Solve the cubic equation $2x^3 - 7x^2 - 42x + 72 = 0$.

3 Write the cubic $3x^3 + 5x^2 + 3x + 5$ in reduced form and use Tartaglia's method to find the real root.

4 Use the method of bisection to find a solution to $x^3 - 5 = 0$ correct to 4 significant figures.

5 Use the Newton–Raphson method to find a positive solution of the following equation, correct to 6 decimal places:

$$\cos 3x = x^2$$

6 Use the modified Newton–Raphson method to find the solution correct to 6 decimal places near to $x = 2$ of the equation

$$x^3 - 6x^2 + 13x - 9 = 0$$

7 Given the table of values

x	$f(x)$
1	0
2	19
3	70
4	171
5	340
6	595

estimate

(a) $f(2\cdot5)$ using the Gregory–Newton forward difference formula

(b) $f(3\cdot4)$ using the Gauss central difference formula

(c) $f(5\cdot6)$ using the Gregory–Newton backward difference formula.

8 Given the table of values

x	$f(x)$
1	4
2	−9
5	−108

use Lagrangian interpolation to estimate the value of $f(2\cdot2)$.

Further problems 1

87

1   Given that $x = \dfrac{-1 - j\sqrt{3}}{2}$ and $x = \dfrac{-1 + j}{\sqrt{2}}$ are two roots of a quartic equation with real coefficients, find the other two roots and hence the quartic equation.

2   Solve the equation $x^3 - 5x^2 - 8x + 12 = 0$, given that the sum of two of the roots is 7.

3   Find the values of the constants p and q such that the function $f(x) = 2x^3 + px^2 + qx + 6$ may be exactly divisible by $(x - 2)(x + 1)$.

4   If $f(x) = 4x^4 + px^3 - 23x^2 + qx + 11$ and when $f(x)$ is divided by $2x^2 + 7x + 3$ the remainder is $3x + 2$, determine the values of p and q.

5   If one root of the equation $x^3 - 2x^2 - 9x + 18 = 0$ is the negative of another, determine the three roots.

6   Solve the equation $x^3 - 7x^2 - 21x + 27 = 0$, given that the roots form a geometric sequence.

7   Form the equation whose roots are those of the equation $x^3 + x^2 + 9x + 9 = 0$ each increased by 2.

8   Form the equation whose roots exceed by 3 the roots of the equation
$$x^3 - 4x^2 + x + 6 = 0.$$

9   If the equation $4x^3 - 4x^2 - 5x + 3 = 0$ is known to have two roots whose sum is 2, solve the equation.

10   Solve the equation $x^3 - 10x^2 + 8x + 64 = 0$, given that the product of two of the roots is the negative of the third.

11   Form the equation whose roots exceed by 2 those of the equation
$$2x^3 - 3x^2 - 11x + 6 = 0.$$

12   If α, β, γ are the roots of the equation $x^3 + px^2 + qx + r = 0$, prove that
$$\alpha^2 + \beta^2 + \gamma^2 = p^2 - 2q.$$

13   Using Tartaglia's solution, find the real root of the equation $2x^3 + 4x - 5 = 0$ giving the result to 4 significant figures.

14   Solve the equation $x^3 - 6x - 4 = 0$.

15   Rewrite the equation $x^3 + 6x^2 + 9x + 4 = 0$ in reduced form and hence determine the three roots.

16   Show that the equation $x^3 + 3x^2 - 4x - 6 = 0$ has a root between $x = 1$ and $x = 2$, and use the Newton–Raphson iterative method to evaluate this root to 4 significant figures.

17   Find the real root of the equations:
(a) $x^3 + 4x + 3 = 0$ (b) $5x^3 + 2x - 1 = 0$.

▶

18 Solve the following equations:
   (a) $x^3 - 5x + 1 = 0$ (b) $x^3 + 2x - 3 = 0$
   (c) $x^3 - 4x + 1 = 0$.

19 Express the following in reduced form and determine the roots:
   (a) $x^3 + 6x^2 + 9x + 5 = 0$
   (b) $8x^3 + 20x^2 + 6x - 9 = 0$
   (c) $4x^3 - 9x^2 + 42x - 10 = 0$.

20 Use the Newton–Raphson iterative method to solve the following.
   (a) Show that a root of the equation $x^3 + 3x^2 + 5x + 9 = 0$ occurs between $x = -2$ and $x = -3$. Evaluate the root to four significant figures.
   (b) Show graphically that the equation $e^{2x} = 25x - 10$ has two real roots and find the larger root correct to four significant figures.
   (c) Verify that the equation $x - \cos x = 0$ has a root near to $x = 0.8$ and determine the root correct to three significant figures.
   (d) Obtain graphically an approximate root of the equation $2 \ln x = 3 - x$. Evaluate the root correct to four significant figures.
   (e) Verify that the equation $x^4 + 5x - 20 = 0$ has a root at approximately $x = 1.8$. Determine the root correct to five significant figures.
   (f) Show that the equation $x + 3 \sin x = 2$ has a root between $x = 0.4$ and $x = 0.6$. Evaluate the root correct to five significant figures.
   (g) The equation $2 \cos x = e^x - 1$ has a real root between $x = 0.8$ and $x = 0.9$. Evaluate the root correct to four significant figures.
   (h) The equation $20x^3 - 22x^2 + 5x - 1 = 0$ has a root at approximately $x = 0.6$. Determine the value of the root correct to four significant figures.

21 A polynomial function is defined by the following set of function values

x	2	4	6	8	10
$y = f(x)$	$-7 \cdot 00$	$9 \cdot 00$	$97 \cdot 0$	305	681

   Find
   (a) $f(4 \cdot 8)$ using the Gregory–Newton forward difference formula
   (b) $f(7 \cdot 2)$ using the Gauss central difference formula
   (c) $f(8 \cdot 5)$ using the Gregory–Newton backward difference formula.

22 For the function $f(x)$

x	4	5	6	7	8	9	10
$f(x)$	-10	12	56	128	234	380	572

   Find
   (a) $f(4 \cdot 5)$ and $f(6 \cdot 4)$ using the Gregory–Newton forward difference formula
   (b) $f(7 \cdot 1)$ and $f(8 \cdot 9)$ using the Gregory–Newton backward difference formula. ▶

23

x	2	4	6	8	10	12
$f(x)$	-9	35	231	675	1463	2691

For the function defined in the table above, evaluate (a) $f(2.6)$ and (b) $f(7.2)$.

24 A function $f(x)$ is defined by the following table

x	-4	-2	0	2	4	6	8
$f(x)$	277	51	1	-17	-147	-533	-1319

Find

(a) $f(-3)$ and $f(1.6)$ using the Gregory–Newton forward difference formula

(b) $f(0.2)$ and $f(3.1)$ using the Gauss central difference formula

(c) $f(4.4)$ and $f(7)$ using the Gregory–Newton backward difference formula.

25 Given the table of values

x	$f(x)$
-1	-2.71828
3	-0.04979
5	-0.00674

use Lagrangian interpolation to find the value of $f(3.4)$.

26 Given the table of values

x	$f(x)$
6	0.801153
7.2	-0.82236
9	-0.73922
13	0.994808

use Lagrangian interpolation to find the value of $f(8)$.

27 Given the table of values

x	$f(x)$
-2	-2.63906
0	-2.48491
5	-1.94591
6	-1.79176

use Lagrangian interpolation to find the values of

(a) $f(-0.8)$ (b) $f(0.8)$ (c) $f(5.5)$.

Laplace transforms 1

Learning outcomes

When you have completed this Programme you will be able to:

- Obtain the Laplace transforms of simple standard expressions
- Use the first shift theorem to find the Laplace transform of a simple expression multiplied by an exponential
- Find the Laplace transform of a simple expression multiplied or divided by a variable
- Use partial fractions to find the inverse Laplace transform
- Use the 'cover up' rule
- Use the Laplace transforms of derivatives to solve differential equations
- Use the Laplace transform to solve simultaneous differential equations

Prerequisite: Engineering Mathematics (Sixth Edition)
Programme 26   Introduction to Laplace transforms

Introduction

The solution of a linear, ordinary differential equation with constant coefficients such as the second-order equation **1**

$$af''(t) + bf'(t) + cf(t) = g(t)$$

can be solved by first obtaining the general form for the expression $f(t)$. This general form will contain a number of integration constants whose values can be found by applying the appropriate boundary conditions (see *Engineering Mathematics, Sixth Edition*, Programme 25). A more systematic way of solving such equations is to use the Laplace transform which converts the differential equation into an algebraic equation and has the added advantage of incorporating the boundary conditions from the beginning. Furthermore, in situations where $f(t)$ represents a function with discontinuities, the Laplace transform method can succeed where other methods fail.

Laplace transform techniques also provide powerful tools in numerous fields of technology such as Control Theory where a knowledge of the system transfer function is essential and where the Laplace transform comes into its own. Let us see what it is all about. (For a more detailed introduction see *Engineering Mathematics, Sixth Edition*, Programme 26.)

Laplace transforms

The Laplace transform of an expression $f(t)$ is denoted by $L\{f(t)\}$ and is defined as the semi-infinite integral

$$L\{f(t)\} = \int_{t=0}^{\infty} f(t)e^{-st}\, dt \tag{1}$$

The parameter s is assumed to be positive and large enough to ensure that the integral converges. In more advanced applications s may be complex and in such cases the real part of s must be positive and large enough to ensure convergence.

In determining the transform of an expression, you will appreciate that the limits of the integral are substituted for t, so that the result will be an expression in s. Therefore

$$L\{f(t)\} = \int_{t=0}^{\infty} f(t)e^{-st}\, dt = F(s)$$

Make a note of this general definition: then we can apply it

2

So we have $L\{f(t)\} = \int_0^\infty f(t)e^{-st}\mathrm{d}t = F(s)$

Example 1

To find the Laplace transform of $f(t) = a$ (constant).

$$L\{a\} = \int_0^\infty ae^{-st}\mathrm{d}t = a\left[\frac{e^{-st}}{-s}\right]_0^\infty = -\frac{a}{s}\left[e^{-st}\right]_0^\infty$$

$$= -\frac{a}{s}\{0-1\} = \frac{a}{s}$$

$$\therefore\ L\{a\} = \frac{a}{s}\qquad (s>0) \tag{2}$$

Example 2

To find the Laplace transform of $f(t) = e^{at}$ (a constant). As with all cases, we multiply $f(t)$ by e^{-st} and integrate between $t = 0$ and $t = \infty$.

$$\therefore\ L\{e^{at}\} = \int_0^\infty e^{at}e^{-st}\,\mathrm{d}t = \int_0^\infty e^{-(s-a)t}\,\mathrm{d}t$$

$$= \ldots\ldots\ldots$$

Finish it off.

3

$$\boxed{L\{e^{at}\} = \frac{1}{s-a}}$$

Because

$$L\{e^{at}\} = \int_0^\infty e^{at}e^{-st}\,\mathrm{d}t = \int_0^\infty e^{-(s-a)t}\,\mathrm{d}t = \left[\frac{e^{-(s-a)t}}{-(s-a)}\right]_0^\infty$$

$$= -\frac{1}{s-a}\{0-1\} = \frac{1}{s-a}$$

$$\therefore\ L\{e^{at}\} = \frac{1}{s-a}\qquad (s>a) \tag{3}$$

So we already have two standard transforms

$$L\{a\} = \frac{a}{s}\quad \text{and}\quad L\{e^{at}\} = \frac{1}{s-a}$$

$$\therefore\quad L\{4\} = \ldots\ldots\ldots\ldots;\qquad L\{e^{4t}\} = \ldots\ldots\ldots\ldots$$

$$L\{-5\} = \ldots\ldots\ldots\ldots;\qquad L\{e^{-2t}\} = \ldots\ldots\ldots\ldots$$

4

$$L\{4\} = \frac{4}{s}; \qquad L\{e^{4t}\} = \frac{1}{s-4}$$

$$L\{-5\} = -\frac{5}{s}; \qquad L\{e^{-2t}\} = \frac{1}{s+2}$$

Note that, as we said earlier, the Laplace transform is always an expression in *s*.

Now for some more examples

Example 3

5

To find the Laplace transform of $f(t) = \sin at$. We could, of course, apply the definition and evaluate

$$L\{\sin at\} = \int_0^\infty \sin at \cdot e^{-st}\, dt$$

using integration by parts.

However, it is much shorter if we use the fact that

$$e^{j\theta} = \cos\theta + j\sin\theta$$

so that $\sin\theta$ is the imaginary part of $e^{j\theta}$, written $\mathscr{I}(e^{j\theta})$.

The function $\sin at$ can therefore be written $\mathscr{I}(e^{jat})$ so that

$$L\{\sin at\} = L\{\mathscr{I}(e^{jat})\} = \mathscr{I}\int_0^\infty e^{jat} e^{-st}\, dt = \mathscr{I}\int_0^\infty e^{-(s-ja)t}\, dt$$

$$= \mathscr{I}\left\{\left[\frac{e^{-(s-ja)t}}{-(s-ja)}\right]_0^\infty\right\} = \mathscr{I}\left\{-\frac{1}{(s-ja)}[0-1]\right\}$$

$$= \mathscr{I}\left\{\frac{1}{s-ja}\right\}$$

We can rationalise the denominator by multiplying top and bottom by

....

6

$$s + ja$$

$$\therefore\ L\{\sin at\} = \mathscr{I}\left\{\frac{s+ja}{s^2+a^2}\right\} = \frac{a}{s^2+a^2}$$

$$\therefore\ L\{\sin at\} = \frac{a}{s^2+a^2} \tag{4}$$

We can use the same method to determine $L\{\cos at\}$ since $\cos at$ is the real part of e^{jat}, written $\mathfrak{R}(e^{jat})$.

Then $L\{\cos at\} = \ldots\ldots\ldots\ldots$

7

$$L\{\cos at\} = \frac{s}{s^2 + a^2} \qquad (5)$$

Because

$$L\{\cos at\} = \Re\left\{\frac{s + ja}{s^2 + a^2}\right\} = \frac{s}{s^2 + a^2}$$

Recapping then: $L\{1\} = \ldots\ldots\ldots;$ $L\{e^{3t}\} = \ldots\ldots\ldots$

$L\{\sin 2t\} = \ldots\ldots\ldots;$ $L\{\cos 4t\} = \ldots\ldots\ldots$

8

$$L\{1\} = \frac{1}{s}; \qquad L\{e^{3t}\} = \frac{1}{s - 3}$$

$$L\{\sin 2t\} = \frac{2}{s^2 + 4}; \qquad L\{\cos 4t\} = \frac{s}{s^2 + 16}$$

Example 4

To find the transform of $f(t) = t^n$ where n is a positive integer.

By the definition $L\{t^n\} = \displaystyle\int_0^\infty t^n e^{-st} \, dt.$

Integrating by parts

$$L\{t^n\} = \left[t^n \left(\frac{e^{-st}}{-s}\right)\right]_0^\infty + \frac{n}{s}\int_0^\infty e^{-st} t^{n-1} \, dt$$

$$= -\frac{1}{s}\left[t^n e^{-st}\right]_0^\infty + \frac{n}{s}\int_0^\infty t^{n-1} e^{-st} \, dt$$

We said earlier that in a product such as $t^n e^{-st}$ the numerical value of s is large enough to make the product converge to zero as $t \to \infty$

$$\therefore \left[t^n e^{-st}\right]_0^\infty = 0 - 0 = 0$$

$$\therefore L\{t^n\} = \frac{n}{s}\int_0^\infty t^{n-1} e^{-st} \, dt \qquad (6)$$

You will notice that $\displaystyle\int_0^\infty t^{n-1} e^{-st} \, dt$ is identical to $\displaystyle\int_0^\infty t^n e^{-st} \, dt$ except that n is replaced by $(n - 1)$.

$$\therefore \text{ If } I_n = \int_0^\infty t^n e^{-st} \, dt, \text{ then } I_{n-1} = \int_0^\infty t^{n-1} e^{-st} \, dt$$

and the result (6) becomes $I_n = \dfrac{n}{s}.I_{n-1}$ $\qquad (7)$

This is a reduction formula, and if we now replace n by $(n - 1)$ we get

$$I_{n-1} = \ldots\ldots\ldots$$

$$\boxed{I_{n-1} = \frac{n-1}{s}.I_{n-2}}$$

9

If we replace n by $(n-1)$ again in this last result, we have

$$I_{n-2} = \frac{n-2}{s}.I_{n-3}$$

So $I_n = \displaystyle\int_0^\infty t^n e^{-st}\,\mathrm{d}t = \frac{n}{s}.I_{n-1}$

$$= \frac{n}{s}.\frac{n-1}{s}.I_{n-2}$$

$$= \frac{n}{s}.\frac{n-1}{s}.\frac{n-2}{s}.I_{n-3} \quad \text{etc.}$$

$$= \ldots\ldots\ldots \text{ (next line)}$$

$$\boxed{I_n = \frac{n}{s}.\frac{n-1}{s}.\frac{n-2}{s}.\frac{n-3}{s}.I_{n-4}}$$

10

So finally, we have

$$I_n = \frac{n}{s}.\frac{n-1}{s}.\frac{n-2}{s}.\frac{n-3}{s}\ldots\frac{n-(n-1)}{s}.I_0$$

But $\quad I_0 = L\{t^0\} = L\{1\} = \dfrac{1}{s}$

$$\therefore\ I_n = \frac{n(n-1)(n-2)(n-3)\ldots(3)(2)(1)}{s^{n+1}} = \frac{n!}{s^{n+1}}$$

$$\therefore\ L\{t^n\} = \frac{n!}{s^{n+1}} \tag{8}$$

$$\therefore\ L\{t\} = \frac{1}{s^2};\quad L\{t^2\} = \frac{2}{s^3};\quad L\{t^3\} = \frac{6}{s^4}$$

and with $n = 0$, since $0! = 1$, the general result includes $L\{1\} = \dfrac{1}{s}$ which we have already established.

Example 5

Laplace transforms of $f(t) = \sinh at$ and $f(t) = \cosh at$.

Starting from the exponential definitions of $\sinh at$ and $\cosh at$, i.e.

$$\sinh at = \tfrac{1}{2}(e^{at} - e^{-at}) \quad \text{and} \quad \cosh at = \tfrac{1}{2}(e^{at} + e^{-at})$$

we proceed as follows, recalling that the Laplace transform is a linear transformation so that:

$$L\{af(t) + bg(t)\} = aL\{f(t)\} + bL\{g(t)\} \quad \text{where } a \text{ and } b \text{ are constants}$$

[Refer: *Engineering Mathematics, Sixth Edition*, Programme 26, Frame 11.]

(a) $f(t) = \sinh t$. $L\{\sinh t\} = L\left\{\dfrac{1}{2}e^{at} - \dfrac{1}{2}e^{-at}\right\}$

$$= \dfrac{1}{2}L\{e^{at}\} - \dfrac{1}{2}L\{e^{-at}\}$$

$$= \dotsb$$

Complete it

11

$$\boxed{L\{\sinh t\} = \dfrac{a}{s^2 - a^2}}$$

Because

$$\dfrac{1}{2}L\{e^{at}\} - \dfrac{1}{2}L\{e^{-at}\} = \dfrac{1}{2}\dfrac{1}{s-a} - \dfrac{1}{2}\dfrac{1}{s+a}$$

$$= \dfrac{1}{2}\left(\dfrac{(s+a) - (s-a)}{s^2 - a^2}\right)$$

$$= \dfrac{a}{s^2 - a^2} \tag{9}$$

(b) $f(t) = \cosh t$. Proceeding in the same way

$$L\{\cosh t\} = \dotsb$$

Next frame

12

$$\boxed{L\{\cosh t\} = \dfrac{s}{s^2 - a^2}}$$

$$L\{\cosh t\} = L\left\{\dfrac{1}{2}e^{at} + \dfrac{1}{2}e^{-at}\right\}$$

$$= \dfrac{1}{2}L\{e^{at}\} + \dfrac{1}{2}L\{e^{-at}\}$$

$$= \dfrac{1}{2}\dfrac{1}{s-a} + \dfrac{1}{2}\dfrac{1}{s+a}$$

$$= \dfrac{1}{2}\left(\dfrac{(s+a) + (s-a)}{s^2 - a^2}\right)$$

$$= \dfrac{s}{s^2 - a^2} \tag{10}$$

So we have accumulated several standard results:

$$L\{a\} = \dfrac{a}{s}; \qquad L\{e^{at}\} = \dfrac{1}{s-a}; \qquad L\{t^n\} = \dfrac{n!}{s^{n+1}}$$

$$L\{\sin at\} = \dfrac{a}{s^2 + a^2}; \qquad L\{\cos at\} = \dfrac{s}{s^2 + a^2}$$

$$L\{\sinh at\} = \dfrac{a}{s^2 - a^2}; \qquad L\{\cosh at\} = \dfrac{s}{s^2 - a^2}$$

Make a note of this list if you have not already done so: it forms the basis of much that is to follow.

The Laplace transform is a linear transform, by which is meant that:

13

(1) *The transform of a sum (or difference) of expressions is the sum (or difference) of the individual transforms. That is*

$$L\{f(t) \pm g(t)\} = L\{f(t)\} \pm L\{g(t)\}$$

(2) *The transform of an expression that is multiplied by a constant is the constant multiplied by the transform of the expression. That is*

$$L\{kf(t)\} = kL\{f(t)\}$$

Note: Two transforms must **not** be multiplied together to form the transform of a product of expressions – we shall see later that the product of two transforms is the transform of the *convolution* of two expressions.

Example 6

(a) $L\{2e^{-t} + t\} = L\{2e^{-t}\} + L\{t\}$
$$= 2L\{e^{-t}\} + L\{t\}$$
$$= \frac{2}{s+1} + \frac{1}{s^2} = \frac{2s^2 + s + 1}{s^2(s+1)}$$

(b) $L\{2\sin 3t + \cos 3t\} = 2L\{\sin 3t\} + L\{\cos 3t\}$
$$= 2.\frac{3}{s^2+9} + \frac{s}{s^2+9} = \frac{s+6}{s^2+9}$$

(c) $L\{4e^{2t} + 3\cosh 4t\} = 4L\{e^{2t}\} + 3L\{\cosh 4t\}$
$$= 4.\frac{1}{s-2} + 3.\frac{s}{s^2-16} = \frac{4}{s-2} + \frac{3s}{s^2-16}$$
$$= \frac{7s^2 - 6s - 64}{(s-2)(s^2-16)}$$

So 1. $L\{2\sin 3t + 4\sinh 3t\} = \ldots\ldots\ldots$
    2. $L\{5e^{4t} + \cosh 2t\} = \ldots\ldots\ldots$
    3. $L\{t^3 + 2t^2 - 4t + 1\} = \ldots\ldots\ldots$

14

1. $\dfrac{18(s^2+3)}{s^4-81}$;	2. $\dfrac{6s^2 - 4s - 20}{(s-4)(s^2-4)}$;	3. $\dfrac{1}{s^4}\{s^3 - 4s^2 + 4s + 6\}$

The working is straightforward.

1. $L\{2\sin 3t + 4\sinh 3t\} = 2.\dfrac{3}{s^2+9} + 4.\dfrac{3}{s^2-9}$
$$= \frac{6}{s^2+9} + \frac{12}{s^2-9} = \frac{18(s^2+3)}{s^4-81}$$

2. $L\{5e^{4t} + \cosh 2t\} = \dfrac{5}{s-4} + \dfrac{s}{s^2-4} = \dfrac{6s^2 - 4s - 20}{(s-4)(s^2-4)}$

▶

3. $L\{t^3 + 2t^2 - 4t + 1\} = \dfrac{3!}{s^4} + 2.\dfrac{2!}{s^3} - 4.\dfrac{1!}{s^2} + \dfrac{1}{s}$

$$= \dfrac{1}{s^4}\{s^3 - 4s^2 + 4s + 6\}$$

We have been building up a list of standard transforms of simple expressions. Before we leave this part of the work, there are three important and useful theorems which enable us to deal with rather more complicated expressions.

---

15   Theorem 1   The first shift theorem

The first shift theorem states that if $L\{f(t)\} = F(s)$ then

$$L\{e^{-at}f(t)\} = F(s+a)$$

Because $L\{e^{-at}f(t)\} = \displaystyle\int_{t=0}^{\infty} e^{-at}f(t)e^{-st}\,dt = \int_{t=0}^{\infty} f(t)e^{-(s+a)t}\,dt = F(s+a)$

That is

$$L\{e^{-at}f(t)\} = F(s+a)$$

The transform $L\{e^{-at}f(t)\}$ is thus the same as $L\{f(t)\}$ with s everywhere in the result replaced by $(s+a)$.

For example $L\{\sin 2t\} = \dfrac{2}{s^2 + 4}$

then $L\{e^{-3t}\sin 2t\} = \dfrac{2}{(s+3)^2 + 4} = \dfrac{2}{s^2 + 6s + 13}$

Similarly, $L\{t^2\} = \dfrac{2}{s^3}$ $\therefore$ $L\{t^2 e^{4t}\} = \ldots\ldots\ldots\ldots$

16

$$\boxed{\dfrac{2}{(s-4)^3}}$$

Because $L\{t^2\} = \dfrac{2}{s^3}$. $\therefore$ $L\{t^2 e^{4t}\}$ is the same with s replaced by $(s-4)$.

$\therefore$ $L\{t^2 e^{4t}\} = \dfrac{2}{(s-4)^3}$

Here is a short exercise by way of practice.

Exercise

Determine the following.

1. $L\{e^{-2t}\cosh 3t\}$ 4. $L\{e^{2t}\cos t\}$
2. $L\{2e^{3t}\sin 3t\}$ 5. $L\{e^{3t}\sinh 2t\}$
3. $L\{4te^{-t}\}$ 6. $L\{t^3 e^{-4t}\}$

Complete all six and then check with the results in the next frame

Here they are.

17

1. $L\{\cosh 3t\} = \dfrac{s}{s^2 - 9}$

$\therefore L\{e^{-2t}\cosh 3t\} = \dfrac{s+2}{(s+2)^2 - 9}$

$= \dfrac{s+2}{s^2 + 4s - 5}$

2. $L\{\sin 3t\} = \dfrac{3}{s^2 + 9}$

$\therefore L\{2e^{3t}\sin 3t\} = \dfrac{6}{(s-3)^2 + 9}$

$= \dfrac{6}{s^2 - 6s + 18}$

3. $L\{4t\} = 4.\dfrac{1}{s^2}$

$\therefore L\{4te^{-t}\} = \dfrac{4}{(s+1)^2}$

4. $L\{\cos t\} = \dfrac{s}{s^2 + 1}$

$\therefore L\{e^{2t}\cos t\} = \dfrac{s-2}{(s-2)^2 + 1}$

$= \dfrac{s-2}{s^2 - 4s + 5}$

5. $L\{\sinh 2t\} = \dfrac{2}{s^2 - 4}$

$\therefore L\{e^{3t}\sinh 2t\} = \dfrac{2}{(s-3)^2 - 4}$

$= \dfrac{2}{s^2 - 6s + 5}$

6. $L\{t^3\} = \dfrac{3!}{s^4}$

$\therefore L\{t^3 e^{-4t}\} = \dfrac{6}{(s+4)^4}$

Now let us deal with the next theorem

Theorem 2   Multiplying by *t* and *tⁿ*

18

If $L\{f(t)\} = F(s)$ then $L\{tf(t)\} = -F'(s)$

Because $L\{tf(t)\} = \displaystyle\int_{t=0}^{\infty} tf(t)e^{-st}\,\mathrm{d}t = \int_{t=0}^{\infty} f(t)\left(-\dfrac{\mathrm{d}e^{-st}}{\mathrm{d}s}\right)\mathrm{d}t$

$= -\dfrac{\mathrm{d}}{\mathrm{d}s}\displaystyle\int_{t=0}^{\infty} f(t)e^{-st}\,\mathrm{d}t = -F'(s)$

That is

$L\{tf(t)\} = -F'(s)$

For example, $\qquad L\{\sin 2t\} = \dfrac{2}{s^2 + 4}$

$\therefore L\{t\sin 2t\} = -\dfrac{\mathrm{d}}{\mathrm{d}s}\left(\dfrac{2}{s^2 + 4}\right) = \dfrac{4s}{(s^2 + 4)^2}$

and similarly, $\qquad L\{t\cosh 3t\} = \ldots\ldots\ldots\ldots$

19

$$\frac{s^2 + 9}{(s^2 - 9)^2}$$

Because $L\{t \cosh 3t\} = -\dfrac{d}{ds}\left(\dfrac{s}{s^2 - 9}\right) = -\dfrac{(s^2 - 9) - s(2s)}{(s^2 - 9)^2} = \dfrac{s^2 + 9}{(s^2 - 9)^2}$

We could, if necessary, take this a stage further and find $L\{t^2 \cosh 3t\}$

$$L\{t^2 \cosh 3t\} = L\{t(t \cosh 3t)\} = -\frac{d}{ds}\left\{\frac{s^2 + 9}{(s^2 - 9)^2}\right\}$$

$$= \frac{2s(s^2 + 27)}{(s^2 - 9)^3}$$

Likewise, starting with $L\{\sin 4t\} = \dfrac{4}{s^2 + 16}$

$L\{t \sin 4t\} = \ldots\ldots\ldots$ and $L\{t^2 \sin 4t\} = \ldots\ldots\ldots$

20

$$\frac{8s}{(s^2 + 16)^2}; \quad \frac{8(3s^2 - 16)}{(s^2 + 16)^3}$$

applying $L\{tf(t)\} = -\dfrac{d}{ds}\{F(s)\}$ in each case.

Theorem 2 obviously extends the range of functions that we can deal with. So, in general, if $L\{f(t)\} = F(s)$, then

$$L\{t^n f(t)\} = (-1)^n \frac{d^n}{ds^n}\{F(s)\}$$

Make a note of this in your record book

Theorem 3   Dividing by *t*

21

If $L\{f(t)\} = F(s)$ then $L\left\{\dfrac{f(t)}{t}\right\} = \displaystyle\int_{\sigma=s}^{\infty} F(\sigma)\,d\sigma$

provided $\displaystyle Lim_{t \to 0}\left(\dfrac{f(t)}{t}\right)$ exists. To demonstrate this we start from the right-hand side of the result

$$\int_{\sigma=s}^{\infty} F(\sigma)\,d\sigma = \int_{\sigma=s}^{\infty}\left\{\int_{t=0}^{\infty} f(t)e^{-\sigma t}\,dt\right\}d\sigma$$

$$= \int_{t=0}^{\infty}\int_{\sigma=s}^{\infty} f(t)e^{-\sigma t}\,d\sigma\,dt$$

$$= \int_{t=0}^{\infty} f(t)\left\{\int_{\sigma=s}^{\infty} e^{-\sigma t}\,d\sigma\right\}dt$$

$$= \int_{t=0}^{\infty} f(t)\frac{e^{-st}}{t}\,dt$$

$$= L\left\{\frac{f(t)}{t}\right\}$$

Notice the dummy variable σ. The end result is an expression in s which comes from the lower limit of the integral so the variable of integration, which is absorbed during the process of integration, is changed to σ. Notice also that we interchange the order of integration.

This rule is somewhat restricted in use, since it is applicable only if $\displaystyle Lim_{t \to 0}\left(\dfrac{f(t)}{t}\right)$ exists. In indeterminate cases, we use L'Hôpital's rule to find out. Let's try a couple of examples.

Example 1

22

Determine $L\left\{\dfrac{\sin at}{t}\right\}$

First we test $\displaystyle Lim_{t \to 0}\left\{\dfrac{\sin at}{t}\right\} = \left\{\dfrac{0}{0}\right\}$ which gives the indeterminate form of $\dfrac{0}{0}$. So, by L'Hôpital's rule, we differentiate top and bottom separately and substitute $t = 0$ in the result to ascertain the limit of the new expression.

$\displaystyle Lim_{t \to 0}\left\{\dfrac{\sin at}{t}\right\} = Lim_{t \to 0}\left\{\dfrac{a\cos at}{1}\right\} = a$, that is, the limit exists and the theorem can therefore be applied.

So $L\{\sin at\} = \dfrac{a}{s^2 + a^2}$, therefore $L\left\{\dfrac{\sin at}{t}\right\} = \displaystyle\int_{s}^{\infty} \frac{a}{\sigma^2 + a^2}\,d\sigma$

$$= \left[\arctan\left(\frac{\sigma}{a}\right)\right]_{s}^{\infty}$$

$$= \frac{\pi}{2} - \arctan\left(\frac{s}{a}\right)$$

$$= \arctan\left(\frac{a}{s}\right)$$

▶

Notice that $\arctan\left(\dfrac{a}{s}\right) + \arctan\left(\dfrac{s}{a}\right) = \dfrac{\pi}{2}$, as can be seen from the figure

Example 2

Determine $L\left\{\dfrac{1 - \cos 2t}{t}\right\}$

First we test whether $\underset{t \to 0}{Lim}\left\{\dfrac{1 - \cos 2t}{t}\right\}$ exists. Result? . . . .

23

$$\boxed{\text{the limit exists}}$$

$$\underset{t \to 0}{Lim}\left\{\dfrac{1 - \cos 2t}{t}\right\} = \dfrac{1 - 1}{0} = \dfrac{0}{0} = ?$$

$$\underset{t \to 0}{Lim}\left\{\dfrac{1 - \cos 2t}{t}\right\} = \underset{t \to 0}{Lim}\left\{\dfrac{2 \sin 2t}{1}\right\} = \dfrac{0}{1} = 0 \quad \therefore \text{ limit exists.}$$

$$L\{1 - \cos 2t\} = \dfrac{1}{s} - \dfrac{s}{s^2 + 4}$$

Then, by Theorem 3

$$L\left\{\dfrac{1 - \cos 2t}{t}\right\} = \int_{\sigma = s}^{\infty} \left\{\dfrac{1}{\sigma} - \dfrac{\sigma}{\sigma^2 + 4}\right\} d\sigma$$

$$= \left[\ln \sigma - \dfrac{1}{2}\ln(\sigma^2 + 4)\right]_{\sigma = s}^{\infty} = \dfrac{1}{2}\left[\ln\left(\dfrac{\sigma^2}{\sigma^2 + 4}\right)\right]_{\sigma = s}^{\infty}$$

When $\sigma \to \infty$, $\ln\left(\dfrac{\sigma^2}{\sigma^2 + 4}\right) \to \ln 1 = 0$

Therefore, $L\left\{\dfrac{1 - \cos 2t}{t}\right\} = \ldots\ldots\ldots\ldots$

Complete it

24

$$\boxed{\ln \sqrt{\dfrac{s^2 + 4}{s^2}}}$$

Because

$$L\left\{\dfrac{1 - \cos 2t}{t}\right\} = -\dfrac{1}{2}\ln\left(\dfrac{s^2}{s^2 + 4}\right) = \ln\left(\dfrac{s^2}{s^2 + 4}\right)^{-1/2}$$

$$= \ln \sqrt{\dfrac{s^2 + 4}{s^2}}$$

Let us pause here for a while and take stock, for we have met a number of results important in the future work.

▶

1 *Standard transforms*

$f(t)$	$L\{f(t)\} = F(s)$
a	$\dfrac{a}{s}$
e^{at}	$\dfrac{1}{s-a}$
$\sin at$	$\dfrac{a}{s^2+a^2}$
$\cos at$	$\dfrac{s}{s^2+a^2}$
$\sinh at$	$\dfrac{a}{s^2-a^2}$
$\cosh at$	$\dfrac{s}{s^2-a^2}$
t^n	$\dfrac{n!}{s^{n+1}}$

(n a positive integer)

2 *Theorem 1* The first shift theorem

If $L\{f(t)\} = F(s)$, then $L\{e^{-at}f(t)\} = F(s+a)$

3 *Theorem 2* Multiplying by t

If $L\{f(t)\} = F(s)$, then $L\{tf(t)\} = -\dfrac{\mathrm{d}}{\mathrm{d}s}\{F(s)\}$

4 *Theorem 3* Dividing by t

If $L\{f(t)\} = F(s)$, then $L\left\{\dfrac{f(t)}{t}\right\} = \displaystyle\int_{\sigma=s}^{\infty} F(\sigma)\,\mathrm{d}\sigma$

provided $\underset{t\to0}{Lim}\left\{\dfrac{f(t)}{t}\right\}$ exists.

Now let us work through a short revision exercise, so move on

Exercise   25

Determine the Laplace transforms of the following expressions.

1	$\sin 3t$	6	$t\cosh 4t$
2	$\cos 2t$	7	$t^2 - 3t + 4$
3	e^{4t}	8	$\dfrac{e^{3t}-1}{t}$
4	$6t^2$	9	$e^{3t}\cos 4t$
5	$\sinh 3t$	10	$t^2\sin t$

Complete the whole set and then check results with the next frame

26

Here are the results.

1 $\dfrac{3}{s^2 + 9}$

6 $\dfrac{s^2 + 16}{(s^2 - 16)^2}$

2 $\dfrac{s}{s^2 + 4}$

7 $\dfrac{1}{s^3}(4s^2 - 3s + 2)$

3 $\dfrac{1}{s - 4}$

8 $\ln\left(\dfrac{s}{s - 3}\right)$

4 $\dfrac{12}{s^3}$

9 $\dfrac{s - 3}{s^2 - 6s + 25}$

5 $\dfrac{3}{s^2 - 9}$

10 $\dfrac{6s^2 - 2}{(s^2 + 1)^3}$

It is just a case of applying the standard tranforms and the three theorems.

Now on to the next piece of work

Inverse transforms

27

Here we have the reverse process, i.e. given a Laplace transform, we have to find the function of t to which it belongs.

For example, we know that $\dfrac{a}{s^2 + a^2}$ is the Laplace transform of $\sin at$, so we can now write $L^{-1}\left\{\dfrac{a}{s^2 + a^2}\right\} = \sin at$, the symbol L^{-1} indicating the inverse transform and **not** a reciprocal.

∴   (a) $L^{-1}\left\{\dfrac{1}{s - 2}\right\} = \ldots\ldots\ldots$;   (c) $L^{-1}\left\{\dfrac{4}{s}\right\} = \ldots\ldots\ldots$

(b) $L^{-1}\left\{\dfrac{s}{s^2 + 25}\right\} \ldots\ldots\ldots$;   (d) $L^{-1}\left\{\dfrac{12}{s^2 - 9}\right\} = \ldots\ldots\ldots$

28

> (a) $L^{-1}\left\{\dfrac{1}{s - 2}\right\} = e^{2t}$;   (c) $L^{-1}\left\{\dfrac{4}{s}\right\} = 4$
>
> (b) $L^{-1}\left\{\dfrac{s}{s^2 + 25}\right\} = \cos 5t$;   (d) $L^{-1}\left\{\dfrac{12}{s^2 - 9}\right\} = 4\sinh 3t$

Therefore, given a transform, we can write down the corresponding expression in t, provided we can recognise it from our table of transforms.

▶

But what about $L^{-1}\left\{\dfrac{3s+1}{s^2-s-6}\right\}$? This certainly did not appear in our list of standard transforms.

In considering $L^{-1}\left\{\dfrac{3s+1}{s^2-s-6}\right\}$, it happens that we can write $\dfrac{3s+1}{s^2-s-6}$ as the sum of two simpler functions $\dfrac{1}{s+2}+\dfrac{2}{s-3}$ which, of course, makes all the difference, since we can now proceed

$$L^{-1}\left\{\frac{3s+1}{s^2-s-6}\right\}=L^{-1}\left\{\frac{1}{s+2}+\frac{2}{s-3}\right\}$$

which we immediately recognise as ....

$$\boxed{e^{-2t}+2e^{3t}}$$

29

The two simpler expressions $\dfrac{1}{s+2}$ and $\dfrac{2}{s-3}$ are called the *partial fractions* of $\dfrac{3s+1}{s^2-s-6}$, and the ability to represent a complicated algebraic fraction in terms of its partial fractions is the key to much of this work. Let us take a closer look at the rules.

Rules of partial fractions

1. The numerator must be of lower degree than the denominator. This is usually the case in Laplace transforms. If it is not, then we first divide out.

2. Factorise the denominator into its prime factors. These determine the shapes of the partial fractions.

3. A linear factor $(s+a)$ gives a partial fraction $\dfrac{A}{s+a}$ where A is a constant to be determined.

4. A repeated factor $(s+a)^2$ gives $\dfrac{A}{(s+a)}+\dfrac{B}{(s+a)^2}$.

5. Similarly $(s+a)^3$ gives $\dfrac{A}{(s+a)}+\dfrac{B}{(s+a)^2}+\dfrac{C}{(s+a)^3}$.

6. A quadratic factor (s^2+ps+q) gives $\dfrac{Ps+Q}{s^2+ps+q}$.

7. Repeated quadratic factors $(s^2+ps+q)^2$ give

$$\frac{Ps+Q}{s^2+ps+q}+\frac{Rs+T}{(s^2+ps+q)^2}.$$

So $\dfrac{s-19}{(s+2)(s-5)}$ has partial fractions of the form ....

30

$$\frac{A}{s+2} + \frac{B}{s-5}$$

and $\dfrac{3s^2 - 4s + 11}{(s+3)(s-2)^2}$ has partial fractions of the form ....

Be careful of the repeated factor.

31

$$\frac{A}{s+3} + \frac{B}{(s-2)} + \frac{C}{(s-2)^2}$$

Let us work through the various steps with an example.

Example 1

To determine $L^{-1}\left\{\dfrac{5s+1}{s^2 - s - 12}\right\}$.

(a) First we check that the numerator is of lower degree than the denominator. In fact, this is so.

(b) Factorise the denominator $\dfrac{5s+1}{s^2 - s - 12} = \dfrac{5s+1}{(s-4)(s+3)}$.

(c) Then the partial fractions are of the form ....

32

$$\frac{A}{s-4} + \frac{B}{s+3}$$

We therefore have the identity

$$\frac{5s+1}{s^2 - s - 12} \equiv \frac{A}{s-4} + \frac{B}{s+3}$$

If we multiply through both sides by the denominator $s^2 - s - 12 \equiv (s-4)(s+3)$ we have

$$5s+1 \equiv A(s+3) + B(s-4)$$

This is also an identity and true for any value of s we care to substitute – our job is now to find the values of A and B.

We now substitute convenient values for s

(a) Let $(s-4) = 0$, i.e. $s = 4$ $\therefore$ $21 = A(7) + B(0)$ $\therefore$ $A = 3$

(b) Let $(s+3) = 0$, i.e. $s = -3$ and we get ....

$$\boxed{B = 2}$$

$$\therefore \quad \frac{5s + 1}{s^2 - s - 12} \equiv \frac{3}{s - 4} + \frac{2}{s + 3}$$

$$\therefore \quad L^{-1}\left\{\frac{5s + 1}{s^2 - s - 12}\right\} = \dots\dots\dots$$

$$\boxed{3e^{4t} + 2e^{-3t}}$$

Example 2

Determine $L^{-1}\left\{\dfrac{9s - 8}{s^2 - 2s}\right\}$.

Working as before, $f(t) = \dots\dots\dots$

$$\boxed{4 + 5e^{2t}}$$

Because

$$L\{f(t)\} = \frac{9s - 8}{s^2 - 2s}.$$

(a) Numerator of first degree; denonominator of second degree. Therefore rule satisfied.

(b) $\dfrac{9s - 8}{s(s - 2)} \equiv \dfrac{A}{s} + \dfrac{B}{s - 2}.$

(c) Multiply by $s(s - 2)$. $\therefore$ $9s - 8 = A(s - 2) + Bs.$

(d) Put $s = 0$. $-8 = A(-2) + B(0)$ $\therefore$ $A = 4.$

(e) Put $s - 2 = 0$, i.e. $s = 2$. $10 = A(0) + B(2)$ $\therefore$ $B = 5.$

$$\therefore \quad f(t) = L^{-1}\left\{\frac{4}{s} + \frac{5}{s - 2}\right\} = 4 + 5e^{2t}$$

Example 3

Express $F(s) = \dfrac{s^2 - 15s + 41}{(s + 2)(s - 3)^2}$ in partial fractions and hence determine its inverse transform.

$\dfrac{s^2 - 15s + 41}{(s + 2)(s - 3)^2}$ has partial fractions of the form $\dots\dots\dots$

36

$$\frac{A}{s+2} + \frac{B}{s-3} + \frac{C}{(s-3)^2}$$

Now we multiply throughout by $(s+2)(s-3)^2$ and get

$$s^2 - 15s + 41 \equiv A(s-3)^2 + B(s+2)(s-3) + C(s+2)$$

Putting $(s-3) = 0$ and then $(s+2) = 0$ we obtain ....

37

$$\boxed{A = 3 \text{ and } C = 1}$$

Now that we have run out of 'crafty' substitutions, we equate coefficients of the highest power of s on each side, i.e. the coefficients of s^2. This gives

....

38

$$\boxed{1 = A + B \quad \therefore \ 1 = 3 + B \quad \therefore \ B = -2}$$

So $\dfrac{s^2 - 15s + 41}{(s+2)(s-3)^2} = \dfrac{3}{s+2} - \dfrac{2}{s-3} + \dfrac{1}{(s-3)^2}$

Now $L^{-1}\left\{\dfrac{3}{s+2}\right\} = ...$ and $L^{-1}\left\{\dfrac{2}{s-3}\right\} = ...$

39

$$\boxed{3e^{-2t} \text{ and } 2e^{3t}}$$

But what about $L^{-1}\left\{\dfrac{1}{(s-3)^2}\right\}$?

We remember that $L^{-1}\left\{\dfrac{1}{s^2}\right\} = ...$

40

$$\boxed{t}$$

and that by Theorem 1, if $L\{f(t)\} = F(s)$ then $L\{e^{-at}f(t)\} = F(s+a)$.

$\therefore \ \dfrac{1}{(s-3)^2}$ is like $\dfrac{1}{s^2}$ with s replaced by $(s-3)$ i.e. $a = -3$.

$$\therefore \ L^{-1}\left\{\frac{1}{(s-3)^2}\right\} = te^{3t}$$

$$\therefore \ L^{-1}\left\{\frac{s^2 - 15s + 41}{(s+2)(s-3)^2}\right\} = 3e^{-2t} - 2e^{3t} + te^{3t}$$

Example 4

Determine $L^{-1}\left\{\dfrac{4s^2 - 5s + 6}{(s+1)(s^2+4)}\right\}$.

Notice that this time we have a quadratic factor in the denominator

$$\frac{4s^2 - 5s + 6}{(s+1)(s^2+4)} \equiv \frac{A}{s+1} + \frac{Bs+C}{s^2+4}$$

$$\therefore\ 4s^2 - 5s + 6 \equiv A(s^2+4) + (Bs+C)(s+1).$$

(a) Putting $(s+1) = 0$, i.e. $s = -1$, $15 = 5A$ $\therefore\ A = 3$

(b) Equate coefficients of highest power, i.e. s^2

     $4 = A + B$ $\therefore\ 4 = 3 + B$ $\therefore\ B = 1$

(c) We now equate the lowest power on each side, i.e. the constant term

     $6 = 4A + C$ $\therefore\ 6 = 12 + C$ $\therefore\ C = -6$

Now you can finish it off. $f(t) = \ldots\ldots\ldots\ldots$

$$\boxed{f(t) = 3e^{-t} + \cos 2t - 3\sin 2t}$$

41

Because

$$L\{f(t)\} = \frac{3}{s+1} + \frac{s}{s^2+4} - \frac{6}{s^2+4}$$

$$\therefore\ f(t) = 3e^{-t} + \cos 2t - 3\sin 2t$$

Example 5

Determine $L^{-1}\left\{\dfrac{s^2 - 9s - 7}{(s^2+2s+2)(s-1)}\right\}$

Again we have a quadratic factor in the denominator that does not have simple factors

$$\frac{s^2 - 9s - 7}{(s^2+2s+2)(s-1)} \equiv \frac{A}{s-1} + \frac{Bs+C}{s^2+2s+2}$$

$$\therefore\ s^2 - 9s - 7 \equiv A(s^2+2s+2) + (Bs+C)(s-1)$$

(a) Putting $s - 1 = 0$, that is $s = 1$, $-15 = 5A$ $\therefore\ A = -3$

(b) Equate coefficients of highest power, that is s^2

     $1 = A + B$ $\therefore\ 1 = -3 + B$ $\therefore\ B = 4$

(c) We now equate the lowest power on each side, that is the constant term

     $-7 = 2A - C$ $\therefore\ -7 = -6 - C$ $\therefore\ C = 1$

So that $L\{f(t)\} = \dfrac{4s+1}{s^2+2s+2} - \dfrac{3}{s-1}.$

▶

Here we have a denominator with no simple factors. We therefore complete the square and use the First Shift Theorem [Refer: Frames 15–17].

$$L\{f(t)\} = \frac{4s + 1}{s^2 + 2s + 2} - \frac{3}{s - 1}$$

$$= \frac{4s + 1}{(s + 1)^2 + 1} - \frac{3}{s - 1}$$

$$= \frac{4(s + 1) - 3}{(s + 1)^2 + 1} - \frac{3}{s - 1}$$

$$= \frac{4(s + 1)}{(s + 1)^2 + 1} - \frac{3}{(s + 1)^2 + 1} - \frac{3}{s - 1}$$

By the First Shift Theorem $f(t) = 4e^{-t} \cos t - 3e^{-t} \sin t - 3e^{t}$

Now you try one.

Given $L\{f(t)\} = \dfrac{2s^2 + s - 3}{(s^2 + 4s + 5)(s + 1)}$ then $f(t) = \ldots\ldots\ldots$

Next frame

42

$$\boxed{f(t) = 3e^{-2t} \cos t - 4e^{-2t} \sin t - e^{-t}}$$

Because

$$\frac{2s^2 + s - 3}{(s^2 + 4s + 5)(s + 1)} \equiv \frac{A}{s + 1} + \frac{Bs + C}{s^2 + 4s + 5}$$

$$\therefore \ 2s^2 + s - 3 \equiv A(s^2 + 4s + 5) + (Bs + C)(s + 1)$$

(a) Putting $s + 1 = 0$, that is $s = -1$, $-2 = 2A$ $\therefore A = -1$

(b) Equate coefficients of highest power, that is s^2

$$2 = A + B \quad \therefore \ 2 = -1 + B \quad B = 3$$

(c) We now equate the lowest power on each side, that is the constant term

$$-3 = 5A + C \quad \therefore \ -3 = -5 + C \quad \therefore \ C = 2$$

So that $L\{f(t)\} = \dfrac{3s + 2}{s^2 + 4s + 5} - \dfrac{1}{s + 1}$

$$= \frac{3s + 2}{(s + 2)^2 + 1} - \frac{1}{s + 1}$$

$$= \frac{3(s + 2) - 4}{(s + 2)^2 + 1} - \frac{1}{s + 1}$$

$$= \frac{3(s + 2)}{(s + 2)^2 + 1} - \frac{4}{(s + 2)^2 + 1} - \frac{1}{s + 1}$$

By the First Shift Theorem $f(t) = 3e^{-2t} \cos t - 4e^{-2t} \sin t - e^{-t}$

The 'cover up' rule

While we can always find A, B, C, etc., there are many cases where we can use the 'cover up' methods and write down the values of the constant coefficients almost on sight. However, this method only works when the denominator of the original fraction has non-repeated, linear factors. The following examples illustrate the method.

43

Example 1

We know that $F(s) = \dfrac{9s - 8}{s(s - 2)}$ has partial fractions of the form $\dfrac{A}{s} + \dfrac{B}{s - 2}$. By the 'cover up' rule, the constant A, that is the coefficient of $\dfrac{1}{s}$, is found by temporarily covering up the factor s in the denominator of $F(s)$ and finding the limiting value of what remains when s (the factor covered up) tends to zero.

Therefore $A = $ coefficient of $\dfrac{1}{s} = \underset{s \to 0}{Lim} \left\{ \dfrac{9s - 8}{s - 2} \right\} = 4$. That is $A = 4$.

Similarly, B, the coefficient of $\dfrac{1}{s - 2}$, is obtained by covering up the factor $(s - 2)$ in the denominator of $F(s)$ and finding the limiting value of what remains when $(s - 2) \to 0$, that is $s \to 2$.

Therefore $B = $ coefficient of $\dfrac{1}{s - 2} = \underset{s \to 2}{Lim} \left\{ \dfrac{9s - 8}{s} \right\} = 5$. That is $B = 5$. So that

$$\frac{9s - 8}{s(s - 2)} = \frac{4}{s} + \frac{5}{s - 2}$$

Another example

Example 2

44

$$F(s) = \frac{s + 17}{(s - 1)(s + 2)(s - 3)} \equiv \frac{A}{s - 1} + \frac{B}{s + 2} + \frac{C}{s - 3}.$$

A: cover up $(s - 1)$ in $F(s)$ and find

$$\underset{s \to 1}{Lim} \left\{ \frac{s + 17}{(s + 2)(s - 3)} \right\} = \frac{18}{-6} \quad \therefore A = -3$$

Similarly

    B: .... $\therefore B = $ ....

    C: .... $\therefore C = $ ....

45

$$B = \underset{s \to -2}{Lim} \left\{ \frac{s + 17}{(s - 1)(s - 3)} \right\} = \frac{15}{(-3)(-5)} = 1 \qquad \therefore \ B = 1$$

$$C = \underset{s \to 3}{Lim} \left\{ \frac{s + 17}{(s - 1)(s + 2)} \right\} = \frac{20}{(2)(5)} = 2 \qquad \therefore \ C = 2$$

$$\therefore \ F(s) = \frac{1}{s + 2} + \frac{2}{s - 3} - \frac{3}{s - 1}$$

So $f(t) = e^{-2t} + 2e^{3t} - 3e^t$

Every entry in our table of standard transforms gives rise to a corresponding entry in a similar table of inverse transforms. Let us tabulate such a list.

46 ## Table of inverse transforms

$F(s)$	$f(t)$	
$\dfrac{a}{s}$	a	
$\dfrac{1}{s + a}$	e^{-at}	
$\dfrac{n!}{s^{n+1}}$	t^n	(n a positive integer)
$\dfrac{1}{s^n}$	$\dfrac{t^{n-1}}{(n - 1)!}$	(n a positive integer)
$\dfrac{a}{s^2 + a^2}$	$\sin at$	
$\dfrac{s}{s^2 + a^2}$	$\cos at$	
$\dfrac{a}{s^2 - a^2}$	$\sinh at$	
$\dfrac{s}{s^2 - a^2}$	$\cosh at$	

Theorem 1

The first shift theorem can be stated as follows.

If $F(s)$ is the Laplace transform of $f(t)$ then $F(s + a)$ is the Laplace transform of $e^{-at}f(t)$.

Here is a short revision exercise.

▶

Exercise

1 Find the inverse transforms of

(a) $\dfrac{1}{2s-3}$; (b) $\dfrac{5}{(s-4)^3}$; (c) $\dfrac{3s+4}{s^2+9}$.

2 Express in partial fractions

(a) $\dfrac{22s+16}{(s+1)(s-2)(s+3)}$; (b) $\dfrac{s^2-11s+6}{(s+1)(s-2)^2}$.

3 Determine

(a) $L^{-1}\left\{\dfrac{4s^2-17s-24}{s(s+3)(s-4)}\right\}$; (b) $L^{-1}\left\{\dfrac{5s^2-4s-7}{(s-3)(s^2+4)}\right\}$;

(c) $L^{-1}\left\{\dfrac{4s^2-21s+30}{(s^2-6s+13)(s-4)}\right\}$.

---

1 (a) $\dfrac{1}{2}e^{3t/2}$; (b) $\dfrac{5}{2}t^2e^{4t}$; (c) $3\cos 3t+\dfrac{4}{3}\sin 3t$

2 (b) $\dfrac{1}{s+1}+\dfrac{4}{s-2}-\dfrac{5}{s+3}$; (b) $\dfrac{2}{s+1}-\dfrac{1}{s-2}-\dfrac{4}{(s-2)^2}$

3 (a) $2+3e^{-3t}-e^{4t}$; (b) $2e^{3t}+3\cos 2t+\dfrac{5}{2}\sin 2t$

(c) $2e^{4t}+2e^{3t}\cos 2t+\dfrac{5}{2}e^{3t}\sin 2t$

Solution of differential equations by Laplace transforms

To solve a differential equation by Laplace transforms, we go through four distinct stages **47**

(a) Rewrite the equation in terms of Laplace transforms.

(b) Insert the given initial conditions.

(c) Rearrange the equation algebraically to give the transform of the solution.

(d) Determine the inverse transform to obtain the particular solution.

We have spent some time finding the transforms of a variety of functions of t and the inverse transforms of functions of s, i.e. we have largely covered steps (a) and (d) of the above list. However, to write a differential equation in Laplace transforms, we must obtain the transforms of the first and second derivatives of $f(t)$, that is the transforms of $f'(t)$ and $f''(t)$.

▶

Transforms of derivatives

By definition $L\{f'(t)\} = \int_0^\infty e^{-st}f'(t)\,dt.$

Integrating by parts

$$L\{f'(t)\} = \left[e^{-st}f(t)\right]_0^\infty - \int_0^\infty f(t)\{-se^{-st}\}\,dt$$

When $t \to \infty$, $e^{-st}f(t) \to \dots\dots\dots\dots$

48

Because s is positive and large enough to ensure that e^{-st} decays faster than any possible growth of $f(t)$.

$\therefore\ L\{f'(t)\} = -f(0) + sL\{f(t)\}$

Replacing $f(t)$ by $f'(t)$ gives

$$L\{f''(t)\} = \dots\dots\dots\dots$$

49

$$\boxed{L\{f''(t)\} = s^2F(s) - sf(0) - f'(0)}$$

Because

$$L\{f'(t)\} = -f(0) + sL\{f(t)\}$$

so$\qquad\ L\{f''(t)\} = -f'(0) + sL\{f'(t)\}$

$$= -f'(0) + s(-f(0) + sL\{f(t)\})$$

Writing$\qquad L\{f(t)\} = F(s)$ as usual, we have

$$L\{f(t)\} = F(s)$$

$$L\{f'(t)\} = sF(s) - f(0)$$

$$L\{f''(t)\} = s^2F(s) - sf(0) - f'(0)$$

We can see a pattern emerging

$$L\{f'''(t)\} = \dots\dots\dots\dots$$

$$L\{f'''(t)\} = s^3 F(s) - s^2 f(0) - sf'0) - f''(0)$$

50

Alternative notation

We make the working neater by adopting the following notation.

Let $x = f(t)$ and at $t = 0$, we write

$x = x_0 \qquad$ i.e. $f(0) = x_0$

$\dfrac{\mathrm{d}x}{\mathrm{d}t} = x_1 \qquad$ i.e. $f'(0) = x_1$

$\dfrac{\mathrm{d}^2 x}{\mathrm{d}t^2} = x_2 \qquad$ i.e. $f''(0) = x_2$ etc.

$\therefore \quad \dfrac{\mathrm{d}^n x}{\mathrm{d}t^n} = x_n \qquad$ i.e. $f^n(0) = x_n$

Also we denote the Laplace transform of x by $\bar{x}$,

i.e. $\quad \bar{x} = L\{x\} = L\{f(t)\} = F(s)$.

So, using the 'dot' notation for derivatives, the previous results can be written ....

51

$$L\{x\} = \bar{x}$$
$$L\{\dot{x}\} = s\bar{x} - x_0$$
$$L\{\ddot{x}\} = s^2\bar{x} - sx_0 - x_1$$
$$L\{\dddot{x}\} = s^3\bar{x} - s^2 x_0 - sx_1 - x_2$$

In each case, the subscript indicates the order of the derivative, i.e. $x_n = $ the value of $\dfrac{\mathrm{d}^n x}{\mathrm{d}t^n}$ at $t = 0$.

Notice the pattern of the results.

$$L\{\dddot{x}\} = \ldots\ldots\ldots$$

52

$$L\{\ddddot{x}\} = s^4\bar{x} - s^3 x_0 - s^2 x_1 - sx_2 - x_3$$

Now, at long last, we can start solving differential equations.

▶

Solution of first-order differential equations

Example 1

Solve the equation $\dfrac{dx}{dt} - 2x = 4$ given that at $t = 0$, $x = 1$.

We go through the four stages.

(a) Rewrite the equation in Laplace transforms, using the last notation

$$L\{x\} = \bar{x}; \quad L\{\dot{x}\} = \ldots\ldots\ldots$$
$$L\{4\} = \ldots\ldots\ldots$$

53

$$\boxed{L\{\dot{x}\} = s\bar{x} - x_0; \quad L\{4\} = \frac{4}{s}}$$

Then the equation becomes $\quad (s\bar{x} - x_0) - 2\bar{x} = \dfrac{4}{s}$

(b) Insert the initial condition that at $t = 0$, $x = 1$, i.e. $x_0 = 1$

$$\therefore \ s\bar{x} - 1 - 2\bar{x} = \frac{4}{s}$$

(c) Now we rearrange this to give an expression for $\bar{x}$

$$\bar{x} = \ldots\ldots\ldots$$

54

$$\boxed{\bar{x} = \frac{s+4}{s(s-2)}}$$

(d) Finally, we take inverse transforms to obtain x.

$\dfrac{s+4}{s(s-2)}$ in partial fractions gives $\ldots\ldots\ldots$

55

$$\boxed{\frac{3}{s-2} - \frac{2}{s}}$$

Because

$$\frac{s+4}{s(s-2)} \equiv \frac{A}{s} + \frac{B}{s-2} \qquad\qquad \therefore \ s+4 = A(s-2) + Bs$$

(1) Put $(s-2) = 0$, i.e. $s = 2$ $\qquad\qquad \therefore \ 6 = B(2) \qquad \therefore \ B = 3$

(2) Put $s = 0$ $\qquad\qquad\qquad\qquad\qquad \therefore \ 4 = A(-2) \quad \therefore \ A = -2$

$$\therefore \ \bar{x} = \frac{s+4}{s(s-2)} = \frac{3}{s-2} - \frac{2}{s}$$

Therefore, taking inverse transforms

$$x = L^{-1}\left\{\frac{s+4}{s(s-2)}\right\} = L^{-1}\left\{\frac{3}{s-2} - \frac{2}{s}\right\} = \ldots\ldots\ldots$$

$$\boxed{x = 3e^{2t} - 2}$$ **56**

This solution should now be substituted back into the differential equation to verify that it is, indeed, correct.

Example 2

Solve the equation $\dfrac{dx}{dt} + 2x = 10e^{3t}$ given that at $t = 0$, $x = 6$.

(a) Convert the equations to Laplace transforms, i.e.

. . . .

$$\boxed{(s\bar{x} - x_0) + 2\bar{x} = \frac{10}{s - 3}}$$ **57**

(b) Insert the initial condition, $x_0 = 6$

$$s\bar{x} - 6 + 2\bar{x} = \frac{10}{s - 3}$$

(c) Rearrange to obtain $\bar{x} = $ . . . .

$$\boxed{\bar{x} = \frac{6s - 8}{(s + 2)(s - 3)}}$$ **58**

(d) Taking inverse transforms to obtain x

$$x = L^{-1}\left\{ \frac{6s - 8}{(s + 2)(s - 3)} \right\} = \dots\dots\dots$$

Complete the solution

$$\boxed{x = 4e^{-2t} + 2e^{3t}}$$ **59**

Because

$$\frac{6s - 8}{(s + 2)(s - 3)} \equiv \frac{A}{s + 2} + \frac{B}{s - 3}$$
$$\therefore \ 6s - 8 = A(s - 3) + B(s + 2)$$

(1) Put $(s - 3) = 0$, i.e. $s = 3$ $\quad \therefore \quad 10 = B(5) \quad \therefore B = 2$
(2) Put $(s + 2) = 0$, i.e. $s = -2.$ $\quad \therefore \quad -20 = A(-5) \quad \therefore A = 4$

$$\therefore \ \bar{x} = \frac{6s - 8}{(s + 2)(s - 3)} = \frac{4}{s + 2} + \frac{2}{s - 3}$$
$$\therefore \ x = L^{-1}\left\{ \frac{4}{s + 2} + \frac{2}{s - 3} \right\} = 4e^{-2t} + 2e^{3t}$$

▶

Example 3

Solve the equation $\dfrac{\mathrm{d}x}{\mathrm{d}t} - 4x = 2e^{2t} + e^{4t}$, given that at $t = 0$, $x = 0$.

Work this through the four steps in the same way as before and complete it on your own.

$$x = \ldots\ldots\ldots\ldots$$

60

$$\boxed{x = e^{4t} - e^{2t} + te^{4t}}$$

The working is quite standard.

$$\frac{\mathrm{d}x}{\mathrm{d}t} - 4x = 2e^{2t} + e^{4t}$$

(a) $(s\bar{x} - x_0) - 4\bar{x} = \dfrac{2}{s-2} + \dfrac{1}{s-4}$

(b) $x_0 = 0 \quad \therefore \ s\bar{x} - 4\bar{x} = \dfrac{2}{s-2} + \dfrac{1}{s-4}$

(c) $\therefore \ \bar{x} = \dfrac{2}{(s-2)(s-4)} + \dfrac{1}{(s-4)^2}$

(d) $\dfrac{2}{(s-2)(s-4)} \equiv \dfrac{A}{s-2} + \dfrac{B}{s-4} \quad \therefore \ 2 = A(s-4) + B(s-2)$

   Putting $(s-2) = 0$, i.e. $s = 2$ $\quad \therefore \ 2 = A(-2) \quad \therefore \ A = -1$

   Putting $(s-4) = 0$, i.e. $s = 4$ $\quad \therefore \ 2 = B(2) \quad\quad \therefore \ B = 1$

   $\therefore \ \bar{x} = \dfrac{1}{s-4} - \dfrac{1}{s-2} + \dfrac{1}{(s-4)^2}$

   $\therefore \ x = e^{4t} - e^{2t} + te^{4t}$

Now on to the next frame

---

61 **Solution of second-order differential equations**

The method is, in effect, the same as before, going through the same four distinct stages.

Example 1

Solve the equation $\dfrac{\mathrm{d}^2 x}{\mathrm{d}t^2} - 3\dfrac{\mathrm{d}x}{\mathrm{d}t} + 2x = 2e^{3t}$, given that at $t = 0$, $x = 5$ and $\dfrac{\mathrm{d}x}{\mathrm{d}t} = 7$.

(a) We rewrite the equation in terms of its transforms, remembering that

$$L\{x\} = \bar{x}$$
$$L\{\dot{x}\} = s\bar{x} - x_0$$
$$L\{\ddot{x}\} = s^2\bar{x} - sx_0 - x_1$$

The equation becomes $\ldots\ldots\ldots\ldots$

$$\left(s^2\bar{x} - sx_0 - x_1\right) - 3(s\bar{x} - x_0) + 2\bar{x} = \frac{2}{s-3}$$

62

(b) Insert the initial conditions. In this case $x_0 = 5$ and $x_1 = 7$

$$\therefore \ (s^2\bar{x} - 5s - 7) - 3(s\bar{x} - 5) + 2\bar{x} = \frac{2}{s-3}$$

(c) Rearrange to obtain $\bar{x} = \ldots\ldots\ldots$

$$\bar{x} = \frac{5s^2 - 23s + 26}{(s-1)(s-2)(s-3)}$$

63

Because

$$s^2\bar{x} - 5s - 7 - 3s\bar{x} + 15 + 2\bar{x} = \frac{2}{s-3}$$

$$\left(s^2 - 3s + 2\right)\bar{x} - 5s + 8 = \frac{2}{s-3}$$

$$(s-1)(s-2)\bar{x} = \frac{2}{s-3} + 5s - 8 = \frac{2 + 5s^2 - 23s + 24}{s-3}$$

$$\therefore \ \bar{x} = \frac{5s^2 - 23s + 26}{(s-1)(s-2)(s-3)}$$

(d) Now for partial fractions

$$\frac{5s^2 - 23s + 26}{(s-1)(s-2)(s-3)} = \frac{A}{s-1} + \frac{B}{s-2} + \frac{C}{s-3}$$

$$\therefore \ 5s^2 - 23s + 26 = A(s-2)(s-3) + B(s-1)(s-3) + C(s-1)(s-2)$$

So that $A = \ldots\ldots\ldots; \quad B = \ldots\ldots\ldots; \quad C = \ldots\ldots\ldots$

$$A = 4; \quad B = 0; \quad C = 1$$

64

$$\therefore \ \bar{x} = \frac{4}{s-1} + \frac{1}{s-3}$$

$$\therefore \ x = \ldots\ldots\ldots$$

$$x = 4e^t + e^{3t}$$

65

As you see, the Laplace transform method can be considerably shorter than the classical method which requires

(a) determination of the complementary function

(b) determination of a particular integral

(c) obtaining the general solution, before

(d) arriving at the particular solution by substitution of the initial conditions in the general solution.

▶

Here is another example.

Example 2

Solve $\dfrac{d^2x}{dt^2} - 4x = 24\cos 2t$ given that at $t = 0$, $x = 3$ and $\dfrac{dx}{dt} = 4$.

(a) In Laplace transforms . . . .

66

$$\boxed{(s^2\bar{x} - sx_0 - x_1) - 4\bar{x} = \frac{24s}{s^2 + 4}}$$

(b) Insert initial condition, i.e. $x_0 = 3$; $x_1 = 4$

$$s^2\bar{x} - 3s - 4 - 4\bar{x} = \frac{24s}{s^2 + 4}$$

$$\therefore \ (s^2 - 4)\bar{x} = 3s + 4 + \frac{24s}{s^2 + 4}$$

$$= \frac{3s^3 + 4s^2 + 36s + 16}{s^2 + 4}$$

(c) $\bar{x} = \dfrac{3s^3 + 4s^2 + 36s + 16}{(s^2 + 4)(s - 2)(s + 2)}$

Expressed in partial fractions, this becomes

. . . .

67

$$\boxed{\frac{3s^3 + 4s^2 + 36s + 16}{(s^2 + 4)(s - 2)(s + 2)} \equiv \frac{As + B}{s^2 + 4} + \frac{C}{s - 2} + \frac{D}{s + 2}}$$

$\therefore \ 3s^3 + 4s^2 + 36s + 16 \equiv (As + B)(s - 2)(s + 2) + C(s^2 + 4)(s + 2)$
$$+ D(s^2 + 4)(s - 2)$$

Putting $(s - 2) = 0$, i.e. $s = 2$, gives $\quad C = 4$

Putting $(s + 2) = 0$, i.e. $s = -2$, gives $\quad D = 2$

Equating coefficients of s^3 and also the constant terms gives $A = -3$ and $B = 0$.

$$\therefore \ \bar{x} = \frac{3s^3 + 4s^2 + 36s + 16}{(s^2 + 4)(s - 2)(s + 2)} = \frac{4}{s - 2} + \frac{2}{s + 2} - \frac{3s}{s^2 + 4}$$

$$\therefore \ x = \ldots \ldots \ldots \ldots$$

68

$$\boxed{x = 4e^{2t} + 2e^{-2t} - 3\cos 2t}$$

Now let us solve another equation, this time using the 'cover up' rule.

▶

Example 3

Solve $\ddot{x} + 5\dot{x} + 6x = 4t$, given that at $t = 0$, $x = 0$ and $\dot{x} = 0$.

As usual we begin $(s^2\bar{x} - sx_0 - x_1) + 5(s\bar{x} - x_0) + 6\bar{x} = \dfrac{4}{s^2}$

$$x_0 = 0;\ x_1 = 0 \quad \therefore\ (s^2 + 5s + 6)\bar{x} = \dfrac{4}{s^2}$$

$$\therefore\ \bar{x} = \dfrac{4}{s^2(s+2)(s+3)}$$

The s^2 in the denominator can be awkward, so we introduce a useful trick and detach one factor s outside the main expression, thus

$$\bar{x} = \dfrac{1}{s}\left\{\dfrac{4}{s(s+2)(s+3)}\right\} = \dfrac{1}{s}\left\{\dfrac{A}{s} + \dfrac{B}{s+2} + \dfrac{C}{s+3}\right\}$$

Applying the 'cover up' rule to the expressions within the brackets

$$\bar{x} = \dfrac{1}{s}\left\{\dfrac{4}{6}\cdot\dfrac{1}{s} - \dfrac{2}{(s+2)} + \dfrac{4}{3}\cdot\dfrac{1}{s+3}\right\}$$

Now we bring the external $\dfrac{1}{s}$ back into the fold

$$\bar{x} = \dfrac{2}{3}\cdot\dfrac{1}{s^2} - \dfrac{2}{s(s+2)} + \dfrac{4}{3}\cdot\dfrac{1}{s(s+3)}$$

and the second and third terms can be expressed in simple partial fractions so that

$$\bar{x} = \ldots\ldots\ldots$$

$$\boxed{\bar{x} = \dfrac{2}{3}\cdot\dfrac{1}{s^2} - \dfrac{1}{s} + \dfrac{1}{s+2} + \dfrac{4}{9}\cdot\dfrac{1}{s} - \dfrac{4}{9}\cdot\dfrac{1}{s+3}}$$

69

which can now be simplified into

$$\bar{x} = \dfrac{2}{3}\cdot\dfrac{1}{s^2} - \dfrac{5}{9}\cdot\dfrac{1}{s} + \dfrac{1}{s+2} - \dfrac{4}{9}\cdot\dfrac{1}{s+3}$$

$$\therefore\ x = \ldots\ldots\ldots$$

$$\boxed{x = \dfrac{2}{3}t - \dfrac{5}{9} + e^{-2t} - \dfrac{4}{9}e^{-3t}}$$

70

There are times when a quadratic coefficient of $\bar{x}$ cannot be expressed in simple linear factors. In that case, we merely complete the square converting the expression into $(s \pm k)^2 \pm a^2$. Let us see such an example.

▶

Example 4

Solve $\ddot{x} - 2\dot{x} + 10x = e^{2t}$, given that at $t = 0$, $x = 0$ and $\dot{x} = 1$.

We find the expression for $\bar{x}$ as before.

$$\bar{x} = \ldots\ldots\ldots$$

71

$$\bar{x} = \frac{s - 1}{(s - 2)(s^2 - 2s + 10)}$$

Because

$$(s^2\bar{x} - sx_0 - x_1) - 2(s\bar{x} - x_0) + 10\bar{x} = \frac{1}{s - 2}$$

$$x_0 = 0;\ x_1 = 1 \qquad \therefore\ s^2\bar{x} - 1 - 2s\bar{x} + 10\bar{x} = \frac{1}{s - 2}$$

$$\therefore\ (s^2 - 2s + 10)\bar{x} = 1 + \frac{1}{s - 2} = \frac{s - 1}{s - 2}$$

$$\therefore\ \bar{x} = \frac{s - 1}{(s - 2)(s^2 - 2s + 10)}$$

Expressing this in partial fractions

$$\bar{x} = \ldots\ldots\ldots \qquad \text{Evaluate the coefficients.}$$

72

$$\bar{x} = \frac{1}{10}\left\{ \frac{1}{s - 2} - \frac{s - 10}{s^2 - 2s + 10} \right\}$$

Because

$$\frac{s - 1}{(s - 2)(s^2 - 2s + 10)} \equiv \frac{A}{(s - 2)} + \frac{Bs + C}{s^2 - 2s + 10}$$

$$\therefore\ s - 1 = A(s^2 - 2s + 10) + (s - 2)(Bs + C)$$

Put $(s - 2) = 0$, i.e. $s = 2$ $1 = A(4 - 4 + 10)$ $\therefore\ A = \dfrac{1}{10}$

$[s^2]$ $0 = A + B$ $\therefore\ B = -\dfrac{1}{10}$

$[\text{CT}]$ $-1 = 10A - 2C$ $\therefore\ 2C = 2$ $\therefore\ C = 1$

$$\therefore\ \bar{x} = \frac{1}{10}\left\{ \frac{1}{s - 2} - \frac{s - 10}{s^2 - 2s + 10} \right\}$$

Now we have to find the inverse transforms to obtain x. The first term $\dfrac{1}{s - 2}$ is easy enough, but what of $\dfrac{s - 10}{s^2 - 2s + 10}$? The denominator will not factorise into simple linear factors; therefore we complete the square in the denominator and write it as

$$\frac{s - 10}{s^2 - 2s + 10} = \frac{s - 10}{(s - 1)^2 + 9}$$

▶

and then we improve this still further and write it in the form $\dfrac{(s-1)-9}{(s-1)^2+9}$. We are quite happy with this, for $\dfrac{s-1}{(s-1)^2+9}$ is merely $\dfrac{s}{s^2+9}$ with s replaced by $(s-1)$, which indicates an extra factor e^t in the final function of t (Theorem 1).

So $\quad \bar{x} = \dfrac{1}{10}\left\{\dfrac{1}{s-2} - \dfrac{s-1}{(s-1)^2+9} + \dfrac{9}{(s-1)^2+9}\right\}$

$\therefore\ x = \ldots\ldots\ldots\ldots$

$$x = \frac{1}{10}\left\{e^{2t} - e^t\cos 3t + 3e^t\sin 3t\right\}$$

73

Just try one more like this one

Example 5

74

Solve $\ddot{x} + \dot{x} + x = e^{-t}$ given that at $t = 0$, $x = 0$ and $\dot{x} = 1$. We find the expression for $\bar{x}$ as before.

$$\bar{x} = \ldots\ldots\ldots\ldots$$

$$\bar{x} = \frac{s+2}{(s+1)(s^2+s+1)}$$

75

Because $\left(s^2\bar{x} - sx_0 - x_1\right) + \left(s\bar{x} - x_0\right) + \bar{x} = \dfrac{1}{s+1}$ where $x_0 = 0$ and $x_1 = 1$ so that

$$s^2\bar{x} - 1 + s\bar{x} + \bar{x} = \frac{1}{s+1}$$

therefore

$$\bar{x}\left(s^2 + s + 1\right) = 1 + \frac{1}{s+1} = \frac{s+2}{s+1}$$

giving

$$\bar{x} = \frac{s+2}{(s+1)(s^2+s+1)}$$

Expressing this in partial fractions

$$\bar{x} = \ldots\ldots\ldots\ldots$$

Evaluate the coefficients

76

$$\bar{x} = \frac{1}{s+1} - \frac{s-1}{s^2+s+1}$$

Because

$$\bar{x} = \frac{s+2}{(s+1)(s^2+s+1)} = \frac{A}{s+1} + \frac{Bs+C}{s^2+s+1}$$

so that

$$s+2 = A(s^2+s+1) + (Bs+C)(s+1)$$

Put $s+1 = 0$, that is $s = -1$ then

$$1 = A(1-1+1) \quad \text{so that } A = 1$$
$$[s^2] \quad 0 = A+B \qquad \text{so that } B = -1$$
$$[\text{CT}] \quad 2 = A+C \qquad \text{so that } C = 1$$

Therefore

$$\bar{x} = \frac{1}{s+1} - \frac{s-1}{s^2+s+1}$$

Completing the squares in the second term gives

$$\frac{s-1}{s^2+s+1} = \ldots\ldots\ldots\ldots$$

77

$$\frac{s-1}{s^2+s+1} = \frac{s+\frac{1}{2}}{\left(s+\frac{1}{2}\right)^2+\left(\frac{\sqrt{3}}{2}\right)^2} - \frac{\sqrt{3}\times\frac{\sqrt{3}}{2}}{\left(s+\frac{1}{2}\right)^2+\left(\frac{\sqrt{3}}{2}\right)^2}$$

Because

$$\frac{s-1}{s^2+s+1} = \frac{s-1}{\left(s+\frac{1}{2}\right)^2+\frac{3}{4}}$$

$$= \frac{s+\frac{1}{2}-\frac{3}{2}}{\left(s+\frac{1}{2}\right)^2+\left(\frac{\sqrt{3}}{2}\right)^2}$$

$$= \frac{s+\frac{1}{2}}{\left(s+\frac{1}{2}\right)^2+\left(\frac{\sqrt{3}}{2}\right)^2} - \frac{\sqrt{3}\times\frac{\sqrt{3}}{2}}{\left(s+\frac{1}{2}\right)^2+\left(\frac{\sqrt{3}}{2}\right)^2}$$

so that

$$\bar{x} = \ldots\ldots\ldots\ldots$$

$$\bar{x} = \frac{1}{s+1} - \frac{s+\frac{1}{2}}{\left(s+\frac{1}{2}\right)^2 + \left(\frac{\sqrt{3}}{2}\right)^2} + \frac{\sqrt{3} \times \frac{\sqrt{3}}{2}}{\left(s+\frac{1}{2}\right)^2 + \left(\frac{\sqrt{3}}{2}\right)^2}$$

and so $x = \ldots\ldots\ldots$

$$x = e^{-t} - e^{-t/2}\cos\frac{\sqrt{3}t}{2} + \sqrt{3}e^{-t/2}\sin\frac{\sqrt{3}t}{2}$$

Before we leave this topic, the same general approach can be employed for solving simultaneous differential equations.

Let us see an example in the next frame

Simultaneous differential equations

Example 1

Solve the pair of simultaneous equations

$$\dot{y} - x = e^t$$
$$\dot{x} + y = e^{-t}$$

given that at $t = 0$, $x = 0$ and $y = 0$.

(a) We first express both equations in Laplace transforms.

$$(s\bar{y} - y_0) - \bar{x} = \frac{1}{s-1}$$

$$(s\bar{x} - x_0) + \bar{y} = \frac{1}{s+1}$$

(b) Then we insert the initial conditions, $x_0 = 0$ and $y_0 = 0$.

$$\left. \begin{array}{l} \therefore \quad s\bar{y} - \bar{x} = \dfrac{1}{s-1} \\[2mm] s\bar{x} + \bar{y} = \dfrac{1}{s+1} \end{array} \right\} \tag{1}$$

(c) We now solve these for $\bar{x}$ and $\bar{y}$ by the normal algebraic method. Eliminating $\bar{y}$ we have

$$s\bar{y} - \bar{x} = \frac{1}{s-1}$$

$$s\bar{y} + s^2\bar{x} = \frac{s}{s+1}$$

$$\therefore \quad (s^2 + 1)\bar{x} = \frac{2}{s+1} - \frac{1}{s-1} = \frac{s^2 - 2s - 1}{(s+1)(s-1)}$$

$$\therefore \quad \bar{x} = \frac{s^2 - 2s - 1}{(s-1)(s+1)(s^2+1)}$$

Representing this in partial fractions gives $\ldots\ldots\ldots\ldots$

81

$$\bar{x} = -\frac{1}{2}\cdot\frac{1}{s-1} - \frac{1}{2}\cdot\frac{1}{s+1} + \frac{s}{s^2+1} + \frac{1}{s^2+1}$$

Because

$$\bar{x} = \frac{s^2 - 2s - 1}{(s-1)(s+1)(s^2+1)} \equiv \frac{A}{s-1} + \frac{B}{s+1} + \frac{Cs+D}{s^2+1}$$

$$\therefore\ s^2 - 2s - 1 = A(s+1)(s^2+1) + B(s-1)(s^2+1)$$
$$+ (s-1)(s+1)(Cs+D)$$

Putting $s = 1$ and $s = -1$ gives $A = -\frac{1}{2}$ and $B = -\frac{1}{2}$.

Comparing coefficients of s^3 and the constant terms gives $C = 1$ and $D = 1$.

$$\therefore\ \bar{x} = \frac{1}{2}\cdot\frac{1}{s-1} - \frac{1}{2}\cdot\frac{1}{s+1} + \frac{s+1}{s^2+1}$$

$$\therefore\ x = \ldots\ldots\ldots\ldots$$

82

$$x = -\tfrac{1}{2}e^t - \tfrac{1}{2}e^{-t} + \cos t + \sin t$$

We now revert to equations (1) and eliminate $\bar{x}$ to obtain $\bar{y}$ and hence y, in the same way. Do this on your own.

$$y = \ldots\ldots\ldots\ldots$$

83

$$y = \tfrac{1}{2}e^t + \tfrac{1}{2}e^{-t} - \cos t + \sin t$$

Here is the working.

$$\left.\begin{array}{r} s^2\bar{y} - s\bar{x} = \dfrac{s}{s-1} \\[2mm] \bar{y} + s\bar{x} = \dfrac{1}{s+1} \end{array}\right\}$$

$$\therefore\ (s^2+1)\bar{y} = \frac{s}{s-1} + \frac{1}{s+1} = \frac{s^2 + 2s - 1}{(s-1)(s+1)}$$

$$\therefore\ \bar{y} = \frac{s^2 + 2s - 1}{(s-1)(s+1)(s^2+1)} \equiv \frac{A}{s-1} + \frac{B}{s+1} + \frac{Cs+D}{s^2+1}$$

$$\therefore\ s^2 + 2s - 1 = A(s+1)(s^2+1) + B(s-1)(s^2+1)$$
$$+ (s-1)(s+1)(Cs+D)$$

Putting $s = 1$ and $s = -1$ gives $A = \frac{1}{2}$ and $B = \frac{1}{2}$.

Equating coefficients of s^3 and the constant terms gives $C = -1$ and $D = 1$.

$$\therefore\ \bar{y} = \frac{1}{2}\cdot\frac{1}{s-1} + \frac{1}{2}\cdot\frac{1}{s+1} - \frac{s}{s^2+1} + \frac{1}{s^2+1}$$

$$\therefore\ y = \frac{1}{2}e^t + \frac{1}{2}e^{-t} - \cos t + \sin t$$

▶

So the results are

$$x = -\frac{1}{2}\left(e^t + e^{-t}\right) + \sin t + \cos t = \sin t + \cos t - \cosh t$$

$$y = \frac{1}{2}\left(e^t + e^{-t}\right) + \sin t - \cos t = \sin t - \cos t + \cosh t$$

$$\therefore\ x = \sin t + \cos t - \cosh t; \quad y = \sin t - \cos t + \cosh t$$

Simultaneous equations are all solved in much the same way. Here is another.

Example 2

Solve the equations

$$2\dot{y} - 6y + 3x = 0$$
$$3\dot{x} - 3x - 2y = 0$$

given that at $t = 0$, $x = 1$ and $y = 3$.

Expressing these in Laplace transforms, we have

....

84

$$2(s\bar{y} - y_0) - 6\bar{y} + 3\bar{x} = 0$$
$$3(s\bar{x} - x_0) - 3\bar{x} - 2\bar{y} = 0$$

Then we insert the initial conditions and simplify, obtaining

....

85

$$3\bar{x} + (2s - 6)\bar{y} = 6 \qquad (1)$$
$$(3s - 3)\bar{x} - 2\bar{y} = 3 \qquad (2)$$

(a) *To find $\bar{x}$*

(1) $\qquad\qquad\qquad 3\bar{x} + (2s - 6)\bar{y} = 6$

(2) $\times (s - 3) \qquad (s - 3)(3s - 3)\bar{x} - (2s - 6)\bar{y} = 3(s - 3)$

Adding, $\qquad\qquad [(s - 3)(3s - 3) + 3]\bar{x} = 3s - 9 + 6$

$\qquad\qquad\qquad\qquad \therefore\ (3s^2 - 12s + 12)\bar{x} = 3s - 3$

$\qquad\qquad\qquad\qquad\qquad (s^2 - 4s + 4)\bar{x} = s - 1$

$$\therefore\ \bar{x} = \frac{s - 1}{(s - 2)^2} \equiv \frac{A}{s - 2} + \frac{B}{(s - 2)^2} = \frac{A(s - 2) + B}{(s - 2)^2}$$

$$\therefore\ s - 1 = A(s - 2) + B \quad \text{giving} \quad A = 1 \quad \text{and} \quad B = 1$$

$$\therefore\ \bar{x} = \frac{1}{s - 2} + \frac{1}{(s - 2)^2} \quad \therefore\ x = e^{2t} + te^{2t}$$

(b) Going back to equations (1) and (2), we can find y.

$$y = \ldots\ldots\ldots\ldots$$

86

$$y = \tfrac{1}{2}\left\{6e^{2t} + 3te^{2t}\right\}$$

Because, eliminating $\bar{x}$ we get

$$\bar{y} = \frac{6s - 9}{2(s - 2)^2} \equiv \frac{1}{2}\left\{\frac{A}{s - 2} + \frac{B}{(s - 2)^2}\right\} = \frac{1}{2}\left\{\frac{A(s - 2) + B}{(s - 2)^2}\right\}$$

$$\therefore\ 6s - 9 = A(s - 2) + B \qquad \therefore\ A = 6;\quad B = 3$$

$$\therefore\ \bar{y} = \frac{1}{2}\left\{\frac{6}{s - 2} + \frac{3}{(s - 2)^2}\right\} \qquad \therefore\ y = \tfrac{1}{2}\left\{6e^{2t} + 3te^{2t}\right\}$$

Simultaneous second-order equations are solved in like manner. Again, with all these solutions it is a worthwhile exercise to substitute the solution back into the differential equation to verify that the solution is correct.

87

Example 3

If x and y are functions of t, solve the equations

$$\ddot{x} + 2x - y = 0$$
$$\ddot{y} + 2y - x = 0$$

given that at $t = 0$, $x_0 = 4$; $y_0 = 2$; $x_1 = 0$; $y_1 = 0$.

We start off as usual with $\qquad (s^2\bar{x} - sx_0 - x_1) + 2\bar{x} - \bar{y} = 0$

and $\qquad\qquad\qquad\qquad\qquad (s^2\bar{y} - sy_0 - y_1) + 2\bar{y} - \bar{x} = 0$

Inserting the initial conditions, we have

$$s^2\bar{x} - 4s + 2\bar{x} - \bar{y} = 0$$
$$s^2\bar{y} - 2s + 2\bar{y} - \bar{x} = 0$$

Simplifying these we can eliminate $\bar{y}$ to obtain $\bar{x}$ and hence x.

$$x = \ldots\ldots\ldots\ldots$$

88

$$x = 3\cos t + \cos\left(\sqrt{3}t\right)$$

Because

$$(s^2 + 2)\bar{x} - \bar{y} = 4s \tag{1}$$
$$-\bar{x} + (s^2 + 2)\bar{y} = 2s \tag{2}$$

Eliminating $\bar{y}$ and simplifying gives

$$\bar{x} = \frac{4s^3 + 10s}{(s^2 + 1)(s^2 + 3)}$$

$$\therefore\ \bar{x} = \frac{4s^3 + 10s}{(s^2 + 1)(s^2 + 3)} \equiv \frac{As + B}{s^2 + 1} + \frac{Cs + D}{s^2 + 3}$$

$$\therefore\ 4s^3 + 10s = (s^2 + 3)(As + B) + (s^2 + 1)(Cs + D)$$

▶

Equating coefficients of like powers of s

$[s^3]$ $4 = A + C$ $\therefore A + C = 4$

$[CT]$ $0 = 3B + D$ $\therefore 3B + D = 0$

     Putting $s = 1$, $14 = 4A + 4B + 2C = 2D$ $\therefore 2A + 2B + C + D = 7$

     Putting $s = -1$ $-14 = -4A + 4B - 2C + 2D$ $\therefore 2A - 2B + C - D = 7$

Putting $C = 4 - A$ and $D = -3B$ in the last two leads to

    $A = \dots\dots\dots\dots;$ $B = \dots\dots\dots\dots;$

    $C = \dots\dots\dots\dots;$ $D = \dots\dots\dots\dots$

$$\boxed{A = 3; \quad B = 0; \quad C = 1; \quad D = 0}$$ **89**

$$\therefore \bar{x} = \frac{3s}{s^2 + 1} + \frac{s}{s^2 + 3}$$

$$\therefore x = \dots\dots\dots\dots$$

$$\boxed{x = 3\cos t + \cos\left(\sqrt{3}t\right)}$$ **90**

To find y we could return to equations (1) and (2) and repeat the process, eliminating $\bar{x}$ so as to obtain $\bar{y}$ and hence y.

But always keep an eye on the original equations, the first of which is

$$\ddot{x} + 2x - y = 0$$

Therefore, in this particular case, $y = \ddot{x} + 2x$.

So all we have to do is to differentiate x twice and substitute

$$x = 3\cos t + \cos\left(\sqrt{3}t\right)$$

$$\dot{x} = -3\sin t - \sqrt{3}\sin\left(\sqrt{3}t\right)$$

$$\ddot{x} = -3\cos t - 3\cos\left(\sqrt{3}t\right)$$

$$\therefore y = -3\cos t - 3\cos\left(\sqrt{3}t\right) + 6\cos t + 2\cos\left(\sqrt{3}t\right)$$

$$\therefore y = 3\cos t - \cos\left(\sqrt{3}t\right)$$

which is a good deal quicker.

So, as we have seen, the method of solving differential equations by Laplace transforms follows a general routine.

(a) Express the equation in Laplace transforms

(b) Insert the initial conditions

(c) Simplify to obtain the transform of the solution

(d) Rewrite the final transform in partial fractions

(e) Determine the inverse transforms

and, by now, you are fully aware of the importance of *partial fractions*!   ▶

That brings us to the end of this particular Programme. We shall continue our study of Laplace transforms in the next Programme. Meanwhile, be sure you are familiar with the items listed in the **Revision summary** that follows, and respond to the questions in the **Can you?** checklist. You will then have no difficulty with the **Test exercise** and the **Further problems** provide additional practice.

 # Revision summary 2

91

1 *Laplace transform* $L\{f(t)\} = \int_0^\infty f(t)e^{-st}\,dt = F(s)$.

2 *Table of transforms*

$f(t)$	$L\{f(t)\} = F(s)$
a	$\dfrac{a}{s}$
e^{at}	$\dfrac{1}{s-a}$
$\sin at$	$\dfrac{a}{s^2 + a^2}$
$\cos at$	$\dfrac{s}{s^2 + a^2}$
$\sinh at$	$\dfrac{a}{s^2 - a^2}$
$\cosh at$	$\dfrac{s}{s^2 - a^2}$
t^n	$\dfrac{n!}{s^{n+1}}$

(n a positive integer)

3 *Linearity of the Laplace transform*

(a) The transform of a sum (or difference) of expressions is the sum (or difference) of the individual transforms. That is

$$L\{f(t) \pm g(t)\} = L\{f(t)\} \pm L\{g(t)\}$$

(b) The transform of an expression that is multiplied by a constant is the constant multiplied by the transform of the expression. That is

$$L\{kf(t)\} = kL\{f(t)\}$$

4 *Theorem 1* First shift theorem

If $L\{f(t)\} = F(s)$, then $L\{e^{-at}f(t)\} = F(s+a)$.

5   *Theorem 2*   Multiplying by t

If $L\{f(t)\} = F(s)$, then $L\{tf(t)\} = -\dfrac{d}{ds}\{F(s)\}$.

6   *Theorem 3*   Dividing by t

If $L\{f(t)\} = F(s)$, then $L\left\{\dfrac{f(t)}{t}\right\} = \displaystyle\int_{s}^{\infty} F(\sigma)\, d\sigma$

provided that $\displaystyle\lim_{t\to 0}\left\{\dfrac{f(t)}{t}\right\}$ exists.

7   *Inverse transform*

If $L\{f(t)\} = F(s)$, then $L^{-1}\{F(s)\} = f(t)$.

8   *Rules of partial fractions*
   (a) The numerator must be of lower degree than the denominator. If not, divide out.
   (b) Factorise the denominator into its prime factors.
   (c) A linear factor $(s + a)$ gives a partial fraction $\dfrac{A}{s + a}$ where A is a constant to be determined.
   (d) A repeated factor $(s + a)^2$ gives $\dfrac{A}{s + a} + \dfrac{B}{(s + a)^2}$.
   (e) Similarly $(s + a)^3$ gives $\dfrac{A}{s + a} + \dfrac{B}{(s + a)^2} + \dfrac{C}{(s + a)^3}$.
   (f) A quadratic factor $(s^2 + ps + q)$ gives $\dfrac{Ps + Q}{s^2 + ps + q}$.
   (g) A repeated quadratic factor $(s^2 + ps + q)^2$ gives
   $$\dfrac{Ps + Q}{s^2 + ps + q} + \dfrac{Rs + T}{(s^2 + ps + q)^2}.$$

9   *The 'cover up' rule*
   The 'cover up' rule often enables the values of the constant coefficients to be written down almost on sight. However, this method only works when the denominator of the original fraction has non-repeated, linear factors.

▶

10 *Table of inverse transforms*

$F(s)$	$f(t)$	
$\dfrac{a}{s}$	a	
$\dfrac{1}{s+a}$	e^{-at}	
$\dfrac{n!}{s^{n+1}}$	t^n	(*n* a positive integer)
$\dfrac{1}{s^n}$	$\dfrac{t^{n-1}}{(n-1)!}$	
$\dfrac{a}{s^2+a^2}$	$\sin at$	
$\dfrac{s}{s^2+a^2}$	$\cos at$	
$\dfrac{a}{s^2-a^2}$	$\sinh at$	
$\dfrac{s}{s^2-a^2}$	$\cosh at$	

By the first shift theorem

If $F(s)$ is the Laplace transform of $f(t)$

then $F(s+a)$ is the Laplace transform of $e^{-at}f(t)$.

11 *Laplace transforms of derivatives*

$$L\{x\} = \bar{x}$$

$$L\left\{\frac{\mathrm{d}x}{\mathrm{d}t}\right\} = L\{\dot{x}\} = s\bar{x} - x_0$$

$$L\left\{\frac{\mathrm{d}^2x}{\mathrm{d}t^2}\right\} = L\{\ddot{x}\} = s\bar{x} - sx_0 - x_1 \text{ etc.}$$

where x_0 = value of x at $t = 0$

$\qquad x_1$ = value of $\dfrac{\mathrm{d}x}{\mathrm{d}t}$ at $t = 0$, etc.

12 *Solution of differential equations*
(a) Rewrite the equation in terms of Laplace transforms.
(b) Insert the given initial conditions.
(c) Rearrange the equation algebraically to give the transform of the solution.
(d) Express the transform in standard forms by partial fractions.
(e) Determine the inverse transforms to obtain the particular solution.

▶

13 *Simultaneous differential equations*
   Convert the simultaneous differential equations into simultaneous algebraic equations by taking the Laplace transform of each equation in turn. Insert the initial values. Solve the simultaneous algebraic equations in the usual manner and take the inverse Laplace transform of the algebraic solutions to find the solutions to the simultaneous differential equations.

Can you?

Checklist 2 92

Check this list before and after you try the end of Programme test.

On a scale of 1 to 5 how confident are you that you can: **Frames**

- Obtain the Laplace transforms of simple standard expressions?   1 to 14
  Yes ☐ ☐ ☐ ☐ ☐ *No*

- Use the first shift theorem to find the Laplace transform of a simple expression multiplied by an exponential?   15 to 17
  Yes ☐ ☐ ☐ ☐ ☐ *No*

- Find the Laplace transform of a simple expression multiplied or divided by a variable?   18 to 26
  Yes ☐ ☐ ☐ ☐ ☐ *No*

- Use partial fractions to find the inverse Laplace transform?   27 to 42
  Yes ☐ ☐ ☐ ☐ ☐ *No*

- Use the 'cover up' rule?   43 to 46
  Yes ☐ ☐ ☐ ☐ ☐ *No*

- Use the Laplace transforms of derivatives to solve differential equations?   47 to 79
  Yes ☐ ☐ ☐ ☐ ☐ *No*

- Use the Laplace transform to solve simultaneous differential equations?   80 to 90
  Yes ☐ ☐ ☐ ☐ ☐ *No*

Test exercise 2

1 Determine the Laplace transforms of the following functions.

   (a) $3e^{-4t} - 5e^{4t}$ (b) $\sin 4t + \cos 4t$ (c) $t^3 + 2t^2 - t + 4$

   (d) $e^{-2t}\cos 5t$ (e) $t\sin 3t$ (f) $\dfrac{e^{-t} - e^{-2t}}{t}$.

2 Determine the inverse transforms of the following.

   (a) $\dfrac{s-5}{(s-3)(s-4)}$ (b) $\dfrac{s^2 + 3s - 7}{(s-1)(s^2+2)}$

   (c) $\dfrac{s^2 - 3s - 4}{(s-3)(s-1)^2}$ (d) $\dfrac{2s^2 - 6s - 1}{(s-3)(s^2 - 2s + 5)}$.

3 Solve the following equations by Laplace transforms.

   (a) $\dfrac{dx}{dt} + 3x = e^{-2t}$ given that $x = 2$ when $t = 0$

   (b) $3\dot{x} - 6x = \sin 2t$ given that $x = 1$ when $t = 0$

   (c) $\ddot{x} - 7\dot{x} + 12x = 2$ given that at $t = 0$, $x = 1$ and $\dot{x} = 5$

   (d) $\ddot{x} - 2\dot{x} + x = te^t$ given that at $t = 0$, $x = 1$ and $\dot{x} = 0$.

4 Solve the following pair of simultaneous equations where x and y are functions of t and given that at $t = 0$, $x = 4$ and $y = -1$.

$$\dot{x} + \dot{y} + x + 2y = e^{-3t}$$
$$\dot{x} + 3x + 5y = 5e^{-2t}$$

 Further problems 2

1 Determine the Laplace transforms of the following functions.

   (a) $e^{4t}\cos 2t$ (b) $t\sin 2t$ (c) $t^3 + 4t^2 + 5$

   (d) $e^{3t}(t^2 + 4)$ (e) $t^2\cos t$ (f) $\dfrac{\sinh 2t}{t}$.

2 Determine the inverse transforms of the following.

   (a) $\dfrac{2s-6}{(s-2)(s-4)}$ (b) $\dfrac{5s-8}{s(s-4)}$ (c) $\dfrac{s^2 - 2s + 3}{(s-2)^3}$

   (d) $\dfrac{2 - 11s}{(s-2)(s^2 + 2s + 2)}$ (e) $\dfrac{s}{(s^2+1)(s^2+4)}$ (f) $\dfrac{s-5}{s^2 + 4s + 20}$.

In Questions 3 to 11, solve the equations by Laplace transforms.

3   $\dot{x} - 4x = 8$ at $t = 0$, $x = 2$.

4   $3\dot{x} - 4x = \sin 2t$ at $t = 0$, $x = \frac{1}{3}$.

5   $\ddot{x} - 2\dot{x} + x = 2(t + \sin t)$ at $t = 0$, $x = 6$, $\dot{x} = 5$.

6   $\ddot{x} - 6\dot{x} + 8x = e^{3t}$ at $t = 0$, $x = 0$, $\dot{x} = 2$.

7   $\ddot{x} + 9x = \cos 2t$ at $t = 0$, $x = 1$, $\dot{x} = 3$.

8   $\ddot{x} - 2\dot{x} + 5x = e^{2t}$ at $t = 0$, $x = 0$, $\dot{x} = 1$.

9   $\ddot{x} + 4\dot{x} + 4x = t^2 + e^{-2t}$ at $t = 0$, $x = \frac{1}{2}$, $\dot{x} = 0$.

10   $\ddot{x} + 8\dot{x} + 32x = 32 \sin 4t$ at $t = 0$, $x = \dot{x} = 0$.

11   $\ddot{x} + 25x = 10(\cos 5t - 2 \sin 5t)$ at $t = 0$, $x = 1$, $\dot{x} = 2$.

In Questions 12 to 17, solve the pairs of simultaneous equations by Laplace transforms.

12   $\left.\begin{array}{l} \dot{y} + 3x = e^{-2t} \\ \dot{x} - 3y = e^{2t} \end{array}\right\}$ at $t = 0$, $x = y = 0$.

13   $\left.\begin{array}{l} 4\dot{x} - 2\dot{y} + 10x - 5y = 0 \\ \dot{y} - 18x + 15y = 10 \end{array}\right\}$ at $t = 0$, $y = 4$, $x = 2$.

14   $\left.\begin{array}{l} \dot{x} - 2\dot{y} - 3x + 6y = 12 \\ 3\dot{y} + 5x + 2y = 16 \end{array}\right\}$ at $t = 0$, $x = 12$, $y = 8$.

15   $\left.\begin{array}{l} 2\dot{x} + 3\dot{y} + 7x = 14t + 7 \\ 5\dot{x} - 3\dot{y} + 4x + 6y = 14t - 14 \end{array}\right\}$ at $t = 0$, $x = y = 0$.

16   $\left.\begin{array}{l} 2\dot{x} + 2x + 3\dot{y} + 6y = 56e^t - 3e^{-t} \\ \dot{x} - 2x - \dot{y} - 3y = -21e^t - 7e^{-t} \end{array}\right\}$ at $t = 0$, $x = 8$, $y = 3$.

17   $\left.\begin{array}{l} \ddot{x} - \ddot{y} + x - y = 5e^{2t} \\ 2\dot{x} - \dot{y} + y = 0 \end{array}\right\}$ at $t = 0$, $x = 1$, $y = 2$, $\dot{x} = 0$, $\dot{y} = 2$.

18   Find an expression for x in terms of t, given that

$$\ddot{y} - \dot{x} + 2x = 10 \sin 2t$$
$$\dot{y} + 2y + x = 0 \qquad \text{and when } t = 0, \ x = y = 0.$$

19   If $\ddot{x} + 8x + 2y = 24 \cos 4t$

and $4\ddot{y} + 2x + 5y = 0$

and at $t = 0$, $x = y = 0$, $\dot{x} = 1$, $\dot{y} = 2$, determine an expression for y in terms of t.

20   Solve completely, the pair of simultaneous equations

$$5\ddot{x} + 12\ddot{y} + 6x = 0$$
$$5\ddot{x} + 16\ddot{y} + 6y = 0$$

given that, at $t = 0$, $x = \frac{7}{4}$, $y = 1$, $\dot{x} = 0$, $\dot{y} = 0$.

Laplace transforms 2

Learning outcomes

When you have completed this Programme you will be able to:

- Use the Heaviside unit step function to 'switch' expressions on and off
- Obtain the Laplace transform of expressions involving the Heaviside unit step function
- Solve linear, constant coefficient ordinary differential equations with piecewise continuous right-hand sides
- Understand what is meant by the convolution of two functions and use the convolution theorem to find the inverse transform of a product of transforms

Introduction

In the previous Programme, we dealt with the Laplace transforms of continuous functions of t. In practical applications, it is convenient to have a function which, in effect, 'switches on' or 'switches off' a given term at pre-described values of t. This we can do with the *Heaviside unit step function*.

Heaviside unit step function

Consider a function that maintains a zero value for all values of t up to $t = c$ and a unit value for $t = c$ and all values of $t \geq c$.

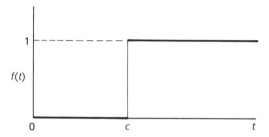

$$f(t) = 0 \quad \text{for } t < c$$
$$f(t) = 1 \quad \text{for } t \geq c$$

This function is the *Heaviside unit step function* and is denoted by

$$f(t) = u(t - c)$$

where the c indicates the value of t at which the function changes from a value of 0 to a value of 1.

Thus, the function

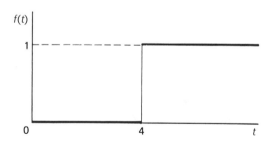

is denoted by $f(t) = \ldots\ldots\ldots\ldots$

$$\boxed{f(t) = u(t - 4)}$$

Similarly, the graph of $f(t) = 2u(t - 3)$ is

$$\ldots\ldots\ldots\ldots$$

3

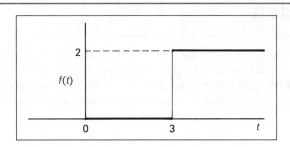

So $u(t - c)$ has just two values

for $t < c$, $u(t - c) = \ldots\ldots\ldots$
for $t \geq c$, $u(t - c) = \ldots\ldots\ldots$

4

$$t < c,\ u(t - c) = 0; \quad t \geq c,\ u(t - c) = 1$$

Unit step at the origin

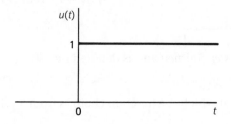

If the unit step occurs at the origin, then $c = 0$ and $f(t) = u(t - c)$ becomes

$$f(t) = u(t)$$

i.e. $u(t) = 0$ for $t < 0$
$u(t) = 1$ for $t \geq 0$.

Effect of the unit step function

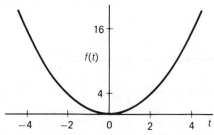

The graph of $f(t) = t^2$ is, of course, as shown.

Remembering the definition of $u(t - c)$, the graph of

$$f(t) = u(t - 2) \cdot t^2 \text{ is}$$

$\ldots\ldots\ldots$

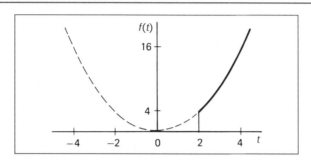

5

For $t < 2$, $u(t - 2) = 0$ $\therefore$ $u(t - 2) \cdot t^2 = 0 \cdot t^2 = 0$

$t \geq 2$, $u(t - 2) = 1$ $\therefore$ $u(t - 2) \cdot t^2 = 1 \cdot t^2 = t^2$

So the function $u(t - 2)$ suppresses the function t^2 for all values of t up to $t = 2$ and 'switches on' the function t^2 at $t = 2$.

Now we can sketch the graphs of the following functions.

(a) $f(t) = \sin t$ for $0 < t < 2\pi$

(b) $f(t) = u(t - \pi/4) \cdot \sin t$ for $0 < t < 2\pi$.

These give ....

and ....

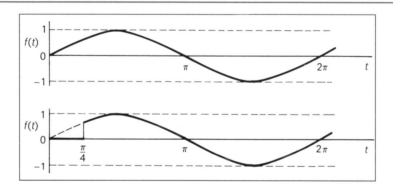

6

That is, the graph of $f(t) = u(t - \pi/4) \cdot \sin t$ is the graph of $f(t) = \sin t$ but suppressed for all values prior to $t = \pi/4$.

If we sketch the graph of $f(t) = \sin(t - \pi/4)$ we have

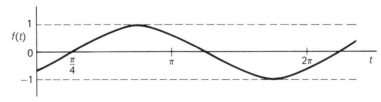

Since $u(t - c)$ has the effect of suppressing a function for $t < c$, then the graph of $f(t) = u(t - \pi/4) \cdot \sin(t - \pi/4)$ is

....

7

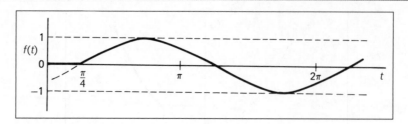

That is, the graph of $f(t) = u(t - \pi/4) \cdot \sin(t - \pi/4)$ is the graph of $f(t) = \sin t$ $(t > 0)$, shifted $\pi/4$ units along the t-axis.

In general, the graph of $f(t) = u(t - c) \cdot \sin(t - c)$ is the graph of $f(t) = \sin t$ $(t > 0)$, shifted along the t-axis through an interval of c units.

Similarly, for $t > 0$, sketch the graphs of

(a) $f(t) = e^{-t}$

(b) $f(t) = u(t - c) \cdot e^{-t}$

(c) $f(t) = u(t - c) \cdot e^{-(t-c)}$

(d) $f(t) = e^{-t}\{u(t - 1) - u(t - 2)\}$.

Arrange the graphs under each other to show the important differences.

8

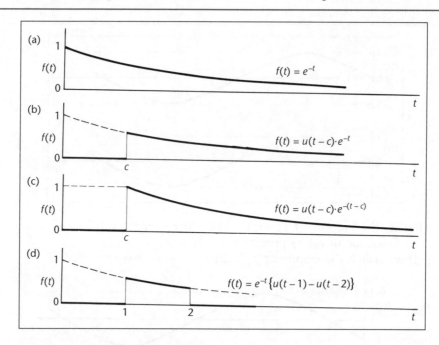

In (a), we have the graph of $f(t) = e^{-t}$

In (b), the same graph is suppressed prior to $t = c$

In (c), the graph of $f(t) = e^{-t}$ is shifted c units along the t-axis

In (d), the graph of $f(t) = e^{-t}$ is turned on at $t = 1$ and off at $t = 2$.

Laplace transform of $u(t - c)$

$$L\{u(t - c)\} = \frac{e^{-cs}}{s}$$

Because

$$L\{u(t - c)\} = \int_0^\infty e^{-st} u(t - c)\, dt$$

but

$$e^{-st}u(t - c) = \begin{cases} 0 & \text{for } 0 < t < c \\ e^{-st} & \text{for } t \geq c \end{cases}$$

so that

$$L\{u(t - c)\} = \int_0^\infty e^{-st}u(t - c)\, dt = \int_c^\infty e^{-st}\, dt$$

$$= \left[\frac{e^{-st}}{-s}\right]_c^\infty = \frac{e^{-sc}}{s}$$

Therefore, the Laplace transform of the unit step at the origin is

$$L\{u(t)\} = \ldots\ldots\ldots\ldots$$

$$\boxed{\dfrac{1}{s}}$$

9

Because $c = 0$.

So $\qquad L\{u(t - c)\} = \dfrac{e^{-cs}}{s}$

and $\qquad L\{u(t)\} = \dfrac{1}{s}$.

Also from the definition of $u(t)$:

$$L(1) = L\{1 \cdot u(t)\}$$
$$L(t) = L\{t \cdot u(t)\}$$
$$L\{f(t)\} = L\{f(t) \cdot u(t)\}$$

Make a note of these results: we shall be using them

As we have seen, the unit step function $u(t - c)$ is often combined with other functions of t, so we now consider the Laplace transform of $u(t - c) \cdot f(t - c)$.

10

▶

Laplace transform of $u(t-c) \cdot f(t-c)$
(the second shift theorem)

$$L\{u(t-c) \cdot (f(t-c)\} = e^{-cs}L\{f(t)\} = e^{-cs}F(s)$$

Because

$$L\{u(t-c) \cdot f(t-c)\} = \int_0^\infty e^{-st}u(t-c) \cdot f(t-c)\,dt$$

$$\text{but } e^{-st}u(t-c) = \begin{cases} 0 & \text{for } 0 < t < c \\ e^{-st} & \text{for } t \geq c \end{cases}$$

so that

$$L\{u(t-c) \cdot f(t-c)\} = \int_c^\infty e^{-st}f(t-c)\,dt$$

We now make the substitution $t - c = v$ so that $t = c + v$ and $dt = dv$. Also for the limits, when $t = c$, $v = 0$ and when $t \to \infty$, $v \to \infty$. Therefore

$$L\{u(t-c) \cdot f(t-c)\} = \int_0^\infty e^{-s(c+v)}f(v)\,dv$$

$$= e^{-cs}\int_0^\infty e^{-sv}f(v)\,dv$$

Now $\int_0^\infty e^{-sv}f(v)\,dv$ has exactly the same value as $\int_0^\infty e^{-st}f(t)\,dt$ which is, of course, the Laplace transform of $f(t)$. Therefore

$$L\{u(t-c) \cdot f(t-c)\} = e^{-cs}L\{f(t)\} = e^{-cs}F(s)$$

11

$$\boxed{L\{u(t-c) \cdot f(t-c)\} = e^{-cs} \cdot F(s) \quad \text{where } F(s) = L\{f(t)\}}$$

So $L\left\{u(t-4) \cdot (t-4)^2\right\} = e^{-4s} \cdot F(s) \quad \text{where } F(s) = L\{t^2\}$

$$= e^{-4s}\left(\frac{2!}{s^3}\right) = \frac{2e^{-4s}}{s^3}$$

Note that $F(s)$ is the transform of t^2 and *not* of $(t-4)^2$.
  In the same way:

$$L\{u(t-3) \cdot \sin(t-3)\} = \ldots\ldots\ldots\ldots$$

12

$$\boxed{\dfrac{e^{-3s}}{s^2+1}}$$

Because $L\{u(t-3) \cdot \sin(t-3)\} = e^{-3s} \cdot F(s) \quad \text{where } F(s) = L\{\sin t\} = \dfrac{1}{s^2+1}$

$$\therefore \ L\{u(t-3) \cdot \sin(t-3)\} = e^{-3s}\left(\frac{1}{s^2+1}\right)$$

▶

So now do these in the same way.

(a) $L\left\{u(t-2)\cdot(t-2)^3\right\}$ $\qquad = \ldots\ldots\ldots$

(b) $L\{u(t-1)\cdot\sin 3(t-1)\}$ $\qquad = \ldots\ldots\ldots$

(c) $L\left\{u(t-5)\cdot e^{(t-5)}\right\}$ $\qquad = \ldots\ldots\ldots$

(d) $L\{u(t-\pi/2)\cdot\cos 2(t-\pi/2)\} = \ldots\ldots\ldots$

Here they are   **13**

(a) $L\left\{u(t-2)\cdot(t-2)^3\right\} = e^{-2s}\cdot F(s)$ where $F(s) = L\{t^3\}$

$$= e^{-2s}\left(\frac{3!}{s^4}\right) = \frac{6e^{-2s}}{s^4}$$

(b) $L\{u(t-1)\cdot\sin 3(t-1)\} = e^{-s}\cdot F(s)$ where $F(s) = L\{\sin 3t\}$

$$= e^{-s}\left(\frac{3}{s^2+9}\right) = \frac{3e^{-s}}{s^2+9}$$

(c) $L\left\{u(t-5)\cdot e^{(t-5)}\right\} = e^{-5s}\cdot F(s)$ where $F(s) = L\{e^t\}$

$$= e^{-5s}\left(\frac{1}{s-1}\right) = \frac{e^{-5s}}{s-1}$$

(d) $L\{u(t-\pi/2)\cdot\cos 2(t-\pi/2)\} = e^{-\pi s/2}\cdot F(s)$ where $F(s) = L\{\cos 2t\}$

$$= e^{-\pi s/2}\left(\frac{s}{s^2+4}\right) = \frac{s\cdot e^{-\pi s/2}}{s^2+4}$$

So $L\{u(t-c)\cdot f(t-c)\} = e^{-cs}\cdot F(s)$ where $F(s) = L\{f(t)\}$.

Written in reverse, this becomes

$\qquad$ If $F(s) = L\{f(t)\}$, then $e^{-cs}\cdot F(s) = L\{u(t-c)\cdot f(t-c)\}$

where c is real and positive.

This is known as the *second shift theorem*.

Make a note of it: then we will use it

$\boxed{\text{If } F(s) = L\{f(t)\}, \text{ then } e^{-cs}\cdot F(s) = L\{u(t-c)\cdot f(t-c)\}}$ **14**

This is useful in finding inverse transforms, as we shall now see.

▶

Example 1

Find the function whose transform is $\dfrac{e^{-4s}}{s^2}$.

The numerator corresponds to e^{-cs} where $c = 4$ and therefore indicates $u(t-4)$.

Then $\dfrac{1}{s^2} = F(s) = L\{t\} \quad \therefore \ f(t) = t$.

$$\therefore \ L^{-1}\left\{\dfrac{e^{-4s}}{s^2}\right\} = u(t-4)\cdot(t-4)$$

Remember that in writing the final result, $f(t)$ is replaced by

. . . .

15

$$\boxed{f(t-c)}$$

Example 2

Determine $L^{-1}\left\{\dfrac{6e^{-2s}}{s^2+4}\right\}$.

The numerator contains e^{-2s} and therefore indicates . . . .

16

$$\boxed{u(t-2)}$$

The remainder of the transform, i.e. $\dfrac{6}{s^2+4}$, can be written as $3\left(\dfrac{2}{s^2+4}\right)$

$$\therefore \ \dfrac{6}{s^2+4} = F(s) = L\{\ldots\ldots\ldots\}$$

17

$$\boxed{L\{3\sin 2t\}}$$

$$\therefore \ L^{-1}\left\{\dfrac{6e^{-2s}}{s^2+4}\right\} = \ldots\ldots\ldots$$

18

$$\boxed{3u(t-2)\cdot\sin 2(t-2)}$$

Because

$$L^{-1}\left\{\dfrac{6e^{-2s}}{s^2+4}\right\} = u(t-2)\cdot f(t-2) \quad \text{where } f(t) = L^{-1}\left\{\dfrac{6}{s^2+4}\right\}$$

$$= u(t-2)\cdot 3\sin 2(t-2)$$

Example 3

Determine $L^{-1}\left\{\dfrac{s\cdot e^{-s}}{s^2+9}\right\}$.

This, in similar manner, is . . . .

$$\boxed{u(t-1) \cdot \cos 3(t-1)}$$

19

Because the numerator contains e^{-s} which indicates $u(t-1)$.

Also $\dfrac{s}{s^2 + 9} = F(s) = L\{\cos 3t\}$

$\therefore f(t) = \cos 3t \quad \therefore f(t-1) = \cos 3(t-1)$.

$$\therefore L^{-1}\left\{\frac{s \cdot e^{-s}}{s^2 + 9}\right\} = u(t-1) \cdot \cos 3(t-1)$$

Remember that, having obtained $f(t)$, the result contains $f(t-c)$.
Here is a short exercise by way of practice.

Exercise

Determine the inverse transforms of the following.

(a) $\dfrac{2e^{-5s}}{s^3}$ 　　　　　　　　(d) $\dfrac{2s \cdot e^{-3s}}{s^2 - 16}$

(b) $\dfrac{3e^{-2s}}{s^2 - 1}$ 　　　　　　　　(e) $\dfrac{5e^{-s}}{s}$

(c) $\dfrac{8e^{-4s}}{s^2 + 4}$ 　　　　　　　　(f) $\dfrac{s \cdot e^{-s/2}}{s^2 + 2}$

Results – all very straightforward.

20

(a) $u(t-5) \cdot (t-5)^2$

(b) $3u(t-2) \cdot \sinh(t-2)$

(c) $4u(t-4) \cdot \sin 2(t-4)$

(d) $2u(t-3) \cdot \cosh 4(t-3)$

(e) $5u(t-1)$

(f) $u(t-1/2) \cdot \cos\sqrt{2}(t-1/2)$.

Before looking at a more interesting example, let us collect our results together as far as we have gone.

The main points are

21

(a) $u(t-c) = 0 \quad\quad 0 < t < c$
$\quad\quad\quad\quad\ = 1 \quad\quad\quad t \geq c$ 　　　　　　　(1)

(b) $L\{u(t-c)\} = \dfrac{e^{-cs}}{s}$

$\quad\ L\{u(t)\} = \dfrac{1}{s}$ 　　　　　　　　(2)

(c) $L\{u(t-c) \cdot f(t-c)\} = e^{-cs} \cdot F(s) \quad$ where $F(s) = L\{f(t)\}$ 　　(3)

(d) If $F(s) = L\{f(t)\}$, then $e^{-cs} \cdot F(s) = L\{u(t-c)\} \cdot f(t-c)\}$ 　　(4)

Now let us apply these to some further examples.

▶

Example 1

Determine the expression $f(t)$ for which

$$L\{f(t)\} = \frac{3}{s} - \frac{4e^{-s}}{s^2} + \frac{5e^{-2s}}{s^2}$$

We take each term in turn and find its inverse transform.

(a) $L^{-1}\left\{\dfrac{3}{s}\right\} = 3L^{-1}\left\{\dfrac{1}{s}\right\} = 3$ i.e. $3u(t)$

(b) $L^{-1}\left\{\dfrac{4e^{-s}}{s^2}\right\} = u(t-1) \cdot 4(t-1)$

(c) $L^{-1}\left\{\dfrac{5e^{-2s}}{s^2}\right\} = \ldots\ldots\ldots$

22

$$\boxed{u(t-2) \cdot 5(t-2)}$$

So we have $L^{-1}\left\{\dfrac{3}{s}\right\} = 3u(t)$

$$L^{-1}\left\{\dfrac{4e^{-s}}{s^2}\right\} = u(t-1) \cdot 4(t-1)$$

$$L^{-1}\left\{\dfrac{5e^{-2s}}{s^2}\right\} = u(t-2) \cdot 5(t-2)$$

$$\therefore\ F(t) = 3u(t) - u(t-1) \cdot 4(t-1) + u(t-2) \cdot 5(t-2)$$

To sketch the graph of $f(t)$ we consider the values of the function within the three sections $0 < t < 1$, $1 < t < 2$, and $2 < t$.

Between $t = 0$ and $t = 1$, $f(t) = \ldots\ldots\ldots$

23

$$\boxed{f(t) = 3}$$

Because in this interval, $u(t) = 1$, but $u(t-1) = 0$ and $u(t-2) = 0$. In the same way, between $t = 1$ and $t = 2$, $f(t) = \ldots\ldots\ldots$

24

$$\boxed{f(t) = 7 - 4t}$$

Because between $t = 1$ and $t = 2$, $u(t) = 1$, $u(t-1) = 1$, but $u(t-2) = 0$.

$$\therefore\ f(t) = 3 - 4(t-1) + 0 = 3 - 4t + 4 = 7 - 4t$$

Similarly, for $t > 2$, $f(t) = \ldots\ldots\ldots$

$$\boxed{f(t) = t - 3}$$

25

Because for $t > 2$, $u(t) = 1$, $u(t - 1) = 1$ and $u(t - 2) = 1$

$\therefore\ f(t) = 3 - 4(t - 1) + 5(t - 2)$
$\quad\quad\quad = 3 - 4t + 4 + 5t - 10 = t - 3$

So, collecting the results together, we have

for $\ \ 0 < t < 1$, $\ \ f(t) = 3$

$\quad\quad 1 < t < 2$, $\ \ f(t) = 7 - 4t \quad (t = 1, f(t) = 3; t = 2, f(t) = -1)$

$\quad\quad 2 < t$, $\ \ \ \ \ \ \ f(t) = t - 3 \quad (t = 2, f(t) = -1; t = 3, f(t) = 0)$

Using these facts we can sketch the graph of $f(t)$, which is ....

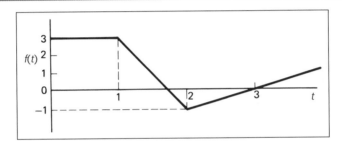

26

Here is another.

Example 2

Determine the expression $f(t) = L^{-1}\left\{\dfrac{2}{s} + \dfrac{3e^{-s}}{s^2} - \dfrac{3e^{-3s}}{s^2}\right\}$ and sketch the graph of $f(t)$.

First we express the inverse transform of each term in terms of the unit step function.

This gives ....

27

$$L^{-1}\left\{\frac{2}{s}\right\} = 2u(t); \quad L^{-1}\left\{\frac{3e^{-s}}{s^2}\right\} = u(t - 1) \cdot 3(t - 1)$$

$$L^{-1}\left\{\frac{3e^{-3s}}{s^2}\right\} = u(t - 3) \cdot 3(t - 3)$$

$\therefore\ f(t) = 2u(t) + u(t - 1) \cdot 3(t - 1) - u(t - 3) \cdot 3(t - 3)$

So there are 'break points', i.e. changes of function, at $t = 1$ and $t = 3$, and we investigate $f(t)$ within the three intervals.

$\quad\quad 0 < t < 1 \quad\quad f(t) = ...$

$\quad\quad 1 < t < 3 \quad\quad f(t) = ...$

$\quad\quad 3 < t \quad\quad\quad\ f(t) = ...$

28

$$0 < t < 1, f(t) = 2; \quad 1 < t < 3, f(t) = 3t - 1; \quad 3 < t, f(t) = 8$$

Because with

$0 < t < 1, \quad u(t) = 1$, but $u(t - 1) = u(t - 3) = 0 \qquad \therefore f(t) = 2$

$1 < t < 3, \quad u(t) = 1, u(t - 1) = 1$, but $u(t - 3) = 0$

$\qquad \therefore f(t) = 2 + 3(t - 1) = 3t - 1 \qquad\qquad \therefore f(t) = 3t - 1$

$3 < t, \qquad u(t) = 1, u(t - 1) = 1, u(t - 3) = 1$

$\qquad \therefore f(t) = 2 + 3t - 3 - 3t + 9 \qquad\qquad \therefore f(t) = 8$

Therefore, the graph of $f(t)$ is ....

29

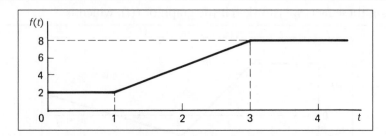

Between the break points, $f(t) = 3t - 1$ $\qquad \begin{cases} t = 1, f(t) = 2 \\ t = 3, f(t) = 8 \end{cases}$

Now move on for the next example

30   **Example 3**

If $f(t) = L^{-1}\left\{ \dfrac{\left(1 - e^{-2s}\right)\left(1 + e^{-4s}\right)}{s^2} \right\}$, determine $f(t)$ and sketch the graph of the function.

Although at first sight this looks more complicated, we simply multiply out the numerator and proceed as before.

$$f(t) = L^{-1}\left\{ \frac{1 - e^{-2s} + e^{-4s} - e^{-6s}}{s^2} \right\}$$

$$= L^{-1}\left\{ \frac{1}{s^2} - \frac{e^{-2s}}{s^2} + \frac{e^{-4s}}{s^2} - \frac{e^{-6s}}{s^2} \right\}$$

We now write down the inverse transform of each term in terms of the unit function, so that

$$f(t) = \ldots\ldots\ldots$$

$$f(t) = u(t) \cdot t - u(t-2) \cdot (t-2) + u(t-4) \cdot (t-4) - u(t-6) \cdot (t-6)$$

31

and we can see there are break points at $t = 2$, $t = 4$, $t = 6$.

For $0 < t < 2$, $f(t) = t - 0 + 0 - 0$ $f(t) = t$
      $2 < t < 4$, $f(t) = t - (t-2) + 0 - 0$ $f(t) = 2$
      $4 < t < 6$, $f(t) = t - (t-2) + (t-4) - 0$ $f(t) = t - 2$
      $6 < t$, $f(t) = t - (t-2) + (t-4) - (t-6)$ $f(t) = 4$

The second and fourth components are constant, but before sketching the graph of the function, we check the values of $f(t) = t$ and $f(t) = t - 2$ at the relevant break points.

$f(t) = t$.   At $t = 0$, $f(t) = 0$;   at $t = 2$, $f(t) = 2$
$f(t) = t - 2$.   At $t = 4$, $f(t) = 2$;   at $t = 6$, $f(t) = 4$.

So the graph of the function is . . . .

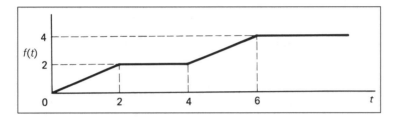

32

It is always wise to calculate the function values at break points, since discontinuities, or jumps, sometimes occur.

On to the next frame

Now for one in reverse.

33

Example 4

A function $f(t)$ is defined by

$f(t) = 4$ for $0 < t < 2$
      $= 2t - 3$ for $2 < t$.

Sketch the graph of the function and determine its Laplace transform.

We see that for $t = 0$ to $t = 2$, $f(t) = 4$.

34

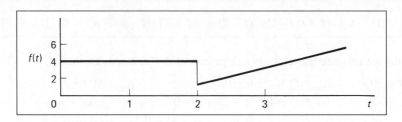

Notice the discontinuity at $t = 2$.

Expressing the function in unit step form:

$$f(t) = 4u(t) - 4u(t-2) + u(t-2) \cdot (2t-3)$$

Note that the second term cancels $f(t) = 4$ at $t = 2$ and that the third switches on $f(t) = 2t - 3$ at $t = 2$.

Before we can express this in Laplace transforms, $(2t - 3)$ in the third term must be written as a function of $(t - 2)$ to correspond to $u(t-2)$. Therefore, we write $2t - 3$ as $2(t-2) + 1$.

Then $f(t) = 4u(t) - 4u(t-2) + u(t-2) \cdot \{2(t-2) + 1\}$

$$= 4u(t) - 4u(t-2) + u(t-2) \cdot 2(t-2) + u(t-2)$$

$$= 4u(t) - 3u(t-2) + u(t-2) \cdot 2(t-2)$$

$$\therefore \ L\{f(t)\} = \ldots\ldots\ldots\ldots$$

35

$$\boxed{L\{f(t)\} = \frac{4}{s} - \frac{3e^{-2s}}{s} + \frac{2e^{-2s}}{s^2}}$$

Here is one for you to work through in much the same way.

Example 5

A function is defined by $\quad f(t) = 6 \qquad\quad 0 < t < 1$

$$= 8 - 2t \quad 1 < t < 3$$

$$= 4 \qquad\quad 3 < t.$$

Sketch the graph and find the Laplace transform of the function.

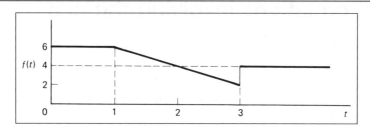

36

Expressing this in unit step form we have

$$f(t) = 6u(t) - 6u(t-1) + u(t-1) \cdot (8 - 2t)$$
$$- u(t-3) \cdot (8 - 2t) + u(t-3) \cdot 4$$

where the second term switches off the first function $f(t) = 6$ at $t = 1$ and the third term switches on the second function $f(t) = 8 - 2t$, which in turn is switched off by the fourth term at $t = 3$ and replaced by $f(t) = 4$ in the fifth term.

Before we can write down the transforms of the third and fourth terms, we must express $f(t) = 8 - 2t$ in terms of $(t-1)$ and $(t-3)$ respectively.

$$8 - 2t = 6 + 2 - 2t = 6 - 2(t-1)$$
$$8 - 2t = 2 + 6 - 2t = 2 - 2(t-3)$$
$$\therefore \; f(t) = 6u(t) - 6u(t-1) + u(t-1) \cdot \{6 - 2(t-1)\}$$
$$- u(t-3) \cdot \{2 - 2(t-3)\} + 4u(t-3)$$
$$= 6u(t) - 6u(t-1) + 6u(t-1)$$
$$- u(t-1) \cdot 2(t-1) - 2u(t-3)$$
$$+ u(t-3) \cdot 2(t-3) + 4u(t-3)$$

which simplifies finally to $f(t) = \ldots\ldots\ldots$

$$f(t) = 6u(t) - u(t-1) \cdot 2(t-1) + u(t-3) \cdot 2(t-3) + 2u(t-3)$$

37

from which $L\{f(t)\} = \ldots\ldots\ldots$

$$L\{f(t)\} = \frac{6}{s} - \frac{2e^{-s}}{s^2} + \frac{2e^{-3s}}{s^2} + \frac{2e^{-3s}}{s}$$

38

Note that, in building up the function in unit step form

(a) to 'switch on' a function $f(t)$ at $t = c$, we add the term $u(t-c) \cdot f(t-c)$

(b) to 'switch off' a function $f(t)$ at $t = c$, we subtract $u(t-c) \cdot f(t-c)$.

Next we shall look at some differential equations that use what we have done so far in the Programme.

Next frame

Differential equations involving the unit step function

39 We can now use the work on the unit step function to solve constant coefficient differential equations with a piecewise continuous right-hand side. To solve the differential equation:

$$f'(t) + 3f(t) = u(t - 1) \text{ where } f(0) = 0$$

we start by taking the Laplace transform of both sides to find that

$$L\{f'(t) + 3f(t)\} = L\{u(t - 1)\} \text{ that is } sF(s) - f(0) + 3F(s) = \frac{e^{-s}}{s} \text{ so that}$$

$$(s + 3)F(s) = \frac{e^{-s}}{s} \text{ giving } F(s) = \frac{e^{-s}}{s(s + 3)}$$

$$= \frac{e^{-s}}{3}\left\{\frac{1}{s} - \frac{1}{s + 3}\right\}$$

Therefore $f(t) = L^{-1}\left\{\frac{e^{-s}}{3s}\right\} - L^{-1}\left\{\frac{e^{-s}}{3(s + 3)}\right\}$

$$= \frac{1}{3}\left[L^{-1}\left\{\frac{e^{-s}}{s}\right\} - L^{-1}\left\{\frac{e^{-s}}{s + 3}\right\}\right]$$

$$= \frac{1}{3}\left[u(t - 1) - u(t - 1)e^{-3(t-1)}\right]$$

$$= \frac{u(t - 1)}{3}\left(1 - e^{-3(t-1)}\right)$$

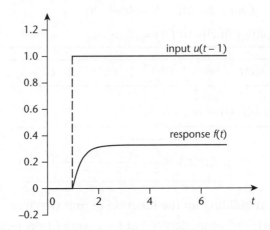

You try one. The solution of the equation $5f'(t) - f(t) = u(t - 4)$ where $f(0) = 0$ is:

$$f(t) = \dots\dots\dots$$

Next frame

$$u(t-4)\left(\exp\frac{(t-4)}{5}-1\right)$$

Because

Taking the Laplace transform of both sides we find that

$$L\{5f'(t)-f(t)\}=L\{u(t-4)\}\text{ that is }5sF(s)-f(0)-F(s)=\frac{e^{-4s}}{s}\text{ so that}$$

$$(5s-1)F(s)=\frac{e^{-4s}}{s}\text{ giving }F(s)=\frac{e^{-4s}}{s(5s-1)}$$

$$=e^{-4s}\left\{\frac{5}{5s-1}-\frac{1}{s}\right\}$$

Therefore $f(t)=L^{-1}\left\{\frac{5e^{-4s}}{5s-1}\right\}-L^{-1}\left\{\frac{e^{-4s}}{s}\right\}$

$$=\left[L^{-1}\left\{\frac{e^{-4s}}{s-\frac{1}{5}}\right\}-L^{-1}\left\{\frac{e^{-4s}}{s}\right\}\right]$$

$$=\left[u(t-4)e^{(t-4)/5}-u(t-4)\right]$$

$$=u(t-4)\left(\frac{\exp(t-4)}{5}-1\right)$$

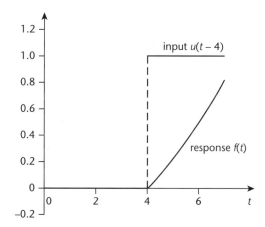

And another. The solution of the equation

$$f''(t)+5f'(t)+6f(t)=u(t-2)\sin(t-2)$$

where $f'(0)=f(0)=0$ is:

$$f(t)=\dots\dots\dots$$

Next frame

41

$$f(t) = \frac{u(t-2)}{10}\left[\sin(t-2) - \cos(t-2) + 2e^{-2(t-2)} - e^{-3(t-2)}\right]$$

Because

Taking the Laplace transform of both sides we find that

$$L\{f''(t) + 5f'(t) + 6f(t)\} = L\{u(t-2)\sin(t-2)\}$$

that is $s^2F(s) - sf(0) - f'(0) + 5F(s) - 5f(0) + 6F(s) = \dfrac{e^{-2s}}{s^2+1}$

so that $(s^2 + 5s + 6)F(s) = \dfrac{e^{-2s}}{s^2+1}$

giving $F(s) = \dfrac{e^{-2s}}{(s^2+1)(s+2)(s+3)}$

$$= \frac{1}{10}\frac{e^{-2s}}{s^2+1} + \frac{1}{10}\frac{se^{-2s}}{s^2+1} + \frac{1}{5}\frac{e^{-2s}}{s+2} - \frac{1}{10}\frac{e^{-2s}}{s+3}$$

Therefore $f(t) = \dfrac{u(t-2)}{10}\left[\sin(t-2) + \cos(t-2) + 2e^{-2(t-2)} - e^{-3(t-2)}\right]$

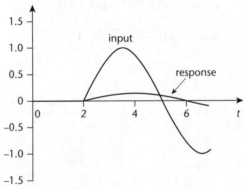

And just one more to make sure. The differential equation

$$f''(t) + f'(t) + f(t) = g(t) \text{ where } f'(0) = f(0) = 0$$

and where the graph of $g(t)$ is as shown

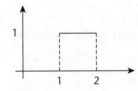

Next frame

42

$$f(t) = u(t-1)e^{-(t-1)/2}\left(1 - \cos\frac{\sqrt{3}(t-1)}{2} - \frac{1}{\sqrt{3}}\sin\frac{\sqrt{3}(t-1)}{2}\right)$$

$$- u(t-2)e^{-(t-2)/2}\left(1 - \cos\frac{\sqrt{3}(t-2)}{2} - \frac{1}{\sqrt{3}}\sin\frac{\sqrt{3}(t-2)}{2}\right)$$

Because

The graph is a single square pulse of width 1 between $t = 1$ and $t = 2$ This is described algebraically as the unit step turned on at $t = 1$ and then turned off at $t = 2$ That is

$$g(t) = u(t-1) - u(t-2)$$

the differential equation then becomes

$$f''(t) + f'(t) + f(t) = u(t-1) - u(t-2) \text{ where } f'(0) = f(0) = 0$$

Taking the Laplace transform of both sides we find that

$$L\{f''(t) + f'(t) + f(t)\} = L\{u(t-1) - u(t-2)\} \text{ that is}$$

$$s^2 F(s) - sf(0) - f'(0) + F(s) - f(0) + F(s) = \frac{e^{-s}}{s} - \frac{e^{-2s}}{s} \text{ so that}$$

$$(s^2 + s + 1)F(s) = (e^{-s} - e^{-2s})\frac{1}{s} \text{ and so } F(s) = (e^{-s} - e^{-2s})\frac{1}{s(s^2 + s + 1)}$$

Separating the factors in the denominator we find that

$$F(s) = (e^{-s} - e^{-2s})\left\{\frac{1}{s} - \frac{s+1}{s^2+s+1}\right\}$$

$$= (e^{-s} - e^{-2s})\left\{\frac{1}{s} - \frac{s+1}{\left(s+\frac{1}{2}\right)^2 + \left(\frac{\sqrt{3}}{2}\right)^2}\right\}$$

$$= (e^{-s} - e^{-2s})\left\{\frac{1}{s} - \frac{s+\frac{1}{2}}{\left(s+\frac{1}{2}\right)^2 + \left(\frac{\sqrt{3}}{2}\right)^2} - \frac{\frac{1}{2}}{\left(s+\frac{1}{2}\right)^2 + \left(\frac{\sqrt{3}}{2}\right)^2}\right\}$$

$$= (e^{-s} - e^{-2s})\left\{\frac{1}{s} - \frac{s+\frac{1}{2}}{\left(s+\frac{1}{2}\right)^2 + \left(\frac{\sqrt{3}}{2}\right)^2} - \frac{1}{\sqrt{3}}\frac{\frac{\sqrt{3}}{2}}{\left(s+\frac{1}{2}\right)^2 + \left(\frac{\sqrt{3}}{2}\right)^2}\right\}$$

Taking the inverse Laplace transforms we see that

$$f(t) = u(t-1)e^{-(t-1)/2}\left(1 - \cos\frac{\sqrt{3}(t-1)}{2} - \frac{1}{\sqrt{3}}\sin\frac{\sqrt{3}(t-1)}{2}\right)$$

$$- u(t-2)e^{-(t-2)/2}\left(1 - \cos\frac{\sqrt{3}(t-2)}{2} - \frac{1}{\sqrt{3}}\sin\frac{\sqrt{3}(t-2)}{2}\right)$$

Graph overleaf

▶

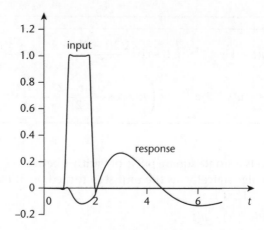

And now a slight digression. In Frame 13 of the previous Programme it was stated that two Laplace transforms must not be multiplied together to form the transform of a product of expressions. This is because the product of two transformations is not the transform of a product of expressions but rather, the transform of the *convolution* of the two expressions. We shall now see what is meant by convolution.

Move to the next frame

Convolution

43

The convolution of the two functions $f(t)$ and $g(t)$ is defined as:

$$c(t) = f(t) * g(t) = \int_{x=-\infty}^{\infty} f(x)g(t-x)\,dx$$

where the $*$ denotes the operation of convolution. You will notice that this is a function t as denoted by $c(t)$ and exactly what is happening here does require an explanation.

We wish to end up with a function depending on the variable t so we set the variable of integration to be x. Next, against the same set of coordinate axes we draw the graphs of the functions $f(x)$ and $g(-x)$ where the graph of $g(-x)$ is simply the graph of $g(x)$ reflected in the vertical axis. If this reflected graph were to be moved along the horizontal axis to the point where its leading edge was at $x = t$ then in its new position it would be the graph of $g(t - x)$.

As an example consider the rectangular function:

$$f(t) = \begin{cases} 2 & 0 \le t \le 3 \\ 0 & \text{otherwise} \end{cases} \qquad \text{with graph}$$

and the rectangular function:

$$g(t) = \begin{cases} 1 & 0 \le t \le 2 \\ 0 & \text{otherwise} \end{cases} \qquad \text{with graph}$$

The second function reflected in the vertical is:

$$g(-t) = \begin{cases} 1 & -2 \le t \le 0 \\ 0 & \text{otherwise} \end{cases} \qquad \text{with graph}$$

The convolution integral is then:

$$c(t) = f(t) * g(t) = \int_{x=-\infty}^{\infty} f(x)g(t-x)\,\mathrm{d}x$$

with the following graphical configuration:

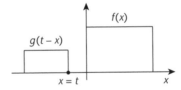

In this position $t < 0$ and $f(x)g(t-x) = 0$ for all values of x. This means that the value of the convolution integral is also zero.

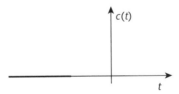

As t becomes positive the non-zero parts of the graphs of $g(t-x)$ and $f(x)$ overlap and so the integrand $f(x)g(t-x)$ becomes non-zero as does the resulting convolution integral. As t increases further so the range of values of x for which the integrand is non-zero also increases.

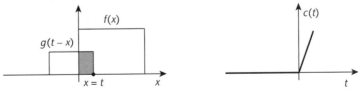

Here, $c(t) = f(t) * g(t)$

$$= \int_{x=0}^{t} 2 \times 1 \, \mathrm{d}x \quad \text{provided } 0 \le t \le 2$$

$$= 2t$$

▶

Eventually the range of values of x for which the integrand is non-zero reaches its maximum and remains there as t increases further. During this stage the convolution integral assumes a constant value.

Here, $c(t) = f(t) * g(t)$

$$= \int_{x=t-2}^{t} 2 \times 1 \, dx \quad \text{provided } 2 \le t \le 3$$
$$= 2t - 2(t-2)$$
$$= 4$$

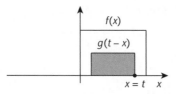 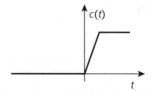

At some point the range of values of x for which the integrand is non-zero decreases.

Here, $c(t) = f(t) * g(t)$

$$= \int_{x=t-2}^{3} 2 \times 1 \, dx \quad \text{provided } 3 \le t$$
$$= 6 - 2(t-2)$$
$$= 10 - 2t$$

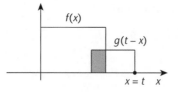 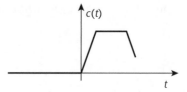

Finally, as t increases further there comes a point where $f(x)g(t-x) = 0$ for all values of x greater than t.

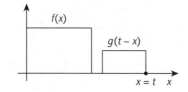 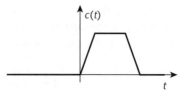

In conclusion:

$$c(t) = \begin{cases} 2t & 0 \le t < 2 \\ 4 & 2 \le t < 3 \\ 10 - 2t & 3 \le t \le 5 \\ 0 & \text{otherwise} \end{cases}$$

▶

Now you try one. Given the rectangular function

$$f(t) = \begin{cases} 1 & -1 \leq t \leq 1 \\ 0 & \text{otherwise} \end{cases}$$

and the truncated exponential

$$g(t) = \begin{cases} e^{-t} & 0 \leq t \leq 1 \\ 0 & \text{otherwise} \end{cases}$$

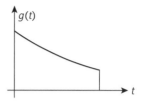

the convolution of these two functions is

$$c(t) = \ldots\ldots\ldots\ldots$$

The answer is in the next frame

44

$$c(t) = \begin{cases} 1 - e^{-1-t} & -1 \leq t < 0 \\ 1 - e^{-1} & 0 \leq t < 1 \\ e^{1-t} - e^{-1} & 1 \leq t \leq 2 \\ 0 & \text{otherwise} \end{cases}$$

Because:

$$c(t) = f(t) * g(t)$$
$$= \int_{x=-\infty}^{\infty} f(x)g(t-x)\,\mathrm{d}x$$
$$= \int_{x=-1}^{1} f(x)g(t-x)\,\mathrm{d}x$$

The evaluation of this integral is separated into three parts and it is always advisable to draw small sketches of the configurations to help in deciding the limits of the integrals.

(a) $-1 \leq t < 0$:

$$c(t) = f(t) * g(t)$$
$$= \int_{x=-1}^{1} f(x)g(t-x)\,\mathrm{d}x$$
$$= \int_{x=-1}^{t} 1 \times e^{-(t-x)}\,\mathrm{d}x$$
$$= \int_{x=-1}^{t} e^{x-t}\,\mathrm{d}x$$
$$= \left[e^{x-t} \right]_{-1}^{t}$$
$$= e^{0} - e^{-1-t} = 1 - e^{-1-t}$$

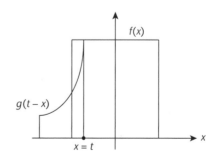

▶

(b) $0 \le t < 1$:

$$c(t) = f(t) * g(t)$$

$$= \int_{x=-1}^{1} f(x)g(t-x)\,\mathrm{d}x$$

$$= \int_{x=t-1}^{t} 1 \times e^{-(t-x)}\,\mathrm{d}x$$

$$= \int_{x=t-1}^{t} e^{x-t}\,\mathrm{d}x$$

$$= \left[e^{x-t}\right]_{t-1}^{t} = e^{0} - e^{t-1-t} = 1 - e^{-1}$$

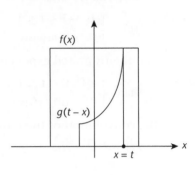

(c) $1 \le t < 2$:

$$c(t) = f(t) * g(t)$$

$$= \int_{x=-1}^{1} f(x)g(t-x)\,\mathrm{d}x$$

$$= \int_{x=t-1}^{1} 1 \times e^{-(t-x)}\,\mathrm{d}x$$

$$= \int_{x=t-1}^{1} e^{x-t}\,\mathrm{d}x$$

$$= \left[e^{x-t}\right]_{t-1}^{1} = e^{1-t} - e^{-1}$$

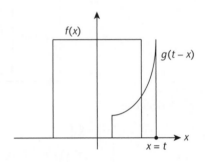

$$\text{Giving } c(t) = \begin{cases} 1 - e^{-1-t} & -1 \le t < 0 \\ 1 - e^{-1} & 0 \le t < 1 \\ e^{1-t} - e^{-1} & 1 \le t \le 2 \\ 0 & \text{otherwise} \end{cases}$$

with the following graph.

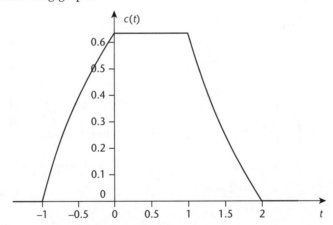

We can now return to our main theme, namely the properties of the Laplace transform

Next frame

The convolution theorem

The convolution theorem concerns the product of a pair of Laplace transforms.   **45**

Given two functions $f(t)$ and $g(t)$ where

$f(t) = 0$ and $g(t) = 0$ when $t < 0$

and their respective Laplace transforms $F(s)$ and $G(s)$ then the Laplace transform of the convolution of $f(t)$ and $g(t)$ is equal to the product of their Laplace transforms. That is:

$L\{f(t) * g(t)\} = L\{f(t)\}L\{g(t)\}$

Because $f(t)$ and $g(t)$ are interchangeable in this equation, that is:

$L\{f(t) * g(t)\} = L\{f(t)\}L\{g(t)\} = L\{g(t)\}L\{f(t)\} = L\{g(t) * f(t)\}$

we see that convolution is a commutative operation; $f(t) * g(t) = g(t) * f(t)$. Furthermore, because $f(t) = g(t) = 0$ when $t < 0$ this means that

$$F(s)G(s) = L\left\{\int_0^t f(t-x)g(x)\,dx\right\} = L\left\{\int_0^t g(t-x)f(x)\,dx\right\}$$

Notice that the upper limit is t because $f(t-x) = 0$ and $g(t-x) = 0$ for $t - x < 0$, that is for $x > t$. For example using the convolution theorem to evaluate $L^{-1}\left\{\dfrac{1}{s^2(s-3)}\right\}$, we see that if

$F(s) = \dfrac{1}{s^2} = L\{f(t)\}$ then $f(t) = t$ and

$G(s) = \dfrac{1}{s-3} = L\{g(t)\}$ then $g(t) = e^{3t}$.

As a result $F(s)G(s) = L\{f(t) * g(t)\}$ so that

$$L^{-1}\left\{\frac{1}{s^2(s-3)}\right\} = f(t) * g(t)$$

$$= \int_{-\infty}^{\infty} f(x)g(t-x)\,dx$$

Since convolution is a commutative operation the choice is made of which expression is represented by $f(x)$ and which by $g(t-x)$ so as to result in the simplest integral. Therefore

$$L^{-1}\left\{\frac{1}{s^2(s-3)}\right\} = \int_0^t xe^{3(t-x)}\,dx$$

$$= e^{3t}\int_0^t xe^{-3x}\,dx = \left(-\frac{t}{3} - \frac{1}{9} + \frac{e^{3t}}{9}\right)$$

which is in agreement with the partial fraction procedure. Now you try one.

By the convolution theorem $L^{-1}\left\{\dfrac{s}{(s^2+1)^2}\right\} = \ldots\ldots\ldots$

The answer is in the next frame

46

$$\boxed{\dfrac{t \sin t}{4}}$$

Because:

$$\frac{s}{(s^2 + 1)^2} = \frac{1}{(s^2 + 1)} \times \frac{s}{(s^2 + 1)}$$

$$= F(s) \times G(s) \quad \text{so that } f(t) = \sin t \text{ and } g(t) = \cos t$$

$$L^{-1}\{F(s)G(s)\} = f(t) * g(t)$$

$$= \int_0^t \sin(t - x) \cos x \, dx$$

$$= \int_0^t \{(\sin t \cos x - \sin x \cos t) \cos x\} \, dx$$

$$= \sin t \int_0^t \cos^2 x \, dx - \cos t \int_0^t \sin x \cos x \, dx$$

$$= \sin t \int_0^t \left[\frac{\cos 2x + 1}{2}\right] dx - \cos t \int_0^t \frac{\sin 2x}{2} \, dx$$

$$= \frac{\sin t}{2} \left[\frac{\sin 2x}{2} + x\right]_0^t - \frac{\cos t}{2} \left[-\frac{\cos 2x}{2}\right]_0^t$$

$$= \frac{1}{4}\{\sin t \sin 2t + t \sin t + \cos t \cos 2t - \cos t\}$$

$$= \frac{t \sin t}{4}$$

You have now reached the end of this Programme and this brings you to the **Revision summary** and the **Can you?** checklist. Following that is the **Test exercise**. Work through this *at your own pace*. A set of **Further problems** provides additional valuable practice.

 # Revision summary 3

47   1   *Heaviside unit step function: $u(t - c)$*

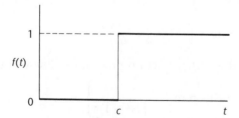

$$f(t) = 0 \quad 0 < t < c$$
$$= 1 \quad c < t$$

▶

2 *Suppression and shift*

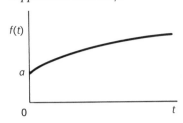 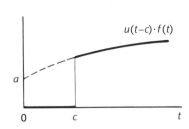

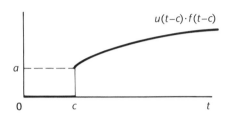

3 *Laplace transform of $u(t - c)$*

$$L\{u(t - c)\} = \frac{e^{-cs}}{s}; \quad L\{u(t)\} = \frac{1}{s}.$$

4 *Laplace transform of $u(t - c) \cdot f(t - c)$*

$$L\{u(t - c) \cdot f(t - c)\} = e^{-cs} \cdot F(s) \quad \text{where } F(s) = L\{f(t)\}.$$

5 *Second shift theorem*
If $F(s) = L\{f(t)\}$, then $e^{-cs} \cdot F(s) = L\{u(t - c) \cdot f(t - c)\}$ where c is real and positive.

6 *Convolution theorem*
The convolution of two expressions $f(t)$ and $g(t)$ is denoted as $f(t) * g(t)$ and is defined as the definite integral

$$f(t) * g(t) = \int_{x=-\infty}^{t} f(t - x)g(x) \, dx$$

Also, convolution is a commutative operation. That is

$$f(t) * g(t) = \int_{x=-\infty}^{\infty} f(t - x)g(x) \, dx = g(t) * f(t) = \int_{x=-\infty}^{\infty} g(t - x)f(x) \, dx$$

The convolution theorem states that if $L\{f(t)\} = F(s)$ and $L\{g(t)\} = G(s)$ then

$$L\{f(t) * g(t)\} = L\{f(t)\}L\{g(t)\}$$
$$= F(s)G(s)$$

so that

$$L^{-1}\{F(s)G(s)\} = f(t) * g(t)$$

Can you?

48 **Checklist 3**

Check this list before and after you try the end of Programme test.

On a scale of 1 to 5 how confident are you that you can: **Frames**

- Use the Heaviside unit step function to 'switch' expressions on
  and off?   [1] to [8]
  Yes ☐   ☐   ☐   ☐   ☐   *No*

- Obtain the Laplace transform of expressions involving the
  Heaviside unit step function?   [8] to [38]
  Yes ☐   ☐   ☐   ☐   ☐   *No*

- Solve linear, constant coefficient ordinary differential
  equations with piecewise continuous right-hand sides   [39] to [42]
  Yes ☐   ☐   ☐   ☐   ☐   *No*

- Understand what is meant by the convolution of two
  functions and use the convolution theorem to find the inverse
  transform of a product of transforms   [43] to [46]
  Yes ☐   ☐   ☐   ☐   ☐   *No*

 # Test exercise 3

49 1 In each of the following cases, sketch the graph of the function and find its
     Laplace transform.

   (a) $f(t) = 3t \quad 0 \leq t < 2$
   $\qquad = 6 \quad\ 2 \leq t$

   (b) $f(t) = e^{-2t} \quad 0 \leq t < 3$
   $\qquad = 0 \qquad 3 \leq t$

   (c) $f(t) = t^2 \quad 0 \leq t < 2$
   $\qquad = 2 \quad\ 2 \leq t < 3$
   $\qquad = 4 \quad\ 3 \leq t$

   (d) $f(t) = \sin 2t \quad 0 \leq t < \pi$
   $\qquad = 0 \qquad\ \pi \leq t.$

   2 Determine the function $f(t)$ whose transform $F(s)$ is

   $$F(s) = \frac{1}{s}\left\{2 - 5e^{-s} + 8e^{-3s}\right\}.$$

   Sketch the graph of the function between $t = 0$ and $t = 4$.

   ▶

3   If $f(t) = L^{-1}\left\{\dfrac{(1 + 3e^{-2s})(1 - e^{-3s})}{s^2}\right\}$, determine $f(t)$ and sketch the graph of the function.

4   Determine the function $f(t)$ for which
$$f(t) = L^{-1}\left\{\frac{2(1 - e^{-s})}{s(1 - e^{-3s})}\right\}.$$
Sketch the waveform and express the function in analytical form.

5   Solve the differential equation
$$f'(t) - f(t) = u(t)e^{-t} - u(t - 1)e^{-(t-1)} \text{ where } f(0) = 0$$

6   Use the convolution theorem to find $L^{-1}\left\{\dfrac{2s}{(s^2 - 16)^2}\right\}$ where $f(0) = 0$.

Further problems 3

1   If $L\{f(t)\} = \dfrac{1}{s^2}\left\{3s + 2e^{-2s} - 2e^{-5s}\right\}$, determine $f(t)$.

50

2   If $f(t) = L^{-1}\left\{\dfrac{(1 - e^{-s})(1 + e^{-2s})}{s^2}\right\}$, find $f(t)$ in terms of the unit step function.

3   A function $f(t)$ is defined by
$$\begin{aligned} f(t) &= 4 & 0 \le t < 3 \\ &= 2t + 1 & 3 \le t. \end{aligned}$$
Sketch the graph of the function and determine its Laplace transform.

4   Express in terms of the Heaviside unit step function
(a) $\begin{aligned} f(t) &= t^2 & 0 \le t < 3 \\ &= 5t & 3 \le t. \end{aligned}$

(b) $\begin{aligned} f(t) &= \cos t & 0 \le t < \pi \\ &= \cos 2t & \pi \le t < 2\pi \\ &= \cos 3t & 2\pi \le t. \end{aligned}$

5   A function $f(t)$ is defined by
$$\begin{aligned} f(t) &= 0 & 0 \le t < 2 \\ &= t + 1 & 2 \le t < 3 \\ &= 0 & 3 \le t. \end{aligned}$$
Determine $L\{f(t)\}$.

6   A function $f(t)$ is defined by
$$\begin{aligned} f(t) &= t^2 & 0 \le t < 2 \\ &= 4 & 2 \le t < 5 \\ &= 0 & 5 \le t. \end{aligned}$$
Determine (a) the function in terms of the unit step function
(b) the Laplace transform of $f(t)$.

▶

7 Solve the differential equations

(a) $f'(t) + 2f(t) = tu(t) - (t-1)u(t-1)$ where $f(0) = 0$

(b) $f''(t) - 4f'(t) + 4f(t) = u(t) - u(t-2)$ where $f'(0) = f(0) = 0$

(c) $f''(t) - f(t) = u(t)\sin 3t - (t-4)\sin 3(t-4)$ where $f'(0) = f(0) = 0$

(d) $f''(t) + f'(t) + f(t) = (t-1)u(t-1)$ where $f'(0) = f(0) = 0$

8 Determine the inverse Laplace transforms of each of the following

(a) $\dfrac{1}{(s+1)(s^2+1)}$ (b) $\dfrac{1}{s^3(s^2-2)}$ (c) $\dfrac{1}{(3s^2-4)(2s^2-1)}$

9 Show that

(a) $u(t) - u(t) = tu(t)$

(b) $tu(t) * e^t u(t) = (e^t - t - 1)$ where $u(t)$ is the Heaviside unit step function.

Laplace transforms 3

Learning outcomes

When you have completed this Programme you will be able to:

- Find the Laplace transforms of periodic functions
- Obtain the inverse Laplace transforms of transforms of periodic functions
- Describe and use the unit impulse to evaluate integrals
- Obtain the Laplace transform of the unit impulse
- Use the Laplace transform to solve differential equations involving the unit impulse
- Solve the equation and describe the behaviour of an harmonic oscillator

Laplace transforms of periodic functions

Periodic functions

Let $f(t)$ represent a periodic function with period T so that $f(t + nT) = f(t)$ with a graph of the following form

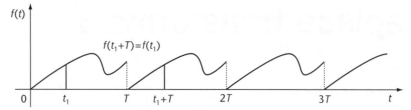

If we describe the first cycle by $\bar{f}(t)$ then

$$\bar{f}(t) = \begin{cases} f(t) & \text{for } 0 \le t < T \\ 0 & \text{otherwise} \end{cases}$$

The second cycle is identical to the first cycle except that it is shifted by T units of time along the t-axis. Therefore the second cycle can be described in terms of the Heaviside unit step function as $\bar{f}(t - T)u(t - T)$. That is

$$\bar{f}(t - T)u(t - T) = \begin{cases} f(t) & \text{for } T \le t < 2T \\ 0 & \text{otherwise} \end{cases}$$

By this reasoning the periodic function $f(t)$ is represented by

$$f(t) = \bar{f}(t)u(t) + \dots\dots\dots$$

$$\boxed{f(t) = \bar{f}(t)u(t) + \bar{f}(t - T)u(t - T) + \bar{f}(t - 2T)u(t - 2T) + \cdots}$$

Because

$u(t)$ switches on $\bar{f}(t)$ at time $t = 0$, $u(t - T)$ switches on $\bar{f}(t - T)$ at time $t = T$ and $u(t - 2T)$ switches on $\bar{f}(t - 2T)$ at time $t = 2T$, etc.

Consider now the Laplace transform of $\bar{f}(t)$. By definition

$$L\{\bar{f}(t)\} = \int_0^\infty e^{-st}\bar{f}(t)\,dt = \int_0^T e^{-st}f(t)\,dt = \bar{F}(s)$$

because for $t > T$, $\bar{f}(t) = 0$ and so the semi-infinite integral becomes an integral just over the period of $f(t)$. Using the second shift theorem (see Frame 10 of Programme 3), the Laplace transform of $f(t)$ is

$$L\{f(t)\} = L\{\bar{f}(t)u(t)\} + L\{\bar{f}(t - T)u(t - T)\}$$
$$+ L\{\bar{f}(t - 2T)u(t - 2T)\} + \cdots$$

That is

$$L\{f(t)\} = \dots\dots\dots$$

$$L\{f(t)\} = \bar{F}(s) + e^{-sT}\bar{F}(s) + e^{-2sT}\bar{F}(s) + \cdots$$

3

Because

$$L\{\bar{f}(t)u(t-c)\} = e^{-sc}L\{\bar{f}(t)\} \text{ by the second shift theorem.}$$

We can factor out $\bar{F}(s)$ and write $L\{f(t)\}$ as

$$L\{f(t)\} = \left(1 + e^{-sT} + e^{-2sT} + \ldots\right)\bar{F}(s)$$

Now, do you remember the series $1 + x + x^2 + x^3 + \ldots$? This can be written in closed form as

$$1 + x + x^2 + x^3 + \ldots = \ldots\ldots\ldots$$

$$1 + x + x^2 + x^3 + \ldots = \frac{1}{1-x}$$

4

Because

$$\frac{1}{1-x} = (1-x)^{-1} = 1 + x + x^2 + x^3 + \ldots$$

either by the binomial theorem or by performing the long division.
    So, if we let $x = e^{-sT}$ then

$$1 + e^{-sT} + e^{-2sT} + \ldots = \ldots\ldots\ldots$$

$$1 + e^{-sT} + e^{-2sT} + \ldots = \frac{1}{1 - e^{-sT}}$$

5

And so the Laplace transform of $f(t)$ is given as

$$L\{f(t)\} = \left(1 + e^{-sT} + e^{-2sT} + \ldots\right)\bar{F}(s) = \ldots\ldots\ldots \text{ where } \bar{F}(s) = \ldots\ldots\ldots$$

$$L\{f(t)\} = \frac{1}{(1 - e^{-sT})}\bar{F}(s) \text{ where } \bar{F}(s) = \int_0^T e^{-st}f(t)\,dt$$

6

Note that we integrate $e^{-st}f(t)$ over one cycle, that is from $t = 0$ to $t = T$, and not from $t = 0$ to $t = \infty$ as we did previously.

This is an important result. Make a note of it – then we shall apply it

▶

Example 1

Find the Laplace transform of the function $f(t)$ defined by

$$f(t) = 3 \quad 0 < t < 2 \atop = 0 \quad 2 < t < 4 \Big\} \quad f(t+4) = f(t)$$

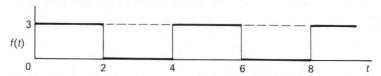

The expression for $L\{f(t)\}$ is

....   (do not evaluate it yet)

7

$$L\{f(t)\} = \frac{1}{1 - e^{-4s}} \int_0^4 e^{-st} \cdot f(t)\, dt$$

Because the period $= 4$, i.e. $T = 4$.

The function $f(t) = 3$ for $0 < t < 2$ and $f(t) = 0$ for $2 < t < 4$.

$$\therefore \ L\{f(t)\} = \frac{1}{1 - e^{-4s}} \int_0^2 e^{-st} \cdot 3\, dt = \ldots\ldots\ldots$$

8

$$L\{f(t)\} = \frac{3}{s(1 + e^{-2s})}$$

Because

$$L\{f(t)\} = \frac{3}{1 - e^{-4s}} \left[\frac{e^{-st}}{-s}\right]_0^2 = \frac{3}{1 - e^{-4s}} \left\{\left(\frac{e^{-2s}}{-s}\right) - \left(\frac{1}{-s}\right)\right\}$$

$$= \frac{3}{1 - e^{-4s}} \left\{\frac{1 - e^{-2s}}{s}\right\} = \frac{3}{s(1 + e^{-2s})}$$

That is all there is to it. Now for another, so move on

---

9   **Example 2**

Find the Laplace transform of the periodic function defined by

$$f(t) = t/2 \qquad 0 < t < 3$$
$$f(t+3) = f(t)$$

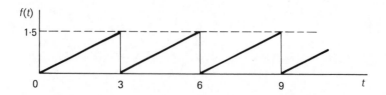

Because in this case, period $= 3$, i.e. $T = 3$.

$$\therefore \ L\{f(t)\} = \frac{1}{1 - e^{-Ts}} \int_0^T e^{-st} \cdot f(t)\,dt$$

$$= \frac{1}{1 - e^{-3s}} \int_0^3 e^{-st} \cdot \left(\frac{t}{2}\right) dt$$

$$\therefore \ 2(1 - e^{-3s})L\{f(t)\} = \int_0^3 t \cdot e^{-st}\,dt$$

Integrating by parts and simplifying the result gives

$$L\{f(t)\} = \ldots\ldots\ldots\ldots$$

$$\boxed{L\{f(t)\} = \frac{1}{2s^2}\left\{1 - \frac{3s}{e^{3s} - 1}\right\}}$$

10

Because

$$2(1 - e^{-3s})L\{f(t)\} = \int_0^3 te^{-st}\,dt$$

$$= \left[t\left(\frac{e^{-st}}{-s}\right)\right]_0^3 + \frac{1}{s}\int_0^3 e^{-st}\,dt$$

$$= -\frac{3e^{-3s}}{s} + \frac{1}{s}\left[\frac{e^{-st}}{-s}\right]_0^3$$

$$= -\frac{3e^{-3s}}{s} - \frac{e^{-3s}}{s^2} + \frac{1}{s^2}$$

$$\therefore \ L\{f(t)\} = \frac{1}{2s^2}\left\{1 - \frac{3se^{-3s}}{1 - e^{-3s}}\right\}$$

$$= \frac{1}{2s^2}\left\{1 - \frac{3s}{e^{3s} - 1}\right\}$$

Example 3

Sketch the graph of the function

$$f(t) = e^t \qquad 0 < t < 5$$
$$f(t + 5) = f(t)$$

and determine its Laplace transform.

First we sketch the graph of $f(t)$, which is $\ldots\ldots\ldots\ldots$

11

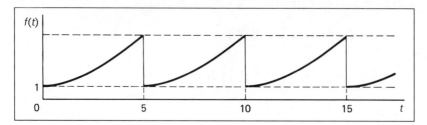

Clearly, period $= 5$ $\therefore$ $T = 5$

$$L\{f(t)\} = \frac{1}{1 - e^{-Ts}} \int_0^T e^{-st} \cdot f(t)\, dt \quad \text{gives}$$

$$L\{f(t)\} = \ldots\ldots\ldots\ldots$$

Complete the working

12

$$\boxed{L\{f(t)\} = \frac{1 - e^{-5(s-1)}}{(s-1)(1 - e^{-5s})}}$$

Because

$$L\{f(t)\} = \frac{1}{1 - e^{-5s}} \int_0^5 e^{-st} \cdot e^t\, dt$$

$$\therefore (1 - e^{-5s})L\{f(t)\} = \int_0^5 e^{-(s-1)t}\, dt$$

$$= \left[\frac{e^{-(s-1)t}}{-(s-1)}\right]_0^5 = \frac{1}{s-1}\left\{1 - e^{-5(s-1)}\right\}$$

$$\therefore L\{f(t)\} = \frac{1 - e^{-5(s-1)}}{(s-1)(1 - e^{-5s})}$$

All very straightforward.

Example 4

Determine the Laplace transform of the half-wave rectifier output waveform defined by

$$\left.\begin{array}{ll} f(t) = 8\sin t & 0 < t < \pi \\ = 0 & \pi < t < 2\pi \end{array}\right\} \quad f(t + 2\pi) = f(t)$$

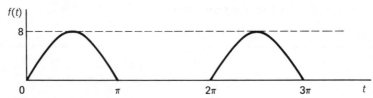

Here the period is 2π i.e. $T = 2\pi$.

In general, for a periodic function of period T

$$L\{f(t)\} = \ldots\ldots\ldots\ldots$$

13

$$L\{f(t)\} = \frac{1}{1 - e^{-Ts}} \int_0^T e^{-st} \cdot f(t)\, dt$$

So, for this example

$$L\{f(t)\} = \frac{1}{1 - e^{-2\pi s}} \int_0^{2\pi} e^{-st} \cdot f(t)\, dt$$

$$\therefore \ (1 - e^{-2\pi s})L\{f(t)\} = \int_0^{\pi} e^{-st} \cdot 8 \sin t\, dt$$

Writing $\sin t$ as the imaginary part of e^{jt}, i.e. $\sin t \equiv \mathscr{I} e^{jt}$,

$$(1 - e^{-2\pi s})L\{f(t)\} = 8\mathscr{I} \int_0^{\pi} e^{-st} \cdot e^{jt}\, dt$$

$$= 8\mathscr{I} \int_0^{\pi} e^{-(s-j)t}\, dt$$

and this you can finish off in the usual manner, giving

$$L\{f(t)\} = \ldots\ldots\ldots\ldots$$

14

$$L\{f(t)\} = \frac{8}{(s^2 + 1)(1 - e^{-\pi s})}$$

Because

$$(1 - e^{-2\pi s})L\{f(t)\} = 8 \cdot \mathscr{I} \int_0^{\pi} e^{-(s-j)t}\, dt$$

$$= 8 \cdot \mathscr{I} \left[\frac{e^{-(s-j)t}}{-(s-j)}\right]_0^{\pi}$$

$$= \mathscr{I}\left\{\frac{-8}{s-j}\left[e^{-(s-j)\pi} - 1\right]\right\}$$

$$= 8 \cdot \mathscr{I}\left\{\frac{1}{s-j}\left[1 - e^{-s\pi}e^{j\pi}\right]\right\}$$

But $e^{j\pi} = \cos \pi + j \sin \pi = -1$.

$$\therefore \ (1 - e^{-2\pi s})L\{f(t)\} = 8 \cdot \mathscr{I}\left\{\frac{1}{s-j}(1 + e^{-s\pi})\right\}$$

$$= 8 \cdot \mathscr{I}\left\{\frac{s+j}{s^2 + 1}(1 + e^{-\pi s})\right\} = 8\left\{\frac{1 + e^{-\pi s}}{s^2 + 1}\right\}$$

$$\therefore \ L\{f(t)\} = \frac{1}{1 - e^{-2\pi s}} \times 8\left\{\frac{1 + e^{-\pi s}}{s^2 + 1}\right\}$$

$$= \frac{8}{(1 - e^{-\pi s})(s^2 + 1)}$$

Now let us consider the corresponding inverse transforms when periodic functions are involved.

15   Inverse transforms

Finding inverse transforms of functions of *s* which are transforms of periodic functions is not as straightforward as in earlier examples, for the transforms result from integration over one cycle and not from $t = 0$ to $t = \infty$. Hence we have no simple table of inverse transforms upon which to draw.

However, all difficulties can be surmounted and an example will show how we deal with this particular problem.

Example 1

Determine the inverse transform

$$L^{-1}\left\{\frac{2 + e^{-2s} - 3e^{-s}}{s(1 - e^{-2s})}\right\}$$

The first thing we see is the factor $(1 - e^{-2s})$ in the denominator, which suggests a periodic function of period 2 units, i.e. $\dfrac{1}{1 - e^{-Ts}}$ where $T = 2$.

The key to the solution is to write $(1 - e^{-2s})$ in the denominator as $(1 - e^{-2s})^{-1}$ in the numerator and to expand this as a binomial series.

We remember that $(1 - x)^{-1} = \ldots\ldots\ldots\ldots$

16

$$\boxed{(1 - x)^{-1} = 1 + x + x^2 + x^3 + \ldots}$$

$$\therefore\ (1 - e^{-2s})^{-1} = 1 + (e^{-2s}) + (e^{-2s})^2 + (e^{-2s})^3 + \ldots$$

$$= 1 + e^{-2s} + e^{-4s} + e^{-6s} + \ldots$$

$$\therefore\ L\{f(t)\} = \frac{2 + e^{-2s} - 3e^{-s}}{s(1 - e^{-2s})} = \frac{1}{s}\left(2 + e^{-2s} - 3e^{-s}\right)\left(1 - e^{-2s}\right)^{-1}$$

$$= \frac{1}{s}\left(2 + e^{-2s} - 3e^{-s}\right)\left(1 + e^{-2s} + e^{-4s} + e^{-6s} + e^{-8s} + \ldots\right)$$

We now multiply the second series by each term of the first in turn and collect up like terms, giving

$$L\{f(t)\} = \frac{1}{s}\left\{\begin{array}{llll} 2 & +2e^{-2s} & +2e^{-4s} & +2e^{-6s}\ \ldots \\ & +\ e^{-2s} & +\ e^{-4s} & +\ e^{-6s}\ \ldots \\ -3e^{-s} & -3e^{-3s} & -3e^{-5s} & \ldots \end{array}\right\}$$

$$= \ldots\ldots\ldots\ldots$$

17

$$L\{f(t)\} = \frac{1}{s}\{2 - 3e^{-s} + 3e^{-2s} - 3e^{-3s} + 3e^{-4s} - 3e^{-5s} + \ldots\}$$

Each term is of the form $\dfrac{e^{-cs}}{s}$, so, expressing $f(t)$ in unit step form, we have

$$f(t) = \ldots\ldots\ldots\ldots$$

18

$$f(t) = 2u(t) - 3u(t-1) + 3u(t-2) - 3u(t-3) + 3u(t-4)\ldots$$

and from this we can sketch the waveform, which is therefore

$$\ldots\ldots\ldots\ldots$$

19

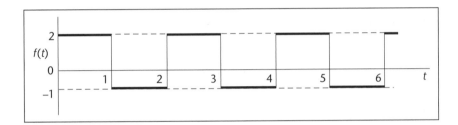

We can finally define this periodic function in analytical terms.

$$f(t) = \ldots\ldots\ldots\ldots$$

20

$$\left.\begin{array}{ll} f(t) = 2 & 0 < t < 1 \\ \quad = -1 & 1 < t < 2 \end{array}\right\} \ f(t+2) = f(t)$$

The key to the whole process is thus to $\ldots\ldots\ldots\ldots$

21

express $(1 - e^{-Ts})$ in the denominator as $(1 - e^{-Ts})^{-1}$ in the numerator and to expand this as a binomial series.

We do this by making use of the basic series

$$(1 - x)^{-1} = \ldots\ldots\ldots\ldots$$

22

$$\boxed{(1-x)^{-1} = 1 + x + x^2 + x^3 + x^4 + \dots}$$

Example 2

Determine $L^{-1}\left\{\dfrac{3(1-e^{-s})}{s(1-e^{-3s})}\right\}$ and sketch the resulting waveform of $f(t)$.

$$L\{f(t)\} = \frac{3}{s}(1 - e^{-s})(1 - e^{-3s})^{-1}$$

$$= \dots\dots\dots\dots \qquad \text{(next step)}$$

23

$$\boxed{L\{f(t)\} = \frac{3}{s}(1 - e^{-s})(1 + e^{-3s} + e^{-6s} + e^{-9s} + \dots)}$$

which multiplied out gives

$$L\{f(t)\} = \frac{3}{s}(1 - e^{-s} + e^{-3s} - e^{-4s} + e^{-6s} - e^{-7s} + \dots)$$

$$= \frac{3}{s} - \frac{3e^{-s}}{s} + \frac{3e^{-3s}}{s} - \frac{3e^{-4s}}{s} + \frac{3e^{-6s}}{s} - \dots$$

And in unit step form, this gives

$$f(t) = \dots\dots\dots\dots$$

24

$$\boxed{f(t) = 3u(t) - 3u(t-1) + 3u(t-3) - 3u(t-4) + \dots}$$

The waveform is thus $\dots\dots\dots\dots$

25

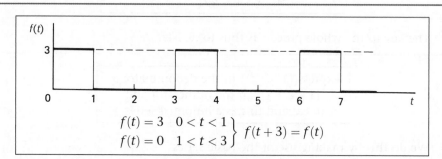

$$\begin{aligned} f(t) &= 3 \quad 0 < t < 1 \\ f(t) &= 0 \quad 1 < t < 3 \end{aligned}\Bigg\} \; f(t+3) = f(t)$$

And now, one more. They are all done in the same way

Example 3 26

If $L\{f(t)\} = \dfrac{1}{2s^2} - \dfrac{2e^{-4s}}{s(1 - e^{-4s})}$, determine $f(t)$ and sketch the waveform.

The first term is easy enough. In unit step form $L^{-1}\left\{\dfrac{1}{2s^2}\right\} = \dfrac{t}{2} \cdot u(t)$

From the second term

$$\dfrac{2e^{-4s}}{s(1 - e^{-4s})} = \dfrac{2}{s}\left\{e^{-4s}(1 - e^{-4s})^{-1}\right\}$$

$$= \dfrac{2}{s}\left\{e^{-4s}(1 + e^{-4s} + e^{-8s} + e^{-12s} + \ldots)\right\}$$

$$= \dfrac{2e^{-4s}}{s} + \dfrac{2e^{-8s}}{s} + \dfrac{2e^{-12s}}{s} + \dfrac{2e^{-16s}}{s} + \ldots$$

$$\therefore\ f(t) = \ldots\ldots\ldots\ldots \quad \text{(in unit step form)}$$

27

$$f(t) = \dfrac{t}{2} \cdot u(t) - 2u(t - 4) - 2u(t - 8) - 2u(t - 12) - \ldots$$

Now we have to draw the waveform. Consider the function terms up to each break point in turn.

$0 < t < 4 \quad f(t) = \dfrac{t}{2} \qquad\qquad f(0) = 0;\ \ f(4) = 2$

$4 < t < 8 \quad f(t) = \dfrac{t}{2} - 2 \qquad\quad f(4) = 0;\ \ f(8) = 2$

$8 < t < 12 \ \ f(t) = \dfrac{t}{2} - 2 - 2 \ \ f(8) = 0;\ \ f(12) = 2 \ \text{etc.}$

So the waveform is $\ldots\ldots\ldots\ldots$

28

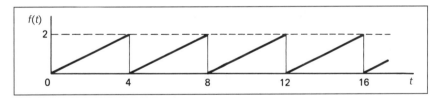

Expressed analytically, we finally have

$$f(t) = \dfrac{t}{2} \quad 0 < t < 4, \quad f(t + 4) = f(t)$$

The Dirac delta – the unit impulse

29

So far we have dealt with a number of standard Laplace transforms and then the Heaviside unit step function with some of its applications. We now come to consider an entity that is different from any of the functions we have used before because it is not a proper function. Rather than being defined by its inputs and corresponding outputs it is defined by its effect on other functions. If $f(t)$ represents a function then the Dirac delta $\delta(t)$ is defined by the integral

$$\int_{-\infty}^{\infty} f(t)\delta(t-a)\,\mathrm{d}t = f(a)$$

$\delta(t)$ is often referred to as the **Dirac delta function** even though it is not a function in the conventional sense of being completely defined in terms of its outputs for the corresponding inputs. The nearest that can be achieved in defining it in function terms is

$$\delta(t) = \begin{cases} 0 & t \neq 0 \\ \text{undefined} & t = 0 \end{cases}$$

From the definition, if $f(t) = 1$ then

$$\int_{-\infty}^{\infty} \delta(t-a)\,\mathrm{d}t = \dots\dots\dots$$

30

$$\boxed{\int_{-\infty}^{\infty} \delta(t-a)\,\mathrm{d}t = 1}$$

Because

$$\int_{-\infty}^{\infty} f(t)\delta(t-a)\,\mathrm{d}t = f(a) \text{ and } f(t) = 1 \text{ so } f(a) = 1, \text{ therefore}$$

$$\int_{-\infty}^{\infty} \delta(t-a)\,\mathrm{d}t = 1 \text{ hence the name } \textit{unit impulse}.$$

Also, if $p < a < q$ then

$$\int_{p}^{q} \delta(t-a)\,\mathrm{d}t = \dots\dots\dots$$

31

$$\int_p^q \delta(t-a)\,dt = 1$$

Because

$$\int_{-\infty}^{\infty} \delta(t-a)\,dt = \int_{-\infty}^{p} \delta(t-a)\,dt + \int_p^q \delta(t-a)\,dt + \int_q^{\infty} \delta(t-a)\,dt$$

$$= 0 + \int_p^q \delta(t-a)\,dt + 0 \qquad \begin{array}{l} \text{since } \delta(t-a) = 0 \\ \text{for } -\infty < t \le p \\ \text{and } q \le t < \infty \end{array}$$

$$= 1$$

So that $\quad \int_p^q \delta(t-a)\,dt = 1$

Graphical representation

Graphically the Dirac delta or unit impulse $\delta(t-a)$ is represented by the horizontal axis with a vertical line of infinite length at $t = a$ and where the infinite nature of the line is indicated by an arrow-head.

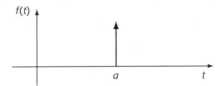

32

So far, then, we have

(a) $\displaystyle\int_p^q \delta(t-a)\,dt = 1$

(b) $\displaystyle\int_p^q f(t)\cdot\delta(t-a)\,dt = f(a)$

provided, in each case, that $p < a < q$.

Example 1

To evaluate $\displaystyle\int_1^3 (t^2 + 4)\cdot\delta(t-2)\,dt$.

The factor $\delta(t-2)$ shows that the impulse occurs at $t = 2$, i.e. $a = 2$.

$$f(t) = t^2 + 4 \qquad \therefore\ f(a) = f(2) = 4 + 4 = 8$$

$$\therefore\ \int_1^3 (t^2 + 4)\cdot\delta(t-2)\,dt = f(2) = 8$$

▶

Example 2

To evaluate $\int_0^\pi \cos 6t \cdot \delta(t - \pi/2)\, dt$.

$$\int_0^\pi \cos 6t \cdot \delta(t - \pi/2)\, dt = f(\pi/2) = \cos 3\pi = -1$$

and in the same way

(a) $\int_0^6 5 \cdot \delta(t - 3)\, dt = \ldots\ldots\ldots$

(b) $\int_2^5 e^{-2t} \cdot \delta(t - 4)\, dt = \ldots\ldots\ldots$

(c) $\int_0^\infty (3t^2 - 4t + 5) \cdot \delta(t - 2)\, dt = \ldots\ldots\ldots$

33

> (a) $\int_0^6 5 \cdot \delta(t - 3)\, dt = 5 \times 1 = 5$
>
> (b) $\int_2^5 e^{-2t} \cdot \delta(t - 4)\, dt = f(4) = \left[e^{-2t}\right]_{t=4} = e^{-8}$
>
> (c) $\int_0^\infty (3t^2 - 4t + 5) \cdot \delta(t - 2)\, dt = 12 - 8 + 5 = 9$

Nothing could be easier. It all rests on the fact that, provided $p < a < q$

$$\int_p^q f(t) \cdot \delta(t - a)\, dt = \ldots\ldots\ldots$$

34

$$\boxed{f(a)}$$

Now let us consider the Laplace transform of $\delta(t - a)$.

On then to the next frame

---

35 Laplace transform of $\delta(t-a)$

We have already shown that

$$\int_p^q f(t) \cdot \delta(t - a)\, dt = f(a) \qquad p < a < q$$

Therefore, if $p = 0$ and $q = \infty$

$$\int_0^\infty f(t) \cdot \delta(t - a)\, dt = f(a)$$

Hence, if $f(t) = e^{-st}$, this becomes

$$\int_0^\infty e^{-st} \cdot \delta(t - a)\, dt = L\{\delta(t - a)\} = \ldots\ldots\ldots$$

$$\boxed{e^{-as}}$$ **36**

i.e. the value of $f(t)$, i.e. e^{-st}, at $t = a$.

$$L\{\delta(t-a)\} = e^{-as}$$

It follows from this that the Laplace transform of the impulse function at the origin is . . . .

$$\boxed{1}$$ **37**

Because

For $a = 0$, $L\{\delta(t-a)\} = L\{\delta(t)\} = e^0 = 1$

$$\therefore \ L\{\delta(t)\} = 1$$

Finally, let us deal with the more general case of $L\{f(t) \cdot \delta(t-a)\}$. We have

$$L\{f(t) \cdot \delta(t-a)\} = \int_0^\infty e^{-st} \cdot f(t) \cdot \delta(t-a) \, dt.$$

Now the integrand $e^{-st} \cdot f(t) \cdot \delta(t-a) = 0$ for all values of t except at $t = a$ at which point $e^{-st} = e^{-as}$, and $f(t) = f(a)$.

$$\therefore \ L\{f(t) \cdot \delta(t-a)\} = f(a) \cdot e^{-as} \int_0^\infty \delta(t-a) \, dt$$

$$= f(a) \cdot e^{-as}(1)$$

$$\therefore \ L\{f(t) \cdot \delta(t-a)\} = f(a)e^{-as}$$

Another important result to note. Then let us deal with some examples

---

We have $L\{f(t) \cdot \delta(t-a)\} = f(a) \cdot e^{-as}$ **38**

Therefore

(a) $L\{6 \cdot \delta(t-4)\}$ $a = 4$, $\therefore \ L\{6 \cdot \delta(t-4)\} = 6e^{-4s}$

(b) $L\{t^3 \cdot \delta(t-2)\}$ $a = 2$, $\therefore \ L\{t^3 \cdot \delta(t-2)\} = 8e^{-2s}$

Similarly

(c) $L\{\sin 3t \cdot \delta(t - \pi/2)\} = $ . . . .

$$\boxed{-e^{-\pi s/2}}$$ **39**

Because

$$L\{\sin 3t \cdot \delta(t - \pi/2)\} = \left[\sin 3t\right]_{t=\pi/2} \cdot e^{-\pi s/2} = -e^{-\pi s/2}$$

and

(d) $L\{\cosh 2t \cdot \delta(t)\} = $ . . . .

40

$\boxed{1}$

Because

$$L\{\cosh 2t \cdot \delta(t)\} = \left[\cosh 2t\right]_{t=0} \cdot e^0 = \cosh 0 \cdot (1) = 1$$

So our main conclusions so far are as follows.

(1) $\displaystyle\int_p^q \delta(t-a)\,dt = \ldots\ldots\ldots$ provided $\ldots\ldots\ldots$

(2) $\displaystyle\int_p^q f(t)\cdot\delta(t-a)\,dt = \ldots\ldots\ldots$ provided $\ldots\ldots\ldots$

(3) $L\{\delta(t-a)\} = \ldots\ldots\ldots$

(4) $L\{\delta(t)\} = \ldots\ldots\ldots$

(5) $L\{f(t)\cdot\delta(t-a)\} = \ldots\ldots\ldots$

41

> (1) $\displaystyle\int_p^q \delta(t-a)\,dt = 1$ provided $p < a < q$
>
> (2) $\displaystyle\int_p^q f(t)\cdot\delta(t-a)\,dt = f(a)$ provided $p < a < q$
>
> (3) $L\{\delta(t-a)\} = e^{-as}$
>
> (4) $L\{\delta(t)\} = 1$
>
> (5) $L\{f(t)\cdot\delta(t-a)\} = f(a)\cdot e^{-as}$

Just check that you have noted this important list – the basis of all work on the Dirac delta function.

Now for one further example on this section

Example

Impulses of 1, 4, 7 units occur at $t=1$, $t=3$ and $t=4$ respectively, in the directions shown.

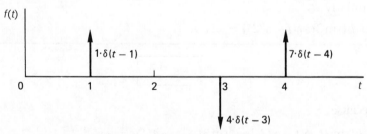

Write down an expression for $f(t)$ and determine its Laplace transform.

We have $f(t) = 1\cdot\delta(t-1) - 4\cdot\delta(t-3) + 7\cdot\delta(t-4)$.

Then $L\{f(t)\} = \ldots\ldots\ldots$

$$L\{f(t)\} = e^{-s} - 4e^{-3s} + 7e^{-4s}$$

and that is all there is to that.

The derivative of the unit step function

One further consideration is interesting.

Consider some function $f(t)$ that is zero outside some finite interval $[a, b]$ of the real line. That is, $f(t) = 0$ for $t < a$ and $t > b$, then

$$\int_{-\infty}^{\infty} [u(t)f(t)]' \, dt = [u(t)f(t)]_{-\infty}^{\infty} = 0$$

where $u(t)$ is the unit step function and $f(t)$ is zero at the limits. Now

$$\int_{-\infty}^{\infty} [u(t)f(t)]' \, dt = \int_{-\infty}^{\infty} u'(t)f(t) \, dt + \int_{-\infty}^{\infty} u(t)f'(t) \, dt$$

and so

$$\int_{-\infty}^{\infty} u'(t)f(t) \, dt = -\int_{-\infty}^{\infty} u(t)f'(t) \, dt$$

This means that

$$\int_{-\infty}^{\infty} u'(t)f(t) \, dt = -\int_{-\infty}^{\infty} u(t)f'(t) \, dt$$

$$= -\int_{0}^{\infty} f'(t) \, dt \qquad \text{Because the unit step is zero for negative } t$$

$$= -\left[f(t)\right]_{0}^{\infty}$$

$$= -f(\infty) + f(0)$$

$$= f(0) \qquad \text{Because } f(\infty) = 0 \text{ by definition}$$

$$= \int_{-\infty}^{\infty} \delta(t)f(t) \, dt \qquad \text{By the definition of the Dirac delta}$$

and so $u'(t) = \delta(t)$ – *the unit impulse is equal to the derivative of the unit step function.*

Differential equations involving the unit impulse

43

Example 1

A system has the equation of motion

$$\ddot{x} + 6\dot{x} + 8x = g(t)$$

where $g(t)$ is an impulse of 4 units applied at $t = 5$. At $t = 0$, $x = 0$ and $\dot{x} = 3$. Determine an expression for the displacement x in terms of t.

The impulse of 4 units is applied at $t = 5$. $\therefore$ $g(t) = 4 \cdot \delta(t - 5)$.

$\therefore$ $\ddot{x} + 6\dot{x} + 8x = 4 \cdot \delta(t - 5)$ At $t = 0$, $x = 0$, $\dot{x} = 3$.

Taking Laplace transforms this differential equation becomes

. . . .

44

$$\boxed{(s^2\bar{x} - sx_0 - x_1) + 6(s\bar{x} - x_0) + 8\bar{x} = 4e^{-5s}}$$

Now $x_0 = 0$; $x_1 = 3$

$\therefore$ $s^2\bar{x} - 3 + 6s\bar{x} + 8\bar{x} = 4e^{-5s}$

$\therefore$ $(s^2 + 6s + 8)\bar{x} = 3 + 4e^{-5s}$

$\therefore$ $\bar{x} = (3 + 4e^{-5s}) \dfrac{1}{(s+2)(s+4)}$

Writing $\dfrac{1}{(s+2)(s+4)}$ in partial fractions, we get

$$\bar{x} = \ldots\ldots\ldots$$

45

$$\boxed{\bar{x} = (3 + 4e^{-5s})\left\{ \frac{1}{2} \cdot \frac{1}{s+2} - \frac{1}{2} \cdot \frac{1}{s+4} \right\}}$$

$$\therefore \bar{x} = \frac{3}{2}\left\{ \frac{1}{s+2} - \frac{1}{s+4} \right\} + 2\left\{ \frac{e^{-5s}}{s+2} - \frac{e^{-5s}}{s+4} \right\}$$

Taking inverse transforms

$$x = \frac{3}{2}\left\{ e^{-2t} - e^{-4t} \right\} + 2\left\{ e^{-2(t-5)} \cdot u(t-5) - e^{-4(t-5)} \cdot u(t-5) \right\}$$

$$= \frac{3}{2}\left\{ e^{-2t} - e^{-4t} \right\} + 2\left\{ e^{-2t} \cdot e^{10} \cdot u(t-5) - e^{-4t} \cdot e^{20} \cdot u(t-5) \right\}$$

which simplifies to $x = \ldots\ldots\ldots$

46

$$\boxed{x = e^{-2t}\left\{ \frac{3}{2} + 2e^{10} \cdot u(t-5) \right\} - e^{-4t}\left\{ \frac{3}{2} + 2e^{20} \cdot u(t-5) \right\}}$$

▶

Example 2

Solve the equation $\ddot{x} + 4\dot{x} + 13x = 2 \cdot \delta(t)$ where, at $t = 0$, $x = 2$ and $\dot{x} = 0$.

$$\ddot{x} + 4\dot{x} + 13x = 2 \cdot \delta(t) \qquad x_0 = 2; \; x_1 = 0$$

Expressing in Laplace transforms, we have

. . . .

$$\boxed{(s^2\bar{x} - sx_0 - x_1) + 4(s\bar{x} - x_0) + 13\bar{x} = 2 \cdot (1)}$$

47

Inserting the initial conditions and simplifying,

$$\bar{x} = \ldots\ldots\ldots\ldots$$

$$\boxed{\bar{x} = (2s + 10)\frac{1}{s^2 + 4s + 13}}$$

48

Rearranging the denominator by completing the square, this can be written

$$\bar{x} = (2s + 10)\frac{1}{(s + 2)^2 + 9}$$

$\therefore \; x = \ldots\ldots\ldots\ldots$

$$\boxed{x = 2e^{-2t}\{\cos 3t + \sin 3t\}}$$

49

Because

$$\bar{x} = \frac{2(s + 2)}{(s + 2)^2 + 9} + \frac{6}{(s + 2)^2 + 9}$$

$\therefore \; x = 2e^{-2t}\cos 3t + 2e^{-2t}\sin 3t$

$\therefore \; x = 2e^{-2t}\{\cos 3t + \sin 3t\}$

Now for one further example for you to work through on your own.

So move on

Example 3

50

The equation of motion of a system is

$$\ddot{x} + 5\dot{x} + 4x = g(t) \text{ where } g(t) = 3 \cdot \delta(t - 2).$$

At $t = 0$, $x = 2$ and $\dot{x} = -2$. Determine an expression for the displacement x in terms of t.

We have $\ddot{x} + 5\dot{x} + 4x = 3 \cdot \delta(t - 2)$ with $x_0 = 2$ and $x_1 = -2$.

As before, you can express this in Laplace transforms, substitute the initial conditions, simplify to obtain an expression for x and finally take inverse transforms to determine the required expression for x.

Work right through it carefully. It is good revision and there are no snags.

$$x = \ldots\ldots\ldots\ldots$$

51

$$x = e^{-t}\{2 + e^2 \cdot u(t-2)\} - e^8 \cdot e^{-4t} \cdot u(t-2)$$

Here is the working for you to check.

$\ddot{x} + 5\dot{x} + 4x = 3 \cdot \delta(t-2)$ with $x_0 = 2$ and $x_1 = -2$

$(s^2\bar{x} - sx_0 - x_1) + 5(s\bar{x} - x_0) + 4\bar{x} = 3e^{-2s}$

$s^2\bar{x} - 2s + 2 + 5s\bar{x} - 10 + 4\bar{x} = 3e^{-2s}$

$(s^2 + 5s + 4)\bar{x} - 2s - 8 = 3e^{-2s}$

$\therefore \quad (s+1)(s+4)\bar{x} = 2s + 8 + 3e^{-2s}$

$\therefore \quad \bar{x} = \dfrac{2(s+4)}{(s+1)(s+4)} + e^{-2s} \cdot \dfrac{3}{(s+1)(s+4)}$

$\quad = \dfrac{2}{s+1} + e^{-2s}\left\{\dfrac{1}{s+1} - \dfrac{1}{s+4}\right\}$

$\therefore \quad \bar{x} = \dfrac{2}{s+1} + \dfrac{e^{-2s}}{s+1} - \dfrac{e^{-2s}}{s+4}$

$\therefore \quad x = 2e^{-t} + u(t-2) \cdot e^{-(t-2)} - u(t-2) \cdot e^{-4(t-2)}$

$\quad = 2e^{-t} + u(t-2) \cdot e^2 \cdot e^{-t} - u(t-2) \cdot e^8 \cdot e^{-4t}$

$x = e^{-t}\{2 + e^2 \cdot u(t-2)\} - e^8 \cdot e^{-4t} \cdot u(t-2)$

Harmonic oscillators

52

If the position of a system at time t is described by the expression $f(t)$ where $f(t)$ satisfies the differential equation

$af''(t) + bf(t) = 0$, $f(0) = \alpha$ and $f'(0) = \beta$

(and where a and b have the same sign)

then, taking Laplace transforms of both sides gives

$L\{af''(t) + bf(t)\} = L\{0\}$

That is

$a[s^2F(s) - s\alpha - \beta] + b[F(s)] = 0$

Collecting like terms gives

$(as^2 + b)F(s) = a(s\alpha + \beta)$

giving

$F(s) = \dfrac{a(s\alpha + \beta)}{as^2 + b}$

Therefore $F(s) = \dfrac{s\alpha}{s^2 + (b/a)} + \dfrac{\beta}{s^2 + (b/a)}$ and so

$$f(t) = \alpha \cos \sqrt{\dfrac{b}{a}} t + \beta \sqrt{\dfrac{a}{b}} \sin \sqrt{\dfrac{b}{a}} t$$

The system executes *simple harmonic, oscillatory motion with natural frequency* $\sqrt{\dfrac{b}{a}}$ radians per unit of time and with period $\dfrac{2\pi}{\sqrt{b/a}} = 2\pi \sqrt{\dfrac{a}{b}}$. It is called an **harmonic oscillator**. Let's try some examples.

Example 1

Find the solution to the harmonic oscillator

$$f''(t) + 16f(t) = 0 \text{ where } f(0) = 1 \text{ and } f'(0) = 0$$

Taking Laplace transforms gives

$$F(s) = \ldots\ldots\ldots\ldots$$

$$F(s) = \dfrac{s}{s^2 + 16}$$

53

Because

Taking Laplace transforms $L\{f''(t) + 16f(t)\} = L\{0\}$.

That is $s^2 F(s) - s + 16F(s) = 0$ and so

$$F(s) = \dfrac{s}{s^2 + 16}$$

This means that

$$f(t) = \ldots\ldots\ldots\ldots$$

$$f(t) = \cos 4t$$

54

Because

$$F(s) = \dfrac{s}{s^2 + 16} = \dfrac{s}{s^2 + 4^2} \text{ so } f(t) = \cos 4t \text{ from the Table of Laplace}$$
transforms on page 68.

The motion of this system is then periodic with frequency 4 radians per unit of time and with period $2\pi/4 = \pi/2$ units of time.

▶

Example 2

The frequency and period of the harmonic oscillator whose position $f(t)$ satisfies the differential equation

$$5f''(t) + 10f(t) = 0 \text{ where } f(0) = 0 \text{ and } f'(0) = 4$$

is given as

frequency .... radians per unit of time
and period .... units of time

55

> frequency $\sqrt{2}$ and period $\sqrt{2}\pi$

Because

Taking Laplace transforms gives

$$L\{5f''(t) + 10f(t)\} = L\{0\} \text{ that is } 5s^2F(s) - 4 + 10F(s) = 0 \text{ so that}$$
$$F(s) = \frac{4}{5s^2 + 10} = \frac{4/5}{s^2 + 2}$$

and from the Table of Laplace transforms on page 68

$$f(t) = \frac{2\sqrt{2}}{5}\sin\sqrt{2}t$$

This is periodic with frequency $\sqrt{2}$ radians per unit of time and period $2\pi/\sqrt{2} = \sqrt{2}\pi$ units of time.

Notice that the amplitude of the motion is $\dfrac{2\sqrt{2}}{5}$.

---

56 Damped motion

Consider the equation

$$5f''(t) + 5f'(t) + 10f(t) = 0 \text{ where } f(0) = 0 \text{ and } f'(0) = 4$$

This is the same as the last equation in Frame 54 with an extra term added, namely $5f'(t)$. This term describes a particular effect on the system as you will see from the solution.

Solving the differential equation gives

$$f(t) = \ldots\ldots\ldots$$

57

$$f(t) = \frac{8}{5\sqrt{7}}e^{-t/2}\sin\left(\sqrt{7}t/2\right)$$

Because

Taking Laplace transforms gives

$$L\{5f''(t) + 5f'(t) + 10f(t)\} = L\{0\} \text{ that is}$$
$$5(s^2F(s) - 4) + 5sF(s) + 10F(s) = 0$$

so that

$$F(s) = \frac{20}{5s^2 + 5s + 10} = \frac{4}{s^2 + s + 2} = \frac{4}{(s + 1/2)^2 + (\sqrt{7}/2)^2}$$

and from the Table of Laplace transforms on page 68

$$f(t) = \frac{8}{\sqrt{7}}e^{-t/2}\sin\left(\sqrt{7}t/2\right)$$

This is periodic with frequency 1 radian per unit of time and period 2π units of time but with an amplitude that is decreasing with time. The graph of this function is as follows

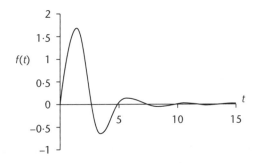

The effect of the $5f'(t)$ in the differential equation is to introduce **damping** into the oscillatory motion so causing the oscillations to decay. Let's try another example.

Example 3

Consider the equation

$$5f''(t) + f'(t) + 10f(t) = 0 \text{ where } f(0) = 0 \text{ and } f'(0) = 4$$

This equation is again similar to the previous equation but with a smaller damping term of $f'(t)$ instead of $5f'(t)$. Then here

$$f(t) = \ldots\ldots\ldots$$

58

$$f(t) = \frac{4}{\sqrt{1\cdot99}} e^{-0.1t} \sin \sqrt{1\cdot99}t$$

Because

Taking Laplace transforms gives

$$L\{5f''(t) + f'(t) + 10f(t)\} = L\{0\} \text{ that is}$$

$$5(s^2F(s) - 4) + sF(s) + 10F(s) = 0$$

so that

$$F(s) = \frac{20}{5s^2 + 1s + 10} = \frac{4}{s^2 + 0\cdot2s + 2} = \frac{4}{(s + 0\cdot1)^2 + 1\cdot99}$$

and from the Table of Laplace transforms on page 68

$$f(t) = \frac{4}{\sqrt{1\cdot99}} e^{-0.1t} \sin \sqrt{1\cdot99}t$$

This is periodic with frequency $\sqrt{1\cdot99}$ radians per unit of time and period $2\pi/\sqrt{1\cdot99}$ units of time and with an amplitude that is decreasing with time. The graph of this function is as follows

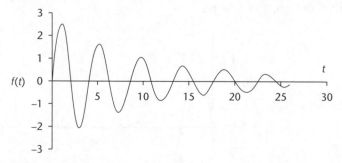

Again, the effect of the $f'(t)$ in the differential equation is to introduce damping into the oscillatory motion so causing it to decay. Also because the coefficient of $f'(t)$ is smaller in this example, the damping is less severe.

Forced harmonic motion with damping

59

The equation

$$f''(t) + f'(t) + f(t) = e^t \text{ where } f(0) = 0 \text{ and } f'(0) = 0$$

we know would represent damped harmonic motion were it not for the exponential on the right-hand side. To see the effect of the exponential we solve the equation.

Taking Laplace transforms we see that

$$F(s) = \ldots\ldots\ldots\ldots$$

60

$$F(s) = \frac{1}{(s-1)(s^2+s+1)}$$

Because

$$L\{f''(t) + f'(t) + f(t)\} = L\{e^t\} \text{ that is } (s^2+s+1)F(s) = \frac{1}{s-1} \text{ so}$$

$$F(s) = \frac{1}{(s-1)(s^2+s+1)}$$

Separating into partial fractions gives

$$F(s) = \ldots\ldots\ldots\ldots$$

61

$$F(s) = \frac{1}{3(s-1)} - \frac{s+2}{3(s^2+s+1)}$$

Because

$$\frac{1}{(s-1)(s^2+s+1)} = \frac{A}{(s-1)} + \frac{Bs+C}{(s^2+s+1)}$$

$$= \frac{A(s^2+s+1) + (Bs+C)(s-1)}{(s-1)(s^2+s+1)}$$

Equating numerators and then comparing coefficients of powers of s gives

$$1 = A(s^2+s+1) + (Bs+C)(s-1)$$

[s^2]: $0 = A + B$ (1) So (2) + (3): $1 = 2A - B$

[s]: $0 = A - B + C$ (2) $2 \times$ (1): $0 = 2A + 2B$

[CT]: $1 = A - C$ (3) Therefore: $-1 = 3B$

so $B = -1/3 = -A$ and $C = -2/3$

Thus $F(s) = \dfrac{1}{(s-1)(s^2+s+1)} = \dfrac{1}{3(s-1)} - \dfrac{s+2}{3(s^2+s+1)}$

Consequently $f(t) = \ldots\ldots\ldots\ldots$

62

$$f(t) = \frac{e^t}{3} - \frac{1}{3}e^{-t/2}\left(\cos\frac{\sqrt{3}}{2}t + \sqrt{3}\sin\frac{\sqrt{3}}{2}t\right)$$

Because

$$F(s) = \frac{1}{3(s-1)} - \frac{s+2}{3(s^2+s+1)}$$

$$= \frac{1}{3(s-1)} - \frac{s+\frac{1}{2}}{3\left(\left(s+\frac{1}{2}\right)^2+\frac{3}{4}\right)} - \frac{\frac{3}{2}}{3\left(\left(s+\frac{1}{2}\right)^2+\frac{3}{4}\right)}$$

So

$$f(t) = \frac{e^t}{3} - \frac{1}{3}e^{-t/2}\left(\cos\frac{\sqrt{3}}{2}t + \sqrt{3}\sin\frac{\sqrt{3}}{2}t\right)$$

from the Table of Laplace transforms on page 68.

▶

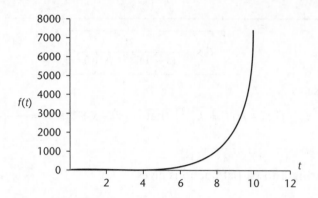

Notice that the term $\frac{1}{3}e^{-t/2}\left(\cos\frac{\sqrt{3}}{2}t + \sqrt{3}\sin\frac{\sqrt{3}}{2}t\right)$ represents damped harmonic motion and is called the **transient** term whereas the term $\frac{e^t}{3}$ represents a
steady-state term, so called because as the transient term decays the steady-
state term remains the dominant part of the solution. The steady-state
solution is a direct consequence of the term on the right-hand side of the
differential equation.

Try another one for yourself. The transient and steady-state terms of the
system described by the differential equation

$$f''(t) + 2f'(t) + 5f(t) = e^{2t} \text{ where } f(0) = 0 \text{ and } f'(0) = 1$$

are   Transient term . . . .   Steady-state term . . . .

63

$$-\frac{1}{13}e^{-t}\cos 2t + \frac{5}{13}e^{-t}\sin 2t,\ \frac{1}{13}e^{2t}$$

Because

Taking Laplace transforms, $L\{f''(t) + 2f'(t) + 5f(t)\} = L\{e^{2t}\}$. That is

$$[s^2F(s) - 1] + 2sF(s) + 5F(s) = \frac{1}{s-2}, \text{ that is}$$

$$(s^2 + 2s + 5)F(s) = 1 + \frac{1}{s-2} = \frac{s-1}{s-2}$$

So that $F(s) = \dfrac{s-1}{(s-2)(s^2 + 2s + 5)} = \dfrac{A}{s-2} + \dfrac{Bs+C}{s^2 + 2s + 5}$. Hence

$s - 1 = A(s^2 + 2s + 5) + (Bs + C)(s - 2)$. Equating powers of s gives

$[s^2]$: $0 = A + B$
$[s]$: $1 = 2A - 2B + C$
$[CT]$: $-1 = 5A - 2C$

Solving these three equations gives $A = 1/13$, $B = -1/13$ and $C = 9/13$ so that

$$F(s) = \frac{1}{13(s-2)} - \frac{s-9}{13(s^2 + 2s + 5)}$$

$$= \frac{1}{13(s-2)} - \frac{s-9}{13\left((s+1)^2 + 2^2\right)}. \text{ That is}$$

$$F(s) = \frac{1}{13(s-2)} - \frac{s+1}{13\left((s+1)^2 + 2^2\right)} + \frac{10}{13\left((s+1)^2 + 2^2\right)}$$

Therefore

$$f(t) = \frac{1}{13}e^{2t} - \frac{1}{13}e^{-t}\cos 2t + \frac{5}{13}e^{-t}\sin 2t$$

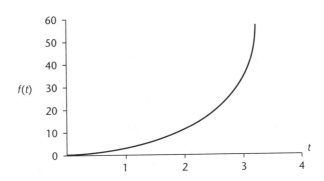

Next frame

Resonance

64

These differential equations with a function on the right-hand side are called **inhomogeneous differential equations**. They represent systems whose behaviour $f(t)$ is dictated by the structure of the left-hand side and the **forcing function** on the right-hand side. If an external force is applied to an undamped harmonic oscillator with a vibrational frequency equal to the oscillator's natural frequency [Frame 52] the oscillator will be set in motion and vibrate in sympathy at its natural frequency. This is called **resonance**. If the applied force is maintained unabated the oscillator will continue to resonate but with an increasing amplitude. An example will illustrate this.

The differential equation

$$f''(t) + f(t) = 0 \text{ where } f(0) = 0 \text{ and } f'(0) = 1$$

represents an undamped, unforced system with behaviour

$$f(t) = \dots\dots\dots$$

65

$$f(t) = \sin t$$

Because

Taking the Laplace transform of both sides of the equation gives

$$L\{f''(t) + f(t)\} = L\{0\} \text{ that is } s^2 F(s) - 1 + F(s) = 0 \text{ so that}$$

$$F(s) = \frac{1}{s^2 + 1} \text{ giving } f(t) = \sin t$$

If the forcing term $-2\sin t$ is applied to the right-hand side of the equation it has the same period as the natural frequency of the system being forced and so resonance will set in. The differential equation to solve is then

$$f''(t) + f(t) = -2\sin t \text{ where } f(0) = 0 \text{ and } f'(0) = 1$$

This has the solution $f(t) = \ldots\ldots\ldots\ldots$

66

$$f(t) = t\cos t$$

Because

Taking the Laplace transform of both sides of the equation gives

$$L\{f''(t) + f(t)\} = L\{-2\sin t\} \text{ that is } s^2 F(s) - 1 + F(s) = -\frac{2}{s^2 + 1}$$

so that $F(s) = \dfrac{1}{s^2 + 1} - \dfrac{2}{(s^2 + 1)^2}$ giving $F(s) = \dfrac{s^2 - 1}{(s^2 + 1)^2}$. Now, the

Laplace transform of $\cos t$ is $\dfrac{s}{s^2 + 1}$ and $\left(\dfrac{s}{s^2 + 1}\right)' = -\dfrac{s^2 - 1}{(s^2 + 1)^2}$.

Therefore $f(t) = t\cos t$

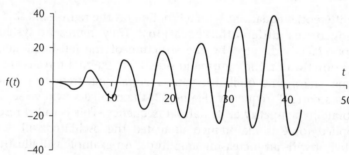

The system undergoes periodic behaviour with an increasing amplitude.

You have now reached the end of this Programme and this brings you to the **Revision summary** and the **Can you?** checklist. Following that is the **Test exercise**. Work through this *at your own pace*. A set of **Further problems** provides additional valuable practice.

Revision summary 4

67

1   *Periodic functions*

$$f(t) = f(t + nT) \qquad n = 1, 2, 3, \dots \qquad \text{Period} = T.$$

2   *Laplace transform of a periodic function with period T*

$$L\{f(t)\} = \frac{1}{1 - e^{-Ts}} \int_0^T e^{-st} \cdot f(t)\, dt.$$

3   *Inverse transforms involving periodic functions*

$$\text{e.g.} \quad L^{-1}\left\{ \frac{1 + 2e^{-3s} - 3e^{-2s}}{s(1 - e^{-3s})} \right\}$$

Expand $(1 - e^{-3s})^{-1}$ as a binomial series, like

$$(1 - x)^{-1} = 1 + x + x^2 + x^3 + \dots$$

Multiply out and take inverse transforms of each term in turn.

4   *Dirac delta function* or unit impulse function

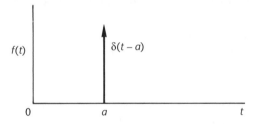

$$\begin{aligned} \delta(t - a) &= 0 & t \neq a \\ &= \infty & t = a. \end{aligned}$$

5   *Delta function at the origin*

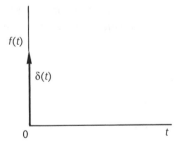

$$a = 0 \quad \therefore \ \delta(t) = 0 \qquad t \neq 0$$
$$= \infty \qquad t = 0.$$

6   *Area of pulse = 1*

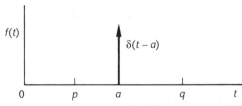

$$\therefore \ \int_p^q \delta(t - a)\, dt = 1$$
$$p < a < q$$

▶

7 *Integration of the impulse function*

$$\int_p^q f(t) \cdot \delta(t-a)\,\mathrm{d}t = f(a) \qquad p < a < q$$

8 *Laplace transform of $\delta(t-a)$*

$L\{\delta(t-a)\} = e^{-as}$

$L\{\delta(t)\} = 1$ because $a = 0$

$L\{f(t) \cdot \delta(t-a)\} = f(a) \cdot e^{-as}.$

9 *Harmonic oscillators*

The equation of $af''(t) + bf(t) = 0$, $f(0) = \alpha$ and $f'(0) = \beta$, where a and b are of the same sign, represents a system undergoing simple harmonic motion and is referred to as an harmonic oscillator. The system oscillates with a frequency of $\sqrt{\dfrac{b}{a}}$ radians per unit of time and with period $\dfrac{2\pi}{\sqrt{b/a}} = 2\pi\sqrt{\dfrac{a}{b}}$ units of time. If a first derivative term is added to the left-hand side of the equation then, provided all three coefficients have the same sign, the system will undergo damped harmonic motion.

10 *Forced harmonic motion*

Forced harmonic motion is achieved by the existence of a term on the right-hand side of the equation giving rise to transient and steady-state parts of the solution.

11 *Resonance*

If an external force is applied to an undamped harmonic oscillator with a vibrational frequency equal to the oscillator's natural frequency the oscillator will be set in motion and vibrate in sympathy at its natural frequency. This is called **resonance**. If the applied force is maintained unabated the oscillator will continue to resonate but with an increasing amplitude.

 Can you?

68 **Checklist 4**

Check this list before and after you try the end of Programme test.

On a scale of 1 to 5, how confident are you that you can: **Frames**

- Find the Laplace transforms of periodic functions?

  Yes   ☐   ☐   ☐   ☐   ☐   *No*   1 to 14

- Obtain the inverse Laplace transforms of transforms of periodic functions?

  Yes   ☐   ☐   ☐   ☐   ☐   *No*   15 to 28

▶

- Describe and use the unit impulse to evaluate integrals?
  Yes ☐ ☐ ☐ ☐ ☐ *No*

- Obtain the Laplace transform of the unit impulse?
  Yes ☐ ☐ ☐ ☐ ☐ *No*

- Use the Laplace transform to solve differential equations
  involving the unit impulse?
  Yes ☐ ☐ ☐ ☐ ☐ *No*

- Solve the equation and describe the behaviour of an harmonic
  oscillator?
  Yes ☐ ☐ ☐ ☐ ☐ *No*

 # Test exercise 4

---

1 Determine the Laplace transform of the periodic function shown. **69**

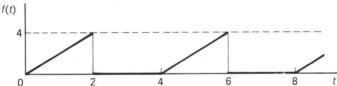

2 Evaluate

(a) $\int_0^4 e^{-3t} \cdot \delta(t-2)\,dt$ (b) $\int_0^\infty \sin 3t \cdot \delta(t-\pi)\,dt$ (c) $\int_1^3 (2t^2+3) \cdot \delta(t-2)\,dt$.

3 Determine (a) $L\{4 \cdot \delta(t-3)\}$, (b) $L\{e^{-3t} \cdot \delta(t-2)\}$.

4 Sketch the graph of $f(t) = 3 \cdot \delta(t) + 4 \cdot \delta(t-2) - 3 \cdot \delta(t-4)$ and determine its
  Laplace transform.

5 Solve the equation $\ddot{x} + 6\dot{x} + 10x = 7 \cdot \delta(t)$ given that, at $t = 0$, $x = -1$ and $\dot{x} = 0$.

6 The equation of motion of a system is
  $$\ddot{x} + 3\dot{x} + 2x = 3 \cdot \delta(t-4).$$
  At $t = 0$, $x = 2$ and $\dot{x} = -4$. Determine an expression for the displacement x in
  terms of t.

7 Find the frequency, periodic time and solution for each of the following
  harmonic oscillators.
  (a) $f''(t) + f(t) = 0$ given that $f(0) = 0$ and $f'(0) = 1$
  (b) $6f''(t) + 2f'(t) + 9f(t) = 0$ given that $f(0) = 0$ and $f'(0) = 3$.

8 Find the transient and steady-state solutions of the forced harmonic oscillator
  $f''(t) + 2f'(t) + 3f(t) = 4e^{5t}$ given that $f(0) = -2$ and $f'(0) = 6$.

Further problems 4

70   **1** If $f(t) = a\sin t \quad \begin{aligned} 0 \le t < \pi \\ = 0 \quad\quad \pi \le t < 2\pi \end{aligned} \Big\} \; f(t + 2\pi) = f(t),$

prove that $L\{f(t)\} = \dfrac{a}{(s^2 + 1)(1 - e^{-\pi s})}$.

2 If $f(t) = a\sin t \quad 0 \le t < \pi \quad f(t + \pi) = f(t)$, determine $L\{f(t)\}$.

3 Find the Laplace transforms of the following periodic functions.

(a) $f(t) = t \quad\quad 0 \le t < T \quad\quad f(t + T) = f(t)$

(b) $f(t) = e^t \quad\quad 0 \le t < 2\pi \quad\quad f(t + 2\pi) = f(t)$

(c) $f(t) = t \quad\quad 0 \le t < 1 \atop = 0 \quad\quad 1 \le t < 2 \Big\} \quad f(t + 2) = f(t)$

(d) $f(t) = t^2 \quad\quad 0 \le t < 2 \atop = 4 \quad\quad 2 \le t < 3 \Big\} \quad f(t + 3) = f(t)$

4 A mass M is attached to a spring of stiffness $\omega^2 M$ and is set in motion at $t = 0$ by an impulsive force P. The equation of motion is

$$M\ddot{x} + M\omega^2 x = P \cdot \delta(t).$$

Obtain an expression for x in terms of t.

5 An impulsive voltage E is applied at $t = 0$ to a series circuit containing inductance L and capacitance C. Initially, the current and charge are zero. The current i at time t is given by

$$L\frac{di}{dt} + \frac{q}{C} = E \cdot \delta(t)$$

where q is the instantaneous value of the charge on the capacitor. Since $i = \dfrac{dq}{dt}$, determine an expression for the current i in the circuit at time t.

6 A system has the equation of motion

$$\ddot{x} + 5\dot{x} + 6x = F(t)$$

where, at $t = 0$, $x = 0$ and $\dot{x} = 2$. If $F(t)$ is an impulse of 20 units applied at $t = 4$, determine an expression for x in terms of t.

7 Find the frequency, periodic time and solution for each of the following harmonic oscillators.

(a) $12f''(t) + f(t) = 0$ given that $f(0) = -1$ and $f'(0) = 2$

(b) $f''(t) + 12f(t) = 0$ given that $f(0) = 2$ and $f'(0) = -1$.

8 Solve for each of the following harmonic oscillators.

(a) $4{\cdot}6f''(t) + 2{\cdot}2f(t) = 0$ given that $f(0) = 1{\cdot}6$ and $f'(0) = -3{\cdot}1$

(b) $\sqrt{2}f''(t) + \sqrt{3}f(t) = 0$ given that $f(0) = 0$ and $f'(0) = \pi$.

9 Find the transient and steady-state solutions of the forced harmonic oscillator

$$4f''(t) + 3f'(t) + 2f(t) = e^t$$

given that $f(0) = 0$ and $f'(0) = 6$.

Difference equations and the *Z* transform

Learning outcomes

When you have completed this Programme you will be able to:

- Convert the descriptive prescription of the output form of a sequence into a recursive description and recognise the importance of initial terms

- Recognise a difference equation, determine its order and generate its terms from a recursive description

- Obtain the solution to a difference equation as a sum of the homogeneous solution and the particular solution

- Define the *Z* transform of a sequence and derive transforms of specified sequences

- Make reference to a table of standard *Z* transforms

- Recognise the *Z* transform as being a linear transform and so obtain the transform of linear combinations of standard sequences

- Apply the first and second shift theorems, the translation theorem, the initial and final value theorems and the derivative theorem

- Use partial fractions to derive the inverse transforms

- Use the *Z* transform to solve linear, constant coefficient difference equations

- Create a sequence by sampling a continuous function and demonstrate the relationship between the Laplace and the *Z* transform

Introduction

1

The Laplace transform deals with continuous functions and can be used to solve many differential equations that arise in science and engineering. There are occasions, however, when we have to deal with discrete functions – *sequences* – and their associated **difference equations**. For example, the central processing unit of your computer can only handle information in the form of pulses of electricity. This information transmission is called **digital** transmission. There are, however, times when information is fed into the computer in the form of a continuously varying signal called an **analogue** signal. For instance, a mouse can be moved about the flat surface of your desk in a continuous manner but the central processing unit will only recognise position on the screen to the nearest pixel. The analogue signal coming from the mouse needs to be converted into a digital signal for recognition by the computer's central processing unit. This conversion of a signal from analogue to digital is achieved by a device called a **demodulator** that *samples* the analogue signal at regular intervals of time and outputs the sampled values as the digital signal – as a sequence of numbers. The *Z* transform, which is allied to the Laplace transform, deals with such sequences and the recurrence relations – or difference equations – that arise.

Sequences

2

Any function f of a single real variable whose input is restricted to integer values n has an output $f(n)$ in the form of a discrete sequence of numbers. Accordingly, such a function is called a **sequence**. For example, the function defined by the prescription

$f(n) = 5n - 2$ where n is an integer ≥ 1

is a sequence. The first three output values corresponding to the successive input values 1, 2 and 3 are

$f(1) = 5 \times 1 - 2 = 3$
$f(2) = 5 \times 2 - 2 = 8$
$f(3) = 5 \times 3 - 2 = 13$

Each output value $f(n)$ of the sequence is called a **term** of the sequence. An alternative way of describing the terms of this sequence can be found from the following consideration:

$f(1) = 5 \times 1 - 2 = 3$
$f(2) = 5 \times 2 - 2 = 3 + 5$
$f(3) = 5 \times 3 - 2 = 8 + 5$

The value of any term is the value of the previous term plus 5 and provided we know that the first term is 3 we can compute any other term. A process such as ▶

this one that repeatedly uses known values to compute an unknown value is called a **recursive process**. That is:

$$f(n+1) = 5(n+1) - 2$$
$$= [5n - 2] + 5$$
$$= f(n) + 5 \quad \text{so that } f(n+1) = f(n) + 5 \text{ where } f(1) = 3.$$

This description, in which each term of the sequence is seen to depend upon another term of the same sequence, is called a *recursive* description and can make the computing of the terms of the sequence more efficient and very amenable to a spreadsheet implementation. In this particular example we simply start with 3 and just add 5 to each preceding term to get:

3, 8, 13, 18, 23, 28, 33, 38, ...

Notice that without the initial term $f(1) = 3$ the recursive description would be of little worth because we would not know how to start the sequence.

Find the recursive description and compute the first four terms of each of the following sequences:

(a) $f(n) = 7n + 4$ where n is an integer ≥ 1

(b) $f(n) = 8 - 2n$ where n is an integer ≥ 0

(c) $f(n) = 4^n$ where n is an integer ≥ -3
   [slightly different involving multiplication rather than addition]

The answers are in the following frame

$$f(n+1) = f(n) + 7 \text{ where } f(1) = 11: \; 11, \, 18, \, 25, \, 32$$
$$f(n+1) = f(n) - 2 \text{ where } f(0) = 8: \; 8, \, 6, \, 2, \, 4$$
$$f(n+1) = 4f(n) \text{ where } f(-3) = \frac{1}{64}: \; \frac{1}{64}, \, \frac{1}{16}, \, \frac{1}{4}, \, 1$$

3

Because:

(a) $f(n+1) = 7(n+1) + 4$
$$= [7n + 4] + 7$$
$$= f(n) + 7 \text{ so } f(n+1) = f(n) + 7 \text{ where } f(1) = 11$$
   giving the first four terms as 11, 18, 25, 32

(b) $f(n+1) = 8 - 2(n+1)$
$$= [8 - 2n] - 2$$
$$= f(n) - 2 \text{ so } f(n+1) = f(n) - 2 \text{ where } f(0) = 8$$
   giving the first four terms as 8, 6, 4, 2

(c) $f(n+1) = 4^{n+1}$
$$= 4[4^n]$$
$$= 4f(n) \text{ so } f(n+1) = 4f(n) \text{ where } f(-3) = 4^{-3} = \frac{1}{64}$$

giving the first four terms as $\dfrac{1}{64}, \, \dfrac{1}{16}, \, \dfrac{1}{4}, \, 1$

Next frame

Difference equations

4

The recursive equation $f(n+1) = f(n) + 5$ can also be written as

$$f(n+1) - f(n) = 5$$

and is an example of a *first order, constant coefficient, linear difference equation* or *linear recurrence relation*. It is linear because there are no products of terms such as $f(n) \times f(m)$ and it is first order because $f(n+1)$ is just one term away from $f(n)$. The order of a difference equation is taken from the maximum number of terms between any pair of terms so that, for example:

(a) $f(n+2) + 2f(n) = 3n^4 + 2$ is a *second* order difference equation because $f(n+2)$ is two terms away from $f(n)$

and

(b) $-3f(n+3) - f(n+2) + 5f(n-1) + 4f(n-2) = -6n^2 \cos n\pi$ is a *fifth* order difference because $f(n+3)$ is five terms away from $f(n-2)$

So the order of the difference equation

$$89f(n-3) + 17f(n+1) - 3f(n-2) + 5f(n+5) = 13n^2 - 2n^4$$

is ....

The answers are in the following frame

5

$$\boxed{8}$$

Because:

$f(n+5)$ is 8 terms away from $f(n-3)$.

In order to generate the terms of the sequence from the recursive description it is necessary to have as many initial terms as the order of the difference equation. For example if we are given the second order difference equation with a single initial term:

$f(n+2) + 2f(n) = 3n + 2$ where $f(1) = 1$

then by substituting into the difference equation we see that:

$f(1+2) + 2f(1) = 3 \times 1 + 2$ that is $f(3) = 5 - 2f(1) = 3$ and
$f(2+2) + 2f(2) = 3 \times 2 + 2$ that is $f(4) = 8 - 2f(2) = ?$ and
$f(3+2) + 2f(3) = 3 \times 3 + 2$ that is $f(5) = 11 - 2f(3) = 5$ and
$f(4+2) + 2f(4) = 3 \times 4 + 2$ that is $f(6) = 16 - 2f(4) = ?$

With the single initial term given we can find all those terms of the sequence that correspond to an odd value of n but unless we are given the value of a term that corresponds to an even value of n, a second initial term, we cannot find any of the other terms of the sequence.

The order and the number of initial conditions necessary to generate the terms of the sequence $f(n)$ from the difference equation:

$$f(n+4) - 3f(n+2) + 5f(n-3) = 2nu(n) \text{ are } \dots\dots \text{ and } \dots\dots$$

The answers are in the following frame

6

$$\boxed{7 \text{ and } 7}$$

Because:

$f(n+4)$ is 7 terms away from $f(n-3)$ and the number of initial conditions required to recover the terms of the sequence from a recursive description is equal to the order of the equation.

For example, if a sequence has terms that satisfy the second order difference equation

$f(n+2) - 3f(n+1) + 2f(n) = 1$ where $f(0) = 0$ and $f(1) = 1$

then the first five terms of the sequence are

$$0, 1, \dots\dots\dots, \dots\dots\dots, \dots\dots\dots$$

Next frame

7

$$\boxed{0, 1, 4, 11, 26}$$

Because:

Since $f(n+2) - 3f(n+1) + 2f(n) = 1$ where $f(0) = 0$ and $f(1) = 1$ then

$f(2) - 3f(1) + 2f(0) = 1$ that is $f(2) - 3 \times 1 + 2 \times 0 = 1$ and so $f(2) = 4$
$f(3) - 3f(2) + 2f(1) = 1$ that is $f(3) - 3 \times 4 + 2 \times 1 = 1$ and so $f(3) = 11$
$f(4) - 3f(3) + 2f(2) = 1$ that is $f(4) - 3 \times 11 + 2 \times 4 = 1$ and so $f(4) = 26$

Try another yourself. The first six terms of the sequence that satisfies the second-order difference equation

$$f(n+2) - f(n) = 1 \text{ where } f(0) = 0 \text{ and } f(1) = -1 \text{ are}$$
$$0, 1, \dots\dots\dots, \dots\dots\dots, \dots\dots\dots, \dots\dots\dots$$

Next frame

8

$$\boxed{0, -1, 1, 0, 2, 1}$$

Because:

Since $f(n+2) - f(n) = 1$ where $f(0) = 0$ and $f(1) = -1$ then

$f(2) - f(0) = 1$ that is $f(2) - 0 = 1$ and so $f(2) = 1$
$f(3) - f(1) = 1$ that is $f(3) - f(-1) = 1$ and so $f(2) = 0$
$f(4) - f(2) = 1$ that is $f(4) - 1 = 1$ and so $f(4) = 2$
$f(5) - f(3) = 1$ that is $f(5) - 0 = 1$ and so $f(5) = 1$

They are all done the same way.

Move on to the next frame

Solving difference equations

9

We have seen how the prescription for a sequence such as:

$f(n) = 5n - 2$ where n is an integer ≥ 1

can be manipulated to create the difference equation

$f(n + 1) - f(n) = 5$ where $f(1) = 3$

What we wish to be able to do now is to reverse this process. That is, given the difference equation we wish to find the prescription for the sequence which is the *solution to the difference equation*. In their most general form these linear difference equations can be written as:

$$a_n f(n) + a_{n-1} f(n - 1) + \ldots + a_{n-k} f(n - k) = b_m g(m) + b_{m-1} g(m - 1) + \ldots$$
$$+ b_{m-l} g(m - l)$$

where the sequence f on the left is unknown and the sequence g on the right along with all the a and b coefficients are known. It is not dissimilar in structure to an ordinary differential equation and we shall find that the process of finding the solution is similar as well [Ref: *Engineering Mathematics, Sixth Edition*]. Read on.

Move on to the next frame

---

10 **Solution by inspection**

The solution to a constant coefficient, linear recursive difference equation is analogous to the solution of a constant coefficient, linear differential equation. It is of the form

$f(n) = f_h(n) + f_p(n)$

where $f_h(n)$ is the solution to the homogeneous equation:

$a_n f(n) + a_{n-1} f(n - 1) + \ldots + a_{n-k} f(n - k) = 0$

and $f_p(n)$ is a particular solution of the inhomogeneous difference equation. Also, for a complete solution, an nth order difference equation must be accompanied by n initial terms.

For example to solve the second order difference equation:

$f(n + 2) - 7f(n + 1) + 12f(n) = 1$ for $n \geq 0$ where $f(0) = 0$ and $f(1) = 1$

we first consider the homogeneous difference equation

$f(n + 2) - 7f(n + 1) + 12f(n) = 0$

and assume a solution of the form

$f_h(n) = Kw^n$

so that the equation becomes:

$$\ldots\ldots\ldots\ldots = 0$$

Next frame

$$\boxed{Kw^n\{w^2 - 7w + 12\} = 0}$$ **11**

Because:

Substitution of $f(n) = Kw^n$ into $f(n+2) - 7f(n+1) + 12f(n) = 0$ yields

$$Kw^{n+2} - 7Kw^{n+1} + 12Kw^n = K(w^{n+2} - 7w^{n+1} + 12w^n)$$
$$= Kw^n\{w^2 - 7w + 12\}$$
$$= 0$$

This is called the *characteristic equation* of the difference equation and it has roots given from:

$w^2 - 7w + 12 = (w - 3)(w - 4) = 0$. That is $w = 3$ or $w = 4$ therefore:

$f_h(n) = A \times 3^n + B \times 4^n$ where A and B are constants

To find $f_p(n)$ we assume a form of $f_p(n) = C_1 n + C_2$ where C_1 and C_2 are constants. Substitution yields:

$$C_1 = \ldots\ldots\ldots \text{ and } C_2 = \ldots\ldots\ldots$$

Next frame

$$\boxed{C_1 = 0 \text{ and } C_2 = \frac{1}{6}}$$ **12**

Because:

Substituting $f(n) = C_1 n + C_2$ into $f(n+2) - 7f(n+1) + 12f(n) = 1$ yields

$(C_1(n+2) + C_2) - 7(C_1(n+1) + C_2) + 12(C_1 n + C_2) = 1$, that is

$6C_1 n - 5C_1 + 6C_2 = 1$ so that $C_1 = 0$ and $C_2 = \dfrac{1}{6}$.

Therefore $f_p(n) = \dfrac{1}{6}$

The complete solution is then:

$$f(n) = A \times 3^n + B \times 4^n + \frac{1}{6}$$

From the initial terms we have:

$$f(0) = 0: \quad A \times 3^0 + B \times 4^0 + \frac{1}{6} = 0 \text{ that is } A + B = -\frac{1}{6}$$
$$f(1) = 1: \quad A \times 3^1 + B \times 4^1 + \frac{1}{6} = 1 \text{ that is } 3A + 4B = \frac{5}{6}$$

From these two equations we see that:

$$A = \ldots\ldots\ldots \text{ and } B = \ldots\ldots\ldots \text{ and hence } f(n) = \ldots\ldots\ldots$$

Next frame

13

$$A = -\frac{3}{2}, \ B = \frac{4}{3}, \ f(n) = -\frac{1}{6}\left(3^{n+2} - 2 \times 4^{n+1} - 1\right)$$

Because:

Since $3A + 4B = \dfrac{5}{6}$ and $A + B = -\dfrac{1}{6}$ so that $3A + 3B = -\dfrac{3}{6}$ and $4A + 4B = -\dfrac{4}{6}$

then $A = -\dfrac{3}{2}$ and $B = \dfrac{4}{3}$ and hence

$$\begin{aligned}
f(n) &= -\frac{3}{2} \times 3^n + \frac{4}{3} \times 4^n + \frac{1}{6} \\
&= -\frac{3^{n+1}}{2} + \frac{4^{n+1}}{3} + \frac{1}{6} \\
&= -\frac{3^{n+2}}{6} + \frac{2 \times 4^{n+1}}{6} + \frac{1}{6} \\
&= -\frac{1}{6}\left(3^{n+2} - 2 \times 4^{n+1} - 1\right)
\end{aligned}$$

Just to re-cap to make sure you are clear of what we have done here. The sequence f with terms

$$f(n) = -\frac{1}{6}\left(3^{n+2} - 2 \times 4^{n+1} - 1\right)$$

is the solution of the difference equation:

$$f(n+2) - 7f(n+1) + 12f(n) = 1 \text{ where } f(0) = 0, \ f(1) = 1$$

As a check:

$$\begin{aligned}
&f(n+2) - 7f(n+1) + 12f(n) \\
&= -\frac{1}{6}\left(3^{n+4} - 2 \times 4^{n+3} - 1\right) - 7\left[-\frac{1}{6}\left(3^{n+3} - 2 \times 4^{n+2} - 1\right)\right] \\
&\quad + 12\left[-\frac{1}{6}\left(3^{n+2} - 2 \times 4^{n+1} - 1\right)\right] \\
&= -\frac{1}{6}\left(81 \times 3^n - 128 \times 4^n - 1\right) + \frac{7}{6}\left(27 \times 3^n - 32 \times 4^n - 1\right) \\
&\quad - 2\left(9 \times 3^n - 8 \times 4^n - 1\right) \\
&= 3^n\left(-\frac{81}{6} + \frac{189}{6} - 18\right) + 4^n\left(\frac{128}{6} - \frac{224}{6} + 16\right) + \left(\frac{1}{6} - \frac{7}{6} + 2\right) \\
&= 1
\end{aligned}$$

Now you try one. If you follow the route given here you will find that it is quite straightforward

The difference equation:

$$f(n+2) + 3f(n+1) - 10f(n) = 4$$

where $f(0) = 1$ and $f(1) = 0$ has solution $f(n) = \ldots\ldots\ldots\ldots$

The answer is in the following frame

$$\frac{1}{21} \left(2^3 \times (-5)^n + 3^3 \times 2^n - 14 \right)$$

Because:

$f(n) = f_h(n) + f_p(n)$ and assuming $f_h(n) = Kw^n$ we arrive at the characteristic equation:

$w^2 + 3w - 10 = (w + 5)(w - 2)$ with roots $w = -5$ and $w = 2$ therefore:
$f_h(n) = A \times (-5)^n + B \times 2^n$

To find $f_p(n)$ we assume a form of $f_p(n) = C_1 n + C_2$ then substitution yields:

$(C_1(n + 2) + C_2) + 3(C_1(n + 1) + C_2) - 10(C_1 n + C_2) = 4$ that is,

$- 6C_1 n + 5C_1 - 6C_2 = 4$ so that $C_1 = 0$ and $C_2 = -\dfrac{2}{3} = f_p(n)$

The complete solution is then:

$$f(n) = A \times (-5)^n + B \times 2^n - \frac{2}{3}$$

From the initial terms we find that:

$$f(0) = A \times (-5)^0 + B \times 2^0 - \frac{2}{3} = 1 \text{ that is } A + B = \frac{5}{3}$$

$$f(1) = A \times (-5)^1 + B \times 2^1 - \frac{2}{3} = 0 \text{ that is } - 5A + 2B = \frac{2}{3}$$

From these two equations we find that:

$$A = \frac{8}{21} \text{ and } B = \frac{9}{7} \text{ therefore:}$$

$$f(n) = \frac{8}{21} \times (-5)^n + \frac{9}{7} \times 2^n - \frac{2}{3}$$

$$= \frac{1}{21} \left(2^3 \times (-5)^n + 3^3 \times 2^n - 14 \right)$$

Move on to the next frame

The particular solution

15

To find the particular solution of a difference equation we make an assumption about a certain form for the solution and apply it to the difference equation. The form assumed depends upon the form of the right-hand side of the equation and a sample of these are listed in the following table:

$g(n)$	Particular solution
Polynomial term n^m	$C_m n^m + C_{m-1} n^{m-1} + \ldots + C_1 n + C_0$
Exponential a^n	Ca^n
$a^n \cos bn$, $a^n \sin bn$	$a^n(C_1 \cos bn + C_2 \sin bn)$

▶

For example, the solution to the difference equation $f(n+1) - 3f(n) = g(n)$ where $f(0) = 1$ and

(a) $g(n) = n^2$

(b) $g(n) = 2^n$

(c) $g(n) = \cos 2n$ is . . . .

The answers are in the next frame

16

> (a) $f(n) = \dfrac{3^{n+1} - (n^2 + n + 1)}{2}$
>
> (b) $f(n) = 2(3^n - 2^{n-1})$
>
> (c) $f(n) = 3^n + \dfrac{\cos 2 - 3}{10 - 6\cos 2}[\cos 2n - 3^n] + \dfrac{\sin 2}{10 - 6\cos 2}\sin 2n$

Because

The homogeneous equation $f(n+1) - 3f(n) = 0$ has solution $f_h(n) = Kw^n$: giving the characteristic equation $Kw^n(w - 3) = 0$ so that $w = 3$ and $f_h(n) = K \times 3^n$. The particular solution is:

(a) $f_p(n) = Cn^2 + Dn + e$. Substitution into the inhomogeneous equation yields

$$C(n+1)^2 + D(n+1) + E - 3(Cn^2 + Dn + E) = n^2$$

so that

$$C(n^2 + 2n + 1) + D(n+1) + E - 3(Cn^2 + Dn + E)$$
$$= n^2(-2C) + n(2C - 2D) + (C + D - 2E)$$
$$= n^2$$

Hence

$$C = D = E = -\frac{1}{2}$$

therefore

$$f_p(n) = -\frac{n^2 + n + 1}{2} \quad \text{and} \quad f(n) = K \times 3^n - \frac{n^2 + n + 1}{2}.$$

Applying the boundary condition $f(0) = 1$ we see that $f(0) = K - 1/2 = 1$ giving $K = 3/2$ so that:

$$f(n) = \frac{3^{n+1} - (n^2 + n + 1)}{2}$$

Check: $f(0) = \dfrac{3^1 - 1}{2} = 1$

▶

(b) $f_p(n) = C \times 2^n$. Substitution into the inhomogeneous equation yields

$$C \times 2^{n+1} - 3C \times 2^n = 2^n \text{ so that}$$

$$
\begin{aligned}
C \times 2^{n+1} - 3C \times 2^n &= (2C - 3C) \times 2^n \\
&= -C \times 2^n \\
&= 2^n
\end{aligned}
$$

Hence $C = -1$ therefore $f_p(n) = -2^n$ and $f(n) = K \times 3^n - 2^n$.

Applying the boundary condition $f(0) = 1$ we see that $f(0) = K - 1 = 1$ giving $K = 2$ so that:

$$f(n) = 2(3^n - 2^{n-1})$$

Check: $f(0) = 2(1 - 2^{-1}) = 1$

(c) $f_p(n) = A \cos 2n + B \sin 2n$. Substitution into the inhomogeneous equation yields

$$A \cos 2[n+1] + B \sin 2[n+1] - 3A \cos 2n - 3B \sin 2n = \cos 2n \text{ so that}$$

$$A(\cos 2n \cos 2 - \sin 2n \sin 2) + (B \sin 2n \cos 2 + \sin 2 \cos 2n)$$
$$- 3A \cos 2n - 3B \sin 2n = \cos 2n.$$

Hence

$$\{A \cos 2 + B \sin 2 - 3A\} \cos 2n + \{-A \sin 2 + B \cos 2 - 3B\} \sin 2n = \cos 2n$$

so that

$$A(\cos 2 - 3) + B \sin 2 = 1$$
$$-A \sin 2 + B(\cos 2 - 3) = 0$$

Multiplying the first equation by $\sin 2$ and the second by $\cos 2 - 3$ gives

$$A(\cos 2 - 3) \sin 2 + B \sin^2 2 = \sin 2$$
$$-A(\cos 2 - 3) \sin 2 + B(\cos 2 - 3)^2 = 0$$

and multiplying the second equation by $\sin 2$ and the first by $\cos 2 - 3$ gives

$$A(\cos 2 - 3)^2 + B \sin 2(\cos 2 - 3) = \cos 2 - 3$$
$$-A \sin^2 2 + B \sin 2(\cos 2 - 3) = 0$$

so that

$$A\left\{(\cos 2 - 3)^2 + \sin^2 2\right\} = A\{\cos^2 2 + \sin^2 2 - 6 \cos 2 + 9\} = \cos 2 - 3$$

and

$$B\left\{(\cos 2 - 3)^2 + \sin^2 2\right\} = B\{\cos^2 2 + \sin^2 2 - 6 \cos 2 + 9\} = \sin 2$$

therefore

$$A = \frac{\cos 2 - 3}{10 - 6 \cos 2} \text{ and } B = \frac{\sin 2}{10 - 6 \cos 2}$$

Therefore

$$f_p(n) = \frac{\cos 2 - 3}{10 - 6 \cos 2} \cos 2n + \frac{\sin 2}{10 - 6 \cos 2} \sin 2n \text{ and}$$

$$f(n) = K \times 3^n + \frac{\cos 2 - 3}{10 - 62} \cos 2n + \frac{\sin 2}{10 - 6 \cos 2} \sin 2n.$$

▶

Applying the boundary condition $f(0) = 1$ we see that

$$f(0) = K + \frac{\cos 2 - 3}{10 - 6\cos 2} = 1 \text{ giving } K = 1 - \frac{\cos 2 - 3}{10 - 6\cos 2} \text{ so that:}$$

$$f(n) = 3^n + \frac{\cos 2 - 3}{10 - 6\cos 2}[\cos 2n - 3^n] + \frac{\sin 2}{10 - 6\cos 2}\sin 2n$$

Check: $f(0) = 3^0 + \dfrac{\cos 2 - 3}{10 - 6\cos 2}[\cos 0 - 3^0] + \dfrac{\sin 2}{10 - 6\cos 2}\sin 0 = 1$

Solving linear, constant coefficient **difference** equations in this way is quite straightforward and quite analogous to the method used for solving linear, constant coefficient **differential** equations for particularly simple equations. As soon as the inhomogeneous equation becomes in any way complicated the algebraic manipulation becomes very labour intensive. Fortunately, there is a simpler way out. Just as we can solve constant coefficient linear **differential** equations by using the Laplace transform and simultaneously incorporating the boundary conditions so we can solve constant coefficient **difference** equations using a transform called the **Z transform** again, simultaneously incorporating the initial conditions. Read on.

Next frame

The Z transform

17

We have already seen that the Laplace transform of the piecewise continuous function $f(t)$ is given as

$$L\{f(t)\} = \int_{t=0}^{\infty} f(t)e^{-st}\, \mathrm{d}t$$

$$= \int_{t=0}^{\infty} \frac{f(t)}{e^{st}}\, \mathrm{d}t$$

This is, in fact, a single-sided Laplace transform and is a special case of what is called the **bilateral Laplace transform** where the integration ranges from minus infinity to plus infinity:

$$L\{f(t)\} = \int_{t=-\infty}^{\infty} f(t)e^{-st}\, \mathrm{d}t$$

$$= \int_{t=-\infty}^{\infty} \frac{f(t)}{e^{st}}\, \mathrm{d}t$$

The bilateral transform is identical to the familiar single-sided transform when $f(t) = 0$ for $t < 0$ The equivalent transform for a function that is not piecewise continuous but discrete is:

$$Z\{f(n)\} = \sum_{n=-\infty}^{\infty} \frac{f(n)}{z^n}$$

$$= F(z) \quad \text{where } n \text{ is an integer}$$

▶

This is called the **Z transform** of the sequence. For example, the sequence $\dots, 3^{-2}, 3^{-1}, 3^0, 3^1, 3^2, \dots$ has a general term of the form $f(n) = 3^n$ and its Z transform is:

$$Z\{f(n)\} = \sum_{n=-\infty}^{\infty} \frac{f(n)}{z^n}$$

$$= \sum_{n=-\infty}^{\infty} \frac{3^n}{z^n}$$

$$= \sum_{n=-\infty}^{\infty} \left(\frac{3}{z}\right)^n$$

$$= \dots + \left(\frac{3}{z}\right)^{-1} + \left(\frac{3}{z}\right)^0 + \left(\frac{3}{z}\right)^1 + \left(\frac{3}{z}\right)^2 + \dots$$

Using this definition the Z transform of the sequence $f(n) = 1$ is $\dots\dots\dots\dots$

The answer is in the following frame

18

$$\dots + z^2 + z + 1 + \frac{1}{z} + \frac{1}{z^2} + \dots$$

Because:

$$Z\{f(n)\} = \sum_{n=-\infty}^{\infty} \frac{f(n)}{z^n}$$

$$= \sum_{n=-\infty}^{\infty} \frac{1}{z^n}$$

$$= \dots + \frac{1}{z^{-2}} + \frac{1}{z^{-1}} + \frac{1}{z^0} + \frac{1}{z^1} + \frac{1}{z^2} + \dots$$

$$= \dots + z^2 + z + 1 + \frac{1}{z} + \frac{1}{z^2} + \dots$$

It is noticeable that this sum does not converge for any value of z because $\dots + z^2 + z + 1$ converges to $\frac{1}{1-z}$ only if $|z| < 1$ and diverges if $|z| \geq 1$

and $\frac{1}{z} + \frac{1}{z^2} + \frac{1}{z^3} + \dots = \frac{1}{z}\left(1 + \frac{1}{z^2} + \frac{1}{z^3} + \dots\right)$ converges to $\frac{1}{z}\left(\frac{1}{1-1/z}\right)$ only if $\left|\frac{1}{z}\right| \leq 1$, that is $|z| > 1$ and diverges if $|z| \leq 1$. Since $|z| \leq 1$ or $|z| > 1$ the sum must necessarily diverge.

▶

For a Z transform to have any worth it must converge and we need to know the values of z that ensures this. As a first step we shall avoid doubly infinite sequences and only concern ourselves with those sequences for which $f(n) = 0$ for $n < 0$. For example, the Z transform $F(z)$ of the **discrete unit step function** $u(n)$ where:

$$u(n) = \begin{cases} 1 & n \geq 0 \\ 0 & n < 0 \end{cases} \quad \text{is}$$

$$F(z) = Z\{u(n)\}$$

$$= \sum_{n=-\infty}^{\infty} \frac{u(n)}{z^n}$$

$$= \sum_{n=0}^{\infty} \frac{1}{z^n}$$

$$= \frac{1}{z^0} + \frac{1}{z^1} + \frac{1}{z^2} + \frac{1}{z^3} + \frac{1}{z^4} + \cdots$$

$$= 1 + \frac{1}{z} + \frac{1}{z^2} + \frac{1}{z^3} + \frac{1}{z^4} + \cdots$$

$$= \frac{1}{1 - \dfrac{1}{z}} \qquad \text{recall that } (1-x)^{-1} = 1 + x + x^2 + x^3 + \dots \text{ provided } |x| < 1$$

$$= \frac{z}{z-1} \qquad \text{provided } \left|\frac{1}{z}\right| < 1, \text{ that is } |z| > 1$$

Therefore, the sequence $f(n) = \begin{cases} a^n & n \geq 0 \\ 0 & n < 0 \end{cases}$, which can be written as $f(n) = a^n u(n)$ has the Z transform

$$F(z) = \ldots\ldots\ldots \text{ provided } |z| > \ldots\ldots\ldots$$

The answer is in the following frame

19

$$\frac{z}{z-a} \quad \text{provided } |z| > |a|$$

Because:

Since $f(n) = a^n u(n)$ then

$$F(z) = Z\{f(n)\}$$

$$= \sum_{n=-\infty}^{\infty} \frac{f(n)}{z^n}$$

$$= \sum_{n=-\infty}^{\infty} \frac{a^n u(n)}{z^n}$$

$$= \sum_{n=0}^{\infty} \frac{a^n}{z^n}$$

$$= \sum_{n=0}^{\infty} \left(\frac{a}{z}\right)^n$$

$$= \left(\frac{a}{z}\right)^0 + \left(\frac{a}{z}\right)^1 + \left(\frac{a}{z}\right)^2 + \left(\frac{a}{z}\right)^3 + \left(\frac{a}{z}\right)^4 + \ldots$$

$$= 1 + \left(\frac{a}{z}\right) + \left(\frac{a}{z}\right)^2 + \left(\frac{a}{z}\right)^3 + \left(\frac{a}{z}\right)^4 + \ldots$$

$$= \frac{1}{1 - \frac{a}{z}} \quad \text{provided } \left|\frac{a}{z}\right| < 1$$

$$= \frac{z}{z-a} \quad \text{provided } |z| > |a|$$

Let's try another. The sequence $f(n) = nu(n)$ has the Z transform

$$F(z) = \ldots\ldots\ldots\ldots$$

The answer is in the following frame

20

$$\frac{1}{z} + \frac{2}{z^2} + \frac{3}{z^3} + \frac{4}{z^4} + \ldots$$

Because:

$$F(z) = Z\{f(n)\}$$

$$= \sum_{n=-\infty}^{\infty} \frac{f(n)}{z^n}$$

$$= \sum_{n=-\infty}^{\infty} \frac{nu(n)}{z^n}$$

$$= \sum_{n=0}^{\infty} \frac{n}{z^n}$$

$$= \frac{0}{z^0} + \frac{1}{z^1} + \frac{2}{z^2} + \frac{3}{z^3} + \frac{4}{z^4} + \ldots$$

$$= \frac{1}{z} + \frac{2}{z^2} + \frac{3}{z^3} + \frac{4}{z^4} + \ldots$$

▶

By comparing this sequence with the derivative of the series representation of $(1-x)^{-1}$, this sequence can be written as a rational expression in z as:

$$F(z) = \ldots\ldots\ldots \text{ provided } |z| \ldots\ldots\ldots$$

The answer is in the following frame

21

$$\boxed{\dfrac{z}{(z-1)^2} \text{ provided } |z| > 1}$$

Because:

$$(1-x)^{-1} = 1 + x + x^2 + x^3 + x^4 \ldots \quad \text{provided } |x| < 1$$

and by differentiating both sides

$$\frac{1}{(1-x)^2} = 1 + 2x + 3x^2 + 4x^3 + \ldots \quad \text{provided } |x| < 1.$$

Comparing this with

$$
\begin{aligned}
F(z) &= \frac{1}{z} + \frac{2}{z^2} + \frac{3}{z^3} + \frac{4}{z^4} + \ldots \\
&= \frac{1}{z}\left(1 + \frac{2}{z} + \frac{3}{z^2} + \frac{4}{z^3} + \frac{5}{z^4} + \ldots\right) \\
&= \frac{1}{z}\left[\frac{1}{(1 - 1/z)^2}\right] \quad \text{provided } \left|\frac{1}{z}\right| < 1 \text{ that is provided } |z| > 1
\end{aligned}
$$

So, multiplying numerator and denominator by z^2

$$F(z) = \frac{z}{(z-1)^2} \quad \text{provided } |z| > 1$$

And another example. The Z transform of the **discrete unit impulse**:

$$\delta(n) = \begin{cases} 1 & n = 0 \\ 0 & \text{otherwise} \end{cases} \quad \text{is } F(z) = \ldots\ldots\ldots$$

The answer is in the following frame

22

$$\boxed{1}$$

Because:

$$
\begin{aligned}
F(z) &= Z\{\delta(n)\} \\
&= \sum_{n=0}^{\infty} \frac{\delta(n)}{z^n} \\
&= \frac{1}{z^0} + \frac{0}{z^1} + \frac{0}{z^2} + \ldots \\
&= 1
\end{aligned}
$$

Next frame

Table of *Z* transforms

We list the results that we have obtained so far as well as some additional ones **23** for future reference.

Sequence	Transform $F(z)$	Permitted values of z
$\delta(n) = \{1, 0, 0, \ldots\}$	1	All values of z
$u(n) = \{1, 1, 1, \ldots\}$	$\dfrac{z}{z-1}$	$\lvert z \rvert > 1$
$n\,u(n) = \{0, 1, 2, 3, \ldots\}$	$\dfrac{z}{(z-1)^2}$	$\lvert z \rvert > 1$
$n^2 u(n) = \{0, 1, 4, 9, \ldots\}$	$\dfrac{z(z+1)}{(z-1)^3}$	$\lvert z \rvert > 1$
$n^3 u(n) = \{0, 1, 8, 27, \ldots\}$	$\dfrac{z(z^2 + 4z + 1)}{(z-1)^4}$	$\lvert z \rvert > 1$
$a^n u(n) = \{1, a, a^2, a^3, \ldots\}$	$\dfrac{z}{(z-a)}$	$\lvert z \rvert > \lvert a \rvert$
$n\,a^n u(n) = \{0, a, 2a^2, 3a^3, \ldots\}$	$\dfrac{az}{(z-a)^2}$	$\lvert z \rvert > \lvert a \rvert$

Next frame

Properties of *Z* transforms

1 *Linearity*   **24**

The Z transform is a linear transform. That is, if a and b are constants then

$$Z(af(n) + bg(n)) = aZ\{f(n)\} + bZ\{g(n)\}$$

For example, the Z transform of the sequence $n\,u(n)$ is $Z\{n\,u(n)\} = \ldots\ldots\ldots$
and the Z transform of the sequence $e^{-2n}u(n)$ is $Z\{e^{-2n}u(n)\} = \ldots\ldots\ldots$

25

$$\boxed{Z\{n\,u(n)\} = \frac{z}{(z-1)^2} \text{ and } Z\{e^{-2n}u(n)\} = \frac{z}{z - e^{-2}}}$$

Because

$$Z\{n\,u(n)\} = \frac{z}{(z-1)^2} \text{ from the table and, also from the table,}$$

$$Z\{a^n u(n)\} = \frac{z}{z-a} \text{ so when } a = e^{-2},$$

$$Z\{e^{-2n}u(n)\} = \frac{z}{z - e^{-2}}$$

Consequently, the Z transform of $(3n - 5e^{-2n})u(n)$ is $\ldots\ldots\ldots$

26

$$\boxed{\dfrac{-5z^3 + 13z^2 - z(3e^{-2} + 5)}{(z-1)^2(z-e^{-2})}}$$

Because

$$Z\{(3n - 5e^{-2n})u(n)\} = 3Z\{n\,u(n)\} - 5Z\{e^{-2n}u(n)\}$$

$$= \frac{3z}{(z-1)^2} - \frac{5z}{(z-e^{-2})}$$

$$= \frac{3z(z-e^{-2}) - 5z(z-1)^2}{(z-1)^2(z-e^{-2})}$$

$$= \frac{3z^2 - 3ze^{-2} - 5z^3 + 10z^2 - 5z}{(z-1)^2(z-e^{-2})}$$

$$= \frac{-5z^3 + 13z^2 - z(3e^{-2} + 5)}{(z-1)^2(z-e^{-2})}$$

2 *First shift theorem* (shifting to the left)

If $Z\{f(n)\} = F(z)$ then

$$Z\{f(n+m)\} = z^m F(z) - \left[z^m f(0) + z^{m-1}f(1) + \ldots + zf(m-1)\right]$$

is the Z transform of the sequence that has been shifted by m places to the left.

For example

$$Z\{f(n+1)\} = zF(z) - zf(0)$$
$$Z\{f(n+2)\} = z^2 F(z) - z^2 f(0) - zf(1)$$

These will be used later when solving difference equations. Note the similarity between these results and the Laplace transforms for the first and second derivatives for continuous functions.

For example, given that $Z\{4^n u(n)\} = \dfrac{z}{z-4}$ then

$$Z\{4^{n+3}u(n)\} = \ldots\ldots\ldots\ldots$$

$$\boxed{\dfrac{64z}{z-4}}$$ **27**

Because

$$Z\{f(n+m)\} = z^m F(z) - \left[z^m f(0) + z^{m-1} f(1) + \ldots + zf(m-1)\right] \text{ so}$$

$$Z\{4^{n+3} u(n)\} = z^3 Z\{4^n u(n)\} - \left[z^3 4^0 + z^2 4^1 + z 4^2\right] \text{ where } Z\{4^n u(n)\} = \dfrac{z}{z-4}$$

$$= z^3 \dfrac{z}{z-4} - \left[z^3 + 4z^2 + 16z\right]$$

$$= \dfrac{z^4}{z-4} - \left[z^3 + 4z^2 + 16z\right]$$

$$= \dfrac{z^4 - \left(z^3 + 4z^2 + 16z\right)(z-4)}{z-4}$$

$$= \dfrac{z^4 - \left(z^4 - 64z\right)}{z-4}$$

$$= \dfrac{64z}{z-4}$$

In this way we have derived the Z transform of the sequence $\{64, 256, 1024, \ldots\}$ by shifting the sequence $\{1, 4, 16, 64, 256, \ldots\}$ three places to the left and losing the first three terms.

Try another. Given that $Z\{n u(n)\} = \dfrac{z}{(z-1)^2}$ then

$$Z\{(n+1)u(n)\} = \ldots\ldots\ldots\ldots$$

$$\boxed{\dfrac{z^2}{(z-1)^2}}$$ **28**

Because

$$Z\{x_{k+m}\} = z^m F(z) - \left[z^m x_0 + z^{m-1} x_1 + \ldots + zx_{m-1}\right] \text{ so}$$

$$Z\{k+1\} = z\dfrac{z}{(z-1)^2} - [z \times 0]$$

$$= \dfrac{z^2}{(z-1)^2}$$

3 *Second shift theorem* (shifting to the right)

If $Z\{f(n)\} = F(z)$ then
$$Z\{f(n-m)\} = z^{-m} F(z)$$
the Z transform of the sequence that has been shifted by m places to the right.

For example, given that $Z\{f(n)u(n)\} = \dfrac{z}{z-1}$ then

$$Z\{f(n-3)u(n-3)\} = \ldots\ldots\ldots\ldots$$

29

$$\boxed{\dfrac{1}{z^2(z-1)}}$$

Because

$$Z\{f(n-m)\} = z^{-m}F(z) \text{ so}$$

$$Z\{f(n-3)u(n-3)\} = z^{-3}\dfrac{z}{z-1}$$

$$= \dfrac{1}{z^2(z-1)}$$

In this way we have derived the Z transform of the sequence $\{0, 0, 0, 1, 1, 1, \ldots\}$ by shifting the sequence $\{1, 1, 1, 1, \ldots\}$ three places to the right and defining the first three terms as zeros.

Try this one. The sequence $\{f(n)u(n)\}$ with Z transform

$$Z\{f(n)u(n)\} = \dfrac{1}{(z-a)}, \text{ where } a \text{ is a constant, is } \{\ldots\ldots\ldots\}$$

30

$$\boxed{f(n) = a^{n-1}u(n-1)}$$

Because

From the table of transforms the nearest transform to the one in question is $\dfrac{z}{(z-a)}$ which is the Z transform of $\{a^n u(n)\}$. Now

$$\dfrac{1}{(z-a)} = \dfrac{1}{z} \times \dfrac{z}{(z-a)}$$

$$= z^{-1}F(z) \quad \text{where } F(z) = Z\{a^n u(n)\}$$

and so

$$\dfrac{1}{(z-a)} = Z\{a^{n-1}u(n-1)\}$$

which is the Z transform of $a^n u(n)$, shifted one place to the right.

4 Translation

If the sequence $f(n)$ has the Z transform $Z\{f(n)\} = F(z)$ then the sequence $a^n f(n)$ has the Z transform $Z\{a^n f(n)\} = F(a^{-1}z)$.

For example, $Z\{nu(n)\} = \dfrac{z}{(z-1)^2}$ so that $Z\{2^n nu(n)\} = \ldots\ldots\ldots$

31

$$\boxed{\frac{2z}{(z-2)^2}}$$

Because

Since $Z\{n\,u(n)\} = \dfrac{z}{(z-1)^2} = F(z)$ then by the translation property

$$Z\{2^n n\,u(n)\} = F(2^{-1}z)$$

$$= \frac{2^{-1}z}{(2^{-1}z - 1)^2}$$

$$= \frac{2z}{(z-2)^2}$$

5 *Final value theorem*

For the sequence $f(n)$ with Z transform $F(z)$

$$\underset{n\to\infty}{Lim}\, f(n) = \underset{z\to 1}{Lim}\left\{\left(\frac{z-1}{z}\right)F(z)\right\} \text{ provided that } \underset{n\to\infty}{Lim}\, f(n) \text{ exists.}$$

For example, the sequence $f(n) = \left(\frac{1}{2}\right)^n u(n)$ has the Z transform

$$F(z) = \frac{z}{z - \frac{1}{2}} = \frac{2z}{2z - 1}.$$

Now

$$\underset{z\to 1}{Lim}\left\{\left(\frac{z-1}{z}\right)F(z)\right\} = \underset{z\to 1}{Lim}\left\{\frac{2(z-1)}{2z-1}\right\} = 0$$

and

$$\underset{n\to\infty}{Lim}\left\{\left(\frac{1}{2}\right)^n u(n)\right\} = 0 \text{ which confirms the final value theorem.}$$

Using the final value theorem the final value of the sequence with the Z transform

$$F(z) = \frac{10z^2 + 2z}{(z-1)(5z-1)^2} \text{ is } \ldots\ldots\ldots\ldots$$

32

$$\boxed{0{\cdot}75}$$

Because

$$\underset{z\to 1}{Lim}\left\{\left(\frac{z-1}{z}\right)F(z)\right\} = \underset{z\to 1}{Lim}\left\{\left(\frac{z-1}{z}\right)\frac{10z^2 + 2z}{(z-1)(5z-1)^2}\right\}$$

$$= \underset{z\to 1}{Lim}\left\{\frac{10z + 2}{(5z-1)^2}\right\}$$

$$= \frac{12}{16}$$

$$= 0{\cdot}75$$

▶

6 *The initial value theorem*

For the sequence $f(n)$ with Z transform $F(z)$

$$f(0) = \underset{z \to \infty}{Lim} \{F(z)\}$$

For example, the sequence $f(n) = a^n u(n)$ has the Z transform $F(z) = \dfrac{z}{z - a}$ and

$$\underset{z \to \infty}{Lim} F(z) = \underset{z \to \infty}{Lim} \dfrac{z}{z - a} = \underset{z \to \infty}{Lim} \dfrac{1}{1} = 1$$

by L'Hôpital's rule. Furthermore $f(0) = a^0 = 1$, so demonstrating the validity of the theorem.

7 *The derivative of the transform*

If $Z\{f(n)\} = F(z)$ then $-zF'(z) = Z\{nf(n)\}$

This is easily proved.

$$F(z) = \sum_{n=0}^{\infty} f(n)z^{-n} \text{ and so } F'(z) = \sum_{n=0}^{\infty} f(n)(-n)z^{-n-1} = -\frac{1}{z}\sum_{n=0}^{\infty} f(n)n\, z^{-n}$$

$$= -\frac{1}{z}Z\{nf(n)\}$$

and so $-zF'(z) = Z\{nf(n)\}$

For example, the sequence $f(n) = a^n u(n)$ has the Z transform $F(z) = \dfrac{z}{z - a}$ and

so the sequence $na^n u(n)$ has Z transform

$$Z\{na^n u(n)\} = -zF'(z) = \ldots\ldots\ldots\ldots$$

33

$$Z\{na^n u(n)\} = \frac{az}{(z - a)^2}$$

Because

$$-zF'(z) = -z\left(\frac{z}{z - a}\right)' = -z\left(\frac{z - a - z}{(z - a)^2}\right) = \frac{az}{(z - a)^2}$$

Notice that this is in agreement with the Table of transforms in Frame 23.

Next frame

---

34 Summary

We now summarise the properties that we have just discussed.

Linearity

$$Z\{af(n) + bg(n)\} = aZ\{f(n)\} + bZ\{g(n)\}$$

▶

If $Z\{f(n)\} = F(z)$ then:

Shifting to the left

$$Z\{f(n+m)\} = z^m F(z) - \left[z^m f(0) + z^{m-1} f(1) + \ldots + z f(m-1) \right]$$

Shifting to the right

$$Z\{f(n-m)\} = z^{-m} F(z)$$

Translation

$$Z\{a^n f(n)\} = F\left(a^{-1} z\right)$$

Final value theorem

$$\underset{n \to \infty}{Lim}\, f(n) = \underset{z \to 1}{Lim} \left\{ \left(\frac{z-1}{z} \right) F(z) \right\} \text{ provided } \underset{n \to \infty}{Lim}\, f(n) \text{ exists}$$

Initial value theorem

$$f(0) = \underset{z \to \infty}{Lim}\, \{F(z)\}$$

Derivative of the transform

$$-z F'(z) = Z\{n f(n)\}$$

Inverse transforms

If the sequence $f(n)$ has Z transform $Z\{f(n)\} = F(z)$, the inverse transform is defined as

35

$$Z^{-1} F(z) = f(n)$$

There are many times when, given the Z transform of a sequence, it is not possible to immediately read off the sequence from the Table of transforms. Instead some manipulation may be required and, as with Laplace transforms, very often this involves using partial fractions.

Example

The sequence $f(n)$ has Z transform $F(z) = \dfrac{z}{z^2 - 5z + 6}$. To find the inverse transform, and hence the sequence, we recognise that the denominator can be factorised and separated into partial fractions as

$$F(z) = \ldots\ldots\ldots\ldots$$

36

$$F(z) = \frac{3}{z-3} - \frac{2}{z-2}$$

Because

$$F(z) = \frac{z}{z^2 - 5z + 6}$$

$$= \frac{z}{(z-2)(z-3)}$$

$$= \frac{A}{z-2} + \frac{B}{z-3}$$

$$= \frac{A(z-3) + B(z-2)}{(z-2)(z-3)}$$

Equating numerators gives $z = A(z-3) + B(z-2)$, giving $A + B = 1$ and $-3A - 2B = 0$. From these two equations we find that $A = -2$ and $B = 3$. So

$$F(z) = \frac{3}{z-3} - \frac{2}{z-2}$$

The nearest Z transform in the table to either of these two partial fractions is $Z\{a^n u(n)\} = \dfrac{z}{z-a}$. Therefore if we write

$$F(z) = \frac{3}{z-3} - \frac{2}{z-2}$$

$$= \frac{3}{z} \times \frac{z}{z-3} - \frac{2}{z} \times \frac{z}{z-2}$$

so $Z^{-1}F(z) = \ldots\ldots\ldots$

37

$$Z^{-1}F(z) = (3^n - 2^n)u(n)$$

Because

$$F(z) = \frac{3}{z} \times \frac{z}{z-3} - \frac{2}{z} \times \frac{z}{z-2}$$

$$= 3 \times z^{-1}Z\{3^n u(n)\} - 2 \times z^{-1}Z\{2^n u(n)\}$$

and so

$$Z^{-1}F(z) = 3 \times 3^{n-1}u(n-1) - 2 \times 2^{n-1}u(n-1) \text{ by the second shift theorem}$$

$$= 3^n u(n) - 2^n u(n)$$

So $f(n) = (3^n - 2^n)u(n)$.

There is a simpler way of doing this without employing the second shift theorem. Recognising that z appears in the numerator of $F(z)$, we consider instead the partial fraction breakdown of $\dfrac{F(z)}{z}$

$$\frac{F(z)}{z} = \ldots\ldots\ldots$$

$$\boxed{\dfrac{1}{z-3} - \dfrac{1}{z-2}}$$

Because

$$\frac{F(z)}{z} = \frac{1}{z} \times \frac{z}{z^2 - 5z + 6}$$

$$= \frac{1}{z^2 - 5z + 6}$$

$$= \frac{1}{(z-2)(z-3)}$$

$$= \frac{A}{z-2} + \frac{B}{z-3}$$

$$= \frac{A(z-3) + B(z-2)}{(z-2)(z-3)}$$

Equating numerators gives $1 = A(z-3) + B(z-2)$, giving

[z]: $\qquad A + B = 0$

[CT]: $\qquad -3A - 2B = 1$ with solution $A = -1$ and $B = 1$. So that

$$\frac{F(z)}{z} = \frac{1}{z-3} - \frac{1}{z-2} \text{ that is}$$

$$F(z) = \frac{z}{z-3} - \frac{z}{z-2}$$

$$= Z\{3^n u(n)\} - Z\{2^n u(n)\} \text{ and so}$$

$$Z^{-1}F(z) = 3^n u(n) - 2^n u(n)$$

$$= (3^n - 2^n)u(n)$$

Thus the use of the second shift theorem is avoided.

So try one yourself. The sequence $f(n)$ has Z transform

$$F(z) = \frac{5z}{(z^2 - 4z + 4)(z + 2)}$$

therefore $f(n) = \ldots\ldots\ldots\ldots$

39

$$\boxed{\frac{5}{16}\left[(2n-1)2^n + (-2)^n\right]u(n)}$$

Because

$$\frac{F(z)}{z} = \frac{1}{z} \times \frac{5z}{(z^2 - 4z + 4)(z + 2)}$$

$$= \frac{5}{(z-2)^2(z+2)}$$

$$= \frac{A}{(z-2)^2} + \frac{B}{z-2} + \frac{C}{z+2}$$

$$= \frac{A(z+2) + B(z-2)(z+2) + C(z-2)^2}{(z-2)^2(z+2)}$$

Equating numerators gives $5 = A(z+2) + B(z^2 - 4) + C(z^2 - 4z + 4)$, giving

$[z^2]$: $B + C = 0$

$[z]$: $A - 4C = 0$

$[CT]$: $2A - 4B + 4C = 5$

with solution $A = 5/4$, $B = -5/16$ and $C = 5/16$, so

$$\frac{F(z)}{z} = \frac{5/4}{(z-2)^2} - \frac{5/16}{z-2} + \frac{5/16}{z+2} \text{ giving}$$

$$F(z) = \frac{5}{8} \times \frac{2z}{(z-2)^2} - \frac{5}{16} \times \frac{z}{z-2} + \frac{5}{16} \times \frac{z}{z+2} \text{ and so}$$

$$Z^{-1}F(z) = \frac{5}{8} \times n2^n u(n) - \frac{5}{16} \times 2^n u(n) + \frac{5}{16} \times (-2)^n u(n)$$

$$= \frac{5}{16}\left[(2n-1)2^n + (-2)^n\right]u(n)$$

Move on to the next frame

Solving difference equations

40

If a sequence satisfies a difference equation with given initial terms then the general term of the sequence can be found by using the Z transform. For example, to solve the difference equation

$$f(n+2) - 5f(n+1) + 6f(n) = 1 \text{ where } f(0) = 0 \text{ and } f(1) = 1$$

we begin by taking the Z transform of both sides of the equation to give:

$$Z\{f(n+2) - 5f(n+1) + 6f(n)\} = Z\{1\} \text{ that is}$$
$$Z\{f(n+2)\} - 5Z\{f(n+1)\} + 6Z\{f(n)\} = Z\{1\}$$

Using the first shift theorem where $Z\{f(n)\} = F(z)$ this then becomes

$$\left(z^2 F(z) - z^2 f(0) - z f(1)\right) - \left(5z F(z) - z f(0)\right) + 6F(z) = \frac{z}{z-1}$$

▶

Collecting like terms and substituting for the initial terms $f(0) = 0$ and $f(1) = 1$ gives

$$(z^2 - 5z + 6)F(z) - z = \frac{z}{z-1} \quad \text{so} \quad (z^2 - 5z + 6)F(z) = z + \frac{z}{z-1} = \frac{z^2}{z-1} \quad \text{that is}$$

$$F(z) = \frac{z^2}{(z-1)(z^2 - 5z + 6)} = \frac{z^2}{(z-1)(z-2)(z-3)} \quad \text{and so}$$

$$\frac{F(z)}{z} = \frac{z}{(z-1)(z-2)(z-3)}$$

This has the partial fraction breakdown

$$\frac{F(z)}{z} = \frac{\cdots\cdots}{z-1} + \frac{\cdots\cdots}{z-2} + \frac{\cdots\cdots}{z-3}$$

The answer is in the next frame

41

$$F(z) = \frac{1/2}{z-1} - \frac{4}{z-2} + \frac{9/2}{z-3}$$

Because:

Letting $\dfrac{z}{(z-1)(z-2)(z-3)} = \dfrac{A}{z-1} + \dfrac{B}{z-2} + \dfrac{C}{z-3}$

$$= \frac{A(z-2)(z-3) + B(z-1)(z-3) + C(z-1)(z-2)}{(z-1)(z-2)(z-3)}$$

and so $z = A(z-2)(z-3) + B(z-1)(z-3) + C(z-1)(z-2)$.

Taking $z = 1$, 2 and 3 in turn yields $A = 1/2$, $B = -2$ and $C = 3/2$.

Consequently,

$$F(z) = \frac{1}{2}\left(\frac{z}{z-1}\right) - 2\left(\frac{z}{z-2}\right) + \frac{3}{2}\left(\frac{z}{z-3}\right) \quad \text{and so } f(n) = \cdots\cdots\cdots$$

The answer is in the next frame

42

$$f(n) = \left(\frac{1}{2} - 2^{n+1} + \frac{3^{n+1}}{2}\right)u(n)$$

Because:

$$f(n) = Z^{-1}\{F(z)\}$$

$$= Z^{-1}\left\{\frac{1}{2}\left(\frac{z}{z-1}\right) - 2\left(\frac{z}{z-2}\right) + \frac{3}{2}\left(\frac{z}{z-3}\right)\right\}$$

$$= \frac{1}{2}Z^{-1}\left\{\frac{z}{z-1}\right\} - 2Z^{-1}\left\{\frac{z}{z-2}\right\} + \frac{3}{2}Z^{-1}\left\{\frac{z}{z-3}\right\}$$

$$= \frac{1}{2}u(n) - 2 \times 2^n u(n) + \frac{3}{2} \times 3^n u(n)$$

$$= \left(\frac{1}{2} - 2^{n+1} + \frac{3^{n+1}}{2}\right)u(n)$$

▶

Try one yourself.

The solution of the second order difference equation

$$f(n+2) - f(n) = 1 \text{ where } f(0) = 0 \text{ and } f(1) = -1 \text{ is } f(n) = \ldots\ldots\ldots$$

Next frame

43

$$f(n) = \left(\frac{1}{4}(2n - 3) + \frac{3}{4}(-1)^n \right) u(n)$$

Because:

Taking the Z transform of the difference equation gives

$Z\{f(n+2) - f(n)\} = Z\{1\}$. That is $Z\{f(n+2)\} - Z\{f(n)\} = Z\{1\}$ so that

$$\left(z^2 F(z) - z^2 f(0) - zf(1) \right) - F(z) = \frac{z}{z-1}.$$

Substituting $f(0) = 0$ and $f(1) = -1$ gives

$$F(z) = \ldots\ldots\ldots$$

Next frame

44

$$F(z) = \frac{-z^2 + 2z}{(z+1)(z-1)^2}$$

Because:

$$\left(z^2 F(z) - z^2 f(0) - zf(1) \right) - F(z) = \frac{z}{z-1} \text{ becomes}$$

$$\left(z^2 F(z) + z \right) - F(z) = \frac{z}{z-1} \text{ so that}$$

$$(z^2 - 1)F(z) = -z + \frac{z}{z-1} = \frac{-z^2 + 2z}{z-1} \text{ and so}$$

$$F(z) = \frac{-z^2 + 2z}{(z^2 - 1)(z - 1)} = \frac{-z^2 + 2z}{(z+1)(z-1)^2}$$

Therefore

$$\frac{F(z)}{z} = \frac{\ldots\ldots}{(z-1)^2} + \frac{\ldots\ldots}{z-1} + \frac{\ldots\ldots}{z+1}$$

Next frame

45

$$\boxed{\frac{F(z)}{z} = \frac{1/2}{(z-1)^2} - \frac{3/4}{z-1} + \frac{3/4}{z+1}}$$

Because:

$$\frac{F(z)}{z} = \frac{-z+2}{(z+1)(z-1)^2}$$

$$= \frac{A}{(z-1)^2} + \frac{B}{z-1} + \frac{C}{z+1}$$

$$= \frac{A(z+1) + B(z+1)(z-1) + C(z-1)^2}{(z+1)(z-1)^2} \quad \text{giving}$$

$$-z+2 = A(z+1) + B(z+1)(z-1) + C(z-1)^2$$

and hence $A = \dfrac{1}{2}$, $B = -\dfrac{3}{4}$ and $C = \dfrac{3}{4}$.

From this we conclude that:

$$f(n) = Z^{-1}\{F(z)\}$$

$$= \frac{1}{2}Z^{-1}\left\{\frac{z}{(z-1)^2}\right\} - \frac{3}{4}Z^{-1}\left\{\frac{z}{z-1}\right\} + \frac{3}{4}Z^{-1}\left\{\frac{z}{z+1}\right\}$$

$$= \frac{1}{2}n\,u(n) - \frac{3}{4}u(n) + \frac{3}{4}(-1)^n u(n)$$

$$= \left(\frac{1}{4}(2n-3) + \frac{3}{4}(-1)^n\right)u(n)$$

Move on to the next frame

Sampling

46

If a continuous function $f(t)$ of time t progresses from $t = 0$ onwards and is measured at every time interval T then what will result is the sequence of values

$$\{f(kT)\} = \{f(0), f(T), f(2T), f(3T), \ldots\}$$

A new, piecewise continuous function $f^*(t)$ can then be created from the sequence of sampled values such that

$$f^*(t) = \begin{cases} f(kT) & \text{if } t = kT \\ 0 & \text{otherwise} \end{cases}$$

▶

The graph of this new function consists of a series of spikes at the regular intervals $t = kT$

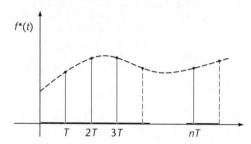

This function can alternatively be described in terms of the delta function $\delta(t)$ as

$$f^*(t) = f(0)\delta(t) + f(T)\delta(t - T) + f(2T)\delta(t - 2T) + f(3T)\delta(t - 3T) + \ldots$$

$$= \sum_{k=0}^{\infty} f(kT)\delta(t - kT)$$

The Laplace transform of $f^*(t)$ is then given as

$$F^*(s) = L\{f^*(t)\}$$

$$= \int_0^\infty \{f(0)\delta(t) + f(T)\delta(t - T) + f(2T)\delta(t - 2T) + \ldots\}e^{-st}\, dt$$

$$= f(0) + f(T)e^{-sT} + f(2T)e^{-2sT} + f(3T)e^{-3sT} + \ldots$$

$$= \sum_{k=0}^{\infty} f(kT)e^{-ksT}$$

Define a new variable $z = e^{sT}$ and we see that

$$L\{f^*(t)\} = \sum_{k=0}^{\infty} f(kT)z^{-k} = \sum_{k=0}^{\infty} \frac{f(kT)}{z^k}$$

which is the Z transform of the sequence $\{f(kT)\}$.

Example 1

The function $f(t) = e^{-at}$ is sampled every interval of T.

     The Z transform of the sampled function is then $\ldots\ldots\ldots\ldots$

47

$$\boxed{F(z) = \frac{z}{z - e^{-aT}}}$$

Because

Defining $f^*(t) = \sum_{k=0}^{\infty} f(kT)\delta(t - kT) = \sum_{k=0}^{\infty} e^{-akT}\delta(t - kT)$ then the Laplace transform of $f^*(t)$ is given as

$$F^*(s) = \sum_{k=0}^{\infty} e^{-kaT}e^{-ksT}$$

▶

This means that the Z transform of $\{f(kT)\}$ is

$$F(z) = \sum_{k=0}^{\infty} \frac{e^{-kaT}}{z^k} = \frac{1}{1 - \dfrac{e^{-aT}}{z}} = \frac{z}{z - e^{-aT}}$$

Notice that this agrees with the Z transform of the sequence $b^n u(n)$ $\left(\text{which is } \dfrac{z}{z-b}\right)$ when b is replaced by e^{-aT}.

Try another.

Example 2

The function $f(t) = t$ is sampled every interval of T.

The Z transform of the sampled function is then ....

$$\boxed{F(z) = \frac{Tz}{(z-1)^2}}$$

48

Because

The Z transform of $\{f(kT)\}$ is $F(z) = \sum_{k=0}^{\infty} \dfrac{f(kT)}{z^k}$. Here $f(kT) = kT$ and so

$$F(z) = \sum_{k=0}^{\infty} \frac{kT}{z^k}$$

$$= T\left(\frac{1}{z} + \frac{2}{z^2} + \frac{3}{z^3} + \cdots\right)$$

$$= \frac{T}{z}\left(1 + 2z^{-1} + 3z^{-2} + 4z^{-3} + \cdots\right)$$

$$= -Tz\frac{d}{dz}\left(1 + z^{-1} + z^{-2} + z^{-3} + \cdots\right\}$$

$$= -Tz\frac{d}{dz}\left(1 - \frac{1}{z}\right)^{-1} = \frac{T}{z}\left(1 - \frac{1}{z}\right)^{-2} = \frac{Tz}{(z-1)^2}$$

Example 3

The function $f(t) = \cos t$ is sampled every interval of T.

The Z transform of the sampled function is then ....

49

$$F(z) = \frac{z(z - \cos T)}{z^2 - 2\cos T + 1}$$

Because

$$f(T) = \cos T = \frac{e^{jT} + e^{-jT}}{2} \text{ and the } Z \text{ transform of } \{e^{-kaT}\} \text{ is}$$

$$F(z) = \frac{z}{z - e^{-aT}}.$$

Therefore the Z transform of $\dfrac{e^{jT} + e^{-jT}}{2}$ is

$$\frac{1}{2}\left(\frac{z}{z - e^{-jT}} + \frac{z}{z - e^{jT}}\right) = \frac{1}{2}\left(\frac{z(z - e^{jT}) + z(z - e^{-jT})}{(z - e^{-jT})(z - e^{jT})}\right)$$

$$= \frac{1}{2}\left(\frac{2z^2 - z(e^{jT} + e^{-jT})}{z^2 - [e^{jT} + e^{-jT}]z + 1}\right)$$

$$= \frac{z(z - \cos T)}{z^2 - 2z\cos T + 1}$$

And that is the end of the Programme on Z transforms. All that remain are the **Revision summary** and the **Can you?** checklist. Read through these closely and make sure that you understand all the workings of this Programme. Then try the **Test exercise**; there is no need to hurry, take your time and work through the questions carefully. The **Further problems** then provide a valuable collection of additional exercises for you to try.

 Revision summary 5

50

1 *Sequences*
   Any function f whose input is restricted to integer values n has an output $f(n)$ in the form of a discrete sequence of numbers. A sequence can be defined by a prescription for the nth term. Alternatively, it can be defined recursively where terms are defined by the values of previous terms. A recursively defined sequence requires one or more initial terms to start the process of evaluating successive terms.

2 *Difference equations*
   The equation that recursively defines a sequence is called a difference equation. A linear, constant coefficient difference equation consists of a sum of general terms of the sequence, each multiplied by a constant. The order of a difference equation is the maximum number of terms between any pair of terms in the equation.

▶

3 *Solving difference equations*

In analogy with linear constant coefficient inhomogeneous differential equations, a linear constant coefficient inhomogeneous difference equation can be solved by first finding the inhomogeneous solution in terms of unknown constants, adding this to the particular solution and then applying the initial terms to find the values of the unknown constants.

4 *Z transform*

The *Z* transform of the sequence $f(n)$ is

$$Z\{f(n)\} = \sum_{n=-\infty}^{\infty} \frac{f(n)}{z^n} = F(z) \text{ where the value of } z \text{ is chosen to}$$
$$\text{ensure that the sum converges.}$$

$f(n)$ and $Z\{f(n)\}$ form a *Z* transform pair.

5 *Table of Z transforms*

Sequence	Transform $F(z)$	Permitted values of z
$\delta(n) = \{1, 0, 0, \ldots\}$	1	All values of z
$u(n) = \{1, 1, 1, \ldots\}$	$\dfrac{z}{z-1}$	$\|z\| > 1$
$n\,u(n) = \{0, 1, 2, 3, \ldots\}$	$\dfrac{z}{(z-1)^2}$	$\|z\| > 1$
$n^2 u(n) = \{0, 1, 4, 9, \ldots\}$	$\dfrac{z(z+1)}{(z-1)^3}$	$\|z\| > 1$
$n^3 u(n) = \{0, 1, 8, 27, \ldots\}$	$\dfrac{z(z^2 + 4z + 1)}{(z-1)^4}$	$\|z\| > 1$
$a^n u(n) = \{1, a, a^2, a^3, \ldots\}$	$\dfrac{z}{(z-a)}$	$\|z\| > \|a\|$
$n\,a^n u(n) = \{0, a, 2a^2, 3a^3, \ldots\}$	$\dfrac{az}{(z-a)^2}$	$\|z\| > \|a\|$

6 *Linearity*

The *Z* transform is a linear transform. That is, if a and b are constants then

$$Z\{af(n) + bg(n)\} = aZ\{f(n)\} + bZ\{g(n)\}.$$

7 *First shift theorem* (shifting to the left)

If $Z\{f(n)\} = F(z)$ then

$$Z\{f(n+m)\} = z^m F(z) - \left[z^m f(0) + z^{m-1} f(1) + \ldots + z f(m-1)\right]$$

the *Z* transform of the sequence that has been shifted by m places to the left.

8 *Second shift theorem* (shifting to the right)

If $Z\{f(n)\} = F(z)$ then

$$Z\{n-m\} = z^{-m} F(z)$$

the *Z* transform of the sequence that has been shifted by m places to the right.

▶

9 *Translation*

If the sequence $f(n)$ has the Z transform $Z\{f(n)\} = F(z)$ then the sequence $a^n f(n)$ has the Z transform $Z\{a^n f(n)\} = F(a^{-1}z)$.

10 *Final value theorem*

For the sequence $f(n)$ with Z transform $F(z)$

$$\underset{n \to \infty}{Lim}\, f(n) = \underset{z \to 1}{Lim} \left\{ \left(\frac{z-1}{z} \right) F(z) \right\} \text{ provided that } \underset{n \to \infty}{Lim}\, f(n) \text{ exists.}$$

11 *The initial value theorem*

For the sequence $f(n)$ with Z transform $F(z)$

$$f(0) = \underset{z \to \infty}{Lim}\, \{F(z)\}.$$

12 *The derivative of the transform*

If $Z\{f(n)\} = F(z)$ then $-zF'(z) = Z\{nf(n)\}$.

13 *Inverse transformations*

If the sequence $f(n)$ has Z transform $Z\{f(n)\} = F(z)$, the inverse transform is defined as

$$Z^{-1}F(z) = f(n).$$

15 *Solving difference equations*

If a sequence $f(n)$ satisfies a difference equation with given initial conditions then the general term of the sequence can be found by using the Z transform where $Z\{f(n)\} = F(z)$. This is referred to as *solving the difference equation*.

15 *Sampling*

If a continuous function $f(t)$ is sampled at equal intervals, the resulting sequence has a Z transform that is related to the Laplace transform of the piecewise function created from the sequence of sample values.

$$L\{f^*(t)\} = \sum_{k=0}^{\infty} f(kT)z^{-k} = \sum_{k=0}^{\infty} \frac{f(kT)}{z^k} = Z\{f(kT)\}$$

where

$$\{f(kT)\} = \{f(0), f(T), f(2T), f(3T), \ldots\},$$
$$f^*(t) = \begin{cases} f(kT) & \text{if } t = k \\ 0 & \text{otherwise} \end{cases}$$

and

$$z = e^{sT}.$$

Can you?

Checklist 5

51

Check this list before and after you try the end of Programme test.

On a scale of 1 to 5 how confident are you that you can: **Frames**

- Convert the descriptive prescription of the output form of a sequence into a recursive description and recognise the importance of initial terms?

  Yes ☐ ☐ ☐ ☐ ☐ *No*   1 to 3

- Recognise a difference equation, determine its order and generate its terms from a recursive description?

  Yes ☐ ☐ ☐ ☐ ☐ *No*   4 to 8

- Obtain the solution to a difference equation as a sum of the homogeneous solution and the particular solution?

  Yes ☐ ☐ ☐ ☐ ☐ *No*   9 to 16

- Define the Z transform of a sequence and derive transforms of specified sequences?

  Yes ☐ ☐ ☐ ☐ ☐ *No*   17 to 22

- Make reference to a table of standard Z transforms?

  Yes ☐ ☐ ☐ ☐ ☐ *No*   23

- Recognise the Z transform as being a linear transform and so obtain the transform of linear combinations of standard sequences?

  Yes ☐ ☐ ☐ ☐ ☐ *No*   24 to 26

- Apply the first and second shift theorems, the translation theorem, the initial and final value theorems and the derivative theorem?

  Yes ☐ ☐ ☐ ☐ ☐ *No*   26 to 34

- Use partial fractions to derive the inverse transforms

  Yes ☐ ☐ ☐ ☐ ☐ *No*   35 to 39

- Use the Z transform to solve linear, constant coefficient difference equations?

  Yes ☐ ☐ ☐ ☐ ☐ *No*   40 to 45

- Create a sequence by sampling a continuous function and demonstrate the relationship between the Laplace and the Z transform?

  Yes ☐ ☐ ☐ ☐ ☐ *No*   46 to 49

Test exercise 5

52 1 Find a recursive description corresponding to each of the following prescriptions for the output of a sequence:

(a) $f(n) = 5n - 9$ where n is an integer ≥ 1

(b) $f(n) = 23 - 4n$ where n is an integer ≥ 0

(c) $f(n) = 3^{-n}$ where n is an integer ≥ -2

2 Determine the order and find the first six terms of each of the following sequences:

(a) $f(n+3) - f(n) = 5n$ where $f(0) = 1$, $f(1) = -1$ and $f(2) = 3$

(b) $f(n+1) - 5f(n) + 6f(n-1) = 2n$ where $f(-1) = 0$ and $f(0) = 1$

(c) $f(n+2) - f(n+1) + 12f(n) = 3u(n)$ where $f(0) = -2$ and $f(1) = 5$

3 Obtain the solution to the following difference equation in the form of a sum of homogeneous and particular solutions:

$$f(n+1) - 5f(n) + 6f(n-1) = 2n \text{ where } f(-1) = 0 \text{ and } f(0) = 1$$

Check that your answer is in agreement with the answer to 2(b).

4 Find the Z transform of each of the sequences with output:

(a) $f(n) = (-1)^n u(n)$

(b) $f(n) = (4n - 2a^n)u(n)$

(c) $f(n) = (n-3)u(n)$

(d) $f(n) = (5^{n+2})u(n)$

5 Find the inverse Z transform of

$$F(z) = \frac{z^2(z-3)}{(z^2 - 2z + 1)(z-2)}.$$

6 Solve the difference equation

$$f(n+2) - 4f(n+1) + 4f(n) = 3 \text{ where } f(0) = 1 \text{ and } f(1) = 0.$$

7 The function $f(t) = \sin t$ is sampled at equal intervals of $t = T$. Find the Z transform of the resulting sequence of values.

 Further problems 5

53 1 Find the Z transform of $f(n) = (-a)^n$ where $a > 0$.

2 Solve each of the following difference equations in the form of the homogeneous solution plus the particular solution:

(a) $f(n+2) + 5f(n+1) + 6f(n) = 1$ where $f(0) = 0$ and $f(1) = 1$

(b) $3f(n+2) - 7f(n+1) + 2f(n) = n$ where $f(0) = 1$ and $f(1) = 0$

(c) $f(n+2) - 9f(n) = 2n^2$ where $f(0) = 1$ and $f(1) = 1$.

3 Given that $a(n+1) = b(n)$ and that $b(n+1) = c(n)$ where $c(n) = f(n) - g(n)$, show that $f(n+2) + f(n) = g(n)$ and solve for $f(n)$ when $g(n) = \delta(n)$, the unit impulse sequence where $f(0) = 0$ and $f(1) = 1$.

4 If $p(n+1) = q(n)$

$q(n+1) = r(n)$

$r(n) = f(n) - \alpha q(n) - \beta p(n)$

where α and β are constants, show that

$p(n+2) + \alpha p(n+1) + \beta p(n) = f(n)$

Solve this recurrence relation when $f(0) = 1$, $f(1) = 0$ for

(a) $\alpha = 4$, $\beta = 4$ and $f(n) = \delta(n)$, the unit impulse sequence

(b) $\alpha = 4$, $\beta = 4$ and $f(n) = u(n)$ the unit step sequence.

5 Find the Z transform of each of the following sequences.

(a) $\{1, 0, 1, 0, 1, 0, \ldots\}$

(b) $\{0, 1, 0, 1, 0, 1, \ldots\}$

(c) $\{1, 0, 1, 1, 0, 0, 0, 1\}$

(d) $\{1, 1, 1, 0, 0, 0, 1, 1\}$

(e) $\{0, 0, 0, 1, 1, 1, 0, 0, 0, 1, 1\}$

(f) $\{1, 1, 0, 0, 0, 1, 1\}$

Note that the last four of these are finite sequences.

6 Find the inverse transform of

(a) $F(z) = \dfrac{z}{(z+1)(z+2)(z+3)}$

(b) $F(z) = \dfrac{z^2}{(z+1)(z+2)(z+3)}$

(c) $F(z) = \dfrac{z(3z+1)}{(z-2)(z-3)}$

(d) $F(z) = \dfrac{z^2}{2 - 3z + z^2}$.

7 Given

$$F(z) = \frac{3z^2}{z^2 - z + 1}$$

show that

$Z^{-1}F(z) = \{3, 3, -3, -3, \ldots\}$.

Hint: Use long division on $F(z)$.

8 Given

$$F(z) = \left(1 + \frac{2}{z}\right)^{-3}$$

show that

$Z^{-1}F(z) = \{1, -6, 24, -48, \ldots\}$.

Hint: Use the binomial theorem on $F(z)$.

▶

9 Find the final value of the sequence $f(n)$ with Z transform

$$F(z) = \frac{4z^2 - z}{2z^2 - 3z + 1}.$$

10 What is the initial value of the sequence whose Z transform is given by

$$F(z) = \frac{2z^2 - z + 1}{5 - 3z - 7z^2}?$$

11 Given the sequence of n terms $f(k)$ for $0 \le k \le n - 1$ with Z transform $F_n(z)$, show that the Z transform of the sequence formed by continually repeating the terms $f(k)$ is given as

$$F(z) = \frac{F_n(z)}{1 - z^{-n}}.$$

12 Using the result of Question 11, show that the Z transform of the sequence obtained by continually repeating the three term sequence $\{1, 0, -1\}$ is

$$F(z) = \frac{z^2}{z^2 + 1}.$$

13 Find the Z transforms of the sequence of values obtained when $f(t)$ is sampled at regular intervals of $t = T$ where
  (a) $f(t) = \sinh t$
  (b) $f(t) = \cosh at$
  (c) $f(t) = e^{-at} \cosh bt$.

14 Solve each of the following difference equations using the Z tranform
  (a) $f(n + 2) + 5f(n + 1) + 6f(n) = 1$ where $f(0) = 0$ and $f(1) = 1$
  (b) $f(n + 2) - 7f(n + 1) + 2f(n) = n$ where $f(0) = 1$ and $f(1) = 0$
  (c) $3f(n + 2) - 9f(n) = 2$ where $f(0) = 1$ and $f(1) = 1$
  (d) $f(n + 2) + 2f(n + 1) - 15f(n) = -4n$ where $f(0) = 0$ and $f(1) = 1$

15 If $f(n + 1) = 3(n + 1)f(n)$ show that $f(n + 1) = 3^{n+1}(n + 1)!f(0)$

16 Show that the difference equation $g(n + 2) - g(n + 1) - 6g(n) = 0$ can be derived from the coupled difference equation

$$f(n + 1) = g(n)$$
$$g(n + 1) = g(n) + 6f(n)$$

  Find $f(n)$ and $g(n)$ given that $f(1) = 0$ and $g(1) = 1$.

17 Show that $f(n) = n!u(n)$ satisfies the difference equation
  $$f(n + 1) - (n + 1)f(n) = \delta(n + 1).$$

18 Use the derivative property to find the Z transform of $f(n) = 3^n n u(n - 3)$.

19 Solve the equation for the Fibonacci sequence:
  $$f(n + 2) = f(n + 1) + f(n) \text{ where } f(0) = 0, f(1) = 1 \text{ and } n \ge 0$$

Introduction to invariant linear systems

Learning outcomes

When you have completed this Programme you will be able to:

- Recognise a system as a process whereby an input (either continuous or discrete) is converted to an output, also called the response of the system
- Distinguish between linear and non-linear systems and recognise time-invariant and shift-invariant systems
- Determine the zero-input response and the zero-state response
- Appreciate why zero valued boundary conditions give rise to a time-invariant system
- Demonstrate that the response of a continuous, linear, time-invariant system to an arbitrary input is the convolution of the input with response of the system to a unit impulse
- Understand the role of the exponential function with respect to a linear, time-invariant system
- Use the convolution theorem to find the response of a continuous, linear, time-invariant system to an arbitrary input
- Derive the system transfer function of a constant coefficient linear differential equation and use it to solve the equation
- Demonstrate that the response of a discrete, linear, shift-invariant system to an arbitrary input is the convolution sum of the input with response of the system to a unit impulse
- Understand the role of the exponential function with respect to a discrete linear, shift-invariant system
- Derive the system transfer function of a constant coefficient linear difference equation and use it to solve the equation
- Derive the constant coefficient difference equation from knowledge of its unit impulse response.

Prerequisites: Advanced Engineering Mathematics (Fifth Edition)
Programme 3 Laplace transforms 2, Programme 4 Laplace transforms 3 and **Programme 5 Difference equations and the *Z* transform**

Invariant linear systems

1 Systems

A **system** is a **process** that is capable of accepting an **input**, processing the input and producing an **output**, also called the **response** of the system. In *Engineering Mathematics, Sixth Edition* a function was described as an example of a system where the input was a number x which was processed by the function f to produce a number output $f(x)$ as shown in the box diagram:

$$\xrightarrow{\quad x \quad} \boxed{\quad f \quad} \xrightarrow{\quad f(x) \quad}$$

For example the function f with input x and output $f(x) = \sin x$ can be represented as:

$$\xrightarrow{\quad x \quad} \boxed{\quad f \quad} \xrightarrow{\quad f(x) = \sin x \quad}$$

This system description simply links the input number to the output number via the function. *How* the function performs the process of evaluating the sine of the input number is not accounted for in this description it is just accepted that the function f can do it.

In this Programme we are going to extend this application of a system to one that will accept an expression as input, process the expression and produce another expression as output. For continuous inputs the box diagram for this system will be:

$$\xrightarrow{\quad x(t) \quad} \boxed{\quad L \quad} \xrightarrow{\quad y(t) \quad}$$

that is $y(t) = L\{x(t)\}$ or $L\{x(t)\} = y(t)$

Here the input and output expressions involve the parameter t. In what follows we shall take this to represent the variable time but it can represent whatever variable is appropriate to the problem in hand. For a discrete system the box diagram is:

$$\xrightarrow{\quad x[n] \quad} \boxed{\quad L \quad} \xrightarrow{\quad y[n] \quad}$$

that is $y[n] = L\{x[n]\}$ or $L\{x[n]\} = y[n]$

Here the input and output expressions involve the discrete integer parameter n. For the purposes of this Programme the integer parameter is placed within square braces to indicate the discrete nature as opposed to round braces used to indicate a continuous nature. That is:

$$x_1, x_2, x_3, \ldots, x_n, \ldots \text{ or } x[1], x[2], x[3], \ldots, x[n], \ldots$$

Just as a system can be used to describe the input–output relationship linking two numbers so a system can be used to describe the input–output relationship linking two expressions. What we need to look for now are ▶

input–output relationships linking two expressions that can be described by a system. Further, just as there are many different types of relationship there are many different types of system. The specific type of system we shall be interested in is an invariant linear system (but we get ahead of ourselves).

Move to the next frame

Input-response relationships

2

Many physical situations in science and engineering can be described by a linear, constant coefficient, ordinary differential equation of the type met in the previous Programmes. Their method of solution may differ depending on the structure of the differential equation but the desire to obtain the solution is common to all. Take for instance the particularly simple first order differential equation

$$\frac{dy_1(t)}{dt} = 2t \text{ where } y_1(0) = 0$$

By integrating this equation:

$$\int \frac{dy_1(t)}{dt} \, dt = \int 2t \, dt \text{ that is } \int dy_1(t) = y_1(t) = 2\frac{t^2}{2} + C = t^2 + C$$

and applying the boundary condition $y_1(0) = 0 = 0 + C$ we arrive at the solution $y_1(t) = t^2$.

The same equation, but with a different right-hand side,

$$\frac{dy_2(t)}{dt} = 4t^3 \text{ where } y_2(0) = 0 \text{ has solution } \cdots\cdots\cdots$$

The answer is in the next frame

$$\boxed{t^4}$$

3

Because:

By integrating this equation:

$$\int \frac{dy_2(t)}{dt} \, dt = \int 4t^3 \, dt \text{ that is } \int dy_2(t) = y_2(t) = 4\frac{t^4}{4} + C' = t^4 + C'$$

and applying the boundary condition $y_2(0) = 0 = 0 + C'$ we arrive at the solution $y_2(t) = t^4$.

The general form of this simple equation can be given as:

$$\frac{dy(t)}{dt} = x(t) \text{ where } y(0) = 0$$

and in both cases we insert the specific expression $x(t)$ in the right-hand side and then manipulate the equation to obtain the solution $y(t)$. It is this commonality of procedure that merits further study.

In each case, the method used to find the solution can be represented by a *system* where the differential equation specifies the relationship between the ▶

input and the output. The *process* is that of integration and evaluating the integration constant, the *input* is the term on the right-hand side and the *output* or the *system response* is what we are trying to find, the solution to the differential equation. We can use a box diagram to represent the system:

In the first box diagram $2t$ is input and t^2 is the response and in the second box diagram $4t^3$ is input and t^4 is the response The process L is the same for each differential equation; what differs are the respective inputs and their corresponding responses.

The response of the differential equation $\dfrac{dy(t)}{dt} = x(t)$ to the input $x(t) = \sin t$ where $y(0) = 0$ is

$$y(t) = \dots\dots\dots$$

The answer is in the next frame

4

$$\boxed{-\cos t + 1}$$

Because:

In the differential equation $\dfrac{dy(t)}{dt} = x(t)$, $y(t)$ is the response to the input $x(t) = \sin t$ so that

$$\int \frac{dy(t)}{dt}\, dt = \int \sin t\, dt.$$

That is

$$\int dy(t) = y(t) = -\cos t + A \text{ where } A \text{ is the integration constant.}$$

and applying the boundary condition $y(0) = 0 = -1 + A$ we arrive at the solution $y(t) = -\cos t + 1$.

Move to the next frame

5 Linear systems

Systems that are **linear** are of particular interest because many problems in science and engineering can be posed as linear systems. A system $y(t) = L\{x(t)\}$ is *linear* if sums and scalar multiples are preserved, that is if

$$L\{x_1(t) + x_2(t)\} = L\{x_1(t)\} + L\{x_2(t)\}$$

and

$$L\{\alpha x(t)\} = \alpha L\{x(t)\} \text{ where } \alpha \text{ is a constant.}$$

In particular, by choosing $\alpha = 0$ then $L\{0\} = 0$ which shows that if nothing is put into a linear system nothing will come out – **zero input yields zero output**.

▶

These two properties can be combined. $y(t) = L\{x(t)\}$ is a linear system if:

$L\{ax_1(t) + bx_2(t)\} = aL\{x_1(t)\} + bL\{x_2(t)\}$ where a and b are constants.

For the discrete case, the system is linear if:

$L\{ax_1[n] + bx_2[n]\} = aL\{x_1[n]\} + bL\{x_2[n]\}$ where a and b are constants.

For example, consider the system in which the output is 5 times the input. That is:

$$y(t) = L\{x(t)\} = 5x(t)$$

To show that this is a linear system we consider two distinct inputs $x_1(t)$ and $x_2(t)$ and their respective responses $y_1(t) = 5x_1(t)$ and $y_2(t) = 5x_2(t)$. We also consider the linear combination of the inputs $x(t) = ax_1(t) + bx_2(t)$ where a and b are constants and where $y(t)$ is the corresponding response. Then:

$$
\begin{aligned}
y(t) &= L\{x(t)\} \\
&= L\{ax_1(t) + bx_2(t)\} \\
&= 5[ax_1(t) + bx_2(t)] \qquad \text{the response is 5 times the input} \\
&= 5ax_1(t) + 5bx_2(t) \\
&= ay_1(t) + by_2(t) \\
&= aL\{x_1(t)\} + bL\{x_2(t)\}
\end{aligned}
$$

that is

$$L\{ax_1(t) + bx_2(t)\} = aL\{x_1(t)\} + bL\{x_2(t)\}$$

Therefore the system is linear. On the other hand, the discrete system whose output is the square of the input, that is:

$$y[n] = L\{x[n]\} = x^2[n]$$

is not linear because, if $y_1[n] = x_1{}^2[n]$, $y_2[n] = x_2{}^2[n]$ and $x[n] = ax_1[n] + bx_2[n]$ where a and b are constants and where $y[n]$ is the corresponding response, then:

$$
\begin{aligned}
y[n] &= L\{x[n]\} \\
&= L\{ax_1[n] + bx_2[n]\} \\
&= [ax_1[n] + bx_2[n]]^2 \qquad \text{the response is the square of the input} \\
&= a^2x_1{}^2[n] + 2abx_1[n]x_2[n] + b^2x_2{}^2[n]
\end{aligned}
$$

However, $aL\{x_1[n]\} + bL\{x_2[n]\} = ax_1{}^2[n] + bx_2{}^2[n]$ therefore

$$L\{ax_1[n] + bx_2[n]\} \neq aL\{x_1[n]\} + bL\{x_2[n]\} \text{ and so the system is not linear.}$$

A system that is not linear is called a *non-linear* system. So, which of the following represent a linear system and which a non-linear system?

(a) $y(t) = L\{x(t)\} = x(t)\sin pt$

(b) $y(t) = L\{x(t)\} = e^{x(t)}$

(c) $y[n] = L\{x[n]\} = x[n] + 4x[n-1]$

(d) $y[n] = L\{x[n]\} = \cos(x[n])$

The answers are in the following frame

6

$$\boxed{\begin{array}{ll} \text{(a) linear} & \text{(b) non-linear} \\ \text{(c) linear} & \text{(d) non-linear} \end{array}}$$

Because:

(a) $y(t) = L\{x(t)\} = x(t)\sin pt$

so if $y_1(t) = L\{x_1(t)\} = x_1(t)\sin pt$, $y_2(t) = L\{x_2(t)\} = x_2(t)\sin pt$
and $x(t) = ax_1(t) + bx_2(t)$

where a and b are constants and where $y(t)$ is the corresponding response. Then:

$$\begin{aligned}
y(t) &= L\{x(t)\} \\
&= L\{ax_1(t) + bx_2(t)\} \\
&= [ax_1(t) + bx_2(t)]\sin pt \qquad \text{the response is the input times } \sin pt \\
&= ax_1(t)\sin pt + bx_2(t)\sin pt \\
\\
&= aL\{x_1(t)\} + bL\{x_2(t)\} \qquad \text{and so sums and scalar products are} \\
& \hspace{10.5em} \text{preserved}
\end{aligned}$$

The system is linear.

(b) $y(t) = L\{x(t)\} = e^{x(t)}$ so if $y_1(t) = L\{x_1(t)\} = e^{x_1(t)}$, $y_2(t) = L\{x_2(t)\} = e^{x_2(t)}$
and $x(t) = ax_1(t) + bx_2(t)$ where a and b are constants and where $y(t)$ is the corresponding response. Then:

$$\begin{aligned}
y(t) &= L\{x(t)\} \\
&= L\{ax_1(t) + bx_2(t)\} \\
&= e^{ax_1(t) + bx_2(t)} \qquad \text{the response is } e \text{ to the power of the input} \\
&= e^{ax_1(t)} \times e^{bx_2(t)} \\
&= L\{ax_1(t)\} \times L\{bx_2(t)\} \\
&\neq L\{ax_1(t)\} + L\{bx_2(t)\}
\end{aligned}$$

Therefore $y(t) = L\{x(t)\} = e^{x(t)}$ is not a linear system – it is a non-linear system

(c) $y[n] = L\{x[n]\} = x[n] + 4x[n-1]$ so if $y_1[n] = L\{x_1[n]\} = x_1[n] + 4x_1[n-1]$, $y_2[n] = \{x_2[n]\} = x_1[n] + 4x_2[n-1]$ and $x[n] = ax_1[n] + bx_2[n]$ where a and b are constants and where $y[n]$ is the corresponding response. Then:

$$\begin{aligned}
y[n] &= L\{x[n]\} \\
&= L\{ax_1[n] + bx_2[n]\} \\
&= (ax_1[n] + bx_2[n]) + 4(ax_1[n-1] + bx_2[n-1]) \\
&= a(x_1[n] + 4x_1[n-1]) + b(x_2[n] + 4x_2[n]) \\
\\
&= aL\{x_1[n]\} + bL\{x_2[n]\} \qquad \text{and so sums and scalar products are} \\
& \hspace{11em} \text{preserved}
\end{aligned}$$

The system is linear.

$\blacktriangleright$

(d) $y[n] = L\{x[n]\} = \cos(x[n])$ so if $y_1[n] = L\{x_1[n]\} = \cos(x_1[n])$, $y_1[n] = L\{x_1[n]\}$ $= \cos(x_2[n])$ and $x[n] = ax_1[n] + bx_2[n]$ where a and b are constants and where $y[n]$ is the corresponding response. Then:

$$
\begin{aligned}
y[n] &= L\{x[n]\} \\
&= L\{ax_1[n] + bx_2[n]\} \\
&= \cos(ax_1[n] + bx_2[n]) \\
&\neq L\{ax_1[n]\} + L\{bx_2[n]\}
\end{aligned}
$$

because $L\{ax_1[n]\} + L\{bx_2[n]\} = \cos(ax_1[n]) + \cos(bx_2[n])$. The system is non-linear.

Move to the next frame

Time-invariance of a continuous system

7

Consider the plot of the input to and the corresponding response of an arbitrary continuous system

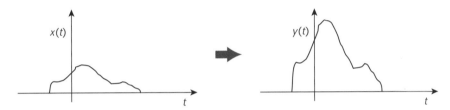

If this response pattern is retained but shifted wholesale through t_0 when the input is similarly shifted through t_0 then the system is said to be **time-invariant**. In other words it does not matter when we activate the system, we always get the same response for the same input; the response will be the same on Tuesday as it was on Monday.

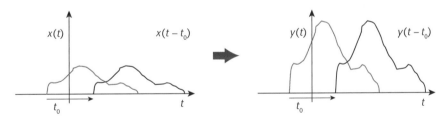

That is, a system is said to be **time-invariant** if:

$$L\{x(t)\} = y(t) \text{ and } L\{x(t - t_0)\} = y(t - t_0) \text{ where } t_0 \text{ is a constant}$$

For example, if

$$L\{x(t)\} = y(t) = \int_{-\infty}^{t} x(\tau)\,d\tau \text{ then if } x_1(t) = x(t - t_0)$$

$$L\{x(t - t_0)\} = L\{x_1(t)\}$$

$$= \int_{-\infty}^{t} x_1(\tau)\,d\tau$$

$$= \int_{-\infty}^{t} x(\tau - t_0)\,d\tau \quad \text{because } x_1(t) = x(t - t_0)$$

$$= \int_{-\infty}^{t - t_0} x(s)\,ds \qquad \text{where } s = \tau - t_0 \text{ so that } ds = d\tau \text{ and when}$$
$$t = 1,\ s = t - t_0$$

$$= y(t - t_0)$$

Therefore:

$$L\{x(t)\} = y(t) \text{ and } L\{x(t - t_0)\} = y(t - t_0)$$

so the system is time-invariant.

Which of the following are time-invariant?
(a) $L\{x(t)\} = y(t) = x(t)\sin pt$
(b) $L\{x(t)\} = y(t) = e^{x(t)}$

The answers are in the next frame

8

> (a) not time-invariant
> (b) time-invariant

Because:

(a) $L\{x(t)\} = y(t) = x(t)\sin pt$ so if $x_1(t) = x(t - t_0)$ then

$$L\{x(t - t_0)\} = L\{x_1(t)\}$$
$$= x_1(t)\sin pt$$
$$= x(t - t_0)\sin pt \text{ because } x_1(t) = x(t - t_0)$$

However, $y(t - t_0) = x(t - t_0)\sin p(t - t_0) \neq L\{x(t - t_0)\}$ so that L is not time-invariant.

(b) $L\{x(t)\} = y(t) = e^{x(t)}$ so let $x_1(t) = x(t - t_0)$ then

$$L\{x(t - t_0)\} = L\{x_1(t)\}$$
$$= e^{x_1(t)}$$
$$= e^{x(t - t_0)}$$
$$= y(t - t_0)$$

Therefore $L\{x(t)\} = y(t)$ and $L\{x(t - t_0)\} = y(t - t_0)$ so that L is time-invariant.

Next frame

Shift-invariance of a discrete system

9

A discrete system is said to be **shift-invariant** if:

$L\{x[n]\} = y[n]$ and $L\{x[n - n_0]\} = y[n - n_0]$

For example, if

$L\{x[n]\} = y[n] = x[n + 3]$ then if $x_1[m] = x[m - n_0]$ we see that

$L\{x[n - n_0]\} = L\{x_1[n]\}$

$\qquad = x_1[n + 3]$

$\qquad = x[n + 3 - n_0] \qquad$ here $m = n + 3$

$\qquad = x[n - n_0 + 3]$

$\qquad = y[n - n_0]$

Therefore $L\{x[n]\} = y[n]$ and $L\{x[n - n_0]\} = y[n - n_0]$ so the system is shift-invariant.

Which of the following are shift-invariant:

(a) $L\{x[n]\} = y[n] = x[n - 1] + x[n - 2]$

(b) $L\{x[n]\} = y[n] = nx[n]$

The answers are in the next frame

> (a) shift-invariant
> (b) not shift-invariant

10

Because:

(a) $L\{x[n]\} = y[n] = x[n - 1] + x[n - 2]$ so if $x_1[m] = x[m - n_0]$ then

$L\{x[n - n_0]\} = L\{x_1[n]\}$

$\qquad = x_1[n - 1] + x_1[n - 2]$

$\qquad = x[n - 1 - n_0] + x[n - 2 - n_0]$ here $m = n - 1$

$\qquad = x[n - n_0 - 1] + x[n - n_0 - 2]$

$\qquad = y[n - n_0]$

Therefore $L\{x[n]\} = y[n]$ and $L\{x[n - n_0]\} = y[n - n_0]$ and so L is shift-invariant

(b) $L\{x[n]\} = y[n] = nx[n]$ so if $x_1[m] = x[m - n_0]$ then

$L\{x[n - n_0]\} = L\{x_1[n]\}$

$\qquad = nx_1[n] \qquad$ here $m = n$

$\qquad = nx[n - n_0]$

$\qquad = y_1[n]$

However, $y[n - n_0] = (n - n_0)x[n - n_0] \neq y_1[n]$ and so L is not shift-invariant.

Move to the next frame

Differential equations

11 The general *n*th-order equation

Linear, constant coefficient differential equations define a linear system so let us refresh our memory.

The general *n*th-order, linear, constant coefficient, inhomogeneous differential equation:

$$a_n \frac{d^n y(t)}{dt^n} + a_{n-1} \frac{d^{n-1} y(t)}{dt^{n-1}} + \ldots + a_0 y(t)$$
$$= b_m \frac{d^m x(t)}{dt^m} + b_{m-1} \frac{d^{m-1} x(t)}{dt^{m-1}} + \ldots + b_0 x(t)$$

coupled with the values of the *n* boundary conditions:

$$\frac{d^n y(t)}{dt^n}\bigg|_{t=t_0}, \frac{d^{n-1} y(t)}{dt^{n-1}}\bigg|_{t=t_0}, \ldots, y(t_0)$$

describes the input-response relationship of a continuous linear system with input $x(t)$ and response $y(t)$.

Such an equation has a solution in the form $y(t) = y_h(t) + y_p(t)$ where $y_h(t)$ is the complementary function solution to the homogeneous equation

$$a_n \frac{d^n y(t)}{dt^n} + a_{n-1} \frac{d^{n-1} y(t)}{dt^{n-1}} + \ldots + a_0 y(t) = 0$$

and $y_p(t)$ is a particular integral or particular solution to the inhomogeneous equation. [Refer to Page 1101, Frame 23ff of *Engineering Mathematics, Sixth Edition*.] The procedure for solving such an equation is:

(i) Find the homogeneous solution $y_h(t)$ in terms of unknown integration constants
(ii) Find the particular solution $y_p(t)$ and form the complete solution $y(t) = y_h(t) + y_p(t)$
(iii) Apply the boundary conditions to find the values of the unknown integration constants in $y_h(t)$.

The solution can alternatively be written as $y(t) = y_{zi}(t) + y_{zs}(t)$ where $y_{zi}(t)$ is called the *zero-input response* and $y_{zs}(t)$ is called the *zero-state response*. The zero-input response of the equation depends only on the initial conditions and is independent of the input. It is obtained by solving the homogeneous equation **and applying the boundary conditions**. The zero-state response depends only on the input and is independent of the initial conditions, it is obtained by solving the inhomogeneous equation **but with all the boundary conditions equated to zero**.

Here the procedure is:

(i) Find the homogeneous solution $y_h(t)$ in terms of unknown integration constants
(ii) Find the particular solution $y_p(t)$ and form the complete solution $y(t) = y_h(t) + y_p(t)$

▶

(iii) Equate the boundary conditions to zero and then find the values of the unknown integration constants in $y(t)$. This is then the zero-state response $y_{zs}(t)$

(iv) Apply the original boundary conditions to find the values of the unknown integration constants in $y_h(t)$. This is then the zero-input solution $y_{zi}(t)$.

Move to the next frame for an example

Zero-input response and zero-state response

12

Consider the first-order, linear, constant coefficient, inhomogeneous differential equation:

$$\frac{dy(t)}{dt} + 4y(t) = tu(t)$$

where $y(0) = y_0$ is the boundary condition and $y(t) = 0$ for $t < 0$.

Homogeneous solution $y_h(t)$

From $\frac{dy_h(t)}{dt} + 4y_h(t) = 0$ the auxiliary equation is $m + 4 = 0$ and so $m = -4$ and $y_h(t) = Ae^{-4t}u(t)$.

Note the use of the Heaviside unit step function which acts as a switch to ensure that $y_h(t) = 0$ for $t < 0$ [Refer to Frame 4 of Programme 3].

Particular solution $y_p(t)$

Based on the form of the right-hand side the form of the particular solution is $y_p(t) = Ct + D$ (C, D are constants). Substituting $y_p(t) = Ct + D$ into $\frac{dy(t)}{dt} + 4y(t) = tu(t)$ gives:

$$C + 4C + 4D = t \text{ for } t \geq 0$$

from which we can see that $4C = 1$ and $C + 4D = 0$.

Therefore

$$C = 1/4 \text{ and } D = -1/16 \text{ so that}$$

$$y_p(t) = \left(\frac{t}{4} - \frac{1}{16}\right)u(t) \text{ and } y(t) = \left(Ae^{-4t} + \frac{t}{4} - \frac{1}{16}\right)u(t)$$

Again, the Heaviside unit step function ensures that $y_p(t) = 0$ and $y(t) = 0$ for $t < 0$.

Zero-state solution

To obtain the zero-state response we now equate the boundary condition to zero and find the value of the integration constant A. That is, we write $y(0) = y_0 = 0$ and find that:

$$y(0) = A - \frac{1}{16} = 0 \text{ so that } A = \frac{1}{16} \text{ and } y_{zs}(t) = \frac{1}{16}\left(e^{-4t} + 4t - 1\right)u(t)$$

▶

Zero-input solution

To obtain the zero-input response we apply the original boundary condition $y(0) = y_0$ to $y_h(t) = Ae^{-4t}u(t)$ to give:

$$y_{zi}(t) = y_0 e^{-4t} u(t)$$

The complete solution is the sum of the zero-state and zero-input responses. That is:

$$y(t) = \frac{1}{16} \left([16y_0 + 1]e^{-4t} + 4t - 1 \right) u(t)$$

So, the zero-input and zero-state responses of each of the differential equations:

(a) $\dfrac{dy(t)}{dt} - y(t) = e^{-t}u(t)$ where $y(0) = 1$ and $y(t) = 0$ for $t < 0$

(b) $\dfrac{d^2y(t)}{dt^2} - 5\dfrac{dy(t)}{dt} + 6y(t) = tu(t)$ where $y(0) = 1$ and $\left.\dfrac{dy(t)}{dt}\right|_{t=0} = -1$ and
   $y(t) = 0$ for $t < 0$

are ....

The answers are in the next frame

13

(a) $y_{zi}(t) = e^t u(t)$

$\quad y_{zs}(t) = u(t) \sinh t$

(b) $y_{zi}(t) = \left(4e^{2t} - 3e^{3t} \right) u(t)$

$\quad y_{zs}(t) = \left(-9e^{2t} + 4e^{3t} + 6t + 5 \right) \dfrac{u(t)}{36}$

Because:

(a) $\dfrac{dy(t)}{dt} - y(t) = e^{-t}u(t)$ where $y(0) = 1$ and $y(t) = 0$ for $t < 0$

Homogeneous solution $y_h(t)$

From $\dfrac{dy_h(t)}{dt} - y_h(t) = 0$ the auxiliary equation is $m - 1 = 0$ and so $m = 1$ and $y_h(t) = Ae^t u(t)$.

Particular solution $y_p(t)$

Based on the form of the right-hand side the form of the particular solution is $y_p(t) = Be^{-t}$ (B constant). Substituting $y_p(t) = Be^{-t}$ into $\dfrac{dy(t)}{dt} - y(t) = e^{-t}u(t)$ gives:

$$-Be^{-t} - Be^{-t} = e^{-t} \text{ for } t \geq 0 \text{ from which we can see that } B = -\frac{1}{2}.$$

Therefore $y_p(t) = -\dfrac{1}{2}e^{-t}u(t)$ and $y(t) = \left(Ae^t - \dfrac{1}{2}e^{-t} \right) u(t)$

▶

Zero-state solution

To obtain the zero-state response we now equate the boundary condition to zero and find the value of the integration constant A. That is, we write $y(0) = 0$ and find that:

$$y(0) = A - \frac{1}{2} = 0 \text{ so that } A = \frac{1}{2} \text{ and } y_{zs}(t) = \left(\frac{1}{2}e^t - \frac{1}{2}e^{-t}\right)u(t) = u(t)\sinh t$$

Zero-input solution

To obtain the zero-input response we apply the original boundary condition $y(0) = 1$ to $y_h(t) = Ae^t u(t)$ to give:

$$y_{zi}(t) = e^t u(t)$$

The complete solution is the sum of the zero-state and zero-input responses. That is:

$$y(t) = (e^t + \sinh t)u(t)$$

(b) $\dfrac{d^2y}{dt^2} - 5\dfrac{dy(t)}{dt} + 6y(t) = tu(t)$ where $y(0) = 1$ and $\left.\dfrac{dy}{dt}\right|_{t=0} = -1$

Homogeneous solution $y_h(t)$

From $\dfrac{d^2y(t)}{dt^2} - 5\dfrac{dy(t)}{dt} + 6y(t) = 0$ the auxiliary equation is

$$m^2 - 5m + 6 = (m-2)(m-3) = 0$$

and so $m = 2, 3$ and $y_h(t) = \left(Ae^{2t} + Be^{3t}\right)u(t)$.

Particular solution $y_p(t)$

Based on the form of the right-hand side the form of the particular solution is

$$y_p(t) = Ct + D \quad (C, D \text{ constants}).$$

Substituting $y_p(t) = Ct + D$ into the differential equation

$$\frac{d^2y(t)}{dt^2} - 5\frac{dy(t)}{dt} + 6y(t) = tu(t) \text{ gives:}$$
$$-5C + 6(Ct + D) = t \text{ for } t \geq 0$$

from which we can see that $6C = 1$ and $-5C + 6D = 0$
Therefore

$$C = 1/6 \text{ and } D = 5/36 \text{ so that } y_p(t) = \left(\frac{t}{6} + \frac{5}{36}\right)u(t) \text{ giving}$$

$$y(t) = \left(Ae^{2t} + Be^{3t} + \frac{t}{6} + \frac{5}{36}\right)u(t) \text{ and } y'(t) = 2Ae^{2t} + 3Be^{3t} + \frac{1}{6}$$

▶

Zero-state solution

To obtain the zero-state response we now equate the boundary conditions to zero and find the value of the integration constants A and B. That is, we write $y(0) = 0$ and $y'(0) = 0$ to find that:

$$y(0) = A + B + \frac{5}{36} \text{ and } y'(0) = 2A + 3B + \frac{1}{6} = 0 \text{ giving } A = -\frac{9}{36} \text{ and } B = \frac{4}{36}$$

so that:

$$y_{zs}(t) = \left(-9e^{2t} + 4e^{3t} + 6t + 5\right)\frac{u(t)}{36}$$

Zero-input solution

To obtain the zero-input response we apply the original boundary conditions $y(0) = 1$ and $y'(0) = -1$ to $y_h(t) = \left(Ae^{2t} + Be^{3t}\right)u(t)$ to give:

$$y_{zi}(t) = \left(4e^{2t} - 3e^{3t}\right)u(t)$$

The complete solution is the sum of the zero-state and zero-input responses. That is:

$$y(t) = \left(135e^{2t} - 104e^{3t} + 6t + 5\right)\frac{u(t)}{36}$$

Move to the next frame

---

14 Zero-input, zero-response

We have already seen that a system $y(t) = L\{x(t)\}$ is *linear* if sums and scalar multiples are preserved, that is if

$$L\{x_1(t) + x_2(t)\} = L\{x_1(t)\} + L\{x_2(t)\} \text{ and } L\{\alpha x(t)\} = \alpha L\{x(t)\}$$

where α is a constant.

In particular, by choosing $\alpha = 0$ then $L\{0\} = 0$ which shows that if nothing is put into a linear system nothing will come out – **zero input yields zero output**.

In the previous frame we considered two systems described by the differential equations:

(a) $\dfrac{dy(t)}{dt} - y(t) = e^{-t}u(t)$ where $y(0) = 1$

(b) $\dfrac{d^2y(t)}{dt^2} - 5\dfrac{dy(t)}{dt} + 6y(t) = tu(t)$ where $y(0) = 1$ and $\left.\dfrac{dy(t)}{dt}\right|_{t=0} = -1$

Do these equations give rise to linear or non-linear systems?

(a) ....

(b) ....

The answers are in the following frame

15

| (a) Non-linear   (b) Non-linear |

Because:

(a) $y_{zi}(t) = e^t u(t)$ so that the zero-input response is not zero. Therefore this differential equation does not give rise to a linear system but to a non-linear system.

(b) $y_{zi}(t) = (4e^{2t} - 3e^{3t})u(t)$ so that the zero-input response is not zero. Therefore this differential equation does not give rise to a linear system but to a non-linear system.

Is it possible for a differential equation to give rise to a linear system? To answer this question we now re-cast these two differential equations with general boundary conditions. Firstly,

$$\frac{dy(t)}{dt} - y(t) = e^{-t}u(t) \text{ where } y(0) = A$$

Here, for $t < 0$ the equation becomes, with zero-input:

$$\frac{dy(t)}{dt} - y(t) = 0 \text{ where } y(0) = A$$

With solution being the zero-input response $y_{zi}(t) = Ae^t$ and this can only be zero if $A = 0$. The differential equation only gives rise to a linear system if the value of the boundary condition is zero.

Now you try. The conditions that the differential equation:

$$\frac{d^2y(t)}{dt^2} - 5\frac{dy}{dt} + 6y(t) = tu(t) \text{ where } y(0) = K_1 \text{ and } \left.\frac{dy(t)}{dt}\right|_{t=0} = K_2$$

gives rise to a linear system are . . . .

The answer is in the next frame

16

| $K_1 = 0$ and $K_2 = 0$ |

Because:

Here, for $t < 0$ the equation becomes, with zero-input:

$$\frac{d^2y(t)}{dt^2} - 5\frac{dy(t)}{dt} + 6y(t) = 0 \text{ where } y(0) = K_1 \text{ and } \left.\frac{dy(t)}{dt}\right|_{t=0} = K_2$$

with solution being the zero-input response $y_{zi}(t) = Ae^{2t} + Be^{3t}$. Applying the boundary conditions:

$$y(0) = K_1: \quad A + B = K_1$$

$$\left.\frac{dy(t)}{dt}\right|_{t=0} = K_2: \quad 2A + 3B = K_2$$

with solution $A = 3K_1 - K_2$ and $B = K_2 - 2K_1$ giving

$$y_{zi}(t) = (3K_1 - K_2)e^{2t} + (K_2 - 2K_1)e^{3t}$$

▶

This can only be zero if:

$3K_1 - K_2 = 0$

$-2K_1 + K_2 = 0$ that is if $K_1 = 0$ and $K_2 = 0$

The differential equation only gives rise to a linear system if the values of the boundary conditions are zero.

This is a general property – a constant coefficient, linear differential equation only gives rise to a linear system if the values of all the boundary conditions are zero. This is an important fact to be remembered.

Next we look at these differential equations and time-invariance.

Move to the next frame

17   Time-invariance

Consider the differential equation in Frame 12, namely:

$$\frac{dy(t)}{dt} + 4y(t) = tu(t) \text{ where } y(0) = y_0 \text{ and } y(t) = 0 \text{ for } t < 0$$

with solution

$$y(t) = \frac{1}{16}\big([16y_0 + 1]e^{-4t} + 4t - 1\big)u(t).$$

If we re-visit this equation but this time change the boundary condition to $y(0) = 0$ the solution is the zero-state solution:

$$y_{zs}(t) = \frac{1}{16}\big(e^{-4t} + 4t - 1\big)u(t)$$

If we now delay the input by 3 units so that the input becomes $(t-3)u(t-3)$ the differential equation becomes:

$$\frac{dy(t)}{dt} + 4y(t) = (t-3)u(t-3) \text{ where } y(3) = 0 \text{ and } y(t) = 0 \text{ for } t < 3$$

The homogeneous solution is again $y_h(t) = Ae^{-4t}u(t)$ and the particular solution has the form $y_p(t) = Ct + D$. However, substituting into the differential equation we now find that

$C + 4Ct + 4D = t - 3$ for $t \geq 3$ from which we find that

$4C = 1$ and $C + 4D = -3$.

Therefore:

$$C = 1/4, D = -\frac{13}{16} = -\frac{12}{16} - \frac{1}{16} = -\frac{3}{4} - \frac{1}{16} \text{ and the particular solution is}$$

$$y_p(t) = \left(\frac{t-3}{4} - \frac{1}{16}\right)u(t-3) \text{ so that } y(t) = \left(Ae^{-4t} + \frac{t-3}{4} - \frac{1}{16}\right)u(t-3).$$

Applying the boundary condition $y(3) = 0$:

$$y(3) = Ae^{-12} - \frac{1}{16} = \frac{1}{16} \text{ that is } A = \frac{2e^{12}}{16} \text{ giving the solution to}$$

$$\frac{dy(t)}{dt} + 4y(t) = (t - 3)u(t - 3) \text{ where } y(3) = 0 \text{ as}$$

$$y(t) = \frac{1}{16}\left(2e^{-4(t-3)} + 4(t - 3) - 1\right)u(t - 3)$$

This is the same solution but delayed by the same amount as the input. Consequently, the system is not only linear but it is also time-invariant. **Indeed, the zero values of the boundary conditions ensure that the general constant coefficient, linear differential equation gives rise to a system that is not only linear but also time-invariant.**

Move to the next frame

Responses of a continuous system

Impulse response

18

We shall soon see that any continuous, linear, time-invariant system has the important property that its response to *any input* can be found from knowing its response to the unit impulse $\delta(t)$ – a property that can be exploited to solve such differential equations as considered here [refer to Programme 4].

When the input to a linear, time-invariant system is the unit impulse $\delta(t)$ the response is denoted by $h(t)$ and is referred to as the **impulse response**. That is:

$$h(t) = L\{\delta(t)\}$$

and, because the system is time-invariant

$$h(t - t_0) = L\{\delta(t - t_0)\}$$

Arbitrary input

From the properties of the unit impulse $\delta(t)$ we can express an arbitrary input $x(t)$ in terms of the unit impulse $\delta(t)$ as:

$$x(t) = \int_{-\infty}^{\infty} x(\tau)\delta(t - \tau)\,d\tau$$

so that if the response to this arbitrary input is $y(t)$ then:

$$y(t) = L\{x(t)\}$$
$$= L\left\{\int_{-\infty}^{\infty} x(\tau)\delta(t - \tau)\,d\tau\right\}$$

▶

Because the variable τ inside the integral is the variable of integration and the operator L is acting on t and not on τ the operator can be moved inside the integral. (Recall that an integral is a limit of a sum and, for linear systems, sums are preserved.) Therefore:

$$L\left\{\int_{-\infty}^{\infty} x(\tau)\delta(t-\tau)\,d\tau\right\} = \int_{-\infty}^{\infty} x(\tau)L\{\delta(t-\tau)\}d\tau$$

$$= \int_{-\infty}^{\infty} x(\tau)h(t-\tau)d\tau$$

because the system is given as time invariant.

This is a remarkable result so we shall look at it very closely. We have just found that:

$$L\{x(t)\} = \int_{-\infty}^{\infty} x(\tau)h(t-\tau)d\tau.$$

You have seen integrals like this one before, can you recall?

This integral is the .... between the input to the system and the impulse response of the system.

The answer is in the next frame

---

19 $\boxed{\text{convolution}}$

Because:

The convolution between $x(t)$ and $h(t)$ is obtained by first reversing $h(t)$ to form $h(-t)$, changing the variable to the dummy variable τ of the integral to form $x(\tau)$ and $h(-\tau)$, advancing $h(-\tau)$ by t to form $h(-\tau+t)=h(t-\tau)$, taking the product of this with $x(\tau)$ and integrating with respect to τ to form [refer to page 112, Frame 43]:

$$y(t) = \int_{-\infty}^{\infty} x(\tau)h(t-\tau)d\tau = x(t) * h(t)$$

Aside: It is also worth remembering that convolution is a commutative operation. That is:

$$y(t) = x(t) * h(t) = \int_{-\infty}^{\infty} x(\tau)h(t-\tau)d\tau \text{ and}$$

$$y(t) = h(t) * x(t) = \int_{-\infty}^{\infty} h(\tau)x(t-\tau)d\tau$$

Bear this fact in mind as we shall make use of it little later on.

This is a most important result because it tells us that **if we know the impulse response of a continuous, linear, time-invariant system then we can find the response of the system to any input simply by evaluating** *the convolution of the input with the system impulse response*.

As an example consider the response to a unit step input $u(t)$ at time $t = 0$ of a system that has an impulse response:

$$h(t) = u(t)e^{-kt} \quad k > 0$$

The graphs of the input and the impulse response are:

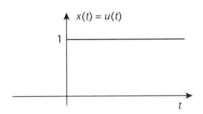
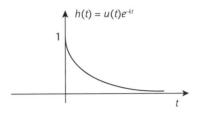

To evaluate the convolution $y(t) = x(t) * h(t)$ we first need to change t to the variable of integration τ to form $x(\tau)$ and $h(\tau)$. We then require $h(\tau)$ to be advanced by t to form $h(\tau - t)$ where $t > 0$. Next we flip $h(\tau - t)$ about the vertical to form $h(-[\tau - t]) = h(t - \tau)$ to overlap with the unit step.

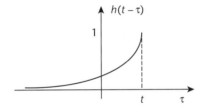
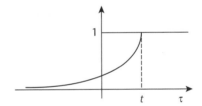

The only non-zero overlap inside the convolution integral is then between the values $\tau = 0$ and $\tau = t$ provided $t > 0$. If $t < 0$ then there is no overlap at all with the unit step function so that:

$$y(t) = x(t) * h(t)$$

$$= \int_{-\infty}^{\infty} x(\tau)h(t - \tau)d\tau$$

$$= \int_{-\infty}^{0} x(\tau)h(t - \tau)d\tau + \int_{0}^{t} x(\tau)h(t - \tau)d\tau + \int_{t}^{\infty} x(\tau)h(t - \tau)d\tau$$

$$= 0 + \int_{0}^{t} x(\tau)h(t - \tau)d\tau + 0 \qquad x(\tau) = 0 \text{ for } \tau < 0 \text{ and } h(t - \tau) = 0 \text{ for } \tau > t$$

Now $x(\tau) = u(\tau)$ and $h(t - \tau) = u(t - \tau)e^{-k(t-\tau)}$ so that

$$y(t) = \int_{0}^{t} u(\tau)u(t - \tau)e^{-k(t-\tau)}\, d\tau$$

$$= \int_{0}^{t} u(t - \tau)e^{-k(t-\tau)}\, d\tau$$

$$= \int_{t}^{0} u(x)e^{-k(x)}d(-x) \quad \text{where } x = t - \tau \text{ so } x = t \text{ when } \tau = 0 \text{ and } x = 0$$
$$\text{when } \tau = t$$

Furthermore, $\tau = t - x$ so $d\tau = d(-x) = -dx$ so that

$$y(t) = -\int_t^0 u(x)e^{-kx}\,dx$$

$$= \int_0^t u(x)e^{-kx}\,dx$$

$$= \int_0^t e^{-kx}\,dx$$

$$= \left[\frac{e^{-kx}}{-k}\right]_0^t = \frac{1}{k}\left(1 - e^{-(t-\tau)}\right)$$

The response can then be written as $y(t) = \frac{1}{k}\left(1 - e^{-kt}\right)u(t)$

You try one. The response to the input $x(t) = u(t - a)$ of a system with impulse response $h(t) = u(t)e^{-bt}$ where $a > 0$ and $b > 0$ is ....

The answer is in the next frame

20

$$y(t) = \frac{1}{b}\left(1 - e^{-b(t-a)}\right)u(t - a)$$

Because:

The graphs of the input and the impulse response are:

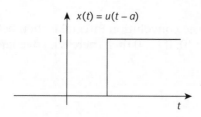
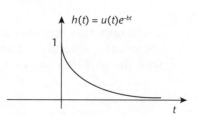

To evaluate the convolution $y(t) = x(t) * h(t)$ we first need to change t to the variable of integration τ to form $x(\tau)$ and $h(\tau)$. We then require $h(\tau)$ to be advanced by t to form $h(t - \tau)$ where $t > 0$. Next we flip $h(t - \tau)$ about the vertical to form $h(-[\tau - t])$ to overlap with the unit step. If $t < a$ there is no overlap with the unit step function.

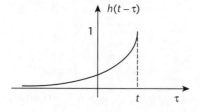
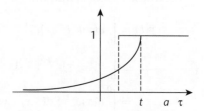

Provided $t > a$ the only non-zero overlap inside the convolution integral is over the interval a to t so that:

$$y(t) = x(t) * h(t)$$

$$= \int_{-\infty}^{\infty} x(\tau)h(t-\tau)\,d\tau$$

$$= \int_{-\infty}^{a} x(\tau)h(t-\tau)\,d\tau + \int_{a}^{t} x(\tau)h(t-\tau)\,d\tau + \int_{t}^{\infty} x(\tau)h(t-\tau)\,d\tau$$

$$= 0 + \int_{a}^{t} x(\tau)h(t-\tau)\,d\tau + 0 \qquad \begin{array}{l} x(\tau) = 0 \text{ for } \tau < a \text{ and } h(t-\tau) = 0 \text{ for } \\ \tau > t \end{array}$$

$$= \int_{a}^{t} u(\tau-a)u(t-\tau)e^{-b(t-\tau)}\,d\tau$$

$$= \int_{a}^{t} u(t-\tau)e^{-b(t-\tau)}\,d\tau$$

$$= \int_{t-a}^{0} u(x)e^{-b(x)}\,d(-x) \qquad \text{where } x = t - \tau$$

$$= \int_{0}^{t-a} u(x)e^{-bx}\,dx$$

$$= \left[\frac{e^{-bx}}{-b}\right]_{0}^{t-a}$$

$$= \frac{1}{b}\left(1 - e^{-b(t-a)}\right) \text{ for } t > a$$

The response can then be written as $y(t) = \dfrac{1}{b}\left(1 - e^{-b(t-a)}\right)u(t-a)$

Next frame

Exponential response

21

When the input to a continuous, linear, time-invariant system with impulse response $h(t)$ is the exponential $x(t) = Ae^{st}$ (A and s being constants) the system response is given as:

$$y(t) = Ae^{st} * h(t) \text{ or, because convolution is commutative, } y(t) = h(t) * Ae^{st}$$

That is:

$$y(t) = A\int_{-\infty}^{\infty} e^{s\tau}h(t-\tau)\,d\tau \text{ or } y(t) = A\int_{-\infty}^{\infty} h(\tau)e^{s(t-\tau)}\,d\tau$$

Taking the latter form of the convolution of the input with the impulse response we see that:

$$y(t) = A\int_{-\infty}^{\infty} h(\tau)e^{s(t-\tau)}\,d\tau$$

$$= Ae^{st}\int_{-\infty}^{\infty} h(\tau)e^{-s\tau}\,d\tau$$

$$= Ae^{st}H(s)$$

where $H(s) = \displaystyle\int_{-\infty}^{\infty} h(\tau)e^{-st}\,d\tau.$

▶

Notice that if $h(t) = 0$ for $t < 0$ then:

$$H(s) = \int_0^\infty h(\tau)e^{-s\tau}\,d\tau \text{ so that } H(s) \text{ is the } \ldots\ldots\ldots\ldots \text{ transform of } h(t)$$

The answer is in the next frame

22

$\boxed{\text{Laplace}}$

Because:

Recalling from Frame 1 on page 47 the Laplace transform $F(s)$ of function $f(t)$ is defined as:

$$F(s) = \int_0^\infty f(t)e^{-st}\,dt$$

The expression $H(s)$ is called the system's **transfer function**. Furthermore, because the system response is simply a scaled version of the input, the scaling factor being $H(s)$ we can say that:

$$L\{Ae^{st}\} = H(s)(Ae^{st})$$

which tells us that Ae^{st} is an *eigenfunction* of the operator L with corresponding *eigenvalue* $H(s)$ (refer to *Engineering Mathematics, Sixth Edition*, page 578). The converse is also true. The eigenfunctions of a linear, time-invariant system are exponential functions, which places the exponential function in a special position with respect to linear time-invariant systems as we shall appreciate later. For example, the transfer function corresponding to the input $5e^{-st}$ of the continuous, linear, time-invariant system with impulse response:

$$h(t) = \begin{cases} e^{-6t} & t \geq 0 \\ 0 & t < 0 \end{cases}$$

is:

$$H(s) = 5\int_0^\infty e^{-6t}e^{-s\tau}\,d\tau$$

$$= 5\int_0^\infty e^{-(6+s)t}\,d\tau$$

$$= 5\left[\frac{e^{-(6+s)t}}{-(6+s)}\right]_0^\infty$$

$$= 5\left(0 - \frac{1}{-(6+s)}\right)$$

$$= \frac{5}{s+6} \qquad \text{provided } s > -6$$

If $s \leq -6$ then $s + 6 \leq 0$ and the integral diverges.

So the transfer function corresponding to the input $3e^{-st}u(t-2)$ of the continuous, linear, time-invariant system with impulse response

$$h(t) = e^{-t}u(t) \text{ is } \ldots\ldots\ldots\ldots$$

The answer is in the next frame

23

$$\boxed{\dfrac{3e^{-2(1+s)}}{s+1} \quad \text{provided } s > -1}$$

Because:

$$H(s) = 3\int_0^\infty e^{-st} u(t-2)u(t)e^{-\tau}\,d\tau$$

$$= 3\int_2^\infty e^{-(1+s)t}\,d\tau$$

$$= 3\left[\dfrac{e^{-(1+s)t}}{-(1+s)}\right]_2^\infty$$

$$= 3\left(0 - \dfrac{e^{-(1+s)2}}{-(1+s)}\right)$$

$$= \dfrac{3e^{-2(1+s)}}{s+1} \quad \text{provided } s > -1$$

If $s \le -1$ then $s + 1 \le 0$ and the integral diverges.

Move to the next frame

24

The transfer function

We have seen that the response $y(t)$ of a continuous, linear, time-invariant system to an input $x(t)$ is given in terms of the system's unit impulse response $h(t)$ as the convolution:

$$y(t) = x(t) * h(t)$$

We have also seen that provided $h(t) = 0$ for $t < 0$ then the system's transfer function $H(s)$ is the Laplace transform of $h(t)$. That is:

$$H(s) = \int_0^\infty h(t)e^{-st}\,dt$$

Referring now to Frame 45 of page 117 we see that the convolution theorem states that the Laplace transform of a convolution of two functions is equal to the product of their respective Laplace transforms. Therefore, if

$$Y(s) = \int_0^\infty y(t)e^{-st}\,dt \text{ and } X(s) = \int_0^\infty x(t)e^{-st}\,dt \text{ then } Y(s) = X(s)H(s)$$

For example to find the response of a time-invariant linear system with impulse response $h(t) = u(t)e^{-t}$ to an input $x(t) = u(t) - u(t-1)$ all we need do is:

(a) Find the Laplace transforms of $h(t) = u(t)e^{-t}$ and $x(t) = u(t) - u(t-1)$

$$\text{These are } H(s) = \dfrac{e^{-s}}{s+1} \text{ and } X(s) = \dfrac{1}{s} - \dfrac{e^{-s}}{s}$$

▶

(b) Obtain $Y(s) = X(s)H(s)$

This is $Y(s) = \dfrac{e^{-s}}{s(s+1)} - \dfrac{e^{-2s}}{s(s+1)}$

$$= (e^{-s} - e^{-2s})\left\{\dfrac{1}{s} - \dfrac{1}{s+1}\right\}$$

(c) Take the inverse Laplace transform

$$Y(s) = \left\{\dfrac{e^{-s}}{s} - \dfrac{e^{-s}}{s+1}\right\} - \left\{\dfrac{e^{-2s}}{s} - \dfrac{e^{-2s}}{s+1}\right\}$$

so $y(t) = \left(u(t-1) - u(t-1)e^{-(t-1)}\right) - \left(u(t-2) - u(t-2)e^{-(t-2)}\right)$

$$= u(t-1)\left(1 - e^{-(t-1)}\right) - u(t-2)\left(1 - e^{-(t-2)}\right)$$

You try one. The response of a time-invariant linear system with impulse response $h(t) = u(t-1)$ to an input $x(t) = u(t)\sin t - u(t-1)\sin(t-1)$ is

$$y(t) = \ldots\ldots\ldots$$

The answer is in the next frame

25

$$\boxed{u(t-1)(1 - \cos(t-1)) - u(t-2)(1 - \cos(t-2))}$$

Because:

(a) The Laplace transforms of

$$h(t) = u(t-1) \text{ and } x(t) = u(t)\sin t - u(t-1)\sin(t-1)$$

are:

$$H(s) = \dfrac{e^{-s}}{s} \text{ and } X(s) = \dfrac{1}{s^2+1} - \dfrac{e^{-s}}{s^2+1}$$

(b) The Laplace transform of $y(t)$ is:

$$Y(s) = X(s)H(s)$$

$$= \dfrac{e^{-s}}{s(s^2+1)} - \dfrac{e^{-2s}}{s(s^2+1)}$$

$$= (e^{-s} - e^{-2s})\left\{\dfrac{1}{s} - \dfrac{s}{s^2+1}\right\}$$

(c) Then the inverse Laplace transform is

$$\mathscr{L}^{-1}\{Y(s)\} = \mathscr{L}^{-1}\left\{\dfrac{e^{-s}}{s} - \dfrac{se^{-2s}}{s^2+1}\right\} - \mathscr{L}^{-1}\left\{\dfrac{e^{-2s}}{s} - \dfrac{se^{-2s}}{s^2+1}\right\}$$

so $y(t) = (u(t-1) - u(t)\cos(t-1)) - (u(t-2) - u(t-2)\cos(t-2))$

$$= u(t-1)(1 - \cos(t-1)) - u(t-2)(1 - \cos(t-2))$$

▶

In summary, to find the response of a continuous, linear, time-invariant system all we need to do is take the inverse Laplace transform of the product of the Laplace transform of the input and the transfer function – the transfer function being the Laplace transform of the unit impulse response.

$$y(t) = \mathcal{L}^{-1}\{X(s)H(s)\} \text{ where } H(s) = \mathcal{L}\{h(t)\}$$

So, given the input all we need do to proceed is to determine the impulse response. However, for those differential equations that give rise to continuous, linear, time-invariant systems we have a much simpler way of determining the transfer function.

Next frame

Differential equations

26

To solve the equation

$$y''(t) - 5y'(t) + 6y(t) = x(t) \text{ where } y(0) = 0,\ y'(0) = 0 \text{ and } y''(0) = 0$$

we note that the differential equation gives rise to a continuous, linear, time-invariant system. That is

$$y(t) = L\{x(t)\}$$

Accordingly, the exponential function Ae^{st} is an eigenfunction of the system whose corresponding eigenvalue is the system's transfer function $H(s)$. That is:

if $x(t) = Ae^{st}$ then $y(t) = Ae^{st}H(s)$

Substituting these into the differential equation, we then see that:

$$\left(Ae^{st}H(s)\right)'' - 5\left(Ae^{st}H(s)\right)' + 6\left(Ae^{st}H(s)\right) = Ae^{st}$$
$$\{s^2 - 5s + 6\}H(s)Ae^{st} = Ae^{st}$$
$$H(s) = \frac{1}{s^2 - 5s + 6}$$

$H(s)$ is the Laplace transform of the left-hand side of the differential equation and now you see the importance of all the boundary conditions having a value of zero. If they did not then their non-zero values would be automatically incorporated into the Laplace transform so giving a different expression to this one here.

Now if, for example, the input is $x(t) = e^{-5t}u(t)$ then its Laplace transform is

$X(s) = \dfrac{1}{s+5}$ giving the Laplace transform of the system's response $y(t)$ as:

$$Y(s) = X(s)H(s)$$

$$= \frac{1}{(s+5)(s-2)(s-3)}$$

$$= \frac{P}{s+5} + \frac{Q}{s-2} + \frac{R}{s-3}$$

Taking inverse Laplace transforms we see that $y(t) = Pe^{-5t} + Qe^{2t} + Re^{3t}$ where the values of P, Q and R can be found – the usual partial fractions procedure giving the solution to the differential equation as:

$$y(t) = \frac{e^{-5t}}{56} - \frac{e^{2t}}{7} + \frac{e^{3t}}{8}$$

Therefore, the solution to

$$y''(t) + 3y'(t) - 28y(t) = e^{-t}u(t) \text{ where } y(0) = 0, \ y'(0) = 0 \text{ and } y''(0) = 0 \text{ is}$$

$$y(t) = \ldots\ldots\ldots\ldots$$

The answer is in the next frame

27

$$\boxed{y(t) - \frac{e^{-t}}{30} + \frac{e^{-7t}}{66} + \frac{e^{4t}}{55}}$$

Because:

The auxiliary equation is $s^2 + 3s - 28 = 0$ so that $H(s) = \dfrac{1}{s^2 + 3s - 28}$

The Laplace transform of the input $x(t) = e^{-t}u(t)$ is $X(s) = \dfrac{1}{s+1}$

The Laplace transform of the response $y(t)$ is then

$$Y(s) = \frac{1}{s+1} \times \frac{1}{s^2 + 3s - 28}$$

$$= \frac{1}{(s+1)(s+7)(s-4)}$$

$$= \frac{P}{s+1} + \frac{Q}{s+7} + \frac{R}{s-4}$$

which gives the response as $y(t) = Pe^{-t} + Qe^{-7t} + Re^{4t}$. Now

$$\frac{1}{(s+1)(s+7)(s-4)} = \frac{P}{s+1} + \frac{Q}{s+7} + \frac{R}{s-4}$$

$$= \frac{P(s+7)(s-4) + Q(s+1)(s-4) + R(s+1)(s+7)}{(s+1)(s+7)(s-4)}$$

$$= \frac{(P+Q+R)s^2 + (3P - 3Q + 8R)s + (-28P - 4Q + 7R)}{(s+1)(s+7)(s-4)}$$

Therefore

$$\begin{array}{c} P + Q + R = 0 \\ 3P - 3Q + 8R = 0 \text{ that is} \\ -28P - 4Q + 7R = 1 \end{array} \quad \begin{pmatrix} 1 & 1 & 1 \\ 3 & -3 & 8 \\ -28 & -4 & 7 \end{pmatrix} \begin{pmatrix} P \\ Q \\ R \end{pmatrix} = \begin{pmatrix} 0 \\ 0 \\ 1 \end{pmatrix}$$

giving $\begin{pmatrix} P \\ Q \\ R \end{pmatrix} = \begin{pmatrix} -1/30 \\ 1/66 \\ 1/55 \end{pmatrix}$

Therefore $y(t) = -\dfrac{e^{-t}}{30} + \dfrac{e^{-7t}}{66} + \dfrac{e^{4t}}{55}$

▶

And just one more. To find the solution to

$$y''(t) - 4y(t) = [u(t) - u(t-1)]t \text{ where } y(0) = 0, \ y'(0) = 0 \text{ and } y''(0) = 0$$

we first need to arrange the input into a form that is Laplace transformable. In other words, we need to convert

$u(t-1)t$ into a form involving $u(t-1)(t-1)$ that is:

$$[u(t) - u(t-1)]t = \ldots\ldots\ldots\ldots$$

The answer is in the next frame

$$\boxed{[u(t) - u(t-1)]t = u(t)t - u(t-1)(t-1) - u(t-1)}$$

28

Because

$$u(t-1)(t-1) = u(t-1)t - u(t-1) \text{ therefore}$$
$$[u(t) - u(t-1)]t = u(t)t - u(t-1)(t-1) - u(t-1)$$

The differential equation then becomes:

$$y''(t) - 4y(t) = u(t)t - u(t-1)(t-1) - u(t-1)$$

where $y(0) = 0$, $y'(0) = 0$ and $y''(0) = 0$.

The solution is then

$$y(t) = \ldots\ldots\ldots\ldots$$

The answer is in the next frame

$$\boxed{\frac{u(t)}{16}\left\{-4t + e^{2t} - e^{-2t}\right\} - \frac{u(t-1)}{16}\left\{-4t + 3e^{2(t-1)} + e^{-2(t-1)}\right\}}$$

29

Because

Taking the Laplace transform of the left-hand side tells us that

$$H(s) = \frac{1}{s^2 - 4} = \frac{1}{4}\left\{\frac{1}{s-2} - \frac{1}{s+2}\right\}$$

Taking the Laplace transform of the right hand side we get

$$X(s) = \frac{1}{s^2} - \frac{e^{-s}}{s^2} - \frac{e^{-s}}{s}$$

Therefore

$$Y(s) = \left(\frac{1}{s^2} - \frac{e^{-s}}{s^2} - \frac{e^{-s}}{s}\right)\left(\frac{1}{4}\left\{\frac{1}{s-2} - \frac{1}{s+2}\right\}\right)$$

▶

Breaking into partial fractions:

$$\frac{1}{s^2(s+2)} = \frac{1}{4}\left\{\frac{2}{s^2} - \frac{1}{s} + \frac{1}{s+2}\right\}$$

$$\frac{1}{s(s+2)} = \frac{1}{2}\left\{\frac{1}{s} - \frac{1}{s+2}\right\}$$

$$\frac{1}{s^2(s-2)} = \frac{1}{4}\left\{-\frac{2}{s^2} - \frac{1}{s} + \frac{1}{s-2}\right\}$$

$$\frac{1}{s(s-2)} = \frac{1}{2}\left\{\frac{1}{s-2} - \frac{1}{s}\right\}$$

Therefore

$$Y(s) = \left(\frac{1}{s^2} - \frac{e^{-s}}{s^2} - \frac{e^{-s}}{s}\right)\left(\frac{1}{4}\left\{\frac{1}{s-2} - \frac{1}{s+2}\right\}\right)$$

$$= \frac{(1-e^{-s})}{16}\left\{-\frac{4}{s^2} + \frac{1}{s-2} - \frac{1}{s+2}\right\} - \frac{e^{-s}}{8}\left\{-\frac{2}{s} + \frac{1}{s-2} + \frac{1}{s+2}\right\}$$

Giving

$$y(t) = \frac{u(t)}{16}\left\{-4t + e^{2t} - e^{-2t}\right\} - \frac{u(t-1)}{16}\left\{-4(t-1) + e^{2(t-1)} - e^{-2(t-1)}\right\}$$

$$\qquad - \frac{u(t-1)}{8}\left\{-2 + e^{2(t-1)} + e^{-2(t-1)}\right\}$$

$$= \frac{u(t)}{16}\left\{-4t + e^{2t} - e^{-2t}\right\} - \frac{u(t-1)}{16}\left\{-4t + 3e^{2(t-1)} + e^{-2(t-1)}\right\}$$

This completes our work on continuous linear systems. Now we shall move on to consider discrete linear systems. Read on.

Responses of a discrete system

30 The discrete unit impulse

The value of the unit impulse $\delta(t)$ in the study of continuous linear systems cannot be overestimated. It permits the system response to any input to be found once the system's response to the unit impulse is known. In the case of discrete systems the equivalent is called the **discrete unit impulse** $\delta[n]$ which is defined as:

$$\delta[n] = \begin{cases} 1 & n = 0 \\ 0 & n \neq 0 \end{cases} \quad \text{where } n \text{ is an integer.}$$

Associated with the discrete unit impulse is the **shifted** discrete unit impulse:

$$\delta[n-k] = \begin{cases} 1 & n = k \\ 0 & n \neq 0 \end{cases}$$

which enables us to select a particular component of an expression $x[n]$ via the equation:

$$x[k] = x[n]\delta[n-k]$$

▶

This is because the right-hand side $x[n]\delta[n-k] = 0$ unless $n = k$. Indeed, any sequence $x[n]$ can be considered as consisting of a collection of scaled and shifted discrete unit impulses. For example, the geometric sequence:

$x[n] = 3^n$ has values $\ldots, 3^{-2}, 3^{-1}, 1, 3, 3^2, 3^3, \ldots$

and can be alternatively written as the sum

$$x[n] = \ldots + 3^{-2}\delta[n-(-2)] + 3^{-1}\delta[n-(-1)] + 1\delta[n-0] + 3\delta[n-1]$$
$$+ 3^2\delta[n-2] + \ldots$$

From this sum any term of the sequence can be selected. For instance:

$$x[2] = \ldots + 3^{-2}\delta[2-(-2)] + 3^{-1}\delta[2-(-1)] + 1\delta[2-0]$$
$$+ 3\delta[2-1] + 3^2[2-2] + \ldots$$
$$= \ldots + 3^{-2}\delta[4] + 3^{-1}\delta[3] + 1\delta[2] + 3\delta[1] + 3^2\delta[0] + \ldots$$
$$= \ldots + 3^{-2} \times 0 + 3^{-1} \times 0 + 1 \times 0 + 3 \times 0 + 3^2 \times 1 + \ldots$$
$$= 3^2$$

For this reason the discrete unit impulse is also referred to as the **unit sample** and it can be used to decompose any sequence into a sum of weighted and shifted unit samples. For example:

$$x[n] = \ldots + x[-2]\delta[n-(-2)] + x[-1]\delta[n-(-1)] + x[0]\delta[n-0] + x[1]\delta[n-1]$$
$$+ x[2]\delta[n-2] + \ldots$$
$$= \sum_{k=-\infty}^{\infty} x[k]\delta[n-k]$$

Note the analogy with the continuous case:

$$f(t) = \int_{-\infty}^{\infty} f(s)\delta(t-s)\,ds$$

When the input to a linear, shift-invariant system is the discrete unit impulse $\delta[n]$ the response is denoted by $h[n]$ and is referred to as the **discrete unit impulse response**. That is:

$$h[n] = L\{\delta[n]\} \text{ and, because it is shift-invariant, } h[n-n_0] = L\{\delta[n-n_0]\}$$

Next frame

Arbitrary input

31

Just like the continuous system a discrete, linear, shift-invariant system has the important property that its response to any input can be found from its response to the discrete unit impulse $\delta[n]$. Recalling the discrete decomposition of a sequence as

$$x[n] = \sum_{k=-\infty}^{\infty} x[k]\delta[n-k]$$

▶

then the response of a linear system to this input is $y[n]$ where:

$$y[n] = L\{x[n]\}$$

$$= L\left\{\sum_{k=-\infty}^{\infty} x[k]\delta[n-k]\right\}$$

$$= \sum_{k=-\infty}^{\infty} L\{x[k]\delta[n-k]\} \quad \text{because } L \text{ is linear and so sums are preserved}$$

$$= \sum_{k=-\infty}^{\infty} x[k]L\{\delta[n-k]\} \quad \text{because } L \text{ is linear and so scalar multiples are preserved}$$

$$= \sum_{k=-\infty}^{\infty} x[k]h[n-k] \quad \text{because } L \text{ is shift-invariant}$$

That is $y[n] = \sum_{k=-\infty}^{\infty} x[k]h[n-k]$ which is referred to as the **convolution sum** of $x[n]$ and $h[n]$, alternatively written as:

$$x[n] * h[n] \quad \text{(also } h[n] * x[n] \text{ since the convolution sum is commutative)}$$

So, by direct analogy with a continuous system, the response of a discrete linear system can be obtained *from the convolution sum of the input with the system's unit impulse response.*

For example, a discrete, linear, shift-invariant system has the unit impulse response:

$$h[n] = u[n-1] \text{ the discrete unit step function where } u[n] = \begin{cases} 1 & n \geq 0 \\ 0 & n < 0 \end{cases}$$

To find the response $y[n]$ to the input $x[n] = 2^n u[n]$ we see that:

$$y[n] = x[n] * h[n]$$

$$= \sum_{k=-\infty}^{\infty} x[k]h[n-k]$$

$$= \sum_{k=-\infty}^{\infty} 2^k u[k]u[n-k-1]$$

$$= \sum_{k=0}^{\infty} 2^k u[n-k-1] \quad \text{since } u[k] = 0 \text{ for } k < 0$$

$$= \sum_{k=0}^{n-1} 2^k \quad \text{since } u[n-k-1] = 0 \text{ for } n-k-1 < 0 \text{ ie } k > n-1$$

$$= 2^n - 1 \quad \text{sum of the first } n \text{ terms of a geometric series with common ratio } 2$$

Try one yourself.

A discrete linear shift-invariant system has a unit impulse response $h[n] = u[n-4]$ and a response to the input $x[n] = nu[n]$ of :

$$y[n] = \ldots\ldots\ldots\ldots$$

The answer is in the next frame

$$y[n] = \frac{1}{2}(n-4)(n-3)$$

32

Because:

$$y[n] = x[n] * h[n]$$

$$= \sum_{k=-\infty}^{\infty} x[k]h[n-k]$$

$$= \sum_{k=-\infty}^{\infty} ku[k]u[n-k-4]$$

$$= \sum_{k=0}^{\infty} ku[n-k-4] \qquad \text{since } ku[k] = 0 \text{ for } k \leq 0$$

$$= \sum_{k=0}^{n-4} k \qquad \text{since } u[n-k-4] = 0 \text{ for } k > n-4$$

$$= \frac{(n-4)(n-5)}{2} \qquad \text{sum of the first } n-4 \text{ integers}$$

Move to the next frame

Exponential response

33

When the input to a discrete, linear, shift-invariant system with impulse response $h[n]$ is the exponential $x[n] = Az^n$ (A being constant) the system response is given as:

$$y[n] = h[n] * x[n]$$

$$= \sum_{k=-\infty}^{\infty} h[k]x[n-k]$$

$$= \sum_{k=-\infty}^{\infty} h[k]Az^{n-k}$$

$$= Az^n \sum_{k=-\infty}^{\infty} h[k]z^{-k}$$

$$= Az^n H[z]$$

where $H[z] = \sum_{k=-\infty}^{\infty} h[k]z^{-k}$ which you will recognise as the Z transform of $h[k]$ [refer to Programme 5].

▶

As in the continuous case we call $H[z]$ the system transfer function. For example, the transfer function $H[z]$ of a discrete, linear, shift-invariant system with impulse response:

$$h[n] = \begin{cases} \left(\frac{1}{5}\right)^n & 0 \le n \le 4 \\ 0 & \text{otherwise} \end{cases}$$

is

$$H(z) = \sum_{k=-\infty}^{\infty} h[k]z^{-k}$$

$$= \sum_{k=0}^{4} \left(\frac{1}{5}\right)^k z^{-k}$$

$$= \left(\frac{1}{5}\right)^0 z^{-0} + \left(\frac{1}{5}\right)^1 z^{-1} + \left(\frac{1}{5}\right)^2 z^{-2} + \left(\frac{1}{5}\right)^3 z^{-3} + \left(\frac{1}{5}\right)^4 z^{-4}$$

$$= 1 + \frac{1}{5z} + \frac{1}{25z^2} + \frac{1}{125z^3} + \frac{1}{625z^4}$$

So, the transfer function $H(z)$ of a discrete, linear, shift-invariant system with impulse response:

$$h[n] = \begin{cases} nu[n] & 0 \le n \le 3 \\ 0 & \text{otherwise} \end{cases}$$

is ....

The answer is in the next frame

34

$$\boxed{H(z) = \frac{1}{z} + \frac{2}{z^2} + \frac{3}{z^3}}$$

Because:

$$H(z) = \sum_{k=-\infty}^{\infty} h[k]z^{-k}$$

$$= \sum_{k=0}^{3} ku[k]z^{-k}$$

$$= 0u[0]z^{-0} + 1u[1]z^{-1} + 2u[2]z^{-2} + 3u[3]z^{-3}$$

$$= \frac{1}{z} + \frac{2}{z^2} + \frac{3}{z^3}$$

Move to the next frame

35 Transfer function

We have seen in Frame 31 that the response $y[n]$ to the input $x[n]$ to a discrete, linear, shift-invariant system with impulse response $h[n]$ is given as the convolution sum:

$$y[n] = h[n] * x[n]$$

$$\sum_{k=-\infty}^{\infty} h[k]x[n-k]$$

▶

So the Z transform of the response $Y(z)$ is given as:

$$Y(z) = \sum_{n=-\infty}^{\infty} y[n]z^{-n}$$

$$= \sum_{n=-\infty}^{\infty} \left(\sum_{k=-\infty}^{\infty} h[k]x[n-k] \right) z^{-n}$$

$$= \sum_{k=-\infty}^{\infty} h[k] \left(\sum_{n=-\infty}^{\infty} x[n-k]z^{-n} \right) \qquad \text{interchanging sums}$$

$$= \sum_{k=-\infty}^{\infty} h[k]z^{-k}X(z) \qquad \text{by the first shift property of the } Z$$

$$= H(z)X(z)$$

where the transfer function $H(z) = \sum_{k=-\infty}^{\infty} h[k]z^{-k}$ is the Z transform of the discrete impulse response $h[k]$. Consequently, for discrete, linear, shift-invariant systems the transfer function (as in the continuous case) completely characterises the system and permits the response to any input to be obtained.

Move to the next frame

Difference equations

36

Given a linear, constant coefficient difference equation it is possible to derive its transfer function and from that the impulse response. For example, consider the difference equation:

$$y[n] = 4y[n-2] + x[n]$$

Taking the Z transform of both sides where $Z\{x[n]\} = X(z)$ and $Z\{y[n]\} = Y(z)$ we see that:

$$Y(z) = (\ldots\ldots\ldots)X(z)$$

Next frame

37

$$\boxed{Y(z) = \frac{1}{(1-4z^{-1})}X(z)}$$

Because

$$Z\{y[n]\} = Z\{4y[n-1] + x[n]\}$$
$$= 4Z\{y[n-1]\} + Z\{x[n]\}$$

That is:

$$Y(z) = 4z^{-1}Y(z) + X(z) \text{ so that } Y(z)(1 - 4z^{-1}) = X(z)$$

and so:

$$Y(z) = \frac{1}{(1-4z^{-1})}X(z)$$

This means that the transfer function is:

$$H(z) = \ldots\ldots\ldots$$

Next frame

38

$$\boxed{\dfrac{z}{z-4}}$$

Because

$$Y(z) = \frac{1}{(1 - 4z^{-1})} X(z)$$
$$= H(z)X(z)$$

Giving:

$$H(z) = \frac{1}{(1 - 4z^{-1})}$$
$$= \frac{z}{z - 4}$$

From this we can now determine the impulse response:

$$h[n] = \dots\dots\dots$$

Next frame

39

$$\boxed{4^n u[n]}$$

Because

$$h[n] = Z^{-1}\{H(z)\}$$
$$= Z^{-1}\left\{\frac{z}{z - 4}\right\}$$
$$= 4^n u[n]$$

Now you try one. The transfer function and hence the impulse response of the difference equation:

$$y[n] = 2y[n-1] + 3y[n-2] + x[n-1] - 2x[n-2]$$

are given as:

$$H(z) = \dots\dots\dots$$
$$h[n] = \dots\dots\dots$$

Next frame

40

$$\boxed{\begin{array}{l} H(z) = \dfrac{3/4}{(z+1)} + \dfrac{1/4}{(z-3)} \\[2mm] h[n] = \left(\dfrac{3}{4}\left\{(-1)^{n-1}\right\} + \dfrac{1}{4}\left\{3^{n-1}\right\}\right)u[n] \end{array}}$$

▶

Because

Taking the Z transform of both sides we see that:

$Z\{y[n]\} = 2Z\{y[n-1]\} + 3Z\{y[n-2]\} + Z\{x[n-1]\} - 2Z\{x[n-2]\}$ that is:

$Y(z) = 2z^{-1}Y(z) + 3z^{-2}Y(z) + z^{-1}X(z) - 2z^{-2}X(z)$ so that:

$Y(z)(1 - 2z^{-1} - 3z^{-2}) = X(z)(z^{-1} - 2z^{-2})$ and so:

$$Y(z) = \frac{(z^{-1} - 2z^{-2})}{(1 - 2z^{-1} - 3z^{-2})} X(z) = H(z)X(z) \text{ giving:}$$

$$H(z) = \frac{(z^{-1} - 2z^{-2})}{(1 - 2z^{-1} - 3z^{-2})}$$

$$= \frac{(z-2)}{(z^2 - 2z - 3)}$$

$$= \frac{(z-2)}{(z+1)(z-3)}$$

$$= \frac{3/4}{(z+1)} + \frac{1/4}{(z-3)}$$

From this we can now determine the impulse response:

$$h[n] = Z^{-1}\{H(z)\}$$

$$= \frac{3}{4}Z^{-1}\left\{z^{-1}\frac{z}{(z+1)}\right\} + \frac{1}{4}Z^{-1}\left\{z^{-1}\frac{z}{(z-3)}\right\}$$

$$= \frac{3}{4}(-1)^{n-1}u[n] + \frac{1}{4}3^{n-1}u[n]$$

Next frame

This procedure can be reversed. For example to find the linear, constant **41** coefficient difference equation whose impulse response is

$$h[n] = 0.5\left((-5)^{n+2}\right)u[n]$$

we proceed as follows. Since the transfer function is the Z transform of the unit impulse then:

$$H(z) = Z\{h[n]\}$$

$$= Z\left\{0.5\left((-5)^{n+2}\right)u[n]\right\}$$

$$= \frac{1}{2}Z\left\{(-5)^{n+2}u[n]\right\}$$

$$= \frac{1}{2} \times z^2 \times \frac{z}{(z+5)}$$

$$= \frac{1}{2}\frac{z^3}{(z+5)}$$

▶

From this we can deduce the input-output relationship:

$$H(z) = \frac{Y(z)}{X(z)}$$

$$= \frac{1}{2}\frac{z^3}{(z+5)}$$

so that:

$$Y(z) = \frac{1}{2}\frac{z^3}{(z+5)}X(z)$$

$$= \frac{1/2}{(z^{-2}+5z^{-3})}X(z)$$

therefore:

$$(z^{-2} + 5z^{-3})Y(z) = \frac{1}{2}X(z)$$

and so the resulting difference equation is:

$$y[n-2] + 5y[n-3] = 0.5x[n] \quad \text{or} \quad y[n+1] + 5y[n] = 0.5x[n+3]$$

You try one now. The difference equation whose unit impulse response is given as:

$$h[n] = 3(2^{n-2})u[n] \text{ is } \ldots\ldots\ldots$$

Next frame

42

$$\boxed{y[n+1] - 2y[n] = 3x[n-1]}$$

Because:

$$H(z) = Z\{h[n]\}$$

$$= Z\{3(2^{n-2})u[n]\}$$

$$= 3Z\{2^{n-2}u[n]\}$$

$$= 3 \times z^{-2} \times \frac{z}{(z-2)}$$

$$= 3\frac{z^{-1}}{(z-2)}$$

From this we can deduce the input-output relationship:

$$H(z) = \frac{Y(z)}{X(z)}$$

$$= 3\frac{z^{-1}}{(z-2)}$$

so that:

$$Y(z) = 3\frac{z^{-1}}{(z-2)}X(z)$$

therefore:

$$(z-2)Y(z) = 3z^{-1}X(z)$$

and so the resulting difference equation is:

$$y[n+1] - 2y[n] = 3x[n-1]$$

▶

And that is the end of the Programme on invariant linear systems. All that remain are the **Revision summary** and the **Can you?** checklist. Read through these thoroughly and make sure you understand all the workings of this Programme. Then try the **Test exercise**; there is no need to hurry, take your time and work through the questions carefully. The **Further problems** then provide a valuable collection of additional exercises for you to try.

 # Revision summary 6

43

1 *Systems*

   A system L is a process capable of accepting an input $x(t)$ and processing the input to produce an output $y(t)$, also called the response of the system. This is written as $y(t) = L\{x(t)\}$.

2 *Linear systems*

   A linear system preserves sums and scalar products. If $y(t) = L\{x(t)\}$ then L is linear if

   $$L\{ax_1(t) + bx_2(t)\} = aL\{x_1(t)\} + bL\{x_2(t)\}$$

3 *Time-invariance*

   A continuous linear system is time-invariant if $y(t) = L\{x(t)\}$ and $y(t \pm t_0) = L\{x(t \pm t_0)\}$

4 *Shift-invariance*

   A discrete linear system is shift-invariant if $y[n] = L\{x[n]\}$ and $y[n \pm n_0] = L\{x[n \pm n_0]\}$

5 *Differential equations*

   The general nth-order, linear, constant coefficient, inhomogeneous differential equation:

   $$a_n \frac{d^n y(t)}{dt^n} + a_{n-1} \frac{d^{n-1} y(t)}{dt^{n-1}} + \ldots + a_0 y(t)$$
   $$= b_m \frac{d^m x(t)}{dt^m} + b_{m-1} \frac{d^{m-1} x(t)}{dt^{m-1}} + \ldots + b_0 x(t)$$

   coupled with the values of the n boundary conditions

   $$\frac{d^n y(t)}{dt^n}\bigg|_{t=t_0}, \frac{d^{n-1} y(t)}{dt^{n-1}}\bigg|_{t=t_0}, \ldots, y(t_0)$$

   describes the input-response relationship of a continuous linear system with input $x(t)$ and response $y(t)$. Such an equation has a solution in the form $y(t) = y_h(t) + y_p(t)$ where $y_h(t)$ is complementary function solution to the homogeneous equation

   $$a_n \frac{d^n y(t)}{dt^n} + a_{n-1} \frac{d^{n-1} y(t)}{dt^{n-1}} + \ldots + a_0 y(t) = 0$$

   and $y_p(t)$ is a particular integral or particular solution to the inhomogeneous equation. The procedure for solving such an equation is:

   ▶

    (i)   Find the homogeneous solution $y_h(t)$ in terms of unknown integration constants

    (ii) Find the particular solution $y_p(t)$ and form the complete solution $y(t) = y_h(t) + y_p(t)$

    (iii) Apply the boundary conditions to find the values of the unknown integration constants in $y_h(t)$.

6   *Zero-input and zero-state*

The solution of the general nth-order, linear, constant coefficient, inhomogeneous differential equation can alternatively be written as $y(t) = y_{zi}(t) + y_{zs}(t)$ where $y_{zi}(t)$ is called the *zero-input response* and $y_{zs}(t)$ is called the *zero-state response*. The zero-input response of the equation depends only on the initial conditions and is independent of the input. It is obtained by solving the homogeneous equation and applying the boundary conditions. The zero-state response depends only on the input and is independent of the initial conditions. It is obtained by solving the inhomogeneous equation but with all the boundary conditions equated to zero.

Here the procedure is:

    (i)   Find the homogeneous solution $y_h(t)$ in terms of unknown integration constants

    (ii) Find the particular solution $y_p(t)$ and form the complete solution $y(t) = y_h(t) + y_p(t)$

    (iii) Equate the boundary conditions to zero and then find the values of the unknown integration constants in $y(t)$. This is then the zero-state response $y_{zs}(t)$.

    (iv) Apply the original boundary conditions to find the values of the unknown integration constants in $y_h(t)$. This is then the zero-input solution $y_{zi}(t)$.

7   *Zero input and zero response*

For a linear time-invariant system zero input yields zero response. This is equivalent to all the boundary conditions having a zero value.

8   *Arbitrary input*

If $h(t)$ is the response of a continuous linear time-invariant system to the unit impulse $\delta(t)$, that is $h(t) = L\{\delta(t)\}$ then the response to an arbitrary input $x(t)$ is the convolution of the input with the unit impulse response. That is:

$$L\{x(t)\} = \int_{-\infty}^{\infty} x(\tau)h(t - \tau)\,d\tau = x(t) * h(t)$$

9   *Exponential response*

The response of a linear, time-invariant system to an exponential input is a scaled exponential. That is $L\{Ae^{st}\} = H(s)(Ae^{st})$. Therefore the exponential is an eigenfunction of the system and the scaling factor $H(s)$ is the eigenvalue. This eigenvalue $H(s)$ is referred to as the system transfer function.

▶

10 *Transfer function*

The transfer function $H(s)$ of a linear, time-invariant system is the Laplace transform of the unit impulse response. That is:

$$H(s) = \int_{-\infty}^{\infty} h(t)e^{-st}\,dt$$

11 *Convolution theorem*

The fact that the response of a continuous linear time-invariant system is the convolution of the input with the unit impulse response enables the use of the convolution theorem as it applies to the Laplace transform:

$$y(t) = x(t) * h(t) \text{ and so } \mathscr{L}\{y(t)\} = \mathscr{L}\{x(t) * h(t)\} = \mathscr{L}\{x(t)\}\mathscr{L}\{h(t)\}$$

That is $Y(s) = X(s)H(s)$, the Laplace transform of the response, is equal to the product of the Laplace transform of the input and the system's transfer function.

12 *Arbitrary input to a discrete system*

If $h[n]$ is the response of a discrete linear shift-invariant system to the discrete unit impulse $\delta[n]$, that is $h[n] = L\{\delta[n]\}$ then the response to an arbitrary input $x[n]$ is the convolution sum of the input with the unit impulse response. That is:

$$L\{x[n]\} = \sum_{k=-\infty}^{\infty} x[k]h[n - k] = x[n] * h[n]$$

13 *Exponential response*

The response of a discrete linear, time-invariant system to an exponential input is a scaled exponential. That is $L\{Az^n\} = H(z)(Az^n)$. Therefore the exponential is an eigenfunction of the system and the scaling factor $H(z)$ is the eigenvalue. This eigenvalue $H(z)$ is referred to as the system transfer function.

14 *Transfer function*

The transfer function $H(z)$ of a discrete linear, shift-invariant system is the Z transform of the unit impulse response. That is:

$$H(z) = \sum_{k=-\infty}^{\infty} h[k]z^{-k}$$

15 *Difference equations*

The transfer function $H(z)$ of a discrete linear system described by a difference equation can be derived by taking the Z transform of the equation. By taking the inverse Z transform of the transfer function the unit impulse response can be found. Alternatively, given the impulse response of a discrete system the corresponding difference equation can be derived.

Can you?

44 **Checklist 6**

Check this list before and after you try the end of Programme test

On a scale of 1 to 5 how confident are you that you can: **Frames**

- Recognise a system as a process whereby an input
  (either continuous or discrete) is converted to an output,
  also called the response of the system?   [1] to [4]
  Yes   ☐   ☐   ☐   ☐   ☐   *No*

- Distinguish between linear and non-linear systems and
  recognise time-invariant and shift-invariant systems?   [5] to [10]
  Yes   ☐   ☐   ☐   ☐   ☐   *No*

- Determine the zero-input response and the zero-state
  response?   [11] to [13]
  Yes   ☐   ☐   ☐   ☐   ☐   *No*

- Appreciate why zero valued boundary conditions give rise
  to a time-invariant system?   [14] to [17]
  Yes   ☐   ☐   ☐   ☐   ☐   *No*

- Demonstrate that the response of a continuous, linear,
  time-invariant system to an arbitrary input is the convolution
  of the input with response of the system to a unit impulse?   [18] to [20]
  Yes   ☐   ☐   ☐   ☐   ☐   *No*

- Understand the role of the exponential function with respect
  to a linear, time-invariant system?   [21] to [23]
  Yes   ☐   ☐   ☐   ☐   ☐   *No*

- Use the convolution theorem to find the response of a
  continuous, linear, time-invariant system to an arbitrary
  input?   [24] to [25]
  Yes   ☐   ☐   ☐   ☐   ☐   *No*

- Derive the system transfer function of a constant coefficient
  linear differential equation and use it to solve the equation?   [26] to [29]
  Yes   ☐   ☐   ☐   ☐   ☐   *No*

- Demonstrate that the response of a discrete, linear,
  shift-invariant system to an arbitrary input is the convolution
  sum of the input with response of the system to a unit
  impulse?   [30] to [32]
  Yes   ☐   ☐   ☐   ☐   ☐   *No*

- Understand the role of the exponential function with respect
  to a discrete linear, shift-invariant system?   [33] to [35]
  Yes   ☐   ☐   ☐   ☐   ☐   *No*

▶

- Derive the system transfer function of a constant coefficient linear difference equation and use it to solve the equation? Yes ☐ ☐ ☐ ☐ ☐ No $\boxed{36}$ to $\boxed{40}$

- Derive the constant coefficient difference equation from knowledge of its unit impulse response? Yes ☐ ☐ ☐ ☐ ☐ No $\boxed{41}$ to $\boxed{42}$

 # Test exercise 6

---

1 Which of the following are linear, non-linear, time-invariant and shift-invariant:

$\boxed{45}$

(a) $y(t) = L\{x(t)\} = -3x(t)$ (e) $y(t) = L\{x(t)\} = tx(t)$

(b) $y[n] = L\{x[n]\} = 2^{x[n-4]}$ (f) $y[n] = L\{x[n]\} = x[n]\delta[n-4]$

(c) $y(t) = L\{x(t)\} = e^{-2t}\sin x(t)$ (g) $y(t) = L\{x(t)\} = \dfrac{x(t)}{4}$

(d) $y[n] = L\{x[n]\} = 2x[n] - \cos x[n]$ (h) $y[n] = L\{x[n]\} = L\{x[n]\} = 4^n x[n]$

2 Find the zero-input response and the zero-state response for each of the following and determine which are time-invariant:

(a) $y'(t) - 3y(t) = t^2 u(t) : y(0) = 2$

(b) $y''(t) - 5y'(t) + 4y(t) = u(t)\sin t : y'(0) = -4, y(0) = 0$

(c) $5y'(t) + 4y(t) = e^{-t}u(t) : y(0) = 0$

(d) $y''(t) + 2y'(t) + y(t) = u(t) : y'(0) = 0, y(0) = 0$

3 A linear, time-invariant system has the impulse response $h(t) = e^{-3t}u(t)$ find the system response to the input $x(t) = u(t) - u(t-3)$.

4 A linear, time-invariant system has the impulse response $h(t) = tu(t-1)$ find the transfer function $H(s)$ and use it to find the response to the input
$x(t) = u(t) - 2u(t-1) + u(t-2)$

5 Given the differential equation
$y''(t) + 3y'(t) - 4y(t) = 30e^{-2t} : y'(0) = 0, y(0) = 0$
find the transfer function and solve the equation.

6 A linear, shift-invariant system has the impulse response $h[n] = nu[n]$ find the system response to the input $x[n] = 4^{-n}u[n]$.

7 Find the impulse response of the difference equation
$y[n+1] - 3y[n] + 2y[n-1] = x[n+1] - x[n]$.

8 A linear, shift-invariant system has the impulse response $h[n] = nu[n]$, find the difference equation.

Further problems

1 For what values of α is the system $y[n] = L\{x[n]\} = x[\alpha n]$ shift-invariant?

2 For what values of a and b is the system $y(t) = L\{x(t)\} = ax(t) + b$ linear?

3 Is the system $y(t) = L\{x(t)\} = \sum_{n=0}^{\infty} x(t)\delta(t - nt_0)$ linear and time-invariant?

4 A linear, time-invariant system has the impulse response $h(t) = e^{-3t}u(t)$ find the system response to the input $x(t) = e^{3t}u(t)$.

5 Is the system $y[n] = L\{x[n]\} = x[n] - x[-n]$ linear and shift-invariant?

6 Is the system $y[n] = L\{x[n]\} = x[n^3]$ linear and shift-invariant?

7 Show that the sequence:
$$x[n] = \begin{cases} 2 & n = 0 \\ 4 & n = 1 \\ 6 & n = 2 \\ 0 & \text{otherwise} \end{cases}$$
can be represented as
$$x[n] = 2\delta[n] + 4\delta[n-1] + 6\delta[n-2] \text{ or as}$$
$$x[n] = 2(u[n] + u[n-1] + u[n-2] - 3u[n-3])$$

8 The sign function $\text{sgn}(x)$ (called the signum function to avoid confusion with the sine function) is defined as:
$$\text{sgn}(x) = \begin{cases} -1 & x < 0 \\ 0 & x = 0 \\ 1 & x > 0 \end{cases} \text{ the discrete form being } \text{sgn}[n] = \begin{cases} -1 & n < 0 \\ 0 & n = 0 \\ 1 & n > 0 \end{cases}$$
The signum function is essentially the sign of a number so that $x = |x|\text{sgn}(x)$ because if $x < 0$ then $x = -|x|$ and if $x > 0$ then $x = |x|$.
    Show that if $x[n] = a^n u[n]$ then the even part of $x[n]$ is:
$$x_e[n] = \frac{1}{2}\left(a^{|n|} + \delta[n]\right)$$
and the odd part is
$$x_o[n] = \frac{1}{2}a^{|n|}\text{sgn}[n].$$

9 Show that the convolution sum of $a[n] = nu[n-1]$ and $b[n] = n^2 u[n]$ is
$$\frac{n^2(n^2 - 1)}{12}.$$

10 Show that if $x[n] = a^n u[n]$ $(a \neq 1)$ and $y[n] = u[n]$ then $x[n] * y[n] = \frac{1 - a^{n+1}}{1 - a}u[n]$.

11 Show that if $p[n] \neq 0$ only for $m_1 \leq n \leq m_2$ and $q[n] \neq 0$ only for $M_1 \leq n \leq M_2$ then $p[n] * q[n] \neq 0$ only for $m_1 + M_1 \leq n \leq m_2 + M_2$.

▶

12 The cross-correlation of two sequences $a[n]$ and $b[n]$ is defined as:

$$a[n] \star b[n] = \sum_{k=-\infty}^{\infty} a[k]b[n+k]$$

Show that if $x[n] = a^n u[n]$ then

$$x[n] \star x[n] = \frac{a^{|n|}}{1-a^2}$$

[the cross-correlation of a sequence with itself is called the autocorrelation of the sequence].

13 A linear, shift-invariant system has the impulse response $h[n] = \left(\frac{1}{4}\right)^n u[n]$ find the system response to the complex input $x[n] = e^{jn\omega_0}u[n]$.

14 Solve the differential equation $y''(t) + 2y'(t) + y(t) = e^{-t}u(t)$: $y(0) = 0$, $y'(0) = 0$.

15 Find the impulse response of the differential equation

$$y'(t) + \frac{1}{a}y(t) = \frac{1}{a}x(t) : y(0) = 0$$

16 Solve the differential equation

$$y'(t) = -\frac{1}{T}y(t) + \frac{G}{T}u(t) : y(0) = 0$$

17 Solve the difference equation $y[n] = \alpha y[n-1] + (1-\alpha)u[n] : y[0] = 0$.

18 Solve the difference equation

$$y[n+1] - y[n] = -\frac{7}{100}(y[n] - 20) : y[0] = 160.$$

19 Solve the difference equation $2y[n] = y[n-1] + y[n+1] : y[0] = 0$, $y[1] = 8$.

20 A continuous, linear, time-invariant system has output $y(t) = tu(t)$ when the input is $x(t) = u(t)$. Find the impulse response of the system and the output when the input is $x(t) = u(t-1)$.

21 A discrete, linear, shift-invariant system has output $y[n] = nu[n]$ when the input is $x[n] = 2^n u[n]$. Find the impulse response of the system and the output when the input is $x[n] = 3^n u[n]$.

Fourier series 1

Learning outcomes

When you have completed this Programme you will be able to:

- Determine the period and amplitude of a periodic function
- Write down the harmonics of a periodic trigonometric function
- Give an analytic description of a non-sinusoidal periodic function
- Evaluate integrals with periodic integrands
- Demonstrate the orthogonality of the trigonometric functions $\sin nx$ and $\cos nx$ for $n = 0, 1, 2, \ldots$
- Describe a periodic function as a Fourier series subject to Dirichlet conditions
- Obtain the Fourier coefficients and hence the Fourier series of a periodic function
- Describe the effects of the harmonics in the construction of the Fourier series
- Find the value of the Fourier series at a point of discontinuity of the periodic function

Prerequisite: Engineering Mathematics (Sixth Edition)
Programmes 15 Integration 1 and 17 Reduction formulas

Introduction

We have seen earlier that many functions can be expressed in the form of **1** infinite series. Problems involving various forms of oscillations are common in fields of modern technology and *Fourier series*, with which we shall now be concerned, enable us to represent a periodic function as an infinite trigonometrical series in sine and cosine terms. One important advantage of a Fourier series is that it can represent a function containing discontinuities, whereas Maclaurin's and Taylor's series require the function to be continuous throughout.

Periodic functions

A function $f(x)$ is said to be *periodic* if its function values repeat at regular intervals of the independent variable. The regular interval between repetitions is the *period* of the oscillations.

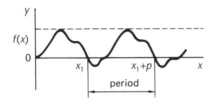

Graphs of $y = A \sin nx$

(a) $y = \sin x$

The obvious example of a periodic function is $y = \sin x$, which goes through its complete range of values while x increases from $0°$ to $360°$. The period is therefore $360°$ or 2π radians and the amplitude, the maximum displacement from the position of rest, is 1.

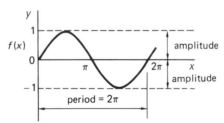

(b) $y = 5 \sin 2x$

The amplitude is 5.
The period is $180°$ and there are thus 2 complete cycles in $360°$.

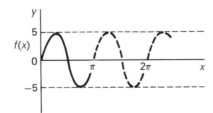

▶

(c) $y = A \sin nx$

Thinking along the same lines, the function $y = A \sin nx$ has
amplitude ....; period ....;
and will have .... complete cycles in 360°.

2

$$\text{amplitude} = A; \text{ period} = \frac{360°}{n} = \frac{2\pi}{n}; n \text{ cycles in } 360°$$

Graphs of $y = A \cos nx$ have the same characteristics.

By way of revising earlier work, then, complete the following short exercise.

Exercise

In each of the following, state (a) the amplitude and (b) the period.

1 $y = 3 \sin 5x$ 5 $y = 5 \cos 4x$

2 $y = 2 \cos 3x$ 6 $y = 2 \sin x$

3 $y = \sin \dfrac{x}{2}$ 7 $y = 3 \cos 6x$

4 $y = 4 \sin 2x$ 8 $y = 6 \sin \dfrac{2x}{3}$

Deal with all eight. They will not take much time.

3

No.	Amplitude	Period	No.	Amplitude	Period
1	3	$2\pi/5$	5	5	$\pi/2$
2	2	$2\pi/3$	6	2	π
3	1	4π	7	3	$\pi/3$
4	4	π	8	6	3π

Harmonics

A function $f(x)$ is sometimes expressed as a series of a number of different sine
components. The component with the largest period is the *first harmonic*, or
fundamental of $f(x)$.

$y = A_1 \sin x$ is the first harmonic or fundamental

$y = A_2 \sin 2x$ is the second harmonic

$y = A_3 \sin 3x$ is the third harmonic, etc.

and in general

$y = A_n \sin nx$ is the .... harmonic, with
amplitude .... and period ....

$$nth \text{ harmonic; amplitude } A_n; \text{ period} = \frac{2\pi}{n}$$

4

Non-sinusoidal periodic functions

Although we introduced the concept of a periodic function via a sine curve, a function can be periodic without being obviously sinusoidal in appearance.

Example

In the following cases, the x-axis carries a scale of t in milliseconds.

(a)
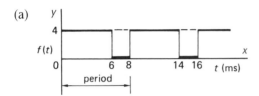
period = 8 ms

(b)
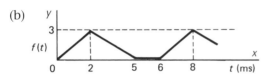
period = ...

(c)
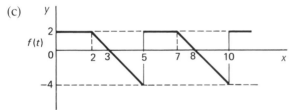
period = ...

(b) period = 6 ms; (c) period = 5 ms

5

Analytic description of a periodic function

A periodic function can be defined analytically in many cases.

Example 1

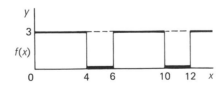

(a) Between $x = 0$ and $x = 4$, $y = 3$, i.e. $f(x) = 3$ $0 < x < 4$
(b) Between $x = 4$ and $x = 6$, $y = 0$, i.e. $f(x) = 0$ $4 < x < 6$

▶

So we could define the function by

$$f(x) = \begin{cases} 3 & 0 < x < 4 \\ 0 & 4 < x < 6 \end{cases}$$

$$f(x+6) = f(x)$$

the last line indicating that ....

6

> the function is periodic with period 6 units

Example 2

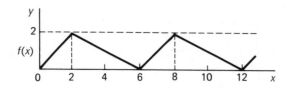

In this case

    (a) Between $x = 0$ and $x = 2$, $y = x$ i.e. $f(x) = x$ $0 < x < 2$

    (b) Between $x = 2$ and $x = 6$, $y = -\dfrac{x}{2} + 3$, i.e. $f(x) = 3 - \dfrac{x}{2}$ $2 < x < 6$

    (c) The period is 6 units   i.e. $f(x+6) = f(x)$.

So we have

$$f(x) = \begin{cases} x & 0 < x < 2 \\ 3 - \dfrac{x}{2} & 2 < x < 6 \end{cases}$$

$$f(x+6) = f(x).$$

Example 3

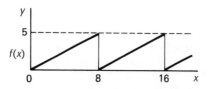

In this case   ....

....

7

> $$f(x) = \frac{5x}{8} \qquad 0 < x < 8$$
>
> $$f(x+8) = f(x)$$

Here is a short exercise.

Exercise

Define analytically the periodic functions shown.

1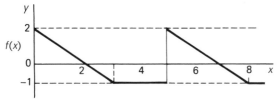

2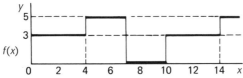

3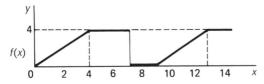

4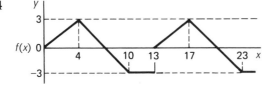

5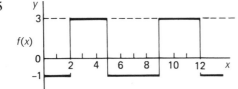

Finish all five and then check the results.

Here are the details.

1

$$f(x) = \begin{cases} 2 - x & 0 < x < 3 \\ -1 & 3 < x < 5 \end{cases}$$

$f(x + 5) = f(x).$

2

$$f(x) = \begin{cases} 3 & 0 < x < 4 \\ 5 & 4 < x < 7 \\ 0 & 7 < x < 10 \end{cases}$$

$f(x + 10) = f(x).$

8

▶

3
$$f(x) = \begin{cases} x & 0 < x < 4 \\ 4 & 4 < x < 7 \\ 0 & 7 < x < 9 \end{cases}$$
$f(x+9) = f(x).$

4
$$f(x) = \begin{cases} \dfrac{3x}{4} & 0 < x < 4 \\ 7 - x & 4 < x < 10 \\ -3 & 10 < x < 13 \end{cases}$$
$f(x+13) = f(x).$

5
$$f(x) = \begin{cases} -1 & 0 < x < 2 \\ 3 & 2 < x < 5 \\ -1 & 5 < x < 7 \end{cases}$$
$f(x+7) = f(x).$

Now we have the same thing in reverse.

Exercise

Sketch the graphs of the following, inserting relevant values.

1
$$f(x) = \begin{cases} 4 & 0 < x < 5 \\ 0 & 5 < x < 8 \end{cases}$$
$f(x+8) = f(x).$

2    $f(x) = 3x - x^2 \qquad 0 < x < 3$
$f(x+3) = f(x).$

3
$$f(x) = \begin{cases} 2\sin x & 0 < x < \pi \\ 0 & \pi < x < 2\pi \end{cases}$$
$f(x+2\pi) = f(x).$

4
$$f(x) = \begin{cases} \dfrac{x}{2} & 0 < x < \pi \\ \pi - \dfrac{x}{2} & \pi < x < 2\pi \end{cases}$$
$f(x+2\pi) = f(x).$

5
$$f(x) = \begin{cases} \dfrac{x^2}{4} & 0 < x < 4 \\ 4 & 4 < x < 6 \\ 0 & 6 < x < 8 \end{cases}$$
$f(x+8) = f(x).$

9   Here they are: check carefully.

1

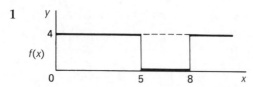

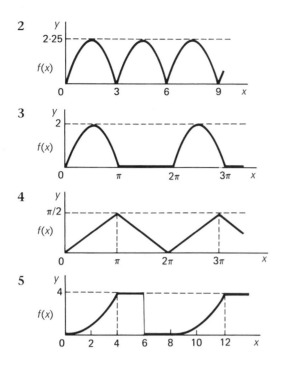

All this is in preparation for what is to come, so let us now consider Fourier series.

Move on then to the next frame

Integrals of periodic functions

Before we proceed we need to consider some specific integrals involving integers m and n. These are integrals over a single period of periodic integrands. You will already know some of these and the others you will easily be able to work out. The integrals that we are concerned with are those of sines, cosines and their combinations where the integration is over a single period from $-\pi$ to π. First, though, we list the integral of the unit constant over the period.

10

1 $\displaystyle\int_{-\pi}^{\pi} dx = \Big[x\Big]_{-\pi}^{\pi} = 2\pi$

2 $\displaystyle\int_{-\pi}^{\pi} \cos nx\, dx = \left[\frac{\sin nx}{n}\right]_{-\pi}^{\pi} \qquad (n \neq 0)$

$\displaystyle\qquad\qquad\qquad = \frac{\sin n\pi}{n} - \frac{\sin(-n\pi)}{n}$

$\displaystyle\qquad\qquad\qquad = 0 \quad \text{because } \sin n\pi = 0$

3 $\displaystyle\int_{-\pi}^{\pi} \sin nx\, dx = \ldots\ldots\ldots\ldots \qquad (n \neq 0)$

▶

11

$$\int_{-\pi}^{\pi} \sin nx \, dx = 0$$

Because

$$\int_{-\pi}^{\pi} \sin nx \, dx = \left[-\frac{\cos nx}{n}\right]_{-\pi}^{\pi} \quad (n \neq 0)$$

$$= -\frac{\cos n\pi}{n} + \frac{\cos(-n\pi)}{n}$$

$$= 0 \quad \text{because } \cos(-x) = \cos x$$

4 $\quad \int_{-\pi}^{\pi} \cos^2 nx \, dx = \int_{-\pi}^{\pi} \frac{\cos 2nx + 1}{2} \, dx \quad \text{because } \cos 2A = 2\cos^2 A - 1$

$$= \left[\frac{\sin 2nx}{4n} + \frac{x}{2}\right]_{-\pi}^{\pi} \quad (n \neq 0)$$

$$= \frac{\sin 2n\pi}{4n} + \frac{\pi}{2} - \frac{\sin(-2n\pi)}{4n} - \frac{(-\pi)}{2}$$

$$= \pi$$

5 $\quad \int_{-\pi}^{\pi} \sin^2 nx \, dx = \ldots\ldots\ldots \quad (n \neq 0)$

12

$$\int_{-\pi}^{\pi} \sin^2 nx \, dx = \pi$$

Because

$$\int_{-\pi}^{\pi} \sin^2 nx \, dx = \int_{-\pi}^{\pi} \frac{1 - \cos 2nx}{2} \, dx \quad \text{because } \cos 2A = 1 - 2\sin^2 A$$

$$= \left[\frac{x}{2} - \frac{\sin 2nx}{4n}\right]_{-\pi}^{\pi} \quad (n \neq 0)$$

$$= \frac{\pi}{2} - \frac{\sin 2n\pi}{4n} - \frac{(-\pi)}{2} + \frac{\sin(-2n\pi)}{4n}$$

$$= \pi$$

6 $\quad \int_{-\pi}^{\pi} \cos mx \cos nx \, dx$

$$= \frac{1}{2} \int_{-\pi}^{\pi} [\cos(m+n)x + \cos(m-n)x] \, dx$$

because $2\cos A \cos B = \cos(A+B) + \cos(A-B)$

$$= \left[\frac{\sin(m+n)x}{m+n} + \frac{\sin(m-n)x}{m-n}\right]_{-\pi}^{\pi} \quad (m \neq n)$$

$$= \frac{\sin(m+n)\pi}{m+n} + \frac{\sin(m-n)\pi}{m-n} - \frac{\sin(m+n)(-\pi)}{m+n} - \frac{\sin(m-n)(-\pi)}{m-n}$$

$$= 0$$

7 $\quad \int_{-\pi}^{\pi} \sin mx \sin nx \, dx = \ldots\ldots\ldots \quad (m \neq n)$

▶

13

$$\boxed{\int_{-\pi}^{\pi} \sin mx \sin nx\, \mathrm{d}x = 0, \quad m \neq n}$$

Because

$$\int_{-\pi}^{\pi} \sin mx \sin nx\, \mathrm{d}x$$

$$= \frac{1}{2} \int_{-\pi}^{\pi} [\cos(m-n)x - \cos(m+n)x]\, \mathrm{d}x$$

because $2 \sin A \sin B = \cos(A-B) - \cos(A+B)$

$$= \left[\frac{\sin(m-n)x}{m-n} - \frac{\sin(m+n)x}{m+n} \right]_{-\pi}^{\pi} \quad (m \neq n)$$

$$= \frac{\sin(m-n)\pi}{m-n} - \frac{\sin(m+n)\pi}{m+n} - \frac{\sin(m-n)(-\pi)}{m-n} + \frac{\sin(m+n)(-\pi)}{m+n}$$

$$= 0$$

8 $\displaystyle\int_{-\pi}^{\pi} \cos mx \sin nx\, \mathrm{d}x \qquad (m \neq n)$

$$= \frac{1}{2} \int_{-\pi}^{\pi} [\sin(m+n)x - \sin(m-n)x]\, \mathrm{d}x$$

because $2 \cos A \sin B = \sin(A+B) - \sin(A-B)$

$$= \frac{1}{2} \left[-\frac{\cos(m+n)x}{m+n} + \frac{\cos(m-n)x}{m-n} \right]_{-\pi}^{\pi} \qquad (m \neq n)$$

$$= \frac{1}{2} \left(-\frac{\cos(m+n)\pi}{m+n} + \frac{\cos(m-n)\pi}{m-n} \right.$$

$$\left. + \frac{\cos(m+n)(-\pi)}{m+n} - \frac{\cos(m-n)(-\pi)}{m-n} \right)$$

$$= 0 \quad \text{because } \cos(-x) = \cos x$$

And finally, when $m = n$

9 $\displaystyle\int_{-\pi}^{\pi} \cos mx \sin mx\, \mathrm{d}x = \ldots\ldots\ldots\ldots$

14

$$\boxed{\int_{-\pi}^{\pi} \cos mx \sin mx \, dx = 0}$$

Because

$$\int_{-\pi}^{\pi} \cos mx \sin mx \, dx$$

$$= \frac{1}{2} \int_{-\pi}^{\pi} \sin 2mx \, dx \quad \text{because} \ \sin 2A = 2 \sin A \cos A$$

$$= \frac{1}{2} \left[-\frac{\cos 2mx}{2m} \right]_{-\pi}^{\pi} \quad (m \neq 0)$$

$$= \frac{1}{2} \left(-\frac{\cos 2m\pi}{2m} + \frac{\cos 2m(-\pi)}{2m} \right)$$

$$= 0 \quad \text{because} \ \cos(-x) = \cos x$$

15   Summary

1   $\displaystyle\int_{-\pi}^{\pi} dx = \Big[x\Big]_{-\pi}^{\pi} = 2\pi$

2   $\displaystyle\int_{-\pi}^{\pi} \cos nx \, dx = 0$

3   $\displaystyle\int_{-\pi}^{\pi} \sin nx \, dx = 0$

4   $\displaystyle\int_{-\pi}^{\pi} \cos mx \cos nx \, dx = \pi \delta_{mn} \quad$ where $\delta_{mn} = \begin{cases} 1 \text{ if } m = n \\ 0 \text{ if } m \neq n \end{cases}$

$(\delta_{mn}$ is called the Kronecker delta)

5   $\displaystyle\int_{-\pi}^{\pi} \sin mx \sin nx \, dx = \pi \delta_{mn}$

6   $\displaystyle\int_{-\pi}^{\pi} \cos mx \sin nx \, dx = 0$

Note that the same results are obtained no matter what the end points of the integrals are, *provided that the interval between them is one period*. So, for example

$$\int_{k}^{k+2\pi} \cos nx \, dx = \left[\frac{\sin nx}{n} \right]_{k}^{k+2\pi} \quad (n \neq 0)$$

$$= \frac{\sin(nk + 2n\pi)}{n} - \frac{\sin nk}{n}$$

$$= 0 \quad \text{because} \ \sin(x + 2n\pi) = \sin x$$

Now to put all these integrals to practical use

Orthogonal functions

16

If two different functions $f(x)$ and $g(x)$ are defined on the interval $a \leq x \leq b$

and $\displaystyle\int_a^b f(x)g(x)\,\mathrm{d}x = 0$

then we say that the two functions are **orthogonal** to each other on the interval $a \leq x \leq b$. In the previous frames we have seen that the trigonometric functions $\sin nx$ and $\cos nx$ where $n = 0, 1, 2, \ldots$ form an infinite collection of periodic functions that are mutually orthogonal on the interval $-\pi \leq x \leq \pi$, indeed on any interval of width 2π. That is

$$\int_{-\pi}^{\pi} \cos mx \cos nx \, \mathrm{d}x = 0 \quad \text{for } m \neq n$$

$$\int_{-\pi}^{\pi} \sin mx \sin nx \, \mathrm{d}x = 0 \quad \text{for } m \neq n$$

and

$$\int_{-\pi}^{\pi} \cos mx \sin nx \, \mathrm{d}x = 0$$

Fourier series

17

Given that certain conditions are satisfied then it is possible to write a periodic function of period 2π as a series expansion of the orthogonal periodic functions just discussed. That is, if $f(x)$ is defined on the interval $-\pi \leq x \leq \pi$ where $f(x + 2n\pi) = f(x)$ then

$$f(x) = \frac{a_0}{2} + \sum_{n=1}^{\infty} (a_n \cos nx + b_n \sin nx)$$

This is the **Fourier series** expansion of $f(x)$ where the a_n and b_n are constants called the *Fourier coefficients*. But how do we find the values of these constants? Quite easily. We make use of the mutual orthogonality of the trigonometric functions in the expansion.

▶

For example, to find a_{10} we multiply $f(x)$ by $\cos 10x$ and integrate over a period. That is

$$\int_{-\pi}^{\pi} f(x) \cos 10x \, dx$$

$$= \int_{-\pi}^{\pi} \left(\frac{a_0}{2} + \sum_{n=1}^{\infty} (a_n \cos nx + b_n \sin nx) \right) \cos 10x \, dx$$

$$= \frac{a_0}{2} \int_{-\pi}^{\pi} \cos 10x \, dx + \sum_{n=1}^{\infty} a_n \int_{-\pi}^{\pi} \cos nx \cos 10x \, dx$$

$$+ \sum_{n=1}^{\infty} b_n \int_{-\pi}^{\pi} \sin nx \cos 10x \, dx$$

$$= \frac{a_0}{2} \times 0 + \sum_{n=1}^{\infty} a_n \pi \delta_{n,10} + \sum_{n=1}^{\infty} b_n \times 0$$

$$= a_0 \pi \times 0 + a_1 \pi \times 0 + \ldots + a_9 \pi \times 0 + a_{10} \pi \times 1 + a_{11} \pi \times 0 + \ldots$$

$$= a_{10} \pi$$

So that

$$a_{10} = \frac{1}{\pi} \int_{-\pi}^{\pi} f(x) \cos 10x \, dx$$

In just the same way $\displaystyle\int_{-\pi}^{\pi} f(x) \cos mx \, dx = \ldots\ldots\ldots$

18

$$\boxed{\int_{-\pi}^{\pi} f(x) \cos mx \, dx = a_m \pi}$$

Because

$$\int_{-\pi}^{\pi} f(x) \cos mx \, dx$$

$$= \int_{-\pi}^{\pi} \left(\frac{a_0}{2} + \sum_{n=1}^{\infty} (a_n \cos nx + b_n \sin nx) \right) \cos mx \, dx$$

$$= \frac{a_0}{2} \int_{-\pi}^{\pi} \cos mx \, dx + \sum_{n=1}^{\infty} a_n \int_{-\pi}^{\pi} \cos nx \cos mx \, dx$$

$$+ \sum_{n=1}^{\infty} b_n \int_{-\pi}^{\pi} \sin nx \cos mx \, dx$$

$$= \frac{a_0}{2} \times 0 + \sum_{n=1}^{\infty} a_n \pi \delta_{n,m} + \sum_{n=1}^{\infty} b_n \times 0$$

$$= a_m \pi$$

and so

$$a_m = \frac{1}{\pi} \int_{-\pi}^{\pi} f(x) \cos mx \, dx$$

Finally

$$\int_{-\pi}^{\pi} f(x) \sin mx \, dx = \ldots\ldots\ldots$$

19

$$\int_{-\pi}^{\pi} f(x) \sin mx \, dx = b_m \pi$$

Because

$$\int_{-\pi}^{\pi} f(x) \sin mx \, dx$$

$$= \int_{-\pi}^{\pi} \left(\frac{a_0}{2} + \sum_{n=1}^{\infty} (a_n \cos nx + b_n \sin nx) \right) \sin mx \, dx$$

$$= \frac{a_0}{2} \int_{-\pi}^{\pi} \sin mx \, dx + \sum_{n=1}^{\infty} a_n \int_{-\pi}^{\pi} \cos nx \sin mx \, dx$$

$$+ \sum_{n=1}^{\infty} b_n \int_{-\pi}^{\pi} \sin nx \sin mx \, dx$$

$$= \frac{a_0}{2} \times 0 + \sum_{n=1}^{\infty} a_n \times 0 + \sum_{n=1}^{\infty} b_n \pi \delta_{n,m}$$

$$= b_m \pi$$

and so

$$b_m = \frac{1}{\pi} \int_{-\pi}^{\pi} f(x) \sin mx \, dx$$

Summary

20

Given that certain conditions are satisfied, if $f(x)$ is defined on the interval $-\pi \le x \le \pi$ and where $f(x + 2n\pi) = f(x)$ then

$$f(x) = \frac{a_0}{2} + \sum_{n=1}^{\infty} (a_n \cos nx + b_n \sin nx)$$

This is the **Fourier series** expansion of $f(x)$ where the a_n and b_n are constants called the *Fourier coefficients* and where

$$a_n = \frac{1}{\pi} \int_{-\pi}^{\pi} f(x) \cos nx \, dx \text{ and } b_n = \frac{1}{\pi} \int_{-\pi}^{\pi} f(x) \sin nx \, dx, \, n = 0, 1, 2, \ldots$$

Look in particular at the constant function $f(x) = c$ which can be considered as a periodic function with any period we wish to choose. Choosing the period to be 2π then

$$a_n = \frac{1}{\pi} \int_{-\pi}^{\pi} c \cos nx \, dx = \frac{c}{\pi} \int_{-\pi}^{\pi} \cos nx \, dx = 2c \delta_{n,0}.$$

That is $a_0 = 2c$ so $c = \dfrac{a_0}{2}$ as expected.

Also

$$b_n = \frac{1}{\pi} \int_{-\pi}^{\pi} c \sin nx \, dx = \frac{c}{\pi} \int_{-\pi}^{\pi} \sin nx \, dx = 0$$

▶

From this we see that we have two choices to represent the Fourier series. We can either write

$$f(x) = \frac{a_0}{2} + \sum_{n=1}^{\infty} \{a_n \cos nx + b_n \sin nx\}$$

where

$$a_n = \frac{1}{\pi} \int_{-\pi}^{\pi} f(x) \cos nx \, dx \text{ and } b_n = \frac{1}{\pi} \int_{-\pi}^{\pi} f(x) \sin nx \, dx$$

or we can write

$$f(x) = \sum_{n=0}^{\infty} \{a_n \cos nx + b_n \sin nx\}$$

where

$$a_0 = \frac{1}{2\pi} \int_{-\pi}^{\pi} f(x) \, dx, \ a_n = \frac{1}{\pi} \int_{-\pi}^{\pi} f(x) \cos nx \, dx \ (n \neq 0)$$

$$\text{and } b_n = \frac{1}{\pi} \int_{-\pi}^{\pi} f(x) \sin nx \, dx.$$

We choose the former and so avoid having a separate integral for a_0.

Dirichlet conditions

21

If a function $f(x)$ is such that

(a) $f(x)$ is defined, single-valued and periodic with period 2π

(b) $f(x)$ and $f'(x)$ have at most a finite number of finite discontinuities over a single period – that is they are *piecewise continuous*

then the series

$$\frac{a_0}{2} + \sum_{n=1}^{\infty} \{a_n \cos nx + b_n \sin nx\}$$

where $a_n = \dfrac{1}{\pi} \displaystyle\int_{-\pi}^{\pi} f(x) \cos nx \, dx$ and $b_n = \dfrac{1}{\pi} \displaystyle\int_{-\pi}^{\pi} f(x) \sin nx \, dx$ converges to $f(x)$

when $(x, f(x))$ is a point of continuity.

The Dirichlet conditions are sufficient for the Fourier series to represent $f(x)$ not only at a point of continuity but, with a slight modification, also at a point of discontinuity, as we shall see later in Frame 36. Also the periodicity of the function need not be restricted to 2π, as we shall see in Programme 8.

Note that these conditions, while being sufficient, are not necessary because there are functions that do not satisfy these conditions which still possess a convergent Fourier series. However, the cases met in science and engineering do generally meet these conditions.

▶

Exercise

If the following functions are defined over the interval $-\pi < x < \pi$ and $f(x + 2\pi) = f(x)$, state whether or not each function can be represented by a Fourier series.

1 $f(x) = x^3$ **4** $f(x) = \dfrac{1}{x - 5}$

2 $f(x) = 4x - 5$ **5** $f(x) = \tan x$

3 $f(x) = \dfrac{2}{x}$ **6** $f(x) = y$ where $x^2 + y^2 = \pi^2$

22

1	Yes	
2	Yes	
3	No: infinite discontinuity at $x = 0$	

4 Yes

5 No: infinite discontinuity at $x = \pi/2$

6 No: two valued

On then

Example 1

23

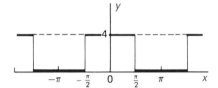

Find the Fourier series for the function shown.

Consider one cycle between $x = -\pi$ and $x = \pi$.

The function can be defined by $f(x) = \begin{cases} 0 & -\pi < x < -\dfrac{\pi}{2} \\ 4 & -\dfrac{\pi}{2} < x < \dfrac{\pi}{2} \\ 0 & \dfrac{\pi}{2} < x < \pi \end{cases}$

$$f(x + 2\pi) = f(x).$$

(a) As before $f(x) = \dfrac{1}{2}a_0 + \sum_{n=1}^{\infty}\{a_n \cos nx + b_n \sin nx\}$

The expression for a_0 is . . . .

24

$$a_0 = \frac{1}{\pi} \int_{-\pi}^{\pi} f(x)\, dx$$

This gives

$$a_0 = \frac{1}{\pi} \left\{ \int_{-\pi}^{-\pi/2} 0\, dx + \int_{-\pi/2}^{\pi/2} 4\, dx + \int_{\pi/2}^{\pi} 0\, dx \right\}$$

$$= \frac{1}{\pi} \left[4x \right]_{-\pi/2}^{\pi/2} \qquad\qquad \therefore\ a_0 = 4$$

(b) *To find a_n*

$$a_n = \frac{1}{\pi} \int_{-\pi}^{\pi} f(x) \cos nx\, dx$$

$$\therefore\ a_n = \frac{1}{\pi} \left\{ \int_{-\pi}^{-\pi/2} (0) \cos nx\, dx + \int_{-\pi/2}^{\pi/2} 4 \cos nx\, dx + \int_{\pi/2}^{\pi} (0) \cos nx\, dx \right\}$$

$$\therefore\ a_n = \ldots\ldots\ldots\ldots$$

25

$$a_n = \frac{8}{\pi n} \sin \frac{n\pi}{2}$$

Then considering different integer values of n, we have

If n is even $a_n = 0$

If $n = 1, 5, 9, \ldots$ $a_n = \dfrac{8}{n\pi}$

If $n = 3, 7, 11, \ldots$ $a_n = -\dfrac{8}{n\pi}$

We keep these in mind while we find b_n.

(c) *To find b_n*

$$b_n = \frac{1}{\pi} \int_{-\pi}^{\pi} f(x) \sin nx\, dx = \ldots\ldots\ldots\ldots$$

26

$$\boxed{b_n = 0}$$

Because we have

$$b_n = \frac{1}{\pi}\left\{\int_{-\pi}^{-\pi/2}(0)\sin nx\,dx + \int_{-\pi/2}^{\pi/2}4\sin nx\,dx + \int_{\pi/2}^{\pi}(0)\sin nx\,dx\right\}$$

$$= \frac{4}{\pi}\int_{-\pi/2}^{\pi/2}\sin nx\,dx = \frac{4}{\pi}\left[\frac{-\cos nx}{n}\right]_{-\pi/2}^{\pi/2}$$

$$= -\frac{4}{n\pi}\left\{\cos\frac{n\pi}{2} - \cos\left(\frac{-n\pi}{2}\right)\right\} = 0 \qquad \therefore\ b_n = 0$$

So with $a_0 = 4$; a_n as stated above; $b_n = 0$; the Fourier series is

$$f(x) = 2 + \frac{8}{\pi}\left\{\cos x - \frac{1}{3}\cos 3x + \frac{1}{5}\cos 5x - \frac{1}{7}\cos 7x + \ldots\right\}$$

In this particular example, there are, in fact, no sine terms.

Example 2

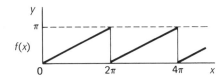

Determine the Fourier series to represent the periodic function shown.

It is more convenient here to take the limits as 0 to 2π.

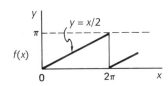

The function can be defined as

$$f(x) = \frac{x}{2} \qquad 0 < x < 2\pi$$

$$f(x + 2\pi) = f(x) \quad \text{i.e. period} = 2\pi.$$

Now to find the coefficients.

(a) $a_0 = \dfrac{1}{\pi}\displaystyle\int_0^{2\pi} f(x)\,dx = \dfrac{1}{\pi}\displaystyle\int_0^{2\pi}\left(\dfrac{x}{2}\right)dx = \dfrac{1}{4\pi}\left[x^2\right]_0^{2\pi}$

$$= \pi \qquad\qquad\qquad \therefore\ a_0 = \pi$$

(b) $a_n = \dfrac{1}{\pi}\displaystyle\int_0^{2\pi} f(x)\cos nx\,dx = \dfrac{1}{\pi}\displaystyle\int_0^{2\pi}\left(\dfrac{x}{2}\right)\cos nx\,dx$

$$= \frac{1}{2\pi}\int_0^{2\pi} x\cos nx\,dx$$

$$= \ldots\ldots\ldots \quad \text{(integrating by parts)}$$

27

$$\boxed{a_n = 0}$$

Because

$$a_n = \frac{1}{2\pi}\int_0^{2\pi} x\cos nx\,dx = \frac{1}{2\pi}\left\{\left[\frac{x\sin nx}{n}\right]_0^{2\pi} - \frac{1}{n}\int_0^{2\pi}\sin nx\,dx\right\}$$

$$= \frac{1}{2\pi}\left\{(0-0) - \frac{1}{n}(0)\right\} = 0 \qquad\qquad \therefore\ a_n = 0$$

(c) $b_n = \dfrac{1}{\pi}\displaystyle\int_0^{2\pi} f(x)\sin nx\,dx = \ldots\ldots\ldots$

28

$$\boxed{b_n = -\frac{1}{n}}$$

Straightforward integration by parts, as for a_n, gives the result stated. So we now have

$$a_0 = \ldots\ldots\ldots; \quad a_n = \ldots\ldots\ldots; \quad b_n = \ldots\ldots\ldots$$

29

$$\boxed{a_0 = \pi; \quad a_n = 0; \quad b_n = -\frac{1}{n}}$$

Now the general expression for a Fourier series is

$$\ldots\ldots\ldots$$

30

$$\boxed{f(x) = \frac{1}{2}a_0 + \sum_{n=1}^{\infty}\{a_n\cos nx + b_n\sin nx\}}$$

Therefore in this case

$$f(x) = \frac{\pi}{2} + \sum_{n=1}^{\infty}\{b_n\sin nx\} \qquad \text{because } a_n = 0$$

$$= \frac{\pi}{2} + \left\{-\frac{1}{1}\sin x - \frac{1}{2}\sin 2x - \frac{1}{3}\sin 3x - \ldots\right\}$$

$$\therefore\ f(x) = \frac{\pi}{2} - \left\{\sin x + \frac{1}{2}\sin 2x + \frac{1}{3}\sin 3x + \ldots\right\}$$

Note that in this example, the series contains a constant term and sine terms only.

▶

Example 3

Find the Fourier series for the function defined by

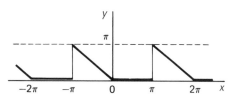

$$f(x) = -x \qquad -\pi < x < 0$$
$$f(x) = 0 \qquad 0 < x < \pi$$
$$f(x + 2\pi) = f(x).$$

The general expressions for a_0, a_n, b_n are

$$a_0 = \dots\dots\dots$$
$$a_n = \dots\dots\dots$$
$$b_n = \dots\dots\dots$$

31

$$a_0 = \frac{1}{\pi} \int_{-\pi}^{\pi} f(x)\, dx$$

$$a_n = \frac{1}{\pi} \int_{-\pi}^{\pi} f(x) \cos nx\, dx$$

$$b_n = \frac{1}{\pi} \int_{-\pi}^{\pi} f(x) \sin nx\, dx$$

With that reminder, in this example $a_0 = \dots\dots\dots$

32

$$a_0 = \frac{\pi}{2}$$

Because

(a) $a_0 = \dfrac{1}{\pi} \int_{-\pi}^{\pi} f(x)\, dx = \dfrac{1}{\pi} \int_{-\pi}^{0} (-x)\, dx = \dfrac{1}{\pi} \left[-\dfrac{x^2}{2} \right]_{-\pi}^{0} = \dfrac{\pi}{2}$

(b) To find a_n

$$a_n = \frac{1}{\pi} \int_{-\pi}^{\pi} f(x) \cos nx\, dx = \dots\dots\dots$$

33

$$a_n = -\frac{2}{\pi n^2} \quad (n \text{ odd}); \quad 0 \quad (n \text{ even})$$

Because

$$a_n = \frac{1}{\pi}\int_{-\pi}^{\pi} f(x)\cos nx\, dx = \frac{1}{\pi}\int_{-\pi}^{0} (-x)\cos nx\, dx$$

$$= -\frac{1}{\pi}\int_{-\pi}^{0} x\cos nx\, dx$$

$$= -\frac{1}{\pi}\left\{\left[x\frac{\sin nx}{n}\right]_{-\pi}^{0} - \frac{1}{n}\int_{-\pi}^{0}\sin nx\, dx\right\}$$

$$= -\frac{1}{\pi}\left\{(0-0) - \frac{1}{n}\left[\frac{-\cos nx}{n}\right]_{-\pi}^{0}\right\} = -\frac{1}{\pi n^2}\{1 - \cos n\pi\}$$

But $\cos n\pi = 1$ (n even) or -1 (n odd)

$$\therefore \ a_n = -\frac{2}{\pi n^2} \quad (n \text{ odd}) \quad \text{or } 0 \quad (n \text{ even})$$

(c) *Now to find b_n.* Working as for a_n, we obtain

$$b_n = \ldots\ldots\ldots\ldots$$

34

$$b_n = -\frac{1}{n} \quad (n \text{ even}) \quad \text{or} \quad \frac{1}{n} \quad (n \text{ odd})$$

Because

$$b_n = \frac{1}{\pi}\int_{-\pi}^{\pi} f(x)\sin nx\, dx = \frac{1}{\pi}\int_{-\pi}^{0} (-x)\sin nx\, dx$$

$$= -\frac{1}{\pi}\int_{-\pi}^{0} x\sin nx\, dx$$

$$= -\frac{1}{\pi}\left\{\left[x\left(\frac{-\cos nx}{n}\right)\right]_{-\pi}^{0} + \frac{1}{n}\int_{-\pi}^{0}\cos nx\, dx\right\}$$

$$= -\frac{1}{\pi}\left\{\frac{\pi\cos n\pi}{n} + \frac{1}{n}\left[\frac{\sin nx}{n}\right]_{-\pi}^{0}\right\} = -\frac{\cos n\pi}{n}$$

$$\therefore \ b_n = -\frac{1}{n} \quad (n \text{ even}); \quad \frac{1}{n} \quad (n \text{ odd})$$

So we have $\quad a_0 = \dfrac{\pi}{2}; \quad a_n = 0 \quad (n \text{ even}) \quad \text{or} \quad -\dfrac{2}{\pi n^2} \quad (n \text{ odd})$

$$b_n = -\frac{1}{n} \quad (n \text{ even}) \quad \text{or} \quad \frac{1}{n} \quad (n \text{ odd})$$

$$\therefore \ f(x) = \ldots\ldots\ldots\ldots$$

Complete the series

35

$$f(x) = \frac{\pi}{4} - \frac{2}{\pi}\left(\cos x + \frac{1}{9}\cos 3x + \frac{1}{25}\cos 5x + \dots\right)$$
$$+ \left(\sin x - \frac{1}{2}\sin 2x + \frac{1}{3}\sin 3x - \frac{1}{4}\sin 4x + \dots\right)$$

It is just a case of substituting $n = 1, 2, 3$, etc.

In this particular example, we have a constant term and both sine and cosine terms.

Effect of harmonics

It is interesting to see just how accurately the Fourier series represents the function with which it is associated. The complete representation requires an infinite number of terms, but we can, at least, see the effect of including the first few terms of the series.

Let us consider the waveform shown. We established earlier in Frames 23–26 that the function

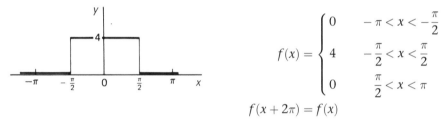

$$f(x) = \begin{cases} 0 & -\pi < x < -\dfrac{\pi}{2} \\ 4 & -\dfrac{\pi}{2} < x < \dfrac{\pi}{2} \\ 0 & \dfrac{\pi}{2} < x < \pi \end{cases}$$

$$f(x + 2\pi) = f(x)$$

gives the Fourier series

$$f(x) = 2 + \frac{8}{\pi}\left\{\cos x - \frac{1}{3}\cos 3x + \frac{1}{5}\cos 5x - \frac{1}{7}\cos 7x + \dots\right\}$$

If we start with just one cosine term, we can then see the effect of including subsequent harmonics. Let us restrict our attention to just the right-hand half of the symmetrical waveform. Detailed plotting of points gives the development on the next page.

▶

1 $f(x) = 2 + \dfrac{8}{\pi}\cos x$

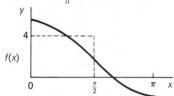

2 $f(x) = 2 + \dfrac{8}{\pi}\left\{\cos x - \dfrac{1}{3}\cos 3x\right\}$

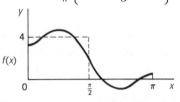

3 $f(x) = 2 + \dfrac{8}{\pi}\left\{\cos x - \dfrac{1}{3}\cos 3x + \dfrac{1}{5}\cos 5x\right\}$

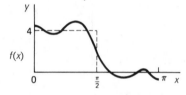

4 $f(x) = 2 + \dfrac{8}{\pi}\left\{\cos x - \dfrac{1}{3}\cos 3x + \dfrac{1}{5}\cos 5x - \dfrac{1}{7}\cos 7x\right\}$

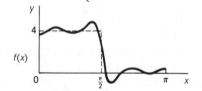

As the number of terms is increased, the graph gradually approaches the shape of the original square waveform. The ripples increase in number and, apart from the one nearest to the step, decrease in amplitude. A perfectly square waveform is unattainable in practice. For practical purposes, the first few terms normally suffice to give an accuracy of acceptable level.

Gibbs' phenomenon

You will notice from the previous diagrams that near the discontinuity, as more terms are taken into acount, the series tends to overshoot on one side and undershoot on the other. This over and undershooting on either side of the discontinuity does not go away as the number of terms in the Fourier series that are taken into account is increased, rather it tends to two spikes on either side of the discontinuity. This effect is called the *Gibbs' phenomenon*.

Sum of a Fourier series at a point of discontinuity 36

$$f(x) = \tfrac{1}{2}a_0 + \sum_{n=1}^{\infty}\{a_n \cos nx + b_n \sin nx\}$$

If $f(x)$ is continuous at $x = x_1$, the series converges to the value $f(x_1)$ as the number of terms included increases to infinity. A particular point of interest occurs at a point of finite discontinuity or 'jump' of the function $y = f(x)$.

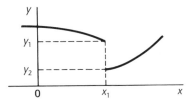

As $x \to x_1$, the expression $f(x)$ approaches y_1 or y_2 depending on the direction of approach.

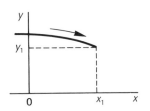

If we approach $x = x_1$ from below that value, the limiting value of $f(x)$ is y_1.

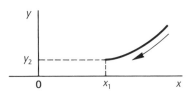

If we approach $x = x_1$ from above that value, the limiting value of $f(x)$ is y_2.

To distinguish between these two values we write

$y_1 = f(x_1 - 0)$ denoting immediately before $x = x_1$
$y_2 = f(x_1 + 0)$ denoting immediately after $x = x_1$.

In fact, if we substitute $x = x_1$ in the Fourier series for $f(x)$, it can be shown that the series converges to the value $\tfrac{1}{2}\{f(x_1 - 0) + f(x_1 + 0)\}$ i.e. $\tfrac{1}{2}(y_1 + y_2)$, the average of y_1 and y_2.

Example

Consider the function

$$f(x) = \begin{cases} 0 & -\pi < x < \pi \\ a & \pi < x < 2\pi \end{cases}$$

$f(x + 2\pi) = f(x)$.

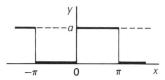

First of all, determine the Fourier series to represent the function. There are no snags.

$$f(x) = \ldots\ldots\ldots\ldots$$

37

$$f(x) = \frac{a}{2} + \frac{2a}{\pi}\{\sin x + \tfrac{1}{3}\sin 3x + \tfrac{1}{5}\sin 5x + \ldots\}$$

Check the working

(a) $a_0 = \dfrac{1}{\pi}\displaystyle\int_{-\pi}^{\pi} f(x)\,dx = \dfrac{1}{\pi}\displaystyle\int_{0}^{\pi} a\,dx = \dfrac{1}{\pi}\Big[ax\Big]_{0}^{\pi} = a$ $\quad \therefore\ a_0 = a$

(b) $a_n = \dfrac{1}{\pi}\displaystyle\int_{-\pi}^{\pi} f(x)\cos nx\,dx = \dfrac{1}{\pi}\displaystyle\int_{0}^{\pi} a\cos nx\,dx$

$\qquad = \dfrac{a}{\pi}\left[\dfrac{\sin nx}{n}\right]_{0}^{\pi} = 0$ $\hspace{3cm} \therefore\ a_n = 0$

(c) $b_n = \dfrac{1}{\pi}\displaystyle\int_{-\pi}^{\pi} f(x)\sin nx\,dx = \dfrac{1}{\pi}\displaystyle\int_{0}^{\pi} a\sin nx\,dx$

$\qquad = \dfrac{a}{\pi}\left[\dfrac{-\cos nx}{n}\right]_{0}^{\pi} = \dfrac{a}{n\pi}(1-\cos n\pi) = \dfrac{a}{n\pi}\left(1-(-1)^n\right)$

and because

$\quad \cos n\pi = 1 \quad (n \text{ even}) \text{ and } -1 \quad (n \text{ odd})$

$\quad b_n = 0 \quad (n \text{ even}); \quad \dfrac{2a}{n\pi} \quad (n \text{ odd})$

$\quad \therefore\ f(x) = \dfrac{1}{2}a_0 + \displaystyle\sum_{n=1}^{\infty} b_n \sin nx$

$\quad \therefore\ f(x) = \dfrac{a}{2} + \dfrac{2a}{\pi}\left\{\sin x + \dfrac{1}{3}\sin 3x + \dfrac{1}{5}\sin 5x + \ldots\right\}$

A finite discontinuity, or 'jump', occurs at $x = 0$. If we substitute $x = 0$ in the series obtained, all the sine terms vanish and we get $f(x) = a/2$, which is, in fact, the average of the two function values at $x = 0$.

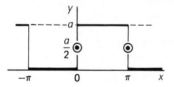

Note also that at $x = \pi$, another finite discontinuity occurs and substituting $x = \pi$ in the series gives the same result.

The **Revision summary** and **Can you?** checklist now follow, after which you will have no trouble with the **Test exercise**. The **Further problems** provide additional practice.

Revision summary 7

38

1 *Graphs of $y = A \sin nx$ and $A \cos nx$*

  Amplitude $= A$; period $= \dfrac{360°}{n} = \dfrac{2\pi}{n}$ radians

2 *Harmonics*

  $y = A_1 \sin x$ is the first harmonic or fundamental
  $y = A_n \sin nx$ is the nth harmonic.

3 *Periodic function*

  $$f(x + P) = f(x) \qquad P = \text{period}$$

4 *Fourier series – functions of period 2π*

  $$f(x) = \tfrac{1}{2}a_0 + a_1 \cos x + a_2 \cos 2x + a_3 \cos 3x + \ldots + a_n \cos nx \ldots$$
  $$+ \, b_1 \sin x + b_2 \sin 2x + b_3 \sin 3x + \ldots + b_n \sin nx \ldots$$
  $$= \tfrac{1}{2}a_0 + \sum_{n=1}^{\infty} \{a_n \cos nx + b_n \sin nx\}$$

5 *Dirichlet conditions*

  (a) The function $f(x)$ must be defined, single-valued and periodic.
  (b) $f(x)$ and $f'(x)$ must be piecewise continuous in the periodic interval.

6 *Fourier coefficients*

  $$a_0 = \frac{1}{\pi} \int_{-\pi}^{\pi} f(x)\, dx$$

  $$a_n = \frac{1}{\pi} \int_{-\pi}^{\pi} f(x) \cos nx\, dx$$

  $$b_n = \frac{1}{\pi} \int_{-\pi}^{\pi} f(x) \sin nx\, dx$$

  where, in each case, $n = 1, 2, 3, \ldots$

7 *Sum of Fourier series at a finite discontinuity*

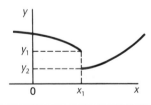

  At $x = x_1$, series for $f(x)$ converges to the value

  $$\tfrac{1}{2}\{f(x_1 - 0) + f(x_1 + 0)\} = \tfrac{1}{2}(y_1 + y_2)$$

Can you?

39 Checklist 7

Check this list before and after you try the end of Programme test.

On a scale of 1 to 5 how confident are you that you can: **Frames**

- Determine the period and amplitude of a periodic function?   [1] and [2]
  Yes ☐ ☐ ☐ ☐ ☐ *No*

- Write down the harmonics of a periodic trigonometric
  function?   [3]
  Yes ☐ ☐ ☐ ☐ ☐ *No*

- Give an analytic description of a non-sinusoidal periodic
  function?   [4] to [9]
  Yes ☐ ☐ ☐ ☐ ☐ *No*

- Evaluate integrals with periodic integrands?   [10] to [15]
  Yes ☐ ☐ ☐ ☐ ☐ *No*

- Demonstrate the orthogonality of the trigonometric functions
  $\sin nx$ and $\cos nx$ for $n = 0, 1, 2, \ldots$?   [16] to [20]
  Yes ☐ ☐ ☐ ☐ ☐ *No*

- Describe a periodic function as a Fourier series subject to the
  Dirichlet conditions?   [21] and [22]
  Yes ☐ ☐ ☐ ☐ ☐ *No*

- Obtain the Fourier coefficients and hence the Fourier series of a
  periodic function?   [23] to [25]
  Yes ☐ ☐ ☐ ☐ ☐ *No*

- Describe the effects of the harmonics in the construction of the
  Fourier series?   [35]
  Yes ☐ ☐ ☐ ☐ ☐ *No*

- Find the value of the Fourier series at a point of discontinuity
  of the periodic function?   [36] and [37]
  Yes ☐ ☐ ☐ ☐ ☐ *No*

Test exercise 7

1 What is the amplitude and the period of the function with output **40**

$$f(x) = \sqrt{2}\cos\frac{3x}{4}?$$

2 Give an analytic description of the function with the following graph:

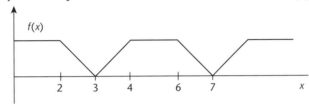

3 Draw the graph of:

$$f(x) = \begin{cases} x & 0 \le x < 1 \\ 1 & 1 \le x < 3 \\ (x - 4^2) & 3 \le x < 4 \end{cases}$$

$$f(x + 4) = f(x) \quad \text{for } 0 \le x \le 8$$

4 If $f(x)$ is defined in the interval $-\pi \le x < \pi$ and $f(x + 2\pi) = f(x)$, state whether or not each of the following functions can be represented by a Fourier series.

(a) $f(x) = x^4$ (d) $f(x) = e^{2x}$

(b) $f(x) = 3 - 2x$ (e) $f(x) = \text{cosec } x$

(c) $f(x) = \dfrac{1}{x}$ (f) $f(x) = \pm\sqrt{4x}$

5 Determine the Fourier series for the function defined by

$$f(x) = 2x \quad 0 \le x \le 2\pi$$
$$f(x + 2\pi) = f(x)$$

6 What is the value at $x = 4$ of the Fourier series for the function defined by

$$f(x) = \begin{cases} 2 & 0 \le x < 2 \\ 4 & 2 \le x < 4 \\ -2 & 4 \le x < 2\pi \end{cases}$$

$$f(x + 2\pi) = f(x)$$

Further problems 7

41   **1** For each of the following graphs give the analytical description of the function
drawn.

(a)

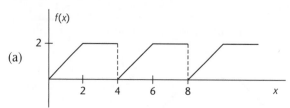

(b)

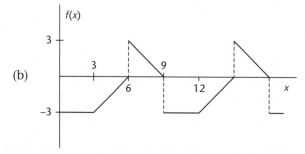

(c)

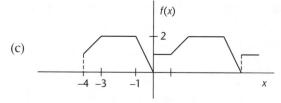

2 Draw the graph of

(a) $(x) = \begin{cases} \sin x & 0 \le x < \pi/2 \\ \cos x & \pi/2 \le x < \pi \end{cases}$

$f(x + \pi) = f(x)$ for $-2\pi \le x \le 2\pi$

(b) $f(x) = \begin{cases} \cos x & 0 \le x < \pi/2 \\ \sin x & \pi/2 \le x < \pi \end{cases}$

$f(x + \pi) = f(x)$ for $-2\pi \le x \le 2\pi$

(c) $f(x) = \begin{cases} x^2 & 0 \le x < 2 \\ 6 - x & 2 \le x < 10 \end{cases}$

$f(x + 10) = f(x)$ for $-20 \le x \le 20$

(d) $f(x) = \begin{cases} x^3 & 0 \le x < 2 \\ 8 & 2 \le x < 3 \\ (5 - x)^3 & 3 \le x < 5 \end{cases}$

$f(x + 5)f(x)$ for $-10 \le x \le 10$

▶

(e) $\quad f(x) = \begin{cases} (x+4)^2 & -4 \le x < -2 \\ 4 - 2x & -2 \le x < 0 \end{cases}$

$\quad f(x+4) = f(x) \qquad$ for $-8 \le x \le 8$

3 A periodic function $f(x)$ is defined by

$$f(x) = 1 - \frac{x}{\pi} \qquad 0 \le x < 2\pi$$

$\quad f(x+2\pi) = f(x)$

Determine the Fourier series up to and including the third harmonic.

4 A function is defined by

$$f(x) = \begin{cases} \pi + x & -\pi \le x < 0 \\ \pi - x & 0 \le x < \pi \end{cases}$$

$\quad f(x+2\pi) = f(x)$

Obtain the Fourier series.

5 A periodic function is defined by

$$f(x) = \begin{cases} A \sin x & 0 \le x < \pi \\ -A \sin x & \pi \le x < 2\pi \end{cases}$$

$\quad f(x+2\pi) = f(x)$

Obtain the Fourier series up to and including the fourth harmonic.

6 A function is defined by

$$f(x) = \begin{cases} 0 & -\pi \le x < 0 \\ x & 0 \le x < \pi \end{cases}$$

$\quad f(x+2\pi) = f(x)$

Obtain the Fourier series.

7 A function is defined by

$$f(x) = \begin{cases} \cos x & -\pi \le x < 0 \\ 0 & 0 \le x < \pi \end{cases}$$

$\quad f(x+2\pi) = f(x)$

Obtain the Fourier series.

8 A function is defined by

$$f(x) = x^2 \qquad -\pi \le x < \pi$$

$\quad f(x+2\pi) = f(x)$

Obtain the Fourier series

9 A function is defined by

$$f(x) = 7 - \frac{3x}{\pi} \qquad -\pi \le x < \pi$$

$\quad f(x+2\pi) = f(x)$

Obtain the Fourier series up to the fourth harmonic.

▶

10   A function is defined by

$$f(x) = \begin{cases} \dfrac{\pi + x}{2} & -\pi \le x < 0 \\ \dfrac{\pi - x}{2} & 0 \le x < \pi \end{cases}$$

$$f(x + 2\pi) = f(x)$$

Obtain the Fourier series.

11   A function is defined by

$$f(x) = x^2 \quad 0 \le x < 2\pi$$
$$f(x + 2\pi) = f(x)$$

Obtain the Fourier series.

12   Given a periodic function $f(x)$ with period 2π and Fourier series

$$f(x) = \frac{a_0}{2} + \sum_{n=1}^{\pi} \{a_n \cos nx + b_n \sin nx\}$$

show that

$$\frac{1}{\pi} \int_{-\pi}^{\pi} [f(x)]^2 \, dx = \frac{a_0^2}{2} + \sum_{n=1}^{\infty} \left\{a_n^2 + b_n^2\right\}$$

13   Given two periodic functions $f(x)$ and $g(x)$ each with period 2π and Fourier series

$$f(x) = \frac{a_0}{2} + \sum_{n=1}^{\infty} \{a_n \cos nx + b_n \sin nx\} \text{ and } g(x) = \frac{p_0}{2} + \sum_{n=1}^{\infty} \{p_n \cos nx + q_n \sin nx\}$$

show that

$$\frac{1}{\pi} \int_{-\pi}^{\pi} f(x)g(x) \, dx = \frac{a_0 p_0}{2} + \sum_{n=1}^{\infty} \{a_n p_n + b_n q_n\}$$

14   Given two periodic functions $f(x)$ and $g(x) = (\pi - x)$ each with period 2π and Fourier series

$$f(x) = \frac{a_0}{2} + \sum_{n=1}^{\infty} \{a_n \cos nx + b_n \sin nx\} \text{ and } g(x) = \frac{p_0}{2} + \sum_{n=1}^{\infty} \{p_n \cos nx + q_n \sin nx\}$$

show that

$$\frac{1}{2\pi} \int_{0}^{2\pi} f(x)(\pi - x) \, dx = \sum_{n=1}^{\infty} \frac{b_n}{n}$$

Fourier series 2

Learning outcomes

When you have completed this Programme you will be able to:

- Obtain the Fourier coefficients of a function with arbitrary period T
- Recognise even and odd functions and their products
- Derive the Fourier series of even and odd functions
- Derive half-range Fourier series
- Recognise the conditions for the Fourier series to contain only odd or only even harmonics
- Explain the geometric significance of the constant term $a_0/2$
- Derive half-range Fourier series with arbitrary period

Functions with periods other than 2π

In Programme 7 we considered functions $f(x)$ with period 2π. In practice, we often encounter functions defined over periodic intervals other than 2π, e.g. from 0 to T, $-\dfrac{T}{2}$ to $\dfrac{T}{2}$, etc.

Functions with period T

If $y = f(x)$ is defined in the range $-\dfrac{T}{2}$ to $\dfrac{T}{2}$, i.e. has a period T, we can convert this to an interval of 2π by changing the units of the independent variable.

In many practical cases involving physical oscillations, the independent variable is time (t) and the periodic interval is normally denoted by T, i.e.

$$f(t + T) = f(t)$$

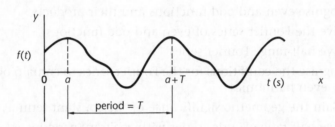

Each cycle is therefore completed in T seconds and the frequency f **hertz** (oscillations per second) of the periodic function is therefore given by $f = \dfrac{1}{T}$.

If the angular velocity, ω radians per second, is defined by $\omega = 2\pi f$, then

$$\omega = \frac{2\pi}{T} \quad \text{and} \quad T = \frac{2\pi}{\omega}.$$

The angle, x radians, at any time t is therefore $x = \omega t$ and the Fourier series to represent the function can be expressed as

$$f(t) = \frac{a_0}{2} + \sum_{n=1}^{\infty} \{a_n \cos n\omega t + b_n \sin n\omega t\}$$

$$= \frac{a_0}{2} + \sum_{n=1}^{\infty} \left\{ a_n \cos \frac{2n\pi t}{T} + b_n \sin \frac{2n\pi t}{T} \right\}$$

Fourier coefficients

With the new variable, the Fourier coefficients become

$$f(t) = \tfrac{1}{2}a_0 + \sum_{n=1}^{\infty}\{a_n \cos n\omega t + b_n \sin n\omega t\}$$

$$a_0 = \frac{2}{T}\int_0^T f(t)\,dt \qquad = \frac{\omega}{\pi}\int_0^{2\pi/\omega} f(t)\,dt$$

$$a_n = \frac{2}{T}\int_0^T f(t)\cos n\omega t\,dt = \frac{\omega}{\pi}\int_0^{2\pi/\omega} f(t)\cos n\omega t\,dt$$

$$b_n = \frac{2}{T}\int_0^T f(t)\sin n\omega t\,dt = \frac{\omega}{\pi}\int_0^{2\pi/\omega} f(t)\sin n\omega t\,dt.$$

We can see that there is very little difference between these expressions and those that have gone before. The limits can, of course, be 0 to T, $-\dfrac{T}{2}$ to $\dfrac{T}{2}$, $-\dfrac{\pi}{\omega}$ to $\dfrac{\pi}{\omega}$, 0 to $\dfrac{2\pi}{\omega}$ etc. as is convenient, so long as they cover a complete period.

Example

Determine the Fourier series for a periodic function defined by

$$f(t) = \begin{cases} 2(1+t) & -1 < t < 0 \\ 0 & 0 < t < 1 \end{cases}$$

$$f(t+2) = f(t)$$

The first step is to sketch the waveform which is ....

3

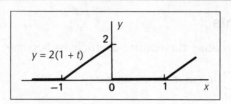

We have

$$f(t) = \frac{a_0}{2} + \sum_{n=1}^{\infty} \left\{ a_n \cos\frac{2n\pi t}{T} + b_n \sin\frac{2n\pi t}{T} \right\}$$

$$= \frac{a_0}{2} + \sum_{n=1}^{\infty} \{ a_n \cos n\pi t + b_n \sin n\pi t \} \quad \text{because } T = 2$$

Therefore

$$a_0 = \frac{2}{T} \int_{-T/2}^{T/2} f(t)\, dt = \int_{-1}^{1} f(t)\, dt = \int_{-1}^{0} 2(1+t)\, dt + \int_{0}^{1} (0)\, dt$$

$$= \left[2t + t^2 \right]_{-1}^{0} = 1$$

and

$$a_n = \frac{2}{T} \int_{-T/2}^{T/2} f(t) \cos n\pi t\, dt = \int_{-1}^{1} f(t) \cos n\pi t\, dt$$

$$= \int_{-1}^{0} 2(1+t)\ \cos n\pi t\, dt = \ldots\ldots\ldots$$

4

$$\boxed{a_n = 0\ (n \text{ even}); \quad a_n = \frac{4}{n^2\pi^2}\ (n \text{ odd})}$$

Because

$$a_n = \int_{-1}^{0} 2(1+t) \cos n\pi t\, dt$$

$$= 2 \left\{ \left[(1+t)\frac{\sin n\pi t}{n\pi} \right]_{-1}^{0} - \frac{1}{n\pi} \int_{-1}^{0} \sin n\pi t\, dt \right\}$$

$$= 2 \left\{ (0-0) - \frac{1}{n\pi} \left[-\frac{\cos n\pi t}{n\pi} \right]_{-1}^{0} \right\} = \frac{2}{n^2\pi^2}(1 - \cos n\pi)$$

$$= \frac{2}{n^2\pi^2}(1 - (-1)^n)$$

so that

$$a_n = 0 \quad (n \text{ even}), \quad a_n = \frac{4}{n^2\pi^2} \quad (n \text{ odd})$$

Now for b_n

$$b_n = \frac{2}{T} \int_{-T/2}^{T/2} f(t) \sin\frac{2n\pi t}{T}\, dt = \ldots\ldots\ldots$$

$$b_n = -\frac{2}{n\pi}$$

5

Because

$$b_n = \int_{-1}^{0} 2(1+t) \sin n\pi t\, dt$$

$$= 2\left\{ \left[(1+t)\frac{-\cos n\pi t}{n\pi} \right]_{-1}^{0} + \frac{1}{n\pi}\int_{-1}^{0} \cos n\pi t\, dt \right\}$$

$$= 2\left\{ -\frac{1}{n\pi} + \left[\frac{\sin n\pi t}{n\pi} \right]_{-1}^{0} \right\} = -\frac{2}{n\pi} + \frac{2}{n^2\pi^2}(\sin n\pi) = -\frac{2}{n\pi}$$

So the first few terms of the series give

$$f(t) = \ldots\ldots\ldots$$

6

$$f(t) = \frac{1}{2} + \frac{4}{\pi^2}\left\{ \cos \pi t + \frac{1}{9}\cos 3\pi t + \frac{1}{25}\cos 5\pi t + \ldots \right\}$$

$$-\frac{2}{\pi}\left\{ \sin \pi t + \frac{1}{2}\sin 2\pi t + \frac{1}{3}\sin 3\pi t + \frac{1}{4}\sin 4\pi t + \ldots \right\}$$

The Fourier series

$$f(t) = \frac{a_0}{2} + \sum_{n=1}^{\infty}\{a_n \cos n\omega t + b_n \sin n\omega t\}$$

can also be written in the form

$$f(t) = \frac{A_0}{2} + \sum_{n=1}^{\infty} B_n \sin(n\omega t + \phi_n)$$

Comparing these two expressions we see that $A_0 = a_0$, $B_n \sin\phi_n = a_n$ and $B_n \cos\phi_n = b_n$. From this it follows that

$$B_n = \ldots\ldots\ldots \text{ and } \phi_n = \ldots\ldots\ldots$$

7

$$B_n = \sqrt{a_n^2 + b_n^2}; \quad \phi_n = \arctan\left(\frac{a_n}{b_n}\right)$$

So

$B_1 \sin(\omega t + \phi_1)$ is the first harmonic or fundamental (lowest frequency)

$B_2 \sin(2\omega t + \phi_2)$ is the second harmonic (frequency twice that of the fundamental)

$B_n \sin(n\omega t + \phi_n)$ is the nth harmonic (frequency n times that of the fundamental).

And for the series to converge, the values of B_n must eventually decrease with higher-order harmonics, i.e. $B_n \to 0$ as $n \to \infty$.

8   Odd and even functions

(a) *Even functions*

A function $f(x)$ is said to be *even* if

$$f(-x) = f(x)$$

i.e. the function value for a particular negative value of x is the same as that for the corresponding positive value of x. The graph of an even function is therefore *reflection symmetrical about the y-axis*.

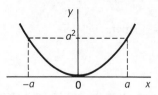

$y = f(x) = x^2$ is an even function because

$$f(-2) = 4 = f(2)$$
$$f(-3) = 9 = f(3) \quad \text{etc.}$$

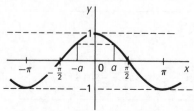

$y = f(x) = \cos x$ is an even function because

$$\cos(-x) = \cos x$$
$$f(-a) = \cos a = f(a).$$

(b) *Odd functions*

A function $f(x)$ is said to be *odd* if

$$f(-x) = -f(x)$$

i.e. the function value for a particular negative value of x is numerically equal to that for the corresponding positive value of x but opposite in sign. If the graph of an odd function is rotated about the origin through $180°$ it coincides with the original graph. We say it is *symmetrical about the origin*.

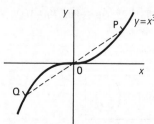

$y = f(x) = x^3$ is an odd function because

$$f(-2) = -8 = -f(2)$$
$$f(-5) = -125 = -f(5) \quad \text{etc.}$$

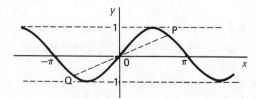

$y = f(x) = \sin x$ is an odd function because

$$\sin(-x) = -\sin x$$
$$f(-a) = -f(a).$$

So, for an even function $f(-x) = f(x)$, symmetrical about the y-axis

for an odd function $f(-x) = -f(x)$, symmetrical about the origin.

▶

Example 1

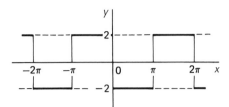

$f(x)$ shown by the waveform is therefore an .... function because it is ....

| odd; symmetrical about the origin, i.e. $f(-x) = -f(x)$ | **9** |

Example 2

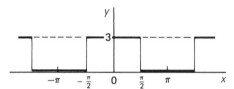

Hence the waveform of $y = f(x)$ depicts an .... function, because it is ....

| even; symmetrical about the y-axis, i.e. $f(-x) = f(x)$ | **10** |

Example 3

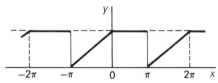

In this case, the waveform shows a function that is .... because ....

| neither even nor odd; not symmetrical about either the y-axis or the origin | **11** |

▶

Exercise

State whether each of the following functions is odd, even, or neither.

1

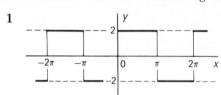

2

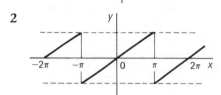

3

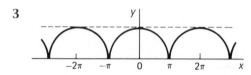

4

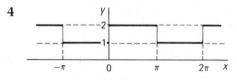

5

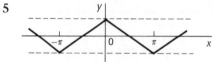

6

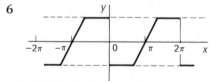

12

1	odd	**2**	odd	**3**	even
4	neither	**5**	even	**6**	odd

We shall shortly see that a knowledge of odd and even functions can save a lot of unnecessary calculation.

First, however, let us consider products of odd and even functions in the next frame

Products of odd and even functions   **13**

The rules closely resemble the elementary rules of signs.

$$(\text{even}) \times (\text{even}) = (\text{even}) \quad \text{like} \quad (+) \times (+) = (+)$$
$$(\text{odd}) \times (\text{odd}) = (\text{even}) \quad\quad\quad (-) \times (-) = (+)$$
$$(\text{odd}) \times (\text{even}) = (\text{odd}) \quad\quad\quad (-) \times (+) = (-).$$

The results can easily be proved.

(a) *Two even functions*

Let $F(x) = f(x)g(x)$ where $f(x)$ and $g(x)$ are even functions.

Then $F(-x) = f(-x)g(-x) = f(x)g(x)$ since $f(x)$ and $g(x)$ are even

$\therefore F(-x) = F(x) \quad$ i.e. $F(x)$ is even

(b) *Two odd functions*

Let $F(x) = f(x)g(x)$ where $f(x)$ and $g(x)$ are odd functions.

Then $F(-x) = f(-x)g(-x)$

$$= \{-f(x)\}\{-g(x)\} \text{ since } f(x) \text{ and } g(x) \text{ are odd}$$

$$= f(x)g(x) = F(x)$$

$\therefore F(-x) = F(x) \quad$ i.e. $F(x)$ is even

Finally

(c) *One odd and one even function*

Let $F(x) = f(x)g(x)$ where $f(x)$ is odd and $g(x)$ even.

Then $F(-x) = f(-x)g(-x) = -f(x)g(x) = -F(x)$

$\therefore F(-x) = -F(x) \quad$ i.e. $F(x)$ is odd

So if $f(x)$ and $g(x)$ are both even, then $f(x)g(x)$ is even
and if $f(x)$ and $g(x)$ are both odd, then $f(x)g(x)$ is even
but if either $f(x)$ or $g(x)$ is even and the other odd, then $f(x)g(x)$ is odd.

Now for a short exercise, so move on

Exercise   **14**

State whether each of the following products is odd, even, or neither.

1	$x^2 \sin 2x$	6	$(2x + 3) \sin 4x$
2	$x^3 \cos x$	7	$\sin^2 x \cos 3x$
3	$\cos 2x \cos 3x$	8	$x^3 e^x$
4	$x \sin nx$	9	$(x^4 + 4) \sin 2x$
5	$3 \sin x \cos 4x$	10	$\dfrac{1}{x+2} \cosh x$

Finish all ten and then check with the next frame

15

1	odd (E)(O) = (O)	6	neither (N)(O) = (N)
2	odd (O)(E) = (O)	7	even (E)(E) = (E)
3	even (E)(E) = (E)	8	neither (O)(N) = (N)
4	even (O)(O) = (E)	9	odd (E)(O) = (O)
5	odd (O)(E) = (O)	10	neither (N)(E) = (N)

Two useful facts emerge from odd and even functions. Thinking in terms of areas under the graphs

(a)

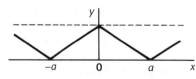

For an *even* function

$$\int_{-a}^{a} f(x)\, dx = 2\int_{0}^{a} f(x)\, dx$$

(b)

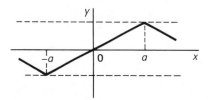

For an *odd* function

$$\int_{-a}^{a} f(x)\, dx = 0$$

We can now look at two important theorems concerning odd and even functions.

Theorem 1

If $f(x)$ is defined over the interval $-\pi < x < \pi$ and $f(x)$ is *even*, then the Fourier series for $f(x)$ contains *cosine terms* only. Included in this is a_0 which may be regarded as $a_n \cos nx$ with $n = 0$.

Proof: Since $f(x)$ is even, $\int_{-\pi}^{0} f(x)\, dx = \int_{0}^{\pi} f(x)\, dx$

(a) $a_0 = \dfrac{1}{\pi}\int_{-\pi}^{\pi} f(x)\, dx = \dfrac{2}{\pi}\int_{0}^{\pi} f(x)\, dx \qquad \therefore\ a_0 = \dfrac{2}{\pi}\int_{0}^{\pi} f(x)\, dx$

(b) $a_n = \dfrac{1}{\pi}\int_{-\pi}^{\pi} f(x)\cos nx\, dx.$

But $f(x)\cos nx$ is the product of two even functions and therefore itself even.

$$\therefore\ a_n = \ldots\ldots\ldots\ldots$$

16

$$a_n = \frac{2}{\pi} \int_0^\pi f(x) \cos nx \, dx$$

Because as the integrand is even,

$$a_n = \frac{1}{\pi} \int_{-\pi}^\pi f(x) \cos nx \, dx = \frac{2}{\pi} \int_0^\pi f(x) \cos nx \, dx.$$

(c) $b_n = \frac{1}{\pi} \int_{-\pi}^\pi f(x) \sin nx \, dx$

Arguing along similar lines, this gives $b_n = \ldots\ldots\ldots$

17

$$b_n = 0$$

Because, since $f(x) \sin nx$ is the product of an even function and an odd function, it is itself odd.

$$\therefore \ b_n = \frac{1}{\pi} \int_{-\pi}^\pi f(x) \sin nx \, dx = 0. \ \therefore \ b_n = 0$$

Therefore, there are no sine terms in the Fourier series for $f(x)$.

Now for an example.

Example

The waveform shown is symmetrical about the y-axis. The function is therefore even and there will be no sine terms in the series.

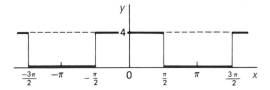

$$\therefore \ f(x) = \frac{1}{2} a_0 + \sum_{n=1}^\infty a_n \cos nx$$

(a) $a_0 = \dfrac{1}{\pi} \displaystyle\int_{-\pi}^\pi f(x) \, dx = \dfrac{2}{\pi} \displaystyle\int_0^\pi f(x) \, dx = \dfrac{2}{\pi} \displaystyle\int_0^{\pi/2} 4 \, dx = \dfrac{2}{\pi} \left[4x \right]_0^{\pi/2} = 4$

(b) $a_n = \dfrac{1}{\pi} \displaystyle\int_{-\pi}^\pi f(x) \cos nx \, dx = \dfrac{2}{\pi} \displaystyle\int_0^\pi f(x) \cos nx \, dx$

$\qquad = \ldots\ldots\ldots$ Finish the integration.

18

$$a_n = 0 \quad (n \text{ even}); \quad a_n = \frac{8}{\pi n} \quad (n = 1,\ 5,\ 9,\ \ldots);$$

$$a_n = -\frac{8}{\pi n} \quad (n = 3,\ 7,\ 11,\ \ldots)$$

Because

$$a_n = \frac{2}{\pi} \int_0^{\pi} f(x) \cos nx \, dx = \frac{2}{\pi} \int_0^{\pi/2} 4 \cos nx \, dx$$

$$= \frac{8}{\pi} \left[\frac{\sin nx}{n} \right]_0^{\pi/2} = \frac{8}{\pi n} \sin \frac{n\pi}{2}$$

But $\sin \dfrac{n\pi}{2} = 0 \quad$ for n even

$$= 1 \quad \text{for } n = 1,\ 5,\ 9,\ \ldots$$

$$= -1 \quad \text{for } n = 3,\ 7,\ 11,\ \ldots \quad \text{Hence the result stated.}$$

(c) We know that $b_n = 0$, because $f(x)$ is an even function. Therefore, the required series is

$$f(x) = \ldots\ldots\ldots$$

19

$$f(x) = 2 + \frac{8}{\pi} \left\{ \cos x - \frac{1}{3} \cos 3x + \frac{1}{5} \cos 5x - \frac{1}{7} \cos 7x + \ldots \right\}$$

If you care to look back to Example 1 in Frame 23 of Programme 7, you will see how much time and effort we have saved by not having to evaluate b_n.

A similar theorem applies to odd functions.

Theorem 2

If $f(x)$ is an *odd* function defined over the interval $-\pi < x < \pi$, then the Fourier series for $f(x)$ contains *sine terms* only.

Proof: Since $f(x)$ is an odd function, $\displaystyle\int_{-\pi}^{0} f(x) \, dx = -\int_0^{\pi} f(x) \, dx.$

(a) $a_0 = \dfrac{1}{\pi} \displaystyle\int_{-\pi}^{\pi} f(x) \, dx.$ But $f(x)$ is odd $\therefore \ a_0 = 0$

(b) $a_n = \dfrac{1}{\pi} \displaystyle\int_{-\pi}^{\pi} f(x) \cos nx \, dx$

   Remembering that $f(x)$ is odd and $\cos nx$ is even, the product $f(x) \cos nx$ is

   $\ldots\ldots\ldots\ldots$

20

$$\boxed{\text{odd}}$$

$$\therefore\ a_n = \frac{1}{\pi}\int_{-\pi}^{\pi} f(x)\cos nx\,dx = \frac{1}{\pi}\int_{-\pi}^{\pi} (\text{odd function})\,dx = 0$$

$$\therefore\ a_n = 0 \quad (\text{including } a_0 = 0)$$

Now for b_n we have

(c) $b_n = \dfrac{1}{\pi}\displaystyle\int_{-\pi}^{\pi} f(x)\sin nx\,dx$ and because $f(x)$ and $\sin nx$ are each odd,

the product $f(x)\sin nx$ is ....

21

$$\boxed{\text{even}}$$

Then $b_n = \dfrac{1}{\pi}\displaystyle\int_{-\pi}^{\pi} f(x)\sin nx\,dx = \dfrac{1}{\pi}\displaystyle\int_{-\pi}^{\pi}(\text{even function})\,dx$

$$= \frac{2}{\pi}\int_{0}^{\pi} f(x)\sin nx\,dx$$

$$\therefore\ b_n = \frac{2}{\pi}\int_{0}^{\pi} f(x)\sin nx\,dx$$

So, if $f(x)$ is odd, $a_0 = 0;\quad a_n = 0;\quad b_n = \dfrac{2}{\pi}\displaystyle\int_{0}^{\pi} f(x)\sin nx\,dx$

i.e. the Fourier series contains sine terms only.

Example

Consider the function shown.

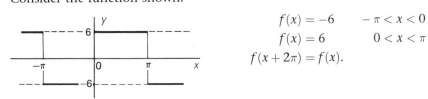

$$f(x) = -6 \qquad -\pi < x < 0$$
$$f(x) = 6 \qquad\quad 0 < x < \pi$$
$$f(x + 2\pi) = f(x).$$

Before we do any evaluation, we can see that this is .... and therefore
....

22

$$\boxed{\text{an odd function; \quad sine terms only, i.e. } a_0 = 0 \text{ and } a_n = 0}$$

$$b_n = \frac{1}{\pi}\int_{-\pi}^{\pi} f(x)\sin nx\,dx. \qquad \begin{array}{l} f(x)\sin nx \text{ is a product of two odd} \\ \text{functions and is therefore even.} \end{array}$$

$$\therefore\ b_n = \frac{2}{\pi}\int_{0}^{\pi} f(x)\sin nx\,dx = \ \ldots\ldots\ldots\ldots$$

23

$$b_n = 0 \quad (n \text{ even}) \quad \text{or} \quad \frac{24}{\pi n} \quad (n \text{ odd})$$

Because

$$b_n = \frac{2}{\pi}\int_0^{\pi} 6\sin nx\,dx = \frac{12}{\pi}\left[\frac{-\cos nx}{n}\right]_0^{\pi} = \frac{12}{\pi n}(1-\cos n\pi).$$

Hence the result stated above.

So the series is $f(x) = \dots\dots\dots$

24

$$f(x) = \frac{24}{\pi}\left\{\sin x + \frac{1}{3}\sin 3x + \frac{1}{5}\sin 5x + \dots\right\}$$

Because $\cos n\pi = (-1)^n$.

Of course, if $f(x)$ is neither an odd nor an even function, then we must obtain expressions for a_0, a_n and b_n in full.

One more example

Example

Determine the Fourier series for the function shown.

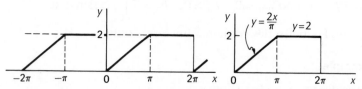

This is neither odd nor even. Therefore we must find a_0, a_n and b_n.

$$f(x) = \frac{1}{2}a_0 + \sum_{n=1}^{\infty}\{a_n\cos nx + b_n\sin nx\}$$

(a) $a_0 = \dfrac{1}{\pi}\displaystyle\int_0^{2\pi} f(x)\,dx = \dfrac{1}{\pi}\left\{\int_0^{\pi}\dfrac{2}{\pi}x\,dx + \int_{\pi}^{2\pi} 2\,dx\right\}$

$$= \frac{1}{\pi}\left\{\left[\frac{x^2}{\pi}\right]_0^{\pi} + \Big[2x\Big]_{\pi}^{2\pi}\right\} = \frac{1}{\pi}\{\pi + 4\pi - 2\pi\} = 3 \qquad \therefore\ a_0 = 3$$

(b) $a_n = \dfrac{1}{\pi}\displaystyle\int_0^{2\pi} f(x)\cos nx\,dx = \dfrac{1}{\pi}\left\{\int_0^{\pi}\left(\dfrac{2}{\pi}x\right)\cos nx\,dx + \int_{\pi}^{2\pi} 2\cos nx\,dx\right\}$

$$= \frac{2}{\pi}\left\{\frac{1}{\pi}\left[\frac{x\sin nx}{n}\right]_0^{\pi} - \frac{1}{\pi n}\int_0^{\pi}\sin nx\,dx + \int_{\pi}^{2\pi}\cos nx\,dx\right\}$$

$$= \dots\dots\dots$$

Finish it off

$$\boxed{a_n = 0 \quad (n \text{ even}); \quad a_n = \frac{-4}{\pi^2 n^2} \quad (n \text{ odd})}$$

25

Because

$$a_n = \frac{2}{\pi}\left\{ \frac{1}{\pi}(0 - 0) - \frac{1}{\pi n}\left[-\frac{\cos nx}{n} \right]_0^\pi + \left[\frac{\sin nx}{n} \right]_\pi^{2\pi} \right\}$$

$$= \frac{2}{\pi}\left\{ -\frac{1}{\pi n^2}(-(-1)^n + 1) + (0 - 0) \right\}$$

$$= -\frac{2}{\pi^2 n^2}(1 - (-1)^n)$$

and so

$$a_n = 0 \quad (n \text{ even}) \text{ and } a_n = -\frac{4}{\pi^2 n^2} \quad (n \text{ odd})$$

(c) To find b_n, we proceed in the same general manner

$$b_n = \dots\dots\dots$$

Complete it on your own

$$\boxed{b_n = -\frac{2}{\pi n}}$$

26

Here is the working.

$$b_n = \frac{1}{\pi}\int_0^{2\pi} f(x)\sin nx\,dx = \frac{1}{\pi}\left\{ \int_0^\pi \left(\frac{2}{\pi}x\right)\sin nx\,dx + \int_\pi^{2\pi} 2\,\sin nx\,dx \right\}$$

$$= \frac{2}{\pi}\left\{ \frac{1}{\pi}\left[\frac{-x\cos nx}{n} \right]_0^\pi + \frac{1}{\pi n}\int_0^\pi \cos nx\,dx + \int_\pi^{2\pi}\sin nx\,dx \right\}$$

$$= \frac{2}{\pi}\left\{ \frac{1}{\pi n}(-\pi\cos n\pi) + \frac{1}{\pi n}\left[\frac{\sin nx}{n} \right]_0^\pi + \left[\frac{-\cos nx}{n} \right]_\pi^{2\pi} \right\}$$

$$= \frac{2}{\pi}\left\{ -\frac{1}{n}\cos n\pi + (0 - 0) - \frac{1}{n}(\cos 2\pi n - \cos n\pi) \right\}$$

$$= \frac{2}{\pi}\left\{ -\frac{1}{n}\cos 2n\pi \right\} = -\frac{2}{\pi n}\cos 2n\pi$$

But $\cos 2n\pi = 1$. $\therefore\ b_n = -\frac{2}{\pi n}$

So the required series is $f(x) = \dots\dots\dots$

27

$$f(x) = \frac{3}{2} - \frac{4}{\pi^2}\left\{\cos x + \frac{1}{9}\cos 3x + \frac{1}{25}\cos 5x + \dots\right\}$$
$$- \frac{2}{\pi}\left\{\sin x + \frac{1}{2}\sin 2x + \frac{1}{3}\sin 3x + \frac{1}{4}\sin 4x \dots\right\}$$

At this stage, let us take stock of our findings so far.

If a function $f(x)$ is defined over the range $-\pi$ to π, or any other periodic interval of 2π, then the Fourier series for $f(x)$ is of the form

$$f(x) = \frac{1}{2}a_0 + \sum_{n=1}^{\infty}\{a_n\cos nx + b_n\sin nx\}$$

where $a_0 = \dfrac{1}{\pi}\displaystyle\int_{-\pi}^{\pi} f(x)\,dx$

$a_n = \dfrac{1}{\pi}\displaystyle\int_{-\pi}^{\pi} f(x)\cos nx\,dx$

$b_n = \dfrac{1}{\pi}\displaystyle\int_{-\pi}^{\pi} f(x)\sin nx\,dx$

We also know that

(a) if $f(x)$ is an *even* function, the series will contain *no sine terms*

(b) if $f(x)$ is an *odd* function, the series will contain *only sine terms*

(c) if $f(x)$ is *neither odd nor even*, the series will, in general, contain a constant term, cosine terms and sine terms.

28 Half-range series

Sometimes a function of period 2π is defined over the range 0 to π, instead of the normal $-\pi$ to π, or 0 to 2π. We then have a choice of how to proceed.

For example, if we are told that between $x = 0$ and $x = \pi$, $f(x) = 2x$, then, since the period is 2π, we have no evidence of how the function behaves between $x = -\pi$ and $x = 0$.

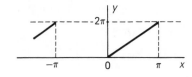

(a)

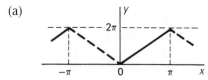

If the waveform were as shown in (a), the function would be an even function, symmetrical about the y-axis and the series would have *only cosine terms* (including possibly a_0).

(b)

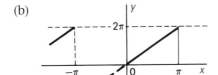

On the other hand, if the waveform were as shown in (b), the function would be odd, being symmetrical about the origin and the series would have *only sine terms*.

(c)

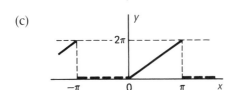

Of course, if we choose something quite different for the waveform between $x = -\pi$ and $x = 0$, then $f(x)$ will be neither odd nor even and the series will then contain ....

| both sine and cosine terms (including a_0) | **29** |

In each case, we are making an assumption on how the function behaves between $x = -\pi$ and $x = 0$, and the resulting Fourier series will therefore apply only to $f(x)$ between $x = 0$ and $x = \pi$ for which it is defined. For this reason, such series are called *half-range series*.

Example 1

A function $f(x)$ is defined by

$$f(x) = 2x \qquad 0 < x < \pi$$
$$f(x + 2\pi) = f(x).$$

Obtain a half-range cosine series to represent the function.

To obtain a cosine series, i.e. a series with no sine terms, we need an .... function.

| even | **30** |

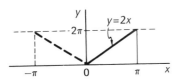

Therefore, we assume the waveform between $x = -\pi$ and $x = 0$ to be as shown, making the total graph symmetrical about the y-axis.

Now we can find expressions for the Fourier coefficients as usual.

$$a_0 = ...$$

31

$$\boxed{a_0 = 2\pi}$$

Because

$$a_0 = \frac{2}{\pi}\int_0^{\pi} f(x)\,dx = \frac{2}{\pi}\int_0^{\pi} 2x\,dx = \frac{2}{\pi}\left[x^2\right]_0^{\pi} = 2\pi \qquad \therefore\ a_0 = 2\pi$$

Then we need a_n which is . . . .

32

$$\boxed{a_n = 0 \quad (n\text{ even}) \qquad = -\frac{8}{\pi n^2} \quad (n\text{ odd})}$$

Because

$$a_n = \frac{2}{\pi}\int_0^{\pi} 2x\cos nx\,dx = \frac{4}{\pi}\int_0^{\pi} x\cos nx\,dx$$

$$= \frac{4}{\pi}\left\{\left[\frac{x\sin nx}{n}\right]_0^{\pi} - \frac{1}{n}\int_0^{\pi}\sin nx\,dx\right\}$$

$$= \frac{4}{\pi}\left\{(0-0) - \frac{1}{n}\left[\frac{-\cos nx}{n}\right]_0^{\pi}\right\} = \frac{4}{\pi n^2}(\cos n\pi - 1)$$

$$\cos n\pi = 1 \quad (n\text{ even}) \qquad = -1 \quad (n\text{ odd})$$

$$\therefore\ a_n = 0 \quad (n\text{ even}) \quad \text{and} \quad a_n = -\frac{8}{\pi n^2} \quad (n\text{ odd})$$

All that now remains is b_n which is . . . .

33

$$\boxed{\text{zero, since } f(x) \text{ is an even function, i.e. } b_n = 0}$$

So $a_0 = 2\pi$, $\quad a_n = 0 \quad (n\text{ even})$ or $\quad -\dfrac{8}{\pi n^2} \quad (n\text{ odd})$, $\quad b_n = 0$.

Therefore $f(x) = $ . . . .

34

$$\boxed{f(x) = \pi - \frac{8}{\pi}\left\{\cos x + \frac{1}{9}\cos 3x + \frac{1}{25}\cos 5x + \ldots\right\}}$$

Let us look at a further example, so move on to the next frame

Example 2   **35**

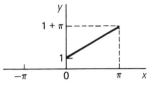

Determine a half-range sine series to represent the function $f(x)$ defined by

$$f(x) = 1 + x \qquad 0 < x < \pi$$
$$f(x + 2\pi) = f(x).$$

We choose the waveform between $x = -\pi$ and $x = 0$ so that the graph is symmetrical about the origin. The function is then an odd function and the series will contain only sine terms.

$$\therefore\ a_0 = 0 \quad \text{and} \quad a_n = 0$$

b_n can now easily be determined and the required series obtained.

$$f(x) = \dots\dots\dots$$

                                                                                  36

$$\boxed{\begin{aligned} f(x) = \left(\frac{4}{\pi} + 2\right)&\left\{\sin x + \frac{1}{3}\sin 3x + \frac{1}{5}\sin 5x + \dots\right\} \\ -2&\left\{\frac{1}{2}\sin 2x + \frac{1}{4}\sin 4x + \frac{1}{6}\sin 6x + \dots\right\} \end{aligned}}$$

Check the working.

$$b_n = \frac{2}{\pi}\int_0^\pi (1 + x)\sin nx\, dx = \frac{2}{\pi}\left\{\left[(1 + x)\frac{-\cos nx}{n}\right]_0^\pi + \frac{1}{n}\int_0^\pi \cos nx\, dx\right\}$$

$$= \frac{2}{\pi}\left\{-\frac{1 + \pi}{n}\cos n\pi + \frac{1}{n} + \frac{1}{n}\left[\frac{\sin nx}{n}\right]_0^\pi\right\}$$

$$= \frac{2}{\pi}\left\{\frac{1}{n} - \frac{1 + \pi}{n}\cos n\pi\right\} = \frac{2}{\pi n}\{1 - (1 + \pi)\cos n\pi\}$$

$$\cos n\pi = 1 \quad (n\text{ even}) \qquad = -1 \quad (n\text{ odd})$$

$$\therefore\ b_n = -\frac{2}{n} \quad (n\text{ even}) \qquad = \frac{4 + 2\pi}{\pi n} \quad (n\text{ odd})$$

Substituting in the general expression $f(x) = \displaystyle\sum_{n=1}^{\infty} b_n \sin nx$ we have

$$f(x) = \frac{4 + 2\pi}{\pi}\left\{\sin x + \frac{1}{3}\sin 3x + \frac{1}{5}\sin 5x + \dots\right\}$$

$$- 2\left\{\frac{1}{2}\sin 2x + \frac{1}{4}\sin 4x + \frac{1}{6}\sin 6x + \dots\right\}$$

So a knowledge of odd and even functions and of half-range series saves a deal of unnecessary work on occasions.

Now let us consider the presence of odd or even harmonics, so move on

37   Series containing only odd harmonics or only even harmonics

$$f(x) = \tfrac{1}{2}a_0 + a_1 \cos x + a_2 \cos 2x + a_3 \cos 3x + \ldots$$
$$+ b_1 \sin x + b_2 \sin 2x + b_3 \sin 3x + \ldots$$

If we replace x by $(x + \pi)$, this becomes

$$f(x + \pi) = \tfrac{1}{2}a_0 + \sum_{n=1}^{\infty} \{a_n \cos n(x + \pi) + b_n \sin n(x + \pi)\}$$

Now $\cos(nx + n\pi) = \cos nx \cos n\pi - \sin nx \sin n\pi$.

But for $n = 1, 2, 3, \ldots$ $\sin n\pi = 0$

$\therefore\ \cos n(x + \pi) = \cos nx \cos n\pi$

Also for $n = 1, 2, 3, \ldots$ $\cos n\pi = 1$ (n even) $= -1$ (n odd).

$\therefore\ \cos n(x + \pi) = \cos nx$ (n even) $= -\cos nx$ (n odd)   (1)

Similarly, $\sin(nx + n\pi) = \sin nx \cos n\pi + \cos nx \sin n\pi$.

Therefore, as before

$\sin n(x + \pi) = \sin nx$ (n even) $= -\sin nx$ (n odd)   (2)

$\therefore\ f(x + \pi) = \tfrac{1}{2}a_0 - a_1 \cos x + a_2 \cos 2x - a_3 \cos 3x + \ldots$
$$- b_1 \sin x + b_2 \sin 2x - b_3 \sin 3x + \ldots$$

But $f(x) = \tfrac{1}{2}a_0 + a_1 \cos x + a_2 \cos 2x + a_3 \cos 3x + \ldots$
$$+ b_1 \sin x + b_2 \sin 2x + b_3 \sin 3x + \ldots$$

If $f(x) = f(x + \pi)$, these two series are equal and the odd harmonics that you see differ in sign must be zero.

$\therefore\ f(x) = f(x + \pi) = \tfrac{1}{2}a_0 + a_2 \cos 2x + a_4 \cos 4x + \ldots$
$$+ b_2 \sin 2x + b_4 \sin 4x + \ldots$$

$\therefore\ $ If $f(x) = f(x + \pi)$, the Fourier series for $f(x)$ contains even harmonics only.

Similarly, from the same two series above

if $f(x) = -f(x + \pi)$, the Fourier series for $f(x)$ contains odd harmonics only.

$\therefore\ f(x) = a_1 \cos x + a_3 \cos 3x + \ldots + b_1 \sin x + b_3 \sin 3x + \ldots$

Make a note of these two results: you will find them useful

38   Example 1

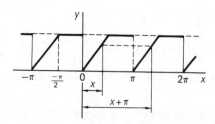

Here $f(x) = f(x + \pi)$

Therefore, the series contains

. . . .

39

even harmonics only

Example 2

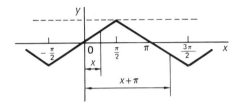

Here we see that $f(x) = -f(x + \pi)$.

Therefore, the series contains

. . . .

40

odd harmonics only

Now we can apply our knowledge to date to the following exercise.

Exercise

From each of the following waveforms, we can describe the nature of the terms in the relevant Fourier series.

1

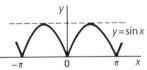

2

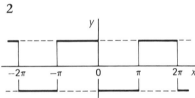

3

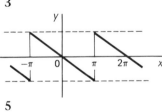

4

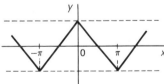

5

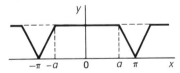

6

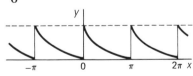

41

> 1   cosine terms ($+a_0$) only; even harmonics only
> 2   sine terms only; odd harmonics only
> 3   sine terms only; all harmonics
> 4   cosine terms ($+a_0$) only; odd harmonics only
> 5   cosine terms ($+a_0$) only; all harmonics
> 6 a_0, sine and cosine terms; even harmonics only.

On we go

42   Significance of the constant term $\frac{1}{2}a_0$

We might, at this point, note that the effect of the constant term $\frac{1}{2}a_0$ is to raise, or lower, the whole waveform on the y-axis.

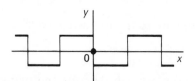 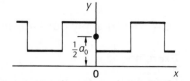

In electrical applications to alternating currents, the constant term $\frac{1}{2}a_0$ of the Fourier series indicates the d.c. component.

For example, from Frames 58–61 we found that the odd square wave

$$f(x) = \begin{cases} -6 & -\pi < x < 0 \\ 6 & 0 < x < \pi \end{cases} \qquad f(x + 2\pi) = f(x)$$

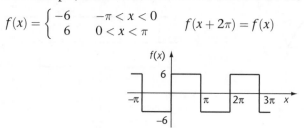

has the Fourier series expansion

$$f(x) = \frac{24}{\pi}\left\{\sin x + \frac{1}{3}\sin 3x + \frac{1}{5}\sin 5x + \dots\right\}$$

The function $g(x) = 2 + f(x)$ has the Fourier series expansion

$$g(x) = 2 + \frac{24}{\pi}\left\{\sin x + \frac{1}{3}\sin 3x + \frac{1}{5}\sin 5x + \dots\right\}$$

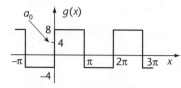

Here $a_0/2 = 2$ – the amount by which the graph of the original function has been raised.

Half-range series with arbitrary period

43

We now extend the work on half-range sine and cosine series to functions with arbitrary period.

(a) *Even function*   Half-range cosine series

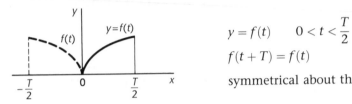

$$y = f(t) \qquad 0 < t < \frac{T}{2}$$

$$f(t + T) = f(t)$$

symmetrical about the y-axis.

With an even function, we know that $b_n = 0$

$$\therefore\ f(t) = \frac{1}{2}a_0 + \sum_{n=1}^{\infty} a_n \cos n\omega t$$

where
$$a_0 = \frac{4}{T}\int_0^{T/2} f(t)\,\mathrm{d}t$$

and
$$a_n = \frac{4}{T}\int_0^{T/2} f(t)\cos n\omega t\,\mathrm{d}t$$

(b) *Odd function*   Half-range sine series

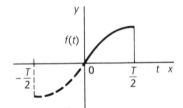

$$y = f(t) \qquad 0 < t < \frac{T}{2}$$

$$f(t + T) = f(t)$$

symmetrical about the origin.

$\therefore\ a_0 = 0$ and $a_n = 0$

Then $f(t) = \dots\dots\dots$

and $b_n = \dots\dots\dots$

44

$$f(t) = \sum_{n=1}^{\infty} b_n \sin n\omega t; \quad b_n = \frac{4}{T}\int_0^{T/2} f(t)\sin n\omega t\,\mathrm{d}t$$

Now for an example or two.

So move on

45

Example 1

A function $f(t)$ is defined by $f(t) = 4 - t$, $\quad 0 < t < 4$.

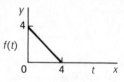

We have to form a half-range cosine series to represent the function in this interval.

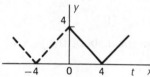

First we form an even function, i.e. symmetrical about the y-axis.

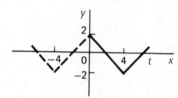

Now for a useful little trick. If we lower the waveform 2 units, i.e. to its 'average' position, balanced above and below the x-axis, then in this new position $\frac{1}{2}a_0 = 0$ and we have been saved one set of calculations.

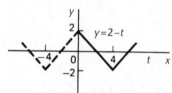

The function is now $y = f_1(t) = 2 - t$ and, for the moment $\frac{1}{2}a_0 = 0$. Also, being an even function $b_n = 0$. All we need to do is to evaluate a_n.

So $a_n = \dfrac{4}{T} \displaystyle\int_0^{T/2} f_1(t) \cos n\omega t \, dt = \dfrac{4}{8} \displaystyle\int_0^4 (2 - t) \cos n\omega t \, dt$

$$= \ldots \ldots \ldots$$

46

$$\boxed{a_n = 0 \quad (n \text{ even}) \qquad = \dfrac{1}{n^2 \omega^2} \quad (n \text{ odd})}$$

Simple integration by parts gives

$$a_n = \frac{1}{2}\left\{ -\frac{2 \sin 4n\omega}{n\omega} - \frac{1}{n^2\omega^2}(\cos 4n\omega - 1) \right\}$$

But $\omega = \dfrac{2\pi}{T} = \dfrac{2\pi}{8} = \dfrac{\pi}{4}$

$$a_n = \frac{1}{2}\left\{ -\frac{2 \sin n\pi}{n\omega} - \frac{1}{n^2\omega^2}(\cos n\pi - 1) \right\} \qquad n = 1, 2, 3, \ldots$$

$$\sin n\pi = 0; \quad \cos n\pi = 1 \quad (n \text{ even}); \quad \cos n\pi = -1 \quad (n \text{ odd})$$

$$\therefore \ a_n = 0 \quad (n \text{ even}) \quad \text{and} \quad a_n = \frac{1}{n^2\omega^2} \quad (n \text{ odd})$$

$$\therefore \ f_1(t) = \ldots \ldots \ldots$$

47

$$f_1(t) = \frac{1}{\omega^2}\left\{\cos\omega t + \frac{1}{9}\cos 3\omega t + \frac{1}{25}\cos 5\omega t + \dots\right\}$$

Now if we finally lift the waveform back to its original position by restoring the 2 units (i.e. $\frac{1}{2}a_0 = 2$), the original function is regained with $f(t) = f_1(t) + 2$.

$$\therefore \ f(t) = 2 + \frac{1}{\omega^2}\left\{\cos\omega t + \frac{1}{9}\cos 3\omega t + \frac{1}{25}\cos 5\omega t + \dots\right\}$$

where $\omega = \dfrac{\pi}{4}$.

Example 2

A function $f(t)$ is defined by $\quad f(t) = 3 + t \qquad 0 < t < 2$
$$f(t + 4) = f(t).$$

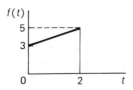

Obtain the half-range sine series for the function in this range.

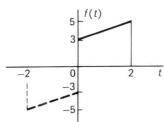

Sine series required. Therefore, we form an odd function, symmetrical about the origin

$$a_0 = 0; \ a_n = 0; \ T = 4$$

$$f(t) = \sum_{n=1}^{\infty} b_n \sin n\omega t$$

$$\therefore \ b_n = \frac{4}{T}\int_0^2 f(t)\ \sin n\omega t\ \mathrm{d}t = \int_0^2 (3 + t)\sin n\omega t\ \mathrm{d}t$$

This you can easily evaluate and then, putting $n = 1, 2, 3, \dots$ obtain the series
$$f(t) = \dots\dots\dots$$

48

$$f(t) = \frac{2}{\omega} \left\{ 4 \sin \omega t - \frac{1}{2} \sin 2\omega t + \frac{4}{3} \sin 3\omega t - \frac{1}{4} \sin 4\omega t \dots \right\}$$

Because

Straightforward integration by parts gives

$$b_n = \frac{1}{n\omega}(3 - 5 \cos 2n\omega) + \frac{1}{n^2\omega^2}(\sin 2n\omega)$$

But $T = \frac{2\pi}{\omega}$ $\therefore$ $\omega = \frac{2\pi}{T} = \frac{\pi}{2}$

$$\therefore \quad b_n = \frac{1}{n\omega}(3 - 5 \cos n\pi) + \frac{1}{n^2\omega^2} \sin n\pi = \begin{cases} -\dfrac{2}{n\omega} & (n \text{ even}) \\[2mm] \dfrac{8}{n\omega} & (n \text{ odd}) \end{cases}$$

Therefore

$$f(t) = \frac{2}{\omega} \left\{ 4 \sin \omega t - \frac{1}{2} \sin 2\omega t + \frac{4}{3} \sin 3\omega t - \frac{1}{4} \sin 4\omega t \dots \right\}$$

And that just about brings this particular Programme to an end. Fourier series have wide applications so it is very worthwhile paying considerable attention to them.

The **Revison summary** and **Can you?** checklist now follow, after which you will have no trouble with the **Test exercise**. The **Further problems** provide additional practice.

 # Revision summary 8

49

1   *Functions with period T*

$$f(t) = \tfrac{1}{2}a_0 + \sum_{n=1}^{\infty}\{a_n \cos n\omega t + b_n \sin n\omega t\}$$

$$a_0 = \frac{2}{T}\int_0^T f(t)\,dt \qquad = \frac{\omega}{\pi}\int_0^{2\pi/\omega} f(t)\,dt$$

$$a_n = \frac{2}{T}\int_0^T f(t)\cos n\omega t\,dt = \frac{\omega}{\pi}\int_0^{2\pi/\omega} f(t)\cos n\omega t\,dt$$

$$b_n = \frac{2}{T}\int_0^T f(t)\sin n\omega t\,dt = \frac{\omega}{\pi}\int_0^{2\pi/\omega} f(t)\sin n\omega t\,dt.$$

where $\omega = \dfrac{2\pi}{T}$ so that $T = \dfrac{2\pi}{\omega}$

2   *Odd and even functions*
   (a) Even function: $f(-x) = f(x)$; symmetrical about the y-axis.
   (b) Odd function: $f(-x) = -f(x)$; symmetrical about the origin.

Product of odd and even functions

$$(\text{even}) \times (\text{even}) = (\text{even})$$
$$(\text{odd}) \times (\text{odd}) = (\text{even})$$
$$(\text{odd}) \times (\text{even}) = (\text{odd})$$

3   *Sine series and cosine series*
    If $f(x)$ is *even*, the series contains *cosine terms only* (including a_0)
    If $f(x)$ is *odd*, the series contains *sine terms only*.

4   *Half-range series*
    A function defined over the domain $0 \le x \le \pi$ can be extended into either
    an odd function or an even function with period 2π.

5   *Odd and even harmonics*
    If $f(x + \pi) = f(x)$, the Fourier series for $f(x)$ contains *even harmonics only*
    If $f(x + \pi) = -f(x)$, the Fourier series for $f(x)$ contains *odd harmonics only*

6   *Significance of the constant term*
    The effect of the constant term $a_0/2$ is to raise or lower the waveform on
    the vertical axis.

7   *Half-range series with arbitrary period T*

    Even function   *Odd function*

$$f(t) = \frac{1}{2}a_0 + \sum_{n=1}^{\infty}\{a_n \cos n\omega t\} \qquad f(t) = \sum_{n=1}^{\infty}\{b_b \sin n\omega t\}$$

$$a_0 = \frac{4}{T}\int_0^T f(t)\,dt \qquad\qquad b_n = \frac{4}{T}\int_0^T f(t)\sin n\omega t\,dt$$

$$a_n = \frac{4}{T}\int_0^T f(t)\cos n\omega t\,dt$$

where $\omega = \dfrac{2\pi}{T}$ that is $T = \dfrac{2\pi}{\omega}$

 Can you?

50

Checklist 8

Check this list before and after you try the end of Programme test.

On a scale of 1 to 5 how confident are you that you can: **Frames**

- Obtain the Fourier coefficients of a function with arbitrary
  period T?
  Yes   ☐   ☐   ☐   ☐   ☐   *No*

- Recognise even and odd functions and their products?
  Yes   ☐   ☐   ☐   ☐   ☐   *No*

▶

- Derive the Fourier series of even and odd functions?
  Yes ☐ ☐ ☐ ☐ ☐ *No*   15 to 27

- Derive half-range Fourier series?
  Yes ☐ ☐ ☐ ☐ ☐ *No*   28 to 36

- Recognise the conditions for the Fourier series to contain only odd or only even harmonics?
  Yes ☐ ☐ ☐ ☐ ☐ *No*   37 to 41

- Explain the geometric significance of the constant term $a_0/2$?
  Yes ☐ ☐ ☐ ☐ ☐ *No*   42

- Derive half-range Fourier series with arbitrary period?
  Yes ☐ ☐ ☐ ☐ ☐ *No*   43 to 48

 # Test exercise 8

51

1   Given the function
$$f(t) = t^2 \qquad 0 \le t < 2$$
$$f(t+2) = f(t)$$
obtain the Fourier series and determine the value of the series when $t = 2$.

2   State whether each of the following products is odd, even, or neither.
- (a) $x^3 \cos 2x$
- (b) $x^2 \sin 3x$
- (c) $\sin 2x \sin 3x$
- (d) $x^2 e^{2x}$
- (e) $(x+5)\cos 2x$
- (f) $\sin^2 x \cos x$.

3   A function $f(x)$ is defined by $f(x) = \pi - x \qquad 0 < x < \pi$
$$f(x+2\pi) = f(x).$$
Express the function
- (a)   as a half-range cosine series
- (b)   as a half-range sine series.

4   Comment on the nature of the terms in the Fourier series for the following functions.

(a)

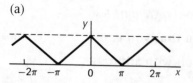

(b)

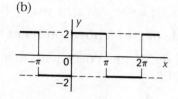

(c)

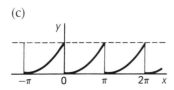

(d)

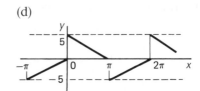

5 A function $f(t)$ is defined by

$$f(t) = \begin{cases} 0 & -2 < t < 0 \\ t & 0 < t < 2 \end{cases}$$

$f(t+4) = f(t)$.

Determine its Fourier series.

 Further problems 8

1 Determine the Fourier series representation of the function $f(t)$ defined by

$$f(t) = \begin{cases} 3 & -2 < t < 0 \\ -5 & 0 < t < 2 \end{cases}$$

$f(t+4) = f(t)$.

52

2 Determine the half-range cosine series for the function $f(x) = \sin x$ defined in the range $0 < x < \pi$.

3 Determine the Fourier series to represent a half-wave rectifier output current, i amperes, defined by

$$i = f(t) = \begin{cases} A \sin \omega t & 0 < t < \dfrac{T}{2} \\ 0 & \dfrac{T}{2} < t < T \end{cases}$$

$f(t+T) = f(t)$.

4 A function $f(x)$ is defined by

$$f(x) = \begin{cases} a & 0 < x < \dfrac{\pi}{3} \\ 0 & \dfrac{\pi}{3} < x < \dfrac{2\pi}{3} \\ -a & \dfrac{2\pi}{3} < x < \pi \end{cases}$$

$f(x+\pi) = f(x)$.

Obtain the Fourier series to represent the function.

5 If $f(x)$ is defined by $f(x) = x(\pi - x)$ $0 < x < \pi$, express the function as
   (a) a half-range cosine series
   (b) a half-range sine series.

6 Determine the Fourier cosine series to represent the function $f(x)$ where

$$f(x) = \begin{cases} \cos x & 0 < x < \dfrac{\pi}{2} \\ 0 & \dfrac{\pi}{2} < x < \pi \end{cases}$$

$f(x + 2\pi) = f(x)$.

7 If

$$f(x) = \begin{cases} 0 & 0 < x < \dfrac{\pi}{2} \\ \cos x & \dfrac{\pi}{2} < x < \pi \end{cases} \qquad f(x + 2\pi) = f(x),$$

obtain the Fourier cosine series for $f(x)$ in the range $x = 0$ to $x = \pi$.

8 A function $f(x)$ is defined over the interval $0 < x < \pi$ by

$$f(x) = \begin{cases} x & 0 < x < \dfrac{\pi}{2} \\ \pi - x & \dfrac{\pi}{2} < x < \pi \end{cases}$$

For the range $x = 0$ to $x = \pi$, determine the Fourier sine series.

9 A function $f(t)$ is defined by

$$f(t) = \begin{cases} -1 & -1 < t < 0 \\ 2t & 0 < t < 1 \end{cases}$$

$f(t + 2) = f(t)$.

Obtain the Fourier series up to and including the third harmonic.

10 A function $f(t)$ is defined by

$$f(t) = 1 - t^2 \qquad -1 < t < 1$$
$$f(t + 2) = f(t).$$

Determine its Fourier series.

11 Determine the Fourier series for a periodic function such that

$$f(t) = \begin{cases} 1 & -2 < t < -1 \\ 0 & -1 < t < 1 \\ -1 & 1 < t < 2 \end{cases}$$

$f(t + 4) = f(t)$.

12 Determine the Fourier series for the function $f(t)$ defined by

$$f(t) = \begin{cases} 0 & -2 < t < 0 \\ \dfrac{3t}{4} & 0 < t < 4 \end{cases}$$

$f(t + 6) = f(t)$.

Introduction to the Fourier transform

Learning outcomes

When you have completed this Programme you will be able to:

- Convert a trigonometric Fourier series into a doubly infinite sum of complex exponentials
- Derive the complex Fourier series of a function that satisfies Dirichlet's conditions
- Recognise the function $\text{sinc}\,(t)$
- Separate a discrete complex spectrum into an amplitude spectrum and a phase spectrum
- State Fourier's integral theorem in terms of complex exponentials
- Define and derive the Fourier transform of a function satisfying Dirichlet's conditions
- Separate a continuous complex spectrum into an amplitude spectrum and a phase spectrum
- Recognise the functions $\Pi_a(t)$ and $\Lambda_a(t)$ and derive their Fourier transforms along with those of the Dirac delta and the Heaviside unit step
- Recognise alternative forms of the function–transform pair
- Reproduce a collection of properties of the Fourier transform
- Evaluate the convolution of two functions and describe its Fourier transform
- Derive the Fourier sine and cosine transformations

Complex Fourier series

1 Introduction

In the previous Programme we saw how a periodic function can be represented by an infinite sum of periodic, trigonometric harmonics. Each harmonic has a definite frequency which is an integer multiple of the fundamental frequency. A non-periodic function can be similarly represented, not as a sum but as an integral over a continuous range of frequencies. Before we do this, however, we shall convert the infinite Fourier series in terms of sines and cosines into a doubly infinite series involving complex exponentials.

Complex exponentials

Recall the exponential form of a complex number and its relationship to the polar form, namely

$$z = r(\cos\theta + j\sin\theta) = re^{j\theta}$$

From this equation we can see that

$$\cos\theta + j\sin\theta = e^{j\theta}$$

and so

$$\cos(-\theta) + j\sin(-\theta) = e^{-j\theta} = \cos\theta - j\sin\theta$$

Using these two equations we can find the complex exponential form of the trigonometric functions as

$$\cos\theta = \ldots\ldots\ldots \quad \text{and} \quad \sin\theta = \ldots\ldots\ldots$$

2

$$\boxed{\cos\theta = \frac{e^{j\theta} + e^{-j\theta}}{2} \quad \text{and} \sin\theta = \frac{e^{j\theta} - e^{-j\theta}}{2j}}$$

Because

$$\cos\theta + j\sin\theta = e^{j\theta} \text{ and } \cos\theta - j\sin\theta = e^{-j\theta}$$

so adding these two equations gives

$$2\cos\theta = e^{j\theta} + e^{-j\theta} \text{ that is } \cos\theta = \frac{e^{j\theta} + e^{-j\theta}}{2} \tag{1}$$

and subtracting the two equations gives

$$2j\sin\theta = e^{j\theta} - e^{-j\theta} \text{ that is } \sin\theta = \frac{e^{j\theta} - e^{-j\theta}}{2j} \tag{2}$$

These two equations permit us to develop an alternative representation of a Fourier series.

▶

In the previous Programme we found that the Fourier series of the piecewise continuous function $f(t)$ with piecewise continuous derivative and where $f(t + T) = f(t)$ is given as

$$f(t) = \frac{a_0}{2} + \sum_{n=1}^{\infty} (a_n \cos n\omega_0 t + b_n \sin n\omega_0 t) \tag{3}$$

where $\omega_0 = \dfrac{2\pi}{T}$ and where $a_n = \dfrac{2}{T} \displaystyle\int_{-T/2}^{T/2} f(t) \cos n\omega_0 t \, dt$

and $b_n = \dfrac{2}{T} \displaystyle\int_{-T/2}^{T/2} f(t) \sin n\omega_0 t \, dt$

Now, if we substitute the right-hand sides of equations (1) and (2) into equation (3) we obtain

$$f(t) = \frac{a_0}{2} + \sum_{n=1}^{\infty} \left(\left\{ \dots\dots\dots \right\} e^{jn\omega_0 t} + \left\{ \dots\dots\dots \right\} e^{-jn\omega_0 t} \right)$$

3

$$f(t) = \frac{a_0}{2} + \sum_{n=1}^{\infty} \left(\left\{ \frac{a_n - jb_n}{2} \right\} e^{jn\omega_0 t} + \left\{ \frac{a_n + jb_n}{2} \right\} e^{-jn\omega_0 t} \right)$$

Because

$$f(t) = \frac{a_0}{2} + \sum_{n=1}^{\infty} (a_n \cos n\omega_0 t + b_n \sin n\omega_0 t)$$

$$= \frac{a_0}{2} + \sum_{n=1}^{\infty} \left(a_n \frac{e^{jn\omega_0 t} + e^{-jn\omega_0 t}}{2} + b_n \frac{e^{jn\omega_0 t} - e^{-jn\omega_0 t}}{2j} \right)$$

$$= \frac{a_0}{2} + \sum_{n=1}^{\infty} \left(\left\{ \frac{a_n + b_n/j}{2} \right\} e^{jn\omega_0 t} + \left\{ \frac{a_n - b_n/j}{2} \right\} e^{-jn\omega_0 t} \right)$$

$$= \frac{a_0}{2} + \sum_{n=1}^{\infty} \left(\left\{ \frac{a_n - jb_n}{2} \right\} e^{jn\omega_0 t} + \left\{ \frac{a_n + jb_n}{2} \right\} e^{-jn\omega_0 t} \right)$$

In the next frame we shall make some notational changes to simplify this expression

4

If we now define $c_n = \dfrac{a_n - jb_n}{2}$ so that the complex conjugate of c_n is

$c_n^* = \dfrac{a_n + jb_n}{2}$ we can write this sum as

$$f(t) = c_0 + \sum_{n=1}^{\infty}\left(c_n e^{jn\omega_0 t} + c_n^* e^{-jn\omega_0 t}\right)$$

Note that we have taken $b_0 = 0$. There is no problem about this. There is no term $\sin 0\omega_0 t$ in the Fourier series and so $b_0 = 0$

$$= c_0 + \sum_{n=1}^{\infty} c_n e^{jn\omega_0 t} + \sum_{n=1}^{\infty} c_n^* e^{-jn\omega_0 t}$$

$$= c_0 + \sum_{n=1}^{\infty} c_n e^{jn\omega_0 t} + \sum_{n=1}^{\infty} c_{-n} e^{-jn\omega_0 t}$$

For notational convenience we denote c_n^* by c_{-n}. This means that $a_{-n} = a_n$ and $b_{-n} = -b_n$

$$= c_0 + \sum_{n=1}^{\infty} c_n e^{jn\omega_0 t} + \sum_{n=-1}^{-\infty} c_n e^{jn\omega_0 t}$$

As n ranges from 1 to ∞ so $-n$ ranges from -1 to $-\infty$

$$= \sum_{n=-\infty}^{-1} c_n e^{jn\omega_0 t} + c_0 + \sum_{n=1}^{\infty} c_n e^{jn\omega_0 t}$$

Notice the reversed order of summation in the first sum

$$= \sum_{n=-\infty}^{\infty} c_n e^{jn\omega_0 t}$$

Combining all three terms into the *doubly infinite sum*

where $c_n = \dfrac{a_n - jb_n}{2} = \dfrac{2}{2T}\displaystyle\int_{-T/2}^{T/2} f(t)(\cos n\omega_0 t - j\sin n\omega_0 t)\,\mathrm{d}t$. That is

$$c_n = \frac{1}{T}\int_{-T/2}^{T/2} f(t) e^{-jn\omega_0 t}\,\mathrm{d}t.$$

In the next frame we shall look at some examples

5

Example 1

To find the complex Fourier series for the function

$$f(t) = \begin{cases} 0 & -T/2 < t < -a/2 \\ 1 & -a/2 < t < a/2 \qquad \text{where } f(t+T) = f(t) \\ 0 & a/2 < t < T/2 \end{cases}$$

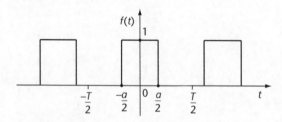

we proceed as on the next page.

$$f(t) = \sum_{n=-\infty}^{\infty} c_n e^{jn\omega_0 t} \qquad \text{where } \omega_0 = \frac{2\pi}{T} \text{ and}$$

$$c_n = \frac{1}{T} \int_{-T/2}^{T/2} f(t) e^{-jn\omega_0 t}$$

$$= \frac{1}{T} \int_{-a/2}^{a/2} e^{-jn\omega_0 t} \, dt \qquad \text{Because } f(t) = 1 \text{ for } -a/2 < t < a/2$$

$$= \frac{1}{T} \left[\frac{e^{-jn\omega_0 t}}{-jn\omega_0} \right]_{-a/2}^{a/2} \qquad \text{Provided } n \neq 0$$

$$= \left(\frac{e^{-jn\omega_0 a/2} - e^{jn\omega_0 a/2}}{-j2n\pi} \right) \qquad \text{Since } \omega_0 = \frac{2\pi}{T}$$

$$= \frac{\sin n\omega_0 a/2}{n\pi} \qquad \text{Recall that } \sin\theta = \frac{e^{j\theta} - e^{-j\theta}}{2j}$$

$$= \frac{\sin n\pi a/T}{n\pi} \qquad \text{Since } \omega_0 = \frac{2\pi}{T}$$

$$= \frac{a}{T} \left(\frac{\sin n\pi a/T}{n\pi a/T} \right) \qquad \text{Provided } n \neq 0$$

When $n = 0$

$$c_0 = \frac{1}{T} \int_{-T/2}^{T/2} f(t) \, dt = \frac{1}{T} \int_{-a/2}^{a/2} dt = \frac{a}{T}$$

Therefore

$$f(t) = \frac{a}{T} + \sum_{\substack{n=-\infty \\ n \neq 0}}^{\infty} \frac{a}{T} \left(\frac{\sin n\pi a/T}{n\pi a/T} \right) e^{jn\omega_0 t}$$

In the next frame we shall look at the same function
retarded by half the width of the peak

Example 2

6

To find the complex Fourier series for the function

$$f(t) = \begin{cases} 1 & 0 < t < a \\ 0 & a < t < T \end{cases} \qquad \text{where } f(t+T) = f(t)$$

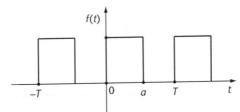

We find that, for $n \neq 0$,

$$c_n = \ldots \ldots \ldots \ldots$$

7

$$c_n = e^{-jn\pi a/T} \frac{a}{T} \left(\frac{\sin n\pi a/T}{n\pi a/T} \right)$$

Because

$$f(t) = \sum_{n=-\infty}^{\infty} c_n e^{jn\omega_0 t} \qquad\qquad \text{where } \omega = \frac{2\pi}{T} \text{ and}$$

$$c_n = \frac{1}{T} \int_{-T/2}^{T/2} f(t) e^{-jn\omega_0 t} \, dt$$

$$= \frac{1}{T} \int_0^a e^{-jn\omega_0 t} \, dt$$

$$= \frac{1}{T} \left[\frac{e^{-jn\omega_0 t}}{-jn\omega_0} \right]_0^a \qquad\qquad \text{Provided } n \neq 0$$

$$= \left(\frac{e^{-jn\omega_0 a} - 1}{-j2n\pi} \right)$$

$$= e^{-jn\omega_0 a/2} \left(\frac{e^{-jn\omega_0 a/2} - e^{jn\omega_0 a/2}}{-j2n\pi} \right)$$

$$= e^{-jn\pi a/T} \frac{a}{T} \left(\frac{\sin n\pi a/T}{n\pi a/T} \right) \qquad\qquad \text{Provided } n \neq 0$$

To finish

$$c_0 = \ldots\ldots\ldots\ldots$$

8

$$c_0 = \frac{a}{T}$$

Because

$$c_0 = \frac{1}{T} \int_{-T/2}^{T/2} f(t) \, dt$$

$$= \frac{1}{T} \int_0^a dt = \frac{a}{T}$$

Therefore

$$f(t) = \frac{a}{T} + \sum_{\substack{n=-\infty \\ n \neq 0}}^{\infty} e^{-jn\pi a/T} \frac{a}{T} \left(\frac{\sin n\pi a/T}{n\pi a/T} \right) e^{jn\omega_0 t}$$

Next frame

9

Before we move on, consider the expression $\dfrac{\sin n\pi a/T}{n\pi a/T}$ that occurs in both of these examples. This is an example of a commonly occurring expression $\dfrac{\sin x}{x}$ which has the special name $\operatorname{sinc}(x)$. Notice that $\operatorname{sinc}(0)$ is not defined. However, because $\underset{x\to 0}{Lim}\operatorname{sinc}(x) = \underset{x\to 0}{Lim}\dfrac{\sin x}{x} = 1$ we define $\operatorname{sinc}(0) = 1$.

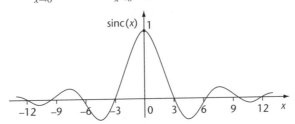

This means that c_0 can be incorporated into the summations so the solutions to Examples 1 and 2 become

$$f(t) = \sum_{n=-\infty}^{\infty} (a/T)\operatorname{sinc}(n\pi a/T)e^{jn\omega_0 t}$$

$$f(t) = \sum_{n=-\infty}^{\infty} (a/T)e^{-jn\pi a/T}\operatorname{sinc}(n\pi a/T)e^{jn\omega_0 t} \quad \text{respectively.}$$

Now let's compare these two results

Complex spectra

10

The coefficients c_n in Example 1 of Frames 5 and 9 are real numbers, namely

$$c_n = \frac{a}{T}\operatorname{sinc}(n\pi a/T)$$

whereas in Example 2 or Frames 8 and 9 they are complex numbers, namely,

$$c_n = \frac{a}{T}\operatorname{sinc}(n\pi a/T)e^{-jn\pi aT}$$

In general, the c_n are complex numbers and can be written as

$$c_n = |c_n|e^{j\phi_n} \text{ where, in Example 2}$$

$$|c_n| = \frac{a}{T}|\operatorname{sinc}(n\pi a/T)| = \frac{a}{T}\left|\frac{\sin n\pi a/T}{n\pi a/T}\right| \text{ for } n \neq 0 \text{ and } c_0 = \frac{a}{T} \text{ and}$$

$$\phi_n = -n\pi a/T.$$

These complex coefficients constitute a **discrete complex spectrum** where c_n represents the *spectral coefficient* of the nth harmonic. Each spectral coefficient couples an **amplitude spectrum** value $|c_n|$ and a **phase spectrum** value ϕ_n. The amplitude spectrum tells us the magnitude of each of the harmonic components and has, for both examples, the graph shown on the next page. ▶

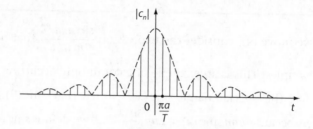

The phase spectrum $\phi_n = -n\pi a/T$ tells us the phase of each harmonic relative to the fundamental harmonic frequency ω_0.

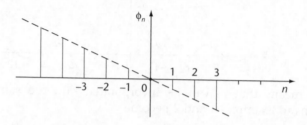

The phase spectrum of the first example is zero for all n and tells us that each harmonic is in phase with the fundamental harmonic. The phase spectrum of the second example, which is a retarded form of the first example, tells us that the nth harmonic is shifted out of phase from the fundamental harmonic by $-n\omega_0 a/2$.

Next frame

The two domains

11   A periodic waveform and its spectrum are described in different terms. The waveform is described in terms of behaviour in time whereas the spectrum is described in terms of behaviour relative to frequency. Thus time and frequency form two domains of definition of our functions and whatever information can be gleaned from within one domain can equally be gleaned from within the other. For example, the *power content* of a periodic function $f(t)$ of period T is defined in the time domain as the mean square value of $f(t)$

$$\frac{1}{T}\int_{-T/2}^{T/2} (f(t))^2 \, dt$$

Within the frequency domain the power content is given as

. . . .

12

$$\boxed{\sum_{n=-\infty}^{\infty} |c_n|^2}$$

Because

$$\frac{1}{T}\int_{-T/2}^{T/2} (f(t))^2 \, dt = \frac{1}{T}\int_{-T/2}^{T/2} \left(\sum_{n=-\infty}^{\infty} c_n e^{jn\omega_0 t}\right) f(t) \, dt$$

$$= \sum_{n=-\infty}^{\infty} c_n \frac{1}{T}\int_{-T/2}^{T/2} f(t) e^{jn\omega_0 t} \, dt$$

$$= \sum_{n=-\infty}^{\infty} c_n \frac{1}{T}\int_{-T/2}^{T/2} f(t) e^{-j(-n)\omega_0 t} \, dt$$

$$= \sum_{n=-\infty}^{\infty} c_n c_{-n} = \sum_{n=-\infty}^{\infty} c_n c_n^*$$

$$= \sum_{n=-\infty}^{\infty} |c_n|^2$$

So the power content can be obtained from either domain.

Next frame

Continuous spectra

13

Of interest in the analysis of periodic functions is the behaviour of the Fourier series as the period increases without limit. Consider Example 1 from Frame 5

$$f(t) = \begin{cases} 0 & -T/2 < t < -a/2 \\ 1 & -a/2 < t < a/2 \\ 0 & a/2 < t < T/2 \end{cases} \qquad \text{where } f(t+T) = f(t)$$

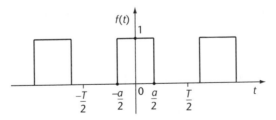

which has the Fourier series

$$f(t) = \sum_{n=-\infty}^{\infty} c_n e^{jn\omega_0 t} \text{ where } \omega_0 = \frac{2\pi}{T} \text{ and where } c_n = \left(\frac{a}{T}\right)\frac{\sin\left(\frac{n\pi a}{T}\right)}{\frac{n\pi a}{T}}$$

As the period increases the separation between the pulses increases and in the limit as $T \to \infty$ only a .... remains and the resulting function is no longer ....

14

only a single pulse remains and the resulting
function is no longer periodic

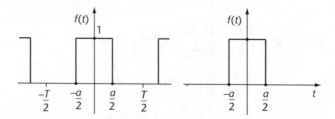

In the Fourier series the distance between neighbouring harmonics in the
the complex spectra is the fundamental frequency $\omega_0 = \dfrac{2\pi}{T}$ and, in the limit as
$T \to \infty$, so $\omega_0 \to 0$. This means that as the period increases the space between
lines in the spectrum decreases so the spectrum lines come closer together and
in the limit merge into a continuous spectrum. That is, for large T

$$n\omega_0 = n\delta\omega \text{ and as } T \to \infty \text{ so } n\delta\omega \to \omega$$

where ω is the continuous frequency variable. To see the effect of this on the
general form of the Fourier series we start with

$$f(t) = \sum_{n=-\infty}^{\infty} c_n e^{jn\omega_0 t} \text{ where } \omega_0 = \frac{2\pi}{T}$$

and where $c_n = \dfrac{1}{T} \displaystyle\int_{-T/2}^{T/2} f(t)e^{-jn\omega_0 t} \, dt$

Substituting the integral form of c_n into the sum gives

$$f(t) = \sum_{n=-\infty}^{\infty} \left[\frac{1}{T} \int_{-T/2}^{T/2} f(u)e^{-jn\omega_0 u} \, du \right] e^{jn\omega_0 t}$$

where u is a dummy variable in place of the variable t.

Now, $\omega_0 = \dfrac{2\pi}{T}$ and so

$$f(t) = \sum_{n=-\infty}^{\infty} \left[\frac{1}{2\pi} \int_{-T/2}^{T/2} f(u)e^{-jn\omega_0 u} \, du \right] \omega_0 e^{jn\omega_0 t}$$

If T is large then $\omega_0 = \delta\omega$ and

$$f(t) = \sum_{n=-\infty}^{\infty} \left[\frac{1}{2\pi} \int_{-T/2}^{T/2} f(u)e^{-jn\delta\omega u} \, du \right] e^{jn\delta\omega t} \delta\omega$$

In the limit as $T \to \infty$ so $n\delta\omega \to \omega$, the sum becomes an integral and $\delta\omega$ becomes the differential $d\omega$ giving

$$f(t) = \int_{\omega=-\infty}^{\infty} \left[\frac{1}{2\pi} \int_{u=-\infty}^{\infty} f(u)e^{-j\omega u} \, du \right] e^{j\omega t} \, d\omega$$

$$= \frac{1}{\sqrt{2\pi}} \int_{\omega=-\infty}^{\infty} \left[\frac{1}{\sqrt{2\pi}} \int_{u=-\infty}^{\infty} f(u)e^{-j\omega u} \, du \right] e^{j\omega t} \, d\omega$$

$$= \frac{1}{\sqrt{2\pi}} \int_{\omega=-\infty}^{\infty} F(\omega)e^{j\omega t} \, d\omega \quad \text{where } F(\omega) = \frac{1}{\sqrt{2\pi}} \int_{u=-\infty}^{\infty} f(u)e^{-j\omega u} \, du$$

These two integrals form the conclusion of *Fourier's integral theorem*.

Next frame

Fourier's integral theorem

Given function $f(t)$ with derivative $f'(t)$ where

15

(a) $f(t)$ and $f'(t)$ are piecewise continuous in every finite interval

(b) $f(t)$ is absolutely integrable in $(\infty, -\infty)$, that is $\int_{-\infty}^{\infty} |f(t)| \, dt$ is finite

then

$$f(t) = \frac{1}{\sqrt{2\pi}} \int_{-\infty}^{\infty} F(\omega)e^{j\omega t} \, d\omega \text{ where } F(\omega) = \frac{1}{\sqrt{2\pi}} \int_{-\infty}^{\infty} f(t)e^{-j\omega t} \, dt$$

The discrete harmonic values $n\omega_0$ of the periodic function are now replaced by the continuous harmonic variable ω and the discrete spectra $c_n = |c_n|e^{j\phi_n}$ are replaced by the *continuous spectra* $F(\omega) = |F(\omega)|e^{j\phi(\omega)}$. $F(\omega)$ is referred to as the **Fourier transform** of $f(t)$ and can also be written as $\mathscr{F}(f(t))$. Deriving the Fourier transform of a function is then a matter of applying the second of these two integrals. The expressions $f(t)$ and $F(\omega)$ form a Fourier transform pair where $f(t)$ can be referred to as the inverse Fourier transform of $F(\omega)$. That is, $f(t) = \mathscr{F}^{-1}[F(\omega)]$.

Next frame

16

Example 3

Find the Fourier transform of

$$f(t) = \begin{cases} 0 & t < -a/2 \\ 1 & -a/2 < t < a/2 \\ 0 & a/2 < t \end{cases}$$

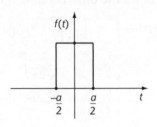

$$F(\omega) = \frac{1}{\sqrt{2\pi}} \int_{-a/2}^{a/2} e^{-j\omega t}\, dt$$

$$= \frac{1}{\sqrt{2\pi}} \left[\frac{e^{-j\omega t}}{-j\omega} \right]_{-a/2}^{a/2}$$

$$= \frac{1}{\sqrt{2\pi}} \left(\frac{e^{-j\omega a/2} - e^{j\omega a/2}}{-j\omega} \right)$$

$$= \sqrt{\frac{2}{\pi}} \left(\frac{e^{-j\omega a/2} - e^{j\omega a/2}}{-2j\omega} \right)$$

$$= \sqrt{\frac{2}{\pi}} \frac{\sin \omega a/2}{\omega}$$

$$= \frac{a}{\sqrt{2\pi}} \frac{\sin \omega a/2}{\omega a/2}$$

$$= \frac{a}{\sqrt{2\pi}} \operatorname{sinc}(\omega a/2)$$

A plot of $F(\omega)$ produces the *continuous amplitude spectrum* of $f(t)$

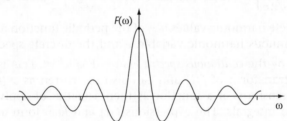

Notice the similarity between the plots of $F(\omega)$ and the discrete spectrum of Frame 10. The lines in the discrete spectrum have merged to form a continuous spectrum while retaining the envelope of the discrete spectrum.

Now you try one

17

Example 4

The function of the previous example time delayed by $t = a/2$ units is

$$f(t) = \begin{cases} 1 & 0 < t < a \\ 0 & \text{otherwise} \end{cases}$$

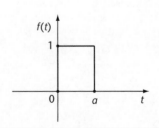

And has the Fourier transform

$$F(\omega) = \ldots\ldots\ldots\ldots$$

18

$$F(\omega) = \frac{ae^{-j\omega a/2}}{\sqrt{2\pi}} \operatorname{sinc}(\omega a/2)$$

Because

$$F(\omega) = \frac{1}{\sqrt{2\pi}} \int_0^a e^{-j\omega t}\, \mathrm{d}t$$

$$= \frac{1}{\sqrt{2\pi}} \left[\frac{e^{-j\omega t}}{-j\omega}\right]_0^a$$

$$= \frac{1}{\sqrt{2\pi}} \left(\frac{e^{-j\omega a} - 1}{-j\omega}\right)$$

$$= \frac{2}{\sqrt{2\pi}} e^{-j\omega a/2} \left(\frac{e^{-j\omega a/2} - e^{j\omega a/2}}{-2j\omega}\right)$$

$$= \frac{2}{\sqrt{2\pi}} e^{-j\omega a/2} \left(\frac{\sin \omega a/2}{\omega}\right)$$

$$= \frac{a}{\sqrt{2\pi}} e^{-j\omega a/2} \left(\frac{\sin \omega a/2}{\omega a/2}\right)$$

$$= \frac{ae^{-j\omega a/2}}{\sqrt{2\pi}} \operatorname{sinc}(\omega a/2)$$

Here $F(\omega)$ is a complex function so we write $F(\omega) = |F(\omega)|e^{j\phi(\omega)}$ where $|F(\omega)| = \left(a/\sqrt{2\pi}\right)\operatorname{sinc}(\omega a/2)|$ is the *continuous amplitude spectrum* and $\phi(\omega) = -a\omega/2$ is the *continuous phase spectrum*.

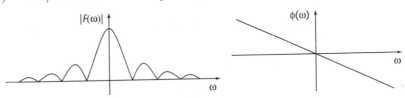

Again, notice the similarity between the plots of $\phi(\omega)$ and the discrete phase spectrum of Frame 10. The lines in the discrete spectrum have merged to form a continuous spectrum while retaining the envelope of the discrete spectrum.

Next frame

Some special functions and their transforms

19 **Even functions**

If $f(t)$ is an even function then

$$f(-t) = f(t) \text{ and } f(t) = \frac{1}{\sqrt{2\pi}} \int_{-\infty}^{\infty} F(\omega)e^{j\omega t} \, d\omega$$

where

$$F(\omega) = \dots\dots\dots \int_{0}^{\infty} f(t) \dots\dots\dots \, dt$$

20

$$\boxed{F(\omega) = \sqrt{\frac{2}{\pi}} \int_{0}^{\infty} f(t) \cos \omega t \, dt}$$

Because

$$F(\omega) = \frac{1}{\sqrt{2\pi}} \int_{-\infty}^{\infty} f(t)e^{-j\omega t} \, dt$$

$$= \frac{1}{\sqrt{2\pi}} \int_{-\infty}^{0} f(t)e^{-j\omega t} \, dt + \frac{1}{\sqrt{2\pi}} \int_{0}^{\infty} f(t)e^{-j\omega t} \, dt$$

$$= -\frac{1}{\sqrt{2\pi}} \int_{0}^{-\infty} f(t)e^{-j\omega t} \, dt + \frac{1}{\sqrt{2\pi}} \int_{0}^{\infty} f(t)e^{-j\omega t} \, dt$$

reversing the limits on the first integral

$$= -\frac{1}{\sqrt{2\pi}} \int_{0}^{\infty} f(-t)e^{j\omega t} \, d(-t) + \frac{1}{\sqrt{2\pi}} \int_{0}^{\infty} f(t)e^{-j\omega t} \, dt$$

changing the variable of integration in the first integral
from t to $-t$

$$= \frac{1}{\sqrt{2\pi}} \int_{0}^{\infty} f(t)\left[e^{j\omega t} + e^{-j\omega t}\right] d(t)$$

$$= \frac{2}{\sqrt{2\pi}} \int_{0}^{\infty} f(t) \cos \omega t \, dt = \sqrt{\frac{2}{\pi}} \int_{0}^{\infty} f(t) \cos \omega t \, dt$$

Notice that if $f(t)$ is even then $F(\omega)$ is real.

Odd functions

If $f(t)$ is an odd function then

$$f(-t) = -f(t) \text{ and } f(t) = \frac{1}{\sqrt{2\pi}} \int_{-\infty}^{\infty} F(\omega)e^{j\omega t} \, d\omega$$

where

$$F(\omega) = \dots\dots\dots \int_{0}^{\infty} f(t) \dots\dots\dots \, dt$$

$$F(\omega) = -j\sqrt{\frac{2}{\pi}} \int_0^\infty f(t) \sin \omega t \, dt$$

Because

$$F(\omega) = \frac{1}{\sqrt{2\pi}} \int_{-\infty}^{\infty} f(t) e^{-j\omega t} \, dt$$

$$= \frac{1}{\sqrt{2\pi}} \int_{-\infty}^{0} f(t) e^{-j\omega t} \, dt + \frac{1}{\sqrt{2\pi}} \int_0^\infty f(t) e^{-j\omega t} \, dt$$

$$= -\frac{1}{\sqrt{2\pi}} \int_0^{-\infty} f(t) e^{-j\omega t} \, dt + \frac{1}{\sqrt{2\pi}} \int_0^\infty f(t) e^{-j\omega t} \, dt$$

reversing the limits on the first integral

$$= -\frac{1}{\sqrt{2\pi}} \int_0^\infty f(-t) e^{j\omega t} \, d(-t) + \frac{1}{\sqrt{2\pi}} \int_0^\infty f(t) e^{-j\omega t} \, dt$$

changing the variable of integration in the first integral
from t to $-t$

$$= \frac{1}{\sqrt{2\pi}} \int_0^\infty f(t) \left[-e^{j\omega t} + e^{-j\omega t} \right] \, dt$$

$$= \frac{-2j}{\sqrt{2\pi}} \int_0^\infty f(t) \sin \omega t \, dt$$

$$= -j\sqrt{\frac{2}{\pi}} \int_0^\infty f(t) \sin \omega t \, dt$$

Notice that if $f(t)$ is odd then $F(\omega)$ is imaginary. An example will show the converse of these two results.

Example

Given that $\mathscr{F}f(t) = F(\omega) = A(\omega) + jB(\omega)$ where $A(\omega)$ and $B(\omega)$ are real functions of ω, then if

(a) $A(\omega) \neq 0$ and $B(\omega) = 0$ then $f(t)$ is an .... function
(b) $A(\omega) = 0$ and $B(\omega) \neq 0$ then $f(t)$ is an .... function

22

> (a) $A(\omega) \neq 0$ and $B(\omega) = 0$ then $f(t)$ is an even function
>
> (b) $A(\omega) = 0$ and $B(\omega) \neq 0$ then $f(t)$ is an odd function

Because

The Fourier transform is given as

$$
\begin{aligned}
F(\omega) &= \frac{1}{\sqrt{2\pi}} \int_{-\infty}^{\infty} f(t) e^{-j\omega t}\, dt \\
&= \frac{1}{\sqrt{2\pi}} \int_{-\infty}^{\infty} f(t)[\cos\omega t - j\sin\omega t]\, dt \\
&= \frac{1}{\sqrt{2\pi}} \int_{-\infty}^{\infty} f(t)\cos\omega t\, dt - j\frac{1}{\sqrt{2\pi}} \int_{-\infty}^{\infty} f(t)\sin\omega t\, dt \\
&= A(\omega) + jB(\omega)
\end{aligned}
$$

(a) If $\displaystyle\int_{-\infty}^{\infty} f(t)\sin\omega t\, dt = 0$ then $f(t)\sin\omega t$ is odd. But $\sin\omega t$ is odd, so $f(t)$ must be even.

(b) If $\displaystyle\int_{-\infty}^{\infty} f(t)\cos\omega t\, dt = 0$ then $f(t)\cos\omega t$ is odd. But $\cos\omega t$ is even, so $f(t)$ must be odd.

Top-hat function

This function is a special form of the function met in Example 3 in Frame 16, and is defined by

$$
f(t) = \begin{cases}
0 & t < -a/2 \\
1/a & -a/2 < t < a/2 \\
0 & a/2 < t
\end{cases}
$$

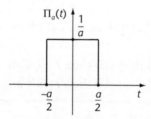

It is, because of its shape, referred to as the *top-hat* function and is denoted by the symbol $\Pi_a(t)$. It is a special form of the function in Example 3 because it has a unit area – width $\times$ height $= a \times (1/a) = 1$, or

$$
\int_{-\infty}^{\infty} \Pi_a(t)\, dt = \int_{-a/2}^{a/2} (1/a)\, dt = \left[\frac{t}{a}\right]_{-a/2}^{a/2} = 1
$$

The Fourier transform of the top-hat function is

$$
F(\omega) = \ldots\ldots\ldots\ldots
$$

23

$$F(\omega) = \frac{1}{\sqrt{2\pi}} \operatorname{sinc}(\omega a/2)$$

Because

$$
\begin{aligned}
F(\omega) &= \frac{1}{\sqrt{2\pi}} \int_{-\infty}^{\infty} \Pi_a(t) e^{-j\omega t}\, dt \\
&= \frac{1}{\sqrt{2\pi}} \int_{-a/2}^{a/2} (1/a) e^{-j\omega t}\, dt \\
&= \frac{1}{a\sqrt{2\pi}} \int_{-a/2}^{a/2} e^{-j\omega t}\, dt \\
&= \frac{1}{\sqrt{2\pi}} \operatorname{sinc}(\omega a/2)
\end{aligned}
$$

This function is useful in that it can be used to select any segment of any function, so acting as a filter. For example

$$\pi \Pi_\pi(t) \sin t$$

selects the segment of $\sin t$ between $\pm\pi/2$ and reduces the rest to zero.

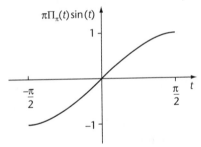

So $\pi \Pi_\pi(t - \pi) \cos t$ selects the segment of $\cos t$ between

.... and ....

24

$$\boxed{\pi/2 \text{ and } 3\pi/2}$$

Because

$$\Pi_\pi(t - \pi) = \begin{cases} 0 & t - \pi < -\pi/2 \\ 1/\pi & -\pi/2 < t - \pi < \pi/2 \\ 0 & \pi/2 < t - \pi \end{cases}$$

that is

$$\Pi_\pi(t - \pi) = \begin{cases} 0 & t < \pi/2 \\ 1/\pi & \pi/2 < t < 3\pi/2 \\ 0 & \pi/2 < t \end{cases}$$

and so

$$\pi\Pi_\pi(t - \pi)\cos t = \begin{cases} \cos t & \pi/2 < t < 3\pi/2 \\ 0 & \text{otherwise} \end{cases}$$

selects the segment of $\cos t$ between $\pi/2$ and $3\pi/2$.

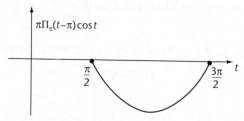

The Dirac delta (refer to Programme 4, Frames 29*ff*)

In science and technology we often need to use the notion of a force that acts for a very brief interval of time. To simulate this mathematically we can use the unit-area pulse – the top-hat function. If we take the duration of this pulse to decrease while at the same time retaining a unit-area then in the limit we are led to the notion of the Dirac delta. That is

$$\underset{a\to0}{Lim} \int_{-\infty}^{\infty} \{\Pi_a(t)\}\, \mathrm{d}t = \underset{a\to0}{Lim}\, 1 = 1$$

Here as $a \to 0$ the width of the top-hat decreases as the height increases but all the while retaining the area beneath the top-hat as unity. It is this limit that we can use to justify the integral definition of the Dirac delta because

$$\underset{a\to0}{Lim} \int_{-\infty}^{\infty} \{\Pi_a(t)\}\, \mathrm{d}t = \int_{-\infty}^{\infty} \underset{a\to0}{Lim}\, \{\Pi_a(t)\}\, \mathrm{d}t = \int_{-\infty}^{\infty} \delta(t)\, \mathrm{d}t = 1$$

and it is also in this sense that we accept the validity of the integral

$$\int_{-\infty}^{\infty} f(t)\delta(t - t_0)\, \mathrm{d}t = f(t_0)$$

because, like the top-hat function, it selects only that part of $f(t)$ over which it is non-zero, namely at $t = t_0$.

So if $f(t) = \delta(t)$ then $F(\omega) = \ldots\ldots\ldots\ldots$

$$\boxed{\dfrac{1}{\sqrt{2\pi}}} \qquad \textbf{25}$$

Because

$$F(\omega) = \frac{1}{\sqrt{2\pi}} \int_{-\infty}^{\infty} \delta(t)e^{-j\omega t}\, \mathrm{d}t$$

$$= \frac{e^{-j\omega 0}}{\sqrt{2\pi}} \qquad \text{because } \delta(t) = \delta(t-0)$$

$$= \frac{1}{\sqrt{2\pi}}$$

Try another.

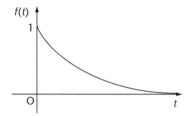

The truncated exponential function

$$f(t) = \begin{cases} e^{-at} & t > 0 \\ 0 & t < 0 \end{cases}$$

where $a > 0$ can be also expressed in the form $f(t) = e^{-at}u(t)$ and has the Fourier transform

$$F(\omega) = \ldots\ldots\ldots$$

$$\boxed{F(\omega) = \dfrac{1}{\sqrt{2\pi}(a + j\omega)}} \qquad \textbf{26}$$

Because

$$F(\omega) = \frac{1}{\sqrt{2\pi}} \int_{-\infty}^{\infty} e^{-at}u(t)e^{-j\omega t}\, \mathrm{d}t$$

$$= \frac{1}{\sqrt{2\pi}} \int_{0}^{\infty} e^{-(a+j\omega)t}\, \mathrm{d}t$$

$$= \frac{1}{\sqrt{2\pi}(a + j\omega)}$$

▶

The triangle function

$$\Lambda_a(t) = \begin{cases} (a+t)/a^2 & -a < t < 0 \\ (a-t)/a^2 & 0 < t < a \\ 0 & |t| > a \end{cases} \qquad \text{Notice that this also has unit area}$$

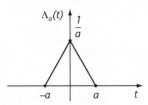

The Fourier transform of $\Lambda_1(t)$

is $F(\omega) = \ldots\ldots\ldots$

27

$$\boxed{F(\omega) = \frac{1}{\sqrt{2\pi}}\operatorname{sinc}^2(\omega/2)}$$

Because

$$F(\omega) = \frac{1}{\sqrt{2\pi}} \int_{-\infty}^{\infty} \Lambda_1(t)e^{-j\omega t}\,\mathrm{d}t$$

$$= \frac{1}{\sqrt{2\pi}} \int_{-1}^{0} (1+t)e^{-j\omega t}\,\mathrm{d}t + \frac{1}{\sqrt{2\pi}} \int_{0}^{1} (1-t)e^{-j\omega t}\,\mathrm{d}t$$

$$= -\frac{1}{\sqrt{2\pi}} \int_{1}^{0} (1-t)e^{j\omega t}\,\mathrm{d}t + \frac{1}{\sqrt{2\pi}} \int_{0}^{1} (1-t)e^{-j\omega t}\,\mathrm{d}t$$

changing the variable of integration in the first integral from t to $-t$

$$= \frac{2}{\sqrt{2\pi}} \int_{0}^{1} (1-t)\cos\omega t\,\mathrm{d}t \qquad \text{and integration by parts yields}$$

$$= \frac{2}{\sqrt{2\pi}} \left(\frac{1}{2}\frac{\sin^2(\omega/2)}{(\omega/2)^2}\right)$$

$$= \frac{1}{\sqrt{2\pi}}\operatorname{sinc}^2(\omega/2)$$

Alternative forms

28

It should be noted that there are a number of alternative forms for the Fourier transform – each dealing with a different location for the constant 2π. Other forms are

$$f(t) = \int_{-\infty}^{\infty} F(\omega)e^{j\omega t}\,\mathrm{d}\omega \text{ where } F(\omega) = \frac{1}{2\pi}\int_{-\infty}^{\infty} f(t)e^{-j\omega t}\,\mathrm{d}t$$

or

$$f(t) = \frac{1}{2\pi}\int_{-\infty}^{\infty} F(\omega)e^{j\omega t}\,\mathrm{d}\omega \text{ where } F(\omega) = \int_{-\infty}^{\infty} f(t)e^{-j\omega t}\,\mathrm{d}t$$

▶

or, by absorbing the 2π in the exponential by defining $\omega = 2\pi\nu$

$$f(t) = \int_{-\infty}^{\infty} F(\nu)e^{j2\pi\nu t}\,d\nu \text{ where } F(\nu) = \int_{-\infty}^{\infty} f(t)e^{-j2\pi\nu t}\,dt$$

We shall remain with our original form because it has the simplest exponential factor and we do not need to remember which integral has the constant in front of it and which does not.

Next frame

Properties of the Fourier transform

We now list a number of properties of the Fourier transform that are useful in their manipulation.

29

Linearity

If the Fourier transforms $\mathscr{F}(f_1(t)) = F_1(\omega)$ and $\mathscr{F}(f_2(t)) = F_2(\omega)$ then

$$\mathscr{F}(\alpha_1 f_1(t) + \alpha_2 f_2(t)) = \alpha_1\mathscr{F}(f_1(t)) + \alpha_2\mathscr{F}(f_2(t)) = \alpha_1 F_1(\omega) + \alpha_2 F_2(\omega)$$

where α_1 and α_2 are constants.

Example

The Fourier transform of $f(t) = 2\Pi_2(t) - 6\Lambda_2(t)$ is

$$F(\omega) = \ldots\ldots\ldots$$

30

$$\boxed{\sqrt{\frac{2}{\pi}}\text{sinc}\,(\omega)(1 - 3\text{sinc}\,(\omega))}$$

Because

If $f(t) = 2\Pi_2(t)$ then $F(\omega) = \dfrac{1}{\sqrt{2\pi}}\text{sinc}\,(\omega)$ and if $f(t) = \Lambda_2(t)$ then

$F(\omega) = \dfrac{1}{\sqrt{2\pi}}\text{sinc}^2(\omega)$. Since $f(t) = 2\Pi_2(t) - 6\Lambda_2(t)$ then

$$F(\omega) = \frac{2}{\sqrt{2\pi}}\text{sinc}\,(\omega) - \frac{6}{\sqrt{2\pi}}\text{sinc}^2(\omega)$$

$$= \sqrt{\frac{2}{\pi}}\text{sinc}\,(\omega)(1 - 3\text{sinc}\,(\omega))$$

▶

Time shifting

If $\mathscr{F}(f(t)) = F(\omega)$ then $\mathscr{F}(f(t - t_0)) = e^{j\omega t_0} F(\omega)$

Example

The Fourier transform of $\Pi_2(t)$ is $\dfrac{1}{\sqrt{2\pi}}\operatorname{sinc}(\omega)$ so, by the time shifting property, the Fourier transform of

$$\Pi_2(t - 5) \text{ is } \ldots\ldots\ldots\ldots \quad \text{and of } \Pi_2(t + 3) \text{ is } \ldots\ldots\ldots\ldots$$

31

$$\frac{e^{j5\omega}}{\sqrt{2\pi}}\operatorname{sinc}(\omega) \quad \text{and} \quad \frac{e^{-j3\omega}}{\sqrt{2\pi}}\operatorname{sinc}(\omega)$$

Frequency shifting

If $\mathscr{F}(f(t)) = F(\omega)$ then $\mathscr{F}\left(f(t)e^{j\omega_0 t}\right) = F(\omega - \omega_0)$

Example

If the Fourier transform of $f(t)$ is $F(\omega)$ then the transform of $f(t)\cos 4t$ is

$$\ldots\ldots\ldots\ldots$$

32

$$\frac{1}{2}(F(\omega + 4) + F(\omega - 4))$$

Because

$$f(t)\cos 4t = f(t)\frac{e^{j4t} + e^{-j4t}}{2}$$

$$= \frac{1}{2}f(t)e^{j4t} + \frac{1}{2}f(t)e^{-j4t}$$

$$= \frac{1}{2}\left(f(t)e^{j4t} + f(t)e^{-j4t}\right)$$

and so the Fourier transform is $\dfrac{1}{2}(F(\omega - 4) + F(\omega + 4))$ by the linearity and the frequency shifting properties.

Time scaling

If $\mathscr{F}(f(t)) = F(\omega)$ then

$$\mathscr{F}(f(kt)) = \frac{1}{|k|}F\left(\frac{\omega}{k}\right)$$

So, for example, given $f(t) = \Pi_a(t)$ with Fourier transform $F(\omega)$, if $f(t)$ is shrunk to half its width then $F(\omega)$ is stretched to twice its width but shrunk to half its height.

▶

Example

If $F(\omega)$ is the Fourier transform of $f(t)$ then the Fourier transform of $f(-t)$ is

. . . .

$$\boxed{F(-\omega)}$$

33

Because

$$|k|^{-1}F(\omega/k) = \frac{1}{\sqrt{2\pi}}\int_{-\infty}^{\infty} f(kt)e^{-j\omega t}\,\mathrm{d}t \text{ and when } k = -1 \text{ then}$$

$$|-1|^{-1}F(\omega/[-1]) = \frac{1}{\sqrt{2\pi}}\int_{-\infty}^{\infty} f(-t)e^{-j\omega t}\,\mathrm{d}t = F(-\omega)$$

Symmetry

If $\mathscr{F}(f(t)) = F(\omega)$ then $\mathscr{F}(F(t)) = f(-\omega)$

Example

The Fourier transform of $f(t) = \Pi_2(t)$ is $F(\omega) = \dfrac{1}{\sqrt{2\pi}}\operatorname{sinc}(\omega)$, so the Fourier transform of

$$F(t) = \frac{1}{\sqrt{2\pi}}\operatorname{sinc}(t) \text{ is } \ldots\ldots\ldots\ldots$$

$$\boxed{f(-\omega) = -\Pi_2(\omega)}$$

34

Because

The Fourier transform of $F(t) = \dfrac{1}{\sqrt{2\pi}}\operatorname{sinc}(t)$

is $f(-\omega) = \Pi_2(-\omega) = -\Pi_2(\omega)$

Try one yourself.

Example

The Fourier transform of the unit constant function $f(t) = 1$ is

$$\mathscr{F}[1] = \ldots\ldots\ldots\ldots$$

$$\boxed{\sqrt{2\pi}\delta(\omega)}$$

35

Because

$$\mathscr{F}[\delta(t)] = \frac{1}{\sqrt{2\pi}} \text{ so } \mathscr{F}\left[\frac{1}{\sqrt{2\pi}}\right] = \delta(\omega), \text{ therefore } \mathscr{F}[1] = \sqrt{2\pi}\delta(\omega)$$

▶

Differentiation

If $f(t) \to 0$ as $t \to \pm\infty$ and if $\mathscr{F}(f(t)) = F(\omega)$ then

$$\mathscr{F}(f'(t)) = \ldots\ldots\ldots\ldots$$

36

$$\boxed{j\omega F(\omega)}$$

Because

$$\mathscr{F}[f'(t)] = \frac{1}{\sqrt{2\pi}} \int_{-\infty}^{\infty} f'(t)e^{-j\omega t}\, dt$$

$$= \frac{1}{\sqrt{2\pi}} \left[f(t)e^{-j\omega t} \right]_{-\infty}^{\infty} + \frac{j\omega}{\sqrt{2\pi}} \int_{-\infty}^{\infty} f(t)e^{-j\omega t}\, dt$$

$$= 0 + j\omega F(\omega)$$

In general, if $f(t) \to 0$ as $t \to \pm\infty$ and if $\mathscr{F}(f(t)) = F(\omega)$ then

If $\mathscr{F}(f(t)) = F(\omega)$ then $\mathscr{F}\left(f^{(n)}(t) \right) = (j\omega)^n F(\omega)$

where the superscript (n) indicates the nth derivative.

Example

The differential equation for unforced and undamped harmonic motion is of the form $mf''(t) + kf(t) = 0$. If we take the Fourier transform of this equation we immediately find that the permitted frequencies of oscillation are

$$\omega = \ldots\ldots\ldots\ldots$$

37

$$\boxed{\omega = \pm\sqrt{\frac{k}{m}}}$$

Because

If $F(\omega)$ is the Fourier transform of $f(t)$ then taking the Fourier transform of both sides of the equation $mf''(t) + kf(t) = 0$ gives by the differentiation property

$$m(j\omega)^2 F(\omega) + kF(\omega) = (-m\omega^2 + k)F(\omega) = 0$$

so if $F(\omega) \neq 0$ then $m\omega^2 = k$ and so the permitted frequencies are

$$\omega = \pm\sqrt{\frac{k}{m}}.$$

Next frame

The Heaviside unit step function

38

The Heaviside unit step function is defined as $u(t)$ where

$$u(t) = \begin{cases} 0 & t < 0 \\ 1 & t > 0 \end{cases}$$

If we follow the definition of the Fourier transform we find that

$$F(\omega) = \frac{1}{\sqrt{2\pi}} \int_{-\infty}^{\infty} u(t) e^{-j\omega t} \, dt$$

So that $F(\omega) = \ldots\ldots\ldots$

39

$$\boxed{F(\omega) = \frac{1}{\sqrt{2\pi} j\omega} - \left\{ 1 - \underset{t \to \infty}{Lim} \left[e^{-j\omega t} \right] \right\}}$$

Because

$$F(\omega) = \frac{1}{\sqrt{2\pi}} \int_{-\infty}^{\infty} u(t) e^{-j\omega t} \, dt$$

$$= \frac{1}{\sqrt{2\pi}} \int_{0}^{\infty} e^{-j\omega t} \, dt$$

$$= \frac{1}{\sqrt{2\pi} j\omega} - \left\{ 1 - \underset{t \to \infty}{Lim} \left[e^{-j\omega t} \right] \right\}$$

Because $e^{-j\omega t} = \cos \omega t - j \sin \omega t$ we cannot say what happens to the exponential as $t \to \infty$. So how do we resolve the problem?

Next frame

40

Let $\mathscr{F} u(t) = F(\omega)$ and so, by the scaling property, $\mathscr{F} u(-t) = F(-\omega)$. Now, $u(t) + u(-t) = 1$, therefore $\mathscr{F}[u(t)] + \mathscr{F} u[(-t)] = \mathscr{F}[1]$. That is, from Frame 35

$$F(\omega) + F(-\omega) = \sqrt{2\pi} \delta(\omega)$$

We now assume that $F(\omega)$ consists of a combination of the Dirac delta and an arbitrary function $G(\omega)$

$$F(\omega) = \alpha \delta(\omega) + G(\omega) \text{ so that}$$

$$F(\omega) + F(-\omega) = \alpha \delta(\omega) + G(\omega) + \alpha \delta(-\omega) + G(-\omega)$$

$$= 2\alpha \delta(\omega) + G(\omega) + G(-\omega) \qquad \text{since } \delta(-\omega) = \delta(\omega)$$

$$= \sqrt{2\pi} \delta(\omega)$$

Therefore $\alpha = \sqrt{\dfrac{\pi}{2}}$ and $G(\omega) + G(-\omega) = 0$. That is, $G(\omega) = -G(-\omega)$.

Consequently $\mathscr{F}[u(t)] = F(\omega) = \sqrt{\dfrac{\pi}{2}} \delta(\omega) + G(\omega)$.

▶

Now, $\mathscr{F}[u'(t)] = j\omega F(\omega) = j\omega\left\{\sqrt{\frac{\pi}{2}}\delta(\omega) + G(\omega)\right\}$ and since $u'(t) = \delta(t)$ then

$$\mathscr{F}[u'(t)] = \mathscr{F}[\delta(t)] = \frac{1}{\sqrt{2\pi}} \text{ giving } j\omega\left\{\sqrt{\frac{\pi}{2}}\delta(\omega) + G(\omega)\right\} = \frac{1}{\sqrt{2\pi}}$$

Since $\omega\delta(\omega) = 0$, then $j\omega G(\omega) = \frac{1}{\sqrt{2\pi}}$ and so $G(\omega) = \frac{1}{j\omega\sqrt{2\pi}}$ thereby giving

$$\mathscr{F}[u(t)] = \frac{1}{\sqrt{2\pi}}\left\{\pi\delta(\omega) + \frac{1}{j\omega}\right\}$$

The next property deals with the Fourier transform of a **product of functions** but before we go any further we need to recap what is meant by the **convolution of two functions**.

Next frame

Convolution

41

You will recall that from Programme 3, Frame 43 onwards we defined the convolution of two functions $f(t)$ and $g(t)$ as

$$f(t) * g(t) = \int_{-\infty}^{\infty} f(x)g(t-x)\,dx = h(t)$$

where $*$ denotes the operation of convolution. As a refresher consider the convolution $f(t) * g(t)$ where

$$f(t) = u(t) \text{ and } g(t) = \begin{cases} \sec^2 t & |t| < \pi/4 \\ 0 & \text{otherwise} \end{cases}$$

where $u(t)$ is the Heaviside function

$$\text{then } h(t) = f(t) * g(t) = \dots\dots\dots$$

42

$$\boxed{\frac{1 + \tan^2 t}{1 + \tan t}}$$

Because

$$\int_{-\infty}^{\infty} f(x)g(t-x)\,dx = \int_{-\infty}^{\infty} u(x)g(t-x)\,dx$$

$$= \int_{0}^{\pi/4} \sec^2(t-x)\,dx \quad \begin{array}{l}\text{because } u(t) = 0 \text{ for } t < 0 \\ \text{and } g(t) = 0 \text{ for } t > \pi/4\end{array}$$

$$= \left[-\tan(t-x)\right]_0^{\pi/4}$$

$$= \{-\tan(t - \pi/4) + \tan t\}$$

$$= -\frac{\tan t - 1}{1 + \tan t} + \tan t = \frac{1 + \tan^2 t}{1 + \tan t}$$

Next frame

The convolution theorem

If $F(\omega)$ and $G(\omega)$ are the Fourier transforms of $f(t)$ and $g(t)$ respectively then **43**

(a) The Fourier transform of the convolution of $f(t)$ and $g(t)$ is equal to the product of the individual Fourier transforms. That is

$$\mathscr{F}[f(t) * g(t)] = \sqrt{2\pi}F(\omega)G(\omega) \text{ and so}$$

$$\mathscr{F}^{-1}[F(\omega)G(\omega)] = \frac{1}{\sqrt{2\pi}}[f(t) * g(t)]$$

(b) The Fourier transform of the product $f(t)g(t)$ is equal to the convolution of the individual Fourier transforms. That is

$$\mathscr{F}[f(t)g(t)] = \frac{1}{\sqrt{2\pi}}F(\omega) * G(\omega) \text{ and so}$$

$$\mathscr{F}^{-1}[F(\omega) * G(\omega)] = \sqrt{2\pi}f(t)g(t)$$

These provide useful methods of finding inverse transforms.

Example

To find the inverse transform of

$$F(\omega) = \frac{1}{2\pi(a + j\omega)^2} = \frac{1}{\sqrt{2\pi}(a + j\omega)} \times \frac{1}{\sqrt{2\pi}(a + j\omega)} \text{ where } a > 0$$

we note that if $F_1(\omega) = \dfrac{1}{\sqrt{2\pi}(a + j\omega)}$ then from Frame 26

$$f_1(t) = \mathscr{F}^{-1}[F_1(\omega)] = \ldots\ldots\ldots\ldots$$

$$\boxed{f_1(t) = e^{-at}u(t)}$$

44

Now, because

$$F(\omega) = F_1(\omega)F_1(\omega)$$

then

$$f(t) = \mathscr{F}^{-1}[F(\omega)] = \mathscr{F}^{-1}[F_1(\omega)F_1(\omega)] = \frac{1}{\sqrt{2\pi}}[f_1(t) * f_1(t)]$$

$$= \frac{1}{\sqrt{2\pi}}\int_{-\infty}^{\infty} f_1(x)f_1(t - x)\,dx$$

$$= \frac{1}{\sqrt{2\pi}}\int_{-\infty}^{\infty} e^{-ax}u(x)e^{-a(t-x)}u(t - x)\,dx$$

$$= \frac{e^{-at}}{\sqrt{2\pi}}\int_{-\infty}^{\infty} e^{-ax}u(x)e^{ax}u(t - x)\,dx$$

$$= \frac{e^{-at}}{\sqrt{2\pi}}\int_{-\infty}^{\infty} u(x)u(t - x)\,dx$$

▶

Now, $u(x)u(t-x) = 0$ when $x < 0$ or when $t - x < 0$, that is when $x > t$.

Therefore, $u(x)u(t-x) = \begin{cases} 1 & \text{if } 0 < x < t \\ 0 & \text{otherwise} \end{cases}$ so

$f(t) = \dfrac{e^{-at}}{\sqrt{2\pi}} \displaystyle\int_0^t dx$

$= \begin{cases} \dfrac{te^{-at}}{\sqrt{2\pi}} & \text{if } t > 0 \\ 0 & \text{if } t < 0 \end{cases}$ that is, $f(t) = \dfrac{te^{-at}}{\sqrt{2\pi}} u(t)$

Now you try one.

The inverse Fourier transform of $F(\omega) = \dfrac{5}{6 + 5j\omega - \omega^2}$ is

$$f(t) = \dots\dots\dots$$

45

$$\boxed{f(t) = \sqrt{50\pi}\left[e^{-2t} - e^{-3t}\right]u(t)}$$

Because

$$F(\omega) = \frac{5}{6 + 5j\omega - \omega^2}$$

$$= \frac{5}{(2 + j\omega)(3 + j\omega)}$$

Let $F_1(\omega) = \dfrac{1}{\sqrt{2\pi}(2 + j\omega)}$ so that $f_1(t) = e^{-2t}u(t)$ and

$F_2(\omega) = \dfrac{1}{\sqrt{2\pi}(3 + j\omega)}$ so that $f_2(t) = e^{-3t}u(t)$ so that

$$F(\omega) = 10\pi[F_1(\omega)F_2(\omega)]$$

By the convolution theorem

$$f(t) = \frac{10\pi}{\sqrt{2\pi}}[f_1(t) * f_2(t)]$$

$$= \sqrt{50\pi} \int_{-\infty}^{\infty} f_1(x)f_2(t-x)\, dx$$

$$= \sqrt{50\pi} \int_{-\infty}^{\infty} e^{-2x}u(x)e^{-3(t-x)}u(t-x)\, dx$$

$$= \sqrt{50\pi}e^{-3t} \int_{-\infty}^{\infty} e^{x}u(x)u(t-x)\, dx$$

$$= \sqrt{50\pi}e^{-3t} \int_0^t e^{x}\, dx \text{ since } u(x)u(t-x) = \begin{cases} 1 & \text{if } 0 < x < t \\ 0 & \text{otherwise} \end{cases}$$

$$= \sqrt{50\pi}e^{-3t}\left[e^{t} - 1\right]u(t) \text{ since } \int_0^t e^{x}\, dx = \begin{cases} e^{t} - 1 & \text{if } t > 0 \\ 0 & \text{if } t < 0 \end{cases}$$

$$= \sqrt{50\pi}\left[e^{-2t} - e^{-3t}\right]u(t)$$

Move to the next frame

Fourier cosine and sine transforms

46

Given that

$$f(t) = \frac{1}{\sqrt{2\pi}} \int_{-\infty}^{\infty} F(\omega) e^{j\omega t} \, d\omega \quad \text{where}$$

$$F(\omega) = \frac{1}{\sqrt{2\pi}} \int_{-\infty}^{\infty} f(t) e^{-j\omega t} \, dt$$

$$= \frac{1}{\sqrt{2\pi}} \int_{-\infty}^{\infty} f(t)(\cos \omega t + j \sin \omega t) \, dt$$

if $f(t)$ is an even function so that $f(-t) = f(t)$ then

$$F(\omega) = \frac{1}{\sqrt{2\pi}} \int_{-\infty}^{\infty} f(t)(\cos \omega t + j \sin \omega t) \, dt$$

$$= \frac{1}{\sqrt{2\pi}} \int_{-\infty}^{\infty} f(t) \cos \omega t \, dt \quad \text{since } f(t) \sin \omega t \text{ is odd}$$

$$= \sqrt{\frac{2}{\pi}} \int_{0}^{\infty} f(t) \cos \omega t \, dt$$

This is referred to as the **Fourier cosine transformation** and is denoted by $F_c(\omega)$. That is

$$F_c(\omega) = \sqrt{\frac{2}{\pi}} \int_{0}^{\infty} f(t) \cos \omega t \, dt$$

Similarly if $f(t)$ is an odd function so that $f(-t) = -f(t)$ then

$$F(\omega) = \frac{1}{\sqrt{2\pi}} \int_{-\infty}^{\infty} f(t)(\cos \omega t + j \sin \omega t) \, dt$$

$$= \frac{j}{\sqrt{2\pi}} \int_{-\infty}^{\infty} f(t) \sin \omega t \, dt \quad \text{since } f(t) \cos \omega t \text{ is odd}$$

$$= j\sqrt{\frac{2}{\pi}} \int_{0}^{\infty} f(t) \sin \omega t \, dt$$

This gives rise to the **Fourier sine transformation**, denoted by $F_s(\omega)$ where

$$F_s(\omega) = \sqrt{\frac{2}{\pi}} \int_{0}^{\infty} f(t) \sin \omega t \, dt$$

Example 1

The Fourier cosine transformation of $f(t) = \begin{cases} 1 & \text{if } 0 < t < a \\ 0 & \text{if } t \geq a \end{cases}$ is

$$F_c(\omega) = \ldots\ldots\ldots\ldots$$

47

$$F_c(\omega) = \sqrt{\frac{2}{\pi}} a \, \text{sinc}\,(\omega a)$$

Because

$$F_c(\omega) = \sqrt{\frac{2}{\pi}} \int_0^\infty f(t) \cos \omega t \, dt$$

$$= \sqrt{\frac{2}{\pi}} \int_0^a \cos \omega t \, dt$$

$$= \sqrt{\frac{2}{\pi}} \left[\frac{\sin \omega t}{\omega} \right]_0^a$$

$$= \sqrt{\frac{2}{\pi}} \left(\frac{\sin \omega a}{\omega} \right) = \sqrt{\frac{2}{\pi}} a \text{sinc}\,(\omega a)$$

Example 2

The Fourier sine transformation of $f(t) = \begin{cases} 1 & \text{if } 0 < t < a \\ 0 & \text{if } t \geq a \end{cases}$ is

$$F_s(\omega) = \ldots\ldots\ldots\ldots$$

48

$$F_s(\omega) = \sqrt{\frac{2}{\pi}} 2a^2 \omega \, \text{sinc}^2(\omega a)$$

Because

$$F_s(\omega) = \sqrt{\frac{2}{\pi}} \int_0^\infty f(t) \sin \omega t \, dt$$

$$= \sqrt{\frac{2}{\pi}} \int_0^a \sin \omega t \, dt$$

$$= \sqrt{\frac{2}{\pi}} \left[-\frac{\cos \omega t}{\omega} \right]_0^a$$

$$= \sqrt{\frac{2}{\pi}} \left(\frac{1 - \cos \omega a}{\omega} \right) = \sqrt{\frac{2}{\pi}} \left(\frac{2 \sin^2 \omega a}{\omega} \right) = \sqrt{\frac{2}{\pi}} 2a^2 \omega \, \text{sinc}^2(\omega a)$$

The Fourier cosine and sine transforms are useful when $f(t)$ is only defined for $t \geq 0$ and where an extension can be added to $f(t)$ for $t < 0$ that makes the extended $f(t)$ into an even or odd function respectively.

Table of transforms

		49
$f(t) = \begin{cases} 1 & \text{if } -a/2 < t < a/2 \\ 0 & \text{otherwise} \end{cases}$	$F(\omega) = \dfrac{a}{\sqrt{2\pi}}\, \text{sinc}\,(\omega a/2)$	
$f(t) = \begin{cases} 1 & \text{if } 0 < t < a \\ 0 & \text{otherwise} \end{cases}$	$F(\omega) = \dfrac{ae^{-j\omega a/2}}{\sqrt{2\pi}}\, \text{sinc}\,(\omega a/2)$	
$\Pi_a(t) = \begin{cases} 1/a & \text{if } -a/2 < t < a/2 \\ 0 & \text{otherwise} \end{cases}$	$F(\omega) = \dfrac{1}{\sqrt{2\pi}}\, \text{sinc}\,(\omega a/2)$	
$f(t) = u(t)$	$F(\omega) = \dfrac{1}{\sqrt{2\pi}}\left\{ \pi\delta(\omega) + \dfrac{1}{j\omega}\right\}$	
$f(t) = e^{-at}u(t)$	$F(\omega) = \dfrac{1}{\sqrt{2\pi}}\left\{ \pi\delta(\omega + a) + \dfrac{1}{j\omega}\right\}$	
$f(t) = te^{-at}u(t)$	$F(\omega) = \dfrac{1}{\sqrt{2\pi}(a+j\omega)^2}$	
$f(t) = \delta(t)$	$F(\omega) = \dfrac{1}{\sqrt{2\pi}}$	

The main points of the Programme are listed in the **Revision summary** that follows. Read it in conjunction with the **Can you?** checklist and refer back to the relevant parts of the Programme, if necessary. You will then have no trouble with the **Test exercise** and the **Further problems** provide valuable additional practice.

Revision summary 9

1 *Complex Fourier series* **50**
The Fourier series of the piecewise continuous function $f(t)$ with piecewise continuous derivative and where $f(t + T) = f(t)$ is given as

$$f(t) = \sum_{n=-\infty}^{\infty} c_n e^{jn\omega_0 t}$$

where $c_n = \dfrac{1}{T}\int_{-T/2}^{T/2} f(t)e^{-jn\omega_0 t}\, dt.$

2 *Discrete complex spectra*
The c_n are complex numbers and can be written as

$$c_n = |c_n|e^{j\phi_n}$$

These complex coefficients constitute a discrete complex spectrum where c_n represents the *spectral coefficient* of the nth harmonic. Each spectral coefficient couples an amplitude spectrum value $|c_n|$ and a phase spectrum value ϕ_n.

3 *Fourier's integral theorem*

If (a) $f(t)$ and $f'(t)$ are piecewise continuous in every finite interval

(b) $f(t)$ is absolutely integrable in $(\infty, -\infty)$, that is $\displaystyle\int_{-\infty}^{\infty} |f(t)|\,dt$ is finite then

$$f(t) = \frac{1}{\sqrt{2\pi}} \int_{-\infty}^{\infty} F(\omega)e^{j\omega t}\,d\omega \text{ where } F(\omega) = \frac{1}{\sqrt{2\pi}} \int_{-\infty}^{\infty} f(t)e^{-j\omega t}\,dt.$$

4 *Continuous complex spectra*

The Fourier transform $F(\omega)$ is a complex function so we write $F(\omega) = |F(\omega)|e^{j\phi(\omega)}$ where $|F(\omega)| = \left|\left(a/\sqrt{2\pi}\right) \operatorname{sinc}(\omega a/2)\right|$ is the *continuous amplitude spectrum* and $\phi(\omega) = -a\omega/2$ is the *continuous phase spectrum*.

5 *Transforms of special functions*

Top-hat function

$$\Pi_a(t) = \begin{cases} 1/a & -a/2 < t < a/2 \\ 0 & \text{otherwise} \end{cases}$$

with Fourier transform $F(\omega) = \dfrac{1}{\sqrt{2\pi}} \operatorname{sinc}(\omega a/2)$.

The Dirac delta

If $f(t) = \delta(t)$ then $F(\omega) = \dfrac{1}{\sqrt{2\pi}}$.

The Heaviside unit step function

$$u(t) = \begin{cases} 0 & t < 0 \\ 1 & t > 0 \end{cases} \text{ has the Fourier transform}$$

$$F(\omega) = \frac{1}{\sqrt{2\pi}} \left\{ \pi\delta(\omega) + \frac{1}{j\omega} \right\}.$$

The triangle function

$$\Lambda(t) = \begin{cases} 0 & |t| > 1 \\ 1 & |t| < 1 \end{cases} \text{ has the Fourier transform}$$

$$F(\omega) = \frac{1}{\sqrt{2\pi}} \operatorname{sinc}^2(\omega/2).$$

6 *Alternative forms*

There are a number of alternative forms for the Fourier transform – each dealing with a different location for the constant 2π. Other forms are

$$f(t) = \int_{-\infty}^{\infty} F(\omega)e^{j\omega t}\,d\omega \text{ where } F(\omega) = \frac{1}{2\pi} \int_{-\infty}^{\infty} f(t)e^{-j\omega t}\,dt \text{ or}$$

$$f(t) = \frac{1}{2\pi} \int_{-\infty}^{\infty} F(\omega)e^{j\omega t}\,d\omega \text{ where } F(\omega) = \int_{-\infty}^{\infty} f(t)e^{-j\omega t}\,dt \text{ or}$$

$$f(t) = \int_{-\infty}^{\infty} F(\omega)e^{j2\pi\omega t}\,d\omega \text{ where } F(\omega) = \int_{-\infty}^{\infty} f(t)e^{-j2\pi\omega t}\,dt.$$

▶

7 *Properties of the Fourier transform*

Time shifting

If $F(\omega) = \dfrac{1}{\sqrt{2\pi}} \displaystyle\int_{-\infty}^{\infty} f(t)e^{-j\omega t}\,dt$ then

$$e^{j\omega t_0}F(\omega) = \dfrac{1}{\sqrt{2\pi}} \int_{-\infty}^{\infty} f(t-t_0)e^{-j\omega t}\,dt.$$

Linearity

If $F_1(\omega)$ and $F_2(\omega)$ are the Fourier transforms of $f_1(t)$ and $f_2(t)$ respectively then $\alpha_1 F_1(\omega) + \alpha_2 F_2(\omega)$ is the Fourier transform of $\alpha_1 f_1(t) + \alpha_2 f_2(t)$ where α_1 and α_2 are constants.

Frequency shifting

If $F(\omega)$ is the Fourier transform of $f(t)$ then the Fourier transform of $f(t)e^{-j\omega_0 t}$ is $F(\omega - \omega_0)$.

Time scaling

If $F(\omega) = \dfrac{1}{\sqrt{2\pi}} \displaystyle\int_{-\infty}^{\infty} f(t)e^{-j\omega t}\,dt$ then

$$|k|^{-1}F(\omega/k) = \dfrac{1}{\sqrt{2\pi}} \int_{-\infty}^{\infty} f(kt)e^{-j\omega t}\,dt.$$

Symmetry

If $F(\omega)$ is the Fourier transform of $f(t)$ then the Fourier transform of $F(t)$ is $f(-\omega)$.

Differentiation

If $F(\omega) = \dfrac{1}{\sqrt{2\pi}} \displaystyle\int_{-\infty}^{\infty} f(t)e^{-j\omega t}\,dt$ then

$$(j\omega)^n F(\omega) = \dfrac{1}{\sqrt{2\pi}} \int_{-\infty}^{\infty} f^{(n)}(t)e^{-j\omega t}\,dt \text{ and}$$

$$F^{(n)}(\omega) = \dfrac{1}{\sqrt{2\pi}} \int_{-\infty}^{\infty} (-j\omega)^n f(t)e^{-j\omega t}\,dt.$$

8 *Convolution*

The convolution of two functions $f(t)$ and $g(t)$ is defined as

$$f(t) * g(t) = \int_{-\infty}^{\infty} f(x)g(t-x)\,dx = h(t).$$

The convolution theorem

If $F(\omega)$ and $G(\omega)$ are the Fourier transforms of $f(t)$ and $g(t)$ respectively then

(a) The Fourier transform of the convolution of $f(t)$ and $g(t)$ is equal to the product of the individual Fourier transforms. That is

$$\mathscr{F}[f(t) * g(t)] = \sqrt{2\pi}F(\omega)G(\omega).$$

(b) The Fourier transform of the product $f(t)g(t)$ is equal to the convolution of the individual Fourier transforms. That is

$$\mathscr{F}[f(t)g(t)] = \dfrac{1}{\sqrt{2\pi}}F(\omega) * G(\omega).$$

▶

9 *Fourier cosine and sine transforms*

Given that $f(t) = \dfrac{1}{\sqrt{2\pi}} \displaystyle\int_{-\infty}^{\infty} F(\omega)e^{j\omega t}\, dt$ where

$$F(\omega) = \frac{1}{\sqrt{2\pi}} \int_{-\infty}^{\infty} f(t)e^{-j\omega t}\, dt$$

where $f(t)$ is even then

$$f(t) = \sqrt{\frac{2}{\pi}} \int_{0}^{\infty} F_c(\omega)\cos \omega t\, d\omega \text{ where } F_c(\omega) = \sqrt{\frac{2}{\pi}} \int_{0}^{\infty} f(t)\cos \omega t\, dt$$

and where $F_c(\omega)$ is called the *Fourier cosine transformation*. This transformation is useful when $f(t)$ is defined only for $t \geq 0$ and where an extension can be added to $f(t)$ for $t < 0$ that makes the extended $f(t)$ into an even function.

If $f(t)$ is odd then

$$f(t) = \sqrt{\frac{2}{\pi}} \int_{0}^{\infty} F_s(\omega)\sin \omega t\, d\omega \text{ where } F_s(\omega) = \sqrt{\frac{2}{\pi}} \int_{0}^{\infty} f(t)\sin \omega t\, dt$$

and where $F_s(\omega)$ is called the *Fourier sine transformation*. This transformation is useful when $f(t)$ is defined only for $t \geq 0$ and where an extension can be added to $f(t)$ for $t < 0$ that makes the extended $f(t)$ into an odd function.

 Can you?

51 Checklist 9

Check this list before and after you try the end of Programme test.

On a scale of 1 to 5 how confident are you that you can: **Frames**

• Convert a trigonometric Fourier series into a doubly infinite sum of complex exponentials?   ☐ 1 ☐ to ☐ 4 ☐

  Yes ☐ ☐ ☐ ☐ ☐ *No*

• Derive the complex Fourier series of a function that satisfies Dirichlet's conditions?   ☐ 5 ☐ to ☐ 8 ☐

  Yes ☐ ☐ ☐ ☐ ☐ *No*

• Recognise the function sinc (t)?   ☐ 9 ☐

  Yes ☐ ☐ ☐ ☐ ☐ *No*

• Separate a discrete complex spectrum into an amplitude spectrum and a phase spectrum?   ☐ 10 ☐ to ☐ 12 ☐

  Yes ☐ ☐ ☐ ☐ ☐ *No*

▶

- State Fourier's integral theorem in terms of complex exponentials?
  Yes ☐ ☐ ☐ ☐ ☐ *No*
  [13] to [15]

- Define and derive the Fourier transform of a function satisfying Dirichlet's conditions?
  Yes ☐ ☐ ☐ ☐ ☐ *No*
  [16] and [17]

- Separate a continuous complex spectrum into an amplitude spectrum and a phase spectrum?
  Yes ☐ ☐ ☐ ☐ ☐ *No*
  [18]

- Recognise the functions $\Pi_a(t)$ and $\Lambda_a(t)$ and derive their Fourier transforms along with those of the Dirac delta and the Heaviside unit step?
  Yes ☐ ☐ ☐ ☐ ☐ *No*
  [19] to [27]

- Recognise alternative forms of the function–transform pair?
  Yes ☐ ☐ ☐ ☐ ☐ *No*
  [28]

- Reproduce a collection of properties of the Fourier transform?
  Yes ☐ ☐ ☐ ☐ ☐ *No*
  [29] to [40]

- Evaluate the convolution of two functions and describe its Fourier transform?
  Yes ☐ ☐ ☐ ☐ ☐ *No*
  [41] to [45]

- Derive the Fourier sine and cosine transformations?
  Yes ☐ ☐ ☐ ☐ ☐ *No*
  [46] to [48]

Test exercise 9

1 Find the complex Fourier series of the sawtooth wave
$$f(t) = t, \quad 0 < t < 1 \quad \text{and where} \quad f(t+1) = f(t).$$

52

2 Find the Fourier transform of
$$f(t) = \begin{cases} e^{-at} & |t| < 1 \\ 0 & \text{otherwise} \end{cases} \qquad a > 0$$

3 Given that the Dirac delta $\delta(t)$ has the Fourier transform $F(\omega) = \dfrac{1}{\sqrt{2\pi}}$, show, by considering the inverse Fourier transform, that
$$\delta(t) = \frac{1}{2\pi} \int_{-\infty}^{\infty} e^{-j\omega t}\, d\omega = \frac{1}{\pi} \int_{0}^{\infty} \cos \omega t\, d\omega.$$

4 If $f(t)$ and $F(\omega)$ form a Fourier transform pair, find the Fourier transform of $f(t) \sin \omega_0 t$ where ω_0 is a constant.

▶

5 Find the inverse transform of $F(\omega) = \dfrac{6}{\omega^2 + 5j\omega - 4}$.

6 Show that

(a) $u(t) * u(t) = t\,u(t)$

(b) $t\,u(t) * e^t u(t) = (e^t - t - 1)u(t)$

where $u(t)$ is the unit step function.

7 Find the Fourier sine and cosine transformations of $f(t) = e^{-kt}$ for $t > 0$ and $k > 0$.

Further problems 9

53 1 By comparing the trigonometric Fourier series of a periodic function with its complex exponential counterpart show that

$$|c_n| = \frac{1}{2}\sqrt{a_n^2 + b_n^2} \text{ and } \phi_n = \arctan\left\{-\frac{b_n}{a_n}\right\} \text{ where } c_n = |c_n|e^{j\phi_n}.$$

2 Prove Parseval's identity for the periodic function with period T

$$\frac{1}{T}\int_{-T/2}^{T/2} \{f(t)\}^2\,dt = \sum_{n=-\infty}^{\infty} |c_n|^2 = \frac{a_0^2}{4} + \frac{1}{2}\sum_{n=-\infty}^{\infty}(a_n^2 + b_n^2)$$

and show that $\displaystyle\sum_{n=1}^{\infty}\frac{1}{n^2} = \frac{\pi^2}{6}$.

3 Draw the graph and find the complex Fourier series of the rectified sine wave
$$f(t) = \sin \pi t, \quad 0 < t < 1 \quad \text{where } f(t+1) = f(t).$$

4 Draw the graph and find the complex Fourier series of the rectified cosine wave
$$f(t) = \cos \pi t, \quad -1/2 < t < 1/2 \quad \text{where } f(t+1) = f(t).$$

5 Draw the graph and find the complex Fourier series of
$$f(t) = e^{\pi t}, \quad 0 < t < 2 \quad \text{where } f(t+2) = f(t).$$

6 Draw the graph and find the complex Fourier series of the sawtooth wave
$$f(t) = -\frac{t}{T} + \frac{1}{2}, \quad 0 < t < T \quad \text{where } f(t+T) = f(t).$$

7 If $f_1(t) = \displaystyle\sum_{n=-\infty}^{\infty} c_n e^{jn\omega_0 t}$ and $f_2(t) = \displaystyle\sum_{n=-\infty}^{\infty} d_n e^{jn\omega_0 t}$ where $\omega_0 = 2\pi/T$, show that the convolution

$$f_1(t) * f_2(t) = \sum_{n=-\infty}^{\infty} c_n d_n e^{jn\omega_0 t}.$$

8 Find the Fourier transform of
$$f(t) = \begin{cases} \cosh t & \text{for } |t| < 1 \\ 0 & \text{for } |t| > 1 \end{cases}$$

▶

9 Find the Fourier transform of
$$f(t) = \begin{cases} \sinh t & \text{for } |t| < 1 \\ 0 & \text{for } |t| > 1. \end{cases}$$

10 Find the Fourier transform of
$$f(t) = \begin{cases} \sin \pi t & \text{for } 0 < t < 1 \\ 0 & \text{otherwise.} \end{cases}$$

11 Find the Fourier transform of
$$f(t) = \begin{cases} \cos \pi t & \text{for } |t| < 1/2 \\ 0 & \text{otherwise.} \end{cases}$$

12 Draw the graph and find the Fourier transform of
$$f(t) = e^{-a|t|}, \quad a > 0.$$

13 Given that
$$f(t) = \begin{cases} 1 & \text{for } -1 < t < 0 \\ -1 & \text{for } 0 < t < 1 \\ 0 & \text{otherwise} \end{cases}$$
(a) Draw the graph of $f(t)$
(b) Express $f(t)$ in terms of the Heaviside unit step function
(c) Find the Fourier transform of $f(t)$.

14 Draw the graph and find the Fourier transform of
$$f(t) = (u(t) - u(t - \pi)) \cos kt.$$

15 Show that if $f(t)$ is real then the corresponding Fourier transform $F(\omega) = |F(\omega)|e^{j\phi(\omega)}$ is such that $|F(\omega)|$ is even and $\phi(\omega)$ is odd.

16 Show that if the Fourier transform of a real function is real then $f(t)$ is even, and if the Fourier transform of a real function is imaginary then $f(t)$ is odd.

17 Defining the squared modulus of the Fourier transform $|F(\omega)|^2 = F(\omega)F^*(\omega)$ where $F^*(\omega)$ is the complex conjugate of $F(\omega)$, prove Parseval's theorem
$$\int_{-\infty}^{\infty} [f(t)]^2 \, dt = \int_{-\infty}^{\infty} |F(\omega)|^2 \, dt.$$

18 Show that the convolution of a top-hat function with itself is the triangle function. That is
$$\Pi_a(t) * \Pi_a(t) = \Lambda_a(t).$$

19 Show that $\text{sinc}(t) * \text{sinc}(t) = \text{sinc}(t)$.

20 Find the Fourier sine and cosine transforms of
$$f(t) = \begin{cases} e^{at} & \text{for } |t| < 1 \\ 0 & \text{otherwise.} \end{cases}$$

21 Find the Fourier sine and cosine transforms of
$$f(t) = \begin{cases} \cosh t & \text{for } |t| < 1 \\ 0 & \text{otherwise.} \end{cases}$$

Power series solutions of ordinary differential equations 1

Learning outcomes

When you have completed this Programme you will be able to:

- Obtain the nth derivative of the exponential, circular and hyperbolic functions
- Apply the Leibnitz theorem to derive the nth derivative of a product of expressions
- Use the Leibnitz–Maclaurin method of obtaining a series solution to a second-order homogeneous differential equation with constant coefficients
- Solve Cauchy–Euler equi-dimensional equations

Prerequisite: Engineering Mathematics (Fifth Edition)
Programmes 13 Series 1, 14 Series 2 and 25 Second-order differential equations

Higher derivatives

1

If $y = \sin x$ $\quad \dfrac{dy}{dx} = \cos x = \sin\left(x + \dfrac{\pi}{2}\right)$

$$\dfrac{d^2y}{dx^2} = -\sin x = \sin(x + \pi) = \sin\left(x + \dfrac{2\pi}{2}\right)$$

$$\dfrac{d^3y}{dx^3} = -\cos x = \sin\left(x + \dfrac{3\pi}{2}\right) \quad \text{etc.}$$

We see a pattern developing. In general $\quad \dfrac{d^n y}{dx^n} = \sin\left(x + \dfrac{n\pi}{2}\right)$. Before we go further, we introduce a shorthand notation for the *n*th derivative of *y* as $y^{(n)} = \dfrac{d^n y}{dx^n}$. Note, however, we still use the 'prime' notation y', y'' and y''' to represent the first, second and third derivatives respectively.

The results above can therefore be written

If $y = \sin x$ $\quad \therefore \ y' = \cos x = \sin\left(x + \dfrac{\pi}{2}\right)$

$$y'' = -\sin x = \sin\left(x + \dfrac{2\pi}{2}\right)$$

$$y''' = -\cos x = \sin\left(x + \dfrac{3\pi}{2}\right)$$

and, in general, $y^{(n)} = \sin\left(x + \dfrac{n\pi}{2}\right)$

It is therefore possible to write down any particular derivative of $\sin x$ without calculating all the previous derivatives. For example

$$\dfrac{d^7 y}{dx^7} = y^{(7)} = \sin\left(x + \dfrac{7\pi}{2}\right) = -\cos x$$

Similarly, starting with $y = \cos x$, we can determine an expression for the *n*th derivative of *y* which is ....

2

$$\boxed{y^{(n)} = \cos\left(x + \dfrac{n\pi}{2}\right)}$$

Because

$y = \cos x$ $\quad \therefore \ y' = -\sin x = \cos\left(x + \dfrac{\pi}{2}\right)$

$$y'' = -\cos x = \cos\left(x + \dfrac{2\pi}{2}\right)$$

$$y''' = \sin x \quad = \cos\left(x + \dfrac{3\pi}{2}\right) \quad \text{etc.}$$

$$\therefore \ y^{(n)} = \cos\left(x + \dfrac{n\pi}{2}\right)$$

Many of the standard functions can be treated in a similar manner.

For example, if $y = e^{ax}$, then $y^{(n)} = \ldots\ldots\ldots$

3

$$y^{(n)} = a^n e^{ax}$$

Because

$$y = e^{ax}, \quad y' = ae^{ax}, \quad y'' = a^2 e^{ax}, \quad y''' = a^3 e^{ax}, \quad \text{etc.}$$

In general, $y^{(n)} = a^n e^{ax}$.

With no great effort, we can now write down expressions for the following

If $y = \sin ax, \quad y^{(n)} = \dots\dots\dots$

If $y = \cos ax, \quad y^{(n)} = \dots\dots\dots$

4

$$y = \sin ax, \quad y^{(n)} = a^n \sin\left(ax + \frac{n\pi}{2}\right)$$

$$y = \cos ax, \quad y^{(n)} = a^n \cos\left(ax + \frac{n\pi}{2}\right)$$

Now one more.

If $y = \ln x, \quad y^{(n)} = \dots\dots\dots$

5

$$y^{(n)} = (-1)^{n-1} \cdot \frac{(n-1)!}{x^n}$$

Because

$$y = \ln x \quad \therefore \quad y' = \frac{1}{x}$$

$$y'' = -\frac{1}{x^2}$$

$$y''' = \frac{2}{x^3}$$

$$y^{(4)} = -\frac{3!}{x^4} \qquad \therefore \quad y^{(n)} = (-1)^{n-1} \cdot \frac{(n-1)!}{x^n}$$

We already know that, if $y = \ln x, \quad \dfrac{dy}{dx} = y' = \dfrac{1}{x} = x^{-1}$.

Therefore, if the result obtained for $y^{(n)}$ is to be valid for $n = 1$, then

$$y' = (-1)^0 \cdot \frac{0!}{x} = \frac{0!}{x}$$

But $y' = x^{-1}$ $\qquad \therefore \quad 0! = \dots\dots\dots$

$$\boxed{0! = 1}$$

6

Now let us consider the derivatives of $\sinh ax$ and $\cosh ax$.

Next frame

If $y = \sinh ax$, $\quad y' = a \cosh ax$

$$y'' = a^2 \sinh ax$$
$$y''' = a^3 \cosh ax \qquad \text{etc.}$$

7

Because $\sinh ax$ is not periodic, we cannot proceed as we did with $\sin ax$. We need to find a general statement for $y^{(n)}$ containing terms in $\sinh ax$ and in $\cosh ax$, such that, when n is even, the term in $\cosh ax$ disappears and, when n is odd, the term in $\sinh ax$ disappears.

This we can do by writing $y^{(n)}$ in the form

$$y^{(n)} = \frac{a^n}{2}\left\{[1 + (-1)^n]\sinh ax + [1 - (-1)^n]\cosh ax\right\}$$

In very much the same way, we can determine the nth derivative of $y = \cosh ax$ as ....

$$\boxed{y^{(n)} = \frac{a^n}{2}\left\{[1 - (-1)^n]\sinh ax + [1 + (-1)^n]\cosh ax\right\}}$$

8

Finally, let us deal with $y = x^a$.

$$y = x^a \quad \therefore \quad y' = ax^{a-1}$$
$$y'' = a(a-1)x^{a-2}$$
$$y''' = a(a-1)(a-2)x^{a-3}$$
$$\cdots\cdots\cdots\cdots$$
$$\therefore \quad y^{(n)} = a(a-1)(a-2)\ldots(a-n+1)x^{a-n}$$
$$\therefore \quad y^{(n)} = \frac{a!}{(a-n)!}x^{a-n} \qquad (a \text{ is a positive integer})$$

So, collecting our results together, we have

$y = x^a$	$y^{(n)} = \dfrac{a!}{(a-n)!}x^{a-n}$
$y = e^{ax}$	$y^{(n)} = a^n e^{ax}$
$y = \sin ax$	$y^{(n)} = a^n \sin\left(ax + \dfrac{n\pi}{2}\right)$
$y = \cos ax$	$y^{(n)} = a^n \cos\left(ax + \dfrac{n\pi}{2}\right)$
$y = \sinh ax$	$y^{(n)} = \dfrac{a^n}{2}\left\{[1 + (-1)^n]\sinh ax + [1 - (-1)^n]\cosh ax\right\}$
$y = \cosh ax$	$y^{(n)} = \dfrac{a^n}{2}\left\{[1 - (-1)^n]\sinh ax + [1 + (-1)^n]\cosh ax\right\}$

Make a note of these, as a set, and then move on to the next frame

9   **Exercise**

Determine the following derivatives

1   $y = \sin 4x$ $\qquad$ $y^{(5)} = \ldots\ldots\ldots$

2   $y = e^{x/2}$ $\qquad$ $y^{(8)} = \ldots\ldots\ldots$

3   $y = \cosh 3x$ $\qquad$ $y^{(12)} = \ldots\ldots\ldots$

4   $y = \cos(x\sqrt{2})$ $\qquad$ $y^{(10)} = \ldots\ldots\ldots$

5   $y = x^8$ $\qquad$ $y^{(6)} = \ldots\ldots\ldots$

6   $y = \sinh 2x$ $\qquad$ $y^{(7)} = \ldots\ldots\ldots$

Finish them all; then check with the next frame

10   Here are the solutions

1   $y^{(5)} = 4^5 \sin\left(4x + \dfrac{5\pi}{2}\right) = 1024 \sin\left(4x + \dfrac{\pi}{2}\right) = 1024 \cos 4x$

2   $y^{(8)} = \left(\dfrac{1}{2}\right)^8 e^{x/2} = \dfrac{1}{256} e^{x/2} = e^{x/2}/256$

3   $y^{(12)} = \dfrac{3^{12}}{2}\{0 \sinh 3x + 2 \cosh 3x\} = 3^{12} \cosh 3x$

4   $y^{(10)} = (\sqrt{2})^{10} \cos\left(x\sqrt{2} + \dfrac{10\pi}{2}\right)$

$\qquad = 32 \cos(x\sqrt{2} + 5\pi) = -32 \cos(x\sqrt{2})$

5   $y^{(6)} = \dfrac{8!}{2!} x^2 = 20\,160\, x^2$

6   $y^{(7)} = \dfrac{2^7}{2}\left\{[1 + (-1)^7]\sinh 2x + [1 - (-1)^7]\cosh 2x\right\}$

$\qquad = 2^7 \cosh 2x$

Leibnitz theorem – *n* th derivative of a product of two functions

If $y = uv$, where u and v are functions of x, then

$$y' = uv' + vu' \quad \text{where} \quad v' = \frac{dv}{dx} \quad \text{and} \quad u' = \frac{du}{dx}$$

and $\quad y'' = uv'' + v'u' + vu'' + u'v' = u''v + 2u'v' + uv''$

If we differentiate the last result and collect like terms, we obtain

$$y''' = \ldots\ldots\ldots$$

$$y''' = u'''v + 3u''v' + 3u'v'' + uv'''$$

11

A further stage of differentiation would give

$$y^{(4)} = u^{(4)}v + 4u^{(3)}v^{(1)} + 6u^{(2)}v^{(2)} + 4u^{(1)}v^{(3)} + uv^{(4)}$$

These results can therefore be written

$$y = uv$$
$$y' = u'v + uv'$$
$$y'' = u''v + 2u'v' + uv''$$
$$y''' = u'''v + 3u''v' + 3u'v'' + uv'''$$
$$y^{(4)} = u^{(4)}v + 4u^{(3)}v^{(1)} + 6u^{(2)}v^{(2)} + 4u^{(1)}v^{(3)} + uv^{(4)}$$

Notice that in each case

(a) the superscript of u decreases regularly by 1

(b) the superscript of v increases regularly by 1

(c) the numerical coefficients are the normal binomial coefficients.

Indeed, $(uv)^{(n)}$ can be obtained by expanding $(u+v)^{(n)}$ using the binomial theorem where the 'powers' are interpreted as derivatives. So the expression for the nth derivative can therefore be written as

$$y^{(n)} = u^{(n)}v + nu^{(n-1)}v^{(1)} + \frac{n(n-1)}{1 \times 2}u^{(n-2)}v^{(2)}$$
$$+ \frac{n(n-1)(n-2)}{1 \times 2 \times 3}u^{(n-3)}v^{(3)} + \dots$$
$$= u^{(n)}v + nu^{(n-1)}v^{(1)} + \frac{n(n-1)}{2!}u^{(n-2)}v^{(2)}$$
$$+ \frac{n(n-1)(n-2)}{3!}u^{(n-3)}v^{(3)} + \dots$$

i.e. $y^{(n)} = u^{(n)}v + {}^nC_1 u^{(n-1)}v^{(1)} + {}^nC_2 u^{(n-2)}v^{(2)} + \dots$
$$+ {}^nC_{n-1}u^{(1)}v^{(n-1)} + uv^{(n)}$$

where ${}^nC_r = \dfrac{n!}{r!(n-r)!}$

If $y = uv$ $\qquad\qquad y^{(n)} = \displaystyle\sum_{r=0}^{n} {}^nC_r u^{(n-r)}v^{(r)}$ where $u^{(0)} \equiv u$

This is the *Leibnitz theorem*. We shall certainly be using it often in the work ahead, so make a note of it for future reference. Then we can see it in use.

12   Choice of function for *u* and *v*

For the product $y = uv$ the function taken as

(a) u is the one whose nth derivative can readily be obtained

(b) v is the one whose derivatives reduce to zero after a small number of stages of differentiation.

Example 1

To find $y^{(n)}$ when $y = x^3 e^{2x}$.

Here we choose $v = x^3$ – whose fourth derivative is zero

$\qquad\qquad\qquad u = e^{2x}$ – because we know that the nth derivative

$\qquad\qquad\qquad u^{(n)} = \ldots\ldots\ldots\ldots$

13

$$u^{(n)} = 2^n e^{2x}$$

Using the Leibnitz theorem:

$$y^{(n)} = u^{(n)}v + nu^{(n-1)}v^{(1)} + \frac{n(n-1)}{2!}\,u^{(n-2)}v^{(2)}$$

$$+ \frac{n(n-1)(n-2)}{3!}\,u^{(n-3)}v^{(3)} + \ldots$$

$v = x^3;\quad v^{(1)} = 3x^2;\quad v^{(2)} = 6x;\quad v^{(3)} = 6;\quad v^{(4)} = 0$

$u = e^{2x};\quad u^{(n)} = 2^n e^{2x}$

$$\therefore\ y^{(n)} = \ldots\ldots\ldots\ldots$$

14

$$y^{(n)} = e^{2x}\,2^{n-3}\{8x^3 + 12nx^2 + n(n-1)\,6x + n(n-1)(n-2)\}$$

Example 2

If $x^2 y'' + xy' + y = 0$, show that

$$x^2 y^{(n+2)} + (2n+1)\,xy^{(n+1)} + (n^2 + 1)\,y^{(n)} = 0.$$

We take the given equation $x^2 y'' + xy' + y = 0$ and differentiate n times, treating each term in turn.

$\qquad\qquad\qquad$ If $w = x^2 y''\qquad w^{(n)} = \ldots\ldots\ldots\ldots$

$\qquad\qquad\qquad$ If $w = xy'\qquad\ \ w^{(n)} = \ldots\ldots\ldots\ldots$

$\qquad\qquad\qquad$ If $w = y\qquad\qquad w^{(n)} = \ldots\ldots\ldots\ldots$

15

$$w = x^2 y'' \qquad \therefore \ w^{(n)} = y^{(n+2)} x^2 + ny^{(n+1)} 2x + \frac{n(n-1)}{2!} y^{(n)} 2 + 0\ldots$$

$$w = xy' \qquad \therefore \ w^{(n)} = y^{(n+1)} x + ny^{(n)} 1 + 0 + \ldots$$

$$w = y \qquad \therefore \ w^{(n)} = y^{(n)}$$

Then $\left[x^2 y'' + xy' + y\right]^{(n)} = 0$ becomes

$$\ldots\ldots\ldots\ldots$$

16

$$x^2 y^{(n+2)} + (2n+1)xy^{(n+1)} + (n^2 + 1)y^{(n)} = 0$$

which is what we had to show.

Example 3

Differentiate n times

$$(1 + x^2)y'' + 2xy' - 5y = 0.$$

The result $\ldots\ldots\ldots\ldots$

17

$$(1 + x^2)\,y^{(n+2)} + 2(n+1)\,xy^{(n+1)} + (n^2 + n - 5)\,y^{(n)} = 0$$

Because, by the Leibnitz theorem

$$\left\{ y^{(n+2)}(1 + x^2) + ny^{(n+1)} 2x + \frac{n(n-1)}{2!} y^{(n)} 2 \right\}$$

$$+ 2\left\{ xy^{(n+1)} + ny^{(n)} \cdot 1 \right\} - 5y^{(n)} = 0$$

$$(1 + x^2)\,y^{(n+2)} + 2(n+1)\,xy^{(n+1)} + \{n(n-1) + 2n - 5\}\,y^{(n)} = 0$$

$$(1 + x^2)\,y^{(n+2)} + 2(n+1)\,xy^{(n+1)} + (n^2 + n - 5)\,y^{(n)} = 0$$

We shall be using the Leibnitz theorem in the rest of this Programme, so let us move on to see some of its applications.

Power series solutions

18

Second-order linear differential equations with constant coefficients of the form $a\dfrac{d^2 y}{dx^2} + b\dfrac{dy}{dx} + cy = 0$ can be solved by algebraic methods giving solutions in terms of the normal elementary functions such as exponentials, trigonometric and polynomial functions.

▶

In general, equations of the form $\dfrac{d^2y}{dx^2} + P(x)\dfrac{dy}{dx} + Q(x)y = 0$, where $P(x)$ and $Q(x)$ are functions of x, cannot be solved in this way. However, it is often possible to obtain solutions in the form of infinite series of powers of x – and the next section of work investigates one of the methods which make this possible.

Leibnitz–Maclaurin method

As the title suggests, for this we need to be familiar with the Leibnitz theorem and with Maclaurin's series.

The *Leibnitz theorem* states that, if $y = uv$, where u and v are functions of x, then

$$y^{(n)} = \ldots\ldots\ldots\ldots$$

19

$$y^{(n)} = u^{(n)}v + nu^{(n-1)}v^{(1)} + \frac{n(n-1)}{2!}u^{(n-2)}v^{(2)} + \ldots$$
$$+ \frac{n(n-1)\ldots(n-r+1)}{r!}u^{(n-r)}v^{(r)} + \ldots + uv^{(n)}$$

where $u^{(r)}$ and $v^{(r)}$ denote $\dfrac{d^r u}{dx^r}$ and $\dfrac{d^r v}{dx^r}$ respectively.

Maclaurin's series for $y = f(x)$ can be stated as

$$y = \ldots\ldots\ldots\ldots$$

20

$$y = (y)_0 + x(y')_0 + \frac{x^2}{2!}(y'')_0 + \ldots + \frac{x^n}{n!}(y^{(n)})_0 + \ldots$$

where $(y^{(n)})_0$ denotes the value of the nth derivative of y at $x = 0$.

On to the next frame

---

21   **Example 1**

Find the power series solution of the equation

$$x\frac{d^2y}{dx^2} + \frac{dy}{dx} + xy = 1.$$

The equation can be written

$$xy'' + y' + xy = 1$$

In the first product term xy'', treat y'' as u and x as v. Then, differentiating the equation n times by the Leibnitz theorem, gives

$$\ldots\ldots\ldots\ldots$$

$$\left(xy^{(n+2)} + n \cdot 1 \cdot y^{(n+1)}\right) + y^{(n+1)} + \left(xy^{(n)} + n \cdot 1 \cdot y^{(n-1)}\right) = 0$$

$$\text{i.e.} \quad xy^{(n+2)} + (n+1)y^{(n+1)} + xy^{(n)} + ny^{(n-1)} = 0$$

22

At $x = 0$, this becomes

$$(n+1)\left(y^{(n+1)}\right)_0 + n\left(y^{(n-1)}\right)_0 = 0$$

$$\therefore \ \left(y^{(n+1)}\right)_0 = -\frac{n}{n+1}\left(y^{(n-1)}\right)_0 \qquad n \geq 1$$

This relationship is called a *recurrence relation*.

We can now substitute $n = 1, 2, 3, \ldots$ and get a set of relationships between the various coefficients.

$n = 1 \qquad (y'')_0 = -\frac{1}{2}(y)_0$

$n = 2 \qquad (y''')_0 = -\frac{2}{3}(y')_0$

$n = 3 \qquad (y^{(4)})_0 = -\frac{3}{4}(y'')_0 = \left(-\frac{3}{4}\right)\left(-\frac{1}{2}\right)(y)_0$

Continuing in the same way,

$(y^{(5)})_0 = \ldots\ldots\ldots$

$(y^{(6)})_0 = \ldots\ldots\ldots$

$(y^{(7)})_0 = \ldots\ldots\ldots$

$(y^{(8)})_0 = \ldots\ldots\ldots$

23

$n = 4 \qquad (y^{(5)})_0 = -\frac{4}{5}(y^{(3)})_0 = \left(-\frac{4}{5}\right)\left(-\frac{2}{3}\right)(y^{(1)})_0$

$n = 5 \qquad (y^{(6)})_0 = -\frac{5}{6}(y^{(4)})_0 = \left(-\frac{5}{6}\right)\left(-\frac{3}{4}\right)\left(-\frac{1}{2}\right)(y)_0$

$n = 6 \qquad (y^{(7)})_0 = -\frac{6}{7}(y^{(5)})_0 = \left(-\frac{6}{7}\right)\left(-\frac{4}{5}\right)\left(-\frac{2}{3}\right)(y^{(1)})_0$

$n = 7 \qquad (y^{(8)})_0 = -\frac{7}{8}(y^{(6)})_0 = \left(-\frac{7}{8}\right)\left(-\frac{5}{6}\right)\left(-\frac{3}{4}\right)\left(-\frac{1}{2}\right)(y)_0$

Notice that, by this means, the values of all the derivatives at $x = 0$ can be expressed in terms of $(y)_0$ and $(y')_0$.

If we now substitute these values for $(y^{(r)})_0$ in the Maclaurin series

$$y = (y)_0 + x(y')_0 + \frac{x^2}{2!}(y'')_0 + \frac{x^3}{3!}(y''')_0 + \ldots + \frac{x^r}{r!}(y^{(r)})_0 + \ldots$$

we obtain $\ldots\ldots\ldots$

24

$$y = (y)_0 + x(y')_0 + \frac{x^2}{2!}\left(-\frac{1}{2}\right)(y)_0 + \frac{x^3}{3!}\left(-\frac{2}{3}\right)(y')_0$$

$$+ \frac{x^4}{4!}\left(-\frac{3}{4}\right)\left(-\frac{1}{2}\right)(y)_0 + \frac{x^5}{5!}\left(-\frac{4}{5}\right)\left(-\frac{2}{3}\right)(y')_0$$

$$+ \frac{x^6}{6!}\left(-\frac{5}{6}\right)\left(-\frac{3}{4}\right)\left(-\frac{1}{2}\right)(y)_0 + \ldots\ldots\ldots$$

Simplifying, this gives

$$y = (y)_0\left\{1 - \frac{x^2}{2^2} + \frac{x^4}{2^2 \times 4^2} - \frac{x^6}{2^2 \times 4^2 \times 6^2} + \cdots\right\}$$

$$+ (y')_0\left\{x - \frac{x^3}{3^2} + \frac{x^5}{3^2 \times 5^2} + \cdots\right\}$$

The values of $(y)_0$ and $(y')_0$ provide the two arbitrary constants for the second-order equation and are obtained from the given initial conditions.

For example, if at $x = 0$, $y = 2$ and $\dfrac{dy}{dx} = 1$, then the relevant particular solution is ....

25

$$y = 2\left\{1 - \frac{x^2}{2^2} + \frac{x^4}{2^2 \times 4^2} - \frac{x^6}{2^2 \times 4^2 \times 6^2} + \cdots\right\}$$

$$+ \left\{x - \frac{x^3}{3^2} + \frac{x^5}{3^2 \times 5^2} + \cdots\right\}$$

Because at $x = 0$, $y = 2$ i.e. $(y)_0 = 2$

$$\frac{dy}{dx} = 1 \quad \text{i.e. } (y')_0 = 1.$$

To be a valid solution, the series obtained must converge. Application of the ratio test will normally indicate any restrictions on the values that x may have.

The Leibnitz–Maclaurin (power series) method therefore involves the following main steps:

(a) Differentiate the given equation n times, using the Leibnitz theorem.

(b) Rearrange the result to obtain the recurrence relation at $x = 0$.

(c) Determine the values of the derivatives at $x = 0$, usually in terms of $(y)_0$ and $(y')_0$.

(d) Substitute in the Maclaurin expansion for $y = f(x)$.

(e) Simplify the result where possible and apply boundary conditions if provided.

That is all there is to it. Let us go through the various steps with another example.

▶

Example 2

Determine a series solution of the equation

$$\frac{d^2y}{dx^2} + x\frac{dy}{dx} + y = 0.$$

The equation can be written $y'' + xy' + y = 0$

(a) Differentiate n times using the Leibnitz theorem, which gives

. . . .

$$y^{(n+2)} + xy^{(n+1)} + (n+1)y^{(n)} = 0$$

26

Because $y'' + xy' + y = 0$

$$\therefore\ y^{(n+2)} + \left\{xy^{(n+1)} + n \cdot 1 \cdot y^{(n)}\right\} + y^{(n)} = 0$$

$$\therefore\ y^{(n+2)} + xy^{(n+1)} + (n+1)y^{(n)} = 0.$$

(b) Determine the recurrence relation at $x = 0$, which is

. . . .

$$y^{(n+2)} = -(n+1)\,y^{(n)}$$

27

(c) Now taking $n = 0$, 1, 2, 3, 4, 5, determine the derivatives at $x = 0$ in terms of $(y)_0$ and $(y')_0$. List them, as we did before, in table form.

28

$$
\begin{aligned}
n = 0 \quad & (y'')_0 = -(y)_0 & &= -(y)_0 \\
1 \quad & (y''')_0 = -2\,(y')_0 & &= -2\,(y')_0 \\
2 \quad & (y^{(4)})_0 = -3\,(y'')_0 = (-3)[-(y)_0] & &= 3\,(y)_0 \\
3 \quad & (y^{(5)})_0 = -4\,(y''')_0 = (-4)[-2(y')_0] & &= 2 \times 4\,(y')_0 \\
4 \quad & (y^{(6)})_0 = -5\,(y^{(4)})_0 = (-5)[-3(y'')_0] & &= -3 \times 5\,(y)_0 \\
5 \quad & (y^{(7)})_0 = -6\,(y^{(5)})_0 = (-6)[-4(y''')_0] & &= -2 \times 4 \times 6\,(y')_0
\end{aligned}
$$

(d) Substitute these expressions for the derivatives in terms of $(y)_0$ and $(y')_0$ in Maclaurin's expansion

$$y = (y)_0 + x\,(y')_0 + \frac{x^2}{2!}\,(y'')_0 + \frac{x^3}{3!}\,(y''')_0 + \frac{x^4}{4!}\,(y^{(4)})_0 + \dots$$

Then $y = \dots\dots\dots$

29

$$y = (y)_0 + x(y')_0 + \frac{x^2}{2!}(-y)_0 + \frac{x^3}{3!}(-2y')_0 + \frac{x^4}{4!}(3y)_0 + \frac{x^5}{5!}(8y')_0$$
$$+ \frac{x^6}{6!}(-15y)_0 + \frac{x^7}{7!}(-48y')_0 + \ldots$$

Collecting now the terms in $(y)_0$ and $(y')_0$, we finally obtain

$$y = (y)_0 \left\{ 1 - \frac{x^2}{2} + \frac{x^4}{2 \times 4} - \frac{x^6}{2 \times 4 \times 6} + \ldots \right\}$$
$$+ (y')_0 \left\{ x - \frac{x^3}{3} + \frac{x^5}{3 \times 5} - \frac{x^7}{3 \times 5 \times 7} + \ldots \right\}$$

They are all done in very much the same way. Here is another.

Example 3

Solve the equation $\dfrac{d^2y}{dx^2} + \dfrac{dy}{dx} + 2xy = 0$ given that at $x = 0$, $y = 0$ and
$\dfrac{dy}{dx} = 1$.

First write the equation as $y'' + y' + 2xy = 0$, differentiate n times by the Leibnitz theorem and obtain the recurrence relation at $x = 0$, which is

. . . .

30

$$y^{(n+2)} = -\left\{ y^{(n+1)} + 2ny^{(n-1)} \right\} \quad n \geq 1$$

Because $y'' + y' + 2xy = 0$

$$\therefore \ y^{(n+2)} + y^{(n+1)} + 2xy^{(n)} + n2y^{(n-1)} = 0$$

At $x = 0$, $\qquad\qquad\qquad y^{(n+2)} + y^{(n+1)} + 2ny^{(n-1)} = 0$

$$\therefore \ y^{(n+2)} = -\left\{ y^{(n+1)} + 2ny^{(n-1)} \right\}$$

Since we have a term in $y^{(n-1)}$, then n must start at 1 to give $(y)_0$. Therefore the recurrence relation applies for $n \geq 1$.

We now take $n = 1, 2, 3, \ldots$ to obtain the relationships between the coefficients up to $(y^{(6)})_0$. Complete the table and check with the next frame.

31

$$
\begin{aligned}
n = 1 &\quad (y^{(3)})_0 = -\{(y^{(2)})_0 + 2(y)_0\} \\
n = 2 &\quad (y^{(4)})_0 = -\{(y^{(3)})_0 + 4(y')_0\} \\
n = 3 &\quad (y^{(5)})_0 = -\{(y^{(4)})_0 + 6(y^{(2)})_0\} \\
n = 4 &\quad (y^{(6)})_0 = -\{(y^{(5)})_0 + 8(y^{(3)})_0\}
\end{aligned}
$$

We therefore have expressions for $(y''')_0, (y^{(4)})_0, (y^{(5)})_0, (y^{(6)})_0$, but what about $(y'')_0$?

If we refer to the initial conditions, we know that at $x = 0$, $y = 0$ and $y' = 1$. $\therefore$ $(y)_0 = 0$ and $(y')_0 = 1$.

We can find $(y'')_0$ by reference to the given equation itself, because

$$y'' + y' + 2xy = 0$$

Therefore, at $x = 0$, $\quad (y'')_0 + (y')_0 = 0$ $\quad \therefore$ $(y'')_0 = -(y')_0 = -1$.

So now we have $\quad (y)_0 = 0$

$$
\begin{aligned}
(y')_0 &= 1 \\
(y'')_0 &= -1 \\
(y''')_0 &= -\{(y'')_0 + 2(y)_0\} &= -\{(-1) + 0\} = 1 \\
(y^{(4)})_0 &= -\{(y''')_0 + 4(y')_0\} &= -\{1 + 4\} &= -5 \\
(y^{(5)})_0 &= -\{(y^{(4)})_0 + 6(y'')_0\} &= -\{(-5) - 6\} = 11 \\
(y^{(6)})_0 &= -\{(y^{(5)})_0 + 8(y''')_0\} &= -\{11 + 8\} &= -19
\end{aligned}
$$

The required series solution is therefore

$$y = \dots\dots\dots$$

32

$$y = x - \frac{x^2}{2!} + \frac{x^3}{3!} - \frac{5x^4}{4!} + \frac{11x^5}{5!} - \frac{19x^6}{6!} + \dots$$

Because

$$
\begin{aligned}
y &= (y)_0 + x(y')_0 + \frac{x^2}{2!}(y'')_0 + \frac{x^3}{3!}(y''')_0 + \frac{x^4}{4!}(y^{(4)})_0 + \dots \\
&= 0 + x(1) + \frac{x^2}{2!}(-1) + \frac{x^3}{3!}(1) + \frac{x^4}{4!}(-5) + \frac{x^5}{5!}(11) + \frac{x^6}{6!}(-19) \\
\therefore \ y &= x - \frac{x^2}{2!} + \frac{x^3}{3!} - \frac{5x^4}{4!} + \frac{11x^5}{5!} - \frac{19x^6}{6!} + \dots
\end{aligned}
$$

33

One more of the same kind.

Example 4

Determine the general series solution of the equation

$$(x^2 + 1)y'' + xy' - 4y = 0$$

As usual, establish the recurrence relation at $x = 0$, which is

$$\dots\dots\dots\dots$$

34

$$y^{(n+2)} = (4 - n^2)y^{(n)}$$

Because

$$(x^2 + 1)y'' + xy' - 4y = 0 \quad \text{therefore}$$

$$\left\{ (x^2 + 1)y^{(n+2)} + 2xny^{(n+1)} + 2\frac{n(n-1)}{2!}y^{(n)} \right\} + \left\{ xy^{(n+1)} + ny^{(n)} \right\} - 4y^{(n)} = 0$$

At $x = 0$, this becomes

$$y^{(n+2)} + n(n-1)y^{(n)} + ny^{(n)} - 4y^{(n)} = 0 \quad \text{that is} \quad y^{(n+2)} = (4 - n^2)y^{(n)}$$

Then, starting with $n = 0$, determine expressions for $(y^{(n)})_0$ as far as $n = 7$.

They are . . . .

35

$n = 0$	$(y'')_0 = 4(y)_0$	$= 4(y)_0$
$n = 1$	$(y''')_0 = 3(y')_0$	$= 3(y')_0$
$n = 2$	$(y^{(4)})_0 = 0$	$= 0$
$n = 3$	$(y^{(5)})_0 = -5(y''')_0$	$= -15(y')_0$
$n = 4$	$(y^{(6)})_0 = -12(y^{(4)})_0 = 0$	
$n = 5$	$(y^{(7)})_0 = -21(y^{(5)})_0 = (-21)(-15)(y')_0$	

Now substitute in Maclaurin's expansion and simplify the result.

$$y = \ldots \ldots \ldots \ldots$$

36

$$y = A(1 + 2x^2) + B\left\{ x + \frac{x^3}{2} - \frac{x^5}{8} + \frac{x^7}{16} + \cdots \right\}$$

Because

$$y = (y)_0 + x(y')_0 + \frac{x^2}{2!}(y'')_0 + \frac{x^3}{3!}(y''')_0 + \frac{x^4}{4!}(y^{(4)})_0 + \cdots$$

$$= (y)_0 + x(y')_0 + \frac{x^2}{2!}4(y)_0 + \frac{x^3}{3!}3(y')_0 + \frac{x^4}{4!}(0) + \frac{x^5}{5!}(-15)(y')_0 + \text{etc.}$$

$$= (y)_0\{1 + 2x^2\} + (y')_0 \left\{ x + \frac{x^3}{2} - \frac{x^5}{8} + \frac{x^7}{16} + \cdots \right\}$$

Putting $(y)_0 = A$ and $(y')_0 = B$, we have the result stated.

Now to something slightly different

Cauchy–Euler equi-dimensional equations

Closely allied to some of the equations that we have been looking at are the Cauchy–Euler equi-dimensional differential equations. The general nth order equation is an inhomogeneous equation with the structure

$$a_n x^n y^{(n)}(x) + a_{n-1} x^{n-1} y^{(n-1)}(x) + \ldots + a_1 x y'(x) + a_0 y(x) = g(x)$$

where $a_0, \ldots, a_n$ are constants and where the nth derivative has a coefficient containing the nth power of x – hence the name *equi-dimensional*. To simplify matters here we shall only deal with second order equations but the method employed can be quite easily extended to higher orders. We shall just look at Cauchy–Euler equations of the form:

$$ax^2 y''(x) + bxy'(x) + cy(x) = g(x)$$

where $x > 0$.

Solutions

Just like the solution to a linear, constant coefficient ordinary differential equation, the solution to $ax^2 y''(x) + bxy'(x) + cy(x) = g(x)$ is in two parts:

$$y(x) = y_h(x) + y_p(x)$$

-- the homogeneous solution $y_h(x)$ where $ax^2 y_h''(x) + bxy_h'(x) + cy_h(x) = 0$

-- the particular solution $y_p(x)$ whose form depends on the form of $g(x)$.

We shall proceed by example.

Example 1

To solve the Cauchy–Euler differential equation $x^2 y''(x) - 4xy'(x) + 6y(x) = 0$ where $y(1) = 1$ and $y(2) = 0$ we first solve the homogeneous equation.

The homogeneous solution

$x^2 y_h''(x) - 4xy_h'(x) + 6y_h(x) = 0$. We assume a solution of the form:

$y_h(x) = Kx^n$ so that

$y_h'(x) = nKx^{n-1}$ and

$y_h''(x) = n(n-1)Kx^{n-2}$.

Substituting into the homogeneous equation gives:

$$\begin{aligned} x^2 y_h''(x) - 4xy_h'(x) + 6y_h(x) &= Kx^n(n(n-1) - 4n + 6) \\ &= Kx^n(n^2 - 5n + 6) \\ &= Kx^n(n-3)(n-2) = 0 \end{aligned}$$

Therefore $n = 3$ or $n = 2$ so that $y_h(x) = Ax^3 + Bx^2$ (A, B constants).

▶

The particular solution

$x^2y''(x) - 4xy'(x) + 6y(x) = x$. Since the right-hand side of the inhomogeneous equation is $g(x) = x$ we assume a form for the inhomogeneous solution of $y_p(x) = Cx + D$ (C, D constants) just as we did for the linear constant coefficient case. This means that that $y_p'(x) = C$ and $y_p''(x) = 0$. Substituting into the inhomogeneous equation gives:

$$-4Cx + 6(Cx + D) = 2Cx + 6D = x$$

Then $C = 1/2$ and $D = 0$ giving $y_p(x) = \dfrac{x}{2}$.

The complete solution

Adding the homogeneous solution to the particular solution gives

$$y(x) = Ax^3 + Bx^2 + \frac{x}{2}.$$

Finally, since $y(1) = 1$ and $y(2) = 0$ we see that:

$$y(1) = A + B + \frac{1}{2} = 1 \quad \text{so that} \quad A + B = \frac{1}{2}$$

$$y(2) = 8A + 4B + 1 = 0 \quad \text{so that} \quad 8A + 4B = -1 \text{ so } A = -\frac{3}{4} \text{ and } B = \frac{5}{4}$$

giving

$$y(x) = \frac{2x + 5x^2 - 3x^3}{4}$$

Check: $y(1) = \dfrac{2 + 5 - 3}{4} = 1$ and $y(2) = \dfrac{4 + 20 - 24}{4} = 0$

Now you try one.

Example 2

Given the Cauchy–Euler equation $x^2y''(x) - 8xy'(x) + 20y(x) = 3x$ where $y(1) = 0$ and $y(3) = 102$ we first consider the homogeneous equation

$$x^2y_h''(x) - 8xy_h'(x) + 20y_h(x) = 0.$$

Assuming a solution of the form $y_h(x) = Kx^n$ we find that:

$$y_h(x) = \ldots\ldots\ldots\ldots$$

38

$$\boxed{y_h(x) = Ax^5 + Bx^4}$$

Because

Since $y_h(x) = Kx^n$ then $y_h'(x) = nKx^{n-1}$ and $y_h''(x) = n(n-1)Kx^{n-2}$ so substituting into $x^2y_h''(x) - 8xy_h'(x) + 20y_h(x) = 0$ gives:

$$n(n-1) - 8n + 20 = n^2 - 9n + 20$$
$$= (n-5)(n-4)$$
$$= 0 \quad \text{so that } n = 5 \text{ or } n = 4.$$

That is $y_h(x) = Ax^5 + Bx^4$ (A, B constants). ▶

The particular solution $y_p(x)$ satisfies the equation

$$x^2 y''(x) - 8xy'(x) + 20y(x) = 3x$$

so that, assuming a form $y_p(x) = Cx + D,$

$$y_p(x) = \ldots\ldots\ldots\ldots \quad \text{and so} \quad y(x) = \ldots\ldots\ldots\ldots$$

$$\boxed{y_p(x) = \frac{x}{4} \quad \text{and} \quad y(x) = Ax^5 + Bx^4 + \frac{x}{4}}$$

39

Because

Assuming a form $y_p(x) = Cx + D$ so that $y_p'(x) = C$ and $y_p''(x) = 0$ and substituting into the inhomogeneous equation gives the equation

$$-8Cx + 20(Cx + D) = 12Cx + 20D = 3x \text{ so that } C = 1/4 \text{ and } D = 0$$

therefore

$$y_p(x) = \frac{x}{4} \text{ and } y(x) = Ax^5 + Bx^4 + \frac{x}{4}.$$

Applying the boundary conditions $y(1) = 0$ and $y(3) = 102$ we find the complete solution to be:

$$y(x) = \ldots\ldots\ldots\ldots$$

$$\boxed{y(x) = \frac{3x^5 - 4x^4 + x}{4}}$$

40

Because

$$y(1) = 0 \quad \text{so that } y(1) = A + B + \frac{1}{4} = 0 \text{ that is } A + B = -\frac{1}{4}$$

$$y(3) = 102 \text{ so that } y(3) = 243A + 81B + \frac{3}{4} = 102 \text{ that is } 3A + B = \frac{5}{4}$$

therefore $A = \dfrac{3}{4}$ and $B = -1$ so that

$$y(x) = \frac{3x^5 - 4x^4 + x}{4}$$

Check: $y(1) = \dfrac{3 - 4 + 1}{4}$ and $y(3) = \dfrac{729 - 324 + 3}{4} = 102$

And just to make sure try another, but this time try and obtain the complete solution yourself.

Example 3

The Cauchy–Euler equation $x^2 y''(x) + xy'(x) - y(x) = 6x^2 + 8x^3$, where $y(1) = 3$, $y(2) = 40$ and $y_p(x)$ is of the form $Cx^2 + Dx^3$, has solution

$$y(x) = \ldots\ldots\ldots\ldots$$

41

$$y(x) = \frac{1}{x}(x^4 + 2x^3 + 16x^2 - 16)$$

Because

We first consider the homogeneous equation $x^2 y_h''(x) + x y_h'(x) - y_h(x) = 0$.
Assuming a solution of the form $y_h(x) = Kx^n$ we find that

$$n(n-1) + n - 1 = n^2 - 1$$
$$= (n+1)(n-1)$$
$$= 0 \qquad \text{so that } n = 1 \text{ or } n = -1.$$

That is $y_h(x) = Ax + Bx^{-1}$ (A, B constants). Since the particular solution is of the form $Cx^2 + Dx^3$ then $y_p(x) = Cx^2 + Dx^3$, $y_p'(x) = 2Cx + 3Dx^2$ and $y_p''(x) = 2C + 6Dx$. Substituting in the equation $x^2 y''(x) + xy'(x) - y(x) = 6x^2 + 8x^3$ results in the equation:

$$x^2(2C + 6Dx) + x(2Cx + 3Dx^2) - (Cx^2 + Dx^3) = 6x^2 + 8x^3.$$

That is:

$$x^2[3C] + x^3[8D] = 6x^2 + 8x^3 \text{ so that } C = 2, D = 1, y_p(x) = 2x^2 + x^3$$
therefore $y(x) = Ax + Bx^{-1} + 2x^2 + x^3$.

Applying the boundary conditions $y(1) = 3$, $y(2) = 40$

$$y(1) = A + B + 2 + 1 = 3 \qquad \text{so that } A + B = 0$$
$$y(2) = 2A + \frac{B}{2} + 8 + 8 = 40 \quad \text{so that } 2A + \frac{B}{2} = 24 \text{ giving } A = 16, B = -16$$

Therefore

$$y(x) = 16x - 16x^{-1} + 2x^2 + x^3$$
$$= \frac{1}{x}(x^4 + 2x^3 + 16x^2 - 16)$$
Check: $y(1) = 1(1 + 2 + 16 - 16) = 3$
$$y(2) = \frac{1}{2}(16 + 16 + 64 - 16) = 40$$

And that completes the work of this Programme. The main points that we have covered in this Programme are listed in the **Revision summary** that follows. Read this in conjunction with the **Can you?** check list and note any sections that may need further attention: refer back to the relevant parts of the Programme, if necessary. There will then be no trouble with the **Test exercise**. The set of **Further problems** provides an opportunity for further practice.

Revision summary 10

1 *Higher derivatives*

42

y	$y^{(n)}$
x^a	$\dfrac{a!}{(a-n)!}x^{a-n}$
e^{ax}	$a^n e^{ax}$
$\sin ax$	$a^n \sin\left(ax + \dfrac{n\pi}{2}\right)$
$\cos ax$	$a^n \cos\left(ax + \dfrac{n\pi}{2}\right)$
$\sinh ax$	$\dfrac{a^n}{2}\left\{[1+(-1)^n]\sinh ax + [1-(-1)^n]\cosh ax\right\}$
$\cosh ax$	$\dfrac{a^n}{2}\left\{[1-(-1)^n]\sinh ax + [1+(-1)^n]\cosh ax\right\}$

2 *Leibnitz theorem* — *n*th derivative of a product of functions.
If $y = uv$

$$y^{(n)} = u^{(n)}v + nu^{(n-1)}v^{(1)} + \frac{n(n-1)}{2!}u^{(n-2)}v^{(2)}$$
$$+ \frac{n(n-1)(n-2)}{3!}u^{(n-3)}v^{(3)} + \dots$$
$$\dots + \frac{n(n-1)(n-2)\dots(n-r+1)}{r!}u^{(n-r)}v^{(r)} + \dots$$

i.e. $y^{(n)} = \displaystyle\sum_{r=0}^{\infty} {}^nC_r u^{(n-r)}v^{(r)}$.

$(uv)^{(n)}$ can be obtained by expanding $(u+v)^{(n)}$ using the binomial theorem where the 'powers' are interpreted as derivatives.

3 *Power series solution of second-order differential equations*
Leibnitz–Maclaurin method
(1) Differentiate the equation *n* times by the Leibnitz theorem.
(2) Put $x = 0$ to establish a recurrence relation.
(3) Substitute $n = 1, 2, 3, \dots$ to obtain $y', y'', y''', \dots$ at $x = 0$.
(4) Substitute in Maclaurin's series and simplify where possible.

▶

4 *Cauchy–Euler equi-dimensional equations*

The second order Cauchy–Euler equi-dimensional equation has the structure

$$ax^2 y''(x) + bxy'(x) + cy(x) = g(x)$$

where the coefficient of the nth derivative contains x^n. The solution consists of the sum of a homogeneous solution $y_h(x)$ and a particular solution $y_p(x)$. The homogeneous solution is assumed to be of the form $y_h(x) = Kx^n$ and substitution into the homogeneous equation results in as many values of n as the degree of the equation. The form of the particular solution depends upon the form of the right-hand side of the equation $g(x)$.

 # Can you?

---

43 **Checklist 10**

Check this list before and after you try the end of Programme test.

On a scale of 1 to 5 how confident are you that you can: **Frames**

- Obtain the nth derivative of the exponential and circular and hyperbolic functions?   1 to 9
  Yes   ☐   ☐   ☐   ☐   ☐   *No*

- Apply the Leibnitz theorem to derive the nth derivative of a product of expressions?   10 to 17
  Yes   ☐   ☐   ☐   ☐   ☐   *No*

- Apply the Leibnitz–Maclaurin method of obtaining a series solution to a second-order homogeneous differential equation with constant coefficients?   18 to 36
  Yes   ☐   ☐   ☐   ☐   ☐   *No*

- Solve Cauchy–Euler equi-dimensional equations?   37 to 43
  Yes   ☐   ☐   ☐   ☐   ☐   *No*

Test exercise 10

1　If $y = e^{x^2+x}$, show that $y'' = y'(2x+1) + 2y$ and hence prove that
　$y^{(n+2)} = (2x+1)y^{(n+1)} + 2(n+1)y^{(n)}$.

44

2　Obtain a power series solution of the equation
　$(1+x^2)y'' - 3xy' - 5y = 0$
　up to and including the term in x^6.

3　Solve each of the following
　(a)　$x^2y''(x) + 2xy'(x) - 2y(x) = 0$
　(b)　$2x^2y''(x) + 5xy'(x) - 9y(x) = x^2$
　(c)　$x^2y''(x) - xy'(x) + y(x) = 3x^2 - 2x^3$ where $y(1) = y(2) = 4$

 # Further problems 10

(a) *Use the Leibnitz theorem* for the following.

45

1　If $y = x^3 e^{4x}$, determine $y^{(5)}$.

2　Find the nth derivative of $y = x^3 e^{-x}$ for $n > 3$.

3　If $y = x^3(2x+1)^2$, find $y^{(4)}$.

4　Find the 6th derivative of $y = x^4 \cos x$.

5　If $y = e^{-x} \sin x$, obtain an expression for $y^{(4)}$.

6　Determine $y^{(3)}$ when $y = x^4 \ln x$.

7　If $x^2 y'' + xy' + y = 0$, show that
　$x^2 y^{(n+2)} + (2n+1)xy^{(n+1)} + (n^2+1)y^{(n)} = 0$.

8　If $y = (2x - \pi)^4 \sin\left(\dfrac{x}{2}\right)$, evaluate $y^{(6)}$ when $x = \pi/2$.

9　If $y = e^{-x} \cos x$, show that $y^{(4)} + 4y = 0$.

10　Find the $(2n)$th derivative of (a) $y = x^2 \sinh x$
　　　　　　　　　　　　　　　　　(b) $y = x^3 \cosh x$.

11　If $y = (x^3 + 3x^2)e^{2x}$, determine an expression for $y^{(6)}$.

12　Find the nth derivative of $y = e^{-ax} \cos ax$ and hence determine $y^{(3)}$.

13　If $y = \dfrac{\sin x}{1 - x^2}$, show that
　（a）　$(1 - x^2)y'' - 4xy' - (1 + x^2)y = 0$
　（b）　$y^{(n+2)} - (n^2 + 3n + 1)y^{(n)} - n(n-1)y^{(n-2)} = 0$ at $x = 0$.

▶

(b) *Use the Leibnitz–Maclaurin method* to determine series solutions for the following.

14 $(1 + x^2)y'' + xy' - 9y = 0.$

15 $(x + 1)y'' + (x - 1)y' - 2y = 0.$

16 $(1 - x^2)y'' - 7xy' - 9y = 0.$

17 $(1 - x^2)y'' - 2xy' + 2y = 0.$

18 $xy'' + y' + 2xy = 0.$

(c) *Solve the following Cauchy–Euler equi-dimensional equations.*

19 $x^2y''(x) - 5xy'(x) + 8y(x) = x^3$ where $y(1) = -3$ and $y(2) = -4$

20 $6x^2y''(x) + 19xy'(x) + 6y(x) = 99x^3 - 56x^2$ where $y(1) = 1 - 4\sqrt{2}$ and $y(8) = 448$

21 $x^2y''(x) - 2y(x) = 4x^3$ where $y(1) = 2$ and $y(2) = -2$

22 $x^2y''(x) + xy(x) - y(x) = 3x^2$ where $y(1) = -1$ and $y(3) = \dfrac{17}{3}$

Power series solutions of ordinary differential equations 2

Learning outcomes

When you have completed this Programme you will be able to:

- Apply Frobenius' method of obtaining a series solution to a second-order homogeneous ordinary differential equation by first differentiating the assumed series several times

- Substitute into the differential equation and equate coefficients of corresponding powers

- Derive the indicial equation

- Distinguish between the possible four distinct outcomes arising from the indicial equation

Introduction

1

In the previous Programme we established the solutions of second order ordinary differential equations as power series in integer powers of x. Such solutions are not always possible and a more general method is to assume a trial solution of the form

$$y = x^c \{a_0 + a_1 x + a_2 x^2 + a_3 x^3 + \ldots + a_r x^r + \ldots\}$$

where a_0 is the first coefficient that is not zero.

The type of equation that can be solved by this method is of the form

$$y'' + Py' + Qy = 0$$

where P and Q are functions of x.

However, certain conditions have to be satisfied.

(a) If the functions P and Q are such that both are finite when x is put equal to zero, $x = 0$ is called an *ordinary point* of the equation.

(b) If xP and $x^2 Q$ remain finite at $x = 0$, then $x = 0$ is called a *regular singular point* of the equation.

In both of these cases, the method of Frobenius can be applied.

(c) If, however, P and Q do not satisfy either of these conditions stated in (a) or (b), then $x = 0$ is called an *irregular singular point* of the equation and the method of Frobenius cannot be applied.

Solution of differential equations by the method of Frobenius

To solve a given equation, we have to find the coefficients $a_0, a_1, a_2, \ldots$ and also the index c in the trial solution. Basically, the steps in the method are as follows

(a) Differentiate the trial series as required.

(b) Substitute the results in the given differential equation.

(c) Equate coefficients of corresponding powers of x on each side of the equation.

The following examples will demonstrate the method – so move on

2

Example 1

Find a series solution for the equation

$$2x \frac{d^2 y}{dx^2} + \frac{dy}{dx} + y = 0.$$

The equation can be written as $2xy'' + y' + y = 0$.

Assume a solution of the form

$$y = x^c \{a_0 + a_1 x + a_2 x^2 + a_3 x^3 + \ldots + a_r x^r + \ldots\} \qquad a_0 \neq 0.$$

$$\therefore \ y = a_0 x^c + a_1 x^{c+1} + a_2 x^{c+2} + \ldots + a_r x^{c+r} + \ldots$$

Differentiating term by term, we get

$$y' = \ldots \ldots \ldots$$

$$y' = a_0 c x^{c-1} + a_1(c+1)x^c + a_2(c+2)x^{c+1} + \ldots + a_r(c+r)x^{c+r-1} + \ldots$$

3

Repeating the process one stage further, we have

$$y'' = \ldots\ldots\ldots\ldots \quad \text{(give yourself plenty of room)}$$

$$y'' = a_0 c(c-1)x^{c-2} + a_1 c(c+1)x^{c-1} + a_2(c+1)(c+2)x^c + \ldots$$
$$+ a_r(c+r-1)(c+r)x^{c+r-2} + \ldots$$

4

So far, we have $2xy'' + y' + y = 0$

$$y = a_0 x^c + a_1 x^{c+1} + a_2 x^{c+2} + \ldots + a_r x^{c+r} + \ldots$$
$$y' = a_0 c x^{c-1} + a_1(c+1)x^c + a_2(c+2)x^{c+1} + \ldots$$
$$+ a_r(c+r)x^{c+r-1} + \ldots$$
$$y'' = a_0 c(c-1)x^{c-2} + a_1 c(c+1)x^{c-1} + a_2(c+1)(c+2)x^c + \ldots$$
$$+ a_r(c+r-1)(c+r)x^{c+r-2} + \ldots$$

Considering each term of the equation in turn

$$2xy'' = 2a_0 c(c-1)x^{c-1} + 2a_1 c(c+1)x^c + 2a_2(c+1)(c+2)x^{c+1}$$
$$+ \ldots + 2a_r(c+r-1)(c+r)x^{c+r-1} + \ldots$$
$$y' = a_0 c x^{c-1} + a_1(c+1)x^c + a_2(c+2)x^{c+1} + \ldots$$
$$+ a_r(c+r)x^{c+r-1} + \ldots$$
$$y = a_0 x^c + a_1 x^{c+1} + \ldots + a_r x^{c+r} + \ldots$$

Adding these three lines to form the left-hand side of the equation, we can equate the total coefficient of each power of x to zero, since the right-hand side is zero.

$$\left[x^{c-1}\right] \text{ gives } \ldots\ldots\ldots\ldots$$

$$\left[x^{c-1}\right]: \quad 2a_0 c(c-1) + a_0 c = 0$$
$$\therefore \ a_0 c(2c-1) = 0$$

5

So, $\left[x^{c-1}\right]$ gives $\quad a_0 c(2c-1) = 0$ (1)

Similarly, $\left[x^c\right]$ gives $\ldots\ldots\ldots\ldots$

6

$$2a_1c\,(c+1) + a_1(c+1) + a_0 = 0$$

Simplifying, this becomes

$$a_1(2c^2 + 3c + 1) + a_0 = 0$$

i.e. $a_1(c+1)(2c+1) + a_0 = 0$ (2)

Also $\left[x^{c+1}\right]$ gives ....

7

$$2a_2(c+1)(c+2) + a_2(c+2) + a_1 = 0$$

and this simplifies straight away to

$$a_2(c+2)(2c+3) + a_1 = 0$$ (3)

Note that the coefficient of x^c involves all three lines of the expressions and, from then on, a general relationship can be obtained for x^{c+r}, $r \geq 0$.

In the expression for $2xy''$ and y' we have terms in x^{c+r-1}. If we replace r by $(r+1)$, we shall obtain the corresponding terms in x^{c+r}.

In the series for $2xy''$,   this is $2a_{r+1}(c+r)(c+r+1)x^{c+r}$
In the series for y',   this is $a_{r+1}(c+r+1)x^{c+r}$
In the series for y,   this is $a_r x^{c+r}$

Therefore, equating the total coefficient of x^{c+r} to zero, we have

....

8

$$2a_{r+1}(c+r)(c+r+1) + a_{r+1}(c+r+1) + a_r = 0$$

and this tidies up to

$$a_{r+1}\{(c+r+1)(2c+2r+1)\} + a_r = 0$$ (4)

Make a note of results (1), (2), (3) and (4): we shall return to them in due course.

Then move on

9 **Indicial equation**

Equation (1), formed from the coefficient of the lowest power of x, that is x^{c-1}, is called the *indicial equation* from which the values of c can be obtained. In the present example $a_0c(2c-1) = 0$

$$\therefore\ c = \ldots\ldots\ldots\ldots$$

10

$$c = 0 \text{ or } \frac{1}{2}, \text{ since } a_0 \neq 0, \text{ by definition}$$

Both values of c are valid, so that we have two possible solutions of the given equation. We will consider each in turn.

(a) *Using $c = 0$*

(2) gives $a_1(1)(1) + a_0 = 0$ $\therefore a_1 = -a_0$

Similarly

(3) gives ....

11

$$a_2(2)(3) + a_1 = 0$$

$a_1 = -a_0$ and $a_2 = -\dfrac{a_1}{2 \times 3} = \dfrac{a_0}{2 \times 3}$

and from (4) $a_{r+1} = \dfrac{-a_r}{(r+1)(2r+1)}$ $r \geq 0$

From the combined series, the term in x^c and all subsequent terms involve all three lines and the coefficient of the general term can be used.

So we have $a_1 = -a_0$ and $a_{r+1} = \dfrac{-a_r}{(r+1)(2r+1)}$ for $r = 0, 1, 2, \ldots$

$$\therefore a_2 = \frac{-a_1}{2 \times 3} = \frac{a_0}{2 \times 3}$$

$$a_2 = \frac{-a_2}{3 \times 5} = \frac{-a_0}{(2 \times 3)(3 \times 5)}$$

$$a_4 = \frac{-a_3}{4 \times 7} = \frac{a_0}{(2 \times 3 \times 4)(3 \times 5 \times 7)} \qquad \text{etc.}$$

$$\therefore y = x^0 \left\{ a_0 - a_0 x + \frac{a_0}{(2 \times 3)} x^2 - \frac{a_0}{(2 \times 3)(3 \times 5)} x^3 + \cdots \right\}$$

$$\therefore y = a_0 \left\{ 1 - x + \frac{x^2}{(2)(3)} - \frac{x^3}{(2 \times 3)(3 \times 5)} + \frac{x^4}{(2 \times 3 \times 4)(3 \times 5 \times 7)} + \cdots \right\}$$

Now we go through the same steps using our second value for c, i.e. $c = \frac{1}{2}$.

Next frame

(b) *Using $c = \frac{1}{2}$*

12

Our equations relating the coefficients were

$a_0 c(2c - 1) = 0$ which gave $c = 0$ or $c = \frac{1}{2}$ (1)

$a_1(c + 1)(2c + 1) + a_0 = 0$ (2)

$a_2(c + 2)(2c + 3) + a_1 = 0$ (3)

$a_{r+1}(c + r + 1)(2c + 2r + 1) + a_r = 0$ (4)

Putting $c = \frac{1}{2}$ in (2) gives ....

13

$$a_1 = -\frac{a_0}{3}$$

Similarly (3) gives $\qquad a_2 = -\frac{a_1}{10} = \frac{a_0}{3 \times 10}$

and from the general relationship, (4), we have ....

14

$$a_{r+1} = \frac{-a_r}{(r+1)(2r+3)}$$

So $\quad a_1 = -\dfrac{a_0}{3}$

$$a_2 = -\frac{a_1}{2 \times 5} = \frac{a_0}{(1 \times 2)(3 \times 5)}$$

$$a_3 = -\frac{a_2}{3 \times 7} = \frac{-a_0}{(1 \times 2 \times 3)(3 \times 5 \times 7)}$$

$$a_4 = -\frac{a_3}{4 \times 9} = \frac{a_0}{(1 \times 2 \times 3 \times 4)(3 \times 5 \times 7 \times 9)} \quad \text{etc.}$$

$$y = x^c \{a_0 + a_1 x + a_2 x^2 + a_3 x^3 + \ldots + a_r x^r + \ldots\}$$

i.e. $\quad y = \ldots\ldots\ldots\ldots$

15

$$y = x^{\frac{1}{2}} \left\{ a_0 - \frac{a_0}{3}x + \frac{a_0}{(1 \times 2)(3 \times 5)}x^2 - \frac{a_0}{(1 \times 2 \times 3)(3 \times 5 \times 7)}x^3 + \ldots \right\}$$

i.e. $\quad y = a_0 x^{\frac{1}{2}} \left\{ 1 - \frac{x}{(1 \times 3)} + \frac{x^2}{(1 \times 2)(3 \times 5)} - \frac{x^3}{(1 \times 2 \times 3)(3 \times 5 \times 7)} + \ldots \right\}$

Since a_0 is an arbitrary (non-zero) constant in each solution, its values may well be different, A and B say. If we denote the first solution by $u(x)$ and the second by $v(x)$, then

$$u = A \left\{ 1 - x + \frac{x^2}{(2 \times 3)} - \frac{x^3}{(2 \times 3)(3 \times 5)} + \frac{x^4}{(2 \times 3 \times 4)(3 \times 5 \times 7)} + \ldots \right\}$$

and

$$v = B x^{\frac{1}{2}} \left\{ 1 - \frac{x}{(1 \times 3)} + \frac{x^2}{(1 \times 2)(3 \times 5)} - \frac{x^3}{(1 \times 2 \times 3)(3 \times 5 \times 7)} + \ldots \right\}$$

The general solution $y = u + v$ is therefore ....

16

$$y = A \left\{ 1 - x + \frac{x^2}{(2 \times 3)} - \frac{x^3}{(2 \times 3)(3 \times 5)} + \ldots \right\} + B x^{\frac{1}{2}} \left\{ 1 - \frac{x}{(1 \times 3)} \right.$$

$$\left. + \frac{x^2}{(1 \times 2)(3 \times 5)} - \frac{x^3}{(1 \times 2 \times 3)(3 \times 5 \times 7)} + \ldots \right\}$$

▶

The method may seem somewhat lengthy, but we have set it out in detail. It is a straightforward routine. Here is another example with the same steps.

Example 2

Find the series solution for the equation

$$3x^2 y'' - xy' + y - xy = 0.$$

We proceed in just the same way as in the previous example.

Assume $\quad\quad y = x^c\{a_0 + a_1 x + a_2 x^2 + a_3 x^3 + \ldots + a_r x^r + \ldots\}$

i.e. $\quad\quad y = a_0 x^c + a_1 x^{c+1} + a_2 x^{c+2} + \ldots + a_r x^{c+r} + \ldots$

$\quad\quad \therefore\ y' = a_0 c x^{c-1} + a_1(c+1)x^c + a_2(c+2)x^{c+1} + \ldots$

$\quad\quad\quad\quad + a_r(c+r)x^{c+r-1} + \ldots$

and $\quad\quad y'' = \ldots\ldots\ldots\ldots$

17

$$y'' = a_0 c(c-1)x^{c-2} + a_1(c+1)c x^{c-1} + a_2(c+2)(c+1)x^c + \ldots$$
$$+ a_r(c+r)(c+r-1)x^{c+r-2} + \ldots$$

Now we build up the terms in the given equation.

$$3x^2 y'' = 3a_0 c(c-1)x^c + 3a_1(c+1)c x^{c+1} + 3a_2(c+2)(c+1)x^{c+2} + \ldots$$
$$+ 3a_r(c+r)(c+r-1)x^{c+r} + \ldots$$

$$-xy' = -a_0 c x^c - a_1(c+1)x^{c+1} - a_2(c+2)x^{c+2} - \ldots - a_r(c+r)x^{c+r} - \ldots$$

$$y = a_0 x^c + a_1 x^{c+1} + a_2 x^{c+2} + \ldots + a_r x^{c+r} + \ldots$$

$$-xy = -a_0 x^{c+1} - a_1 x^{c+2} - \ldots - a_r x^{c+r+1} \ldots$$

The *indicial equation*, i.e. equating the coefficient of the lowest power of x to zero, gives the values of c. Thus, in this case

$$c = \ldots\ldots\ldots\ldots$$

18

$$c = 1 \ \text{ or } \ \frac{1}{3}$$

Because the lowest power is x^c and the coefficient of x^c equated to zero gives

$$3a_0 c(c-1) - a_0 c + a_0 = 0$$

$\therefore\ a_0(3c^2 - 4c + 1) = 0 \quad \therefore\ (3c-1)(c-1) = 0$ since $a_0 \neq 0$

$\therefore\ c = 1 \ $ or $\ \dfrac{1}{3}$

The coefficient of the general term, i.e. x^{c+r} gives

$$3a_r(c+r)(c+r-1) - a_r(c+r) + a_r - a_{r-1} = 0$$

$\therefore\ a_r = \ldots\ldots\ldots\ldots$

19

$$a_r = \frac{a_{r-1}}{3(c+r)^2 - 4(c+r) + 1} = \frac{a_{r-1}}{(c+r-1)(3c+3r-1)}$$

(a) *Using* $c = 1$ the recurrence relation becomes

$$a_r = \frac{a_{r-1}}{r(3r+2)}$$

$$\therefore r = 1 \quad a_1 = \frac{a_0}{1 \times 5}$$

$$r = 2 \quad a_2 = \frac{a_1}{2 \times 8} = \frac{a_0}{(1 \times 2)(5 \times 8)}$$

$$r = 3 \quad a_3 = \frac{a_2}{3 \times 11} = \frac{a_0}{(1 \times 2 \times 3)(5 \times 8 \times 11)}$$

Our first solution is therefore

$$y = \ldots\ldots\ldots$$

20

$$y = x^1 \left\{ a_0 + \frac{a_0 x}{(1 \times 5)} + \frac{a_0 x^2}{(1 \times 2)(5 \times 8)} + \frac{a_0 x^3}{(1 \times 2 \times 3)(5 \times 8 \times 11)} + \cdots \right\}$$

$$\therefore y = Ax \left\{ 1 + \frac{x}{(1 \times 5)} + \frac{x^2}{(1 \times 2)(5 \times 8)} + \frac{x^3}{(1 \times 2 \times 3)(5 \times 8 \times 11)} + \cdots \right\}$$

(b) *For the second solution*, we put $c = \frac{1}{3}$. The recurrence relation then becomes

$$a_r = \ldots\ldots\ldots$$

21

$$a_r = \frac{a_{r-1}}{r(3r-2)}$$

Therefore we can now determine the coefficients for $r = 1, 2, 3, \ldots$ and complete the second solution.

$$y = \ldots\ldots\ldots$$

22

$$y = Bx^{\frac{1}{3}}\left\{1 + x + \frac{x^2}{(2 \times 4)} + \frac{x^3}{(2 \times 3)(4 \times 7)}\right.$$
$$\left. + \frac{x^4}{(2 \times 3 \times 4)(4 \times 7 \times 10)} + \dots\right\}$$

Because

$$a_1 = \frac{a_0}{1 \times 1}; \qquad a_2 = \frac{a_1}{2 \times 4} = \frac{a_0}{(1 \times 2)(2 \times 4)}$$

$$a_3 = \frac{a_2}{3 \times 7} = \frac{a_0}{(2 \times 3)(4 \times 7)}$$

$$a_4 = \frac{a_3}{4 \times 10} = \frac{a_0}{(2 \times 3 \times 4)(4 \times 7 \times 10)}$$

$$\therefore \ y = a_0 x^{\frac{1}{3}}\left\{1 + x + \frac{x^2}{(2 \times 4)} + \frac{x^3}{(2 \times 3)(4 \times 7)}\right.$$
$$\left. + \frac{x^4}{(2 \times 3 \times 4)(4 \times 7 \times 10)} + \dots\right\}$$

Therefore, the general solution is

$$y = \dots\dots\dots$$

23

$$y = Ax\left\{1 + \frac{x}{(1 \times 5)} + \frac{x^2}{(1 \times 2)(5 \times 8)} + \frac{x^3}{(1 \times 2 \times 3)(5 \times 8 \times 11)} + \dots\right\}$$
$$+ Bx^{\frac{1}{3}}\left\{1 + x + \frac{x^2}{(2 \times 4)} + \frac{x^3}{(2 \times 3)(4 \times 7)} + \frac{x^4}{(2 \times 3 \times 4)(4 \times 7 \times 10)} + \dots\right\}$$

Example 3

Find the series solution for the equation

$$\frac{d^2y}{dx^2} - y = 0 \quad \text{i.e.} \quad y'' - y = 0.$$

As usual, we start off with the assumed solution

$$y = x^c\{a_0 + a_1 x + a_2 x^2 + \dots + a_r x^r + \dots\}$$

i.e. $y = a_0 x^c + a_1 x^{c+1} + a_2 x^{c+2} + \dots + a_r x^{c+r} + \dots$

$\therefore \ y' = a_0 c x^{c-1} + a_1(c+1)x^c + a_2(c+2)x^{c+1} + \dots$
$$+ a_r(c+r)x^{c+r-1} + \dots$$

$y'' = a_0 c(c-1)x^{c-2} + a_1(c+1)c x^{c-1} + a_2(c+2)(c+1)x^c + \dots$
$$+ a_r(c+r)(c+r-1)x^{c+r-2} + \dots$$

These three expansions are required regularly, so make a note of them

24

Now we build up the terms in the left-hand side of the equation.

$$y'' = a_0c(c-1)x^{c-2} + a_1(c+1)cx^{c-1} + a_2(c+2)(c+1)x^c + \ldots$$
$$+ a_r(c+r)(c+r-1)x^{c+r-2} + \ldots$$
$$y = a_0x^c + a_1x^{c+1} + \ldots + a_rx^{c+r} + \ldots$$

The term in x^{c+r} in the first of these expansions is

. . . .

25

$$\boxed{a_{r+2}(c+r+2)(c+r+1)x^{c+r}}$$

Because replacing r by $(r+2)$ in $a_r(c+r)(c+r+1)x^{c+r-2}$ gives this result.

Then $y'' - y = \ldots \ldots \ldots$

26

$$\boxed{\begin{aligned} y'' - y &= a_0c(c-1)x^{c-2} + a_1(c+1)cx^{c-1} + [a_2(c+2)(c+1) - a_0]x^c \\ &\quad + \ldots + [a_{r+2}(c+r+2)(c+r+1) - a_r]x^{c+r} + \ldots \end{aligned}}$$

We now equate each coefficient in turn to zero, since the right-hand side of the equation is zero. The coefficient of the lowest power of x gives the *indicial equation* from which we obtain the values of c.

So, in this case, $c = \ldots \ldots \ldots$

27

$$\boxed{c = 0 \quad \text{or} \quad 1}$$

For the term in x^{c-1}, we have

$[x^{c-1}]$: $a_1(c+1)c = 0$.

With $c = 1$, $a_1 = 0$.

But with $c = 0$, a_1 is indeterminate, because any value of a_1 combined with the zero value of c would make the product zero.

$[x^c]$: $a_2(c+2)(c+1) - a_0 = 0$ $\therefore a_2 = \dfrac{a_0}{(c+1)(c+2)}$

For the general term

$[x^{c+r}]$:   . . . .

28

$$a_{r+2} = \frac{a_r}{(c+r+1)(c+r+2)}$$

Because $a_{r+2}(c+r+2)(c+r+1) - a_r = 0$. Hence the result above.

From the indicial equation, $c = 0$ or $c = 1$.

(a) When $c = 0$ a_1 is indeterminate

$$a_2 = \frac{a_0}{2}$$

In general $a_{r+2} = \dfrac{a_r}{(r+1)(r+2)}$

$r = 1$ $\therefore a_3 = \dfrac{a_1}{2 \times 3}$

$r = 2$ $a_4 = \dfrac{a_2}{3 \times 4} = \dfrac{a_0}{4!}$

Therefore, one solution is . . . .

29

$$y = x^0 \left\{ a_0 + a_1 x + \frac{a_0}{2!} x^2 + \frac{a_1}{3!} x^3 + \frac{a_0}{4!} x^4 \ldots \right\}$$

i.e. $y = a_0 \left\{ 1 + \dfrac{x^2}{2!} + \dfrac{x^4}{4!} + \ldots \right\} + a_1 \left\{ x + \dfrac{x^3}{3!} + \dfrac{x^5}{5!} + \ldots \right\}$

a_0 and a_1 are arbitrary constants depending on the boundary conditions.

$$\therefore y = A \left\{ 1 + \frac{x^2}{2!} + \frac{x^4}{4!} + \ldots \right\} + B \left\{ x + \frac{x^3}{3!} + \frac{x^5}{5!} + \ldots \right\}$$

Notice that these two series are the Maclaurin series expansions of the hyperbolic functions, so that

$$y = A \cosh x + B \sinh x$$

It is not very often the case that the series solution is so easily expressible in terms of known functions.

(b) Similarly,

when $c = 1$ $a_1 = 0$

$$a_2 = \frac{a_0}{2 \times 3}$$

$$a_{r+2} = \ldots \ldots \ldots \ldots$$

30

$$a_{r+2} = \frac{a_r}{(r+2)(r+3)}$$

$$\therefore \ a_1 = 0$$

$$a_2 = \frac{a_0}{3!}$$

$r = 1 \qquad a_3 = \frac{a_1}{3 \times 4} = 0$

$r = 2 \qquad a_4 = \frac{a_2}{4 \times 5} = \frac{a_0}{5!}$

$r = 3 \qquad a_5 = \frac{a_3}{5 \times 6} = 0 \quad$ etc.

A second solution with $c = 1$ is therefore

$$y = \dots\dots\dots$$

31

$$y = a_0\left\{x + \frac{x^3}{3!} + \frac{x^5}{5!} + \dots\right\}$$

and, because a_0 is an arbitrary constant

$$y = C\left\{x + \frac{x^3}{3!} + \frac{x^5}{5!} + \frac{x^7}{7!} + \dots\right\}$$

Note: This is not, in fact, a separate solution, since it already forms the second series in the solution for $c = 0$ obtained previously. Therefore, the first solution, with its two arbitrary constants, A and B, gives the general solution. This happens when the two values of c differ by an integer.

Make a note of the following:
 If the two values of c, i.e. c_1 and c_2, differ by an integer, and if $c = c_1$ results in a_1 being indeterminate, then this value of c gives the general solution. The solution resulting from $c = c_2$ is then merely a multiple of one of the series forming the first solution.

Our last problem was an example of this.
So far, we have met two distinct cases concerning the two roots $c = c_1$ and $c = c_2$ of the indicial equation.

(a) *If c_1 and c_2 differ by a quantity NOT an integer* then two independent solutions, $y = u(x)$ and $y = v(x)$, are obtained. The general solution is then $y = Au + Bv$.

(b) *If c_1 and c_2 differ by an integer*, i.e. $c_2 = c_1 + n$, and if one coefficient (a_r) is indeterminate when $c = c_1$, the complete general solution is given by using this value of c. Using $c = c_1 + n$ gives a series which is a simple multiple of one of the series in the first solution.

Make a note of these two points in your record book. Then move on

---

There is a third category to be added to (a) and (b) above.

32

(c) If the roots $c = c_1$ and $c = c_1 + n$ of the indicial equation differ by an integer and one coefficient (a_r) becomes infinite when $c = c_1$, the series is rewritten with a_0 replaced by $k(c - c_1)$.

Putting $c = c_1$ in the rewritten series and that of its derivative with respect to c gives two independent solutions.

Add this to the previous two. Then we will see how it works in practice

---

Example 4

33

Find the series solution of the equation

$$xy'' + (2 + x)y' - 2y = 0.$$

Using $y = x^c(a_0 + a_1 x + a_2 x^2 + a_3 x^3 + \ldots + a_r x^r + \ldots)$ and its first two derivatives, the expansions for

$$xy'' = \ldots\ldots\ldots$$
$$2y' = \ldots\ldots\ldots$$
$$xy' = \ldots\ldots\ldots$$
$$-2y = \ldots\ldots\ldots$$

Method as before.

34

$$xy'' = a_0 c(c - 1)x^{c-1} + a_1(c + 1)cx^c + a_2(c + 2)(c + 1)x^{c+1} + \ldots$$
$$+ a_r(c + r)(c + r - 1)x^{c+r-1} + \ldots$$
$$2y' = 2a_0 c x^{c-1} + 2a_1(c + 1)x^c + 2a_2(c + 2)x^{c+1} + 2a_3(c + 3)x^{c+2}$$
$$+ \ldots + 2a_r(c + r)x^{c+r-1} + \ldots$$
$$xy' = a_0 c x^c + a_1(c + 1)x^{c+1} + a_2(c + 2)c^{c+2} + \ldots$$
$$+ a_r(c + r)x^{c+r} + \ldots$$
$$-2y = -2a_0 x^c - 2a_1 x^{c+1} - 2a_2 x^{c+2} - 2a_3 x^{c+3} - \ldots$$
$$- 2a_r x^{c+r} - \ldots$$

From which, the indicial equation is $\ldots\ldots\ldots$

35

$$\boxed{a_0(c^2 + c) = 0}$$

i.e. equating the coefficient of the lowest power of x, (x^{c-1}), to zero.

$$a_0 \neq 0 \quad \therefore c = 0 \text{ or } -1$$

Also, from the expansions, the total coefficient of x^c gives

$$a_1 = \ldots\ldots\ldots$$

36

$$a_1 = \frac{-a_0(c-2)}{(c+1)(c+2)}$$

From the terms in x^c, all four expansions are involved, so we can form the recurrence relation from the coefficient of x^{c+r}.

$$a_{r+1} = \ldots\ldots\ldots$$

37

$$a_{r+1} = \frac{-a_r(c+r-2)}{(c+r+1)(c+r+2)}$$

Because

$$a_{r+1}(c+r+1)(c+r) + 2a_{r+1}(c+r+1) + a_r(c+r) - 2a_r = 0$$
$$a_{r+1}(c+r+1)(c+r+2) + a_r(c+r-2) = 0$$
$$\therefore\; a_{r+1} = \frac{-a_r(c+r-2)}{(c+r+1)(c+r+2)} \qquad r \geq 0$$

$$\therefore\; a_2 = \ldots\ldots\ldots$$

38

$$a_2 = \frac{a_0(c-1)(c-2)}{(c+1)(c+2)^2(c+3)}$$

and, from the recurrence relation, when $r = 2$

$$a_3 = \ldots\ldots\ldots$$

39

$$a_3 = \frac{-a_0 c(c-1)(c-2)}{(c+1)(c+2)^2(c+3)^2(c+4)}$$

$$\therefore\; y = a_0 x^c \left\{ 1 - \frac{c-2}{(c+1)(c+2)}x + \frac{(c-1)(c-2)}{(c+1)(c+2)^2(c+3)}x^2 \right.$$
$$\left. - \frac{c(c-1)(c-2)}{(c+1)(c+2)^2(c+3)^2(c+4)}x^3 + \ldots \right\}$$

From the indicial equation above, the values of c are 0 and -1.

Putting c = 0, we have one solution

$$y = u = \ldots\ldots\ldots$$

40

$$y = u = a_0 \left\{ 1 + x + \frac{x^2}{6} \right\}$$

Note that coefficients after the x^2 term are zero, because of the factor c in the numerator.

Putting $c = -1$, we soon find that . . . .

41

coefficients become infinite, because of
the factor $(c + 1)$ in the denominator.

Therefore, we substitute $a_0 = k(c - c_1) = k(c - [-1]) = k(c + 1)$.

$$\therefore \ y = k(c+1)x^c \left\{ 1 - \frac{c-2}{(c+1)(c+2)}x + \frac{(c-1)(c-2)}{(c+1)(c+2)^2(c+3)}x^2 \right.$$
$$\left. - \frac{c(c-1)(c-2)}{(c+1)(c+2)^2(c+3)^2(c+4)}x^3 + \dots \right\}$$

$$= kx^c \left\{ (c+1) - \frac{c-2}{c+2}x + \frac{(c-1)(c-2)}{(c+2)^2(c+3)}x^2 \right.$$
$$\left. - \frac{c(c-1)(c-2)}{(c+2)^2(c+3)^2(c+4)}x^3 + \dots \right\}$$

Now, putting $c = -1$:

$$y = \ \dots \dots \dots \dots$$

42

$$y = kx^{-1} \left\{ 3x + 3x^2 + \frac{x^3}{2} \right\}$$

All subsequent terms are zero, since the numerators all contain a factor $(c + 1)$.

$$\therefore \ y = v = \left\{ 3 + 3x + \frac{x^2}{2} \right\}$$

is a solution.

A solution is also given by $\dfrac{\partial y}{\partial c} = 0$.

So, starting from

$$y = kx^c \left\{ (c+1) - \frac{c-2}{c+2}x + \frac{(c-1)(c-2)}{(c+2)^2(c+3)}x^2 \right.$$

$$\left. - \frac{c(c-1)(c-2)}{(c+2)^2(c+3)^2(c+4)}x^3 + \ldots \right\}$$

$$\frac{\partial y}{\partial c} = kx^c \ln x \left\{ (c+1) - \frac{c-2}{c+2}x + \frac{(c-1)(c-2)}{(c+2)^2(c+3)}x^2 \right.$$

$$\left. - \frac{c(c-1)(c-2)}{(c+2)^2(c+3)^2(c+4)}x^3 + \ldots \right\}$$

$$+ kx^c \frac{\partial}{\partial c} \left\{ (c+1) - \frac{c-2}{c+2}x + \frac{(c-1)(c-2)}{(c+2)^2(c+3)}x^2 - \ldots \right\}$$

We now have to determine the partial derivative of each term.

$$\frac{\partial}{\partial c}(c+1) = 1$$

$$\frac{\partial}{\partial c}\left\{\frac{c-2}{c+2}\right\} = \ldots\ldots\ldots$$

43

$$\boxed{\frac{\partial}{\partial c}\left\{\frac{c-2}{c+2}\right\} = \frac{4}{(c+2)^2}}$$

Now we have to differentiate $\dfrac{(c-1)(c-2)}{(c+2)^2(c+3)}$

Let $t = \dfrac{(c-1)(c-2)}{(c+2)^2(c+3)}$

$$\therefore \ \ln t = \ln(c-1) + \ln(c-2) - 2\ln(c+2) - \ln(c+3)$$

$$\therefore \ \frac{1}{t}\frac{\partial t}{\partial c} = \frac{1}{c-1} + \frac{1}{c-2} - \frac{2}{c+2} - \frac{1}{c+3}$$

$$\therefore \ \frac{\partial t}{\partial c} = \frac{(c-1)(c-2)}{(c+2)^2(c+3)}\left\{\frac{1}{c-1} + \frac{1}{c-2} - \frac{2}{c+2} - \frac{1}{c+3}\right\}$$

$$\therefore \ \text{when } c = -1, \quad \frac{\partial}{\partial c}(c+1) = 1$$

$$\frac{\partial}{\partial c}\left\{\frac{c-2}{c+2}\right\} = 4$$

$$\frac{\partial}{\partial c}\left\{\frac{(c-1)(c-2)}{(c+2)^2(c+3)}\right\} = \ldots\ldots\ldots$$

$$\boxed{-10}$$ **44**

Therefore, when $c = -1$:

$$\frac{\partial y}{\partial c} = kx^{-1}\ln x \left\{ 0 + 3x + 3x^2 + \frac{x^3}{2} + \dots \right\}$$
$$+ kx^{-1}\{1 - 4x - 10x^2 + \dots\}$$

$\therefore$ Another solution is

$$y = w = C \left\{ \ln x \left(3 + 3x + \frac{x^2}{2} + \dots \right) + x^{-1}(1 - 4x - 10x^2 + \dots) \right\}$$

Now we have a problem, for we seem to have three separate series solutions for a second-order differential equation.

(a) $y = u = A \left(1 + x + \dfrac{x^2}{6} \right)$

(b) $y = v = B \left(3 + 3x + \dfrac{x^2}{2} \right)$

(c) $y = w = C \left\{ \ln x \left(3 + 3x + \dfrac{x^2}{2} + \dots \right) + x^{-1}(1 - 4x - 10x^3 + \dots) \right\}$

But (b) is clearly a simple multiple of (a) and thus not a distinct solution. So finally, we have just (a) and (c).

i.e. $y = u = A \left(1 + x + \dfrac{x^2}{6} \right)$

and $y = w = B \left\{ \ln x \left(3 + 3x + \dfrac{x^2}{2} + \dots \right) + x^{-1}(1 - 4x - 10x^3 + \dots) \right\}$

The complete solution is then $y = u + w$

In general if $c_1 - c_2 = n$ where n is a non-zero integer the solution is of the form:

$$y = (1 + k\ln x)x^{c_1}\{a_0 + a_1 x + a_2 x^2 + \dots\} + x^{c_2}\{b_0 + b_1 x + b_2 x^2 + \dots\}$$

> *Finally we have just one more variation to the list in Frames 31 and 32,*
> *so move on*

Example 5   **45**

Solve the equation $xy'' + y' - xy = 0$.

Start off as before and build up expansions for the terms in the left-hand side of the equation.

$$xy'' = \dots\dots\dots$$
$$y' = \dots\dots\dots$$
$$-xy = \dots\dots\dots$$

46

$$xy'' = a_0c(c-1)x^{c-1} + a_1(c+1)cx^c + a_2(c+2)(c+1)x^{c+1} + \ldots$$
$$+ a_r(c+r)(c+r-1)x^{c+r-1} + \ldots$$
$$y' = a_0cx^{c-1} + a_1(c+1)x^c + a_2(c+2)x^{c+1} + \ldots$$
$$+ a_r(c+r)x^{c+r-1} + \ldots$$
$$-xy = \qquad -a_0x^{c+1} - a_1x^{c+2} - \ldots$$
$$- a_rx^{c+r+1} - \ldots$$

The indicial equation, therefore, gives $c = \ldots\ldots\ldots\ldots$

47

$$\boxed{c = 0 \quad \text{(twice)}}$$

Because $a_0\{c(c-1)+c\} = 0 \quad a_0 \neq 0 \quad \therefore \ c^2 = 0 \quad \therefore \ c = 0 \ \text{(twice)}$

Coefficient of x^c gives $\ldots\ldots\ldots\ldots$

48

$$\boxed{a_1 = 0}$$

$[x^c]$: $\qquad a_1(c^2 + c + c + 1) = 0 \quad \therefore \ a_1(c+1)^2 = 0 \quad \therefore \ a_1 = 0$
$[x^{c+1}]$:

$[x^{c+r-1}]$: $\quad a_r\{(c+r)(c+r-1) + (c+r)\} - a_{r-2} = 0$
$$\therefore \ a_r(c+r)^2 = a_{r-2} \quad \therefore \ a_r = \frac{a_{r-2}}{(c+r)^2}$$

$$\therefore \ y = \ldots\ldots\ldots\ldots$$

49

$$\boxed{y = x^c\left\{a_0 + \frac{a_0}{(c+2)^2}x^2 + \frac{a_0}{(c+2)^2(c+4)^2}x^4 + \ldots\right\}}$$

i.e. $\quad y = a_0x^c\left\{1 + \frac{x^2}{(c+2)^2} + \frac{x^4}{(c+2)^2(c+4)^2} + \ldots\right\}$

$\therefore$ When $c = 0$

$$y = u = A\left\{1 + \frac{x^2}{2^2} + \frac{x^4}{2^2 \times 4^2} + \ldots\right\} \qquad (1)$$

This is one solution. Another is given by $v = \dfrac{\partial y}{\partial c}$

$$\frac{\partial y}{\partial c} = a_0 x^c \ln x \left\{ 1 + \frac{x^2}{(c+2)^2} + \frac{x^4}{(c+2)^2(c+4)^2} + \cdots \right\}$$

$$+ a_0 x^c \frac{\partial}{\partial c} \left\{ 1 + \frac{x^2}{(c+2)^2} + \frac{x^4}{(c+2)^2(c+4)^2} + \cdots \right\}$$

Now $\quad \dfrac{\partial}{\partial c}(1) = 0; \quad \dfrac{\partial}{\partial c} \left\{ \dfrac{1}{(c+2)^2} \right\} = \dfrac{-2}{(c+2)^3}$

Let $t = \dfrac{1}{(c+2)^2(c+4)^2} \qquad \therefore \quad \ln t = -2\ln(c+2) - 2\ln(c+4)$

$$\therefore \quad \frac{1}{t}\frac{\partial t}{\partial c} = \frac{-2}{c+2} - \frac{2}{c+4} \qquad \therefore \quad \frac{\partial t}{\partial c} = \frac{-2}{(c+2)^2(c+4)^2}\left\{ \frac{1}{c+2} + \frac{1}{c+4} \right\}$$

$$\therefore \quad \frac{\partial y}{\partial c} = a_0 x^c \ln x \left\{ 1 + \frac{x^2}{(c+2)^2} + \frac{x^4}{(c+2)^2(c+4)^2} + \cdots \right\}$$

$$+ a_0 x^c \left\{ 0 - \frac{2x^2}{(c+2)^3} - \frac{4x^4(c+3)}{(c+2)^3(c+4)^3} + \cdots \right\}$$

$\therefore$ When $c = 0$

$$y = v = \ldots\ldots\ldots\ldots$$

$$\boxed{ y = v = B\left\{ \ln x \left(1 + \frac{x^2}{2^2} + \frac{x^4}{2^2 \times 4^2} + \cdots \right) - \frac{x^2}{2^2} - \frac{3x^4}{2^3 \times 4^2} + \cdots \right\} }$$

(2)

50

So our two solutions are $y = u$ (at 1) and $y = v$ (at 2). The complete solution is therefore $y = u + v$.

In general if $c_1 = c_2 = c$ the solution is of the form

$$y = (1 + k\ln x)x^c\{a_0 + a_1 x + a_2 x^2 + \cdots\} + x^c\{b_1 x + b_2 x^2 + \cdots\}$$

The main points that we have covered in this Programme are listed in the **Revision summary** that follows. Read this in conjunction with the **Can you?** checklist and note any sections that may need further attention: refer back to the relevant parts of the Programme, if necessary. There will then be no trouble with the **Test exercise**. The set of **Further examples** provides an opportunity for further practice

Revision summary 11

51

Frobenius' method

Assume a series solution of the form

$$y = x^c\{a_0 + a_1x + a_2x^2 + \ldots + a_rx^r + \ldots\} \quad a_0 \neq 0$$

(1) Differentiate the assumed series to find y' and y''.
(2) Substitute in the equation.
(3) Equate coefficients of corresponding powers of x on each side of the equation – usually written with zero on the right-hand side.
(4) Coefficient of the lowest power of x gives the *indicial equation* from which values of c are obtained, $c = c_1$ and $c = c_2$.

Case 1: c_1 and c_2 differ by a quantity *not an integer*. Substitute $c = c_1$ and $c = c_2$ in the series for y.

Case 2: c_1 and c_2 differ by an *integer* and make a coefficient *indeterminate* when $c = c_1$. Substitute $c = c_1$ to obtain the complete solution.

Case 3: c_1 and c_2 ($c_1 < c_2$) differ by an *integer* and make a coefficient *infinite* when $c = c_1$. Replace a_0 by $k(c - c_1)$. Two independent solutions are then obtained by putting $c = c_1$ in the new series for y and for $\dfrac{\partial y}{\partial c}$.

In general if $c_1 - c_2 = n$ where n is a non-zero integer, the solution is of the form

$$y = (1 + k\ln x)x^{c_1}\{a_0 + a_1x + a_2x^2 + \ldots\} + x^{c_2}\{b_0 + b_1x + b_2x^2 + \ldots\}$$

Case 4: c_1 and c_2 are *equal*. Substitute $c = c_1$ in the series for y and for $\dfrac{\partial y}{\partial c}$. Make the substitution after differentiating.

In general if $c_1 = c_2 = c$, the solution is of the form

$$y = (1 + k\ln x)x^c\{a_0 + a_1x + a_2x^2 + \ldots\} + x^c\{b_1x + b_2x^2 + \ldots\}$$

Can you?

52

Checklist 11

Check this item before and after you try the end of Programme test

On a scale of 1 to 5 how confident are you that you can: **Frames**

- Apply Frobenius' method of obtaining a series solution to a second-order homogeneous ordinary differential equation by first differentiating the assumed series several times?   [1] to [4]

  Yes ☐ ☐ ☐ ☐ ☐   *No*

- Substitute into the differential equation and equate coefficients of corresponding powers?   [5] to [8]

  Yes ☐ ☐ ☐ ☐ ☐   *No*

- Derive the indicial equation?   [9]

  Yes ☐ ☐ ☐ ☐ ☐   *No*

- Distinguish between the possible four distinct outcomes arising from the indicial equation?   [10] to [50]

  Yes ☐ ☐ ☐ ☐ ☐   *No*

 Test exercise 11

53

1   Determine a series solution for each of the following

    (a) $3xy'' + 2y' + y = 0$

    (b) $y'' + x^2 y = 0$

    (c) $xy'' + 3y' - y = 0$

 Further problems 11

54

1   Use the method of Frobenius to obtain a series solution for each of the following

    (a) $3xy'' + y' - y = 0$

    (b) $y'' + y = 0$

    (c) $y'' - xy = 0$

    (d) $3xy'' + 4y' + y = 0$

    (e) $y'' - xy' + y = 0$

    (f) $xy'' - 3y' + y = 0$

    (g) $xy'' + y' - 3y = 0$

Power series solutions of ordinary differential equations 3

Learning outcomes

When you have completed this Programme you will be able to:

- Apply Frobenius' method to Bessel's equation to derive Bessel functions of the first kind
- Apply Frobenius' method to Legendre's equation to derive Legendre polynomials
- Use Rodrigue's formula to derive Legendre polynomials and the generating function to obtain some of their properties
- Recognise a Sturm–Liouville system and the orthogonality properties of its eigenfunctions
- Write a polynomial in *x* as a finite series of Legendre polynomials

Introduction

A common feature of certain differential equations is that they appear in a multiplicity of guises in the application of mathematics to problems in physics and engineering. For example, Bessel's equation appears in the study of electromagnetic radiation, heat conduction, vibrational modes of a membrane and signal processing to name but a few. Many of these equations have solutions (called *special functions*) in the form of infinite series that are accessible by the method of Frobenius and in this Programme we shall consider two of these equations, namely Bessel's equation and Legendre's equation.

Move to the next frame

1

A second-order differential equation that occurs frequently in branches of technology is of the form

$$x^2 y'' + xy' + (x^2 - v^2)y = 0$$

where v is a real constant.

Starting with $y = x^c(a_0 + a_1 x + a_2 x^2 + a_3 x^3 + \ldots + a_r x^r + \ldots)$ and proceeding as before with the Frobenius method of Programme 11, we obtain

$$c = \pm v \quad \text{and} \quad a_1 = 0$$

The recurrence relation is $\quad a_r = \dfrac{a_{r-2}}{v^2 - (c + r)^2} \quad$ for $r \geq 2$.

It follows that $a_1 = a_3 = a_5 = a_7 = \ldots = 0$

and that $\quad a_2 = \ldots\ldots\ldots\ldots; \quad a_4 = \ldots\ldots\ldots\ldots; \quad a_6 = \ldots\ldots\ldots\ldots$

2

$$a_2 = \frac{a_0}{v^2 - (c + 2)^2}; \quad a_4 = \frac{a_0}{\left[v^2 - (c + 2)^2\right]\left[v^2 - (c + 4)^2\right]};$$

$$a_6 = \frac{a_0}{\left[v^2 - (c + 2)^2\right]\left[v^2 - (c + 4)^2\right]\left[v^2 - (c + 6)^2\right]}$$

$\therefore$ When $c = +v \quad a_2 = \ldots\ldots\ldots\ldots; \quad a_4 = \ldots\ldots\ldots\ldots$

$\qquad\qquad\qquad a_6 = \ldots\ldots\ldots\ldots; \quad a_r = \ldots\ldots\ldots\ldots$

3

4

$$a_2 = \frac{-a_0}{2^2(v+1)}; \qquad a_4 = \frac{a_0}{2^4 \times 2(v+1)(v+2)}$$

$$a_6 = \frac{-a_0}{2^6 \times 3!(v+1)(v+2)(v+3)}$$

$$a_r = \frac{(-1)^{r/2}a_0}{2^r \times (r/2)!(v+1)(v+2)\ldots(v+r/2)} \quad \text{for } r \text{ even}$$

The resulting series solution is therefore

$$y = u = \ldots\ldots\ldots\ldots$$

5

$$y = u = Ax^v\left\{1 - \frac{x^2}{2^2(v+1)} + \frac{x^4}{2^4 \times 2!(v+1)(v+2)}\right.$$
$$\left. - \frac{x^6}{2^6 \times 3!(v+1)(v+2)(v+3)} + \ldots\right\}$$

This is valid provided v is not a negative integer.

Similarly, *when* $c = -v$

$$y = w = Bx^{-v}\left\{1 + \frac{x^2}{2^2(v-1)} + \frac{x^4}{2^4 \times 2!(v-1)(v-2)}\right.$$
$$\left. + \frac{x^6}{2^6 \times 3!(v-1)(v-2)(v-3)} + \ldots\right\}$$

This is valid provided v is not a positive integer.

Except for these two restrictions, the complete solution of Bessel's equation is therefore $y = u + w$ with the two arbitrary constants A and B.

6 Bessel functions

It is convenient to present the two results obtained above in terms of the *gamma function* $\Gamma(x)$ where the letter Γ is the capital Greek gamma. We shall deal with gamma functions in more detail in Programme 16 but for now we only require two simple properties of gamma functions, namely

For $x > 0$, $\Gamma(x+1) = x\Gamma(x)$ and $\Gamma(1) = 1$

What happens when $x \leq 0$ or what $\Gamma(x)$ looks like in terms of a general variable x does not matter for now. What is important is that for $x > 0$ these simple properties give rise to the following equations:

$$\Gamma(x+1) = x\Gamma(x)$$
$$\Gamma(x+2) = (x+1)\Gamma(x+1) = (x+1)x\Gamma(x)$$
$$\Gamma(x+3) = (x+2)\Gamma(x+2) = (x+2)(x+1)x\Gamma(x), \quad \text{etc.}$$

▶

Then if $x = 1$

$$\Gamma(1 + 1) = 1 + \Gamma(1) = 1$$
$$\Gamma(1 + 2) = (1 + 1)\Gamma(1 + 1) = 2 \times 1 \times \Gamma(1) = 2 \times 1$$
$$\Gamma(1 + 3) = (1 + 2)\Gamma(1 + 2) = 3 \times 2 \times 1$$
$$\Gamma(1 + 4) = \ldots\ldots\ldots\ldots$$

Next frame

| $4 \times 3 \times 2 \times 1 = 4!$ | **7** |

Because:

$$\Gamma(1 + 4) = (1 + 3)\Gamma(1 + 3) = 4 \times 3 \times 2 \times 1 = 4!$$

And so, if $x = 1$ and n is a positive integer

$$\Gamma(1 + n) = \ldots\ldots\ldots\ldots$$

The answer is in the next frame

| $n \times (n - 1) \times (\ldots) \times 2 \times 1 = n!$ | **8** |

Because:

$$\Gamma(1 + n) = (1 + [n - 1]) \times \Gamma(1 + [n - 1])$$
$$= n \times \Gamma(1 + [n - 1])$$
$$= n \times (1 + [n - 2]) \times \Gamma(1 + [n - 2])$$
$$= n \times (n - 1) \times \Gamma(1 + [n - 2])$$
$$= \ldots\ldots\ldots\ldots$$
$$= n \times (n - 1) \times (\ldots) \times \Gamma(1 + [n - n])$$
$$= n \times (n - 1) \times (\ldots) \times \Gamma(1)$$
$$= n \times (n - 1) \times (\ldots) \times 1$$
$$= n!$$

If now, in Frame 3, we assign to the arbitrary constant a_0 the value $\dfrac{1}{2^v \Gamma(v + 1)}$,

then we have, for $c = v$

$$a_2 = \frac{a_0}{v^2 - (c + 2)^2} = \frac{a_0}{(v - c - 2)(v + c + 2)} = \frac{a_0}{-2(2v + 2)}$$
$$= \frac{-1}{2^2(v + 1)} \cdot \frac{1}{2^v \Gamma(v + 1)} = \frac{-1}{2^{v+2}(1!)\Gamma(v + 2)}$$

Similarly $\qquad a_4 = \ldots\ldots\ldots\ldots$

9

$$a_4 = \frac{1}{2^{v+4}(2!)\Gamma(v+3)}$$

Because

$$a_4 = \frac{a_2}{v^2 - (c+4)^2} = \frac{a_2}{(v-c-4)(v+c+4)} = \frac{a_2}{-4(2v+4)}$$

$$= \frac{-1}{2^3(v+2)} \cdot \frac{-1}{2^{v+2}(1!)\Gamma(v+2)} = \frac{1}{2^{v+4}(2!)\Gamma(v+3)}$$

and $\quad a_6 = \ldots\ldots\ldots$

10

$$a_6 = \frac{-1}{2^{v+6}(3!)\Gamma(v+4)}$$

We can see the pattern taking shape.

$$a_r = \frac{(-1)^{r/2}}{2^{v+r}\left(\frac{r}{2}!\right)\Gamma\left(v+\frac{r}{2}+1\right)} \quad \text{for } r \text{ even.} \quad \therefore \text{ Put } r = 2k$$

The result then becomes

$$a_{2k} = \ldots\ldots\ldots$$

11

$$a_{2k} = \frac{(-1)^k}{2^{v+2k}(k!)\Gamma(v+k+1)} \quad k = 1, 2, 3, \ldots$$

Therefore, we can write the new form of the series for y as

$$y = x^v\left\{\frac{1}{2^v\Gamma(v+1)} - \frac{x^2}{2^{v+2}(1!)\Gamma(v+2)} + \frac{x^4}{2^{v+4}(2!)\Gamma(v+3)} - \cdots\right\}$$

This is called the *Bessel function of the first kind of order v* and is denoted by $J_v(x)$.

$$\therefore J_v(x) = \left(\frac{x}{2}\right)^v\left\{\frac{1}{\Gamma(v+1)} - \frac{x^2}{2^2(1!)\Gamma(v+2)} + \frac{x^4}{2^4(2!)\Gamma(v+3)} - \cdots\right\}$$

This is valid provided v is not $\ldots\ldots\ldots$

12

a negative integer

– otherwise some of the terms would become infinite.

If we take the other value for c, i.e. $c = -v$, the corresponding result becomes

$$J_{-v}(x) = \ldots\ldots\ldots$$

13

$$J_{-v}(x) = \left(\frac{x}{2}\right)^{-v} \left\{ \frac{1}{\Gamma(1-v)} - \frac{x^2}{2(1!)\Gamma(2-v)} + \frac{x^4}{2^2(2!)\Gamma(3-v)} - \cdots \right\}$$

provided that v is not a positive integer.

In general terms

$$J_v(x) = \left(\frac{x}{2}\right)^{v} \sum_{k=0}^{\infty} \frac{(-1)^k x^{2k}}{2^{2k}(k!)\Gamma(v+k+1)}$$

$$J_{-v}(x) = \left(\frac{x}{2}\right)^{-v} \sum_{k=0}^{\infty} \frac{(-1)^k x^{2k}}{2^{2k}(k!)\Gamma(k-v+1)}$$

The convergence of the series for all values of x can be established by the normal ratio test.

$J_v(x)$ and $J_{-v}(x)$ are two independent solutions of the original equation. Hence, the complete solution is

$$y = AJ_v(x) + BJ_{-v}(x)$$

where A and B are constants.

Make a note of the expressions for $J_v(x)$ and $J_{-v}(x)$.
Then on to the next frame

Some Bessel functions are commonly used and are worthy of special mention. **14**
This arises when v is a positive integer, denoted by n.

$$\therefore \ J_n(x) = \left(\frac{x}{2}\right)^{n} \sum_{k=0}^{\infty} \frac{(-1)^k x^{2k}}{2^{2k}(k!)\Gamma(n+k+1)}$$

From our work on gamma functions, $\Gamma(k+1) = k!$ for $k = 0,\ 1,\ 2,\ldots$

$$\therefore \ \Gamma(n+k+1) = (n+k)!$$

and the result above then becomes

$$J_n(x) = \ldots\ldots\ldots\ldots$$

15

$$J_n(x) = \left(\frac{x}{2}\right)^{n} \sum_{k=0}^{\infty} \frac{(-1)^k x^{2k}}{2^{2k}(k!)(n+k)!}$$

We have seen that $J_v(x)$ and $J_{-v}(x)$ are two solutions of Bessel's equation. When v and $-v$ are not integers, the two solutions are independent of each other. Then $y = AJ_v(x) + BJ_{-v}(x)$.

When, however, $v = n$ (integer), then $J_n(x)$ and $J_{-n}(x)$ are not independent, but are related by $J_{-n}(x) = (-1)^n J_n(x)$. This can be shown by referring once again to our knowledge of gamma functions.

$$\Gamma(x+1) = x\Gamma(x) \quad \therefore \ \Gamma(x) = \frac{\Gamma(x+1)}{x}$$

and for negative integral values of x, or zero, $\Gamma(x)$ is infinite.

▶

From the previous result:

$$J_{-v}(x) = \left(\frac{x}{2}\right)^{-v} \sum_{k=0}^{\infty} \frac{(-1)^k x^{2k}}{2^{2k}(k!)\Gamma(k-v+1)} \qquad k = 0, 1, 2, \ldots$$

Let us consider the gamma function $\Gamma(k-v+1)$ in the denominator and let v approach closely to a positive integer n.

Then $\qquad \Gamma(k-v+1) \to \Gamma(k-n+1)$.

When $k-n+1 \le 0$, i.e. when $k \le (n-1)$, then $\Gamma(k-n+1)$ is infinite.

The first finite value of $\Gamma(k-n+1)$ occurs for $k = n$.

When values of $\Gamma(k-v+1)$ are infinite the coefficients of $J_{-v}(x)$ are

. . . .

16

$\boxed{\text{zero}}$

The series, therefore, starts at $k = n$

$$\therefore J_{-n}(x) = \left(\frac{x}{2}\right)^{-n} \sum_{k=n}^{\infty} \frac{(-1)^k x^{2k}}{2^{2k}(k!)\Gamma(k-n+1)}$$

$$= \sum_{k=n}^{\infty} \frac{(-1)^k x^{2k-n}}{2^{2k-n}(k!)\Gamma(k-n+1)} \qquad \text{Put } k = p+n$$

$$= \sum_{p=0}^{\infty} \frac{(-1)^{p+n} x^{2p+n}}{2^{2p+n}(k!)(k-n)!}$$

$$= (-1)^n \sum_{p=0}^{\infty} \frac{(-1)^p x^{2p+n}}{2^{2p+n}(p!)(p+n)!}$$

$$= (-1)^n \left(\frac{x}{2}\right)^n \sum_{p=0}^{\infty} \frac{(-1)^p x^{2p}}{2^{2p}(p!)(p+n)!}$$

$$= (-1)^n \left(\frac{x}{2}\right)^n \sum_{k=0}^{\infty} \frac{(-1)^k x^{2k}}{2^{2k}(k!)(k+n)!}$$

$$\therefore J_{-n}(x) = (-1)^n J_n(x)$$

So, after all that, the series for $J_n(x) = \ldots \ldots \ldots \ldots$

17

$$\boxed{J_n(x) = \left(\frac{x}{2}\right)^n \left\{\frac{1}{n!} - \frac{1}{(n+1)!}\left(\frac{x}{2}\right)^2 + \frac{1}{(2!)(n+2)!}\left(\frac{x}{2}\right)^4 - \ldots \ldots \ldots\right\}}$$

From this we obtain two commonly used functions

$$J_0(x) = \ldots \ldots \ldots \ldots$$

<table>
<tr><td>

$$J_0(x) = 1 - \frac{1}{(1!)^2}\left(\frac{x}{2}\right)^2 + \frac{1}{(2!)^2}\left(\frac{x}{2}\right)^4 - \frac{1}{(3!)^2}\left(\frac{x}{2}\right)^6 + \cdots$$

</td><td>

18

</td></tr>
</table>

and $\qquad\qquad J_1(x) = \ldots\ldots\ldots$

<table>
<tr><td>

$$J_1(x) = \frac{x}{2}\left\{1 - \frac{1}{(1!)(2!)}\left(\frac{x}{2}\right)^2 + \frac{1}{(2!)(3!)}\left(\frac{x}{2}\right)^4 + \cdots\right\}$$

</td><td>

19

</td></tr>
</table>

Bessel functions for a range of values of n and x are tabulated in published lists of mathematical data. Of these, $J_0(x)$ and $J_1(x)$ are most commonly used.

Graphs of Bessel functions $J_0(x)$ and $J_1(x)$ **20**

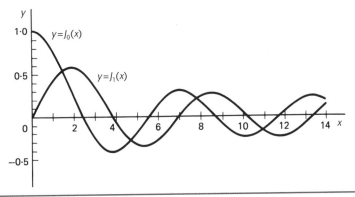

Legendre's equation

21

Another equation of special interest in engineering applications is Legendre's equation of the form

$$(1 - x^2)y'' - 2xy' + k(k+1)y = 0$$

where k is a real constant.

This may be solved by the Frobenius method as before. In this case, the indicial equation gives $c = 0$ and $c = 1$, and the two corresponding solutions are

(a) $c = 0$: $y = a_0\left\{1 - \dfrac{k(k+1)}{2!}x^2 + \dfrac{k(k-2)(k+1)(k+3)}{4!}x^4 - \cdots\right\}$

(b) $c = 1$: $y = a_1\left\{x - \dfrac{(k-1)(k+2)}{3!}x^3\right.$

$$\left. + \frac{(k-1)(k-3)(k+2)(k+4)}{5!}x^5 - \cdots\right\}$$

where a_0 and a_1 are the usual arbitrary constants

▶

Legendre polynomials

When k is an integer (n), one of the solution series terminates after a finite number of terms. The resulting polynomial in x, denoted by $P_n(x)$, is called a *Legendre polynomial*, with a_0 or a_1 being chosen so that the polynomial has unit value when $x = 1$.

For example $P_2(x) = \ldots\ldots\ldots$

22

$$P_2(x) = \frac{1}{2}(3x^2 - 1)$$

Because, in $P_2(x)$, $n = k = 2$

$$\therefore \ y = a_0\left\{1 - \frac{2 \times 3}{2!}x^2 + 0 + 0 + \ldots\right\}$$

$$= a_0\{1 - 3x^2\}$$

The constant a_0 is then chosen to make $y = 1$ when $x = 1$

i.e. $1 = a_0(1 - 3)$ $\therefore \ a_0 = -\frac{1}{2}$

$$\therefore \ P_2(x) = -\tfrac{1}{2}(1 - 3x^2) = \tfrac{1}{2}(3x^2 - 1)$$

Similarly $P_3(x) = \ldots\ldots\ldots$

23

$$P_3(x) = \frac{1}{2}(5x^3 - 3x)$$

Here $n = k = 3$

$$\therefore \ y = a_1\left\{x - \frac{2 \times 5}{3!}x^3 + 0 + 0 + \ldots\right\}$$

$$= a_1\left\{x - \frac{5x^3}{3}\right\}$$

$y = 1$ when $x = 1$ $\therefore \ a_1\left(1 - \frac{5}{3}\right) = 1$ $\therefore \ a_1 = -\frac{3}{2}$

$$\therefore \ P_3(x) = -\frac{3}{2}\left(x - \frac{5x^3}{3}\right) = \frac{1}{2}(5x^3 - 3x)$$

---

24   **Rodrigue's formula and the generating function**

Legendre polynomials can be derived by using *Rodrigue's formula*

$$P_n(x) = \frac{1}{2^n n!}\frac{\mathrm{d}^n}{\mathrm{d}x^n}(x^2 - 1)^n$$

so using this formula

$$P_4(x) = \ldots\ldots\ldots$$

25

$$P_4(x) = \frac{1}{8}\left(35x^4 - 30x^2 + 3\right)$$

Because

$$P_4(x) = \frac{1}{2^4 4!}\frac{d^4}{dx^4}\left(x^2 - 1\right)^4$$

$$= \frac{1}{384}\frac{d^4}{dx^4}\left(x^8 - 4x^6 + 6x^4 - 4x^2 + 1\right)$$

$$= \frac{1}{384}\frac{d^3}{dx^3}\left(8x^7 - 24x^5 + 24x^3 - 8x\right)$$

$$= \frac{1}{384}\frac{d^2}{dx^2}\left(56x^6 - 120x^4 + 72x^2 - 8\right)$$

$$= \frac{1}{384}\frac{d}{dx}\left(336x^5 - 480x^3 + 144x\right)$$

$$= \frac{1}{384}\left(1680x^4 - 1440x^2 + 144\right)$$

$$= \frac{1}{8}\left(35x^4 - 30x^2 + 3\right)$$

In addition to Rodrigue's formula, the function

$$\frac{1}{\sqrt{1 - 2xt + t^2}} = \sum_{n=0}^{\infty} P_n(x)t^n, \qquad |t| < 1$$

is called the *generating function* for Legendre polynomials and can be used to obtain some of their properties. For example using this generating function we find that

$$P_n(1) = \dots\dots\dots$$

26

$$P_n(1) = 1$$

Because

When $x = 1$ the generating function becomes

$$\frac{1}{\sqrt{1 - 2t + t^2}} = \sum_{n=0}^{\infty} P_n(1)t^n, \qquad |t| < 1$$

Noting that $\dfrac{1}{\sqrt{1 - 2t + t^2}} = \dfrac{1}{\sqrt{(1-t)^2}} = \dfrac{1}{1 - t} = (1 - t)^{-1}$, the left-hand side

is expanded by the binomial theorem to give

$$(1 - t)^{-1} = 1 + t + t^2 + t^3 + \dots = \sum_{n=0}^{\infty} t^n.$$

Therefore $\displaystyle\sum_{n=0}^{\infty} t^n = \sum_{n=0}^{\infty} P_n(1)t^n$ and so $P_n(1) = 1$

By a similar reasoning

$$P_n(-1) = \dots\dots\dots$$

27

$$\boxed{P_n(-1) = (-1)^n}$$

Because

When $x = -1$ the generating function becomes

$$\frac{1}{\sqrt{1 + 2t + t^2}} = \sum_{n=0}^{\infty} P_n(-1)t^n$$

Noting that $\dfrac{1}{\sqrt{1 + 2t + t^2}} = \dfrac{1}{\sqrt{(1+t)^2}} = \dfrac{1}{1+t} = (1+t)^{-1}$, the left-hand side

is expanded by the binomial theorem to give

$$(1+t)^{-1} = 1 - t + t^2 - t^3 + \ldots = \sum_{n=0}^{\infty} (-1)^n t^n. \text{ Therefore}$$

$$\sum_{n=0}^{\infty} (-1)^n t^n = \sum_{n=0}^{\infty} P_n(-1)t^n \text{ and so } P_n(-1) = (-1)^n$$

Legendre's equation, whose solutions are expressed in terms of Legendre polynomials, is an example of a particular class of differential equations referred to as Sturm–Liouville systems. In the following frames we shall look at such systems more closely.

So on to the next frame

Sturm–Liouville systems

28   A boundary value problem that is described by a differential equation of the general form

$$(p(x)y')' + (q(x) + \lambda r(x))y = 0 \quad \text{for} \quad a \le x \le b \text{ and } r(x) > 0$$

where the boundary conditions can be written in the form

$$\alpha_1 y(a) + \alpha_2 y'(a) = 0 \quad \text{and} \quad \beta_1 y(b) + \beta_2 y'(b) = 0$$

is called a **Sturm–Liouville** system. Solutions of such a system are in the form of an infinite sequence of *eigenfunctions* y_n, each corresponding to an *eigenvalue* λ_n of the system for $n = 0, 1, 2, \ldots$.

For example, consider the differential equation

$$y'' + \lambda y = 0 \quad \text{for} \quad 0 \le x \le 5$$

where here, $a = 0$ and $b = 5$. Also

$$y(0) = 0 \quad \text{and} \quad y(5) = 0$$

By comparing this equation with the general form given above we can see that

$$p(x) = \ldots\ldots\ldots\ldots; \quad q(x) = \ldots\ldots\ldots\ldots; \quad r(x) = \ldots\ldots\ldots\ldots;$$

$$\alpha_2 = \ldots\ldots\ldots\ldots; \quad \beta_2 = \ldots\ldots\ldots\ldots$$

$$\boxed{p(x) = 1; \quad q(x) = 0; \quad r(x) = 1; \quad \alpha_2 = 0; \quad \beta_2 = 0}$$ **29**

Because

By performing the differentiation on the left-hand term of

$$(p(x)y')' + (q(x) + \lambda r(x))y = 0$$

we find that the differential equation can be written as

$$p(x)y'' + p'(x)y' + (q(x) + \lambda r(x))y = 0$$

By inspection, comparing this form with the differential equation $y'' + \lambda y = 0$ it is easily seen that $p(x) = 1$, $q(x) = 0$, $r(x) = 1$ and comparing boundary conditions gives $\alpha_2 = 0$ and $\beta_2 = 0$.

To solve the equation $y'' + \lambda y = 0$ we use the auxiliary equation $m^2 + \lambda = 0$ which has solutions $m = \pm j\sqrt{\lambda}$ (refer to *Engineering Mathematics (Sixth Edition)*, page 1093, Frame 5ff). This means that the solution can be written in the form

$$y = A \sin \ldots\ldots\ldots + B \cos \ldots\ldots\ldots$$

$$\boxed{y = A \sin \sqrt{\lambda}x + B \cos \sqrt{\lambda}x}$$ **30**

Because

When the solutions to the auxiliary equation are of the form $m = \alpha \pm j\beta$ the solution to the differential equation is of the form

$$y = e^{\alpha x}(A \sin \beta x + B \cos \beta x) \quad \text{and here} \quad \alpha = 0 \quad \text{and} \quad \beta = \sqrt{\lambda}$$

Applying the boundary condition $y(0) = 0$ then $B = \ldots\ldots\ldots$

$$\boxed{B = 0}$$ **31**

Because

$y = A \sin \sqrt{\lambda}x + B \cos \sqrt{\lambda}x$ and so $y(0) = A \sin 0 + B \cos 0 = B = 0$. Therefore $y = A \sin \sqrt{\lambda}x$

Applying the boundary condition $y(5) = 0$ then

$$\lambda = \ldots\ldots\ldots$$

$$\boxed{\lambda = \frac{n^2\pi^2}{25}}$$ **32**

Because

$y = A \sin \sqrt{\lambda}x$ therefore $y(5) = A \sin \sqrt{\lambda}5 = 0$. If $A = 0$ the solution reduces to the trivial solution $y = 0$. For a non-trivial solution $\sin \sqrt{\lambda}5 = 0$ and so $\sqrt{\lambda}5 = n\pi$, $n = 0, 1, 2, 3, \ldots$. This means that

$$\sqrt{\lambda} = \frac{n\pi}{5} \quad \text{and so} \quad \lambda = \frac{n^2\pi^2}{25}$$

▶

There is an infinity of eigenvalues, the nth eigenvalue being denoted by λ_n where $\lambda_n = \dfrac{n^2\pi^2}{25}$ and to each eigenvalue there is an eigenvector solution $y_n = A_n \sin\dfrac{n\pi x}{5}$.

---

33 Orthogonality

If two different functions $f(x)$ and $g(x)$ are defined on the interval $a \le x \le b$ and

$$\int_a^b f(x)g(x)\,dx = 0$$

then we say that the two functions are mutually **orthogonal**. If, on the other hand, a third function $w(x) > 0$ exists such that

$$\int_a^b f(x)g(x)w(x)\,dx = 0$$

then we say that $f(x)$ and $g(x)$ are mutually orthogonal *with respect to the weight function* $w(x)$.

One important property of the solutions to a Sturm–Liouville system is that the solutions are all mutually orthogonal with respect to the weight function $r(x)$. For instance, in the previous example the individual solutions were given as

$$y_n = A_n \sin\frac{n\pi x}{5} \quad \text{where} \quad r(x) = 1$$

and so if $m \ne n$

$$\int_0^5 y_m(x)y_n(x)r(x)\,dx = \ldots\ldots\ldots\ldots$$

34

$$\boxed{\int_0^5 y_m(x)y_n(x)r(x)\,dx = 0}$$

Because

$$\int_0^5 y_m(x)y_n(x)r(x)\,dx = \int_0^5 A_m \sin\frac{m\pi x}{5} A_n \sin\frac{n\pi x}{5}\,dx \qquad \text{where } r(x) = 1$$

$$= A_m A_n \int_0^5 \sin\frac{m\pi x}{5}\sin\frac{n\pi x}{5}\,dx$$

$$= \frac{A_m A_n}{2} \int_0^5 \left(\cos\frac{(m-n)\pi x}{5} - \cos\frac{(m+n)\pi x}{5}\right)dx$$

$$= \frac{A_m A_n}{2}\left[-\frac{5}{(m-n)\pi}\sin\frac{(m-n)\pi x}{5}\right.$$

$$\left.+ \frac{5}{(m+n)\pi}\sin\frac{(m+n)\pi x}{5}\right]_0^5 \qquad \text{provided } m \ne n$$

$$= 0$$

Summary   **35**

1   A Sturm–Liouville system is a differential equation of the form

$$p(x)y'' + p'(x)y' + (q(x) + \lambda r(x))y = 0 \quad \text{for}$$
$$a \le x \le b \text{ and } r(x) > 0$$

where the boundary conditions can be written in the form

$$\alpha_1 y(a) + \alpha_2 y'(a) = 0 \quad \text{and} \quad \beta_1 y(b) + \beta_2 y'(b) = 0$$

2   Solutions y_n to a Sturm–Liouville system are called eigenvectors, each corresponding to an eigenvalue λ_n for $n = 0, 1, 2, \ldots$

3   The solutions y_n are mutually orthogonal with respect to the weighting $r(x)$. That is

$$\int_a^b y_m(x)y_n(x)r(x)\,\mathrm{d}x = 0 \qquad (m \neq n)$$

Keep going

Legendre's equation revisited   **36**

The equation $(1 - x^2)y'' - 2xy' + n(n+1)y = 0$ is Legendre's equation and has Legendre polynomials as solutions. That is

$$y_n = P_n(x) \qquad \text{where } P_n(1) = 1 \text{ and } P_n(-1) = (-1)^n$$

This equation is an example of a Sturm–Liouville system

$$p(x)y'' + p'(x)y' + (q(x) + \lambda r(x))y = 0$$

with boundary conditions

$$\alpha_1 y(a) + \alpha_2 y'(a) = 0 \quad \text{and} \quad \beta_1 y(b) + \beta_2 y'(b) = 0 \quad \text{where}$$
$$p(x) = \ldots\ldots\ldots\ldots; \quad q(x) = \ldots\ldots\ldots\ldots; \quad r(x) = \ldots\ldots\ldots\ldots;$$
$$\alpha_1, \alpha_2 = \ldots\ldots\ldots\ldots; \quad \beta_1, \beta_2 = \ldots\ldots\ldots\ldots$$

$p(x) = 1 - x^2;$ $q(x) = 0;$ $r(x) = 1;$ $\alpha_1, \alpha_2 = 1, 0;$ $\beta_1, \beta_2 = 1, 0$

37

Consequently, Legendre polynomials are mutually orthogonal. That is, if $m \neq n$

$$\int_{-1}^{1} P_m(x)P_n(x)\,\mathrm{d}x = \ldots\ldots\ldots\ldots$$

$\displaystyle\int_{-1}^{1} P_m(x)P_n(x)\,\mathrm{d}x = 0$

38

▶

Polynomials as a finite series of Legendre polynomials

Many differential equations cannot be solved by the normal analytical means and solution by power series provides a powerful tool in many situations. Furthermore, any polynomial can be written as a finite series of Legendre polynomials.

Example 1

Show that $f(x) = x^2$ can be written as a series of Legendre polynomials.

Assume that

$$f(x) = x^2 = \sum_{n=0}^{\infty} a_n P_n(x), \text{ then}$$

$$x^2 = a_0 P_0(x) + a_1 P_1(x) + a_2 P_2(x) + \dots$$

$$= a_0(1) + a_1(x) + a_2 \frac{3x^2 - 1}{2} + a_3 \frac{5x^3 - 3x}{2} + \dots$$

Since the left-hand side is a polynomial of degree 2 then any Legendre polynomial on the right-hand side containing powers of x greater than 2 must be excluded. Therefore $a_3 = a_4 = \dots = 0$, so that

$$x^2 = a_0 - \frac{a_2}{2} + a_1 x + \frac{3}{2} a_2 x^2 \quad \text{giving} \quad a_2 = \frac{2}{3}, \ a_1 = 0, \ a_0 - \frac{a_2}{2} = 0$$

therefore $a_0 = \frac{1}{3}$, and

$$x^2 = \frac{1}{3} P_0(x) + \frac{2}{3} P_2(x)$$

Now you try one

39

Example 2

The polynomial $1 + x + x^3$ can be written as a series of Legendre polynomials in the form

$$1 + x + x^3 = \dots\dots\dots$$

40

$$\boxed{1 + x + x^3 = P_0(x) + \frac{8}{5} P_1(x) + \frac{2}{5} P_3(x)}$$

Because

$$1 + x + x^3 = a_0 P_0(x) + a_1 P_1(x) + a_2 P_2(x) + \dots$$

$$= a_0(1) + a_1(x) + a_2 \frac{3x^2 - 1}{2} + a_3 \frac{5x^3 - 3x}{2} + \dots$$

Since the left-hand side is a polynomial of degree 3 then any Legendre polynomial on the right-hand side containing powers of x greater than 3 must be excluded. Therefore $a_4 = a_5 = \ldots = 0$, so that

$$1 + x + x^3 = a_0 - \frac{a_2}{2} + \left(a_1 - \frac{3}{2}a_3\right)x + \frac{3}{2}a_2 x^2 + \frac{5}{2}a_3 x^3$$

This gives $a_3 = \frac{2}{5}$, $a_2 = 0$, $a_1 - \frac{3}{2}a_3 = 1$, $a_0 - \frac{a_2}{2} = 1$ therefore $a_0 = 1$, and $a_1 = \frac{8}{5}$ so

$$1 + x + x^3 = P_0(x) + \frac{8}{5}P_1(x) + \frac{2}{5}P_3(x)$$

As usual, the main points that we have covered in this Programme are listed in the **Revision summary** that follows. Read this in conjunction with the **Can you?** checklist and note any sections that may need further attention: refer back to the relevant parts of the Programme, if necessary. There will then be no trouble with the **Test exercise**. The set of **Further problems** provides an opportunity for further practice.

 # Revision summary 12

1 *Bessel's equation*   **41**

$$x^2 y'' + xy' + (x^2 - v^2) = 0$$

where v is a real constant.

Bessel functions: Express the two solutions obtained in terms of gamma functions.

$$J_v(x) = \left(\frac{x}{2}\right)^v \left\{ \frac{1}{\Gamma(v+1)} - \frac{x^2}{2^2(1!)\Gamma(v+2)} + \frac{x^4}{2^4(2!)\Gamma(v+3)} - \cdots \right\}$$

This is the *Bessel function of the first kind of order v* – valid for v not a negative integer.

Also $J_{-v}(x) = \left(\frac{x}{2}\right)^{-v} \left\{ \frac{1}{\Gamma(1-v)} - \frac{x^2}{2(1!)\Gamma(2-v)} + \frac{x^4}{2^2(2!)\Gamma(3-v)} - \cdots \right\}$

provided that v is not a positive integer.

Complete solution is therefore $y = AJ_v(x) + BJ_{-v}(x)$.

▶

When $v = n$ (an integer) $J_{-n}(x) = (-1)^n J_n(x)$

$$J_n(x) = \left(\frac{x}{2}\right)^n \left\{ \frac{1}{n!} - \frac{1}{(n+1)!} \left(\frac{x}{2}\right)^2 + \frac{1}{(2!)(n+2)!} \left(\frac{x}{2}\right)^4 \right.$$
$$\left. - \frac{1}{(3!)(n+3)!} \left(\frac{x}{2}\right)^6 + \cdots \right\}$$

In particular

$$J_0(x) = 1 - \frac{1}{(1!)^2} \left(\frac{x}{2}\right)^2 + \frac{1}{(2!)^2} \left(\frac{x}{2}\right)^4 - \frac{1}{(3!)^2} \left(\frac{x}{2}\right)^6 + \cdots$$

and

$$J_1(x) = \frac{x}{2} \left\{ 1 - \frac{1}{(1!)(2!)} \left(\frac{x}{2}\right)^2 + \frac{1}{(2!)(3!)} \left(\frac{x}{2}\right)^4 - \frac{1}{(3!)(4!)} \left(\frac{x}{2}\right)^6 + \cdots \right\}$$

2 *Legendre's equation*

$$(1 - x^2)y'' - 2xy' + k(k+1)y = 0$$

where k is a real constant.

Solution by Frobenius gives

$$c = 0: \quad y = a_0 \left\{ 1 - \frac{k(k+1)}{2!} x^2 + \frac{k(k-2)(k+1)(k+3)}{4!} x^4 - \cdots \right\}$$

$$c = 1: \quad y = a_1 \left\{ x - \frac{(k-1)(k+2)}{3!} x^3 \right.$$
$$\left. + \frac{(k-1)(k-3)(k+2)(k+4)}{5!} x^5 - \cdots \right\}$$

When k *is an integer*, one series terminates. The resulting polynomial in x, $P_n(x)$, is a *Legendre polynomial*, with a_0 or a_1 being chosen so that the polynomial has unit value when $x = 1$.

3 *Rodrigue's formula*

$$P_n(x) = \frac{1}{2^n n!} \frac{d^n}{dx^n} \left(x^2 - 1\right)^n$$

Generating function

$$\frac{1}{\sqrt{1 - 2xt + t^2}} = \sum_{n=0}^{\infty} P_n(x) t^n$$

4 *Sturm–Liouville systems*

$(p(x)y')' + (q(x) + \lambda r(x))y = 0$ for $a \le x \le b$ and $r(x) > 0$ with boundary conditions $\alpha_1 y(a) + \alpha_2 y'(a) = 0$ and $\beta_1 y(b) + \beta_2 y'(b) = 0$

Solutions y_n to a Sturm–Liouville system are called eigenvectors, each corresponding to an eigenvalue λ_n for $n = 0, 1, 2, \ldots$

▶

5 *Orthogonality*

If two different functions $f(x)$ and $g(x)$ are defined on the interval $a \le x \le b$ and

$$\int_a^b f(x)g(x)\,\mathrm{d}x = 0$$

then the two functions are **orthogonal** to each other. If a function $w(x) > 0$ exists such that

$$\int_a^b f(x)g(x)w(x)\,\mathrm{d}x = 0$$

then $f(x)$ and $g(x)$ are orthogonal to each other *with respect to the weight function $w(x)$.*

The solutions of a Sturm–Liouville system y_n are mutually orthogonal with respect to the weighting $r(x)$. That is

$$\int_a^b y_m(x)y_n(x)r(x)\,\mathrm{d}x = 0 \qquad (m \ne n)$$

6 *Legendre polynomials are mutually orthogonal*

If $m \ne n$ then

$$\int_{-1}^1 P_m(x)P_n(x)\,\mathrm{d}x = 0$$

The orthogonality of the Legendre polynomials permits any polynomial to be written as a finite series of Legendre polynomials.

 # Can you?

Checklist 12

42

Check this list before and after you try the end of Programme test.

On a scale of 1 to 5 how confident are you that you can: **Frames**

- Apply Frobenius' method to Bessel's equation to derive Bessel functions of the first kind?
  Yes ☐ ☐ ☐ ☐ ☐ *No*
  1 to 20

- Apply Frobenius' method to Legendre's equation to derive Legendre polynomials?
  Yes ☐ ☐ ☐ ☐ ☐ *No*
  21 to 23

- Use Rodrigue's formula to derive Legendre polynomials and the generating function to obtain some of their properties?
  Yes ☐ ☐ ☐ ☐ ☐ *No*
  24 to 27

▶

- Recognise a Sturm–Liouville system and the orthogonality
  properties of its eigenfunctions?   28 to 37
  Yes   ☐   ☐   ☐   ☐   ☐   *No*

- Write a polynomial in x as a finite series of Legendre
  polynomials?   38 to 40
  Yes   ☐   ☐   ☐   ☐   ☐   *No*

 ## Test exercise 12

---

43 1 Use Rodrigue's formula $P_n(x) = \dfrac{1}{2^n n!} \dfrac{d^n}{dx^n} (x^2 - 1)^n$ to derive the Legendre

polynomials $P_2(x)$ and $P_3(x)$, and show that $P_2(x)$ and $P_3(x)$ are orthogonal
on $(-1, 1)$.

2 Write $f(x) = 1 - 2x^2$ as a series of Legendre polynomials.

 ## Further problems 12

---

44 1 Verify that $y'' + \lambda y = 0$ where $y'(0) = 0$ and $y(2) = 0$ is a Sturm–Liouville
system. Find the eigenvalues and eigenfunctions of the system and prove that
they are orthogonal in $(0, 2)$.

2 Series solutions of the equation $y'' - 2xy' + 2ny = 0$ are known as Hermite
polynomials, $H_n(x)$, where

$$H_n(x) = (-1)^n e^{x^2} \frac{d^n}{dx^n} \left(e^{-x^2} \right)$$

Derive the first four Hermite polynomials and show that they are orthogonal
with respect to the weight e^{-x^2} in $(-\infty, \infty)$.

3 Series solutions of the equation $xy'' + (1 - x)y' + ny = 0$ are known as Laguerre
polynomials, $L_n(x)$, where

$$L_n(x) = e^x \frac{d^n}{dx^n} (x^n e^{-x})$$

Derive the first four Laguerre polynomials and show that they are orthogonal
with respect to the weight e^{-x} in $(0, \infty)$.

4 Given the generating function for Laguerre polynomials $L_n(x)$ as

$$\frac{e^{-xt/(1-t)}}{1 - t} = \sum_{n=0}^{\infty} \frac{L_n(x)}{n!} t^n$$

show that $L_n(0) = n!$

▶

5   Given the generating function for Hermite polynomials $H_n(x)$ as

$$e^{2tx-t^2} = \sum_{n=0}^{\infty} \frac{H_n(x)}{n!} t^n$$

show that $H_{2n+1}(0) = 0$.

6   Given the generating function for Legendre polynomials $P_n(x)$ as

$$\frac{1}{\sqrt{1-2xt+t^2}} = \sum_{n=0}^{\infty} P_n(x)t^n$$

show that $P_{2n+1}(0) = 0$.

Numerical solutions of ordinary differential equations

Learning outcomes

When you have completed this Programme you will be able to:

- Derive a form of Taylor's series from Maclaurin's series and from it describe a function increment as a series of first and higher-order derivatives of the function

- Describe and apply by means of a spreadsheet the Euler method, the Euler–Cauchy method and the Runge–Kutta method for first-order differential equations

- Describe and apply by means of a spreadsheet the Euler second-order method and the Runge–Kutta method for second-order ordinary differential equations

- Describe and apply by means of a spreadsheet a simple predictor–corrector method.

Prerequisite: Engineering Mathematics (Sixth Edition)
Programme F.4 (Using a spreadsheet)

Introduction

The range of differential equations that can be solved by straightforward analytical methods is relatively restricted. Even solution in series may not always be satisfactory, either because of the slow convergence of the resulting series or because of the involved manipulation in repeated stages of differentiation.

In such cases, where a differential equation and known boundary conditions are given, an approximate solution is often obtainable by the application of numerical methods, where a numerical solution is obtained at discrete values of the independent variable.

The solution of differential equations by numerical methods is a wide subject. The present Programme introduces some of the simpler methods, which nevertheless are of practical use.

Taylor's series

Let us start off by briefly revising the fundamentals of Maclaurin's and Taylor's series.

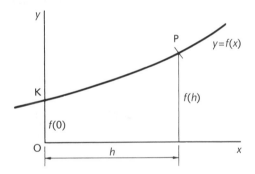

Maclaurin's series for $f(x)$ is

$$f(x) = f(0) + xf'(0) + \frac{x^2}{2!}f''(0) + \ldots + \frac{x^n}{n!}f^n(0) + \ldots \tag{1}$$

and expresses the function $f(x)$ in terms of its successive derivatives at $x = 0$, i.e. at the point K.

Therefore, at P, $\qquad f(h) = \ldots\ldots\ldots\ldots$

2

$$f(h) = f(0) + hf'(0) + \frac{h^2}{2!}f''(0) + \ldots + \frac{h^n}{n!}f^n(0) + \ldots$$

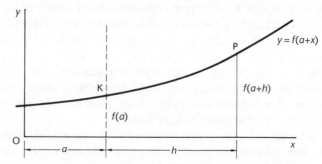

If the y-axis and origin are moved a units to the left, the equation of the same curve relative to the new axes becomes $y = f(a + x)$ and the function value at K is $f(a)$.

At P, $f(a + h) = f(a) + hf'(a) + \frac{h^2}{2!}f''(a) + \ldots + \frac{h^n}{n!}f^n(a) + \ldots$

This is one common form of Taylor's series.

Make a note of it and then move on

3 **Function increment**

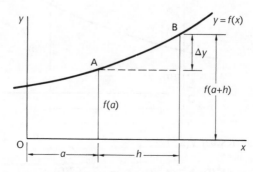

If we know the function value $f(a)$ at A, i.e. at $x = a$, we can apply Taylor's series to determine the function value at a neighbouring point B, i.e. at $x = a + h$.

$$f(a + h) = f(a) + hf'(a) + \frac{h^2}{2!}f''(a) + \frac{h^3}{3!}f'''(a) + \ldots \qquad (3)$$

The *function increment* from A to B $= \Delta y = f(a + h) - f(a)$

i.e. $f(a + h) = f(a) + \Delta y$

where $\Delta y = hf'(a) + \frac{h^2}{2!}f''(a) + \frac{h^3}{3!}f'''(a) + \ldots$

▶

This entails evaluation of an infinite number of derivatives at $x = a$: in practice an approximation is accepted by restricting the number of terms that are used in the series.

This approximation of Taylor's series forms the basis of several numerical methods, some of which we shall now introduce. It should be noted that these early examples have been selected because exact solutions can also be found. The purpose of this is to enable a comparison between the results obtained by a particular method with those obtained from an exact solution, and so to demonstrate the accuracy of the method.

On then to the next frame

First-order differential equations

Numerical solution of $\dfrac{dy}{dx} = f(x, y)$ with the initial condition that, at $x = x_0$, $y = y_0$.

4

Euler's method

The simplest of the numerical methods for solving first-order differential equations is *Euler's method*, in which the Taylor's series

$$f(a + h) = f(a) + hf'(a) \;\left[\; + \frac{h^2}{2!}f''(a) + \frac{h^3}{3!}f'''(a) + \ldots \right.$$

is truncated after the second term to give

$$f(a + h) \approx f(a) + hf'(a) \tag{4}$$

This is a severe approximation, but in practice the 'approximately equals' sign is replaced by the normal 'equals' sign, in the knowledge that the result we obtain will necessarily differ to some extent from the function value we seek. With this in mind, we write

$$f(a + h) = f(a) + hf'(a)$$

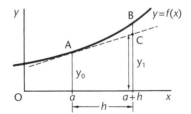

If h is the interval between two near ordinates and if we denote $f(a)$ by y_0, then the relationship

$$f(a + h) = f(a) + hf'(a)$$

becomes

$$y_1 = y_0 + h(y')_0 \tag{5}$$

Hence, knowing y_0, h and $(y')_0$, we can compute y_1, an approximate value for the function value at B.

Make a note of result (5): we shall be using it quite a lot.

Then move on for an example

5

Example 1

Given that $\dfrac{dy}{dx} = 2(1+x) - y$ with the initial condition that at $x = 2$, $y = 5$, we can find an approximate value of y at $x = 2 \cdot 2$, as follows.

We have $y' = 2(1+x) - y$ with $x_0 = 2$, $y_0 = 5$

$$\therefore \ (y')_0 = \dots\dots\dots$$

6

$$\boxed{(y')_0 = 1}$$

We obtain this by substituting x_0 and y_0 in the given equation:

$$(y')_0 = 2(1 + x_0) - y_0 = 2(1 + 2) - 5 \quad \therefore \ (y')_0 = 1$$

So we have $x_0 = 2$; $y_0 = 5$; $(y')_0 = 1$; $x_1 = 2 \cdot 2$; $h = 0 \cdot 2$.

By Euler's relationship:

$$y_1 = y_0 + h(y')_0 \quad \therefore \ y_1 = \dots\dots\dots$$

7

$$\boxed{y_1 = 5 \cdot 2}$$

Because

$$y_1 = y_0 + h(y')_0 = 5 + (0 \cdot 2)1 = 5 \cdot 2$$

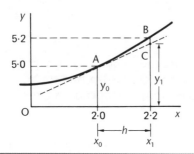

At B, $x_1 = 2 \cdot 2$; $y_1 = 5 \cdot 2$;   and

$$(y')_1 = \dots\dots\dots$$

$$\boxed{(y')_1 = 1\cdot2}$$ 8

$$(y')_1 = 2(1 + x_1) - y_1 = 2(1 + 2\cdot2) - 5\cdot2 = 1\cdot2$$

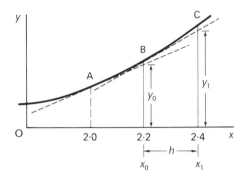

If we take the values of x, y and y' that we have just found for the point B and treat these as new starter values x_0, y_0, $(y')_0$, we can repeat the process and find values corresponding to the point C.

At B, $x_0 = 2\cdot2$; $y_0 = 5\cdot2$; $(y')_0 = 1\cdot2$; $x_1 = 2\cdot4$.

Then at C: $y_1 = \ldots\ldots\ldots$; $(y')_1 = \ldots\ldots\ldots$

$$\boxed{y_1 = 5\cdot44;\quad (y')_1 = 1\cdot36}$$ 9

$$y_1 = y_0 + h(y')_0 = 5\cdot2 + (0\cdot2)1\cdot2 = 5\cdot44$$
$$(y')_1 = 2(1 + x_1) - y_1 = 2(1 + 2\cdot4) - 5\cdot44 = 1\cdot36$$

So we could continue in a step-by-step method. At each stage, the determined values of x_1, y_1 and $(y')_1$ become the new starter values x_0, y_0 and $(y')_0$ for the next stage.

Our results so far can be tabulated thus

x_0	y_0	$(y')_0$	x_1	y_1	$(y')_1$
2·0	5·0	1·0	2·2	5·2	1·2
2·2	5·2	1·2	2·4	5·44	1·36
2·4	5·44	1·36			

Continue the table with a constant interval of $h = 0\cdot2$. The third row can be completed to give

$$x_1 = \ldots\ldots\ldots;\quad y_1 = \ldots\ldots\ldots;\quad (y')_1 = \ldots\ldots\ldots$$

10

$$x_1 = 2 \cdot 6; \quad y_1 = 5 \cdot 712; \quad (y')_1 = 1 \cdot 488$$

Because

$$x_1 = x_0 + h = 2 \cdot 4 + 0 \cdot 2 = 2 \cdot 6$$
$$y_1 = y_0 + h(y')_0 = 5 \cdot 44 + (0 \cdot 2)1 \cdot 36 = 5 \cdot 712$$
$$(y')_1 = 2(1 + x_1) - y_1 = 2(1 + 2 \cdot 6) - 5 \cdot 712 = 1 \cdot 488$$

Now you can continue in the same way and complete the table for

$$x = 2 \cdot 0, \ 2 \cdot 2, \ 2 \cdot 4, \ 2 \cdot 6, \ 2 \cdot 8, \ 3 \cdot 0$$

Finish it off and compare results with the next frame

11   Here is the result.

x_0	y_0	$(y')_0$		x_1	y_1	$(y')_1$
2·0	5·0	1·0		2·2	5·2	1·2
2·2	5·2	1·2		2·4	5·44	1·36
2·4	5·44	1·36		2·6	5·712	1·488
2·6	5·712	1·488		2·8	6·009 6	1·590 4
2·8	6·009 6	1·590 4		3·0	6·327 68	1·672 32
3·0	6·327 68	1·672 32				

In practice, we do not, in fact, enter the values in the right-hand half of the table, but write them in directly as new starter values in the left-hand section of the table.

x_0	y_0	$(y')_0$
2·0	5·0	1·0
2·2	5·2	1·2
2·4	5·44	1·36
2·6	5·712	1·488
2·8	6·009 6	1·590 4
3·0	6·327 68	1·672 32

The particular solution is given by the values of y against x and a graph of the function can be drawn.

Draw the graph of the function carefully on graph paper.

Graph of the solution of $\dfrac{dy}{dx} = 2(1 + x) - y$ with $y = 5$ at $x = 2$. **12**

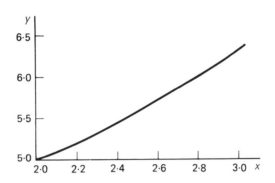

It is an advantage to plot the points step-by-step as the results are built up. In **13** that way, one can check that there is a smooth progression and that no apparent errors in the calculations occur at any one stage.

The differential equation $\dfrac{dy}{dx} = 2(1 + x) - y$ can be solved by the integration factor method (see *Engineering Mathematics, Sixth Edition*, Programme 24) to give the solution

$$y = 2x + e^{2-x}$$

and in the following table we compare our results with the actual values to determine the errors.

x	y (Euler)	y (actual)	Absolute error
2·0	5·0	5·0	0
2·2	5·2	5·218 731	0·018 731
2·4	5·44	5·470 320	0·030 320
2·6	5·712	5·748 812	0·036 812
2·8	6·009 6	6·049 329	0·039 729
3·0	6·327 68	6·367 879	0·040 199

The errors involved in the process are shown. These errors are due mainly to

....

14

> the fact that Taylor's series was truncated after the second term

By now you will appreciate the amount of arithmetic manipulation involved in solving these differential equations – a large amount of which is repetitive. To avoid the tedium and to make the computations more efficient we shall resort to the use of a spreadsheet. If the use of a spreadsheet is a totally new experience to you then you are referred to Programme F.4 of *Engineering Mathematics, Sixth Edition*, where the spreadsheet is introduced as a tool for constructing graphs of functions. If you have a limited knowledge then you will be able to follow the text from here. The spreadsheet we shall be using here is Microsoft Excel, though all commercial spreadsheets possess the equivalent functionality. Alternatively, an iteration process can be used in any computer algebra package such as *Derive*, *Maple* or *Mathematica*.

Open your spreadsheet and in cell A1 enter the letter *n* and press **Enter**. In this first column we are going to enter the iteration numbers. In cell A2 enter the number 0 and press **Enter**. Place the cell highlight in cell A2 and highlight the block of cells A2 to A12 by holding down the mouse button and wiping the highlight down to cell A12. Click the **Edit** command on the Command bar and point at **Fill** from the drop-down menu. Select **Series** from the next drop-down menu and accept the default **Step value** of 1 by clicking **OK** in the Series window.

The cells A3 to A12 fill with . . . .

15

> The numbers 1 to 10

In cell B1 enter the letter *x* – this column is going to contain the successive *x*-values for which the *y*-value is going to be enumerated. In cell B2 enter the number 2 – the initial *x*-value. We now could fill the column in much the same way as we filled the first column, but we have a better way.

Place the cell highlight in cell F1 and enter the number 0·2 – this is the value of *h*, the increment in *x*. Now place the cell highlight in cell B3 and enter the formula

= B2 + F1 followed by **Enter**   (uppercase or lowercase, it does not matter)

The number 2·2 appears in cell B3. Place the cell highlight in cell B3, click the **Edit** command and select **Copy** from the drop-down menu. You have now copied the contents of cell B3 to the clipboard. Now place the cell highlight in cell B4 and highlight the block of cells from B4 to B12. Click the **Edit** command again but this time select **Paste** from the drop-down menu.

The cells B4 to B12 fill with the numbers . . . .

| The numbers 2·4 to 4·0 in intervals of 0·2 | **16** |

How has this happened? When you typed in the cell reference B2 into the formula in cell B3, the spreadsheet understood this to mean *the contents of the cell immediately above current cell B3*. When the formula is copied into cell B4 it means *the contents of the cell immediately above current cell B4*. Entered in this way the address B2 is a *relative address*. On the other hand, when you typed in F1 the spreadsheet understood this to mean the contents of cell F1 and that meaning remains when it is copied – the dollar signs indicate an *absolute address*. So as you move down the column the contents of a cell contain the contents of the cell immediately above it plus the contents of cell F1. You will shortly see the advantages of all this.

For now, place the cell highlight in cell C1 and enter the letter y – this column is going to contain the computed y-values against the corresponding x-values in column B. Place the cell highlight in cell C2 and enter the number 5 – the initial y-value. Before we can compute the y-values in column C we need to be able to tabulate the values of y' – the derivatives of y. Place the cell highlight in cell D1 and enter y' – this column will contain the values of the derivatives of y against the corresponding x-values. Cell D2 will contain the initial value of y' which can be computed from the equation

$$y' = 2(1 + x) - y$$

When $x = x_0 = 2$ and $y = y_0 = 5$ then

$$y_0' = 2(1 + x_0) - y_0 = 2(1 + 2) - 5 = 1$$

so place the cell highlight in cell D2 and enter the formula

$$= 2 * (1 + B2) - C2 \qquad \text{(B2 contains } x_0 \text{ and C2 contains } y_0)$$

The number 1 appears in cell D2. We need to copy this formula down the y' column. Place the cell highlight in cell D2, click **Edit** and select **Copy**. Now place the cell highlight in cell D3 and highlight the block of cells D3 to D12. Click the **Edit** command again and select **Paste**.

The cells D3 to D12 fill with . . . .

17

| The numbers 6·4 to 10·0 in intervals of 0·4 |

Because the cells in the C2 column are currently empty, these values are just $2 * (1 + B2) - 0$.

Now, to compute the *y*-values we use the equation $y_1 = y_0 + h(y')_0$. Place the cell highlight in cell C3 and enter the formula

$$= C2 + \$F\$1 * D2 \qquad \text{(C2 contains } y_0, \text{ F1 contains } h \text{ and D2 contains } (y')_0)$$

and the number 5·2 appears. That is, $y_1 = 5 + (0·2)(1) = 5·2$. This now completes the sequence of operations required to find y_1. To find the values of $y_2 = y(x_2) = y(2·4)$ this sequence is repeated and, to ensure this, all that remains is to copy the formula in cell C3 into cells C4 to C12. So do this to reveal the following display

n	x	y	y'	0·2
0	2	5	1	
1	2·2	5·2	1·2	
2	2·4	5·44	1·36	
3	2·6	5·712	1·488	
4	2·8	6·0096	1·5904	
5	3	6·32768	1·67232	
6	3·2	6·662144	1·737856	
7	3·4	7·0097152	1·7902848	
8	3·6	7·36777216	1·83222784	
9	3·8	7·734217728	1·865782272	
10	4	8·107374182	1·892625818	

Now that was a lot easier than all that arithmetic manipulation by hand, wasn't it? We can tidy this display up by using the **Format** command and by using the various options on the tool bars to change the column widths and to display the numbers in a regular format of 10 decimal places to produce a display that is easier to read.

Next frame

18

n	x	y	y'	$h = 0.2$
0	2·0	5·0000000000	1·0000000000	
1	2·2	5·2000000000	1·2000000000	
2	2·4	5·4400000000	1·3600000000	
3	2·6	5·7120000000	1·4880000000	
4	2·8	6·0096000000	1·5904000000	
5	3·0	6·3276800000	1·6723200000	
6	3·2	6·6621440000	1·7378560000	
7	3·4	7·0097152000	1·7902848000	
8	3·6	7·3677721600	1·8322278400	
9	3·8	7·7342177280	1·8657822720	
10	4·0	8·1073741824	1·8926258176	

Notice that we have added *h=* in cell E1 and justified it to the right and then justified the number 0·2 in F2 to the left so that together they read as an equation. The advantage of isolating the step value 0·2 in cell F1, as we have done, is that we can change the value and immediately see the effects on the calculations. For example, if the contents of F1 are changed to 0·1 the display changes automatically to

n	x	y	y'	$h = 0.1$
0	2·0	5·0000000000	1·0000000000	
1	2·1	5·1000000000	1·1000000000	
2	2·2	5·2100000000	1·1900000000	
3	2·3	5·3290000000	1·2710000000	
4	2·4	5·4561000000	1·3439000000	
5	2·5	5·5904900000	1·4095100000	
6	2·6	5·7314410000	1·4685590000	
7	2·7	5·8782969000	1·5217031000	
8	2·8	6·0304672100	1·5695327900	
9	2·9	6·1874204890	1·6125795110	
10	3·0	6·3486784401	1·6513215599	

Notice that the different values of *h* produce different corresponding values in the tables. For example, for $h = 0.2$ we find that $y(3.0) = 6.327\,680\,0000$ whereas for $h = 0.1$ we have $y(3.0) = 6.348\,678\,4401$. The smaller the value of *h* then, the smaller the errors in the calculation – we shall see this demonstrated explicitly in the next frame.

Go to the next frame

19 The exact value and the errors

The differential equation

$$y' = 2(1 + x) - y$$

can be solved using the integration factor method (see *Engineering Mathematics, Sixth Edition*, Programme 24) to give the solution

$$y = 2x + e^{2-x}$$

We can programme this into the spreadsheet to compare the exact solution with the solution obtained numerically and compute the actual errors. Place the cell highlight in cell E1 and highlight cells E1 and F1. Click **Insert** on the Command bar and select **Columns**. Immediately two new columns appear. Notice that the numbers in the display do not change despite the fact that the *h*-value of 0·2 has moved from F1 to H1 – all the formulas in the spreadsheet will have automatically adjusted themselves. You can check this by highlighting a cell with a formula in it to see the change.

In cell E1 enter the word **Exact** and in cell F1 enter **Errors (%)**. In cell E2 enter the right-hand side of the equation $y = 2x + e^{2-x}$ by using the formula

$$= 2 * B2 + EXP(2 - B2) \quad \text{(the EXP stands for the exponential function)}$$

and copy this into the block of cells E3 to E12. In cell F2 enter the formula for the error

$$= (E2 - C2) * 100/E2 \quad \text{(the error as a percentage of the exact value)}$$

and copy this into the block of cells F3 to F12 to produce the following display

n	x	y	y'	Exact	Errors (%)	h=0·2
0	2·0	5·0000000000	1·0000000000	5·0000000000	0·00	
1	2·2	5·2000000000	1·2000000000	5·2187307531	0·36	
2	2·4	5·4400000000	1·3600000000	5·4703200460	0·55	
3	2·6	5·7120000000	1·4880000000	5·7488116361	0·64	
4	2·8	6·0096000000	1·5904000000	6·0493289641	0·66	
5	3·0	6·3276800000	1·6723200000	6·3678794412	0·63	
6	3·2	6·6621440000	1·7378560000	6·7011942119	0·58	
7	3·4	7·0097152000	1·7902848000	7·0465969639	0·52	
8	3·6	7·3677721600	1·8322278400	7·4018965180	0·46	
9	3·8	7·7342177280	1·8657822720	7·7652988882	0·40	
10	4·0	8·1073741824	1·8926258176	8·1353352832	0·34	

▶

Change the value of *h* to 0·1 and produce the following display

n	x	y	y′	Exact	Errors *h* = 0·1 (%)
0	2·0	5·0000000000	1·0000000000	5·0000000000	0·00
1	2·1	5·1000000000	1·1000000000	5·1048374180	0·09
2	2·2	5·2100000000	1·1900000000	5·2187307531	0·17
3	2·3	5·3290000000	1·2710000000	5·3408182207	0·22
4	2·4	5·4561000000	1·3439000000	5·4703200460	0·26
5	2·5	5·5904900000	1·4095100000	5·6065306597	0·29
6	2·6	5·7314410000	1·4685590000	5·7488116361	0·30
7	2·7	5·8782969000	1·5217031000	5·8965853038	0·31
8	2·8	6·0304672100	1·5695327900	6·0493289641	0·31
9	2·9	6·1874204890	1·6125795110	6·2065696597	0·31
10	3·0	6·3486784401	1·6513215599	6·3678794412	0·30

When $h = 0·2$ the error in $y(3·0)$ is 0·63% whereas when $h = 0·1$ the error in $y(3·0)$ is 0·30%.

The smaller the value of *h* the ....

| smaller the error | **20** |

Having completed your first spreadsheet you can now use it as a template for similar problems.

To avoid losing the work that you have already done, save your spreadsheet under some suitable name. When that is complete, highlight all the cells from A1 to G12 and copy them onto the clipboard using the **Edit-Copy** sequence of commands. Now click the **Sheet 2** tab at the bottom of your spreadsheet to reveal a blank worksheet. Place the cell highlight in cell A1, click **Edit** and select **Paste**. The entire contents of **Sheet 1** are now copied to **Sheet 2** in readiness for editing to accommodate a new problem.

So let's look at another example.

Example 2

Obtain a numerical solution of the equation

$$\frac{dy}{dx} = 1 + x - y$$

with the initial condition that $y = 2$ at $x = 1$, for the range $x = 1·0(0·2)3·0$, that is from $x = 1·0$ to $x = 3·0$ with step length $x = 0·2$.

As initial conditions, we have

$$x_0 = \ldots\ldots\ldots\ldots \text{ and } y_0 = \ldots\ldots\ldots\ldots$$

21

$$x_0 = 1, \quad y_0 = 2$$

Because

$x_0 = 1$ and $y_0 = 2$ are given initial conditions.

These values can now be inserted into the
spreadsheet in cells . . . .

22

$$x_1 = 1 \text{ in B2}, \quad y_0 = 2 \text{ in C2}$$

Notice how the numbers in column B have changed to accommodate the new sequence of x-values. The contents of the cells in column C do not need to be changed as they refer to the equation

$$y_1 = y_0 + h(y')_0$$

which is the same in this spreadsheet as it was in the previous spreadsheet. The contents of column D do have to be changed because they currently refer to the equation to be solved in the previous problem. The equation to be solved here is

$$y' = 1 + x - y$$

so in cell D3 the contents need to be changed to . . . .

23

$$= 1 + B2 - C2$$

This formula must then be copied into cells C3 to C12. Finally, the **Exact** column needs to be amended to reflect the exact solution to this equation, which is again found by using the integration factor method as

$$y = x + e^{1-x}$$

So, in E2, enter the formula . . . .

24

$$= B2 + EXP(1 - B2)$$

This formula needs to be copied into cells E3 to E12. This completes the editing of the spreadsheet to reflect the new problem to give the display

n	x	y	y′	Exact	Errors (%)	h=0·2
0	1·0	2·0000000000	0·0000000000	2·0000000000	0·00	
1	1·2	2·0000000000	0·2000000000	2·0187307531	0·93	
2	1·4	2·0400000000	0·3600000000	2·0703200460	1·46	
3	1·6	2·1120000000	0·4880000000	2·1488116361	1·71	
4	1·8	2·2096000000	0·5904000000	2·2493289641	1·77	
5	2·0	2·3276800000	0·6723200000	2·3678794412	1·70	
6	2·2	2·4621440000	0·7378560000	2·5011942119	1·56	
7	2·4	2·6097152000	0·7902848000	2·6465969639	1·39	
8	2·6	2·7677721600	0·8322278400	2·8018965180	1·22	
9	2·8	2·9342177280	0·8657822720	2·9652988882	1·05	
10	3·0	3·1073741824	0·8926258176	3·1353352832	0·89	

A plot of the graph of y against x for both the computed value and the exact value looks as follows

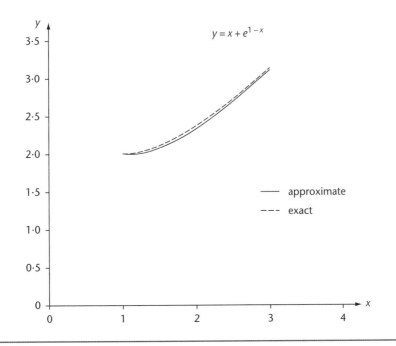

25   Graphical interpretation of Euler's method

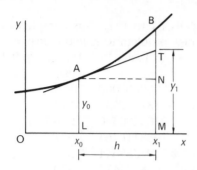

If AT is the tangent to the curve at A,

$$\text{then } \frac{NT}{AN} = \left[\frac{dy}{dx}\right]_{x=x_0} = (y')_0$$

$$\frac{NT}{h} = (y')_0 \qquad \therefore \ NT = h(y')_0$$

$$\therefore \ \text{At } x = x_1, \ MT = y_0 + h(y')_0$$

By Euler's relationship, $y_1 = y_0 + h(y')_0$ i.e. MT.

The difference between the calculated value of y, i.e. MT, and the actual value of the function y, i.e. MB, at $x = x_1$, is indicated by TB. This error can be considerable, depending on the curvature of the graph and the size of the interval h. It is inherent to the method and corresponds to the truncation of the Taylor's series after the second term.

  Euler's method, then

(a) is simple in procedure

(b) is lacking in accuracy, especially away from the starter values of the initial conditions

(c) is of use only for very small values of the interval h.

In spite of its practical limitations, it is the foundation of several more sophisticated methods and hence it is worthy of note.
  Here is one more example to work on your own.

Example 3

Obtain the solution of $\dfrac{dy}{dx} = x + y$ with the initial condition that $y = 1$ at $x = 0$, for the range $x = 0(0{\cdot}1)0{\cdot}5$.

By using a previously constructed spreadsheet as a template, the solution is

....

The function values are given in the next frame

n	x	y	y′	Exact	Errors (%)	h = 0·1
0	0·0	1·0000000000	1·0000000000	1·0000000000	0·00	
1	0·1	1·1000000000	1·2000000000	1·1103418362	0·93	
2	0·2	1·2200000000	1·4200000000	1·2428055163	1·84	
3	0·3	1·3620000000	1·6620000000	1·3997176152	2·69	
4	0·4	1·5282000000	1·9282000000	1·5836493953	3·50	
5	0·5	1·7210200000	2·2210200000	1·7974425414	4·25	
6	0·6	1·9431220000	2·5431220000	2·0442376008	4·95	
7	0·7	2·1974342000	2·8974342000	2·3275054149	5·59	
8	0·8	2·4871776200	3·2871776200	2·6510818570	6·18	
9	0·9	2·8158953820	3·7158953820	3·0192062223	6·73	
10	1·0	3·1874849202	4·1874849202	3·4365636569	7·25	

Because

The initial conditions are entered as

    0 in cell B2   (the initial x-value)
    1 in cell C2   (the initial y-value)
    0·1 in cell H1 (the x step length)

The formulas are entered as

    = B2 + C2 in cell D2, copied into cells D3 to D12
                (the successive y'-values)
    = C2 + \$H\$1 * D2 in cell C3 copied into cells C4 to C12
                (the successive y-values)

The exact solution found by using the integration factor method is $y = 2e^x - x - 1$ and so

    = 2 * EXP(B2) − B2 − 1 is entered into cell E2 and copied into cells E3 to E12

Notice how the errrors here are significant, which is very evident from the graphs of the computed values and the exact values.

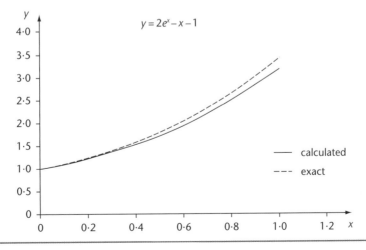

27 The Euler–Cauchy method – or the improved Euler method

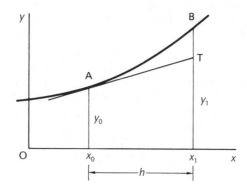

In Euler's method, we use the slope $(y')_0$ at A (x_0, y_0) across the whole interval h to obtain an approximate value of y_1 at B. TB is the resulting error in the result.

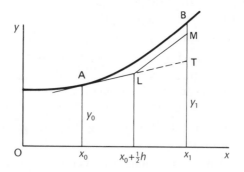

In the Euler–Cauchy method, we use the slope at A (x_0, y_0) across half the interval and then continue with a line whose slope approximates to the slope of the curve at x_1.

Let $\bar{\bar{y}}_1$ be the y-value of the point at T.

The error (MB) in the result is now considerably less than the error (TB) associated with the basic Euler method and the calculated results will accordingly be of greater accuracy.

Euler–Cauchy calculations

28

The steps in the Euler–Cauchy method are as follows.

1 We start with the given equation $y' = f(x, y)$ with the initial condition that at $x = x_0$, $y = y_0$. We have to determine function values for $x = x_0(h)x_n$.

2 From the equation and the initial condition we obtain $(y')_0 = f(x_0, y_0)$.

3 Knowing x_0, y_0, $(y')_0$ and h, we then evaluate

(a) $x_1 = x_0 + h$

(b) the auxiliary value of y, denoted by $\bar{\bar{y}}$ where
$\bar{\bar{y}}_1 = y_0 + h(y')_0$. This is the same step as in Euler's method.

(c) Then $y_1 = y_0 + \frac{1}{2}h\{(y')_0 + f(x_1, \bar{\bar{y}}_1)\}$

Note that $f(x_1, y_1)$ is the right-hand side of the given equation with x and y replaced by the calculated values of x_1 and $\bar{\bar{y}}_1$.

(d) Finally $(y')_1 = f(x_1, y_1)$.

We have thus evaluated x_1, y_1 and $(y')_1$.

The whole process is then repeated, the calculated values of x_1, y_1 and $(y')_1$ becoming the starter values x_0, y_0, $(y')_0$ for the next stage.

Make a note of the relationships above. We shall be using them quite often.

Then on to the next frame for an example of their use

Example 1

29

Apply the Euler–Cauchy method to solve the equation

$y' = x + y$

with the initial condition that at $x = 0$, $y = 1$, for the range $x = 0(0\cdot1)1\cdot0$.

We proceed as before by copying our template solution to a new worksheet. Before we continue we need to decide what the entries are going to be in our spreadsheet.

1 We are going to have to enter new initial conditions, so

Enter 0 in cell B2   that is $x_0 = 0$
Enter 1 in cell C2   that is $y_0 = 1$
Enter 0·1 in cell H1   this is the x step length

2 The equation to be solved is $y' = x + y$, so enter the formula

$= B2 + C2$ in cell D2 and copy the contents of D2 into cells D3 to D12

3 The Euler–Cauchy method tells us that

$$y_1 = y_0 + \frac{1}{2}h\{(y')_0 + f(x, \bar{\bar{y}}_1)\}$$

where $\bar{\bar{y}}_1 = y_0 + h(y')_0$ so that

$$f(x_1, \bar{\bar{y}}_1) = x_1 + \bar{\bar{y}}_1 = x_1 + y_0 + h(y')_0$$

Therefore $y_1 = \ldots\ldots\ldots\ldots$

30

$$y_1 = y_0 + \frac{1}{2}h\{x_1 + y_0 + (1+h)(y')_0\}$$

Because

By replacing $f(x_1, \bar{\bar{y}}_1)$ with $x_1 + y_0 + h(y')_0$ in the expression

$$y_1 = y_0 + \frac{1}{2}h\{(y')_0 + f(x_1, \bar{\bar{y}}_1)\}$$

we find that

$$y_1 = y_0 + \frac{1}{2}h\{(y')_0 + x_1 + y_0 + h(y')_0\}$$
$$= y_0 + \frac{1}{2}h\{x_1 + y_0 + (1+h)(y')_0\}$$

In cell C3 enter the formula ....

31

$$= C2 + (0\cdot5) * \$H\$1 * (B3 + C2 + (1 + \$H\$1) * D2)$$

Because

y_0 is in cell C2, h is in cell H1, x_1 is in cell B3 and $(y')_0$ is in cell D2.

Copy the contents of cell C3 into cells C4 to C12.

4 Finally, for comparison purposes, the exact solution of this equation is $y = 2e^x - x - 1$ and this is

entered into E2 by the formula ....
and copied into cells ....

32

$$= 2 * EXP(B2) - B2 - 1 \text{ and copied into cells E3 to E12}$$

The resulting display looks as follows

n	x	y	y'	Exact	Errors (%)	h = 0·1
0	0·0	1·0000000000	1·0000000000	1·0000000000	0·00	
1	0·1	1·1100000000	1·2100000000	1·1103418362	0·03	
2	0·2	1·2420500000	1·4420500000	1·2428055163	0·06	
3	0·3	1·3984652500	1·6984652500	1·3997176152	0·09	
4	0·4	1·5818041013	1·9818041013	1·5836493953	0·12	
5	0·5	1·7948935319	2·2948935319	1·7974425414	0·14	
6	0·6	2·0408573527	2·6408573527	2·0442376008	0·17	
7	0·7	2·3231473748	3·0231473748	2·3275054149	0·19	
8	0·8	2·6455778491	3·4455778491	2·6510818570	0·21	
9	0·9	3·0123635233	3·9123635233	3·0192062223	0·23	
10	1·0	3·4281616932	4·4281616932	3·4365636569	0·24	

▶

Comparing these results with the same equation being solved by the Euler method demonstrates how much more accurate the Euler–Cauchy method is, as can be seen from the following table of comparative errors

x	Euler	Euler–Cauchy
0·0	0·00	0·00
0·1	0·93	0·03
0·2	1·84	0·06
0·3	2·69	0·09
0·4	3·50	0·12
0·5	4·25	0·14
0·6	4·95	0·17
0·7	5·59	0·19
0·8	6·18	0·21
0·9	6·73	0·23
1·0	7·25	0·24

Next frame

33

Now for another example, but before that, complete the following without reference to your notes – if possible. In the Euler–Cauchy method the relevant relationships are

$$x_1 = \ldots\ldots\ldots$$
$$\bar{\bar{y}}_1 = \ldots\ldots\ldots$$
$$y_1 = \ldots\ldots\ldots$$
$$(y')_1 = \ldots\ldots\ldots$$

Next frame

34

$$x_1 = x_0 + h$$
$$\bar{\bar{y}}_1 = y_0 + h(y')_0$$
$$y_1 = y_0 + \frac{1}{2}h\{(y')_0 + f(x_1, \bar{\bar{y}}_1)\}$$
$$(y')_1 = f(x_1, y_1)$$

Example 2

Determine a numerical solution of the equation $y' = 2(1 + x) - y$ with the initial condition that $y = 5$ when $x = 2$, for the range 2·0(0·2)4·0. Try this one yourself.

The exact solution is given as $y = 2x + e^{-2x}$
and the final display of results is $\ldots\ldots\ldots\ldots$

35

n	x	y	y'	Exact	Errors (%)	h = 0·2
0	2·0	5·0000000000	1·0000000000	5·0000000000	0·00	
1	2·2	5·2200000000	1·1800000000	5·2187307531	−0·02	
2	2·4	5·4724000000	1·3276000000	5·4703200460	−0·04	
3	2·6	5·7513680000	1·4486320000	5·7488116361	−0·04	
4	2·8	6·0521217600	1·5478782400	6·0493289641	−0·05	
5	3·0	6·3707398432	1·6292601568	6·3678794412	−0·04	
6	3·2	6·7040066714	1·6959933286	6·7011942119	−0·04	
7	3·4	7·0492854706	1·7507145294	7·0465969639	−0·04	
8	3·6	7·4044140859	1·7955859141	7·4018965180	−0·03	
9	3·8	7·7676195504	1·8323804496	7·7652988882	−0·03	
10	4·0	8·1374480313	1·8625519687	8·1353352832	−0·03	

Because

1 The initial conditions are entered as

Enter 2 in cell B2 (that is $x_0 = 2$); enter 5 in cell C2 (that is $y_0 = 5$)
Enter 0·2 in cell H1 (this is the x step length)

2 The equation to be solved is $y' = 2(1 + x) - y$, so enter the formula

$= 2 * (1 + B2) - C2$ in cell D2 and copy the contents of D2 into cells D3 to D12

3 The Euler–Cauchy method tells us that

$$y_1 = y_0 + \frac{1}{2}h\{(y')_0 + f(x_1, \bar{\bar{y}}_1)\} \quad \text{where } \bar{\bar{y}}_1 = y_0 + h(y')_0 \text{ so that}$$

$$f(x_1, \bar{\bar{y}}_1) = 2(1 + x_1) - \bar{\bar{y}}_1 = 2(1 + x_1) - y_0 - h(y')_0 \text{ therefore}$$

$$y_1 = y_0 + \frac{1}{2}h\{(y')_0 + 2(1 + x_1) - y_0 - h(y')_0\} \text{ that is}$$

$$y_1 = y_0 + \frac{1}{2}h\{2(1 + x_1) - y_0 + (1 - h)(y')_0\}$$

This is accommodated by the formula in C3 (copied into cells C4 to C12)

$$= C2 + (0·5) * \$H\$1 * (2 * (1 + B3) - C2 + (1 - \$H\$1) * D2)$$

4 Finally the exact solution $y = 2x + e^{-2x}$ is entered into cell E2 as $= 2 * B2 + EXP(-2 * B2)$ and copied into cells E3 to E12.

Refer to Frame 19 for a comparison of errors between this method and the Euler method. Then another example for you to try just to make sure you are clear about the processes involved.

Next frame

Example 3

36

Solve the equation $y' = y^2 + xy$ with initial condition that at $x = 1$, $y = 1$, for the range $x = 1{\cdot}0(0{\cdot}1)1{\cdot}7$. Use the Euler–Cauchy method and work to 6 places of decimals.

The solution is ....

37

n	x	y	y'	$h = 0{\cdot}1$
0	1·0	1·000000	2·000000	
1	1·1	1·238000	2·894444	
2	1·2	1·591023	4·440583	
3	1·3	2·152410	7·431004	
4	1·4	3·145846	14·300528	
5	1·5	5·251007	35·449581	
6	1·6	11·595613	153·011211	
7	1·7	57·704110	3427·861242	

Because

1 The initial conditions are entered as

Enter 1 in cell B2 (that is $x_0 = 1$); enter 1 in cell C2 (that is $y_0 = 1$)
Enter 0·1 in cell H1 (this is the x step length)

2 The equation to be solved is $y' = y^2 + xy$, so

Enter the formula $= C2\wedge 2 + B2 * C2$ in cell D2 and copy the contents of D2 into cells D3 to D9. Note that $C2\wedge 2 = C2 * C2$ – the 'hat' indicates raising to a power.

3 The Euler-Cauchey method tell us that

$$y_1 = y_0 + \frac{1}{2}h\{(y')_0 + f(x_1, \bar{\bar{y}}_1)\} \quad \text{where } \bar{\bar{y}}_1 = y_0 + h(y')_0 \text{ so that}$$

$$f(x_1, \bar{\bar{y}}_1) = \bar{\bar{y}}_1^2 + x_1\bar{\bar{y}}_1 = (y_0 + h(y')_0)^2 + x_1(y_0 + h(y')_0) \text{ therefore}$$

$$y_1 = y_0 + \frac{1}{2}h\{(y')_0 + (y_0 + h(y')_0)^2 + x_1(y_0 + h(y')_0)\}$$

This is accommodated by the formula in C3 (copied into cells C4 to C9)

$$= C2 + (0{\cdot}5) * \$F\$1 * (D2 + (C2 + \$F\$1 * D2))\wedge 2 + B3 * (C2 + \$F\$1 * D2))$$

The table shows that as x increases, the computed values of y and its derivative increase dramatically. This is an indication that the exact solution increases without bound near to the larger values of x considered, so bringing the accuracy of these computed values into question. This emphasises the importance of checking every method against a known solution so as to form some idea of the method's accuracy. However, all numerical methods produce significant accuracies whenever the exact solution diverges in this way.

38 Runge–Kutta method

The Runge–Kutta method for solving first-order differential equations is widely used and affords a high degree of accuracy. It is a further step-by-step process where a table of function values for a range of values of x is accumulated. Several intermediate calculations are required at each stage, but these are straightforward and present little difficulty.

In general terms, the method is as follows.

To solve $y' = f(x, y)$ with initial condition $y = y_0$ at $x = x_0$, for a range of values of $x = x_0(h)x_n$.

Starting as usual with $x = x_0$, $y = y_0$, $y' = (y')_0$ and h, we have

$$x_1 = x_0 + h$$

Finding y_1 requires four intermediate calculations

$$k_1 = hf(x_0, y_0) = h(y')_0$$
$$k_2 = hf(x_0 + \tfrac{1}{2}h, \ y_0 + \tfrac{1}{2}k_1)$$
$$k_3 = hf(x_0 + \tfrac{1}{2}h, \ y_0 + \tfrac{1}{2}k_2)$$
$$k_4 = hf(x_0 + h, \ y_0 + k_3)$$

The increment Δy_0 in the y-values from $x = x_0$ to $x = x_1$ is then

$$\Delta y_0 = \tfrac{1}{6}\{k_1 + 2k_2 + 2k_3 + k_4\}$$

and finally $y_1 = y_0 + \Delta y_0.$

We shall be using these repeatedly, so make a note of them for future reference. Then let us see an example

39 Example 1

Find the numerical solution of $y' = x + y$ using the Runge–Kutta method with $y = 1$ and $x = 0$ for values in the range $x = 0(0{\cdot}1)1{\cdot}0$.

We shall proceed with the solution of this differential equation using a spreadsheet in much the same manner as before. However, we are going to require a different structure in order to accommodate the four variables k_i for $i = 1, 2, 3, 4$. The structure we shall use is headed by

	A	B	C	D	E	F	G	H	I
1	n	x	k1	k2	k3	k4	y	y'	h =

where the value of h is held in cell J1.

1 Enter the values 0 to 10 in column A from A2 to A12 using the **Edit-Fill-Series** sequence of commands. These are the iteration numbers.
2 Enter the x step value of $0{\cdot}1$ in cell J1.
3 Enter the initial value of x in cell B2 as 0 and in B3 enter the formula $= \text{B2} + \$\text{J}\$1$. Now copy the contents of B3 into cells B4 to B12.
4 Enter the initial value of y in cell G2 as 1.

▶

We can now progressively enter the table of values from the left.

5   $k_1 = hf(x_0, y_0) = h(y')_0$ – the y'-values are in column H, so in cell C2 enter the formula $= \$J\$1 * H2$. Copy the contents of C2 into cells C3 to C12.

6   $k_2 = hf\left(x_0 + \frac{1}{2}h, y_0 + \frac{1}{2}k_1\right) = h\left(x_0 + \frac{1}{2}h + y_0 + \frac{1}{2}k_1\right)$, so in cell D2 enter the formula $= \$J\$1 * (B2 + 0.5 * \$J\$1 + G2 + 0.5 * C2)$. Copy the contents of D2 into cells D3 to D12.

7   $k_3 = hf\left(x_0 + \frac{1}{2}h, y_0 + \frac{1}{2}k_2\right) = h\left(x_0 + \frac{1}{2}h + y_0 + \frac{1}{2}k_2\right)$, so in cell E2 enter the formula $= \$J\$1 * (B2 + 0.5 * \$J\$1 + G2 + 0.5 * D2)$. Copy the contents of E2 into cells E3 to E12.

8   $k_4 = hf(x_0 + h, y_0 + k_3) = h(x_0 + h + y_0 + k_3)$, so in cell F2 enter the formula $= \$J\$1 * (B2 + \$J\$1 + G2 + E2)$. Copy the contents of F2 into cells F3 to F12.

9   $y_1 = y_0 + \frac{1}{6}\{k_1 + 2k_2 + 2k_3 + k_4\}$, so in cell G3 enter the formula $= G2 + (1/6) * (C2 + 2 * D2 + 2 * E2 + F2)$. Copy the contents of G3 into cells G4 to G12.

10   $y' = x + y$, so in H2 enter the formula $= B2 + G2$. Copy the contents of H2 into cells H3 to H12.

The results are displayed in the next frame

n	x	k1	k2	k3	k4	y	y'	$h = 0.1$	**40**
0	0.0	0.1000000	0.1100000	0.1105000	0.1210500	1.0000000	1.0000000		
1	0.1	0.1210342	0.1320859	0.1326385	0.1442980	1.1103417	1.2103417		
2	0.2	0.1442805	0.1564945	0.1571052	0.1699910	1.2428051	1.4428051		
3	0.3	0.1699717	0.1834703	0.1841452	0.1983862	1.3997170	1.6997170		
4	0.4	0.1983648	0.2132831	0.2140290	0.2297677	1.5836485	1.9836485		
5	0.5	0.2297441	0.2462313	0.2470557	0.2644497	1.7974413	2.2974413		
6	0.6	0.2644236	0.2826448	0.2835558	0.3027792	2.0442359	2.6442359		
7	0.7	0.3027503	0.3228878	0.3238947	0.3451398	2.3275033	3.0275033		
8	0.8	0.3451079	0.3673633	0.3684761	0.3919555	2.6510791	3.4510791		
9	0.9	0.3919203	0.4165163	0.4177461	0.4436949	3.0192028	3.9192028		
10	1.0	0.4436559	0.4708387	0.4721979	0.5008757	3.4365595	4.4365595		

with the following errors

n	x	Exact	Error (%)	Error (%)
0	0.0	1.0000000	0.0000000	0.00
1	0.1	1.1103418	0.0000153	0.93
2	0.2	1.2428055	0.0000301	1.84
3	0.3	1.3997176	0.0000444	2.69
4	0.4	1.5836494	0.0000578	3.50
5	0.5	1.7974425	0.0000703	4.25
6	0.6	2.0442376	0.0000820	4.95
7	0.7	2.3275054	0.0000929	5.59
8	0.8	2.6510819	0.0001030	6.18
9	0.9	3.0192062	0.0001124	6.73
10	1.0	3.4365637	0.0001213	7.25

The column to the far right contains the errors using the Euler method and, as you can see, the Runge–Kutta method provides a significant improvement in accuracy.

▶

Now, without reference to your notes, complete the following expressions for

$$k_1 = \ldots\ldots\ldots$$
$$k_2 = \ldots\ldots\ldots$$
$$k_3 = \ldots\ldots\ldots$$
$$k_4 = \ldots\ldots\ldots$$
$$\Delta y_0 = \ldots\ldots\ldots$$
$$y_1 = \ldots\ldots\ldots$$

It speeds up your working if you can remember them.

41

$$k_1 = h(y')_0$$
$$k_2 = hf\left(x_0 + \tfrac{1}{2}h,\ y_0 + \tfrac{1}{2}k_1\right)$$
$$k_3 = hf\left(x_0 + \tfrac{1}{2}h,\ y_0 + \tfrac{1}{2}k_2\right)$$
$$k_4 = hf\left(x_0 + h,\ y_0 + k_3\right)$$
$$\Delta y_0 = \tfrac{1}{6}(k_1 + 2k_2 + 2k_3 + k_4)$$
$$y_1 = y_0 + \Delta y_0$$

With those in mind, let us move on to a further example. Next frame

42   **Example 2**

Solve $y' = \sqrt{x^2 + y}$ for $x = 0(0 \cdot 2)2 \cdot 0$ given that at $x = 0$, $y = 0 \cdot 8$.

Using the spreadsheet for the previous example as a template for this example. The solution is $\ldots\ldots\ldots$

43

n	x	k1	k2	k3	k4	y	y'	h = 0·2
0	0·0	0·1788854	0·1896779	0·1902460	0·2030021	0·8000000	0·8944272	
1	0·2	0·2030063	0·2174206	0·2180825	0·2339548	0·9902892	1·0150316	
2	0·4	0·2339473	0·2510185	0·2516977	0·2698134	1·2082838	1·1697366	
3	0·6	0·2698011	0·2887709	0·2894271	0·3091435	1·4598160	1·3490055	
4	0·8	0·3091304	0·3294604	0·3300769	0·3509482	1·7490394	1·5456518	
5	1·0	0·3509358	0·3722562	0·3728285	0·3945492	2·0788983	1·7546790	
6	1·2	0·3945381	0·4165946	0·4171237	0·4394829	2·4515074	1·9726904	
7	1·4	0·4394732	0·4620889	0·4625781	0·4854274	2·8684170	2·1973659	
8	1·6	0·4854190	0·5084682	0·5089213	0·5321545	3·3307894	2·4270948	
9	1·8	0·5321472	0·5555390	0·5559599	0·5794989	3·8395148	2·6607358	
10	2·0	0·5794925	0·6031595	0·6035518	0·6273385	4·3952888	2·8974625	

Because

1   The initial conditions are entered as $x_0 = 0$ and $y_0 = 0 \cdot 8$. The x step length is entered as $0 \cdot 2$

2   The formula for the variable k_1 remains the same as $= \$J\$1 * H2$

▶

3   The formula for the variable k_2 is changed to
    $= \$J\$1*(((B2+0\cdot5*\$J\$1)^\wedge2+G2+0\cdot5*C2)^\wedge0\cdot5)$
4   The formula for the variable k_3 is changed to
    $= \$J\$1*(((B2+0\cdot5*\$J\$1)^\wedge2+G2+0\cdot5*D2)^\wedge0\cdot5)$
5   The formula for the variable k_4 is changed to
    $= \$J\$1*(((B2+\$J\$1)^\wedge2+G2+E2)^\wedge0\cdot5)$
6   The formula for y remains the same as
    $= G2+\{1/6\}*(C2+2*D2+2*E2+F2)$
7   The formula for y' is changed to $= (B2^\wedge2+G2)^\wedge0\cdot5$

That is it. Now move on to the next frame where we make a new start and apply similar methods to the solution of second-order differential equations by numerical methods.

Second-order differential equations

Euler second-order method

44

The first method we will deal with is really an extension of the Euler method for the first-order equations and is a direct application of a truncated form of Taylor's series. We anticipate, therefore, that the method will be relatively easy, but the results will not be accurate to a high degree.

Taylor's series:

$$f(x+h) = f(x) + hf'(x) + \frac{h^2}{2!}f''(x) + \frac{h^3}{3!}f'''(x) + \dots$$

Differentiating term by term with respect to x, we obtain

$$f'(x+h) = f'(x) + hf''(x) + \frac{h^2}{2!}f'''(x) + \frac{h^3}{3!}f''''(x) + \dots$$

If we neglect terms in $f'''(x)$ and subsequent terms in each of these two series, we have the approximations

$$f(x+h) \approx \dots\dots\dots\dots$$
$$f'(x+h) \approx \dots\dots\dots\dots$$

45

$$f(x+h) \approx f(x) + hf'(x) + \frac{h^2}{2!}f''(x)$$
$$f'(x+h) \approx f'(x) + hf''(x)$$

Although these are approximations, in practice we tend to write them with the 'equals' sign. Therefore, at $x = a$, these become

$$\dots\dots\dots\dots\dots\dots$$

and $$\dots\dots\dots\dots\dots\dots$$

46

$$f(a+h) = f(a) + hf'(a) + \frac{h^2}{2!}f''(a)$$
$$f'(a+h) = f'(a) + hf''(a)$$

and these, with the notation we have previously used, can be written

$$y_1 = y_0 + h(y')_0 + \frac{h^2}{2!}(y'')_0$$
$$(y')_1 = (y')_0 + h(y'')_0$$

Thus, if x_0, y_0, $(y')_0$ and $(y'')_0$ are known, we can find an approximate value of y_1 at $x_1 = x_0 + h$.

Make a note of these two relationships: then we can apply them.

47

Example

Solve the equation $y'' = xy' + y$ for $x = 0(0\cdot2)2\cdot0$ given that at $x = 0$, $y = 1$ and $y' = 0$.

We shall set about finding the numerical solution to this equation as we have done previously by using a spreadsheet. The headings for the sheet will be

	A	B	C	D	E	F	G	H
1	n	x	y	y'	y''	Exact	Errors (%)	h=

The entries will then be

1 Column A contains the iteration number from 0 in A2 to 10 in A12.
2 Cell I1 contains the x step length which is $0\cdot2$.
3 Column B contains the successive x-values from $0\cdot0$ to $2\cdot0$ in steps of $0\cdot2$. The initial value of $x_0 = 0$ is entered into cell B2 and the formula $= B2 + \$I\$1$ is entered into cell B3 and copied into cells B4 to B12.
4 Column C contains the computed y-values. The initial value of $y_0 = 1$ is entered into cell C2 and the equation

$$y_1 = y_0 + h(y')_0 + \frac{h^2}{2!}(y'')_0$$

is represented in cell C3 by the formula

$$= C2 + \$I\$1 * D2 + (\$I\$1^\wedge 2) * E2/2$$

copied into cells C4 to C12.
5 Column D contains the computed y'-values. The initial value of $(y')_0 = 0$ is entered into cell D2 and the equation

$$(y')_1 = (y')_0 + h(y'')_0$$

is represented in cell D3 by the formula $= D2 + \$I\$1 * E2$ copied into cells D4 to D12.
6 Column E contains the y''-values which are obtained from the equation $y'' = xy' + y$ which is represented in cell E2 by the formula $= B2 * D2 + C2$ copied into cells E3 to E12.

▶

7 Column F contains the values obtained from the exact solution which can be shown to be $y = e^{x^2/2}$. This is represented in cell F2 by the formula = EXP((B2^2)/2) copied into cells F3 to F12.

8 Column G contains the percentage errors. In cell G2 enter the formula = (F2 − C2) * 100/F2 copied into cells G3 to G12.

Your spreadsheet should now look like the one on the next page (with the appropriate formatting to make it easier to read).

n	x	y	y'	y''	Exact	Errors (%)	$h = 0.2$
0	0·0	1·0000000	0·0000000	1·0000000	1·0000000	0·00	
1	0·2	1·0200000	0·2000000	1·0600000	1·0202013	0·02	
2	0·4	1·0812000	0·4120000	1·2460000	1·0832871	0·19	
3	0·6	1·1885200	0·6612000	1·5852400	1·1972174	0·73	
4	0·8	1·3524648	0·9782480	2·1350632	1·3771278	1·79	
5	1·0	1·5908157	1·4052606	2·9960763	1·6487213	3·51	
6	1·2	1·9317893	2·0044759	4·3371604	2·0544332	5·97	
7	1·4	2·4194277	2·8719080	6·4400989	2·6644562	9·20	
8	1·6	3·1226113	4·1599278	9·7784957	3·5966397	13·18	
9	1·8	4·1501667	6·1156269	15·1582952	5·0530903	17·87	
10	2·0	5·6764580	9·1472859	23·9710299	7·3890561	23·18	

You will notice that the errors are significant and grow dramatically as the value of x increases. The main cause of errors is ....

| the truncation of the Taylor's series on which the method is based | **48** |

A greater degree of accuracy can be obtained by using the Runge–Kutta method for second-order differential equations, which is an extension of the method we have already used for first-order equations. As before, more intermediate calculations are required, but the reliability of results reflects the extra work involved.

Runge–Kutta method for second-order differential equations

Starting with the given equation $y'' = f(x, y, y')$ and initial conditions that at $x = x_0$, $y = y_0$ and $y' = (y')_0$, we can obtain the value of y_1 at $x_1 = x_0 + h$ as follows.

(a) We evaluate

$$k_1 = \tfrac{1}{2}h^2 f\{x_0, y_0, (y')_0\} = \tfrac{1}{2}h^2(y'')_0$$

$$k_2 = \tfrac{1}{2}h^2 f\left\{x_0 + \tfrac{1}{2}h, y_0 + \tfrac{1}{2}h(y')_0 + \tfrac{1}{4}k_1, (y')_0 + \frac{k_1}{h}\right\}$$

$$k_3 = \tfrac{1}{2}h^2 f\left\{x_0 + \tfrac{1}{2}h, y_0 + \tfrac{1}{2}h(y')_0 + \tfrac{1}{4}k_1, (y')_0 + \frac{k_2}{h}\right\}$$

$$k_4 = \tfrac{1}{2}h^2 f\left\{x_0 + h, y_0 + h(y')_0 + k_3, (y')_0 + \frac{2k_3}{h}\right\}$$

▶

(b) From these results, we then determine

$$P = \tfrac{1}{3}\{k_1 + k_2 + k_3\}$$
$$Q = \tfrac{1}{3}\{k_1 + 2k_2 + 2k_3 + k_4\}$$

(c) Finally, we have

$$x_1 = x_0 + h$$
$$y_1 = y_0 + h(y')_0 + P$$
$$(y')_1 = (y')_0 + \frac{Q}{h}$$

It is not as complicated as it looks at first sight. Copy down this list of relationships for reference when dealing with some examples that follow.

Then move on

49

Note the following

1   Four evaluations for k are required to determine a single new point on the solution curve.
2   The method is self-starting in that no preliminary calculations are required. The equation and initial conditions are sufficient to provide the next point on the curve.
3   As with the Runge–Kutta method for first-order equations, the method contains no self-correcting element or indication of any error involved.

Example 1

Use the Runge–Kutta method to solve the equation $y'' = xy' + y$ for $x = 0 \cdot 0 (0 \cdot 2) 2 \cdot 0$ given that at $x = 0$, $y = 1$ and $y' = 0$.

This is the same problem that we have just encountered and in due course we shall compare results. As expected, we shall use a spreadsheet to derive the solution. The headings for the sheet this time will be

	A	B	C	D	E	F	G	H	I	J	K	L
1	n	x	k1	k2	k3	k4	P	Q	y	y′	y″	h =

The entries will then be

1   Column A contains the iteration number from 0 in A2 to 10 in A12.
2   Cell M1 contains the x step length which is $0 \cdot 2$.
3   Column B contains the successive x-values from $0 \cdot 0$ to $2 \cdot 0$ in steps of $0 \cdot 2$. The initial value of $x_0 = 0$ is entered into cell B2 and the formula $= B2 + \$M\$1$ is entered into cell B3 and copied into cells B4 to B12.
4   Column C contains the computed k_1-values and the equation $k_1 = \tfrac{1}{2}h^2(y'')_0$ is represented in cell C2 by the formula . . . .

$$= (0.5) * (\$M\$1\wedge 2) * K2$$

50

The contents of cell C2 are then copied into cells C3 to C12.

5 Column D contains the computed k_2-values and the equation

$$k_2 = \tfrac{1}{2}h^2 f\left(x_0 + \tfrac{1}{2}h, \; y_0 + \tfrac{1}{2}h(y')_0 + \tfrac{1}{4}k_1, \; (y')_0 + k_1/h\right)$$
$$= \tfrac{1}{2}h^2\left((x_0 + \tfrac{1}{2}h)((y')_0 + k_1/h) + y_0 + \tfrac{1}{2}h(y')_0 + \tfrac{1}{4}k_1\right)$$

is represented in cell D2 by the formula ....

$$=(0.5) * (\$M\$1\wedge 2) * ((B2 + 0.5 * \$M\$1) * (J2 + C2/\$M\$1)$$
$$+ I2 + 0.5 * \$M\$1 * J2 + 0.25 * C2)$$

51

The contents of cell D2 are then copied into cells D3 to D12.

6 Column E contains the computed k_3-values and the equation

$$k_3 = \tfrac{1}{2}h^2 f\left(x_0 + \tfrac{1}{2}h, \; y_0 + \tfrac{1}{2}h(y')_0 + \tfrac{1}{4}k_1, \; (y')_0 + k_2/h\right)$$
$$= \tfrac{1}{2}h^2\left((x_0 + \tfrac{1}{2}h)((y')_0 + k_2/h) + y_0 + \tfrac{1}{2}h(y')_0 + \tfrac{1}{4}k_1\right)$$

is represented in cell E2 by the formula ....

$$=(0.5) * (\$M\$1\wedge 2) * ((B2 + 0.5 * \$M\$1) * (J2 + D2/\$M\$1)$$
$$+ I2 + 0.5 * \$M\$1 * J2 + 0.25 * C2)$$

52

The contents of cell E2 are then copied into cells E3 to E12.

7 Column F contains the computed k_4-values and the equation

$$k_4 = \tfrac{1}{2}h^2 f\left(x_0 + h, \; y_0 + h(y')_0 + k_3, \; (y')_0 + 2k_3/h\right)$$
$$= \tfrac{1}{2}h^2\left((x_0 + h)((y')_0 + 2k_3/h) + y_0 + h(y')_0 + k_3\right)$$

is represented in cell F2 by the formula ....

$$=(0.5) * (\$M\$1\wedge 2) * ((B2 + \$M\$1) * (J2 + 2 * E2/\$M\$1)$$
$$+ I2 + \$M\$1 * J2 + E2)$$

53

The contents of cell F2 are then copied into cells F3 to F12.

8 Column G contains the computed P-values and the equation
$P = \tfrac{1}{3}(k_1 + k_2 + k_3)$ is represented in cell G2 by the formula ....

54

$$=(1/3)*(C2+D2+E2)$$

The contents of cell G2 are then copied into cells G3 to G12.

9 Column H contains the computed Q-values and the equation $Q = \frac{1}{3}(k_1 + 2k_2 + 2k_3 + k_4)$ is represented in cell H2 by the formula

. . . .

55

$$=(1/3)*(C2+2*D2+2*E2+F2)$$

The contents of cell H2 are then copied into cells H3 to H12.

10 Column I contains the computed y-values. The initial value of $y_0 = 1$ is entered into cell I2 and the equation

$$y_1 = y_0 + h(y')_0 + P$$

is represented in cell I3 by the formula . . . .

56

$$= I2 + \$M\$1 * J2 + G2$$

The contents of cell I3 are then copied into cells I4 to I12.

11 Column J contains the computed y'-values. The initial value of $(y')_0 = 0$ is entered into cell J2 and the equation $(y')_1 = (y')_0 + Q/h$ is represented in cell J3 by the formula . . . .

57

$$= J2 + H2/\$M\$1$$

The contents of cell J3 are then copied into cells J4 to J12.

12 Column K contains the y''-values which are obtained from the equation $y'' = xy' + y$ which is represented in cell K2 by the formula . . . .

58

$$= B2 * J2 + I2$$

The contents of cell K2 are then copied into cells K3 to K12 and the final spreadsheet looks like the following

n	x	k1	k2	k3	k4	P	Q	y	y'	y''	h=0·2
0	0·0	0·0200000	0·0203000	0·0203030	0·0212182	0·0202010	0·0408081	1·0000000	0·0000000	1·0000000	
1	0·2	0·0212202	0·0227790	0·0228258	0·0251351	0·0222750	0·0458550	1·0202010	0·2040403	1·0610091	
2	0·4	0·0251322	0·0282477	0·0284035	0·0325752	0·0272612	0·0570033	1·0832841	0·4333153	1·2566102	
3	0·6	0·0325641	0·0378798	0·0382519	0·0451961	0·0362319	0·0766745	1·1972083	0·7183318	1·6282074	
4	0·8	0·0451694	0·0538673	0·0546501	0·0660061	0·0512289	0·1094035	1·3771066	1·1017045	2·2584702	
5	1·0	0·0659480	0·0801269	0·0816865	0·1003762	0·0759205	0·1633170	1·6486764	1·6487218	3·2973982	
6	1·2	0·1002542	0·1236497	0·1266912	0·1579840	0·1168650	0·2529733	2·0543413	2·4653068	5·0127095	
7	1·4	0·1577302	0·1970991	0·2030044	0·2565931	0·1859446	0·4048434	2·6642677	3·7301734	7·8865105	
8	1·6	0·2560654	0·3238945	0·3354254	0·4295622	0·3051284	0·6680891	3·5962469	5·7543906	12·8032719	
9	1·8	0·4284592	0·5483881	0·5711745	0·7411112	0·5160073	1·1362318	5·0522535	9·0948361	21·4229585	
10	2·0	0·7387844	0·9567270	1·0024949	1·3181400	0·8993354	1·9917894	7·3872279	14·7759954	36·9392186	

▶

The errors have been dramatically reduced, as can be seen from the following table in comparison with those in Frame 47.

n	x	Exact	Error (%)
0	0·0	1·0000000	0·00
1	0·2	1·0202013	0·00
2	0·4	1·0832871	0·00
3	0·6	1·1972174	0·00
4	0·8	1·3771278	0·00
5	1·0	1·6487213	0·00
6	1·2	2·0544332	0·00
7	1·4	2·6644562	0·01
8	1·6	3·5966397	0·01
9	1·8	5·0530903	0·02
10	2·0	7·3890561	0·02

Next frame

59

Now here is one for you to do entirely on your own. The method is exactly the same as before and there are no snags. Use the spreadsheet that you created for the previous example as a template for this one.

Example 2

Solve the equation

$$y'' = x - y^2$$

for $x = 0·0(0·2)2·0$ where at $x = 0$, $y = 0$ and $y' = 0$.

When you have finished, check the results with the next frame

60

n	x	k1	k2	k3	k4	P	Q	y	y′	y″	h=0·2
0	0·0	0·0000000	0·0020000	0·0020000	0·0039999	0·0013333	0·0040000	0·0000000	0·0000000	0·0000000	
1	0·2	0·0040000	0·0059996	0·0059996	0·0079974	0·0053331	0·0119986	0·0013333	0·0199999	0·1999982	
2	0·4	0·0079977	0·0099915	0·0099915	0·0119731	0·0093269	0·0199789	0·0106664	0·0799930	0·3998862	
3	0·6	0·0119741	0·0139351	0·0139351	0·0158524	0·0132814	0·0278556	0·0359919	0·1798875	0·5987046	
4	0·8	0·0158546	0·0177065	0·0177065	0·0194436	0·0170892	0·0353748	0·0852508	0·3191655	0·7927323	
5	1·0	0·0194477	0·0210264	0·0210264	0·0223594	0·0205002	0·0419709	0·1661731	0·4960396	0·9723865	
6	1·2	0·0223654	0·0233782	0·0233782	0·0239421	0·0230406	0·0466068	0·2858812	0·7058940	1·1182719	
7	1·4	0·0239482	0·0239504	0·0239504	0·0232394	0·0239497	0·0476631	0·4501006	0·9389280	1·1974094	
8	1·6	0·0232395	0·0216639	0·0216639	0·0191107	0·0221891	0·0430019	0·6618359	1·1772436	1·1619732	
9	1·8	0·0190914	0·0153806	0·0153806	0·0105578	0·0166175	0·0303905	0·9194737	1·3922530	0·9545681	
10	2·0	0·0104978	0·0043750	0·0043750	−0·0026809	0·0064159	0·0084390	1·2145418	1·5442053	0·5248882	

Because

The only items that need amending from the previous spreadsheet are the references to the actual differential equation. Consequently

The formula in D2 for k_2 now reads as

$$= (0·5) * (\$M\$1\char94 2) * (B2 + 0·5 * \$M\$1 - (I2 + 0·5 * \$M\$1 * J2 + 0·25 * C2)\char94 2)$$

The formula in E2 for k_3 now reads as

$$= 0·5 * (\$M\$1\char94 2) * (B2 + 0·5 * \$M\$1 - (I2 + 0·5 * \$M\$1 * J2 + 0·25 * C2)\char94 2)$$

The formula in F2 for k_4 now reads as

$= 0.5*(\$M\$1^\wedge2)*(B2+\$M\$1 - (I2+\$M\$1*J2+E2)^\wedge2)$

The formula in K2 for y'' now reads as

$= B2 - I2^\wedge2$

Predictor–corrector methods

61

So far, all the methods that we have used for the numerical solution of differential equations have been *single-step* methods. By this is meant that, given the differential equation $y' = f(x, y)$, a set of starting values (x_0 and y_0) and a step length (h), we can then find the value of y_1. The values of x_1 and y_1 become the starting values for the next iteration and so the procedure goes on, one step at a time. More accurate methods employ a *multi-step* procedure where, instead of starting with just a single set of initial values, we use a collection of previously calculated values.

A very simple multi-step method is given by the equations

$$\bar{y}_1 = y_0 + hf(x_0, y_0)$$
$$y_1 = y_0 + \tfrac{1}{2}h\big(f(x_0, y_0) + f(x_1, \bar{y}_1)\big)$$

Here we calculate $\bar{y}_1$ first from the given initial conditions x_0 and y_0. We call this equation the *predictor* because it gives $\bar{y}_1$ as a first estimate of y_1. Using $\bar{y}_1$ in the second equation then gives a more accurate value for y_1. We call this equation the *corrector*.

An even better pair of predictor–corrector equations is given by

$$\bar{y}_{i+1} = y_i + \tfrac{1}{2}h(3f(x_i, y_i) - f(x_{i-1}, y_{i-1}))$$
$$y_{i+1} = y_i + \tfrac{1}{2}h\big(f(x_i, y_i) + f(x_{i+1}, \bar{y}_{i+1})\big) \quad \text{for } i = 0, 1, 2, 3, \ldots$$

Here, in order to use the predictor for the first time when $i = 0$ we need to know the value of $f(x_{0-1}, y_{0-1}) = f(x_{-1}, y_{-1})$, which we do not. Instead we shall use the equation $\bar{y}_1 = y_0 + hf(x_0, y_0)$ when $i = 0$.

In the next frame we shall look at an example

62

Example

Solve the equation $y' = x + y$ for $x = 0.0(0.1)1.0$ where $y = 1$ when $x = 0$.

We have solved this equation before in Frame 32 using the Euler–Cauchy method and have viewed the accuracy of this method when compared with the exact solution. Here we shall see that this predictor–corrector method is even more accurate. Set up the following heading on your spreadsheet

	A	B	C	D	E	F	G
1	n	x	y*	y	Exact	Errors (%)	h=

▶

As usual, column A contains the iteration numbers 0 to 10 in cells A2 to A12 and column B contains the *x*-values stepped according to the step length $h = 0.1$ which is in cell H1. The initial value of $y = 1$ must be entered into cell D2.

Column C contains the predictor values given by the equations

$$\bar{y}_1 = y_0 + hf(x_0, y_0)$$
$$\bar{y}_{i+1} = y_i + \tfrac{1}{2}h(3f(x_i, y_i) - f(x_{i-1}, y_{i-1})) \quad \text{for } i > 0$$

To accommodate these equations in cell C3 enter the formula . . . .

$$= D2 + \$H\$1 * (B2 + D2)$$

63

And in cell C4 enter the formula . . . .

$$= D3 + 0.5 * \$H\$1 * (3 * B3 + 3 * D3 - B2 - D2)$$

64

And copy into cells C5 to C12.

Column D contains the corrector values given by the equation

$$y_{i+1} = y_i + \tfrac{1}{2}h(f(x_i, y_i) + f(x_{i+1}, \bar{y}_{i+1}))$$

To accommodate this equation in cell D3 enter the formula . . . .

$$= D2 + 0.5 * \$H\$1 * (B2 + D2 + B3 + C3)$$

65

And copy into cells D4 to D12.

We have seen that the exact solution to this equation is $2e^x - x - 1$, so this can be programmed into the sheet entering the formula

$= 2 * \text{EXP(B2)} - \text{B2} - 1$ in cell E2 and then copying it into cells E3 to E12.

The final table looks as follows

n	x	y*	y	Exact	Error (%)	h=0·1
0	0·0		1·0000000	1·0000000	0·00	
1	0·1	1·1000000	1·1100000	1·1103418	0·03	
2	0·2	1·2415000	1·2425750	1·2428055	0·02	
3	0·3	1·3984613	1·3996268	1·3997176	0·01	
4	0·4	1·5824421	1·5837303	1·5836494	−0·01	
5	0·5	1·7963085	1·7977322	1·7974425	−0·02	
6	0·6	2·0432055	2·0447791	2·0442376	−0·03	
7	0·7	2·3266093	2·3283485	2·3275054	−0·04	
8	0·8	2·6503618	2·6522840	2·6510819	−0·05	
9	0·9	3·0187092	3·0208337	3·0192062	−0·05	
10	1·0	3·4363445	3·4386926	3·4365637	−0·06	

▶

Here the errors are significantly reduced, as seen from the comparisons below.

1	2	3
0·00	0·00	0·00
0·93	0·03	0·03
1·84	0·06	0·02
2·69	0·09	0·01
3·50	0·12	−0·01
4·25	0·14	−0·02
4·95	0·17	−0·03
5·59	0·19	−0·04
6·18	0·21	−0·05
6·73	0·23	−0·05
7·25	0·24	−0·06

Here **1** refers to Euler, **2** refers to Euler–Cauchy and **3** refers to the predictor–corrector method just used.

And that is it. There are many other more sophisticated methods for the solution of ordinary differential equations by numerical methods and a detailed study of these is a course in itself. The methods we have used give an introduction to the processes and are practical in application.

The **Revision summary** and **Can You?** checklist now follow as usual. Check them carefully and refer back to the Programme for any points that may need further brushing up. Then you will be ready for the **Test exercise**, and the **Further problems** provide further practice.

 Revision summary 13

66

1 *Taylor's series*

$$f(a + h) = f(a) + hf'(a) + \frac{h}{2!}f''(a) + \frac{h}{3!}f'''(a) + \dots$$

2 *Solution of first-order differential equations*
   Equation $y' = f(x, y)$ with $y = y_0$ at $x = x_0$ for $x_0(h)x_n$.

   (a) *Euler's method*

   $$y_1 = y_0 + h(y')_0.$$

   (b) *Euler–Cauchy method*

   $$x_1 = x_0 + h$$
   $$\bar{\bar{y}}_1 = y_0 + h(y')_0$$
   $$y_1 = y_0 + \tfrac{1}{2}h\{(y')_0 + f(x_1, \bar{\bar{y}}_1)\}$$
   $$(y')_1 = f(x_1, y_1).$$

▶

(c) *Runge–Kutta method*

$$x_1 = x_0 + h$$
$$k_1 = hf(x_0, y_0) = h(y')_0$$
$$k_2 = hf(x_0 + \tfrac{1}{2}h, y_0 + \tfrac{1}{2}k_1)$$
$$k_3 = hf(x_0 + \tfrac{1}{2}h, y_0 + \tfrac{1}{2}k_2)$$
$$k_4 = hf(x_0 + h, y_0 + k_3)$$
$$\Delta y_0 = \tfrac{1}{6}(k_1 + 2k_2 + 2k_3 + k_4)$$
$$y_1 = y_0 + \Delta y_0$$
$$(y')_1 = f(x_1, y_1).$$

3 *Solution of second-order differential equations*

Equation $y'' = f(x, y, y')$ with $y = y_0$ and $y' = (y')_0$ at $x = x_0$ for $x = x_0(h)x_n$.

(a) *Euler's second-order method*

$$y_1 = y_0 + h(y')_0 + \frac{h^2}{2!}(y'')_0$$
$$(y')_1 = (y')_0 + h(y'')_0.$$

(b) *Runge–Kutta method*

$$x_1 = x_0 + h$$
$$k_1 = \tfrac{1}{2}h^2 f\{x_0, y_0, (y')_0\} = \tfrac{1}{2}h^2(y'')_0$$
$$k_2 = \tfrac{1}{2}h^2 f\left\{x_0 + \tfrac{1}{2}h, y_0 + \tfrac{1}{2}h(y')_0 + \tfrac{1}{4}k_1, (y')_0 + \frac{k_1}{h}\right\}$$
$$k_3 = \tfrac{1}{2}h^2 f\left\{x_0 + \tfrac{1}{2}h, y_0 + \tfrac{1}{2}h(y')_0 + \tfrac{1}{4}k_1, (y')_0 + \frac{k_2}{h}\right\}$$
$$k_4 = \tfrac{1}{2}h^2 f\left\{x_0 + h, y_0 + h(y')_0 + k_3, (y')_0 + \frac{2k_3}{h}\right\}$$
$$P = \tfrac{1}{3}(k_1 + k_2 + k_3)$$
$$Q = \tfrac{1}{3}(k_1 + 2k_2 + 2k_3 + k_4)$$
$$y_1 = y_0 + h(y')_0 + P$$
$$(y')_1 = (y')_0 + \frac{Q}{h}$$
$$(y'')_1 = f\{x_1, y_1, (y')_1\}.$$

4 *Predictor–corrector*

Equation $y' = f(x, y)$ with $y = y_0$ and $y' = (y')_0$ at $x = x_0$ for $x = x_0(h)x_n$, then

Predictor

$$\bar{y}_{i+1} = y_i + \tfrac{1}{2}h(3f(x_i, y_i) - f(x_{i-1}, y_{i-1})) \quad \text{for } i = 1, 2, 3, \ldots$$
$$\bar{y}_1 = y_0 + hf(x_0, y_0) \quad \text{for } i = 0$$

Corrector

$$y_{i+1} = y_i + \tfrac{1}{2}h\big(f(x_i, y_i) + f(x_{i+1}, \bar{y}_{i+1})\big) \quad \text{for } i = 0, 1, 2, 3, \ldots$$

Can you?

67 **Checklist 13**

Check this list before and after you try the end of Programme test.

On a scale of 1 to 5 how confident are you that you can: **Frames**

- Derive a form of Taylor's series from Maclaurin's series and
  from it describe a function increment as a series of first and
  higher-order derivatives of the function?   [1] to [3]
  Yes   ☐   ☐   ☐   ☐   ☐   *No*

- Describe and apply by means of a spreadsheet the Euler
  method, the Euler–Cauchy method and the Runge–Kutta
  method for first-order differential equations?   [4] to [43]
  Yes   ☐   ☐   ☐   ☐   ☐   *No*

- Describe and apply by means of a spreadsheet the Euler
  second-order method and the Runge–Kutta method for
  second-order ordinary differential equations?   [44] to [60]
  Yes   ☐   ☐   ☐   ☐   ☐   *No*

- Describe and apply by means of a spreadsheet a simple
  predictor–corrector method?   [61] to [65]
  Yes   ☐   ☐   ☐   ☐   ☐   *No*

 # Test exercise 13

68 1 Apply Euler's method to solve the equation
$$\frac{dy}{dx} = 1 + xy \quad \text{for} \quad x = 0(0\cdot1)0\cdot5$$
given that at $x = 0$, $y = 1$.

2 The equation $\dfrac{dy}{dx} = x^2 - 2y$ is subject to the initial condition $y = 0$ at $x = 1$. Use
the Euler–Cauchy method to obtain function values for $x = 1\cdot0(0\cdot2)2\cdot0$.

3 Using the Runge–Kutta method, solve the equation
$$\frac{dy}{dx} = 1 + y - x \quad \text{for} \quad x = 0(0\cdot1)0\cdot5$$
given that $y = 1$ when $x = 0$.

4 Apply Euler's second-order method to solve the equation
$$y'' = y - x \quad \text{for} \quad x = 2\cdot0(0\cdot1)2\cdot5$$
given that at $x = 2$, $y = 3$ and $y' = 0$.

▶

5 Use the Runge–Kutta method to solve the equation
$$y'' = (y'/x) + y \quad \text{for} \quad x = 1 \cdot 0(0 \cdot 1)1 \cdot 5$$
given the initial conditions that at $x = 1 \cdot 0$, $y = 0$ and $y' = 1 \cdot 0$.

6 Use the predictor–corrector method in the text to solve the equation
$$y' = 1 + xy \quad \text{for} \quad x = 0(0 \cdot 1)1$$
given that $x = 0$ when $y = 0$.

 # Further problems 13

Solve the following differential equations by the methods indicated. **69**

Euler's method

1	$y' = 2x - y$	$x = 0, y = 1$	$x = 0(0 \cdot 2)1 \cdot 0$
2	$y' = 2x + y^2$	$x = 0, y = 1 \cdot 4$	$x = 0(0 \cdot 1)0 \cdot 5$

Euler–Cauchy method

3	$y' = 2 - y/x$	$x = 1, y = 2$	$x = 1 \cdot 0(0 \cdot 2)2 \cdot 0$
4	$y' = x^2 - 2x + y$	$x = 0, y = 0 \cdot 5$	$x = 0(0 \cdot 1)0 \cdot 5$
5	$y' = (y - x^2)^{\frac{1}{2}}$	$x = 0, y = 1$	$x = 0(0 \cdot 1)0 \cdot 5$
6	$y' = \dfrac{x + y}{xy}$	$x = 1, y = 1$	$x = 1 \cdot 0(0 \cdot 1)1 \cdot 5$
7	$y' = y \sin x + \cos x$	$x = 0, y = 0$	$x = 0(0 \cdot 1)0 \cdot 5$

Runge Kutta method

8	$y' = 2x - y$	$x = 0, y = 1$	$x = 0(0 \cdot 2)1 \cdot 0$
9	$y' = x - y^2$	$x = 0, y = 1$	$x = 0(0 \cdot 1)0 \cdot 5$
10	$y' = y^2 - xy$	$x = 0, y = 0 \cdot 4$	$x = 0(0 \cdot 2)1 \cdot 0$
11	$y' = \sqrt{2x + y}$	$x = 1, y = 2$	$x = 1 \cdot 0(0 \cdot 2)2 \cdot 0$
12	$y' = 1 - x^3/y$	$x = 0, y = 1$	$x = 0(0 \cdot 2)1 \cdot 0$
13	$y' = \dfrac{y - x}{y + x}$	$x = 0, y = 1$	$x = 0(0 \cdot 2)1 \cdot 0$

Euler second-order method

14	$y'' = (x + 1)y' + y$	$x = 0, y = 1, y' = 1$	$x = 0(0 \cdot 1)0 \cdot 5$
15	$y'' = 2(xy' - 4y)$	$x = 0, y = 3, y' = 0$	$x = 0(0 \cdot 1)0 \cdot 5$

Runge–Kutta second-order method

16 $y'' = x - y - xy'$ $x = 0, y = 0, y' = 1$ $x = 0(0 \cdot 2)1 \cdot 0$

17 $y'' = (1 - x)y' - y$ $x = 0, y = 1, y' = 1$ $x = 0(0 \cdot 2)1 \cdot 0$

18 $y'' = 1 + x - y^2$ $x = 0, y = 2, y' = 1$ $x = 0(0 \cdot 1)0 \cdot 5$

19 $y'' = (x + 2)y - 2y'$ $x = 0, y = 1, y' = 0$ $x = 0(0 \cdot 2)1 \cdot 0$

20 $y'' = \dfrac{y - xy'}{x^2}$ $x = 1, y = 0, y' = 1$ $x = 1 \cdot 0(0 \cdot 2)2 \cdot 0$

Predictor–corrector

21 $y' = 2 - y/x$ $x = 1, y = 2$ $x = 1 \cdot 0(0 \cdot 2)2 \cdot 0$

22 $y' = 2x - y$ $x = 0, y = 1$ $x = 0 \cdot 0(0 \cdot 2)1 \cdot 0$

23 $y' = \sqrt{2x + y}$ $x = 1, y = 2$ $x = 1 \cdot 0(0 \cdot 2)2 \cdot 0$

Partial differentiation

Learning outcomes

When you have completed this Programme you will be able to:

- Derive the expression for a small increment in an expression of two real variables using Taylor's theorem
- Apply the notion of small increments in expressions in two and three real variables to a variety of problems
- Determine the rate of change with respect to time of an expression involving two or three real variables
- Differentiate implicit functions
- Determine first and second derivatives involving change of variables in expressions of two real variables
- Use the Jacobian to obtain the derivatives of inverse functions of two real variables
- Locate and identify maxima, minima and saddle points of functions of two real variables
- Solve problems where the independent variables are constrained by using the method of Lagrange undetermined multipliers for functions of two and three real variables

Prerequisite: Engineering Mathematics (Sixth Edition)
Programmes 9 Differentiation applications 2, 10 Partial differentiation 1 and 11 Partial differentiation 2

Small increments

1   Taylor's theorem for one independent variable

Taylor's theorem expands $f(x + h)$ in terms of $f(x)$, powers of h and successive derivatives of $f(x)$, and can be stated as

$$f(x + h) = f(x) + hf'(x) + \frac{h^2}{2!}f''(x) + \ldots + \frac{h^n}{n!}f^{(n)}(x) + \ldots$$

where $f^{(n)}(x)$ denotes the nth derivative of $f(x)$. You will also, no doubt, remember that, by putting $x = 0$ in the result and then letting $h = x$, we obtain Maclaurin's series

$$f(h) = f(0) + hf'(0) + \frac{h^2}{2!}f''(0) + \ldots + \frac{h^n}{n!}f^{(n)}(0) + \ldots$$

Taylor's theorem for two independent variables

If we consider $z = f(x, y)$ where z is a function of two independent variables x and y, then, in general, increases in x and y will produce a combined increase in z.

So, if $z = f(x, y)$ then $z + \delta z = f(x + h, \ y + k)$

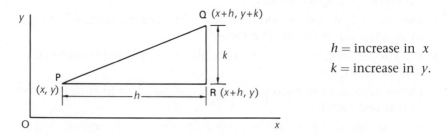

$h = $ increase in x

$k = $ increase in y.

For R: $f(x + h, \ y) = f(x, \ y) + hf_x(x, \ y) + \dfrac{h^2}{2!}f_{xx}(x, \ y) + \ldots$ (1)

where $f_x(x, \ y)$ denotes $\dfrac{\partial}{\partial x} f(x, \ y)$; $f_{xx}(x, \ y)$ denotes $\dfrac{\partial^2}{\partial x^2}f(x, \ y)$ etc.

From R to Q: $(x + h)$ is constant; y changes to $(y + k)$

$$\therefore \ f(x + h, \ y + k) = f(x + h, \ y) + kf_y(x + h, \ y) + \frac{k^2}{2!}f_{yy}(x + h, \ y) + \ldots \quad (2)$$

To express (2) in terms of $f(x, y)$ we can substitute result (1) for the first term $f(x + h, y)$ and similar expressions which we shall obtain for $f_y(x + h, y)$, $f_{yy}(x + h, y)$ and so on.

If we differentiate (1) with respect to y, we have

$$f_y(x + h, \ y) = \ldots\ldots\ldots\ldots$$

2

$$f_y(x+h,\ y) = f_y(x,\ y) + hf_{yx}(x,\ y) + \frac{h^2}{2!}f_{yxx}(x,\ y) + \dots$$

If we now differentiate this result again with respect to y

$$f_{yy}(x+h,\ y) = \dots\dots\dots$$

3

$$f_{yy}(x+h,y) = f_{yy}(x,y) + hf_{yyx}(x,y) + \frac{h^2}{2!}f_{yyxx}(x,y) + \dots$$

Then our previous expansion (2), i.e.

$$f(x+h,y+k) = f(x+h,y) + kf_y(x+h,y) + \frac{k^2}{2!}f_{yy}(x+h,y) + \dots$$

now becomes

$$f(x+h,\ y+k) = f(x,\ y) + hf_x(x,\ y) + \frac{h^2}{2!}f_{xx}(x,\ y) + \dots$$

$$+ k\left\{f_y(x,\ y) + hf_{yx}(x,\ y) + \frac{h^2}{2!}f_{yxx}(x,\ y) + \dots\right\}$$

$$+ \frac{k^2}{2!}\left\{f_{yy}(x,\ y) + hf_{yyx}(x,\ y) + \frac{h^2}{2!}f_{yyxx}(x,\ y) + \dots\right\}$$

$$+ \dots$$

Rearranging the terms by collecting together all the first derivatives, and then all the second derivatives, and so on, we get

$$f(x+h,\ y+k) = \dots\dots\dots$$

4

$$f(x+h,y+k) = f(x,y) + \{hf_x(x,y) + kf_y(x,y)\}$$

$$+ \frac{1}{2!}\{h^2f_{xx}(x,y) + 2hkf_{xy}(x,y) + k^2f_{yy}(x,y)\} + \dots$$

This is Taylor's theorem for two independent variables.

▶

Small increments

If $z = f(x, y)$, $h = \delta x$, $k = \delta y$, then Taylor's theorem can be written as

$$z + \delta z = z + \left\{ h\frac{\partial z}{\partial x} + k\frac{\partial z}{\partial y} \right\} + \frac{1}{2!}\left\{ h^2\frac{\partial^2 z}{\partial x^2} + 2hk\frac{\partial^2 z}{\partial y\, \partial x} + k^2\frac{\partial^2 z}{\partial y^2} \right\} + \cdots$$

Subtracting z from each side:

$$\delta z = \frac{\partial z}{\partial x}\delta x + \frac{\partial z}{\partial y}\delta y + \frac{1}{2!}\left\{ \frac{\partial^2 z}{\partial x^2}(\delta x)^2 + 2\frac{\partial^2 z}{\partial y\, \partial x}(\delta x\, \delta y) + \frac{\partial^2 z}{\partial y^2}(\delta y)^2 \right\} + \cdots$$

Since δx and δy are small, the expression in the brackets is of the next order of smallness and can be discarded for our purposes. Therefore, we arrive at the result

If $z = f(x, y)$ then $\delta z = \dfrac{\partial z}{\partial x}\delta x + \dfrac{\partial z}{\partial y}\delta y$

As already explained above, this result is, in fact, an approximation since the smaller terms in the series have been neglected. For practical purposes, however, the result can be used as stated. **Be sure to make a note of the result, for it is the foundation of much that follows.**

5

$$z = f(x, y); \qquad \delta z = \frac{\partial z}{\partial x}\delta x + \frac{\partial z}{\partial y}\delta y$$

The following diagram illustrates the result.

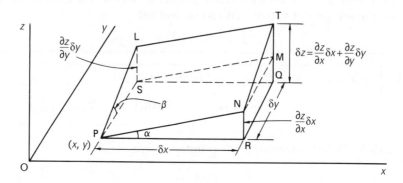

$\dfrac{\partial z}{\partial x}$ is the slope of PN $\quad \therefore \quad \text{RN} = \dfrac{\partial z}{\partial x}\delta x = \text{QM}$

$\dfrac{\partial z}{\partial y}$ is the slope of PL $\quad \therefore \quad \text{SL} = \dfrac{\partial z}{\partial y}\delta y = \text{MT}$

$\text{QT} = \text{QM} + \text{MT} \quad \therefore \quad \delta z = \dfrac{\partial z}{\partial x}\delta x + \dfrac{\partial z}{\partial y}\delta y$

This is the total increment of $z = f(x, y)$ from P to Q.

▶

It is worth noting at this stage that the result can be extended to the case of three independent variables, i.e. if $u = f(x, y, z)$

$$\delta u = \frac{\partial u}{\partial x} \delta x + \frac{\partial u}{\partial y} \delta y + \frac{\partial u}{\partial z} \delta z$$

One or two straightforward applications will lay the foundations for future development.

Example

A rectangular box has sides measured as 30 mm, 40 mm and 60 mm. If these measurements are liable to be in error by ± 0.5 mm, ± 0.8 mm and ± 1.0 mm respectively, calculate the length of the diagonal of the box and the maximum possible error in the result.

First build up an expression for the diagonal d in terms of the sides, a, b and c.

$$d = $$

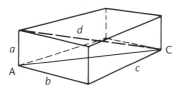

$$\boxed{d = \sqrt{a^2 + b^2 + c^2}}$$

6

Because

$$d^2 = a^2 + AC^2 = a^2 + b^2 + c^2 \text{ and so } d = \sqrt{a^2 + b^2 + c^2}$$

Then $\delta d = \dfrac{\partial d}{\partial a} \delta a + \dfrac{\partial d}{\partial b} \delta b + \dfrac{\partial d}{\partial c} \delta c$

We now determine the partial differential coefficients and obtain an expression for δd, but all in terms of a, b and c. Do not yet insert numerical values.

$$\delta d = \ldots\ldots\ldots$$

$$\boxed{\delta d = \frac{1}{\sqrt{a^2 + b^2 + c^2}} \{ a\delta a + b\delta b + c\delta c \}}$$

7

Now, substituting the values $a = 30$, $b = 40$, $c = 60$

$$\delta a = \pm 0.5, \ \delta b = \pm 0.8, \ \delta c = \pm 1.0$$

the calculated length of the diagonal $= \ldots\ldots\ldots$

the maximum possible error $= \ldots\ldots\ldots$

8

$$\boxed{\text{diagonal} = \sqrt{a^2 + b^2 + c^2} = 78 \cdot 10 \text{ mm} \\ \text{maximum error} = \pm 1 \cdot 37 \text{ mm}}$$

Because

$$\delta d = \frac{1}{78 \cdot 10}\{30(\pm 0 \cdot 5) + 40(\pm 0 \cdot 8) + 60(\pm 1 \cdot 0)\}$$

Greatest error when the signs are the same

$$\therefore \ \delta d = \frac{1}{78 \cdot 10}\{\pm (15 + 32 + 60)\} = \pm 1 \cdot 37 \text{ mm}$$

Rates of change

If $z = f(x, y)$, then we have seen that $\delta z = \dfrac{\partial z}{\partial x}\delta x + \dfrac{\partial z}{\partial y}\delta y$

Dividing through by δt: $\dfrac{\delta z}{\delta t} = \dfrac{\partial z}{\partial x}\dfrac{\delta x}{\delta t} + \dfrac{\partial z}{\partial y}\dfrac{\delta y}{\delta t}$

Then if $\delta t \to 0$: $\dfrac{dz}{dt} = \dfrac{\partial z}{\partial x}\dfrac{dx}{dt} + \dfrac{\partial z}{\partial y}\dfrac{dy}{dt}$

Note the result. Then on to an example.

Example

The base radius r of a right circular cone is increasing at the rate of $1 \cdot 5$ mm/s while the perpendicular height is decreasing at $6 \cdot 0$ mm/s. Determine the rate at which the volume V is changing when $r = 12$ mm and $h = 24$ mm.

Find an expression for $\dfrac{dV}{dt}$ in terms of r and h which is $\ldots\ldots\ldots\ldots$

9

$$\boxed{\dfrac{dV}{dt} = \dfrac{2\pi rh}{3}\cdot\dfrac{dr}{dt} + \dfrac{\pi r^2}{3}\cdot\dfrac{dh}{dt}}$$

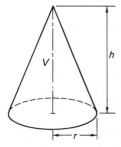

$$V = \frac{1}{3}\pi r^2 h; \quad \frac{dV}{dt} = \frac{\partial V}{\partial r}\cdot\frac{dr}{dt} + \frac{\partial V}{\partial h}\cdot\frac{dh}{dt}$$

$$\frac{\partial V}{\partial r} = \frac{2\pi rh}{3}; \quad \frac{\partial V}{\partial h} = \frac{\pi r^2}{3}$$

$$\therefore \ \frac{dV}{dt} = \frac{2\pi rh}{3}\cdot\frac{dr}{dt} + \frac{\pi r^2}{3}\cdot\frac{dh}{dt}$$

▶

Finally, we insert the numerical values:

$$r = 12; \quad h = 24; \quad \frac{dr}{dt} = 1\cdot5; \quad \frac{dh}{dt} = -6\cdot0 \quad (h \text{ is decreasing})$$

$$\frac{dV}{dt} = 288\pi - 288\pi = 0$$

$\therefore$ At the instant when $r = 12$ mm and $h = 24$ mm,

the volume is unchanging.

Implicit functions

The same initial result, $\delta z = \dfrac{\partial z}{\partial x}\delta x + \dfrac{\partial z}{\partial y}\delta y$ enables us to determine the derivative of an implicit function $f(x, y) = 0$, i.e. in a case where y is not defined explicitly in terms of x.

If $f(x, y) = 0$ is an implicit function, we let $z = f(x, y)$.

Then, as before:
$$\delta z = \frac{\partial z}{\partial x}\delta x + \frac{\partial z}{\partial y}\delta y$$

Dividing through by δx:
$$\frac{\delta z}{\delta x} = \frac{\partial z}{\partial x} + \frac{\partial z}{\partial y}\cdot\frac{\delta y}{\delta x}$$

Then, if $\delta x \to 0$:
$$\frac{dz}{dx} = \frac{\partial z}{\partial x} + \frac{\partial z}{\partial y}\cdot\frac{dy}{dx}$$

But $z = 0$ $\therefore \dfrac{dz}{dx} = 0$
$$\therefore \frac{\partial z}{\partial x} + \frac{\partial z}{\partial y}\cdot\frac{dy}{dx} = 0$$

$$\therefore \frac{dy}{dx} = -\left(\frac{\partial z}{\partial x}\Big/\frac{\partial z}{\partial y}\right)$$

So, if $x^2 - xy - y^2 = 0$, $\dfrac{dy}{dx} = \ldots\ldots\ldots\ldots$

10

$$\boxed{\frac{dy}{dx} = \frac{2x - y}{x + 2y}}$$

Putting $z = x^2 - xy - y^2$, $\dfrac{\partial z}{\partial x} = 2x - y$ and $\dfrac{\partial z}{\partial y} = -x - 2y$

The rest follows immediately.

Now on to the next frame

11

The work so far, important though it is, is largely by way of revision of the more basic ideas of partial differentiation. We now extend these same ideas to further applications.

▶

Change of variables

If $z = f(x, y)$ and x and y are themselves functions of two new independent variables, u and v, then we need expressions for $\dfrac{\partial z}{\partial u}$ and $\dfrac{\partial z}{\partial v}$.

Yet again, we start from the result we established at the beginning of this Programme.

$$\text{If } z = f(x, y) \quad \text{then} \quad \delta z = \frac{\partial z}{\partial x}\delta x + \frac{\partial z}{\partial y}\delta y$$

Dividing in turn by δu and δv:

$$\frac{\delta z}{\delta u} = \frac{\partial z}{\partial x}\cdot\frac{\delta x}{\delta u} + \frac{\partial z}{\partial y}\cdot\frac{\delta y}{\delta u}$$

$$\frac{\delta z}{\delta v} = \frac{\partial z}{\partial x}\cdot\frac{\delta x}{\delta v} + \frac{\partial z}{\partial y}\cdot\frac{\delta y}{\delta v}$$

Then, as $\delta u \to 0$ and $\delta v \to 0$, these become

$$\frac{\partial z}{\partial u} = \frac{\partial z}{\partial x}\cdot\frac{\partial x}{\partial u} + \frac{\partial z}{\partial y}\cdot\frac{\partial y}{\partial u}$$

$$\frac{\partial z}{\partial v} = \frac{\partial z}{\partial x}\cdot\frac{\partial x}{\partial v} + \frac{\partial z}{\partial y}\cdot\frac{\partial y}{\partial v}$$

Example 1

If $z = x^2 - y^2$ and $x = r\cos\theta$ and $y = r\sin\theta$, then

$$\frac{\partial z}{\partial r} = \frac{\partial z}{\partial x}\cdot\frac{\partial x}{\partial r} + \frac{\partial z}{\partial y}\cdot\frac{\partial y}{\partial r}$$

$$\text{and} \quad \frac{\partial z}{\partial \theta} = \frac{\partial z}{\partial x}\cdot\frac{\partial x}{\partial \theta} + \frac{\partial z}{\partial y}\cdot\frac{\partial y}{\partial \theta}$$

We now need the various partial derivatives

$$\frac{\partial z}{\partial x} = \ldots\ldots\ldots; \quad \frac{\partial x}{\partial r} = \ldots\ldots\ldots; \quad \frac{\partial y}{\partial r} = \ldots\ldots\ldots$$

$$\frac{\partial z}{\partial y} = \ldots\ldots\ldots; \quad \frac{\partial x}{\partial \theta} = \ldots\ldots\ldots; \quad \frac{\partial y}{\partial \theta} = \ldots\ldots\ldots$$

12

$$\frac{\partial z}{\partial x} = 2x; \quad \frac{\partial x}{\partial r} = \cos\theta; \quad \frac{\partial y}{\partial r} = \sin\theta$$

$$\frac{\partial z}{\partial y} = -2y; \quad \frac{\partial x}{\partial \theta} = -r\sin\theta; \quad \frac{\partial y}{\partial \theta} = r\cos\theta$$

Substituting in the two equations and simplifying:

$$\frac{\partial z}{\partial r} = \ldots\ldots\ldots; \quad \frac{\partial z}{\partial \theta} = \ldots\ldots\ldots$$

$$\frac{\partial z}{\partial r} = 2x \cos\theta - 2y \sin\theta; \quad \frac{\partial z}{\partial \theta} = -(2xr \sin\theta + 2yr \cos\theta)$$

13

Finally, we can express x and y in terms of r and θ as given, so that, after tidying up, we obtain

$$\frac{\partial z}{\partial r} = \ldots\ldots\ldots; \quad \frac{\partial z}{\partial \theta} = \ldots\ldots\ldots$$

$$\frac{\partial z}{\partial r} = 2r(\cos^2\theta - \sin^2\theta); \quad \frac{\partial z}{\partial \theta} = -4r^2 \sin\theta \cos\theta$$

14

Of course, we could express these as

$$\frac{\partial z}{\partial r} = 2r \cos 2\theta \quad \text{and} \quad \frac{\partial z}{\partial \theta} = -2r^2 \sin 2\theta$$

From these results, we can, if necessary, find the second partial derivatives in the normal manner.

$$\frac{\partial^2 z}{\partial r^2} = \frac{\partial}{\partial r}\left(\frac{\partial z}{\partial r}\right) = \frac{\partial}{\partial r}(2r \cos 2\theta) = 2 \cos 2\theta$$

Similarly $\quad \dfrac{\partial^2 z}{\partial \theta^2} = \ldots\ldots\ldots \quad$ and $\quad \dfrac{\partial^2 z}{\partial r \partial \theta} = \ldots\ldots\ldots$

$$\frac{\partial^2 z}{\partial \theta^2} = -4r^2 \cos 2\theta; \quad \frac{\partial^2 z}{\partial r \partial \theta} = -4r \sin 2\theta$$

15

Because

$$\frac{\partial^2 z}{\partial \theta^2} = \frac{\partial}{\partial \theta}\left(\frac{\partial z}{\partial \theta}\right) = \frac{\partial}{\partial \theta}(-2r^2 \sin 2\theta) = -4r^2 \cos 2\theta$$

and $\quad \dfrac{\partial^2 z}{\partial r \partial \theta} = \dfrac{\partial}{\partial r}\left(\dfrac{\partial z}{\partial \theta}\right) = \dfrac{\partial}{\partial r}(-2r^2 \sin 2\theta) = -4r \sin 2\theta$

Example 2

If $z = f(x, y)$ and $x = \frac{1}{2}(u^2 - v^2)$ and $y = uv$, show that

$$u\frac{\partial z}{\partial v} - v\frac{\partial z}{\partial u} = 2\left(x\frac{\partial z}{\partial y} - y\frac{\partial z}{\partial x}\right)$$

Although this is much the same as the previous example, there is, at least, one difference. In this case, we are not told the precise nature of $f(x, y)$. We must remember that z is a function of x and y and, therefore, of u and v. With that in mind, we set off with the usual two equations.

$$\frac{\partial z}{\partial u} = \ldots\ldots\ldots$$

$$\frac{\partial z}{\partial v} = \ldots\ldots\ldots$$

16

$$\frac{\partial z}{\partial u} = \frac{\partial z}{\partial x} \cdot \frac{\partial x}{\partial u} + \frac{\partial z}{\partial y} \cdot \frac{\partial y}{\partial u}$$

$$\frac{\partial z}{\partial v} = \frac{\partial z}{\partial x} \cdot \frac{\partial x}{\partial v} + \frac{\partial z}{\partial y} \cdot \frac{\partial y}{\partial v}$$

From the given information:

$$\frac{\partial x}{\partial u} = \dots\dots\dots\dots; \quad \frac{\partial y}{\partial u} = \dots\dots\dots\dots$$

$$\frac{\partial x}{\partial v} = \dots\dots\dots\dots; \quad \frac{\partial y}{\partial v} = \dots\dots\dots\dots$$

17

$$\frac{\partial x}{\partial u} = u; \quad \frac{\partial y}{\partial u} = v$$

$$\frac{\partial x}{\partial v} = -v; \quad \frac{\partial y}{\partial v} = u$$

Whereupon $\quad \dfrac{\partial z}{\partial u} = \dots\dots\dots\dots$

$$\frac{\partial z}{\partial v} = \dots\dots\dots\dots$$

18

$$\frac{\partial z}{\partial u} = u\frac{\partial z}{\partial x} + v\frac{\partial z}{\partial y}$$

$$\frac{\partial z}{\partial v} = -v\frac{\partial z}{\partial x} + u\frac{\partial z}{\partial y}$$

If we now multiply the first of these by $(-v)$ and the second by u and add the two equations, we get the desired result.

$$-v\frac{\partial z}{\partial u} = -uv\frac{\partial z}{\partial x} - v^2\frac{\partial z}{\partial y}$$

$$u\frac{\partial z}{\partial v} = -uv\frac{\partial z}{\partial x} + u^2\frac{\partial z}{\partial y}$$

Adding $\quad u\dfrac{\partial z}{\partial v} - v\dfrac{\partial z}{\partial u} = -2uv\dfrac{\partial z}{\partial x} + (u^2 - v^2)\dfrac{\partial z}{\partial y}$

$$= -2y\frac{\partial z}{\partial x} + 2x\frac{\partial z}{\partial y}$$

$$\therefore \; u\frac{\partial z}{\partial v} - v\frac{\partial z}{\partial u} = 2\left(x\frac{\partial z}{\partial y} - y\frac{\partial z}{\partial x}\right)$$

▶

With the same given data, i.e.

$$z = f(x, y) \text{ with } x = \frac{1}{2}(u^2 - v^2) \text{ and } y = uv$$

we can now show that $\dfrac{\partial^2 z}{\partial u^2} + \dfrac{\partial^2 z}{\partial v^2} = (u^2 + v^2)\left(\dfrac{\partial^2 z}{\partial x^2} + \dfrac{\partial^2 z}{\partial y^2}\right)$.

In determining the second partial derivatives, keep in mind that z is a function of u and v and that both of these variables also occur in $\dfrac{\partial z}{\partial x}$ and $\dfrac{\partial z}{\partial y}$.

$$\frac{\partial^2 z}{\partial u^2} = \ldots\ldots\ldots$$

19

$$\boxed{\frac{\partial^2 z}{\partial u^2} = \frac{\partial z}{\partial x} + u^2 \frac{\partial^2 z}{\partial x^2} + 2uv \frac{\partial^2 z}{\partial x \partial y} + v^2 \frac{\partial^2 z}{\partial y^2}}$$

Because

$$\frac{\partial}{\partial u} = \left(u\frac{\partial}{\partial x} + v\frac{\partial}{\partial y}\right) \quad \text{and} \quad \frac{\partial}{\partial v} = \left(-v\frac{\partial}{\partial x} + u\frac{\partial}{\partial y}\right)$$

$$\therefore \frac{\partial^2 z}{\partial u^2} = \frac{\partial}{\partial u}\left(u\frac{\partial z}{\partial x} + v\frac{\partial z}{\partial y}\right) = \frac{\partial z}{\partial x} + u\frac{\partial}{\partial u}\left(\frac{\partial z}{\partial x}\right) + v\frac{\partial}{\partial u}\left(\frac{\partial z}{\partial y}\right)$$

$$= \frac{\partial z}{\partial x} + u\left(u\frac{\partial}{\partial x} + v\frac{\partial}{\partial y}\right)\frac{\partial z}{\partial x} + v\left(u\frac{\partial}{\partial x} + v\frac{\partial}{\partial y}\right)\frac{\partial z}{\partial y}$$

$$= \frac{\partial z}{\partial x} + u^2\frac{\partial^2 z}{\partial x^2} + uv\frac{\partial^2 z}{\partial x \partial y} + uv\frac{\partial^2 z}{\partial x \partial y} + v^2\frac{\partial^2 z}{\partial y^2}$$

$$\therefore \frac{\partial^2 z}{\partial u^2} = \frac{\partial z}{\partial x} + u^2\frac{\partial^2 z}{\partial x^2} + 2uv\frac{\partial^2 z}{\partial x \partial y} + v^2\frac{\partial^2 z}{\partial y^2} \qquad (1)$$

Likewise, $\dfrac{\partial^2 z}{\partial v^2} = \dfrac{\partial}{\partial v}\left(\dfrac{\partial z}{\partial v}\right) = \dfrac{\partial}{\partial v}\left(-v\dfrac{\partial z}{\partial x} + u\dfrac{\partial z}{\partial y}\right)$

$$= \ldots\ldots\ldots$$

20

$$\boxed{\frac{\partial^2 z}{\partial v^2} = \frac{\partial z}{\partial x} + v^2 \frac{\partial^2 z}{\partial x^2} - 2uv \frac{\partial^2 z}{\partial x \partial y} + u^2 \frac{\partial^2 z}{\partial y^2}}$$

Because

$$\frac{\partial^2 z}{\partial v^2} = \frac{\partial}{\partial v}\left(-v\frac{\partial z}{\partial x} + u\frac{\partial z}{\partial y}\right) = -\frac{\partial z}{\partial x} + \frac{\partial}{\partial v}\left(\frac{\partial z}{\partial x}\right) + u\frac{\partial}{\partial v}\left(\frac{\partial z}{\partial y}\right)$$

$$= \frac{\partial z}{\partial x} - v\left(-v\frac{\partial}{\partial x} + u\frac{\partial}{\partial y}\right)\frac{\partial z}{\partial x} + u\left(-v\frac{\partial}{\partial x} + u\frac{\partial}{\partial y}\right)\frac{\partial z}{\partial y}$$

$$= \frac{\partial z}{\partial x} + v^2\frac{\partial^2 z}{\partial x^2} - uv\frac{\partial^2 z}{\partial x \partial y} - uv\frac{\partial^2 z}{\partial x \partial y} + u^2\frac{\partial^2 z}{\partial y^2}$$

$$\therefore \frac{\partial^2 z}{\partial v^2} = \frac{\partial z}{\partial x} + v^2\frac{\partial^2 z}{\partial x^2} - 2uv\frac{\partial^2 z}{\partial x \partial y} + u^2\frac{\partial^2 z}{\partial y^2} \qquad (2)$$

Adding together results (1) and (2), we get $\ldots\ldots\ldots\ldots$

21

$$\frac{\partial^2 z}{\partial u^2} + \frac{\partial^2 z}{\partial v^2} = (u^2 + v^2)\left(\frac{\partial^2 z}{\partial x^2} + \frac{\partial^2 z}{\partial y^2}\right)$$

and that is it.

Now, for something slightly different, move on to the next frame

Inverse functions

22

If $z = f(x, y)$ and x and y are functions of two independent variables u and v defined by $u = g(x, y)$ and $v = h(x, y)$, we can theoretically solve these two equations to obtain x and y in terms of u and v. Hence we can determine $\dfrac{\partial x}{\partial u}, \dfrac{\partial x}{\partial v}, \dfrac{\partial y}{\partial u}, \dfrac{\partial y}{\partial v}$ and then $\dfrac{\partial z}{\partial x}$ and $\dfrac{\partial z}{\partial y}$ as required.

In practice, however, the solution of $u = g(x, y)$ and $v = h(x, y)$ may well be difficult or even impossible by normal means. The following example shows how we can get over this difficulty.

Example 1

If $z = f(x, y)$ and $u = e^x \cos y$ and $v = e^{-x} \sin y$, we have to find $\dfrac{\partial x}{\partial u}, \dfrac{\partial x}{\partial v}, \dfrac{\partial y}{\partial u}, \dfrac{\partial y}{\partial v}$.

We start off once again with our standard relationships

$$\delta u = \frac{\partial u}{\partial x}\delta x + \frac{\partial u}{\partial y}\delta y \tag{1}$$

$$\delta v = \frac{\partial v}{\partial x}\delta x + \frac{\partial v}{\partial y}\delta y \tag{2}$$

Now $u = e^x \cos y$ and $v = e^{-x} \sin y$

So $\dfrac{\partial u}{\partial x} = \ldots\ldots\ldots;$ $\dfrac{\partial u}{\partial y} = \ldots\ldots\ldots$

$\dfrac{\partial v}{\partial x} = \ldots\ldots\ldots;$ $\dfrac{\partial v}{\partial y} = \ldots\ldots\ldots$

23

$$\frac{\partial u}{\partial x} = e^x \cos y; \qquad \frac{\partial u}{\partial y} = -e^x \sin y$$

$$\frac{\partial v}{\partial x} = -e^{-x} \sin y; \qquad \frac{\partial v}{\partial y} = e^{-x} \cos y$$

Substituting in equations (1) and (2) above, we have

$$\delta u = e^x \cos y \, \delta x - e^x \sin y \, \delta y \qquad\qquad (3)$$
$$\delta v = -e^{-x} \sin y \, \delta x + e^{-x} \cos y \, \delta y \qquad\qquad (4)$$

Eliminating δy from (3) and (4), we get

$$\delta x = \ldots\ldots\ldots$$

24

$$\delta x = \frac{e^{-x} \cos y}{\cos 2y} \delta u + \frac{e^x \sin y}{\cos 2y} \delta v$$

Because

$(3) \times e^{-x} \cos y:$ $e^{-x} \cos y \, \delta u = \cos^2 y \, \delta x - \sin y \cos y \, \delta y$

$(4) \times e^x \sin y:$ $e^x \sin y \, \delta v = -\sin^2 y \, \delta x + \sin y \cos y \, \delta y$

Adding: $e^{-x} \cos y \, \delta u + e^x \sin y \, \delta v = (\cos^2 y - \sin^2 y) \, \delta x$

$$\therefore \; \delta x = \frac{e^{-x} \cos y}{\cos 2y} \delta u + \frac{e^x \sin y}{\cos 2y} \delta v$$

But $\delta x = \dfrac{\partial x}{\partial u} \delta u + \dfrac{\partial x}{\partial v} \delta v$

$$\therefore \; \frac{\partial x}{\partial u} = \frac{e^{-x} \cos y}{\cos 2y} \quad \text{and} \quad \frac{\partial x}{\partial v} = \frac{e^x \sin y}{\cos 2y}$$

which are, of course, two of the expressions we have to find.
Starting again with equations (3) and (4), we can obtain

$$\delta y = \ldots\ldots\ldots$$

25

$$\delta y = \frac{e^{-x} \sin y}{\cos 2y} \delta u + \frac{e^x \cos y}{\cos 2y} \delta v$$

Because

$(3) \times e^{-x} \sin y:$ $e^{-x} \sin y \, \delta u = \sin y \cos y \, \delta x - \sin^2 y \, \delta y$

$(4) \times e^x \cos y:$ $e^x \cos y \, \delta v = -\sin y \cos y \, \delta x + \cos^2 y \, \delta y$

Adding: $e^{-x} \sin y \, \delta u + e^x \cos y \, \delta v = (\cos^2 y - \sin^2 y) \, \delta y$

$$\therefore \; \delta y = \frac{e^{-x} \sin y}{\cos 2y} \delta u + \frac{e^x \cos y}{\cos 2y} \delta v$$

But, $\delta y = \ldots\ldots\ldots$ Finish it off.

26

$$\delta y = \frac{\partial y}{\partial u} \delta u + \frac{\partial y}{\partial v} \delta v$$

$$\therefore \quad \frac{\partial y}{\partial u} = \frac{e^{-x} \sin y}{\cos 2y} \quad \text{and} \quad \frac{\partial y}{\partial v} = \frac{e^{x} \cos y}{\cos 2y}$$

So, collecting our four results together:

$$\frac{\partial x}{\partial u} = \frac{e^{-x} \cos y}{\cos 2y}; \quad \frac{\partial x}{\partial v} = \frac{e^{x} \sin y}{\cos 2y}$$

$$\frac{\partial y}{\partial u} = \frac{e^{-x} \sin y}{\cos 2y}; \quad \frac{\partial y}{\partial v} = \frac{e^{x} \cos y}{\cos 2y}$$

We can tackle most similar problems in the same way, but it is more efficient to investigate a general case and to streamline the results. Let us do that.

27 **General case**

If $z = f(x, y)$ with $u = g(x, y)$ and $v = h(x, y)$, then we have

$$\delta u = \frac{\partial u}{\partial x} \delta x + \frac{\partial u}{\partial y} \delta y \tag{1}$$

$$\delta v = \frac{\partial v}{\partial x} \delta x + \frac{\partial v}{\partial y} \delta y \tag{2}$$

We now solve these for δx and δy. Eliminating δy, we have

$(1) \times \dfrac{\partial v}{\partial y}:$ $\qquad \dfrac{\partial v}{\partial y} \delta u = \dfrac{\partial v}{\partial y} \cdot \dfrac{\partial u}{\partial x} \delta x + \dfrac{\partial v}{\partial y} \cdot \dfrac{\partial u}{\partial y} \delta y$

$(2) \times \dfrac{\partial u}{\partial y}:$ $\qquad \dfrac{\partial u}{\partial y} \delta v = \dfrac{\partial u}{\partial y} \cdot \dfrac{\partial v}{\partial x} \delta x + \dfrac{\partial u}{\partial y} \cdot \dfrac{\partial v}{\partial y} \delta y$

Subtracting: $\qquad \dfrac{\partial v}{\partial y} \delta u - \dfrac{\partial u}{\partial y} \delta v = \left(\dfrac{\partial u}{\partial x} \cdot \dfrac{\partial v}{\partial y} - \dfrac{\partial v}{\partial x} \cdot \dfrac{\partial u}{\partial y} \right) \delta x$

$$\therefore \quad \delta x = \frac{\dfrac{\partial v}{\partial y} \delta u - \dfrac{\partial u}{\partial y} \delta v}{\dfrac{\partial u}{\partial x} \cdot \dfrac{\partial v}{\partial y} - \dfrac{\partial v}{\partial x} \cdot \dfrac{\partial u}{\partial y}}$$

Starting afresh from (1) and (2) and eliminating δx, we have

$$\delta y = \ldots \ldots \ldots$$

28

$$\delta y = \frac{\dfrac{\partial u}{\partial x}\delta v - \dfrac{\partial v}{\partial x}\delta u}{\dfrac{\partial u}{\partial x}\cdot\dfrac{\partial v}{\partial y} - \dfrac{\partial v}{\partial x}\cdot\dfrac{\partial u}{\partial y}}$$

The two results so far are therefore

$$\delta x = \frac{\dfrac{\partial v}{\partial y}\delta u - \dfrac{\partial u}{\partial y}\delta v}{\dfrac{\partial u}{\partial x}\cdot\dfrac{\partial v}{\partial y} - \dfrac{\partial v}{\partial x}\cdot\dfrac{\partial u}{\partial y}} \quad \text{and} \quad \delta y = \frac{\dfrac{\partial u}{\partial x}\delta v - \dfrac{\partial v}{\partial x}\delta u}{\dfrac{\partial u}{\partial x}\cdot\dfrac{\partial v}{\partial y} - \dfrac{\partial v}{\partial x}\cdot\dfrac{\partial u}{\partial y}}$$

You will notice that the denominator is the same in each case and that it can be expressed in determinant form

$$\frac{\partial u}{\partial x}\cdot\frac{\partial v}{\partial y} - \frac{\partial v}{\partial x}\cdot\frac{\partial u}{\partial y} = \begin{vmatrix} \dfrac{\partial u}{\partial x} & \dfrac{\partial v}{\partial x} \\ \dfrac{\partial u}{\partial y} & \dfrac{\partial v}{\partial y} \end{vmatrix}$$

This determinant is called the *Jacobian* of u, v with respect to x, y and is denoted by the symbol J:

i.e. $\quad J = \begin{vmatrix} \dfrac{\partial u}{\partial x} & \dfrac{\partial v}{\partial x} \\ \dfrac{\partial u}{\partial y} & \dfrac{\partial v}{\partial y} \end{vmatrix}\quad$ and is often written as $\quad \dfrac{\partial(u,v)}{\partial(x,y)}$

So $\quad J = \dfrac{\partial(u,v)}{\partial(x,y)} = \begin{vmatrix} \dfrac{\partial u}{\partial x} & \dfrac{\partial v}{\partial x} \\ \dfrac{\partial u}{\partial y} & \dfrac{\partial v}{\partial y} \end{vmatrix}$

Our last two results can therefore be written

$$\delta x = \ldots\ldots\ldots\ldots; \quad \delta y = \ldots\ldots\ldots\ldots$$

29

$$\delta x = \frac{\dfrac{\partial v}{\partial y}\delta u - \dfrac{\partial u}{\partial y}\delta v}{J} = \frac{\begin{vmatrix} \delta u & \delta v \\ \dfrac{\partial u}{\partial y} & \dfrac{\partial v}{\partial y} \end{vmatrix}}{\begin{vmatrix} \dfrac{\partial u}{\partial x} & \dfrac{\partial v}{\partial x} \\ \dfrac{\partial u}{\partial y} & \dfrac{\partial v}{\partial y} \end{vmatrix}}, \quad \delta y = \frac{\dfrac{\partial u}{\partial x}\delta v - \dfrac{\partial v}{\partial x}\delta u}{J} = \frac{\begin{vmatrix} \dfrac{\partial u}{\partial x} & \dfrac{\partial v}{\partial x} \\ \delta u & \delta v \end{vmatrix}}{\begin{vmatrix} \dfrac{\partial u}{\partial x} & \dfrac{\partial v}{\partial x} \\ \dfrac{\partial u}{\partial y} & \dfrac{\partial v}{\partial y} \end{vmatrix}}$$

▶

We can now get a number of useful relationships.

(a) *If v is kept constant, $\delta v = 0$* $\therefore \ \delta x = \dfrac{\partial v}{\partial y}\delta u \Big/ J$

Dividing by δu and letting $\delta u \to 0$ $\dfrac{\partial x}{\partial u} = \dfrac{\partial v}{\partial y}\Big/ J$

Similarly $\dfrac{\partial y}{\partial u} = -\dfrac{\partial v}{\partial x}\Big/ J$

(b) *If u is kept constant, $\delta u = 0$* $\therefore \ \delta x = -\dfrac{\partial u}{\partial y}\delta v \Big/ J$

Dividing by δv and letting $\delta v \to 0$ $\dfrac{\partial x}{\partial v} = -\dfrac{\partial u}{\partial y}\Big/ J$

Similarly $\dfrac{\partial y}{\partial v} = \dfrac{\partial u}{\partial x}\Big/ J$

So, at this stage, we had better summarise the results.

Summary

If $z = f(x, y)$ and $u = g(x, y)$ and $v = h(x, y)$ then

$$\frac{\partial x}{\partial u} = \frac{\partial v}{\partial y}\Big/ J \qquad \frac{\partial x}{\partial v} = -\frac{\partial u}{\partial y}\Big/ J$$

$$\frac{\partial y}{\partial u} = -\frac{\partial v}{\partial x}\Big/ J \qquad \frac{\partial y}{\partial v} = \frac{\partial u}{\partial x}\Big/ J$$

where, in each case

$$J = \frac{\partial(u, v)}{\partial(x, y)} = \begin{vmatrix} \dfrac{\partial u}{\partial x} & \dfrac{\partial v}{\partial x} \\ \dfrac{\partial u}{\partial y} & \dfrac{\partial v}{\partial y} \end{vmatrix}$$

Let us put this into practice by doing again the same example that we started with (Example 1 in Frame 22), but by the new method. First of all, however, make a note of the important summary listed above for future reference.

Example 1

30

If $z = f(x, y)$ and $u = e^x \cos y$ and $v = e^{-x} \sin y$, find the derivatives $\dfrac{\partial x}{\partial u}, \dfrac{\partial x}{\partial v}, \dfrac{\partial y}{\partial u}, \dfrac{\partial y}{\partial v}$.

$$u = e^x \cos y \qquad\qquad v = e^{-x} \sin y$$

$$\frac{\partial u}{\partial x} = e^x \cos y \qquad\qquad \frac{\partial v}{\partial x} = -e^{-x} \sin y$$

$$\frac{\partial u}{\partial y} = -e^x \sin y \qquad\qquad \frac{\partial v}{\partial y} = e^{-x} \cos y$$

$$J = \frac{\partial(u, v)}{\partial(x, y)} = \begin{vmatrix} \dfrac{\partial u}{\partial x} & \dfrac{\partial v}{\partial x} \\[2mm] \dfrac{\partial u}{\partial y} & \dfrac{\partial v}{\partial y} \end{vmatrix} = \begin{vmatrix} e^x \cos y & -e^{-x} \sin y \\[2mm] -e^x \sin y & e^{-x} \cos y \end{vmatrix}$$

$$= (e^x \cos y)(e^{-x} \cos y) - (-e^x \sin y)(-e^{-x} \sin y)$$

$$= \cos^2 y - \sin^2 y$$

$$= \cos 2y$$

Then $\quad \dfrac{\partial x}{\partial u} = \dfrac{\partial v}{\partial y} \Big/ J = \dfrac{e^{-x} \cos y}{\cos 2y}; \qquad \dfrac{\partial x}{\partial v} = -\dfrac{\partial u}{\partial y} \Big/ J = \dfrac{e^x \sin y}{\cos 2y}$

$\qquad\qquad \dfrac{\partial y}{\partial u} = -\dfrac{\partial v}{\partial x} \Big/ J = \dfrac{e^{-x} \sin y}{\cos 2y}; \qquad \dfrac{\partial y}{\partial v} = \dfrac{\partial u}{\partial x} \Big/ J = \dfrac{e^x \cos y}{\cos 2y}$

which is a lot shorter than our first approach.

Move on for a further example

Example 2

31

If $z = f(x, y)$ with $u = x^2 - y^2$ and $v = xy$, find expressions for $\dfrac{\partial x}{\partial u}, \dfrac{\partial x}{\partial v}, \dfrac{\partial y}{\partial u}, \dfrac{\partial y}{\partial v}$.

First we need

$$\frac{\partial u}{\partial x} = \ldots\ldots\ldots; \quad \frac{\partial u}{\partial y} = \ldots\ldots\ldots; \quad \frac{\partial v}{\partial x} = \ldots\ldots\ldots; \quad \frac{\partial v}{\partial y} = \ldots\ldots\ldots$$

32

$$\boxed{\frac{\partial u}{\partial x} = 2x; \quad \frac{\partial u}{\partial y} = -2y; \quad \frac{\partial v}{\partial x} = y; \quad \frac{\partial v}{\partial y} = x}$$

Then we calculate J which, in this case, is $\ldots\ldots\ldots$

33

$$J = 2(x^2 + y^2)$$

Because

$$J = \frac{\partial(u, v)}{\partial(x, y)} = \begin{vmatrix} \dfrac{\partial u}{\partial x} & \dfrac{\partial v}{\partial x} \\ \dfrac{\partial u}{\partial y} & \dfrac{\partial v}{\partial y} \end{vmatrix} = \begin{vmatrix} 2x & y \\ -2y & x \end{vmatrix} = 2x^2 + 2y^2$$

Finally, we have the four relationships

$$\frac{\partial x}{\partial u} = \frac{\partial v}{\partial y}\bigg/ J = \ldots\ldots\ldots\ldots; \qquad \frac{\partial x}{\partial v} = -\frac{\partial u}{\partial y}\bigg/ J = \ldots\ldots\ldots\ldots$$

$$\frac{\partial y}{\partial u} = -\frac{\partial v}{\partial x}\bigg/ J = \ldots\ldots\ldots\ldots; \qquad \frac{\partial y}{\partial v} = \frac{\partial u}{\partial x}\bigg/ J = \ldots\ldots\ldots\ldots$$

34

$$\frac{\partial x}{\partial u} = \frac{x}{2(x^2 + y^2)}; \qquad \frac{\partial x}{\partial v} = \frac{y}{x^2 + y^2}$$

$$\frac{\partial y}{\partial u} = \frac{-y}{2(x^2 + y^2)}; \qquad \frac{\partial y}{\partial v} = \frac{x}{x^2 + y^2}$$

And that is all there is to it.

If we know the details of the function $z = f(x, y)$ then we can go one stage further and use the results $\dfrac{\partial x}{\partial u}, \dfrac{\partial x}{\partial v}, \dfrac{\partial y}{\partial u}, \dfrac{\partial y}{\partial v}$ to find $\dfrac{\partial z}{\partial u}$ and $\dfrac{\partial z}{\partial v}$.

Let us see this in a further example.

Example 3

If $z = 2x^2 + 3xy + 4y^2$ and $u = x^2 + y^2$ and $v = x + 2y$, determine

(a) $\dfrac{\partial x}{\partial u}, \dfrac{\partial x}{\partial v}, \dfrac{\partial y}{\partial u}, \dfrac{\partial y}{\partial v}$ \qquad (b) $\dfrac{\partial z}{\partial u}$ and $\dfrac{\partial z}{\partial v}$.

Section (a) is just like the previous example. Complete that on your own.

35

$$\boxed{\frac{\partial x}{\partial u} = \frac{1}{2x - y}; \quad \frac{\partial x}{\partial v} = \frac{-y}{2x - y}; \quad \frac{\partial y}{\partial u} = \frac{-1}{2(2x - y)}; \quad \frac{\partial y}{\partial v} = \frac{x}{2x - y}}$$

Because if $u = x^2 + y^2$ and $v = x + 2y$

$$\frac{\partial u}{\partial x} = 2x; \quad \frac{\partial u}{\partial y} = 2y; \quad \frac{\partial v}{\partial x} = 1; \quad \frac{\partial v}{\partial y} = 2$$

$$J = \frac{\partial(u, v)}{\partial(x, y)} = \begin{vmatrix} \dfrac{\partial u}{\partial x} & \dfrac{\partial v}{\partial x} \\ \dfrac{\partial u}{\partial y} & \dfrac{\partial v}{\partial y} \end{vmatrix} = \begin{vmatrix} 2x & 1 \\ 2y & 2 \end{vmatrix} = 4x - 2y = 2(2x - y)$$

Then
$$\frac{\partial x}{\partial u} = \frac{\partial v}{\partial y} \bigg/ J = 2 \bigg/ 2(2x - y) = \frac{1}{2x - y}$$

$$\frac{\partial x}{\partial v} = -\frac{\partial u}{\partial y} \bigg/ J = -2y \bigg/ 2(2x - y) = \frac{-y}{2x - y}$$

$$\frac{\partial y}{\partial u} = -\frac{\partial v}{\partial x} \bigg/ J = -1 \bigg/ 2(2x - y) = \frac{-1}{2(2x - y)}$$

$$\frac{\partial y}{\partial v} = \frac{\partial u}{\partial x} \bigg/ J = 2x \bigg/ 2(2x - y) = \frac{x}{2x - y}$$

$$\therefore \quad \frac{\partial x}{\partial u} = \frac{1}{2x - y}; \quad \frac{\partial x}{\partial v} = \frac{-y}{2x - y}; \quad \frac{\partial y}{\partial u} = \frac{-1}{2(2x - y)}; \quad \frac{\partial y}{\partial v} = \frac{x}{2x - y}$$

Now for part (b).

Since z is also a function of u and v, the expressions for $\dfrac{\partial z}{\partial u}$ and $\dfrac{\partial z}{\partial v}$ are

$$\frac{\partial z}{\partial u} = \dots\dots\dots$$

$$\frac{\partial z}{\partial v} = \dots\dots\dots$$

36

$$\boxed{\begin{aligned} \frac{\partial z}{\partial u} &= \frac{\partial z}{\partial x} \cdot \frac{\partial x}{\partial u} + \frac{\partial z}{\partial y} \cdot \frac{\partial y}{\partial u} \\ \frac{\partial z}{\partial v} &= \frac{\partial z}{\partial x} \cdot \frac{\partial x}{\partial v} + \frac{\partial z}{\partial y} \cdot \frac{\partial y}{\partial v} \end{aligned}}$$

The only remaining items of information we need are the expressions for $\dfrac{\partial z}{\partial x}$ and $\dfrac{\partial z}{\partial y}$ which we obtain from $z = 2x^2 + 3xy + 4y^2$

$$\frac{\partial z}{\partial x} = 4x + 3y \quad \text{and} \quad \frac{\partial z}{\partial y} = 3x + 8y$$

Using these and the previous set of derivatives, we now get

$$\frac{\partial z}{\partial u} = \dots\dots\dots; \quad \frac{\partial z}{\partial v} = \dots\dots\dots$$

37

$$\frac{\partial z}{\partial u} = \frac{5x - 2y}{2(2x - y)}; \quad \frac{\partial z}{\partial v} = \frac{3x^2 + 4xy - 3y^2}{2x - y}$$

Because

$$\frac{\partial z}{\partial u} = \frac{\partial z}{\partial x} \cdot \frac{\partial x}{\partial u} + \frac{\partial z}{\partial y} \cdot \frac{\partial y}{\partial u}$$

$$\therefore \frac{\partial z}{\partial u} = (4x + 3y)\left\{\frac{1}{2x - y}\right\} + (3x + 8y)\left\{\frac{-1}{2(2x - y)}\right\}$$

$$= \frac{5x - 2y}{2(2x - y)} \qquad\qquad \therefore \frac{\partial z}{\partial u} = \frac{5x - 2y}{2(2x - y)}$$

and $\dfrac{\partial z}{\partial v} = \dfrac{\partial z}{\partial x} \cdot \dfrac{\partial x}{\partial v} + \dfrac{\partial z}{\partial y} \cdot \dfrac{\partial y}{\partial v}$

$$\therefore \frac{\partial z}{\partial v} = (4x + 3y)\left\{\frac{-y}{2x - y}\right\} + (3x + 8y)\left\{\frac{x}{2x - y}\right\}$$

$$= \frac{3x^2 + 4xy - 3y^2}{2x - y} \qquad\qquad \therefore \frac{\partial z}{\partial v} = \frac{3x^2 + 4xy - 3y^2}{2x - y}$$

They are all done in the same general way.

Now on to the next topic

Stationary values of a function

38

You will doubtless remember that in earlier work you established the characteristics of *stationary points* on a plane curve and derived the conditions that enable these critical points to be calculated.

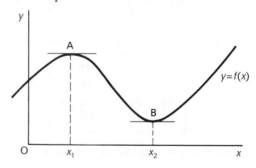

At A and B $\dfrac{dy}{dx} = 0$

For maximum $\dfrac{d^2y}{dx^2}$ is negative $(x = x_1)$

For minimum $\dfrac{d^2y}{dx^2}$ is positive $(x = x_2)$

We now progress to the application of these same considerations to three dimensions, where $z = f(x, y)$. The function is now represented by a surface and stationary values of the function $z = f(x, y)$ occur when the tangent plane to the surface at a point P (a, b) is parallel to the plane $z = 0$, i.e. to the x–y plane.

Let us take a closer look at this.

Maximum and minimum values

39

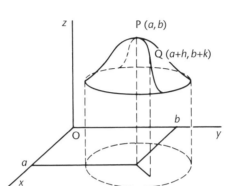

A function $z = f(x, y)$ is said to have *maximum* value at P (a, b) if $f(a, b)$ is greater than the value at a near-by point Q $(a + h, b + k)$ for all values of h and k however small, positive or negative, i.e. in all directions from P.

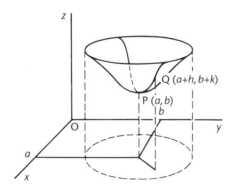

Similarly, $z = f(x, y)$ is said to have a *minimum* value at P (a, b) if $f(a, b)$ is less than the value at a neighbouring point Q $(a + h, b + k)$ in any direction from P.

To establish maximum and minimum values, we must therefore investigate the sign of the value of $f(a + h, b + k) - f(a, b)$.

If $f(a + h, b + k) - f(a, b) < 0$ we have a maximum value at P (a, b).

If $f(a + h, b + k) - f(a, b) > 0$ we have a minimum value at P (a, b).

▶

To pursue this further we turn to the total differential

$$df(x, y) = \frac{\partial f}{\partial x} dx + \frac{\partial f}{\partial y} dy$$

The total differential measures the rise or fall in the tangent plane from the point of its contact with the surface at (x, y) to the point $(x + dx, y + dy)$.

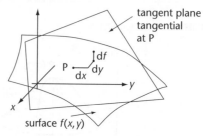

If the point of contact is a maximum or a minimum then

$$\frac{\partial f}{\partial x} = \ldots\ldots\ldots \quad \text{and} \quad \frac{\partial f}{\partial y} = \ldots\ldots\ldots$$

40

$$\boxed{\frac{\partial f}{\partial x} = 0 \quad \text{and} \quad \frac{\partial f}{\partial y} = 0}$$

Because

The tangent plane is parallel with the x–y plane and so the tangent plane neither rises nor falls, so that

$$df(x, y) = \frac{\partial f}{\partial x} dx + \frac{\partial f}{\partial y} dy = 0$$

Also because $dx \neq 0$ and $dy \neq 0$ then $\frac{\partial f}{\partial x} = 0$ and $\frac{\partial f}{\partial y} = 0$.

Notice the logic here. If there is a maximum or a minimum, then $\frac{\partial f}{\partial x} = 0$ and $\frac{\partial f}{\partial y} = 0$. However, just because $\frac{\partial f}{\partial x} = 0$ and $\frac{\partial f}{\partial y} = 0$ at a point does not imply that a maximum or a minimum exists at that point. What we can say is that a *stationary* point exists at that point and, as we shall see later, not all stationary points are maxima or minima.

Example 1

Determine the values of x and y at which the stationary values of

$$f(x, y) = x^2 + xy + y^2 + 5x - 5y + 3$$

occur.

All we need to do is to obtain expressions for $\frac{\partial f}{\partial x}$ and $\frac{\partial f}{\partial y}$, equate each to zero and then solve the pair of simultaneous equations so obtained. In which case

$$x = \ldots\ldots\ldots \quad \text{and} \quad y = \ldots\ldots\ldots$$

$$\boxed{x = -5 \quad \text{and} \quad y = 5}$$

Because

$\dfrac{\partial f}{\partial x} = 2x + y + 5$ and $\dfrac{\partial f}{\partial y} = x + 2y - 5$ giving the pair of simultaneous equations

$$2x + y + 5 = 0 \tag{1}$$
$$x + 2y - 5 = 0 \tag{2}$$

Adding (1) + (2) gives $3x + 3y = 0$, that is $y = -x$

Substitution in (1) gives $x = -5$ and so $y = 5$

Although a stationary value occurs at $(-5, 5)$ we have no evidence as to whether it is a maximum or a minimum value. Let us investigate further.

From the previous definitions

$f(a, b)$ will be a maximum value if $f(a + h, b + k) - f(a, b) < 0$

$f(a, b)$ will be a minimum value if $f(a + h, b + k) - f(a, b) > 0$

Now, from Taylor's theorem

$$f(a + h, b + k) = f(a, b) + h\frac{\partial f}{\partial x} + k\frac{\partial f}{\partial y}$$
$$+ \frac{1}{2!}\left(h^2\frac{\partial^2 f}{\partial x^2} + 2hk\frac{\partial^2 f}{\partial x\partial y} + k^2\frac{\partial^2 f}{\partial y^2}\right) + \dots$$

and we have already seen that at a stationary value $\dfrac{\partial f}{\partial x} = 0$ and $\dfrac{\partial f}{\partial y} = 0$. So, at a stationary point, Taylor's theorem becomes

$$f(a + h, b + k) - f(a, b) = \frac{1}{2!}\left(h^2\frac{\partial^2 f}{\partial x^2} + 2hk\frac{\partial^2 f}{\partial x\partial y} + k^2\frac{\partial^2 f}{\partial y^2}\right) + \dots$$

where subsequent terms are of higher orders of h and k and are neglected.

The expression in the brackets on the right-hand side can be written as

$$\frac{1}{\dfrac{\partial^2 f}{\partial x^2}}\left\{\left(h\frac{\partial^2 f}{\partial x^2} + k\frac{\partial^2 f}{\partial x\partial y}\right)^2 + k^2\left(\frac{\partial^2 f}{\partial x^2}\cdot\frac{\partial^2 f}{\partial y^2} - \left[\frac{\partial^2 f}{\partial x\partial y}\right]^2\right)\right\}$$

Take a moment and expand the brackets and confirm that this is so.

42

So $h^2 \dfrac{\partial^2 f}{\partial x^2} + 2hk \dfrac{\partial^2 f}{\partial x \partial y} + k^2 \dfrac{\partial^2 f}{\partial y^2}$

$$= \dfrac{1}{\dfrac{\partial^2 f}{\partial x^2}} \left\{ \left(h \dfrac{\partial^2 f}{\partial x^2} + k \dfrac{\partial^2 f}{\partial x \partial y} \right)^2 + k^2 \left(\dfrac{\partial^2 f}{\partial x^2} \cdot \dfrac{\partial^2 f}{\partial y^2} - \left[\dfrac{\partial^2 f}{\partial x \partial y} \right]^2 \right) \right\}$$

Now $\left(h \dfrac{\partial^2 f}{\partial x^2} + k \dfrac{\partial^2 f}{\partial x \partial y} \right)^2$, being a square, is always positive and if

$\dfrac{\partial^2 f}{\partial x^2} \cdot \dfrac{\partial^2 f}{\partial y^2} > \left[\dfrac{\partial^2 f}{\partial x \partial y} \right]^2$ the second term will also be positive.

In that case the sign of the whole expression is given by that of $\dfrac{\partial^2 f}{\partial x^2}$ at the front.

Furthermore, if $\dfrac{\partial^2 f}{\partial x^2} \cdot \dfrac{\partial^2 f}{\partial y^2} > \left(\dfrac{\partial^2 f}{\partial x \partial y} \right)^2$, i.e. $\dfrac{\partial^2 f}{\partial x^2} \cdot \dfrac{\partial^2 f}{\partial y^2} - \left(\dfrac{\partial^2 f}{\partial x \partial y} \right)^2 > 0$, this can be so only if $\dfrac{\partial^2 f}{\partial x^2}$ and $\dfrac{\partial^2 f}{\partial y^2}$ have the same sign. Therefore,

for $f(a, b)$ to be a maximum, $\dfrac{\partial^2 f}{\partial x^2}$ and $\dfrac{\partial^2 f}{\partial y^2}$ are both negative

and for $f(a, b)$ to be a minimum, $\dfrac{\partial^2 f}{\partial x^2}$ and $\dfrac{\partial^2 f}{\partial y^2}$ are both positive.

So, to determine whether a known stationary value is a maximum or a minimum value, we must find the second derivatives $\dfrac{\partial^2 f}{\partial x^2}, \dfrac{\partial^2 f}{\partial y^2}$ and $\dfrac{\partial^2 f}{\partial x \partial y}$.

Then

(a) If $\dfrac{\partial^2 f}{\partial x^2} \cdot \dfrac{\partial^2 f}{\partial y^2} - \left(\dfrac{\partial^2 f}{\partial x \partial y} \right)^2 > 0$, the stationary value is a true maximum or minimum value.

(b) In that case

(1) if $\dfrac{\partial^2 f}{\partial x^2}$ and $\dfrac{\partial^2 f}{\partial y^2}$ are both *negative*, $f(a, b)$ is a *maximum*

(2) if $\dfrac{\partial^2 f}{\partial x^2}$ and $\dfrac{\partial^2 f}{\partial y^2}$ are both *positive*, $f(a, b)$ is a *minimum*.

Make a careful note of the conclusions (a) and (b): then let us apply them.

43

Example 2

Investigate further the stationary value of the function

$$z = x^2 + xy + y^2 + 5x - 5y + 3$$

We have already seen that this function has a stationary point at

$$x = \ldots\ldots\ldots; \quad y = \ldots\ldots\ldots$$

$$\boxed{x = -5; \quad y = 5}$$

44

Next, we investigate the value of $\left(\dfrac{\partial^2 z}{\partial x^2}\right)\left(\dfrac{\partial^2 z}{\partial y^2}\right) - \left(\dfrac{\partial^2 z}{\partial x \partial y}\right)^2$. If this is greater than zero at $(-5, 5)$, then either a maximum or a minimum occurs at that point.

Check whether this is so.

$$\boxed{\text{Yes.} \quad \left(\dfrac{\partial^2 z}{\partial x^2}\right)\left(\dfrac{\partial^2 z}{\partial y^2}\right) - \left(\dfrac{\partial^2 z}{\partial x \partial y}\right)^2 > 0}$$

Because

$$\frac{\partial^2 z}{\partial x^2} = 2; \quad \frac{\partial^2 z}{\partial y^2} = 2; \quad \frac{\partial^2 z}{\partial x \partial y} = 1.$$

This confirms that $(-5, 5)$ is either a maximum or a minimum.

To decide which it is, we note that $\dfrac{\partial^2 z}{\partial x^2}$ and $\dfrac{\partial^2 z}{\partial y^2}$ are both *positive*.

$$\therefore \quad \text{at } (-5, 5), \ z \text{ is a } \ldots\ldots\ldots\ldots$$

$$\boxed{\text{minimum}}$$

Of course to find the actual minimum value of z we substitute $x = -5$ and $y = 5$ into the expression for z. That is really all there is to it. Another example.

Example 3

Determine the stationary values, if any, of the function

$$z = x^3 - 6xy + y^3$$

The four steps in the routine are:

(a) Find $\dfrac{\partial z}{\partial x}$ and $\dfrac{\partial z}{\partial y}$ and solve the equations $\dfrac{\partial z}{\partial x} = 0$ and $\dfrac{\partial z}{\partial y} = 0$.

(b) Determine whether $\left(\dfrac{\partial^2 z}{\partial x^2}\right)\left(\dfrac{\partial^2 z}{\partial y^2}\right) - \left(\dfrac{\partial^2 z}{\partial x \partial y}\right)^2 > 0$.

(c) If so, note the sign of $\dfrac{\partial^2 z}{\partial x^2}$ and $\dfrac{\partial^2 z}{\partial y^2}$ to distinguish between max. and min.

(d) Evaluate the maximum or minimum value of z.

In this example, stationary values occur at $\ldots\ldots\ldots\ldots$

47

$$z = 0 \text{ at } (0, 0) \quad \text{and} \quad z = -8 \text{ at } (2, 2)$$

Because

$$z = x^3 - 6xy + y^3 \quad \therefore \quad \frac{\partial z}{\partial x} = 3x^2 - 6y \qquad \frac{\partial z}{\partial y} = -6x + 3y^2$$

$$\frac{\partial z}{\partial x} = 0 \text{ and } \frac{\partial z}{\partial y} = 0 \quad \therefore \quad x^2 - 2y = 0 \text{ and } -2x + y^2 = 0$$

A possible stationary point exists when $x^2 - 2y = 0$ and $-2x + y^2 = 0$. From the first equation $y = x^2/2$ and substitution into $-2x + y^2 = 0$ gives $-2x + x^4/4 = 0$.

That is $x^4 - 8x = x(x^3 - 8) = 0$ and so $x = 0$ or $x = 2$.

When $x = 0$ then $y = 0$ and when $x = 2$ then $y = 2$.

$\therefore$ There are stationary values at $(0, 0)$ and $(2, 2)$

Next we determine whether $\left(\dfrac{\partial^2 z}{\partial x^2}\right)\left(\dfrac{\partial^2 z}{\partial y^2}\right) - \left(\dfrac{\partial^2 z}{\partial x \partial y}\right)^2 > 0$

Result . . . .

48

No max. or min. at $(0, 0)$;   Either max. or min. at $(2, 2)$

Because

$$\frac{\partial z}{\partial x} = 3x^2 - 6y \qquad \therefore \quad \frac{\partial^2 z}{\partial x^2} = 6x$$
$$\frac{\partial^2 z}{\partial x \partial y} = -6$$
$$\frac{\partial z}{\partial y} = -6x + 3y^2 \qquad \therefore \quad \frac{\partial^2 z}{\partial y^2} = 6y$$

$\therefore$ at $(0, 0)$ $\left(\dfrac{\partial^2 z}{\partial x^2}\right)\left(\dfrac{\partial^2 z}{\partial y^2}\right) - \left(\dfrac{\partial^2 z}{\partial x \partial y}\right)^2 = (0)(0) - 36 < 0$

$\therefore$ No max. or min. at $(0, 0)$

At $(2, 2)$ $\left(\dfrac{\partial^2 z}{\partial x^2}\right)\left(\dfrac{\partial^2 z}{\partial y^2}\right) - \left(\dfrac{\partial^2 z}{\partial x \partial y}\right)^2 = (12)(12) - 36 > 0$

$\therefore$ Either max. or min. at $(2, 2)$

We see that at $(2, 2)$ both $\dfrac{\partial^2 z}{\partial x^2}$ and $\dfrac{\partial^2 z}{\partial y^2}$ are positive. Therefore the stationary value at $(2, 2)$ is a . . . .

49

minimum

Finally, the minimum value of z is . . . .

50

$$\boxed{-8}$$

Therefore, $z_{min} = -8$ and occurs at (2, 2)

Before doing a further example, let us consider one other aspect of stationary values.

On to a new frame

Saddle point

51

In the previous example, when we substituted the coordinates (0, 0) in the expression $\left(\dfrac{\partial^2 z}{\partial x^2}\right)\left(\dfrac{\partial^2 z}{\partial y^2}\right) - \left(\dfrac{\partial^2 z}{\partial x \partial y}\right)^2$ we found that this did not satisfy the condition that for a maximum or minimum value

$$\left(\frac{\partial^2 z}{\partial x^2}\right)\left(\frac{\partial^2 z}{\partial y^2}\right) - \left(\frac{\partial^2 z}{\partial x \partial y}\right)^2 > 0$$

In fact, if $\dfrac{\partial z}{\partial x} = 0$ and $\dfrac{\partial z}{\partial y} = 0$

and $\left(\dfrac{\partial^2 z}{\partial x^2}\right)\left(\dfrac{\partial^2 z}{\partial y^2}\right) - \left(\dfrac{\partial^2 z}{\partial x \partial y}\right)^2 < 0$

this is an indication of a form of stationary value described as a *saddle point*, as shown at P below.

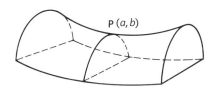

A saddle point is, in effect, a combined maximum and minimum configuration in different directions. Its name is obvious from the shape.

Add this then to the list of conditions for stationary values that we have built up.

At this stage, one naturally asks, what is implied if

52

$$\left(\frac{\partial^2 z}{\partial x^2}\right)\left(\frac{\partial^2 z}{\partial y^2}\right) - \left(\frac{\partial^2 z}{\partial x \partial y}\right)^2 = 0$$

In such a case, further detailed study of the function is necessary.
Now for an example to see it all in practice.

Example 4

Determine the stationary values of $z = 5xy - 4x^2 - y^2 - 2x - y + 5$.
Stationary values (or turning points) occur where

$$\frac{\partial z}{\partial x} = 0 \quad \text{and} \quad \frac{\partial z}{\partial y} = 0, \quad \text{i.e. at} \dots\dots\dots$$

53

$$\boxed{x = 1, \quad y = 2}$$

Because

$$\frac{\partial z}{\partial x} = 5y - 8x - 2 \qquad \frac{\partial z}{\partial y} = 5x - 2y - 1$$

$$\therefore \quad \left.\begin{matrix} 8x - 5y + 2 = 0 \\ 5x - 2y - 1 = 0 \end{matrix}\right\} \text{ gives } x = 1, \, y = 2$$

Therefore, the only stationary value occurs at $(1, 2)$.
Next we substitute these x and y values in

$$\left(\frac{\partial^2 z}{\partial x^2}\right)\left(\frac{\partial^2 z}{\partial y^2}\right) - \left(\frac{\partial^2 z}{\partial x \partial y}\right)^2 \quad \text{and find} \ldots\ldots\ldots\ldots$$

54

$$\boxed{\left(\frac{\partial^2 z}{\partial x^2}\right)\left(\frac{\partial^2 z}{\partial y^2}\right) - \left(\frac{\partial^2 z}{\partial x \partial y}\right)^2 < 0}$$

Because

$$\frac{\partial^2 z}{\partial x^2} = -8; \quad \frac{\partial^2 z}{\partial y^2} = -2; \quad \frac{\partial^2 z}{\partial x \partial y} = 5$$

$$\therefore \quad \left(\frac{\partial^2 z}{\partial x^2}\right)\left(\frac{\partial^2 z}{\partial y^2}\right) - \left(\frac{\partial^2 z}{\partial x \partial y}\right)^2 = (-8)(-2) - 25 = -9 \ \text{ i.e. } < 0$$

The stationary value at $(1, 2)$ is therefore a $\ldots\ldots\ldots\ldots$

55

$$\boxed{\text{saddle point}}$$

Example 5

Determine stationary values of $z = x^3 - 3x + xy^2$ and their nature.

We go through the same routine as before.

First find $\dfrac{\partial z}{\partial x}$ and $\dfrac{\partial z}{\partial y}$ and solve $\dfrac{\partial z}{\partial x} = 0$ and $\dfrac{\partial z}{\partial y} = 0$.

Possible stationary values therefore occur at $\ldots\ldots\ldots\ldots$

$$\boxed{x = 0,\ y = \pm\sqrt{3};\quad x = \pm1,\ y = 0}$$ **56**

Because

$$\frac{\partial z}{\partial x} = 3x^2 - 3 + y^2 \quad \frac{\partial z}{\partial y} = 2xy \quad \therefore\ x = 0\ \text{or}\ y = 0$$

If $x = 0$, $y^2 = 3$ $\therefore\ y = \pm\sqrt{3}$ $x = 0,\ y = \pm\sqrt{3}$

If $y = 0$, $3x^2 = 3$ $\therefore\ x = \pm1$ $x = \pm1,\ y = 0$.

Now we need the second derivatives and the usual tests. Finish if off. The nature of the stationary values:

$(0,\ \sqrt{3})$ ....; $(0,\ -\sqrt{3})$ ....

$(1,\ 0)$ ....; $(-1,\ 0)$ ....

$$\boxed{\begin{array}{ll}(0,\ \sqrt{3})\ \text{saddle point}; & (0,\ -\sqrt{3})\ \text{saddle point}\\(1,\ 0)\ \text{minimum}; & (-1,\ 0)\ \text{maximum}\end{array}}$$ **57**

Because

$$\frac{\partial^2 z}{\partial x^2} = 6x;\quad \frac{\partial^2 z}{\partial y^2} = 2x;\quad \frac{\partial^2 z}{\partial x\partial y} = 2y$$

$$\left(\frac{\partial^2 z}{\partial x^2}\right)\left(\frac{\partial^2 z}{\partial y^2}\right) - \left(\frac{\partial^2 z}{\partial x\partial y}\right)^2$$

$(0,\ \sqrt{3})$	$(0)(0) - 12$	i.e. <0 $\therefore$ saddle point
$(0,\ -\sqrt{3})$	$(0)(0) - 12$	i.e. <0 $\therefore$ saddle point
$(1,\ 0)$	$(6)(2) - 0$	i.e. >0 $\therefore$ minimum
$(-1,\ 0)$	$(-6)(-2) - 0$	i.e. >0 $\therefore$ maximum

and that just about does everything.

Substitution of $(1,\ 0)$ and $(-1,\ 0)$ in $z = x^3 - 3x + xy^2$ gives the minimum and maximum values of z.

$$z_{min} = -2;\quad z_{max} = 2.$$

The value of z at each of the saddle points is zero.

Let's now look at some examples where the second derivative test fails

Example 6   **58**

Determine the stationary values of $z = x^2 - 6xy + 9y^2$.

Here we see that $\dfrac{\partial z}{\partial x} = 2x - 6y,\ \dfrac{\partial z}{\partial y} = -6x + 18y$ and so these two derivatives vanish when

....

59

$$\boxed{y = x/3}$$

Because

$\dfrac{\partial z}{\partial x} = 2x - 6y = 0$ when $2x = 6y$, that is when $y = x/3$ and

$\dfrac{\partial z}{\partial y} = -6x + 18y = 0$ when $6x = 18y$, that is when $y = x/3$, and so there is an

infinity of stationary points lying along the line $y = x/3$.

Now

$$\left(\frac{\partial^2 z}{\partial x^2}\right)\left(\frac{\partial^2 z}{\partial y^2}\right) - \left(\frac{\partial^2 z}{\partial x \partial y}\right)^2 = \ldots\ldots\ldots$$

60

$$\boxed{0}$$

Because

$$\frac{\partial^2 z}{\partial x^2} = 2, \quad \frac{\partial^2 z}{\partial y^2} = 18 \quad \text{and} \quad \frac{\partial^2 z}{\partial x \partial y} = -6 \quad \text{so that}$$

$$\left(\frac{\partial^2 z}{\partial x^2}\right)\left(\frac{\partial^2 z}{\partial y^2}\right) - \left(\frac{\partial^2 z}{\partial x \partial y}\right)^2 = 18 \times 2 - 36 = 0$$

So the second derivative test does not apply and we must look elsewhere to decide the nature of the stationary points.

Since $x^2 - 6xy + 9y^2 = (x - 3y)^2$ then $z \geq 0$ for all values of x and y. Therefore the stationary points are minima – there is an infinity of minimum points along the line $y = x/3$.

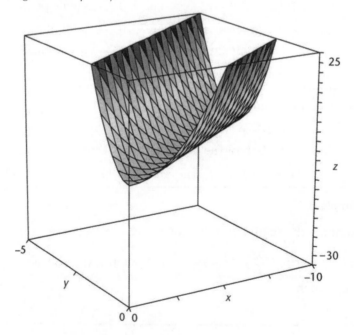

▶

Example 7

Find the stationary points of $z = x^4 - y^3$.

Here we see that $\dfrac{\partial z}{\partial x} = 4x^3$, $\dfrac{\partial z}{\partial y} = -3y^2$ and so these two derivatives vanish when $x = \dots\dots\dots$, $y = \dots\dots\dots$

$$\boxed{x = 0, \quad y = 0}$$

61

Because

$\dfrac{\partial z}{\partial x} = 4x^3 = 0$ when $x = 0$ and $\dfrac{\partial z}{\partial y} = -2y^2 = 0$ when $y = 0$, so there is just one stationary point at $(0, 0)$.

Now, at the stationary point

$$\left(\frac{\partial^2 z}{\partial x^2}\right)\left(\frac{\partial^2 z}{\partial y^2}\right) - \left(\frac{\partial^2 z}{\partial x \partial y}\right)^2 = \dots\dots\dots$$

$$\boxed{0}$$

62

Because

$$\frac{\partial^2 z}{\partial x^2} = 12x^2, \quad \frac{\partial^2 z}{\partial y^2} = -4y \quad \text{and} \quad \frac{\partial^2 z}{\partial x \partial y} = 0 \quad \text{so that at } (0, 0):$$

$$\left(\frac{\partial^2 z}{\partial x^2}\right)\left(\frac{\partial^2 z}{\partial y^2}\right) - \left(\frac{\partial^2 z}{\partial x \partial y}\right)^2 = 0$$

So the second derivative test does not apply. However, in the z–x plane $y = 0$ and so $z = x^4$. This means that the line of intersection of the surface with the z–x plane has a minimum at the origin. In the z–y plane $x = 0$ and so $z = -y^3$. This means that the line of intersection of the surface with the z–y plane has a point of inflexion at the origin.

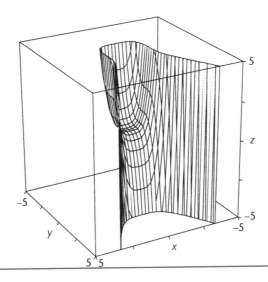

Lagrange undetermined multipliers

63

Closely allied to the problem of locating the stationary points of some function $u = f(x, y)$ is the problem of locating points where $u = f(x, y)$ attains its greatest or its least value (an extremal value) subject to the condition that x and y are related to each other via the equation

$$\phi(x, y) = 0 \qquad (1)$$

The problem can be clarified if we consider it graphically.

The graph of $u = f(x, y)$ is a surface within the (x, y, u) coordinate system. Selecting a plane parallel to the x–y plane on which the value of u is constant, u_k, we see that the surface intersects the plane in a curve given by the equation $f(x, y) = u_k$.

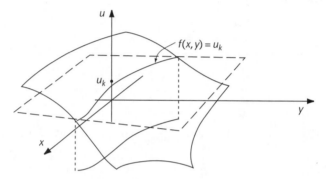

This line of intersection can now be projected onto the x–y plane to form what is known as a *level curve*. Different values of u_k determine different planes (all parallel to the x–y plane), different lines of intersection and hence different level curves. Accordingly, an alternative graphical description of $u = f(x, y)$ is

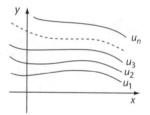

that of a family of level curves in the x–y plane with each member of the family being associated with a particular value of u_k, where we assume that $u_1 < u_2 < u_3 < \ldots < u_n$ or $u_1 > u_2 > u_3 > \ldots > u_n$.

We now superimpose onto this family of level curves the graph of the constraint equation $\phi(x, y) = 0$.

Clearly, in the figure alongside, u_3 is the extremal value of $f(x, y)$ that coincides with $\phi(x, y) = 0$, and at the point P where they meet they share the same tangent line $\mathrm{d}y/\mathrm{d}x$. Now, since $\phi(x, y) = 0$ we see that

$$\frac{\mathrm{d}y}{\mathrm{d}x} = \ldots\ldots\ldots\ldots$$

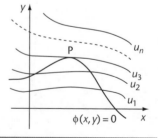

64

$$\boxed{\frac{dy}{dx} = -\frac{\partial\phi/\partial x}{\partial\phi/\partial y}}$$

Because

$$d\phi = \frac{\partial\phi}{\partial x}dx + \frac{\partial\phi}{\partial y}dy = 0 \quad \text{so that} \quad \frac{dy}{dx} = -\frac{\partial\phi/\partial x}{\partial\phi/\partial y}$$

The same tangent can be found from

$$du = \frac{\partial f}{\partial x}dx + \frac{\partial f}{\partial y}dy$$

by equating the differential $du = 0$. Therefore

$$\frac{dy}{dx} = -\frac{\partial f/\partial x}{\partial f/\partial y} = -\frac{\partial\phi/\partial x}{\partial\phi/\partial y}$$

The latter two fractions are equivalent fractions which means that the two numerators and the two denominators each differ by the same multiplicative factor K, enabling us to say that

$$\frac{\partial f}{\partial x} = K\frac{\partial\phi}{\partial x} \quad \text{and} \quad \frac{\partial f}{\partial y} = K\frac{\partial\phi}{\partial y} \quad \text{so that}$$

$$\frac{\partial f}{\partial x} + \lambda\frac{\partial\phi}{\partial x} = 0 \tag{2}$$

$$\frac{\partial f}{\partial y} + \lambda\frac{\partial\phi}{\partial y} = 0 \tag{3}$$

$\lambda = -K$ is called a Lagrange multiplier and equations (2) and (3), coupled with the constraint equation $\phi(x, y) = 0$, give us three relationships from which the values of x and y at the extremal points – and also the value of λ if required – can be found. Quite often the value of λ is not important.

Let us see how it works in a simple example.

Example 1

65

Find the stationary points of the function $u = x^2 + y^2$ subject to the constraint $x^2 + y^2 + 2x - 2y + 1 = 0$.

In this case, $\quad u = x^2 + y^2$

$$\phi = x^2 + y^2 + 2x - 2y + 1$$

We need to know

$$\frac{\partial u}{\partial x} = \ldots\ldots\ldots ; \quad \frac{\partial u}{\partial y} = \ldots\ldots\ldots$$

$$\frac{\partial\phi}{\partial x} = \ldots\ldots\ldots ; \quad \frac{\partial\phi}{\partial y} = \ldots\ldots\ldots$$

66

$$\frac{\partial u}{\partial x} = 2x; \quad \frac{\partial u}{\partial y} = 2y; \quad \frac{\partial \phi}{\partial x} = 2x + 2; \quad \frac{\partial \phi}{\partial y} = 2y - 2$$

Then we form and solve

$$\frac{\partial u}{\partial x} + \lambda \frac{\partial \phi}{\partial x} = 0$$

$$\frac{\partial u}{\partial y} + \lambda \frac{\partial \phi}{\partial y} = 0$$

together with $\qquad \phi = x^2 + y^2 + 2x - 2y + 1 = 0$

which gives $\qquad x = \dots \dots; \quad y = \dots \dots; \quad \lambda = \dots \dots$

67

$$x = -1 \pm \frac{\sqrt{2}}{2}; \quad y = 1 \mp \frac{\sqrt{2}}{2}; \quad \lambda = \sqrt{2} - 1$$

$\dfrac{\partial u}{\partial x} + \lambda \dfrac{\partial \phi}{\partial x} = 0 \quad \therefore \; 2x + \lambda(2x + 2) = 0 \quad \therefore \; x + \lambda(x + 1) = 0$

$\dfrac{\partial u}{\partial y} + \lambda \dfrac{\partial \phi}{\partial y} = 0 \quad \therefore \; 2y + \lambda(2x - 2) = 0 \quad \therefore \; y + \lambda(y - 1) = 0$

$\therefore \; \dfrac{x}{y} = \dfrac{-\lambda(x + 1)}{-\lambda(y - 1)} \quad \therefore \; xy - x = xy + y \quad \therefore \; y = -x$

Substituting this in ϕ

$\qquad x^2 + x^2 + 2x + 2x + 1 = 0 \qquad 2x^2 + 4x + 1 = 0$

$$\therefore \; x = -1 \pm \frac{\sqrt{2}}{2}$$

But $y = -x \qquad\qquad\qquad \therefore \; y = 1 \mp \dfrac{\sqrt{2}}{2}$

To find λ, we have $x + \lambda(x + 1) = 0 \quad \therefore \; \lambda = \sqrt{2} \mp 1$

As we have already said, we do not really need to find the value of λ.

On to the next

Functions with three independent variables   **68**

The argument is very much the same as before.

To find stationary points of the function $u = f(x, y, z)$ (1)

subject to the constraint $\phi(x, y, z) = 0$ (2)

Again we have, at stationary points

$$\frac{\partial u}{\partial x}\delta x + \frac{\partial u}{\partial y}\delta y + \frac{\partial u}{\partial z}\delta z = 0 \tag{3}$$

and since $\phi(x, y, z) = 0$

then $\dfrac{\partial \phi}{\partial x}\delta x + \dfrac{\partial \phi}{\partial y}\delta y + \dfrac{\partial \phi}{\partial z}\delta z = 0$ (4)

Multiplying each term in (4) by λ and adding (4) to (3), we have

$$\cdots\cdots\cdots$$

                                                            69

$$\boxed{\left(\frac{\partial u}{\partial x} + \lambda\frac{\partial \phi}{\partial x}\right)\delta x + \left(\frac{\partial u}{\partial y} + \lambda\frac{\partial \phi}{\partial y}\right)\delta y + \left(\frac{\partial u}{\partial z} + \lambda\frac{\partial \phi}{\partial z}\right)\delta z = 0}$$

from which $\dfrac{\partial u}{\partial x} + \lambda\dfrac{\partial \phi}{\partial x} = 0$ (5)

                       $\dfrac{\partial u}{\partial y} + \lambda\dfrac{\partial \phi}{\partial y} = 0$ (6)

                       $\dfrac{\partial u}{\partial z} + \lambda\dfrac{\partial \phi}{\partial z} = 0$ (7)

Equations (5), (6), (7), together with the constraint (2), provide all the information to determine x, y, z, and, if necessary, λ.

Example 2   **70**

To find the stationary points of the function

$$u = x^2 + 2y^2 + z$$

subject to the constraint $\phi(x, z) = x^2 - z^2 - 2 = 0$.

So $\dfrac{\partial u}{\partial x} = \cdots\cdots$; $\dfrac{\partial u}{\partial y} = \cdots\cdots$; $\dfrac{\partial u}{\partial z} = \cdots\cdots$

        $\dfrac{\partial \phi}{\partial x} = \cdots\cdots$; $\dfrac{\partial \phi}{\partial y} = \cdots\cdots$; $\dfrac{\partial \phi}{\partial z} = \cdots\cdots$

71

$$\frac{\partial u}{\partial x} = 2x; \qquad \frac{\partial u}{\partial y} = 4y; \qquad \frac{\partial u}{\partial z} = 1$$

$$\frac{\partial \phi}{\partial x} = 2x; \qquad \frac{\partial \phi}{\partial y} = 0; \qquad \frac{\partial \phi}{\partial z} = -2z$$

Now compile the equations

$$\frac{\partial u}{\partial x} + \lambda \frac{\partial \phi}{\partial x} = 0; \quad \frac{\partial u}{\partial y} + \lambda \frac{\partial \phi}{\partial y} = 0; \quad \frac{\partial u}{\partial z} + \lambda \frac{\partial \phi}{\partial z} = 0$$

and, together with the constraint $\phi = x^2 - z^2 - 2 = 0$, establish that stationary points occur at ....

72

$$\left(\tfrac{3}{2},\ 0,\ -\tfrac{1}{2}\right) \quad \text{and} \quad \left(-\tfrac{3}{2},\ 0,\ -\tfrac{1}{2}\right)$$

Because

$$\frac{\partial u}{\partial x} + \lambda \frac{\partial \phi}{\partial x} = 0 \quad \therefore\ 2x + \lambda 2x = 0 \quad \therefore\ \lambda = -1$$

$$\frac{\partial u}{\partial y} + \lambda \frac{\partial \phi}{\partial y} = 0 \quad 4y + \lambda(0) = 0 \quad \therefore\ y = 0$$

$$\frac{\partial u}{\partial z} + \lambda \frac{\partial \phi}{\partial z} = 0 \qquad 1 - \lambda 2z = 0 \qquad \therefore\ z = \frac{1}{2\lambda} = -\frac{1}{2}$$

$$\phi = x^2 - z^2 - 2 = 0 \quad \therefore\ x^2 - \tfrac{1}{4} - 2 = 0 \quad \therefore\ x = \pm\tfrac{3}{2}.$$

Therefore, stationary points at $\left(\tfrac{3}{2},\ 0,\ -\tfrac{1}{2}\right)$ and $\left(-\tfrac{3}{2},\ 0,\ -\tfrac{1}{2}\right)$.

The method of Lagrange multipliers does not lend itself easily to give a distinction between the various types of stationary points. In many practical applications, however, whether a result is a maximum or a minimum value will be apparent from the physical consideration of the problem.

Let us finish with one further example.

So move on

73

Example 3

A hot water storage tank is a vertical cylinder surmounted by a hemispherical top of the same diameter. The tank is designed to hold 400 m³ of liquid. Determine the total height and the diameter of the tank if the surface heat loss is to be a minimum.

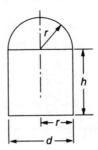

We first write down the function for the total surface area, A.

$$A = \ldots\ldots\ldots\ldots$$

74

$$A = 3\pi r^2 + 2\pi rh$$

Because

The surface area of the hemisphere is $2\pi r^2$, the area of the base of the tank is πr^2 and the area of the cylindrical side is $2\pi rh$, giving a total area of $3\pi r^2 + 2\pi rh$.

This is the function which has to be a minimum. The constraint in this problem is that ....

75

$$\text{the volume is } 400 \text{ m}^3$$

So we have

$$A = 3\pi r^2 + 2\pi rh \tag{1}$$

constraint $\quad V = \pi r^2 h + \dfrac{2}{3}\pi r^3 = 400$

So let $\quad \phi = \pi r^2 h + \dfrac{2}{3}\pi r^3 - 400 = 0 \tag{2}$

We now want $\quad \dfrac{\partial A}{\partial r} = \ldots\ldots\ldots ; \quad \dfrac{\partial A}{\partial h} = \ldots\ldots\ldots$

$$\dfrac{\partial \phi}{\partial r} = \ldots\ldots\ldots ; \quad \dfrac{\partial \phi}{\partial h} = \ldots\ldots\ldots$$

76

$$\dfrac{\partial A}{\partial r} = 6\pi r + 2\pi h; \qquad \dfrac{\partial \phi}{\partial r} = 2\pi rh + 2\pi r^2$$
$$\dfrac{\partial A}{\partial h} = 2\pi r; \qquad \dfrac{\partial \phi}{\partial h} = \pi r^2$$

Now we form $\qquad \dfrac{\partial A}{\partial r} + \lambda\dfrac{\partial \phi}{\partial r} = 0$

and $\qquad \dfrac{\partial A}{\partial h} + \lambda\dfrac{\partial \phi}{\partial h} = 0$

and, with the constraint, $\phi = \pi r^2 h + \dfrac{2}{3}\pi r^3 - 400 = 0$,

we eventually obtain $r = \ldots\ldots\ldots$ and $h = \ldots\ldots\ldots$

Finish it off and hence find the total height and the diameter.

77

$$r = 4\cdot243 \text{ m}; \quad h = 4\cdot243 \text{ m}$$

Check the working:

$$\frac{\partial A}{\partial r} + \lambda\frac{\partial \phi}{\partial r} = 0 \quad \therefore \quad 6\pi r + 2\pi h + \lambda(2\pi rh + 2\pi r^2) = 0 \tag{3}$$

$$\frac{\partial A}{\partial h} + \lambda\frac{\partial \phi}{\partial h} = 0 \quad \therefore \quad 2\pi r + \lambda\pi r^2 = 0 \tag{4}$$

From (4): $\lambda = -\dfrac{2}{r}$ Substitute this in (3)

$$6\pi r + 2\pi h - \frac{2}{r}(2\pi rh + 2\pi r^2) = 0$$

$$\therefore \quad 6r + 2h - 4h - 4r = 0 \quad \therefore \quad h = r$$

Also $\pi r^2 h + \dfrac{2}{3}\pi r^3 = 400 \quad \therefore \quad \dfrac{5}{3}\pi r^3 = 400 \quad \therefore \quad r = 4\cdot243$

$\therefore$ Total height $= h + r = 8\cdot49$ m;   Diameter $= 8\cdot49$ m

That brings us to the end of this particular Programme and to the usual **Revision summary** that follows. Check through the **Can you?** checklist and be sure to revise any section should you feel that is necessary. Then you will find the **Test exercise** straightforward – no tricks. The **Further problems** provide valuable additional practice.

 # Revision summary 14

78

1   *Small increments*

$$z = f(x,y) \quad \delta z = \frac{\partial z}{\partial x}\delta x + \frac{\partial z}{\partial y}\delta y$$

$$u = f(x,y,z) \quad \delta u = \frac{\partial u}{\partial x}\delta x + \frac{\partial u}{\partial y}\delta y + \frac{\partial u}{\partial z}\delta z$$

2   *Rates of change*

$$z = f(x,y) \quad \frac{dz}{dt} = \frac{\partial z}{\partial x}\cdot\frac{dx}{dt} + \frac{\partial z}{\partial y}\cdot\frac{dy}{dt}$$

3   *Implicit functions*

$$z = f(x,y) = 0 \quad \frac{dy}{dx} = -\left(\frac{\partial z}{\partial x}\Big/\frac{\partial z}{\partial y}\right)$$

4   *Change of variables*

$z = f(x,y)$ x and y are functions of u and v

$$\frac{\partial z}{\partial u} = \frac{\partial z}{\partial x}\cdot\frac{\partial x}{\partial u} + \frac{\partial z}{\partial y}\cdot\frac{\partial y}{\partial u}$$

$$\frac{\partial z}{\partial v} = \frac{\partial z}{\partial x}\cdot\frac{\partial x}{\partial v} + \frac{\partial z}{\partial y}\cdot\frac{\partial y}{\partial v}$$

▶

5 *Inverse functions*

$$z = f(x, y) \quad u = g(x, y) \quad v = h(x, y)$$

$$\frac{\partial x}{\partial u} = \frac{\partial v}{\partial y} \bigg/ J; \quad \frac{\partial x}{\partial v} = -\frac{\partial u}{\partial y} \bigg/ J$$

$$\frac{\partial y}{\partial u} = -\frac{\partial v}{\partial x} \bigg/ J; \quad \frac{\partial y}{\partial v} = \frac{\partial u}{\partial x} \bigg/ J$$

where $\quad J = \dfrac{\partial(u, v)}{\partial(x, y)} = \begin{vmatrix} \dfrac{\partial u}{\partial x} & \dfrac{\partial v}{\partial x} \\ \dfrac{\partial u}{\partial y} & \dfrac{\partial v}{\partial y} \end{vmatrix}$

6 *Stationary points*

$z = f(x, y)$ (a) $\dfrac{\partial z}{\partial x} = 0$ and $\dfrac{\partial z}{\partial y} = 0$

(b) $\left(\dfrac{\partial^2 z}{\partial x^2}\right)\left(\dfrac{\partial^2 z}{\partial y^2}\right) - \left(\dfrac{\partial^2 z}{\partial x \partial y}\right)^2 > 0$ for max. or min.

$\left(\dfrac{\partial^2 z}{\partial x^2}\right)\left(\dfrac{\partial^2 z}{\partial y^2}\right) - \left(\dfrac{\partial^2 z}{\partial x \partial y}\right)^2 < 0$ for saddle point

$\left(\dfrac{\partial^2 z}{\partial x^2}\right)\left(\dfrac{\partial^2 z}{\partial y^2}\right) - \left(\dfrac{\partial^2 z}{\partial x \partial y}\right)^2 = 0$ no decision without further information

(c) $\dfrac{\partial^2 z}{\partial x^2}$ and $\dfrac{\partial^2 z}{\partial y^2}$ both *negative* for maximum

$\dfrac{\partial^2 z}{\partial x^2}$ and $\dfrac{\partial^2 z}{\partial y^2}$ both *positive* for minimum.

7 *Lagrange multipliers*
Two independent variables
$u = f(x, y)$ with constraint $\phi(x, y) = 0$

Solve $\quad \dfrac{\partial u}{\partial x} + \lambda \dfrac{\partial \phi}{\partial x} = 0$

$\dfrac{\partial u}{\partial y} + \lambda \dfrac{\partial \phi}{\partial y} = 0$

with $\quad \phi(x, y) = 0.$

Three independent variables
$u = f(x, y, z)$ with constraint $\phi(x, y, z) = 0$

Solve $\quad \dfrac{\partial u}{\partial x} + \lambda \dfrac{\partial \phi}{\partial x} = 0$

$\dfrac{\partial u}{\partial y} + \lambda \dfrac{\partial \phi}{\partial y} = 0$

$\dfrac{\partial u}{\partial z} + \lambda \dfrac{\partial \phi}{\partial z} = 0$

with $\quad \phi(x, y, z) = 0.$

Can you?

79 **Checklist 14**

Check this list before and after you try the end of Programme test.

On a scale of 1 to 5 how confident are you that you can: **Frames**

- Derive the expression for a small increment in an expression of
  two real variables using Taylor's theorem?   ⎡ 1 ⎤ to ⎡ 4 ⎤
  Yes   ☐   ☐   ☐   ☐   ☐   *No*

- Apply the notion of small increments in expressions in two
  and three real variables to a variety of problems?   ⎡ 5 ⎤ to ⎡ 7 ⎤
  Yes   ☐   ☐   ☐   ☐   ☐   *No*

- Determine the rate of change with respect to time of an
  expression involving two or three real variables?   ⎡ 8 ⎤ and ⎡ 9 ⎤
  Yes   ☐   ☐   ☐   ☐   ☐   *No*

- Differentiate implicit functions?   ⎡ 9 ⎤ and ⎡ 10 ⎤
  Yes   ☐   ☐   ☐   ☐   ☐   *No*

- Determine first and second derivatives involving change of
  variables in expressions of two real variables?   ⎡ 11 ⎤ to ⎡ 21 ⎤
  Yes   ☐   ☐   ☐   ☐   ☐   *No*

- Use the Jacobian to obtain the derivatives of inverse functions
  of two real variables?   ⎡ 22 ⎤ to ⎡ 37 ⎤
  Yes   ☐   ☐   ☐   ☐   ☐   *No*

- Locate and identify maxima, minima and saddle points of
  functions of two real variables?   ⎡ 38 ⎤ to ⎡ 57 ⎤
  Yes   ☐   ☐   ☐   ☐   ☐   *No*

- Solve problems where the independent variables are
  constrained by using the method of Lagrange undetermined
  multipliers for functions of two and three real variables?   ⎡ 58 ⎤ to ⎡ 77 ⎤
  Yes   ☐   ☐   ☐   ☐   ☐   *No*

Test exercise 14

1 If $z = \dfrac{xy}{x - y}$, show that **80**

(a) $x\dfrac{\partial z}{\partial x} + y\dfrac{\partial z}{\partial y} = z$

(b) $x^2\dfrac{\partial^2 z}{\partial x^2} - y^2\dfrac{\partial^2 z}{\partial y^2} = 0$

(c) $z\dfrac{\partial^2 z}{\partial x \partial y} = 2\dfrac{\partial z}{\partial x} \cdot \dfrac{\partial z}{\partial y}$.

2 Two sides of a triangular plate are measured as 125 mm and 160 mm, each to the nearest millimetre. The included angle is quoted as $60° \pm 1°$. Calculate the length of the remaining side and the maximum possible error in the result.

3 If $z = \left(x^2 - y^2\right)^{1/2}$ and x is increasing at 3·5 m/s, determine at what rate y must change in order that z shall be neither increasing nor decreasing at the instant when $x = 5$ m and $y = 3$ m.

4 If $2x^2 + 4xy + 3y^2 = 1$, obtain expressions for $\dfrac{dy}{dx}$ and $\dfrac{d^2 y}{dx^2}$.

5 If $u = x^2 + y^2$ and $v = 4xy$, determine

$\dfrac{\partial x}{\partial u}, \dfrac{\partial x}{\partial v}, \dfrac{\partial y}{\partial u}, \dfrac{\partial y}{\partial v}$.

6 Determine the position and nature of the stationary points of the functions:

(a) $z = 2x^2 y^2 + 4xy^2 - 4y^3 + 16y + 5$

(b) $z = 4 - 25x^2 + 20xy - 4y^2$.

7 A rectangular storage tank is to have a capacity of 1·0 m³. If the tank is closed and the top is made of metal half as thick as the sides and base, use Lagrange's method of undetermined multipliers to determine the dimensions of the tank for the total amount of metal used in its construction to be a minimum.

8 Use Lagrange's method of undetermined multipliers to obtain the stationary values of $u = x^2 + y^2 + z^2$ subject to the constraint $\phi = 3x - 2y + z - 4$.

Further problems 14

1 If $z = 2x^2 - 3y$ with $u = x^2 \sin y$ and $v = 2y \cos x$, determine expressions for $\dfrac{\partial z}{\partial u}$ and $\dfrac{\partial z}{\partial v}$.

2 If $u = x^2 + e^{-3y}$ and $v = 2x + e^{3y}$, determine $\dfrac{\partial x}{\partial u}, \dfrac{\partial x}{\partial v}, \dfrac{\partial y}{\partial u}, \dfrac{\partial y}{\partial v}$.

3 If $z = f(x, y)$ where $x = uv$ and $y = u^2 - v^2$, show that

(a) $2x\dfrac{\partial z}{\partial x} + 2y\dfrac{\partial z}{\partial y} = u\dfrac{\partial z}{\partial u} + v\dfrac{\partial z}{\partial v}$

(b) $2\dfrac{\partial z}{\partial y} = \dfrac{1}{u^2 + v^2}\left\{ u\dfrac{\partial z}{\partial u} - v\dfrac{\partial z}{\partial v} \right\}$.

4 If $V = f(x, y)$ and $x = r\cos\theta$ and $y = r\sin\theta$, show that

$$\frac{\partial^2 V}{\partial x^2} + \frac{\partial^2 V}{\partial y^2} = \frac{\partial^2 V}{\partial r^2} + \frac{1}{r}\frac{\partial V}{\partial r} + \frac{1}{r^2}\frac{\partial^2 V}{\partial \theta^2}.$$

5 If $z = \cosh 2x \sin 3y$ and $u = e^x(1 + y^2)$ and $v = 2ye^{-x}$, determine expressions for $\dfrac{\partial x}{\partial u}, \dfrac{\partial x}{\partial v}, \dfrac{\partial y}{\partial u}, \dfrac{\partial y}{\partial v}$, and hence find $\dfrac{\partial z}{\partial u}$ and $\dfrac{\partial z}{\partial v}$.

6 If $z = f(u, v)$ where $u = \frac{1}{2}(x^2 - y^2)$ and $v = xy$, prove that

$$\frac{\partial^2 z}{\partial x^2} - \frac{\partial^2 z}{\partial y^2} = 2u\left(\frac{\partial^2 z}{\partial u^2} - \frac{\partial^2 z}{\partial v^2} \right) + 4v\frac{\partial^2 z}{\partial u \partial v} + 2\frac{\partial z}{\partial u}.$$

7 Locate the stationary points of the following functions. Determine the nature of the points and calculate the critical function values.

(a) $z = y^2 + xy + x^2 + 4y - 4x + 5$

(b) $z = y^2 + xy + 2x + 3y + 6$

(c) $z = 3xy - 6y^2 - 3x^2 + 6y + 6x + 7$.

8 Find the stationary points of the function

$$z = (x^2 + y^2)^2 - 8(x^2 - y^2)$$

and determine their nature.

9 Verify that the function $z = (x + y - 1)/(x^2 + 2y^2 + 2)$ has stationary values at $(2, 1)$ and $(-\frac{2}{3}, -\frac{1}{3})$ and determine their nature.

10 Locate stationary points of the function

$$z = 4x^2 + 10xy + 4y^2 - x^2y^2$$

and determine their nature.

11 Find the stationary points of the following functions and determine their nature.

(a) $z = x(x^2 - 3) + 3y(x - 1)^2 + 18y^2(2y - 3)$

(b) $z = x^2y^2 - x^2 - y^2$.

▶

12 Find the stationary points of the following functions and determine their nature.

(a) $z = (x - y)(x^2 + xy + y^2)$

(b) $z = 6 - x^2 + 8xy - 16y^2$

(c) $z = \cos(x^2 + y^2)$.

13 A metal channel is formed by turning up the sides of width x of a rectangular sheet of metal through an angle θ. If the sheet is 200 mm wide, determine the values of x and θ for which the cross-section of the channel will be a maximum.

14 A container is in the form of a right circular cylinder of length l and diameter d, with equal conical ends of the same diameter and height h. If V is the fixed volume of the container, find the dimensions l, h and d for minimum surface area.

15 A solid consists of a cylinder of length l and diameter d, surmounted at one end by a cone of vertex angle 2θ and base diameter d, and at the other end by a hemisphere of the same diameter. If the volume V of the solid is 50 cm^3, determine the dimensions l, d and θ so that the total surface area shall be a minimum.

16 A rectangular solid of maximum volume is to be cut from a solid sphere of radius r. Determine the dimensions of the solid so formed and its volume.

17 Use Lagrange's method of undetermined multipliers to obtain the stationary values of the following functions u, subject in each case to the constraint ϕ.

(a) $u = x^2 y^2 z^2 \qquad \phi = x^2 + y^2 + z^2 - 4 = 0$

(b) $u = x^2 + y^2 \qquad \phi = 4x^2 + 6xy + 4y^2 = 9$.

Partial differential equations

Learning outcomes

When you have completed this Programme you will be able to:

- Summarise the introductory methods of solving ordinary differential equations
- Solve partial differential equations that are amenable to solution by direct integration
- Apply initial and boundary conditions
- Solve the one-dimensional wave and heat equations by separating the variables and obtaining eigenfunctions and corresponding eigenvalues
- Solve the two-dimensional Laplace equation in Cartesian coordinates
- Recognise the need for alternative coordinate systems and solve the two-dimensional Laplace equation in plane polar coordinates

Prerequisite: Engineering Mathematics (Sixth Edition)
Programmes 24 First-order differential equations and 25 Second-order differential equations

Introduction

The formation of ordinary linear differential equations and their solution by various methods were covered in some detail in Programmes 24 and 25 of *Engineering Mathematics (Sixth Edition)*, and reference to these before undertaking the new work of this Programme could be beneficial – especially Programme 25 which dealt with second-order equations. Working through the Test exercise of that Programme would provide worthwhile revision.

The main results obtained are listed here for convenience and easy reference.

1 Equations of the form $a\dfrac{d^2y}{dx^2} + b\dfrac{dy}{dx} + cy = 0$

Auxiliary equation $am^2 + bm + c = 0$. Solutions depend on the roots of this equation.

(a) Real and different roots: $m = m_1$ and $m = m_2$

$$y = Ae^{m_1 x} + Be^{m_2 x} \tag{1}$$

(b) Real and equal roots: $m = m_1$ (twice)

$$y = e^{m_1 x}(A + Bx) \tag{2}$$

(c) Complex roots: $m = \alpha \pm j\beta$

$$y = e^{\alpha x}(A\cos\beta x + B\sin\beta x) \tag{3}$$

2 Equations of the form $\dfrac{d^2y}{dx^2} \pm n^2 y = 0$

(a) $\dfrac{d^2y}{dx^2} + n^2 y = 0$ $\therefore m^2 + n^2 = 0$ $\therefore m^2 = -n^2$ $\therefore m = \pm jn$

$$\text{Solution } y = A\cos nx + B\sin nx \tag{4}$$

(b) $\dfrac{d^2y}{dx^2} - n^2 y = 0$ $\therefore m^2 - n^2 = 0$ $\therefore m^2 = n^2$ $\therefore m = \pm n$

$$\left. \begin{aligned} &\text{Solution } y = A\cosh nx + B\sinh nx \\ &\text{or } y = Ae^{nx} + Be^{-nx} \\ &\text{or } y = A\sinh n(x + \phi) \end{aligned} \right\} \tag{5}$$

In each case, A and B are arbitrary constants depending on the initial conditions, and in the last form ϕ is an arbitrary constant.

Partial differential equations

2

A partial differential equation is a relationship between a dependent variable u and two or more independent variables $(x, y, t, \ldots)$ and partial derivatives of u with respect to these independent variables. The solution is therefore of the form $u = f(x, y, t, \ldots)$.

Solution by direct integration

The simplest form of partial differential equation is such that a solution can be determined by direct partial integration.

Example 1

Solve the equation $\dfrac{\partial^2 u}{\partial x^2} = 12x^2(t+1)$ given that at $x = 0$, $u = \cos 2t$ and $\dfrac{\partial u}{\partial x} = \sin t$. Notice that the boundary conditions are functions of t and not just constants. $\dfrac{\partial^2 u}{\partial x^2} = 12x^2(t+1)$. Integrating partially with respect to x, we have

$\dfrac{\partial u}{\partial x} = 4x^3(t+1) + \phi(t)$ where the arbitrary function $\phi(t)$ takes the place of the normal arbitrary constant in ordinary integration. Integrating partially again with respect to x gives

$$u = \ldots\ldots\ldots$$

3

$$\boxed{u = x^4(t+1) + x\phi(t) + \theta(t)}$$

where $\theta(t)$ is a second arbitrary function.

To find the two arbitrary functions $\phi(t)$ and $\theta(t)$, we apply the given initial conditions that at $x = 0$, $\dfrac{\partial u}{\partial x} = \sin t$ and $u = \cos 2t$. Substituting these in the relevant equations gives

$$\phi(t) = \ldots\ldots\ldots\ldots; \quad \theta(t) = \ldots\ldots\ldots\ldots$$

4

$$\boxed{\phi(t) = \sin t; \quad \theta(t) = \cos 2t}$$

Because

$$u = x^4(t+1) + x\sin t + \cos 2t$$

Example 2

Solve the equation $\dfrac{\partial^2 u}{\partial x \partial y} = \sin(x+y)$, given that at $y = 0$, $\dfrac{\partial u}{\partial x} = 1$ and at $x = 0$, $u = (y-1)^2$.

In just the same way as before, $u = \ldots\ldots\ldots\ldots$

$$\boxed{u = -\sin(x+y) + x + \sin x + (y-1)^2}$$

5

Because

$$\frac{\partial^2 u}{\partial x \partial y} = \sin(x+y) \quad \therefore \quad \frac{\partial u}{\partial x} = -\cos(x+y) + \phi(x).$$

At $y = 0$, $\dfrac{\partial u}{\partial x} = 1$ $\therefore$ $1 = -\cos x + \phi(x)$ $\therefore$ $\phi(x) = 1 + \cos x$

$$\therefore \quad \frac{\partial u}{\partial x} = -\cos(x+y) + 1 + \cos x$$

Integrating again partially, this time with respect to x, we have

$$u = -\sin(x+y) + x + \sin x + \theta(y)$$

But at $x = 0$, $u = (y-1)^2$. $\therefore$ $(y-1)^2 = -\sin y + \theta(y)$

$$\therefore \quad \theta(y) = (y-1)^2 + \sin y$$

$$\therefore \quad u = -\sin(x+y) + x + \sin x + \sin y + (y-1)^2$$

Initial conditions and boundary conditions

As with any differential equation, the arbitrary constants or arbitrary functions in any particular case are determined from the additional information given concerning the variables of the equation. These extra facts are called the *initial conditions* or, more generally, the *boundary conditions* since they do not always refer to zero values of the independent variables.

Example 3

Solve the equation $\dfrac{\partial^2 u}{\partial x \partial y} = \sin x \cos y$, subject to the boundary conditions that

at $y = \dfrac{\pi}{2}$, $\dfrac{\partial u}{\partial x} = 2x$ and at $x = \pi$, $u = 2\sin y$.

Work through it: it is easy enough. $u = \dots\dots\dots$

$$\boxed{u = x^2 + \cos x(1 - \sin y) + \sin y + 1 - \pi^2}$$

6

Because

$$\frac{\partial^2 u}{\partial x \partial y} = \sin x \cos y \qquad\qquad \therefore \quad \frac{\partial u}{\partial x} = \sin x \sin y + \phi(x)$$

But $\dfrac{\partial u}{\partial x} = 2x$ at $y = \dfrac{\pi}{2}$ $\qquad\qquad \therefore$ $\phi(x) = 2x - \sin x$

$$\therefore \quad \frac{\partial u}{\partial x} = 2x - \sin x(1 - \sin y) \qquad \therefore \quad u = x^2 + \cos x(1 - \sin y) + \theta(y)$$

But $u = 2\sin y$ at $x = \pi$ $\therefore$ $\theta(y) = 1 - \pi^2 + \sin y$

$$u = x^2 + \cos x(1 - \sin y) + \sin y + 1 - \pi^2$$

On to the next frame

7

Before we take a closer look at some of the more important partial differential equations occurring in branches of technology, let us recall the fact that if $u = u_1$, $u = u_2$, $u = u_3, \ldots$ are different solutions of a linear partial differential equation, so also is the *linear combination*

$$u = c_1 u_1 + c_2 u_2 + c_3 u_3 + \ldots$$

where c_1, c_2, $c_3, \ldots$ are arbitrary constants.

There are many types of partial differential equations, some requiring special treatment in their solution. In this Programme we are concerned with a restricted number of such equations that occur in branches of science and technology, which can be solved by the method of separating the variables, and which also link up with the work we have done on Fourier series techniques.

Let us make a new start

8 **The wave equation**

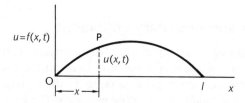

Consider a perfectly flexible elastic string stretched between two points at $x = 0$ and $x = l$ with uniform tension T. If the string is displaced slightly from its initial position of rest and released, with the end points remaining fixed, then the string will vibrate. The position of any point P in the string will then depend on its distance from one end and on the instant in time. Its displacement u at any time t can thus be expressed as $u = f(x, t)$ where x is its distance from the left-hand end.

The equation of motion is given by $\dfrac{\partial^2 u}{\partial x^2} = \dfrac{1}{c^2} \cdot \dfrac{\partial^2 u}{\partial t^2}$, where $c^2 = \dfrac{T}{\rho}$ in which T is the tension in the string and ρ the mass per unit length of the string. The displacement of the string is regarded as small so that T and ρ remain constant.

Now let us deal with the solution of this equation.

On to the next frame

Solution of the wave equation

9

The new equation $\dfrac{\partial^2 u}{\partial x^2} = \dfrac{1}{c^2} \cdot \dfrac{\partial^2 u}{\partial t^2}$ has a solution $u(x, t)$.

Boundary conditions:

(a) The string is fixed at both ends, i.e. at $x = 0$ and at $x = l$ for all values of time t. Therefore $u(x, t)$ becomes

$$\left.\begin{array}{l} u(0,\ t) = 0 \\ u(l,\ t) = 0 \end{array}\right\} \quad \text{for all values of } t \geq 0$$

Initial conditions:

(b) If the initial deflection of P at $t = 0$ is denoted by $f(x)$, then

$$u(x,\ 0) = f(x)$$

(c) Let the initial velocity of P be $g(x)$, then

$$\left[\frac{\partial u}{\partial t}\right]_{t=0} = g(x)$$

So now we have listed all the information available from the question. Next we turn to solving the equation.

Solution by separating the variables

We assume a trial solution of the form $u(x,\ t) = X(x)T(t)$ where

$X(x)$ is a function of x only

$T(t)$ is a function of t only.

If we simplify the symbols to $u = XT$ and denote derivatives with respect to their own independent variables by primes, we have

$$u = XT \quad \therefore \quad \frac{\partial u}{\partial x} = X'T \quad \text{and} \quad \frac{\partial^2 u}{\partial x^2} = X''T$$

$$\frac{\partial u}{\partial t} = XT' \quad \text{and} \quad \frac{\partial^2 u}{\partial t^2} = XT''$$

The wave equation $\dfrac{\partial^2 u}{\partial x^2} = \dfrac{1}{c^2} \cdot \dfrac{\partial^2 u}{\partial t^2}$ can then be written as

. . . .

10

$$X''T = \frac{1}{c^2}XT''$$

and this can be transposed into $\dfrac{X''}{X} = \dfrac{1}{c^2} \cdot \dfrac{T''}{T}$

Notice that the left-hand side expression involves functions of x only and that the right-hand side expression involves functions of t only. Therefore, if these two expressions are to be equal for all values of the separate variables, then both expressions must be equal to

. . . .

11

| a constant |

Denote this arbitrary constant by k. Then we have

$$\frac{X''}{X} = k \quad \text{and} \quad \frac{1}{c^2} \cdot \frac{T''}{T} = k$$

$$\therefore \ X'' - kX = 0 \quad \text{and} \quad T'' - c^2kT = 0$$

Let us consider the first of these two equations for different values of k.

(1) *If* $k = 0$, $X'' = 0$ $\therefore X' = a$ $\therefore X = ax + b$.

But $X = 0$ at $x = 0$ $\therefore b = 0$ $\therefore X = ax$
and $X = 0$ at $x = l$ $\therefore a = 0$ $\left.\right\} \therefore a = b = 0$

$\therefore X = 0$ which is not oscillatory as the problem requires it to be.

(2) *If* k *is positive*, let $k = p^2$ $\therefore X'' - p^2X = 0$.

The auxiliary equation is therefore $m^2 - p^2 = 0$ $\therefore m^2 = p^2$

$$m = \pm p$$

$\therefore X = Ae^{px} + Be^{-px}$

But $X = 0$ at $x = 0$ $\therefore 0 = A + B$ $\therefore B = -A$

and $X = 0$ at $x = l$ $\therefore 0 = Ae^{pl} - Ae^{-pl}$ $\therefore 0 = A(e^{pl} - e^{-pl})$

$$\therefore A = 0 \quad \therefore A = B = 0$$

Here again $X = 0$ which is not oscillatory.

(3) *If* k *is negative*, let $k = -p^2$ $\therefore X'' + p^2X = 0$.

This is one of the standard equations listed at the beginning of the Programme and gives a solution

$$X = A\cos px + B\sin px \tag{1}$$

which fits the requirements.

The second equation $T'' - c^2kT = 0$ therefore now becomes

. . . .

$$\boxed{T'' + c^2 p^2 T = 0}$$

12

because the same value for k must apply. This equation is of the same form as before and gives the solution

$$T = C \cos cpt + D \sin cpt \qquad (2)$$

So our suggested solution $u = XT$ now becomes

$$u(x, t) = (A \cos px + B \sin px)(C \cos cpt + D \sin cpt)$$

and, if we put $cp = \lambda \quad \therefore \ p = \dfrac{\lambda}{c}$, this becomes

$$u(x, t) = \left(A \cos \frac{\lambda}{c}x + B \sin \frac{\lambda}{c}x\right)(C \cos \lambda t + D \sin \lambda t) \qquad (3)$$

where A, B, C, D are arbitrary constants.

The result, of course, must satisfy the set of boundary conditions which we now turn to.

(a) $u = 0$ when $x = 0$ for all values of t. From this, we get

. . . .

$$\boxed{A = 0}$$

13

Because, substituting $u = 0$ and $x = 0$ in result (3) above

$$0 = A(C \cos \lambda t + D \sin \lambda t) \ \text{ for all } t \quad \therefore \ A = 0$$

$$\therefore \ u(x, t) = B \sin \frac{\lambda}{c}x(C \cos \lambda t + D \sin \lambda t)$$

(b) $u = 0$ when $x = l$ for all $t \quad \therefore \ 0 = B \ \sin \dfrac{\lambda l}{c}(C \cos \lambda t + D \sin \lambda t)$

Now $B \neq 0$ or $u(x, t)$ would be identically zero. $\therefore \ \sin \dfrac{\lambda l}{c} = 0$.

$$\therefore \ \frac{\lambda l}{c} = n\pi \text{ where } n = 1,\ 2,\ 3, \ldots \quad \therefore \ \lambda = \frac{nc\pi}{l} \text{ for } n = 1,\ 2,\ 3, \ldots$$

Note that we exclude $n = 0$ since this would also make $u(x, t)$ identically zero.

▶

As we can see, there is an infinite set of values of λ and each separate value gives a particular solution for $u(x,t)$. The values of λ are called the *eigenvalues* and each corresponding solution the *eigenfunction*.

Putting $n = 1, 2, 3, \ldots$ we therefore have

	Eigenvalues	Eigenfunctions
n	$\lambda = \dfrac{nc\pi}{l}$	$u(x,t) = B\sin\dfrac{\lambda x}{c}\{C\cos\lambda t + D\sin\lambda t\}$
1	$\lambda_1 = \dfrac{c\pi}{l}$	$u_1 = \sin\dfrac{\pi x}{l}\left\{C_1\cos\dfrac{c\pi t}{l} + D_1\sin\dfrac{c\pi t}{l}\right\}$
2	$\lambda_2 = \dfrac{2c\pi}{l}$	$u_2 = \sin\dfrac{2\pi x}{l}\left\{C_2\cos\dfrac{2c\pi t}{l} + D_2\sin\dfrac{2c\pi t}{l}\right\}$
3	$\lambda_3 = \dfrac{3c\pi}{l}$	$u_3 = \sin\dfrac{3\pi x}{l}\left\{C_3\cos\dfrac{3c\pi t}{l} + D_3\sin\dfrac{3c\pi t}{l}\right\}$
$\vdots$	$\vdots$	$\vdots$
r	$\lambda_r = \dfrac{rc\pi}{l}$	$u_r = \sin\dfrac{r\pi x}{l}\left\{C_r\cos\dfrac{rc\pi t}{l} + D_r\sin\dfrac{rc\pi t}{l}\right\}$

Note that the constant B has been absorbed into the constants C and D so that $BC = C_n$ and $BD = D_n$, where $C_1, C_2, C_3, \ldots$ and $D_1, D_2, D_3, \ldots$ are arbitrary constants.

Since the original wave equation is linear in form, we have already noted that if $u = u_1, u = u_2, u = u_3 \ldots$ are particular solutions, a more general solution is

$$\ldots\ldots\ldots\ldots$$

14

$$\boxed{u = u_1 + u_2 + u_3 + \ldots}$$

The more general solution is therefore

$$u(x,t) = \sum_{r=1}^{\infty} u_r = \sum_{r=1}^{\infty}\left\{\sin\frac{r\pi x}{l}\left(C_r\cos\frac{rc\pi t}{l} + D_r\sin\frac{rc\pi t}{l}\right)\right\} \qquad (4)$$

We still have to find C_r and D_r and for this we use the initial conditions which we have not yet taken into account.

(c) At $t = 0$, $u(x, 0) = f(x)$ for $0 \le x \le l$

Therefore from (4), $u(x, 0) = f(x) = \displaystyle\sum_{r=1}^{\infty} C_r \sin\frac{r\pi x}{l}$.

(d) Also at $t = 0$, $\left[\dfrac{\partial u}{\partial t}\right]_{t=0} = g(x)$ for $0 \le x \le l$

We therefore differentiate (4) with respect to t and put $t = 0$, which gives

$$\ldots\ldots\ldots\ldots$$

15

$$g(x) = \frac{c\pi}{l} \sum_{r=1}^{\infty} D_r \, r \sin\frac{r\pi x}{l}$$

Because

$$\frac{\partial u}{\partial t} = \sum_{r=1}^{\infty} \sin\frac{r\pi x}{l} \left\{ -C_r \frac{rc\pi}{l} \sin\frac{rc\pi t}{l} + D_r \frac{rc\pi}{l} \cos\frac{rc\pi t}{l} \right\}$$

$\therefore$ With $t = 0$, $\quad \dfrac{\partial u}{\partial t} = g(x) = \displaystyle\sum_{r=1}^{\infty} D_r \frac{rc\pi}{l} \sin\frac{r\pi x}{l}$

$$\therefore \; g(x) = \frac{c\pi}{l} \sum_{r=1}^{\infty} D_r \, r \sin\frac{r\pi x}{l}$$

Finally we can draw on our knowledge of Fourier series techniques to determine the coefficients C_r and D_r.

$C_r = 2 \times$ mean value of $f(x) \sin\dfrac{r\pi x}{l}$ between $x = 0$ and $x = l$

$$\therefore \; C_r = \frac{2}{l} \int_0^l f(x) \sin\frac{r\pi x}{l} \, dx \qquad r = 1, \, 2, \, 3, \dots$$

and $D_r \dfrac{rc\pi}{l} = 2 \times$ mean value of $g(x) \sin\dfrac{r\pi x}{l}$ between $x = 0$ and $x = l$

$$\therefore \; D_r = \frac{2}{rc\pi} \int_0^l g(x) \sin\frac{r\pi x}{l} \, dx \qquad r = 1, \, 2, \, 3, \dots$$

The general solution (4) then becomes

$$u(x,\,t) = \sum_{r=1}^{\infty} \left\{ \left[\frac{2}{l} \int_0^l f(w) \sin\frac{r\pi w}{l} \, dw \right] \cos\frac{rc\pi t}{l} \sin\frac{r\pi x}{l} \right.$$
$$\left. + \left[\frac{2}{rc\pi} \int_0^l g(w) \sin\frac{r\pi w}{l} \, dw \right] \sin\frac{rc\pi t}{l} \sin\frac{r\pi x}{l} \right\} \qquad (5)$$

Notice that the variable of integration has been changed from x to w because we wish to use the variable x in the final expression for $u(x,t)$. The value of a definite integral depends only on the limit points of the integral and we are free to use any symbol that we desire for the variable of integration – we call such a variable a *dummy variable*.

At first sight, the solution seems very involved, but it can be analysed into a definite sequence of logical steps. Given the equation and relevant initial and boundary conditions, we go through the following stages.

(a) Assume a solution of the form $u = XT$ and express the equation in terms of X and T and their derivatives.

(b) Transpose the equation by separation of the variables and equate each side to a constant, so obtaining two separate equations, one in x and the other in t.

(c) Choose $k = -p^2$ to give an oscillatory solution.

▶

(d) The two solutions are of the form

$$X = A\cos px + B\sin px$$
$$T = C\cos cpt + D\sin cpt$$

Then $u(x,t) = \{A\cos px + B\sin px\}\{C\cos cpt + D\sin cpt\}$.

(e) Putting $cp = \lambda$, i.e. $p = \dfrac{\lambda}{c}$, this becomes

$$u(x,t) = \left\{A\cos\frac{\lambda}{c}x + B\sin\frac{\lambda}{c}x\right\}\{C\cos\lambda t + D\sin\lambda t\}.$$

(f) Apply boundary conditions to determine A and B.

(g) List the eigenvalues and eigenfunctions for $n = 1, 2, 3, \ldots$ and determine the general solution as an infinite sum.

(h) Apply the remaining initial or boundary conditions.

(i) Determine the coefficients C_r and D_r by Fourier series techniques.

Make a list of these steps: then we can follow them with an example.

---

16   **Example**

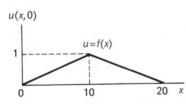

A stretched string of length 20 cm is set oscillating by displacing its mid-point a distance 1 cm from its rest position and releasing it with zero initial velocity.

Solve the wave equation $\dfrac{\partial^2 u}{\partial x^2} = \dfrac{1}{c^2}\cdot\dfrac{\partial^2 u}{\partial t^2}$

where $c^2 = 1$ to determine the resulting motion, $u(x,t)$.

First we make a list of the boundary conditions from the data given in the question.

$$u(0,\ t) = \ldots\ldots\ldots\ldots;\quad u(20,\ t) = \ldots\ldots\ldots\ldots$$
$$u(x,\ 0) = \ldots\ldots\ldots\ldots$$
$$\ldots\ldots\ldots\ldots$$
$$\left[\frac{\partial u}{\partial t}\right]_{t=0} = \ldots\ldots\ldots\ldots$$

17

$$u(0,t) = 0;\qquad u(20,t) = 0 \quad \text{(fixed end points)}$$
$$u(x,0) = f(x) = \frac{x}{10} \qquad 0 \le x \le 10$$
$$= \frac{20-x}{10} \qquad 10 \le x \le 20$$
$$\left[\frac{\partial u}{\partial t}\right]_{t=0} = 0 \quad \text{(zero initial velocity)}$$

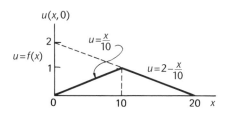

Now we can apply our sequence of operations which we listed.

So move on

(a) Assume a solution $u = XT$ where X is a function of x only and T is a **18**

function of t only. Then the equation $\dfrac{\partial^2 u}{\partial x^2} = \dfrac{\partial^2 u}{\partial t^2}$ (since $c^2 = 1$) becomes

....

$$\boxed{X''T = XT''}$$ **19**

Because

$$u = XT \qquad \therefore \quad \frac{\partial u}{\partial x} = X'T \qquad \frac{\partial^2 u}{\partial x^2} = X''T$$

$$\text{and} \qquad \frac{\partial u}{\partial t} = XT' \qquad \frac{\partial^2 u}{\partial t^2} = XT''$$

$$\frac{\partial^2 u}{\partial x^2} = \frac{\partial^2 u}{\partial t^2} \qquad \therefore \quad X''T = XT''$$

(b) Next we rearrange the equation to separate the variables, giving

....

$$\boxed{\dfrac{X''}{X} = \dfrac{T''}{T}}$$ **20**

(c) Since the two sides are equal for all values of the variables, each must be equal to a constant k and to give an oscillatory solution we put $k = -p^2$. The two separate equations then are written

.... and ....

$$\boxed{X'' + p^2 X = 0 \quad \text{and} \quad T'' + p^2 T = 0}$$ **21**

(d) These have solution $X = $ ....
$$T = \text{............}$$
so that $u(x,\, t) = $ ....

22

$$X = A\cos px + B\sin px; \quad T = C\cos pt + D\sin pt$$
$$\therefore \ u(x,\,t) = \{A\cos px + B\sin px\}\{C\cos pt + D\sin pt\}$$

(e) We normally now put $cp = \lambda$, but in this case $c = 1$ $\therefore$ $p = \lambda$ and
$$u(x,\,t) = \ldots\ldots\ldots\ldots$$

23

$$u(x,\,t) = \{A\cos \lambda x + B\sin \lambda x\}\{C\cos \lambda t + D\sin \lambda t\}$$

(f) Now we determine A and B from the boundary conditions.

(1) $u(0,\,t) = 0$ $\quad \therefore$ $0 = A(C\cos \lambda t + D\sin \lambda t)$ $\quad \therefore$ $A = 0$

$\quad \therefore$ $u(x,\,t) = B\sin \lambda x(C\cos \lambda t + D\sin \lambda t)$

(2) $u(20,\,t) = 0$ $\quad \therefore$ $0 = B\sin 20\lambda(C\cos \lambda t + D\sin \lambda t)$

$\quad B \neq 0$ or u would be identically zero. $\quad \therefore$ $\sin 20\lambda = 0$.

$\quad \therefore$ $20\lambda = n\pi$ $\quad \therefore$ $\lambda = \dfrac{n\pi}{20}$

$\quad \therefore$ $u(x,\,t) = \sin\dfrac{n\pi}{20}x\left\{P\cos\dfrac{n\pi}{20}t + Q\sin\dfrac{n\pi}{20}t\right\}$

where $P = B \times C$ and $Q = B \times D$.

(g) The next step is to list the eigenvalues and eigenfunctions.

n	Eigenvalues $\lambda = \dfrac{n\pi}{20}$	Eigenfunctions $u(x,t) = \sin \lambda x\{P\cos \lambda t + Q\sin \lambda t\}$
1	$\lambda_1 = \dfrac{\pi}{20}$	$u_1 = \sin\dfrac{\pi x}{20}\left\{P_1\cos\dfrac{\pi t}{20} + Q_1\sin\dfrac{\pi t}{20}\right\}$
2	$\lambda_2 = \dfrac{2\pi}{20}$	$u_2 = \sin\dfrac{2\pi x}{20}\left\{P_2\cos\dfrac{2\pi t}{20} + Q_2\sin\dfrac{2\pi t}{20}\right\}$
3	$\lambda_3 = \dfrac{3\pi}{20}$	$u_3 = \sin\dfrac{3\pi x}{20}\left\{P_3\cos\dfrac{3\pi t}{20} + Q_3\sin\dfrac{3\pi t}{20}\right\}$
$\vdots$	$\vdots$	$\vdots$
r	$\lambda_r = \dfrac{r\pi}{20}$	$u_r = \sin\dfrac{r\pi x}{20}\left\{P_r\cos\dfrac{r\pi t}{20} + Q_r\sin\dfrac{r\pi t}{20}\right\}$

$$u = u_1 + u_2 + u_3 + \ldots \quad \therefore \ u(x,\,t) = \sum_{r=1}^{\infty}\sin\dfrac{r\pi x}{20}\left\{P_r\cos\dfrac{r\pi t}{20} + Q_r\sin\dfrac{r\pi t}{20}\right\}$$

▶

(h) Now we apply the remaining initial conditions

(1) $\quad u(x, 0) = f(x) = \dfrac{x}{10} \qquad\qquad 0 \le x \le 10$

$\qquad\qquad\qquad\quad = \dfrac{20 - x}{10} \qquad 10 \le x \le 20$

$\quad$ Also $u(x, 0) = \ldots\ldots\ldots$

$$u(x,\ 0) = \sum_{r=1}^{\infty} P_r \sin \frac{r\pi x}{20}$$

24

Then $P_r = 2 \times$ mean value of $f(x)\ \sin \dfrac{r\pi x}{20}$ between $x = 0$ and $x = 20$

$$= \frac{2}{20} \int_0^{20} f(x)\ \sin \frac{r\pi x}{20}\, dx$$

$$\therefore\ 10P_r = \int_0^{10} \frac{x}{10} \sin \frac{r\pi x}{20}\, dx + \int_{10}^{20} \frac{20 - x}{10} \sin \frac{r\pi x}{20}\, dx$$

$$= \qquad I_1 \qquad + \qquad I_2$$

$$I_1 = \int_0^{10} \frac{x}{10} \sin \frac{r\pi x}{20}\, dx = \ldots\ldots\ldots$$

$$I_1 = -\frac{20}{r\pi} \cos \frac{r\pi}{2} + \frac{40}{r^2\pi^2} \sin \frac{r\pi}{2}$$

25

Using integration by parts

$$I_2 = \int_{10}^{20} \frac{20 - x}{10} \sin \frac{r\pi x}{20}\, dx = \ldots\ldots\ldots$$

$$I_2 = \frac{20}{r\pi} \cos \frac{r\pi}{2} - \frac{40}{r^2\pi^2} \left(\sin r\pi - \sin \frac{r\pi}{2} \right)$$

26

Then $10\, P_r = -\dfrac{20}{r\pi} \cos \dfrac{r\pi}{2} + \dfrac{40}{r^2\pi^2} \sin \dfrac{r\pi}{2} + \dfrac{20}{r\pi} \cos \dfrac{r\pi}{2} - \dfrac{40}{r^2\pi^2} \left(\sin r\pi - \sin \dfrac{r\pi}{2} \right)$

$$\therefore\ \text{For } r = 1,\ 2,\ 3, \ldots\ P_r = \frac{8}{r^2\pi^2} \sin \frac{r\pi}{2}$$

$$\therefore\ u(x, t) = \sum_{r=1}^{\infty} \sin \frac{r\pi x}{20} \left\{ \frac{8}{r^2\pi^2} \sin \frac{r\pi}{2} \cos \frac{r\pi t}{20} + Q_r \sin \frac{r\pi t}{20} \right\}$$

(2) Also at $t = 0$, $\dfrac{\partial u}{\partial t} = 0$.

$$\frac{\partial u}{\partial t} = \ldots\ldots\ldots$$

27

$$\frac{\partial u(x,t)}{\partial t} = \sum_{r=1}^{\infty} \sin\frac{r\pi x}{20}\left\{ \left(\frac{8}{r^2\pi^2}\sin\frac{r\pi}{2}\right)\left(-\frac{r\pi}{20}\sin\frac{r\pi t}{20}\right) \right.$$
$$\left. + Q_r\frac{r\pi}{20}\cos\frac{r\pi t}{20}\right\}$$

$\therefore$ At $t = 0,\qquad 0 = \sum_{r=1}^{\infty}\sin\frac{r\pi x}{20}Q_r\frac{r\pi}{20}\qquad \therefore\ Q_r = 0$

So finally we have $\qquad u(x,t) = \ldots\ldots\ldots$

28

$$u(x,t) = \frac{8}{\pi^2}\sum_{r=1}^{\infty}\frac{1}{r^2}\sin\frac{r\pi x}{20}\sin\frac{r\pi}{2}\cos\frac{r\pi t}{20}$$

And that is it.

Now let us turn to a slightly different equation, but one for which the method of solution is very much along the same lines.

The heat conduction equation for a uniform finite bar

The conduction of heat in a uniform bar depends on the initial distribution of temperature and on the physical properties of the bar, i.e. the thermal conductivity and specific heat of the material, and the mass per unit length of the bar.

With a uniform bar insulated except at its ends, any heat flow is along the bar and, at any instant, the temperature u at a point P is a function of its distance x from one end and of the time t.

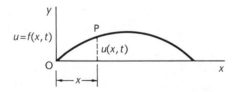

The one-dimensional heat equation is then of the form

$$\frac{\partial^2 u}{\partial x^2} = \frac{1}{c^2}\cdot\frac{\partial u}{\partial t}\tag{1}$$

where $c^2 = \dfrac{k}{\sigma\rho}$ in which $k =$ thermal conductivity of the material; $\sigma =$ specific heat of the material; $\rho =$ mass per unit length of the bar.

You will already have noticed that the heat equation differs from the wave equation only in the fact that the right-hand side contains the first partial derivative instead of the second. It is not surprising therefore that the method of solution is very much like that of our previous examples.

Solutions of the heat conduction equation

Consider the case where

(a) the bar extends from $x = 0$ to $x = l$
(b) the temperature of the ends of the bar is maintained at zero
(c) the initial temperature distribution along the bar is defined by $f(x)$.

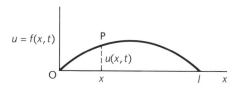

The boundary conditions can be expressed as . . . .

29

$$u(0, t) = 0 \quad \text{and} \quad u(l, t) = 0 \quad \text{for all } t \geq 0$$
$$u(x, 0) = f(x) \quad \text{for } 0 \leq x \leq l$$

As before, we assume a solution of the form $u(x, t) = X(x)T(t)$ where

X is a function of x only

T is a function of t only.

Then, starting with $u = XT$ we can write the equation $\dfrac{\partial^2 u}{\partial x^2} = \dfrac{1}{c^2} \cdot \dfrac{\partial u}{\partial t}$ in terms of X and T, and separating the variables, we obtain

. . . .

30

$$\frac{X''}{X} = \frac{1}{c^2} \cdot \frac{T'}{T}$$

Arguing as before, since the left-hand side is a function of x only and the right-hand side a function of t only, for these to be equal each side must equal the same constant. Let this be $(-p^2)$ as before.

$$\therefore \frac{X''}{X} = -p^2 \quad \therefore X'' + p^2 X = 0 \quad \text{giving } X = A \cos px + B \sin px$$

and $\dfrac{1}{c^2} \cdot \dfrac{T'}{T} = -p^2 \quad \therefore T' + p^2 c^2 T = 0 \quad \text{giving } T = \ldots \ldots \ldots \ldots$

31

$$\boxed{T = Ce^{-p^2c^2t}}$$

Because

$$\frac{T'}{T} = -p^2c^2 \quad \therefore \quad \ln T = -p^2c^2t + c_1 \quad \therefore \quad T = Ce^{-p^2c^2t}$$

$$u(x,t) = XT = \{A\cos px + B\sin px\}\,Ce^{-p^2c^2t}$$

$$\therefore \quad u(x,t) = \{P\cos px + Q\sin px\}e^{-p^2c^2t} \qquad \text{where } P = AC \text{ and } Q = BC$$

Now put $pc = \lambda \quad \therefore \quad p = \dfrac{\lambda}{c}$

$$\therefore \quad u(x,t) = \left\{P\cos\frac{\lambda}{c}x + Q\sin\frac{\lambda}{c}x\right\}e^{-\lambda^2 t}$$

Applying the boundary condition $u(0,t) = 0$ gives

$$\dots\dots\dots \quad \text{and} \quad \dots\dots\dots$$

32

$$\boxed{P = 0 \quad \text{and} \quad u(x,t) = Qe^{-\lambda^2 t}\sin\frac{\lambda}{c}x}$$

Also $u(l,t) = 0$ and from this we get

$$\dots\dots\dots\dots$$

33

$$\boxed{\lambda = \frac{nc\pi}{l} \quad \text{for} \quad n = 1, 2, 3, \dots}$$

Because

if $u = 0$ when $x = l$, $\qquad 0 = Qe^{-\lambda^2 t}\sin\dfrac{\lambda l}{c}$

$Q \neq 0$ or $u(x, t)$ would be identically zero $\quad\therefore\quad \sin\dfrac{\lambda l}{c} = 0$

$$\therefore \quad \frac{\lambda l}{c} = n\pi \qquad \therefore \quad \lambda = \frac{nc\pi}{l} \qquad n = 1, 2, 3, \dots$$

▶

Now we can compile the table of eigenfunctions.

n	$\lambda = \dfrac{nc\pi}{l}$	$u(x,t) = Q e^{-\lambda^2 t} \sin\dfrac{n\pi x}{l}$
1	$\lambda_1 = \dfrac{c\pi}{l}$	$u_1 = Q_1 e^{-\lambda_1^2 t} \sin\dfrac{\pi x}{l}$
2	$\lambda_2 = \dfrac{2c\pi}{l}$	$u_2 = Q_2 e^{-\lambda_2^2 t} \sin\dfrac{2\pi x}{l}$
3	$\lambda_3 = \dfrac{3c\pi}{l}$	$u_3 = Q_3\, e^{-\lambda_3^2 t} \sin\dfrac{3\pi x}{l}$
$\vdots$	$\vdots$	$\vdots$
r	$\lambda_r = \dfrac{rc\pi}{l}$	$u_r = Q_r\, e^{-\lambda_r^2 t} \sin\dfrac{r\pi x}{l}$

$$u = u_1 + u_2 + u_3 + \ldots$$

$$\therefore\ u(x,t) = \sum_{r=1}^{\infty}\left\{ Q_r\, e^{-\lambda_r^2 t} \sin\dfrac{r\pi x}{l}\right\}$$

The remaining boundary condition still to be applied is that when

$$t = 0,\ \ u(x,0) = f(x)\quad 0 \le x \le l$$

This gives $f(x) = \ldots\ldots\ldots\ldots$

$$\boxed{f(x) = \sum_{r=1}^{\infty}\left\{ Q_r \sin\dfrac{r\pi x}{l}\right\}}$$

34

and from our knowledge of Fourier series techniques:

$$Q_r = \ldots\ldots\ldots\ldots$$

$$\boxed{Q_r = 2 \times \text{mean value of } f(x)\sin\dfrac{r\pi x}{l} \text{ from } x = 0 \text{ to } x = l}$$

35

$$\therefore\ Q_r = \dfrac{2}{l}\int_0^l f(x)\sin\dfrac{r\pi x}{l}\,dx \text{ and the final solution becomes}$$

$$u(x,t) = \dfrac{2}{l}\sum_{r=1}^{\infty}\left\{\left[\int_0^l f(w)\sin\dfrac{r\pi w}{l}\,dw\right] e^{-\lambda_r^2 t}\sin\dfrac{r\pi x}{l}\right\}$$

where $\lambda_r = \dfrac{rc\pi}{l}\qquad r = 1,\,2,\,3,\,\ldots$

Now on to the next frame for an example

36   **Example**

A bar of length 2 m is fully insulated along its sides. It is initially at a uniform temperature of 10°C and at $t = 0$ the ends are plunged into ice and maintained at a temperature of 0°C. Determine an expression for the temperature at a point P a distance x from one end at any subsequent time t seconds after $t = 0$.

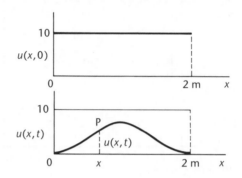

We have the heat equation $\dfrac{\partial^2 u}{\partial x^2} = \dfrac{1}{c^2} \cdot \dfrac{\partial u}{\partial t}$ with the boundary conditions

............; ....; and ....

37                   $$u(0,\, t) = 0; \quad u(2,\, t) = 0; \quad u(x,\, 0) = 10$$

Assuming a solution of the form $u = XT$, we know that this gives for

this equation $X = A \cos px + B \sin px$

and $T = C e^{-p^2 c^2 t}$

so that the general solution is

$u(x, t) = \{P \cos px + Q \sin px\} e^{-p^2 c^2 t}$

If we now write $pc = \lambda, \quad p = \dfrac{\lambda}{c}$ and the solution becomes

$u(x, t) = \left\{ P \cos \dfrac{\lambda}{c} x + Q \sin \dfrac{\lambda}{c} x \right\} e^{-\lambda^2 t}$

Applying the first two of the boundary conditions gives us

....

38

$$P = 0 \quad \text{and} \quad u(x, t) = \left\{ Q \sin \frac{n\pi x}{2} \right\} e^{-\lambda^2 t}$$

Because

$$u(0, t) = 0 \quad \therefore \quad 0 = P e^{-\lambda^2 t} \quad \therefore \quad P = 0$$

$$\therefore \quad u(x, t) = \left\{ Q \sin \frac{\lambda}{c} x \right\} e^{-\lambda^2 t}$$

Also $u(2, t) = 0 \qquad \therefore \quad 0 = \left\{ Q \sin \frac{2\lambda}{c} \right\} e^{-\lambda^2 t}$

$$Q \neq 0 \quad \therefore \quad \sin \frac{2\lambda}{c} = 0 \quad \therefore \quad \frac{2\lambda}{c} = n\pi \quad \therefore \quad \lambda = \frac{n c \pi}{2} \quad n = 1, 2, 3, \ldots$$

$$\therefore \quad u(x, t) = \left\{ Q \sin \frac{n\pi x}{2} \right\} e^{-\lambda^2 t}$$

There is, of course, an infinite number of such solutions with different values of n. We can write the solution so far therefore as

$$u(x, t) = \ldots\ldots\ldots\ldots$$

39

$$u(x, t) = \sum_{r=1}^{\infty} Q_r \sin \frac{r\pi x}{2} e^{-\lambda_r^2 t}$$

Finally, there is the remaining initial condition that at $t = 0$, $u = 10$.

$$\therefore \quad u(x, 0) = f(x) = 10 \quad \therefore \quad 10 = \sum_{r=1}^{\infty} Q_r \sin \frac{r\pi x}{2}$$

where $Q_r = 2 \times$ mean value of $10 \sin \frac{r\pi x}{2}$ from $x = 0$ to $x = 2$.

$$\therefore \quad Q_r = \ldots\ldots\ldots\ldots$$

40

$$0 \ (r \text{ even}); \quad \frac{40}{\pi r} \ (r \text{ odd})$$

Because

$$Q_r = \frac{2}{2} \int_0^2 10 \sin \frac{r\pi x}{2} \, dx = 10 \int_0^2 \sin \frac{r\pi x}{2} \, dx$$

$$= -\frac{20}{\pi r} \left[\cos \frac{r\pi x}{2} \right]_0^2 = \frac{20}{\pi r} \{1 - \cos r\pi\}$$

$$= 0 \ (r \text{ even}) \quad \text{and} \quad \frac{40}{r\pi} \ (r \text{ odd})$$

Therefore the required solution is

$$u(x, t) = \ldots\ldots\ldots\ldots$$

41

$$u(x, t) = \frac{40}{\pi} \sum_{r \,(\text{odd})=1}^{\infty} \frac{1}{r} \sin \frac{r\pi x}{2} e^{-\lambda_r^2 t} \qquad r = 1, 3, 5, \ldots$$

where $\lambda_r = \dfrac{rc\pi}{2}$

By now you will appreciate that the approach to all these problems is very much the same, as indeed it still is with the next important equation.

Laplace's equation

The Laplace equation concerns the distribution of a field, e.g. temperature, potential, etc., over a plane area subject to certain boundary conditions.

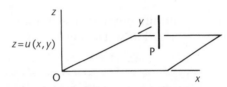

The potential at a point P in a plane can be indicated by an ordinate axis and is a function of its position, i.e. $z = u(x, y)$ where $u(x, y)$ is the solution of the Laplace two-dimensional equation $\dfrac{\partial^2 u}{\partial x^2} + \dfrac{\partial^2 u}{\partial y^2} = 0$.

Let us consider the situation in the next frame

42 Solution of the Laplace equation

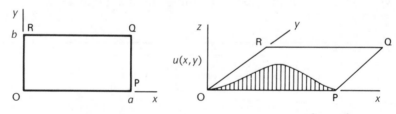

We are required to determine a solution of the equation $\dfrac{\partial^2 u}{\partial x^2} + \dfrac{\partial^2 u}{\partial y^2} = 0$ for the rectangle bounded by the lines $x = 0$, $y = 0$, $x = a$, $y = b$, subject to the following boundary conditions

$\quad u = 0 \qquad$ when $\quad x = 0 \quad 0 \le y \le b$

$\quad u = 0 \qquad$ when $\quad x = a \quad 0 \le y \le b$

$\quad u = 0 \qquad$ when $\quad y = b \quad 0 \le x \le a$

$\quad u = f(x) \quad$ when $\quad y = 0 \quad 0 \le x \le a$

$\quad$ i.e. $u(0, y) = 0$ and $u(a, y) = 0$ for $0 \le y \le b$

$\quad$ and $u(x, b) = 0$ and $u(x, 0) = f(x)$ for $0 \le x \le a$.

▶

The solution $z = u(x, y)$ will give the potential at any point within the rectangle OPQR.

We start off, as usual, by assuming a solution of the form $u(x, y) = X(x)Y(y)$ where X is a function of x only and Y is a function of y only. We now express the equation in terms of X and Y and separate the variables to give

. . . .

$$\boxed{\frac{X''}{X} = -\frac{Y''}{Y}}$$

43

Because

$$u = XY \qquad \therefore \quad \frac{\partial u}{\partial x} = X'Y \quad \text{and} \quad \frac{\partial^2 u}{\partial x^2} = X''Y$$

$$\frac{\partial u}{\partial y} = XY' \quad \text{and} \quad \frac{\partial^2 u}{\partial y^2} = XY''$$

The equation is then $X''Y = -XY'' \qquad \therefore \quad \dfrac{X''}{X} = -\dfrac{Y''}{Y}$

Putting each side equal to a constant $(-p^2)$ gives two equations

$$X'' + p^2 X = 0 \quad \text{and} \quad Y'' - p^2 Y = 0$$

$X'' + p^2 X = 0$ has a solution $X = $ . . . .

$$\boxed{X = A \cos px + B \sin px}$$

44

In the introduction to this Programme we said that the equation $Y'' - p^2 Y = 0$ has a solution of the form $Y = C \cosh py + D \sinh py$ which can also be expressed as $Y = E \sinh p(y + \phi)$.

$$\therefore \quad u(x, y) = \{A \cos px + B \sin px\} E \sinh p(y + \phi)$$
$$\therefore \quad u(x, y) = \{P \cos px + Q \sin px\} \sinh p(y + \phi)$$

Now we apply the first of the boundary conditions.

$$u(0, y) = 0 \quad \therefore \quad 0 = P \sinh p(y + \phi) \quad \therefore \quad P = 0$$
$$\therefore \quad u(x, y) = Q \sin px \sinh p(y + \phi)$$

From the second boundary condition, we have

$$u(a, y) = 0 \quad \therefore \quad 0 = Q \sin pa \sinh p(y + \phi) \quad \therefore \quad \sin pa = 0$$
$$\therefore \quad pa = n\pi \qquad \text{for } n = 1, 2, 3, \dots$$

If we write $\lambda = p$ then $\lambda = \dfrac{n\pi}{a}$ and $u(x, y) = Q \sin \lambda x \sinh \lambda(y + \phi)$

Now from the third condition

$$u(x, b) = 0 \text{ from which we have } \dots \dots \dots \dots$$

45

$$u(x, y) = Q \sin \lambda x \sinh \lambda(b - y)$$

Because

$$0 = Q \sin \lambda x \sinh \lambda(b + \phi) \quad \therefore \quad \sinh \lambda(b + \phi) = 0 \quad \therefore \quad \phi = -b.$$

$$\therefore \quad u(x, y) = Q \sin \lambda x \sinh \lambda(y - b)$$

$$\sinh \lambda(y - b) = -\sinh \lambda(b - y) \quad \therefore \quad u(x, y) = Q \sin \lambda x \sinh \lambda(b - y),$$

the minus sign being absorbed in the symbol Q whose value has yet to be found. Now $\lambda = \dfrac{n\pi}{a}$ with $n = 1, 2, 3, \ldots$ and there is therefore an infinite number of values for λ and hence an infinite number of solutions for $u(x, y)$. Therefore, again using $u = u_1 + u_2 + u_3 + \ldots$ we have

$$u(x, y) = \ldots\ldots\ldots\ldots$$

46

$$u(x, y) = \sum_{r=1}^{\infty} Q_r \sin \lambda_r x \sinh \lambda_r(b - y)$$

Now there remains the fourth boundary condition to be applied.

$$u(x, 0) = f(x) \quad \therefore \quad f(x) = \sum_{r=1}^{\infty} Q_r \sin \lambda_r x \ \sinh \lambda_r b$$

$$\therefore \quad Q_r \sinh \lambda_r b = 2 \times \text{ mean value of } f(x) \sin \lambda_r x \text{ from } x = 0 \text{ to } x = a$$

$$= \frac{2}{a} \int_0^a f(x) \sin \lambda_r x \ \mathrm{d}x$$

$$= \frac{2}{a} \int_0^a f(x) \sin \frac{r\pi x}{a} \mathrm{d}x$$

from which the coefficients Q_r can be found.

Let us work through an example with numerical values.

Example

Determine a solution $u(x, y)$ of the Laplace equation $\dfrac{\partial^2 u}{\partial x^2} + \dfrac{\partial^2 u}{\partial y^2} = 0$ subject to the following boundary conditions

$$u = 0 \text{ when } x = 0; \quad u = 0 \text{ when } x = \pi$$
$$u \to 0 \text{ when } y \to \infty; \quad u = 3 \text{ when } y = 0$$

As always, we begin with $u(x, y) = X(x)Y(y)$, rewrite the equation in terms of X and Y and separate the variables. The equation then becomes

$$\ldots\ldots\ldots\ldots$$

$$\boxed{\dfrac{X''}{X} = -\dfrac{Y''}{Y}}$$

47

Equating each side to $-p^2$, we have $X'' + p^2 X = 0$ and $Y'' - p^2 Y = 0$.

$$X'' + p^2 X = 0 \text{ has a solution } \ldots\ldots\ldots\ldots$$

$$\boxed{X = A \cos px + B \sin px}$$

48

The solution of $Y'' - p^2 Y = 0$ can be stated in three different forms

$$Y = C \cosh py + D \sinh py; \quad Y = Ce^{py} + De^{-py}; \quad Y = C \sinh p(y + \phi)$$

On this occasion, we will use the second one

$$Y = Ce^{py} + De^{-py}$$

Then $u(x, y) = \{A \cos px + B \sin px\}\{Ce^{py} + De^{-py}\}$

Application of the first boundary condition $u(0, y) = 0$ gives

$$\ldots\ldots\ldots\ldots \text{ and } \ldots\ldots\ldots\ldots$$

$$\boxed{A = 0 \text{ and } u(x, y) = \sin px\{Pe^{py} + Qe^{-py}\}}$$

49

Because

$$0 = A\{Ce^{py} + De^{-py}\} \qquad \therefore A = 0$$

and $u(x, y) = B \sin px\{Ce^{py} + De^{-py}\} = \sin px\{Pe^{py} + Qe^{-py}\}$.

The second boundary condition $u(\pi, y) = 0$ then gives

$$\ldots\ldots\ldots\ldots$$

$$\boxed{u(x, y) = \sin nx\ \{Pe^{ny} + Qe^{-ny}\} \qquad n = 1, 2, 3, \ldots}$$

50

Because

$$u = 0 \text{ when } x = \pi \ \therefore\ 0 = \sin p\pi\{Pe^{py} + Qe^{-py}\}$$
$$\therefore\ \sin p\pi = 0 \ \therefore\ p\pi = n\pi \ \therefore\ p = n \qquad n = 1, 2, 3, \ldots$$
$$\therefore\ u(x, y) = \sin nx\ \{Pe^{ny} + Qe^{-ny}\}$$

The third condition is that $u \to 0$ as $y \to \infty$.

▶

Because $e^{-ny} \to 0$ as $y \to \infty$ then $0 = \sin nx \{Pe^{ny}\}$, so that $P = 0$

$\therefore \ u(x, y) = Qe^{-ny} \sin nx$

But n can have an infinite number of values giving an infinite number of solutions

$u_1 = Q_1 e^{-y} \sin x$

$u_2 = Q_2 e^{-2y} \sin 2x$

$u_3 = Q_3 e^{-3y} \sin 3x$

$\vdots \qquad \vdots$

$u_r = Q_r e^{-ry} \sin rx$

So the solution at this stage can be written as

$$u(x, y) = \ldots\ldots\ldots\ldots$$

51

$$u(x, y) = \sum_{r=1}^{\infty} Q_r e^{-ry} \sin rx$$

Now we turn to the final boundary condition that $u = 3$ when $y = 0$.

$\therefore \ 3 = \displaystyle\sum_{r=1}^{\infty} Q_r \sin rx$ from which we obtain

$Q_r = \ldots\ldots\ldots\ldots$

52

$$Q_r = 0 \ (r \text{ even}); \quad Q_r = \frac{12}{r\pi} \ (r \text{ odd})$$

Because

$$Q_r = 2 \times \text{ mean value of } 3\sin rx \text{ between } x = 0 \text{ and } x = \pi$$

$$= \frac{2}{\pi} \int_0^{\pi} 3 \sin rx \, dx = \frac{6}{\pi} \left[-\frac{\cos rx}{r} \right]_0^{\pi} = \frac{6}{r\pi}(1 - \cos r\pi)$$

$$\therefore \ Q_r = 0 \ (r \text{ even}) \text{ and } \frac{12}{r\pi} \ (r \text{ odd})$$

$$\therefore \ u(x, y) = \sum_{r\,(\text{odd})=1}^{\infty} \frac{12}{r\pi} e^{-ry} \sin rx \qquad r = 1, 3, 5, \ldots$$

$$\therefore \ u(x, y) = \frac{12}{\pi} \left\{ e^{-y} \sin x + \frac{1}{3} e^{-3y} \sin 3x + \frac{1}{5} e^{-5y} \sin 5x + \ldots \right\}$$

Laplace's equation in plane polar coordinates

53

Laplace's equation

$$\frac{\partial^2 u(x, y)}{\partial x^2} + \frac{\partial^2 u(x, y)}{\partial y^2} = 0$$

is often referred to as the *potential equation* because such physical entities as the electrostatic and gravitational potentials can be shown to satisfy it. It is an equation that is commonly met in science and engineering. Solving this equation inside a region of the *x–y* plane subject to some specified condition applied to $u(x, y)$ on the boundary of the region is known as a *Dirichlet problem*. To solve this Dirichlet problem we proceed, as we have seen, by separating the variables to find the general solution and then matching up the general solution to the boundary conditions to find the specific solution. However, the process of finding the specific solution from the general solution is very sensitive to the shape of the boundary, and difficulties can arise if the symmetries of the boundary do not match the symmetries of the coordinate system used. For example, if the region under consideration is bounded by the circle

$$x^2 + y^2 = a^2$$

employing Cartesian coordinates will create difficulties when we come to match up the general solution in Cartesians to the boundary conditions on the circular boundary. To avoid such difficulties we choose a coordinate system that has the same symmetries as the boundary where the coordinate symmetries are exhibited when we let one variable vary while keeping all the others constant. The Cartesian coordinate system (x, y) produces straight lines $x = $ constant as y varies and $y = $ constant as x varies. The plane polar coordinate system (r, θ), on the other hand, produces circles $r = $ constant when θ varies and so is suitable for dealing with circular boundaries in the plane.

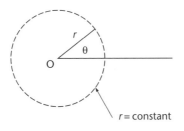

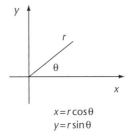

Before we attempt to find the solution we must pose the problem *from the beginning* in terms of the coordinates that are appropriate to the boundary conditions. This means, of course, that Laplace's equation must also be given in the same coordinates. To convert Laplace's equation from its current form in Cartesians (x, y) to a new form in plane polar coordinates (r, θ) where

$$x = r\cos\theta \quad \text{and} \quad y = r\sin\theta$$

requires manipulations using Frame 11 onwards of Programme 14. We shall not go into this here, suffice it to say that in plane polar coordinates Laplace's equation is

$$\frac{\partial^2 v(r, \theta)}{\partial r^2} + \frac{1}{r}\frac{\partial v(r, \theta)}{\partial r} + \frac{1}{r^2}\frac{\partial^2 v(r, \theta)}{\partial \theta^2} = 0$$

where $v(r, \theta)$ is the expression obtained by changing the coordinates in $u(x, y)$ using $x = r\cos\theta$ and $y = r\sin\theta$.

We shall now pose the problem anew in the next frame

54 The problem

Find the solution to

$$\frac{\partial^2 v(r, \theta)}{\partial r^2} + \frac{1}{r}\frac{\partial v(r, \theta)}{\partial r} + \frac{1}{r^2}\frac{\partial^2 v(r, \theta)}{\partial \theta^2} = 0$$

in the circular region $x^2 + y^2 = a^2$ (that is, for $0 \leq r \leq a$) of the plane where

1   $v(r, \theta)$ is finite for $0 \leq r \leq a$ and for all θ
2   $v(a, \theta) = f(\theta)$ – the condition on the boundary of the circular region
3   θ is unbounded but $v(r, \theta + 2\pi) = v(r, \theta)$ for $0 \leq r \leq a$. That is, though θ can take any finite value, the value of $v(r, \theta)$ repeats itself as θ winds round every 2π.

Separating the variables

The variables are r and θ and we assume they are separable and write $v(r, \theta) = R(r)\Theta(\theta)$. This form is then substituted into Laplace's equation and the entire equation multiplied by $\dfrac{r^2}{R(r)\Theta(\theta)}$ to obtain

$$\ldots\ldots\ldots\ldots = 0$$

$$\boxed{\frac{r^2}{R(r)}\frac{\mathrm{d}^2 R(r)}{\mathrm{d}r^2} + \frac{r}{R(r)}\frac{\mathrm{d}R(r)}{\mathrm{d}r} + \frac{1}{\Theta(\theta)}\frac{\mathrm{d}^2 \Theta(\theta)}{\mathrm{d}\theta^2} = 0}$$

55

Because

Substituting $R(r)\Theta(\theta)$ for $v(r, \theta)$ gives

$$\frac{\partial^2 R(r)\Theta(\theta)}{\partial r^2} + \frac{1}{r}\frac{\partial R(r)\Theta(\theta)}{\partial r} + \frac{1}{r^2}\frac{\partial^2 R(r)\Theta(\theta)}{\partial \theta^2} = 0$$

That is

$$\Theta(\theta)\frac{\mathrm{d}^2 R(r)}{\mathrm{d}r^2} + \frac{\Theta(\theta)}{r}\frac{\mathrm{d}R(r)}{\mathrm{d}r} + \frac{R(r)}{r^2}\frac{\mathrm{d}^2 \Theta(\theta)}{\mathrm{d}\theta^2} = 0$$

Multiplying the entire equation by $\dfrac{r^2}{R(r)\Theta(\theta)}$ then gives

$$\frac{r^2}{R(r)}\frac{\mathrm{d}^2 R(r)}{\mathrm{d}r^2} + \frac{r}{R(r)}\frac{\mathrm{d}R(r)}{\mathrm{d}r} + \frac{1}{\Theta(\theta)}\frac{\mathrm{d}^2 \Theta(\theta)}{\mathrm{d}\theta^2} = 0$$

From this result we can say that

$$\frac{r^2}{R(r)}\frac{\mathrm{d}^2 R(r)}{\mathrm{d}r^2} + \frac{r}{R(r)}\frac{\mathrm{d}R(r)}{\mathrm{d}r} = -\frac{1}{\Theta(\theta)}\frac{\mathrm{d}^2 \Theta(\theta)}{\mathrm{d}\theta^2} = k$$

which gives rise to the two uncoupled, second-order ordinary differential equations

$$\frac{r^2}{R(r)}\frac{\mathrm{d}^2 R(r)}{\mathrm{d}r^2} + \frac{r}{R(r)}\frac{\mathrm{d}R(r)}{\mathrm{d}r} = k \quad \text{so that}$$

$$r^2\frac{\mathrm{d}^2 R(r)}{\mathrm{d}r^2} + r\frac{\mathrm{d}R(r)}{\mathrm{d}r} = kR(r) \tag{1}$$

and

$$\frac{1}{\Theta(\theta)}\frac{\mathrm{d}^2 \Theta(\theta)}{\mathrm{d}\theta^2} = -k \quad \text{so that} \quad \frac{\mathrm{d}^2 \Theta(\theta)}{\mathrm{d}\theta^2} = -k\Theta(\theta) \tag{2}$$

The general solution to equation (2) for $k > 0$ is

. . . .

$$\boxed{\Theta_n(\theta) = a_n \cos n\theta + b_n \sin n\theta \quad \text{where } n = 1, 2, \ldots}$$

56

Because

To solve $\dfrac{\mathrm{d}^2 \Theta(\theta)}{\mathrm{d}\theta^2} = -k\Theta(\theta)$, that is $\dfrac{\mathrm{d}^2 \Theta(\theta)}{\mathrm{d}\theta^2} + k\Theta(\theta) = 0$ we use the auxiliary equation $m^2 + k = 0$ with solutions $m = \pm j\sqrt{k}$. This gives the solution, periodic with period 2π as

$$\Theta(\theta) = A \cos \sqrt{k}\theta + B \sin \sqrt{k}\theta \tag{3}$$

▶

provided $k > 0$ so that m is pure imaginary. If $k < 0$ then non-periodic solutions would result which would be physically incorrect. To ensure periodicity, that is to ensure that $k > 0$ write $k = n^2$, $n = 1, 2, \ldots$.

$\Theta_n(\theta) = a_n \cos n\theta + b_n \sin n\theta$ is a solution to equation (2).

We shall look at the case $n = 0$ later.
Substituting $k = n^2$ into equation (1) then gives

$$r^2 \frac{d^2 R(r)}{dr^2} + r \frac{dR(r)}{dr} = n^2 R(r) \tag{4}$$

As a trial solution to equation (4) let $R(r) = pr^q$. Substitution into (4) gives

$$q = \ldots\ldots\ldots\ldots$$

57

$$\boxed{q = \pm n \ \text{where} \ n = 1, 2, \ldots}$$

Because

$$r \frac{dR(r)}{dr} = r \frac{d(pr^q)}{dr} = rpqr^{q-1} = pqr^q. \ \text{Similarly} \ r^2 \frac{d^2 R(r)}{dr^2} = pq(q-1)r^q.$$

Therefore, substitution into $r^2 \dfrac{d^2 R(r)}{dr^2} + r \dfrac{dR(r)}{dr} = n^2 R(r)$ gives

$[q(q-1) + q]pr^q = n^2 pr^q$ and so $[q^2 - n^2]pr^q = 0$

giving $q = \pm n$ where $n = 1, 2, \ldots$

Therefore, a solution to equation (4) is

$R_n(r) = c_n r^n + d_n r^{-n}$ provided $n \neq 0$. The case $n = 0$ is special.

---

58 Summary

To summarise the results so far, we have started to solve Laplace's equation

$$\frac{\partial^2 v(r, \theta)}{\partial r^2} + \frac{1}{r} \frac{\partial v(r, \theta)}{\partial r} + \frac{1}{r^2} \frac{\partial^2 v(r, \theta)}{\partial \theta^2} = 0$$

in the circular region $x^2 + y^2 = a^2$ (that is, for $0 \leq r \leq a$) of the plane where

1. $v(r, \theta)$ is finite for $0 \leq r \leq a$ and for all θ
2. $v(a, \theta) = f(\theta)$
3. θ is unbounded but $v(r, \theta + 2\pi) = v(r, \theta)$ for $0 \leq r \leq a$.

We have found that, assuming $v(r, \theta) = R(r)\Theta(\theta)$ then, provided $n \neq 0$

$\Theta_n(\theta) = a_n \cos n\theta + b_n \sin n\theta$
$R_n(r) = c_n r^n + d_n r^{-n}$

So that

$$v_n(r, \theta) = R_n(r)\Theta_n(\theta) = (c_n r^n + d_n r^{-n})(a_n \cos n\theta + b_n \sin n\theta)$$

If we now apply the boundary condition **1** we find that

$$d_n = \ldots\ldots\ldots\ldots$$

$$\boxed{d_n = 0}$$

59

Because

$v(r, \theta)$ is finite for $0 \leq r \leq a$. In particular, the solution is finite when $r = 0$ and so we cannot have a term of the form r^{-n}. Accordingly $d_n = 0$, so omitting the r^{-n} term the solution then becomes

$$v_n(r, \theta) = c_n r^n (a_n \cos n\theta + b_n \sin n\theta)$$

There is an infinite number of such solutions (eigenfunctions), one for each eigenvalue n. The complete solution to Laplace's equation is then a linear combination of all these eigenfunctions. That is

$$v(r, \theta) = \sum_{n=1}^{\infty} e_n v_n(r, \theta) = \sum_{n=1}^{\infty} r^n (A_n \cos n\theta + B_n \sin n\theta)$$

And now for the $n = 0$ case

The *n* = 0 case

60

When $n = 0$ then $k = 0$ and equation (1) becomes

$$r^2 \frac{d^2 R(r)}{dr^2} + r \frac{dR(r)}{dr} = 0$$

and if we let $S(r) = \dfrac{dR(r)}{dr}$ then this equation becomes

$$r^2 \frac{dS(r)}{dr} + rS(r) = 0, \text{ that is } r\left[r\frac{dS(r)}{dr} + S(r)\right] = 0 \text{ and so}$$

$$r\frac{dS(r)}{dr} + S(r) = \frac{d[rS(r)]}{dr} = 0$$

This has the solution

$$rS(r) = \alpha \text{ (constant) and so } S(r) = \frac{dR(r)}{dr} = \frac{\alpha}{r}$$

$$\text{giving } R(r) = \alpha \ln r + \beta \tag{5}$$

When $n = 0$ then $k = 0$ and equation (2) becomes

$$\frac{d^2 \Theta(\theta)}{d\theta^2} = 0 \text{ with solution } \Theta(\theta) = \gamma\theta + \delta \tag{6}$$

Applying the boundary conditions to the solutions (5) and (6) gives

$$\alpha = \ldots\ldots\ldots\ldots \text{ and } \gamma = \ldots\ldots\ldots\ldots$$

61

$$\boxed{\alpha = 0 \text{ and } \gamma = 0}$$

Because

(a) $v(r, \theta)$ is finite for $0 \le r \le a$, in particular when $r = 0$, and so $\alpha = 0$

(b) $v(r, \theta + 2\pi) = v(r, \theta)$. That is, though θ can take any finite value, the value of $v(r, \theta)$ repeats itself as θ winds round every 2π and this means that $\gamma = 0$.

So, when $n = 0$ the solution is $v_0(r, \theta) = $ constant. We therefore write the complete solution as

$$v(r, \theta) = \frac{A_0}{2} + \sum_{n=1}^{\infty} r^n (A_n \cos n\theta + B_n \sin n\theta)$$

where the constant is taken to be in the form $\dfrac{A_0}{2}$.

Applying the condition on the boundary where $v(a, \theta) = f(\theta)$ we see that

$$f(\theta) = \frac{A_0}{2} + \sum_{n=1}^{\infty} a^n (A_n \cos n\theta + B_n \sin n\theta)$$

which is a Fourier series and hence the form of the constant term being taken as $\dfrac{A_0}{2}$.

The Fourier coefficients are then

$$A_n = \frac{1}{2\pi a^n} \int_0^{2\pi} f(\theta) \cos n\theta \, d\theta \quad \text{and} \quad B_n = \frac{1}{2\pi a^n} \int_0^{2\pi} f(\theta) \sin n\theta \, d\theta$$

Example

Solve Laplace's equation

$$\frac{\partial^2 v(r, \theta)}{\partial r^2} + \frac{1}{r} \frac{\partial v(r, \theta)}{\partial r} + \frac{1}{r^2} \frac{\partial^2 v(r, \theta)}{\partial \theta^2} = 0$$

in the circular region $x^2 + y^2 = a^2$ of the plane where

1 $v(r, \theta)$ is finite for $0 \le r \le a$ and for all θ
2 $v(a, \theta) = \sin \theta$
3 $v(r, \theta + 2\pi) = v(r, \theta)$ for $0 \le r \le a$.

The solution, as we have seen, is

$$v(r, \theta) = \frac{A_0}{2} + \sum_{n=1}^{\infty} r^n (A_n \cos n\theta + B_n \sin n\theta) \quad \text{where}$$

$$A_n = \dots\dots\dots \quad \text{and} \quad B_n = \dots\dots\dots$$

62

$$A_n = 0 \quad \text{and} \quad B_n = \frac{1}{2a^n}\delta_{1,n}$$

Because

$$A_n = \frac{1}{2\pi a^n}\int_0^{2\pi} f(\theta)\cos n\theta\,d\theta = \frac{1}{2\pi a^n}\int_0^{2\pi}\sin\theta\cos n\theta\,d\theta = 0 \quad \text{and}$$

$$B_n = \frac{1}{2\pi a^n}\int_0^{2\pi} f(\theta)\sin n\theta\,d\theta = \frac{1}{2\pi a^n}\int_0^{2\pi}\sin\theta\sin n\theta\,d\theta = \frac{1}{2\pi a^n}\pi\delta_{1,n}$$

where $\delta_{1,n}$ is the Kronecker delta

That is, $B_1 = \dfrac{1}{2a}$, $B_n = 0$ for $n = 2, 3, \ldots$. The complete solution is then

$$v(r,\,\theta) = \frac{r}{a}\sin\theta$$

Notice that all three conditions in Frame 61 are satisfied by this solution, that is

1   $v(r,\,\theta) = \dfrac{r}{a}\sin\theta$ is finite for $0 \le r \le a$ and for all θ

2   $v(a,\,\theta) = \dfrac{a}{a}\sin\theta = \sin\theta$

3   $v(r,\,\theta + 2\pi) = \dfrac{r}{a}\sin(\theta + 2\pi) = \dfrac{r}{a}\sin\theta = v(r,\,\theta)$ for $0 \le r \le a$.

That covers the main steps in the method of solving linear, second-order partial differential equations applied specifically to the wave equation, the heat conduction equation and Laplace's equation. The same approach can be made with other similar equations.

The **Revision summary** and the **Can you?** checklist now follow, then the **Test exercise** with problems like those we have considered. Although the solutions take rather more steps than with other forms of equations, the method is straightforward and follows a clear pattern. The **Further problems** give additional practice.

Revision summary 15

63

1 *Ordinary second-order linear differential equations*

  (a) Equation of the form $a\dfrac{d^2y}{dx^2} + b\dfrac{dy}{dx} + cy = 0$

  Auxiliary equation $am^2 + bm + c = 0$

  (1) Real and different roots: $m = m_1$ and $m = m_2$

     $y = Ae^{m_1 x} + Be^{m_2 x}$

  (2) Real and equal roots: $m = m_1$ (twice)

     $y = e^{m_1 x}(A + Bx)$

  (3) Complex roots: $m = \alpha \pm j\beta$

     $y = e^{\alpha x}\{A \cos \beta x + B \sin \beta x\}.$

  (b) Equations of the form $\dfrac{d^2y}{dx^2} \pm n^2 y = 0$

  (1) $\dfrac{d^2y}{dx^2} + n^2 y = 0;$ $y = A \cos nx + B \sin nx$

  (2) $\dfrac{d^2y}{dx^2} - n^2 y = 0;$ $y = A \cosh nx + B \sinh nx$

         or $y = Ae^{nx} + Be^{-nx}$

         or $y = A \sinh n(x + \phi).$

2 *Partial differential equations*   Solution $u = f(x, y, t, \ldots)$
  Linear equations: If $u = u_1$, $u = u_2$, $u = u_3$, $\ldots$ are solutions, so also is

  $$u = u_1 + u_2 + u_3 + \ldots + u_r + \ldots = \sum_{r=1}^{\infty} u_r.$$

  (a) *Wave equation* – transverse vibrations of an elastic string

     $$\frac{\partial^2 u}{\partial x^2} = \frac{1}{c^2} \cdot \frac{\partial^2 u}{\partial t^2}$$

  (b) *Heat conduction equation* – heat flow in uniform finite bar

     $$\frac{\partial^2 u}{\partial x^2} = \frac{1}{c^2} \cdot \frac{\partial u}{\partial t} \qquad \text{where } c^2 = \frac{k}{\sigma \rho}$$

     k = thermal conductivity of material
     σ = specific heat of the material
     ρ = mass per unit length of bar.

  (c) *Laplace equation* – distribution of a field over a plane area

     $$\frac{\partial^2 u}{\partial x^2} + \frac{\partial^2 u}{\partial y^2} = 0.$$

▶

3 *Separating the variables*
   Let $u(x, y) = X(x)Y(y)$ where $X(x)$ is a function of x only
   and $Y(y)$ is a function of y only.

   $$\text{Then} \quad \frac{\partial u}{\partial x} = X'Y; \qquad \frac{\partial^2 u}{\partial x^2} = X''Y$$

   $$\frac{\partial u}{\partial y} = XY'; \qquad \frac{\partial^2 u}{\partial y^2} = XY''$$

   Substitute in the given partial differential equation and form separate differential equations to give $X(x)$ and $Y(y)$ by introducing a common constant $(-p^2)$. Determine arbitrary functions by use of the initial and boundary conditions.

4 *Laplace's equation in plane polar coordinates*

   $$\frac{\partial^2 v(r, \theta)}{\partial r^2} + \frac{1}{r} \frac{\partial v(r, \theta)}{\partial r} + \frac{1}{r^2} \frac{\partial^2 v(r, \theta)}{\partial \theta^2} = 0$$

   Separating the variables by $v(r, \theta) = R(r)\Theta(\theta)$ produces two uncoupled, second-order ordinary differential equations

   $$r^2 \frac{d^2 R(r)}{dr^2} + r \frac{dR(r)}{dr} = kR(r)$$

   $$\text{and} \quad \frac{d^2 \Theta(\theta)}{d\theta^2} = -k\Theta(\theta)$$

   These two ordinary differential equations can then be solved under the application of appropriate boundary conditions.

 # Can you?

Checklist 15

64

Check this list before and after you try the end of Programme test.

On a scale of 1 to 5 how confident are you that you can: | **Frames**

• Summarise the introductory methods of solving ordinary differential equations?

Yes ☐ ☐ ☐ ☐ ☐ *No* | 1

• Solve partial differential equations that are amenable to solution by direct integration?

Yes ☐ ☐ ☐ ☐ ☐ *No* | 2 to 7

• Apply initial and boundary conditions?

Yes ☐ ☐ ☐ ☐ ☐ *No* | 5 to 7

▶

- Solve the one-dimensional wave and heat equations by separating the variables and obtaining eigenfunctions and corresponding eigenvalues?

  Yes ☐ ☐ ☐ ☐ ☐ No   8 to 40

- Solve the two-dimensional Laplace equation in Cartesian coordinates?

  Yes ☐ ☐ ☐ ☐ ☐ No   41 to 52

- Recognise the need for alternative coordinate systems and solve the two-dimensional Laplace equation in plane polar coordinates?

  Yes ☐ ☐ ☐ ☐ ☐ No   53 to 62

 Test exercise 15

---

65   1   Solve the following equations

(a) $\dfrac{\partial^2 u}{\partial x^2} = 24x^2(t - 2)$, given that at $x = 0$, $u = e^{2t}$ and $\dfrac{\partial u}{\partial x} = 4t$.

(b) $\dfrac{\partial^2 u}{\partial x \partial y} = 4e^y \cos 2x$, given that at $y = 0$, $\dfrac{\partial u}{\partial x} = \cos x$

   and at $x = \pi$, $u = y^2$.

2   A perfectly elastic string is stretched between two points 10 cm apart. Its centre point is displaced 2 cm from its position of rest at right angles to the original direction of the string and then released with zero velocity. Applying the equation $\dfrac{\partial^2 u}{\partial x^2} = \dfrac{1}{c^2} \cdot \dfrac{\partial^2 u}{\partial t^2}$ with $c^2 = 1$, determine the subsequent motion $u(x, t)$.

3   One end A of an insulated metal bar AB of length 2 m is kept at 0°C while the other end B is maintained at 50°C until a steady state of temperature along the bar is achieved. At $t = 0$, the end B is suddenly reduced to 0°C and kept at that temperature. Using the heat conduction equation $\dfrac{\partial^2 u}{\partial x^2} = \dfrac{1}{c^2} \cdot \dfrac{\partial u}{\partial t}$, determine an expression for the temperature at any point in the bar distance x from A at any time t.

4   A square plate is bounded by the lines $x = 0$, $y = 0$, $x = 2$, $y = 2$. Apply the Laplace equation $\dfrac{\partial^2 u}{\partial x^2} + \dfrac{\partial^2 u}{\partial y^2} = 0$ to determine the potential distribution $u(x, y)$ over the plate, subject to the following boundary conditions.

   $u = 0$ when $x = 0$ $0 \le y \le 2$
   $u = 0$ when $x = 2$ $0 \le y \le 2$
   $u = 0$ when $y = 0$ $0 \le x \le 2$
   $u = 5$ when $y = 2$ $0 \le x \le 2$.

▶

5   Solve Laplace's equation in plane polar coordinates

$$\frac{\partial^2 v(r,\theta)}{\partial r^2} + \frac{1}{r}\frac{\partial v(r,\theta)}{\partial r} + \frac{1}{r^2}\frac{\partial^2 v(r,\theta)}{\partial \theta^2} = 0$$

in the circular region $x^2 + y^2 = 1$ of the plane where

(1) $v(r,\theta)$ is finite for $0 \le r \le 1$ and for all θ
(2) $v(1,\theta) = 5\cos 3\theta$
(3) $v(r,\theta+2\pi) = v(r,\theta)$ for $0 \le r \le 1$.

 Further problems 15

1   Show that the equation $\dfrac{\partial^2 u}{\partial x^2} - \dfrac{1}{c^2}\cdot\dfrac{\partial^2 u}{\partial t^2} = 0$ is satisfied by   **66**

$u = f(x+ct) + F(x-ct)$ where f and F are arbitrary functions.

2   If $\dfrac{\partial^2 u}{\partial x^2} = \dfrac{1}{c^2}\cdot\dfrac{\partial^2 u}{\partial t^2}$ and $c = 3$, determine the solution $u = f(x,t)$ subject to the
boundary conditions

$u(0,t) = 0$ and $u(2,t) = 0$ for $t \ge 0$

$u(x,0) = x(2-x)$ and $\left[\dfrac{\partial u}{\partial t}\right]_{t=0} = 0 \qquad 0 \le x \le 2.$

3   The centre point of a perfectly elastic string stretched between two points A
and B, 4 m apart, is deflected a distance 0·01 m from its position of rest
perpendicular to AB and released initially with zero velocity. Apply the wave
equation $\dfrac{\partial^2 u}{\partial x^2} = \dfrac{1}{c^2}\cdot\dfrac{\partial^2 u}{\partial t^2}$ where $c = 10$ to determine the subsequent motion of a
point P distant x from A at time t.

4   An elastic string is stretched between two points 10 cm apart. A point P on the
string 2 cm from the left-hand end, i.e. the origin, is drawn aside 1 cm from its
position of rest and released with zero velocity. Solve the one-dimensional
wave equation to determine the displacement of any point at any instant.

5   An insulated uniform metal bar, 10 units long, has the temperature of its ends
maintained at 0°C and at $t = 0$ the temperature distribution $f(x)$ along the bar
is defined by $f(x) = x(10-x)$. Solve the heat conduction equation $\dfrac{\partial^2 u}{\partial x^2} = \dfrac{1}{c^2}\cdot\dfrac{\partial u}{\partial t}$
with $c^2 = 4$ to determine the temperature u of any point in the bar at time t.

6   The ends of an insulated rod AB, 10 units long, are maintained at 0°C. At $t = 0$,
the temperature within the rod rises uniformly from each end reaching 2°C at
the mid-point of AB. Determine an expression for the temperature $u(x,t)$ at any
point in the rod, distant x from the left-hand end at any subsequent time t.

▶

7   A rectangular plate OPQR is bounded by the lines $x = 0$, $y = 0$, $x = 4$, $y = 2$. Determine the potential distribution $u(x, y)$ over the rectangle using the Laplace equation $\dfrac{\partial^2 u}{\partial x^2} + \dfrac{\partial^2 u}{\partial y^2} = 0$, subject to the following boundary conditions

$u(0, y) = 0$ $0 \leq y \leq 2$
$u(4, y) = 0$ $0 \leq y \leq 2$
$u(x, 2) = 0$ $0 \leq x \leq 4$
$u(x, 0) = x(4 - x)$ $0 \leq x \leq 4.$

8   Two sides AB and AD of a rectangular plate ABCD lie along the x and y axes respectively. The remaining two sides are the lines $x = 5$ and $y = 2$. The sides BC, CD and DA are maintained at zero temperature. The temperature distribution along AB is defined by $f(x) = x(x - 5)$. Determine an expression for the steady-state temperature at any point in the plate.

9   Solve Laplace's equation in plane polar coordinates
$$\frac{\partial^2 v(r, \theta)}{\partial r^2} + \frac{1}{r}\frac{\partial v(r, \theta)}{\partial r} + \frac{1}{r^2}\frac{\partial^2 v(r, \theta)}{\partial \theta^2} = 0$$
in the circular region $x^2 + y^2 = 1$ of the plane where
   (1) $v(r, \theta)$ is finite for $0 \leq r \leq 1$ and for all θ
   (2) $v(1, \theta) = \sin 2\theta - 4\cos\theta$
   (3) $v(r, \theta + 2\pi) = v(r, \theta)$ for $0 \leq r \leq 1$.

10   Solve Laplace's equation in plane polar coordinates
$$\frac{\partial^2 v(r, \theta)}{\partial r^2} + \frac{1}{r}\frac{\partial v(r, \theta)}{\partial r} + \frac{1}{r^2}\frac{\partial^2 v(r, \theta)}{\partial \theta^2} = 0$$
in the circular region $x^2 + y^2 = 1$ of the plane where
   (1) $v(r, \theta)$ is finite for $0 \leq r \leq 1$ and for all θ
   (2) $v(1, \theta) = 3\sin^2\theta$
   (3) $v(r, \theta + 2\pi) = v(r, \theta)$ for $0 \leq r \leq 1$.

Matrix algebra

Learning outcomes

When you have completed this Programme you will be able to:

- Determine whether a matrix is singular or non-singular
- Determine the rank of a matrix
- Determine the consistency of a set of linear equations and hence demonstrate the uniqueness of their solution
- Obtain the solution of a set of simultaneous linear equations by using matrix inversion, by row transformation, by Gaussian elimination, by triangular decomposition and by using an electronic spreadsheet
- Use matrices to represent transformations between coordinate systems

Prerequisite: Engineering Mathematics (Sixth Edition)
Programmes 4 Determinants and 5 Matrices

Singular and non-singular matrices

1   Every square matrix **A** has associated with it a number called the determinant of **A** and denoted by |**A**|. If |**A**| $\neq$ 0 then **A** is called a *non-singular* matrix. Otherwise if |**A**| = 0, then **A** is called a *singular* matrix.

Example 1

Is $\mathbf{A} = \begin{pmatrix} 1 & 2 & 8 \\ 4 & 7 & 6 \\ 9 & 5 & 3 \end{pmatrix}$ singular or non-singular?

$$|\mathbf{A}| = \begin{vmatrix} 1 & 2 & 8 \\ 4 & 7 & 6 \\ 9 & 5 & 3 \end{vmatrix}$$

$$= 1 \begin{vmatrix} 7 & 6 \\ 5 & 3 \end{vmatrix} - 2 \begin{vmatrix} 4 & 6 \\ 9 & 3 \end{vmatrix} + 8 \begin{vmatrix} 4 & 7 \\ 9 & 5 \end{vmatrix}$$

$$= (21 - 30) - 2(12 - 54) + 8(20 - 63)$$

$$= -9 + 84 - 344$$

$$= -269$$

Because |**A**| $\neq$ 0 then **A** is non-singular.

Example 2

Is $\mathbf{A} = \begin{pmatrix} 3 & 9 & 2 \\ 1 & 5 & 6 \\ 2 & 7 & 4 \end{pmatrix}$ singular or non-singular?

A is . . . .

2
$$\boxed{\text{singular}}$$

Because

$$|\mathbf{A}| = \begin{vmatrix} 3 & 9 & 2 \\ 1 & 5 & 6 \\ 2 & 7 & 4 \end{vmatrix}$$

$$= 3(20 - 42) - 9(4 - 12) + 2(7 - 10)$$

$$= -66 + 72 - 6$$

$$= 0$$

Because |**A**| = 0 then |**A**| is singular.

▶

Exercise

Determine whether each of the following is singular or non-singular.

1 $|\mathbf{A}| = \begin{pmatrix} 4 & 5 \\ 2 & 3 \end{pmatrix}$ 2 $|\mathbf{B}| = \begin{pmatrix} 3 & -4 \\ -6 & 8 \end{pmatrix}$

3 $|\mathbf{C}| = \begin{pmatrix} 4 & 1 & -2 \\ 1 & 7 & 3 \\ 5 & 8 & 1 \end{pmatrix}$ 4 $|\mathbf{D}| = \begin{pmatrix} 3 & 2 & 4 \\ 5 & 1 & 6 \\ 2 & 0 & 3 \end{pmatrix}$

---

1 non-singular	2 singular
3 singular	4 non-singular

3

Because

Straightforward evaluation of the relevant determinants gives

1 $|\mathbf{A}| = 2$ 2 $|\mathbf{B}| = 0$

3 $|\mathbf{C}| = 0$ 4 $|\mathbf{D}| = -5$

Closely related to the notion of the singularity or otherwise of a square matrix is the notion of **rank** of a general $n \times m$ matrix.

Rank of a matrix

The rank of an $n \times m$ matrix $\mathbf{A}$ is the order of the largest square, non-singular sub-matrix. That is, the largest square sub-matrix whose determinant is non-zero. If $n = m$, so making $\mathbf{A}$ itself square, then this sub-matrix could be the matrix $\mathbf{A}$ itself.

Example

To find the rank of the matrix $\mathbf{A} = \begin{pmatrix} 3 & 4 & 5 \\ 1 & 2 & 3 \\ 4 & 5 & 6 \end{pmatrix}$ we note that

$$|\mathbf{A}| = \begin{vmatrix} 3 & 4 & 5 \\ 1 & 2 & 3 \\ 4 & 5 & 6 \end{vmatrix} = \ldots\ldots\ldots\ldots$$

4

$$\boxed{0}$$

Because

$$|A| = \begin{vmatrix} 3 & 4 & 5 \\ 1 & 2 & 3 \\ 4 & 5 & 6 \end{vmatrix}$$

$$= 3(12 - 15) - 4(6 - 12) + 5(5 - 8)$$
$$= -9 + 24 - 15 = 0$$

Therefore we can say that the rank of **A** is ....

5

$$\boxed{\text{not } 3}$$

Because

$|A| = 0$ and therefore **A** is singular.

Now try a sub-matrix of order 2.

$$\begin{vmatrix} 3 & 4 \\ 1 & 2 \end{vmatrix} = 6 - 4 = -2 \neq 0. \text{ Therefore the rank of } A \text{ is } \ldots\ldots\ldots$$

6

$$\boxed{2}$$

Because

The largest square, non-singular sub-matrix of **A** has order 2 therefore **A** has rank 2.

This method of finding the rank of a matrix can be a very hit and miss affair and a better, more systematic method is to use **elementary operations** and the notion of an **equivalent matrix**.

Next frame

Elementary operations and equivalent matrices

7

Each of the following row operations on matrix **A** produces a *row equivalent matrix* **B**, where the order and rank of **B** is the same as that of **A**. We write **A** ~ **B**.

1 Interchanging two rows
2 Multiplying each element of a row by the same non-zero scalar quantity
3 Adding or subtracting corresponding elements from those of another row

are operations called *elementary row operations*. There is a corresponding set of three *elementary column operations* that can be used to form *column equivalent matrices*.

▶

Example 1

Given $\mathbf{A} = \begin{pmatrix} 3 & 4 & 5 \\ 1 & 2 & 3 \\ 4 & 5 & 6 \end{pmatrix}$ then

$\begin{pmatrix} 3 & 4 & 5 \\ 1 & 2 & 3 \\ 4 & 5 & 6 \end{pmatrix} \sim \begin{pmatrix} 0 & -2 & -4 \\ 1 & 2 & 3 \\ 4 & 5 & 6 \end{pmatrix}$ by subtracting 3 times each element of row 2 from row 1

$\sim \begin{pmatrix} 0 & -2 & -4 \\ 1 & 2 & 3 \\ 0 & -3 & -6 \end{pmatrix}$ by subtracting 4 times each element of row 2 from row 3

$\sim \begin{pmatrix} 0 & -3 & -6 \\ 1 & 2 & 3 \\ 0 & -3 & -6 \end{pmatrix}$ by multiplying each element of row 1 by 3/2

$\sim \begin{pmatrix} 0 & -3 & -6 \\ 1 & 2 & 3 \\ 0 & 0 & 0 \end{pmatrix}$ by subtracting corresponding elements of row 1 from row 3

$= \mathbf{B}$

The row of zeros in matrix **B** means that its determinant is zero and so its rank is not 3. The largest sub-matrix with non-zero determinant has order 2 and so the rank of **B** is 2. Because matrix **B** is row equivalent to matrix **A** we can say that the rank of **A** is also 2.

Example 2

Determine the rank of $\mathbf{A} = \begin{pmatrix} 1 & 2 & 8 \\ 4 & 7 & 6 \\ 9 & 5 & 3 \end{pmatrix}$

By taking 4 times the elements of row 1 from row 2 we obtain the equivalent matrix . . . .

$$\begin{pmatrix} 1 & 2 & 8 \\ 0 & -1 & -26 \\ 9 & 5 & 3 \end{pmatrix}$$

8

By taking 9 times the elements of row 1 from row 3 we obtain the equivalent matrix . . . .

$$\begin{pmatrix} 1 & 2 & 8 \\ 0 & -1 & -26 \\ 0 & -13 & -69 \end{pmatrix}$$

9

By multiplying the elements of row 2 by -13 we obtain the equivalent matrix
. . . .

10

$$\begin{pmatrix} 1 & 2 & 8 \\ 0 & 13 & 338 \\ 0 & -13 & -69 \end{pmatrix}$$

By adding corresponding elements of row 2 to row 3 we obtain the equivalent matrix . . . .

11

$$\begin{pmatrix} 1 & 2 & 8 \\ 0 & 13 & 269 \\ 0 & 0 & -69 \end{pmatrix}$$

Because all the elements below the main diagonal of this matrix are zero we call the matrix an *upper triangular matrix*. By inspection we can see that the determinant of this triangular matrix is non-zero, being the product of its three diagonal elements $1 \times 13 \times (-69) = -897$. Therefore its rank is 3 and so the rank of matrix **A** is also 3.

Try another one for yourself.

Example 3

The rank of $\mathbf{A} = \begin{pmatrix} 3 & 9 & 2 \\ 1 & 5 & 6 \\ 2 & 7 & 4 \end{pmatrix}$ is . . . .

12

$$\boxed{2}$$

Because

$$\mathbf{A} = \begin{pmatrix} 3 & 9 & 2 \\ 1 & 5 & 6 \\ 2 & 7 & 4 \end{pmatrix} \sim \begin{pmatrix} 0 & -6 & -16 \\ 1 & 5 & 6 \\ 2 & 7 & 4 \end{pmatrix}$$ Subtracting 3 times row 2 from row 1

$$\sim \begin{pmatrix} 0 & -6 & -16 \\ 1 & 5 & 6 \\ 0 & -3 & -8 \end{pmatrix}$$ Subtracting 2 times row 2 from row 3

$$\sim \begin{pmatrix} 0 & 3 & 8 \\ 1 & 5 & 6 \\ 0 & -3 & -8 \end{pmatrix}$$ Multiplying row 1 by $-1/2$

$$\sim \begin{pmatrix} 0 & 3 & 8 \\ 1 & 5 & 6 \\ 0 & 0 & 0 \end{pmatrix}$$ Adding row 1 to row 3

$$\sim \begin{pmatrix} 1 & 5 & 6 \\ 0 & 3 & 8 \\ 0 & 0 & 0 \end{pmatrix}$$ Interchanging rows 1 and 2

▶

and $\begin{vmatrix} 1 & 5 & 6 \\ 0 & 3 & 8 \\ 0 & 0 & 0 \end{vmatrix} = 0$. So the rank of this matrix is not 3. The largest

square sub-matrix of this matrix with non-zero determinant is, by inspection, of order 2 and so the rank of this matrix, and hence the rank of the equivalent matrix **A** is 2.

Finally try a non-square matrix.

Example 4

The rank of $\mathbf{A} = \begin{pmatrix} 2 & 2 & 3 & 1 \\ 0 & 8 & 2 & 4 \\ 1 & 7 & 3 & 2 \end{pmatrix}$ is ....

$$\boxed{3}$$

13

Because

$$\mathbf{A} = \begin{pmatrix} 2 & 2 & 3 & 1 \\ 0 & 8 & 2 & 4 \\ 1 & 7 & 3 & 2 \end{pmatrix} \sim \begin{pmatrix} 0 & -12 & -3 & -3 \\ 0 & 8 & 2 & 4 \\ 1 & 7 & 3 & 2 \end{pmatrix} \quad \text{Subtracting 2 times row 3 from row 1}$$

$$\sim \begin{pmatrix} 0 & -8 & -2 & -2 \\ 0 & 8 & 2 & 4 \\ 1 & 7 & 3 & 2 \end{pmatrix} \quad \text{Multiplying row 1 by 2/3}$$

$$\sim \begin{pmatrix} 0 & 0 & 0 & 2 \\ 0 & 8 & 2 & 4 \\ 1 & 7 & 3 & 2 \end{pmatrix} \quad \text{Adding row 2 to row 1}$$

It is possible to find a 3×3 sub-matrix of this matrix that has non-zero determinant, namely

$$\begin{pmatrix} 0 & 0 & 2 \\ 8 & 2 & 4 \\ 7 & 3 & 2 \end{pmatrix} \text{ where } \begin{vmatrix} 0 & 0 & 2 \\ 8 & 2 & 4 \\ 7 & 3 & 2 \end{vmatrix} = 2(24 - 14) = 20.$$

Consequently, this matrix and hence matrix **A** has rank 3.

Consistency of a set of equations

14

In solving sets of simultaneous equations, we can express the equations in matrix form. For example

$$a_{11}x_1 + a_{12}x_2 + a_{13}x_3 = b_1$$
$$a_{21}x_1 + a_{22}x_2 + a_{23}x_3 = b_2$$
$$a_{31}x_1 + a_{32}x_2 + a_{33}x_3 = b_3$$

can be written in the form

$$\begin{pmatrix} a_{11} & a_{12} & a_{13} \\ a_{21} & a_{22} & a_{23} \\ a_{31} & a_{32} & a_{33} \end{pmatrix} \begin{pmatrix} x_1 \\ x_2 \\ x_3 \end{pmatrix} = \begin{pmatrix} b_1 \\ b_2 \\ b_3 \end{pmatrix}$$

i.e. $\qquad \mathbf{Ax} = \mathbf{b}$

The set of three equations is said to be *consistent* if solutions for x_1, x_2, x_3 exist and *inconsistent* if no such solutions can be found.

In practice, we can solve the equations by operating on the *augmented coefficient matrix*, i.e. we write the constant terms as a fourth column of the coefficient matrix to form $\mathbf{A_b}$.

$$\mathbf{A_b} = \begin{pmatrix} a_{11} & a_{12} & a_{13} & b_1 \\ a_{21} & a_{22} & a_{23} & b_2 \\ a_{31} & a_{32} & a_{33} & b_3 \end{pmatrix}$$

which, of course, is a (3×4) matrix.

The general test for consistency is then:

A set of n simultaneous equations in n unknowns is consistent if the rank of the coefficient matrix $\mathbf{A}$ is equal to the rank of the augmented matrix $\mathbf{A_b}$.

If the rank of $\mathbf{A}$ is less than the rank of $\mathbf{A_b}$, then the equations are inconsistent and have no solution.

Make a note of this test. It can save time in working

15

Example

If $\begin{pmatrix} 1 & 3 \\ 2 & 6 \end{pmatrix} \begin{pmatrix} x_1 \\ x_2 \end{pmatrix} = \begin{pmatrix} 4 \\ 5 \end{pmatrix}$ then

$\mathbf{A} = \begin{pmatrix} 1 & 3 \\ 2 & 6 \end{pmatrix}$ and $\mathbf{A_b} = \begin{pmatrix} 1 & 3 & 4 \\ 2 & 6 & 5 \end{pmatrix}$

Rank of $\mathbf{A}$: $\quad \begin{vmatrix} 1 & 3 \\ 2 & 6 \end{vmatrix} = 6 - 6 = 0 \qquad \therefore$ rank of $\mathbf{A} = 1$

Rank of $\mathbf{A_b}$: $\quad \begin{vmatrix} 1 & 3 \\ 2 & 6 \end{vmatrix} = 0$ as before

but $\qquad \begin{vmatrix} 3 & 4 \\ 6 & 5 \end{vmatrix} = 15 - 24 = -9 \quad \therefore$ rank of $\mathbf{A_b} = 2$

In this case, rank of $\mathbf{A} <$ rank of $\mathbf{A_b}$ $\quad \therefore \quad$ . . . .

$$\boxed{\text{no solution exists}}$$ **16**

Remember that, for consistency,

$$\text{rank of } \mathbf{A} = \ldots\ldots\ldots$$

$$\boxed{\text{rank of } \mathbf{A}_b}$$ **17**

Uniqueness of solutions

1 With a set of n equations in n unknowns, the equations are consistent if the coefficient matrix $\mathbf{A}$ and the augmented matrix $\mathbf{A}_b$ are each of rank n. There is then a *unique* solution for the n equations.

   Note that if the rank of $\mathbf{A} = n$ then $\mathbf{A}$ is a non-singular sub-matrix of $\mathbf{A}_b$ and so the rank of $\mathbf{A}_b = n$ also. Therefore there is no need to test for the rank of $\mathbf{A}_b$ in this case.

2 If the rank of $\mathbf{A}$ and that of $\mathbf{A}_b$ is m, where $m < n$, then the matrix $\mathbf{A}$ is singular, i.e. $|\mathbf{A}| = 0$, and there will be an *infinite number* of solutions for the equations.

3 As we have already seen, if the rank of $\mathbf{A} <$ the rank of $\mathbf{A}_b$, then *no solution* exists.

Copy these up in your record book; they are important

Writing the results in a slightly different way: **18**

   With a set of n equations in n unknowns, checking the rank of the coefficient matrix $\mathbf{A}$ and that of the augmented matrix $\mathbf{A}_b$ enables us to see whether

(a) a unique solution exists

$$\text{rank } \mathbf{A} = \text{rank } \mathbf{A}_b = n$$

(b) an infinite number of solutions exist

$$\text{rank } \mathbf{A} = \text{rank } \mathbf{A}_b = m < n$$

(c) no solution exists

$$\text{rank } \mathbf{A} < \text{rank } \mathbf{A}_b$$

Example

$$\begin{pmatrix} -4 & 5 \\ -8 & 10 \end{pmatrix}\begin{pmatrix} x_1 \\ x_2 \end{pmatrix} = \begin{pmatrix} -3 \\ -6 \end{pmatrix}$$

Finding the rank of $\mathbf{A}$ and of $\mathbf{A}_b$ leads us to the conclusion that

$$\ldots\ldots\ldots$$

19

there is an infinite
number of solutions

Because

$$\mathbf{A} = \begin{pmatrix} -4 & 5 \\ -8 & 10 \end{pmatrix} \text{ and } \mathbf{A_b} = \begin{pmatrix} -4 & 5 & -3 \\ -8 & 10 & -6 \end{pmatrix}$$

Rank of $\mathbf{A}$: $\begin{vmatrix} -4 & 5 \\ -8 & 10 \end{vmatrix} = -40 + 40 = 0$ ∴ Rank of $\mathbf{A} = 1$

Rank of $\mathbf{A_b}$: $\begin{vmatrix} -4 & 5 \\ -8 & 10 \end{vmatrix} = 0;$ $\begin{vmatrix} 5 & -3 \\ 10 & -6 \end{vmatrix} = 0;$ $\begin{vmatrix} -4 & -3 \\ -8 & -6 \end{vmatrix} = 0$

∴ Rank of $\mathbf{A_b} = 1$

∴ Rank of $\mathbf{A} = $ rank of $\mathbf{A_b} = 1$

But there are two equations in two unknowns, i.e. $n = 2$

∴ Rank of $\mathbf{A} = $ rank of $\mathbf{A_b} = 1 < n$

∴ Infinite number of solutions.

The solutions can be written as x_1 arbitrary and $x_2 = \dfrac{4x_1 - 3}{5}$.

You will recall that, for a unique solution of n equations in n unknowns

. . . .

20

rank $\mathbf{A} = $ rank $\mathbf{A_b} = n$

Now for some examples for you to try. In each of the following cases, apply
the rank tests to determine the nature of the solutions. Do not solve the sets of
equations.

Example 1

$$\begin{pmatrix} 1 & 2 & -1 \\ 3 & 4 & 2 \\ 1 & 4 & 3 \end{pmatrix} \begin{pmatrix} x_1 \\ x_2 \\ x_3 \end{pmatrix} = \begin{pmatrix} 1 \\ -2 \\ 3 \end{pmatrix}$$

$$\mathbf{A} = \begin{pmatrix} 1 & 2 & -1 \\ 3 & 4 & 2 \\ 1 & 4 & 3 \end{pmatrix} \text{ and } \mathbf{A_b} = \begin{pmatrix} 1 & 2 & -1 & 1 \\ 3 & 4 & 2 & -2 \\ 1 & 4 & 3 & 3 \end{pmatrix}$$

Finish it off and we find that . . . .

$$\boxed{\text{a unique solution exists}}$$

Because

$$n = 3; \quad \text{rank of } \mathbf{A} = 3; \quad \text{rank of } \mathbf{A}_b = 3.$$
$$\therefore \quad \text{rank of } \mathbf{A} = \text{rank of } \mathbf{A}_b = 3 = n \quad \therefore \quad \text{Solution unique}$$

And this one.

Example 2

$$\begin{pmatrix} 2 & -1 & 7 \\ 4 & 2 & 2 \\ 3 & 1 & 3 \end{pmatrix} \begin{pmatrix} x_1 \\ x_2 \\ x_3 \end{pmatrix} = \begin{pmatrix} 2 \\ 5 \\ 1 \end{pmatrix}$$

This time we find that . . . .

$$\boxed{\text{no solution is possible}}$$

Because

$$\mathbf{A} = \begin{pmatrix} 2 & -1 & 7 \\ 4 & 2 & 2 \\ 3 & 1 & 3 \end{pmatrix}; \qquad \mathbf{A}_b = \begin{pmatrix} 2 & -1 & 7 & 2 \\ 4 & 2 & 2 & 5 \\ 3 & 1 & 3 & 1 \end{pmatrix}$$

$$n = 3; \qquad \text{rank of } \mathbf{A} = 2; \qquad \text{rank of } \mathbf{A}_b = 3$$

$$\therefore \quad \text{rank of } \mathbf{A} < \text{rank of } \mathbf{A}_b$$

$$\therefore \quad \text{No solution exists}$$

and finally

Example 3

$$\begin{pmatrix} 1 & 2 & -3 \\ 1 & 3 & 4 \\ 2 & 5 & 1 \end{pmatrix} \begin{pmatrix} x_1 \\ x_2 \\ x_3 \end{pmatrix} = \begin{pmatrix} 1 \\ 2 \\ 3 \end{pmatrix}$$

In this case, we find that . . . .

23

$$\boxed{\text{infinite number of solutions possible}}$$

Because

$$A = \begin{pmatrix} 1 & 2 & -3 \\ 1 & 3 & 4 \\ 2 & 5 & 1 \end{pmatrix} \quad \text{and} \quad A_b = \begin{pmatrix} 1 & 2 & -3 & 1 \\ 1 & 3 & 4 & 2 \\ 2 & 5 & 1 & 3 \end{pmatrix}$$

Rank of **A**:

$$A = \begin{pmatrix} 1 & 2 & -3 \\ 1 & 3 & 4 \\ 2 & 5 & 1 \end{pmatrix} \sim \begin{pmatrix} 1 & 2 & -3 \\ 0 & 1 & 7 \\ 2 & 5 & 1 \end{pmatrix} \quad \text{Subtracting row 1 from row 2}$$

$$\sim \begin{pmatrix} 1 & 2 & -3 \\ 0 & 1 & 7 \\ 0 & 1 & 7 \end{pmatrix} \quad \text{Subtracting 2 times row 1 from row 2}$$

$$\sim \begin{pmatrix} 1 & 2 & -3 \\ 0 & 1 & 7 \\ 0 & 0 & 0 \end{pmatrix} \quad \text{Subtracting row 2 from row 3}$$

and so rank of **A** is 2 by inspection.

Rank of **A**$_b$:

$$A_b = \begin{pmatrix} 1 & 2 & -3 & 1 \\ 1 & 3 & 4 & 2 \\ 2 & 5 & 1 & 3 \end{pmatrix} \sim \begin{pmatrix} 1 & 2 & -3 & 1 \\ 0 & 1 & 7 & 1 \\ 2 & 5 & 1 & 3 \end{pmatrix} \quad \text{Subtracting row 1 from row 2}$$

$$\sim \begin{pmatrix} 1 & 2 & -3 & 1 \\ 0 & 1 & 7 & 1 \\ 0 & 1 & 7 & 1 \end{pmatrix} \quad \begin{array}{l} \text{Subtracting 2 times row 1} \\ \text{from row 2} \end{array}$$

$$\sim \begin{pmatrix} 1 & 2 & -3 & 1 \\ 0 & 1 & 7 & 1 \\ 0 & 0 & 0 & 0 \end{pmatrix} \quad \text{Subtracting row 2 from row 3}$$

and so rank of **A**$_b$ is 2 by inspection.

Therefore rank of **A** = rank of **A**$_b$ = 2 < n (that is 3), therefore there is an infinite number of solutions.

Now let us move on to a new section of the work

Solution of sets of equations

1 *Inverse method*

24

Let us work through an example by way of explanation.

Example 1

To solve
$$3x_1 + 2x_2 - x_3 = 4$$
$$2x_1 - x_2 + 2x_3 = 10$$
$$x_1 - 3x_2 - 4x_3 = 5.$$

We first write this in matrix form, which is ....

25

$$\begin{pmatrix} 3 & 2 & -1 \\ 2 & -1 & 2 \\ 1 & -3 & -4 \end{pmatrix} \begin{pmatrix} x_1 \\ x_2 \\ x_3 \end{pmatrix} = \begin{pmatrix} 4 \\ 10 \\ 5 \end{pmatrix}$$

Then if $\mathbf{A} = \begin{pmatrix} 3 & 2 & -1 \\ 2 & -1 & 2 \\ 1 & -3 & -4 \end{pmatrix}$ then $\begin{pmatrix} x_1 \\ x_2 \\ x_3 \end{pmatrix} = \mathbf{A}^{-1} \begin{pmatrix} 4 \\ 10 \\ 5 \end{pmatrix}$

where $\mathbf{A}^{-1}$ is the *inverse* of $\mathbf{A}$.

To find $\mathbf{A}^{-1}$

(a) Form the determinant of $\mathbf{A}$ and evaluate it.

$$|\mathbf{A}| = \begin{vmatrix} 3 & 2 & -1 \\ 2 & -1 & 2 \\ 1 & -3 & -4 \end{vmatrix} = 3(4+6) - 2(-8-2) - 1(-6+1) = 55$$

(b) Form a new matrix $\mathbf{C}$ consisting of the cofactors of the elements in $\mathbf{A}$.

The cofactor of any one element is its minor together with its 'place sign'

i.e. $\mathbf{C} = \begin{pmatrix} A_{11} & A_{12} & A_{13} \\ A_{21} & A_{22} & A_{23} \\ A_{31} & A_{32} & A_{33} \end{pmatrix}$

where A_{11} is the cofactor of a_{11} in $\mathbf{A}$.

$$A_{11} = \begin{vmatrix} -1 & 2 \\ -3 & -4 \end{vmatrix} = 10; \quad A_{12} = -\begin{vmatrix} 2 & 2 \\ 1 & -4 \end{vmatrix} = 10;$$

$$A_{13} = \begin{vmatrix} 2 & -1 \\ 1 & -3 \end{vmatrix} = -5$$

$$A_{21} = -\begin{vmatrix} 2 & -1 \\ -3 & -4 \end{vmatrix} = 11; \quad A_{22} = \begin{vmatrix} 3 & -1 \\ 1 & -4 \end{vmatrix} = -11;$$

$$A_{23} = -\begin{vmatrix} 3 & 2 \\ 1 & -3 \end{vmatrix} = 11$$

$$A_{31} = \ldots\ldots\ldots\ldots; \qquad A_{32} = \ldots\ldots\ldots\ldots; \qquad A_{33} = \ldots\ldots\ldots\ldots$$

26

$$A_{31} = \begin{vmatrix} 2 & -1 \\ -1 & 2 \end{vmatrix} = 3; \quad A_{32} = -\begin{vmatrix} 3 & -1 \\ 2 & 2 \end{vmatrix} = -8; \quad A_{33} = \begin{vmatrix} 3 & 2 \\ 2 & -1 \end{vmatrix} = -7$$

So $\mathbf{C} = \begin{pmatrix} 10 & 10 & -5 \\ 11 & -11 & 11 \\ 3 & -8 & -7 \end{pmatrix}$

We now write the transpose of **C**, i.e. $\mathbf{C}^{\mathrm{T}}$ in which we write rows as columns and columns as rows.

$$\mathbf{C}^{\mathrm{T}} = \ldots\ldots\ldots$$

27

$$\mathbf{C}^{\mathrm{T}} = \begin{pmatrix} 10 & 11 & 3 \\ 10 & -11 & -8 \\ -5 & 11 & -7 \end{pmatrix}$$

This is called the *adjoint* (adj) of the original matrix **A**

i.e. $\operatorname{adj} \mathbf{A} = \mathbf{C}^{\mathrm{T}}$

Then the inverse of **A**, i.e. $\mathbf{A}^{-1}$ is given by

$$\mathbf{A}^{-1} = \frac{1}{|\mathbf{A}|} \times \mathbf{C}^{\mathrm{T}} = \frac{1}{55} \begin{pmatrix} 10 & 11 & 3 \\ 10 & -11 & -8 \\ -5 & 11 & -7 \end{pmatrix}$$

As a check that all the calculations have been done correctly and without error, the product of matrix **A** with its adjoint should be equal to the unit matrix multiplied by the determinant of **A**. That is

$$\mathbf{A} \times \operatorname{adj} \mathbf{A} = \det \mathbf{A} \times \mathbf{I}$$

For this case

$$\mathbf{A} \times \operatorname{adj} \mathbf{A} = \begin{pmatrix} 3 & 2 & -1 \\ 2 & -1 & 2 \\ 1 & -3 & -4 \end{pmatrix} \begin{pmatrix} 10 & 11 & 3 \\ 10 & -11 & -8 \\ -5 & 11 & -7 \end{pmatrix}$$

$$= \begin{pmatrix} 55 & 0 & 0 \\ 0 & 55 & 0 \\ 0 & 0 & 55 \end{pmatrix}$$

$$= \det \mathbf{A} \times \mathbf{I}$$

Thus all is well. We can now continue to find the solution.

So $\begin{pmatrix} x_1 \\ x_2 \\ x_3 \end{pmatrix} = \mathbf{A}^{-1} \begin{pmatrix} 4 \\ 10 \\ 5 \end{pmatrix}$ becomes

$$\begin{pmatrix} x_1 \\ x_2 \\ x_3 \end{pmatrix} = \frac{1}{55} \begin{pmatrix} 10 & 11 & 3 \\ 10 & -11 & -8 \\ -5 & 11 & -7 \end{pmatrix} \begin{pmatrix} 4 \\ 10 \\ 5 \end{pmatrix} = \ldots\ldots\ldots$$

$$\boxed{x_1 = 3; \quad x_2 = -2; \quad x_3 = 1}$$ **28**

Because

$$\begin{pmatrix} x_1 \\ x_2 \\ x_3 \end{pmatrix} = \frac{1}{55} \begin{pmatrix} 10 & 11 & 3 \\ 10 & -11 & -8 \\ -5 & 11 & -7 \end{pmatrix} \begin{pmatrix} 4 \\ 10 \\ 5 \end{pmatrix}$$

$$= \frac{1}{55} \begin{pmatrix} 40 & +110 & +15 \\ 40 & -110 & -40 \\ -20 & +110 & -35 \end{pmatrix} = \begin{pmatrix} 3 \\ -2 \\ 1 \end{pmatrix}$$

$$\therefore \ x_1 = 3; \quad x_2 = -2; \quad x_3 = 1$$

The method is the same every time.

To solve $\mathbf{Ax} = \mathbf{b}$ $\qquad \mathbf{x} = \mathbf{A}^{-1}\mathbf{b}$

To find $\mathbf{A}^{-1}$

(1) Evaluate $|\mathbf{A}|$
   If $|\mathbf{A}| \neq 0$ then proceed to (2)
   If $|\mathbf{A}| = 0$ then there is no inverse and hence no unique solution. Later we shall discover how to determine whether there is an infinity of solutions or none.

(2) Form $\mathbf{C}$, the matrix of cofactors of $\mathbf{A}$

(3) Write $\mathbf{C}^{\mathrm{T}}$, the transpose of $\mathbf{C}$

(4) Then $\mathbf{A}^{-1} = \dfrac{1}{|\mathbf{A}|} \times \mathbf{C}^{\mathrm{T}}$.

Now apply the method to Example 2.

Example 2

$$4x_1 + 5x_2 + \ x_3 = 2$$
$$x_1 - 2x_2 - 3x_3 = 7$$
$$3x_1 - \ x_2 - 2x_3 = 1.$$
$$x_1 = \ldots\ldots\ldots; \qquad x_2 = \ldots\ldots\ldots; \qquad x_3 = \ldots\ldots\ldots$$

29

$$x_1 = -2; \quad x_2 = 3; \quad x_3 = -5$$

Here is the complete working.

$$\mathbf{A} = \begin{pmatrix} 4 & 5 & 1 \\ 1 & -2 & -3 \\ 3 & -1 & -2 \end{pmatrix} \quad \therefore \; |\mathbf{A}| = \begin{vmatrix} 4 & 5 & 1 \\ 1 & -2 & -3 \\ 3 & -1 & -2 \end{vmatrix} = -26$$

$$\mathbf{C} = \begin{pmatrix} A_{11} & A_{12} & A_{13} \\ A_{21} & A_{22} & A_{23} \\ A_{31} & A_{32} & A_{33} \end{pmatrix}$$

$$A_{11} = \begin{vmatrix} -2 & -3 \\ -1 & -2 \end{vmatrix} = 1 \qquad A_{12} = -\begin{vmatrix} 1 & -3 \\ 3 & -2 \end{vmatrix} = -7 \qquad A_{13} = \begin{vmatrix} 1 & -2 \\ 3 & -1 \end{vmatrix} = 5$$

$$A_{21} = -\begin{vmatrix} 5 & 1 \\ -1 & -2 \end{vmatrix} = 9 \qquad A_{22} = \begin{vmatrix} 4 & 1 \\ 3 & -2 \end{vmatrix} = -11 \qquad A_{23} = -\begin{vmatrix} 4 & 5 \\ 3 & -1 \end{vmatrix} = 19$$

$$A_{31} = \begin{vmatrix} 5 & 1 \\ -2 & -3 \end{vmatrix} = -13 \qquad A_{32} = -\begin{vmatrix} 4 & 1 \\ 1 & -3 \end{vmatrix} = 13 \qquad A_{33} = \begin{vmatrix} 4 & 5 \\ 1 & -2 \end{vmatrix} = -13$$

$$\therefore \; \mathbf{C} = \begin{pmatrix} 1 & -7 & 5 \\ 9 & -11 & 19 \\ -13 & 13 & -13 \end{pmatrix} \qquad \therefore \; \mathbf{C}^{\mathrm{T}} = \begin{pmatrix} 1 & 9 & -13 \\ -7 & -11 & 13 \\ 5 & 19 & -13 \end{pmatrix}$$

$$\mathbf{A}^{-1} = \frac{1}{|\mathbf{A}|} \times \mathbf{C}^{\mathrm{T}} = -\frac{1}{26} \begin{pmatrix} 1 & 9 & -13 \\ -7 & -11 & 13 \\ 5 & 19 & -13 \end{pmatrix}$$

$$\begin{pmatrix} x_1 \\ x_2 \\ x_3 \end{pmatrix} = \mathbf{A}^{-1} \begin{pmatrix} 2 \\ 7 \\ 1 \end{pmatrix} = -\frac{1}{26} \begin{pmatrix} 1 & 9 & -13 \\ -7 & -11 & 13 \\ 5 & 19 & -13 \end{pmatrix} \begin{pmatrix} 2 \\ 7 \\ 1 \end{pmatrix}$$

$$= -\frac{1}{26} \begin{pmatrix} 2 & +63 & -13 \\ -14 & -77 & +13 \\ 10 & +133 & -13 \end{pmatrix}$$

$$= -\frac{1}{26} \begin{pmatrix} 52 \\ -78 \\ 130 \end{pmatrix} = -\begin{pmatrix} 2 \\ -3 \\ 5 \end{pmatrix}$$

$$\therefore \; x_1 = -2; \quad x_2 = 3; \quad x_3 = -5$$

With a set of four equations with four unknowns, the method becomes somewhat tedious as there are then sixteen cofactors to be evaluated and each one is a third-order determinant! There are, however, other methods that can be applied – so let us see method 2.

2 *Row transformation method* 30

Elementary row transformations that can be applied are as follows

(a) Interchange any two rows.

(b) Multiply (or divide) every element in a row by a non-zero scalar (constant) k.

(c) Add to (or subtract from) all the elements of any row k times the corresponding elements of any other row.

Equivalent matrices

Two matrices, **A** and **B**, are said to be equivalent if **B** can be obtained from **A** by a sequence of elementary transformations.

Solutions of equations

The method is best described by working through a typical example.

Example 1

Solve
$$2x_1 + x_2 + x_3 = 5$$
$$x_1 + 3x_2 + 2x_3 = 1$$
$$3x_1 - 2x_2 - 4x_3 = -4.$$

This can be written
$$\begin{pmatrix} 2 & 1 & 1 \\ 1 & 3 & 2 \\ 3 & -2 & -4 \end{pmatrix} \begin{pmatrix} x_1 \\ x_2 \\ x_3 \end{pmatrix} = \begin{pmatrix} 5 \\ 1 \\ -4 \end{pmatrix}$$

and for convenience we introduce the unit matrix

$$\begin{pmatrix} 2 & 1 & 1 \\ 1 & 3 & 2 \\ 3 & -2 & -4 \end{pmatrix} \begin{pmatrix} x_1 \\ x_2 \\ x_3 \end{pmatrix} = \begin{pmatrix} 1 & 0 & 0 \\ 0 & 1 & 0 \\ 0 & 0 & 1 \end{pmatrix} \begin{pmatrix} 5 \\ 1 \\ -4 \end{pmatrix}$$

where $\begin{pmatrix} 1 & 0 & 0 \\ 0 & 1 & 0 \\ 0 & 0 & 1 \end{pmatrix}$ may be regarded as the coefficient of $\begin{pmatrix} 5 \\ 1 \\ -4 \end{pmatrix}$

We then form the combined coefficient matrix

$$\begin{pmatrix} 2 & 1 & 1 & 1 & 0 & 0 \\ 1 & 3 & 2 & 0 & 1 & 0 \\ 3 & -2 & -4 & 0 & 0 & 1 \end{pmatrix}$$

and work on this matrix from now on.

On then to the next frame

31   The rest of the working is mainly concerned with applying row transformations to convert the left-hand half of the matrix to a unit matrix and the right-
hand side to the inverse, eventually obtaining

$$\begin{pmatrix} 1 & 0 & 0 & a & b & c \\ 0 & 1 & 0 & d & e & f \\ 0 & 0 & 1 & g & h & i \end{pmatrix}$$

with $a, b, c, \ldots g, h, i$ being evaluated in the process.

The following notation will be helpful to denote the transformation used:

(1) ~ (2) denotes 'interchange rows 1 and 2'

(3) − 2(1) denotes 'subtract twice row 1 from row 3', etc.

So off we go.

$$(1) \sim (2) \quad \begin{pmatrix} 1 & 3 & 2 & 0 & 1 & 0 \\ 2 & 1 & 1 & 1 & 0 & 0 \\ 3 & -2 & -4 & 0 & 0 & 1 \end{pmatrix}$$

$$\begin{matrix}(2) - 2(1) \\ (3) - 3(1)\end{matrix} \quad \begin{pmatrix} 1 & 3 & 2 & 0 & 1 & 0 \\ 0 & -5 & -3 & 1 & -2 & 0 \\ 0 & -11 & -10 & 0 & -3 & 1 \end{pmatrix}$$

$$(3) - 2(2) \quad \begin{pmatrix} 1 & 3 & 2 & 0 & 1 & 0 \\ 0 & -5 & -3 & 1 & -2 & 0 \\ 0 & -1 & -4 & -2 & 1 & 1 \end{pmatrix}$$

$$-(2) \sim -(3) \quad \begin{pmatrix} 1 & 3 & 2 & 0 & 1 & 0 \\ 0 & 1 & 4 & 2 & -1 & -1 \\ 0 & 5 & 3 & -1 & 2 & 0 \end{pmatrix}$$

$$(3) - 5(2) \quad \begin{pmatrix} 1 & 3 & 2 & 0 & 1 & 0 \\ 0 & 1 & 4 & 2 & -1 & -1 \\ 0 & 0 & -17 & -11 & 7 & 5 \end{pmatrix}$$

$$\begin{matrix}(1) - 3(2) \\ (3) \div (-17)\end{matrix} \quad \begin{pmatrix} 1 & 0 & -10 & -6 & 4 & 3 \\ 0 & 1 & 4 & 2 & -1 & -1 \\ 0 & 0 & 1 & 11/17 & -7/17 & -5/17 \end{pmatrix}$$

$$\begin{matrix}(1) + 10(3) \\ (2) - 4(3)\end{matrix} \quad \begin{pmatrix} 1 & 0 & 0 & 8/17 & -2/17 & 1/17 \\ 0 & 1 & 0 & -10/17 & 11/17 & 3/17 \\ 0 & 0 & 1 & 11/17 & -7/17 & -5/17 \end{pmatrix}$$

We now have

$$\begin{pmatrix} 1 & 0 & 0 \\ 0 & 1 & 0 \\ 0 & 0 & 1 \end{pmatrix} \begin{pmatrix} x_1 \\ x_2 \\ x_3 \end{pmatrix} = \frac{1}{17} \begin{pmatrix} 8 & -2 & 1 \\ -10 & 11 & 3 \\ 11 & -7 & -5 \end{pmatrix} \begin{pmatrix} 5 \\ 1 \\ -4 \end{pmatrix}$$

$$\therefore x_1 = \ldots\ldots\ldots; \quad x_2 = \ldots\ldots\ldots; \quad x_3 = \ldots\ldots\ldots$$

$$x_1 = 2; \qquad x_2 = -3; \qquad x_3 = 4$$ **32**

$$
\begin{pmatrix} x_1 \\ x_2 \\ x_3 \end{pmatrix} = \frac{1}{17} \begin{pmatrix} 40 & -2 & -4 \\ -50 & +11 & -12 \\ 55 & -7 & +20 \end{pmatrix} = \frac{1}{17} \begin{pmatrix} 34 \\ -51 \\ 68 \end{pmatrix} = \begin{pmatrix} 2 \\ -3 \\ 4 \end{pmatrix}
$$

$$x_1 = 2; \qquad x_2 = -3; \qquad x_3 = 4$$

Of course, there is no set pattern of how to carry out the row transformations. It depends on one's ingenuity and every case is different. Here is a further example.

Example 2

$$2x_1 - x_2 - 3x_3 = 1$$
$$x_1 + 2x_2 + x_3 = 3$$
$$2x_1 - 2x_2 - 5x_3 = 2.$$

First write the set of equations in matrix form – with the unit matrix included. This gives . . . .

$$
\begin{pmatrix} 2 & -1 & -3 \\ 1 & 2 & 1 \\ 2 & -2 & -5 \end{pmatrix} \begin{pmatrix} x_1 \\ x_2 \\ x_3 \end{pmatrix} = \begin{pmatrix} 1 & 0 & 0 \\ 0 & 1 & 0 \\ 0 & 0 & 1 \end{pmatrix} \begin{pmatrix} 1 \\ 3 \\ 2 \end{pmatrix}
$$ **33**

The combined coefficient matrix is now . . . .

$$
\begin{pmatrix} 2 & -1 & -3 & 1 & 0 & 0 \\ 1 & 2 & 1 & 0 & 1 & 0 \\ 2 & -2 & -5 & 0 & 0 & 1 \end{pmatrix}
$$ **34**

If we start off by interchanging the top two rows, we obtain a 1 at the beginning of the top row which is a help.

$$(1) \sim (2) \quad \begin{pmatrix} 1 & 2 & 1 & 0 & 1 & 0 \\ 2 & -1 & -3 & 1 & 0 & 0 \\ 2 & -2 & -5 & 0 & 0 & 1 \end{pmatrix}$$

Now, if we subtract $2 \times$ row 1 from row 2

and $\qquad\qquad 2 \times$ row 1 from row 3, we get

. . . .

35

$$\begin{pmatrix} 1 & 2 & 1 & 0 & 1 & 0 \\ 0 & -5 & -5 & 1 & -2 & 0 \\ 0 & -6 & -7 & 0 & -2 & 1 \end{pmatrix}$$

Continuing with the same line of reasoning, we then have

$(2)-(3)$
$$\begin{pmatrix} 1 & 2 & 1 & 0 & 1 & 0 \\ 0 & 1 & 2 & 1 & 0 & -1 \\ 0 & -6 & -7 & 0 & -2 & 1 \end{pmatrix}$$

$(3)+6(2)$
$$\begin{pmatrix} 1 & 2 & 1 & 0 & 1 & 0 \\ 0 & 1 & 2 & 1 & 0 & -1 \\ 0 & 0 & 5 & 6 & -2 & -5 \end{pmatrix}$$

$(1)-2(2)$
$(3)\div 5$
$$\begin{pmatrix} 1 & 0 & -3 & -2 & 1 & 2 \\ 0 & 1 & 2 & 1 & 0 & -1 \\ 0 & 0 & 1 & \frac{6}{5} & -\frac{2}{5} & -1 \end{pmatrix}$$
Notice the three diagonal 1s appearing at the left-hand end

What do you suggest we should do now?

....

36

| Add three times row 3 to row 1 and subtract twice row 3 from row 2 |

Right. That gives

$(1)+3(3)$
$(2)-3(3)$
$$\begin{pmatrix} 1 & 0 & 0 & \frac{8}{5} & -\frac{1}{5} & -1 \\ 0 & 1 & 0 & -\frac{7}{5} & \frac{4}{5} & 1 \\ 0 & 0 & 1 & \frac{6}{5} & -\frac{2}{5} & -1 \end{pmatrix}$$

$$\therefore \begin{pmatrix} 1 & 0 & 0 \\ 0 & 1 & 0 \\ 0 & 0 & 1 \end{pmatrix}\begin{pmatrix} x_1 \\ x_2 \\ x_3 \end{pmatrix} = \frac{1}{5}\begin{pmatrix} 8 & -1 & -5 \\ -7 & 4 & 5 \\ 6 & -2 & -5 \end{pmatrix}\begin{pmatrix} 1 \\ 3 \\ 2 \end{pmatrix}$$

Now you can finish it off.

$$x_1 = \ldots\ldots\ldots ; \quad x_2 = \ldots\ldots\ldots ; \quad x_3 = \ldots\ldots\ldots$$

37

| $x_1 = -1; \quad x_2 = 3; \quad x_3 = -2$ |

Because

$$\begin{pmatrix} x_1 \\ x_2 \\ x_3 \end{pmatrix} = \frac{1}{5}\begin{pmatrix} 8 - 3 - 10 \\ -7 + 12 + 10 \\ 6 - 6 - 10 \end{pmatrix} = \frac{1}{5}\begin{pmatrix} -5 \\ 15 \\ -10 \end{pmatrix} = \begin{pmatrix} -1 \\ 3 \\ -2 \end{pmatrix}$$

Let us now look at a somewhat similar method with rather fewer steps involved.

So move on

3 Gaussian elimination method

38

Once again we will demonstrate the method by a typical example.

Example 1

$2x_1 - 3x_2 + 2x_3 = 9$

$3x_1 + 2x_2 - x_3 = 4$

$x_1 - 4x_2 + 2x_3 = 6.$

We start off as usual

$$\begin{pmatrix} 2 & -3 & 2 \\ 3 & 2 & -1 \\ 1 & -4 & 2 \end{pmatrix} \begin{pmatrix} x_1 \\ x_2 \\ x_3 \end{pmatrix} = \begin{pmatrix} 9 \\ 4 \\ 6 \end{pmatrix}$$

We then form the *augmented coefficient matrix* by including the constants as an extra column on the right-hand side of the matrix

$$\begin{pmatrix} 2 & -3 & 2 & \vdots & 9 \\ 3 & 2 & -1 & \vdots & 4 \\ 1 & -4 & 2 & \vdots & 6 \end{pmatrix}$$

Now we operate on the rows to convert the first three columns into an upper triangular matrix

$(1) \sim (3)$ $\begin{pmatrix} 1 & -4 & 2 & 6 \\ 3 & 2 & -1 & 4 \\ 2 & -3 & 2 & 9 \end{pmatrix}$ $\qquad$ $(2) \sim (3)$ $\begin{pmatrix} 1 & -4 & 2 & 6 \\ 2 & -3 & 2 & 9 \\ 3 & 2 & -1 & 4 \end{pmatrix}$

$\begin{matrix}(2) - 2(1) \\ (3) - 3(1)\end{matrix}$ $\begin{pmatrix} 1 & -4 & 2 & 6 \\ 0 & 5 & -2 & -3 \\ 0 & 14 & -7 & -14 \end{pmatrix}$ $\qquad$ $\begin{matrix}(2) \div 5 \\ (3) \div 7\end{matrix}$ $\begin{pmatrix} 1 & -4 & 2 & 6 \\ 0 & 1 & -\frac{2}{5} & -\frac{3}{5} \\ 0 & 2 & -1 & -2 \end{pmatrix}$

$(3) - 2(2)$ $\begin{pmatrix} 1 & -4 & 2 & 6 \\ 0 & 1 & -\frac{2}{5} & -\frac{3}{5} \\ 0 & 0 & -\frac{1}{5} & -\frac{4}{5} \end{pmatrix}$ $\qquad$ $(3) \times (-5)$ $\begin{pmatrix} 1 & -4 & 2 & 6 \\ 0 & 1 & -\frac{2}{5} & -\frac{3}{5} \\ 0 & 0 & 1 & 4 \end{pmatrix}$

The first three columns now form an upper triangular matrix which has been our purpose. If we now detach the fourth column back to its original position on the right-hand side of the matrix equation, we have

. . . .

39

$$\begin{pmatrix} 1 & -4 & 2 \\ 0 & 1 & -\frac{2}{5} \\ 0 & 0 & 1 \end{pmatrix} \begin{pmatrix} x_1 \\ x_2 \\ x_3 \end{pmatrix} = \begin{pmatrix} 6 \\ -\frac{3}{5} \\ 4 \end{pmatrix}$$

Expanding from the bottom row, working upwards

$$x_3 = 4 \qquad\qquad\qquad \therefore x_3 = 4$$
$$x_2 - \tfrac{2}{5}x_3 = -\tfrac{3}{5} \quad \therefore x_2 = -\tfrac{3}{5} + \tfrac{8}{5} = 1 \quad \therefore x_2 = 1$$
$$x_1 - 4x_2 + 2x_3 = 6 \quad \therefore x_1 - 4 + 8 = 6 \qquad \therefore x_1 = 2$$
$$\therefore x_1 = 2; \quad x_2 = 1; \quad x_3 = 4$$

It is a very useful method and entails fewer tedious steps, and can be used to solve efficiently higher-order sets of equations and non-square systems. It can also solve a sequence of problems with the same coefficient matrix $\mathbf{A}$ by using the augmented matrix $(\mathbf{A}\mathbf{b}_1\mathbf{b}_2 \dots \mathbf{b}_n)$.

Example 2

$$x_1 + 3x_2 - 2x_3 + x_4 = -1$$
$$2x_1 - 2x_2 + x_3 - 2x_4 = 1$$
$$x_1 + x_2 - 3x_3 + x_4 = 6$$
$$3x_1 - x_2 + 2x_3 - x_4 = 3.$$

First we write this in matrix form and compile the augmented matrix which is

$$\dots\dots\dots\dots$$

40

$$\left(\begin{array}{cccc|c} 1 & 3 & -2 & 1 & -1 \\ 2 & -2 & 1 & -2 & 1 \\ 1 & 1 & -3 & 1 & 6 \\ 3 & -1 & 2 & -1 & 3 \end{array}\right)$$

Next we operate on rows to convert the left-hand side to an upper triangular matrix. There is no set way of doing this. Use any trickery to save yourself unnecessary work.

So now you can go ahead and complete the transformations and obtain

$$x_1 = \dots\dots\dots; \quad x_2 = \dots\dots\dots$$
$$x_3 = \dots\dots\dots; \quad x_4 = \dots\dots\dots$$

$$\boxed{x_1 = 2; \quad x_2 = -3; \quad x_3 = -1; \quad x_4 = 4}$$

41

Here is one way. You may well have taken quite a different route.

$$\begin{pmatrix} 1 & 3 & -2 & 1 & \vdots & -1 \\ 2 & -2 & 1 & -2 & \vdots & 1 \\ 1 & 1 & -3 & 1 & \vdots & 6 \\ 3 & -1 & 2 & -1 & \vdots & 3 \end{pmatrix}$$

$$\begin{matrix} (2) - 2(1) \\ (3) - (1) \\ (4) - [(1) + (2)] \end{matrix} \begin{pmatrix} 1 & 3 & -2 & 1 & \vdots & -1 \\ 0 & -8 & 5 & -4 & \vdots & 3 \\ 0 & -2 & -1 & 0 & \vdots & 7 \\ 0 & -2 & 3 & 0 & \vdots & 3 \end{pmatrix}$$

$$\begin{matrix} (2) - 4(4) \\ (3) - (4) \end{matrix} \begin{pmatrix} 1 & 3 & -2 & 1 & \vdots & -1 \\ 0 & 0 & -7 & -4 & \vdots & -9 \\ 0 & 0 & -4 & 0 & \vdots & 4 \\ 0 & -2 & 3 & 0 & \vdots & 3 \end{pmatrix}$$

$$\begin{matrix} (2) \sim (4) \\ (3) \div 4 \end{matrix} \begin{pmatrix} 1 & 3 & -2 & 1 & \vdots & -1 \\ 0 & -2 & 3 & 0 & \vdots & 3 \\ 0 & 0 & -1 & 0 & \vdots & 1 \\ 0 & 0 & -7 & -4 & \vdots & -9 \end{pmatrix}$$

$$(4) - 7(3) \begin{pmatrix} 1 & 3 & -2 & 1 & \vdots & -1 \\ 0 & -2 & 3 & 0 & \vdots & 3 \\ 0 & 0 & -1 & 0 & \vdots & 1 \\ 0 & 0 & 0 & -4 & \vdots & -16 \end{pmatrix}$$

Returning the right-hand column to its original position

$$\begin{pmatrix} 1 & 3 & -2 & 1 \\ 0 & -2 & 3 & 0 \\ 0 & 0 & -1 & 0 \\ 0 & 0 & 0 & -4 \end{pmatrix} \begin{pmatrix} x_1 \\ x_2 \\ x_3 \\ x_4 \end{pmatrix} = \begin{pmatrix} -1 \\ 3 \\ 1 \\ -16 \end{pmatrix}$$

Expanding from the bottom row, we have

$-4x_4 = -16$ $\qquad\qquad\qquad\qquad\qquad\qquad \therefore\ x_4 = 4$

$-x_3 = 1$ $\qquad\qquad\qquad\qquad\qquad\qquad\qquad \therefore\ x_3 = -1$

$-2x_2 + 3x_3 = 3 \quad \therefore\ -2x_2 = 6$ $\qquad\qquad\qquad \therefore\ x_2 = -3$

$x_1 + 3x_2 - 2x_3 + x_4 = -1 \quad \therefore\ x_1 - 9 + 2 + 4 = -1 \quad \therefore\ x_1 = 2$

$$\therefore\ x_1 = 2; \quad x_2 = -3; \quad x_3 = -1; \quad x_4 = 4$$

42   We still have a further method for solving sets of simultaneous equations.

4 *Triangular decomposition method*

A square matrix A can usually be written as a product of a lower-triangular matrix L and an upper-triangular matrix U, where $A = LU$.

For example, if $A = \begin{pmatrix} 1 & 2 & 3 \\ 3 & 5 & 8 \\ 4 & 9 & 10 \end{pmatrix}$, A can be expressed as

$$A = LU = \begin{pmatrix} l_{11} & 0 & 0 \\ l_{21} & l_{22} & 0 \\ l_{31} & l_{32} & l_{33} \end{pmatrix} \begin{pmatrix} u_{11} & u_{12} & u_{13} \\ 0 & u_{22} & u_{23} \\ 0 & 0 & u_{33} \end{pmatrix}$$

$$\qquad\qquad (L) \qquad\qquad\qquad (U)$$

$$= \begin{pmatrix} l_{11}u_{11} & l_{11}u_{12} & l_{11}u_{13} \\ l_{21}u_{11} & l_{21}u_{12} + l_{22}u_{22} & l_{21}u_{13} + l_{22}u_{23} \\ l_{31}u_{11} & l_{31}u_{12} + l_{32}u_{22} & l_{31}u_{13} + l_{32}u_{23} + l_{33}u_{33} \end{pmatrix}$$

Note that, in L and U, elements occur in the major diagonal in each case. These are related in the product and whatever values we choose to put for u_{11}, u_{22}, $u_{33} \ldots$ then the corresponding values of l_{11}, l_{22}, $l_{33} \ldots$ will be determined – and vice versa.

For convenience, we put $u_{11} = u_{22} = u_{33} \ldots = 1$

Then $A = LU = \begin{pmatrix} l_{11} & l_{11}u_{12} & l_{11}u_{13} \\ l_{21} & l_{21}u_{12} + l_{22} & l_{21}u_{13} + l_{22}u_{23} \\ l_{31} & l_{31}u_{12} + l_{32} & l_{31}u_{13} + l_{32}u_{23} + l_{33} \end{pmatrix}$

In our example, $A = \begin{pmatrix} 1 & 2 & 3 \\ 3 & 5 & 8 \\ 4 & 9 & 10 \end{pmatrix}$

$\therefore l_{11} = 1; \qquad l_{11}u_{12} = 2 \quad \therefore u_{12} = 2; \qquad l_{11}u_{13} = 3 \quad \therefore u_{13} = 3$

$\quad l_{21} = 3; \qquad$ Similarly $l_{22} = \ldots\ldots\ldots\ldots;$ $\qquad\qquad u_{23} = \ldots\ldots\ldots\ldots$

$\quad l_{31} = 4; \qquad\qquad\quad l_{32} = \ldots\ldots\ldots\ldots;$ $\qquad\qquad l_{33} = \ldots\ldots\ldots\ldots$

43 $\boxed{l_{22} = -1; \quad u_{23} = 1; \quad l_{32} = 1; \quad l_{33} = -3}$

Because

$\qquad l_{21}u_{12} + l_{22}u_{22} = 5$ that is $3 \times 2 + l_{22} \times 1 = 5$ and so $l_{22} = -1$

$\qquad l_{21}u_{13} + l_{22}u_{23} = 8$ that is $3 \times 3 + (-1) \times u_{23} = 8$ and so $u_{23} = 1$

$\qquad l_{31}u_{12} + l_{32}u_{22} = 9$ that is $4 \times 2 + l_{32} \times 1 = 9$ and so $l_{32} = 1$

$\qquad l_{31}u_{13} + l_{32}u_{23} + l_{33}u_{33} = 10$ that is $4 \times 3 + 1 \times 1 + l_{33} \times 1 = 10$

$\qquad\qquad\qquad\qquad\qquad\qquad\qquad\qquad\qquad\qquad\qquad$ and so $l_{33} = -3$

Now we substitute all these values back into the upper and lower triangular matrices and obtain $A = LU = \ldots\ldots\ldots\ldots$

44

$$A = LU = \begin{pmatrix} 1 & 0 & 0 \\ 3 & -1 & 0 \\ 4 & 1 & -3 \end{pmatrix} \begin{pmatrix} 1 & 2 & 3 \\ 0 & 1 & 1 \\ 0 & 0 & 1 \end{pmatrix}$$

We have thus expressed the given matrix **A** as the product of lower and upper triangular matrices. Let us now see how we use them.

Example 1

$$x_1 + 2x_2 + 3x_3 = 16$$
$$3x_1 + 5x_2 + 8x_3 = 43$$
$$4x_1 + 9x_2 + 10x_3 = 57.$$

i.e. $\begin{pmatrix} 1 & 2 & 3 \\ 3 & 5 & 8 \\ 4 & 9 & 10 \end{pmatrix} \begin{pmatrix} x_1 \\ x_2 \\ x_3 \end{pmatrix} = \begin{pmatrix} 16 \\ 43 \\ 57 \end{pmatrix}$ i.e. $\mathbf{Ax} = \mathbf{b}$.

We have seen above that **A** can be written as **LU** where

$$\mathbf{A} = \mathbf{LU} = \begin{pmatrix} 1 & 0 & 0 \\ 3 & -1 & 0 \\ 4 & 1 & -3 \end{pmatrix} \begin{pmatrix} 1 & 2 & 3 \\ 0 & 1 & 1 \\ 0 & 0 & 1 \end{pmatrix}$$

To solve $\mathbf{Ax} = \mathbf{b}$, we have $\mathbf{LUx} = \mathbf{b}$ i.e. $\mathbf{L}(\mathbf{Ux}) = \mathbf{b}$

Putting $\mathbf{Ux} = \mathbf{y}$, we solve $\mathbf{Ly} = \mathbf{b}$ to obtain $\mathbf{y}$

and then $\mathbf{Ux} = \mathbf{y}$ to obtain $\mathbf{x}$.

(a) Solving $\mathbf{Ly} = \mathbf{b}$ $\begin{pmatrix} 1 & 0 & 0 \\ 3 & -1 & 0 \\ 4 & 1 & -3 \end{pmatrix} \begin{pmatrix} y_1 \\ y_2 \\ y_3 \end{pmatrix} = \begin{pmatrix} 16 \\ 43 \\ 57 \end{pmatrix}$

Expanding from the top $y_1 = 16$; $3y_1 - y_2 = 43$ $\therefore y_2 = 5$; and
$4y_1 + y_2 - 3y_3 = 57$ $\therefore$ $64 + 5 - 3y_3 = 57$ $\therefore$ $y_3 = 4$

$$\therefore \begin{pmatrix} y_1 \\ y_2 \\ y_3 \end{pmatrix} = \begin{pmatrix} 16 \\ 5 \\ 4 \end{pmatrix}$$

(b) Solving $\mathbf{Ux} = \mathbf{y}$ $\begin{pmatrix} 1 & 2 & 3 \\ 0 & 1 & 1 \\ 0 & 0 & 1 \end{pmatrix} \begin{pmatrix} x_1 \\ x_2 \\ x_3 \end{pmatrix} = \begin{pmatrix} 16 \\ 5 \\ 4 \end{pmatrix}$

Expanding from the bottom, we then have

$$x_1 = \ldots\ldots\ldots\ldots; \qquad x_2 = \ldots\ldots\ldots\ldots; \qquad x_3 = \ldots\ldots\ldots\ldots$$

45

$$x_1 = 2; \quad x_2 = 1; \quad x_3 = 4$$

Note:

1   If $l_{ii} = 0$, then either decomposition is not possible, or, if **A** is singular, i.e. $|\mathbf{A}| = 0$, there is an infinite number of possible decompositions.

2   Instead of putting $u_{11} = u_{22} = u_{33} \ldots = 1$, we could have used the alternative substitution $l_{11} = l_{22} = l_{33} \ldots = 1$ and obtained values of u_{11}, u_{22}, $u_{33} \ldots$ etc. The working is as before.

3   One advantage of employing **LU** decomposition over Gaussian elimination is in the solution of a sequence of problems in which the same coefficient matrix occurs.

Now for another example.

46   **Example 2**

$$x_1 + 3x_2 + 2x_3 = 19$$

$$2x_1 + x_2 + x_3 = 13$$

$$4x_1 + 2x_2 + 3x_3 = 31.$$

$$\therefore \quad \begin{pmatrix} 1 & 3 & 2 \\ 2 & 1 & 1 \\ 4 & 2 & 3 \end{pmatrix} \begin{pmatrix} x_1 \\ x_2 \\ x_3 \end{pmatrix} = \begin{pmatrix} 19 \\ 13 \\ 31 \end{pmatrix} \quad \text{i.e. } \mathbf{Ax} = \mathbf{b}$$

$$\mathbf{A} = \mathbf{LU} = \begin{pmatrix} l_{11} & 0 & 0 \\ l_{21} & l_{22} & 0 \\ l_{31} & l_{32} & l_{33} \end{pmatrix} \begin{pmatrix} 1 & u_{12} & u_{13} \\ 0 & 1 & u_{23} \\ 0 & 0 & 1 \end{pmatrix}$$

$$= \begin{pmatrix} l_{11} & l_{11}u_{12} & l_{11}u_{13} \\ l_{21} & l_{21}u_{12} + l_{22} & l_{21}u_{13} + l_{22}u_{23} \\ l_{31} & l_{31}u_{12} + l_{32} & l_{31}u_{13} + l_{32}u_{23} + l_{33} \end{pmatrix}$$

$$= \begin{pmatrix} 1 & 3 & 2 \\ 2 & 1 & 1 \\ 4 & 2 & 3 \end{pmatrix}$$

▶

Now we have to find the values of the various elements. The usual order of doing this is shown by the diagram.

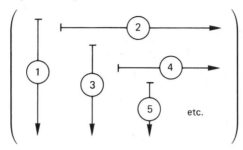

That is, first we can write down values for l_{11}, l_{21}, l_{31} from the left-hand column; then follow this by finding u_{12}, u_{13} from the top row; and proceed for the others.

So, completing the two triangular matrices, we have

$$A = LU = \ldots\ldots\ldots$$

47

$$A = LU = \begin{pmatrix} 1 & 0 & 0 \\ 2 & -5 & 0 \\ 4 & -10 & 1 \end{pmatrix} \begin{pmatrix} 1 & 3 & 2 \\ 0 & 1 & \frac{3}{5} \\ 0 & 0 & 1 \end{pmatrix}$$

As we stated before: $\mathbf{Ax} = \mathbf{b}$; $\mathbf{L(Ux)} = \mathbf{b}$. Put $\mathbf{Ux} = \mathbf{y}$

then   (a) solve $\mathbf{Ly} = \mathbf{b}$ to obtain $\mathbf{y}$

and   (b) solve $\mathbf{Ux} = \mathbf{y}$ to obtain $\mathbf{x}$.

Solving $\mathbf{Ly} = \mathbf{b}$ gives $\begin{pmatrix} y_1 \\ y_2 \\ y_3 \end{pmatrix} = \begin{pmatrix} \cdots \\ \cdots \\ \cdots \end{pmatrix}$

48

$$\begin{pmatrix} y_1 \\ y_2 \\ y_3 \end{pmatrix} = \begin{pmatrix} 19 \\ 5 \\ 5 \end{pmatrix}$$

Because

$$\begin{pmatrix} 1 & 0 & 0 \\ 2 & -5 & 0 \\ 4 & -10 & 1 \end{pmatrix} \begin{pmatrix} y_1 \\ y_2 \\ y_3 \end{pmatrix} = \begin{pmatrix} 19 \\ 13 \\ 31 \end{pmatrix}$$

Expanding from the top gives

$y_1 = 19$; $y_2 = 5$; $y_3 = 5$.

(b) Now solve $\mathbf{Ux} = \mathbf{y}$ from which $x_1 = \ldots\ldots\ldots$; $x_2 = \ldots\ldots\ldots$;
$x_3 = \ldots\ldots\ldots$

49

$$x_1 = 3; \quad x_2 = 2; \quad x_3 = 5$$

Because we have

$$\mathbf{Ux} = y$$

i.e.
$$\begin{pmatrix} 1 & 3 & 2 \\ 0 & 1 & \frac{3}{5} \\ 0 & 0 & 1 \end{pmatrix} \begin{pmatrix} x_1 \\ x_2 \\ x_3 \end{pmatrix} = \begin{pmatrix} 19 \\ 5 \\ 5 \end{pmatrix}$$

Expanding from the bottom $x_3 = 5$; $x_2 + \dfrac{3}{5}x_3 = 5$ $\therefore x_2 = 2$

and $x_1 + 3x_2 + 2x_3 = 19$ $\therefore x_1 + 6 + 10 = 19$ $\therefore x_1 = 3$

$$\therefore x_1 = 3; \quad x_2 = 2; \quad x_3 = 5$$

We can of course apply the same method to a set of four equations.

Example 3

$$\begin{aligned} x_1 + 2x_2 - x_3 + 3x_4 &= 9 \\ 2x_1 - x_2 + 3x_3 + 2x_4 &= 23 \\ 3x_1 + 3x_2 + x_3 + x_4 &= 5 \\ 4x_1 + 5x_2 - 2x_3 + 2x_4 &= -2. \end{aligned}$$

i.e.
$$\begin{pmatrix} 1 & 2 & -1 & 3 \\ 2 & -1 & 3 & 2 \\ 3 & 3 & 1 & 1 \\ 4 & 5 & -2 & 2 \end{pmatrix} \begin{pmatrix} x_1 \\ x_2 \\ x_3 \\ x_4 \end{pmatrix} = \begin{pmatrix} 9 \\ 23 \\ 5 \\ -2 \end{pmatrix}$$ i.e. $\mathbf{Ax} = \mathbf{b}$

$$\mathbf{A} = \mathbf{LU} = \begin{pmatrix} l_{11} & 0 & 0 & 0 \\ l_{21} & l_{22} & 0 & 0 \\ l_{31} & l_{32} & l_{33} & 0 \\ l_{41} & l_{42} & l_{43} & l_{44} \end{pmatrix} \begin{pmatrix} 1 & u_{12} & u_{13} & u_{14} \\ 0 & 1 & u_{23} & u_{24} \\ 0 & 0 & 1 & u_{34} \\ 0 & 0 & 0 & 1 \end{pmatrix} = \begin{pmatrix} 1 & 2 & -1 & 3 \\ 2 & -1 & 3 & 2 \\ 3 & 3 & 1 & 1 \\ 4 & 5 & -2 & 2 \end{pmatrix}$$

$$\mathbf{A} = \begin{pmatrix} l_{11} & l_{11}u_{12} & l_{11}u_{13} & l_{11}u_{14} \\ l_{21} & l_{21}u_{12} + l_{22} & l_{21}u_{13} + l_{22}u_{23} & l_{21}u_{14} + l_{22}u_{24} \\ l_{31} & l_{31}u_{12} + l_{32} & l_{31}u_{13} + l_{32}u_{23} + l_{33} & l_{31}u_{14} + l_{32}u_{24} + l_{33}u_{34} \\ l_{41} & l_{41}u_{12} + l_{42} & l_{41}u_{13} + l_{42}u_{23} + l_{43} & l_{41}u_{14} + l_{42}u_{24} + l_{43}u_{34} + l_{44} \end{pmatrix}$$

Now we have to find the values of the individual elements. It is easy enough if we follow the order indicated in the diagram earlier. So the two triangular matrices are

$$\mathbf{A} = \mathbf{LU} = (\ldots\ldots\ldots\ldots)(\ldots\ldots\ldots\ldots)$$

$$\boxed{A = LU = \begin{pmatrix} 1 & 0 & 0 & 0 \\ 2 & -5 & 0 & 0 \\ 3 & -3 & 1 & 0 \\ 4 & -3 & -1 & -\frac{66}{5} \end{pmatrix} \begin{pmatrix} 1 & 2 & -1 & 3 \\ 0 & 1 & -1 & \frac{4}{5} \\ 0 & 0 & 1 & -\frac{28}{5} \\ 0 & 0 & 0 & 1 \end{pmatrix}}$$

50

As usual $\mathbf{Ax} = \mathbf{b}$; $\mathbf{L(Ux)} = \mathbf{b}$. Put $\mathbf{Ux} = \mathbf{y}$ $\therefore$ $\mathbf{Ly} = \mathbf{b}$

(a) Solving $\mathbf{Ly} = \mathbf{b}$

$$\begin{pmatrix} 1 & 0 & 0 & 0 \\ 2 & -5 & 0 & 0 \\ 3 & -3 & 1 & 0 \\ 4 & -3 & -1 & -\frac{66}{5} \end{pmatrix} \begin{pmatrix} y_1 \\ y_2 \\ y_3 \\ y_4 \end{pmatrix} = \begin{pmatrix} 9 \\ 23 \\ 5 \\ -2 \end{pmatrix}$$

$$\therefore \begin{pmatrix} y_1 \\ y_2 \\ y_3 \\ y_4 \end{pmatrix} = \begin{pmatrix} \cdots \\ \cdots \\ \cdots \\ \cdots \end{pmatrix}$$

$$\boxed{\begin{pmatrix} y_1 \\ y_2 \\ y_3 \\ y_4 \end{pmatrix} = \begin{pmatrix} 9 \\ -1 \\ -25 \\ 5 \end{pmatrix}}$$

51

(b) Solving $\mathbf{Ux} = \mathbf{y}$

$$\begin{pmatrix} 1 & 2 & -1 & 3 \\ 0 & 1 & -1 & \frac{4}{5} \\ 0 & 0 & 1 & -\frac{28}{5} \\ 0 & 0 & 0 & 1 \end{pmatrix} \begin{pmatrix} x_1 \\ x_2 \\ x_3 \\ x_4 \end{pmatrix} = \begin{pmatrix} 9 \\ -1 \\ -25 \\ 5 \end{pmatrix}$$

which finally gives

$$x_1 = \ldots\ldots\ldots\ldots; \quad x_2 = \ldots\ldots\ldots\ldots$$
$$x_3 = \ldots\ldots\ldots\ldots; \quad x_4 = \ldots\ldots\ldots\ldots$$

52 $x_1 = 1; \quad x_2 = -2; \quad x_3 = 3; \quad x_4 = 5$

5 *Using an electronic spreadsheet*

The four methods just considered for multiplying and inverting matrices clearly demonstrate the effects of certain properties of matrices and their algebraic manipulation. For this reason alone these methods are invaluable for providing a facility and familiarity with matrix algebra. However, by far the most efficient way of proceeding to multiply and invert small matrices with a numerical content is to use an electronic spreadsheet. This not only provides a faster method but also provides a method that is not prone human arithmetic error!

The spreadsheet used to demonstrate the method is the *Excel 2010* spreadsheet provided by Microsoft. There are two functions in particular that we shall use, namely:

MINVERSE(array)   for obtaining the inverse of a matrix
MMULT(array1, array2)   for multiplying two matrices together

Their use is quite straightforward. We start with an example we have done before in Frame 25 to solve the matrix equation:

$$\begin{pmatrix} 3 & 2 & -1 \\ 2 & -1 & 2 \\ 1 & -3 & -4 \end{pmatrix} \begin{pmatrix} x_1 \\ x_2 \\ x_3 \end{pmatrix} = \begin{pmatrix} 4 \\ 10 \\ 5 \end{pmatrix}$$

The solution is given as:

$$\begin{pmatrix} x_1 \\ x_2 \\ x_3 \end{pmatrix} = \begin{pmatrix} 3 & 2 & -1 \\ 2 & -1 & 2 \\ 1 & -3 & -4 \end{pmatrix}^{-1} \begin{pmatrix} 4 \\ 10 \\ 5 \end{pmatrix}$$

where $\begin{pmatrix} 3 & 2 & -1 \\ 2 & -1 & 2 \\ 1 & -3 & -4 \end{pmatrix}^{-1}$ is the inverse of matrix $\begin{pmatrix} 3 & 2 & -1 \\ 2 & -1 & 2 \\ 1 & -3 & -4 \end{pmatrix}$.

Open up a new blank worksheet and enter the elements of the matrix

$\begin{pmatrix} 3 & 2 & -1 \\ 2 & -1 & 2 \\ 1 & -3 & -4 \end{pmatrix}$ into cells A1 to C3.

Now highlight the empty cells A5 to C7 this is an empty 3×3 array ready to take the inverse matrix. With these cells highlighted type in:

=MINVERSE(A1:C3)   **and wait!**

You may be tempted just to press **Enter** but don't; you must press **Ctrl-Shift-Enter** (hold down the **Ctrl** and **Shift** keys together and then press **Enter**).

And there, in the allotted array appear the numbers . . . .

Next frame

$$\begin{array}{ccc} 0.181818182 & 0.2 & 0.054545455 \\ 0.181818182 & -0.2 & -0.145454545 \\ -0.090909091 & 0.2 & -0.127272727 \end{array}$$

53

This is the inverse matrix. We now need the 3×1 matrix so in cells E1 to E3 enter the numbers:

4
10
5

Now place the cursor in cell A9 and highlight the three cells A9 to A11. With these three cells highlighted type in:

=MMULT(A5:C7,E1:E3)

and press **Ctrl-Shift-Enter**.

The result is ....

Next frame

$$\begin{array}{c} 3 \\ -2 \\ 1 \end{array}$$

54

Because:

$$\begin{pmatrix} x_1 \\ x_2 \\ x_3 \end{pmatrix} = \begin{pmatrix} 3 & 2 & -1 \\ 2 & -1 & 2 \\ 1 & -3 & 4 \end{pmatrix}^{-1} \begin{pmatrix} 4 \\ 10 \\ 5 \end{pmatrix}$$

$$= \begin{pmatrix} 0.1818\ldots & 0.2 & 0.0545\ldots \\ 0.1818\ldots & -0.2 & -0.1454\ldots \\ -0.0909\ldots & 0.2 & -0.1272\ldots \end{pmatrix} \begin{pmatrix} 4 \\ 10 \\ 5 \end{pmatrix}$$

$$= \begin{pmatrix} 3 \\ -2 \\ 1 \end{pmatrix}$$

Try one yourself. The solution of the set of equations:

$$2x_1 - x_2 - 3x_3 = 1$$
$$x_1 + 2x_2 + x_3 = 3$$
$$2x_1 - 2x_2 - 5x_3 = 2 \text{ is } x_1 = \ldots\ldots\ldots, x_2 = \ldots\ldots\ldots, x_3 = \ldots\ldots\ldots$$

The answer is in the next frame

55

$$\boxed{x_1 = -1,\ x_2 = 3,\ x_3 = -2}$$

Because:

$$2x_1 - x_2 - 3x_3 = 1$$
$$x_1 + 2x_2 + x_3 = 3$$
$$2x_1 - 2x_2 - 5x_3 = 2$$

can be written in matrix form as

$$\begin{pmatrix} 2 & -1 & -3 \\ 1 & 2 & 1 \\ 2 & -2 & -5 \end{pmatrix} \begin{pmatrix} x_1 \\ x_2 \\ x_3 \end{pmatrix} = \begin{pmatrix} 1 \\ 3 \\ 2 \end{pmatrix}$$

Entering the 3×3 array on the left in cells A1 to C3 and the 3×1 array on the right in cells E1 to E3 we then highlight cells A5 to C7 and type in the instruction:

=MINVERSE(A1:C3)

and press **Ctrl-Shift-Enter** to reveal the display . . . .

Next frame

56

$$\begin{array}{rrr} 1.6 & -0.2 & -1 \\ -1.4 & 0.8 & 1 \\ 1.2 & -0.4 & -1 \end{array}$$

This is the inverse matrix. Next, we multiply this array by the array in cells E1 to E3 so highlight the cells A9 to A11 and type in the formula

=MMULT(A5:C7,E1:E3)

Press **Ctrl-Shift-Enter** to reveal the result

-1

$\quad 3$ giving the solution to the three equations as $x_1 = -1,\ x_2 = 3,\ x_3 = -2$

-2

Now try this one and see how much easier the whole process is as the number of equations increases. The solution to the set of equations:

$$2x - 3y + z + 4w = 13$$
$$x + 2y - 3z + w = 25$$
$$-3x - y + 4z - 2w = -34$$
$$x + y + z + w = 6$$

is

$$x = \ldots\ldots\ldots\ldots,\ y = \ldots\ldots\ldots\ldots,\ z = \ldots\ldots\ldots\ldots,\ w = \ldots\ldots\ldots\ldots$$

The answer is in the next frame

$$\boxed{x = 1,\ y = 3,\ y = 3,\ z = -4,\ w = 6}$$ **57**

Because:

$$2x - 3y + z + 4w = 13$$
$$x + 2y - 3z + w = 25$$
$$-3x - y + 4z - 2w = -34$$
$$x + y + z + w = 6$$

can be written in matrix form as

$$\begin{pmatrix} 2 & -3 & 1 & 4 \\ 1 & 2 & -3 & 1 \\ -3 & -1 & 4 & -2 \\ 1 & 1 & 1 & 1 \end{pmatrix} \begin{pmatrix} x \\ y \\ z \\ w \end{pmatrix} = \begin{pmatrix} 13 \\ 25 \\ -34 \\ 6 \end{pmatrix}$$

Entering the 4×4 array on the left in cells A1 to D4 and the 4×1 array on the right in cells F1 to F4 we then highlight cells A6 to D9 and type in the instruction:

=MINVERSE(A1:D4)

and press **Ctrl-Shift-Enter** to reveal the display . . . .

Next frame

58

-0196078431	-0.764705882	-0.607843137	0.333333333
-0.078431373	0.294117647	0.156862745	0.333333333
-0.019607843	-0.176470588	0.039215686	0.333333333
0.294117647	0.647058824	0.411764706	0

This is the inverse matrix. Next, we multiply this array by the array in cells F1 to F4 so highlight the cells A11 to A14 and type in the formula

=MMULT(A6:D9,F1:F4)

Press **Ctrl-Shift-Enter** to reveal the result

$$1$$
$$3$$
$$-4$$
$$6$$

giving the solution to the three equations as $x = 1,\ y = 3,\ z = -1,\ w = 6$

▶

We can even combine the two processes of taking the inverse and performing the multiplication into one formula. For example, to solve the matrix equation:

$$\begin{pmatrix} 2 & 1 & 1 \\ 1 & 3 & 2 \\ 3 & -2 & -4 \end{pmatrix} \begin{pmatrix} x_1 \\ x_2 \\ x_3 \end{pmatrix} = \begin{pmatrix} 5 \\ 1 \\ -4 \end{pmatrix}$$

Enter the 3×3 array on the left in cells A1 to C3 and the 3×1 array on the right in cells E1 to E3. We then highlight cells A5 to A7 and type in the instruction:

=MMULT(MINVERSE(A1:C3),E1:E3)

and press **Ctrl-Shift-Enter** to reveal the display ....

Next frame

59

$$\begin{array}{c} 2 \\ -3 \\ 4 \end{array}$$

giving the solution $x_1 = 2$, $x_2 = -3$, $x_3 = 4$.

Comparison of methods

Inverse method

This is an elementary method but it is very inefficient when the number of equations to solve increases beyond three.

Row transformation method

An efficient method but each case is different and relies on ingenuity to see the way forward.

Gaussian elimination method

The most efficient method and should be used in most cases. It must be used when there is a singular or non-square system.

Triangular decomposition method

An alternative to Gaussian elimination in some cases and by far the most efficient method of all for very large matrices.

Spreadsheet method

Whilst a spreadsheet cannot be used for matrices with algebraic content, for numerical content it provides an efficient method for small matrices.

Now let us proceed to something rather different,
so move on to the next frame for a new start

Matrix transformation

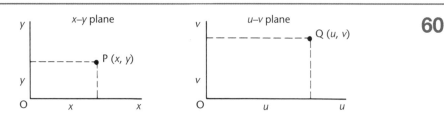

If for every point Q (u, v) in the u–v plane there is a corresponding point P (x, y) in the x–y plane, then there is a relationship between the two sets of coordinates. In the simple case of scaling the coordinate where

$$u = ax \text{ and } v = by$$

we have a *linear transformation* and we can combine these in matrix form

$$\begin{pmatrix} u \\ v \end{pmatrix} = \begin{pmatrix} a & 0 \\ 0 & b \end{pmatrix} \begin{pmatrix} x \\ y \end{pmatrix}$$

The matrix $\begin{pmatrix} a & 0 \\ 0 & b \end{pmatrix}$ then provides the transformation between the vector $\begin{pmatrix} x \\ y \end{pmatrix}$ in one set of coordinates and the vector $\begin{pmatrix} u \\ v \end{pmatrix}$ in the other set of coordinates.

Similarly, if we solve the two equations for x and y, we have

$$x = \frac{1}{a}u \text{ and } y = \frac{1}{b}v$$

$$\therefore \begin{pmatrix} x \\ y \end{pmatrix} = \begin{pmatrix} 1/a & 0 \\ 0 & 1/b \end{pmatrix} \begin{pmatrix} u \\ v \end{pmatrix}$$

which allows us to transform back from the u–v plane coordinates to the x–y plane coordinates.

Now for an example.

▶

Example

If $X = \begin{pmatrix} x \\ y \end{pmatrix} = \begin{pmatrix} 2 \\ 1 \end{pmatrix}$ with the transformation $T = \begin{pmatrix} -2 & 0 \\ 2 & 1 \end{pmatrix}$ determine

$U = \begin{pmatrix} u \\ v \end{pmatrix} = TX$ and show the positions on the *x–y* and *u–v* planes.

In this case

$$\begin{pmatrix} u \\ v \end{pmatrix} = \begin{pmatrix} -2 & 0 \\ 2 & 1 \end{pmatrix} \begin{pmatrix} 2 \\ 1 \end{pmatrix} = \begin{pmatrix} -4 \\ 5 \end{pmatrix}$$

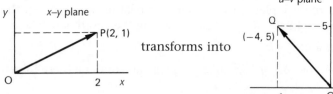

transforms into

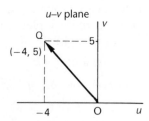

If **T** is non-singular and **U = TX** then **X = T⁻¹U** and since

$$T = \begin{pmatrix} -2 & 0 \\ 2 & 1 \end{pmatrix} \text{ then } T^{-1} = \ldots\ldots\ldots$$

61

$$T^{-1} = \begin{pmatrix} -1/2 & 0 \\ 1 & 1 \end{pmatrix}$$

There are several ways of finding the inverse of a matrix. One method is as follows.

$$T = \begin{pmatrix} -2 & 0 \\ 2 & 1 \end{pmatrix}$$

$$\begin{pmatrix} -2 & 0 & | & 1 & 0 \\ 2 & 1 & | & 0 & 1 \end{pmatrix} \sim \begin{pmatrix} -2 & 0 & | & 1 & 0 \\ 0 & 1 & | & 1 & 1 \end{pmatrix}$$

$$\sim \begin{pmatrix} 1 & 0 & | & -1/2 & 0 \\ 0 & 1 & | & 1 & 1 \end{pmatrix}$$

$$\therefore T^{-1} = \begin{pmatrix} -1/2 & 0 \\ 1 & 1 \end{pmatrix}$$

So we have **U = TX** ∴ **X = T⁻¹U**

$$\therefore \begin{pmatrix} x \\ y \end{pmatrix} = \begin{pmatrix} -1/2 & 0 \\ 1 & 1 \end{pmatrix} \begin{pmatrix} u \\ v \end{pmatrix}$$

Hence a vector $\begin{pmatrix} 1 \\ 4 \end{pmatrix}$ in the *u–v* plane transforms into $\begin{pmatrix} x \\ y \end{pmatrix}$ in the *x–y*

plane where $\begin{pmatrix} x \\ y \end{pmatrix} = \ldots\ldots\ldots$

62

$$\left(\begin{array}{c} x \\ y \end{array}\right) = \left(\begin{array}{c} -1/2 \\ 5 \end{array}\right)$$

$$\left(\begin{array}{c} x \\ y \end{array}\right) = \left(\begin{array}{cc} -1/2 & 0 \\ 1 & 1 \end{array}\right)\left(\begin{array}{c} 1 \\ 4 \end{array}\right) = \left(\begin{array}{c} -1/2 \\ 5 \end{array}\right)$$

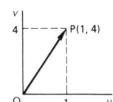

transforms into

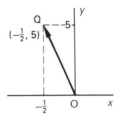

Rotation of axes

A more interesting case occurs with a degree of rotation between the two sets of coordinate axes.

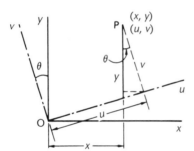

Let P be the point (x, y) in the x–y plane and the point (u, v) in the u–v plane.

Let θ be the angle of rotation between the two systems. From the diagram we can see that

$$\left.\begin{array}{l} x = u\cos\theta - v\sin\theta \\ y = u\sin\theta + v\cos\theta \end{array}\right\} \tag{1}$$

In matrix form, this becomes $\left(\begin{array}{c} x \\ y \end{array}\right) = \left(\begin{array}{cc} \cos\theta & -\sin\theta \\ \sin\theta & \cos\theta \end{array}\right)\left(\begin{array}{c} u \\ v \end{array}\right)$

which enables us to transform from the u–v plane coordinates to the corresponding x–y plane coordinates.

Make a note of this and then move on

63

If we solve equations (1) for u and v, we have

$$x \sin \theta = u \sin \theta \cos \theta - v \sin^2 \theta$$
$$y \cos \theta = u \sin \theta \cos \theta + v \cos^2 \theta$$
$$\therefore \ y \cos \theta - x \sin \theta = v(\cos^2 \theta + \sin^2 \theta) = v$$

Also
$$x \cos \theta = u \cos^2 \theta - v \sin \theta \cos \theta$$
$$y \sin \theta = u \sin^2 \theta + v \sin \theta \cos \theta$$
$$\therefore \ x \cos \theta + y \sin \theta = u(\cos^2 \theta + \sin^2 \theta) = u$$

So
$$u = x \cos \theta + y \sin \theta$$
$$v = -x \sin \theta + y \cos \theta$$

and written in matrix form, this is . . . .

64

$$\boxed{\begin{pmatrix} u \\ v \end{pmatrix} = \begin{pmatrix} \cos \theta & \sin \theta \\ -\sin \theta & \cos \theta \end{pmatrix} \begin{pmatrix} x \\ y \end{pmatrix}}$$

So we have

$$\begin{pmatrix} x \\ y \end{pmatrix} = \begin{pmatrix} \cos \theta & -\sin \theta \\ \sin \theta & \cos \theta \end{pmatrix} \begin{pmatrix} u \\ v \end{pmatrix}$$

and
$$\begin{pmatrix} u \\ v \end{pmatrix} = \begin{pmatrix} \cos \theta & \sin \theta \\ -\sin \theta & \cos \theta \end{pmatrix} \begin{pmatrix} x \\ y \end{pmatrix}$$

i.e. $\mathbf{X} = \mathbf{TU}$ and $\mathbf{U} = \mathbf{T^{-1}X}$

where $\mathbf{T}$ is the matrix of transformation and the equations provide a linear transformation between the two sets of coordinates.

Example

If the u–v plane axes rotate through $30°$ in an anticlockwise manner from the x–y plane axes, determine the (u, v) coordinates of a point whose (x, y) coordinates are $x = 2$, $y = 3$ in the x–y plane.

This is a straightforward application of the results above.

So $\quad \begin{pmatrix} u \\ v \end{pmatrix} = $ . . . .

$$\boxed{\begin{pmatrix} u \\ v \end{pmatrix} = \begin{pmatrix} \sqrt{3} + 3/2 \\ -1 + 3\sqrt{3}/2 \end{pmatrix} = \begin{pmatrix} 3\cdot23 \\ 1\cdot60 \end{pmatrix}}$$

65

Because

$$\begin{pmatrix} u \\ v \end{pmatrix} = \begin{pmatrix} \cos\theta & \sin\theta \\ -\sin\theta & \cos\theta \end{pmatrix} \begin{pmatrix} 2 \\ 3 \end{pmatrix} \qquad \cos\theta = \sqrt{3}/2$$
$$\sin\theta = 1/2$$

$$= \begin{pmatrix} \sqrt{3}/2 & 1/2 \\ -1/2 & \sqrt{3}/2 \end{pmatrix} \begin{pmatrix} 2 \\ 3 \end{pmatrix}$$

$$= \begin{pmatrix} \sqrt{3} + 3/2 \\ -1 + 3\sqrt{3}/2 \end{pmatrix} = \begin{pmatrix} 3\cdot23 \\ 1\cdot60 \end{pmatrix}$$

As usual, the Programme ends with the **Revision summary**, to be read in conjunction with the **Can you?** checklist. Go back to the relevant part of the Programme for any points on which you are unsure. The **Test exercise** should then be straightforward and the **Further problems** give valuable additional practice.

Revision summary 16

1 *Singular* square matrix: $|\mathbf{A}| = 0$
   Non-singular square matrix $|\mathbf{A}| \neq 0$.

66

2 *Rank of a matrix* – order of the largest non-zero determinant that can be formed from the elements of the matrix.

3 *Elementary operations and equivalent matrices*
   Each of the following row operations on matrix **A** produces a *row equivalent matrix* **B** where the order and rank of **B** are the same as those of **A**. We write **A** ~ **B**.

   (1) Interchanging two rows
   (2) Multiplying each element of a row by the same non-zero scalar quantity
   (3) Adding or subtracting corresponding elements from those of another row.

   These operations are called *elementary row operations*. There is a corresponding set of three *elementary column operations* that can be used to form *column equivalent matrices*.

▶

4   *Consistency* of a set of n equations in n unknowns with coefficient matrix $\mathbf{A}$ and augmented matrix $\mathbf{A_b}$.

   (a) Consistent if rank of $\mathbf{A} = $ rank of $\mathbf{A_b}$

   (b) Inconsistent if rank of $\mathbf{A} < $ rank of $\mathbf{A_b}$.

5   *Uniqueness of solutions – n equations with n unknowns.*

   (a) rank of $\mathbf{A} = $ rank of $\mathbf{A_b} = n$ *unique solutions*

   (b) rank of $\mathbf{A} = $ rank of $\mathbf{A_b} = m < n$ *infinite number of solutions*

   (c) rank of $\mathbf{A} < $ rank of $\mathbf{A_b}$ *no solution*

6   *Solution of sets of equations*

   (a) *Inverse matrix method* $\mathbf{Ax} = \mathbf{b}$; $\mathbf{x} = \mathbf{A}^{-1}\mathbf{b}$

   To find $\mathbf{A}^{-1}$

   (1) evaluate $|\mathbf{A}|$

   (2) form $\mathbf{C}$, the matrix of cofactors of $\mathbf{A}$

   (3) write $\mathbf{C}^{\mathrm{T}}$, the transpose of $\mathbf{A}$

   (4) $\mathbf{A}^{-1} = \dfrac{1}{|\mathbf{A}|} \times \mathbf{C}^{\mathrm{T}}$.

   (b) *Row transformation method* $\mathbf{Ax} = \mathbf{b}$; $\mathbf{Ax} = \mathbf{Ib}$

   (1) form the combined coefficient matrix $[\mathbf{A}|\mathbf{I}]$

   (2) row transformations to convert to $[\mathbf{I}|\mathbf{A}^{-1}]$

   (3) then solve $\mathbf{x} = \mathbf{A}^{-1}\mathbf{b}$.

   (c) *Gaussian elimination method* $\mathbf{Ax} = \mathbf{b}$

   (1) form augmented matrix $[\mathbf{A}|\mathbf{b}]$

   (2) operate on rows to convert to $[\mathbf{U}|\mathbf{b}']$ where $\mathbf{U}$ is the upper-triangular matrix.

   (3) expand from bottom row to obtain $\mathbf{x}$.

   (d) *Triangular decomposition method* $\mathbf{Ax} = \mathbf{b}$

   Write $\mathbf{A}$ as the product of upper and lower triangular matrices.

   $\mathbf{A} = \mathbf{LU}$, $\mathbf{L}(\mathbf{Ux}) = \mathbf{b}$. Put $\mathbf{Ux} = \mathbf{y}$ $\therefore$ $\mathbf{Ly} = \mathbf{b}$

   (1) solve $\mathbf{Ly} = \mathbf{b}$ to obtain $\mathbf{y}$
   (2) solve $\mathbf{Ux} = \mathbf{y}$ to obtain $\mathbf{x}$.

   (e) *Using an electronic spreadsheet*

   The spreadsheet used to demonstrate the method is the *Excel* spreadsheet provided by Microsoft. The two functions used are:

   MINVERSE(array)   for obtaining the inverse of a matrix
   MMULT(array1, array2)   for multiplying two matrices together

7 *Matrix transformation*

(a) **U = TX**, where **T** is a transformation matrix, transforms a vector in the *x–y* plane to a corresponding vector in the *u–v* plane. Similarly, **X = T⁻¹U** converts a vector in the *u–v* plane to a corresponding vector in the *x–y* plane.

(b) *Rotation of axes*

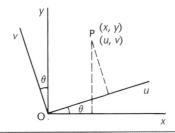

$$\begin{pmatrix} u \\ v \end{pmatrix} = \begin{pmatrix} \cos\theta & \sin\theta \\ -\sin\theta & \cos\theta \end{pmatrix}\begin{pmatrix} x \\ y \end{pmatrix}$$

$$\begin{pmatrix} x \\ y \end{pmatrix} = \begin{pmatrix} \cos\theta & -\sin\theta \\ \sin\theta & \cos\theta \end{pmatrix}\begin{pmatrix} u \\ v \end{pmatrix}$$

 # Can you?

Checklist 16

67

Check this list before and after you try the end of Programme test.

On a scale of 1 to 5 how confident are you that you can: **Frames**

- Determine whether a matrix is singular or non-singular?   [1] to [3]
  Yes ☐ ☐ ☐ ☐ ☐   *No*

- Determine the rank of a matrix?   [3] to [13]
  Yes ☐ ☐ ☐ ☐ ☐   *No*

- Determine the consistency of a set of linear equations and hence demonstrate the uniqueness of their solution?   [14] to [23]
  Yes ☐ ☐ ☐ ☐ ☐   *No*

- Obtain the solution of a set of simultaneous linear equations by using matrix inversion, by row transformation, by Gaussian elimination, by triangular decomposition and by using a spreadsheet?   [24] to [59]
  Yes ☐ ☐ ☐ ☐ ☐   *No*

- Use matrices to represent transformations between coordinate systems?   [60] to [65]
  Yes ☐ ☐ ☐ ☐ ☐   *No*

Test exercise 16

68

1   Determine the rank of **A** and of $\mathbf{A_b}$ for the following sets of equations and hence determine the nature of the solutions. Do *not* solve the equations.

(a) $x_1 + 3x_2 - 2x_3 = 6$ (b) $x_1 + 2x_2 - 4x_3 = 3$
    $4x_1 + 5x_2 + 2x_3 = 3$ $x_1 + 2x_2 + 3x_3 = -4$
    $x_1 + 3x_2 + 4x_3 = 7$ $2x_1 + 4x_2 + x_3 = -3.$

2   If $\mathbf{Ax} = \mathbf{b}$ where $\mathbf{A} = \begin{pmatrix} 2 & 3 & -2 \\ 3 & 5 & -4 \\ 1 & 2 & -3 \end{pmatrix}$ and $\mathbf{b} = \begin{pmatrix} 4 \\ 10 \\ 9 \end{pmatrix}$, determine $\mathbf{A}^{-1}$ and hence solve the set of equations.

3   Given that $3x_1 + 2x_2 + x_3 = 1$
                $x_1 - x_2 + 3x_3 = 5$
                $2x_1 + 5x_2 - 2x_3 = 0$

    apply the method of row transformation to obtain the value of x_1, x_2, x_3.

4   By the method of Gaussian elimination, solve the equations $\mathbf{Ax} = \mathbf{b}$, where

$$\mathbf{A} = \begin{pmatrix} 1 & -2 & -4 \\ 2 & 1 & -3 \\ 1 & 3 & 2 \end{pmatrix} \text{ and } \mathbf{b} = \begin{pmatrix} -3 \\ 4 \\ 5 \end{pmatrix}.$$

5   If $\mathbf{Ax} = \mathbf{b}$ where $\mathbf{A} = \begin{pmatrix} 1 & -2 & 1 \\ 3 & 1 & -2 \\ 5 & 3 & 3 \end{pmatrix}$ and $\mathbf{b} = \begin{pmatrix} 7 \\ -3 \\ 5 \end{pmatrix}$, express **A** as the product

    $\mathbf{A} = \mathbf{LU}$ where **L** and **U** are lower and upper-triangular matrices and hence determine the values of x_1, x_2, x_3.

6   Use an electronic spreadsheet to solve the set of equations:
    $3a - 2b + 4c - d + e = 32$
    $-a + 3b - 2c + 5d + 3e = -3$
    $a - b + c - d + e = 12$
    $4a - 6b + 2c + 8d - e = 20$
    $a - 5b - 7c + 2d - 3e = -40$

7   (a) Determine the vector in the *u–v* plane formed by $\mathbf{U} = \mathbf{TX}$, where the transformation matrix is $\mathbf{T} = \begin{pmatrix} -2 & 1 \\ 3 & 4 \end{pmatrix}$ and $\mathbf{X} = \begin{pmatrix} 3 \\ -2 \end{pmatrix}$ is a vector in the *x–y* plane.

    (b) The coordinate axes in the *x–y* plane and in the *u–v* plane have the same origin O, but OU is inclined to OX at an angle of 60° in an anticlockwise manner. Transform a vector $\mathbf{X} = \begin{pmatrix} 4 \\ 6 \end{pmatrix}$ in the *x–y* plane into the corresponding vector in the *u–v* plane.

Further problems 16

69

1 If $\mathbf{Ax = b}$ where $\mathbf{A} = \begin{pmatrix} 5 & 2 & 3 \\ 3 & -2 & -2 \\ 4 & 3 & 1 \end{pmatrix}$ and $\mathbf{b} = \begin{pmatrix} 6 \\ 5 \\ -5 \end{pmatrix}$, determine $\mathbf{A}^{-1}$ and

hence solve the set of equations.

2 Apply the method of row transformation to solve the following sets of equations.

(a) $\quad x_1 - 3x_2 - 2x_3 = 8$
$\qquad 2x_1 + 2x_2 + x_3 = 4$
$\qquad 3x_1 - 4x_2 + 2x_3 = -3$

(b) $\quad x_1 - 3x_2 + 2x_3 = 8$
$\qquad 2x_1 - x_2 + x_3 = 9$
$\qquad 3x_1 + 2x_2 + 3x_3 = 5.$

3 Solve the following sets of equations by Gaussian elimination.

(a) $\quad x_1 - 2x_2 - x_3 + 3x_4 = 4$
$\qquad 2x_1 + x_2 + x_3 - 4x_4 = 3$
$\qquad 3x_1 - x_2 - 2x_3 + 2x_4 = 6$
$\qquad x_1 + 3x_2 - x_3 + x_4 = 8$

(b) $\quad 2x_1 + 3x_2 - 2x_3 + 2x_4 = 2$
$\qquad 4x_1 + 2x_2 - 3x_3 - x_4 = 6$
$\qquad x_1 - x_2 + 4x_3 - 2x_4 = 7$
$\qquad 3x_1 + 2x_2 + x_3 - x_4 = 5$

(c) $\quad x_1 + 2x_2 + 5x_3 + x_4 = 4$
$\qquad 3x_1 - 4x_2 + 3x_3 - 2x_4 = 7$
$\qquad 4x_1 + 3x_2 + 2x_3 - x_4 = 1$
$\qquad x_1 - 2x_2 - 4x_3 - x_4 = 2.$

4 Using the method of triangular decomposition, solve the following sets of equations.

(a) $\begin{pmatrix} 1 & 4 & -1 \\ 4 & 2 & 3 \\ 7 & -3 & 2 \end{pmatrix} \begin{pmatrix} x_1 \\ x_2 \\ x_3 \end{pmatrix} = \begin{pmatrix} -2 \\ -1 \\ -18 \end{pmatrix}$

(b) $\begin{pmatrix} 1 & -2 & 3 \\ 2 & 1 & -5 \\ 6 & -3 & 2 \end{pmatrix} \begin{pmatrix} x_1 \\ x_2 \\ x_3 \end{pmatrix} = \begin{pmatrix} -2 \\ 17 \\ 22 \end{pmatrix}$

(c) $\begin{pmatrix} 1 & -2 & 3 & -1 \\ 3 & 1 & -3 & 2 \\ 5 & 3 & 2 & 3 \\ 2 & -4 & -2 & 4 \end{pmatrix} \begin{pmatrix} x_1 \\ x_2 \\ x_3 \\ x_4 \end{pmatrix} = \begin{pmatrix} -3 \\ 14 \\ 21 \\ -10 \end{pmatrix}.$

5 Use an electronic spreadsheet to solve all the equations in questions 2, 3 and 4. [Hint: For all those sets of three equations you only need a single template just change the numbers. The same applies for all those sets of four equations.

▶

6   Invert the matrix $\mathbf{A} = \begin{pmatrix} 8 & 10 & 7 \\ 5 & 9 & 4 \\ 9 & 11 & 8 \end{pmatrix}$ and hence solve the equations

$$8I_1 + 10I_2 + 7I_3 = 0$$
$$5I_1 + 9I_2 + 4I_3 = -9$$
$$9I_1 + 11I_2 + 8I_3 = 1.$$

7   If $\mathbf{A} = \begin{pmatrix} 1 & 2 & 3 \\ 4 & 6 & 7 \\ 5 & 8 & 9 \end{pmatrix}$ and $\mathbf{B} = \begin{pmatrix} -2 & 6 & -4 \\ -1 & -6 & 5 \\ 2 & 2 & -2 \end{pmatrix}$, verify that $\mathbf{AB} = k\mathbf{I}$ where $\mathbf{I}$ is a

unit matrix and k is a constant. Hence solve the equations

$$x_1 + 2x_2 + 3x_3 = 2$$
$$4x_1 + 6x_2 + 7x_3 = 2$$
$$5x_1 + 8x_2 + 9x_3 = 3.$$

Systems of ordinary differential equations

Learning outcomes

When you have completed this Programme you will be able to:

- Obtain the eigenvalues and corresponding eigenvectors of a square matrix
- Demonstrate the validity of the Cayley–Hamilton theorem
- Solve systems of first-order ordinary differential equations using eigenvalue and eigenvector methods
- Construct the modal matrix from the eigenvectors of a matrix and the spectral matrix from the eigenvalues
- Solve systems of second-order ordinary differential equations using diagonalisation

Eigenvalues and eigenvectors

1 Introduction

Matrices commonly appear in technological problems, for example those involving coupled oscillations and vibrations, and give rise to equations of the form

$$\mathbf{A}\mathbf{x} = \lambda\mathbf{x}$$

where $\mathbf{A} = (a_{ij})$ is a square matrix, $\mathbf{x} = (x_i)$ is a column matrix and λ is a scalar quantity, that is a number.

For non-trivial solutions, that is for $\mathbf{x} \neq \mathbf{0}$, the values of λ are called the *eigenvalues*, *characteristic values* or *latent roots* of the matrix $\mathbf{A}$ and the corresponding solutions of the given equations $\mathbf{A}\mathbf{x} = \lambda\mathbf{x}$ are called the *eigenvectors*, or *characteristic vectors* of $\mathbf{A}$ (refer to *Engineering Mathematics, Sixth Edition*, pages 578ff).

The set of equations

$$\begin{pmatrix} a_{11} & a_{12} & \cdots & a_{1n} \\ a_{21} & a_{22} & \cdots & a_{2n} \\ \vdots & \vdots & & \vdots \\ a_{n1} & a_{n2} & \cdots & a_{nn} \end{pmatrix} \begin{pmatrix} x_1 \\ x_2 \\ \vdots \\ x_n \end{pmatrix} = \lambda \begin{pmatrix} x_1 \\ x_2 \\ \vdots \\ x_n \end{pmatrix}$$

then simplifies to

$$\begin{pmatrix} (a_{11} - \lambda) & a_{12} & \cdots & a_{1n} \\ a_{21} & (a_{22} - \lambda) & \cdots & a_{2n} \\ \vdots & \vdots & & \vdots \\ a_{n1} & a_{n2} & & (a_{nn} - \lambda) \end{pmatrix} \begin{pmatrix} x_1 \\ x_2 \\ \vdots \\ x_n \end{pmatrix} = \begin{pmatrix} 0 \\ 0 \\ \vdots \\ 0 \end{pmatrix}$$

That is, $\mathbf{A}\mathbf{x} = \lambda\mathbf{x}$ becomes $\mathbf{A}\mathbf{x} - \lambda\mathbf{x} = \mathbf{0}$

$$\text{i.e.} \quad (\mathbf{A} - \lambda\mathbf{I})\mathbf{x} = \mathbf{0}$$

the unit matrix $\mathbf{I}$ being introduced since we can subtract only a matrix from another matrix.

For this set of homogeneous linear equations (right-hand side constant terms all zero) to have non-trivial solutions

$$|\mathbf{A} - \lambda\mathbf{I}| \text{ must be zero}$$

This is called the *characteristic determinant* of $\mathbf{A}$ and $|\mathbf{A} - \lambda\mathbf{I}| = 0$ is the *characteristic equation*, the solution of which gives the values of λ , i.e. the eigenvalues of $\mathbf{A}$.

Example 1

2

Find the eigenvalues and corresponding eigenvectors of

$$\mathbf{Ax} = \lambda\mathbf{x} \text{ where } \mathbf{A} = \begin{pmatrix} 2 & 3 \\ 4 & 1 \end{pmatrix}.$$

The characteristic equation is $|\mathbf{A} - \lambda\mathbf{I}| = 0$

i.e. $\begin{vmatrix} 2 - \lambda & 3 \\ 4 & 1 - \lambda \end{vmatrix} = 0$, which, when expanded, gives

$$\lambda_1 = \ldots\ldots\ldots \quad \text{and} \quad \lambda_2 = \ldots\ldots\ldots$$

$$\boxed{\lambda_1 = -2 \quad \text{and} \quad \lambda_2 = 5}$$

3

Because

$$(2 - \lambda)(1 - \lambda) - 12 = 0 \quad \therefore \; 2 - 3\lambda + \lambda^2 - 12 = 0$$

$$\lambda^2 - 3\lambda - 10 = 0 \quad (\lambda - 5)(\lambda + 2) = 0 \quad \therefore \; \lambda = -2 \text{ or } 5$$

Now we substitute each value of λ in turn in the equation

$$(\mathbf{A} - \lambda\mathbf{I})\mathbf{x} = 0$$

With $\lambda = -2$

$$\left\{ \begin{pmatrix} 2 & 3 \\ 4 & 1 \end{pmatrix} - (-2)\begin{pmatrix} 1 & 0 \\ 0 & 1 \end{pmatrix} \right\} \begin{pmatrix} x_1 \\ x_2 \end{pmatrix} = \begin{pmatrix} 0 \\ 0 \end{pmatrix}$$

$$\left\{ \begin{pmatrix} 2 & 3 \\ 4 & 1 \end{pmatrix} + \begin{pmatrix} 2 & 0 \\ 0 & 2 \end{pmatrix} \right\} \begin{pmatrix} x_1 \\ x_2 \end{pmatrix} = \begin{pmatrix} 0 \\ 0 \end{pmatrix}$$

$$\begin{pmatrix} 4 & 3 \\ 4 & 3 \end{pmatrix} \begin{pmatrix} x_1 \\ x_2 \end{pmatrix} = \begin{pmatrix} 0 \\ 0 \end{pmatrix}$$

Multiplying out the left-hand side, we get

$$\ldots\ldots\ldots\ldots$$

$$\boxed{4x_1 + 3x_2 = 0}$$

4

from which we get $x_2 = -\frac{4}{3}x_1$ i.e. not specific values for x_1 and x_2, but a relationship between them. Whatever value we assign to x_1 we obtain a corresponding value of x_2.

$$\mathbf{x}_1 = \begin{pmatrix} x_1 \\ x_2 \end{pmatrix} = \begin{pmatrix} 3 \\ -4 \end{pmatrix} \text{ or } \begin{pmatrix} 6 \\ -8 \end{pmatrix} \text{ or } \begin{pmatrix} 9 \\ -12 \end{pmatrix}, \text{ etc.}$$

The most convenient way to do this is to choose $x_1 = 1$ and then scale x_1 to obtain integer elements. So here we find for $x_1 = 1$ then $x_2 = -4/3$ so x_1 is of the form

$$\begin{pmatrix} 1 \\ -\dfrac{4}{3} \end{pmatrix}$$

▶

This is now scaled up by multiplying by 3 to give

$$\mathbf{x}_1 = \alpha \begin{pmatrix} 3 \\ -4 \end{pmatrix} \text{ where } \alpha \text{ is a constant multiplier.}$$

The simplest result, with $\alpha = 1$, is the one normally quoted.

$$\therefore \text{ for } \lambda_1 = -2, \quad \mathbf{x}_1 = \begin{pmatrix} 3 \\ -4 \end{pmatrix}$$

Similarly, for $\lambda_2 = 5$, the corresponding eigenvector is ....

5

$$\boxed{\mathbf{x}_2 = \begin{pmatrix} 1 \\ 1 \end{pmatrix}}$$

Because, with $\lambda_2 = 5$, $(\mathbf{A} - \lambda \mathbf{I})\mathbf{x} = \mathbf{0}$ becomes

$$\left\{ \begin{pmatrix} 2 & 3 \\ 4 & 1 \end{pmatrix} - 5 \begin{pmatrix} 1 & 0 \\ 0 & 1 \end{pmatrix} \right\} \begin{pmatrix} x_1 \\ x_2 \end{pmatrix} = \begin{pmatrix} 0 \\ 0 \end{pmatrix}$$

$$\left\{ \begin{pmatrix} 2 & 3 \\ 4 & 1 \end{pmatrix} - \begin{pmatrix} 5 & 0 \\ 0 & 5 \end{pmatrix} \right\} \begin{pmatrix} x_1 \\ x_2 \end{pmatrix} = \begin{pmatrix} 0 \\ 0 \end{pmatrix}$$

$$\begin{pmatrix} -3 & 3 \\ 4 & -4 \end{pmatrix} \begin{pmatrix} x_1 \\ x_2 \end{pmatrix} = \begin{pmatrix} 0 \\ 0 \end{pmatrix}$$

$\therefore \quad -3x_1 + 3x_2 = 0 \quad \text{i.e. } x_2 = x_1$

$\therefore \quad$ with $\lambda_2 = 5$, the corresponding eigenvector is $\mathbf{x}_2 = \beta \begin{pmatrix} 1 \\ 1 \end{pmatrix}$

Again, taking $\beta = 1$, for $\lambda_2 = 5$, $\quad \mathbf{x}_2 = \begin{pmatrix} 1 \\ 1 \end{pmatrix}$

So the required eigenvectors are

$$\mathbf{x}_1 = \begin{pmatrix} 3 \\ -4 \end{pmatrix} \quad \text{corresponding to } \lambda_1 = -2$$

$$\mathbf{x}_2 = \begin{pmatrix} 1 \\ 1 \end{pmatrix} \quad \text{corresponding to } \lambda_2 = 5.$$

Example 2

Determine the eigenvalues and corresponding eigenvectors of

$$\mathbf{A}\mathbf{x} = \lambda \mathbf{x} \text{ where } \mathbf{A} = \begin{pmatrix} 3 & 10 \\ 2 & 4 \end{pmatrix}.$$

The characteristic equation is $|\mathbf{A} - \lambda \mathbf{I}| = 0$, which in this case can be written as

....

$$\begin{vmatrix} 3 - \lambda & 10 \\ 2 & 4 - \lambda \end{vmatrix} = 0$$

Expanding the determinant and solving the equation gives

$$\lambda_1 = \ldots\ldots\ldots\ldots; \quad \lambda_2 = \ldots\ldots\ldots\ldots$$

$$\lambda_1 = -1; \quad \lambda_2 = 8$$

Because the equation is $(3 - \lambda)(4 - \lambda) - 20 = 0 \quad \therefore \lambda^2 - 7\lambda - 8 = 0$

$\therefore (\lambda + 1)(\lambda - 8) = 0 \quad \therefore \lambda = -1$ or 8

(a) With $\lambda_1 = -1$, we solve $(\mathbf{A} - \lambda \mathbf{I})\mathbf{x} = \mathbf{0}$ to obtain an eigenvector, which is

$$\ldots\ldots\ldots\ldots$$

$$\mathbf{x_1} = \begin{pmatrix} 5 \\ -2 \end{pmatrix}$$

Because

$$\mathbf{A} = \begin{pmatrix} 3 & 10 \\ 2 & 4 \end{pmatrix} \quad \therefore \quad \left\{ \begin{pmatrix} 3 & 10 \\ 2 & 4 \end{pmatrix} - (-1) \begin{pmatrix} 1 & 0 \\ 0 & 1 \end{pmatrix} \right\} \begin{pmatrix} x_1 \\ x_2 \end{pmatrix} = \begin{pmatrix} 0 \\ 0 \end{pmatrix}$$

$$\left\{ \begin{pmatrix} 3 & 10 \\ 2 & 4 \end{pmatrix} + \begin{pmatrix} 1 & 0 \\ 0 & 1 \end{pmatrix} \right\} \begin{pmatrix} x_1 \\ x_2 \end{pmatrix} = \begin{pmatrix} 0 \\ 0 \end{pmatrix}$$

$$\begin{pmatrix} 4 & 10 \\ 2 & 5 \end{pmatrix} \begin{pmatrix} x_1 \\ x_2 \end{pmatrix} = \begin{pmatrix} 0 \\ 0 \end{pmatrix}$$

$\therefore \ 4x_1 + 10x_2 = 0 \quad \therefore \ x_2 = -\dfrac{2}{5}x_1 \quad \mathbf{x_1} = \alpha \begin{pmatrix} 5 \\ -2 \end{pmatrix}$

$\therefore$ with $\alpha = 1 \quad \lambda_1 = -1$ and $\mathbf{x_1} = \begin{pmatrix} 5 \\ -2 \end{pmatrix}$

(b) In the same way the corresponding eigenvector $\mathbf{x_2}$ for $\lambda_2 = 8$ is

$$\ldots\ldots\ldots\ldots$$

9

$$\mathbf{x}_2 = \begin{pmatrix} 2 \\ 1 \end{pmatrix}$$

Because

$$\left\{ \begin{pmatrix} 3 & 10 \\ 2 & 4 \end{pmatrix} - 8 \begin{pmatrix} 1 & 0 \\ 0 & 1 \end{pmatrix} \right\} \begin{pmatrix} x_1 \\ x_2 \end{pmatrix} = \begin{pmatrix} 0 \\ 0 \end{pmatrix}$$

$$\left\{ \begin{pmatrix} 3 & 10 \\ 2 & 4 \end{pmatrix} - \begin{pmatrix} 8 & 0 \\ 0 & 8 \end{pmatrix} \right\} \begin{pmatrix} x_1 \\ x_2 \end{pmatrix} = \begin{pmatrix} 0 \\ 0 \end{pmatrix}$$

$$\begin{pmatrix} -5 & 10 \\ 2 & -4 \end{pmatrix} \begin{pmatrix} x_1 \\ x_2 \end{pmatrix} = \begin{pmatrix} 0 \\ 0 \end{pmatrix}$$

$$\therefore \ -5x_1 + 10x_2 = 0 \quad \therefore \ x_2 = \frac{1}{2}x_1 \quad \mathbf{x}_2 = \beta \begin{pmatrix} 2 \\ 1 \end{pmatrix}$$

$$\therefore \ \text{with } \beta = 1, \quad \lambda_2 = 8 \ \text{ and } \ \mathbf{x}_2 = \begin{pmatrix} 2 \\ 1 \end{pmatrix}$$

The same basic method can similarly be applied to third-order sets of equations.

Example 3

Determine the eigenvalues and eigenvectors of $\mathbf{Ax} = \lambda\mathbf{x}$ where

$$\mathbf{A} = \begin{pmatrix} 1 & 0 & 4 \\ 0 & 2 & 0 \\ 3 & 1 & -3 \end{pmatrix}.$$

As before, we have $(\mathbf{A} - \lambda\mathbf{I})\mathbf{x} = \mathbf{0}$ with characteristic equation $|\mathbf{A} - \lambda\mathbf{I}| = 0$.

i.e.
$$\begin{vmatrix} 1 - \lambda & 0 & 4 \\ 0 & 2 - \lambda & 0 \\ 3 & 1 & -3 - \lambda \end{vmatrix} = 0$$

Expanding this we have

$$\lambda_1 = \ldots\ldots\ldots\ldots; \quad \lambda_2 = \ldots\ldots\ldots\ldots; \quad \lambda_3 = \ldots\ldots\ldots\ldots$$

$$\boxed{\lambda_1 = 2; \quad \lambda_2 = 3; \quad \lambda_3 = -5}$$ **10**

Because

$(1 - \lambda)\{(2 - \lambda)(-3 - \lambda) - 0\} + 4\{0 - 3(2 - \lambda)\} = 0$

$(1 - \lambda)(2 - \lambda)(-3 - \lambda) - 12(2 - \lambda) = 0$

$\therefore \quad (2 - \lambda)\{(1 - \lambda)(-3 - \lambda) - 12\} = 0$

$\therefore \quad \lambda = 2 \quad \text{or} \quad \lambda^2 + 2\lambda - 15 = 0 \quad \therefore \quad (\lambda - 3)(\lambda + 5) = 0$

$\therefore \quad \lambda = 2, 3, \text{ or } -5$

(a) With $\lambda_1 = 2$, $(\mathbf{A} - \lambda\mathbf{I})\mathbf{x} = \mathbf{0}$ becomes

$$\left\{ \begin{pmatrix} 1 & 0 & 4 \\ 0 & 2 & 0 \\ 3 & 1 & -3 \end{pmatrix} - 2 \begin{pmatrix} 1 & 0 & 0 \\ 0 & 1 & 0 \\ 0 & 0 & 1 \end{pmatrix} \right\} \begin{pmatrix} x_1 \\ x_2 \\ x_3 \end{pmatrix} = \begin{pmatrix} 0 \\ 0 \\ 0 \end{pmatrix}$$

$$\left\{ \begin{pmatrix} 1 & 0 & 4 \\ 0 & 2 & 0 \\ 3 & 1 & -3 \end{pmatrix} - \begin{pmatrix} 2 & 0 & 0 \\ 0 & 2 & 0 \\ 0 & 0 & 2 \end{pmatrix} \right\} \begin{pmatrix} x_1 \\ x_2 \\ x_3 \end{pmatrix} = \begin{pmatrix} 0 \\ 0 \\ 0 \end{pmatrix}$$

$$\therefore \quad \begin{pmatrix} -1 & 0 & 4 \\ 0 & 0 & 0 \\ 3 & 1 & -5 \end{pmatrix} \begin{pmatrix} x_1 \\ x_2 \\ x_3 \end{pmatrix} = \begin{pmatrix} 0 \\ 0 \\ 0 \end{pmatrix}$$

from which a corresponding eigenvector $\mathbf{x}_1$ is ....

$$\boxed{\mathbf{x}_1 = \begin{pmatrix} 4 \\ -7 \\ 1 \end{pmatrix}}$$ **11**

Because we have $-x_1 + 4x_3 = 0 \qquad \therefore \quad x_3 = \frac{1}{4}x_1$

$3x_1 + x_2 - 5x_3 = 0 \quad \therefore \quad 3x_1 + x_2 - \frac{5}{4}x_1 = 0 \quad \therefore \quad x_2 = -\frac{7}{4}x_1$

$\therefore \quad x_1, x_2, x_3$ are in the ratio $1 : -\dfrac{7}{4} : \dfrac{1}{4}$ i.e. $4 : -7 : 1 \quad \therefore \quad \mathbf{x}_1 = \begin{pmatrix} 4 \\ -7 \\ 1 \end{pmatrix}$

(b) Similarly for $\lambda_2 = 3$, $(\mathbf{A} - \lambda\mathbf{I})\mathbf{x} = \mathbf{0}$

$$\left\{ \begin{pmatrix} 1 & 0 & 4 \\ 0 & 2 & 0 \\ 3 & 1 & -3 \end{pmatrix} - 3 \begin{pmatrix} 1 & 0 & 0 \\ 0 & 1 & 0 \\ 0 & 0 & 1 \end{pmatrix} \right\} \begin{pmatrix} x_1 \\ x_2 \\ x_3 \end{pmatrix} = \begin{pmatrix} 0 \\ 0 \\ 0 \end{pmatrix}$$

from which a corresponding eigenvector is

$$\mathbf{x}_2 = ...$$

12

$$\mathbf{x}_2 = \begin{pmatrix} 2 \\ 0 \\ 1 \end{pmatrix}$$

Because

$$\left\{ \begin{pmatrix} 1 & 0 & 4 \\ 0 & 2 & 0 \\ 3 & 1 & -3 \end{pmatrix} - \begin{pmatrix} 3 & 0 & 0 \\ 0 & 3 & 0 \\ 0 & 0 & 3 \end{pmatrix} \right\} \begin{pmatrix} x_1 \\ x_2 \\ x_3 \end{pmatrix} = \begin{pmatrix} 0 \\ 0 \\ 0 \end{pmatrix}$$

$$\begin{pmatrix} -2 & 0 & 4 \\ 0 & -1 & 0 \\ 3 & 1 & -6 \end{pmatrix} \begin{pmatrix} x_1 \\ x_2 \\ x_3 \end{pmatrix} = \begin{pmatrix} 0 \\ 0 \\ 0 \end{pmatrix}$$

$$\therefore -2x_1 + 4x_3 = 0 \quad \therefore x_3 = \tfrac{1}{2}x_1$$

Also $\quad -x_2 = 0 \quad \therefore x_2 = 0 \qquad \therefore \mathbf{x}_2 = \begin{pmatrix} 2 \\ 0 \\ 1 \end{pmatrix}$

(c) All that now remains is $\lambda_3 = -5$. A corresponding eigenvector $\mathbf{x}_3$ is

$$\mathbf{x}_3 = \ldots\ldots\ldots\ldots$$

Finish it on your own. Method just the same as before.

13

$$\mathbf{x}_3 = \begin{pmatrix} 2 \\ 0 \\ -3 \end{pmatrix}$$

Check the working.

$$\mathbf{A} = \begin{pmatrix} 1 & 0 & 4 \\ 0 & 2 & 0 \\ 3 & 1 & -3 \end{pmatrix} \quad \text{and} \quad \lambda_3 = -5 \quad \text{with} \quad (\mathbf{A} - \lambda\mathbf{I})\mathbf{x} = \mathbf{0}.$$

$$\left\{ \begin{pmatrix} 1 & 0 & 4 \\ 0 & 2 & 0 \\ 3 & 1 & -3 \end{pmatrix} + 5 \begin{pmatrix} 1 & 0 & 0 \\ 0 & 1 & 0 \\ 0 & 0 & 1 \end{pmatrix} \right\} \begin{pmatrix} x_1 \\ x_2 \\ x_3 \end{pmatrix} = \begin{pmatrix} 0 \\ 0 \\ 0 \end{pmatrix}$$

$$\begin{pmatrix} 6 & 0 & 4 \\ 0 & 7 & 0 \\ 3 & 1 & 2 \end{pmatrix} \begin{pmatrix} x_1 \\ x_2 \\ x_3 \end{pmatrix} = \begin{pmatrix} 0 \\ 0 \\ 0 \end{pmatrix}$$

$$\therefore 6x_1 + 4x_3 = 0 \quad \therefore x_3 = -\tfrac{3}{2}x_1$$

$$7x_2 = 0 \quad \therefore x_2 = 0 \qquad \therefore \mathbf{x}_3 = \begin{pmatrix} 2 \\ 0 \\ -3 \end{pmatrix}$$

▶

Collecting the results together, we finally have

$$\lambda_1 = 2, \quad \mathbf{x_1} = \begin{pmatrix} 4 \\ -7 \\ 1 \end{pmatrix}; \quad \lambda_2 = 3, \quad \mathbf{x_2} = \begin{pmatrix} 2 \\ 0 \\ 1 \end{pmatrix}; \quad \lambda_3 = -5, \quad \mathbf{x_3} = \begin{pmatrix} 2 \\ 0 \\ -3 \end{pmatrix}$$

Cayley–Hamilton theorem

The Cayley–Hamilton theorem states that every square matrix satisfies its characteristic equation. For example the matrix

14

$$\mathbf{A} = \begin{pmatrix} 2 & 3 \\ 4 & 1 \end{pmatrix}$$

of Frame 54 has the characteristic equation

$$\lambda^2 - 3\lambda - 10 = 0$$

and so the Cayley–Hamilton theorem tells us that

$$\mathbf{A}^2 - 3\mathbf{A} - 10\mathbf{I} = \mathbf{0}$$

To verify this we note that

$$\mathbf{A}^2 = \begin{pmatrix} 2 & 3 \\ 4 & 1 \end{pmatrix}\begin{pmatrix} 2 & 3 \\ 4 & 1 \end{pmatrix} = \begin{pmatrix} 16 & 9 \\ 12 & 13 \end{pmatrix} \quad \text{so that}$$

$$\mathbf{A}^2 - 3\mathbf{A} - 10\mathbf{I} = \begin{pmatrix} 16 & 9 \\ 12 & 13 \end{pmatrix} - 3\begin{pmatrix} 2 & 3 \\ 4 & 1 \end{pmatrix} - 10\begin{pmatrix} 1 & 0 \\ 0 & 1 \end{pmatrix}$$

$$= \begin{pmatrix} 16 & 9 \\ 12 & 13 \end{pmatrix} - \begin{pmatrix} 6 & 9 \\ 12 & 3 \end{pmatrix} - \begin{pmatrix} 10 & 0 \\ 0 & 10 \end{pmatrix} = \begin{pmatrix} 0 & 0 \\ 0 & 0 \end{pmatrix}$$

You try one. Verify that the matrix $\mathbf{A} = \begin{pmatrix} 3 & 10 \\ 2 & 4 \end{pmatrix}$ of Frame 5 with the characteristic equation

$$\lambda^2 - 7\lambda - 8 = 0$$

satisfies the Cayley–Hamilton theorem, that is . . . .

15

$$\boxed{\mathbf{A}^2 - 7\mathbf{A} - 8\mathbf{I} = 0}$$

Because

$$\mathbf{A}^2 = \begin{pmatrix} 3 & 10 \\ 2 & 4 \end{pmatrix}\begin{pmatrix} 3 & 10 \\ 2 & 4 \end{pmatrix} = \begin{pmatrix} 29 & 70 \\ 14 & 36 \end{pmatrix} \text{ so that}$$

$$\mathbf{A}^2 - 7\mathbf{A} - 8\mathbf{I} = \begin{pmatrix} 29 & 70 \\ 14 & 36 \end{pmatrix} - 7\begin{pmatrix} 3 & 10 \\ 2 & 4 \end{pmatrix} - 8\begin{pmatrix} 1 & 0 \\ 0 & 1 \end{pmatrix}$$

$$= \begin{pmatrix} 29 & 70 \\ 14 & 36 \end{pmatrix} - \begin{pmatrix} 21 & 70 \\ 14 & 28 \end{pmatrix} - \begin{pmatrix} 8 & 0 \\ 0 & 8 \end{pmatrix} = \begin{pmatrix} 0 & 0 \\ 0 & 0 \end{pmatrix}$$

Now on to something different

Systems of first-order ordinary differential equations

16

Matrix methods involving eigenvalues and their associated eigenvectors can be used to solve systems of coupled differential equations, though we shall only consider cases where the relevant eigenvalues are distinct. We proceed by example.

Example 1

Consider the system of two coupled ordinary differential equations

$$f_1'(x) = 2f_1(x) + 3f_2(x)$$
$$f_2'(x) = 4f_1(x) + f_2(x)$$
where $f_1(0) = 2$ and $f_2(0) = 1$

These can be written in matrix form as . . . .

17

$$\boxed{\begin{pmatrix} f_1'(x) \\ f_2'(x) \end{pmatrix} = \begin{pmatrix} 2 & 3 \\ 4 & 1 \end{pmatrix}\begin{pmatrix} f_1(x) \\ f_2(x) \end{pmatrix}}$$

That is

$$\mathbf{F}'(x) = \mathbf{A}\mathbf{F}(x)$$

where $\mathbf{F}(x) = \begin{pmatrix} f_1(x) \\ f_2(x) \end{pmatrix}$, $\mathbf{F}'(x) = \begin{pmatrix} f_1'(x) \\ f_2'(x) \end{pmatrix}$ and $\mathbf{A} = \begin{pmatrix} 2 & 3 \\ 4 & 1 \end{pmatrix}$ and where

$\mathbf{F}(0) = \begin{pmatrix} f_1(0) \\ f_2(0) \end{pmatrix} = \begin{pmatrix} 2 \\ 1 \end{pmatrix}$ are the boundary conditions in matrix form.

▶

The matrix differential equation $\mathbf{F}'(x) = \mathbf{AF}(x)$ is similar in form to the single differential equation $f'(x) = af(x)$ (a constant) which has solution $f(x) = \alpha e^{ax}$ (α constant), so to solve the matrix equation we try a solution of the form

$\mathbf{F}(x) = \mathbf{C}e^{kx}$ where the number k and the constants c_1 and c_2 of the

matrix $\mathbf{C} = \begin{pmatrix} c_1 \\ c_2 \end{pmatrix}$ are to be determined.

Substituting $\mathbf{F}(x) = \mathbf{C}e^{kx}$ into the matrix equation $\mathbf{F}'(x) = \mathbf{AF}(x)$ gives

. . . .

$$k\mathbf{C}e^{kx} = \mathbf{AC}e^{kx}$$

18

Because

$\mathbf{F}(x) = \mathbf{C}e^{kx}$ so $\mathbf{F}'(x) = k\mathbf{C}e^{kx}$. Since $\mathbf{F}'(x) = \mathbf{AF}(x)$ then $k\mathbf{C}e^{kx} = \mathbf{AC}e^{kx}$

Dividing both sides by e^{kx} gives

$k\mathbf{C} = \mathbf{AC}$ that is $\mathbf{AC} = k\mathbf{C}$.

So, from Frame 1, k is an *eigenvalue* of $\mathbf{A}$ and $\mathbf{C}$ is the corresponding *eigenvector*. Therefore, we must first find the eigenvalues of $\mathbf{A}$ and for this matrix these have been found earlier in Frames 2 to 5. They are

$\lambda = -2$ (and so $k = -2$) with corresponding eigenvector $\begin{pmatrix} 3 \\ -4 \end{pmatrix}$

$\lambda = 5$ (and so $k = 5$) with corresponding eigenvector $\begin{pmatrix} 1 \\ 1 \end{pmatrix}$

To each eigenvalue the matrix $\mathbf{F}(x) = \mathbf{C}e^{kx}$ is a solution. The complete solution to $\mathbf{F}' = \mathbf{AF}$ is then

$$\mathbf{F}_1(x) = \begin{pmatrix} \cdots \\ \cdots \end{pmatrix} a_1 e^{\cdots} \quad \text{and} \quad \mathbf{F}_2(x) = \begin{pmatrix} \cdots \\ \cdots \end{pmatrix} a_2 e^{\cdots}$$

$$\mathbf{F}_1(x) = \begin{pmatrix} 3 \\ -4 \end{pmatrix} a_1 e^{-2x} \quad \text{and} \quad \mathbf{F}_2(x) = \begin{pmatrix} 1 \\ 1 \end{pmatrix} a_2 e^{5x}$$

19

Because

$\mathbf{F}(x) = \mathbf{C}e^{kx}$ is the solution corresponding to the eigenvalue k with associated eigenvector $\mathbf{C}$.

The complete solution to the equation $\mathbf{F}'(x) = \mathbf{AF}(x)$ is then a combination of these two solutions in the form

$$\mathbf{F}(x) = A\begin{pmatrix} 3 \\ -4 \end{pmatrix} e^{-2x} + B\begin{pmatrix} 1 \\ 1 \end{pmatrix} e^{5x}$$

Applying the boundary conditions gives $\mathbf{F}(0) = \ldots\ldots\ldots\ldots$

20

$$\boxed{\mathbf{F}(0) = \begin{pmatrix} 3A + B \\ -4A + B \end{pmatrix} = \begin{pmatrix} 2 \\ 1 \end{pmatrix}}$$

Because

$$\mathbf{F}(x) = A\begin{pmatrix} 3 \\ -4 \end{pmatrix}e^{-2x} + B\begin{pmatrix} 1 \\ 1 \end{pmatrix}e^{5x} \text{ and so } \mathbf{F}(0) = A\begin{pmatrix} 3 \\ -4 \end{pmatrix} + B\begin{pmatrix} 1 \\ 1 \end{pmatrix}$$

$$= \begin{pmatrix} 3A + B \\ -4A + B \end{pmatrix} = \begin{pmatrix} 2 \\ 1 \end{pmatrix}$$

Therefore

$$\begin{matrix} 3A + B = 2 \\ -4A + B = 1 \end{matrix} \text{ with solution } A = 1/7 \text{ and } B = 11/7,$$

giving the final solution as $\mathbf{F}(x) = \ldots\ldots\ldots$

21

$$\boxed{\mathbf{F}(x) = \begin{pmatrix} 3/7 \\ -4/7 \end{pmatrix}e^{-2x} + \begin{pmatrix} 11/7 \\ 11/7 \end{pmatrix}e^{5x}}$$

Summary

To solve an equation of the form

$$\mathbf{F}'(x) = \mathbf{AF}(x)$$

1   Find the eigenvalues $\lambda_1, \lambda_2, \ldots, \lambda_n$ of $\mathbf{A}$ (assuming they are all distinct)
2   Find the associated eigenvectors $\mathbf{C}_1, \mathbf{C}_2, \ldots, \mathbf{C}_n$
3   Write the solution of the equation as $\mathbf{F}(x) = \sum_{r=1}^{n}(A_r e^{\lambda_r x})\mathbf{C}_r$ and use the boundary conditions to find the values of a_r for $r = 1, 2, \ldots, n$.

Now you try one.

Next frame

22   **Example 2**

The system of two coupled ordinary differential equations

$$\begin{matrix} f_1'(x) = 3f_1(x) + 10f_2(x) \\ f_2'(x) = 2f_1(x) + 4f_2(x) \end{matrix} \text{ where } f_1(0) = 0 \text{ and } f_2(0) = 1$$

has the solution (refer to Frames 5 to 9)

$$f_1(x) = \ldots\ldots\ldots$$
$$f_2(x) = \ldots\ldots\ldots$$

23

$$f_1(x) = -\frac{10}{9}e^{-x} + \frac{10}{9}e^{8x}$$

$$f_2(x) = \frac{4}{9}e^{-x} + \frac{5}{9}e^{8x}$$

Because

$$f_1'(x) = 3f_1(x) + 10f_2(x)$$
$$f_2'(x) = 2f_1(x) + 4f_2(x)$$

can be written in matrix form as

. . . .

24

$$\begin{pmatrix} f_1'(x) \\ f_2'(x) \end{pmatrix} = \begin{pmatrix} 3 & 10 \\ 2 & 4 \end{pmatrix} \begin{pmatrix} f_1(x) \\ f_2(x) \end{pmatrix}$$

That is

$$\mathbf{F}'(x) = \mathbf{A}\mathbf{F}(x)$$

where $\mathbf{F}(x) = \begin{pmatrix} f_1(x) \\ f_2(x) \end{pmatrix}$, $\mathbf{F}'(x) = \begin{pmatrix} f_1'(x) \\ f_2'(x) \end{pmatrix}$ and $\mathbf{A} = \begin{pmatrix} 3 & 10 \\ 2 & 4 \end{pmatrix}$ and where

$\mathbf{F}(0) = \begin{pmatrix} 0 \\ 1 \end{pmatrix}$.

To solve the matrix equation we first need the eigenvalues and associated eigenvectors of the matrix **A**. These have already been found in Frames 5 to 9 and they are

$\lambda = -1$ with corresponding eigenvector $\begin{pmatrix} 5 \\ -2 \end{pmatrix}$

$\lambda = 8$ with corresponding eigenvector $\begin{pmatrix} 2 \\ 1 \end{pmatrix}$

The complete solution of $\mathbf{F}' = \mathbf{A}\mathbf{F}$ is then

$$\mathbf{F}(x) = A\begin{pmatrix} 5 \\ -2 \end{pmatrix}e^{-x} + B\begin{pmatrix} 2 \\ 1 \end{pmatrix}e^{8x}$$

That is $f_1(x) = \ldots\ldots\ldots\ldots$
$f_2(x) = \ldots\ldots\ldots\ldots$

25

$$\boxed{\begin{aligned} f_1(x) &= 5Ae^{-x} + 2Be^{8x} \\ f_2(x) &= -2Ae^{-x} + Be^{8x} \end{aligned}}$$

Because

$$\mathbf{F}(x) = \begin{pmatrix} f_1(x) \\ f_2(x) \end{pmatrix} = A\begin{pmatrix} 5 \\ -2 \end{pmatrix}e^{-x} + B\begin{pmatrix} 2 \\ 1 \end{pmatrix}e^{8x}$$

and so

$$f_1(x) = 5Ae^{-x} + 2Be^{8x}$$
$$f_2(x) = -2Ae^{-x} + Be^{8x}$$

Applying the boundary conditions, we find

$$\mathbf{F}(0) = \begin{pmatrix} 0 \\ 1 \end{pmatrix} = \begin{pmatrix} \dots A + \dots B \\ \dots A + \dots B \end{pmatrix}$$

26

$$\boxed{\mathbf{F}(0) = \begin{pmatrix} 0 \\ 1 \end{pmatrix} = \begin{pmatrix} f_1(0) \\ f_2(0) \end{pmatrix} = \begin{pmatrix} 5A + 2B \\ -2A + B \end{pmatrix}}$$

Because

The boundary conditions are $f_1(0) = 0$ and $f_2(0) = 1$ therefore

$$\mathbf{F}(0) = \begin{pmatrix} 0 \\ 1 \end{pmatrix} = \begin{pmatrix} f_1(0) \\ f_2(0) \end{pmatrix} = \begin{pmatrix} 5A + 2B \\ -2A + B \end{pmatrix}$$

This gives the pair of simultaneous equations

$$\begin{aligned} 5A + 2B &= 0 \\ -2A + B &= 1 \end{aligned}$$ which have solution

$$A = \dots\dots\dots \quad \text{and} \quad B = \dots\dots\dots$$

27

$$\boxed{A = -2/9 \quad \text{and} \quad B = 5/9}$$

This gives the complete solution as

$$\mathbf{F}(x) = \begin{pmatrix} f_1(x) \\ f_2(x) \end{pmatrix} = \begin{pmatrix} -10/9 \\ 4/9 \end{pmatrix}e^{-x} + \begin{pmatrix} 10/9 \\ 5/9 \end{pmatrix}e^{8x}$$

$$f_1(x) = -\frac{10}{9}e^{-x} + \frac{10}{9}e^{8x}$$

$$f_2(x) = \frac{4}{9}e^{-x} + \frac{5}{9}e^{8x}$$

Diagonalisation of a matrix

Modal matrix

28

We have already discussed the eigenvalues and eigenvectors of a matrix $\mathbf{A}$ of order n. In this section we shall assume that all the eigenvalues are distinct. If the n eigenvectors $\mathbf{x}_i$ are arranged as columns of a square matrix, the *modal matrix* of $\mathbf{A}$, denoted by $\mathbf{M}$, is formed

i.e. $\mathbf{M} = (\mathbf{x}_1, \mathbf{x}_2, \mathbf{x}_3, \ldots, \mathbf{x}_n)$

For example, we have seen earlier that if

$$\mathbf{A} = \begin{pmatrix} 1 & 0 & 4 \\ 0 & 2 & 0 \\ 3 & 1 & -3 \end{pmatrix} \text{ then } \lambda_1 = 2, \lambda_2 = 3, \lambda_3 = -5$$

and the corresponding eigenvectors are

$$\mathbf{x}_1 = \begin{pmatrix} 4 \\ -7 \\ 1 \end{pmatrix}, \quad \mathbf{x}_2 = \begin{pmatrix} 2 \\ 0 \\ 1 \end{pmatrix}, \quad \mathbf{x}_3 = \begin{pmatrix} 2 \\ 0 \\ -3 \end{pmatrix}$$

Then the modal matrix $\mathbf{M} = \begin{pmatrix} 4 & 2 & 2 \\ -7 & 0 & 0 \\ 1 & 1 & -3 \end{pmatrix}$

Spectral matrix

Also, we define the *spectral matrix* of $\mathbf{A}$, i.e. $\mathbf{S}$, as a diagonal matrix with the eigenvalues only on the main diagonal

i.e. $\mathbf{S} = \begin{pmatrix} \lambda_1 & 0 & 0 & \ldots & 0 \\ 0 & \lambda_2 & 0 & \ldots & 0 \\ \vdots & \vdots & \vdots & & \vdots \\ 0 & 0 & 0 & \ldots & \lambda_n \end{pmatrix}$

So, in the example above, $\mathbf{S} = \ldots\ldots\ldots$

29

$$\boxed{\mathbf{S} = \begin{pmatrix} 2 & 0 & 0 \\ 0 & 3 & 0 \\ 0 & 0 & -5 \end{pmatrix}}$$

Note that the eigenvalues of $\mathbf{S}$ and $\mathbf{A}$ are the same.

So, if $\mathbf{A} = \begin{pmatrix} 5 & -6 & 1 \\ 1 & 1 & 0 \\ 3 & 0 & 1 \end{pmatrix}$ has eigenvalues $\lambda = 1, 2, 4$ and

corresponding eigenvectors $\begin{pmatrix} 0 \\ 1 \\ 6 \end{pmatrix}, \begin{pmatrix} 1 \\ 1 \\ 3 \end{pmatrix}, \begin{pmatrix} 3 \\ 1 \\ 3 \end{pmatrix}$

then $\mathbf{M} = \ldots\ldots\ldots$ and $\mathbf{S} = \ldots\ldots\ldots$

30

$$\mathbf{M} = \begin{pmatrix} 0 & 1 & 3 \\ 1 & 1 & 1 \\ 6 & 3 & 3 \end{pmatrix}; \quad \mathbf{S} = \begin{pmatrix} 1 & 0 & 0 \\ 0 & 2 & 0 \\ 0 & 0 & 4 \end{pmatrix}$$

Now how are these connected? Let us investigate.

The eigenvectors $\mathbf{x}$ arranged in the modal matrix satisfy the original equation

$$\mathbf{Ax} = \lambda\mathbf{x}$$

Also $\mathbf{M} = (\mathbf{x}_1 \quad \mathbf{x}_2 \quad \ldots \quad \mathbf{x}_n)$

Then $\mathbf{AM} = \mathbf{A}(\mathbf{x}_1 \quad \mathbf{x}_2 \quad \ldots \quad \mathbf{x}_n)$

$$= (\mathbf{Ax}_1 \quad \mathbf{Ax}_2 \quad \ldots \quad \mathbf{Ax}_n)$$

$$= (\lambda_1\mathbf{x}_1 \quad \lambda_2\mathbf{x}_2 \quad \ldots \quad \lambda_n\mathbf{x}_n) \quad \text{since } \mathbf{Ax} = \lambda\mathbf{x}$$

Now $\mathbf{S} = \begin{pmatrix} \lambda_1 & 0 & \ldots & 0 \\ 0 & \lambda_2 & \ldots & 0 \\ \vdots & \vdots & & \vdots \\ 0 & 0 & \ldots & \lambda_n \end{pmatrix}$ $\therefore (\lambda_1\mathbf{x}_1 \quad \lambda_2\mathbf{x}_2 \quad \ldots \quad \lambda_n\mathbf{x}_n) = \mathbf{MS}$

$$\therefore \mathbf{AM} = \mathbf{MS}$$

If we now pre-multiply both sides by $\mathbf{M}^{-1}$ we have

$$\mathbf{M}^{-1}\mathbf{AM} = \mathbf{M}^{-1}\mathbf{MS} \quad \text{But } \mathbf{M}^{-1}\mathbf{M} = \mathbf{I}$$

$$\therefore \mathbf{M}^{-1}\mathbf{AM} = \mathbf{S}$$

Make a note of this result. Then we will consider an example

31   **Example 1**

From the results of a previous example in Frame 13, if

$$\mathbf{A} = \begin{pmatrix} 1 & 0 & 4 \\ 0 & 2 & 0 \\ 3 & 1 & -3 \end{pmatrix} \text{ then } \lambda_1 = 2, \ \lambda_2 = 3, \ \lambda_3 = -5 \text{ and}$$

$$\mathbf{x}_1 = \begin{pmatrix} 4 \\ -7 \\ 1 \end{pmatrix}, \quad \mathbf{x}_2 = \begin{pmatrix} 2 \\ 0 \\ 1 \end{pmatrix}, \quad \mathbf{x}_3 = \begin{pmatrix} 2 \\ 0 \\ -3 \end{pmatrix}.$$

Also $\mathbf{M} = \begin{pmatrix} 4 & 2 & 2 \\ -7 & 0 & 0 \\ 1 & 1 & -3 \end{pmatrix}.$

We can find $\mathbf{M}^{-1}$ by any of the methods we have established previously.

$$\mathbf{M}^{-1} = \ldots\ldots\ldots\ldots$$

$$\mathbf{M}^{-1} = \begin{pmatrix} 0 & -1/7 & 0 \\ 3/8 & 1/4 & 1/4 \\ 1/8 & 1/28 & -1/4 \end{pmatrix}$$

Here is one way of determining the inverse. You may have done it by another.

$$\begin{pmatrix} 4 & 2 & 2 & | & 1 & 0 & 0 \\ -7 & 0 & 0 & | & 0 & 1 & 0 \\ 1 & 1 & -3 & | & 0 & 0 & 1 \end{pmatrix} \sim \begin{pmatrix} 7 & 0 & 0 & | & 0 & -1 & 0 \\ 1 & 1 & -3 & | & 0 & 0 & 1 \\ 4 & 2 & 2 & | & 1 & 0 & 0 \end{pmatrix}$$

$$\sim \begin{pmatrix} 1 & 0 & 0 & | & 0 & -1/7 & 0 \\ 0 & 1 & -3 & | & 0 & 1/7 & 1 \\ 0 & 2 & 2 & | & 1 & 4/7 & 0 \end{pmatrix} \sim \begin{pmatrix} 1 & 0 & 0 & | & 0 & -1/7 & 0 \\ 0 & 1 & -3 & | & 0 & 1/7 & 1 \\ 0 & 0 & 8 & | & 1 & 2/7 & -2 \end{pmatrix}$$

$$\sim \begin{pmatrix} 1 & 0 & 0 & | & 0 & -1/7 & 0 \\ 0 & 1 & -3 & | & 0 & 1/7 & 1 \\ 0 & 0 & 1 & | & 1/8 & 1/28 & -1/4 \end{pmatrix}$$

$$\sim \begin{pmatrix} 1 & 0 & 0 & | & 0 & -1/7 & 0 \\ 0 & 1 & 0 & | & 3/8 & 7/28 & 1/4 \\ 0 & 0 & 1 & | & 1/8 & 1/28 & -1/4 \end{pmatrix}$$

$$\therefore \ \mathbf{M}^{-1} = \begin{pmatrix} 0 & -1/7 & 0 \\ 3/8 & 1/4 & 1/4 \\ 1/8 & 1/28 & -1/4 \end{pmatrix}$$

So now $\mathbf{A} = \begin{pmatrix} 1 & 0 & 4 \\ 0 & 2 & 0 \\ 3 & 1 & -3 \end{pmatrix}$ and $\mathbf{M} = \begin{pmatrix} 4 & 2 & 2 \\ -7 & 0 & 0 \\ 1 & 1 & -3 \end{pmatrix}$

$$\therefore \ \mathbf{AM} = \begin{pmatrix} 1 & 0 & 4 \\ 0 & 2 & 0 \\ 3 & 1 & -3 \end{pmatrix} \begin{pmatrix} 4 & 2 & 2 \\ -7 & 0 & 0 \\ 1 & 1 & -3 \end{pmatrix} = \begin{pmatrix} 8 & 6 & -10 \\ -14 & 0 & 0 \\ 2 & 3 & 15 \end{pmatrix}$$

Then $\mathbf{M}^{-1}\mathbf{AM} = \begin{pmatrix} 0 & -1/7 & 0 \\ 3/8 & 1/4 & 1/4 \\ 1/8 & 1/28 & -1/4 \end{pmatrix} \begin{pmatrix} 8 & 6 & -10 \\ -14 & 0 & 0 \\ 2 & 3 & 15 \end{pmatrix}$

$$= \ldots\ldots\ldots\ldots$$

33

$$M^{-1}AM = \begin{pmatrix} 2 & 0 & 0 \\ 0 & 3 & 0 \\ 0 & 0 & -5 \end{pmatrix}$$

So we have transformed the original matrix A into a diagonal matrix and notice that the elements on the main diagonal are, in fact, the eigenvalues of A

i.e. $M^{-1}AM = S$

Therefore, let us list a few relevant facts

1 $M^{-1}AM$ transforms the square matrix A into a diagonal matrix S.
2 A square matrix A of order n can be so transformed if the matrix has n independent eigenvectors.
3 A matrix A always has n linearly independent eigenvectors if it has n distinct eigenvalues or if it is a symmetric matrix.
4 If the matrix has repeated eigenvalues and is not symmetric, it may or may not have n linearly independent eigenvectors.

Now here is one straightforward example with which to finish.

Example 2

If $A = \begin{pmatrix} -6 & 5 \\ 4 & 2 \end{pmatrix}$, $M = \ldots\ldots\ldots\ldots$; $M^{-1} = \ldots\ldots\ldots\ldots$;

and hence $M^{-1}AM = \ldots\ldots\ldots\ldots$

Work through it entirely on your own:

(1) Determine the eigenvalues and corresponding eigenvectors.

(2) Hence form the matrix M.

(3) Determine M^{-1}, the inverse of M.

(4) Finally form the matrix products AM and $M^{-1}(AM)$.

34

$$M = \begin{pmatrix} 1 & 5 \\ 2 & -2 \end{pmatrix}; \quad M^{-1} = \begin{pmatrix} 1/6 & 5/12 \\ 1/6 & -1/12 \end{pmatrix}; \quad M^{-1}AM = \begin{pmatrix} 4 & 0 \\ 0 & -8 \end{pmatrix}$$

Here is the working. See whether you agree.

$$A = \begin{pmatrix} -6 & 5 \\ 4 & 2 \end{pmatrix} \qquad \therefore \begin{vmatrix} -6 - \lambda & 5 \\ 4 & 2 - \lambda \end{vmatrix} = 0$$

$(-6 - \lambda)(2 - \lambda) - 20 = 0 \qquad \therefore \lambda^2 + 4\lambda - 32 = 0$

$(\lambda - 4)(\lambda + 8) = 0 \qquad \therefore \lambda = 4 \text{ or } -8$

▶

(a) $\lambda_1 = 4$ $\left\{ \begin{pmatrix} -6 & 5 \\ 4 & 2 \end{pmatrix} - \begin{pmatrix} 4 & 0 \\ 0 & 4 \end{pmatrix} \right\} \begin{pmatrix} x_1 \\ x_2 \end{pmatrix} = \begin{pmatrix} 0 \\ 0 \end{pmatrix}$

$$\begin{pmatrix} -10 & 5 \\ 4 & -2 \end{pmatrix} \begin{pmatrix} x_1 \\ x_2 \end{pmatrix} = \begin{pmatrix} 0 \\ 0 \end{pmatrix}$$

$\therefore \ -10x_1 + 5x_2 = 0 \quad \therefore \ x_2 = 2x_1 \quad \mathbf{x}_1 = \begin{pmatrix} 1 \\ 2 \end{pmatrix}$

(b) $\lambda_2 = -8$ $\left\{ \begin{pmatrix} -6 & 5 \\ 4 & 2 \end{pmatrix} + \begin{pmatrix} 8 & 0 \\ 0 & 8 \end{pmatrix} \right\} \begin{pmatrix} x_1 \\ x_2 \end{pmatrix} = \begin{pmatrix} 0 \\ 0 \end{pmatrix}$

$$\begin{pmatrix} 2 & 5 \\ 4 & 10 \end{pmatrix} \begin{pmatrix} x_1 \\ x_2 \end{pmatrix} = \begin{pmatrix} 0 \\ 0 \end{pmatrix}$$

$\therefore \ 2x_1 + 5x_2 = 0 \quad \therefore \ x_2 = -\frac{2}{5}x_1 \quad \therefore \ \mathbf{x}_2 = \begin{pmatrix} 5 \\ -2 \end{pmatrix}$

$\therefore \ \mathbf{M} = \begin{pmatrix} 1 & 5 \\ 2 & -2 \end{pmatrix}$

To find $\mathbf{M}^{-1}$ $\begin{pmatrix} 1 & 5 & | & 1 & 0 \\ 2 & -2 & | & 0 & 1 \end{pmatrix}$

Operating on rows, we have

$$\begin{pmatrix} 0 & 5 & | & 1 & 0 \\ 0 & -12 & | & -2 & 1 \end{pmatrix} = \begin{pmatrix} 1 & 5 & | & 1 & 0 \\ 0 & 1 & | & 1/6 & -1/12 \end{pmatrix}$$
$$= \begin{pmatrix} 1 & 0 & | & 1/6 & 5/12 \\ 0 & 1 & | & 1/6 & -1/12 \end{pmatrix}$$

$$\therefore \ \mathbf{M}^{-1} = \begin{pmatrix} 1/6 & 5/12 \\ 1/6 & -1/12 \end{pmatrix}$$

$$\therefore \ \mathbf{AM} = \begin{pmatrix} -6 & 5 \\ 4 & 2 \end{pmatrix} \begin{pmatrix} 1 & 5 \\ 2 & -2 \end{pmatrix} = \begin{pmatrix} 4 & -40 \\ 8 & 16 \end{pmatrix}$$

$$\therefore \ \mathbf{M}^{-1}\mathbf{AM} = \begin{pmatrix} 1/6 & 5/12 \\ 1/6 & -1/12 \end{pmatrix} \begin{pmatrix} 4 & -40 \\ 8 & 16 \end{pmatrix} = \begin{pmatrix} 4 & 0 \\ 0 & -8 \end{pmatrix}$$

$$\therefore \ \mathbf{M}^{-1}\mathbf{AM} = \begin{pmatrix} 4 & 0 \\ 0 & -8 \end{pmatrix}$$

Systems of second-order differential equations

The process of uncoupling a system of differential equations to obtain their solution can be achieved by diagonalising the matrix of coefficients. For simplicity we shall only consider second-order equations and again, we proceed by example.

Example 1

Consider the system of coupled second-order differential equations

$$f_1''(x) = 2f_1(x) + 3f_2(x)$$
$$f_2''(x) = 4f_1(x) + f_2(x)$$

where $f_1(0) = 2$, $f_2(0) = 1$, $f_1'(0) = 4$ and $f_2'(0) = 3$

These can be written in matrix form as

. . . .

$$\begin{pmatrix} f_1''(x) \\ f_2''(x) \end{pmatrix} = \begin{pmatrix} 2 & 3 \\ 4 & 1 \end{pmatrix} \begin{pmatrix} f_1(x) \\ f_2(x) \end{pmatrix}$$

That is

$$\mathbf{F}''(x) = \mathbf{A}\mathbf{F}(x)$$

where $\mathbf{F}(x) = \begin{pmatrix} f_1(x) \\ f_2(x) \end{pmatrix}$, $\mathbf{F}''(x) = \begin{pmatrix} f_1''(x) \\ f_2''(x) \end{pmatrix}$ and $\mathbf{A} = \begin{pmatrix} 2 & 3 \\ 4 & 1 \end{pmatrix}$ and where

$\mathbf{F}(0) = \begin{pmatrix} f_1(0) \\ f_2(0) \end{pmatrix} = \begin{pmatrix} 2 \\ 1 \end{pmatrix}$ and $\mathbf{F}'(0) = \begin{pmatrix} f_1'(0) \\ f_2'(0) \end{pmatrix} = \begin{pmatrix} 4 \\ 3 \end{pmatrix}$ are the boundary

conditions in matrix form.

The matrix differential equation $\mathbf{F}''(x) = \mathbf{A}\mathbf{F}(x)$ is similar in form to the single differential equation $f''(x) = af(x)$ (a constant) which has solution $f(x) = \alpha e^{\sqrt{a}x} + \beta e^{-\sqrt{a}x}$ (α, β constants), so to solve the matrix equation we try a solution of this form. We already know from Frames 2 to 5 that the eigenvalues and eigenvectors of matrix $\mathbf{A}$ are

$\lambda = -2$ with corresponding eigenvector $\begin{pmatrix} 3 \\ -4 \end{pmatrix}$

$\lambda = 5$ with corresponding eigenvector $\begin{pmatrix} 1 \\ 1 \end{pmatrix}$

The modal matrix of $\mathbf{A}$ is the matrix $\mathbf{M}$ and the spectral matrix of $\mathbf{A}$ is the matrix $\mathbf{S}$ where

$$\mathbf{M} = \begin{pmatrix} \cdots & \cdots \\ \cdots & \cdots \end{pmatrix} \text{ and } \mathbf{S} = \begin{pmatrix} \cdots & \cdots \\ \cdots & \cdots \end{pmatrix}$$

37

$$M = \begin{pmatrix} 3 & 1 \\ -4 & 1 \end{pmatrix} \text{ and } S = \begin{pmatrix} -2 & 0 \\ 0 & 5 \end{pmatrix}$$

Because

The modal matrix is formed from the eigenvectors of **A**. That is

$$M = \begin{pmatrix} 3 & 1 \\ -4 & 1 \end{pmatrix} \text{ where the two eigenvectors are } \begin{pmatrix} 3 \\ -4 \end{pmatrix} \text{ and } \begin{pmatrix} 1 \\ 1 \end{pmatrix}$$

The spectral matrix is formed from the eigenvalues of **A**. That is

$$S = \begin{pmatrix} -2 & 0 \\ 0 & 5 \end{pmatrix} \text{ where the two eigenvalues are } -2 \text{ and } 5$$

If we now define the matrix $G(x)$ by the equation $F(x) = MG(x)$, then differentiating gives

$$F''(x) = [MG(x)]'' = MG''(x) \text{ where}$$
$$F''(x) = AF(x) = AMG(x)$$

and so, from Frame 85, $M^{-1}MG''(x) = G''(x) = M^{-1}AMG(x) = SG(x)$. That is

$$G''(x) = SG(x)$$

Therefore, in component terms

$$G''(x) = \begin{pmatrix} g_1''(x) \\ g_2''(x) \end{pmatrix} = SG(x) = \begin{pmatrix} -2 & 0 \\ 0 & 5 \end{pmatrix} \begin{pmatrix} g_1(x) \\ g_2(x) \end{pmatrix}$$

and so

$$g_1''(x) = \dots g_1(x) \text{ with solution } g_1(x) = k_{11}e^{\dots x} + k_{12}e^{-\dots x}$$
$$g_2''(x) = \dots g_2(x) \text{ with solution } g_2(x) = k_{21}e^{\dots x} + k_{22}e^{-\dots x}$$

38

$$g_1''(x) = -2g_1(x) \text{ with solution } g_1(x) = k_{11}e^{j\sqrt{2}x} + k_{12}e^{-j\sqrt{2}x}$$
$$g_2''(x) = 5g_2(x) \text{ with solution } g_2(x) = k_{21}e^{\sqrt{5}x} + k_{22}e^{-\sqrt{5}x}$$

Now, $F(x) = MG(x)$ so

$$F(x) = \begin{pmatrix} f_1(x) \\ f_2(x) \end{pmatrix} = \begin{pmatrix} \dots \\ \dots \end{pmatrix}$$

39

$$\mathbf{F}(x) = \begin{pmatrix} f_1(x) \\ f_2(x) \end{pmatrix} = \begin{pmatrix} 3k_{11}e^{j\sqrt{2}x} + 3k_{12}e^{-j\sqrt{2}x} + k_{21}e^{\sqrt{5}x} + k_{22}e^{-\sqrt{5}x} \\ -4k_{11}e^{j\sqrt{2}x} - 4k_{12}e^{-j\sqrt{2}x} + k_{21}e^{\sqrt{5}x} + k_{22}e^{-\sqrt{5}x} \end{pmatrix}$$

Because

$$\mathbf{F}(x) = \begin{pmatrix} f_1(x) \\ f_2(x) \end{pmatrix} = \mathbf{MG}(x) = \begin{pmatrix} 3 & 1 \\ -4 & 1 \end{pmatrix} \begin{pmatrix} k_{11}e^{j\sqrt{2}x} + k_{12}e^{-j\sqrt{2}x} \\ k_{21}e^{\sqrt{5}x} + k_{22}e^{-\sqrt{5}x} \end{pmatrix}$$

and so

$$f_1(x) = 3k_{11}e^{j\sqrt{2}x} + 3k_{12}e^{-j\sqrt{2}x} + k_{21}e^{j\sqrt{5}x} + k_{22}e^{-\sqrt{5}x}$$

and

$$f_2(x) = -4k_{11}e^{j\sqrt{2}x} - 4k_{12}e^{-j\sqrt{2}x} + k_{21}e^{\sqrt{5}x} + k_{22}e^{-\sqrt{5}x}$$

This solution can be written in terms of circular and hyperbolic trigonometric expressions as

$$\mathbf{F}(x) = \begin{pmatrix} \cdots & \cdots \\ \cdots & \cdots \end{pmatrix} \begin{pmatrix} P\cos\ldots x + Q\sin\ldots x \\ R\cosh\ldots x + S\sinh\ldots x \end{pmatrix}$$

40

$$\mathbf{F}(x) = \begin{pmatrix} 3 & 1 \\ -4 & 1 \end{pmatrix} \begin{pmatrix} P\cos\sqrt{2}x + Q\sin\sqrt{2}x \\ R\cosh\sqrt{5}x + S\sinh\sqrt{5}x \end{pmatrix}$$

Because

$$3k_{11}e^{j\sqrt{2}x} + 3k_{12}e^{-j\sqrt{2}x}$$
$$= 3k_{11}\left(\cos\sqrt{2}x + j\sin\sqrt{2}x\right) + 3k_{12}\left(\cos\sqrt{2}x - j\sin\sqrt{2}x\right)$$
$$= P\cos\sqrt{2}x + Q\sin\sqrt{2}x$$

where $P = 3k_{11} + 3k_{12}$ and $Q = (3k_{11} - 3k_{12})j$

and

$$k_{21}e^{\sqrt{5}x} + k_{22}e^{-\sqrt{5}x}$$
$$= k_{21}\left(\cosh\sqrt{5}x + \sinh\sqrt{5}x\right) + k_{22}\left(\cosh\sqrt{5}x - \sinh\sqrt{5}x\right)$$
$$= R\cosh\sqrt{5}x + S\sinh\sqrt{5}x \text{ where } R = k_{21} + k_{22} \text{ and } S = k_{21} - k_{22}$$

Therefore

$$\mathbf{F}(x) = \begin{pmatrix} \cdots \\ \cdots \end{pmatrix}$$

$$\boxed{\mathbf{F}(x) = \begin{pmatrix} 3P\cos\sqrt{2}x + 3Q\sin\sqrt{2}x + R\cosh\sqrt{5}x + S\sinh\sqrt{5}x \\ -4P\cos\sqrt{2}x - 4Q\sin\sqrt{2}x + R\cosh\sqrt{5}x + S\sinh\sqrt{5}x \end{pmatrix}}$$

41

That is

$$f_1(x) = \ldots\ldots\ldots$$
$$f_2(x) = \ldots\ldots\ldots$$

$$\boxed{\begin{aligned} f_1(x) &= 3P\cos\sqrt{2}x + 3Q\sin\sqrt{2}x + R\cosh\sqrt{5}x + S\sinh\sqrt{5}x \\ f_2(x) &= -4P\cos\sqrt{2}x - 4Q\sin\sqrt{2}x + R\cosh\sqrt{5}x + S\sinh\sqrt{5}x \end{aligned}}$$

42

Because

$$\mathbf{F}(x) = \begin{pmatrix} f_1(x) \\ f_2(x) \end{pmatrix}$$

and so

$$f_1(x) = 3P\cos\sqrt{2}x + 3Q\sin\sqrt{2}x + R\cosh\sqrt{5}x + S\sinh\sqrt{5}x$$
$$f_2(x) = -4P\cos\sqrt{2}x - 4Q\sin\sqrt{2}x + R\cosh\sqrt{5}x + S\sinh\sqrt{5}x$$

Applying the boundary conditions, we find

$$\mathbf{F}(0) = \begin{pmatrix} 2 \\ 1 \end{pmatrix} = \begin{pmatrix} \ldots P + \ldots R \\ \ldots P + \ldots R \end{pmatrix} \text{ and } \mathbf{F}'(0) = \begin{pmatrix} 4 \\ 3 \end{pmatrix} = \begin{pmatrix} \ldots Q + \ldots S \\ \ldots Q + \ldots S \end{pmatrix}$$

$$\boxed{\mathbf{F}(0) = \begin{pmatrix} 2 \\ 1 \end{pmatrix} = \begin{pmatrix} 3P + R \\ -4P + R \end{pmatrix} \text{ and } \mathbf{F}'(0) = \begin{pmatrix} 4 \\ 3 \end{pmatrix} = \begin{pmatrix} 3\sqrt{2}Q + \sqrt{5}S \\ -4\sqrt{2}Q + \sqrt{5}S \end{pmatrix}}$$

43

Because

$$f_1(0) = 2, \; f_2(0) = 1, \; f_1'(0) = 4 \text{ and } f_2'(0) = 3 \text{ and so}$$

$$\mathbf{F}(0) = \begin{pmatrix} 2 \\ 1 \end{pmatrix} = \begin{pmatrix} f_1(0) \\ f_2(0) \end{pmatrix} = \begin{pmatrix} 3P + R \\ -4P + R \end{pmatrix} \text{ and}$$

$$\mathbf{F}'(0) = \begin{pmatrix} 4 \\ 3 \end{pmatrix} = \begin{pmatrix} f_1'(0) \\ f_2'(0) \end{pmatrix} = \begin{pmatrix} 3\sqrt{2}Q + \sqrt{5}S \\ -4\sqrt{2}Q + \sqrt{5}S \end{pmatrix}$$

This gives the two sets of simultaneous equations

$$\begin{aligned} 3P + R &= 2 \\ -4P + R &= 1 \end{aligned} \quad \text{and} \quad \begin{aligned} 3\sqrt{2}Q + \sqrt{5}S &= 4 \\ -4\sqrt{2}Q + \sqrt{5}S &= 3 \end{aligned} \quad \text{which have solution}$$

$$P = \ldots\ldots\ldots, \quad R = \ldots\ldots\ldots,$$
$$Q = \ldots\ldots\ldots \text{ and } S = \ldots\ldots\ldots$$

44

$$P = 1/7, \ R = 11/7, \ Q = 1/\left(7\sqrt{2}\right) \text{ and } S = 25/\left(7\sqrt{5}\right)$$

This gives the complete solution as

$$f_1(x) = \frac{3}{7}\cos\sqrt{2}x + \frac{3}{7\sqrt{2}}\sin\sqrt{2}x + \frac{11}{7}\cosh\sqrt{5}x + \frac{25}{7\sqrt{5}}\sinh\sqrt{5}x$$

$$f_2(x) = -\frac{4}{7}\cos\sqrt{2}x - \frac{4}{7\sqrt{2}}\sin\sqrt{2}x + \frac{11}{7}\cosh\sqrt{5}x + \frac{25}{7\sqrt{5}}\sinh\sqrt{5}x$$

This method is quite straightforwardly extended to three or more such coupled differential equations.

Summary

To solve the system of coupled second-order differential equations

$$\mathbf{F}''(x) = \mathbf{AF}(x)$$

1 Find the eigenvalues and eigenvectors of matrix $\mathbf{A}$ and construct the modal matrix $\mathbf{M}$ and the diagonal spectral matrix $\mathbf{S}$
2 Solve the equation $\mathbf{G}'(x) = \mathbf{SG}(x)$
   (note that even though $\mathbf{M}^{-1}$ is used there was no need to calculate it)
3 Apply $\mathbf{F}(x) = \mathbf{MG}(x)$ to find $\mathbf{F}(x)$.

Try one yourself.

Next frame

45   ### Example 2

The system of coupled second-order differential equations (refer to Frames 5 to 9)

$$f_1''(x) = 3f_1(x) + 10f_2(x)$$

$$f_2''(x) = 2f_1(x) + 4f_2(x)$$

where $f_1(0) = 0$, $f_2(0) = 1$, $f_1'(0) = 1$ and $f_2'(0) = 0$

has the solution

$$f_1(x) = \ldots\ldots\ldots\ldots$$
$$f_2(x) = \ldots\ldots\ldots\ldots$$

$$f_1(x) = 10\cos x + \frac{5}{9}\sin x + 10\cosh 2\sqrt{2}x + \frac{2}{9\sqrt{2}}\sinh 2\sqrt{2}x$$

$$f_2(x) = -4\cos x - \frac{2}{9}\sin x + 5\cosh 2\sqrt{2}x + \frac{1}{9\sqrt{2}}\sinh 2\sqrt{2}x$$

Because

$$f_1''(x) = 3f_1(x) + 10f_2(x)$$
$$f_2''(x) = 2f_1(x) + 4f_2(x)$$

can be written in matrix form as

. . . .

$$\begin{pmatrix} f_1''(x) \\ f_2''(x) \end{pmatrix} = \begin{pmatrix} 3 & 10 \\ 2 & 4 \end{pmatrix}\begin{pmatrix} f_1(x) \\ f_2(x) \end{pmatrix}$$

That is

$$\mathbf{F}''(x) = \mathbf{AF}(x)$$

where $\mathbf{F}(x) = \begin{pmatrix} f_1(x) \\ f_2(x) \end{pmatrix}$, $\mathbf{F}''(x) = \begin{pmatrix} f_1''(x) \\ f_2''(x) \end{pmatrix}$ and $\mathbf{A} = \begin{pmatrix} 3 & 10 \\ 2 & 4 \end{pmatrix}$

and where $\mathbf{F}(0) = \begin{pmatrix} 0 \\ 1 \end{pmatrix}$ and $\mathbf{F}'(0) = \begin{pmatrix} f_1'(0) \\ f_2'(0) \end{pmatrix} = \begin{pmatrix} 1 \\ 0 \end{pmatrix}$.

To solve the matrix equation we first need the eigenvalues and associated eigenvectors of the matrix **A**. These have already been found in Frames 5 to 9 and they are

$\lambda = -1$ with corresponding eigenvector $\begin{pmatrix} 5 \\ -2 \end{pmatrix}$

$\lambda = 8$ with corresponding eigenvector $\begin{pmatrix} 2 \\ 1 \end{pmatrix}$

The complete solution of $\mathbf{F}'' = \mathbf{AF}$ is then

$$\mathbf{F}(x) = (P\cos x + Q\sin x)\begin{pmatrix} 5 \\ -2 \end{pmatrix} + \left(R\cosh 2\sqrt{2}x + S\sinh 2\sqrt{2}x\right)\begin{pmatrix} 2 \\ 1 \end{pmatrix}$$

$$= \begin{pmatrix} 5P\cos x + 5Q\sin x + 2R\cosh 2\sqrt{2}x + 2S\sinh 2\sqrt{2}x \\ -2P\cos x - 2Q\sin x + R\cosh 2\sqrt{2}x + S\sinh 2\sqrt{2}x \end{pmatrix}$$

That is

$$f_1(x) = \ldots\ldots\ldots$$
$$f_2(x) = \ldots\ldots\ldots$$

48

$$f_1(x) = 5P\cos x + 5Q\sin x + 2R\cosh 2\sqrt{2}x + 2S\sinh 2\sqrt{2}x$$
$$f_2(x) = -2P\cos x - 2Q\sin x + R\cosh 2\sqrt{2}x + S\sinh 2\sqrt{2}x$$

Because

$$\mathbf{F}(x) = \begin{pmatrix} f_1(x) \\ f_2(x) \end{pmatrix}$$

and so

$$f_1(x) = 5P\cos x + 5Q\sin x + 2R\cosh 2\sqrt{2}x + 2S\sinh 2\sqrt{2}x$$
$$f_2(x) = -2P\cos x - 2Q\sin x + R\cosh 2\sqrt{2}x + S\sinh 2\sqrt{2}x$$

Applying the boundary conditions, we find

$$\mathbf{F}(0) = \begin{pmatrix} 0 \\ 1 \end{pmatrix} = \begin{pmatrix} \dots P + \dots R \\ \dots P + \dots R \end{pmatrix} \text{ and } \mathbf{F}'(0) = \begin{pmatrix} 1 \\ 0 \end{pmatrix} = \begin{pmatrix} \dots Q + \dots S \\ \dots Q + \dots S \end{pmatrix}$$

49

$$\mathbf{F}(0) = \begin{pmatrix} 0 \\ 1 \end{pmatrix} = \begin{pmatrix} 5P + 2R \\ -2P + R \end{pmatrix} \text{ and } \mathbf{F}'(0) = \begin{pmatrix} 1 \\ 0 \end{pmatrix} = \begin{pmatrix} 5Q + 4\sqrt{2}S \\ -2Q + 2\sqrt{2}S \end{pmatrix}$$

Because

The boundary conditions are $f_1(0) = 0$, $f_2(0) = 1$, $f_1'(0) = 1$ and $f_2'(0) = 0$, therefore

$$\mathbf{F}(0) = \begin{pmatrix} 0 \\ 1 \end{pmatrix} = \begin{pmatrix} f_1(0) \\ f_2(0) \end{pmatrix} = \begin{pmatrix} 5P + 2R \\ -2P + R \end{pmatrix} \text{ and }$$

$$\mathbf{F}'(0) = \begin{pmatrix} 1 \\ 0 \end{pmatrix} = \begin{pmatrix} f_1'(0) \\ f_2'(0) \end{pmatrix} = \begin{pmatrix} 5Q + 4\sqrt{2}S \\ -2Q + 2\sqrt{2}S \end{pmatrix}$$

This gives the two sets of simultaneous equations

$$5P + 2R = 0 \quad \text{and} \quad 5Q + 4\sqrt{2}S = 1 \quad \text{which have solution}$$
$$-2P + R = 1 \quad \quad\quad -2Q + 2\sqrt{2}S = 0$$

$$P = \dots\dots\dots, \quad R = \dots\dots\dots,$$
$$Q = \dots\dots\dots \text{ and } S = \dots\dots\dots$$

50

$$P = -2/9, \quad R = 5/9, \quad Q = 1/9 \text{ and } S = 1/\left(9\sqrt{2}\right)$$

This gives the complete solution as

$$f_1(x) = -\frac{10}{9}\cos x + \frac{5}{9}\sin x + \frac{10}{9}\cosh 2\sqrt{2}x + \frac{2}{9\sqrt{2}}\sinh 2\sqrt{2}x$$

$$f_2(x) = \frac{4}{9}\cos x - \frac{2}{9}\sin x + \frac{5}{9}\cosh 2\sqrt{2}x + \frac{1}{9\sqrt{2}}\sinh 2\sqrt{2}x$$

As usual, the Programme ends with the **Revision summary**, to be read in conjunction with the **Can you?** checklist. Go back to the relevant part of the Programme for any points on which you are unsure. The **Test exercise** should then be straightforward and the **Further problems** give valuable additional practice.

 # Revision summary 17

51

1 *Eigenvalues and eigenvectors* $\mathbf{Ax} = \lambda \mathbf{x}$

Sets of equations of form $\mathbf{Ax} = \lambda \mathbf{x}$, where $\mathbf{A} = $ coefficient matrix, $\mathbf{x} = $ column matrix, $\lambda = $ scalar quantity.

Equations become $(\mathbf{A} - \lambda \mathbf{I})\mathbf{x} = \mathbf{0}$.

For non-trivial solutions, $|\mathbf{A} - \lambda \mathbf{I}| = 0$ is the *characteristic equation* and gives values of λ i.e. the *eigenvalues*.

Substitution of each eigenvalue gives a corresponding *eigenvector*.

2 *Cayley–Hamilton theorem*

Every square matrix satisfies its own characteristic equation.

3 *Solving systems of first-order ordinary differential equations*

To solve the system of coupled first-order differential equations

$$\mathbf{F}'(x) = \mathbf{AF}(x)$$

(a) Find the eigenvalues and eigenvectors of matrix $\mathbf{A}$ and construct the modal matrix $\mathbf{M}$ and the diagonal spectral matrix $\mathbf{S}$

(b) Solve the equation $\mathbf{G}'(x) = \mathbf{SG}(x)$

(c) Apply $\mathbf{F}(x) = \mathbf{MG}(x)$ to find $\mathbf{F}(x)$.

4 *Diagonalisation of a matrix*

Modal matrix of $\mathbf{A}$

If $\mathbf{A}$ has distinct eigenvalues $\mathbf{M} = (\mathbf{x}_1, \mathbf{x}_2, \ldots, \mathbf{x}_n)$, where $\mathbf{x}_1, \mathbf{x}_2, \ldots, \mathbf{x}_n$ are eigenvectors of $\mathbf{A}$, then $\mathbf{M}^{-1}\mathbf{AM} = \mathbf{S}$ where $\mathbf{S}$ is the *spectral matrix* of $\mathbf{A}$

$$\text{and} \quad \mathbf{S} = \begin{pmatrix} \lambda_1 & 0 & \cdots & 0 \\ 0 & \lambda_2 & \cdots & 0 \\ \cdot & \cdot & & \cdot \\ \cdot & \cdot & & \cdot \\ \cdot & \cdot & & \cdot \\ 0 & 0 & \cdots & \lambda_n \end{pmatrix}$$

$\lambda_1, \lambda_2, \ldots, \lambda_n$ are the eigenvalues of $\mathbf{A}$.

5   *Solving systems of second-order ordinary differential equations*
To solve an equation of the form

$$\mathbf{F}''(x) = \mathbf{A}\mathbf{F}(x)$$

(a) Find the eigenvalues $\lambda_1, \lambda_2, \ldots, \lambda_n$ of $\mathbf{A}$

(b) Assuming the eigenvectors are all distinct, find the associated eigenvectors $\mathbf{C}_1, \mathbf{C}_2, \ldots, \mathbf{C}_n$

(c) Write the solution of the equation as

$$\mathbf{F}(x) = \sum_{r=1}^{n} \left(a_r e^{\sqrt{\lambda_r}x} + b_r e^{-\sqrt{\lambda_r}x} \right) \mathbf{C}_r$$

and use the boundary conditions to find the values of a_r and b_r for $r = 1, 2, \ldots, n$.

 # Can you?

52   Checklist 17

Check this list before and after you try the end of Programme test.

On a scale of 1 to 5 how confident are you that you can: **Frames**

- Obtain the eigenvalues and corresponding eigenvectors of a square matrix?   ☐ 1 ☐ to ☐ 13
  Yes   ☐   ☐   ☐   ☐   ☐   *No*

- Demonstrate the validity of the Cayley–Hamilton theorem?   ☐ 14 ☐ and ☐ 15
  Yes   ☐   ☐   ☐   ☐   ☐   *No*

- Solve systems of first-order ordinary differential equations using eigenvalue and eigenvector methods?   ☐ 16 ☐ to ☐ 27
  Yes   ☐   ☐   ☐   ☐   ☐   *No*

- Construct the modal matrix from the eigenvectors of a matrix and the spectral matrix from the eigenvalues?   ☐ 28 ☐ to ☐ 34
  Yes   ☐   ☐   ☐   ☐   ☐   *No*

- Solve systems of second-order ordinary differential equations using diagonalisation?   ☐ 35 ☐ to ☐ 50
  Yes   ☐   ☐   ☐   ☐   ☐   *No*

Test exercise 17

1 Determine the eigenvalues and corresponding eigenvectors of $\mathbf{A}\mathbf{x} = \lambda\mathbf{x}$ where **53**

$$\mathbf{A} = \begin{pmatrix} 1 & 3 & 0 \\ 1 & 2 & 1 \\ -2 & 1 & -1 \end{pmatrix}.$$

2 If $\mathbf{x}_1$ and $\mathbf{x}_2$ are eigenvectors of $\mathbf{A}\mathbf{x} = \lambda\mathbf{x}$ where $\mathbf{A} = \begin{pmatrix} 3 & 2 \\ 4 & 1 \end{pmatrix}$ determine

(a) $\mathbf{M} = (\mathbf{x}_1\mathbf{x}_2)$

(b) $\mathbf{M}^{-1}$

(c) $\mathbf{M}^{-1}\mathbf{A}\mathbf{M}.$

3 Solve the system of first-order differential equations
$$\begin{aligned} f_1'(x) &= 5f_1(x) - 2f_2(x) \\ f_2'(x) &= -f_1(x) + 4f_2(x) \end{aligned} \quad \text{where } f_1(0) = -3 \text{ and } f_2(0) = 2.$$

4 Solve the system of second-order differential equations
$$\begin{aligned} f_1''(x) &= f_1(x) + 6f_2(x) \\ f_2''(x) &= 3f_1(x) - 2f_2(x) \end{aligned} \quad \text{where } f_1(0) = 1, \ f_2(0) = 0, \ f_1'(0) = 2, \ f_2'(0) = -1.$$

 # Further problems 17

1 Solve each of the following systems of first-order differential equations. **54**

(a) $f_1'(x) = 2f_1(x) - 5f_2(x)$
$f_2'(x) = f_1(x) - 4f_2(x)$
where $f_1(0) = 1$ and $f_2(0) = 0$

(b) $f_1'(x) = -5f_1(x) + 9f_2(x)$
$f_2'(x) = f_1(x) + 3f_2(x)$
where $f_1(0) = 0$ and $f_2(0) = -2$

(c) $f_1'(x) = 5f_1(x) - 6f_2(x) + f_3(x)$
$f_2'(x) = f_1(x) + f_2(x)$
$f_3'(x) = 3f_1(x) + f_3(x)$
where $f_1(0) = 1$, $f_2(0) = 0$ and $f_3(0) = 2$

(d) $f_1'(x) = 4f_1(x) + 10f_2(x) - 8f_3(x)$
$f_2'(x) = f_1(x) + 2f_2(x) + f_3(x)$
$f_3'(x) = -f_1(x) + 2f_2(x) + 3f_3(x)$
where $f_1(0) = 4$, $f_2(0) = -2$ and $f_3(0) = -1.$

▶

2   If $A = \begin{pmatrix} 1 & 3 & 0 \\ 3 & 10 & -3 \\ 0 & -3 & 9 \end{pmatrix}$, determine the three eigenvalues λ_1, λ_2, λ_3 of A and

   verify that if $M = \begin{pmatrix} -9 & 1 & 1 \\ 3 & 2 & 4 \\ 1 & 3 & -3 \end{pmatrix}$ then $M^{-1}AM = S$, where S is a diagonal

   matrix with elements λ_1, λ_2, λ_3.

3   Solve each of the following systems of second-order differential equations.

   (a) $f_1''(x) = 4f_1(x) + 3f_2(x)$
   $f_2''(x) = 2f_1(x) + 5f_2(x)$
   where $f_1(0) = 0$, $f_2(0) = 1$, $f_1'(0) = 4$ and $f_2'(0) = 1$

   (b) $f_1''(x) = -6f_1(x) + 5f_2(x)$
   $f_2''(x) = 4f_1(x) + 2f_2(x)$
   where $f_1(0) = 0$, $f_2(0) = 1$, $f_1'(0) = 1$ and $f_2'(0) = 0$

   (c) $f_1''(x) = 2f_1(x) + 7f_2(x)$
   $f_2''(x) = f_1(x) + 3f_2(x) + f_3(x)$
   $f_3''(x) = 5f_1(x) + 8f_3(x)$
   where $f_1(0) = 1$, $f_2(0) = 1$, $f_3(0) = 0$, $f_1'(0) = 0$, $f_2'(0) = 0$
   and $f_3'(0) = 1$

   (d) $f_1''(x) = -3f_1(x) + 6f_3(x)$
   $f_2''(x) = 4f_1(x) + 5f_2(x) + 3f_3(x)$
   $f_3''(x) = f_1(x) + 2f_2(x) + f_3(x)$
   where $f_1(0) = 1$, $f_2(0) = 1$, $f_3(0) = 0$, $f_1'(0) = 0$, $f_2'(0) = 0$, $f_3'(0) = 1$.

Numerical solutions of partial differential equations

Learning outcomes

When you have completed this Programme you will be able to:

- Derive the finite difference formulas for the first partial derivatives of a function of two real variables and construct the central finite difference formula to represent a first-order partial differential equation
- Draw a rectangular grid of points overlaid on the domain of a function of two real variables and evaluate the function at the boundary grid points
- Construct the computational molecule for a first-order partial differential equation in two real variables and use the molecule to evaluate the solutions to the equation at the grid points interior to the boundary
- Describe the solution as a set of simultaneous linear equations and use matrices to represent them
- Invert the coefficient matrix and thereby represent the solution to the partial differential equation as a column matrix
- Take account of a boundary condition in the form of the derivative normal to the boundary
- Obtain the central finite difference formulas for the second derivatives of a function of two real variables and construct finite difference formulas for second-order partial differential equations
- Use the forward difference formula for the first time derivatives in partial differential equations involving time and distance
- Use the Crank–Nicolson procedure for a partial differential equation involving a first time derivative
- Appreciate the use of dimensional analysis in the conversion of a partial differential equation modelling a physical system into a dimensionless equation

Introduction

1   The numerical solution of partial differential equations is a large subject and can form the content of a course in itself. Here we shall just introduce the subject by considering the basic methods of solving some first- and second-order partial differential equations that involve functions of two real variables. The approach that is used is to construct finite difference formulas for the first and second partial derivatives and then to construct a finite difference formula that represents an approximation to the differential equation. However, before we move into the realm of functions of two real variables we shall derive the finite difference formulas for the ordinary first derivative of a function of a single real variable.

Next frame

Numerical approximation to derivatives

2   A function of one real variable $f(x)$ has the Taylor series expansion

$$f(x + h) = f(x) + hf'(x) + \frac{h^2}{2!}f''(x) + \frac{h^3}{3!}f'''(x) + \ldots$$

and, equally, replacing h by $-h$

$$f(x - h) = f(x) - hf'(x) + \frac{h^2}{2!}f''(x) - \frac{h^3}{3!}f'''(x) + \ldots$$

From the first equation we can see that by dividing through by h, we have

$$\frac{f(x + h) - f(x)}{h} = f'(x) + \frac{h}{2!}f''(x) + \frac{h^2}{3!}f'''(x) + \ldots$$

and from the second equation

$$\frac{f(x - h) - f(x)}{h} = -f'(x) + \frac{h}{2!}f''(x) - \frac{h^2}{3!}f'''(x) + \ldots$$

If we now neglect terms of the order two and higher we see that

$$f'(x) \approx \frac{f(x + h) - f(x)}{h} \qquad \text{[this is the *forward difference formula* for the first derivative of } f(x)\text{]}$$

and

$$f'(x) \approx \frac{f(x) - f(x - h)}{h} \qquad \text{[this is the *backward difference formula* for the first derivative of } f(x)\text{]}$$

▶

and both of these are accurate up to terms of order two. A more accurate estimate of the derivative can be obtained by subtracting the two Taylor series expansions from each other to get

$$f'(x) \approx \text{............} \quad \text{neglecting terms of the order of ....}$$

$$f'(x) \approx \frac{f(x+h) - f(x-h)}{2h}$$

neglecting terms of the order two and higher

3

Because

$$f(x+h) - f(x-h) = \left(f(x) + hf'(x) + \frac{h^2}{2!}f''(x) + \frac{h^3}{3!}f'''(x) + \cdots \right)$$

$$- \left(f(x) - hf'(x) + \frac{h^2}{2!}f''(x) - \frac{h^3}{3!}f'''(x) + \cdots \right)$$

$$= 2\left(hf'(x) + \frac{h^3}{3!}f'''(x) + \cdots \right)$$

and so

$$\frac{f(x+h) - f(x-h)}{2h} = f'(x) + \frac{h^2}{3!}f'''(x) + \cdots$$

giving

$$f'(x) \approx \frac{f(x+h) - f(x-h)}{2h}$$

neglecting terms of the order two and higher.

The derivative at x is given as the difference between the two values either side of $f(x)$ divided by $2h$.

This is called the *central difference formula* for the derivative of $f(x)$ and because it is the most accurate of the three for small h, it is the one that we shall use in the remainder of the Programme.

Now we need to look at the second derivative. By adding the first two Taylor series expansions in Frame 2 we find that

$$f''(x) \approx \text{............} \quad \text{neglecting terms of the order ....}$$

4

$$f''(x) \approx \frac{f(x+h) - 2f(x) + f(x-h)}{h^2}$$

neglecting terms of the order two and higher

Because

$$f(x+h) + f(x-h) = \left(f(x) + hf'(x) + \frac{h^2}{2!}f''(x) + \frac{h^3}{3!}f'''(x) + \ldots \right)$$

$$+ \left(f(x) - hf'(x) + \frac{h^2}{2!}f''(x) - \frac{h^3}{3!}f'''(x) + \ldots \right)$$

$$= 2\left(f(x) + \frac{h^2}{2!}f''(x) + \frac{h^4}{4!}f^{iv}(x) + \ldots \right)$$

$$= 2f(x) + h^2 f''(x) + \frac{h^4}{12}f^{iv}(x) + \ldots$$

and so

$$\frac{f(x+h) - 2f(x) + f(x-h)}{h^2} = f''(x) + \frac{h^2}{12}f^{iv}(x) + \ldots$$

Therefore

$$f''(x) \approx \frac{f(x+h) - 2f(x) + f(x-h)}{h^2} \qquad \text{neglecting terms of the order two and higher}$$

This is the *central difference formula* for the second derivative and, as you see, it possesses the same level of accuracy as the central difference formula for the first derivative.

Functions of two real variables

A function of two real variables $f(x, y)$ is graphically represented as a surface in three-dimensional space.

5

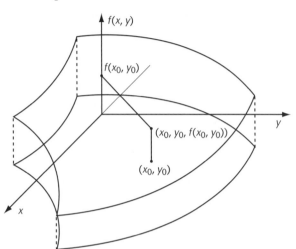

If $f(x, y)$ is single-valued, then to every *domain* point (x, y) there corresponds a single range point $f(x, y)$ and hence a single surface point $(x, y, f(x, y))$. If we know the exact form of $f(x, y)$ then we can compute its value at any domain point (x, y) selected at random. If we do not know the exact form of $f(x, y)$ but we do know that it satisfies a given differential equation then to evaluate $f(x, y)$ numerically we have to be more systematic. What we do is to lay a rectangular grid over the domain and evaluate $f(x, y)$ at the grid points – the points of intersection of the lines parallel with the x-axis and the lines parallel with the y-axis.

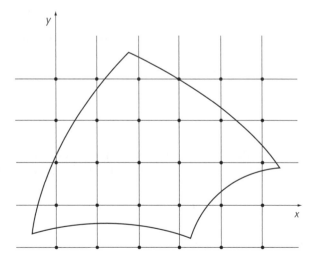

In this Programme we shall be considering functions of two real variables that satisfy given differential equations and whose domains are restricted to being rectangular. This restriction avoids many of the problems that occur with arbitrary domain shapes where the grid lines can cross the domain boundary.

Grid values

6

The rectangular domain of the function is overlaid by a grid whose mesh size is of h units in the x direction and k units in the y direction. We shall denote the value of $f(x, y)$ at the ijth grid point as

$$f_{i,j} \equiv f(ih, jk)$$

The values of the expression $f(x, y)$ are required to be found at the grid points as shown:

$$
\begin{array}{ccccc}
\cdots & \cdots & \cdots & \cdots & \cdots \\
\cdots & f_{i-1,j+1} & f_{i,j+1} & f_{i+1,j+1} & \cdots \\
\cdots & f_{i-1,j} & f_{i,j} & f_{i+1,j} & \cdots \\
\cdots & f_{i-1,j-1} & f_{i,j-1} & f_{i+1,j-1} & \cdots \\
\cdots & \cdots & \cdots & \cdots & \cdots
\end{array}
$$

Notice as you move along the jth row of this table that the value of y is constant at $y_j = y_0 + jk$ for all points on that row. Similarly, as you move up and down the ith column that the value of x is constant at $x_i = x_0 + ih$ for all points in that column. These facts now enable us to define the central difference formulas for the partial derivatives of $f(x, y)$.

The first partial derivative of $f(x, y)$ with respect to the variable x is obtained by differentiating $f(x, y)$ with respect to x whilst keeping the value of the variable y constant. Therefore, as with the ordinary derivative

$$\left. \frac{\partial f(x, y)}{\partial x} \right|_{ij}$$ *is equal to the difference between the two adjacent values of $f(x, y)$ in the x-direction divided by twice the mesh size in the x-direction.*

That is

$$\left. \frac{\partial f(x, y)}{\partial x} \right|_{-ij} = \frac{f_{i+1,j} - f_{i-1,j}}{2h}$$

This is the central difference formula for the partial derivative with respect to x. Similarly, the central difference formula for the partial derivative with respect to y is

$$\left. \frac{\partial f(x, y)}{\partial y} \right|_{ij} = \ldots\ldots\ldots$$

7

$$\left.\frac{\partial f(x, y)}{\partial y}\right|_{ij} = \frac{f_{i,j+1} - f_{i,j-1}}{2k}$$

Because

$$\left.\frac{\partial f(x, y)}{\partial y}\right|_{ij}$$ *is equal to the difference between the two adjacent values of $f(x, y)$ in the y-direction divided by twice the mesh size in the y-direction.*

That is

$$\left.\frac{\partial f(x, y)}{\partial y}\right|_{ij} = \frac{f_{i,j+1} - f_{i,j-1}}{2k}$$

Let's try an example so that we can put all this information together.

Example 1

Find the solution to $3\dfrac{\partial f(x, y)}{\partial x} - 4\dfrac{\partial f(x, y)}{\partial y} = 0$, for $0 \le x \le 1$ and $0 \le y \le 1$ given that the boundary conditions are

$f(x, 0) = 4x + 4$
$f(x, 1) = 4x + 7$
$f(0, y) = 3y + 4$
$f(1, y) = 3y + 8$

for a mesh of size $1/4$ in the x-direction and of size $1/3$ in the y-direction.

Next frame

8

The first thing we must do is to make a reasonable drawing of the domain of the function with the grid overlaid. The domain of $f(x, y)$ is the square of side length 1 as shown in the diagram.

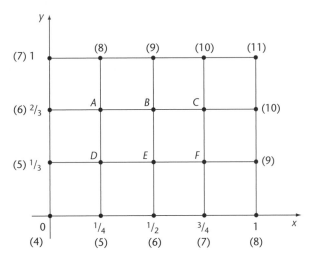

Overlaid on the function domain in the x–y plane is a mesh of grid points. The values of $f(x, y)$ that we can compute directly from the boundary conditions are shown in brackets. For example, from $f(x, 0) = 4x + 4$ we obtain $f(1/4, 0) = 5$, $f(1/2, 0) = 6$, $f(3/4, 0) = 7$ and $f(1, 0) = 8$. From $f(1, y) = 3y + 8$ we obtain $f(1, 0) = 8$, $f(1, 1/3) = 9$, $f(1, 2/3) = 10$ and $f(1, 1) = 11$. Notice that the value found at $f(1, 0) = 8$ using $f(x, 0) = 4x + 4$ is the same as the value found using $f(1, y) = 3y + 8$, as of course it must be. The values of $f(x, y)$ that we have to determine are labelled A to F.

The second part of the procedure is to find the central difference formula that describes the differential equation:

We have $\dfrac{\partial f(x, y)}{\partial x}\bigg|_{ij} = \dfrac{f_{i+1,j} - f_{i-1,j}}{2h} = 2(f_{i+1,j} - f_{i-1,j})$ because $h = 1/4$

$\dfrac{\partial f(x, y)}{\partial y}\bigg|_{ij} = \dfrac{f_{i,j+1} - f_{i,j-1}}{2k} = 1\cdot5(f_{i,j+1} - f_{i,j-1})$ because $k = 1/3$

Therefore

$$3\frac{\partial f(x, y)}{\partial x} - 4\frac{\partial f(x, y)}{\partial y} = 0 \text{ becomes} \quad \ldots\ldots\ldots\ldots$$

9

$$\boxed{6(f_{i+1,j} - f_{i-1,j}) - 6(f_{i,j+1} - f_{i,j-1}) = 0}$$

Because

$3\dfrac{\partial f(x, y)}{\partial x} - 4\dfrac{\partial f(x, y)}{\partial y} = 0$ evaluated at the ijth grid point is

$3\dfrac{\partial f(x, y)}{\partial x}\bigg|_{ij} - 4\dfrac{\partial f(x, y)}{\partial y}\bigg|_{ij} = 0$

which is

$$3 \times 2(f_{i+1,j} - f_{i-1,j}) - 4 \times 1\cdot5(f_{i,j+1} - f_{i,j-1}) = 0, \text{ that is}$$
$$6(f_{i+1,j} - f_{i-1,j}) - 6(f_{i,j+1} - f_{i,j-1}) = 0$$

Computational molecules

10

The value of the first derivative with respect to x at the point (x_i, y_j) on the grid overlaying the function domain is found by evaluating the right-hand side of the equation

$$\frac{\partial f(x, y)}{\partial x}\bigg|_{ij} = \frac{f_{i+1,j} - f_{i-1,j}}{2h} = \frac{-f_{i-1,j} + f_{i+1,j}}{2h}$$

and this process is repeated for every grid point in the function domain. We can construct a graphic template to assist us in this process:

The three circles in a row are used to calculate the contribution of three adjacent row members to the equation. If the circle labelled ij is laid over the ijth grid point then the derivative at that point is given by multiplying the value of the function at the $i-1, j$ grid point (one to the left) by $-1/2h$ and adding the product of the value of the function at the $i+1, j$ grid point (one to the right) by $1/2h$. The number 0 in the centre circle means that we multiply $f_{i,j}$ by zero because it does not enter into the formula. This template is called a *computational molecule*. The horizontal structure reflects the fact that we are evaluating along a row. By a similar reasoning the first derivative with respect to y at the ijth grid point is

$$\frac{\partial f(x, y)}{\partial y}\bigg|_{ij} = \frac{f_{i,j+1} - f_{i,j-1}}{2k}$$

and this is represented by the computational molecule:

The vertical structure reflects the fact that we are evaluating up and down a column.

By combining such computational molecules we can construct a composite molecule that represents the entire differential equation. For example, the partial differential equation

$$a\frac{\partial f(x, y)}{\partial x} + b\frac{\partial f(x, y)}{\partial y} = c$$

evaluated at the *ij*th grid point is

$$a\frac{\partial f(x, y)}{\partial x}\bigg|_{ij} + b\frac{\partial f(x, y)}{\partial y}\bigg|_{ij} = c$$

and is represented by the central difference formula

$$\frac{a}{2h}(f_{i+1,j} - f_{i-1,j}) + \frac{b}{2k}(f_{i,j+1} - f_{i,j-1}) = c$$

which is in turn represented by the composite computational molecule:

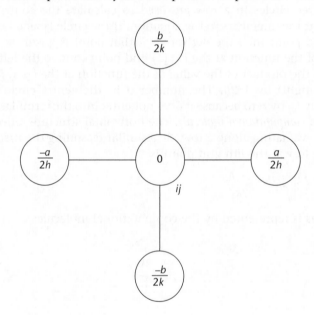

So the equation $3\dfrac{\partial f(x, y)}{\partial x} - 4\dfrac{\partial f(x, y)}{\partial y} = 0$ which is represented by the finite difference formula

$$6(f_{i+1,j} - f_{i-1,j}) - 6(f_{i,j+1} - f_{i,j-1}) = 0$$

has the computational molecule . . . .

11

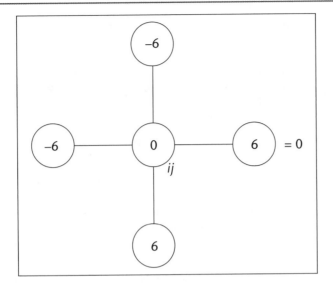

We now place the centre of the molecule, in turn, on each of the grid points at which we need to find the value of $f(x, y)$:

On A	$-36 - 48 + 6B + 6D = 0$
On B	$-6A - 54 + 6C + 6E = 0$
On C	$\dots\dots\dots$
On D	$\dots\dots\dots$
On E	$\dots\dots\dots$
On F	$\dots\dots\dots$

12

On A	$-36 - 48 + 6B + 6D = 0$
On B	$-6A - 54 + 6C + 6E = 0$
On C	$-6B - 60 + 60 + 6F = 0$
On D	$-30 - 6A + 6E + 30 = 0$
On E	$-6D - 6B + 6F + 36 = 0$
On F	$-6E - 6C + 54 + 42 = 0$

We now have six simultaneous linear equations in six unknowns.

These can be written in matrix form as $\dots\dots\dots$

13

$$\begin{pmatrix} 0 & 6 & 0 & 6 & 0 & 0 \\ -6 & 0 & 6 & 0 & 6 & 0 \\ 0 & -6 & 0 & 0 & 0 & 6 \\ -6 & 0 & 0 & 0 & 6 & 0 \\ 0 & -6 & 0 & -6 & 0 & 6 \\ 0 & 0 & -6 & 0 & -6 & 0 \end{pmatrix} \begin{pmatrix} A \\ B \\ C \\ D \\ E \\ F \end{pmatrix} = \begin{pmatrix} 84 \\ 54 \\ 0 \\ 0 \\ -36 \\ -96 \end{pmatrix}$$

That is: $\mathbf{Ax} = \mathbf{b}$ with solution $\mathbf{x} = \mathbf{A}^{-1}\mathbf{b}$

There are many ways to derive the inverse matrix $\mathbf{A}^{-1}$, many of them time consuming and prone to arithmetic error. An efficient method in terms of time and accuracy is to use a spreadsheet, provided of course that the spreadsheet has the appropriate functionality. Here we shall use the *Microsoft Excel* spreadsheet which possesses matrix functions. If your spreadsheet does not have these functions then you are referred to Programme 16, Matrix algebra.

If you do possess the *Microsoft Excel* spreadsheet then follow the instructions in the next frame.

Next frame

14

1 Open your spreadsheet.
2 Place the cell highlight in cell A1 and then enter the values of matrix $\mathbf{A}$ into the cells A1 to F6.
3 Place the cell highlight in cell H1 and then enter the values of matrix $\mathbf{b}$ into the cells H1 to H6.
4 Place the cell highlight in cell A8 and drag the mouse to highlight the block of cells A8 to F13 – this is where the inverse of $\mathbf{A}$ is going to go.
5 With this block of cells highlighted, type the function:

    =MINVERSE(A1:F6) and then press the three keys **Ctrl-Shift-Enter** together

As you type, the function is entered into cell A8 and when you press the **Ctrl-Shift-Enter** keys together the block of cells A8 to F13 fills with entries. This block of cells is the inverse matrix $\mathbf{A}^{-1}$. (*Note*: You must remember to press the three keys **Ctrl-Shift-Enter** together. If you just press **Enter** it will not work.)

    MINVERSE(array) is the *Excel* function that computes the inverse of the square matrix denoted by **array**.

6 Place the cell highlight in cell H8 and drag the mouse to highlight the block of cells H8 to H13 – this is where the solution $\mathbf{x}$ is going to go.
7 With this block of cells highlighted type the function:

    =MMULT(A8:F13, H8:H13) and then press the three keys **Ctrl-Shift-Enter** together

    MMULT(array1,array2) is the *Excel* function that multiplies the two matrices denoted by **array1** and **array2**.

▶

As you type, the function is entered into cell H8 and when you press the **Ctrl-Shift-Enter** keys together the block of cells H8 to H13 fills with entries. This block of cells is the product matrix $\mathbf{A}^{-1}\mathbf{b}$, that is, the solution matrix $\mathbf{x}$.

$$\begin{pmatrix} A \\ B \\ C \\ D \\ E \\ F \end{pmatrix} = \begin{pmatrix} 7 \\ 8 \\ 9 \\ 6 \\ 7 \\ 8 \end{pmatrix}$$

These values are identical to the values found from the exact solution which is $f(x, y) = 4x + 3y + 4$.

Next frame

Summary of procedures

The procedure to solve a first-order partial differential equation requires a number of steps to be completed in a certain order, and the following list describes the sequence:

15

1 Draw the domain of the function with the grid overlaid.
2 On the drawing enter the values of $f(x, y)$ that can be obtained from the boundary conditions.

Put these values in brackets so that they will be easily distinguished from the x- and y-values on the axes.

3 Label the grid points at which $f(x, y)$ is to be evaluated with capital letters.
4 Construct the central difference equation that represents the numerical approximation to the partial differential equation.
5 Construct the computational molecule for this equation.
6 Lay the centre of the molecule on each of the lettered grid points in turn and derive a set of simultaneous linear equations – the unknowns being represented by the letters at the grid points.
7 Write the simultaneous linear equations in matrix form $\mathbf{Ax} = \mathbf{b}$.
8 Find the inverse matrix $\mathbf{A}^{-1}$ and compute the solution $\mathbf{x} = \mathbf{A}^{-1}\mathbf{b}$.

Now try one yourself. Just follow the procedure in order and you should have no problems.

▶

Example 2

The solution to $x\dfrac{\partial f(x, y)}{\partial x} - y\dfrac{\partial f(x, y)}{\partial y} = 0$, for $0 \le x \le 1$ and $0 \le y \le 1$ given that

$f(x, 0) = 2$

$f(x, 1) = x + 2$

$f(0, y) = 2$

$f(1, y) = y + 2$

for a mesh of $1/4$ in the x-direction and $1/3$ in the y-direction is: ....

16

$$\begin{pmatrix} A \\ B \\ C \\ D \\ E \\ F \end{pmatrix} = \begin{pmatrix} 2{\cdot}166\ldots \\ 2{\cdot}33\ldots \\ 2{\cdot}5 \\ 2{\cdot}0833\ldots \\ 2{\cdot}166\ldots \\ 2{\cdot}25 \end{pmatrix} = \begin{pmatrix} 13/6 \\ 7/3 \\ 5/2 \\ 25/12 \\ 13/6 \\ 9/4 \end{pmatrix}$$

Because

The domain of the function $f(x, y)$ with the overlaid grid looks as follows:

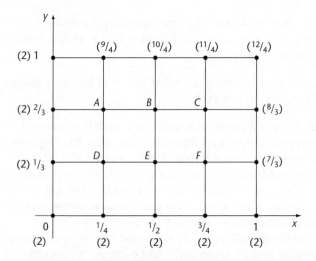

where the numbers at the grid points in brackets are the values of $f(x, y)$ obtained by applying the boundary conditions and the letters $A \ldots F$ represent the values of $f(x, y)$ that we have yet to determine.

The central difference formulas for the two first partial derivatives of $f(x, y)$ are

$$\left.\frac{\partial f(x, y)}{\partial x}\right|_{ij} = \frac{f_{i+1,j} - f_{i-1,j}}{2h} = 2(f_{i+1,j} - f_{i-1,j}) \text{ because } h = 1/4$$

$$\left.\frac{\partial f(x, y)}{\partial y}\right|_{ij} = \frac{f_{i,j+1} - f_{i,j-1}}{2k} = 1.5(f_{i,j+1} - f_{i,j-1}) \text{ because } k = 1/3$$

Therefore

$$x\frac{\partial f(x, y)}{\partial x} - y\frac{\partial f(x, y)}{\partial y} = 0 \text{ becomes } \ldots\ldots\ldots\ldots$$

17

$$\boxed{2(x_i f_{i+1,j} - x_i f_{i-1,j}) - 1.5(y_j f_{i,j+1} - y_j f_{i,j-1}) = 0}$$

Because

$$x\frac{\partial f(x, y)}{\partial x} - y\frac{\partial f(x, y)}{\partial y} = 0$$

is written using the central difference formulas as

$$2x_i(f_{i+1,j} - f_{i-1,j}) - 1.5y_j(f_{i,j+1} - f_{i,j-1})$$
$$= 2(x_i f_{i+1,j} - x_i f_{i-1,j}) - 1.5(y_j f_{i,j+1} - y_j f_{i,j-1}) = 0$$

This has the following computational molecule:

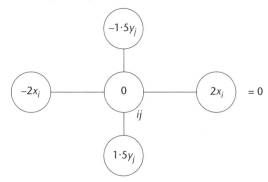

Placing the centre of the molecule, in turn, on each of the grid points that we need to evaluate, we obtain the six simultaneous equations:

On A at $\left(\frac{1}{4}, \frac{2}{3}\right)$: $\quad -2\left(\frac{1}{4}\right)(2) - \frac{3}{2}\left(\frac{2}{3}\right)\left(\frac{9}{4}\right) + 2\left(\frac{1}{4}\right)B + \frac{3}{2}\left(\frac{2}{3}\right)D = 0$

On B at $\left(\frac{1}{2}, \frac{2}{3}\right)$: $\quad -2\left(\frac{1}{2}\right)A - \frac{3}{2}\left(\frac{2}{3}\right)\left(\frac{10}{4}\right) + 2\left(\frac{1}{2}\right)C + \frac{3}{2}\left(\frac{2}{3}\right)E = 0$

On C at $\left(\frac{3}{4}, \frac{2}{3}\right)$: $\quad -2\left(\frac{3}{4}\right)B - \frac{3}{2}\left(\frac{2}{3}\right)\left(\frac{11}{4}\right) + 2\left(\frac{3}{4}\right)\left(\frac{8}{3}\right) + \frac{3}{2}\left(\frac{2}{3}\right)F = 0$

On D at $\left(\frac{1}{4}, \frac{1}{3}\right)$: $\quad -2\left(\frac{1}{4}\right)(2) - \frac{3}{2}\left(\frac{1}{3}\right)A + 2\left(\frac{1}{4}\right)E + \frac{3}{2}\left(\frac{1}{3}\right)(2) = 0$

On E at $\left(\frac{1}{2}, \frac{1}{3}\right)$: $\quad -2\left(\frac{1}{2}\right)D - \frac{3}{2}\left(\frac{1}{3}\right)B + 2\left(\frac{1}{2}\right)F + \frac{3}{2}\left(\frac{1}{3}\right)(2) = 0$

On F at $\left(\frac{3}{4}, \frac{1}{3}\right)$: $\quad -2\left(\frac{3}{4}\right)E - \frac{3}{2}\left(\frac{1}{3}\right)C + 2\left(\frac{3}{4}\right)\left(\frac{7}{3}\right) + \frac{3}{2}\left(\frac{1}{3}\right)(2) = 0$

These six equations can be simplified as $\ldots\ldots\ldots\ldots$

18

On A	$B/2 + D = 13/4$
On B	$-A + C + E = 10/4$
On C	$-3B/2 + F = -5/4$
On D	$-A/2 + E/2 = 0$
On E	$-B/2 - D + F = -1$
On F	$-C/2 - 3E/2 = -9/2$

These six simultaneous linear equations can be expressed in matrix form as
....

19

$$\begin{pmatrix} 0 & 0.5 & 0 & 1 & 0 & 0 \\ -1 & 0 & 1 & 0 & 1 & 0 \\ 0 & -1.5 & 0 & 0 & 0 & 1 \\ -0.5 & 0 & 0 & 0 & 0.5 & 0 \\ 0 & -0.5 & 0 & -1 & 0 & 1 \\ 0 & 0 & -0.5 & 0 & -1.5 & 0 \end{pmatrix} \begin{pmatrix} A \\ B \\ C \\ D \\ E \\ F \end{pmatrix} = \begin{pmatrix} 13/4 \\ 10/4 \\ -5/4 \\ 0 \\ -1 \\ -9/2 \end{pmatrix}$$

That is

$\mathbf{Ax} = \mathbf{b}$ with solution $\mathbf{x} = \mathbf{A}^{-1}\mathbf{b}$

Inverting the matrix $\mathbf{A}$ we find that

$$\begin{pmatrix} A \\ B \\ C \\ D \\ E \\ F \end{pmatrix} = \begin{pmatrix} 2.166\ldots \\ 2.3\ldots \\ 2.5 \\ 2.0833\ldots \\ 2.166\ldots \\ 2.25 \end{pmatrix} = \begin{pmatrix} 13/6 \\ 7/3 \\ 5/2 \\ 25/12 \\ 13/6 \\ 9/4 \end{pmatrix}$$

which is identical to the values found from the exact solution $f(x, y) = xy + 2$.

Next frame

Derivative boundary conditions

20

The process of solving a differential equation, either ordinary or partial, involves using indefinite integration and each time we integrate we produce an integration constant. For a differential equation to have a complete solution, where all the integration constants are evaluated, the differential equation must be accompanied by a set of conditions that are sufficient to do this.

If the differential equation involves time t then it is natural for these conditions to give values of the function and its derivatives at time $t = 0$. Such conditions are known as *initial conditions* and we have met these before when ▶

we studied the Laplace transform, for example. Other conditions, like the conditions we met in the previous two examples, are called *boundary conditions* because they gave the values of the function on the boundary of the function domain. We now consider boundary conditions in the form of derivatives normal to the boundary and this we do in the following example.

Example 3

Find the solution to $4\dfrac{\partial f(x,\,y)}{\partial x} + 2\dfrac{\partial f(x,\,y)}{\partial y} = 3$, for $0 \le x \le 1$ and $0 \le y \le 1$

given that the boundary conditions are

$$f(x,0) = f(x,1) = f(0,y) = 10$$

and $\quad \dfrac{\partial f(x,\,y)}{\partial x}\bigg|_{x=1} = 2$

for a mesh of size $1/3$ in both the x-direction and the y-direction.

Next frame

The domain of $f(x,\,y)$ is the square of side length 1 as shown in the diagram: **21**

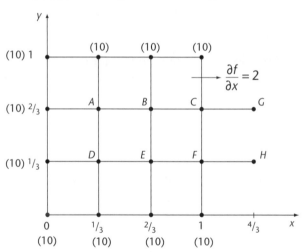

Overlaid on the function domain in the x–y plane is a mesh of grid points. Because the boundary condition relating to the side $x = 1$ is in the form of a derivative normal to the side, *we extend the grid over the boundary of the function domain by adding two additional points outside the domain* and distant $1/3$ from it, as shown in the figure.

The values of $f(x, y)$ that we can compute from the boundary conditions alone are shown in brackets. The values of $f(x, y)$ that we have to determine are labelled A to F and we shall need the two additional points G and H outside the domain of $f(x, y)$ to do this.

The second part of the procedure is to find the central difference formula that describes the differential equation:

We have $\dfrac{\partial f(x,\,y)}{\partial x}\bigg|_{ij} = \dfrac{f_{i+1,j} - f_{i-1,j}}{2h} = 1{\cdot}5\left(f_{i+1,j} - f_{i-1,j}\right)$

$\dfrac{\partial f(x,\,y)}{\partial y}\bigg|_{ij} = \dfrac{f_{i,j+1} - f_{i,j-1}}{2k} = 1{\cdot}5\left(f_{i,j+1} - f_{i,j-1}\right)$

because both h and $k = 1/3$

Therefore

$$4\dfrac{\partial f(x,\,y)}{\partial x} + 2\dfrac{\partial f(x,\,y)}{\partial y} = 3 \text{ becomes } \ldots\ldots\ldots\ldots$$

22

$$\boxed{6\left(f_{i+1,j} - f_{i-1,j}\right) + 3\left(f_{i,j+1} - f_{i,j-1}\right) = 3}$$

Because

$$4\dfrac{\partial f(x,\,y)}{\partial x} + 2\dfrac{\partial f(x,\,y)}{\partial y} = 3 \text{ can be written as}$$

$4 \times 1{\cdot}5\left(f_{i+1,j} - f_{i-1,j}\right) + 2 \times 1{\cdot}5\left(f_{i,j+1} - f_{i,j-1}\right) = 3$, that is

$6\left(f_{i+1,j} - f_{i-1,j}\right) + 3\left(f_{i,j+1} - f_{i,j-1}\right) = 3$

This has the computational molecule ....

23

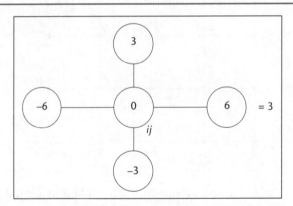

We now place the centre of the molecule, in turn, on each of the grid points that we need to evaluate:

On A	$-60 + 30 + 6B - 3E = 3$
On B	$-6A + 30 + 6C - 3E = 3$
On C	....
On D	....
On E	....
On F	....

24

On A	$-60 + 30 + 6B - 3E = 3$
On B	$-6A + 30 + 6C - 3E = 3$
On C	$-6B + 30 + 6G - 3F = 3$
On D	$-60 + 3A + 6E - 30 = 3$
On E	$-6D + 3B + 6F - 30 = 3$
On F	$-6E + 3C + 6H - 30 = 3$

At the boundary $x = 1$ the boundary condition $\left.\dfrac{\partial f(x, y)}{\partial x}\right|_{x=1} = 2$ can be written using the central difference formula as

$$\left.\frac{\partial f(x, y)}{\partial x}\right|_{\substack{x=1 \\ y=y_j}} = \frac{f_{i+1,j} - f_{i-1,j}}{2h} = 1{\cdot}5(f_{i+1,j} - f_{i-1,j}) = 2$$

which has the computational molecule:

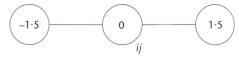

We now place the centre of this molecule, in turn, on each of the grid points C and F to obtain

$$\text{On } C \quad -1{\cdot}5B + 1{\cdot}5G = 2$$
$$\text{On } F \quad \ldots\ldots\ldots\ldots$$

25

On C	$-1{\cdot}5B + 1{\cdot}5G = 2$
On F	$-1{\cdot}5E + 1{\cdot}5H = 2$

We can now use these last two equations either to eliminate the points G and H from the six equations in Frame 24 or to form an 8×8 system. We shall eliminate the points G and H to obtain the six equations, with the constant on the right-hand side as

$$\ldots\ldots\ldots\ldots$$

26

On A	$6B - 3E = 33$
On B	$-6A + 6C - 3E = -27$
On C	$-3F = -35$
On D	$3A + 6E = 93$
On E	$-6D + 3B + 6F = 33$
On F	$3C = 25$

These six simultaneous linear equations can be expressed in matrix form as

$$\ldots\ldots\ldots\ldots$$

27

$$\begin{pmatrix} 0 & 6 & 0 & 0 & -3 & 0 \\ -6 & 0 & 6 & 0 & -3 & 0 \\ 0 & 0 & 0 & 0 & 0 & -3 \\ 3 & 0 & 0 & 0 & 6 & 0 \\ 0 & 3 & 0 & -6 & 0 & 6 \\ 0 & 0 & 3 & 0 & 0 & 0 \end{pmatrix} \begin{pmatrix} A \\ B \\ C \\ D \\ E \\ F \end{pmatrix} = \begin{pmatrix} 33 \\ -27 \\ -35 \\ 93 \\ 33 \\ 25 \end{pmatrix}$$

That is

$\mathbf{Ax} = \mathbf{b}$ with solution $\mathbf{x} = \mathbf{A}^{-1}\mathbf{b}$

Inverting the matrix $\mathbf{A}^{-1}$ we find that $\mathbf{x} = \dots\dots\dots$

28

$$\begin{pmatrix} A \\ B \\ C \\ D \\ E \\ F \end{pmatrix} = \begin{pmatrix} 6 \cdot 777 \dots \\ 11 \cdot 555 \dots \\ 8 \cdot 333 \dots \\ 11 \cdot 9444 \dots \\ 12 \cdot 111 \dots \\ 11 \cdot 666 \ \dots \end{pmatrix} = \begin{pmatrix} 61/9 \\ 104/9 \\ 25/3 \\ 215/18 \\ 109/9 \\ 35/3 \end{pmatrix}$$

Next frame

Second-order partial differential equations

29

The most general form of a second-order partial differential equation is

$$a(x,y)\frac{\partial^2 f}{\partial x^2} + b(x,y)\frac{\partial^2 f}{\partial x \partial y} + c(x,y)\frac{\partial^2 f}{\partial y^2} + d(x,y)\frac{\partial f}{\partial x} + e(x,y)\frac{\partial f}{\partial y} + g(x,y) = 0$$

Three types of equation are of particular interest because they feature so prominently in engineering and science.

Elliptic equations

If $b^2 - 4ac < 0$ the partial differential equation is called an *elliptic* equation. Such equations arise out of steady-state problems as occur in potential or flow theory. Two examples are

Poisson's equation

$$\frac{\partial^2 \phi(x,y)}{\partial x^2} + \frac{\partial^2 \phi(x,y)}{\partial y^2} = g(x,y)$$

Laplace's equation

$$\frac{\partial^2 \phi(x,y)}{\partial x^2} + \frac{\partial^2 \phi(x,y)}{\partial y^2} = 0$$

In both cases $a = 1$, $b = 0$ and $c = 1$ and so $b^2 - 4ac < 0$.

▶

Hyperbolic equations

If $b^2 - 4ac > 0$ the partial differential equation is called an *hyperbolic* equation. Such equations arise out of vibrational and radiative problems as occur in wave mechanics. An example is

The wave equation

$$\frac{\partial^2 \phi(x, t)}{\partial x^2} = \frac{1}{\kappa^2} \frac{\partial^2 \phi(x, t)}{\partial t^2}$$

Here $a = 1$, $b = 0$ and $c = -\dfrac{1}{\kappa^2}$ and so $b^2 - 4ac > 0$.

Parabolic equations

If $b^2 - 4ac = 0$ the partial differential equation is called a *parabolic* equation. Such equations arise out of transient flow problems as occur in conduction or consolidation. An example is

The consolidation (or heat conduction) equation

$$\frac{\partial^2 \phi(x, t)}{\partial x^2} = \frac{1}{\kappa} \frac{\partial \phi(x, t)}{\partial t}$$

Here $a = 1$, $b = 0$ and $c = 0$ and so $b^2 - 4ac = 0$.

In the equations above a, b and c are constant but in the general case they depend on x and y and so a given equation may change from one type to another within the same domain.

Determine whether each of the following equations are elliptic, hyperbolic or parabolic:

(a) $\dfrac{\partial^2 \phi(x, y)}{\partial x^2} + \dfrac{\partial^2 \phi(x, y)}{\partial y^2} + k\phi(x, y) = \Phi(x, y)$

(b) $\dfrac{\partial \phi(r, t)}{\partial t} = a\left[\dfrac{\partial^2 \phi(r, t)}{\partial r^2} + \dfrac{2}{r}\dfrac{\partial \phi(r, t)}{\partial r}\right] + \Phi(r, t)$

(c) $\dfrac{\partial^2 \phi(x, t)}{\partial t^2} + k\dfrac{\partial \phi(x, t)}{\partial t} = p^2 \dfrac{\partial^2 \phi(x, x)}{\partial x^2} + q\phi(x, y)$

(d) $\dfrac{\partial \phi(x, t)}{\partial t} = \dfrac{\partial^2 \phi(x, t)}{\partial x^2} + \phi(x, t)\dfrac{\partial \phi(x, t)}{\partial x}$

(e) $\dfrac{\partial^2}{\partial x}\left(px^n \dfrac{\partial \phi(x, y)}{\partial x}\right) + \dfrac{\partial^2}{\partial y}\left(py^n \dfrac{\partial \phi(x, y)}{\partial y}\right) = r\phi(x, y)$

(f) $\dfrac{\partial^2 \phi(x, t)}{\partial t^2} = p\dfrac{\partial}{\partial x}\left(\phi^n(x, y)\dfrac{\partial \phi(x, y)}{\partial y}\right) + q^m(x, y)$

Next frame

30

(a) Elliptic	(b) Parabolic
(c) Hyperbolic	(d) Parabolic
(e) Elliptic	(f) Hyperbolic

Because:

Comparing with the general equation

$$a(x,y)\frac{\partial^2 f}{\partial x^2} + b(x,y)\frac{\partial^2 f}{\partial x \partial y} + c(x,y)\frac{\partial^2 f}{\partial y^2} + d(x,y)\frac{\partial f}{\partial x} + e(x,y)\frac{\partial f}{\partial y} + g(x,y) = 0$$

(a) $\dfrac{\partial^2 \phi(x,\ y)}{\partial x^2} + \dfrac{\partial^2 \phi(x,\ y)}{\partial y^2} + k\phi(x,\ y) = \Phi(x,\ y)$

Here $b = 0$ so $b^2 - 4ac < 0$, Elliptic. The Helmholtz equation.

(b) $\dfrac{\partial \phi(r,\ t)}{\partial t} = a\left[\dfrac{\partial^2 \phi(r,\ t)}{\partial r^2} + \dfrac{2}{r}\dfrac{\partial \phi(r,\ t)}{\partial r}\right] + \Phi(r,\ t)$

Here $b = c = 0$ so $b^2 - 4ac = 0$. Parabolic. The heat equation with central symmetry.

(c) $\dfrac{\partial^2 \phi(x,\ t)}{\partial t^2} + k\dfrac{\partial \phi(x,\ t)}{\partial t} = p^2 \dfrac{\partial^2 \phi(x,\ x)}{\partial x^2} + q\phi(x,\ y)$

Here $b = 0$, $a > 0$ and $c < 0$ so $b^2 - 4ac > 0$. Hyperbolic. The telegraph equation.

(d) $\dfrac{\partial \phi(x,\ t)}{\partial t} = \dfrac{\partial^2 \phi(x,\ t)}{\partial x^2} + \phi(x,\ t)\dfrac{\partial \phi(x,\ t)}{\partial x}$

Here $b = c = 0$ so $b^2 - 4ac = 0$. Parabolic. Burger's equation.

(e) $\dfrac{\partial^2}{\partial x}\left(px^n \dfrac{\partial \phi(x,\ y)}{\partial x}\right) + \dfrac{\partial^2}{\partial y}\left(py^n \dfrac{\partial \phi(x,\ y)}{\partial y}\right) = r\phi(x,\ y)$

Here $b = 0$ so $b^2 - 4ac < 0$. Elliptic. The anisotropic heat diffusion equation.

(f) $\dfrac{\partial^2 \phi(x,\ t)}{\partial t^2} = p\dfrac{\partial}{\partial x}\left(\phi^n(x,\ y)\dfrac{\partial \phi(x,\ y)}{\partial y}\right) + q^m(x,\ y)$

Here $b = 0$, $a > 0$ and $c < 0$ so $b^2 - 4ac > 0$. Hyperbolic.

Next frame

Second partial derivatives

31

In Frame 4 we found that for a function of a single real variable $f(x)$ the central difference formula approximating the second derivative was

$$f''(x) \approx \frac{f(x-h) - 2f(x) + f(x+h)}{h^2}$$

The second derivative at x is given as the sum of the two adjacent values less twice the value at the point, all divided by h^2.

If we apply this to a function of two real variables $f(x, y)$ and use $f_{i,j} \equiv f(ih, jk)$ to represent the value of $f(x, y)$ at the point (ih, jk) then the central difference formulas for the second partial derivatives with respect to x and y are seen to be ....

$$\left. \frac{\partial^2 f(x, y)}{\partial x^2} \right|_{ij} \approx \frac{f_{i-1,j} - 2f_{i,j} + f_{i+1,j}}{h^2}$$

$$\left. \frac{\partial^2 f(x, y)}{\partial y^2} \right|_{ij} \approx \frac{f_{i,j-1} - 2f_{i,j} + f_{i,j+1}}{k^2}$$

Because

The second derivative at x_i is given as the sum of the two adjacent values on the jth row less twice the value at x_i, all divided by the cell width squared – h^2, and so

$$\left. \frac{\partial^2 f(x, y)}{\partial x^2} \right|_{ij} \approx \frac{f_{i-1,j} - 2f_{i,j} + f_{i+1,j}}{h^2}$$

The second derivative at y_j is given as the sum of the two adjacent values in the jth column less twice the value at y_j, all divided by the cell height squared – k^2, and so

$$\left. \frac{\partial^2 f(x, y)}{\partial y^2} \right|_{ij} \approx \frac{f_{i,j-1} - 2f_{i,j} + f_{i,j+1}}{k^2}$$

We are now ready to consider the construction of central difference formulas for second-order partial differential equations. We shall proceed by example.

▶

Example 4

Given a grid with mesh size $h = k = 1/3$, find a numerical solution to the equation

$$\frac{\partial^2 f(x, y)}{\partial x^2} + \frac{\partial^2 f(x, y)}{\partial y^2} = 0 \text{ for } 0 \le x \le 1, \ 0 \le y \le 1, \text{ given that}$$

$$f(x, 0) = f(x, 1) = 5 - 3x$$

$$f(0, y) = 9y^2 - 9y + 5 \text{ and}$$

$$\left. \frac{\partial f(x, y)}{\partial x} \right|_{x=1} = -6$$

The domain with the grid overlaid is ....

33

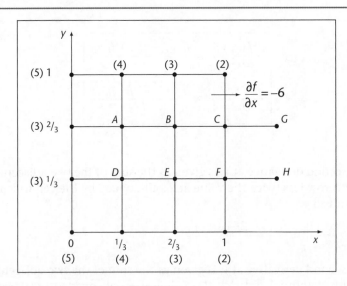

The solution is to be evaluated at the grid points A to F – the external grid points G and H are inserted to accommodate the derivative boundary condition. The numbers in brackets are the values of $f(x, y)$ as found from the boundary conditions.

The central difference formula that represents
the partial differential equation is ....

$$\boxed{9\left(f_{i+1,j} + f_{i,j+1} - 4f_{i,j} + f_{i-1,j} + f_{i,j-1}\right)}$$

34

Because

$$\left.\frac{\partial^2 f(x,\,y)}{\partial x^2}\right|_{ij} + \left.\frac{\partial^2 f(x,\,y)}{\partial y^2}\right|_{ij} \approx \frac{f_{i+1,j} - 2f_{i,j} + f_{i-1,j}}{h^2} + \frac{f_{i,j+1} - 2f_{i,j} + f_{i,j-1}}{k^2}$$

$$= 9\left(f_{i+1,j} - 2f_{i,j} + f_{i-1,j}\right) + 9\left(f_{i,j+1} - 2f_{i,j} + f_{i,j-1}\right)$$

$$= 9\left(f_{i+1,j} + f_{i,j+1} - 4f_{i,j} + f_{i-1,j} + f_{i,j-1}\right)$$

From this we can construct the computational molecule for this differential equation as . . . .

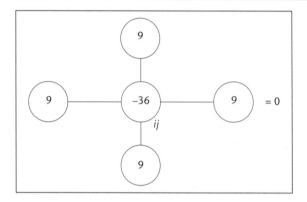

35

If we applied this computational molecule to the grid points A to F then the six simultaneous linear equations that result would all have a common factor of 9 arising from the 9 in the molecule. If we divided every equation by 9 to remove this common factor we would not change the overall validity of the equations. So, to make the computation simpler we divide each term in the computational molecule by 9 and use the resulting molecule:

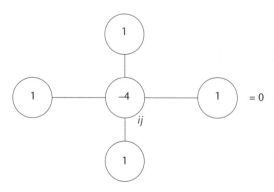

We now proceed as we have done before. Laying the centre of the computational molecule on each grid point in turn gives the six simultaneous linear equations:

On A	$3 + 4 + B + D - 4A = 0$
On B	. . . .
On C	. . . .
On D	. . . .
On E	. . . .
On F	. . . .

36

On A	$3 + 4 + B + D - 4A = 0$
On B	$A + 3 + C + E - 4B = 0$
On C	$B + 2 + G + F - 4C = 0$
On D	$3 + A + E + 4 - 4D = 0$
On E	$D + B + F + 3 - 4E = 0$
On F	$E + C + H + 2 - 4F = 0$

We now apply the derivative boundary condition at the grid points C and F by using the computational molecule for the first partial derivative with respect to x

$$\left.\frac{\partial f(x, y)}{\partial x}\right|_{x=1} = \frac{f_{i+1,j} - f_{i-1,j}}{2h} = \frac{3}{2}(f_{i+1,j} - f_{i-1,j}) = -6$$

This gives . . . .

37

$$\frac{3}{2}(-B + G) = -6$$
$$\frac{3}{2}(-E + H) = -6$$

Because

The computational molecule for the first partial derivative with respect to x is

$$\left.\frac{\partial f(x, y)}{\partial x}\right|_{ij} = \frac{-f_{i-1,j} + f_{i+1,j}}{2h} = \frac{3}{2}(-f_{i-1,j} + f_{i+1,j}) \quad \text{because } h = 1/3$$

Applying this molecule at the boundary points C and F gives the two equations

$$\frac{3}{2}(-B + G) = -6 \text{ so } G = -4 + B$$
$$\frac{3}{2}(-E + H) = -6 \text{ so } H = -4 + E$$

▶

Substitution of these two equations into the first six eliminates the grid points G and H to produce the six equations in six unknowns.

These are written in matrix form as ....

38

$$\begin{pmatrix} -4 & 1 & 0 & 1 & 0 & 0 \\ 1 & -4 & 1 & 0 & 1 & 0 \\ 0 & 2 & -4 & 0 & 0 & 1 \\ 1 & 0 & 0 & -4 & 1 & 0 \\ 0 & 1 & 0 & 1 & -4 & 1 \\ 0 & 0 & 1 & 0 & 2 & -4 \end{pmatrix} \begin{pmatrix} A \\ B \\ C \\ D \\ E \\ F \end{pmatrix} = \begin{pmatrix} -7 \\ -3 \\ 2 \\ -7 \\ -3 \\ 2 \end{pmatrix}$$

which has solution ....

39

$$\begin{pmatrix} A \\ B \\ C \\ D \\ E \\ F \end{pmatrix} = \begin{pmatrix} 28/9 \\ 7/3 \\ 8/9 \\ 28/9 \\ 7/3 \\ 8/9 \end{pmatrix}$$

Next frame

Time-dependent equations

40

Many physical systems have their behaviour modelled by a differential equation. For example, a long thin metal bar of length L, insulated along its length, has its ends maintained at a temperature of $0°C$ and, at time $t = 0$, the temperature distribution is given by

$$T(x, 0) = x^2 - 2xL + L^2$$

The future distribution of temperature $T(x, t)$ can then be found by solving the partial differential equation (the *heat equation*)

$$\frac{\partial^2 T(x, t)}{\partial x^2} = \frac{1}{\kappa} \frac{\partial T(x, t)}{\partial t}$$

subject to the given boundary and initial conditions. The constant $\kappa = \dfrac{K}{\omega}$ is called the *diffusivity* constant where K is the *thermal conductivity* and ω is the *specific heat per unit volume* of the metal that constitutes the rod. Apart from the physical considerations that set up the equation in the first place, the dimensions of κ are $[L^2 T^{-1}]$ and are necessary to balance the dimensions on either side of the equation.

If we wished to solve the heat equation numerically as it stands then we would need to know the value of κ, and this would vary depending upon the specific metal used for the bar. We can overcome this problem by absorbing κ using a process of *dimensional analysis* when we transform the equation into an equation of the form

$$\frac{\partial^2 f(x, t)}{\partial x^2} = \frac{\partial f(x, t)}{\partial t}$$

where the variables x and t are now dimensionless – they are measured in numbers rather than units of distance and time respectively. How this is done we shall leave to the end of the Programme. For now we are interested in numerically solving such dimensionless equations over a rectangular domain of width 1, and as usual we shall proceed by example.

Next frame

41

Example 5

Solve the partial differential equation

$$\frac{\partial^2 f(x, t)}{\partial x^2} = \frac{\partial f(x, t)}{\partial t}$$

for $0 \le x \le 1$ and $t \ge 0$ where

$f(0, t) = 1$

$f(x, 0) = 1 + x$ and

$\left. \dfrac{\partial f(x, t)}{\partial x} \right|_{x=1} = 0$

We now have a change in procedure. Hitherto, the first thing we did was to draw the domain of the function with the grid overlaid. We could do this because we knew the step lengths in the x- and y-directions from the beginning. Here, the first thing we must do is to construct the finite difference formula that will represent the differential equation because its structure will dictate the step lengths. We can immediately write down the central difference formula for the second derivative on the left of this equation.

It is . . . .

42

$$\left. \frac{\partial^2 f(x, t)}{\partial x^2} \right|_{ij} \approx \frac{f_{i-1,j} - 2f_{i,j} + f_{i+1,j}}{h^2}$$

To use a central difference formula for the derivative with respect to t would require a knowledge of $f(x, t)$ for values of $t < 0$ and this we do not possess. Consequently, for the derivative with respect to t we use the *forward* difference formula. Do you remember this one?

It is . . . .

$$\left.\frac{\partial f(x, t)}{\partial t}\right|_{ij} \approx \frac{f_{i,j+1} - f_{i,j}}{k}$$

43

Because

For a function of a single real variable the forward difference formula is given as

$$f'(x) \approx \frac{f(x+h) - f(x)}{h} \quad \text{and so} \quad \left.\frac{\partial f(x, t)}{\partial t}\right|_{ij} \approx \frac{f_{i,j+1} - f_{i,j}}{k}$$

Using these two finite difference formulas we can write down the finite difference representation of the partial differential equation.

The finite difference representation is ....

$$\frac{f_{i-1,j} - 2f_{i,j} + f_{i+1,j}}{h^2} = \frac{f_{i,j+1} - f_{i,j}}{k}$$

44

That is

$$f_{i,j+1} = f_{i,j} + \frac{k}{h^2}(f_{i-1,j} - 2f_{i,j} + f_{i+1,j})$$

It can be shown that there will be no growth of rounding errors when evaluating this equation if $\frac{k}{h^2} \le \frac{1}{2}$.

In compliance with this condition we shall take $h = 0.2$ and $k = 0.02$ so that $\frac{k}{h^2} = \frac{1}{2}$. We shall also restrict ourselves to finding solutions for t ranging from 0 to 0.16.

The finite difference equation then reduces to ....

$$f_{i,j+1} = \frac{1}{2}(f_{i-1,j} + f_{i+1,j})$$

45

Because

$$f_{i,j+1} = f_{i,j} + \frac{k}{h^2}(f_{i-1,j} - 2f_{i,j} + f_{i+1,j}) \quad \text{and so}$$

$$f_{i,j+1} = f_{i,j} + \frac{1}{2}(f_{i-1,j} - 2f_{i,j} + f_{i+1,j}) = \frac{1}{2}(f_{i-1,j} + f_{i+1,j})$$

Notice that this is an equation for stepping forwards in time, so that given the solution is known at $t = 0$ then the solution at $t = k$ can be found from this equation. We can use our spreadsheet to construct the solution from this equation. Open your spreadsheet and

1 Cell A1 enter $t \setminus x$ to represent the fact that the first column will contain the t-values and the first row the x-values.
2 In cells B1 to H1 enter the values of x from 0 to 1·2 in steps of 0·2.

▶

The column headed 1·2 contains grid points outside the domain of $f(x, t)$ to accommodate the derivative boundary condition.

3 In cells A2 to A10 enter the values of t from 0 to 0·16 in steps of 0·02.
4 In cells B2 to B10 enter the value 1 to represent the boundary condition $f(0, t) = 1$.
5 In cell C2 enter the formula $= 1 + C1$ to represent the initial condition $f(x, 0) = 1 + x$. Copy this formula into cells D2 to G2.
6 In cell C3 enter the formula $= 0.5(B2 + D2)$ to represent the finite difference equation

$$f_{i,j+1} = \frac{1}{2}\left(f_{i-1,j} + f_{i+1,j}\right)$$

7 Copy the contents of cell C3 into the block of cells C3 to G10.

Because the derivative boundary condition $\left.\dfrac{\partial f(x,\,t)}{\partial x}\right|_{x=1} = 0$ is represented by the central difference formula $f_{i+1,j} - f_{i-1,j} = 0$, the values of $f(x, t)$ at the external grid points when $x = 1\cdot2$ are equal to the values at the internal grid points when $x = 0\cdot8$.

8 In cell H2 enter the formula $= F2$ and copy this into cells H3 to H10 to produce the following final display:

$t \setminus x$	0·0	0·2	0·4	0·6	0·8	1·0	1·2
0·00	1·00000	1·20000	1·40000	1·60000	1·80000	2·00000	1·80000
0·02	1·00000	1·20000	1·40000	1·60000	1·80000	1·80000	1·80000
0·04	1·00000	1·20000	1·40000	1·60000	1·70000	1·80000	1·70000
0·06	1·00000	1·20000	1·40000	1·55000	1·70000	1·70000	1·70000
0·08	1·00000	1·20000	1·37500	1·55000	1·62500	1·70000	1·62500
0·10	1·00000	1·18750	1·37500	1·50000	1·62500	1·62500	1·62500
0·12	1·00000	1·18750	1·34375	1·50000	1·56250	1·62500	1·56250
0·14	1·00000	1·17188	1·34375	1·45313	1·56250	1·56250	1·56250
0·16	1·00000	1·17188	1·31250	1·45313	1·50781	1·56250	1·50781

If the diffusion equation in Frame 40 to which this solution refers is taken to represent the temperature distribution along a heated rod then this tableau displays how the temperature is changing both in time and spatially along the rod. Notice how, as the heat diffuses through the rod, the temperature changes faster at points that are further away from the end that is maintained at constant temperature.

Try one yourself. Next frame

Example 6

46

The solution of the partial differential equation

$$\frac{\partial^2 f(x, t)}{\partial x^2} = \frac{\partial f(x, t)}{\partial t}$$

for $0 \le x \le 1$ taken in steps of $h = 0 \cdot 2$ and $0 \le t \le 0 \cdot 16$ in steps of $k = 0 \cdot 02$ where

$$f(0, t) = 2, \ f(x, 0) = 2 + x \ \text{ and } \ \left. \frac{\partial f(x, t)}{\partial x} \right|_{x=1} = 0 \cdot 5$$

is ....

47

$t \setminus x$	0·0	0·2	0·4	0·6	0·8	1·0	1·2
0·00	2·00000	2·20000	2·40000	2·60000	2·80000	3·00000	3·00000
0·02	2·00000	2·20000	2·40000	2·60000	2·80000	2·90000	3·00000
0·04	2·00000	2·20000	2·40000	2·60000	2·75000	2·90000	2·95000
0·06	2·00000	2·20000	2·40000	2·57500	2·75000	2·85000	2·95000
0·08	2·00000	2·20000	2·38750	2·57500	2·71250	2·85000	2·91250
0·10	2·00000	2·19375	2·38750	2·55000	2·71250	2·81250	2·91250
0·12	2·00000	2·19375	2·37188	2·55000	2·68125	2·81250	2·88125
0·14	2·00000	2·18594	2·37188	2·52656	2·68125	2·78125	2·88125
0·16	2·00000	2·18594	2·35625	2·52656	2·65391	2·78125	2·85391

Because

$$f_{i,j+1} = f_{i,j} + \frac{k}{h^2}\left(f_{i-1,j} - 2f_{i,j} + f_{i+1,j}\right) \text{ and so}$$

$$f_{i,j+1} = f_{i,j} + \frac{1}{2}\left(f_{i-1,j} - 2f_{i,j} + f_{i+1,j}\right) = \frac{1}{2}\left(f_{i-1,j} + f_{i+1,j}\right)$$

We can use our spreadsheet to construct the solution from this equation. Open your spreadsheet and

1 In cell A1 enter $t \setminus x$ to represent the fact that the first column will contain the t-values and the first row the x-values.
2 In cells B1 to H1 enter the values of x from 0 to 1·2 in steps of 0·2.

The column headed 1·2 contains grid points outside the domain of $f(x, t)$ to accommodate the derivative boundary condition.

3 In cells A2 to A10 enter the values of t from 0 to 0·16 in steps of 0·02.
4 In cells B2 to B10 enter the value 2 to represent the boundary condition $f(0, t) = 2$.
5 In cell C2 enter the formula $= 2 + \text{C1}$ to represent the initial condition $f(x, 0) = 2 + x$. Copy this formula into cells D2 to G2.
6 In cell C3 enter the formula to represent the finite difference equation

$$f_{i,j+1} = \frac{1}{2}\left(f_{i-1,j} + f_{i+1,j}\right)$$

The formula is ....

48

$$= 0.5(B2 + D2)$$

7   Copy the contents of cell C3 into the block of cells C3 to G10.

Because the derivative boundary condition $\left.\dfrac{\partial f(x,\, t)}{\partial x}\right|_{x=1} = 0.5$ is represented by the central difference formula $f_{i+1,j} - f_{i-1,j} = 0.2$, the values of $f(x,\, t)$ at the external grid points when $x = 1.2$ are equal to

The values at the internal grid points when

$$x = \ldots\ldots\ldots\ldots \text{ plus } \ldots\ldots\ldots\ldots$$

49

$$x = 0.8 \text{ plus } 0.2$$

8   In cell H2 enter the formula $= F2 + 0.2$ and copy this into cells H3 to H10 to produce the following display:

$t \setminus x$	0·0	0·2	0·4	0·6	0·8	1·0	1·2
0·00	2·00000	2·20000	2·40000	2·60000	2·80000	3·00000	3·00000
0·02	2·00000	2·20000	2·40000	2·60000	2·80000	2·90000	3·00000
0·04	2·00000	2·20000	2·40000	2·60000	2·75000	2·90000	2·95000
0·06	2·00000	2·20000	2·40000	2·57500	2·75000	2·85000	2·95000
0·08	2·00000	2·20000	2·38750	2·57500	2·71250	2·85000	2·91250
0·10	2·00000	2·19375	2·38750	2·55000	2·71250	2·81250	2·91250
0·12	2·00000	2·19375	2·37188	2·55000	2·68125	2·81250	2·88125
0·14	2·00000	2·18594	2·37188	2·52656	2·68125	2·78125	2·88125
0·16	2·00000	2·18594	2·35625	2·52656	2·65391	2·78125	2·85391

The Crank–Nicolson procedure

50   The forward difference formula that we used for the derivative with respect to time is not as accurate as a central difference formula. However, because we do not possess information about $f(x,\, t)$ for $t < 0$ we were forced to adopt the forward difference formula. To overcome this the Crank–Nicolson procedure makes the assumption that the partial differential equation is satisfied not just at the grid points but also at points in time halfway between two grid points. That is

$$\left.\frac{\partial^2 f(x,\, t)}{\partial x^2}\right|_{i,\,j+1/2} = \left.\frac{\partial f(x,\, t)}{\partial t}\right|_{i,\,j+1/2}$$

▶

We can then derive a central finite difference formula for the time derivative based on this intermediate point

$$\frac{\partial f(x, t)}{\partial t}\bigg|_{i,j+1/2} = \frac{f_{i,j+1} - f_{i,j}}{2(k/2)} = \frac{f_{i,j+1} - f_{i,j}}{k}$$

Here the two grid points either side of the $i, j + 1/2$th point are the i, jth and the $i, j + 1$th, each separated by half the grid step in the time direction. You will note that the outcome is identical to the forward difference taken from the i, jth grid point. However, the finite difference formula that represents the partial differential equation will *not* be the same. For the second derivative with respect to x on the left-hand side of the equation we use a finite difference formula that is the average of the central difference formulas for the i, jth grid point and the $i, j + 1$th grid point. That is

$$\frac{\partial^2 f(x, t)}{\partial x^2}\bigg|_{i,j+1/2} = \frac{1}{2}\left(\frac{f_{i-1,j} - 2f_{i,j} + f_{i+1,j}}{h^2} + \frac{f_{i-1,j+1} - 2f_{i,j+1} + f_{i+1,j+1}}{h^2}\right)$$

The partial differential equation is then represented by the central difference formula . . . .

$$\frac{1}{2}\left(\frac{f_{i-1,j} - 2f_{i,j} + f_{i+1,j}}{h^2} + \frac{f_{i-1,j+1} - 2f_{i,j+1} + f_{i+1,j+1}}{h^2}\right) = \frac{f_{i,j+1} - f_{i,j}}{k}$$

51

That is

$$-f_{i,j+1} + \frac{k}{2h^2}\left(f_{i-1,j+1} - 2f_{i,j+1} + f_{i+1,j+1}\right)$$
$$= -f_{i,j} - \frac{k}{2h^2}\left(f_{i-1,j} - 2f_{i,j} + f_{i+1,j}\right)$$

Unlike the previous case there is now no restriction on the value of $\dfrac{k}{2h^2}$ and different choices of h and k will result in different difference formulas. If we choose $\dfrac{k}{2h^2} = 1$ this difference formula becomes

$$f_{i-1,j+1} - 3f_{i,j+1} + f_{i+1,j+1} = -f_{i-1,j} + f_{i,j} - f_{i+1,j}$$

So we have three unknown quantities on the left-hand side of this equation given in terms of three known quantities on the right. We shall do an example to see exactly how this procedure operates.

Next frame

52

Example 7

Use the Crank–Nicolson procedure to solve the partial differential equation

$$\frac{\partial^2 f(x,\ t)}{\partial x^2} = \frac{\partial f(x,\ t)}{\partial t}$$

for $0 \le x \le 1$ taken in steps of $h = 0.25$ and $0 \le t \le 0.5$ in steps of $k = 0.125$ where:

$$f(0,\ t) = f(1,\ t) = 0$$
$$f(x,\ 0) = x(1 - x)$$

We can use our spreadsheet to construct the solution from this equation. Open your spreadsheet and

1   In cell A1 enter $t \setminus x$ to represent the fact that the first column will contain the t-values and the first row the x-values.
2   In cells B1 to F1 enter the values of x from 0 to 1 in steps of 0.25.
3   In cells A2 to A6 enter the values of t from 0 to 0.5 in steps of 0.125.
4   In cells B2 to B6 enter the value 0 to represent the boundary condition $f(0,\ t) = 0$.
5   In cells F2 to F6 enter the value 0 to represent the boundary condition $f(1,\ t) = 0$.
6   In cell C2 enter the formula $= C1(1 - C1)$ to represent the boundary condition $f(x,\ 0) = x(1 - x)$ and copy into cells D2 to F2.

We now want to know the values that are going to go into the block of cells C3 to E6. We shall work on one row at a time and consider cells C3, D3 and E3 – we shall call these values A, B and C respectively.

Applying the central difference formula for the differential equation

$$f_{i-1,j+1} - 3f_{i,j+1} + f_{i+1,j+1} = -f_{i-1,j} + f_{i,j} - f_{i+1,j}$$

we find that by working along rows 2 and 3

From columns B to D: $0 - 3A + B = -0 + 0.1875 - 0.25$, that is
$$-3A + B = -0.0625$$

From columns C to E: $A - 3B + C = -0.1875 + 0.25 - 0.1875$, that is
$$A - 3B + C = -0.125$$

From columns D to F: $B - 3C + 0 = -0.25 + 0.1875 - 0$, that is
$$B - 3C = -0.0625$$

These equations have solution

$$A = 0.044643, \ B = 0.071429 \text{ and } C = 0.044643$$

Enter these values into cells C3 to E3 respectively and repeat the procedure to find the values in cells C4 to E4.

These are C4: …………, D4: ………… and E4: …………

53

C4: 0·014031, D4: 0·015306 and E4: 0·014031

Because

From columns B to D: $-3A + B = -0.026786$

From columns C to E: $A - 3B + C = -0.017857$

From columns D to F: $B - 3C = -0.026786$

These equations have solution

$A = 0.014031$, $B = 0.015306$ and $C = 0.014031$

This process is repeated until all the appropriate values have been found, giving the following display:

$t \setminus x$	0·00	0·25	0·50	0·75	1·00
0·000	0·000000	0·187500	0·250000	0·187500	0·000000
0·125	0·000000	0·044643	0·071429	0·044643	0·000000
0·250	0·000000	0·014031	0·015306	0·014031	0·000000
0·375	0·000000	0·002369	0·005831	0·002369	0·000000
0·500	0·000000	0·001328	0·000521	0·001328	0·000000

Try one yourself.

Next frame

Example 8

54

Use the Crank–Nicolson procedure to solve the partial differential equation

$$\frac{\partial^2 f(x, t)}{\partial x^2} = \frac{\partial f(x, t)}{\partial t}$$

for $0 \le x \le 1$ taken in steps of $h = 0.2$ and $0 \le t \le 0.2$ in steps of $k = 0.04$ where

$f(0, t) = 2$

$f(1, t) = 1$

$f(x, 0) = 2 - x^2$

The very first thing we must do in solving
this equation numerically is ....

55

Derive the finite difference equation to be used

Because

The Crank–Nicolson procedure tells us that

$$- f_{i,j+1} + \frac{k}{2h^2}\left(f_{i-1,j+1} - 2f_{i,j+1} + f_{i+1,j+1}\right)$$
$$= -f_{i,j} - \frac{k}{2h^2}\left(f_{i-1,j} - 2f_{i,j} + f_{i+1,j}\right)$$

so for each different ratio $\dfrac{k}{2h^2}$ we have a different finite difference formula.

Here we choose $h = 0{\cdot}2$ and $k = 0{\cdot}04$ so that $\dfrac{k}{2h^2} = \dfrac{1}{2}$ and the terms in $f_{i,j}$ do not appear.

This gives the finite difference formula ....

56

$$\boxed{f_{i-1,j+1} - 4f_{i,j+1} + f_{i+1,j+1} = -\left(f_{i-1,j} + f_{i+1,j}\right)}$$

Because

$$- f_{i,j+1} + \frac{k}{2h^2}\left(f_{i-1,j+1} - 2f_{i,j+1} + f_{i+1,j+1}\right)$$
$$= -f_{i,j} - \frac{k}{2h^2}\left(f_{i-1,j} - 2f_{i,j} + f_{i+1,j}\right)$$

and so

$$-f_{i,j+1} + \frac{1}{2}\left(f_{i-1,j+1} - 2f_{i,j+1} + f_{i+1,j+1}\right) = -f_{i,j} - \frac{1}{2}\left(f_{i-1,j} - 2f_{i,j} + f_{i+1,j}\right)$$

that is

$$\frac{1}{2}\left(f_{i-1,j+1} - 4f_{i,j+1} + f_{i+1,j+1}\right) = -\frac{1}{2}\left(f_{i-1,j} + f_{i+1,j}\right)$$

giving

$$\left(f_{i-1,j+1} - 4f_{i,j+1} + f_{i+1,j+1}\right) = -\left(f_{i-1,j} + f_{i+1,j}\right)$$

The complete solution required is ....

$t \setminus x$	0·00	0·20	0·40	0·60	0·80	1·00
0·000	2·000000	1·960000	1·840000	1·640000	1·360000	1·000000
0·040	2·000000	1·901818	1·767273	1·567273	1·301818	1·000000
0·080	2·000000	1·870083	1·713058	1·513058	1·270083	1·000000
0·120	2·000000	1·847483	1·676875	1·476875	1·247483	1·000000
0·160	2·000000	1·832271	1·65221	1·45221	1·232271	1·000000
0·200	2·000000	1·821919	1·635467	1·435467	1·221919	1·000000

Because

Using your spreadsheet to construct the solution from this equation

1 In cell A1 enter $t \setminus x$ to represent the fact that the first column will contain the t-values and the first row the x-values.
2 In cells B1 to G1 enter the values of x from 0 to 1 in steps of 0·2.
3 In cells A2 to A7 enter the values of t from 0 to 0·2 in steps of 0·04.
4 In cells B2 to B7 enter the value 2 to represent the boundary condition $f(0, t) = 2$.
5 In cells G2 to G7 enter the value 1 to represent the boundary condition $f(1, t) = 1$.
6 In cell C2 enter the formula $=2-C1^2$ to represent the boundary condition $f(x,0) = 2 - x^2$ and copy into cells D2 to F2.

We now want to know the values that are going to go into the block of cells C3 to F7. We shall work on one row at a time and consider cells C3, D3, E3 and F3 – we shall call these values A, B, C and D respectively.

Applying the central difference formula for the differential equation

$$f_{i-1,j+1} - 4f_{i,j+1} + f_{i+1,j+1} = -\left(f_{i-1,j} + f_{i+1,j}\right)$$

Then by working along rows 2 and 3

From columns B to D: $2 - 4A + B = -2 - 1·6$, that is
$-4A + B = -5·6$

From columns C to E: ....

From columns D to F: ....

From columns E to G: ....

58

From columns B to D: $-4A + B = -5 \cdot 6$
From columns C to E: $A - 4B + C = -3 \cdot 2$
From columns D to F: $B - 4C + D = -2 \cdot 8$
From columns E to G: $C - 4D = -3 \cdot 4$

Because

From columns B to D: $2 - 4A + B = -2 - 1 \cdot 6$, that is $-4A + B = -5 \cdot 6$
From columns C to E: $A - 4B + C = -1 \cdot 8 - 1 \cdot 4$, that is $A - 4B + C = -3 \cdot 2$
From columns D to F: $B - 4C + D = -1 \cdot 6 - 1 \cdot 2$, that is $B - 4C + D = -2 \cdot 8$
From columns E to G: $C - 4D + 1 = -1 \cdot 4 - 1 \cdot 0$, that is $C - 4D = -3 \cdot 4$

These equations have solution

$$A = \ldots\ldots\ldots, \quad B = \ldots\ldots\ldots,$$
$$C = \ldots\ldots\ldots \quad \text{and} \quad D = \ldots\ldots\ldots$$

59

$A = 1 \cdot 901818$
$B = 1 \cdot 767273$
$C = 1 \cdot 567273$
$D = 1 \cdot 301818$

Enter these values into cells C3 to F3 respectively and repeat the procedure to find the values for cells C4 to F4.

These are   C4: $\ldots\ldots\ldots$,   D4: $\ldots\ldots\ldots$,
E4: $\ldots\ldots\ldots$ and   F4: $\ldots\ldots\ldots$

60

C4: $1 \cdot 870083$
D4: $1 \cdot 713058$
E4: $1 \cdot 513058$
F4: $1 \cdot 270083$

Continuing in this way we find the complete solution as:

$t \backslash x$	0·00	0·20	0·40	0·60	0·80	1·00
0·000	2·000000	1·960000	1·840000	1·640000	1·360000	1·000000
0·040	2·000000	1·901818	1·767273	1·567273	1·301818	1·000000
0·080	2·000000	1·870083	1·713058	1·513058	1·270083	1·000000
0·120	2·000000	1·847483	1·676875	1·476875	1·247483	1·000000
0·160	2·000000	1·832271	1·65221	1·45221	1·232271	1·000000
0·200	2·000000	1·821919	1·635467	1·435467	1·221919	1·000000

Next frame

Dimensional analysis

The equation of Frame 40

$$\frac{\partial^2 T(x, t)}{\partial x^2} = \frac{1}{\kappa} \frac{\partial T(x, t)}{\partial t} \quad \text{for } 0 \leq x \leq L \text{ and } t \geq 0$$

models the temperature distribution $T(x, t)$ along a long thin metal bar of length L. Solutions of this equation will produce values for the temperature distant x along the rod ($0 \leq x \leq L$) at time t. The dimensions of the left- and right-hand sides of this equation due to the derivatives are

$$\left[\frac{\partial^2}{\partial x^2}\right] \equiv [L^{-2}] \quad \text{and} \quad \left[\frac{\partial}{\partial t}\right] \equiv [T^{-1}]$$

To ensure that the dimensions of the left-hand side are the same as the dimensions of the right-hand side we find that the dimensions of $\frac{1}{\kappa}$ are

$$\left[\frac{1}{\kappa}\right] \equiv [L^{-2}T]$$

This then ensures that the equation compares quantities with the same dimension. To solve this equation numerically would require a knowledge of the value of κ which would be different for different problems. To avoid this we transform the equation into a dimensionless form, so ensuring that the variables are measured in numbers and not in any particular dimensional units. We do this as follows.

Define new dimensionless variables as: $X = \dfrac{x}{L}$ (so that $0 \leq X \leq 1$), $\tau = \dfrac{\kappa t}{L^2}$ and define

$$U(X, \tau) = T(x[X], t[\tau])$$

then

$$\frac{\partial T}{\partial t} = \frac{d\tau}{dt} \frac{\partial U}{\partial \tau} = \frac{\kappa}{L^2} \frac{\partial U}{\partial \tau} \quad \text{and}$$

$$\frac{\partial T}{\partial x} = \frac{dX}{dx} \frac{\partial U}{\partial X} = \frac{1}{L} \frac{\partial U}{\partial X}$$

therefore $\dfrac{\partial^2 T}{\partial x^2} = \dfrac{\partial}{\partial x} \dfrac{\partial T}{\partial x} = \dfrac{\partial}{\partial x} \dfrac{1}{L} \dfrac{\partial U}{\partial X} = \dfrac{dX}{dx} \dfrac{1}{L} \dfrac{\partial^2 U}{\partial X^2} = \dfrac{1}{L^2} \dfrac{\partial^2 U}{\partial X^2}$

This means that

$$\frac{\partial^2 T(x, t)}{\partial x^2} = \frac{1}{\kappa^2} \frac{\partial T(x, t)}{\partial t} \quad \text{becomes}$$

$$\frac{1}{L^2} \frac{\partial^2 U(X, \tau)}{\partial X^2} = \frac{1}{\kappa} \frac{\kappa}{L^2} \frac{\partial U(X, \tau)}{\partial \tau} = \frac{1}{L^2} \frac{\partial U(X, \tau)}{\partial \tau}$$

so $\qquad \dfrac{\partial^2 U(X, \tau)}{\partial X^2} = \dfrac{\partial U(X, \tau)}{\partial \tau}$

is the required equation in dimensionless form.

This now completes the work for this Programme. Read through the **Revision summary** that follows and then check your understanding against the **Can you?** checklist. When you are satisified that you do understand the contents of the Programme, try the **Test exercises**. There are no tricks and you should find them quite straightforward. Finally there are some **Further problems** to give additional practice.

 # Revision summary 18

62

1 *Numerical approximation to derivatives of* $f(x)$
   The *forward difference formula*

   $$f'(x) \approx \frac{f(x+h) - f(x)}{h} \quad \text{neglecting terms of the order } h$$

   The *backward difference formula*

   $$f'(x) \approx \frac{f(x) - f(x-h)}{h} \quad \text{neglecting terms of the order } h$$

   The *central difference formulas*

   $$f'(x) \approx \frac{f(x+h) - f(x-h)}{2h} \quad \text{neglecting terms of the order } h^2$$

   $$f''(x) \approx \frac{f(x+h) - 2f(x) + f(x-h)}{h^2} \quad \text{neglecting terms of the order } h^2.$$

2 *Functions of two real variables*
   If $f(x, y)$ is single-valued, then to every domain point (x, y) there corresponds a single range point $f(x, y)$.

   Grid values
   The rectangular domain of the function is overlaid by a grid whose mesh size is of h units in the x-direction and k units in the y-direction. The value of $f(x, y)$ at the *ij*th grid point is denoted by

   $$f_{i,j} \equiv f(x_0 + ih, y_0 + jk)$$

   The values of the expression $f(x, y)$ are required to be found at the grid points

   $$\begin{array}{ccccc}
   \cdots & \cdots & \cdots & \cdots & \cdots \\
   \cdots & f_{i-1,j+1} & f_{i,j+1} & f_{i+1,j+1} & \cdots \\
   \cdots & f_{i-1,j} & f_{i,j} & f_{i+1,j} & \cdots \\
   \cdots & f_{i-1,j-1} & f_{i,j-1} & f_{i+1,j-1} & \cdots \\
   \cdots & \cdots & \cdots & \cdots & \cdots
   \end{array}$$

3 *Central difference formulas for partial derivatives*

   $$\left.\frac{\partial f(x, y)}{\partial x}\right|_{ij} = \frac{f_{i+1,j} - f_{i-1,j}}{2h} \quad \text{and} \quad \left.\frac{\partial f(x, y)}{\partial y}\right|_{ij} = \frac{f_{i,j+1} - f_{i,j-1}}{2k}$$

▶

4 *Computational molecules*

The partial differential equation $a\dfrac{\partial f(x,\,y)}{\partial x}+b\dfrac{\partial f(x,\,y)}{\partial y}=c$, evaluated at

the *ij*th grid point, is $a\dfrac{\partial f(x,\,y)}{\partial x}\bigg|_{ij}+b\dfrac{\partial f(x,\,y)}{\partial y}\bigg|_{ij}=c$ and is by the central

difference formula

$$\frac{a}{2h}\left(f_{i+1,j}-f_{i-1,j}\right)+\frac{b}{2k}\left(f_{i,j+1}-f_{i,j-1}\right)=c$$

which is in turn represented by the composite computational molecule:

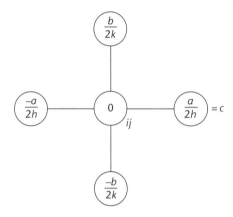

5 *Numerical solutions*

The solutions are in the form of simultaneous linear equations in that they can be written in matrix form as $\mathbf{Ax}=\mathbf{b}$ with solution $\mathbf{x}=\mathbf{A}^{-1}\mathbf{b}$. Using the *Microsoft Excel* spreadsheet the two functions **MINVERSE(array)** and **MMULT(array1, array2)** are employed.

6 *Derivative boundary conditions*

The grid is extended over the boundary of the function domain by adding additional points outside the domain.

7 *Second-order partial differential equations*

The most general form of a second-order partial differential equation is

$$a(x,y)\frac{\partial^2 f}{\partial x^2}+b(x,y)\frac{\partial^2 f}{\partial x\partial y}+c(x,y)\frac{\partial^2 f}{\partial y^2}+d(x,y)\frac{\partial f}{\partial x}+e(x,y)\frac{\partial f}{\partial y}+g(x,y)=0$$

Elliptic equations

If $b^2-4ac<0$ then the partial differential equation is called an *elliptic* equation

Hyperbolic equations

If $b^2-4ac>0$ then the partial differential equation is called an *hyperbolic* equation

Parabolic equations

If $b^2-4ac=0$ then the partial differential equation is called a *parabolic* equation.

▶

8 *Second partial derivatives – central difference formulas*

$$\left.\frac{\partial^2 f(x, y)}{\partial x^2}\right|_{ij} \approx \frac{f_{i-1,j} - 2f_{i,j} + f_{i+1,j}}{h^2}$$

and $\left.\dfrac{\partial^2 f(x, y)}{\partial y^2}\right|_{ij} \approx \dfrac{f_{i,j-1} - 2f_{i,j} + f_{i,j+1}}{k^2}$

9 *Time-dependent equations*

To use a central difference formula for the derivative with respect to *t* would require a knowledge of $f(x, t)$ for values of $t < 0$ and this we do not possess. Consequently, for the derivative with respect to *t* we use the *forward* difference formula

$$\left.\frac{\partial f(x, t)}{\partial t}\right|_{ij} \approx \frac{f_{i,j+1} - f_{i,j}}{k}$$

So the partial differential equation $\dfrac{\partial^2 f(x, t)}{\partial x^2} = \dfrac{\partial f(x, t)}{\partial t}$ becomes

$$f_{i,j+1} = f_{i,j} + \frac{k}{h^2}\left(f_{i-1,j} - 2f_{i,j} + f_{i+1,j}\right)$$

where it can be shown that there will be no growth of rounding errors when evaluating this equation if

$$\frac{k}{h^2} \leq \frac{1}{2}.$$

10 *The Crank–Nicolson procedure*

The Crank–Nicolson procedure makes the assumption that the partial differential equation can be satisfied at points in time halfway between two grid points. That is

$$\left.\frac{\partial^2 f(x, t)}{\partial x^2}\right|_{i,j+1/2} = \left.\frac{\partial f(x, t)}{\partial t}\right|_{i,j+1/2}$$

This gives

$$\left.\frac{\partial f(x, t)}{\partial t}\right|_{i,j+1/2} = \frac{f_{i,j+1} - f_{i,j}}{2(k/2)} = \frac{f_{i,j+1} - f_{i,j}}{k}$$

$$\left.\frac{\partial^2 f(x, t)}{\partial x^2}\right|_{i,j+1/2} = \frac{1}{2}\left(\frac{f_{i-1,j} - 2f_{i,j} + f_{i+1,j}}{h^2} + \frac{f_{i-1,j+1} - 2f_{i,j+1} + f_{i+1,j+1}}{h^2}\right)$$

So that

$$-f_{i,j+1} + \frac{k}{2h^2}\left(f_{i-1,j+1} - 2f_{i,j+1} + f_{i+1,j+1}\right)$$

$$= -f_{i,j} - \frac{k}{2h^2}\left(f_{i-1,j} - 2f_{i,j} + f_{i+1,j}\right)$$

with no restriction on the value of $\dfrac{k}{2h^2}$.

Can you?

Checklist 18

63

Check this list before and after you try the end of Programme test.

On a scale of 1 to 5 how confident are you that you can: **Frames**

- Derive the finite difference formulas for the first partial derivatives of a function of two real variables and construct the central finite difference formula to represent a first-order partial differential equation?

  Yes ☐ ☐ ☐ ☐ ☐ *No*   1 to 4

- Draw a rectangular grid of points overlaid on the domain of a function of two real variables and evaluate the function at the boundary grid points?

  Yes ☐ ☐ ☐ ☐ ☐ *No*   5 to 9

- Construct the computational molecule for a first-order partial differential equation in two real variables and use the molecule to evaluate the solutions to the equation at the grid points interior to the boundary?

  Yes ☐ ☐ ☐ ☐ ☐ *No*   10 and 11

- Describe the solution as a set of simultaneous linear equations and use matrices to represent them?

  Yes ☐ ☐ ☐ ☐ ☐ *No*   12 and 13

- Invert the coefficient matrix and thereby represent the solution to the partial differential equation as a column matrix?

  Yes ☐ ☐ ☐ ☐ ☐ *No*   14 to 19

- Take account of a boundary condition in the form of the derivative normal to the boundary?

  Yes ☐ ☐ ☐ ☐ ☐ *No*   20 to 28

- Obtain the central finite difference formulas for the second derivatives of a function of two real variables and construct finite difference formulas for second-order partial differential equations?

  Yes ☐ ☐ ☐ ☐ ☐ *No*   29 to 39

- Use the forward difference formula for the first time derivatives in partial differential equations involving time and distance?

  Yes ☐ ☐ ☐ ☐ ☐ *No*   40 to 49

▶

- Use the Crank–Nicolson procedure for a partial differential equation involving a first time derivative?
  Yes ☐ ☐ ☐ ☐ ☐ No   50 to 60

- Appreciate the use of dimensional analysis in the conversion of a partial differential equation modelling a physical system into a dimensionless equation?
  Yes ☐ ☐ ☐ ☐ ☐ No   61

 # Test exercise 18

64

1 Solve the following equation numerically.

$$5\frac{\partial f(x, y)}{\partial x} - 4\frac{\partial f(x, y)}{\partial y} = -5$$

for $0 \le x \le 1$ with a step length $h = 1/4$ and $0 \le y \le 1$ with a step length $k = 1/3$ where

$$f(x, 0) = 3x - 4, \ f(x, 1) = 3x + 1, \ f(0, y) = 5y - 4 \text{ and } f(1, y) = 5y - 1$$

2 Solve the following equation numerically.

$$10\frac{\partial f(x, y)}{\partial x} + 8\frac{\partial f(x, y)}{\partial y} = -10$$

for $0 \le x \le 1$ with a step length $h = 1/3$ and $0 \le y \le 1$ with a step length $k = 1/3$ where

$$f(x, 0) = 7x + 5, \ f(x, 1) = 7x - 5, \ f(0, y) = 5 - 10y \text{ and } \left.\frac{\partial f(x, y)}{\partial x}\right|_{x=1} = 7$$

3 Name the type of equation in each of the following.

(a) $2\dfrac{\partial f(x, y)}{\partial x} - 3y\dfrac{\partial f(x, y)}{\partial y} = 4xy$

(b) $\dfrac{\partial f(x, y)}{\partial x} + \dfrac{\partial^2 f(x, y)}{\partial x \partial y} - \dfrac{\partial f(x, y)}{\partial y} = \dfrac{x}{y}$

(c) $\dfrac{\partial^2 f(x, y)}{\partial x^2} - 2\dfrac{\partial^2 f(x, y)}{\partial x \partial y} + \dfrac{\partial^2 f(x, y)}{\partial y^2} = 0$

(d) $\dfrac{\partial}{\partial x}\left[\dfrac{\partial f(x, y)}{\partial x} + \dfrac{\partial f(x, y)}{\partial y}\right] = \dfrac{x^2}{y^3}$

(e) $3\dfrac{\partial^2 f(x, y)}{\partial x^2} - 2\dfrac{\partial^2 f(x, y)}{\partial x \partial y} + \dfrac{\partial^2 f(x, y)}{\partial y^2} = 3xy$

4 Solve the following equation numerically.

$$\frac{\partial^2 f(x, y)}{\partial x^2} + \frac{\partial^2 f(x, y)}{\partial y^2} = -2 \text{ for } 0 \le x \le 1 \text{ and } 0 \le y \le 1$$

with step lengths $h = k = 1/3$ where

$$f(x, 0) = f(x, 1) = x - 2, \ f(0, y) = y^2 - y - 2 \text{ and } \left.\frac{\partial f(x, y)}{\partial x}\right|_{x=1} = 1$$

▶

5 Solve the following equation numerically using the forward difference approximation for the first derivative with respect to time.

$$\frac{\partial^2 f(x, t)}{\partial x^2} = \frac{\partial f(x, t)}{\partial t}$$

for $0 \leq x \leq 1$ with a step length $h = 0.2$ and $0 \leq t \leq 0.2$ with step length $k = 0.02$ where

$$f(x, 0) = x^2, \; f(0, t) = 0 \text{ and } \left.\frac{\partial f(x, t)}{\partial x}\right|_{x=1} = 0.25$$

6 Solve the following equation numerically using the Crank–Nicolson procedure.

$$\frac{\partial^2 f(x, t)}{\partial x^2} = \frac{\partial f(x, t)}{\partial t}$$

for $0 \leq x \leq 1$ with a step length $h = 0.2$ and $0 \leq t \leq 0.2$ with step length $k = 0.04$ where

$$f(x, 0) = x^2 - x + 1 \text{ and } f(0, t) = f(1, t) = 1$$

 # Further problems 18

1 Solve the following equation numerically.

65

$$-2\frac{\partial f(x, y)}{\partial x} + \frac{\partial f(x, y)}{\partial y} = 0$$

for $0 \leq x \leq 1$ with a step length $h = 1/4$ and $0 \leq y \leq 1$ with a step length $k = 1/3$ where

$$f(x, 0) = x - 3, \; f(x, 1) = x - 1, \; f(0, y) = 2y - 3 \text{ and } f(1, y) = 2y - 2$$

2 Solve the following equation numerically.

$$9\frac{\partial f(x, y)}{\partial x} - 7\frac{\partial f(x, y)}{\partial y} = -7$$

for $0 \leq x \leq 1$ with a step length $h = 1/3$ and $0 \leq y \leq 1$ with a step length $k = 1/3$ where

$$f(x, 0) = 7x + 4, \; f(x, 1) = 7x + 14, \; f(0, y) = 10y + 4$$
$$\text{and } f(1, y) = 10y + 11$$

3 Solve the following equation numerically.

$$x\frac{\partial f(x, y)}{\partial x} + (y + 1)\frac{\partial f(x, y)}{\partial y} = 0$$

for $0 \leq x \leq 1$ with a step length $h = 1/3$ and $0 \leq y \leq 1$ with a step length $k = 1/3$ where

$$f(x, 0) = x - 1, \; f(x, 1) = (x - 2)/2, \; f(0, y) = -1$$
$$\text{and } f(1, y) = -y/(y + 1)$$

4   Solve the following equation numerically.

$$\frac{\partial f(x,\, y)}{\partial y} - \frac{\partial f(x,\, y)}{\partial x} = x^2 + y^2$$

for $0 \le x \le 1$ with a step length $h = 1/4$ and $0 \le y \le 1$ with a step length $k = 1/3$ where

$$f(x,\, 0) = 0, \; f(x,\, 1) = x(x-1), \; f(0,\, y) = 0 \text{ and } f(1,\, y) = y(1-y)$$

5   Solve the following equation numerically.

$$3\frac{\partial f(x,\, y)}{\partial x} - 5\frac{\partial f(x,\, y)}{\partial y} = -4$$

for $0 \le x \le 1$ with a step length $h = 1/3$ and $0 \le y \le 1$ with a step length $k = 1/3$ where

$$f(x,\, 0) = 7x + 15, \; f(x,\, 1) = 7x + 20, \; f(0,\, y) = 5y + 15$$

and $\left.\dfrac{\partial f(x,\, y)}{\partial x}\right|_{x=1} = 7$

6   Solve the following equation numerically.

$$11\frac{\partial f(x,\, y)}{\partial x} + 12\frac{\partial f(x,\, y)}{\partial y} = 19$$

for $0 \le x \le 1$ with a step length $h = 1/3$ and $0 \le y \le 1$ with a step length $k = 1/3$ where

$$f(x,\, 0) = 5x + 21, \; f(x,\, 1) = 5x + 18, \; f(0,\, y) = 21 - 3y$$

and $\left.\dfrac{\partial f(x,\, y)}{\partial x}\right|_{x=1} = 5$

7   Solve the following equation numerically.

$$2x\frac{\partial f(x,\, y)}{\partial x} - y\frac{\partial f(x,\, y)}{\partial y} = 8x^2$$

for $0 \le x \le 1$ with a step length $h = 1/3$ and $0 \le y \le 1$ with a step length $k = 1/3$ where

$$f(x,\, 0) = 2x^2 + 4, \; f(x,\, 1) = 2x^2 - 3x + 4, \; f(0,\, y) = 4$$

and $\left.\dfrac{\partial f(x,\, y)}{\partial x}\right|_{x=1} = 4 - 3y^2$

8   Solve the following equation numerically.

$$y\frac{\partial f(x,\, y)}{\partial x} + x\frac{\partial f(x,\, y)}{\partial y} = x^4 - y^4$$

for $0 \le x \le 1$ with a step length $h = 1/3$ and $0 \le y \le 1$ with a step length $k = 1/3$ where

$$f(x,\, 0) = 0, \; f(x,\, 1) = x(x+1)(x-1), \; f(0,\, y) = 0$$

and $\left.\dfrac{\partial f(x,\, y)}{\partial x}\right|_{x=1} = y(3 - y^2)$

▶

9 Solve the following equation numerically.

$$\frac{\partial^2 f(x, y)}{\partial x^2} + \frac{\partial^2 f(x, y)}{\partial y^2} = -4$$

for $0 \le x \le 1$ and $0 \le y \le 1$ with step lengths $h = k = 1/3$ where

$$f(x, 0) = 3x^2,\ f(x, 1) = 3x^2 - 5,\ f(0, y) = -5y^2 \text{ and } \left.\frac{\partial f(x, y)}{\partial x}\right|_{x=1} = 6$$

10 Solve the following equation numerically.

$$\frac{\partial^2 f(x, y)}{\partial x^2} + \frac{\partial^2 f(x, y)}{\partial y^2} = 2(x + y)$$

for $0 \le x \le 1$ and $0 \le y \le 1$ with step lengths $h = k = 1/3$ where

$$f(x, 0) = -1,\ f(x, 1) = x^2 + 3x - 1,\ f(0, y) = -1$$

and $\left.\dfrac{\partial f(x, y)}{\partial x}\right|_{x=1} = y^2 + 4y$

11 Solve the following equation numerically.

$$\frac{\partial^2 f(x, y)}{\partial x^2} + \frac{\partial^2 f(x, y)}{\partial y^2} = (2 - x^2)\cos y$$

for $0 \le x \le 1$ and $0 \le y \le 1$ with step lengths $h = k = 1/3$ where

$$f(x, 0) = x^2,\ f(x, 1) = 0.540302x^2,\ f(0, y) = 0 \text{ and } \left.\frac{\partial f(x, y)}{\partial x}\right|_{x=1} = 2x\cos y$$

12 Solve the following equation numerically.

$$\frac{\partial^2 f(x, y)}{\partial x^2} - \frac{\partial^2 f(x, y)}{\partial y^2} = 4(x - y)$$

for $0 \le x \le 1$ and $0 \le y \le 1$ with step lengths $h = k = 1/3$ where

$$f(x, 0) = x^3,\ f(x, 1) = (x + 1)(x^2 + 1),\ f(0, y) = y^3$$

and $\left.\dfrac{\partial f(x, y)}{\partial x}\right|_{x=1} = y^2 + 2y + 3$

13 Given the central difference formula

$$\left.\frac{\partial^2 f(x, y)}{\partial x \partial y}\right|_{ij} = \frac{1}{4h^2}\left(f_{i-1,j-1} - f_{i+1,j-1} - f_{i-1,j+1} + f_{i+1,j+1}\right)$$

where the step length in both directions is h, construct the computational molecule for this formula.

Solve the equation

$$\frac{\partial^2 f(x, y)}{\partial x \partial y} = 1$$

for $0 \le x \le 1$ and $0 \le y \le 1$ with step lengths $h = 1/3$ where

$$f(x, 0) = 0,\ f(x, 1) = x,\ f(0, y) = 0 \text{ and } \left.\frac{\partial f(x, y)}{\partial x}\right|_{x=1} = y$$

▶

14 Given the central difference formula

$$\left.\frac{\partial^2 f(x, y)}{\partial x \partial y}\right|_{ij} = \frac{1}{4h^2}\left(f_{i-1,j-1} - f_{i+1,j-1} - f_{i-1,j+1} + f_{i+1,j+1}\right)$$

where the step length in both directions is h, construct the computational molecule for this formula.

Solve the equation

$$\frac{\partial^2 f(x, y)}{\partial x \partial y} = 2(x - y)$$

for $0 \leq x \leq 1$ and $0 \leq y \leq 1$ with step lengths $h = 1/3$ where

$$f(x, 0) = 0, \ f(x, 1) = x(x - 1), \ f(0, y) = 0 \text{ and } \left.\frac{\partial f(x, y)}{\partial x}\right|_{x=1} = y(2 - y)$$

15 Solve the following equation numerically using the forward difference approximation for the first derivative with respect to time.

$$\frac{\partial^2 f(x, t)}{\partial x^2} = \frac{\partial f(x, t)}{\partial t}$$

for $0 \leq x \leq 1$ with a step length $h = 0{\cdot}2$ and $0 \leq t \leq 0{\cdot}2$ with a step length $k = 0{\cdot}02$ where

$$f(x, 0) = x(x - 1), \ f(0, t) = 2t \text{ and } \left.\frac{\partial f(x, t)}{\partial x}\right|_{x=1} = 1$$

16 Solve the following equation numerically using the forward difference approximation for the first derivative with respect to time.

$$\frac{\partial^2 f(x, t)}{\partial x^2} = \frac{1}{0{\cdot}1}\frac{\partial f(x, t)}{\partial t}$$

for $0 \leq x \leq 1$ with a step length $h = 0{\cdot}2$ and $0 \leq t \leq 0{\cdot}2$ with a step length $k = 0{\cdot}02$ where

$$f(x, 0) = \sin x, \ f(0, t) = 0 \text{ and } \left.\frac{\partial f(x, t)}{\partial x}\right|_{x=1} = 0{\cdot}54e^{-t/10}$$

17 Solve the following equation numerically using the forward difference approximation for the first derivative with respect to time.

$$\frac{\partial^2 f(x, t)}{\partial x^2} = \frac{\partial f(x, t)}{\partial t}$$

for $0 \leq x \leq 1$ with a step length $h = 0{\cdot}2$ and $0 \leq t \leq 0{\cdot}2$ with a step length $k = 0{\cdot}02$ where

$$f(x, 0) = 3\sin(0{\cdot}64x), \ f(0, t) = 0 \text{ and } \left.\frac{\partial f(x, t)}{\partial x}\right|_{x=1} = 2{\cdot}41e^{-0{\cdot}41t}$$

18 Solve the following equation numerically using the Crank–Nicolson procedure.

$$\frac{\partial^2 f(x, t)}{\partial x^2} = \frac{\partial f(x, t)}{\partial t}$$

for $0 \leq x \leq 1$ with a step length $h = 0{\cdot}2$ and $0 \leq t \leq 0{\cdot}6$ with a step length $k = 0{\cdot}04$ where

$$f(x, 0) = x^2 + x - 1 \text{ and } f(0, t) = 2t - 1, \ f(1, t) = 1 + 2t$$

▶

19 Solve the following equation numerically using the Crank–Nicolson procedure.

$$\frac{\partial^2 f(x,\, t)}{\partial x^2} = \frac{\partial f(x,\, t)}{\partial t}$$

for $0 \leq x \leq 1$ with a step length $h = 0{\cdot}1$ and $0 \leq t \leq 0{\cdot}14$ with a step length $k = 0{\cdot}02$ where

$$f(x,\, 0) = 10x(x - 1) \text{ and } f(0, t) = f(1, t) = 20t$$

20 Solve the following equation numerically using the Crank–Nicolson procedure.

$$\frac{\partial^2 f(x,\, t)}{\partial x^2} = \frac{\partial f(x,\, t)}{\partial t}$$

for $0 \leq x \leq 1$ with a step length $h = 0{\cdot}1$ and $0 \leq t \leq 0{\cdot}6$ with a step length $k = 0{\cdot}04$ where

$$f(x,\, 0) = 100 \sin \pi x \text{ and } f(0, t) = f(1, t) = 0$$

Multiple integration 1

Learning outcomes

When you have completed this Programme you will be able to:

- Evaluate double and triple integrals and apply them to the determination of the areas of plane figures and the volumes of solids
- Understand the role of the differential of a function of two or more real variables
- Determine exact differentials in two real variables and their integrals
- Evaluate the area enclosed by a closed curve by contour integration
- Evaluate line integrals and appreciate their properties
- Evaluate line integrals around closed curves within a simply connected region
- Link line integrals to integrals along the x-axis
- Link line integrals to integrals along a contour given in parametric form
- Discuss the dependence of a line integral between two points on the path of integration
- Determine exact differentials in three real variables and their integrals
- Demonstrate the validity and use of Green's theorem

Prerequisite: Engineering Mathematics (Sixth Edition)
Programme 23   Multiple integrals

Introduction

The introductory work on double and triple integrals was covered in detail in Programme 23 of *Engineering Mathematics (Sixth Edition)* and another look at the main points before launching forth on the current development could well be worth while.

You will no doubt recognise the following.

1   *Double integrals*

$$\int_{y_1}^{y_2} \int_{x_1}^{x_2} f(x,\, y)\, dx\, dy$$

is a double integral and is evaluated from the inside outwards, i.e.

$$\int_{y_1}^{y_2} \boxed{\int_{x_1}^{x_2} f(x,\, y)\, dx} \overset{\textcircled{1}}{} dy \overset{\textcircled{2}}{}$$

A double integral is sometimes expressed in the form

$$\int_{y_1}^{y_2} dy \int_{x_1}^{x_2} f(x,\, y)\, dx$$

in which case, we evaluate from the right-hand end, i.e.

$$\int_{y_1}^{y_2} dy \boxed{\int_{x_1}^{x_2} f(x,\, y)\, dx} \overset{\textcircled{1}}{}$$

then

$$\boxed{\int_{y_1}^{y_2} \int_{x_1}^{x_2} f(x,\, y)\, dx\ dy} \overset{\textcircled{2}}{}$$

2   *Triple integrals*

Triple integrals follow the same procedure.

$$\int_{z_1}^{z_2} \int_{y_1}^{y_2} \int_{x_1}^{x_2} f(x,\, y,\, z)\, dx\, dy\, dz \text{ is evaluated in the order}$$

$$\int_{z_1}^{z_2} \int_{y_1}^{y_2} \boxed{\int_{x_1}^{x_2} f(x,y,z)\, dx} \overset{\textcircled{1}}{} dy \overset{\textcircled{2}}{} dz \overset{\textcircled{3}}{}$$

▶

3 *Applications*

(a) *Areas of plane figures*

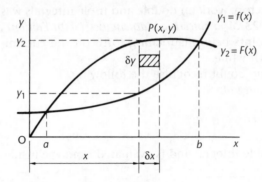

Area of element $\delta A = \delta x \delta y$

Area of strip $\approx \displaystyle\sum_{y=y_1}^{y=y_2} \delta x \delta y$

Area of all such strips $\approx \displaystyle\sum_{x=a}^{x=b} \left\{ \sum_{y=y_1}^{y=y_2} \delta x \delta y \right\}$

If $\delta x \to 0$ and $\delta y \to 0$, $A = \displaystyle\int_a^b \int_{y_1}^{y_2} \mathrm{d}y \, \mathrm{d}x$

(b) *Areas of plane figures bounded by a polar curve $r = f(\theta)$ and radius vectors at $\theta = \theta_1$ and $\theta = \theta_2$*

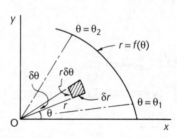

Small arc of circle of radius r, subtending angle $\delta\theta$ at centre.

$\therefore$ Arc $= r\delta\theta$

Area of element $\delta A \approx r\delta\theta \, \delta r$

Area of thin sector $\approx \displaystyle\sum_{r=0}^{r=f(\theta)} r \, \delta\theta \, \delta r$

$\therefore$ Total area of all such sectors $\approx \displaystyle\sum_{\theta=\theta_1}^{\theta=\theta_2} \left\{ \sum_{r=0}^{r=f(\theta)} r \, \delta r \, \delta\theta \right\}$

$\therefore$ If $\delta r \to 0$ and $\delta\theta \to 0$, $A = \displaystyle\int_{\theta_1}^{\theta_2} \int_0^{r=f(\theta)} r \, \mathrm{d}r \, \mathrm{d}\theta$

▶

(c) *Volume of solids*

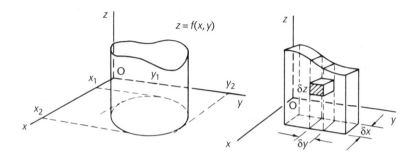

Volume of element $\delta V = \delta x \, \delta y \, \delta z$

Volume of column $\approx \displaystyle\sum_{z=0}^{z=f(x,\ y)} \delta x \, \delta y \, \delta z$

Volume of slice $\approx \displaystyle\sum_{y=y_1}^{y=y_2} \left\{ \sum_{z=0}^{z=f(x,\ y)} \delta x \, \delta y \, \delta z \right\}$

$\therefore$ Total volume $V \approx$ sum of all such slices

i.e. $\quad V \approx \displaystyle\sum_{x=x_1}^{x=x_2} \sum_{y=y_1}^{y=y_2} \sum_{z=0}^{z=f(x,\ y)} \delta x \, \delta y \, \delta z$

Then, if $\delta x \to 0$, $\delta y \to 0$, $\delta z \to 0$,

$$V = \int_{x_1}^{x_2} \int_{y_1}^{y_2} \int_{0}^{z=f(x,\ y)} \mathrm{d}z \, \mathrm{d}y \, \mathrm{d}x$$

If $z = f(x, y)$, this becomes

$$V = \int_{x_1}^{x_2} \int_{y_1}^{y_2} f(x, y) \, \mathrm{d}y \, \mathrm{d}x$$

2

4 *Revision examples* As a means of 'warming up', let us work through one or two straightforward examples on the previous work.

Example 1

Find the area of the plane figure bounded by the curves $y_1 = (x-1)^2$ and $y_2 = 4 - (x-3)^2$.

The first thing, as always, is to sketch the curves – each of which is a parabola – and to determine their points of intersection.

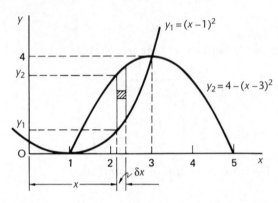

Points of intersection: $(x-1)^2 = 4 - (x-3)^2$

$\qquad x^2 - 2x + 1 = 4 - x^2 + 6x - 9 \quad$ i.e. $x^2 - 4x + 3 = 0$

$\qquad \therefore \ (x-1)(x-3) = 0 \quad \therefore \ x = 1 \quad$ or $\quad x = 3$.

Now we have all the information to determine the required area, which is

. . . .

3

$$\boxed{A = 2\tfrac{2}{3} \text{ square units}}$$

Because

$$A = \int_{x=1}^{x=3} \int_{y_1}^{y_2} dy\, dx = \int_{x=1}^{x=3} \int_{y=(x-1)^2}^{y=4-(x-3)^2} dy\, dx$$

$$= \int_{1}^{3} \{4 - (x-3)^2 - (x-1)^2\}\, dx = -2\int_{1}^{3} (x^2 - 4x + 3)\, dx$$

$$= -2\left[\frac{x^3}{3} - 2x^2 + 3x\right]_{1}^{3} = 2\tfrac{2}{3} \text{ square units}$$

Now for another.

Example 2

A rectangular plate is bounded by the x and y axes and the lines $x = 6$ and $y = 4$. The thickness t of the plate at any point is proportional to the square of the distance of the point from the origin. Determine the total volume of the plate.

First of all draw the figure and build up the appropriate double integral. Do not evaluate it yet. The expression is therefore

$$V = \ldots\ldots\ldots$$

$$\boxed{V = \int_{x=0}^{x=6} \int_{y=0}^{y=4} k(x^2 + y^2)\, \mathrm{d}y\, \mathrm{d}x}$$

4

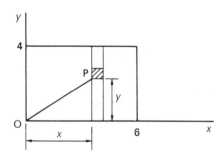

Thickness t of plate at P is

$$t = k\,\mathrm{OP}^2 = k(x^2 + y^2)$$

Element of area $= \delta y\, \delta x$

$\therefore$ Element of volume at P $\approx k(x^2 + y^2)\,\delta y\, \delta x$

$\therefore$ Total volume $V = \displaystyle\int_{x=0}^{x=6} \int_{y=0}^{y=4} k(x^2 + y^2)\, \mathrm{d}y\, \mathrm{d}x$

Now we can evaluate the integral. We start from the inside with

$$\int_{y=2}^{y=4} k(x^2 + y^2)\, \mathrm{d}y,$$

remembering that for this integral (volume of the strip) x is constant.

This gives $\ldots\ldots\ldots\ldots$

$$\boxed{k\left(4x^2 + \frac{64}{3}\right)}$$

5

Because

$$k\int_0^4 (x^2 + y^2)\, \mathrm{d}y = k\left[x^2 y + \frac{y^3}{3}\right]_{y=0}^{y=4} = k\left(4x^2 + \frac{64}{3}\right)$$

Then $\quad V = k\displaystyle\int_0^6 \left(4x^2 + \frac{64}{3}\right)\, \mathrm{d}x = \ldots\ldots\ldots\ldots$

6

$$\boxed{V = 416\,k \text{ cubic units}}$$

That was easy enough. Notice that an alternative interpretation of this problem could be that of a uniform lamina with a variable density $\rho = k(x^2 + y^2)$ at any point (x, y). Now for one in polar coordinates.

Example 3

Express as a double integral the area enclosed by one loop of the curve $r = 3\cos 2\theta$ and evaluate the integral (refer to *Engineering Mathematics (Sixth Edition)*, Programme 23, Frame 10).

Consider the half loop shown.

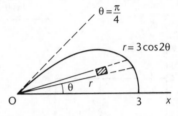

First set up the double integral which is . . . .

7

$$A = \int_{\theta=0}^{\theta=\pi/4} \int_{r=0}^{r=3\cos 2\theta} r\,dr\,d\theta$$

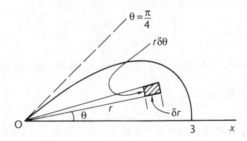

Area of element $= r\,\delta r\,\delta\theta$

$$\therefore \text{ Area of sector} \approx \sum_{r=0}^{r=3\cos 2\theta} r\,\delta r\,\delta\theta$$

$$\therefore \text{ Area of half loop} \approx \sum_{\theta=0}^{\theta=\pi/4} \sum_{r=0}^{r=3\cos 2\theta} r\,\delta r\,\delta\theta$$

If $\delta r \to 0$ and $\delta\theta \to 0$,

$$A = \int_{\theta=0}^{\theta=\pi/4} \int_{r=0}^{r=3\cos 2\theta} r\,dr\,d\theta$$

Now finish it off to find the area of the whole loop, which is

. . . .

$$\boxed{\dfrac{9\pi}{8} \text{ square units}}$$

8

Because

$$A = \int_{\theta=0}^{\theta=\pi/4} \int_{r=0}^{r=3\cos 2\theta} r\,dr\,d\theta$$

$$= \int_0^{\pi/4} \left[\frac{r^2}{2}\right]_0^{3\cos 2\theta} d\theta$$

$$= \frac{9}{2} \int_0^{\pi/4} \cos^2 2\theta\,d\theta$$

$$= \frac{9}{4} \int_0^{\pi/4} (1+\cos 4\theta)\,d\theta$$

$$= \frac{9}{4} \left[\theta + \frac{\sin 4\theta}{4}\right]_0^{\pi/4}$$

$$= \frac{9\pi}{16}$$

This is the area of a half loop.

$$\text{Required area} = \frac{9\pi}{8} \text{ square units}$$

Now here is another.

Example 4

Find the volume of the solid bounded by the planes $z = 0$, $x = 1$, $x = 3$, $y = 1$, $y = 2$ and the surface $z = x^2 y^2$.

As always, we start off by sketching the figure. When you have done that, check the result with the next frame.

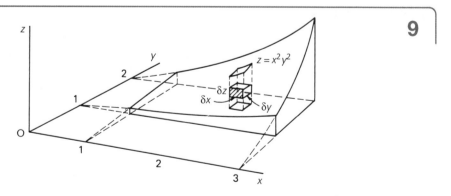

9

We now build up the integral which will give us the volume of the solid.

Element of volume $\delta V = \delta x\, \delta y\, \delta z$

Volume of column $\approx \displaystyle\sum_{z=0}^{z=x^2y^2} \delta x\, \delta y\, \delta z$

Volume of slice $\approx \displaystyle\sum_{y=1}^{y=2} \left\{ \sum_{z=0}^{z=x^2y^2} \delta x\, \delta y\, \delta z \right\}$

Volume of solid $\approx \displaystyle\sum_{x=1}^{x=3} \left\{ \sum_{y=1}^{y=2} \sum_{z=0}^{z=x^2y^2} \delta x\, \delta y\, \delta z \right\}$

When $\delta x \to 0$, $\delta y \to 0$, $\delta z \to 0$,

$$V = \int_{x=1}^{x=3} \int_{y=1}^{y=2} \int_{z=0}^{z=x^2y^2} dz\, dy\, dx$$

Evaluating this, $V = \ldots\ldots\ldots\ldots$

10

$$\boxed{V = 20\tfrac{2}{9} \text{ cubic units}}$$

Because, starting with the innermost integral

$$V = \int_{x=1}^{x=3} \int_{y=1}^{y=2} \left[z \right]_0^{x^2y^2} dy\, dx = \int_1^3 \int_1^2 x^2y^2\, dy\, dx$$

$$= \int_1^3 \left[\frac{x^2\, y^3}{3} \right]_{y=1}^{y=2} dx \qquad = \int_1^3 \frac{7x^2}{3}\, dx = 20\tfrac{2}{9}$$

Now that we have revised the basics, let us move on to something
rather different

Differentials

11

It is convenient in various branches of the calculus to denote small increases in value of a variable by the use of *differentials*. The method is particularly useful in dealing with the effects of small finite changes and shortens the writing of calculus expressions.

We are already familiar with the diagram from which finite changes δy and δx in a function $y = f(x)$ are depicted.

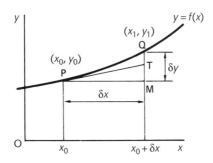

The increase in y from P to Q $= MQ = \delta y = f(x_0 + \delta x) - f(x_0)$

If PT is the tangent at P, then $MQ = MT + TQ$. Also $\dfrac{MT}{\delta x} = f'(x_0)$

$\therefore\ \ MT = f'(x_0)\delta x$

$\therefore\ \ MQ = \delta y = f'(x_0) \cdot \delta x + TQ$

and, if Q is close to P, then $\delta y \approx f'(x_0)\delta x$

We define the differentials dy and dx as finite quantities such that

$\quad dy = f'(x_0)\,dx$

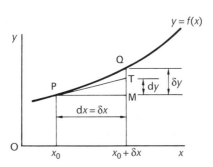

Note that the differentials dy and dx are finite quantities – not necessarily zero – and can therefore exist alone.

Note too that $dx = \delta x$.

From the diagram, we can see that

$\quad \delta y$ is the increase in y as we move from P to Q along the curve.
$\quad dy$ is the increase in y as we move from P to T along the tangent.

As Q approaches P, the difference between δy and dy decreases to zero. The use of differentials simplifies the writing of many relationships and is based on the general statement $dy = f'(x)\,dx$.

▶

For example

(a) $y = x^5$ then $dy = 5x^4\, dx$

(b) $y = \sin 3x$ then $dy = 3\cos 3x\, dx$

(c) $y = e^{4x}$ then $dy = 4\,e^{4x}\, dx$

(d) $y = \cosh 2x$ then $dy = 2\,\sinh 2x\, dx$

Note that when the left-hand side is a differential dy the right-hand side must also contain a differential. Remember therefore to include the 'dx' on the right-hand side.

The product and quotient rules can also be expressed in differentials.

$$\frac{d}{dx}(uv) = u\frac{dv}{dx} + v\frac{du}{dx} \quad \text{becomes} \quad d(uv) = u\,dv + v\,du$$

$$\frac{d}{dx}\left(\frac{u}{v}\right) = \frac{v\dfrac{du}{dx} - u\dfrac{dv}{dx}}{v^2} \quad \text{becomes} \quad d\left(\frac{u}{v}\right) = \frac{v\,du - u\,dv}{v^2}$$

So, if $y = e^{2x}\sin 4x$, $dy = \dots\dots\dots$

and if $y = \dfrac{\cos 2t}{t^2}$ $dy = \dots\dots\dots$

12

$$\boxed{\begin{aligned} &y = e^{2x}\sin 4x, \quad dy = 2e^{2x}(2\cos 4x + \sin 4x)\,dx \\ &y = \frac{\cos 2t}{t^2}, \quad dy = -\frac{2}{t^3}\{t\sin 2t + \cos 2t\}\,dt \end{aligned}}$$

That was easy enough. Let us now consider a function of two independent variables, $z = f(x, y)$.

If $z = f(x, y)$ then $z + \delta z = f(x + \delta x, y + \delta y)$

$$\therefore\ \delta z = f(x + \delta x, y + \delta y) - f(x, y)$$

Expanding δz in terms of δx and δy, gives

$\delta z = A\delta x + B\delta y + \text{higher powers of } \delta x \text{ and } \delta y,$

where A and B are functions of x and y.

If y remains constant, i.e. $\delta y = 0$, then

$$\delta z = A\,\delta x + \text{higher powers of } \delta x \quad \therefore\ \frac{\delta z}{\delta x} \approx A$$

$$\therefore\ \text{If } \delta x \to 0, \text{ then } A = \frac{\partial z}{\partial x}$$

Similarly, if x remains constant, i.e. $\delta x = 0$, then

$$\delta z = B\,\delta y + \text{higher powers of } \delta y \quad \therefore\ \frac{\delta z}{\delta y} \approx B$$

$$\therefore\ \text{If } \delta y \to 0, \text{ then } B = \frac{\partial z}{\partial y}$$

$$\therefore\ \delta z = \frac{\partial z}{\partial x}\delta x + \frac{\partial z}{\partial y}\delta y + \text{higher powers of small quantities}$$

$$\therefore\ \delta z = \frac{\partial z}{\partial x}\delta x + \frac{\partial z}{\partial y}\delta y$$

▶

In terms of differentials, this result can be written

If $z = f(x, y)$. then $dz = \dfrac{\partial z}{\partial x} dx + \dfrac{\partial z}{\partial y} dy$

The result can be extended to functions of more than two independent variables.

If $z = f(x, y, w)$, $dz = \dfrac{\partial z}{\partial x} dx + \dfrac{\partial z}{\partial y} dy + \dfrac{\partial z}{\partial w} dw$

Make a note of these results in differential form as shown.

Exercise

Determine the differential dz for each of the following functions.

1   $z = x^2 + y^2$
2   $z = x^3 \sin 2y$
3   $z = (2x - 1) e^{3y}$
4   $z = x^2 + 2y^2 + 3w^2$
5   $z = x^3 y^2 w.$

Finish all five and then check the results.

13

1   $dz = 2(x\,dx + y\,dy)$
2   $dz = x^2(3\sin 2y\,dx + 2x\cos 2y\,dy)$
3   $dz = e^{3y}\{2\,dx + (6x - 3)dy\}$
4   $dz = 2(x\,dx + 2y\,dy + 3w\,dw)$
5   $dz = x^2 y(3yw\,dx + 2xw\,dy + xy\,dw)$

Now move on

Exact differential

14

We have just established that if $z = f(x, y)$

$dz = \dfrac{\partial z}{\partial x} dx + \dfrac{\partial z}{\partial y} dy$

We now work in reverse.
Any expression $dz = P\,dx + Q\,dy$, where P and Q are functions of x and y, is an *exact differential* if it can be integrated to determine z.

$\therefore\ P = \dfrac{\partial z}{\partial x}$ and $Q = \dfrac{\partial z}{\partial y}$

Now $\dfrac{\partial P}{\partial y} = \dfrac{\partial^2 z}{\partial y\,\partial x}$ and $\dfrac{\partial Q}{\partial x} = \dfrac{\partial^2 z}{\partial x\,\partial y}$ and we know that $\dfrac{\partial^2 z}{\partial y\,\partial x} = \dfrac{\partial^2 z}{\partial x\,\partial y}$.

Therefore, for dz to be an exact differential $\dfrac{\partial P}{\partial y} = \dfrac{\partial Q}{\partial x}$ and this is the test we apply.

▶

Example 1

$dz = (3x^2 + 4y^2)\,dx + 8xy\,dy.$

If we compare the right-hand side with $P\,dx + Q\,dy$, then

$$P = 3x^2 + 4y^2 \quad \therefore \ \frac{\partial P}{\partial y} = 8y$$

$$Q = 8xy \qquad \therefore \ \frac{\partial Q}{\partial x} = 8y$$

$\dfrac{\partial P}{\partial y} = \dfrac{\partial Q}{\partial x} \qquad \therefore$ dz is an exact differential

Similarly, we can test this one.

Example 2

$dz = (1 + 8xy)\,dx + 5x^2\,dy.$

From this we find . . . .

15

$\boxed{dz \text{ is } not \text{ an exact differential}}$

Because $dz = (1 + 8xy)\,dx + 5x^2\,dy$

$$\therefore \ P = 1 + 8xy \quad \therefore \ \frac{\partial P}{\partial y} = 8x$$

$$Q = 5x^2 \qquad \therefore \ \frac{\partial Q}{\partial x} = 10x$$

$\dfrac{\partial P}{\partial y} \neq \dfrac{\partial Q}{\partial x} \quad \therefore$ dz is not an exact differential.

Exercise

Determine whether each of the following is an exact differential.

1   $dz = 4x^3y^3\,dx + 3x^4y^2\,dy$
2   $dz = (4x^3y + 2xy^3)\,dx + (x^4 + 3x^2y^2)\,dy$
3   $dz = (15y^2e^{3x} + 2xy^2)\,dx + (10ye^{3x} + x^2y)\,dy$
4   $dz = (3x^2e^{2y} - 2y^2e^{3x})\,dx + (2x^3e^{2y} - 2ye^{3x})\,dy$
5   $dz = (4y^3\cos 4x + 3x^2\cos 2y)\,dx + (3y^2\sin 4x - 2x^3\sin 2y)\,dy.$

16

$\boxed{\textbf{1 \ Yes} \quad \textbf{2 \ Yes} \quad \textbf{3 \ No} \quad \textbf{4 \ No} \quad \textbf{5 \ Yes}}$

We have just tested whether certain expressions are, in fact, exact differentials – and we said previously that, by definition, an exact differential can be integrated. But how exactly do we go about it? The following examples will show.

▶

Integration of exact differentials

$dz = P\,dx + Q\,dy$ where $P = \dfrac{\partial z}{\partial x}$ and $Q = \dfrac{\partial z}{\partial y}$

$\therefore z = \displaystyle\int P\,dx$ and also $z = \displaystyle\int Q\,dy$

Example 1

$dz = (2xy + 6x)\,dx + (x^2 + 2y^3)\,dy$.

$P = \dfrac{\partial z}{\partial x} = 2xy + 6x \qquad \therefore z = \displaystyle\int (2xy + 6x)\,dx$

$\therefore z = x^2y + 3x^2 + f(y)$ where $f(y)$ is an arbitrary function of y only, and is akin to the constant of integration in a normal integral.

Also $Q = \dfrac{\partial z}{\partial y} = x^2 + 2y^3 \qquad \therefore z = \displaystyle\int (x^2 + 2y^3)\,dy$

$\therefore z = \ldots\ldots\ldots\ldots$

17

$$z = x^2y + \frac{y^4}{2} + F(x) \text{ where } F(x) \text{ is an arbitrary function of } x \text{ only}$$

So the two results tell us

$z = x^2y + 3x^2 + f(y)$ \hfill (1)

and $z = x^2y + \dfrac{y^4}{2} + F(x)$ \hfill (2)

For these two expressions to represent the same function, then

$f(y)$ in (1) must be $\dfrac{y^4}{2}$ already in (2)

and $F(x)$ in (2) must be $3x^2$ already in (1)

$\therefore z = x^2y + 3x^2 + \dfrac{y^4}{2}$

Example 2

Integrate $dz = (8e^{4x} + 2xy^2)\,dx + (4\cos 4y + 2x^2y)\,dy$.

Argue through the working in just the same way, from which we obtain

$z = \ldots\ldots\ldots\ldots$

18

$$z = 2e^{4x} + x^2y^2 + \sin 4y$$

Here it is. $dz = (8e^{4x} + 2xy^2)\,dx + (4\cos 4y + 2x^2y)\,dy$

$$P = \frac{\partial z}{\partial x} = 8e^{4x} + 2xy^2 \qquad \therefore\ z = \int (8e^{4x} + 2xy^2)dx$$

$$\therefore\ z = 2e^{4x} + x^2y^2 + f(y) \tag{1}$$

$$Q = \frac{\partial z}{\partial y} = 4\cos 4y + 2x^2y \quad \therefore\ z = \int (4\cos 4y + 2x^2y)\,dy$$

$$\therefore\ z = \sin 4y + x^2y^2 + F(x) \tag{2}$$

For (1) and (2) to agree, $f(y) = \sin 4y$ and $F(x) = 2e^{4x}$

$$\therefore\ z = 2\,e^{4x} + x^2y^2 + \sin 4y$$

They are all done in the same way, so you will have no difficulty with the short exercise that follows. *On you go.*

Exercise

Integrate the following exact differentials to obtain the function *z*.

1   $dz = (6x^2 + 8xy^3)\,dx + (12x^2y^2 + 12y^3)\,dy$
2   $dz = (3x^2 + 2xy + y^2)\,dx + (x^2 + 2xy + 3y^2)\,dy$
3   $dz = 2(y+1)e^{2x}\,dx + (e^{2x} - 2y)\,dy$
4   $dz = (3y^2\cos 3x - 3\sin 3x)\,dx + (2y\sin 3x + 4)\,dy$
5   $dz = (\sinh y + y\sinh x)dx + (x\cosh y + \cosh x)\,dy.$

Finish all five before checking with the next frame.

19

$$
\begin{array}{ll}
1 & z = 2x^3 + 4x^2y^3 + 3y^4 \\
2 & z = x^3 + x^2y + xy^2 + y^3 \\
3 & z = e^{2x}(1 + y) - y^2 \\
4 & z = y^2\sin 3x + \cos 3x + 4y \\
5 & z = x\sinh y + y\cosh x.
\end{array}
$$

In the last one, of course, we find that the two expressions for *z* agree without any further addition of $f(y)$ or $F(x)$.

We shall be meeting exact differentials again later on, but for the moment let us deal with something different. On then to the next frame

Area enclosed by a closed curve

One of the earliest applications of integration is finding the area of a plane figure bounded by the x-axis, the curve $y = f(x)$ and ordinates at $x = x_1$ and $x = x_2$.

20

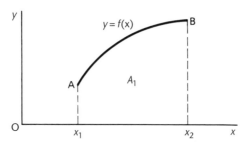

$$A_1 = \int_{x_1}^{x_2} y \, dx = \int_{x_1}^{x_2} f(x) \, dx$$

If points A and B are joined by another curve, $y = F(x)$

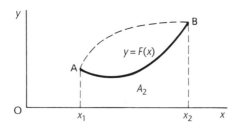

$$A_2 = \int_{x_1}^{x_2} F(x) \, dx$$

Combining the two figures, we have

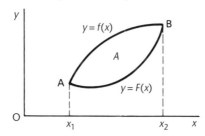

$$A = A_1 - A_2$$

$$\therefore \; A = \int_{x_1}^{x_2} f(x) \, dx - \int_{x_1}^{x_2} F(x) \, dx$$

It is convenient on occasions to arrange the limits so that the integration follows the path round the enclosed area in a regular order.

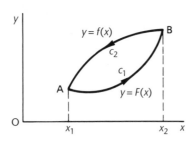

▶

For example

$\int_{x_1}^{x_2} F(x)\,dx$ gives A_2 as before, but integrating from B to A along c_2 with

$y = f(x)$, i.e. $\int_{x_2}^{x_1} f(x)\,dx$, is the integral for A_1 with the sign changed, i.e.

$$\int_{x_2}^{x_1} f(x)\,dx = -\int_{x_1}^{x_2} f(x)\,dx$$

∴ The result $A = A_1 - A_2 = \int_{x_1}^{x_2} f(x)\,dx - \int_{x_1}^{x_2} F(x)\,dx$ becomes

$$A = \ldots\ldots\ldots\ldots$$

21

$$\boxed{A = -\int_{x_1}^{x_2} F(x)\,dx - \int_{x_2}^{x_1} f(x)\,dx}$$

i.e. $$A = -\left\{ \int_{x_1}^{x_2} F(x)\,dx + \int_{x_2}^{x_1} f(x)\,dx \right\}$$

If we proceed round the boundary in an *anticlockwise manner*, the enclosed area is kept on the *left-hand side* and the resulting area is considered *positive*. If we proceed round the boundary in a *clockwise manner*, the enclosed area remains on the *right-hand side* and the resulting area is *negative*.

The final result above can be written in the form

$$A = -\oint y\,dx$$

where the symbol $\oint$ indicates that the integral is to be evaluated round the closed boundary in the positive (i.e. anticlockwise) direction

$$\therefore\ A = -\oint y\,dx = -\left\{ \int_{x_1}^{x_2} F(x)\,dx + \int_{x_2}^{x_1} f(x)\,dx \right\}$$
$$\text{(along } c_1) \quad \text{(along } c_2)$$

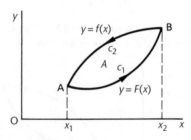

Let us apply this result to a very simple case.

Example 1

Determine the area enclosed by the graphs of $y = x^3$ and $y = 4x$ for $x \geq 0$.

First we need to know the points of intersection. These are

. . . .

$$\boxed{x = 0 \ \text{ and } \ x = 2}$$

22

We integrate in an anticlockwise manner

c_1: $y = x^3$, limits $x = 0$ to $x = 2$

c_2: $y = 4x$, limits $x = 2$ to $x = 0$.

$$A = -\oint y \, \mathrm{d}x = \ldots\ldots\ldots\ldots$$

$$\boxed{A = 4 \ \text{square units}}$$

23

Because

$$A = -\oint y \, \mathrm{d}x = -\left\{ \int_0^2 x^3 \, \mathrm{d}x + \int_2^0 4x \, \mathrm{d}x \right\}$$

$$= -\left\{ \left[\frac{x^4}{4} \right]_0^2 + \left[2x^2 \right]_2^0 \right\} = 4$$

Another example.

Example 2

Find the area of the triangle with vertices $(0, 0)$, $(5, 3)$ and $(2, 6)$.

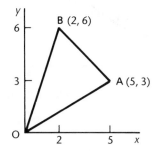

The equation of

OA is . . . .

BA is . . . .

OB is . . . .

24

$$\boxed{\begin{array}{l} \text{OA is } y = \tfrac{3}{5}x \\ \text{BA is } y = 8 - x \\ \text{OB is } y = 3x \end{array}}$$

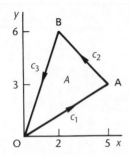

Then $A = -\oint y\,dx$

$= \ldots\ldots\ldots$

Write down the component integrals with appropriate limits.

25

$$\boxed{A = -\oint y\,dx = -\left\{ \int_0^5 \frac{3}{5}x\,dx + \int_5^2 (8-x)\,dx + \int_2^0 3x\,dx \right\}}$$

The limits chosen must progress the integration round the boundary of the figure in an *anticlockwise manner*. Finishing off the integration, we have

$$A = \ldots\ldots\ldots$$

26

$$\boxed{A = 12 \text{ square units}}$$

The actual integration is easy enough.

The work we have just done leads us on to consider line integrals, so let us make a fresh start in the next frame

Line integrals

27

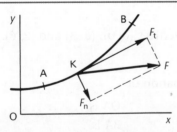 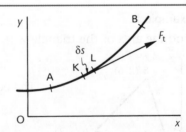

If a field exists in the x–y plane, producing a force F on a particle at K, then F can be resolved into two components

F_t along the tangent to the curve AB at K

F_n along the normal to the curve AB at K.

▶

The work done in moving the particle through a small distance δs from K to L along the curve is then approximately $F_t \, \delta s$. So the total work done in moving a particle along the curve from A to B is given by

. . . .

$$\underset{\delta s \to 0}{Lim} \sum F_t \, \delta s = \int F_t \, ds \text{ from A to B}$$

28

This is normally written $\int_{AB} F_t \, ds$ where A and B are the end points of the curve, or as $\int_c F_t \, ds$ where the curve c connecting A and B is defined.

Such an integral thus formed is called a *line integral* since integration is carried out along the path of the particular curve c joining A and B.

$$\therefore I = \int_{AB} F_t \, ds = \int_c F_t \, ds$$

where c is the curve $y = f(x)$ between A (x_1, y_1) and B (x_2, y_2).

There is in fact an alternative form of the integral which is often useful, so let us also consider that

Alternative form of a line integral

29

It is often more convenient to integrate with respect to x or y than to take arc length as the variable.

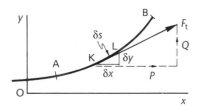

If F_t has a component

    P in the x-direction

    Q in the y-direction

then the work done from K to L can be stated as $P \, \delta x + Q \, \delta y$.

$$\therefore \int_{AB} F_t \, ds = \int_{AB} (P \, dx + Q \, dy)$$

where P and Q are functions of x and y.

In general then, the line integral can be expressed as

$$I = \int_c F_t \, ds = \int_c (P \, dx + Q \, dy)$$

where c is the prescribed curve and F, or P and Q, are functions of x and y.

Make a note of these results – then we will apply them to one or two examples

30

Example 1

Evaluate $\int_c (x + 3y)\,dx$ from A (0, 1) to B (2, 5) along the curve $y = 1 + x^2$.

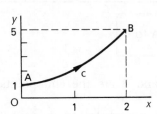

The line integral is of the form

$$\int_c (P\,dx + Q\,dy)$$

where, in this case, $Q = 0$ and c is the curve $y = 1 + x^2$.

It can be converted at once into an ordinary integral by substituting for y and applying the appropriate limits of x.

$$I = \int_c (P\,dx + Q\,dy) = \int_c (x + 3y)\,dx = \int_0^2 (x + 3 + 3x^2)\,dx$$

$$= \left[\frac{x^2}{2} + 3x + x^3\right]_0^2 = 16$$

Now for another, so move on

31

Example 2

Evaluate $I = \int_c (x^2 + y)\,dx + (x - y^2)\,dy$ from A (0, 2) to B (3, 5) along the curve $y = 2 + x$.

$$I = \int_c (P\,dx + Q\,dy)$$

$$P = x^2 + y = x^2 + 2 + x = x^2 + x + 2$$

$$Q = x - y^2 = x - (4 + 4x + x^2)$$

$$= -(x^2 + 3x + 4)$$

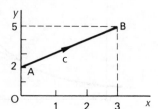

Also $y = 2 + x$ $\quad \therefore$ $dy = dx$ and the limits are $x = 0$ to $x = 3$.

$$\therefore I = \ldots\ldots\ldots\ldots$$

32

$$\boxed{I = -15}$$

Because

$$I = \int_0^3 \{(x^2 + x + 2)\,dx - (x^2 + 3x + 4)\,dx\}$$

$$\int_0^3 -(2x + 2)\,dx = \left[x^2 - 2x\right]_0^3 = -15$$

▶

Here is another.

Example 3

Evaluate $I = \int_c \{(x^2 + 2y)\,dx + xy\,dy\}$ from O (0, 0) to B (1, 4) along the curve $y = 4x^2$.

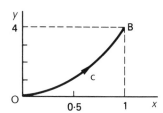

In this case, c is the curve $y = 4x^2$.

$$\therefore \ dy = 8x\,dx$$

Substitute for y in the integral and apply the limits.

$$\text{Then } I = \ldots\ldots\ldots$$

Finish it off: it is quite straightforward.

$$\boxed{I = 9\cdot4}$$
33

Because

$$I = \int_c \{(x^2 + 2y)\,dx + xy\,dy\} \qquad y = 4x^2 \qquad \therefore \ dy = 8x\,dx$$

Also $x^2 + 2y = x^2 + 8x^2 = 9x^2;$ $xy = 4x^3$

$$\therefore \ I = \int_0^1 \{9x^2\,dx + 32x^4\,dx\} = \int_0^1 (9x^2 + 32x^4)\,dx = 9.4$$

They are all done in very much the same way.

Move on for Example 4

Example 4
34

Evaluate $I = \int_c \{(x^2 + 2y)\,dx + xy\,dy\}$ from O (0, 0) to A (1, 0) along line $y = 0$ and then from A (1, 0) to B (1, 4) along the line $x = 1$.

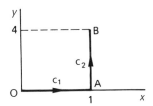

(1) OA: c_1 is the line $y = 0$ $\therefore$ $dy = 0$. Substituting $y = 0$ and $dy = 0$ in the given integral gives

$$I_{OA} = \int_0^1 x^2\,dx = \left[\frac{x^3}{3}\right]_0^1 = \frac{1}{3}$$

(2) AB: Here c_2 is the line $x = 1$ $\therefore$ $dx = 0$

$$\therefore \ I_{AB} = \ldots\ldots\ldots\ldots$$

35

$$\boxed{I_{AB} = 8}$$

Because

$$I_{AB} = \int_0^4 \{(1 + 2y)(0) + y\,dy\}$$

$$= \int_0^4 y\,dy$$

$$= \left[\frac{y^2}{2}\right]_0^4 = 8$$

Then $I = I_{OA} + I_{AB} = \frac{1}{3} + 8 = 8\frac{1}{3}$ $\therefore I = 8\frac{1}{3}$

If we now look back to Examples 3 and 4 just completed, we find that we have evaluated the same integral between the same two end points, but . . . .

36

$$\boxed{\text{along different paths of integration}}$$

If we combine the two diagrams, we have

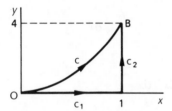

where c is the curve $y = 4x^2$ and $c_1 + c_2$ are the lines $y = 0$ and $x = 1$.

The results obtained were

$$I_c = 9\frac{2}{5} \text{ and } I_{c_1 + c_2} = 8\frac{1}{3}$$

Notice therefore that integration along two distinct paths joining the same two end points does not necessarily give the same results.

37   Let us pause here a moment and list the main properties of line integrals.

Properties of line integrals

1   $\displaystyle\int_c F\,ds = \int_c \{P\,dx + Q\,dy\}$

2   $\displaystyle\int_{AB} F\,ds = -\int_{BA} F\,ds$ and $\int_{AB}\{P\,dx + Q\,dy\} = -\int_{BA}\{P\,dx + Q\,dy\}$

   i.e. the sign of a line integral is reversed when the direction of the integration along the path is reversed.

3   (a) For a path of integration parallel to the y-axis, i.e. $x = k$,

   $$dx = 0. \quad \therefore \int_c P\,dx = 0 \quad \therefore I_c = \int_c Q\,dy.$$

   (b) For a path of integration parallel to the x-axis, i.e. $y = k$,

   $$dy = 0. \quad \therefore \int_c Q\,dy = 0 \quad \therefore I_c = \int_c P\,dx.$$

4 If the path of integration c joining A to B is divided into two parts AK and KB, then $I_c = I_{AB} = I_{AK} + I_{KB}$.

5 In all cases, the function $y = f(x)$ that describes the path of integration involved must be continuous and single-valued – or dealt with as in item **6** below.

6 If the function $y = f(x)$ that describes the path of integration c is not single-valued for part of its extent, the path is divided into two sections.

$y = f_1(x)$ from A to K

$y = f_2(x)$ from K to B.

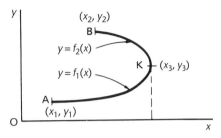

Make a note of this list for future reference and revision

Example **38**

Evaluate $I = \int_c (x + y)\, dx$ from A $(0, 1)$ to B $(0, -1)$ along the semi-circle $x^2 + y^2 = 1$ for $x \geq 0$.

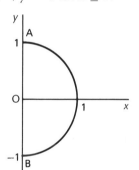

The first thing we notice is that

. . . .

39

| the function $y = f(x)$ that describes the path of integration c is *not* single-valued |

For any value of x, $y = \pm\sqrt{1 - x^2}$. Therefore, we divide c into two parts

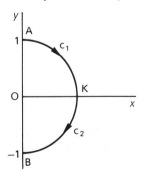

(1) $y = \sqrt{1 - x^2}$ from A to K

(2) $y = -\sqrt{1 - x^2}$ from K to B.

As usual, $I = \int_c (P\, dx + Q\, dy)$ and in this

particular case, $Q = $ . . . .

40

$$Q = 0$$

$$\therefore I = \int_c P\,dx = \int_0^1 \left(x + \sqrt{1 - x^2}\right) dx + \int_1^0 \left(x - \sqrt{1 - x^2}\right) dx$$

$$= \int_0^1 (x + \sqrt{1 - x^2} - x + \sqrt{1 - x^2})\,dx = 2\int_0^1 \sqrt{1 - x^2}\,dx$$

Now substitute $x = \sin\theta$ and finish it off.

$$I = \dots\dots\dots$$

41

$$I = \frac{\pi}{2}$$

Because

$$I = 2\int_0^1 \sqrt{1 - x^2}\,dx \quad x = \sin\theta \quad \therefore\ dx = \cos\theta\,d\theta$$

$$\sqrt{1 - x^2} = \cos\theta$$

Limits: $x = 0,\ \theta = 0;\ x = 1,\ \theta = \dfrac{\pi}{2}$

$$\therefore I = 2\int_0^{\pi/2} \cos^2\theta\,d\theta = \int_0^{\pi/2} (1 + \cos 2\theta)\,d\theta$$

$$= \left[\theta + \frac{\sin 2\theta}{2}\right]_0^{\pi/2}$$

$$= \frac{\pi}{2}$$

Now let us extend this line of development a stage further.

42 Regions enclosed by closed curves

A region is said to be *simply connected* if a path joining A and B can be deformed to coincide with any other line joining A and B without going outside the region.

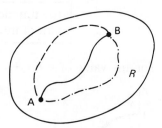

Another definition is that a region is simply connected if any closed path in the region can be contracted to a single point without leaving the region.

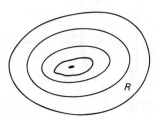

▶

Clearly, this would not be satisfied in the case where the region *R* contains one or more 'holes'.

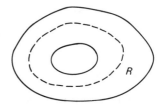

The closed curves involved in problems in this Programme all relate to simply connected regions, so no difficulties will arise.

Line integrals round a closed curve

43

We have already introduced the symbol $\oint$ to indicate that an integral is to be evaluated round a closed curve in the positive (anticlockwise) direction.

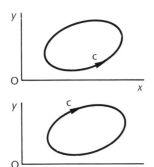

Positive direction (anticlockwise) line integral denoted by $\oint$.

Negative direction (clockwise) line integral denoted by $-\oint$.

With a closed curve, the *y*-values on the path c cannot be single-valued. Therefore, we divide the path into two or more parts and treat each separately.

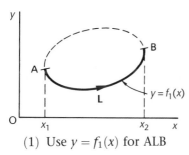 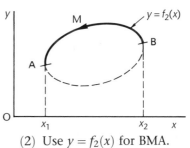

$$(1)\ \text{Use } y = f_1(x) \text{ for ALB} \qquad (2)\ \text{Use } y = f_2(x) \text{ for BMA.}$$

Unless specially required otherwise, we always proceed round the closed curve in an . . . .

44

anticlockwise direction

Example 1

Evaluate the line integral $I = \oint_c (x^2 \, dx - 2xy \, dy)$ where c comprises the three

sides of the triangle joining O $(0, 0)$, A $(1, 0)$ and B $(0, 1)$.

First draw the diagram and mark in c_1, c_2 and c_3, the proposed directions of integration. Do just that.

45

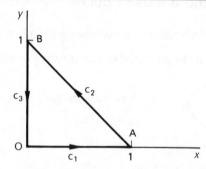

The three sections of the path of integration must be arranged in an anticlockwise manner round the figure. Now we deal with each part separately.

(a) OA: c_1 is the line $y = 0$ $\therefore$ $dy = 0$.

Then $I = \oint (x^2 \, dx - 2xy \, dy)$ for this part becomes

$$I_1 = \int_0^1 x^2 \, dx = \left[\frac{x^3}{3} \right]_0^1 = \frac{1}{3}   \therefore I_1 = \frac{1}{3}$$

(b) AB: c_2 is the line $y = 1 - x$ $\therefore$ $dy = -dx$

$$I_2 = \dots\dots\dots \quad \text{(evaluate it)}$$

46

$$\boxed{I_2 = -\tfrac{2}{3}}$$

Because c_2 is the line $y = 1 - x$ $\therefore$ $dy = -dx$.

$$I_2 = \int_1^0 \{x^2 \, dx + 2x(1 - x) \, dx\} = \int_1^0 (x^2 + 2x - 2x^2) \, dx$$

$$= \int_1^0 (2x - x^2) \, dx = \left[x^2 - \frac{x^3}{3} \right]_1^0 = -\frac{2}{3}   \therefore I_2 = -\frac{2}{3}$$

Note that anticlockwise progression is obtained by arranging the limits in the appropriate order.

Now we have to determine I_3 for BO.

(c) BO: c_3 is the line $x = 0$

$$I_3 = \dots\dots\dots$$

$$\boxed{I_3 = 0}$$ **47**

Because for c_3, $x = 0$ $\therefore$ $dx = 0$ $\therefore$ $I_3 = \int 0\,dy = 0$ $\therefore$ $I_3 = 0$

Finally, $I = I_1 + I_2 + I_3 = \frac{1}{3} - \frac{2}{3} + 0 = -\frac{1}{3}$ $\therefore$ $I = -\frac{1}{3}$

Let us work through another example.

Example 2

Evaluate $\oint_c y\,dx$ when c is the circle $x^2 + y^2 = 4$.

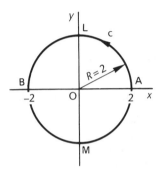

$x^2 + y^2 = 4$ $\therefore$ $y = \pm\sqrt{4 - x^2}$

y is thus not single-valued. Therefore use $y = \sqrt{4 - x^2}$ for ALB between $x = 2$ and $x = -2$ and $y = -\sqrt{4 - x^2}$ for BMA between $x = -2$ and $x = 2$.

$$\therefore I = \int_2^{-2} \sqrt{4 - x^2}\,dx + \int_{-2}^2 \{-\sqrt{4 - x^2}\}\,dx$$

$$= 2\int_2^{-2} \sqrt{4 - x^2}\,dx = -2\int_{-2}^2 \sqrt{4 - x^2}\,dx$$

$$= -4\int_0^2 \sqrt{4 - x^2}\,dx.$$

To evaluate this integral, substitute $x = 2\sin\theta$ and finish it off.

$$I = \dots\dots\dots$$

$$\boxed{I = -4\pi}$$ **48**

Because

$x = 2\sin\theta$ $\therefore$ $dx = 2\cos\theta\,d\theta$ $\therefore$ $\sqrt{4 - x^2} = 2\cos\theta$

limits: $x = 0$, $\theta = 0$; $x = 2$, $\theta = \dfrac{\pi}{2}$

$$\therefore I = -4\int_0^{\pi/2} 2\cos\theta\, 2\cos\theta\,d\theta = -16\int_0^{\pi/2} \cos^2\theta\,d\theta$$

$$= -8\int_0^{\pi/2} (1 + \cos 2\theta)\,d\theta = -8\left[\theta + \frac{\sin 2\theta}{2}\right]_0^{\pi/2} = -4\pi$$

Now for one more

▶

Example 3

Evaluate $I = \oint_c \{xy\,dx + (1+y^2)\,dy\}$ where c is the boundary of the rectangle joining A $(1, 0)$, B $(3, 0)$, C $(3, 2)$ and D $(1, 2)$.

First draw the diagram and insert c_1, c_2, c_3, c_4.

That gives ....

49

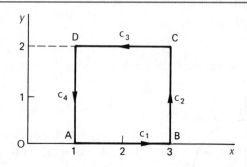

Now evaluate I_1 for AB; I_2 for BC; I_3 for CD; I_4 for DA; and finally I.

Complete the working and then check with the next frame

50

$$I_1 = 0; \quad I_2 = 4\tfrac{2}{3}; \quad I_3 = -8; \quad I_4 = -4\tfrac{2}{3}; \quad I = -8$$

Here is the complete working.

$$I = \oint_c \{xy\,dx + (1+y^2)\,dy\}$$

(a) AB: c_1 is $y = 0$ $\quad \therefore \quad dy = 0$ $\quad \therefore \quad I_1 = 0$

(b) BC: c_2 is $x = 3$ $\quad \therefore \quad dx = 0$

$$\therefore \ I_2 = \int_0^2 (1+y^2)\,dy = \left[y + \frac{y^3}{3}\right]_0^2 = 4\tfrac{2}{3} \qquad \therefore \ I_2 = 4\tfrac{2}{3}$$

(c) CD: c_3 is $y = 2$ $\quad \therefore \quad dy = 0$

$$\therefore \ I_3 = \int_3^1 2x\,dx = \left[x^2\right]_3^1 = -8 \qquad\qquad \therefore \ I_3 = -8$$

(d) DA: c_4 is $x = 1$ $\quad \therefore \quad dx = 0$

$$\therefore \ I_4 = \int_2^0 (1+y^2)\,dy = \left[y + \frac{y^3}{3}\right]_2^0 = -4\tfrac{2}{3} \quad \therefore \ I_4 = -4\tfrac{2}{3}$$

Finally

$$I = I_1 + I_2 + I_3 + I_4$$
$$= 0 + 4\tfrac{2}{3} - 8 - 4\tfrac{2}{3} = -8 \qquad \therefore \ I = -8$$

Remember that, unless we are directed otherwise, we always proceed round the closed boundary in an anticlockwise manner.

On now to the next piece of work

Line integral with respect to arc length

51

We have already established that

$$I = \int_{AB} F_t \, ds = \int_{AB} \{ P \, dx + Q \, dy \}$$

where F_t denoted the tangential force along the curve c at the sample point K (x, y).

The same kind of integral can, of course, relate to any function $f(x, y)$ which is a function of the position of a point on the stated curve, so that

$$I = \int_c f(x, y) \, ds.$$

This can readily be converted into an integral in terms of x. (Refer to *Engineering Mathematics (Sixth Edition)*, Programme 19, Frame 30.)

$$I = \int_c f(x, y) \, ds = \int_c f(x, y) \frac{ds}{dx} \, dx \quad \text{where} \quad \frac{ds}{dx} = \sqrt{1 + \left(\frac{dy}{dx} \right)^2}$$

$$\therefore \ \int_c f(x, y) \, ds = \int_{x_1}^{x_2} f(x, y) \sqrt{1 + \left(\frac{dy}{dx} \right)^2} \, dx \tag{1}$$

Example

Evaluate $I = \int_c (4x + 3xy) \, ds$ where c is the straight line joining O $(0, 0)$ to A $(1, 2)$.

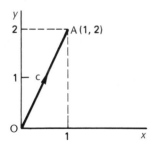

c is the line $y = 2x \quad \therefore \ \dfrac{dy}{dx} = 2$

$$\therefore \ \frac{ds}{dx} = \sqrt{1 + \left(\frac{dy}{dx} \right)^2} = \sqrt{5}$$

$$\therefore \ I = \int_{x=0}^{x=1} (4x + 3xy) \, ds = \int_0^1 (4x + 3xy)(\sqrt{5}) \, dx. \quad \text{But } y = 2x$$

$$\therefore \ I = \ldots\ldots\ldots\ldots$$

52

$$I = 4\sqrt{5}$$

Because

$$I = \int_0^1 (4x + 6x^2)(\sqrt{5})\,dx = 2\sqrt{5}\int_0^1 (2x + 3x^2)\,dx = 4\sqrt{5}$$

Try another.

The path length of the parabola defined by $y = x^2$ betwen the values $x = 0$ and $x = 2$ is given by the integral

$$I = \int_c ds = \ldots\ldots\ldots \quad \text{to 3 dp}$$

53

$$3\cdot393 \text{ to 3 dp}$$

Because

$$I = \int_c ds = \int_{x=0}^2 \sqrt{1 + \left(\frac{dy}{dx}\right)^2}\,dx$$

$$= \int_{x=0}^2 \sqrt{1 + 2x}\,dx$$

Let $u = 1 + 2x$ so that $du = 2dx$ and so

$$I = \int_{u=1}^5 u^{1/2}\frac{du}{2}$$

$$= \frac{1}{2}\left[\frac{2}{3}u^{3/2}\right]_1^5$$

$$= \frac{1}{3}\left(125^{1/2} - 1\right)$$

$$= 3\cdot393 \text{ to 3 dp}$$

54 Parametric equations

When x and y are expressed in parametric form, e.g. $x = f(t)$, $y = g(t)$, then

$$\frac{ds}{dt} = \sqrt{\left(\frac{dx}{dt}\right)^2 + \left(\frac{dy}{dt}\right)^2} \quad \therefore \quad ds = \sqrt{\left(\frac{dx}{dt}\right)^2 + \left(\frac{dy}{dt}\right)^2}\,dt$$

and result (1) above becomes

$$I = \int_c f(x, y)\,ds = \int_{t_1}^{t_2} f(x, y)\sqrt{\left(\frac{dx}{dt}\right)^2 + \left(\frac{dy}{dt}\right)^2}\,dt \tag{2}$$

Make a note of results (1) and (2) for future use

Example 55

Evaluate $I = \oint_c 4xy\,\mathrm{d}s$ where c is defined as the curve $x = \sin t$, $y = \cos t$

between $t = 0$ and $t = \dfrac{\pi}{4}$.

We have $x = \sin t$ $\therefore$ $\dfrac{\mathrm{d}x}{\mathrm{d}t} = \cos t$

$\qquad\qquad y = \cos t$ $\therefore$ $\dfrac{\mathrm{d}y}{\mathrm{d}t} = -\sin t$

$\qquad\qquad\qquad$ $\therefore$ $\dfrac{\mathrm{d}s}{\mathrm{d}t} = \ldots\ldots\ldots\ldots$

$$\boxed{\dfrac{\mathrm{d}s}{\mathrm{d}t} = 1}$$ 56

Because

$$\frac{\mathrm{d}s}{\mathrm{d}t} = \sqrt{\left(\frac{\mathrm{d}x}{\mathrm{d}t}\right)^2 + \left(\frac{\mathrm{d}y}{\mathrm{d}t}\right)^2} = \sqrt{\cos^2 t + \sin^2 t} = 1$$

$$\therefore I = \int_{t_1}^{t_2} f(x,\,y)\sqrt{\left(\frac{\mathrm{d}x}{\mathrm{d}t}\right)^2 + \left(\frac{\mathrm{d}y}{\mathrm{d}t}\right)^2}\,\mathrm{d}t = \int_0^{\pi/4} 4\sin t\,\cos t\,\mathrm{d}t$$

$$= 2\int_0^{\pi/4} \sin 2t\,\mathrm{d}t = -2\left[\frac{\cos 2t}{2}\right]_0^{\pi/4} = 1 \quad \therefore I = 1$$

Dependence of the line integral on the path of integration

We saw earlier in the Programme that integration along two separate paths joining the same two end points does not necessarily give identical results.

With this in mind, let us investigate the following problem.

Example

Evaluate $I = \oint_c \{3x^2y^2\,\mathrm{d}x + 2x^3y\,\mathrm{d}y\}$ between O $(0, 0)$ and A $(2, 4)$

(a) along c_1 i.e. $y = x^2$

(b) along c_2 i.e. $y = 2x$

(c) along c_3 i.e. $x = 0$ from $(0, 0)$ to $(0, 4)$ and $y = 4$ from $(0, 4)$ to $(2, 4)$.

Let us concentrate on section (a).

First we draw the figure and insert relevant information.

$\qquad\qquad\qquad\qquad$ This gives $\ldots\ldots\ldots\ldots$

57

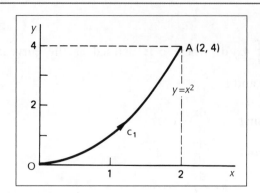

(a) $I = \int_c \{3x^2y^2\,dx + 2x^3y\,dy\}$

The path c_1 is $y = x^2$ $\therefore$ $dy = 2x\,dx$

$$\therefore I_1 = \int_0^2 \{3x^2x^4\,dx + 2x^3x^2 2x\,dx\} = \int_0^2 (3x^6 + 4x^4)\,dx$$

$$= \left[x^7\right]_0^2 = 128 \quad \therefore I_1 = 128$$

(b) In (b), the path of integration changes to c_2, i.e. $y = 2x$

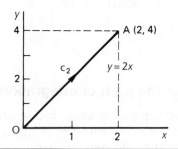

So, in this case,

$I_2 = \ldots\ldots\ldots$

58

$$\boxed{I_2 = 128}$$

Because with c_2, $y = 2x$ $\therefore$ $dy = 2\,dx$

$$\therefore I_2 = \int_0^2 \{3x^2\,4x^2\,dx + 2x^3\,2x2\,dx\} = \int_0^2 20x^4\,dx$$

$$= 4\left[x^5\right]_0^2 = 128 \quad \therefore I_2 = 128$$

(c) In the third case, the path c_3 is split

$x = 0$ from $(0,\,0)$ to $(0,\,4)$

$y = 4$ from $(0,\,4)$ to $(2,\,4)$

Sketch the diagram and determine I_3.

$$I_3 = \ldots\ldots\ldots$$

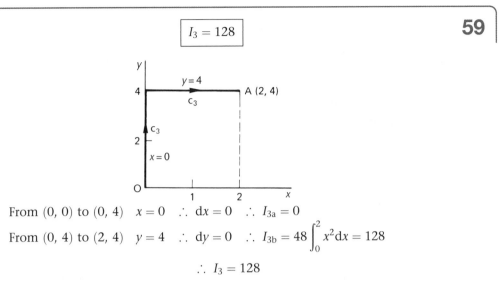

$$\boxed{I_3 = 128}$$

59

From $(0, 0)$ to $(0, 4)$ $x = 0$ $\therefore$ $dx = 0$ $\therefore$ $I_{3a} = 0$

From $(0, 4)$ to $(2, 4)$ $y = 4$ $\therefore$ $dy = 0$ $\therefore$ $I_{3b} = 48 \int_0^2 x^2 dx = 128$

$$\therefore I_3 = 128$$

On to the next frame

60

In the example we have just worked through, we took three different paths and in each case, the line integral produced the same result. It appears, therefore, that in this case, the value of the integral is independent of the path of integration taken.

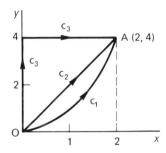

How then does this integral perhaps differ from those of previous cases?

Let us investigate

61

We have been dealing with $I = \int_c \{3x^2y^2\,dx + 2x^3y\,dy\}$

On reflection, we see that the integrand $3x^2y^2\,dx + 2x^3y\,dy$ is of the form $P\,dx + Q\,dy$ which we have met before and that it is, in fact, an *exact differential* of the function $z = x^3y^2$, because

$$\frac{\partial z}{\partial x} = 3x^2y^2 \quad \text{and} \quad \frac{\partial z}{\partial y} = 2x^3y$$

Provided P, Q and their first partial derivatives are finite and continuous at all points inside and on any closed curve, this always happens. If the integrand of the given integral is seen to be an *exact differential*, then the value of the line integral is *independent of the path taken and depends only on the coordinates of the two end points*

Make a note of this. It is important

62

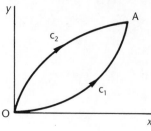

If $I = \int_c \{P \, dx + Q \, dy\}$ and $(P \, dx + Q \, dy)$ is an exact differential, then

$$I_{c_1} = I_{c_2}$$

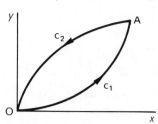

If we reverse the direction of c_2, then

$$I_{c_1} = -I_{c_2}$$

i.e. $I_{c_1} + I_{c_2} = 0$

Hence, *if $(P \, dx + Q \, dy)$ is an exact differential, then the integration taken round a closed curve is zero.*

$\therefore$ If $(P \, dx + Q \, dy)$ is an exact differential, $\oint (P \, dx + Q \, dy) = 0$

63

Example 1

Evaluate $I = \int_c \{3y \, dx + (3x + 2y) \, dy\}$ from A $(1, 2)$ to B $(3, 5)$.

No path is given, so the integrand is probably an exact differential of some function $z = f(x, y)$. In fact $\dfrac{\partial P}{\partial y} = 3 = \dfrac{\partial Q}{\partial x}$.

We have already dealt with the integration of exact differentials, so there is no difficulty. Compare with $I = \int_c \{P \, dx + Q \, dy\}$.

$$P = \frac{\partial z}{\partial x} = 3y \qquad \therefore \ z = \int 3y \, dx = 3xy + f(y) \tag{1}$$

$$Q = \frac{\partial z}{\partial y} = 3x + 2y \quad \therefore \ z = \int (3x + 2y) \, dy = 3xy + y^2 + F(x) \tag{2}$$

For (1) and (2) to agree

$$f(y) = \ldots\ldots\ldots\ldots \quad \text{and} \quad F(x) = \ldots\ldots\ldots\ldots$$

64

$$\boxed{f(y) = y^2; \quad F(x) = 0}$$

Hence $z = 3xy + y^2$

$$\therefore \ I = \int_c \{3y\,dx + (3x + 2y)\,dy\} = \int_{(1,\ 2)}^{(3,\ 5)} d(3xy + y^2)$$

$$= \left[3xy + y^2\right]_{(1,\ 2)}^{(3,\ 5)}$$

$$= (45 + 25) - (6 + 4)$$

$$= 60$$

Example 2

Evaluate $I = \int_c \{(x^2 + ye^x)\,dx + (e^x + y)\,dy\}$ between A $(0, 1)$ and B $(1, 2)$.

As before, compare with $\int_c \{P\,dx + Q\,dy\}$.

$$P = \frac{\partial z}{\partial x} = x^2 + ye^x \quad \therefore \ z = \ldots\ldots\ldots$$

$$Q = \frac{\partial z}{\partial y} = e^x + y \quad \therefore \ z = \ldots\ldots\ldots$$

Continue the working and complete the evaluation.

When you have finished, check the result with the next frame

65

$$\boxed{\begin{aligned} z &= \frac{x^3}{3} + ye^x + f(y) \\ z &= ye^x + \frac{y^2}{2} + F(x) \end{aligned}}$$

For these expressions to agree, $\qquad f(y) = \frac{y^2}{2}; \quad F(x) = \frac{x^3}{3}$

Then $I = \left[\dfrac{x^3}{3} + ye^x + \dfrac{y^2}{2}\right]_{(0,\ 1)}^{(1,\ 2)}$

$$= \frac{5}{6} + 2e$$

So the main points are that, if $(P\,dx + Q\,dy)$ is an exact differential

(a) $I = \int_c (P\,dx + Q\,dy)$ is independent of the path of integration

(b) $I = \int_c (P\,dx + Q\,dy)$ is zero when c is a closed curve.

On to the next frame

66   Exact differentials in three independent variables

A line integral in space naturally involves three independent variables, but the method is very much like that for two independent variables.

$dw = Pdx + Qdy + Rdz$ is an exact differential of $w = f(x, y, z)$

if $\dfrac{\partial P}{\partial y} = \dfrac{\partial Q}{\partial x}$; $\dfrac{\partial P}{\partial z} = \dfrac{\partial R}{\partial x}$; $\dfrac{\partial R}{\partial y} = \dfrac{\partial Q}{\partial z}$

If the test is successful, then

(a) $\displaystyle\int_c (P\,dx + Q\,dy + R\,dz)$ is independent of the path of integration

(b) $\displaystyle\oint_c (P\,dx + Q\,dy + R\,dz)$ is zero when c is a closed curve.

Example

Verify that $dw = (3x^2yz + 6x)dx + (x^3z - 8y)dy + (x^3y + 1)dz$ is an exact differential and hence evaluate $\displaystyle\int_c dw$ from A $(1, 2, 4)$ to B $(2, 1, 3)$.

First check that dw is an exact differential by finding the partial derivatives above, when $P = 3x^2yz + 6x$; $Q = x^3z - 8y$; and $R = x^3y + 1$.

We have ....

67

$$\begin{array}{lll}
\dfrac{\partial P}{\partial y} = 3x^2z; & \dfrac{\partial Q}{\partial x} = 3x^2z & \therefore \ \dfrac{\partial P}{\partial y} = \dfrac{\partial Q}{\partial x} \\[2mm]
\dfrac{\partial P}{\partial z} = 3x^2y; & \dfrac{\partial R}{\partial x} = 3x^2y & \therefore \ \dfrac{\partial P}{\partial z} = \dfrac{\partial R}{\partial x} \\[2mm]
\dfrac{\partial R}{\partial y} = x^3; & \dfrac{\partial Q}{\partial z} = x^3 & \therefore \ \dfrac{\partial R}{\partial y} = \dfrac{\partial Q}{\partial z}
\end{array}$$

$$\therefore \ dw \text{ is an exact differential}$$

Now to find w. $P = \dfrac{\partial z}{\partial x}$; $Q = \dfrac{\partial z}{\partial y}$; $R = \dfrac{\partial w}{\partial z}$

$\therefore \ \dfrac{\partial w}{\partial x} = 3x^2yz + 6x$ $\therefore \ w = \displaystyle\int (3x^2yz + 6x)dx$

$\qquad\qquad\qquad\qquad\qquad = x^3yz + 3x^2 + f(y, z)$

$\dfrac{\partial w}{\partial y} = x^3z - 8y$ $\therefore \ w = \displaystyle\int (x^3z - 8y)\,dy$

$\qquad\qquad\qquad\qquad = x^3zy - 4y^2 + F(x, z)$

$\dfrac{\partial w}{\partial z} = x^3y + 1$ $\therefore \ w = \displaystyle\int (x^3y + 1)\,dz$

$\qquad\qquad\qquad\qquad = x^3yz + z + g(x, y)$

For these three expressions for z to agree

$\qquad f(y, z) = \ldots\ldots\ldots\ldots$; $F(x, z) = \ldots\ldots\ldots\ldots$; $g(x, y) = \ldots\ldots\ldots\ldots$

68

$$f(y,z) = -4y^2; \quad F(x,z) = z; \quad g(x,y) = 3x^2$$

$$\therefore \ w = x^3yz + 3x^2 - 4y^2 + z$$

$$\therefore \ I = \left[x^3yz + 3x^2 - 4y^2 + z \right]_{(1,2,4)}^{(2,1,3)}$$

$$= \ldots\ldots\ldots$$

69

$$I = 36$$

Because

$$I = \left[x^3yz + 3x^2 - 4y^2 + z \right]_{(1,2,4)}^{(2,1,3)}$$

$$= (24 + 12 - 4 + 3) - (8 + 3 - 16 + 4) = 36$$

The extension to line integrals in space is thus quite straightforward.

Finally, we have a theorem that can be very helpful on occasions and which links up with the work we have been doing.

It is important, so let us start a new section

Green's theorem

70

Let P and Q be two functions of x and y that are, along with their first partial derivatives, finite and continuous inside and on the boundary c of a region R in the x–y plane.

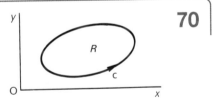

If the first partial derivatives are continuous within the region and on the boundary, then Green's theorem states that

$$\int\int_R \left(\frac{\partial P}{\partial y} - \frac{\partial Q}{\partial x} \right) dx\,dy = -\oint_c (P\,dx + Q\,dy)$$

That is, a double integral over the plane region R can be transformed into a line integral over the boundary c of the region – and the action is reversible.

Let us see how it works.

▶

Example 1

Evaluate $I = \oint_c \{(2x - y)\,dx + (2y + x)\,dy\}$ around the boundary c of the ellipse $x^2 + 9y^2 = 16$.

The integral is of the form $I = \oint_c \{P\,dx + Q\,dy\}$ where

$$P = 2x - y \quad \therefore \quad \frac{\partial P}{\partial y} = -1$$

and $Q = 2y + x \quad \therefore \quad \frac{\partial Q}{\partial x} = 1$.

$$\therefore \ I = -\int_R\int \left(\frac{\partial P}{\partial y} - \frac{\partial Q}{\partial x}\right) dx\,dy$$

$$= -\int_R\int (-1 - 1)\,dx\,dy$$

$$= 2\int_R\int dx\,dy$$

But $\int_R\int dx\,dy$ over any closed region gives ....

71

$$\boxed{\text{the area of the figure}}$$

In this case, then, $I = 2A$ where A is the area of the ellipse

$$x^2 + 9y^2 = 16 \quad \text{i.e.} \quad \frac{x^2}{16} + \frac{9y^2}{16} = 1$$

$$\therefore \ a = 4; \ b = \frac{4}{3}$$

$$\therefore \ A = \pi ab = \frac{16\pi}{3}$$

$$\therefore \ I = 2A = \frac{32\pi}{3}$$

To demonstrate the advantage of Green's theorem, let us work through the next example (a) by the method of line integrals, and (b) by applying Green's theorem.

Example 2

Evaluate $I = \oint_c \{(2x + y)\,dx + (3x - 2y)\,dy\}$ taken in anticlockwise manner round the triangle with vertices at O (0, 0), A (1, 0) and B (1, 2).

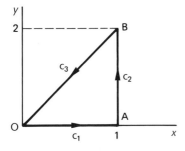

$$I = \oint_c \{(2x + y)\,dx + (3x - 2y)\,dy\}$$

(a) *By the method of line integrals*
There are clearly three stages with c_1, c_2, c_3. Work through the complete evaluation to determine the value of *I*. It will be good revision.

When you have finished, check the result with the solution in the next frame

$$\boxed{I = 2}$$

(a) (1) c_1 is $y = 0$ $\quad \therefore \ dy = 0$

$$\therefore \ I_1 = \int_0^1 2x\,dx = \left[x^2\right]_0^1 = 1 \quad \therefore \ I_1 = 1$$

(2) c_2 is $x = 1$ $\quad \therefore \ dx = 0$

$$\therefore \ I_2 = \int_0^2 (3 - 2y)\,dy = \left[3y - y^2\right]_0^2 = 2 \quad \therefore \ I_2 = 2$$

(3) c_3 is $y = 2x$ $\quad \therefore \ dy = 2\,dx$

$$\therefore \ I_3 = \int_1^0 \{4x\,dx + (3x - 4x)2\,dx\}$$

$$= \int_1^0 2x\,dx = \left[x^2\right]_1^0 = -1 \qquad \therefore \ I_3 = -1$$

$$I = I_1 + I_2 + I_3 = 1 + 2 + (-1) = 2 \quad \therefore \ I = 2$$

Now we will do the same problem by applying Green's theorem, so move on

73

(b) *By Green's theorem*

$$I = \oint_c \{(2x + y)\,dx + (3x - 2y)\,dy\}$$

$$P = 2x + y \quad \therefore \ \frac{\partial P}{\partial y} = 1; \quad Q = 3x - 2y \quad \therefore \ \frac{\partial Q}{\partial x} = 3$$

$$I = -\int_R \int \left(\frac{\partial P}{\partial y} - \frac{\partial Q}{\partial x} \right) dx\,dy$$

Finish it off. $I = \ldots\ldots\ldots\ldots$

74

$$\boxed{I = 2}$$

Because

$$I = -\int_R \int (1 - 3)\,dx\,dy$$

$$= 2\int_R \int dx\,dy = 2A$$

$$= 2 \times \text{ the area of the triangle}$$

$$= 2 \times 1 = 2 \quad \therefore \ I = 2$$

Application of Green's theorem is not always the quickest method. It is useful, however, to have both methods available. If you have not already done so, make a note of Green's theorem.

$$\int_R \int \left(\frac{\partial P}{\partial y} - \frac{\partial Q}{\partial x} \right) dx\,dy = -\oint_c (P\,dx + Q\,dy)$$

75

Example 3

Evaluate the line integral $I = \oint_c \{xy\,dx + (2x - y)\,dy\}$ round the region bounded by the curves $y = x^2$ and $x = y^2$ by the use of Green's theorem.

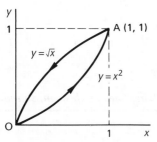

Points of intersection are O (0, 0) and A (1, 1). P and Q are known, so there is no difficulty.

Complete the working.

$$I = \ldots\ldots\ldots\ldots$$

$$\boxed{I = \frac{31}{60}}$$

76

Here is the working.

$$I = \oint_c \{xy\,dx + (2x - y)\,dy\}$$

$$\oint_c \{P\,dx + Q\,dy\} = -\int_R \int \left(\frac{\partial P}{\partial y} - \frac{\partial Q}{\partial x}\right) dx\,dy$$

$$P = xy \quad \therefore \quad \frac{\partial P}{\partial y} = x; \quad Q = 2x - y \quad \therefore \quad \frac{\partial Q}{\partial x} = 2$$

$$I = -\int_R \int (x - 2)\,dx\,dy$$

$$= -\int_0^1 \int_{y=x^2}^{y=\sqrt{x}} (x - 2)\,dy\,dx$$

$$= -\int_0^1 (x - 2)\Big[y\Big]_{x^2}^{\sqrt{x}} dx$$

$$\therefore \quad I = -\int_0^1 (x - 2)(\sqrt{x} - x^2)\,dx$$

$$= -\int_0^1 (x^{3/2} - x^3 - 2x^{1/2} + 2x^2)\,dx$$

$$= -\left[\frac{2}{5}x^{5/2} - \frac{1}{4}x^4 - \frac{4}{3}x^{3/2} + \frac{2}{3}x^3\right]_0^1 = \frac{31}{60}$$

Before we finally leave this section of the work, there is one more result to note.

In the special case when $P = y$ and $Q = -x$

$$\frac{\partial P}{\partial y} = 1 \quad \text{and} \quad \frac{\partial Q}{\partial x} = -1$$

Green's theorem then states

$$\int_R \int \{1 - (-1)\}\,dx\,dy = -\oint_c (P\,dx + Q\,dy)$$

i.e.
$$2\int_R \int dx\,dy = -\oint_c (y\,dx - x\,dy)$$

$$= \oint_c (x\,dy - y\,dx)$$

Therefore, the area of the closed region

$$A = \int_R \int dx\,dy = \frac{1}{2}\oint_c (x\,dy - y\,dx)$$

Note this result in your record book. Then let us see an example

77

Example 1

Determine the area of the figure enclosed by $y = 3x^2$ and $y = 6x$.

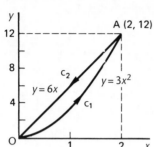

Points of intersection:

$$3x^2 = 6x \quad \therefore \quad x = 0 \text{ or } 2$$

Area $A = \frac{1}{2} \oint_c (x\,dy - y\,dx)$

We evaluate the integral in two parts, i.e.   OA along c_1

and   AO along c_2

$$2A = \int_{c_1 \text{ (along OA)}} (x\,dy - y\,dx) + \int_{c_2 \text{ (along AO)}} (x\,dy - y\,dx) = I_1 + I_2$$

I_1: c_1 is $y = 3x^2$ $\therefore$ $dy = 6x\,dx$

$$\therefore \quad I_1 = \int_0^2 (6x^2\,dx - 3x^2\,dx) = \int_0^2 3x^2\,dx = \left[x^3\right]_0^2 = 8$$

$$\therefore \quad I_1 = 8$$

Similarly, $I_2 = \ldots\ldots\ldots\ldots$

78

$$\boxed{I_2 = 0}$$

Because

c_2 is $y = 6x$ $\therefore$ $dy = 6\,dx$

$$\therefore \quad I_2 = \int_2^0 (6x\,dx - 6x\,dx) = 0 \quad \therefore \quad I_2 = 0$$

$$\therefore \quad I = I_1 + I_2 = 8 + 0 = 8 \quad \therefore \quad A = 4 \text{ square units}$$

Finally, here is one for you to do entirely on your own.

Example 2

Determine the area bounded by the curves $y = 2x^3$, $y = x^3 + 1$ and the axis $x = 0$ for $x \geq 0$.

Complete the working and see if you agree with the working in the next frame

Here it is. **79**

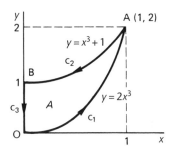

$$y = 2x^3; \quad y = x^3 + 1; \quad x = 0$$

Point of intersection

$$2x^3 = x^3 + 1 \quad \therefore \ x^3 = 1 \quad \therefore \ x = 1$$

Area $\quad A = \dfrac{1}{2} \oint_c (x\,dy - y\,dx)$

$$\therefore \ 2A = \oint_c (x\,dy - y\,dx)$$

(a) OA: c_1 is $y = 2x^3 \quad \therefore \ dy = 6x^2\,dx$

$$\therefore \ I_1 = \int_{c_1} (x\,dy - y\,dx) = \int_0^1 (6x^3\,dx - 2x^3\,dx)$$

$$= \int_0^1 4x^3\,dx = \left[x^4 \right]_0^1 = 1 \qquad \therefore \ I_1 = 1$$

(b) AB: c_2 is $y = x^3 + 1 \quad \therefore \ dy = 3x^2\,dx$

$$\therefore \ I_2 = \int_1^0 \{3x^3\,dx - (x^3 + 1)\,dx\} = \int_1^0 (2x^3 - 1)\,dx$$

$$= \left[\frac{x^4}{2} - x \right]_1^0 = -(\tfrac{1}{2} - 1) = \tfrac{1}{2} \qquad \therefore \ I_2 = \tfrac{1}{2}$$

(c) BO: c_3 is $x = 0 \quad \therefore \ dx = 0$

$$I_3 = \int_{y=1}^{y=0} (x\,dy - y\,dx) = 0 \qquad\qquad \therefore \ I_3 = 0$$

$$\therefore \ 2A = I = I_1 + I_2 + I_3 = 1 + \tfrac{1}{2} + 0 = 1\tfrac{1}{2}$$
$$\therefore \ A = \tfrac{3}{4} \text{ square units}$$

And that brings this Programme to an end. We have covered some important topics, so check down the **Revision summary** and the **Can you?** checklist that follow and revise any part of the text if necessary, before working through the **Test exercise**. The **Further problems** provide an opportunity for additional practice.

Revision summary 19

80

1 *Differentials* d*y* and d*x*

(a)

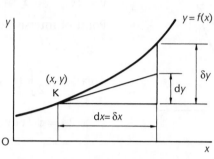

$$dy = f'(x)\,dx$$

(b) If $z = f(x, y)$, $dz = \dfrac{\partial z}{\partial x}\,dx + \dfrac{\partial z}{\partial y}\,dy$

If $z = f(x, y, w)$, $dz = \dfrac{\partial z}{\partial x}\,dx + \dfrac{\partial z}{\partial y}\,dy + \dfrac{\partial z}{\partial w}\,dw.$

(c) $dz = P\,dx + Q\,dy$, where P and Q are functions of x and y, is an *exact differential* if $\dfrac{\partial P}{\partial y} = \dfrac{\partial Q}{\partial x}.$

2 *Line integrals* – definition

$$I = \int_c f(x, y)\,ds = \int_c (P\,dx + Q\,dy)$$

3 *Properties of line integrals*

(a) Sign of line integral is reversed when the direction of integration along the path is reversed.

(b) Path of integration parallel to y-axis, $dx = 0$ $\therefore$ $I_c = \displaystyle\int_c Q\,dy.$

Path of integration parallel to x-axis, $dy = 0$ $\therefore$ $I_c = \displaystyle\int_c P\,dx.$

(c) The y-values on the path of integration must be continuous and single-valued.

4 *Line of integral round a closed curve* $\oint$

Positive direction $\oint$ anticlockwise

Negative direction $\oint$ clockwise, i.e. $\oint = -\oint.$

5 *Line integral related to arc length*

$$I = \int_{AB} F \, ds = \int_{AB} (P \, dx + Q \, dy)$$

$$= \int_{x_1}^{x_2} f(x, y) \sqrt{1 + \left(\frac{dy}{dx}\right)^2} \, dx$$

With parametric equations, x and y in terms of t,

$$I = \int_{c} f(x, y) \, ds = \int_{t_1}^{t_2} f(x, y) \sqrt{\left(\frac{dx}{dt}\right)^2 + \left(\frac{dy}{dt}\right)^2} \, dt$$

6 *Dependence of line integral on path of integration*

In general, the value of the line integral depends on the particular path of integration.

7 *Exact differential*

If $P \, dx + Q \, dy$ is an exact differential where P, Q and their first derivatives are finite and continuous inside the simply connected region R

(a) $\dfrac{\partial P}{\partial y} = \dfrac{\partial Q}{\partial x}$

(b) $I = \displaystyle\int_{c} (P \, dx + Q \, dy)$ is independent of the path of integration where c lies entirely within R

(c) $I = \displaystyle\oint_{c} (P \, dx + Q \, dy)$ is zero when c is a closed curve lying entirely within R.

8 *Exact differentials in three variables*

If $P \, dx + Q \, dy + R \, dz$ is an exact differential where P, Q, R and their first partial derivatives are finite and continuous inside a simply connected region containing path c

(a) $\dfrac{\partial P}{\partial y} = \dfrac{\partial Q}{\partial x}; \quad \dfrac{\partial P}{\partial z} = \dfrac{\partial R}{\partial x}; \quad \dfrac{\partial R}{\partial y} = \dfrac{\partial Q}{\partial z}$

(b) $\displaystyle\int_{c} (P \, dx + Q \, dy + R \, dz)$ is independent of the path of integration

(c) $\displaystyle\oint_{c} (P \, dx + Q \, dy + R \, dz)$ is zero when c is a closed curve.

9 *Green's theorem*

$$\oint_{c} (P \, dx + Q \, dy) = -\int_{R}\int \left\{ \frac{\partial P}{\partial y} - \frac{\partial Q}{\partial x} \right\} dx \, dy$$

and, for a simple closed curve

$$\oint_{c} (x \, dy - y \, dx) = 2 \int_{R}\int dx \, dy = 2A$$

where A is the area of the enclosed figure.

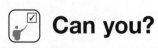 **Can you?**

81
Checklist 19

Check this list before and after you try the end of Programme test.

On a scale of 1 to 5 how confident are you that you can: **Frames**

- Evaluate double and triple integrals and apply them to the
  determination of the areas of plane figures and the volumes of
  solids?   `1` to `10`
  Yes   ☐   · ☐   ☐   ☐   ☐   *No*

- Understand the role of the differential of a function of two or
  more real variables?   `11` to `13`
  Yes   ☐   ☐   ☐   ☐   ☐   *No*

- Determine exact differentials in two real variables and their
  integrals?   `14` to `19`
  Yes   ☐   ☐   ☐   ☐   ☐   *No*

- Evaluate the area enclosed by a closed curve by contour
  integration?   `20` to `26`
  Yes   ☐   ☐   ☐   ☐   ☐   *No*

- Evaluate line integrals and appreciate their properties?   `27` to `41`
  Yes   ☐   ☐   ☐   ☐   ☐   *No*

- Evaluate line integrals around closed curves within a simply
  connected region?   `42` to `50`
  Yes   ☐   ☐   ☐   ☐   ☐   *No*

- Link line integrals to integrals along the *x*-axis?   `51` to `53`
  Yes   ☐   ☐   ☐   ☐   ☐   *No*

- Link line integrals to integrals along a contour given in
  parametric form?   `54` to `56`
  Yes   ☐   ☐   ☐   ☐   ☐   *No*

- Discuss the dependence of a line integral between two points
  on the path of integration?   `56` to `#65`
  Yes   ☐   ☐   ☐   ☐   ☐   *No*

- Determine exact differentials in three real variables and their
  integrals?   `66` to `69`
  Yes   ☐   ☐   ☐   ☐   ☐   *No*

- Demonstrate the validity and use of Green's theorem?   `70` to `79`
  Yes   ☐   ☐   ☐   ☐   ☐   *No*

Test exercise 19

1. Determine the differential dz of each of the following. **82**
   (a) $z = x^4 \cos 3y$; (b) $z = e^{2y} \sin 4x$; (c) $z = x^2 y w^3$.

2. Determine which of the following are exact differentials and integrate where appropriate to determine z.
   (a) $dz = (3x^2 y^4 + 8x)\, dx + (4x^3 y^3 - 15y^2)\, dy$
   (b) $dz = (2x \cos 4y - 6 \sin 3x)dx - 4(x^2 \sin 4y - 2y)\, dy$
   (c) $dz = 3e^{3x}(1 - y)\, dx + (e^{3x} + 3y^2)\, dy$.

3. Calculate the area of the triangle with vertices at O $(0, 0)$, A $(4, 2)$ and B $(1, 5)$.

4. Evaluate the following.
   (a) $I = \int_c \{(x^2 - 3y)\, dx + xy^2\, dy\}$ from A $(1, 2)$ to B $(2, 8)$ along the curve $y = 2x^2$.
   (b) $I = \int_c (2x + y)\, dx$ from A $(0, 1)$ to B $(0, -1)$ along the semicircle $x^2 + y^2 = 1$ for $x \geq 0$.
   (c) $I = \oint_c \{(1 + xy)\, dx + (1 + x^2)\, dy\}$ where c is the boundary of the rectangle joining A $(1, 0)$, B $(4, 0)$, C $(4, 3)$ and D $(1, 3)$.
   (d) $I = \int_c 2xy\, ds$ where c is defined by the parametric equations $x = 4 \cos \theta$, $y = 4 \sin \theta$ between $\theta = 0$ and $\theta = \dfrac{\pi}{3}$.
   (e) $I = \int_c \{(8xy + y^3)\, dx + (4x^2 + 3xy^2)\, dy\}$ from A$(1, 3)$ to B$(2, 1)$.
   (f) $I = \oint_c \{(3x + y)\, dx + (y - 2x)\, dy\}$ round the boundary of the ellipse $x^2 + 4y^2 = 36$.

5. Apply Green's theorem to determine the area of the plane figure bounded by the curves $y = x^3$ and $y = \sqrt{x}$.

6. Verify that $dw = (2xyz + 2z - y^2)dx + (x^2 z - 2yx)dy + (x^2 y + 2x)dz$ is an exact differential and find the value of
   $$\int_c dw \quad \text{where}$$
   (a) c is the straight line joining $(0, 0, 0)$ to $(1, 1, 1)$
   (b) c is the curve of intersection of the unit sphere centred on the origin and the plane $x + y + z = 1$.

Further problems 19

83

1   Show that $I = \int_c \{xy^2w^2\,dx + x^2yw^2\,dy + x^2y^2w\,dw\}$ is independent of the path of integration c and evaluate the integral from A (1, 3, 2) to B (2, 4, 1).

2   Determine whether $dz = 3x^2(x^2 + y^2)\,dx + 2y(x^3 + y^4)\,dy$ is an exact differential. If so, determine z and hence evaluate $\int_c dz$ from A (1, 2) to B (2, 1).

3   Evaluate the line integral $I = \oint_c \left\{ \dfrac{x\,dy - y\,dx}{x^2 + y^2 + 4} \right\}$ where c is the boundary of the segment formed by the arc of the circle $x^2 + y^2 = 4$ and the chord $y = 2 - x$ for $x \geq 0$.

4   Show that
$$I = \int_c \{(3x^2 \sin y + 2 \sin 2x + y^3)\,dx + (x^3 \cos y + 3xy^2)\,dy\}$$
is independent of the path of integration and evaluate it from A $(0, 0)$
to B $\left(\dfrac{\pi}{2}, \pi\right)$.

5   Evaluate the integral $I = \int_c xy\,ds$ where c is defined by the parametric equations
$x = \cos^3 t$, $y = \sin^3 t$ from $t = 0$ to $t = \dfrac{\pi}{2}$.

6   Verify that $dz = \dfrac{x\,dx}{x^2 - y^2} - \dfrac{y\,dy}{x^2 - y^2}$ for $x^2 > y^2$ is an exact differential and evaluate
$z = f(x, y)$ from A (3, 1) to B (5, 3).

7   The parametric equations of a circle, centre (1, 0) and radius 1, can be expressed as $x = 2 \cos^2 \theta$, $y = 2 \cos \theta \sin \theta$.
Evaluate $I = \int_c \{(x + y)\,dx + x^2\,dy\}$ along the semicircle for which $y \geq 0$ from O (0, 0) to A (2, 0).

8   Evaluate $\oint_c \{x^3y^2\,dx + x^2y\,dy\}$ where c is the boundary of the region enclosed by the curve $y = 1 - x^2$, $x = 0$ and $y = 0$ in the first quadrant.

9   Use Green's theorem to evaluate
$$I = \oint_c \{(4x + y)\,dx + (3x - 2y)\,dy\}$$
where c is the boundary of the trapezium with vertices A (0, 1), B (5, 1), C (3, 3) and D (1, 3).

10   Evaluate $I = \int_c \{(3x^2y^2 + 2 \cos 2x - 2xy)\,dx + (2x^3y + 8y - x^2)\,dy\}$
(a) along the curve $y = x^2 - x$ from A (0, 0) to B (2, 2)
(b) round the boundary of the quadrilateral joining the points (1, 0), (3, 1), (2, 3) and (0, 3)

11   Verify that $dw = \dfrac{y}{z}\,dx + \dfrac{x}{z}\,dy - \dfrac{xy}{z^2}\,dz$ is an exact differential and find the value
of $\int_c dw$ where c is the straight line joining (0, 0, 1) to (1, 2, 3) for either region
$z > 0$ or $z < 0$.

Multiple integration 2

Learning outcomes

When you have completed this Programme you will be able to:

- Evaluate double integrals and surface integrals
- Relate three-dimensional Cartesian coordinates to cylindrical and spherical polar forms
- Evaluate volume integrals in Cartesian coordinates and in cylindrical and spherical polar coordinates
- Use the Jacobian to convert integrals given in Cartesian coordinates into general curvilinear coordinates in two and three dimensions

Double integrals

1   Let us start off with an example with which we are already familiar.

Example 1

A solid is enclosed by the planes $z = 0$, $y = 1$, $y = 2$, $x = 0$, $x = 3$ and the surface $z = x + y^2$. We have to determine the volume of the solid so formed.

First take some care in sketching the figure, which is

. . . .

2

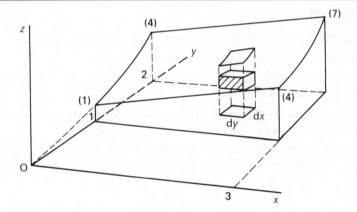

In the plane $y = 1$, $z = x + 1$, i.e. a straight line joining $(0, 1, 1)$ and $(3, 1, 4)$
In the plane $y = 2$, $z = x + 4$, i.e. a straight line joining $(0, 2, 4)$ and $(3, 2, 7)$
In the plane $x = 0$, $z = y^2$, i.e. a parabola joining $(0, 1, 1)$ and $(0, 2, 4)$
In the plane $x = 3$, $z = 3 + y^2$, i.e. a parabola joining $(3, 1, 4)$ and $(3, 2, 7)$.

Consideration like this helps us to visualise the problem and the time involved is well spent.
    Now we can proceed.

The element of volume $\delta v = \delta x\, \delta y\, \delta z$

Then the total volume $V = \displaystyle\iiint dx\, dy\, dz$ between appropriate limits in each case.

We could also have said that the element of area on the $z = 0$ plane

$$\delta a = \delta y \, \delta x$$

and that the volume of the column $\delta v_c = z \, \delta a = z \, \delta x \, \delta y$

Then, since $z = x + y^2$, this becomes $\delta v_c = (x + y^2) \, \delta x \, \delta y$

Summing in the usual way then gives

$$V = \int z \, da$$

$$= \int_R \int (x + y^2) \, dx \, dy$$

where R is the region bounded in the x–y plane.

Now we insert the appropriate limits and complete the integration

$$V = \dots\dots\dots$$

$$\boxed{V = 11\cdot5 \text{ cubic units}}$$

3

Because

$$V = \int_{y=1}^{y=2} \int_{x=0}^{x=3} (x + y^2) \, dx \, dy$$

$$= \int_1^2 \left[\frac{x^2}{2} + xy^2 \right]_{x=0}^{x=3} dy$$

$$= \int_1^2 \left(\frac{9}{2} + 3y^2 \right) dy$$

$$= \left[\frac{9}{2}y + y^3 \right]_1^2$$

$$= 11\cdot5$$

$$\therefore \; V = 11\cdot5 \text{ cubic units}$$

Although we have found a volume, this is, in fact, an example of a *double integral* since the expression for z was a function of position in the x–y plane within the closed region

i.e. $I = \int_R \int f(x, y) \, da$

$$= \int_R \int f(x, y) \, dy \, dx$$

In this particular case, R is the region in the x–y plane bounded by $x = 0$, $x = 3$, $y = 1$, $y = 2$.

▶

Example 2

A triangular thin plate has the dimensions shown and a variable density ρ where $\rho = 1 + x + xy$.

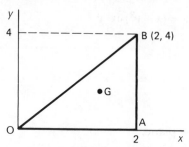

We have to determine

(a) the mass of the plate

(b) the position of its centre of gravity G.

(a) Consider an element of area at the point P(x, y) in the plate

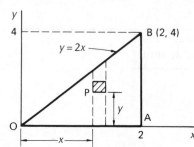

$$\delta a = \delta x\, \delta y$$

The mass δm of the element is then

$$\delta m = \rho\, \delta x\, \delta y$$

$$\therefore \text{ Total mass } M = \int_R \int dm = \int_R \int \rho\, dx\, dy$$

Now we insert the limits and complete the integration, remembering that $\rho = (1 + x + xy)$

$$M = \ldots\ldots\ldots$$

4

$$\boxed{M = 17\frac{1}{3}}$$

Because we have

$$M = \int_R \int \rho\, dx\, dy = \int_{x=0}^{x=2} \int_{y=0}^{y=2x} (1 + x + xy)\, dy\, dx$$

$$= \int_0^2 \left[y + xy + \frac{xy^2}{2} \right]_{y=0}^{y=2x} dx$$

$$= \int_0^2 \{2x + 2x^2 + 2x^3\}\, dx$$

$$= \left[x^2 + \frac{2x^3}{3} + \frac{x^4}{2} \right]_0^2 = 17\frac{1}{3}$$

(b) To find the position of the centre of gravity, we need to know

$$\ldots\ldots\ldots$$

5

the sum of the moments of mass about OY and OX

(1) To find $\bar{x}$, we take moments about OY.

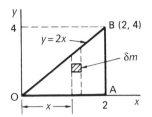

Moment of mass of element about OY

$$= x\,\delta m$$

$$= x(1 + x + xy)\,\delta x\,\delta y$$

∴ Sum of first moments $= \displaystyle\int_R \int (x + x^2 + x^2 y)\,\mathrm{d}x\,\mathrm{d}y$

$$= \ldots\ldots\ldots$$

6

$$26\frac{2}{15}$$

Because sum of first moments $= \displaystyle\int_{x=0}^{x=2} \int_{y=0}^{y=2x} (x + x^2 + x^2 y)\,\mathrm{d}y\,\mathrm{d}x$

$$= \int_0^2 \left[xy + x^2 y + \frac{x^2 y^2}{2} \right]_{y=0}^{y=2x} \mathrm{d}x$$

$$= \int_0^2 \{2x^2 + 2x^3 + 2x^4\}\mathrm{d}x$$

$$= 2\int_0^2 (x^2 + x^3 + x^4)\,\mathrm{d}x$$

$$= 2\left[\frac{x^3}{3} + \frac{x^4}{4} + \frac{x^5}{5} \right]_0^2 = 26\frac{2}{15}$$

Now $M\bar{x} = $ sum of moments ∴ $\bar{x} = \ldots\ldots\ldots$

7

$$\bar{x} = 1{\cdot}508$$

We found previously that $M = 17\dfrac{1}{3}$ ∴ $\left(17\dfrac{1}{3}\right)\bar{x} = 26\dfrac{2}{15}$

which gives $\bar{x} = 1\dfrac{33}{65} = 1{\cdot}508$

(2) To find $\bar{y}$ we proceed in just the same way, this time taking moments about OX. Work right through it on your own.

$$\bar{y} = \ldots\ldots\ldots$$

8

$$\bar{y} = 1{\cdot}754$$

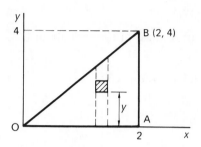

Moment of element of mass δm about OX

$$= y\,\delta m = y(1 + x + xy)\,\delta x\,\delta y$$

$\therefore$ Sum of first moments about OX $= \displaystyle\int_R\!\!\int (y + xy + xy^2)\,\mathrm{d}x\,\mathrm{d}y$

$$= \int_{x=0}^{x=2}\int_{y=0}^{y=2x} (y + xy + xy^2)\,\mathrm{d}y\,\mathrm{d}x$$

$$= \int_0^2 \left[\frac{y^2}{2} + \frac{xy^2}{2} + \frac{xy^3}{3}\right]_{y=0}^{y=2x}\,\mathrm{d}x$$

$$= \int_0^2 \left\{2x^2 + 2x^3 + \frac{8x^4}{3}\right\}\,\mathrm{d}x$$

$$= \left[\frac{2x^3}{3} + \frac{x^4}{2} + \frac{8x^5}{15}\right]_0^2$$

$$= 30\frac{2}{5}$$

$\therefore\; M\bar{y} = 30\dfrac{2}{5}\quad \therefore\; \bar{y} = 30\dfrac{2}{5}\Big/17\dfrac{1}{3} = 1{\cdot}754$

So we finally have:

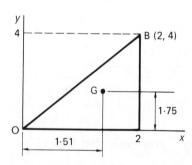

Note that this again referred to a plane figure in the x–y plane.

Now let us move on to something slightly different

Surface integrals

9

When the area over which we integrate is not restricted to the x–y plane, matters become rather more involved, but also more interesting.

If S is a two-sided surface in space and R is its projection on the x–y plane, then the equation of S is of the form $z = f(x, y)$ where f is a single-valued function and continuous throughout R.

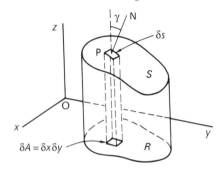

Let δA denote an element of R and δS the corresponding element of area of S at the point $P(x, y, z)$ in S.

Let also $\phi(x, y, z)$ be a function of position on S (e.g. potential) and let γ denote the angle between the outward normal PN to the surface at P and the positive z-axis.

Then $\delta A \approx \delta S \cos \gamma$ i.e. $\delta S \approx \dfrac{\delta A}{\cos \gamma} = \delta A \sec \gamma$ and

$\sum \phi(x, y, z) \delta S$ is the total value of $\phi(x, y, z)$ taken over the surface S.

As $\delta S \to 0$, this sum becomes the integral

$$I = \int_S \phi(x,\ y,\ z)\, \mathrm{d}S$$

and, since $\delta S \approx \delta A \sec \gamma$, the result can be written

$$I = \int_R \int \phi(x,\ y,\ z) \sec \gamma\, \mathrm{d}x\, \mathrm{d}y \qquad \left(\gamma < \frac{\pi}{2}\right)$$

Notice that $\cos \gamma = \hat{\mathbf{n}} \cdot \mathbf{k}$, where $\mathbf{k}$ is the unit vector in the z-direction and $\hat{\mathbf{n}}$ is the unit normal to the surface at P.

With limits inserted for x and y, the integral seems straightforward, except for the factor $\sec \gamma$, which naturally varies over the surface S.

We can, in fact, show that $\sec \gamma = \sqrt{1 + \left(\dfrac{\partial z}{\partial x}\right)^2 + \left(\dfrac{\partial z}{\partial y}\right)^2}$

(see Appendix, page 1065)

Therefore, the *surface integral* of $\phi(x,\ y,\ z)$ over the surface S is given by

$$\text{(a)} \quad I = \int_S \phi(x,\ y,\ z)\, \mathrm{d}S \tag{1}$$

$$\text{or} \quad \text{(b)} \quad I = \int_R \int \phi(x,\ y,\ z) \sqrt{1 + \left(\frac{\partial z}{\partial x}\right)^2 + \left(\frac{\partial z}{\partial y}\right)^2}\, \mathrm{d}x\, \mathrm{d}y \tag{2}$$

where $z = f(x, y)$

▶

Note that, when $\phi(x, y, z) = 1$, then $I = \displaystyle\int_S dS$ gives the area of the surface S.

$$\therefore\ S = \int_S dS = \int_R\!\!\int \sqrt{1 + \left(\frac{\partial z}{\partial x}\right)^2 + \left(\frac{\partial z}{\partial y}\right)^2}\, dx\, dy \qquad (3)$$

Make a note of these three important results.

Then we will apply them to a few examples.

10

Example 1

Find the area of the surface $z = \sqrt{x^2 + y^2}$ over the region bounded by $x^2 + y^2 = 1$.

$$S = \int_R\!\!\int \sqrt{1 + \left(\frac{\partial z}{\partial x}\right)^2 + \left(\frac{\partial z}{\partial y}\right)^2}\, dx\, dy$$

So we now find $\dfrac{\partial z}{\partial x}$ and $\dfrac{\partial z}{\partial y}$ and determine $\sqrt{1 + \left(\dfrac{\partial z}{\partial x}\right)^2 + \left(\dfrac{\partial z}{\partial y}\right)^2}$

which is ....

11

$$\boxed{\sqrt{2}}$$

Because

$$z = (x^2 + y^2)^{1/2} \quad \therefore\ \frac{\partial z}{\partial x} = \frac{1}{2}(x^2 + y^2)^{-1/2}2x = \frac{x}{\sqrt{x^2 + y^2}}$$

$$\frac{\partial z}{\partial y} = \frac{1}{2}(x^2 + y^2)^{-1/2}2y = \frac{y}{\sqrt{x^2 + y^2}}$$

$$\therefore\ 1 + \left(\frac{\partial z}{\partial x}\right)^2 + \left(\frac{\partial z}{\partial y}\right)^2 = 1 + \frac{x^2 + y^2}{x^2 + y^2} = 2$$

$$\therefore\ \sqrt{1 + \left(\frac{\partial z}{\partial x}\right)^2 + \left(\frac{\partial z}{\partial y}\right)^2} = \sqrt{2}$$

$$\therefore\ S = \sqrt{2}\int_R\!\!\int dx\, dy = \sqrt{2} \times \ldots\ldots\ldots\ldots$$

| the area of the region R | **12** |

But R is bounded by $x^2 + y^2 = 1$, i.e. a circle, centre the origin and radius 1.

$\therefore$ area $= \pi$

$$\therefore \ S = \sqrt{2} \int_R \int dx\, dy = \sqrt{2}\pi$$

Example 2

Find the area of the surface S of the paraboloid $z = x^2 + y^2$ cut off by the cone $z = 2\sqrt{x^2 + y^2}$.

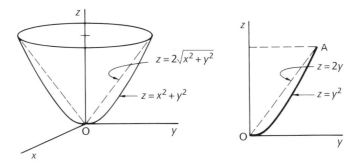

We can find the point of intersection A by considering the y–z plane, i.e. put $x = 0$.

Coordinates of A are ....

| A (2, 4) | **13** |

The projection of the surface S on the x–y plane is

....

| the circle $x^2 + y^2 = 4$ | **14** |

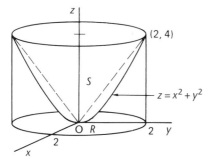

$$S = \int_R \int \sqrt{1 + \left(\frac{\partial z}{\partial x}\right)^2 + \left(\frac{\partial z}{\partial y}\right)^2}\, dx\, dy$$

For this we use the equation of the surface S. The information from the projection R on the x–y plane will later provide the limits of the two stages of integration.

For the time being, then, $S = $ ....

15

$$S = \int_R \int \sqrt{1 + 4x^2 + 4y^2} \, dx \, dy$$

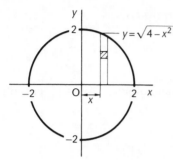

Using Cartesian coordinates, we could integrate with respect to y from $y = 0$ to $y = \sqrt{4 - x^2}$ and then with respect to x from $x = 0$ to $x = 2$. Finally, we should multiply by four to cover all four quadrants.

i.e. $S = 4 \int_{x=0}^{x=2} \int_{y=0}^{y=\sqrt{4-x^2}} \sqrt{1 + 4x^2 + 4y^2} \, dy \, dx$

But how do we carry out the actual integration?

It becomes a lot easier if we use polar coordinates.

The same integral in polar coordinates is . . . .

16

$$S = \int_{\theta=0}^{\theta=2\pi} \int_{r=0}^{r=2} \sqrt{1 + 4r^2} \, r \, dr \, d\theta$$

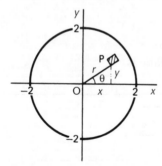

$x = r\cos\theta;$ $y = r\sin\theta$

$x^2 + y^2 = r^2$ $dx \, dy = r \, dr \, d\theta$

(refer to Frame 67)

$$S = \int_{\theta=0}^{\theta=2\pi} \int_{r=0}^{r=2} \sqrt{1 + 4r^2} \, r \, dr \, d\theta$$

$\therefore \ S = \ . . . . $

Finish it off.

$$\boxed{S = 36 \cdot 18 \text{ square units}}$$

17

Because

$$S = \int_{\theta=0}^{\theta=2\pi} \int_{r=0}^{r=2} (1 + 4r^2)^{1/2} r \, dr \, d\theta = \int_0^{2\pi} \left[\frac{1}{12}(1 + 4r^2)^{3/2}\right]_0^2 d\theta$$

$$= \frac{1\!\!Đ}{12} \int_0^{2\pi} \{17^{3/2} - 1\} \, d\theta = 5 \cdot 7577 \Big[\theta\Big]_0^{2\pi} = 36 \cdot 18$$

Now on to Example 3.

Example 3

18

To determine the moment of inertia of a thin spherical shell of radius a about a diameter as axis. The mass per unit area of shell is ρ.

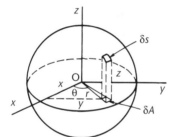

Equation of sphere

$$x^2 + y^2 + z^2 = a^2$$

Mass of element $= m = \rho \, \delta S$

$$I \approx \Sigma mr^2 \approx \Sigma \rho \, \delta S r^2$$

Let us deal with the upper hemisphere

$$\therefore \ I_H = \int_S \rho r^2 \, dS$$

$$= \int_R \int \rho r^2 \sqrt{1 + \left(\frac{\partial z}{\partial x}\right)^2 + \left(\frac{\partial z}{\partial y}\right)^2} \, dx \, dy$$

Now determine the partial derivatives and simplify the integral as far as possible in Cartesian coordinates.

$$I_H = \ldots\ldots\ldots$$

19

$$\boxed{I_H = \int_R \int \rho r^2 \frac{a}{\sqrt{a^2 - x^2 - y^2}} \, dx \, dy}$$

In this particular example, R is, of course, the region bounded by the circle $x^2 + y^2 = a^2$ in the x–y plane.

Converting to polar coordinates

$$x = r\cos\theta; \quad y = r\sin\theta; \quad dx \, dy = r \, dr \, d\theta$$

the integral becomes $I_H = \ldots\ldots\ldots$

20

$$I_H = \rho a \int_{\theta=0}^{\theta=2\pi} \int_{r=0}^{r=a} \frac{r^3}{\sqrt{a^2-r^2}} \, dr \, d\theta$$

Because for $x^2 + y^2 = r^2$:   limits of r: $r = 0$ to $r = a$

limits of θ: $\theta = 0$ to $\theta = 2\pi$

$$I_H = \int_R \int \rho r^2 \frac{a}{\sqrt{a^2-r^2}} r \, dr \, d\theta$$

$$= \rho a \int_{\theta=0}^{\theta=2\pi} \int_{r=0}^{r=a} \frac{r^3}{\sqrt{a^2-r^2}} \, dr \, d\theta$$

First we have to evaluate

$$I_r = \int_0^a \frac{r^3}{\sqrt{a^2-r^2}} \, dr$$

If we substitute $u = a^2 - r^2$ then the integral is evaluated as

$$I_r = \ldots\ldots\ldots\ldots$$

21

$$I_r = \frac{2a^3}{3}$$

Because

When $u = a^2 - r^2$ then $du = -2r \, dr$ so that $r^2 = a^2 - u$ and $r \, dr = -\dfrac{du}{2}$. Therefore

$$I_r = \int_0^a \frac{r^3}{\sqrt{a^2-r^2}} \, dr = \int_{r=0}^a \frac{r^2}{\sqrt{a^2-r^2}} r \, dr$$

$$= -\int_{u=a^2}^0 \frac{a^2-u}{\sqrt{u}} \frac{du}{2}$$

$$= -\frac{a^2}{2} \int_{u=a^2}^0 u^{-1/2} \, du + \frac{1}{2} \int_{u=a^2}^0 u^{1/2} \, du$$

$$= -\frac{a^2}{2} \left[2u^{1/2} \right]_{u=a^2}^0 + \frac{1}{2} \left[\frac{2}{3} u^{3/2} \right]_{u=a^2}^0$$

$$= a^3 - \frac{a^3}{3}$$

$$= \frac{2a^3}{3}$$

Now, to complete I_H we have

$$I_H = \rho a \int_0^{2\pi} \frac{2a^3}{3} \, d\theta$$

$$= \ldots\ldots\ldots\ldots$$

22

$$I_H = \frac{4\pi\rho a^4}{3}$$

Because

$$I_H = \rho a \int_0^{2\pi} \frac{2a^3}{3}\,\mathrm{d}\theta = \frac{2a^4\rho}{3}\Big[\theta\Big]_0^{2\pi} = \frac{4\pi a^4\rho}{3}$$

Therefore, the moment of inertia for the complete spherical shell is

$$I_s = \frac{8\pi a^4\rho}{3}$$

The total mass of the shell $M = 4\pi a^2\rho$ $\therefore\ I = \frac{2Ma^2}{3}$

Now let us turn our attention towards *volume integrals* and in preparation review systems of space coordinates.

Space coordinate systems

23

1   *Cartesian coordinates* (x, y, z) – referred to three coordinate axes OX, OY, OZ at right angles to each other. These are arranged in a *right-handed* manner, i.e. turning from OX to OY gives a right-handed screw action in the positive direction of OZ.

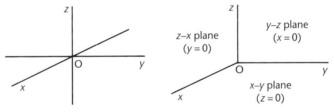

The three coordinate planes, $x = 0$, $y = 0$, $z = 0$, divide the space into eight sections called *octants*. The section containing $x \geq 0$, $y \geq 0$, $z \geq 0$ is called the *first octant*.

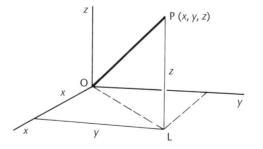

For a point P (x, y, z)

$$OL^2 = x^2 + y^2$$
$$OP^2 = x^2 + y^2 + z^2$$

Note that this is Pythagoras' theorem in three dimensions.

We are all familiar with this system of coordinates.

24

2 *Cylindrical coordinates* (r, θ, z) are useful where an axis of symmetry occurs.

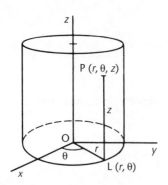

Any point P is considered as having a position on a cylinder. If L is the projection of P on the *x–y* plane, then (r, θ) are the usual polar coordinates of L. The cylindrical coordinates of P then merely require the addition of the *z*-coordinate.

$$r \geq 0$$

Relationship between Cartesian and cylindrical coordinates

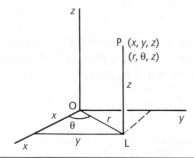

If we consider a combined figure, we can easily relate the two systems.

Expressing each of the following in terms of the alternative system,

$x = \dots\dots\dots$ $r = \dots\dots\dots$

$y = \dots\dots\dots$ $\theta = \dots\dots\dots$

$z = \dots\dots\dots$ $z = \dots\dots\dots$

25

$$\begin{array}{ll} x = r\cos\theta & r = \sqrt{x^2 + y^2} \\ y = r\sin\theta & \theta = \arctan(y/x) \\ z = z & z = z \end{array}$$

So, in cylindrical coordinates, the surface defined by

(1) $r = 5$ is $\dots\dots\dots$

(2) $\theta = \pi/6$ is $\dots\dots\dots$

(3) $z = 4$ is $\dots\dots\dots$

> **26**
>
> (1) $r = 5$ is a right cylinder, radius 5, with OZ as axis.
>
> (2) $\theta = \pi/6$ is a plane through OZ, making an angle $\pi/6$ with OX.
>
> (3) $z = 4$ is a plane parallel to the x–y plane cutting OZ at 4 units above the origin.

So position P (2, 3, 4) in Cartesian coordinates

$$= \ldots\ldots\ldots\ldots \text{ in cylindrical coordinates}$$

and position Q (2·5, $\pi/3$, 6) in cylindrical coordinates

$$= \ldots\ldots\ldots\ldots \text{ in Cartesian coordinates.}$$

> **27**
>
> P (2, 3, 4) = ($\sqrt{13}$, 0·983, 4) in cylindrical coordinates
>
> Q (2·5, $\pi/3$, 6) = (1·25, 2·165, 6) in Cartesian coordinates.

3   *Spherical coordinates* (r, θ, ϕ) are appropriate where a centre of symmetry occurs. The position of a point is considered as being a point on a sphere.

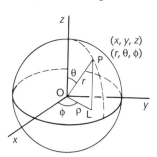

r is the distance of P from the origin and is always taken as positive.

L is the projection of P on the x–y plane

θ is the angle between OP and the positive OZ axis

ϕ is the angle between OL and the OX axis.

Note that (a) ϕ may be regarded as the longitude of P from OX

(b) θ may be regarded as the complement of the latitude of P.

Relationship between Cartesian and spherical coordinates

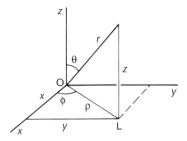

The combined figure shows the connection between the two systems, so

$x = \ldots\ldots\ldots\ldots$ $r = \ldots\ldots\ldots\ldots$

$y = \ldots\ldots\ldots\ldots$ $\theta = \ldots\ldots\ldots\ldots$

$z = \ldots\ldots\ldots\ldots$ $\phi = \ldots\ldots\ldots\ldots$

28

$$x = r \sin\theta \cos\phi \qquad r = \sqrt{x^2 + y^2 + z^2}$$
$$y = r \sin\theta \sin\phi \qquad \theta = \arccos(z/r)$$
$$z = r \cos\theta \qquad\qquad \phi = \arctan(y/x)$$

For the spherical coordinates of any point in space

$$r \geq 0; \quad 0 \leq \theta \leq \pi; \quad 0 \leq \phi \leq 2\pi$$

So, converting Cartesian coordinates (2, 3, 4) to spherical coordinates gives

. . . .

29

$$P\,(r,\,\theta,\,\phi) = (5\cdot385,\,0\cdot734,\,0\cdot983)$$

Because

$$x = 2, y = 3, z = 4$$
$$\therefore\ r = \sqrt{x^2 + y^2 + z^2} = \sqrt{4 + 9 + 16} = \sqrt{29} = 5\cdot385$$
$$\theta = \arccos(z/r) = \arccos(4/\sqrt{29}) = 0\cdot734$$
$$\phi = \arctan(y/x) = \arctan 1\cdot5 = 0\cdot983$$

And, in reverse, spherical coordinates (5, $\pi/4$, $\pi/3$) transform into Cartesian coordinates . . . .

30

$$P\,(x,\,y,\,z) = (1\cdot768,\,3\cdot061,\,3\cdot536)$$

Because

$$x = r \sin\theta \cos\phi = 5 \sin\frac{\pi}{4} \cos\frac{\pi}{3} = 5(0\cdot707)(0\cdot5) \quad = 1\cdot768$$
$$y = r \sin\theta \sin\phi = 5 \sin\frac{\pi}{4} \sin\frac{\pi}{3} = 5(0\cdot707)(0\cdot866) = 3\cdot061$$
$$z = r \cos\theta \qquad\quad = 5 \cos\frac{\pi}{4} \qquad\quad = 5(0\cdot707) \qquad\quad = 3\cdot536.$$

One of the main uses of cylindrical and spherical coordinates occurs in integrals dealing with volumes of solids. In preparation for this, let us consider the next important section of the work.

So move on

Element of volume in space in the three coordinate systems   31

1 Cartesian coordinates

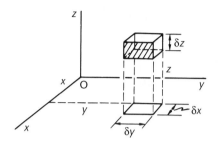

We have already used this many times.

$$\delta v = \delta x\, \delta y\, \delta z$$

2 Cylindrical coordinates

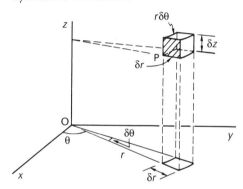

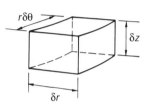

$$\delta v = r\delta\theta\, \delta r\, \delta z$$
$$\therefore\ \delta v = r\, \delta r\, \delta\theta\, \delta z$$

3 Spherical coordinates

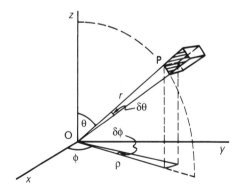

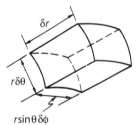

$$\delta v = \delta r\, r\delta\theta\, r\sin\theta\, \delta\phi$$
$$\therefore\ \delta v = r^2 \sin\theta\, \delta r\, \delta\theta\, \delta\phi$$

It is important to make a note of these results, since they are required when we change the variables in various types of integrals. We shall meet them again before long, so be sure of them now.

Volume integrals

32

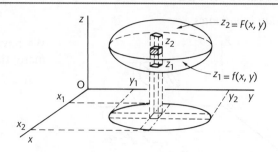

A solid is enclosed by a lower surface $z_1 = f(x, y)$ and an upper surface $z_2 = F(x, y)$.

Then, in general, using Cartesian coordinates, the element of volume is $\delta v = \delta x \, \delta y \, \delta z$.

The approximate value of the total volume V is then found

(a) by summing δv from $z = z_1$ to $z = z_2$ to obtain the volume of the column

(b) by summing all such columns from $y = y_1$ to $y = y_2$ to obtain the volume of the slice

(c) by summing all such slices from $x = x_1$ to $x = x_2$ to obtain the total volume V.

Then, when $\delta x \to 0$, $\delta y \to 0$, $\delta z \to 0$, the summation becomes an integral

$$V = \int_{x=x_1}^{x=x_2} \int_{y=y_1}^{y=y_2} \int_{z=z_1}^{z=z_2} \mathrm{d}z \, \mathrm{d}y \, \mathrm{d}x$$

Example 1

Find the volume of the solid bounded by the planes $z = 0$, $x = 0$, $y = 0$, $x^2 + y^2 = 4$ and $z = 6 - xy$ for $x \geq 0$, $y \geq 0$, $z \geq 0$.

First sketch the figure, so that we can see what we are doing. Take your time over it.

33

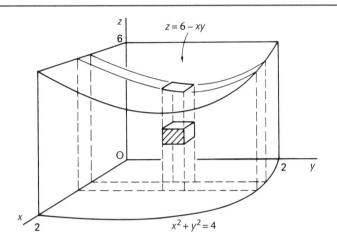

$\delta v = \delta x\, \delta y\, \delta z$

Volume of column $\approx \displaystyle\sum_{z=0}^{z=6-xy} \delta x\, \delta y\, \delta z$

Volume of slice $\approx \displaystyle\sum_{y=0}^{\sqrt{4-x^2}} \left\{ \sum_{z=0}^{6-xy} \delta x\, \delta y\, \delta z \right\}$

Total volume $\approx \displaystyle\sum_{x=0}^{2} \sum_{y=0}^{\sqrt{4-x^2}} \sum_{z=0}^{6-xy} \delta x\, \delta y\, \delta z$

If $\delta x \to 0$, $\delta y \to 0$, $\delta z \to 0$, then

$$V = \int_0^2 \int_0^{\sqrt{4-x^2}} \int_0^{6-xy} \mathrm{d}z\, \mathrm{d}y\, \mathrm{d}x$$

Starting with the innermost integral

$$\int_0^{6-xy} \mathrm{d}z = \left[z \right]_0^{6-xy}$$
$$= 6 - xy$$

Then $\displaystyle\int_0^{\sqrt{4-x^2}} (6 - xy)\, \mathrm{d}y = \ldots\ldots\ldots\ldots$

34

$$6\sqrt{4-x^2} - \frac{x}{2}(4-x^2)$$

Because

$$\int_0^{\sqrt{4-x^2}} (6-xy)\,dy = \left[6y - \frac{xy^2}{2}\right]_{y=0}^{y=\sqrt{4-x^2}}$$

$$= 6\sqrt{4-x^2} - \frac{x}{2}(4-x^2)$$

Then finally $V = \int_0^2 \left\{6(4-x^2)^{1/2} - 2x + \frac{x^3}{2}\right\} dx$

Now we are faced with $\int (4-x^2)^{1/2}\,dx$. You may remember that this is a

standard form $\int \sqrt{a^2 - x^2}\,dx = \frac{1}{2}\left\{x\sqrt{a^2-x^2} + a^2 \arcsin\frac{x}{a}\right\}$.

If not, to evaluate $\int_0^2 \sqrt{4-x^2}\,dx$, put $x = 2\sin\theta$ and proceed from
there.

Finish off the main integral, so that we have

$$V = \dots\dots\dots$$

35

$$\boxed{V = 6\pi - 2 \approx 16{\cdot}8 \text{ cubic units}}$$

Because we had

$$V = \int_0^2 \left\{6(4-x^2)^{1/2} - 2x + \frac{x^3}{2}\right\} dx$$

$$= 3\left[x\sqrt{4-x^2} + 4\arcsin\frac{x}{2}\right]_0^2 - \left[x^2 - \frac{x^4}{8}\right]_0^2$$

$$= 3\{4\ \arcsin 1 - 4\ \arcsin 0\} - 4 + 2$$

$$= 3\{2\pi\} - 2 = 6\pi - 2$$

$$\approx 16{\cdot}8$$

▶

Alternative method

We could, of course, have used cylindrical coordinates in this problem.

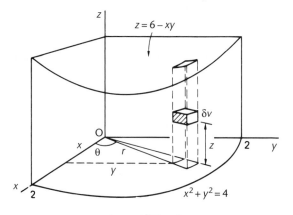

$$\delta v = r \, \delta r \, \delta\theta \, \delta z$$
$$x = r \cos\theta; \ y = r \sin\theta$$
$$\therefore \ z = 6 - xy$$
$$= 6 - r^2 \sin\theta \cos\theta$$
$$= 6 - \frac{r^2}{2} \sin 2\theta$$

$$\therefore \ V = \int_{r=0}^{2} \int_{\theta=0}^{\pi/2} \int_{z=0}^{6-(r^2/2)\sin 2\theta} r \, dr \, d\theta \, dz$$
$$= \int_{\theta=0}^{\pi/2} \int_{r=0}^{2} \int_{z=0}^{6-(r^2/2)\sin 2\theta} dz \, r \, dr \, d\theta$$
$$= \dots\dots\dots$$

Finish it

$$\boxed{V = 6\pi - 2 \text{ (as before)}}$$

36

$$V = \int_{\theta=0}^{\pi/2} \int_{r=0}^{2} \left(6 - \frac{r^2}{2} \sin 2\theta\right) r \, dr \, d\theta$$
$$= \int_{\theta=0}^{\pi/2} \int_{r=0}^{2} \left(6r - \frac{r^3}{2} \sin 2\theta\right) dr \, d\theta$$
$$= \int_{0}^{\pi/2} \left[3r^2 - \frac{r^4}{8} \sin 2\theta\right]_{r=0}^{r=2} d\theta$$
$$= \int_{0}^{\pi/2} (12 - 2\sin 2\theta) \, d\theta$$
$$= \left[12\theta + \cos 2\theta\right]_{0}^{\pi/2}$$
$$= (6\pi - 1) - 1$$
$$\therefore \ V = 6\pi - 2$$

In this case, the use of cylindrical coordinates facilitates the evaluation.

Let us consider another example.

37

Example 2

To find the moment of inertia and radius of gyration of a thick hollow sphere about a diameter as axis. Outer radius $= a$; inner radius $= b$; density of material $= c$.

It is convenient to deal with one-eighth of the sphere in the first octant.

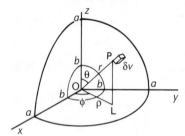

$\therefore$ Total mass of the solid $M_1 = \dfrac{1}{8} M$

$$M_1 = \frac{1}{8} \cdot \frac{4}{3}\pi(a^3 - b^3)c = \frac{\pi}{6}(a^3 - b^3)c$$

Using spherical coordinates, the element of volume

$$\delta v = \ldots\ldots\ldots$$

38

$$\boxed{\delta v = r^2 \sin\theta \, \delta r \, \delta\theta \, \delta\phi}$$

Also the element of mass $m = c\delta v$

Second moment of mass of the element about OZ

$$= m\rho^2 = m(r\sin\theta)^2$$
$$= c\, r^2 \sin\theta \, \delta r \, \delta\theta \, \delta\phi \, r^2 \sin^2\theta$$
$$= c\, r^4 \sin^3\theta \, \delta r \, \delta\theta \, \delta\phi$$

$\therefore$ Total second moment for the solid

$$I_1 \approx \sum_{\phi=0}^{\pi/2} \sum_{\theta=0}^{\pi/2} \sum_{r=b}^{a} c\, r^4 \, \delta r \, \sin^3\theta \, \delta\theta \, \delta\phi$$

Then, as usual, if $\delta r \to 0$, $\delta\theta \to 0$, $\delta\phi \to 0$, we finally obtain

$$I_1 = \int_{\phi=0}^{\pi/2} \int_{\theta=0}^{\pi/2} \int_{r=b}^{a} c\, r^4 \, dr \, \sin^3\theta \, d\theta \, d\phi$$

which you can evaluate without any difficulty and obtain

$$I_1 = \ldots\ldots\ldots$$

$$\boxed{I_1 = \frac{\pi}{15}(a^5 - b^5)c}$$

39

Because

$$I_1 = \int_0^{\pi/2} \int_0^{\pi/2} \left[c\frac{r^5}{5} \right]_b^a \sin^3 \theta \, d\theta \, d\phi$$

$$= \int_0^{\pi/2} \int_0^{\pi/2} \frac{c}{5}(a^5 - b^5) \sin^3 \theta \, d\theta \, d\phi$$

$$= \frac{c}{5}(a^5 - b^5) \int_0^{\pi/2} \int_0^{\pi/2} (1 - \cos^2 \theta) \sin \theta \, d\theta \, d\phi$$

$$= \frac{c}{5}(a^5 - b^5) \int_0^{\pi/2} \left[-\cos\theta + \frac{\cos^3\theta}{3} \right]_0^{\pi/2} d\phi$$

$$= \frac{c}{5}(a^5 - b^5) \int_0^{\pi/2} \left(1 - \frac{1}{3} \right) d\phi$$

$$= \frac{2c}{15}(a^5 - b^5) \left[\phi \right]_0^{\pi/2} = \frac{c\pi}{15}(a^5 - b^5)$$

Therefore, the moment of inertia for the whole sphere I is

$$I = 8I_1 \quad \text{i.e.} \quad I = \frac{8\pi}{15}(a^5 - b^5)c$$

Radius of gyration (k) $\quad Mk^2 = I$

$$\therefore k = \ldots\ldots\ldots\ldots$$

$$\boxed{k = \sqrt{\frac{2}{5}\left(\frac{a^5 - b^5}{a^3 - b^3} \right)}}$$

We had already calculated the total mass $M = \frac{4\pi}{3}(a^3 - b^3)c$ and since

$$I = \frac{8\pi}{15}(a^5 - b^5) \text{ then}$$

$$\frac{4\pi}{3}(a^3 - b^3)ck^2 = \frac{8\pi}{15}(a^5 - b^5)c$$

$$\therefore k^2 = \frac{2}{5}\left(\frac{a^5 - b^5}{a^3 - b^3} \right) \quad \therefore k = \sqrt{\frac{2}{5}\left(\frac{a^5 - b^5}{a^3 - b^3} \right)}$$

We have set the working out in considerable detail, since spherical coordinates may be a new topic. Many of the statements can be streamlined when one is familiar with the system.

Now move on for another example

41   **Example 3**

Find the total mass of a solid sphere of radius a, enclosed by the surface $x^2 + y^2 + z^2 = a^2$ and having variable density c where $c = 1 + r|z|$ and r is the distance of any point from the origin.

   This is a case where spherical coordinates can clearly be used with advantage.

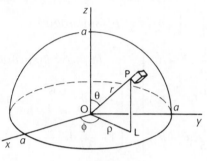

(a) ....

In the element of volume, the three dimensions are

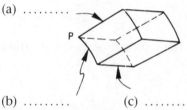

(b) ....   (c) ....

42

$$(a)\ \delta r \quad (b)\ r\,\delta\theta \quad (c)\ \rho\,\delta\phi = r\sin\theta\,\delta\phi$$

so that $\delta v = \ldots\ldots\ldots$

43

$$\delta v = r^2 \sin\theta\,\delta r\,\delta\theta\,\delta\phi$$

Then the mass of the element $= c\,\delta v = (1 + r|z|)\,\delta v$
and $\qquad\qquad\qquad\qquad z = r\cos\theta$
$$\therefore\ m = c\,\delta v = (1 + r^2\cos\theta)\,r^2\sin\theta\,\delta r\,\delta\theta\,\delta\phi$$

Since the density uses $|z| = 1$ we must only consider the region where $\cos\theta \geq 0$ and so we consider the *upper hemisphere* only. The integral for the total mass M_1 is

$$M_1 = \ldots\ldots\ldots$$

Write out the integral and insert the limits.

$$M_1 = \int_{\phi=0}^{\phi=2\pi} \int_{\theta=0}^{\theta=\pi/2} \int_{r=0}^{r=a} (1 + r^2 \cos \theta) r^2 \sin \theta \, dr \, d\theta \, d\phi$$

44

i.e. $M_1 = \int_{\phi=0}^{2\pi} \int_{\theta=0}^{\pi/2} \int_{r=0}^{a} \{r^2 \sin \theta \, dr \, d\theta \, d\phi + r^4 \sin \theta \cos \theta \, dr \, d\theta \, d\phi\}$

$$= \qquad\qquad I_1 \qquad\qquad + \qquad\qquad I_2$$

$$I_1 = \int_0^{2\pi} \int_0^{\pi/2} \int_0^a r^2 \sin \theta \, dr \, d\theta \, d\phi \text{ gives } \ldots\ldots\ldots\ldots$$

Do *not* work it out. You can doubtless recognise what the result would represent.

The volume of the hemisphere

45

Because the integral is simply the summation of elements of volume throughout the region of the hemisphere.

Thus, without more ado, $I_1 = \dfrac{2}{3} \pi a^3$.

Now for I_2.

$$I_2 = \int_0^{2\pi} \int_0^{\pi/2} \int_0^a r^4 \sin \theta \cos \theta \, dr \, d\theta \, d\phi$$

$$= \ldots\ldots\ldots\ldots \text{ Evaluate the triple integral.}$$

$$I_2 = \frac{\pi a^5}{5}$$

46

Because

$$I_2 = \int_0^{2\pi} \int_0^{\pi/2} \frac{a^5}{5} \sin \theta \cos \theta \, d\theta \, d\phi$$

$$= \frac{a^5}{5} \int_0^{2\pi} \left[\frac{\sin^2 \theta}{2} \right]_0^{\pi/2} d\phi$$

$$= \frac{a^5}{10} \int_0^{2\pi} 1 \, d\phi$$

$$= \frac{a^5}{10} \left[\phi \right]_0^{2\pi} = \frac{\pi a^5}{5}$$

$$\therefore I_2 = \frac{\pi a^5}{5}$$

So now finish it off. For the complete sphere

$$M = \ldots\ldots\ldots\ldots$$

47

$$M = \frac{2\pi a^3}{15}(10 + 3a^2)$$

Because

$$M_1 = I_1 + I_2 = \frac{2}{3}\pi a^3 + \frac{\pi a^5}{5} = \frac{\pi a^3}{15}(10 + 3a^2)$$

Then, for the whole sphere, $M = 2M_1 = \dfrac{2\pi a^3}{15}(10 + 3a^2)$

Each problem, then, is tackled in much the same way.

(a) Draw a careful sketch diagram, inserting all relevant information.
(b) Decide on the most appropriate coordinate system to use.
(c) Build up the multiple integral and insert correct limits.
(d) Evaluate the integral.

And now we can apply the general guide lines to a final problem.

Example 4

Determine the volume of the solid bounded by the planes $x = 0$, $y = 0$, $z = x$, $z = 2$ and $y = 4 - x^2$ in the first quadrant.

First we sketch the diagram.

48

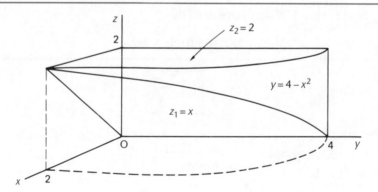

There is no axis of symmetry and no spherical centre. We shall therefore use
.... coordinates.

49

Cartesian

So off you go on your own. There are no snags.

$$V = \ldots\ldots\ldots\ldots$$

50

$$V = 6\frac{2}{3} \text{ cubic units}$$

Here is the complete solution.

$$V \approx \sum_{x=0}^{2} \sum_{y=0}^{4-x^2} \sum_{z=x}^{2} \delta x \, \delta y \, \delta z$$

$$\therefore \; V = \int_{x=0}^{2} \int_{y=0}^{4-x^2} \int_{z=x}^{2} \mathrm{d}z \, \mathrm{d}y \, \mathrm{d}x$$

$$= \int_{0}^{2} \int_{0}^{4-x^2} (2-x) \, \mathrm{d}y \, \mathrm{d}x$$

$$= \int_{0}^{2} \left[2y - xy \right]_{y=0}^{4-x^2} \mathrm{d}x$$

$$= \int_{0}^{2} \{8 - 2x^2 - 4x + x^3\} \mathrm{d}x$$

$$= \left[8x - \frac{2x^3}{3} - 2x^2 + \frac{x^4}{4} \right]_{0}^{2}$$

$$= 6\frac{2}{3}$$

And that is it. Now we move to the next section of work

Change of variables in multiple integrals

51

In Cartesian coordinates, we use the variables (x, y, z); in cylindrical coordinates, we use the variables (r, θ, z); in spherical coordinates, we use the variables (r, θ, ϕ); and we have established relationships connecting these systems of variables, permitting us to transfer from one system to another. These relationships, you will remember, were obtained geometrically in Frames 23 to 30 of this Programme.

There are occasions, however, when it is expedient to make other transformations beside those we have used and it is worth looking at the problem in a rather more general manner.

This we will now do

52   First, however, let us revise a result from an earlier Programme on determinants to find the area of the triangle ABC.

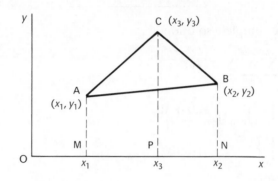

If we arrange the vertices   A (x_1, y_1)
                              B (x_2, y_2)
                              C (x_3, y_3)

in an anticlockwise manner then

area triangle ABC = trapezium AMPC + trapezium CPNB

$\qquad\qquad\qquad$ − trapezium AMNB

$= \frac{1}{2}\{(x_3 - x_1)(y_1 + y_3) + (x_2 - x_3)(y_2 + y_3) - (x_2 - x_1)(y_1 + y_2)\}$

$= \frac{1}{2}\{x_3 y_1 - x_1 y_1 + x_3 y_3 - x_1 y_3 + x_2 y_2 + x_2 y_3 - x_3 y_2 - x_3 y_3$

$\qquad\qquad\qquad\qquad - x_2 y_1 - x_2 y_2 + x_1 y_1 + x_1 y_2\}$

$= \frac{1}{2}\{(x_2 y_3 - x_3 y_2) + (x_3 y_1 - x_1 y_3) + (x_1 y_2 - x_2 y_1)\}$

$= \frac{1}{2}\begin{vmatrix} 1 & 1 & 1 \\ x_1 & x_2 & x_3 \\ y_1 & y_2 & y_3 \end{vmatrix}$

The determinant is positive if the points A, B, C are taken in an anticlockwise manner.

We shall need to use this result in a short while, so keep it in mind.

On to the next frame

Curvilinear coordinates

Consider the double integral $\int_R \int \phi(x, y)\,dA$ where $dA = dx\,dy$ in Cartesian **53** coordinates. Let u and v be two new independent variables defined by $u = F(x, y)$ and $v = G(x, y)$ where these equations can be simultaneously solved to obtain $x = f(u, v)$ and $y = g(u, v)$. Furthermore, these transformation equations are such that every point (x, y) is mapped to a unique point (u, v) and vice versa.

Let us see where this leads us, so on to the next frame

The equation $u = F(x, y)$ will be a family of curves depending on the particular **54** constant value given to u in each case.

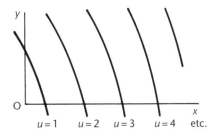

Curves $u = F(x, y)$ for different constant values of u.

Similarly, $v = G(x, y)$ will be a family of curves depending on the particular constant value assigned to v in each case.

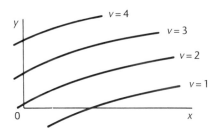

Curves $v = G(x, y)$ for different constant values of v.

These two sets of curves will therefore cover the region R and form a network, and to any point P (x_0, y_0) there will be a pair of curves $u = u_0$ (constant) and $v = v_0$ (constant) that intersect at that point.

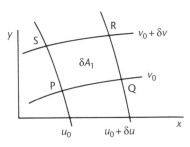

The *u*- and *v*-values relating to any particular point are known as its *curvilinear coordinates* and $x = f(u, v)$ and $y = g(u, v)$ are the *transformation equations* between the two systems.

In the Cartesian coordinates (x, y) system, the element of area $\delta A = \delta x \delta y$ and is the area bounded by the lines $x = x_0$, $x = x_0 + \delta x$, $y = y_0$, and $y = y_0 + \delta y$.

In the new system of *curvilinear coordinates* (u, v) the element of area δA_1 can be taken as that of the figure P, Q, R, S, i.e. the area bounded by the curves $u = u_0$, $u = u_0 + \delta u$, $v = v_0$ and $v = v_0 + \delta v$.

Since δA_1 is small, PQRS may be regarded as a parallelogram

i.e. $\delta A_1 \approx 2 \times$ area of triangle PQS

and this is where we make use of the result previously revised that the area of a triangle ABC with vertices (x_1, y_1), (x_2, y_2), (x_3, y_3) can be expressed in determinant form as

$$\text{Area} = \ldots\ldots\ldots\ldots$$

55

$$\text{Area} = \frac{1}{2} \begin{vmatrix} 1 & 1 & 1 \\ x_1 & x_2 & x_3 \\ y_1 & y_2 & y_3 \end{vmatrix}$$

Before we can apply this, we must find the Cartesian coordinates of P, Q and S in the diagram on page 646 where we omit the subscript $_0$ on the coordinates.

If $x = f(u, v)$, then a small increase δx in x is given by

$$\delta x = \ldots\ldots\ldots\ldots$$

56

$$\delta x = \frac{\partial f}{\partial u} \delta u + \frac{\partial f}{\partial v} \delta v$$

i.e. $\delta x = \dfrac{\partial x}{\partial u} \delta u + \dfrac{\partial x}{\partial v} \delta v$

and, for $y = g(u, v)$

$$\delta y = \ldots\ldots\ldots\ldots$$

57

$$\delta y = \frac{\partial y}{\partial u} \delta u + \frac{\partial y}{\partial v} \delta v$$

Now

(a) P is the point (x, y)

(b) Q corresponds to small changes from P.

$$\delta x = \frac{\partial x}{\partial u} \delta u + \frac{\partial x}{\partial v} \delta v \quad \text{and} \quad \delta y = \frac{\partial y}{\partial u} \delta u + \frac{\partial y}{\partial v} \delta v$$

But along PQ v is constant. $\therefore \delta v = 0$.

$$\therefore \delta x = \frac{\partial x}{\partial u} \delta u \quad \text{and} \quad \delta y = \frac{\partial y}{\partial u} \delta u$$

i.e. Q is the point $\left(x + \frac{\partial x}{\partial u} \delta u, \ y + \frac{\partial y}{\partial u} \delta u \right)$.

(c) Similarly for S, since u is constant along PS $\quad \delta u = 0$ and

$$\therefore \text{ S is the point } \left(x + \frac{\partial x}{\partial v} \delta v, \ y + \frac{\partial y}{\partial v} \delta v \right)$$

So the Cartesian coordinates of P, Q, S are

$$\text{P } (x, y); \quad \text{Q } \left(x + \frac{\partial x}{\partial u} \delta u, \ y + \frac{\partial y}{\partial u} \delta u \right); \quad \text{S } \left(x + \frac{\partial x}{\partial v} \delta v, \ y + \frac{\partial y}{\partial v} \delta v \right)$$

$\therefore$ The determinant for the area PQS is $\ldots \ldots \ldots \ldots$

58

$$\text{Area} = \frac{1}{2} \begin{vmatrix} 1 & 1 & 1 \\ x & x + \dfrac{\partial x}{\partial u} \delta u & x + \dfrac{\partial x}{\partial v} \delta v \\ y & y + \dfrac{\partial y}{\partial u} \delta u & y + \dfrac{\partial y}{\partial v} \delta v \end{vmatrix}$$

Subtracting column 1 from columns 2 and 3 gives

$$\text{Area} = \frac{1}{2} \begin{vmatrix} 1 & 0 & 0 \\ x & \dfrac{\partial x}{\partial u} \delta u & \dfrac{\partial x}{\partial v} \delta v \\ y & \dfrac{\partial y}{\partial u} \delta u & \dfrac{\partial y}{\partial v} \delta v \end{vmatrix}$$

which simplifies immediately to

$\ldots \ldots \ldots \ldots$

59

$$\text{Area} = \frac{1}{2} \begin{vmatrix} \dfrac{\partial x}{\partial u}\delta u & \dfrac{\partial x}{\partial v}\delta v \\ \dfrac{\partial y}{\partial u}\delta u & \dfrac{\partial y}{\partial v}\delta v \end{vmatrix}$$

Then, taking out the factor δu from the first column and the factor δv from the second column, this becomes

$$\text{Area} = \ldots\ldots\ldots\ldots$$

60

$$\frac{1}{2} \begin{vmatrix} \dfrac{\partial x}{\partial u} & \dfrac{\partial x}{\partial v} \\ \dfrac{\partial y}{\partial u} & \dfrac{\partial y}{\partial v} \end{vmatrix} \delta u\delta v$$

The area of the approximate parallelogram is twice the area of the triangle.

$$\therefore \ \text{Area of parallelogram} = \delta A_1 = \begin{vmatrix} \dfrac{\partial x}{\partial u} & \dfrac{\partial x}{\partial v} \\ \dfrac{\partial y}{\partial u} & \dfrac{\partial y}{\partial v} \end{vmatrix} \delta u\,\delta v$$

Expressing this in differentials

$$\mathrm{d}A = \begin{vmatrix} \dfrac{\partial x}{\partial u} & \dfrac{\partial x}{\partial v} \\ \dfrac{\partial y}{\partial u} & \dfrac{\partial y}{\partial v} \end{vmatrix} \mathrm{d}u\,\mathrm{d}v$$

and, for convenience, this is often written

$$\mathrm{d}A = \frac{\partial(x,\,y)}{\partial(u,\,v)}\,\mathrm{d}u\,\mathrm{d}v$$

$\dfrac{\partial(x,\,y)}{\partial(u,\,v)}$ is called the *Jacobian of the transformation* from the Cartesian coordinates $(x,\,y)$ to the curvilinear coordinates $(u,\,v)$.

$$\therefore \ J(u,v) = \frac{\partial(x,\,y)}{\partial(u,\,v)} = \begin{vmatrix} \dfrac{\partial x}{\partial u} & \dfrac{\partial x}{\partial v} \\ \dfrac{\partial y}{\partial u} & \dfrac{\partial y}{\partial v} \end{vmatrix}$$

So, if the transformation equations are

$$x = u(u+v) \quad \text{and} \quad y = uv^2$$
$$J(u,\,v) = \ldots\ldots\ldots\ldots$$

$$\boxed{J(u,v) = uv(4u + v)}$$

61

Because

$$\frac{\partial x}{\partial u} = 2u + v \qquad \frac{\partial x}{\partial v} = u$$

$$\frac{\partial y}{\partial u} = v^2 \qquad \frac{\partial y}{\partial v} = 2uv$$

$$\therefore\ J(u,v) = \begin{vmatrix} 2u + v & u \\ v^2 & 2uv \end{vmatrix} = 4u^2v + 2uv^2 - uv^2$$

$$= 4u^2v + uv^2 = uv(4u + v)$$

Next frame

Sometimes the transformation equations are given the other way round. That is, where u and v are given as expressions in x and y. In such a case $J(u, v)$ can be found using the fact that

62

$$\frac{\partial(x, y)}{\partial(u, v)} = \frac{1}{\left(\dfrac{\partial(u, v)}{\partial(x, y)}\right)}$$

For example, if the transformation equations are given as $u = x^2 + y^2$ and $v = 2xy$ then

$$J(u, v) = \ldots\ldots\ldots\ldots$$

$$\boxed{J(u, v) = \frac{1}{4\sqrt{u^2 - v^2}}}$$

63

Because

$$\frac{\partial(u, v)}{\partial(x, y)} = \begin{vmatrix} \dfrac{\partial u}{\partial x} & \dfrac{\partial u}{\partial y} \\ \dfrac{\partial v}{\partial x} & \dfrac{\partial v}{\partial y} \end{vmatrix} = \begin{vmatrix} 2x & 2y \\ 2y & 2x \end{vmatrix} = 4x^2 - 4y^2$$

and so

$$J(u, v) = \frac{\partial(x, y)}{\partial(u, v)} = \frac{1}{\left(\dfrac{\partial(u, v)}{\partial(x, y)}\right)} = \frac{1}{4(x^2 - y^2)}$$

Now $u - v = x^2 - 2xy + y^2 = (x - y)^2$

and $u + v = x^2 + 2xy + y^2 = (x + y)^2$

and so $x^2 - y^2 = (x - y)(x + y) = \sqrt{u - v}\sqrt{u + v} = \sqrt{u^2 - v^2}$ giving

$$J(u, v) = \frac{1}{4\sqrt{u^2 - v^2}}$$

There is one further point to note in this piece of work, so move on

64

Note: In the transformation, it is possible for the order of the points P, Q, R, S to be reversed with the result that δA may give a negative result when the determinant is evaluated. To ensure a positive element of area, the result is finally written

$$\mathrm{d}A = \left| \frac{\partial(x, y)}{\partial(u, v)} \right| \mathrm{d}u\,\mathrm{d}v$$

where the 'modulus' lines indicate the absolute value of the Jacobian.

Therefore, to rewrite the integral $\int_R \int F(x, y)\,\mathrm{d}x\,\mathrm{d}y$ in terms of the new variables, u and v, where $x = f(u, v)$ and $y = g(u, v)$, we substitute for x and y in $F(x, y)$ and replace $\mathrm{d}x\,\mathrm{d}y$ with $\left| \frac{\partial(x, y)}{\partial(u, v)} \right| \mathrm{d}u\,\mathrm{d}v$.

The integral then becomes

$$\int_R \int F\{f(u, v), g(u, v)\} \left| \frac{\partial(x, y)}{\partial(u, v)} \right| \mathrm{d}u\,\mathrm{d}v$$

Make a note of this result

65

Example 1

Express $I = \int_R \int xy^2\,\mathrm{d}x\,\mathrm{d}y$ in polar coordinates, making the substitutions $x = r\cos\theta$, $y = r\sin\theta$.

$$\frac{\partial x}{\partial r} = \cos\theta \qquad \frac{\partial x}{\partial\theta} = -r\sin\theta$$

$$\frac{\partial y}{\partial r} = \sin\theta \qquad \frac{\partial y}{\partial\theta} = r\cos\theta$$

$$\therefore \ J(r, \theta) = \dots\dots\dots$$

66

$$\boxed{J(r, \theta) = r}$$

$$J(r, \theta) = \begin{vmatrix} \cos\theta & -r\sin\theta \\ \sin\theta & r\cos\theta \end{vmatrix} = r\cos^2\theta + r\sin^2\theta = r$$

Then $I = \int_R \int xy^2\,\mathrm{d}x\,\mathrm{d}y$ \quad becomes $\dots\dots\dots\dots$

$$I = \int_R \int r^3 \sin^2 \theta \cos \theta \, r \, dr \, d\theta$$

67

Because $xy^2 = r \cos \theta \, r^2 \sin^2 \theta = r^3 \sin^2 \theta \cos \theta$

$$\left| \frac{\partial(x, y)}{\partial(r, \theta)} \right| dr \, d\theta = r \, dr \, d\theta$$

$$\therefore \ I = \int_R \int r^3 \sin^2 \theta \cos \theta \, r \, dr \, d\theta = \int_R \int r^4 \sin^2 \theta \cos \theta \, dr \, d\theta$$

Now this one.

Example 2

Express $I = \int_R \int (x^2 + y^2) dx \, dy$ in terms of u and v, given that $x = u^2 - v^2$ and $y = 2uv$.

First of all, the expression for $\dfrac{\partial(x, y)}{\partial(u, v)}$ gives ....

$$4(u^2 + v^2)$$

68

Because

$$x = u^2 - v^2 \qquad \therefore \ \frac{\partial x}{\partial u} = 2u \qquad \frac{\partial x}{\partial v} = -2v$$

$$y = 2uv \qquad \therefore \ \frac{\partial y}{\partial u} = 2v \qquad \frac{\partial y}{\partial v} = 2u$$

$$\therefore \ \frac{\partial(x, y)}{\partial(u, v)} = \begin{vmatrix} \dfrac{\partial x}{\partial u} & \dfrac{\partial y}{\partial u} \\ \dfrac{\partial x}{\partial v} & \dfrac{\partial y}{\partial v} \end{vmatrix} = \begin{vmatrix} 2u & 2v \\ -2v & 2u \end{vmatrix} = 4(u^2 + v^2)$$

Also $x^2 + y^2 = (u^2 - v^2)^2 + (2uv)^2 = u^4 - 2u^2v^2 + v^4 + 4u^2v^2$

$$= u^4 + 2u^2v^2 + v^4 = (u^2 + v^2)^2$$

Then $I = \displaystyle\int_R \int (x^2 + y^2) \, dx \, dy$ becomes $I = $ ....

69

$$I = 4 \int_R \int (u^2 + v^2)^3 \, du \, dv$$

One more.

Example 3

By substituting $x = 2uv$ and $y = u(1 - v)$ where $u > 0$ and $v > 0$, express the integral $I = \int_R \int x^2 y \, dx \, dy$ in terms of u and v.

Complete it: there are no snags. $I = \ldots\ldots\ldots$

70

$$I = 8 \int_R \int u^4 v^2 (1 - v) \, du \, dv$$

Working:

$$x = 2uv \qquad \therefore \frac{\partial x}{\partial u} = 2v \qquad \frac{\partial x}{\partial v} = 2u$$

$$y = u - uv \qquad \frac{\partial y}{\partial u} = 1 - v \qquad \frac{\partial y}{\partial v} = -u$$

$$\therefore J(u, v) = \frac{\partial(x, y)}{\partial(u, v)} = \begin{vmatrix} \dfrac{\partial x}{\partial u} & \dfrac{\partial y}{\partial u} \\ \dfrac{\partial x}{\partial v} & \dfrac{\partial y}{\partial v} \end{vmatrix} = \begin{vmatrix} 2v & 1 - v \\ 2u & -u \end{vmatrix}$$

$$= 2u \begin{vmatrix} v & 1 - v \\ 1 & -1 \end{vmatrix} = 2u \begin{vmatrix} v & 1 \\ 1 & 0 \end{vmatrix} = -2u$$

$$\therefore \left| \frac{\partial(x, y)}{\partial(u, v)} \right| = 2u$$

$$x^2 y = 4u^2 v^2 (u - uv) = 4u^3 v^2 (1 - v)$$

$$\therefore I = \int_R \int 4u^3 v^2 (1 - v) \, 2u \, du \, dv$$

$$I = 8 \int_R \int u^4 v^2 (1 - v) \, du \, dv$$

▶

Transformation in three dimensions

If we extend the previous results to convert variables (x, y, z) to (u, v, w), we proceed in just the same way.

If $x = f(u, v, w); \quad y = g(u, v, w); \quad z = h(u, v, w)$

Then $\quad J(u, v, w) = \dfrac{\partial(x, y, z)}{\partial(u, v, w)} = \begin{vmatrix} \dfrac{\partial x}{\partial u} & \dfrac{\partial y}{\partial u} & \dfrac{\partial z}{\partial u} \\[2mm] \dfrac{\partial x}{\partial v} & \dfrac{\partial y}{\partial v} & \dfrac{\partial z}{\partial v} \\[2mm] \dfrac{\partial x}{\partial w} & \dfrac{\partial y}{\partial w} & \dfrac{\partial z}{\partial w} \end{vmatrix}$

and the element of volume $\mathrm{d}V = \mathrm{d}x\,\mathrm{d}y\,\mathrm{d}z$ becomes

$$\mathrm{d}V = |J(u, v, w)|\,\mathrm{d}u\,\mathrm{d}v\,\mathrm{d}w$$

Also $\displaystyle\iiint F(x, y, z)\,\mathrm{d}x\,\mathrm{d}y\,\mathrm{d}z$ is transformed into

$$\iiint G(u, v, w)\left|\frac{\partial(x, y, z)}{\partial(u, v, w)}\right|\mathrm{d}u\,\mathrm{d}v\,\mathrm{d}w$$

Now for an example, so move on

Example 4

71

To transform a triple integral $I = \displaystyle\iiint F(x, y, z)\,\mathrm{d}x\,\mathrm{d}y\,\mathrm{d}z$ in Cartesian coordinates to spherical coordinates by the transformation equations

$x = r\sin\theta\cos\phi$
$y = r\sin\theta\sin\phi$
$z = r\cos\theta.$

First we need the partial derivatives, from which to build up the Jacobian.

These are ....

72

$$\frac{\partial x}{\partial r} = \sin\theta\cos\phi \qquad \frac{\partial y}{\partial r} = \sin\theta\sin\phi \qquad \frac{\partial z}{\partial r} = \cos\theta$$

$$\frac{\partial x}{\partial \theta} = r\cos\theta\cos\phi \qquad \frac{\partial y}{\partial \theta} = r\cos\theta\sin\phi \qquad \frac{\partial z}{\partial \theta} = -r\sin\theta$$

$$\frac{\partial x}{\partial \phi} = -r\sin\theta\sin\phi \qquad \frac{\partial y}{\partial \phi} = r\sin\theta\cos\phi \qquad \frac{\partial z}{\partial \phi} = 0$$

$$\therefore\ J(r,\theta,\phi) = \begin{vmatrix} \sin\theta\cos\phi & \sin\theta\sin\phi & \cos\theta \\ r\cos\theta\cos\phi & r\cos\theta\sin\phi & -r\sin\theta \\ -r\sin\theta\sin\phi & r\sin\theta\cos\phi & 0 \end{vmatrix}$$

$$= \cos\theta\begin{vmatrix} r\cos\theta\cos\phi & r\cos\theta\sin\phi \\ -r\sin\theta\sin\phi & r\sin\theta\cos\phi \end{vmatrix}$$

$$+ r\sin\theta\begin{vmatrix} \sin\theta\cos\phi & \sin\theta\sin\phi \\ -r\sin\theta\sin\phi & r\sin\theta\cos\phi \end{vmatrix}$$

$$= \dots\dots\dots$$

73

$$\boxed{r^2\sin\theta}$$

Because

$$J(r,\ \theta,\ \phi) = r^2\cos^2\theta\sin\theta\begin{vmatrix} \cos\phi & \sin\phi \\ -\sin\phi & \cos\phi \end{vmatrix}$$

$$+ r^2\sin^3\theta\begin{vmatrix} \cos\phi & \sin\phi \\ -\sin\phi & \cos\phi \end{vmatrix}$$

$$= (r^2\sin^3\theta + r^2\sin\theta\cos^2\theta)\begin{vmatrix} \cos\phi & \sin\phi \\ -\sin\phi & \cos\phi \end{vmatrix}$$

$$= r^2\sin\theta(\sin^2\theta + \cos^2\theta)(\cos^2\phi + \sin^2\phi) = r^2\sin\theta$$

$$\therefore\ I = \iiint G(u,v,w)r^2\sin\theta\,dr\,d\theta\,d\phi$$

which agrees, of course, with the result we had previously obtained by a geometric consideration.

And that is about it. Check carefully down the **Revision summary** and the **Can you?** checklist that now follow, before working through the **Test exercise**. The **Further problems** give additional practice.

Revision summary 20

1 *Surface integrals*

74

$$I = \int_R f(x,\, y)\, \mathrm{d}a = \int_R \int f(x,\, y)\, \mathrm{d}y\, \mathrm{d}x$$

2 *Surface in space*

$$I = \int_S \phi(x, y, z)\, \mathrm{d}S = \int_R \int \phi(x, y, z) \sec \gamma\, \mathrm{d}x\, \mathrm{d}y \qquad (\gamma < \pi/2)$$

$$= \int_R \int \phi(x, y, z) \sqrt{1 + \left(\frac{\partial z}{\partial x}\right)^2 + \left(\frac{\partial z}{\partial y}\right)^2}\, \mathrm{d}x\, \mathrm{d}y$$

3 *Space coordinate systems*

(a) *Cartesian coordinates* $(x,\, y,\, z)$

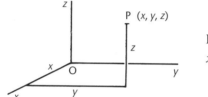

First octant:

$x \geq 0;\ \ y \geq 0;\ \ z \geq 0$

(b) *Cylindrical coordinates* $(r,\, \theta,\, z)$ $\quad r \geq 0$

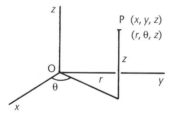

$$\begin{aligned} x &= r \cos \theta & r &= \sqrt{x^2 + y^2} \\ y &= r \sin \theta & \theta &= \arctan(y/x) \\ z &= z & z &= z \end{aligned}$$

(c) *Spherical coordinates* $(r,\, \theta, \phi)$ $\quad r \geq 0$

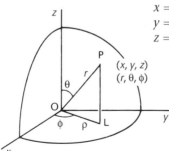

$$\begin{aligned} x &= r \sin \theta \cos \phi & r &= \sqrt{x^2 + y^2 + z^2} \\ y &= r \sin \theta \sin \phi & \theta &= \arccos(z/r) \\ z &= r \cos \theta & \phi &= \arctan(y/x) \end{aligned}$$

▶

4   *Elements of volume*
    (a)   *Cartesian coordinates*

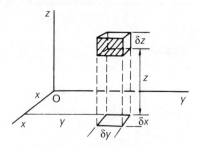

$$\delta v = \delta x \, \delta y \, \delta z$$

    (b)   *Cylindrical coordinates* $r \geq 0$

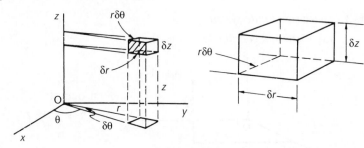

$$\delta v = r \, \delta r \, \delta \theta \, \delta z$$

    (c)   *Spherical coordinates*

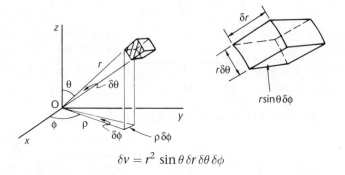

$$\delta v = r^2 \sin \theta \, \delta r \, \delta \theta \, \delta \phi$$

5   *Volume integrals*

$$V = \iiint dz \, dy \, dx$$

$$I = \iiint f(x, y, z) \, dz \, dy \, dx$$

6 *Change of variables* in multiple integrals
   (a) *Double integrals* $x = f(u, v);\quad y = g(u, v)$

$$dA = \left|\frac{\partial(x, y)}{\partial(u, v)}\right| du\, dv; \qquad J(u, v) = \frac{\partial(x, y)}{\partial(u, v)} = \begin{vmatrix} \dfrac{\partial x}{\partial u} & \dfrac{\partial y}{\partial u} \\ \dfrac{\partial x}{\partial v} & \dfrac{\partial y}{\partial v} \end{vmatrix}$$

$$I = \int_R \int F(x, y)\, dx\, dy = \int_R \int F\{f(u, v), g(u, v)\} \left|\frac{\partial(x, y)}{\partial(u, v)}\right| du\, dv$$

   (b) *Triple integrals* $\quad x = f(u, v, w); y = g(u, v, w); z = h(u, v, w)$

$$J(u, v, w) = \frac{\partial(x, y, z)}{\partial(u, v, w)} = \begin{vmatrix} \dfrac{\partial x}{\partial u} & \dfrac{\partial y}{\partial u} & \dfrac{\partial z}{\partial u} \\ \dfrac{\partial x}{\partial v} & \dfrac{\partial y}{\partial v} & \dfrac{\partial z}{\partial v} \\ \dfrac{\partial x}{\partial w} & \dfrac{\partial y}{\partial w} & \dfrac{\partial z}{\partial w} \end{vmatrix}$$

   Then $I = \displaystyle\iiint F(x, y, z)\, dx\, dy\, dz$

$$= \iiint G(u, v, w) \left|\frac{\partial(x, y, z)}{\partial(u, v, w)}\right| du\, dv\, dw$$

 # Can you?

Checklist 20   75

Check this list before and after you try the end of Programme test.

On a scale of 1 to 5, how confident are you that you can: **Frames**

- Evaluate double integrals and surface integrals?   1 to 22
  Yes ☐ ☐ ☐ ☐ ☐   *No*

- Relate three-dimensional Cartesian coordinates to cylindrical and spherical polar forms?   23 to 31
  Yes ☐ ☐ ☐ ☐ ☐   *No*

- Evaluate volume integrals in Cartesian coordinates and in cylindrical and spherical polar coordinates?   32 to 50
  Yes ☐ ☐ ☐ ☐ ☐   *No*

- Use the Jacobian to convert integrals given in Cartesian coordinates into general curvilinear coordinates in two and three dimensions?   51 to 73
  Yes ☐ ☐ ☐ ☐ ☐   *No*

Test exercise 20

76

1   Determine the area of the surface $z = \sqrt{x^2 + y^2}$ over the region bounded by $x^2 + y^2 = 4$.

2   Evaluate the surface integral $I = \int_S \phi \, dS$ where $\phi = \dfrac{1}{\sqrt{x^2 + y^2}}$ over the surface of the sphere $x^2 + y^2 + z^2 = a^2$ in the first octant.

3   (a) Transform the Cartesian coordinates
   (1) (4, 2, 3) to cylindrical coordinates (r, θ, z)
   (2) (3, 1, 5) to spherical coordinates (r, θ, ϕ).

   (b) Express in Cartesian coordinates (x, y, z)
   (1) the cylindrical coordinates $(5, \pi/4, 3)$
   (2) the spherical coordinates $(4, \pi/6, 2)$.

4   Determine the volume of the solid bounded by the plane $z = 0$ and the surfaces $x^2 + y^2 = 4$ and $z = x^2 + y^2 + 1$.

5   Determine the total mass of a solid hemisphere bounded by the plane $z = 0$ and the surface $x^2 + y^2 + z^2 = a^2$ $(z \geq 0)$ if the density at any point is given by $\rho = 1 - z$ $(z < a)$.

6   (a) Express the integral $I = \int_R \int (x - y) \, dx \, dy$ in terms of u and v, where $x = u(1 + v)$ and $y = u - v$.

   (b) Express the triple integral $I = \int \int \int \left(\dfrac{x + z}{y} \right) dx \, dy \, dz$ in terms of u, v, w using the transformation equations
   $$x = u + v + w; \quad y = v^2 w; \quad z = u - w.$$

 Further problems 20

77

1   Evaluate the surface integral $I = \int_S (x^2 + y^2) \, dS$ over the surface of the cone $z^2 = 4(x^2 + y^2)$ between $z = 0$ and $z = 4$.

2   Find the position of the centre of gravity of that part of a thin spherical shell $x^2 + y^2 + z^2 = a^2$ which exists in the first octant.

3   Determine the surface area of the plane $6x + 3y + 4z = 60$ cut off by $x = 0$, $y = 0$, $x = 5$, $y = 8$.

4   Find the surface area of the plane $3x + 2y + 3z = 12$ cut off by the planes $x = 0$, $y = 0$, and the cylinder $x^2 + y^2 = 16$ for $x \geq 0$, $y \geq 0$.

5   Determine the area of the paraboloid $z = 2(x^2 + y^2)$ cut off by the cone $z = \sqrt{x^2 + y^2}$.

6   Find the area of the cone $z^2 = 4(x^2 + y^2)$ which is inside the paraboloid $z = 2(x^2 + y^2)$.

7   Cylinders $x^2 + y^2 = a^2$ and $x^2 + z^2 = a^2$ intersect. Determine the total external surface area of the common portion.

8   Determine the surface area of the sphere $x^2 + y^2 + z^2 = a^2$ cut off by the cylinder $x^2 + y^2 = ax$.

9   A cylinder of radius b, with the z-axis as its axis of symmetry, is removed from a sphere of radius a, $a > b$, with centre at the origin. Calculate the total curved surface area of the ring so formed, including the inner cylindrical surface.

10   Find the volume enclosed by the cylinder $x^2 + y^2 = 9$ and the planes $z = 0$ and $z = 5 - x$.

11   Determine the volume of the solid bounded by the surfaces $y = x^2$, $x = y^2$, $z = 2$ and $x + y + z = 4$.

12   Find the volume of the solid bounded by the plane $z = 0$, the cylinder $x^2 + y^2 = a^2$ and the surface $z = x^2 + y^2$.

13   A solid is bounded by the planes $x = 0$, $y = 0$, $z = 2$, $z = x$ and the surface $x^2 + y^2 = 4$. Determine the volume of the solid.

14   Find the position of the centre of gravity of the part of the solid sphere $x^2 + y^2 + z^2 = a^2$ in the first octant.

15   A solid is bounded by the cone $z = 2\sqrt{x^2 + y^2}$, $z \geq 0$, and the sphere $x^2 + y^2 + (z - a)^2 = 2a^2$. Determine the volume of the solid so formed.

16   Determine the volume enclosed by the ellipsoid $\dfrac{x^2}{a^2} + \dfrac{y^2}{b^2} + \dfrac{z^2}{c^2} = 1$.

17   Find the volume of the solid in the first octant bounded by the planes $x = 0$, $y = 0$, $z = 0$, $z = x + y$ and the surface $x^2 + y^2 = a^2$.

18   Express the integral $\displaystyle\int\int (x^2 + y^2)\,dx\,dy$ in terms of u and v, using the transformations $u = x + y$, $v = x - y$.

19   Determine an expression for the element of volume $dx\,dy\,dz$ in terms of u, v, w using the transformations $x = u(1 - v)$, $y = uv$, $z = uvw$.

20   A solid sphere of radius a has variable density c at any point (x, y, z) given by $c = k(a - z)$ where k is a constant. Determine the position of the centre of gravity of the sphere.

21   Calculate $\displaystyle\int\int x^2 y^2\,dx\,dy$ over the triangular region in the x–y plane with vertices $(0, 0)$, $(1, 1)$, $(1, 2)$.

▶

22 Evaluate the integral $I = \int_0^2 \int_{\sqrt{y(2-y)}}^{\sqrt{4-y^2}} \dfrac{y}{x^2 + y^2} \, dx \, dy$ by transforming to polar coordinates.

23 Evaluate $I = \int_0^1 \int_0^y \dfrac{xy^2}{\sqrt{x^2 + y^2}} \, dx \, dy$.

24 Find the volume bounded by the cylinder $x^2 + y^2 = a^2$, the plane $z = 0$ and the surface $z = x^2 + y^2$. Convert to polar coordinates and show that

$$V = \frac{\pi a^4}{2.}$$

25 By changing the order of integration in the integral

$$I = \int_0^a \int_x^a \frac{y^2 \, dy \, dx}{\sqrt{x^2 + y^2}}$$

show that $I = \tfrac{1}{3} a^3 \ln(1 + \sqrt{2})$.

Integral functions

Learning outcomes

When you have completed this Programme you will be able to:

- Derive the recurrence relation for the gamma function and evaluate the gamma function for certain rational arguments
- Evaluate integrals that require the use of the gamma function in their solution
- Identify the beta function and evaluate integrals that require the use of the beta function in their solution
- Derive the relationship between the gamma function and the beta function
- Use the duplication formula to evaluate the gamma function for half integer arguments
- Recognise the error function and its relation to the Gaussian probability distribution
- Recognise elliptic functions of the first and second kind
- Evaluate integrals that require the use of elliptic functions in their solution
- Use alternative forms of the elliptic functions

Prerequisite: Engineering Mathematics (Sixth Edition)
Programmes 15 Integration 1, 16 Integration 2 and 17 Reduction formulas

Integral functions

1   Some functions are most conveniently defined in the form of integrals and we shall deal with one or two of these in the present Programme.

The gamma function

The gamma function $\Gamma(x)$ is defined by the integral

$$\Gamma(x) = \int_0^\infty t^{x-1} e^{-t} \, dt \tag{1}$$

and is convergent for $x > 0$.

From (1): $\Gamma(x + 1) = \int_0^\infty t^x e^{-t} dt$

Integrating by parts

$$\Gamma(x + 1) = \left[t^x \left(\frac{e^{-t}}{-1} \right) \right]_0^\infty + x \int_0^\infty e^{-t} t^{x-1} \, dt$$

$$= \{0 - 0\} + x\Gamma(x)$$

$$\therefore \ \Gamma(x + 1) = x\Gamma(x) \tag{2}$$

This is a fundamental recurrence relation for gamma functions. It can also be written as $\Gamma(x) = (x - 1)\Gamma(x - 1)$

With it we can derive a number of other results.
   For instance, when $x = n$, a positive integer ≥ 1, then

$$\Gamma(n + 1) = n\Gamma(n) \qquad \text{But} \quad \Gamma(n) = (n - 1)\Gamma(n - 1)$$
$$= n(n - 1)\Gamma(n - 1) \qquad \Gamma(n - 1) = (n - 2)\Gamma(n - 2)$$
$$= n(n - 1)(n - 2)\Gamma(n - 2)$$
$$- - - - - -$$
$$= n(n - 1)(n - 2)(n - 3) \ldots 1\Gamma(1) = n!\Gamma(1)$$

But, from the original definition $\Gamma(1) = \ldots\ldots\ldots\ldots$

2 $$\boxed{\Gamma(1) = 1}$$

Because

$$\Gamma(1) = \int_0^\infty t^0 e^{-t} dt = \left[-e^{-t} \right]_0^\infty = 0 + 1 = 1$$

Therefore, we have $\Gamma(1) = 1$ $\tag{3}$
and $\Gamma(n + 1) = n!$ provided n is a positive integer.

$$\therefore \ \Gamma(7) = \ldots\ldots\ldots\ldots$$

$$\boxed{\Gamma(7) = 720}$$ **3**

Because

$$\Gamma(7) = \Gamma(6 + 1) = 6! = 720.$$

Knowing $\Gamma(7) = 720$, $\Gamma(8) = \ldots\ldots\ldots$ and $\Gamma(9) = \ldots\ldots\ldots$

$$\boxed{\Gamma(8) = 5040; \quad \Gamma(9) = 40\,320}$$ **4**

Because

$$\Gamma(8) = \Gamma(7 + 1) = 7\Gamma(7) = 7(720) = 5040$$
$$\Gamma(9) = \Gamma(8 + 1) = 8\Gamma(8) = 8(5040) = 40\,320$$

We can also use the recurrence relation in reverse

$$\Gamma(x + 1) = x\Gamma(x)$$
$$\therefore \ \Gamma(x) = \frac{\Gamma(x + 1)}{x} \tag{4}$$

For example, given that $\Gamma(7) = 720$, we can determine $\Gamma(6)$

$$\Gamma(6) = \frac{\Gamma(6 + 1)}{6} = \frac{\Gamma(7)}{6} = \frac{720}{6} = 120$$

and then $\Gamma(5) = \ldots\ldots\ldots$

$$\boxed{\Gamma(5) = 24}$$ **5**

$$\Gamma(5) = \frac{\Gamma(5 + 1)}{5} = \frac{\Gamma(6)}{5} = \frac{120}{5} = 24.$$

So far, we have used the original definition

$$\Gamma(x) = \int_0^\infty t^{x-1} e^{-t} \, dt$$

for cases where x is a positive integer n.

What happens when $x = \frac{1}{2}$? We will investigate.

$$\Gamma\left(\tfrac{1}{2}\right) = \int_0^\infty t^{-1/2} e^{-t} \, dt$$

Putting $t = u^2$, $dt = 2u\,du$, then

$$\Gamma\left(\tfrac{1}{2}\right) = \ldots\ldots\ldots$$

6

$$\Gamma(\tfrac{1}{2}) = 2 \int_0^\infty e^{-u^2}\, du$$

Because

$$\Gamma(\tfrac{1}{2}) = \int_0^\infty u^{-1} e^{-u^2} 2u\, du = 2 \int_0^\infty e^{-u^2}\, du.$$

Unfortunately, $\displaystyle\int_0^\infty e^{-u^2}\, du$ cannot easily be determined by normal means. It is, however, important, so we have to find a way of getting round the difficulty.

Evaluation of $\displaystyle\int_0^\infty e^{-x^2}\, dx$

Let $I = \displaystyle\int_0^\infty e^{-x^2}\, dx$, then   also $I = \displaystyle\int_0^\infty e^{-y^2}\, dy$

$$\therefore\ I^2 = \left(\int_0^\infty e^{-x^2}\, dx\right)\left(\int_0^\infty e^{-y^2}\, dy\right) = \int_0^\infty \int_0^\infty e^{-(x^2+y^2)}\, dx\, dy$$

$\delta a = \delta x\, \delta y$ represents an element of area in the x–y plane and the integration with the stated limits covers the whole of the first quadrant.

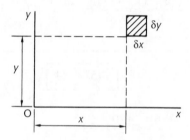

Converting to polar coordinates, the element of area $\delta a = r\, \delta\theta\, \delta r$. Also, $x^2 + y^2 = r^2$

$$\therefore\ e^{-(x^2+y^2)} = e^{-r^2}$$

For the integration to cover the same region as before,

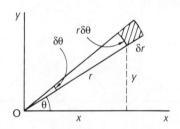

the limits of r are $r = 0$ to $r = \infty$
the limits of θ are $\theta = 0$ to $\theta = \pi/2$.

▶

$$\therefore I^2 = \int_0^{\pi/2} \int_0^\infty e^{-r^2} r \, dr \, d\theta = \int_0^{\pi/2} \left[-\frac{e^{-r^2}}{2} \right]_0^\infty d\theta$$

$$= \int_0^{\pi/2} \left(\frac{1}{2} \right) d\theta = \left[\frac{\theta}{2} \right]_0^{\pi/2} = \frac{\pi}{4}$$

$$\therefore I = \frac{\sqrt{\pi}}{2}$$

$$\therefore \int_0^\infty e^{-x^2} \, dx = \frac{\sqrt{\pi}}{2} \tag{5}$$

This result opens the way for others, so make a note of it and then move on to the next frame

---

Before that diversion, we had established that

7

$$\Gamma\left(\tfrac{1}{2}\right) = 2 \int_0^\infty e^{-u^2} \, du$$

We now know that $\int_0^\infty e^{-u^2} \, du = \frac{\sqrt{\pi}}{2} \qquad \therefore \Gamma\left(\tfrac{1}{2}\right) = \sqrt{\pi}$

From this, using the recurrence relation $\Gamma(x+1) = x\Gamma(x)$, we can obtain the following

$$\Gamma\left(\tfrac{3}{2}\right) = \tfrac{1}{2} \ \Gamma\left(\tfrac{1}{2}\right) = \tfrac{1}{2}(\sqrt{\pi}) \qquad \therefore \Gamma\left(\tfrac{3}{2}\right) = \frac{\sqrt{\pi}}{2}$$

$$\Gamma\left(\tfrac{5}{2}\right) = \tfrac{3}{2} \ \Gamma\left(\tfrac{3}{2}\right) = \tfrac{3}{2}\left(\frac{\sqrt{\pi}}{2} \right) \qquad \therefore \Gamma\left(\tfrac{5}{2}\right) = \frac{3\sqrt{\pi}}{4}$$

$$\Gamma\left(\tfrac{7}{2}\right) = \ldots\ldots\ldots\ldots$$

$$\boxed{\Gamma\left(\tfrac{7}{2}\right) = \frac{15\sqrt{\pi}}{8}}$$

8

Because

$$\Gamma\left(\tfrac{7}{2}\right) = \Gamma\left(\tfrac{5}{2}+1\right) = \tfrac{5}{2} \ \Gamma\left(\tfrac{5}{2}\right) = \frac{5}{2}\left(\frac{3\sqrt{\pi}}{4} \right) = \frac{15\sqrt{\pi}}{8}$$

Using the recurrence relation in reverse, i.e. $\Gamma(x) = \dfrac{\Gamma(x+1)}{x}$, we can also obtain

$$\Gamma\left(-\tfrac{3}{2}\right) = \frac{\Gamma\left(-\tfrac{1}{2}\right)}{-\tfrac{3}{2}} = \frac{\Gamma\left(\tfrac{1}{2}\right)}{\left(-\tfrac{3}{2}\right)\left(-\tfrac{1}{2}\right)} = \tfrac{4}{3}\sqrt{\pi}$$

Negative values of x

Since $\Gamma(x) = \dfrac{\Gamma(x+1)}{x}$, then as $x \to 0$, $\Gamma(x) \to \infty$ $\therefore \Gamma(0) = \infty$.

The same result occurs for all negative integral values of x – which does not follow from the original definition, but which is obtainable from the recurrence relation.

▶

Because at $x = -1$, $\quad \Gamma(-1) = \dfrac{\Gamma(0)}{-1} = \infty$

$x = -2$, $\quad \Gamma(-2) = \dfrac{\Gamma(-1)}{-2} = \infty$ etc.

Also, at $x = -\frac{1}{2}$, $\quad \Gamma\left(-\frac{1}{2}\right) = \dfrac{\Gamma\left(\frac{1}{2}\right)}{-\frac{1}{2}} = -2\sqrt{\pi}$

and at $x = -\frac{3}{2}$, $\quad \Gamma\left(-\frac{3}{2}\right) = \dfrac{\Gamma\left(-\frac{1}{2}\right)}{-\frac{3}{2}} = \dfrac{4}{3}\sqrt{\pi}$

Similarly $\quad \Gamma\left(-\frac{5}{2}\right) = \ldots\ldots\ldots\ldots$

and $\quad \Gamma\left(-\frac{7}{2}\right) = \ldots\ldots\ldots\ldots$

9

$$\boxed{\Gamma\left(-\tfrac{5}{2}\right) = -\dfrac{8}{15}\sqrt{\pi}; \quad \Gamma\left(-\tfrac{7}{2}\right) = \dfrac{16}{105}\sqrt{\pi}}$$

So we have

(a) For n a positive integer

$\Gamma(n+1) = n\Gamma(n) = n!$

$\Gamma(1) = 1; \quad \Gamma(0) = \infty; \quad \Gamma(-n) = \pm\infty$

(b) $\Gamma\left(\frac{1}{2}\right) = \sqrt{\pi}; \qquad\qquad \Gamma\left(-\frac{1}{2}\right) = -2\sqrt{\pi}$

$\Gamma\left(\frac{3}{2}\right) = \dfrac{\sqrt{\pi}}{2}; \qquad\qquad \Gamma\left(-\frac{3}{2}\right) = \dfrac{4}{3}\sqrt{\pi}$

$\Gamma\left(\frac{5}{2}\right) = \dfrac{3\sqrt{\pi}}{4}; \qquad\qquad \Gamma\left(-\frac{5}{2}\right) = -\dfrac{8}{15}\sqrt{\pi}$

$\Gamma\left(\frac{7}{2}\right) = \dfrac{15\sqrt{\pi}}{8}; \qquad\qquad \Gamma\left(-\frac{7}{2}\right) = \dfrac{16}{105}\sqrt{\pi}$

This is quite a useful list. Make a note of it for future use

---

10

Graph of $y = \Gamma(x)$

Values of $\Gamma(x)$ for a range of positive values of x are available in tabulated form in various sets of mathematical tables. These, together with the results established above, enable us to draw the graph of $y = \Gamma(x)$.

x	0	0·5	1·0	1·5	2·0	2·5	3·0	3·5	4·0
$\Gamma(x)$	∞	1·772	1·000	0·886	1·000	1·329	2·000	3·323	6·000

x	−0·5	−1·5	−2·5	−3·5
$\Gamma(x)$	−3·545	2·363	−0·945	0·270

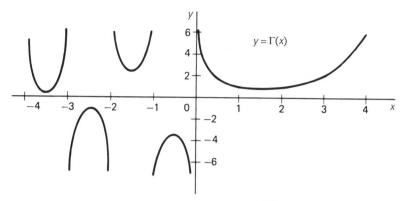

For large n it can be shown that $\Gamma(n+1) \approx \sqrt{2\pi n}\, n^n\, e^{-n}$ which gives rise to Stirling's formula for an approximation to the factorial of a large number

$$n! \approx \sqrt{2\pi n}\, n^n\, e^{-n}$$

Revision **11**

Let us now revise the main points before we move on to some examples.

The definition of $\Gamma(x)$ is that $\Gamma(x) = \dots\dots\dots$

12

$$\boxed{\Gamma(x) = \int_0^\infty t^{x-1}e^{-t}\mathrm{d}t}$$

The recurrence relation states that

$$\Gamma(x+1) = \dots\dots\dots$$

13

$$\boxed{\Gamma(x+1) = x\,\Gamma(x)}$$

When x is a positive integer, i.e. $x = n$, then

$$\Gamma(n+1) = \dots\dots\dots$$

14

$$\boxed{\Gamma(n+1) = n!}$$

Then we have a number of specific results

$$\Gamma(1) = \dots\dots\dots; \quad \Gamma(0) = \dots\dots\dots; \quad \Gamma\left(\tfrac{1}{2}\right) = \dots\dots\dots$$

15

$$\boxed{\Gamma(1) = 1; \quad \Gamma(0) = \infty; \quad \Gamma\left(\tfrac{1}{2}\right) = \sqrt{\pi}}$$

and finally, for all negative integral values of n

$$\Gamma(n) = \dots\dots\dots$$

16

$$\boxed{\Gamma(n) = \pm\infty}$$

Listing them together, we have

$$\Gamma(x) = \int_0^\infty t^{x-1}e^{-t}dt$$

$$\Gamma(x+1) = x\Gamma(x)$$

$$\Gamma(n+1) = n! \qquad \text{for } n \text{ a positive integer}$$

$$\Gamma(1) = 1; \qquad \Gamma(0) = \infty; \quad \Gamma(\tfrac{1}{2}) = \sqrt{\pi}$$

$$\Gamma(n) = \pm\infty \quad \text{for } n \text{ a negative integer.}$$

17   Now for a few examples of evaluation of integrals.

Example 1

Evaluate $\displaystyle\int_0^\infty x^7 e^{-x}dx$.

We recognise this as the standard form of the gamma function

$$\Gamma(x) = \int_0^\infty t^{x-1}e^{-t}dt \qquad \text{with the variables changed.}$$

It is often convenient to write the gamma function as

$$\Gamma(v) = \int_0^\infty x^{v-1}e^{-x}dx$$

Our example then becomes

$$I = \int_0^\infty x^7 e^{-x}dx = \int_0^\infty x^{v-1}e^{-x}\,dx \qquad \text{where } v = \ldots\ldots\ldots$$

18

$$\boxed{v = 8}$$

$$\therefore \ I = \Gamma(v) = \Gamma(8) = \ldots\ldots\ldots\ldots$$

$$\boxed{\Gamma(8) = 7! = 5040}$$

19

i.e. $\displaystyle\int_0^\infty x^7\, e^{-x}\, \mathrm{d}x = \Gamma(8) = 7! = 5040$

Example 2

Evaluate $\displaystyle\int_0^\infty x^3\, e^{-4x}\, \mathrm{d}x$.

If we compare this with $\Gamma(v) = \displaystyle\int_0^\infty x^{v-1}\, e^{-x}\, \mathrm{d}x$, we must reduce the power of e to a single variable, i.e. put $y = 4x$, and we use this substitution to convert the whole integral into the required form.

$$y = 4x \quad \therefore\ \mathrm{d}y = 4\,\mathrm{d}x \qquad \text{Limits remain unchanged.}$$

The integral now becomes . . . .

$$\boxed{I = \int_0^\infty \left(\frac{y}{4}\right)^3 e^{-y}\, \frac{\mathrm{d}y}{4}}$$

20

$$\therefore\ I = \frac{1}{4^4}\int_0^\infty y^3\, e^{-y}\,\mathrm{d}y = \frac{1}{4^4}\Gamma(v) \qquad \text{where } v = \ldots\ldots\ldots$$

$$\boxed{v = 4}$$

21

Because

$$\int_0^\infty y^{v-1}e^{-y}\mathrm{d}y = \int_0^\infty y^3 e^{-y}\mathrm{d}y \qquad \therefore\ v = 4$$

$$\therefore\ I = \frac{1}{4^4}\,\Gamma(4) = \ldots\ldots\ldots\ldots$$

$$\boxed{I = \frac{3}{128}}$$

22

Because

$$I = \frac{1}{256}\Gamma(4) = \frac{1}{256}\,(3!) = \frac{6}{256} = \frac{3}{128}$$

One more.

Example 3

Evaluate $\displaystyle\int_0^\infty x^{1/2}\, e^{-x^2}\, \mathrm{d}x$.

The substitution here is to put . . . .

23

$$\boxed{y = x^2}$$

Work through it as before. When you have completed it, check with the next frame.

24   Here is the working.

$y = x^2$ $\therefore$ $dy = 2x\,dx$ Limits $x = 0, y = 0$; $x = \infty, y = \infty$.

$x = y^{1/2}$ $\therefore$ $x^{1/2} = y^{1/4}$

$\therefore I = \int_0^\infty y^{1/4}\, e^{-y}\, dy/2x = \int_0^\infty \dfrac{y^{1/4}\, e^{-y}\, dy}{2y^{1/2}}$

$\quad = \dfrac{1}{2}\int_0^\infty y^{-1/4}\, e^{-y}\, dy$

$\quad = \dfrac{1}{2}\int_0^\infty y^{\nu-1}\, e^{-y} dy$ where $\nu = \dfrac{3}{4}$ $\therefore I = \dfrac{1}{2}\Gamma\!\left(\dfrac{3}{4}\right)$

From tables, $\Gamma(0{\cdot}75) = 1{\cdot}2254$

$\quad \therefore I = 0{\cdot}613$

Here is part of a table that may be useful.

x	$\Gamma(x)$		x	$\Gamma(x)$
0·25	3·6256		2·75	1·6084
0·50	1·7725		3·00	2·0000
0·75	1·2254		3·25	2·5493
1·00	1·0000		3·50	3·3234
1·25	0·9064		3·75	4·4230
1·50	0·8862		4·00	6·0000
1·75	0·9191		4·25	8·2851
2·00	1·0000		4·50	11·6318
2·25	1·1330		4·75	16·5862
2·50	1·3293		5·00	24·0000

Now we will move on to another set of functions closely related to gamma functions.

Let us start a new frame

The beta function   25

The beta function $B(m, n)$, is defined by

$$B(m, n) = \int_0^1 x^{m-1}(1 - x)^{n-1}\, dx \qquad (1)$$

which converges for $m > 0$ and $n > 0$.

Putting $(1 - x) = u \quad \therefore x = 1 - u \quad \therefore dx = -du$

Limits: when $x = 0$, $u = 1$; when $x = 1$, $u = 0$

$$\therefore \; B(m, n) = -\int_1^0 (1 - u)^{m-1} u^{n-1}\, du = \int_0^1 (1 - u)^{m-1} u^{n-1} du$$

$$= \int_0^1 u^{n-1}(1 - u)^{m-1} du = B(n, m)$$

$$\therefore \; B(m, n) = B(n, m) \qquad (2)$$

Alternative form of the beta function

We had

$$B(m, n) = \int_0^1 x^{m-1}(1 - x)^{n-1}\, dx$$

If we put $x = \sin^2 \theta$, the result then becomes ....

26

$$\boxed{B(m, n) = 2\int_0^{\pi/2} \sin^{2m-1}\theta \; \cos^{2n-1}\theta\, d\theta}$$

Because if $x = \sin^2 \theta$, $dx = 2\sin\theta \; \cos\theta\, d\theta$.

When $x = 0$, $\theta = 0$; when $x = 1$, $\theta = \pi/2$. $1 - x = 1 - \sin^2\theta = \cos^2\theta$

$$\therefore \; B(m, n) = 2\int_0^{\pi/2} \sin^{2m-2}\theta \cos^{2n-2}\theta \sin\theta \; \cos\theta\, d\theta$$

$$\therefore \; B(m, n) = 2\int_0^{\pi/2} \sin^{2m-1}\theta \cos^{2n-1}\theta\, d\theta \qquad (3)$$

Make a note of this result. We shall need to use it later.

27 Reduction formulas

In Programme 17 of *Engineering Mathematics (Sixth Edition)* we established useful reduction formulas relating to integrals of powers of sines and cosines, particularly when the integral limits are 0 and $\pi/2$.

(a) $\displaystyle\int_0^{\pi/2} \sin^n x \, dx = \frac{n-1}{n} \int_0^{\pi/2} \sin^{n-2} x \, dx$ i.e. $S_n = \dfrac{n-1}{n} S_{n-2}$ $\hfill$ (4)

(b) $\displaystyle\int_0^{\pi/2} \cos^n x \, dx = \frac{n-1}{n} \int_0^{\pi/2} \cos^{n-2} x \, dx$ i.e. $C_n = \dfrac{n-1}{n} C_{n-2}$ $\hfill$ (5)

A third reduction formula for products of powers of sines and cosines is

(c) $\displaystyle\int_0^{\pi/2} \sin^m x \, \cos^n x \, dx = \frac{m-1}{m+n} \int_0^{\pi/2} \sin^{m-2} x \, \cos^n x \, dx$

If we denote $\displaystyle\int_0^{\pi/2} \sin^m x \, \cos^n x \, dx$ by $I_{m,n}$, the last result can be written

$$I_{m,n} = \frac{m-1}{m+n} I_{m-2,n} \hfill (6)$$

Alternatively, $\displaystyle\int_0^{\pi/2} \sin^m x \, \cos^n x \, dx$ can be expressed as

$$\frac{n-1}{m+n} \int_0^{\pi/2} \sin^m x \, \cos^{n-2} x \, dx$$

i.e. $\quad I_{m,n} = \dfrac{n-1}{m+n} I_{m,n-2} \hfill (7)$

Now $B(m, n) = 2 \displaystyle\int_0^{\pi/2} \sin^{2m-1}\theta \, \cos^{2n-1}\theta \, d\theta$ and if we apply (6) to the integral, we have

$$\int_0^{\pi/2} \sin^{2m-1}\theta \, \cos^{2n-1}\theta \, d\theta$$

$$= \frac{(2m-1)-1}{(2m-1)+(2n-1)} \int_0^{\pi/2} \sin^{2m-3}\theta \, \cos^{2n-1}\theta \, d\theta$$

$$= \frac{m-1}{m+n-1} \int_0^{\pi/2} \sin^{2m-3}\theta \cos^{2n-1}\theta \, d\theta$$

Now, using (7) with the right-hand integral

$$\int_0^{\pi/2} \sin^{2m-1}\theta \, \cos^{2n-1}\theta \, d\theta$$

$$= \frac{m-1}{m+n-1} \times \frac{(2n-1)-1}{(2m-3)+(2n-1)} \times \int_0^{\pi/2} \sin^{2m-3}\theta \cos^{2n-3}\theta \, d\theta$$

$$= \frac{m-1}{m+n-1} \times \frac{n-1}{m+n-2} \times \int_0^{\pi/2} \sin^{2m-3}\theta \cos^{2n-3}\theta \, d\theta$$

▶

$$\therefore \ B(m, n) = \frac{(m - 1)(n - 1)}{(m + n - 1)(m + n - 2)} \times 2 \int_0^{\pi/2} \sin^{2m-3}\theta \ \cos^{2n-3}\theta \ d\theta$$

$$\text{i.e. } B(m, n) = \frac{(m - 1)(n - 1)}{(m + n - 1)(m + n - 2)} B(m - 1, n - 1) \tag{8}$$

This is obviously a reduction formula for B(m, n) and the process can be repeated as required.

For example B(4, 3) = ...

28

$$B(4, 3) = \frac{(3)(2)\,(2)(1)}{(6)(5)\,(4)(3)} B(2, 1)$$

Because, applying (8)

$$B(4, 3) = \frac{(3)(2)}{(6)(5)} B(3, 2) = \frac{(3)(2)\,(2)(1)}{(6)(5)\,(4)(3)} B(2, 1)$$

Now we must evaluate B(2, 1) for we can go no further in the reduction process, since, from the definition of B(m, n), m and n must be

....

29

$$> 0$$

But $B(2, 1) = 2 \int_0^{\pi/2} \sin^3\theta \ \cos\theta \ d\theta = 2\left[\frac{\sin^4\theta}{4}\right]_0^{\pi/2} = \frac{1}{2}$

$$\therefore \ B(4, 3) = \frac{(3)(2)\,(2)(1)}{(6)(5)\,(4)(3)}\frac{1}{2}$$

$$= \frac{(3)(2)(1) \times (2)(1)}{(6)(5)(4)(3)(2)(1)} = \frac{(3!)(2!)}{(6!)}$$

Similarly, B(5, 3) = ...

30

$$B(5, 3) = \frac{(4!)(2!)}{(7!)}$$

Because

$$B(5, 3) = \frac{(4)(2)}{(7)(6)} B(4, 2) = \frac{(4)(2)\,(3)(1)}{(7)(6)\,(5)(4)} B(3, 1)$$

$$B(3, 1) = 2 \int_0^{\pi/2} \sin^5\theta \ \cos\theta \ d\theta = 2\left[\frac{\sin^6\theta}{6}\right]_0^{\pi/2} = \frac{1}{3}$$

$$\therefore \ B(5, 3) = \frac{(4)(2)\,(3)(1)}{(7)(6)\,(5)(4)}\frac{1}{3}\frac{(2)}{(2)} = \frac{(4!)(2!)}{(7!)}$$

▶

In general $B(m, n) = \dfrac{(m-1)!(n-1)!}{(m+n-1)!}$ (9)

Note that $B(k, 1) = 2\displaystyle\int_0^{\pi/2} \sin^{2k-1}\theta \, \cos\theta \, d\theta$

$= 2\displaystyle\int_0^{\pi/2} \sin^{2k-1}\theta \, d(\sin\theta)$

$= 2\left[\dfrac{\sin^{2k}\theta}{2k}\right]_0^{\pi/2} = \dfrac{1}{k}$

$\therefore \; B(k, 1) = \dfrac{1}{k}$

$\therefore \; B(k, 1) = B(1, k) = \dfrac{1}{k}$ (10)

We can also use the trigonometrical definition (3) to evaluate $B\left(\frac{1}{2}, \frac{1}{2}\right)$

$$B\left(\tfrac{1}{2}, \tfrac{1}{2}\right) = \ldots\ldots\ldots\ldots$$

31

$$\boxed{B\left(\tfrac{1}{2}, \tfrac{1}{2}\right) = \pi}$$

Because

$$B(m, n) = 2\int_0^{\pi/2} \sin^{2m-1}\theta \, \cos^{2n-1}\theta \, d\theta$$

$$\therefore \; B\left(\tfrac{1}{2}, \tfrac{1}{2}\right) = 2\int_0^{\pi/2} \sin^0\theta \, \cos^0\theta \, d\theta$$

$$= 2\int_0^{\pi/2} 1 \, d\theta = 2\Big[\theta\Big]_0^{\pi/2} = \pi \qquad (11)$$

Now let us summarise our various results so far.   *Next frame*

---

32 Revision

$$B(m, n) = \int_0^1 x^{m-1}(1-x)^{n-1} \, dx \quad m > 0, \; n > 0$$

$$B(m, n) = B(n, m)$$

$$B(m, n) = 2\int_0^{\pi/2} \sin^{2m-1}\theta \, \cos^{2n-1}\theta \, d\theta$$

$$B(m, n) = \dfrac{(m-1)(n-1)}{(m+n-1)(m+n-2)}B(m-1, n-1)$$

$$B(m, n) = \dfrac{(m-1)!(n-1)!}{(m+n-1)!} \qquad m \text{ and } n \text{ positive integers}$$

$$B(k, 1) = B(1, k) = \dfrac{1}{k} \qquad \therefore \; B(1, 1) = 1$$

$$B\left(\tfrac{1}{2}, \tfrac{1}{2}\right) = \pi$$

Be sure that you are familiar with all these. We shall be using them all in due course.

Relation between the gamma and beta functions

33

If m and n are positive integers

$$B(m, n) = \frac{(m - 1)!(n - 1)!}{(m + n - 1)!}$$

Also, we have previously established that, for n a positive integer,

$$n! = \Gamma(n + 1)$$

$$\therefore \ (m - 1)! = \Gamma(m) \ \text{ and } \ (n - 1)! = \Gamma(n)$$

and also $(m + n - 1)! = \Gamma(m + n)$

$$\therefore \ B(m, n) = \frac{(m - 1)!(n - 1)!}{(m + n - 1)!} = \frac{\Gamma(m)\Gamma(n)}{\Gamma(m + n)} \tag{12}$$

The relation $B(m, n) = \dfrac{\Gamma(m)\Gamma(n)}{\Gamma(m + n)}$ holds good even when m and n are not necessarily integers.

We will prove this in the next frame, so move on

Proof that $\ B(m, n) = \dfrac{\Gamma(m)\Gamma(n)}{\Gamma(m + n)}$

34

Let $\Gamma(m) = \displaystyle\int_0^\infty x^{m-1} \, e^{-x} \, dx \ $ and $\ \Gamma(n) = \displaystyle\int_0^\infty y^{n-1} \, e^{-y} \, dy$

$$\therefore \ \Gamma(m)\Gamma(n) = \int_0^\infty x^{m-1} \, e^{-x} \, dx \int_0^\infty y^{n-1} \, e^{-y} \, dy$$

$$= \int_0^\infty \int_0^\infty x^{m-1} \, y^{n-1} \, e^{-(x+y)} \, dx \, dy$$

Note that the integration is carried out over the first quadrant of the x–y plane.

Putting $x = u^2$ and $y = v^2$ $\quad dx = 2u \, du \ $ and $\quad dy = 2v \, dv$

$$\therefore \ \Gamma(m)\Gamma(n) = 4 \int_0^\infty \int_0^\infty u^{2m-2} \, v^{2n-2} \, e^{-(u^2+v^2)} uv \, du \, dv$$

$$= 4 \int_0^\infty \int_0^\infty u^{2m-1} \, v^{2n-1} \, e^{-(u^2+v^2)} \, du \, dv$$

If we now convert to polar coordinates,
$u = r\cos\theta;\quad v = r\sin\theta;\quad \mathrm{d}u\,\mathrm{d}v = r\,\mathrm{d}r\,\mathrm{d}\theta$
$u^2 + v^2 = r^2\quad 0 < r < \infty\quad 0 < \theta < \pi/2$

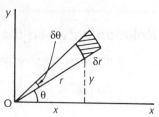

$$\therefore\ \Gamma(m)\Gamma(n) = 4\int_0^{\pi/2}\int_0^\infty r^{2m-1}\cos^{2m-1}\theta\, r^{2n-1}\sin^{2n-1}\theta\ e^{-r^2}\, r\,\mathrm{d}r\,\mathrm{d}\theta$$

$$= 4\int_0^{\pi/2}\int_0^\infty r^{2m+2n-2}\, e^{-r^2}\cos^{2m-1}\theta\,\sin^{2n-1}\theta\, r\,\mathrm{d}r\,\mathrm{d}\theta$$

Then, writing $w = r^2\quad\therefore\ \mathrm{d}w = 2r\,\mathrm{d}r$

$$\Gamma(m)\Gamma(n) = 2\int_0^\infty w^{m+n-1}\, e^{-w}\,\mathrm{d}w\int_0^{\pi/2}\sin^{2n-1}\theta\,\cos^{2m-1}\theta\,\mathrm{d}\theta$$

$$= \Gamma(m+n)\times \mathrm{B}(m,\, n)$$

$$\therefore\ \mathrm{B}(m,\, n) = \frac{\Gamma(m)\Gamma(n)}{\Gamma(m+n)} \tag{13}$$

So $\mathrm{B}(\tfrac{3}{2},\, \tfrac{1}{2}) = \ldots\ldots\ldots\ldots$

35

$$\boxed{\mathrm{B}(\tfrac{3}{2},\, \tfrac{1}{2}) = \frac{\pi}{2}}$$

Because

$$\mathrm{B}(\tfrac{3}{2},\, \tfrac{1}{2}) = \frac{\Gamma(\tfrac{3}{2})\Gamma(\tfrac{1}{2})}{\Gamma(2)} = \frac{\sqrt{\pi}/2 \times \sqrt{\pi}}{1} = \frac{\pi}{2}$$

Now for some examples.

---

36   Application of gamma and beta functions

The use of gamma and beta functions in the evaluation of definite integrals depends largely on the ability to change the variables to express the integral in the basic form of the beta function $\displaystyle\int_0^1 x^{m-1}(1-x)^{n-1}\mathrm{d}x$ or its trigonometrical form $\displaystyle 2\int_0^{\pi/2}\sin^{2m-1}\theta\cos^{2n-1}\theta\,\mathrm{d}\theta$.

Example 1

Evaluate $I = \displaystyle\int_0^1 x^5(1-x)^4\,\mathrm{d}x$.

Compare this with $\mathrm{B}(m,\, n) = \displaystyle\int_0^1 x^{m-1}(1-x)^{n-1}\mathrm{d}x$

Then $m - 1 = 5\quad\therefore\ m = 6$ and $n - 1 = 4\quad\therefore\ n = 5$

$$\therefore\ I = \mathrm{B}(6,\, 5) = \ldots\ldots\ldots\ldots$$

37

$$I = B(6,\ 5) = \frac{5!\ 4!}{10!} = \frac{1}{1260}$$

Example 2

Evaluate $I = \displaystyle\int_0^1 x^4 \sqrt{1-x^2}\,dx$.

Comparing this with $B(m,\ n) = \displaystyle\int_0^1 x^{m-1}(1-x)^{n-1}\,dx$

we see that we have x^2 in the root, instead of a single x.

Therefore, put $x^2 = y$ $\therefore$ $x = y^{\frac{1}{2}}$ $dx = \frac{1}{2}y^{-\frac{1}{2}}\,dy$

The limits remain unchanged. $\therefore$ $I = \ldots\ldots\ldots$

38

$$I = \tfrac{1}{2}B\left(\tfrac{5}{2},\tfrac{3}{2}\right)$$

Because

$$I = \int_0^1 y^2(1-y)^{\frac{1}{2}}\frac{1}{2}y^{-\frac{1}{2}}\,dy = \frac{1}{2}\int_0^1 y^{\frac{3}{2}}(1-y)^{\frac{1}{2}}\,dy$$
$$m - 1 = \tfrac{3}{2} \quad \therefore \quad m = \tfrac{5}{2} \quad \text{and} \quad n - 1 = \tfrac{1}{2} \quad \therefore \quad n = \tfrac{3}{2}$$
$$\therefore \quad I = \tfrac{1}{2}B\left(\tfrac{5}{2},\tfrac{3}{2}\right)$$

Expressing this in gamma functions

$$I = \ldots\ldots\ldots$$

39

$$I = \frac{1}{2}\frac{\Gamma\left(\frac{5}{2}\right)\Gamma\left(\frac{3}{2}\right)}{\Gamma(4)}$$

From our previous work on gamma functions

$$\Gamma\left(\tfrac{3}{2}\right) = \frac{\sqrt{\pi}}{2};\ \ \Gamma\left(\tfrac{5}{2}\right) = \frac{3\sqrt{\pi}}{4};\ \ \Gamma(4) = 3!$$
$$\therefore \quad I = \ldots\ldots\ldots$$

40

$$I = \frac{\pi}{32}$$

Because

$$I = \frac{1}{2} \cdot \frac{(3\sqrt{\pi}/4)(\sqrt{\pi}/2)}{3!} = \frac{\pi}{32}.$$

Now you can work through this one in much the same way. There are no tricks.

Example 3

Evaluate $I = \int_0^3 \dfrac{x^3\,dx}{\sqrt{3-x}}$.

You need to compare this with $B(m, n) = \int_0^1 x^{m-1}(1-x)^{n-1}\,dx$ so bring every-thing up on to the top line and then make the necessary change in the variables. Finish it off and then compare the results with the next frame.

41

$$\boxed{I = \frac{864\sqrt{3}}{35} = 42{\cdot}76}$$

Here is the working; see whether you agree.

$$I = \int_0^3 \frac{x^3\,dx}{\sqrt{3-x}} = \int_0^3 x^3(3-x)^{-\frac{1}{2}}\,dx = 3^{-\frac{1}{2}} \int_0^3 x^3\left(1-\frac{x}{3}\right)^{-\frac{1}{2}}\,dx$$

Put $\dfrac{x}{3} = y$,   i.e. $x = 3y$ $\therefore\ dx = 3\,dy$

Limits: $x = 0,\ y = 0;\quad x = 3,\ y = 1$

$$\therefore\ I = 27\sqrt{3} \int_0^1 y^3(1-y)^{-\frac{1}{2}}\,dy \qquad\qquad \begin{array}{ll} m - 1 = 3 & \therefore\ m = 4 \\ n - 1 = -\frac{1}{2} & \therefore\ n = \frac{1}{2} \end{array}$$

$$\therefore\ I = 27\sqrt{3}\ B\!\left(4, \tfrac{1}{2}\right) = 27\sqrt{3}\ \frac{\Gamma(4)\Gamma\!\left(\tfrac{1}{2}\right)}{\Gamma\left(9/2\right)}$$

Now $\Gamma\!\left(\tfrac{1}{2}\right) = \sqrt{\pi};\ \Gamma(9/2) = \dfrac{105\sqrt{\pi}}{16};\ \Gamma(4) = 3!$

$$\therefore\ I = 27\sqrt{3} \times 6 \times \sqrt{\pi} \times \frac{16}{105\sqrt{\pi}} = \frac{864\sqrt{3}}{35} = 42{\cdot}76$$

Example 4

Evaluate $I = \int_0^{\pi/2} \sin^5\theta \cos^4\theta\,d\theta$.

$$B(m, n) = 2\int_0^{\pi/2} \sin^{2m-1}\theta \cos^{2n-1}\theta\,d\theta$$

$\therefore\ 2m - 1 = 5\quad \therefore\ m = 3;\quad 2n - 1 = 4\quad \therefore\ n = 5/2$

$$\therefore\ I = \tfrac{1}{2}\,B(3,\ 5/2) = \ldots\ldots\ldots\ldots$$

Finish it off

$$\boxed{I = \frac{8}{315}}$$

42

$$I = \frac{1}{2} \, \mathrm{B}(3, 5/2) = \frac{1}{2} \cdot \frac{\Gamma(3)\Gamma(5/2)}{\Gamma(11/2)}$$

$$= \frac{1}{2} \cdot \frac{2!(3\sqrt{\pi})/4}{(945\sqrt{\pi})/32} = \frac{3\sqrt{\pi}}{4} \cdot \frac{32}{945\sqrt{\pi}} = \frac{8}{315}$$

Finally, one more.

Example 5

Evaluate $I = \displaystyle\int_0^{\pi/2} \sqrt{\tan\theta} \; \mathrm{d}\theta$.

Somehow, we need to turn this into the form

$$\mathrm{B}(m, n) = 2\int_0^{\pi/2} \sin^{2m-1}\theta \; \cos^{2n-1}\theta \; \mathrm{d}\theta$$

So off you go; express the result in gamma functions

$$I = \ldots\ldots\ldots$$

$$\boxed{I = \frac{1}{2} \cdot \frac{\Gamma(\frac{3}{4})\Gamma(\frac{1}{4})}{\Gamma(1)}}$$

43

Because

$$I = \int_0^{\pi/2} \sqrt{\tan\theta} \; \mathrm{d}\theta = \int_0^{\pi/2} \sin^{\frac{1}{2}}\theta \cos^{-\frac{1}{2}}\theta \; \mathrm{d}\theta$$

$$\therefore \; 2m - 1 = \frac{1}{2} \quad \therefore \; m = \frac{3}{4}; \quad 2n - 1 = -\frac{1}{2} \quad \therefore \; n = \frac{1}{4}$$

$$\therefore \; I = \frac{1}{2} \, \mathrm{B}\left(\frac{3}{4}, \frac{1}{4}\right) = \frac{1}{2} \cdot \frac{\Gamma(\frac{3}{4})\Gamma(\frac{1}{4})}{\Gamma(1)}$$

and, unless we have appropriate tables to evaluate $\Gamma(\frac{3}{4})$ and $\Gamma(\frac{1}{4})$, we cannot proceed much further. However, we do have such a table in Frame 24 so refer to it to evaluate the integral of our example.

$$I = \ldots\ldots\ldots$$

44

$$\boxed{I = 2\cdot 2214}$$

Because

$$\Gamma(0\cdot 25) = 3\cdot 6256 \quad \text{and} \quad \Gamma(0\cdot 75) = 1\cdot 2254$$

$$\therefore \ I = \frac{1}{2} \cdot \frac{(1\cdot 2254)(3\cdot 6256)}{1\cdot 0000} = 2\cdot 2214$$

Duplication formula for gamma functions

We already know that, when n is a positive integer

$$\Gamma(n) = (n-1)!$$

A useful formula enables us to calculate the gamma functions for values of n halfway between the integers. This is the *duplication formula* which can be stated as

$$\Gamma(n + \tfrac{1}{2}) = \frac{\Gamma(2n)\sqrt{\pi}}{2^{2n-1}\Gamma(n)} \tag{14}$$

Thus, to find $\quad \Gamma(3\cdot 5) \qquad \Gamma(n) = \Gamma(3) = 2!$

$$\Gamma(2n) = \Gamma(6) = 5!$$

$$\therefore \ \Gamma(3\cdot 5) = \Gamma(3 + \tfrac{1}{2}) = \frac{5!\sqrt{\pi}}{2^5\,2!} = 3\cdot 3234$$

The formula is quoted here without proof, but it is useful to have on occasions.

So $\quad \Gamma(6\cdot 5) = \ldots\ldots\ldots\ldots$

45

$$\boxed{\Gamma(6\cdot 5) = 287\cdot 9}$$

$$\Gamma(6\cdot 5) = \Gamma(6 + \tfrac{1}{2}) = \frac{\Gamma(12)\sqrt{\pi}}{2^{11}\Gamma(6)}$$

$$\Gamma(6) = 5!; \ \ \Gamma(12) = 11!; \ \ 2^{11} = 2048$$

$$\therefore \ \Gamma(6\cdot 5) = \frac{11!\sqrt{\pi}}{2048 \times 5!} = 287\cdot 9$$

Now let us consider another function represented by an integral.

On then to the next frame

The error function

The error function erf (x) is defined as

46

$$\text{erf}(x) = \frac{2}{\sqrt{\pi}} \int_0^x e^{-t^2} \, dt$$

and occurs in statistics and various studies in physics and engineering. This integral, for arbitrary x, can only be evaluated numerically and values of erf (x) for various values of x are obtained from tables.

Where the limits of $\int_a^b e^{-t^2} \, dt$ are zero or $\pm\infty$, however, an exact result is possible. We have already considered the integral $I = \int_0^\infty e^{-t^2} \, dt$ in Frame 6 when dealing with gamma functions and we established then that

$$\int_0^\infty e^{-t^2} \, dt =$$

$$\boxed{\int_0^\infty e^{-t^2} \, dt = \frac{1}{2} \Gamma(\tfrac{1}{2}) = \frac{\sqrt{\pi}}{2}}$$

47

Consequently

$$\underset{x \to \infty}{Lim} \, (\text{erf}(x)) = \frac{2}{\sqrt{\pi}} \int_0^\infty e^{-t^2} \, dt = 1$$

By representing the exponential function in the integral by its Maclaurin series we see that

$$\text{erf}(x) = \frac{2}{\sqrt{\pi}} \sum_{n=0}^{\infty} \dots\dots\dots$$

$$\boxed{\text{erf}(x) = \frac{2}{\sqrt{\pi}} \sum_{n=0}^{\infty} \frac{(-1)^n x^{2n+1}}{n!(2n+1)}}$$

48

Because

$$\text{erf}(x) = \frac{2}{\sqrt{\pi}} \int_0^x e^{-t^2} \, dt$$

$$= \frac{2}{\sqrt{\pi}} \int_0^x \left(\sum_{n=0}^{\infty} \frac{(-1)^n t^{2n}}{n!} \right) dt$$

$$= \frac{2}{\sqrt{\pi}} \sum_{n=0}^{\infty} \left(\int_0^x \frac{(-1)^n t^{2n}}{n!} \, dt \right)$$

$$= \frac{2}{\sqrt{\pi}} \sum_{n=0}^{\infty} \frac{(-1)^n x^{2n+1}}{n!(2n+1)}$$

Consequently erf $(-x) = -\text{erf}(x)$ and so erf (x) is an odd function.

▶

The graph of erf(x)

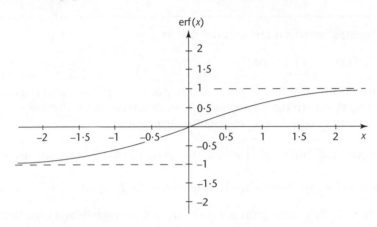

The complementary error function erfc(x)

The complementary error function is defined as

$$\text{erfc}(x) = \frac{2}{\sqrt{\pi}} \int_x^\infty e^{-t^2}\, dt$$

which is related to the error function by the relation

$$\text{erfc}(x) = \dots\dots\dots$$

49

$$\boxed{\text{erfc}(x) = 1 - \text{erf}(x)}$$

Because

$$\text{erfc}(x) = \frac{2}{\sqrt{\pi}} \int_x^\infty e^{-t^2}\, dt$$

$$= \frac{2}{\sqrt{\pi}} \left(\int_0^\infty e^{-t^2}\, dt - \int_0^x e^{-t^2}\, dt \right)$$

$$= 1 - \text{erf}(x)$$

Example 1

In terms of the complementary error function, for $0 < a < b$

$$\int_a^b e^{-t^2}\, dt = \dots\dots\dots$$

$$\frac{\sqrt{\pi}}{2}[\text{erfc}\,(a) - \text{erfc}\,(b)]$$

Because

$$\int_a^b e^{-t^2}\,\mathrm{d}t = \int_0^b e^{-t^2}\,\mathrm{d}t - \int_0^a e^{-t^2}\,\mathrm{d}t$$

$$= \frac{\sqrt{\pi}}{2}\text{erf}\,(b) - \frac{\sqrt{\pi}}{2}\text{erf}\,(a)$$

$$= \frac{\sqrt{\pi}}{2}[1 - \text{erfc}\,(b)] - \frac{\sqrt{\pi}}{2}[1 - \text{erfc}\,(a)]$$

$$= \frac{\sqrt{\pi}}{2}[\text{erfc}\,(a) - \text{erfc}\,(b)]$$

Example 2

In statistics the integral

$$\Phi(x) = \frac{1}{\sqrt{2\pi}}\int_{-\infty}^x e^{-t^2/2}\,\mathrm{d}t$$

is the area beneath the Gaussian or normal probability distribution $\frac{1}{\sqrt{2\pi}}e^{-t^2/2}$ for the values $-\infty < t \le x$.

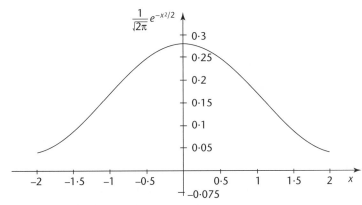

The area beneath the complete Gaussian curve is then

$$\frac{1}{\sqrt{2\pi}}\int_{-\infty}^{\infty} e^{-t^2/2}\,\mathrm{d}t = \ldots\ldots\ldots\ldots$$

51

| 1 |

Because

$$\frac{1}{\sqrt{2\pi}} \int_{-\infty}^{\infty} e^{-t^2/2}\,dt = \frac{1}{\sqrt{2\pi}}\left(2\int_{0}^{\infty} e^{-t^2/2}\,dt\right) \quad \text{because the integrand is even}$$

$$= \sqrt{\frac{2}{\pi}} \int_{0}^{\infty} e^{-t^2/2}\,dt$$

$$= \sqrt{\frac{2}{\pi}} \times \sqrt{2} \int_{0}^{\infty} e^{-u^2}\,du \quad \text{where } u = t/\sqrt{2}$$

$$= 1 \qquad\qquad\qquad \text{from Frame 47}$$

For positive x, $\Phi(x)$ is related to the error function

$$\Phi(x) = \ldots\ldots\ldots$$

$$\boxed{\Phi(x) = \frac{1}{2} + \frac{1}{2}\operatorname{erf}\left(\frac{x}{\sqrt{2}}\right)}$$

52

Because

$$\Phi(x) = \frac{1}{\sqrt{2\pi}} \int_{-\infty}^{x} e^{-t^2/2}\,dt$$

$$= \frac{1}{\sqrt{2\pi}} \int_{-\infty}^{0} e^{-t^2/2}\,dt + \frac{1}{\sqrt{2\pi}} \int_{0}^{x} e^{-t^2/2}\,dt$$

$$= \frac{1}{2} + \frac{1}{\sqrt{2\pi}} \times \sqrt{2} \int_{0}^{x/\sqrt{2}} e^{-u^2}\,du \quad \text{where } u = t/\sqrt{2}$$

$$= \frac{1}{2} + \frac{1}{2}\operatorname{erf}\left(\frac{x}{\sqrt{2}}\right)$$

Now let us consider a new set of integral funtions.

Elliptic functions

53

The use of *elliptic functions* provides a means of evaluating a further range of definite integrals, provided that the integrals can be converted by various appropriate substitutions into certain standard forms.

If an integrand is a rational function of x and of $\sqrt{P(x)}$ where $P(x)$ is a polynomial in x of degree 3 or 4, then the integral is said to be *elliptic*.

For example, $\displaystyle\int_{0}^{1} \frac{dx}{\sqrt{(1-2x^2)(4-3x^2)}}$ is an elliptic integral. The name is derived from such an integral occurring in the determination of the arc length of part of an ellipse.

▶

Standard forms of elliptic functions

(a) *Of the first kind*

$$F(k, \phi) = \int_0^\phi \frac{\mathrm{d}\theta}{\sqrt{1 - k^2 \sin^2 \theta}} \qquad (1)$$

where $0 \le \phi \le \pi/2$ and $0 < k < 1$.

(b) *Of the second kind*

$$E(k, \phi) = \int_0^\phi \sqrt{1 - k^2 \sin^2 \theta}\, \mathrm{d}\theta \qquad (2)$$

where $0 \le \phi \le \dfrac{\pi}{2}$ and $0 < k < 1$.

Make a careful note of these two standard forms: then we can apply them to some examples.

Example 1   **54**

Evaluate $\displaystyle\int_0^{\pi/2} \sqrt{4 - \sin^2 \theta}\, \mathrm{d}\theta$ in terms of an elliptic function.

Taking out a factor 4 to reduce the first term to 1

$$I = 2\int_0^{\pi/2} \sqrt{1 - \frac{1}{4}\sin^2 \theta}\, \mathrm{d}\theta$$

The integral now agrees with the standard form, where $k^2 = \frac{1}{4}$, i.e. $k = \frac{1}{2}$ and $\phi = \pi/2$.

$$\therefore\; I = \ldots\ldots\ldots\ldots$$

$$\boxed{I = 2E(\tfrac{1}{2},\, \pi/2)}$$ **55**

Complete elliptic functions

In each of the cases (1) and (2) listed above, if $\phi = \pi/2$, the integral is said to be *complete* and then

$$F(k, \pi/2) \text{ is denoted by } K(k)$$
and $\quad E(k,\, \pi/2)$ is denoted by $E(k)$.

The method, then, rests on making suitable substitutions in a given integral to transform the integrand into one of the standard forms stated above. For various values of k and ϕ, values of the functions $F(k,\, \phi)$, $E(k,\, \phi)$, $K(k)$ and $E(k)$ are obtainable from published tables. These tables, which are quite extensive, are not reproduced here and so many required values will be given in the text.

Incidentally, the result of Example 1 above, i.e. $I = 2E(\frac{1}{2},\, \pi/2)$ could also be written as

$$I = \ldots\ldots\ldots\ldots$$

56

$$\boxed{I = 2E\left(\tfrac{1}{2}\right)}$$

because, in this case, $\phi = \pi/2$.

From tables, we find that $E\left(\tfrac{1}{2}\right) = 1\cdot4675$ $\therefore$ $I = 2\cdot935$

Example 2

Evaluate $I = \displaystyle\int_0^{\pi/6} \frac{d\theta}{\sqrt{1 - 4\sin^2\theta}}$.

At first sight, this seems to be in standard form, but notice that the value of k^2 is 4, i.e. $k = 2$ – and this does not comply with the requirement that $0 < k < 1$. We therefore proceed as follows.

$$I = \int_0^{\pi/6} \frac{d\theta}{\sqrt{1 - 4\sin^2\theta}}$$

Put $4\sin^2\theta = \sin^2\psi$

i.e. $2\sin\theta = \sin\psi$

$\therefore$ $2\cos\theta\,d\theta = \cos\psi\,d\psi$ $\therefore$ $d\theta = \dfrac{\cos\psi\,d\psi}{2\cos\theta}$

Also, for the new limits, when $\theta = 0$, $\psi = \ldots\ldots\ldots$

and   when $\theta = \pi/6$, $\psi = \ldots\ldots\ldots$

57

$$\boxed{\theta = 0, \ \psi = 0; \quad \theta = \pi/6, \ \psi = \pi/2}$$

$\therefore$ $I = \displaystyle\int_0^{\pi/2} \frac{1}{\sqrt{1 - \sin^2\psi}} \cdot \frac{\cos\psi\,d\psi}{2\,\cos\theta}$

We now transform the $\cos\theta$

$\sin\theta = \tfrac{1}{2}\sin\psi$ $\therefore$ $1 - \cos^2\theta = \tfrac{1}{4}\sin^2\psi$ $\therefore$ $\cos\theta = \sqrt{1 - \tfrac{1}{4}\sin^2\psi}$

$\therefore$ $I = \dfrac{1}{2}\displaystyle\int_0^{\pi/2} \frac{1}{\cos\psi} \cdot \frac{\cos\psi\,d\psi}{\sqrt{1 - \tfrac{1}{4}\sin^2\psi}}$

$= \dfrac{1}{2}\displaystyle\int_0^{\pi/2} \frac{d\psi}{\sqrt{1 - \tfrac{1}{4}\sin^2\psi}}$ which is now in standard form

$\therefore$ $I = \ldots\ldots\ldots$

$$I = \tfrac{1}{2}F\left(\tfrac{1}{2}, \pi/2\right) = \tfrac{1}{2}K\left(\tfrac{1}{2}\right)$$

58

From the appropriate tables, $K\left(\tfrac{1}{2}\right) = 1.6858$ $\therefore I = 0.8429$

Now for another

Example 3

Evaluate $I = \displaystyle\int_0^{\pi/3} \frac{d\theta}{\sqrt{3 - 4\sin^2\theta}}$.

The first step is to ....

take out a factor 3 to reduce the first term to 1

59

$$\therefore I = \frac{1}{\sqrt{3}} \int_0^{\pi/3} \frac{d\theta}{\sqrt{1 - \tfrac{4}{3}\sin^2\theta}}$$

Next, we see that $k^2 > 1$. Therefore, we put ....

$$\tfrac{4}{3}\sin^2\theta = \sin^2\psi$$

60

$$\frac{2}{\sqrt{3}}\sin\theta = \sin\psi \quad \therefore \quad \frac{2}{\sqrt{3}}\cos\theta\, d\theta = \cos\psi\, d\psi \quad \therefore \quad d\theta = \frac{\sqrt{3}\cos\psi\, d\psi}{2\cos\theta}$$

Then, so far, we have $I = $ ....

$$I = \frac{1}{\sqrt{3}} \int_{\theta=0}^{\theta=\pi/3} \frac{1}{\sqrt{1 - \sin^2\psi}} \cdot \frac{\sqrt{3}\cos\psi\, d\psi}{2\cos\theta}$$

61

$$\frac{2}{\sqrt{3}}\sin\theta = \sin\psi$$

Limits: when $\theta = 0$, $\psi = 0$

$$\theta = \frac{\pi}{3}, \quad \frac{2}{\sqrt{3}}\sin\theta = \frac{2}{\sqrt{3}} \cdot \frac{\sqrt{3}}{2} = 1 \quad \therefore \quad \psi = \pi/2$$

Also $\cos\theta = \sqrt{1 - \sin^2\theta} = \sqrt{1 - \tfrac{3}{4}\sin^2\psi}$

$$\therefore I = \text{............}$$

62

$$I = \frac{1}{2} \int_0^{\pi/2} \frac{d\psi}{\sqrt{1 - \frac{3}{4}\sin^2 \psi}}$$

which is now in standard form with $k = \dfrac{\sqrt{3}}{2}$ and $\phi = \pi/2$

$$\therefore \ I = \frac{1}{2} F\left(\frac{\sqrt{3}}{2},\ \pi/2\right) = \frac{1}{2} K\left(\frac{\sqrt{3}}{2}\right)$$

From tables $\quad K\left(\dfrac{\sqrt{3}}{2}\right) = 2 \cdot 1565 \quad \therefore \ I = 1 \cdot 078$

Now, what about this one?

Example 4

Evaluate $I = \displaystyle\int_0^{\pi/2} \frac{d\theta}{\sqrt{1 + 4\sin^2 \theta}}$.

The trouble here is the *plus* sign in the denominator. Were it a minus sign as in Example 2, the integral could be converted into standard form and would present no difficulty.

In this case, the key is to put $\theta = \pi/2 - \psi$, i.e. $\sin \theta = \cos \psi$.

Expressing the integral in terms of ψ, we have

$$I = \ldots\ldots\ldots\ldots$$

63

$$I = \int_{\pi/2}^0 \frac{-d\psi}{\sqrt{5 - 4\sin^2 \psi}}$$

Because

$$\theta = \pi/2 - \psi \quad \therefore \ d\theta = -d\psi$$

$$1 + 4\sin^2 \theta = 1 + 4(1 - \cos^2 \theta) = 5 - 4\cos^2 \theta = 5 - 4\sin^2 \psi$$

Limits: when $\theta = 0$, $\psi = \pi/2$; when $\theta = \pi/2$, $\psi = 0$ and the expression above immediately follows.

Move on

64 So we have $I = \displaystyle\int_{\pi/2}^0 \frac{-d\psi}{\sqrt{5 - 4\sin^2 \psi}}$.

The minus sign in the numerator can be absorbed by $\ldots\ldots\ldots\ldots$

65

$$\therefore\ I = \int_0^{\pi/2} \frac{\mathrm{d}\psi}{\sqrt{5 - 4\sin^2\psi}}$$

Finally, taking out a factor 5 from the denominator, the integral becomes

$$I = \frac{1}{\sqrt{5}} \int_0^{\pi/2} \frac{\mathrm{d}\psi}{\sqrt{1 - \frac{4}{5}\sin^2\psi}}$$

and this can then be written ....

66

$$\boxed{I = \frac{1}{\sqrt{5}} F\left(\frac{2}{\sqrt{5}}, \frac{\pi}{2}\right) = \frac{1}{\sqrt{5}} K\left(\frac{2}{\sqrt{5}}\right)}$$

From tables $\quad K\left(\dfrac{2}{\sqrt{5}}\right) = K(0{\cdot}8944) = 2{\cdot}2435 \quad \therefore\ I = 1{\cdot}003$

Alternative forms of elliptic functions

(a) *Of the first kind*

$$F(k, x) = \int_0^x \frac{\mathrm{d}u}{\sqrt{(1 - u^2)(1 - k^2 u^2)}} \tag{3}$$

where $0 \le x \le 1$ and $0 < k < 1$.

(b) *Of the second kind*

$$E(k, x) = \int_0^x \sqrt{\frac{1 - k^2 u^2}{1 - u^2}}\, \mathrm{d}u \tag{4}$$

where $0 \le x \le 1$ and $0 < k < 1$.

Note these two new forms and then we can deal with a few examples. As before, it is a case of transforming the given integrand into the required form by suitable substitutions.

Example 1

67

Evaluate $I = \displaystyle\int_0^{1/\sqrt{2}} \sqrt{\frac{4 - 3u^2}{1 - u^2}}\, \mathrm{d}u.$

Here we remove a factor 4 from the numerator to reduce the first term to 1.

$$I = 2\int_0^{1/\sqrt{2}} \sqrt{\frac{1 - \frac{3}{4}u^2}{1 - u^2}}\, \mathrm{d}u$$

This is now in standard form with $k = \ldots\ldots\ldots$ and $x = \ldots\ldots\ldots$

68

$$k = \frac{\sqrt{3}}{2}; \quad x = \frac{1}{\sqrt{2}}$$

$$\therefore \ I = 2E\left(\frac{\sqrt{3}}{2}, \frac{1}{\sqrt{2}}\right) = 2(0{\cdot}7282) \text{ from tables}$$

$$\therefore \ I = 1{\cdot}4564$$

Example 2

Evaluate $I = \displaystyle\int_0^{1/2} \frac{du}{\sqrt{5 - 6u^2 + u^4}}$.

Factorising the denominator gives $I = \ldots\ldots\ldots\ldots$

69

$$I = \int_0^{1/2} \frac{du}{\sqrt{(1 - u^2)(5 - u^2)}}$$

Taking out a factor 5

$$I = \frac{1}{\sqrt{5}} \int_0^{1/2} \frac{du}{\sqrt{(1 - u^2)(1 - \frac{1}{5}u^2)}}$$

which is in standard form with $k = 1/\sqrt{5}$ and $x = 1/2$

$$\therefore \ I = \ldots\ldots\ldots\ldots$$

70

$$I = \frac{1}{\sqrt{5}} F\left(\frac{1}{\sqrt{5}}, \frac{1}{2}\right)$$

In some tables, k is quoted as $\sin\theta$, i.e. $\sin\theta = \dfrac{1}{\sqrt{5}}$ $\therefore \ \theta = 26° \ 34'$

and $\quad$ x is quoted as $\sin\phi$, i.e. $\sin\phi = \dfrac{1}{2}$ $\therefore \ \phi = 30°$.

Then $F(1/\sqrt{5}, 1/2) = 0{\cdot}528$

$$\therefore \ I = 0{\cdot}236$$

Now move on for Example 3

Example 3   **71**

Evaluate $I = \int_0^{\sqrt{3}/4} \sqrt{\dfrac{2 - x^2}{1 - 4x^2}}\, dx$.

We have to convert this into the form $\int \sqrt{\dfrac{1 - k^2 u^2}{1 - u^2}}\, du$, so first concentrate on the denominator. Any suggestions?

72

$$\boxed{\text{Put } 4x^2 = u^2 \quad \text{i.e.} \quad 2x = u}$$

$4x^2 = u^2 \quad \therefore\ 2x = u \quad \therefore\ 2\,dx = du$

Limits: when $x = 0$, $u = 0$ and when $x = \sqrt{3}/4$, $u = \sqrt{3}/2$

Also $2 - x^2 = 2 - u^2/4$

The integral now becomes ....

73

$$\boxed{I = \int_0^{\sqrt{3}/2} \sqrt{\frac{2 - u^2/4}{1 - u^2}} \cdot \frac{du}{2}}$$

Finally, taking out the factor 2 in the numerator

$$I = \ldots\ldots\ldots$$

74

$$\boxed{I = \frac{1}{\sqrt{2}} \int_0^{\sqrt{3}/2} \frac{\sqrt{1 - u^2/8}}{1 - u^2}\, du}$$

i.e. $k^2 = \dfrac{1}{8} \quad \therefore\ k = \dfrac{\sqrt{2}}{4} \quad$ and $\quad x = \dfrac{\sqrt{3}}{2}$

$$\text{So} \quad I = \ldots\ldots\ldots$$

75

$$\boxed{I = \frac{1}{\sqrt{2}} E\left(\frac{\sqrt{2}}{4}, \frac{\sqrt{3}}{2}\right)}$$

Then $\sin\theta = \dfrac{\sqrt{2}}{4} \quad \therefore\ \theta = 20° \ 42'$ and $\sin\phi = \dfrac{\sqrt{3}}{2} \quad \therefore\ \phi = 60°$

From tables, $E\left(\dfrac{\sqrt{2}}{4}, \dfrac{\sqrt{3}}{2}\right) = 1{\cdot}029 \quad \therefore\ I = 0{\cdot}728$

So it is all just a question of manipulation to transform the given integral into the required standard forms, and then of reference to the appropriate tables.

▶

The **Revision summary** follows, to be read in conjunction with the **Can you?** checklist, checking with the relevant parts of the Programme any points of which you are unsure. You will then find the **Test exercise** straightforward. Finally the **Further problems** provide additional practice.

 ## Revision summary 21

76

1   *Gamma functions*

    (a) $\Gamma(x) = \displaystyle\int_0^\infty t^{x-1}e^{-t}\,dt \quad x > 0$

       $\Gamma(x+1) = x\Gamma(x)$

    (b) If $x = n$, a positive integer

        $\Gamma(n+1) = n!$

          $\Gamma(1) = 1$

          $\Gamma(0) = \infty \qquad \Gamma(-n) = \pm\infty$

    (c) $\displaystyle\int_0^\infty e^{-x^2}\,dx = \dfrac{\sqrt{\pi}}{2}$

    (d) $\Gamma\left(\tfrac{1}{2}\right) = \sqrt{\pi} \qquad\qquad \Gamma\left(\tfrac{3}{2}\right) = \dfrac{\sqrt{\pi}}{2}$

       $\Gamma\left(\tfrac{5}{2}\right) = \dfrac{3\sqrt{\pi}}{4} \qquad\quad \Gamma\left(\tfrac{7}{2}\right) = \dfrac{15\sqrt{\pi}}{8}$

       $\Gamma\left(-\tfrac{1}{2}\right) = -2\sqrt{\pi} \qquad \Gamma\left(-\tfrac{3}{2}\right) = \dfrac{4\sqrt{\pi}}{3}$

    (e) Duplication formula $\Gamma\left(n+\tfrac{1}{2}\right) = \dfrac{\Gamma(2n)\sqrt{\pi}}{2^{2n-1}\cdot\Gamma(n)}$

2   *Beta functions*

    (a) $B(m, n) = \displaystyle\int_0^1 x^{m-1}(1-x)^{n-1}\,dx \qquad m > 0;\ n > 0$

       $B(m, n) = B(n, m)$

       $B(m, n) = 2\displaystyle\int_0^{\pi/2} \sin^{2m-1}\theta \cos^{2n-1}\theta\,d\theta$

(b) $B(m, n) = \dfrac{(m-1)(n-1)}{(m+n-1)(m+n-2)} B(m-1, n-1)$

$B(k, 1) = B(1, k) = \dfrac{1}{k}$

$B(1, 1) = 1; \qquad B(\tfrac{1}{2}, \tfrac{1}{2}) = \pi$

$B(m, n) = \dfrac{\Gamma(m) \cdot \Gamma(n)}{\Gamma(m+n)}$

(c) m and n positive integers

$B(m, n) = \dfrac{(m-1)!(n-1)!}{(m+n-1)!}$

3 Error function

(a) $\mathrm{erf}\,(x) = \dfrac{2}{\sqrt{\pi}} \displaystyle\int_0^x e^{-t^2}\,\mathrm{d}t$

(b) $\displaystyle\int_0^\infty e^{-x^2}\,\mathrm{d}x = \dfrac{\sqrt{\pi}}{2}$

$\displaystyle\int_{-\infty}^\infty e^{-x^2}\,\mathrm{d}x = \sqrt{\pi}; \qquad \int_{-\infty}^\infty e^{-x^2/2}\,\mathrm{d}x = \sqrt{2\pi}$

Complementary error function

$\mathrm{erfc}\,(x) = \dfrac{2}{\sqrt{\pi}} \displaystyle\int_x^\infty e^{-t^2}\,\mathrm{d}t = 1 - \mathrm{erf}\,(x)$

4 Elliptic functions
(a) *Standard forms*

(1) of the first kind: $F(k, \phi) = \displaystyle\int_0^\phi \dfrac{\mathrm{d}\theta}{\sqrt{1 - k^2 \sin^2 \theta}}$

(2) of the second kind: $E(k, \phi) = \displaystyle\int_0^\phi \sqrt{1 - k^2 \sin^2 \theta}\,\mathrm{d}\theta$

In each case, $\quad 0 \le \phi \le \pi/2; \quad 0 < k < 1.$

(b) *Complete elliptic integrals* $\phi = \dfrac{\pi}{2}$

$F\left(k, \dfrac{\pi}{2}\right) = K\,(k)$

$E\left(k, \dfrac{\pi}{2}\right) = E\,(k)$

(c) *Alternative forms of elliptic functions*

(1) of the first kind: $F(k, x) = \displaystyle\int_0^x \dfrac{\mathrm{d}u}{\sqrt{(1 - u^2)(1 - k^2 u^2)}}$

(2) of the second kind: $E(k, x) = \displaystyle\int_0^x \sqrt{\dfrac{1 - k^2 u^2}{1 - u^2}}\,\mathrm{d}u$

In each case $\quad 0 \le x \le 1; \quad 0 < k < 1.$

Can you?

77 **Checklist 21**

Check this list before and after you try the end of Programme test.

On a scale of 1 to 5 how confident are you that you can: **Frames**

- Derive the recurrence relation for the gamma function and
  evaluate the gamma function for certain rational arguments?   [1] to [16]
  Yes ☐   ☐   ☐   ☐   ☐   *No*

- Evaluate integrals that require the use of the gamma function
  in their solution?   [17] to [24]
  Yes ☐   ☐   ☐   ☐   ☐   *No*

- Identify the beta function and evaluate integrals that require
  the use of the beta function in their solution?   [25] to [32]
  Yes ☐   ☐   ☐   ☐   ☐   *No*

- Derive the relationship between the gamma function and the
  beta function?   [33] to [44]
  Yes ☐   ☐   ☐   ☐   ☐   *No*

- Use the duplication formula to evaluate the gamma function
  for half integer arguments?   [44] and [45]
  Yes ☐   ☐   ☐   ☐   ☐   *No*

- Recognise the error function and its relation to the Gaussian
  probability distribution?   [46] to [52]
  Yes ☐   ☐   ☐   ☐   ☐   *No*

- Recognise elliptic functions of the first and second kind?   [53]
  Yes ☐   ☐   ☐   ☐   ☐   *No*

- Evaluate integrals that require the use of elliptic functions in
  their solution?   [54] to [66]
  Yes ☐   ☐   ☐   ☐   ☐   *No*

- Use alternative forms of the elliptic functions?   [66] to [75]
  Yes ☐   ☐   ☐   ☐   ☐   *No*

Test exercise 21

1 Evaluate (a) $\dfrac{\Gamma(6)}{3\Gamma(4)}$ (b) $\dfrac{\Gamma(1\cdot5)}{\Gamma(2\cdot5)}$ (c) $\dfrac{\Gamma(-\frac{1}{2})}{\Gamma(\frac{1}{2})}$ **78**

    (d) $\displaystyle\int_0^\infty x^5\,e^{-x}\,dx$ (e) $\displaystyle\int_0^\infty x^6\,e^{-4x^2}\,dx$.

2 Determine

    (a) $\displaystyle\int_0^1 x^5(2-x)^4 dx$ (b) $\displaystyle\int_0^{\pi/2} \sin^7\theta\cos^3\theta\,d\theta$ (c) $\displaystyle\int_0^{\pi/8} \sin^2 4\theta\cos^5 4\theta\,d\theta$.

3 Show that

    (a) $\displaystyle\int_{-a}^a e^{-t^2}\,dt = \sqrt{\pi}\,\mathrm{erf}\,(a)$ (b) $\displaystyle\int_0^\infty e^{-k^2t^2}\,dt = \dfrac{\sqrt{\pi}}{2k},\quad k>0$.

4 Evaluate
    (a) $\mathrm{erfc}\,(\infty)$ (b) $\mathrm{erfc}\,(0)$.

5 Express the following in elliptic functions.

    (a) $\displaystyle\int_0^{\pi/4} \dfrac{d\theta}{\sqrt{1-2\sin^2\theta}}$ (b) $\displaystyle\int_0^{\sqrt{3}/2} \dfrac{du}{\sqrt{4-5u^2+u^4}}$.

 Further problems 21

1 Evaluate (a) $\dfrac{\Gamma(5)}{2\Gamma(3)}$; (b) $\dfrac{\Gamma(\frac{1}{2})}{\Gamma(-\frac{1}{2})}$; (c) $\dfrac{\Gamma(2\cdot5)}{\Gamma(3\cdot5)}$; **79**

    (d) $\displaystyle\int_0^\infty x^4 e^{-x}\,dx$; (e) $\displaystyle\int_0^\infty x^8 e^{-2x}\,dx$.

2 Determine (a) $\displaystyle\int_0^\infty x^3 e^{-x}\,dx$; (b) $\displaystyle\int_0^\infty x^4 e^{-3x}\,dx$;

    (c) $\displaystyle\int_0^\infty x^2 e^{-2x^2}\,dx$; (d) $\displaystyle\int_0^\infty \sqrt{x}\cdot e^{-\sqrt{x}}\,dx$.

3 If m and n are positive constants, show that $\displaystyle\int_0^\infty x^m e^{-ax^n}\,dx$ can be expressed in

    the form $\dfrac{1}{n\cdot a^{(m+1)/n}}\Gamma\left(\dfrac{m+1}{n}\right)$.

4 Evaluate the following.

    (a) $\displaystyle\int_0^{1/2} x^4(1-2x)^3 dx$ (b) $\displaystyle\int_0^{1/\sqrt{2}} x^2\sqrt{1-2x^2}\,dx$ (c) $\displaystyle\int_0^{\pi/2} \sin^5\theta\cos^4\theta\,d\theta$

    (d) $\displaystyle\int_0^{\pi/2} \sin\theta\sqrt{\cos^5\theta}\,d\theta$ (e) $\displaystyle\int_0^{\pi/4} \sin^3 2\theta\cos^6 2\theta\,d\theta$ (f) $\displaystyle\int_0^{1/3} x^2\sqrt{1-9x^2}\,dx$.

5 Show that $\dfrac{\mathrm{d}}{\mathrm{d}x}\mathrm{erf}\,(x) = \dfrac{2}{\sqrt{\pi}}e^{-x^2}$.

6 Show that the Laplace transform of the error function is given as

$$F(s) = \int_0^\infty \mathrm{erf}\,(t)\,e^{-st}\,\mathrm{d}t = \frac{e^{-s^2/4}}{s}\,\mathrm{erfc}\left(\frac{s}{2}\right)\ \text{for } s > 0.$$

7 The Fresnel integrals are defined as

$$C(x) = \int_0^x \cos\left(\frac{\pi t^2}{2}\right)\mathrm{d}t \text{ and } S(x) = \int_0^x \sin\left(\frac{\pi t^2}{2}\right)\mathrm{d}t$$

Show that

$$\frac{1}{\sqrt{2j}}\,\mathrm{erf}\left(x\sqrt{\frac{j\pi}{2}}\right) = C(x) - jS(x)$$

8 Express the following in elliptic functions.

(a) $\displaystyle\int_0^{\pi/2}\sqrt{1+4\sin^2\theta}\,\mathrm{d}\theta$ (b) $\displaystyle\int_0^{\pi/2}\frac{\mathrm{d}\theta}{\sqrt{\cos\theta}}$ (c) $\displaystyle\int_0^1\sqrt{\frac{4-x^2}{1-x^2}}\,\mathrm{d}x$

(d) $\displaystyle\int_0^2\frac{\mathrm{d}x}{\sqrt{(9-x^2)(16-x^2)}}$ (e) $\displaystyle\int_0^2\frac{\mathrm{d}x}{\sqrt{(4-x^2)(5-x^2)}}$

(f) $\displaystyle\int_0^{\pi/6}\frac{\mathrm{d}\theta}{\sqrt{\sin^2\theta+2\cos^2\theta}}$ (g) $\displaystyle\int_{\pi/4}^{\pi/3}\frac{\mathrm{d}\theta}{\sqrt{\sin^2\theta+2\cos^2\theta}}$.

9 Using the substitution $x = \tan\theta$ prove that the integral

$$\int_0^1\frac{\mathrm{d}x}{\sqrt{(1+x^2)(1+4x^2)}}$$

can be expressed in the form

$$\frac{1}{2}\int_0^{\pi/4}\frac{\mathrm{d}\theta}{\sqrt{1-\frac{3}{4}\cos^2\theta}}$$

Hence, using $\theta = \dfrac{\pi}{2} - \phi$, evaluate the integral in terms of elliptic functions.

10 Evaluate the following.

(a) $\displaystyle\int_0^{0\cdot5}\frac{\mathrm{d}x}{\sqrt{3-4x^2+x^4}}$ (b) $\displaystyle\int_{0\cdot5}^{1\cdot0}\frac{\mathrm{d}x}{\sqrt{3-4x^2+x^4}}$

(c) $\displaystyle\int_0^{\pi/2}\frac{\mathrm{d}\theta}{\sqrt{25+9\sin^2\theta}}$ (d) $\displaystyle\int_0^{\pi/3}\frac{\mathrm{d}\theta}{\sqrt{4+3\sin^2\theta}}$.

Vector analysis 1

Learning outcomes

When you have completed this Programme you will be able to:

- Obtain the scalar and vector product of two vectors
- Reproduce the relationships between the scalar and vector products of the Cartesian coordinate unit vectors
- Obtain the scalar and vector triple products and appreciate their geometric significance
- Differentiate a vector field and derive a unit vector tangential to the vector field at a point
- Integrate a vector field
- Obtain the gradient of a scalar field, the directional derivative and a unit normal to a surface
- Obtain the divergence of a vector field and recognise a solenoidal vector field
- Obtain the curl of a vector field
- Obtain combinations of div, grad and curl acting on scalar and vector fields as appropriate

Prerequisite: Engineering Mathematics (Sixth Edition)
Programme 6 Vectors

Introduction

1

The initial work on vectors was covered in detail in Programme 6 of *Engineering Mathematics (Sixth Edition)* and, if you are in any doubt, spend some time reviewing that section of the work before proceeding further.

The current Programmes on vector analysis build on these early foundations, so, for quick reference, the essential results of the previous work are summarised in the following list.

Summary of prerequisites

1 A *scalar* quantity has magnitude only; a *vector* quantity has both magnitude and direction.

2 The axes of reference, OX, OY, OZ, form a right-handed set. The symbols **i**, **j**, **k** denote *unit vectors* in the directions OX, OY, OZ, respectively.

   If $\overline{OP} = \mathbf{r} = a_x\mathbf{i} + a_y\mathbf{j} + a_z\mathbf{k}$ then $OP = |\mathbf{r}| = \sqrt{a_x^2 + a_y^2 + a_z^2}$ where $|\mathbf{r}|$ is the modulus of **r**.

3 The *direction cosines* $[l, m, n]$ are the cosines of the angles between the vector **r** and the axes OX, OY, OZ, respectively. For any vector $\mathbf{r} = a_x\mathbf{i} + a_y\mathbf{j} + a_z\mathbf{k}$

$$l = \frac{a_x}{|\mathbf{r}|}; \quad m = \frac{a_y}{|\mathbf{r}|}; \quad n = \frac{a_z}{|\mathbf{r}|}$$

   and $l^2 + m^2 + n^2 = 1$.

4 *Scalar product* ('dot product')
   $\mathbf{A} \cdot \mathbf{B} = AB\cos\theta$ where θ is the angle between **A** and **B** and where A and B are the moduli of **A** and **B**.

   If $\mathbf{A} = a_x\mathbf{i} + a_y\mathbf{j} + a_z\mathbf{k}$ and $\mathbf{B} = b_x\mathbf{i} + b_y\mathbf{j} + b_z\mathbf{k}$ then
   $$\mathbf{A} \cdot \mathbf{B} = a_xb_x + a_yb_y + a_zb_z \quad \text{and} \quad \mathbf{A} \cdot \mathbf{B} = \mathbf{B} \cdot \mathbf{A}$$

5 *Vector product* ('cross product')
   $\mathbf{A} \times \mathbf{B} = AB\sin\theta$ in a direction perpendicular to **A** and **B** so that **A**, **B**, $(\mathbf{A} \times \mathbf{B})$ form a right-handed set.

   Therefore $|\mathbf{A} \times \mathbf{B}| = AB\sin\theta$

   Also $\mathbf{A} \times \mathbf{B} = \begin{vmatrix} \mathbf{i} & \mathbf{j} & \mathbf{k} \\ a_x & a_y & a_z \\ b_x & b_y & b_z \end{vmatrix}$ where $\mathbf{A} \times \mathbf{B} = -\mathbf{B} \times \mathbf{A}$

6 *Angle between two vectors*
   $$\cos\theta = l_1l_2 + m_1m_2 + n_1n_2$$

   where l_1, m_1, n_1 and l_2, m_2, n_2 are the direction cosines of vectors $\mathbf{r}_1$ and $\mathbf{r}_2$ respectively.

   For perpendicular vectors $l_1l_2 + m_1m_2 + n_1n_2 = 0$
   For parallel vectors $l_1l_2 + m_1m_2 + n_1n_2 = 1$.

One or two examples will no doubt help to recall the main points.   ▶

Example 1 Direction cosines

If **i**, **j**, **k** are unit vectors in the directions OX, OY, OZ, respectively, then any position vector $\overline{OP} \ (= \mathbf{r})$ can be represented in the form

$$\overline{OP} = \mathbf{r} = a_x\mathbf{i} + a_y\mathbf{j} + a_z\mathbf{k}.$$

Then $|\mathbf{r}| = \dots\dots\dots\dots$

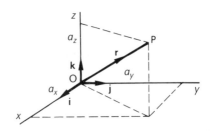

$$\boxed{|\mathbf{r}| = \sqrt{a_x^2 + a_y^2 + a_z^2}}$$

2

The direction of OP is denoted by stating the direction cosines of the angles made by OP and the three coordinate axes.

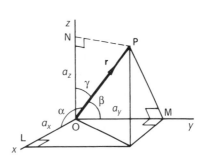

$$l = \cos\alpha = \frac{OL}{OP} = \frac{a_x}{|\mathbf{r}|}$$

$$m = \cos\beta = \frac{OM}{OP} = \frac{a_y}{|\mathbf{r}|}$$

$$n = \cos\gamma = \frac{ON}{OP} = \frac{a_z}{|\mathbf{r}|}$$

$$\therefore \ l, \ m, \ n = \cos\alpha, \ \cos\beta, \ \cos\gamma$$

So, if P is the point (3, 2, 6), then

$$|\mathbf{r}| = \dots\dots\dots\dots;$$
$$l = \dots\dots\dots\dots; \quad m = \dots\dots\dots\dots; \quad n = \dots\dots\dots\dots$$

$$\boxed{\begin{array}{c} |\mathbf{r}| = 7; \\ l = 0\cdot429; \quad m = 0\cdot286; \quad n = 0\cdot857 \end{array}}$$

3

Because

$$(|\mathbf{r}|)^2 = 9 + 4 + 36 = 49 \qquad \therefore \ |\mathbf{r}| = 7$$
$$l = \cos\alpha = \tfrac{3}{7} = 0\cdot4286$$
$$m = \cos\beta = \tfrac{2}{7} = 0\cdot2857$$
$$n = \cos\gamma = \tfrac{6}{7} = 0\cdot8571.$$

▶

Example 2   Angle between two vectors

If the direction cosines of A are l_1, m_1, n_1 and those of B are l_2, m_2, n_2, then the angle between the vectors is given by

$$\cos\theta = l_1 l_2 + m_1 m_2 + n_1 n_2. \tag{1}$$

If $\mathbf{A} = 2\mathbf{i} + 3\mathbf{j} + 4\mathbf{k}$ and $\mathbf{B} = \mathbf{i} - 2\mathbf{j} + 3\mathbf{k}$, we can find the direction cosines of each and hence θ which is ....

4

$$\boxed{\theta = 66°\ 36'}$$

Because

For **A**: $|\mathbf{r}_1| = \sqrt{4+9+16} = \sqrt{29}$

$$\therefore\ l_1 = \frac{2}{\sqrt{29}};\quad m_1 = \frac{3}{\sqrt{29}};\quad n_1 = \frac{4}{\sqrt{29}}$$

For **B**: $|\mathbf{r}_2| = \sqrt{1+4+9} = \sqrt{14}$

$$\therefore\ l_2 = \frac{1}{\sqrt{14}};\quad m_2 = \frac{-2}{\sqrt{14}};\quad n_2 = \frac{3}{\sqrt{14}}$$

Then $\cos\theta = \dfrac{1}{\sqrt{14 \times 29}}\{2 - 6 + 12\} = 0.3970$

$\therefore\ \theta = 66°\ 36'$

Let us now look at the question of scalar and vector products.

On to the next frame

5

Example 3   Scalar product

If A and B are two vectors, the scalar product of A and B is defined as

$$\mathbf{A} \cdot \mathbf{B} = AB\cos\theta \tag{2}$$

where θ is the angle between the two vectors. If $\mathbf{A} \cdot \mathbf{B} = 0$ then $\mathbf{A} \perp \mathbf{B}$.

If we consider the *scalar products of the unit vectors* i, j, k, which are mutually perpendicular, then

$$\mathbf{i} \cdot \mathbf{j} = (1)(1)\cos 90° = 0 \qquad \therefore\ \mathbf{i} \cdot \mathbf{j} = \mathbf{j} \cdot \mathbf{k} = \mathbf{k} \cdot \mathbf{i} = 0$$

and $\mathbf{i} \cdot \mathbf{i} = (1)(1)\cos 0° = 1 \qquad \therefore\ \mathbf{i} \cdot \mathbf{i} = \mathbf{j} \cdot \mathbf{j} = \mathbf{k} \cdot \mathbf{k} = 1.$

In general, if $\mathbf{A} = a_x\mathbf{i} + a_y\mathbf{j} + a_z\mathbf{k}$ and $\mathbf{B} = b_x\mathbf{i} + b_y\mathbf{j} + b_z\mathbf{k}$ then $\mathbf{A} \cdot \mathbf{B} = a_x b_x + a_y b_y + a_z b_z$ which is, of course, a scalar quantity.

So, if $\mathbf{A} = 2\mathbf{i} - 3\mathbf{j} + 4\mathbf{k}$ and $\mathbf{B} = \mathbf{i} + 2\mathbf{j} + 5\mathbf{k}$, then

$$\mathbf{A} \cdot \mathbf{B} = \ldots\ldots\ldots$$

6

$$\boxed{\mathbf{A} \cdot \mathbf{B} = 2 - 6 + 20 = 16}$$

Also, since $\mathbf{A} \cdot \mathbf{B} = AB\cos\theta$, we can determine the angle θ between the vectors. In this case $\theta = \ldots\ldots\ldots$

$$\boxed{\theta = 57° \, 9'}$$

7

$\mathbf{A} = 2\mathbf{i} - 3\mathbf{j} + 4\mathbf{k}$ $\therefore \, A = |\mathbf{A}| = \sqrt{4 + 9 + 16} = \sqrt{29}$

$\mathbf{B} = \mathbf{i} + 2\mathbf{j} + 5\mathbf{k}$ $\therefore \, B = |\mathbf{B}| = \sqrt{1 + 4 + 25} = \sqrt{30}$

We have already found that $\mathbf{A} \cdot \mathbf{B} = 16$ and $\mathbf{A} \cdot \mathbf{B} = AB \cos \theta$

$\therefore \, 16 = \sqrt{29} \, \sqrt{30} \cos \theta$ $\therefore \, \cos \theta = 0 \cdot 5425$ $\therefore \, \theta = 57° \, 9'$

So, the *scalar product* of $\mathbf{A} = a_x\mathbf{i} + a_y\mathbf{j} + a_z\mathbf{k}$ and $\mathbf{B} = b_x\mathbf{i} + b_y\mathbf{j} + b_z\mathbf{k}$

is $\mathbf{A} \cdot \mathbf{B} = a_xb_x + a_yb_y + a_zb_z$

and $\mathbf{A} \cdot \mathbf{B} = AB \cos \theta$ where θ is the angle between the vectors.

It can also be shown that

       (a) $\mathbf{A} \cdot \mathbf{B} = \mathbf{B} \cdot \mathbf{A}$

and   (b) $\mathbf{A} \cdot (\mathbf{B} + \mathbf{C}) = \mathbf{A} \cdot \mathbf{B} + \mathbf{A} \cdot \mathbf{C}$

Make a note of these results

Example 4   Vector product

8

If $\mathbf{A} = a_x\mathbf{i} + a_y\mathbf{j} + a_z\mathbf{k}$ and $\mathbf{B} = b_x\mathbf{i} + b_y\mathbf{j} + b_z\mathbf{k}$ the vector product $\mathbf{A} \times \mathbf{B}$ has magnitude $|\mathbf{A} \times \mathbf{B}| = AB \sin \theta$ in the direction perpendicular to $\mathbf{A}$ and $\mathbf{B}$ such that $\mathbf{A}$, $\mathbf{B}$ and $(\mathbf{A} \times \mathbf{B})$ form a right-handed set.

We can write this as

     $\mathbf{A} \times \mathbf{B} = (AB \sin \theta)\mathbf{n}$ (3)

where $\mathbf{n}$ is defined as a unit vector in the positive normal direction to the plane of $\mathbf{A}$ and $\mathbf{B}$, i.e. forming a right-handed set.

Also $\mathbf{A} \times \mathbf{B} = \begin{vmatrix} \mathbf{i} & \mathbf{j} & \mathbf{k} \\ a_x & a_y & a_z \\ b_x & b_y & b_z \end{vmatrix}$ (4)

If we consider the *vector products of the unit vectors*, $\mathbf{i}$, $\mathbf{j}$, $\mathbf{k}$, then

     $\mathbf{i} \times \mathbf{j} = (1)(1) \sin 90° \mathbf{k} = \mathbf{k}$

     $\mathbf{j} \times \mathbf{k} = \mathbf{i}, \quad \mathbf{k} \times \mathbf{i} = \mathbf{j}$

Note that

     $\mathbf{j} \times \mathbf{i} = -(\mathbf{i} \times \mathbf{j}) = -\mathbf{k}$

     $\mathbf{k} \times \mathbf{j} = -\mathbf{i}, \quad \mathbf{i} \times \mathbf{k} = -\mathbf{j}$

Also

     $\mathbf{i} \times \mathbf{i} = (1)(1) \sin 0° \mathbf{n} = 0$

     $\mathbf{j} \times \mathbf{j} = \mathbf{k} \times \mathbf{k} = 0$

▶

It can also be shown that

(a) $\mathbf{A} \times (\mathbf{B} + \mathbf{C}) = \mathbf{A} \times \mathbf{B} + \mathbf{A} \times \mathbf{C}$

and   (b) $\mathbf{A} \times \mathbf{B} = -(\mathbf{B} \times \mathbf{A})$ (5)

Make a note of these results (3), (4) and (5).

Then, if $\mathbf{A} = 3\mathbf{i} - 2\mathbf{j} + 4\mathbf{k}$ and $\mathbf{B} = 2\mathbf{i} - 3\mathbf{j} - 2\mathbf{k}$

$$\mathbf{A} \times \mathbf{B} = \ldots\ldots\ldots\ldots$$

9

$$\boxed{\mathbf{A} \times \mathbf{B} = 16\mathbf{i} + 14\mathbf{j} - 5\mathbf{k}}$$

We simply evaluate the determinant

$$\mathbf{A} \times \mathbf{B} = \begin{vmatrix} \mathbf{i} & \mathbf{j} & \mathbf{k} \\ 3 & -2 & 4 \\ 2 & -3 & -2 \end{vmatrix}$$

$$= \mathbf{i}(4 + 12) - \mathbf{j}(-6 - 8) + \mathbf{k}(-9 + 4) = 16\mathbf{i} + 14\mathbf{j} - 5\mathbf{k}$$

Move on to the next frame

10

We have seen therefore that

the scalar product of two vectors is a scalar

but that   the vector product of two vectors is a vector.

We know also that $|\mathbf{A} \times \mathbf{B}| = AB \sin \theta$

Therefore, the angle between the vectors $\mathbf{A}$ and $\mathbf{B}$ given in Example 4 is

$$\theta = \ldots\ldots\ldots\ldots$$

11

$$\boxed{\theta = 79^\circ\, 40'}$$

Because

$$\mathbf{A} = 3\mathbf{i} - 2\mathbf{j} + 4\mathbf{k}; \quad \mathbf{B} = 2\mathbf{i} - 3\mathbf{j} - 2\mathbf{k}; \quad \text{and} \quad \mathbf{A} \times \mathbf{B} = 16\mathbf{i} + 14\mathbf{j} - 5\mathbf{k}$$

$$\therefore \quad |\mathbf{A} \times \mathbf{B}| = \sqrt{16^2 + 14^2 + 5^2} = \sqrt{477} = 21 \cdot 84$$

$$A = |\mathbf{A}| = \sqrt{3^2 + 2^2 + 4^2} = \sqrt{29} = 5 \cdot 385$$

$$B = |\mathbf{B}| = \sqrt{2^2 + 3^2 + 2^2} = \sqrt{17} = 4 \cdot 123$$

$$\therefore \quad 21 \cdot 84 = (5 \cdot 385)(4 \cdot 123) \sin \theta$$

$$\therefore \quad \sin \theta = 0 \cdot 9838 \quad \therefore \quad \theta = 79^\circ\, 40'$$

▶

So, to recapitulate:

If $\mathbf{A} = a_x\mathbf{i} + a_y\mathbf{j} + a_z\mathbf{k}$ and $\mathbf{B} = b_x\mathbf{i} + b_y\mathbf{j} + b_z\mathbf{k}$ and θ is the angle between them

(a) Scalar product $= \mathbf{A} \cdot \mathbf{B} = a_xb_x + a_yb_y + a_zb_z$

$$= AB\cos\theta$$

(b) Vector product $= \mathbf{A} \times \mathbf{B} = \begin{vmatrix} \mathbf{i} & \mathbf{j} & \mathbf{k} \\ a_x & a_y & a_z \\ b_x & b_y & b_z \end{vmatrix}$

and $\quad | \mathbf{A} \times \mathbf{B} | = AB\sin\theta.$

Make a note of these fundamental results: we shall certainly need them. Then, in the next frame, we can set off on some new work

Triple products

We now deal with the various products that we form with three vectors.

12

Scalar triple product of three vectors

If $\mathbf{A}, \mathbf{B}, \mathbf{C}$ are three vectors, the scalar formed by the product $\mathbf{A} \cdot (\mathbf{B} \times \mathbf{C})$ is called the scalar triple product.

If $\mathbf{A} = a_x\mathbf{i} + a_y\mathbf{j} + a_z\mathbf{k}; \quad \mathbf{B} = b_x\mathbf{i} + b_y\mathbf{j} + b_z\mathbf{k}; \quad \mathbf{C} = c_x\mathbf{i} + c_y\mathbf{j} + c_z\mathbf{k};$

then $\quad \mathbf{B} \times \mathbf{C} = \begin{vmatrix} \mathbf{i} & \mathbf{j} & \mathbf{k} \\ b_x & b_y & b_z \\ c_x & c_y & c_z \end{vmatrix}$

$$\therefore \; \mathbf{A} \cdot (\mathbf{B} \times \mathbf{C}) = (a_x\mathbf{i} + a_y\mathbf{j} + a_z\mathbf{k}) \cdot \begin{vmatrix} \mathbf{i} & \mathbf{j} & \mathbf{k} \\ b_x & b_y & b_z \\ c_x & c_y & c_z \end{vmatrix}$$

Multiplying the top row by the external bracket and remembering that

$$\mathbf{i} \cdot \mathbf{j} = \mathbf{j} \cdot \mathbf{k} = \mathbf{k} \cdot \mathbf{i} = 0 \quad \text{and} \quad \mathbf{i} \cdot \mathbf{i} = \mathbf{j} \cdot \mathbf{j} = \mathbf{k} \cdot \mathbf{k} = 1$$

we have $\; \mathbf{A} \cdot (\mathbf{B} \times \mathbf{C}) = \begin{vmatrix} a_x & a_y & a_z \\ b_x & b_y & b_z \\ c_x & c_y & c_z \end{vmatrix}$ $\hfill (6)$

Example

If $\; \mathbf{A} = 2\mathbf{i} - 3\mathbf{j} + 4\mathbf{k}; \quad \mathbf{B} = \mathbf{i} - 2\mathbf{j} - 3\mathbf{k}; \quad \mathbf{C} = 2\mathbf{i} + \mathbf{j} + 2\mathbf{k};$

then $\; \mathbf{A} \cdot (\mathbf{B} \times \mathbf{C}) = \begin{vmatrix} 2 & -3 & 4 \\ 1 & -2 & -3 \\ 2 & 1 & 2 \end{vmatrix}$

$$= \ldots\ldots\ldots\ldots$$

13
$$\boxed{\mathbf{A} \cdot (\mathbf{B} \times \mathbf{C}) = 42}$$

Because

$$\mathbf{A} \cdot (\mathbf{B} \times \mathbf{C}) = \begin{vmatrix} 2 & -3 & 4 \\ 1 & -2 & -3 \\ 2 & 1 & 2 \end{vmatrix}$$

$$= 2(-4 + 3) + 3(2 + 6) + 4(1 + 4) = 42$$

As simple as that.

14 Properties of scalar triple products

(a) $\quad \mathbf{B} \cdot (\mathbf{C} \times \mathbf{A}) = \begin{vmatrix} b_x & b_y & b_z \\ c_x & c_y & c_z \\ a_x & a_y & a_z \end{vmatrix} = - \begin{vmatrix} a_x & a_y & a_z \\ c_x & c_y & c_z \\ b_x & b_y & b_z \end{vmatrix}$

since interchanging two rows in a determinant reverses the sign. If we now interchange rows 2 and 3 and again change the sign, we have

$$\mathbf{B} \cdot (\mathbf{C} \times \mathbf{A}) = \begin{vmatrix} a_x & a_y & a_z \\ b_x & b_y & b_z \\ c_x & c_y & c_z \end{vmatrix} = \mathbf{A} \cdot (\mathbf{B} \times \mathbf{C})$$

$$\therefore \ \mathbf{A} \cdot (\mathbf{B} \times \mathbf{C}) = \mathbf{B} \cdot (\mathbf{C} \times \mathbf{A}) = \mathbf{C} \cdot (\mathbf{A} \times \mathbf{B}) \tag{7}$$

i.e. the scalar triple product is unchanged by a cyclic change of the vectors involved.

(b) $\quad \mathbf{B} \cdot (\mathbf{A} \times \mathbf{C}) = \begin{vmatrix} b_x & b_y & b_z \\ a_x & a_y & a_z \\ c_x & c_y & c_z \end{vmatrix} = - \begin{vmatrix} a_x & a_y & a_z \\ b_x & b_y & b_z \\ c_x & c_y & c_z \end{vmatrix}$

$$\therefore \ \mathbf{B} \cdot (\mathbf{A} \times \mathbf{C}) = -\mathbf{A} \cdot (\mathbf{B} \times \mathbf{C}) \tag{8}$$

i.e. a change of vectors not in cyclic order, changes the sign of the scalar triple product.

(c) $\quad \mathbf{A} \cdot (\mathbf{B} \times \mathbf{A}) = \begin{vmatrix} a_x & a_y & a_z \\ b_x & b_y & b_z \\ a_x & a_y & a_z \end{vmatrix} = 0$ since two rows are identical.

$$\therefore \ \mathbf{A} \cdot (\mathbf{B} \times \mathbf{A}) = \mathbf{B} \cdot (\mathbf{C} \times \mathbf{B}) = \mathbf{C} \cdot (\mathbf{A} \times \mathbf{C}) = 0 \tag{9}$$

Example

If $\mathbf{A} = \mathbf{i} + 2\mathbf{j} + 3\mathbf{k}; \quad \mathbf{B} = 2\mathbf{i} - 3\mathbf{j} + \mathbf{k}; \quad \mathbf{C} = 3\mathbf{i} + \mathbf{j} - 2\mathbf{k}$

$$\mathbf{A} \cdot (\mathbf{B} \times \mathbf{C}) = \ldots\ldots\ldots\ldots \qquad \mathbf{C} \cdot (\mathbf{B} \times \mathbf{A}) = \ldots\ldots\ldots\ldots$$

$$\boxed{\mathbf{A} \cdot (\mathbf{B} \times \mathbf{C}) = 52; \quad \mathbf{C} \cdot (\mathbf{A} \times \mathbf{B}) = -52}$$

15

Because

$$\mathbf{A} \cdot (\mathbf{B} \times \mathbf{C}) = \begin{vmatrix} 1 & 2 & 3 \\ 2 & -3 & 1 \\ 3 & 1 & -2 \end{vmatrix} = 1(6 - 1) - 2(-4 - 3) + 3(2 + 9) = 52$$

$\mathbf{C} \cdot (\mathbf{B} \times \mathbf{A})$ is not a cyclic change from the above. Therefore

$$\mathbf{C} \cdot (\mathbf{B} \times \mathbf{A}) = -\mathbf{A} \cdot (\mathbf{B} \times \mathbf{C}) = -52$$

Coplanar vectors

The magnitude of the scalar triple product $|\mathbf{A} \cdot (\mathbf{B} \times \mathbf{C})|$ is equal to the volume of the parallelepiped with three adjacent sides defined by $\mathbf{A}$, $\mathbf{B}$ and $\mathbf{C}$.

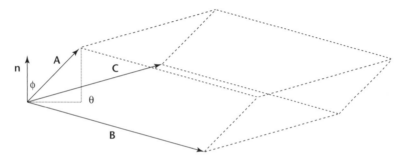

The scalar triple product $\mathbf{A} \cdot (\mathbf{B} \times \mathbf{C}) = \mathbf{A} \cdot (BC \sin \theta \, \mathbf{n}) = ABC \sin \theta \cos \phi$ where $\mathbf{n}$ is a unit vector perpendicular to the plane containing $\mathbf{B}$ and $\mathbf{C}$, θ is the angle between $\mathbf{B}$ and $\mathbf{C}$ and ϕ is the angle between $\mathbf{A}$ and $\mathbf{n}$. Therefore

$$|\mathbf{A} \cdot (\mathbf{B} \times \mathbf{C})| = ABC \, |\sin \theta \cos \phi|$$

Notice that in the figure both θ and ϕ are drawn as acute but in the general case this may not be so. Now, $BC \, |\sin \theta|$ is the area of the parallelogram defined by $\mathbf{B}$ and $\mathbf{C}$. The altitude of the parallelepiped is $A \, |\cos \phi|$ and so $ABC \, |\sin \theta \cos \phi|$ is the *volume* of the parallelepiped with three adjacent sides defined by $\mathbf{A}$, $\mathbf{B}$ and $\mathbf{C}$.

Consequently if $\mathbf{A} \cdot (\mathbf{B} \times \mathbf{C}) = 0$ then the volume of the parallelepiped is zero and the three vectors $\mathbf{A}$, $\mathbf{B}$ and $\mathbf{C}$ are coplanar.

Example 1

Show that $\mathbf{A} = \mathbf{i} + 2\mathbf{j} - 3\mathbf{k}$; $\mathbf{B} = 2\mathbf{i} - \mathbf{j} + 2\mathbf{k}$; and $\mathbf{C} = 3\mathbf{i} + \mathbf{j} - \mathbf{k}$ are coplanar.

We just evaluate $\mathbf{A} \cdot (\mathbf{B} \times \mathbf{C}) = \ldots\ldots\ldots$ and apply the test.

16
$$\boxed{\mathbf{A} \cdot (\mathbf{B} \times \mathbf{C}) = 0}$$

Because

$$\mathbf{A} \cdot (\mathbf{B} \times \mathbf{C}) = \begin{vmatrix} 1 & 2 & -3 \\ 2 & -1 & 2 \\ 3 & 1 & -1 \end{vmatrix} = 1(1 - 2) - 2(-2 - 6) - 3(2 + 3) = 0.$$

Therefore **A, B, C** are coplanar.

Example 2

If $\mathbf{A} = 2\mathbf{i} - \mathbf{j} + 3\mathbf{k}$; $\mathbf{B} = 3\mathbf{i} + 2\mathbf{j} + \mathbf{k}$; $\mathbf{C} = \mathbf{i} + p\mathbf{j} + 4\mathbf{k}$ are coplanar, find the value of p.

The method is clear enough. We merely set up and evaluate the determinant and solve the equation $\mathbf{A} \cdot (\mathbf{B} \times \mathbf{C}) = 0$.

$$p = \ldots\ldots\ldots\ldots$$

17
$$\boxed{p = -3}$$

Because

$$\mathbf{A} \cdot (\mathbf{B} \times \mathbf{C}) = 0 \qquad \therefore \begin{vmatrix} 2 & -1 & 3 \\ 3 & 2 & 1 \\ 1 & p & 4 \end{vmatrix} = 0$$

$$\therefore \ 2(8 - p) + 1(12 - 1) + 3(3p - 2) = 0 \quad \therefore \ 7p = -21 \quad \therefore \ p = -3$$

One more.

Example 3

Determine whether the three vectors $\mathbf{A} = 3\mathbf{i} + 2\mathbf{j} - \mathbf{k}$; $\mathbf{B} = 2\mathbf{i} - \mathbf{j} + 3\mathbf{k}$; $\mathbf{C} = \mathbf{i} - 2\mathbf{j} + 2\mathbf{k}$ are coplanar.

Work through it on your own. The result shows that

$$\ldots\ldots\ldots\ldots$$

18
$$\boxed{\textbf{A, B, C are not coplanar}}$$

Because

$$\text{in this case} \ \ \mathbf{A} \cdot (\mathbf{B} \times \mathbf{C}) = \begin{vmatrix} 3 & 2 & -1 \\ 2 & -1 & 3 \\ 1 & -2 & 2 \end{vmatrix} = 13$$

$$\therefore \ \mathbf{A} \cdot (\mathbf{B} \times \mathbf{C}) \neq 0 \qquad \therefore \ \textbf{A, B, C are not coplanar.}$$

Now on to something different

Vector triple products of three vectors

19

If **A**, **B** and **C** are three vectors, then

$$\left.\begin{array}{l} \mathbf{A} \times (\mathbf{B} \times \mathbf{C}) \\ \text{and} \quad (\mathbf{A} \times \mathbf{B}) \times \mathbf{C} \end{array}\right\} \quad \text{are called the vector triple products.} \qquad (10)$$

Consider $\mathbf{A} \times (\mathbf{B} \times \mathbf{C})$ where $\mathbf{A} = a_x\mathbf{i} + a_y\mathbf{j} + a_z\mathbf{k}$; $\mathbf{B} = b_x\mathbf{i} + b_y\mathbf{j} + b_z\mathbf{k}$ and $\mathbf{C} = c_x\mathbf{i} + c_y\mathbf{j} + c_z\mathbf{k}$.

Then $(\mathbf{B} \times \mathbf{C})$ is a vector perpendicular to the plane of **B** and **C** and $\mathbf{A} \times (\mathbf{B} \times \mathbf{C})$ is a vector perpendicular to the plane containing **A** and $(\mathbf{B} \times \mathbf{C})$, i.e. coplanar with **B** and **C**.

Note that, similarly, $(\mathbf{A} \times \mathbf{B}) \times \mathbf{C}$ is coplanar with **A** and **B** and so in general $\mathbf{A} \times (\mathbf{B} \times \mathbf{C}) \neq (\mathbf{A} \times \mathbf{B}) \times \mathbf{C}$.

Now

$$(\mathbf{B} \times \mathbf{C}) = \begin{vmatrix} \mathbf{i} & \mathbf{j} & \mathbf{k} \\ b_x & b_y & b_z \\ c_x & c_y & c_z \end{vmatrix} = \mathbf{i} \begin{vmatrix} b_y & b_z \\ c_y & c_z \end{vmatrix} - \mathbf{j} \begin{vmatrix} b_x & b_z \\ c_x & c_z \end{vmatrix} + \mathbf{k} \begin{vmatrix} b_x & b_y \\ c_x & c_y \end{vmatrix}$$

$$\text{Then} \quad \mathbf{A} \times (\mathbf{B} \times \mathbf{C}) = \begin{vmatrix} \mathbf{i} & \mathbf{j} & \mathbf{k} \\ a_x & a_y & a_z \\ \begin{vmatrix} b_y & b_z \\ c_y & c_z \end{vmatrix} & -\begin{vmatrix} b_x & b_z \\ c_x & c_z \end{vmatrix} & \begin{vmatrix} b_x & b_y \\ c_x & c_y \end{vmatrix} \end{vmatrix}$$

$$= \begin{vmatrix} \mathbf{i} & \mathbf{j} & \mathbf{k} \\ a_x & a_y & a_z \\ \begin{vmatrix} b_y & b_z \\ c_y & c_z \end{vmatrix} & \begin{vmatrix} b_z & b_x \\ c_z & c_x \end{vmatrix} & \begin{vmatrix} b_x & b_y \\ c_x & c_y \end{vmatrix} \end{vmatrix}$$

In symbolic form, further expansion of the determinant becomes somewhat tedious. However a numerical example will clarify the method.

So make a note of the definition (10) above and then go on to the next frame

Example 1

20

If $\mathbf{A} = 2\mathbf{i} - 3\mathbf{j} + \mathbf{k}$; $\mathbf{B} = \mathbf{i} + 2\mathbf{j} - \mathbf{k}$; $\mathbf{C} = 3\mathbf{i} + \mathbf{j} + 3\mathbf{k}$; determine the vector triple product $\mathbf{A} \times (\mathbf{B} \times \mathbf{C})$.

We start off with $\mathbf{B} \times \mathbf{C} = \ldots\ldots\ldots\ldots$

21

$$\boxed{\mathbf{B} \times \mathbf{C} = 7\mathbf{i} - 6\mathbf{j} - 5\mathbf{k}}$$

Because

$$\mathbf{B} \times \mathbf{C} = \begin{vmatrix} \mathbf{i} & \mathbf{j} & \mathbf{k} \\ 1 & 2 & -1 \\ 3 & 1 & 3 \end{vmatrix} \begin{aligned} &= \mathbf{i}(6+1) - \mathbf{j}(3+3) + \mathbf{k}(1-6) \\ &= 7\mathbf{i} - 6\mathbf{j} - 5\mathbf{k} \end{aligned}$$

Then $\mathbf{A} \times (\mathbf{B} \times \mathbf{C}) = \dots\dots\dots$

22

$$\boxed{\mathbf{A} \times (\mathbf{B} \times \mathbf{C}) = 21\mathbf{i} + 17\mathbf{j} + 9\mathbf{k}}$$

Because

$$\begin{aligned} \mathbf{A} \times (\mathbf{B} \times \mathbf{C}) &= \begin{vmatrix} \mathbf{i} & \mathbf{j} & \mathbf{k} \\ 2 & -3 & 1 \\ 7 & -6 & 5 \end{vmatrix} \\ &= \mathbf{i}(15+6) - \mathbf{j}(-10-7) + \mathbf{k}(-12+21) \\ &= 21\mathbf{i} + 17\mathbf{j} + 9\mathbf{k} \end{aligned}$$

That is fundamental enough. There is, however, an even easier way of determining a vector triple product. It can be proved that

$$\mathbf{A} \times (\mathbf{B} \times \mathbf{C}) = (\mathbf{A} \cdot \mathbf{C})\mathbf{B} - (\mathbf{A} \cdot \mathbf{B})\mathbf{C}$$

and $\quad (\mathbf{A} \times \mathbf{B}) \times \mathbf{C} = (\mathbf{C} \cdot \mathbf{A})\mathbf{B} - (\mathbf{C} \cdot \mathbf{B})\mathbf{A}$
$\hspace{6cm}$ (11)

The proof of this is given in the Appendix. For the moment, make a careful note of the expressions: then we will apply the method to the example we have just completed.

23

$\mathbf{A} = 2\mathbf{i} - 3\mathbf{j} + \mathbf{k}; \quad \mathbf{B} = \mathbf{i} + 2\mathbf{j} - \mathbf{k}; \quad \mathbf{C} = 3\mathbf{i} + \mathbf{j} + 3\mathbf{k}$ and we have

$$\begin{aligned} \mathbf{A} \times (\mathbf{B} \times \mathbf{C}) &= (\mathbf{A} \cdot \mathbf{C})\mathbf{B} - (\mathbf{A} \cdot \mathbf{B})\mathbf{C} \\ &= (6-3+3)(\mathbf{i}+2\mathbf{j}-\mathbf{k}) - (2-6-1)(3\mathbf{i}+\mathbf{j}+3\mathbf{k}) \\ &= 6\,(\mathbf{i}+2\mathbf{j}-\mathbf{k}) + 5(3\mathbf{i}+\mathbf{j}+3\mathbf{k}) \\ &= 21\mathbf{i} + 17\mathbf{j} + 9\mathbf{k} \end{aligned}$$

which is, of course, the result we achieved before.

Here is another.

Example 2

If $\mathbf{A} = 3\mathbf{i} + 2\mathbf{j} - 2\mathbf{k}; \quad \mathbf{B} = 4\mathbf{i} - \mathbf{j} + 3\mathbf{k}; \quad \mathbf{C} = 2\mathbf{i} - 3\mathbf{j} + \mathbf{k}$ determine $(\mathbf{A} \times \mathbf{B}) \times \mathbf{C}$ using the relationship $(\mathbf{A} \times \mathbf{B}) \times \mathbf{C} = (\mathbf{C} \cdot \mathbf{A})\mathbf{B} - (\mathbf{C} \cdot \mathbf{B})\mathbf{A}$.

$$(\mathbf{A} \times \mathbf{B}) \times \mathbf{C} = \dots\dots\dots$$

24

$$\boxed{-50\mathbf{i} - 26\mathbf{j} + 22\mathbf{k}}$$

Because

$$(\mathbf{A} \times \mathbf{B}) \times \mathbf{C} = (\mathbf{C} \cdot \mathbf{A})\mathbf{B} - (\mathbf{C} \cdot \mathbf{B})\mathbf{A}$$
$$= (6 - 6 - 2)(4\mathbf{i} - \mathbf{j} + 3\mathbf{k}) - (8 + 3 + 3)(3\mathbf{i} + 2\mathbf{j} - 2\mathbf{k})$$
$$= -2(4\mathbf{i} - \mathbf{j} + 3\mathbf{k}) - 14(3\mathbf{i} + 2\mathbf{j} - 2\mathbf{k})$$
$$= -50\mathbf{i} - 26\mathbf{j} + 22\mathbf{k}$$

Now one more.

Example 3

If $\mathbf{A} = \mathbf{i} + 3\mathbf{j} + 2\mathbf{k}; \quad \mathbf{B} = 2\mathbf{i} + 5\mathbf{j} - \mathbf{k}; \quad \mathbf{C} = \mathbf{i} + 2\mathbf{j} + 3\mathbf{k}$

$$\mathbf{A} \times (\mathbf{B} \times \mathbf{C}) = \ldots\ldots\ldots$$
$$(\mathbf{A} \times \mathbf{B}) \times \mathbf{C} = \ldots\ldots\ldots$$

Finish them both.

25

$$\boxed{\begin{array}{l} \mathbf{A} \times (\mathbf{B} \times \mathbf{C}) = 11\mathbf{i} + 35\mathbf{j} - 58\mathbf{k} \\ (\mathbf{A} \times \mathbf{B}) \times \mathbf{C} = 17\mathbf{i} + 38\mathbf{j} - 31\mathbf{k} \end{array}}$$

Because

$$\mathbf{A} \times (\mathbf{B} \times \mathbf{C}) = (\mathbf{A} \cdot \mathbf{C})\mathbf{B} - (\mathbf{A} \cdot \mathbf{B})\mathbf{C}$$
$$= (1 + 6 + 6)(2\mathbf{i} + 5\mathbf{j} - \mathbf{k}) - (2 + 15 - 2)(\mathbf{i} + 2\mathbf{j} + 3\mathbf{k})$$
$$= 13(2\mathbf{i} + 5\mathbf{j} - \mathbf{k}) - 15(\mathbf{i} + 2\mathbf{j} + 3\mathbf{k})$$
$$= 11\mathbf{i} + 35\mathbf{j} - 58\mathbf{k}$$

and

$$(\mathbf{A} \times \mathbf{B}) \times \mathbf{C} = (\mathbf{C} \cdot \mathbf{A})\mathbf{B} - (\mathbf{C} \cdot \mathbf{B})\mathbf{A}$$
$$= (1 + 6 + 6)(2\mathbf{i} + 5\mathbf{j} - \mathbf{k}) - (2 + 10 - 3)(\mathbf{i} + 3\mathbf{j} + 2\mathbf{k})$$
$$= 13(2\mathbf{i} + 5\mathbf{j} - \mathbf{k}) - 9(\mathbf{i} + 3\mathbf{j} + 2\mathbf{k}) = 17\mathbf{i} + 38\mathbf{j} - 31\mathbf{k}$$

These two results clearly confirm that

$$\mathbf{A} \times (\mathbf{B} \times \mathbf{C}) \neq (\mathbf{A} \times \mathbf{B}) \times \mathbf{C} \qquad \textit{so beware!}$$

Before we proceed, note the following concerning the unit vectors.

(a) $\quad (\mathbf{i} \times \mathbf{j}) = \mathbf{k}$

$\quad\therefore\ \mathbf{i} \times (\mathbf{i} \times \mathbf{j}) = \mathbf{i} \times \mathbf{k} = -\mathbf{j}$

$\quad\therefore\ \mathbf{i} \times (\mathbf{i} \times \mathbf{j}) = -\mathbf{j}$

(b) $\quad (\mathbf{i} \times \mathbf{i}) \times \mathbf{j} = (0) \times \mathbf{j} = 0$

$\quad\therefore\ (\mathbf{i} \times \mathbf{i}) \times \mathbf{j} = 0$

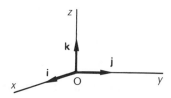

and once again, we see that

$$\mathbf{i} \times (\mathbf{i} \times \mathbf{j}) \neq (\mathbf{i} \times \mathbf{i}) \times \mathbf{j}$$

On to the next

26   Finally, by way of revision:

Example 4

If $\mathbf{A} = 5\mathbf{i} - 2\mathbf{j} + 3\mathbf{k};\ \ \mathbf{B} = 3\mathbf{i} + \mathbf{j} - 2\mathbf{k};\ \ \mathbf{C} = \mathbf{i} - 3\mathbf{j} + 4\mathbf{k};$ determine

(a) the scalar triple product $\mathbf{A} \cdot (\mathbf{B} \times \mathbf{C})$

(b) the vector triple products (1) $\mathbf{A} \times (\mathbf{B} \times \mathbf{C})$
                                            (2) $(\mathbf{A} \times \mathbf{B}) \times \mathbf{C}$.

Finish all these and then check with the next frame

27

> (a) $\mathbf{A} \cdot (\mathbf{B} \times \mathbf{C}) = -12$
>
> (b) (1) $\mathbf{A} \times (\mathbf{B} \times \mathbf{C}) = 62\mathbf{i} + 44\mathbf{j} - 74\mathbf{k}$
>
> (2) $(\mathbf{A} \times \mathbf{B}) \times \mathbf{C} = 109\mathbf{i} + 7\mathbf{j} - 22\mathbf{k}$

Here is the working.

(a) $\mathbf{A} \cdot (\mathbf{B} \times \mathbf{C}) = \begin{vmatrix} 5 & -2 & 3 \\ 3 & 1 & -2 \\ 1 & -3 & 4 \end{vmatrix}$

$\quad = 5(4 - 6) + 2(12 + 2) + 3(-9 - 1) = -12$

(b) (1) $\mathbf{A} \times (\mathbf{B} \times \mathbf{C}) = (\mathbf{A} \cdot \mathbf{C})\mathbf{B} - (\mathbf{A} \cdot \mathbf{B})\mathbf{C}$

$\quad = (5 + 6 + 12)(3\mathbf{i} + \mathbf{j} - 2\mathbf{k})$

$\quad\quad\quad\quad\quad\quad\quad - (15 - 2 - 6)(\mathbf{i} - 3\mathbf{j} + 4\mathbf{k})$

$\quad = 23(3\mathbf{i} + \mathbf{j} - 2\mathbf{k}) - 7(\mathbf{i} - 3\mathbf{j} + 4\mathbf{k})$

$\quad = 62\mathbf{i} + 44\mathbf{j} - 74\mathbf{k}$

(2) $(\mathbf{A} \times \mathbf{B}) \times \mathbf{C} = (\mathbf{C} \cdot \mathbf{A})\mathbf{B} - (\mathbf{C} \cdot \mathbf{B})\mathbf{A}$

$\quad = 23(3\mathbf{i} + \mathbf{j} - 2\mathbf{k}) - (-8)(5\mathbf{i} - 2\mathbf{j} + 3\mathbf{k})$

$\quad = 109\mathbf{i} + 7\mathbf{j} - 22\mathbf{k}$

Let us now move to the next topic

28 **Differentiation of vectors**

In many practical problems, we often deal with vectors that change with time, e.g. velocity, acceleration, etc. If a vector $\mathbf{A}$ depends on a scalar variable t, then $\mathbf{A}$ can be represented as $\mathbf{A}(t)$ and $\mathbf{A}$ is then said to be a function of t.

If $\mathbf{A} = a_x\mathbf{i} + a_y\mathbf{j} + a_z\mathbf{k}$ then a_x, a_y, a_z will also be dependent on the parameter t.

i.e. $\mathbf{A}(t) = a_x(t)\mathbf{i} + a_y(t)\mathbf{j} + a_z(t)\mathbf{k}$

Differentiating with respect to t gives . . . .

$$\frac{d}{dt}\{A(t)\} = i\frac{d}{dt}\{a_x(t)\} + j\frac{d}{dt}\{a_y(t)\} + k\frac{d}{dt}\{a_z(t)\}$$

29

In short $\dfrac{dA}{dt} = i\dfrac{da_x}{dt} + j\dfrac{da_y}{dt} + k\dfrac{da_z}{dt}$.

The independent scalar variable is not, of course, restricted to t. In general, if u is the parameter, then

$$\frac{dA}{du} = \ldots\ldots\ldots$$

$$\frac{dA}{du} = i\frac{da_x}{du} + j\frac{da_y}{du} + k\frac{da_z}{du}$$

30

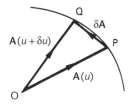

If a position vector $\overline{OP}$ moves to $\overline{OQ}$ when u becomes $u + \delta u$, then as $\delta u \to 0$, the direction of the chord $\overline{PQ}$ becomes that of the tangent to the curve at P, i.e. the direction of $\dfrac{dA}{du}$ is along the tangent to the locus of P.

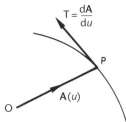

Example 1

If $A = (3u^2 + 4)i + (2u - 5)j + 4u^3k$, then

$$\frac{dA}{du} = \ldots\ldots\ldots$$

$$\frac{dA}{du} = 6ui + 2j + 12u^2k$$

31

If we differentiate this again, we get $\dfrac{d^2A}{du^2} = 6i + 24uk$

When $u = 2$, $\dfrac{dA}{du} = 12i + 2j + 48k$ and $\dfrac{d^2A}{du^2} = 6i + 48k$

Then $\left|\dfrac{dA}{du}\right| = \ldots\ldots\ldots$ and $\left|\dfrac{d^2A}{du^2}\right| = \ldots\ldots\ldots$

32

$$\boxed{\left|\frac{\mathrm{d}A}{\mathrm{d}u}\right| = 49 \cdot 52; \quad \left|\frac{\mathrm{d}^2A}{\mathrm{d}u^2}\right| = 48 \cdot 37}$$

Because

$$\left|\frac{\mathrm{d}A}{\mathrm{d}u}\right| = \{12^2 + 2^2 + 48^2\}^{1/2} = \{2452\}^{1/2} = 49 \cdot 52$$

and $\quad \left|\frac{\mathrm{d}^2A}{\mathrm{d}u^2}\right| = \{6^2 + 48^2\}^{1/2} = \{2340\}^{1/2} = 48 \cdot 37$

Example 2

If $\mathbf{F} = \mathbf{i}\sin 2t + \mathbf{j}e^{3t} + \mathbf{k}(t^3 - 4t)$, then when $t = 1$

$$\frac{\mathrm{d}\mathbf{F}}{\mathrm{d}t} = \ldots\ldots\ldots; \quad \frac{\mathrm{d}^2\mathbf{F}}{\mathrm{d}t^2} = \ldots\ldots\ldots$$

33

$$\boxed{\begin{aligned} \frac{\mathrm{d}\mathbf{F}}{\mathrm{d}t} &= 2\cos 2\mathbf{i} + 3e^3\mathbf{j} - \mathbf{k} \\ \frac{\mathrm{d}^2\mathbf{F}}{\mathrm{d}t^2} &= -4\sin 2\mathbf{i} + 9e^3\mathbf{j} + 6\mathbf{k} \end{aligned}}$$

From these, we could if required find the magnitudes of $\dfrac{\mathrm{d}\mathbf{F}}{\mathrm{d}t}$ and $\dfrac{\mathrm{d}^2\mathbf{F}}{\mathrm{d}t^2}$.

$$\left|\frac{\mathrm{d}\mathbf{F}}{\mathrm{d}t}\right| = \ldots\ldots\ldots; \quad \left|\frac{\mathrm{d}^2\mathbf{F}}{\mathrm{d}t^2}\right| = \ldots\ldots\ldots$$

34

$$\boxed{\left|\frac{\mathrm{d}\mathbf{F}}{\mathrm{d}t}\right| = 60 \cdot 27; \quad \left|\frac{\mathrm{d}^2\mathbf{F}}{\mathrm{d}t^2}\right| = 180 \cdot 9}$$

Because

$$\begin{aligned} \left|\frac{\mathrm{d}\mathbf{F}}{\mathrm{d}t}\right| &= \{(2\cos 2)^2 + 9e^6 + 1\}^{1/2} \\ &= \{0 \cdot 6927 + 3631 + 1\}^{1/2} = 60 \cdot 27 \end{aligned}$$

and $\quad \begin{aligned}[t] \left|\frac{\mathrm{d}^2\mathbf{F}}{\mathrm{d}t^2}\right| &= \{(-4\sin 2)^2 + 81e^6 + 36\}^{1/2} \\ &= \{13 \cdot 23 + 32\,678 + 36\}^{1/2} = 180 \cdot 9 \end{aligned}$

One more example.

Example 3

If $\mathbf{A} = (u+3)\mathbf{i} - (2+u^2)\mathbf{j} + 2u^3\mathbf{k}$, determine

(a) $\dfrac{\mathrm{d}A}{\mathrm{d}u}$ (b) $\dfrac{\mathrm{d}^2A}{\mathrm{d}u^2}$ (c) $\left|\dfrac{\mathrm{d}A}{\mathrm{d}u}\right|$ (d) $\left|\dfrac{\mathrm{d}^2A}{\mathrm{d}u^2}\right|$ at $u = 3$.

Work through all sections and then check with the next frame

Here is the working. $A = (u + 3)i - (2 + u^2)j + 2u^3k$

35

(a) $\dfrac{dA}{du} = i - 2uj + 6u^2k$ At $u = 3$, $\dfrac{dA}{du} = i - 6j + 54k$

(b) $\dfrac{d^2A}{du^2} = -2j + 12uk$ At $u = 3$, $\dfrac{d^2A}{du^2} = -2j + 36k$

(c) $\left|\dfrac{dA}{du}\right| = \{1 + 36 + 2916\}^{1/2} = (2953)^{1/2} = 54.34$

(d) $\left|\dfrac{d^2A}{du^2}\right| = \{4 + 1296\}^{1/2} = (1300)^{1/2} = 36.06$

The next example is of a rather different kind, so move on

Example 4

36

A particle moves in space so that at time t its position is stated as $x = 2t + 3$, $y = t^2 + 3t$, $z = t^3 + 2t^2$. We are required to find the components of its velocity and acceleration in the direction of the vector $2i + 3j + 4k$ when $t = 1$.

First we can write the position as a vector r

$r = (2t + 3)i + (t^2 + 3t)j + (t^3 + 2t^2)k$

Then, at $t = 1$

$$\frac{dr}{dt} = \ldots\ldots\ldots\ldots ; \qquad \frac{d^2r}{dt^2} = \ldots\ldots\ldots\ldots$$

37

$$\frac{dr}{dt} = 2i + 5j + 7k; \qquad \frac{d^2r}{dt^2} = 2j + 10k$$

Because

$$\frac{dr}{dt} = 2i + (2t + 3)j + (3t^2 + 4t)k$$

$$\therefore \text{ At } t = 1, \qquad \frac{dr}{dt} = 2i + 5j + 7k$$

and $\dfrac{d^2r}{dt^2} = 2j + (6t + 4)k$

$$\therefore \text{ At } t = 1, \qquad \frac{d^2r}{dt^2} = 2j + 10k$$

Now, a unit vector parallel to $2i + 3j + 4k$ is $\ldots\ldots\ldots\ldots$

38

$$\boxed{\dfrac{2\mathbf{i}+3\mathbf{j}+4\mathbf{k}}{\sqrt{4+9+16}} = \dfrac{1}{\sqrt{29}}(2\mathbf{i}+3\mathbf{j}+4\mathbf{k})}$$

Denote this unit vector by **I**. Then the component of $\dfrac{d\mathbf{r}}{dt}$ in the direction of **I**

$$= \dfrac{d\mathbf{r}}{dt}\cos\theta$$

$$= \dfrac{d\mathbf{r}}{dt}\cdot\mathbf{I}$$

$$= \dfrac{1}{\sqrt{29}}(2\mathbf{i}+5\mathbf{j}+7\mathbf{k})\cdot(2\mathbf{i}+3\mathbf{j}+4\mathbf{k})$$

$$= \ldots\ldots\ldots$$

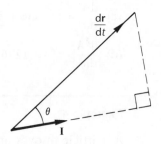

39

$$\boxed{8\cdot73}$$

Because

$$\dfrac{1}{\sqrt{29}}(2\mathbf{i}+5\mathbf{j}+7\mathbf{k})\cdot(2\mathbf{i}+3\mathbf{j}+4\mathbf{k}) = \dfrac{1}{\sqrt{29}}(4+15+28)$$

$$= \dfrac{47}{\sqrt{29}}$$

$$= 8\cdot73$$

Similarly, the component of $\dfrac{d^2\mathbf{r}}{dt^2}$ in the direction of **I** is

$$\ldots\ldots\ldots$$

40

$$\boxed{8\cdot54}$$

Because

$$\dfrac{d^2\mathbf{r}}{dt^2}\cos\theta = \dfrac{d^2\mathbf{r}}{dt^2}\cdot\mathbf{I}$$

$$= \dfrac{1}{\sqrt{29}}(2\mathbf{j}+10\mathbf{k})\cdot(2\mathbf{i}+3\mathbf{j}+4\mathbf{k})$$

$$= \dfrac{1}{\sqrt{29}}(6+40)$$

$$= \dfrac{46}{\sqrt{29}}$$

$$= 8\cdot54$$

▶

Differentiation of sums and products of vectors

If $\mathbf{A} = \mathbf{A}(u)$ and $\mathbf{B} = \mathbf{B}(u)$, then

(a) $\dfrac{d}{du}\{c\mathbf{A}\} = c\dfrac{d\mathbf{A}}{du}$

(b) $\dfrac{d}{du}\{\mathbf{A} + \mathbf{B}\} = \dfrac{d\mathbf{A}}{du} + \dfrac{d\mathbf{B}}{du}$

(c) $\dfrac{d}{du}\{\mathbf{A} \cdot \mathbf{B}\} = \mathbf{A} \cdot \dfrac{d\mathbf{B}}{du} + \dfrac{d\mathbf{A}}{du} \cdot \mathbf{B}$

(d) $\dfrac{d}{du}\{\mathbf{A} \times \mathbf{B}\} = \mathbf{A} \times \dfrac{d\mathbf{B}}{du} + \dfrac{d\mathbf{A}}{du} \times \mathbf{B}$.

These are very much like the normal rules of differentiation.

However, if $\mathbf{A}(u) \cdot \mathbf{A}(u) = a_x^2 + a_y^2 + a_z^2 = |\mathbf{A}|^2 = A^2$ is a constant then

$$\frac{d}{du}\{\mathbf{A}(u) \cdot \mathbf{A}(u)\} = \mathbf{A}(u) \cdot \frac{d}{du}\{\mathbf{A}(u)\} + \mathbf{A}(u) \cdot \frac{d}{du}\{\mathbf{A}(u)\}$$

$$= 2\mathbf{A}(u) \cdot \frac{d}{du}\{\mathbf{A}(u)\} = \frac{d}{du}\{A^2\} = 0$$

Assuming that $\mathbf{A}(u) \neq 0$, then since $\mathbf{A}(u) \cdot \dfrac{d}{du}\{\mathbf{A}(u)\} = \dfrac{d}{du}\{A^2\} = 0$ it follows

that $\mathbf{A}(u)$ and $\dfrac{d}{du}\{\mathbf{A}(u)\}$ are perpendicular vectors because

....

$$\mathbf{A}(u) \cdot \frac{d}{du}\{\mathbf{A}(u)\} = |\,\mathbf{A}(u)\,|\left|\frac{d}{du}\{\mathbf{A}(u)\}\right|\cos\theta = 0$$

$$\therefore \quad \cos\theta = 0 \qquad \therefore \quad \theta = \frac{\pi}{2}$$

41

Now let us deal with unit tangent vectors.

Unit tangent vectors

We have already established in Frame 30 of this Programme that if $\overline{OP}$ is a position vector $\mathbf{A}(u)$ in space, then the direction of the vector denoting $\dfrac{d}{du}\{\mathbf{A}(u)\}$ is

....

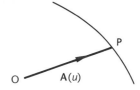

42

parallel to the tangent to the curve at P

Then the unit tangent vector **T** at P can be found from

$$\mathbf{T} = \frac{\dfrac{d}{du}\{\mathbf{A}(u)\}}{\left|\dfrac{d}{du}\{\mathbf{A}(u)\}\right|}$$

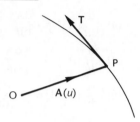

In simpler notation, this becomes:

If $\mathbf{r} = a_x\mathbf{i} + a_y\mathbf{j} + a_z\mathbf{k}$ then the unit tangent vector **T** is given by

$$\mathbf{T} = \frac{d\mathbf{r}/du}{|d\mathbf{r}/du|}$$

Example 1

Determine the unit tangent vector at the point (2, 4, 7) for the curve with parametric equations $x = 2u$; $y = u^2 + 3$; $z = 2u^2 + 5$.

First we see that the point (2, 4, 7) corresponds to $u = 1$.

The vector equation of the curve is

$$\mathbf{r} = a_x\mathbf{i} + a_y\mathbf{j} + a_z\mathbf{k} = 2u\mathbf{i} + (u^2 + 3)\mathbf{j} + (2u^2 + 5)\mathbf{k}$$

$$\therefore \quad \frac{d\mathbf{r}}{du} = \ldots\ldots\ldots\ldots$$

43

$$\frac{d\mathbf{r}}{du} = 2\mathbf{i} + 2u\mathbf{j} + 4u\mathbf{k}$$

and at $u = 1$, $\dfrac{d\mathbf{r}}{du} = 2\mathbf{i} + 2\mathbf{j} + 4\mathbf{k}$

Hence $\left|\dfrac{d\mathbf{r}}{du}\right| = \ldots\ldots\ldots\ldots$ and **T** $= \ldots\ldots\ldots\ldots$

44

$$\left|\frac{d\mathbf{r}}{du}\right| = 2\sqrt{6}; \quad \mathbf{T} = \frac{1}{\sqrt{6}}\{\mathbf{i} + \mathbf{j} + 2\mathbf{k}\}$$

Because

$$\left|\frac{d\mathbf{r}}{du}\right| = \{4 + 4 + 16\}^{1/2} = 24^{1/2} = 2\sqrt{6}$$

$$\mathbf{T} = \frac{\dfrac{d\mathbf{r}}{du}}{\left|\dfrac{d\mathbf{r}}{du}\right|} = \frac{2\mathbf{i} + 2\mathbf{j} + 4\mathbf{k}}{2\sqrt{6}} = \frac{1}{\sqrt{6}}\{\mathbf{i} + \mathbf{j} + 2\mathbf{k}\}$$

Let us do another.

▶

Example 2

Find the unit tangent vector at the point $(2, 0, \pi)$ for the curve with parametric equations $x = 2\sin\theta$; $y = 3\cos\theta$; $z = 2\theta$.

We see that the point $(2, 0, \pi)$ corresponds to $\theta = \pi/2$.

Writing the curve in vector form $\mathbf{r} = \ldots\ldots\ldots$

$$\mathbf{r} = 2\sin\theta\,\mathbf{i} + 3\cos\theta\,\mathbf{j} + 2\theta\,\mathbf{k}$$

45

Then, at $\theta = \pi/2$, $\quad \dfrac{d\mathbf{r}}{d\theta} = \ldots\ldots\ldots$

$$\left|\dfrac{d\mathbf{r}}{d\theta}\right| = \ldots\ldots\ldots$$

$$\mathbf{T} = \ldots\ldots\ldots$$

Finish it off

$$\frac{d\mathbf{r}}{d\theta} = -3\mathbf{j} + 2\mathbf{k}; \quad \left|\frac{d\mathbf{r}}{d\theta}\right| = \sqrt{13}$$

$$\mathbf{T} = \frac{1}{\sqrt{13}}(-3\mathbf{j} + 2\mathbf{k})$$

46

And now

Example 3

Determine the unit tangent vector for the curve

$$x = 3t; \quad y = 2t^2; \quad z = t^2 + t$$

at the point $(6, 8, 6)$.

On your own. $\mathbf{T} = \ldots\ldots\ldots$

$$\mathbf{T} = \frac{1}{\sqrt{98}}(3\mathbf{i} + 8\mathbf{j} + 5\mathbf{k})$$

47

The point $(6, 8, 6)$ corresponds to $t = 2$

$$\mathbf{r} = 3t\mathbf{i} + 2t^2\mathbf{j} + (t^2 + t)\mathbf{k}$$

$$\therefore \quad \frac{d\mathbf{r}}{dt} = 3\mathbf{i} + 4t\mathbf{j} + (2t + 1)\mathbf{k}$$

At $t = 2$, $\mathbf{r} = 6\mathbf{i} + 8\mathbf{j} + 6\mathbf{k}$ and $\dfrac{d\mathbf{r}}{dt} = 3\mathbf{i} + 8\mathbf{j} + 5\mathbf{k}$

$$\therefore \quad \left|\frac{d\mathbf{r}}{dt}\right| = (9 + 64 + 25)^{1/2} = \sqrt{98}$$

$$\therefore \quad \mathbf{T} = \frac{d\mathbf{r}/dt}{|d\mathbf{r}/dt|} = \frac{1}{\sqrt{98}}(3\mathbf{i} + 8\mathbf{j} + 5\mathbf{k})$$

Partial differentiation of vectors

48

If a vector $\mathbf{F}$ is a function of two independent variables u and v, then the rules of differentiation follow the usual pattern.

If $\mathbf{F} = x\mathbf{i} + y\mathbf{j} + z\mathbf{k}$ then x, y, z will also be functions of u and v.

Then
$$\frac{\partial \mathbf{F}}{\partial u} = \frac{\partial x}{\partial u}\mathbf{i} + \frac{\partial y}{\partial u}\mathbf{j} + \frac{\partial z}{\partial u}\mathbf{k}$$

$$\frac{\partial \mathbf{F}}{\partial v} = \frac{\partial x}{\partial v}\mathbf{i} + \frac{\partial y}{\partial v}\mathbf{j} + \frac{\partial z}{\partial v}\mathbf{k}$$

$$\frac{\partial^2 \mathbf{F}}{\partial u^2} = \frac{\partial^2 x}{\partial u^2}\mathbf{i} + \frac{\partial^2 y}{\partial u^2}\mathbf{j} + \frac{\partial^2 z}{\partial u^2}\mathbf{k}$$

$$\frac{\partial^2 \mathbf{F}}{\partial v^2} = \frac{\partial^2 x}{\partial v^2}\mathbf{i} + \frac{\partial^2 y}{\partial v^2}\mathbf{j} + \frac{\partial^2 z}{\partial v^2}\mathbf{k}$$

$$\frac{\partial^2 \mathbf{F}}{\partial u \partial v} = \frac{\partial^2 x}{\partial u \partial v}\mathbf{i} + \frac{\partial^2 y}{\partial u \partial v}\mathbf{j} + \frac{\partial^2 z}{\partial u \partial v}\mathbf{k}$$

and for small finite changes du and dv in u and v, we have

$$d\mathbf{F} = \frac{\partial \mathbf{F}}{\partial u}du + \frac{\partial \mathbf{F}}{\partial v}dv$$

Example

If $\mathbf{F} = 2uv\mathbf{i} + (u^2 - 2v)\mathbf{j} + (u + v^2)\mathbf{k}$

$$\frac{\partial \mathbf{F}}{\partial u} = \ldots\ldots\ldots; \qquad \frac{\partial \mathbf{F}}{\partial v} = \ldots\ldots\ldots$$

$$\frac{\partial^2 \mathbf{F}}{\partial u^2} = \ldots\ldots\ldots; \qquad \frac{\partial^2 \mathbf{F}}{\partial u \partial v} = \ldots\ldots\ldots$$

49

$$\frac{\partial \mathbf{F}}{\partial u} = 2v\mathbf{i} + 2u\mathbf{j} + \mathbf{k}; \qquad \frac{\partial \mathbf{F}}{\partial v} = 2u\mathbf{i} - 2\mathbf{j} + 2v\mathbf{k}$$

$$\frac{\partial^2 \mathbf{F}}{\partial u^2} = 2\mathbf{j}; \qquad \frac{\partial^2 \mathbf{F}}{\partial u \partial v} = 2\mathbf{i}$$

This is straightforward enough.

Integration of vector functions

The process is the reverse of that for differentiation. If a vector $\mathbf{F} = x\mathbf{i} + y\mathbf{j} + z\mathbf{k}$ where $\mathbf{F}$, x, y, z are expressed as functions of u, then

$$\int_a^b \mathbf{F}\, du = \mathbf{i} \int_a^b x\, du + \mathbf{j} \int_a^b y\, du + \mathbf{k} \int_a^b z\, du.$$

▶

Example 1

If $\mathbf{F} = (3t^2 + 4t)\mathbf{i} + (2t - 5)\mathbf{j} + 4t^3\mathbf{k}$, then

$$\int_1^3 \mathbf{F}\,\mathrm{d}t = \mathbf{i}\int_1^3 (3t^2 + 4t)\,\mathrm{d}t + \mathbf{j}\int_1^3 (2t - 5)\,\mathrm{d}t + \mathbf{k}\int_1^3 4t^3\,\mathrm{d}t = \ldots\ldots\ldots\ldots$$

$$\boxed{42\mathbf{i} - 2\mathbf{j} + 80\mathbf{k}}$$

50

Because

$$\int_1^3 \mathbf{F}\,\mathrm{d}t = \left[\mathbf{i}(t^3 + 2t^2) + \mathbf{j}(t^2 - 5t) + \mathbf{k}t^4\right]_1^3$$
$$= (45\mathbf{i} - 6\mathbf{j} + 81\mathbf{k}) - (3\mathbf{i} - 4\mathbf{j} + \mathbf{k}) = 42\mathbf{i} - 2\mathbf{j} + 80\mathbf{k}$$

Here is a slightly different one.

Example 2

If $\quad\mathbf{F} = 3u\mathbf{i} + u^2\mathbf{j} + (u + 2)\mathbf{k}$

and $\quad\mathbf{V} = 2u\mathbf{i} - 3u\mathbf{j} + (u - 2)\mathbf{k}$

evaluate $\displaystyle\int_0^2 (\mathbf{F} \times \mathbf{V})\mathrm{d}u$.

First we must determine $\mathbf{F} \times \mathbf{V}$ in terms of u.

$$\mathbf{F} \times \mathbf{V} = \ldots\ldots\ldots\ldots$$

$$\boxed{\mathbf{F} \times \mathbf{V} = (u^3 + u^2 + 6u)\mathbf{i} - (u^2 - 10u)\mathbf{j} - (2u^3 + 9u^2)\mathbf{k}}$$

51

Because

$$\mathbf{F} \times \mathbf{V} = \begin{vmatrix} \mathbf{i} & \mathbf{j} & \mathbf{k} \\ 3u & u^2 & (u + 2) \\ 2u & -3u & (u - 2) \end{vmatrix}$$

which gives the result above.

Then $\displaystyle\int_0^2 (\mathbf{F} \times \mathbf{V})\,\mathrm{d}u = \ldots\ldots\ldots\ldots$

$$\boxed{\tfrac{4}{3}\{14\mathbf{i} + 13\mathbf{j} - 24\mathbf{k}\}}$$

52

Because

$$\int (\mathbf{F} \times \mathbf{V})\mathrm{d}u = \left(\frac{u^4}{4} + \frac{u^3}{3} + 3u^2\right)\mathbf{i} - \left(\frac{u^3}{3} - 5u^2\right)\mathbf{j} - \left(\frac{u^4}{2} + 3u^3\right)\mathbf{k}$$

$$\therefore \int_0^2 (\mathbf{F} \times \mathbf{V})\mathrm{d}u = (4 + \tfrac{8}{3} + 12)\mathbf{i} - (\tfrac{8}{3} - 20)\mathbf{j} - (8 + 24)\mathbf{k}$$

$$= \tfrac{4}{3}\{14\mathbf{i} + 13\mathbf{j} - 24\mathbf{k}\}$$

▶

Example 3

If $\mathbf{F} = \mathbf{A} \times (\mathbf{B} \times \mathbf{C})$ where

$$\mathbf{A} = 3t^2\mathbf{i} + (2t - 3)\mathbf{j} + 4t\mathbf{k}$$
$$\mathbf{B} = 2\mathbf{i} + 4t\mathbf{j} + 3(1 - t)\mathbf{k}$$
$$\mathbf{C} = 2t\mathbf{i} - 3t^2\mathbf{j} - 2t\mathbf{k}$$

determine $\displaystyle\int_0^1 \mathbf{F}\,dt$.

First we need to find $\mathbf{A} \times (\mathbf{B} \times \mathbf{C})$. The simplest way to do this is to use the relationship

$$\mathbf{A} \times (\mathbf{B} \times \mathbf{C}) = \ldots\ldots\ldots\ldots$$

53

$$\boxed{\mathbf{A} \times (\mathbf{B} \times \mathbf{C}) = (\mathbf{A} \cdot \mathbf{C})\mathbf{B} - (\mathbf{A} \cdot \mathbf{B})\mathbf{C}}$$

So $\quad \mathbf{A} \cdot \mathbf{C} = \ldots\ldots\ldots\ldots$
and $\quad \mathbf{A} \cdot \mathbf{B} = \ldots\ldots\ldots\ldots$

54

$$\boxed{\begin{array}{l} \mathbf{A} \cdot \mathbf{C} = 6t^3 - 6t^3 + 9t^2 - 8t^2 = t^2 \\ \mathbf{A} \cdot \mathbf{B} = 6t^2 + 8t^2 - 12t + 12t - 12t^2 = 2t^2 \end{array}}$$

Then $\mathbf{F} = \mathbf{A} \times (\mathbf{B} \times \mathbf{C})$
$$= t^2\{2\mathbf{i} + 4t\mathbf{j} + 3(1 - t)\mathbf{k}\} - 2t^2\{2t\mathbf{i} - 3t^2\mathbf{j} - 2t\mathbf{k}\}$$

$$\therefore \quad \int_0^1 \mathbf{F}\,dt = \ldots\ldots\ldots\ldots$$

Finish off the simplification and complete the integration.

55

$$\boxed{\tfrac{1}{60}\{-20\mathbf{i} + 132\mathbf{j} + 75\mathbf{k}\}}$$

Because

$$\mathbf{F} = \mathbf{A} \times (\mathbf{B} \times \mathbf{C}) = (2t^2 - 4t^3)\mathbf{i} + (4t^3 + 6t^4)\mathbf{j} + (3t^2 + t^3)\mathbf{k}$$

Integration with respect to t then gives the result stated above.

Now let us move on to the next stage of our development

Scalar and vector fields

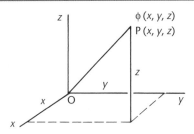

If every point P (x, y, z) of a region R of space has associated with it a scalar quantity $\phi(x, y, z)$, then $\phi(x, y, z)$ is a *scalar function* and a *scalar field* is said to exist in the region R.

56

Examples of scalar fields are temperature, potential, etc.

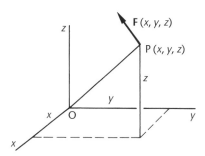

Similarly, if every point P (x, y, z) of a region R has associated with it a vector quantity $\mathbf{F}(x, y, z)$, then $\mathbf{F}(x, y, z)$ is a *vector function* and a *vector field* is said to exist in the region R.

Examples of vector fields are force, velocity, acceleration, etc. $\mathbf{F}(x, y, z)$ can be defined in terms of its components parallel to the coordinate axes, OX, OY, OZ.

That is, $\mathbf{F}(x, y, z) = F_x\mathbf{i} + F_y\mathbf{j} + F_z\mathbf{k}$.

Note these important definitions:
we shall be making good use of them as we proceed

Grad (gradient of a scalar function)

57

If a scalar function $\phi(x, y, z)$ is continuously differentiable with respect to its variables x, y, z, throughout the region, then the *gradient* of ϕ, written *grad ϕ*, is defined as the vector

$$\text{grad } \phi = \frac{\partial \phi}{\partial x}\mathbf{i} + \frac{\partial \phi}{\partial y}\mathbf{j} + \frac{\partial \phi}{\partial z}\mathbf{k} \tag{12}$$

Note that, while ϕ is a scalar function, grad ϕ is a vector function. For example, if ϕ depends upon the position of P and is defined by $\phi = 2x^2yz^3$, then

$$\text{grad } \phi = 4xyz^3\mathbf{i} + 2x^2z^3\mathbf{j} + 6x^2yz^2\mathbf{k}$$

▶

Notation

The expression (12) above can be written

$$\text{grad } \phi = \left\{ \mathbf{i}\frac{\partial}{\partial x} + \mathbf{j}\frac{\partial}{\partial y} + \mathbf{k}\frac{\partial}{\partial z} \right\} \phi$$

where $\left(\mathbf{i}\dfrac{\partial}{\partial x} + \mathbf{j}\dfrac{\partial}{\partial y} + \mathbf{k}\dfrac{\partial}{\partial z} \right)$ is called a *vector differential operator* and is denoted by the symbol ∇ (pronounced 'del' or sometimes 'nabla')

i.e. $\nabla \equiv \left(\mathbf{i}\dfrac{\partial}{\partial x} + \mathbf{j}\dfrac{\partial}{\partial y} + \mathbf{k}\dfrac{\partial}{\partial z} \right)$

Beware! ∇ cannot exist alone: it is an operator and must operate on a stated scalar function $\phi(x, y, z)$.

If **F** is a vector function, ∇**F** has no meaning.

So we have:

$$\nabla \phi = \text{grad } \phi = \left(\mathbf{i}\frac{\partial}{\partial x} + \mathbf{j}\frac{\partial}{\partial y} + \mathbf{k}\frac{\partial}{\partial z} \right) \phi$$

$$= \mathbf{i}\frac{\partial \phi}{\partial x} + \mathbf{j}\frac{\partial \phi}{\partial y} + \mathbf{k}\frac{\partial \phi}{\partial z} \tag{13}$$

Make a note of this definition and then let us see how to use it

---

58

Example 1

If $\phi = x^2yz^3 + xy^2z^2$, determine grad ϕ at the point P (1, 3, 2).

By the definition, $\text{grad } \phi = \nabla \phi = \dfrac{\partial \phi}{\partial x}\mathbf{i} + \dfrac{\partial \phi}{\partial y}\mathbf{j} + \dfrac{\partial \phi}{\partial z}\mathbf{k}.$

All we have to do then is to find the partial derivatives at $x = 1$, $y = 3$, $z = 2$ and insert their values.

$$\therefore \ \nabla \phi = \dots\dots\dots\dots$$

59

$$\boxed{4(21\mathbf{i} + 8\mathbf{j} + 18\mathbf{k})}$$

Because

$$\phi \ = x^2yz^3 + xy^2z^2 \qquad \therefore \ \frac{\partial \phi}{\partial x} = 2xyz^3 + y^2z^2$$

$$\frac{\partial \phi}{\partial y} \ = x^2z^3 + 2xyz^2 \qquad\qquad \frac{\partial \phi}{\partial z} = 3x^2yz^2 + 2xy^2z$$

Then, at (1, 3, 2) $\dfrac{\partial \phi}{\partial x} = 48 + 36 \qquad \therefore \ \dfrac{\partial \phi}{\partial x} = 84$

$$\frac{\partial \phi}{\partial y} = 8 + 24 \qquad \therefore \ \frac{\partial \phi}{\partial y} = 32$$

$$\frac{\partial \phi}{\partial z} = 36 + 36 \qquad \therefore \ \frac{\partial \phi}{\partial z} = 72$$

$$\therefore \ \text{grad } \phi = \nabla \phi = 84\mathbf{i} + 32\mathbf{j} + 72\mathbf{k} = 4(21\mathbf{i} + 8\mathbf{j} + 18\mathbf{k})$$

▶

Example 2

If $\quad \mathbf{A} = x^2 z \mathbf{i} + xy\mathbf{j} + y^2 z\mathbf{k}$

and $\quad \mathbf{B} = yz^2\mathbf{i} + xz\mathbf{j} + x^2 z\mathbf{k}$

determine an expression for grad $(\mathbf{A} \cdot \mathbf{B})$.

This we can soon do since we know that $\mathbf{A} \cdot \mathbf{B}$ is a scalar function of x, y and z.

First then, $\quad \mathbf{A} \cdot \mathbf{B} = \ldots\ldots\ldots$

$$\boxed{\mathbf{A} \cdot \mathbf{B} = x^2 y z^3 + x^2 yz + x^2 y^2 z^2}$$

60

Then $\quad \nabla(\mathbf{A} \cdot \mathbf{B}) = \ldots\ldots\ldots$

$$\boxed{2xyz(z^2 + 1 + yz)\mathbf{i} + x^2 z(z^2 + 1 + 2yz)\mathbf{j} + x^2 y(3z^2 + 1 + 2yz)\mathbf{k}}$$

61

Because

$$\text{if } \phi = \mathbf{A} \cdot \mathbf{B} = (x^2 z\mathbf{i} + xy\mathbf{j} + y^2 z\mathbf{k}) \cdot (yz^2\mathbf{i} + xz\mathbf{j} + x^2 z\mathbf{k})$$

$$= x^2 y z^3 + x^2 yz + x^2 y^2 z^2$$

$$\frac{\partial \phi}{\partial x} = 2xyz^3 + 2xyz + 2xy^2 z^2 = 2xyz(z^2 + 1 + yz)$$

$$\frac{\partial \phi}{\partial y} = x^2 z^3 + x^2 z + 2x^2 yz^2 = x^2 z(z^2 + 1 + 2yz)$$

$$\frac{\partial \phi}{\partial z} = 3x^2 yz^2 + x^2 y + 2x^2 y^2 z = x^2 y(3z^2 + 1 + 2yz)$$

$$\therefore \ \nabla(\mathbf{A} \cdot \mathbf{B}) = 2xyz(z^2 + 1 + yz)\mathbf{i} + x^2 z(z^2 + 1 + 2yz)\mathbf{j}$$
$$+ x^2 y(3z^2 + 1 + 2yz)\mathbf{k}$$

Now let us obtain another useful relationship.

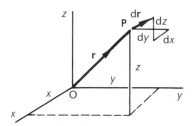

If $\overline{\text{OP}}$ is a position vector $\mathbf{r}$ where $\mathbf{r} = x\mathbf{i} + y\mathbf{j} + z\mathbf{k}$ and $d\mathbf{r}$ is a small displacement corresponding to changes dx, dy, dz in x, y, z respectively, then

$$d\mathbf{r} = dx\,\mathbf{i} + dy\,\mathbf{j} + dz\,\mathbf{k}$$

If $\phi(x, y, z)$ is a scalar function at P, we know that

$$\text{grad}\ \phi = \nabla\phi = \frac{\partial \phi}{\partial x}\mathbf{i} + \frac{\partial \phi}{\partial y}\mathbf{j} + \frac{\partial \phi}{\partial z}\mathbf{k}$$

Then $\quad$ grad $\phi \cdot d\mathbf{r} = \ldots\ldots\ldots$

62

$$\text{grad}\,\phi \cdot \mathbf{dr} = \frac{\partial \phi}{\partial x}\,dx + \frac{\partial \phi}{\partial y}\,dy + \frac{\partial \phi}{\partial z}\,dz$$

Because

$$\text{grad}\,\phi \cdot \mathbf{dr} = \left(\frac{\partial \phi}{\partial x}\mathbf{i} + \frac{\partial \phi}{\partial y}\mathbf{j} + \frac{\partial \phi}{\partial z}\mathbf{k}\right) \cdot (dx\,\mathbf{i} + dy\,\mathbf{j} + dz\,\mathbf{k})$$

$$= \frac{\partial \phi}{\partial x}\,dx + \frac{\partial \phi}{\partial y}\,dy + \frac{\partial \phi}{\partial z}\,dz$$

$$= \text{the total differential } d\phi \text{ of } \phi$$

That is

$$d\phi = \mathbf{dr} \cdot \text{grad}\,\phi \tag{14}$$

This will certainly be useful, so make a note of it

63 Directional derivatives

We have just established that

$$d\phi = \mathbf{dr} \cdot \text{grad}\,\phi$$

If ds is the small element of arc between P $(\mathbf{r})$ and Q $(\mathbf{r} + \mathbf{dr})$ then $ds = |\mathbf{dr}|$

$$\frac{\mathbf{dr}}{ds} = \frac{\mathbf{dr}}{|\mathbf{dr}|}$$

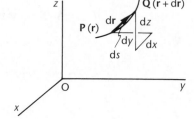

and $\dfrac{\mathbf{dr}}{ds}$ is thus a unit vector in the direction of $\mathbf{dr}$.

$$\therefore \quad \frac{d\phi}{ds} = \frac{\mathbf{dr}}{ds} \cdot \text{grad}\,\phi$$

If we denote the unit vector $\dfrac{\mathbf{dr}}{ds}$ by $\hat{\mathbf{a}}$ then the result becomes

$$\frac{d\phi}{ds} = \hat{\mathbf{a}} \cdot \text{grad}\,\phi$$

$\dfrac{d\phi}{ds}$ is thus the projection of grad ϕ on the unit vector $\hat{\mathbf{a}}$ and is called the *directional derivative* of ϕ in the direction of $\hat{\mathbf{a}}$. It gives the rate of change of ϕ with distance measured in the direction of $\hat{\mathbf{a}}$ and $\dfrac{d\phi}{ds} = \hat{\mathbf{a}} \cdot \text{grad}\,\phi$ will be a maximum when $\hat{\mathbf{a}}$ and grad ϕ have the same direction, since then

$$\hat{\mathbf{a}} \cdot \text{grad}\,\phi = |\,\hat{\mathbf{a}}\,||\,\text{grad}\,\phi\,|\cos\theta \text{ and } \theta \text{ will be zero.}$$

Thus the direction of grad ϕ gives the direction in which the maximum rate of change of ϕ occurs.

Example 1

Find the directional derivative of the function $\phi = x^2 z + 2xy^2 + yz^2$ at the point (1, 2, –1) in the direction of the vector $\mathbf{A} = 2\mathbf{i} + 3\mathbf{j} - 4\mathbf{k}$.

We start off with $\phi = x^2 z + 2xy^2 + yz^2$

$$\therefore \ \nabla\phi = \dots\dots\dots$$

$$\boxed{\nabla\phi = (2xz + 2y^2)\mathbf{i} + (4xy + z^2)\mathbf{j} + (x^2 + 2yz)\mathbf{k}}$$

64

Because

$$\frac{\partial\phi}{\partial x} = 2xz + 2y^2; \qquad \frac{\partial\phi}{\partial y} = 4xy + z^2; \qquad \frac{\partial\phi}{\partial z} = x^2 + 2yz$$

Then, at $(1, 2, -1)$

$$\nabla\phi = (-2 + 8)\mathbf{i} + (8 + 1)\mathbf{j} + (1 - 4)\mathbf{k} = 6\mathbf{i} + 9\mathbf{j} - 3\mathbf{k}$$

Next we have to find the unit vector $\hat{\mathbf{a}}$ where $\mathbf{A} = 2\mathbf{i} + 3\mathbf{j} - 4\mathbf{k}$

$$\hat{\mathbf{a}} = \dots\dots\dots$$

$$\boxed{\hat{\mathbf{a}} = \frac{1}{\sqrt{29}}(2\mathbf{i} + 3\mathbf{j} - 4\mathbf{k})}$$

65

Because

$$\mathbf{A} = 2\mathbf{i} + 3\mathbf{j} - 4\mathbf{k} \quad \therefore \quad |\mathbf{A}| = \sqrt{4 + 9 + 16} = \sqrt{29}$$

$$\hat{\mathbf{a}} = \frac{\mathbf{A}}{|\mathbf{A}|} = \frac{1}{\sqrt{29}}(2\mathbf{i} + 3\mathbf{j} - 4\mathbf{k})$$

So we have $\nabla\phi = 6\mathbf{i} + 9\mathbf{j} - 3\mathbf{k}$ and $\hat{\mathbf{a}} = \frac{1}{\sqrt{29}}(2\mathbf{i} + 3\mathbf{j} - 4\mathbf{k})$

$$\therefore \quad \frac{\mathrm{d}\phi}{\mathrm{d}s} = \hat{\mathbf{a}} \cdot \nabla\phi$$

$$= \dots\dots\dots$$

$$\boxed{\frac{\mathrm{d}\phi}{\mathrm{d}s} = \frac{51}{\sqrt{29}} = 9\cdot47}$$

66

Because

$$\frac{\mathrm{d}\phi}{\mathrm{d}s} = \hat{\mathbf{a}} \cdot \nabla\phi = \frac{1}{\sqrt{29}}(2\mathbf{i} + 3\mathbf{j} - 4\mathbf{k}) \cdot (6\mathbf{i} + 9\mathbf{j} - 3\mathbf{k})$$

$$= \frac{1}{\sqrt{29}}(12 + 27 + 12) = \frac{51}{\sqrt{29}} = 9\cdot47$$

▶

That is all there is to it.

(a) From the given scalar function ϕ, determine $\nabla\phi$.

(b) Find the unit vector $\hat{\mathbf{a}}$ in the direction of the given vector $\mathbf{A}$.

(c) Then $\dfrac{d\phi}{ds} = \hat{\mathbf{a}} \cdot \nabla\phi$.

Example 2

Find the directional derivative of $\phi = x^2 y + y^2 z + z^2 x$ at the point $(1, -1, 2)$ in the direction of the vector $\mathbf{A} = 4\mathbf{i} + 2\mathbf{j} - 5\mathbf{k}$.

Same as before. *Work through it and check the result with the next frame*

67

$$\boxed{\dfrac{d\phi}{ds} = \dfrac{-23}{3\sqrt{5}} = -3{\cdot}43}$$

Because

$$\phi = x^2 y + y^2 z + z^2 x$$
$$\therefore \ \nabla\phi = (2xy + z^2)\mathbf{i} + (x^2 + 2yz)\mathbf{j} + (y^2 + 2zx)\mathbf{k}$$
$$\therefore \ \text{At } (1, -1, 2), \quad \nabla\phi = 2\mathbf{i} - 3\mathbf{j} + 5\mathbf{k}$$
$$\mathbf{A} = 4\mathbf{i} + 2\mathbf{j} - 5\mathbf{k} \quad \therefore \quad |\mathbf{A}| = \sqrt{16 + 4 + 25} = \sqrt{45} = 3\sqrt{5}$$
$$\therefore \ \hat{\mathbf{a}} = \dfrac{1}{3\sqrt{5}} (4\mathbf{i} + 2\mathbf{j} - 5\mathbf{k})$$
$$\therefore \ \dfrac{d\phi}{ds} = \hat{\mathbf{a}} \cdot \nabla\phi = \dfrac{1}{3\sqrt{5}} (4\mathbf{i} + 2\mathbf{j} - 5\mathbf{k}) \cdot (2\mathbf{i} - 3\mathbf{j} + 5\mathbf{k})$$
$$= \dfrac{1}{3\sqrt{5}} (8 - 6 - 25) = \dfrac{-23}{3\sqrt{5}} = -3{\cdot}43$$

Example 3

Find the direction from the point $(1, 1, 0)$ which gives the greatest rate of increase of the function $\phi = (x + 3y)^2 + (2y - z)^2$.

This appears to be different, but it rests on the fact that the greatest rate of increase of ϕ with respect to distance is in

. . . .

68

$$\boxed{\text{the direction of } \nabla\phi}$$

All we need then is to find the vector $\nabla\phi$, which is

. . . .

$$\boxed{\nabla\phi = 4(2\mathbf{i} + 8\mathbf{j} - \mathbf{k})}$$ **69**

Because

$$\phi = (x + 3y)^2 + (2y - z)^2$$

$$\therefore \quad \frac{\partial\phi}{\partial x} = 2(x + 3y); \quad \frac{\partial\phi}{\partial y} = 6(x + 3y) + 4(2y - z); \quad \frac{\partial\phi}{\partial z} = -2(2y - z)$$

$$\therefore \text{ At } (1,\, 1,\, 0), \quad \frac{\partial\phi}{\partial x} = 8; \quad \frac{\partial\phi}{\partial y} = 32; \quad \frac{\partial\phi}{\partial z} = -4$$

$$\therefore \quad \nabla\phi = 8\mathbf{i} + 32\mathbf{j} - 4\mathbf{k} = 4(2\mathbf{i} + 8\mathbf{j} - \mathbf{k})$$

$\therefore$ greatest rate of increase occurs in direction $2\mathbf{i} + 8\mathbf{j} - \mathbf{k}$

So on we go

Unit normal vectors **70**

The equation of $\phi(x,\, y,\, z) = $ constant represents a surface in space. For example, $3x - 4y + 2z = 1$ is the equation of a plane and $x^2 + y^2 + z^2 = 4$ represents a sphere centred on the origin and of radius 2.

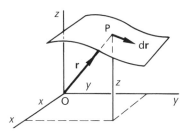

If $\mathbf{dr}$ is a displacement in this surface, then $d\phi = 0$ since ϕ is constant over the surface.

Therefore our previous relationship $\mathbf{dr} \cdot \text{grad } \phi = d\phi$ becomes

$$\mathbf{dr} \cdot \text{grad } \phi = 0$$

for all such small displacements $\mathbf{dr}$ in the surface.

But $\mathbf{dr} \cdot \text{grad } \phi = |\mathbf{dr}| \, |\text{grad } \phi| \cos\theta = 0$.

$\therefore \ \theta = \dfrac{\pi}{2} \quad \therefore \ \text{grad } \phi$ is perpendicular to dr, i.e. grad ϕ is a vector perpendicular to the surface at P, in the direction of maximum rate of change of ϕ. The magnitude of that maximum rate of change is given by $|\text{grad } \phi|$.

The unit vector **N** in the direction of grad ϕ is called the *unit normal vector* at P.

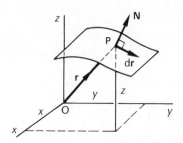

$\therefore$ Unit normal vector

$$N = \frac{\nabla\phi}{|\nabla\phi|} \tag{15}$$

Example 1

Find the unit normal vector to the surface $x^3y + 4xz^2 + xy^2z + 2 = 0$ at the point $(1, 3, -1)$.

$$\text{Vector normal} = \nabla\phi = \dots\dots\dots$$

71

$$\boxed{\nabla\phi = (3x^2y + 4z^2 + y^2z)\mathbf{i} + (x^3 + 2xyz)\mathbf{j} + (8xz + xy^2)\mathbf{k}}$$

Then, at $(1, 3, -1)$, $\nabla\phi = 4\mathbf{i} - 5\mathbf{j} + \mathbf{k}$

and the unit normal at $(1, 3, -1)$ is $\dots\dots\dots\dots$

72

$$\boxed{\frac{1}{\sqrt{42}}\,(4\mathbf{i} - 5\mathbf{j} + \mathbf{k})}$$

Because

$$|\nabla\phi| = \sqrt{16 + 25 + 1} = \sqrt{42}$$

and $N = \dfrac{\nabla\phi}{|\nabla\phi|} = \dfrac{1}{\sqrt{42}}(4\mathbf{i} - 5\mathbf{j} + \mathbf{k})$

One more.

Example 2

Determine the unit normal to the surface

$xyz + x^2y - 5yz - 5 = 0$ at the point $(3, 1, 2)$.

All very straightforward. Complete it.

73

$$\boxed{\text{Unit normal} = \mathbf{N} = \frac{1}{\sqrt{93}}(8\mathbf{i} + 5\mathbf{j} - 2\mathbf{k})}$$

Because

$$\phi = xyz + x^2y - 5yz - 5$$

$$\therefore \ \nabla\phi = (yz + 2xy)\mathbf{i} + (xz + x^2 - 5z)\mathbf{j} + (xy - 5y)\mathbf{k}$$

At $(3, 1, 2)$, $\quad \nabla\phi = 8\mathbf{i} + 5\mathbf{j} - 2\mathbf{k}; \ \ |\nabla\phi| = \sqrt{64 + 25 + 4} = \sqrt{93}$

$$\therefore \ \text{Unit normal} = \mathbf{N} = \frac{\nabla\phi}{|\nabla\phi|} = \frac{1}{\sqrt{93}}(8\mathbf{i} + 5\mathbf{j} - 2\mathbf{k})$$

Collecting our results so far, we have, for $\phi(x, y, z)$ a scalar function

(a) $\text{grad } \phi = \nabla\phi = \dfrac{\partial\phi}{\partial x}\mathbf{i} + \dfrac{\partial\phi}{\partial y}\mathbf{j} + \dfrac{\partial\phi}{\partial z}\mathbf{k}$

(b) $\mathrm{d}\phi = \mathrm{d}\mathbf{r} \cdot \text{grad } \phi$ where $\mathrm{d}\phi = \dfrac{\partial\phi}{\partial x}\mathrm{d}x + \dfrac{\partial\phi}{\partial y}\mathrm{d}y + \dfrac{\partial\phi}{\partial z}\mathrm{d}z$

(c) directional derivative $\dfrac{\mathrm{d}\phi}{\mathrm{d}s} = \hat{\mathbf{a}} \cdot \text{grad } \phi$

(d) unit normal vector $\mathbf{N} = \dfrac{\nabla\phi}{|\nabla\phi|}$.

Copy out this brief summary for future reference. It will help

Grad of sums and products of scalars

74

(a) $\nabla(A + B) = \mathbf{i}\left\{\dfrac{\partial}{\partial x}(A + B)\right\} + \mathbf{j}\left\{\dfrac{\partial}{\partial y}(A + B)\right\} + \mathbf{k}\left\{\dfrac{\partial}{\partial z}(A + B)\right\}$

$\qquad = \left\{\dfrac{\partial A}{\partial x}\mathbf{i} + \dfrac{\partial A}{\partial y}\mathbf{j} + \dfrac{\partial A}{\partial z}\mathbf{k}\right\} + \left\{\dfrac{\partial B}{\partial x}\mathbf{i} + \dfrac{\partial B}{\partial y}\mathbf{j} + \dfrac{\partial B}{\partial z}\mathbf{k}\right\}$

$\therefore \ \nabla(A + B) = \nabla A + \nabla B$

(b) $\nabla(AB) = \mathbf{i}\left\{\dfrac{\partial}{\partial x}(AB)\right\} + \mathbf{j}\left\{\dfrac{\partial}{\partial y}(AB)\right\} + \mathbf{k}\left\{\dfrac{\partial}{\partial z}(AB)\right\}$

$\qquad = \mathbf{i}\left\{A\dfrac{\partial B}{\partial x} + B\dfrac{\partial A}{\partial x}\right\} + \mathbf{j}\left\{A\dfrac{\partial B}{\partial y} + B\dfrac{\partial A}{\partial y}\right\} + \mathbf{k}\left\{A\dfrac{\partial B}{\partial z} + B\dfrac{\partial A}{\partial z}\right\}$

$\qquad = \left\{A\dfrac{\partial B}{\partial x}\mathbf{i} + A\dfrac{\partial B}{\partial y}\mathbf{j} + A\dfrac{\partial B}{\partial z}\mathbf{k}\right\} + \left\{B\dfrac{\partial A}{\partial x}\mathbf{i} + B\dfrac{\partial A}{\partial y}\mathbf{j} + B\dfrac{\partial A}{\partial z}\mathbf{k}\right\}$

$\qquad = A\left\{\dfrac{\partial B}{\partial x}\mathbf{i} + \dfrac{\partial B}{\partial y}\mathbf{j} + \dfrac{\partial B}{\partial z}\mathbf{k}\right\} + B\left\{\dfrac{\partial A}{\partial x}\mathbf{i} + \dfrac{\partial A}{\partial y}\mathbf{j} + \dfrac{\partial A}{\partial z}\mathbf{k}\right\}$

$\therefore \ \nabla(AB) = A(\nabla B) + B(\nabla A)$

Remember that in these results A and B are scalars.

75

Example

If $A = x^2yz + xz^2$ and $B = xy^2z - z^3$, evaluate $\nabla(AB)$ at the point $(2, 1, 3)$.

We know that $\nabla(AB) = A(\nabla B) + B(\nabla A)$

At $(2, 1, 3)$,

$$\nabla B = \ldots\ldots\ldots; \quad \nabla A = \ldots\ldots\ldots$$

76

$$\boxed{\nabla B = 3\mathbf{i} + 12\mathbf{j} - 25\mathbf{k}; \quad \nabla A = 21\mathbf{i} + 12\mathbf{j} + 16\mathbf{k}}$$

$$\nabla B = \frac{\partial B}{\partial x}\mathbf{i} + \frac{\partial B}{\partial y}\mathbf{j} + \frac{\partial B}{\partial z}\mathbf{k} = y^2z\mathbf{i} + 2xyz\mathbf{j} + (xy^2 - 3z^2)\mathbf{k}$$

$$= 3\mathbf{i} + 12\mathbf{j} - 25\mathbf{k} \quad \text{at } (2, 1, 3)$$

$$\nabla A = \frac{\partial A}{\partial x}\mathbf{i} + \frac{\partial A}{\partial y}\mathbf{j} + \frac{\partial A}{\partial z}\mathbf{k} = (2xyz + z^2)\mathbf{i} + x^2z\mathbf{j} + (x^2y + 2xz)\mathbf{k}$$

$$= 21\mathbf{i} + 12\mathbf{j} + 16\mathbf{k} \quad \text{at } (2, 1, 3)$$

Now $\nabla(AB) = A(\nabla B) + B(\nabla A) = \ldots\ldots\ldots$

Finish it

77

$$\boxed{\nabla(AB) = 3(-117\mathbf{i} + 36\mathbf{j} - 362\mathbf{k})}$$

Because

$$\nabla(AB) = A(\nabla B) + B(\nabla A)$$

$$A = x^2yz + xz^2 \quad \therefore \text{ at } (2, 1, 3), \quad A = 12 + 18 = 30$$

$$B = xy^2z - z^3 \quad \therefore \text{ at } (2, 1, 3), \quad B = 6 - 27 = -21$$

$$\therefore \quad \nabla(AB) = 30(3\mathbf{i} + 12\mathbf{j} - 25\mathbf{k}) - 21(21\mathbf{i} + 12\mathbf{j} + 16\mathbf{k})$$

$$= -351\mathbf{i} + 108\mathbf{j} - 1086\mathbf{k}$$

$$= 3(-117\mathbf{i} + 36\mathbf{j} - 362\mathbf{k})$$

So add these to the list of results.

$$\nabla(A + B) = \nabla A + \nabla B$$

$$\nabla(AB) = A(\nabla B) + B(\nabla A)$$

where A and B are scalars.

Now on to the next page

Div (divergence of a vector function)

78

The operator $\nabla \cdot$ (notice the 'dot'; it makes all the difference) can be applied to a vector function $\mathbf{A}(x, y, z)$ to give the *divergence* of $\mathbf{A}$, written in short as *div* $\mathbf{A}$.

If $\mathbf{A} = a_x\mathbf{i} + a_y\mathbf{j} + a_z\mathbf{k}$

$$\text{div } \mathbf{A} = \nabla \cdot \mathbf{A} = \left(\mathbf{i}\frac{\partial}{\partial x} + \mathbf{j}\frac{\partial}{\partial y} + \mathbf{k}\frac{\partial}{\partial z}\right) \cdot \left(a_x\mathbf{i} + a_y\mathbf{j} + a_z\mathbf{k}\right)$$

$$\therefore \quad \text{div } \mathbf{A} = \nabla \cdot \mathbf{A} = \frac{\partial a_x}{\partial x} + \frac{\partial a_y}{\partial y} + \frac{\partial a_z}{\partial z}$$

Note that

(a) the grad operator ∇ acts on a scalar and gives a vector

(b) the div operation $\nabla\cdot$ acts on a vector and gives a scalar.

Example 1

If $\mathbf{A} = x^2y\mathbf{i} - xyz\mathbf{j} + yz^2\mathbf{k}$ then

$$\text{div } \mathbf{A} = \nabla \cdot \mathbf{A} = \ldots\ldots\ldots\ldots$$

$$\boxed{\text{div } \mathbf{A} = \nabla \cdot \mathbf{A} = 2xy - xz + 2yz}$$

79

We simply take the appropriate partial derivatives of the coefficients of $\mathbf{i}$, $\mathbf{j}$ and $\mathbf{k}$. It could hardly be easier.

Example 2

If $\mathbf{A} = 2x^2y\mathbf{i} - 2(xy^2 + y^3z)\mathbf{j} + 3y^2z^2\mathbf{k}$, determine $\nabla \cdot \mathbf{A}$, i.e. div $\mathbf{A}$.

Complete it. $\qquad \nabla \cdot \mathbf{A} = \ldots\ldots\ldots\ldots$

$$\boxed{\nabla \cdot \mathbf{A} = 0}$$

80

Because

$$\mathbf{A} = 2x^2y\mathbf{i} - 2(xy^2 + y^3z)\mathbf{j} + 3y^2z^2\mathbf{k}$$

$$\nabla \cdot \mathbf{A} = \frac{\partial a_x}{\partial x} + \frac{\partial a_y}{\partial y} + \frac{\partial a_z}{\partial z}$$

$$= 4xy - 2(2xy + 3y^2z) + 6y^2z$$

$$= 4xy - 4xy - 6y^2z + 6y^2z = 0$$

Such a vector $\mathbf{A}$ for which $\nabla \cdot \mathbf{A} = 0$ at all points, i.e. for all values of x, y, z, is called a *solenoidal vector*. It is rather a special case.

▶

Curl (curl of a vector function)

The *curl operator* denoted by $\nabla\times$, acts on a vector and gives another vector as a result.

If $A = a_x i + a_y j + a_z k$, then curl $A = \nabla \times A$.

i.e. $\quad$ curl $A = \nabla \times A = \left(i\dfrac{\partial}{\partial x} + j\dfrac{\partial}{\partial y} + k\dfrac{\partial}{\partial z} \right) \times (a_x i + a_y j + a_z k)$

$$= \begin{vmatrix} i & j & k \\ \dfrac{\partial}{\partial x} & \dfrac{\partial}{\partial y} & \dfrac{\partial}{\partial z} \\ a_x & a_y & a_z \end{vmatrix}$$

$$\therefore \ \nabla \times A = i\left(\frac{\partial a_z}{\partial y} - \frac{\partial a_y}{\partial z} \right) + j\left(\frac{\partial a_x}{\partial z} - \frac{\partial a_z}{\partial x} \right) + k\left(\frac{\partial a_y}{\partial x} - \frac{\partial a_x}{\partial y} \right)$$

Curl A is thus a vector function. *It is best remembered in its determinant form, so make a note of it.*

If $\nabla \times A = 0$ then A is said to be *irrotational*.

Then on for an example

81

Example 1

If $A = (y^4 - x^2 z^2)i + (x^2 + y^2)j - x^2 yz k$, determine curl A at the point $(1, 3, -2)$.

$$\text{curl } A = \nabla \times A = \begin{vmatrix} i & j & k \\ \dfrac{\partial}{\partial x} & \dfrac{\partial}{\partial y} & \dfrac{\partial}{\partial z} \\ y^4 - x^2 z^2 & x^2 + y^2 & -x^2 yz \end{vmatrix}$$

Now we expand the determinant

$$\nabla \times A = i\left\{ \frac{\partial}{\partial y}(-x^2 yz) - \frac{\partial}{\partial z}(x^2 + y^2) \right\} - j\left\{ \frac{\partial}{\partial x}(-x^2 yz) - \frac{\partial}{\partial z}(y^4 - x^2 z^2) \right\}$$

$$+ k\left\{ \frac{\partial}{\partial x}(x^2 + y^2) - \frac{\partial}{\partial y}(y^4 - x^2 z^2) \right\}$$

All that now remains is to obtain the partial derivatives and substitute the values of x, y, z.

$$\therefore \ \nabla \times A = \dots\dots\dots\dots$$

82

$$\boxed{2i - 8j - 106k}$$

$$\nabla \times A = i\{-x^2 z\} - j\{-2xyz + 2x^2 z\} + k\{2x - 4y^3\}.$$

$$\therefore \ \text{At } (1, 3, -2), \quad \nabla \times A = i(2) - j(12 - 4) + k(2 - 108)$$

$$= 2i - 8j - 106k$$

▶

Example 2

Determine curl **F** at the point (2, 0, 3) given that

$$\mathbf{F} = ze^{2xy}\mathbf{i} + 2xz\cos y\mathbf{j} + (x+2y)\mathbf{k}.$$

In determinant form, curl $\mathbf{F} = \nabla \times \mathbf{F} = \ldots\ldots\ldots\ldots$

$$\begin{vmatrix} \mathbf{i} & \mathbf{j} & \mathbf{k} \\ \dfrac{\partial}{\partial x} & \dfrac{\partial}{\partial y} & \dfrac{\partial}{\partial z} \\ ze^{2xy} & 2xz\cos y & x+2y \end{vmatrix}$$

83

Now expand the determinant and substitute the values of x, y and z, finally obtaining curl $\mathbf{F} = \ldots\ldots\ldots\ldots$

$$\boxed{\text{curl } \mathbf{F} = \nabla \times \mathbf{F} = -2(\mathbf{i} + 3\mathbf{k})}$$

84

Because

$$\nabla \times \mathbf{F} = \mathbf{i}\{2 - 2x\cos y\} - \mathbf{j}\{1 - e^{2xy}\} + \mathbf{k}\{2z\cos y - 2xze^{2xy}\}$$
$$\therefore \text{ At } (2,\,0,\,3) \qquad \nabla \times \mathbf{F} = \mathbf{i}(2-4) - \mathbf{j}(1-1) + \mathbf{k}(6-12)$$
$$= -2\mathbf{i} - 6\mathbf{k} = -2(\mathbf{i} + 3\mathbf{k})$$

Every one is done in the same way.

Summary of grad, div and curl

(a) *Grad* operator ∇ acts on a *scalar* field to give a *vector* field.

(b) *Div* operator $\nabla\cdot$ acts on a *vector* field to give a *scalar* field.

(c) *Curl* operator $\nabla\times$ acts on a *vector* field to give a *vector* field.

(d) With a *scalar function* $\phi(x,\,y,\,z)$

$$\text{grad }\phi = \nabla\phi = \frac{\partial\phi}{\partial x}\mathbf{i} + \frac{\partial\phi}{\partial y}\mathbf{j} + \frac{\partial\phi}{\partial z}\mathbf{k}$$

(e) With a *vector function* $\mathbf{A} = a_x\mathbf{i} + a_y\mathbf{j} + a_z\mathbf{k}$

$$(1)\ \text{div }\mathbf{A} = \nabla\cdot\mathbf{A} = \frac{\partial a_x}{\partial x} + \frac{\partial a_y}{\partial y} + \frac{\partial a_z}{\partial z}$$

$$(2)\ \text{curl }\mathbf{A} = \nabla\times\mathbf{A} = \begin{vmatrix} \mathbf{i} & \mathbf{j} & \mathbf{k} \\ \dfrac{\partial}{\partial x} & \dfrac{\partial}{\partial y} & \dfrac{\partial}{\partial z} \\ a_x & a_y & a_z \end{vmatrix}$$

Check through that list, just to make sure. We shall need them all

85

By way of revision, here is one further example.

Example 3

If $\phi = x^2y^2 + x^3yz - yz^2$

and $\mathbf{F} = xy^2\mathbf{i} - 2yz\mathbf{j} + xyz\mathbf{k}$

determine for the point P $(1, -1, 2)$,

(a) $\nabla\phi$,   (b) unit normal,   (c) $\nabla \cdot \mathbf{F}$,   (d) $\nabla \times \mathbf{F}$.

Complete all four parts and then check the results with the next frame

86

Here is the working in full. $\phi = x^2y^2 + x^3yz - yz^2$

(a) $\nabla\phi = \dfrac{\partial\phi}{\partial x}\mathbf{i} + \dfrac{\partial\phi}{\partial y}\mathbf{j} + \dfrac{\partial\phi}{\partial z}\mathbf{k}$

$\qquad = (2xy^2 + 3x^2yz)\mathbf{i} + (2x^2y + x^3z - z^2)\mathbf{j} + (x^3y - 2yz)\mathbf{k}$

$\therefore$ At $(1, -1, 2)$ $\nabla\phi = -4\mathbf{i} - 4\mathbf{j} + 3\mathbf{k}$

(b) $\mathbf{N} = \dfrac{\nabla\phi}{|\nabla\phi|}$ $|\nabla\phi| = \sqrt{16 + 16 + 9} = \sqrt{41}$

$\therefore$ $\mathbf{N} = \dfrac{-1}{\sqrt{41}}(4\mathbf{i} + 4\mathbf{j} - 3\mathbf{k})$

(c) $\mathbf{F} = xy^2\mathbf{i} - 2yz\mathbf{j} + xyz\mathbf{k}$ $\nabla \cdot \mathbf{F} = \dfrac{\partial a_x}{\partial x} + \dfrac{\partial a_y}{\partial y} + \dfrac{\partial a_z}{\partial z}$

$\therefore$ $\nabla \cdot \mathbf{F} = y^2 - 2z + xy$

$\therefore$ At $(1, -1, 2)$ $\nabla \cdot \mathbf{F} = 1 - 4 - 1 = -4$ $\therefore$ $\nabla \cdot \mathbf{F} = -4$

(d) $\nabla \times \mathbf{F} = \begin{vmatrix} \mathbf{i} & \mathbf{j} & \mathbf{k} \\ \dfrac{\partial}{\partial x} & \dfrac{\partial}{\partial y} & \dfrac{\partial}{\partial z} \\ xy^2 & -2yz & xyz \end{vmatrix}$

$\therefore$ $\nabla \times \mathbf{F} = \mathbf{i}(xz + 2y) - \mathbf{j}(yz - 0) + \mathbf{k}(0 - 2xy)$

$\qquad = (xz + 2y)\mathbf{i} - yz\mathbf{j} - 2xy\mathbf{k}$

$\therefore$ At $(1, -1, 2)$ $\nabla \times \mathbf{F} = 2\mathbf{j} + 2\mathbf{k}$ $\therefore$ $\nabla \times \mathbf{F} = 2(\mathbf{j} + \mathbf{k})$

Now let us combine some of these operations.

Multiple operations

We can combine the operators grad, div and curl in multiple operations, as in the examples that follow.

Example 1

If $\mathbf{A} = x^2y\mathbf{i} + yz^3\mathbf{j} - zx^3\mathbf{k}$

then $\operatorname{div} \mathbf{A} = \nabla \cdot \mathbf{A} = \left(\dfrac{\partial}{\partial x}\mathbf{i} + \dfrac{\partial}{\partial y}\mathbf{j} + \dfrac{\partial}{\partial z}\mathbf{k} \right) \cdot (x^2y\mathbf{i} + yz^3\mathbf{j} - zx^3\mathbf{k})$

$\qquad\qquad = 2xy + z^3 + x^3 = \phi$ say

Then $\operatorname{grad}(\operatorname{div}\mathbf{A}) = \nabla(\nabla \cdot \mathbf{A}) = \dfrac{\partial \phi}{\partial x}\mathbf{i} + \dfrac{\partial \phi}{\partial y}\mathbf{j} + \dfrac{\partial \phi}{\partial z}\mathbf{k}$

$\qquad\qquad = (2y + 3x^2)\mathbf{i} + (2x)\mathbf{j} + (3z^2)\mathbf{k}$

i.e. $\operatorname{grad}\operatorname{div}\mathbf{A} = \nabla(\nabla \cdot \mathbf{A}) = (2y + 3x^2)\mathbf{i} + 2x\mathbf{j} + 3z^2\mathbf{k}$

Move on for the next example

Example 2

If $\phi = xyz - 2y^2z + x^2z^2$, determine div grad ϕ at the point (2, 4, 1).

First find grad ϕ and then the div of the result.

At (2, 4, 1), div grad $\phi = \nabla \cdot (\nabla \phi) = \ldots\ldots\ldots$

$$\boxed{\operatorname{div} \ \operatorname{grad} \phi = 6}$$

Because we have $\phi = xyz - 2y^2z + x^2z^2$

$\qquad \operatorname{grad} \phi = \nabla\phi = \dfrac{\partial \phi}{\partial x}\mathbf{i} + \dfrac{\partial \phi}{\partial y}\mathbf{j} + \dfrac{\partial \phi}{\partial z}\mathbf{k}$

$\qquad\qquad = (yz + 2xz^2)\mathbf{i} + (xz - 4yz)\mathbf{j} + (xy - 2y^2 + 2x^2z)\mathbf{k}$

$\therefore$ div grad $\phi = \nabla \cdot (\nabla\phi) = 2z^2 - 4z + 2x^2$

$\therefore$ At (2, 4, 1), div grad $\phi = \nabla \cdot (\nabla\phi) = 2 - 4 + 8 = 6$

Example 3

If $\mathbf{F} = x^2yz\mathbf{i} + xyz^2\mathbf{j} + y^2z\mathbf{k}$ determine curl curl $\mathbf{F}$ at the point (2, 1, 1).

Determine an expression for curl $\mathbf{F}$ in the usual way, which will be a vector, and then the curl of the result. Finally substitute values.

$$\operatorname{curl} \ \operatorname{curl} \mathbf{F} = \ldots\ldots\ldots$$

▶

90

$$\boxed{\text{curl curl } \mathbf{F} = \nabla \times (\nabla \times \mathbf{F}) = \mathbf{i} + 2\mathbf{j} + 6\mathbf{k}}$$

Because

$$\text{curl } \mathbf{F} = \begin{vmatrix} \mathbf{i} & \mathbf{j} & \mathbf{k} \\ \dfrac{\partial}{\partial x} & \dfrac{\partial}{\partial y} & \dfrac{\partial}{\partial z} \\ x^2 yz & xyz^2 & y^2 z \end{vmatrix}$$

$$= (2yz - 2xyz)\mathbf{i} + x^2 y\mathbf{j} + (yz^2 - x^2 z)\mathbf{k}$$

Then   curl curl $\mathbf{F} = \begin{vmatrix} \mathbf{i} & \mathbf{j} & \mathbf{k} \\ \dfrac{\partial}{\partial x} & \dfrac{\partial}{\partial y} & \dfrac{\partial}{\partial z} \\ 2yz - 2xyz & x^2 y & yz^2 - x^2 z \end{vmatrix}$

$$= z^2 \mathbf{i} - (-2xz - 2y + 2xy)\mathbf{j} + (2xy - 2z + 2xz)\mathbf{k}$$

$\therefore$ At (2, 1, 1),   curl curl $\mathbf{F} = \nabla \times (\nabla \times \mathbf{F}) = \mathbf{i} + 2\mathbf{j} + 6\mathbf{k}$

91

Remember that grad, div and curl are operators and that they must act on a scalar or vector as appropriate. They cannot exist alone and must be followed by a function.

One or two interesting general results appear.

(a) *Curl grad ϕ* where ϕ is a scalar

$$\text{grad } \phi = \frac{\partial \phi}{\partial x}\mathbf{i} + \frac{\partial \phi}{\partial y}\mathbf{j} + \frac{\partial \phi}{\partial z}\mathbf{k}$$

$$\therefore \text{ curl grad } \phi = \begin{vmatrix} \mathbf{i} & \mathbf{j} & \mathbf{k} \\ \dfrac{\partial}{\partial x} & \dfrac{\partial}{\partial y} & \dfrac{\partial}{\partial z} \\ \dfrac{\partial \phi}{\partial x} & \dfrac{\partial \phi}{\partial y} & \dfrac{\partial \phi}{\partial z} \end{vmatrix}$$

$$= \mathbf{i}\left\{\frac{\partial^2 \phi}{\partial y \partial z} - \frac{\partial^2 \phi}{\partial z \partial y}\right\} - \mathbf{j}\left\{\frac{\partial^2 \phi}{\partial z \partial x} - \frac{\partial^2 \phi}{\partial x \partial z}\right\}$$

$$+ \mathbf{k}\left\{\frac{\partial^2 \phi}{\partial x \partial y} - \frac{\partial^2 \phi}{\partial y \partial x}\right\}$$

$$= 0$$

$$\therefore \text{ curl grad } \phi = \nabla \times (\nabla \phi) = 0$$

(b) *Div curl* **A** where **A** is a vector. $\quad \mathbf{A} = a_x\mathbf{i} + a_y\mathbf{j} + a_z\mathbf{k}$

$$\text{curl } \mathbf{A} = \nabla \times \mathbf{A} = \begin{vmatrix} \mathbf{i} & \mathbf{j} & \mathbf{k} \\ \dfrac{\partial}{\partial x} & \dfrac{\partial}{\partial y} & \dfrac{\partial}{\partial z} \\ a_x & a_y & a_z \end{vmatrix}$$

$$= \mathbf{i}\left(\frac{\partial a_z}{\partial y} - \frac{\partial a_y}{\partial z}\right) - \mathbf{j}\left(\frac{\partial a_z}{\partial x} - \frac{\partial a_x}{\partial z}\right) + \mathbf{k}\left(\frac{\partial a_y}{\partial x} - \frac{\partial a_x}{\partial y}\right)$$

Then div curl $\mathbf{A} = \nabla \cdot (\nabla \times \mathbf{A}) = \left(\mathbf{i}\dfrac{\partial}{\partial x} + \mathbf{j}\dfrac{\partial}{\partial y} + \mathbf{k}\dfrac{\partial}{\partial z}\right) \cdot (\nabla \times \mathbf{A})$

$$= \frac{\partial^2 a_z}{\partial x \partial y} - \frac{\partial^2 a_y}{\partial z \partial x} - \frac{\partial^2 a_z}{\partial x \partial y} + \frac{\partial^2 a_x}{\partial y \partial z} + \frac{\partial^2 a_y}{\partial z \partial x} - \frac{\partial^2 a_x}{\partial y \partial z}$$

$$= 0$$

$\therefore$ div curl $\mathbf{A} = \nabla \cdot (\nabla \times \mathbf{A}) = 0$

(c) *Div grad* ϕ where ϕ is a scalar

$$\text{grad } \phi = \frac{\partial \phi}{\partial x}\mathbf{i} + \frac{\partial \phi}{\partial y}\mathbf{j} + \frac{\partial \phi}{\partial z}\mathbf{k}$$

Then div grad $\phi = \nabla \cdot (\nabla \phi)$

$$= \left(\mathbf{i}\frac{\partial}{\partial x} + \mathbf{j}\frac{\partial}{\partial y} + \mathbf{k}\frac{\partial}{\partial z}\right) \cdot \left(\frac{\partial \phi}{\partial x}\mathbf{i} + \frac{\partial \phi}{\partial y}\mathbf{j} + \frac{\partial \phi}{\partial z}\mathbf{k}\right)$$

$$= \frac{\partial^2 \phi}{\partial x^2} + \frac{\partial^2 \phi}{\partial y^2} + \frac{\partial^2 \phi}{\partial z^2}$$

$\therefore$ div grad $\phi = \nabla \cdot (\nabla \phi) = \dfrac{\partial^2 \phi}{\partial x^2} + \dfrac{\partial^2 \phi}{\partial y^2} + \dfrac{\partial^2 \phi}{\partial z^2}$

$$= \nabla^2 \phi, \text{ the Laplacian of } \phi$$

The operator ∇^2 is called the Laplacian.

So these general results are

(a) curl grad $\phi = \nabla \times (\nabla \phi) = 0$

(b) div curl $\mathbf{A} = \nabla \cdot (\nabla \times \mathbf{A}) = 0$

(c) div grad $\phi = \nabla \cdot (\nabla \phi) = \dfrac{\partial^2 \phi}{\partial x^2} + \dfrac{\partial^2 \phi}{\partial y^2} + \dfrac{\partial^2 \phi}{\partial z^2}.$

That brings us to the end of this particular Programme. We have covered quite a lot of new material, so check carefully through the **Revision summary** and **Can you?** checklist that follow: then you can deal with the **Test exercise**. The **Further problems** provide an opportunity for additional practice.

Revision summary 22

92

If $\mathbf{A} = a_x\mathbf{i} + a_y\mathbf{j} + a_z\mathbf{k}$; $\mathbf{B} = b_x\mathbf{i} + b_y\mathbf{j} + b_z\mathbf{k}$; $\mathbf{C} = c_x\mathbf{i} + c_y\mathbf{j} + c_z\mathbf{k}$; then we have the following relationships.

1 *Scalar product* (dot product) $\qquad \mathbf{A} \cdot \mathbf{B} = AB\cos\theta$

$\qquad \mathbf{A} \cdot \mathbf{B} = \mathbf{B} \cdot \mathbf{A} \quad$ and $\quad \mathbf{A} \cdot (\mathbf{B} + \mathbf{C}) = \mathbf{A} \cdot \mathbf{B} + \mathbf{A} \cdot \mathbf{C}$

If $\mathbf{A} \cdot \mathbf{B} = 0$ and $\mathbf{A}$, $\mathbf{B} \neq \mathbf{0}$ then $\mathbf{A} \perp \mathbf{B}$.

2 *Vector product* (cross product) $\qquad \mathbf{A} \times \mathbf{B} = (AB\sin\theta)\mathbf{n}$

$\mathbf{n}$ = unit normal vector where $\mathbf{A}$, $\mathbf{B}$, $\mathbf{n}$ form a right-handed set.

$$\mathbf{A} \times \mathbf{B} = \begin{vmatrix} \mathbf{i} & \mathbf{j} & \mathbf{k} \\ a_x & a_y & a_z \\ b_x & b_y & b_z \end{vmatrix}$$

$\mathbf{A} \times \mathbf{B} = -(\mathbf{B} \times \mathbf{A})$ and $\mathbf{A} \times (\mathbf{B} + \mathbf{C}) = \mathbf{A} \times \mathbf{B} + \mathbf{A} \times \mathbf{C}$

3 *Unit vectors*

(a) $\mathbf{i} \cdot \mathbf{i} = \mathbf{j} \cdot \mathbf{j} = \mathbf{k} \cdot \mathbf{k} = 1$

$\quad \mathbf{i} \cdot \mathbf{j} = \mathbf{j} \cdot \mathbf{k} = \mathbf{k} \cdot \mathbf{i} = 0.$

(b) $\mathbf{i} \times \mathbf{i} = \mathbf{j} \times \mathbf{j} = \mathbf{k} \times \mathbf{k} = \mathbf{0}$

$\quad \mathbf{i} \times \mathbf{j} = \mathbf{k}, \quad \mathbf{j} \times \mathbf{k} = \mathbf{i}, \quad \mathbf{k} \times \mathbf{i} = \mathbf{j}.$

4 *Scalar triple product* $\qquad \mathbf{A} \cdot (\mathbf{B} \times \mathbf{C})$

$$\mathbf{A} \cdot (\mathbf{B} \times \mathbf{C}) = \begin{vmatrix} a_x & a_y & a_z \\ b_x & b_y & b_z \\ c_x & c_y & c_z \end{vmatrix}$$

$\mathbf{A} \cdot (\mathbf{B} \times \mathbf{C}) = \mathbf{B} \cdot (\mathbf{C} \times \mathbf{A}) = \mathbf{C} \cdot (\mathbf{A} \times \mathbf{B})$

Unchanged by cyclic change of vectors.
Sign reversed by non-cyclic change of vectors.

5 *Coplanar vectors* $\quad \mathbf{A} \cdot (\mathbf{B} \times \mathbf{C}) = 0.$

6 *Vector triple product* $\quad \mathbf{A} \times (\mathbf{B} \times \mathbf{C})$ and $(\mathbf{A} \times \mathbf{B}) \times \mathbf{C}$

$$\mathbf{A} \times (\mathbf{B} \times \mathbf{C}) = (\mathbf{A} \cdot \mathbf{C})\mathbf{B} - (\mathbf{A} \cdot \mathbf{B})\mathbf{C}$$

and $\quad (\mathbf{A} \times \mathbf{B}) \times \mathbf{C} = (\mathbf{C} \cdot \mathbf{A})\mathbf{B} - (\mathbf{C} \cdot \mathbf{B})\mathbf{A}.$

7 *Differentiation of vectors*

If $\mathbf{A}$, a_x, a_y, a_z are functions of u

$$\frac{d\mathbf{A}}{du} = \frac{da_x}{du}\mathbf{i} + \frac{da_y}{du}\mathbf{j} + \frac{da_z}{du}\mathbf{k}$$

8 *Unit tangent vector* $\mathbf{T}$

$$\mathbf{T} = \frac{\dfrac{d\mathbf{A}}{du}}{\left|\dfrac{d\mathbf{A}}{du}\right|}$$

▶

9 *Integration of vectors*

$$\int_a^b \mathbf{A}\, du = \mathbf{i} \int_a^b a_x \, du + \mathbf{j} \int_a^b a_y \, du + \mathbf{k} \int_a^b a_z \, du$$

10 *Grad* (gradient of a scalar function ϕ)

$$\operatorname{grad} \phi = \nabla \phi = \frac{\partial \phi}{\partial x}\mathbf{i} + \frac{\partial \phi}{\partial y}\mathbf{j} + \frac{\partial \phi}{\partial z}\mathbf{k}$$

$$\text{'del'} = \text{operator } \nabla = \left(\mathbf{i}\frac{\partial}{\partial x} + \mathbf{j}\frac{\partial}{\partial y} + \mathbf{k}\frac{\partial}{\partial z} \right)$$

(a) *Directional derivative* $\dfrac{\mathrm{d}\phi}{\mathrm{d}s} = \hat{\mathbf{a}} \cdot \operatorname{grad} \phi = \hat{\mathbf{a}} \cdot \nabla \phi$ where $\hat{\mathbf{a}}$ is a unit vector in a stated direction. Grad ϕ gives the direction for maximum rate of change of ϕ.

(b) *Unit normal vector* $\mathbf{N}$ to surface $\phi(x,\, y,\, z) = \text{constant}$.

$$\mathbf{N} = \frac{\nabla \phi}{|\nabla \phi|}$$

11 *Div* (divergence of a vector function $\mathbf{A}$)

$$\operatorname{div} \mathbf{A} = \nabla \cdot \mathbf{A} = \frac{\partial a_x}{\partial x} + \frac{\partial a_y}{\partial y} + \frac{\partial a_z}{\partial z}$$

If $\nabla \cdot \mathbf{A} = 0$ for all points, $\mathbf{A}$ is a solenoidal vector.

12 *Curl* (curl of a vector function $\mathbf{A}$)

$$\operatorname{curl} \mathbf{A} = \nabla \times \mathbf{A} = \begin{vmatrix} \mathbf{i} & \mathbf{j} & \mathbf{k} \\ \dfrac{\partial}{\partial x} & \dfrac{\partial}{\partial y} & \dfrac{\partial}{\partial z} \\ a_x & a_y & a_z \end{vmatrix}$$

If $\nabla \times \mathbf{A} = 0$ then $\mathbf{A}$ is an irrotational vector.

13 *Operators*
grad (∇) acts on a *scalar* and gives a *vector*
div ($\nabla \cdot$) acts on a *vector* and gives a *scalar*
curl ($\nabla \times$) acts on a *vector* and gives a *vector*.

14 *Multiple operations*
(a) curl grad $\phi = \nabla \times (\nabla \phi) = 0$
(b) div curl $\mathbf{A} = \nabla \cdot (\nabla \times \mathbf{A}) = 0$

(c) div grad $\phi = \nabla \cdot (\nabla \phi) = \dfrac{\partial^2 \phi}{\partial x^2} + \dfrac{\partial^2 \phi}{\partial y^2} + \dfrac{\partial^2 \phi}{\partial z^2}$

$\qquad = \nabla^2 \phi$, the Laplacian of ϕ.

Can you?

93 Checklist 22

Check this list before and after you try the end of Programme test.

On a scale of 1 to 5 how confident are you that you can: **Frames**

- Obtain the scalar and vector product of two vectors?   [1] to [4]
  Yes   ☐   ☐   ☐   ☐   ☐   *No*

- Reproduce the relationships between the scalar and vector
  products of the Cartesian coordinate unit vectors?   [5] to [11]
  Yes   ☐   ☐   ☐   ☐   ☐   *No*

- Obtain the scalar and vector triple products and appreciate
  their geometric significance?   [12] to [27]
  Yes   ☐   ☐   ☐   ☐   ☐   *No*

- Differentiate a vector field and derive a unit vector tangential
  to the vector field at a point?   [28] to [48]
  Yes   ☐   ☐   ☐   ☐   ☐   *No*

- Integrate a vector field?   [49] to [55]
  Yes   ☐   ☐   ☐   ☐   ☐   *No*

- Obtain the gradient of a scalar field, the directional derivative
  and a unit normal to a surface?   [56] to [77]
  Yes   ☐   ☐   ☐   ☐   ☐   *No*

- Obtain the divergence of a vector field and recognise a
  solenoidal vector field?   [78] to [80]
  Yes   ☐   ☐   ☐   ☐   ☐   *No*

- Obtain the curl of a vector field?   [80] to [86]
  Yes   ☐   ☐   ☐   ☐   ☐   *No*

- Obtain combinations of div, grad and curl acting on scalar and
  vector fields as appropriate?   [87] to [91]
  Yes   ☐   ☐   ☐   ☐   ☐   *No*

Test exercise 22

1 Find (a) the scalar product and (b) the vector product of the vectors
$A = 3i - 2j + 4k$ and $B = i + 5j - 2k$.

94

2 If $A = 2i + 3j - 5k$; $B = 3i + j + 2k$; $C = i - j + 3k$; determine
(a) the scalar triple product $A \cdot (B \times C)$
(b) the vector triple product $A \times (B \times C)$.

3 Determine whether the three vectors $A = 2i + 3j + k$; $B = i - 2j + 2k$;
$C = 3i + j + 3k$ are coplanar.

4 If $A = (u^2 + 5)i - (u^2 + 3)j + 2u^3k$, determine
(a) $\dfrac{dA}{du}$; (b) $\dfrac{d^2A}{du^2}$; (c) $\left|\dfrac{dA}{du}\right|$; all at $u = 2$.

5 Determine the unit tangent vector at the point $(2, 4, 3)$ for the curve with
parametric equations
$x = 2u^2$; $y = u + 3$; $z = 4u^2 - u$.

6 If $F = 2i + 4uj + u^2k$ and $G = u^2i - 2uj + 4k$, determine
$$\int_0^2 (F \times G)du.$$

7 Find the directional derivative of the function $\phi = x^2y - 2xz^2 + y^2z$ at the point
$(1, 3, 2)$ in the direction of the vector $A = 3i + 2j - k$.

8 Find the unit normal to the surface $\phi = 2x^3z + x^2y^2 + xyz - 4 = 0$ at the point
$(2, 1, 0)$.

9 If $A = x^2yi + (xy + yz)j + xz^2k$; $B = yzi - 3xzj + 2xyk$; and
$\phi = 3x^2y + xyz - 4y^2z^2 - 3$; determine, at the point $(1, 2, 1)$
(a) $\nabla\phi$; (b) $\nabla \cdot A$; (c) $\nabla \times B$; (d) grad div A; (e) curl curl A.

Further problems 22

1 If $A = 2i + 3j - 4k$; $B = 3i + 5j + 2k$; $C = i - 2j + 3k$; determine $A \cdot (B \times C)$.

95

2 If $A = 2i + j - 3k$; $B = i - 2j + 2k$; $C = 3i + 2j - k$; find $A \times (B \times C)$.

3 If $A = i - 2j + 3k$; $B = 2i + j - 2k$; $C = 3i + 2j + k$; find
(a) $A \times (B \times C)$; (b) $(A \times B) \times C$.

4 If $F = x^2i + (3x + 2)j + \sin xk$, find
(a) $\dfrac{dF}{dx}$; (b) $\dfrac{d^2F}{dx^2}$; (c) $\left|\dfrac{dF}{dx}\right|$; (d) $\dfrac{d}{dx}(F \cdot F)$ at $x = 1$.

5 If $\mathbf{F} = u\mathbf{i} + (1 - u)\mathbf{j} + 3u\mathbf{k}$ and $\mathbf{G} = 2\mathbf{i} - (1 + u)\mathbf{j} - u^2\mathbf{k}$, determine
  (a) $\dfrac{d}{du}(\mathbf{F} \cdot \mathbf{G})$; (b) $\dfrac{d}{du}(\mathbf{F} \times \mathbf{G})$; (c) $\dfrac{d}{du}(\mathbf{F} + \mathbf{G})$.

6 Find the unit normal to the surface $4x^2y^2 - 3xz^2 - 2y^2z + 4 = 0$ at the point $(2, -1, -2)$.

7 Find the unit normal to the surface $2xy^2 + y^2z + x^2z - 11 = 0$ at the point $(-2, 1, 3)$.

8 Determine the unit vector normal to the surface
  $xz^2 + 3xy - 2yz^2 + 1 = 0$ at the point $(1, -2, -1)$.

9 Find the unit normal to the surface $x^2y - 2yz^2 + y^2z = 3$ at the point $(2, -3, 1)$.

10 Determine the directional derivative of $\phi = xe^y + yz^2 + xyz$ at the point $(2, 0, 3)$ in the direction of $\mathbf{A} = 3\mathbf{i} - 2\mathbf{j} + \mathbf{k}$.

11 Find the directional derivative of $\phi = (x + 2y + z)^2 - (x - y - z)^2$ at the point $(2, 1, -1)$ in the direction of $\mathbf{A} = \mathbf{i} - 4\mathbf{j} + 2\mathbf{k}$.

12 Find the scalar triple product of
  (a) $\mathbf{A} = \mathbf{i} + 2\mathbf{j} - 3\mathbf{k}$; $\mathbf{B} = 2\mathbf{i} - \mathbf{j} + 4\mathbf{k}$; $\mathbf{C} = 3\mathbf{i} + \mathbf{j} - 2\mathbf{k}$.
  (b) $\mathbf{A} = 2\mathbf{i} - 3\mathbf{j} + \mathbf{k}$; $\mathbf{B} = 3\mathbf{i} + \mathbf{j} + 2\mathbf{k}$; $\mathbf{C} = \mathbf{i} + 4\mathbf{j} - 2\mathbf{k}$.
  (c) $\mathbf{A} = -2\mathbf{i} + 3\mathbf{j} - 2\mathbf{k}$; $\mathbf{B} = 3\mathbf{i} - \mathbf{j} + 3\mathbf{k}$; $\mathbf{C} = 2\mathbf{i} - 5\mathbf{j} + \mathbf{k}$.

13 Find the vector triple product $\mathbf{A} \times (\mathbf{B} \times \mathbf{C})$ of the following.
  (a) $\mathbf{A} = 3\mathbf{i} + \mathbf{j} - 2\mathbf{k}$; $\mathbf{B} = 2\mathbf{i} + 4\mathbf{j} + 3\mathbf{k}$; $\mathbf{C} = \mathbf{i} - 2\mathbf{j} + \mathbf{k}$.
  (b) $\mathbf{A} = 2\mathbf{i} - \mathbf{j} + 3\mathbf{k}$; $\mathbf{B} = \mathbf{i} + 4\mathbf{j} - 5\mathbf{k}$; $\mathbf{C} = 3\mathbf{i} - 2\mathbf{j} + \mathbf{k}$.
  (c) $\mathbf{A} = 4\mathbf{i} + 2\mathbf{j} - 3\mathbf{k}$; $\mathbf{B} = 2\mathbf{i} - 3\mathbf{j} + 2\mathbf{k}$; $\mathbf{C} = 3\mathbf{i} - 3\mathbf{j} + \mathbf{k}$.

14 If $\mathbf{F} = 4t^3\mathbf{i} - 2t^2\mathbf{j} + 4t\mathbf{k}$, determine when $t = 1$
  (a) $\dfrac{d\mathbf{F}}{dt}$; (b) $\dfrac{d^2\mathbf{F}}{dt^2}$; (c) $\dfrac{d}{dt}(\mathbf{F} \cdot \mathbf{F})$.

15 If $\phi = x^2 \sin z + ze^y$ find, at the point $(1, 3, 2)$, the values of
  (a) $\mathrm{grad}\ \phi$ and (b) $|\ \mathrm{grad}\ \phi|$.

16 Given that $\phi = xy^2 + yz^2 - x^2$, find the derivative of ϕ with respect to distance at the point $(1, 2, -1)$, measured parallel to the vector $2\mathbf{i} - 3\mathbf{j} + 4\mathbf{k}$.

17 Find unit vectors normal to the surfaces $x^2 + y^2 - z^2 + 3 = 0$ and $xy - yz + zx - 10 = 0$ at the point $(3, 2, 4)$ and hence find the angle between the two surfaces at that point.

18 If $\mathbf{r} = (t^2 + 3t)\mathbf{i} - 2 \sin 3t\mathbf{j} + 3e^{2t}\mathbf{k}$, determine
  (a) $\dfrac{d\mathbf{r}}{dt}$; (b) $\dfrac{d^2\mathbf{r}}{dt^2}$; (c) the value of $\left|\dfrac{d^2\mathbf{r}}{dt^2}\right|$ at $t = 0$.

19 (a) Show that curl $(-y\mathbf{i} + x\mathbf{j})$ is a constant vector.
  (b) Show that the vector field $(yz\mathbf{i} + zx\mathbf{j} + xy\mathbf{k})$ has zero divergence and zero curl.

20 If $A = 2xz^2i - xzj + (y + z)k$, find curl curl A.

21 Determine grad ϕ where $\phi = x^2 \cos(2yz - 0.5)$ and obtain its value at the point (1, 3, 1).

22 Determine the value of p such that the three vectors A, B, C are coplanar when
$A = 2i + j + 4k$; $B = 3i + 2j + pk$; $C = i + 4j + 2k$.

23 If $A = pi - 6j - 3k$; $B = 4i + 3j - k$; $C = i - 3j + 2k$

(a) find the values of p for which

(1) A and B are perpendicular to each other

(2) A, B and C are coplanar.

(b) determine a unit vector perpendicular to both A and B when $p = 2$.

Vector analysis 2

Learning outcomes

When you have completed this Programme you will be able to:

- Evaluate the line integral of a scalar and a vector field in Cartesian coordinates
- Evaluate the volume integral of a vector field
- Evaluate the surface integral of a scalar and a vector field
- Determine whether or not a vector field is a conservative vector field
- Apply Gauss' divergence theorem
- Apply Stokes' theorem
- Determine the direction of unit normal vectors to a surface
- Apply Green's theorem in the plane

We dealt in some detail with line, surface and volume integrals in an earlier Programme, when we approached the subject analytically. In many practical problems, it is more convenient to express these integrals in vector form and the methods often lead to more concise working.

1

Line integrals

Let a point P on the curve c joining A and B be denoted by the position vector $\mathbf{r}$ with respect to a fixed origin O.

If Q is a neighbouring point on the curve with position vector $\mathbf{r} + d\mathbf{r}$, then $\overline{PQ} = d\mathbf{r}$.

The curve c can be divided up into many (n) such small arcs, approximating to $d\mathbf{r}_1$, $d\mathbf{r}_2$, $d\mathbf{r}_3 \ldots d\mathbf{r}_p \ldots$ so that

(a)

(b)

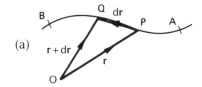

$$\overline{AB} = \sum_{p=1}^{n} d\mathbf{r}_p$$

where $d\mathbf{r}_p$ is a vector representing the element of arc in both magnitude and direction.

Scalar field

If a scalar field V exists for all points on the curve, then $\sum_{p=1}^{n} V \, d\mathbf{r}_p$ with $d\mathbf{r} \to 0$, defines the *line integral* of V along the curve c from A to B,

i.e. line integral $= \displaystyle\int_c V \, d\mathbf{r}$

We can illustrate this integral by erecting a continuous ordinate proportional to V at each point of the curve. $\displaystyle\int_c V \, d\mathbf{r}$ is then represented by the area of the curved surface between the ends A and B of the curve c.

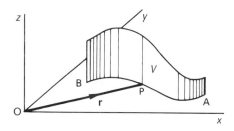

To evaluate a line integral, the integrand is expressed in terms of x, y, z, with $d\mathbf{r} = \ldots\ldots\ldots\ldots$

2

$$\mathbf{dr} = \mathbf{i}\,dx + \mathbf{j}\,dy + \mathbf{k}\,dz$$

In practice, x, y and z are often expressed in terms of parametric equations of a fourth variable (say u), i.e. $x = x(u)$; $y = y(u)$; $z = z(u)$. From these, dx, dy and dz can be written in terms of u and the integral evaluated in terms of this parameter u.

The following examples will show the method.

Example 1

If $V = xy^2z$, evaluate $\int_c V\,\mathbf{dr}$ along the curve c having parametric equations $x = 3u$; $y = 2u^2$; $z = u^3$ between A $(0, 0, 0)$ and B $(3, 2, 1)$.

$$V = xy^2z = (3u)(4u^4)(u^3) = 12u^8$$
$$\mathbf{dr} = \mathbf{i}\,dx + \mathbf{j}\,dy + \mathbf{k}\,dz = \ldots\ldots\ldots$$

3

$$\mathbf{dr} = \mathbf{i}\,3\,du + \mathbf{j}\,4u\,du + \mathbf{k}\,3u^2\,du$$

Because

$x = 3u$, $\therefore$ $dx = 3\,du$
$y = 2u^2$, $\therefore$ $dy = 4u\,du$
$z = u^3$, $\therefore$ $dz = 3u^2\,du$
Limits: A $(0, 0, 0)$ corresponds to $u = \ldots\ldots\ldots$
B $(3, 2, 1)$ corresponds to $u = \ldots\ldots\ldots$

4

$$\text{A}\,(0,0,0) \equiv u = 0 \quad \text{B}\,(3,2,1) \equiv u = 1$$

$$\therefore \int_c V\,\mathbf{dr} = \int_0^1 12u^8\,(\mathbf{i}\,3\,du + \mathbf{j}\,4u\,du + \mathbf{k}\,3u^2\,du)$$

$$= \ldots\ldots\ldots$$

Finish it off

5

$$4\mathbf{i} + \frac{24}{5}\mathbf{j} + \frac{36}{11}\mathbf{k}$$

Because

$$\int_c V\,\mathbf{dr} = 12\int_0^1 (\mathbf{i}\,3u^8\,du + \mathbf{j}\,4u^9\,du + \mathbf{k}\,3u^{10}\,du)$$

which integrates directly to give the result quoted above.

Now for another example.

Example 2   6

If $V = xy + y^2 z$, evaluate $\displaystyle\int_c V\,\mathrm{d}\mathbf{r}$ along the curve c defined by

$x = t^2$; $y = 2t$; $z = t + 5$ between A $(0, 0, 5)$ and B $(4, 4, 7)$.

As before, expressing V and $\mathrm{d}\mathbf{r}$ in terms of the parameter t we have

$$V = \ldots\ldots\ldots\ldots \qquad \mathrm{d}\mathbf{r} = \ldots\ldots\ldots\ldots$$

$$\boxed{V = 6t^3 + 20t^2; \quad \mathrm{d}\mathbf{r} = \mathbf{i}\,2t\,\mathrm{d}t + \mathbf{j}\,2\,\mathrm{d}t + \mathbf{k}\,\mathrm{d}t}$$ 7

Because

$$V = xy + y^2 z = (t^2)(2t) + (4t^2)(t + 5) = 6t^3 + 20t^2.$$

$$\text{Also } \left. \begin{array}{ll} x = t^2 & \mathrm{d}x = 2t\,\mathrm{d}t \\ y = 2t & \mathrm{d}y = 2\,\mathrm{d}t \\ z = t + 5 & \mathrm{d}z = \mathrm{d}t \end{array} \right\} \quad \begin{array}{l} \therefore\ \mathrm{d}\mathbf{r} = \mathbf{i}\,\mathrm{d}x + \mathbf{j}\,\mathrm{d}y + \mathbf{k}\,\mathrm{d}z \\ \qquad = \mathbf{i}\,2t\,\mathrm{d}t + \mathbf{j}\,2\,\mathrm{d}t + \mathbf{k}\,\mathrm{d}t \end{array}$$

$$\therefore\ \int_c V\,\mathrm{d}\mathbf{r} = \int_c (6t^3 + 20t^2)(\mathbf{i}\,2t + \mathbf{j}\,2 + \mathbf{k})\,\mathrm{d}t$$

Limits: A $(0, 0, 5) \equiv t = \ldots\ldots\ldots\ldots$

$\qquad\qquad$ B $(4, 4, 7) \equiv t = \ldots\ldots\ldots\ldots$

$$\boxed{\text{A } (0,0,5) \equiv t = 0; \quad \text{B } (4,4,7) \equiv t = 2}$$ 8

$$\therefore\ \int_c V\,\mathrm{d}\mathbf{r} = \int_0^2 (6t^3 + 20t^2)(\mathbf{i}\,2t + \mathbf{j}\,2 + \mathbf{k})\,\mathrm{d}t$$

$$= \ldots\ldots\ldots\ldots \text{ Complete the integration.}$$

$$\boxed{\dfrac{8}{15}(444\,\mathbf{i} + 290\,\mathbf{j} + 145\,\mathbf{k})}$$ 9

$$\int_c V\,\mathrm{d}\mathbf{r} = 2\int_0^2 \{(6t^4 + 20t^3)\mathbf{i} + (6t^3 + 20t^2)\mathbf{j} + (3t^3 + 10t^2)\mathbf{k}\}\,\mathrm{d}t$$

The actual integration is simple enough and gives the result shown. All line integrals in scalar fields are done in the same way.

10 Vector field

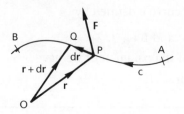

If a vector field **F** exists for all points of the curve c, then for each element of arc we can form the scalar product **F** · **dr**. Summing these products for all elements of arc, we have $\sum_{p=1}^{n} \mathbf{F} \cdot \mathbf{dr}_p$

Then, if $\mathbf{dr}_p \to 0$, the sum becomes the integral $\int_c \mathbf{F} \cdot \mathbf{dr}$,

i.e. the line integral of **F** from A to B along the stated curve

$$= \int_c \mathbf{F} \cdot \mathbf{dr}$$

In this case, since **F** · **dr** is a scalar product, then the line integral is a scalar.

To evaluate the line integral, **F** and **dr** are expressed in terms of x, y, z and the curve in parametric form. We have

$$\mathbf{F} = F_x\mathbf{i} + F_y\mathbf{j} + F_z\mathbf{k}$$

and $\quad \mathbf{dr} = \mathbf{i}\,dx + \mathbf{j}\,dy + \mathbf{k}\,dz$

Then $\quad \mathbf{F} \cdot \mathbf{dr} = (F_x\mathbf{i} + F_y\mathbf{j} + F_z\mathbf{k}) \cdot (\mathbf{i}\,dx + \mathbf{j}\,dy + \mathbf{k}\,dz)$
$$= F_x\,dx + F_y\,dy + F_z\,dz$$

$$\therefore \quad \int_c \mathbf{F} \cdot \mathbf{dr} = \int_c F_x\,dx + \int_c F_y\,dy + \int_c F_z\,dz$$

Now for an example to show it in operation.

Example 1

If $\mathbf{F} = x^2y\mathbf{i} + xz\mathbf{j} - 2yz\mathbf{k}$, evaluate $\int_c \mathbf{F} \cdot \mathbf{dr}$ between A (0, 0, 0) and B (4, 2, 1) along the curve having parametric equations $x = 4t$; $y = 2t^2$; $z = t^3$.

Expressing everything in terms of the parameter t, we have

$$\mathbf{F} = \dots\dots\dots\dots$$
$$dx = \dots\dots\dots\dots; \quad dy = \dots\dots\dots\dots; \quad dz = \dots\dots\dots\dots$$

11

$$\boxed{\begin{array}{l} \mathbf{F} = 32t^4\,\mathbf{i} + 4t^4\,\mathbf{j} - 4t^5\,\mathbf{k} \\ dx = 4\,dt; \quad dy = 4t\,dt; \quad dz = 3t^2\,dt \end{array}}$$

Because

$$\begin{array}{lll} x^2y = (16t^2)(2t^2) = 32t^4 & x = 4t & \therefore \ dx = 4\,dt \\ xz = (4t)(t^3) = 4t^4 & y = 2t^2 & \therefore \ dy = 4t\,dt \\ 2yz = (4t^2)(t^3) = 4t^5 & z = t^3 & \therefore \ dz = 3t^2\,dt \end{array}$$

Then $\displaystyle\int \mathbf{F}\cdot d\mathbf{r} = \int (32t^4\mathbf{i} + 4t^4\mathbf{j} - 4t^5\mathbf{k}) \cdot (\mathbf{i}\,4\,dt + \mathbf{j}\,4t\,dt + \mathbf{k}\,3t^2\,dt)$

$$= \int (128t^4 + 16t^5 - 12t^7)\,dt$$

Limits: A $(0, 0, 0) \equiv t = \ldots\ldots\ldots\ldots;$ B $(4, 2, 1) \equiv t = \ldots\ldots\ldots\ldots$

12

$$\boxed{A \equiv t = 0; \quad B \equiv t = 1}$$

$$\therefore \ \int_c \mathbf{F}\cdot d\mathbf{r} = \int_0^1 (128t^4 + 16t^5 - 12t^7)\,dt = \ldots\ldots\ldots\ldots$$

13

$$\boxed{\dfrac{128}{5} + \dfrac{8}{3} - \dfrac{3}{2} = \dfrac{803}{30} = 26{\cdot}77}$$

If the vector field $\mathbf{F}$ is a *force field*, then the line integral $\displaystyle\int_c \mathbf{F}\cdot d\mathbf{r}$ represents the work done in moving a unit particle along the prescribed curve c from A to B. Now for another example.

Example 2

If $\mathbf{F} = x^2y\mathbf{i} + 2yz\mathbf{j} + 3z^2x\mathbf{k}$, evaluate $\displaystyle\int_c \mathbf{F}\cdot d\mathbf{r}$ between A $(0, 0, 0)$ and B $(1, 2, 3)$

(a) along the straight lines c_1 from $(0, 0, 0)$ to $(1, 0, 0)$
                    then c_2 from $(1, 0, 0)$ to $(1, 2, 0)$
                    and c_3 from $(1, 2, 0)$ to $(1, 2, 3)$
(b) along the straight line c_4 joining $(0, 0, 0)$ to $(1, 2, 3)$.

As before, we first obtain an expression for $\mathbf{F}\cdot d\mathbf{r}$ which is

$$\ldots\ldots\ldots\ldots$$

14

$$\boxed{\mathbf{F} \cdot d\mathbf{r} = x^2 y\, dx + 2yz\, dy + 3z^2 x\, dz}$$

Because

$$\mathbf{F} \cdot d\mathbf{r} = (x^2 y\, \mathbf{i} + 2yz\, \mathbf{j} + 3z^2 x\, \mathbf{k}) \cdot (\mathbf{i}\, dx + \mathbf{j}\, dy + \mathbf{k}\, dz)$$

$$\therefore \int \mathbf{F} \cdot d\mathbf{r} = \int x^2 y\, dx + \int 2yz\, dy + \int 3z^2 x\, dz$$

(a) Here the integration is made in three sections, along c_1, c_2 and c_3.

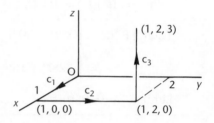

(1) c_1: $y = 0$, $z = 0$, $dy = 0$, $dz = 0$

$$\therefore \int_{c_1} \mathbf{F} \cdot d\mathbf{r} = 0 + 0 + 0 = 0$$

(2) c_2: The conditions along c_2 are

. . . .

15

$$\boxed{c_2: \quad x = 1, \quad z = 0, \quad dx = 0, \quad dz = 0}$$

$$\therefore \int_{c_2} \mathbf{F} \cdot d\mathbf{r} = 0 + 0 + 0 = 0$$

(3) c_3: $x = 1$, $y = 2$, $dx = 0$, $dy = 0$

$$\therefore \int_{c_3} \mathbf{F} \cdot d\mathbf{r} = \ldots \ldots \ldots$$

16

$$\boxed{27}$$

Because

$$\int_{c_3} \mathbf{F} \cdot d\mathbf{r} = 0 + 0 + \int_0^3 3z^2\, dz = 27$$

Summing the three partial results

$$\int_{(0,\,0,\,0)}^{(1,\,2,\,3)} \mathbf{F} \cdot d\mathbf{r} = 0 + 0 + 27 = 27 \quad \therefore \int_{c_1 + c_2 + c_3} \mathbf{F} \cdot d\mathbf{r} = 27$$

▶

(b) If t is taken as the parameter, the
parametric equations of c are

$$x = \ldots\ldots\ldots$$

$$y = \ldots\ldots\ldots$$

$$z = \ldots\ldots\ldots$$

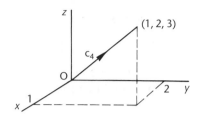

$$\boxed{x = t; \quad y = 2t; \quad z = 3t}$$

17

and the limits of t are $\ldots\ldots\ldots\ldots$

$$\boxed{t = 0 \quad \text{and} \quad t = 1}$$

18

As in Example 1, we now express everything in terms of t and complete the
integral, finally getting

$$\int_{c_4} \mathbf{F} \cdot d\mathbf{r} = \ldots\ldots\ldots$$

$$\boxed{\int_{c_4} \mathbf{F} \cdot d\mathbf{r} = \frac{115}{4} = 28 \cdot 75}$$

19

Because

$$\mathbf{F} = 2t^3\mathbf{i} + 12t^2\mathbf{j} + 27t^3\mathbf{k}$$
$$d\mathbf{r} = \mathbf{i}\,dx + \mathbf{j}\,dy + \mathbf{k}\,dz = \mathbf{i}\,dt + \mathbf{j}\,2\,dt + \mathbf{k}\,3\,dt$$

$$\therefore \int_{c_4} \mathbf{F} \cdot d\mathbf{r} = \int_0^1 (2t^3\mathbf{i} + 12t^2\mathbf{j} + 27t^3\mathbf{k}) \cdot (\mathbf{i} + 2\mathbf{j} + 3\mathbf{k})\,dt$$

$$= \int_0^1 (2t^3 + 24t^2 + 81t^3)\,dt = \int_0^1 (83t^3 + 24t^2)\,dt$$

$$= \left[83\frac{t^4}{4} + 8t^3\right]_0^1 = \frac{115}{4} = 28 \cdot 75$$

So the value of the line integral depends on the path taken between the two
end points A and B

(a) $\int \mathbf{F} \cdot d\mathbf{r}$ via c_1, c_2 and $c_3 = 27$

(b) $\int \mathbf{F} \cdot d\mathbf{r}$ via c_4 $= 28 \cdot 75$

We shall refer to this topic later.

One further example on your own. The working is just the same as before.

▶

Example 3

If $\mathbf{F} = x^2y^2\mathbf{i} + y^3z\mathbf{j} + z^2\mathbf{k}$, evaluate $\displaystyle\int_c \mathbf{F}\cdot d\mathbf{r}$ along the curve $x = 2u^2$, $y = 3u$, $z = u^3$ between A $(2, -3, -1)$ and B $(2, 3, 1)$. Proceed as before. You will have no difficulty.

$$\int_c \mathbf{F}\cdot d\mathbf{r} = \ldots\ldots\ldots$$

20

$$\boxed{\int_c \mathbf{F}\cdot d\mathbf{r} = \frac{500}{21} = 23{\cdot}8}$$

Here is the working for you to check.

$$x = 2u^2 \quad y = 3u \quad z = u^3$$
$$x^2y^2 = (4u^4)(9u^2) = 36\,u^6 \qquad dx = 4u\,du$$
$$y^3z = (27u^3)(u^3) = 27u^6 \qquad dy = 3\,du$$
$$z^2 = u^6 \qquad dz = 3u^2\,du$$

Limits: A $(2, -3, -1)$ corresponds to $u = -1$
                B $(2, 3, 1)$ corresponds to $u = 1$

$$\therefore \int_c \mathbf{F}\cdot d\mathbf{r} = \int_{-1}^{1} (x^2y^2\mathbf{i} + y^3z\mathbf{j} + z^2\mathbf{k})\cdot(\mathbf{i}\,dx + \mathbf{j}\,dy + \mathbf{k}\,dz)$$

$$= \int_{-1}^{1} (36u^6\mathbf{i} + 27u^6\mathbf{j} + u^6\mathbf{k})\cdot(\mathbf{i}\,4u\,du + \mathbf{j}\,3\,du + \mathbf{k}\,3u^2\,du)$$

$$= \int_{-1}^{1} (144u^7 + 81u^6 + 3u^8)\,du$$

$$= \left[18u^8 + \frac{81u^7}{7} + \frac{u^9}{3}\right]_{-1}^{1} = \frac{500}{21} = 23{\cdot}8$$

Now on to the next section

Volume integrals

21

If V is a closed region bounded by a surface $\mathbf{S}$ and $\mathbf{F}$ is a vector field at each point of V and on its boundary surface $\mathbf{S}$, then $\displaystyle\int_V \mathbf{F}\,dV$ is the *volume integral* of $\mathbf{F}$ throughout the region.

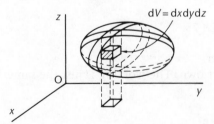

$dV = dx\,dy\,dz$

$$\int_V \mathbf{F}\,dV = \int_{x_1}^{x_2}\int_{y_1}^{y_2}\int_{z_1}^{z_2} \mathbf{F}\,dz\,dy\,dx$$

▶

Example 1

Evaluate $\displaystyle\int_V \mathbf{F}\,dV$ where V is the region bounded by the planes $x = 0$, $x = 2$, $y = 0$, $y = 3$, $z = 0$, $z = 4$, and $\mathbf{F} = xy\mathbf{i} + z\mathbf{j} - x^2\mathbf{k}$.

We start, as in most cases, by sketching the diagram, which is

. . . .

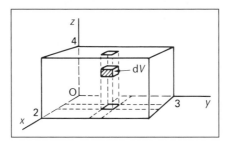

22

Then $\mathbf{F} = xy\,\mathbf{i} + z\,\mathbf{j} - x^2\,\mathbf{k}$ and $dV = dx\,dy\,dz$

$$\therefore \int_V \mathbf{F}\,dV = \int_0^4 \int_0^3 \int_0^2 (xy\mathbf{i} + z\mathbf{j} - x^2\mathbf{k})\,dx\,dy\,dz$$

$$= \int_0^4 \int_0^3 \left[\frac{x^2 y}{2}\mathbf{i} + xz\mathbf{j} - \frac{x^3}{3}\mathbf{k}\right]_{x=0}^{x=2} dy\,dz$$

$$= \int_0^4 \int_0^3 \left(2y\mathbf{i} + 2z\mathbf{j} - \frac{8}{3}\mathbf{k}\right) dy\,dz$$

$$= \ldots\ldots\ldots\ldots \text{ Complete the integral.}$$

$$\boxed{\int_V \mathbf{F}\,dV = 4(9\mathbf{i} + 12\mathbf{j} - 8\mathbf{k})}$$

23

Because

$$\int_V \mathbf{F}\,dV = \int_0^4 \left[y^2\mathbf{i} + 2yz\mathbf{j} - \frac{8}{3}y\mathbf{k}\right]_{y=0}^{y=3} dz$$

$$= \int_0^4 (9\mathbf{i} + 6z\mathbf{j} - 8\mathbf{k})\,dz$$

$$= \left[9z\mathbf{i} + 3z^2\mathbf{j} - 8z\mathbf{k}\right]_0^4$$

$$= 36\mathbf{i} + 48\mathbf{j} - 32\mathbf{k}$$

$$= 4(9\mathbf{i} + 12\mathbf{j} - 8\mathbf{k})$$

Now another.

Example 2

Evaluate $\displaystyle\int_V \mathbf{F}\,dV$ where V is the region bounded by the planes $x = 0$, $y = 0$,
$z = 0$ and $2x + y + z = 2$, and $\mathbf{F} = 2z\mathbf{i} + y\mathbf{k}$.

To sketch the surface $2x + y + z = 2$, note that

when $z = 0$, $2x + y = 2$ i.e. $y = 2 - 2x$

when $y = 0$, $2x + z = 2$ i.e. $z = 2 - 2x$

when $x = 0$, $y + z = 2$ i.e. $z = 2 - y$

Inserting these in the planes $x = 0$, $y = 0$, $z = 0$ will help.

The diagram is therefore

. . . .

24

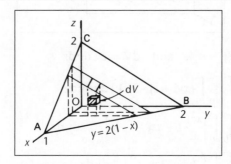

So $2x + y + z = 2$ cuts the axes at A $(1, 0, 0)$; B $(0, 2, 0)$; C $(0, 0, 2)$.

Also $\mathbf{F} = 2z\mathbf{i} + y\mathbf{k}$; $z = 2 - 2x - y = 2(1 - x) - y$

$$\therefore \int_V \mathbf{F}\,dV = \int_0^1 \int_0^{2(1-x)} \int_0^{2(1-x)-y} (2z\mathbf{i} + y\mathbf{k})\,dz\,dy\,dx$$

$$= \int_0^1 \int_0^{2(1-x)} \left[z^2\mathbf{i} + yz\mathbf{k} \right]_{z=0}^{z=2(1-x)-y} dy\,dx$$

$$= \int_0^1 \int_0^{2(1-x)} \{[4(1-x)^2 - 4(1-x)y + y^2]\mathbf{i}$$
$$+ [2(1-x)y - y^2]\mathbf{k}\}\,dy\,dx$$

$$= \int_0^1 \left[\left\{ 4(1-x)^2 y - 2(1-x)y^2 + \frac{y^3}{3} \right\} \mathbf{i} \right.$$
$$\left. + \left\{ (1-x)y^2 - \frac{y^3}{3} \right\} \mathbf{k} \right]_{y=0}^{2(1-x)} dx$$

$$= \ldots \ldots \ldots$$

Finish the last stage

$$\int_V \mathbf{F}\,dV = \frac{1}{3}(2\mathbf{i} + \mathbf{k})$$

25

Because

$$\int_V \mathbf{F}\,dV = \int_0^1 \left\{ \frac{8}{3}(1-x)^3\mathbf{i} + \frac{4}{3}(1-x)^3\mathbf{k} \right\} dx$$

$$= \left[-\frac{2}{3}(1-x)^4\mathbf{i} - \frac{1}{3}(1-x)^4\mathbf{k} \right]_0^1 = \frac{1}{3}(2\mathbf{i} + \mathbf{k})$$

And now one more, slightly different.

Example 3

Evaluate $\displaystyle\int_V \mathbf{F}\,dV$ where $\mathbf{F} = 2\mathbf{i} + 2z\mathbf{j} + y\mathbf{k}$ and V is the region bounded by the planes $z = 0$, $z = 4$ and the surface $x^2 + y^2 = 9$.

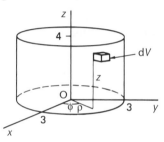

It will be convenient to use cylindrical polar coordinates (ρ, ϕ, z) so the relevant transformations are

$$x = \ldots\ldots\ldots ; \qquad y = \ldots\ldots\ldots$$

$$z = \ldots\ldots\ldots ; \qquad dV = \ldots\ldots\ldots$$

$$\boxed{\begin{aligned} &x = \rho\cos\phi; \qquad y = \rho\sin\phi \\ &z = z; \qquad dV = \rho\,d\rho\,d\phi\,dz \end{aligned}}$$

26

Then $\displaystyle\int_V \mathbf{F}\,dV = \iiint_V (2\mathbf{i} + 2z\mathbf{j} + y\mathbf{k})\,dx\,dy\,dz$.

Changing into cylindrical polar coordinates with appropriate change of limits this becomes

$$\int_V \mathbf{F}\,dV = \int_{\phi=0}^{2\pi}\int_{\rho=0}^{3}\int_{z=0}^{4}(2\mathbf{i} + 2z\mathbf{j} + \rho\sin\phi\,\mathbf{k})\,dz\,\rho\,d\rho\,d\phi$$

$$= \int_{\phi=0}^{2\pi}\int_{\rho=0}^{3}\left[2z\mathbf{i} + z^2\mathbf{j} + \rho\sin\phi\,z\mathbf{k} \right]_{z=0}^{4}\rho\,d\rho\,d\phi$$

$$= \int_{0}^{2\pi}\int_{0}^{3}(8\mathbf{i} + 16\mathbf{j} + 4\rho\sin\phi\,\mathbf{k})\,d\rho\,d\phi$$

$$= 4\int_{0}^{2\pi}\int_{0}^{3}(2\rho\mathbf{i} + 4\rho\mathbf{j} + \rho^2\sin\phi\,\mathbf{k})\,d\rho\,d\phi$$

Completing the working, we finally get

$$\int_V \mathbf{F}\,dV = \ldots\ldots\ldots$$

27

$$\boxed{72\pi(\mathbf{i} + 2\mathbf{j})}$$

Because

$$\int_V \mathbf{F}\,dV = 4\int_0^{2\pi}\left[\rho^2\mathbf{i} + 2\rho^2\mathbf{j} + \frac{\rho^3}{3}\sin\phi\,\mathbf{k}\right]_0^3 d\phi$$

$$= 4\int_0^{2\pi}(9\mathbf{i} + 18\mathbf{j} + 9\sin\phi\,\mathbf{k})\,d\phi$$

$$= 36\int_0^{2\pi}(\mathbf{i} + 2\mathbf{j} + \sin\phi\,\mathbf{k})\,d\phi$$

$$= 36\left[\phi\mathbf{i} + 2\phi\mathbf{j} - \cos\phi\,\mathbf{k}\right]_0^{2\pi}$$

$$= 36\{(2\pi\mathbf{i} + 4\pi\mathbf{j} - \mathbf{k}) - (-\mathbf{k})\}$$

$$= 72\pi(\mathbf{i} + 2\mathbf{j})$$

You will, of course, remember that in appropriate cases, the use of cylindrical polar coordinates or spherical polar coordinates often simplifies the subsequent calculations. So keep them in mind.

Now let us turn to surface integrals – in the next frame

Surface integrals

28

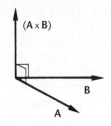

The vector product of two vectors **A** and **B** has magnitude $|\mathbf{A} \times \mathbf{B}| = AB\sin\theta$ at right angles to the plane of **A** and **B** to form a right-handed set.

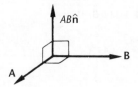

If $\theta = \dfrac{\pi}{2}$, then $|\mathbf{A} \times \mathbf{B}| = AB$ in the direction of the normal. Therefore, if $\hat{\mathbf{n}}$ is a unit normal then

$$\mathbf{A} \times \mathbf{B} = |\mathbf{A}|\,|\mathbf{B}|\hat{\mathbf{n}} = AB\,\hat{\mathbf{n}}$$

▶

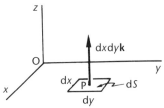

If P (x, y) is a point in the x–y plane, the element of area $dx\,dy$ has a vector area $dS = (i\,dx) \times (j\,dy)$.

i.e. $dS = dx\,dy(i \times j) = dx\,dy\,k$

i.e. a vector of magnitude $dx\,dy$ acting in the direction of k and referred to as the *vector area*.

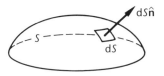

For a general surface S in space, each element of surface dS has a *vector area* dS such that $dS = dS\,\hat{n}$.

You will remember we established previously that for a surface S given by the equation $\phi(x, y, z) = $ constant, the unit normal $\hat{n}$ is given by

$$\hat{n} = \frac{\text{grad}\,\phi}{|\text{grad}\,\phi|} = \frac{\nabla\phi}{|\nabla\phi|}$$

Let us see how we can apply these results to the following examples.

Scalar fields   29

Example 1

A scalar field $V = xyz$ exists over the curved surface S defined by $x^2 + y^2 = 4$ between the planes $z = 0$ and $z = 3$ in the first octant. Evaluate $\int_S V\,dS$ over this surface.

We have $V = xyz$ S: $x^2 + y^2 - 4 = 0$, $z = 0$ to $z = 3$

$$dS = \hat{n}\,dS \qquad \text{where} \quad \hat{n} = \frac{\nabla\phi}{|\nabla\phi|}$$

Now $\nabla\phi = \dfrac{\partial\phi}{\partial x}i + \dfrac{\partial\phi}{\partial y}j + \dfrac{\partial\phi}{\partial z}k = 2xi + 2yj$ and

$$|\nabla\phi| = \sqrt{4x^2 + 4y^2} = 2\sqrt{x^2 + y^2} = 2\sqrt{4} = 4$$

Therefore

$$\hat{n} = \frac{\nabla\phi}{|\nabla\phi|} = \frac{xi + yj}{2} \quad \text{so that} \quad dS = \hat{n}\,dS = \frac{xi + yj}{2}\,dS$$

$$\therefore \quad \int_S V\,dS = \int_S V\,\hat{n}\,dS$$

$$= \frac{1}{2}\int_S xyz(xi + yj)\,dS$$

$$= \frac{1}{2}\int_S (x^2yzi + xy^2zj)\,dS \tag{1}$$

▶

We have to evaluate this integral over the prescribed surface.

Changing to cylindrical coordinates with $\rho = 2$

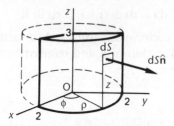

$x = \ldots\ldots\ldots\ldots;$ $y = \ldots\ldots\ldots\ldots$

$z = \ldots\ldots\ldots\ldots;$ $dS = \ldots\ldots\ldots\ldots$

30

$$\boxed{\begin{array}{ll} x = 2\cos\phi; & y = 2\sin\phi \\ z = z; & dS = 2\,d\phi\,dz \end{array}}$$

$\therefore\ x^2 yz = (4\cos^2\phi)(2\sin\phi)(z)$

$\qquad = 8\cos^2\phi\sin\phi\,z$

$\quad xy^2 z = (2\cos\phi)(4\sin^2\phi)(z)$

$\qquad = 8\cos\phi\sin^2\phi\,z$

Then result (1) above becomes

$$\int_S V\,dS = \frac{1}{2}\int_0^{\pi/2}\int_0^3 (8\cos^2\phi\sin\phi\,z\mathbf{i} + 8\cos\phi\sin^2\phi\,z\mathbf{j})\,2\,dz\,d\phi$$

$$= 4\int_0^{\pi/2}\int_0^3 (\cos^2\phi\sin\phi\,\mathbf{i} + \cos\phi\sin^2\phi\,\mathbf{j})\,2z\,dz\,d\phi$$

$$= 4\int_0^{\pi/2} (\cos^2\phi\sin\phi\,\mathbf{i} + \cos\phi\sin^2\phi\,\mathbf{j})\,9\,d\phi$$

and this eventually gives

$$\int_S V\,dS = \ldots\ldots\ldots\ldots$$

31

$$\boxed{\int_S V\,dS = 12(\mathbf{i}+\mathbf{j})}$$

Because

$$\int_S V\,dS = 36\left[-\frac{\cos^3\phi}{3}\mathbf{i} + \frac{\sin^3\phi}{3}\mathbf{j}\right]_0^{\pi/2} = 12(\mathbf{i}+\mathbf{j})$$

▶

Example 2

A scalar field $V = x + y + z$ exists over the surface S defined by $2x + 2y + z = 2$ bounded by $x = 0$, $y = 0$, $z = 0$ in the first octant.

Evaluate $\displaystyle\int_S V\,\mathrm{d}\mathbf{S}$ over this surface.

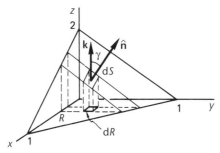

$$S: \quad 2x + 2y + z = 2$$
$$x = 0 \qquad z = 2 - 2y$$
$$y = 0 \qquad z = 2 - 2x$$
$$z = 0 \qquad y = 1 - x$$

$$\mathrm{d}\mathbf{S} = \hat{\mathbf{n}}\,\mathrm{d}S \qquad \text{where} \quad \hat{\mathbf{n}} = \frac{\nabla\phi}{|\nabla\phi|}$$

Now $\nabla\phi = \dfrac{\partial\phi}{\partial x}\mathbf{i} + \dfrac{\partial\phi}{\partial y}\mathbf{j} + \dfrac{\partial\phi}{\partial z}\mathbf{k} = 2\mathbf{i} + 2\mathbf{j} + \mathbf{k}$ and

$$|\nabla\phi| = \sqrt{4 + 4 + 1} = \sqrt{9} = 3$$

Therefore

$$\hat{\mathbf{n}} = \frac{\nabla\phi}{|\nabla\phi|} = \frac{2\mathbf{i} + 2\mathbf{j} + \mathbf{k}}{3} \quad \text{so that} \quad \mathrm{d}\mathbf{S} = \hat{\mathbf{n}}\,\mathrm{d}S = \frac{1}{3}(2\mathbf{i} + 2\mathbf{j} + \mathbf{k})\,\mathrm{d}S$$

If we now project $\mathrm{d}S$ onto the x–y plane, $\mathrm{d}R = \mathrm{d}S\cos\gamma$

$$\cos\gamma = \hat{\mathbf{n}} \cdot \mathbf{k} = \frac{1}{3}(2\mathbf{i} + 2\mathbf{j} + \mathbf{k}) \cdot (\mathbf{k}) = \frac{1}{3}$$

$$\therefore \ \mathrm{d}R = \frac{1}{3}\mathrm{d}S \qquad \therefore \ \mathrm{d}S = 3\mathrm{d}R = 3\,\mathrm{d}x\mathrm{d}y$$

$$\therefore \ \int_S V\,\mathrm{d}\mathbf{S} = \int_S V\hat{\mathbf{n}}\,\mathrm{d}S = \int_S\!\!\int (x + y + z)\frac{1}{3}(2\mathbf{i} + 2\mathbf{j} + \mathbf{k})3\,\mathrm{d}x\,\mathrm{d}y$$

But $z = 2 - 2x - 2y$

$$\therefore \ \int_S V\,\mathrm{d}\mathbf{S} = \int_{x=0}^{1}\int_{y=0}^{1-x} (2 - x - y)(2\mathbf{i} + 2\mathbf{j} + \mathbf{k})\,\mathrm{d}y\,\mathrm{d}x$$

$$= \ldots\ldots\ldots\ldots$$

32

$$\boxed{\frac{2}{3}(2\mathbf{i} + 2\mathbf{j} + \mathbf{k})}$$

Because

$$
\begin{aligned}
\int_S V \, \mathrm{d}S &= \int_0^1 \left[2y - xy - \frac{y^2}{2} \right]_0^{1-x} (2\mathbf{i} + 2\mathbf{j} + \mathbf{k}) \, \mathrm{d}x \\
&= \left[\frac{3}{2}x - x^2 + \frac{x^3}{6} \right]_0^1 (2\mathbf{i} + 2\mathbf{j} + \mathbf{k}) \\
&= \frac{2}{3}(2\mathbf{i} + 2\mathbf{j} + \mathbf{k})
\end{aligned}
$$

33 Vector fields

Example 1

A vector field $\mathbf{F} = y\mathbf{i} + 2\mathbf{j} + \mathbf{k}$ exists over a surface S defined by $x^2 + y^2 + z^2 = 9$ bounded by $x = 0$, $y = 0$, $z = 0$ in the first octant. Evaluate $\displaystyle\int_S \mathbf{F} \cdot \mathrm{d}\mathbf{S}$ over the surface indicated.

$$\mathrm{d}\mathbf{S} = \hat{\mathbf{n}} \, \mathrm{d}S \qquad \text{where} \quad \hat{\mathbf{n}} = \frac{\nabla \phi}{|\nabla \phi|} \quad \text{where} \quad \phi = x^2 + y^2 + z^2 - 9 = 0$$

Now $\nabla\phi = \dfrac{\partial\phi}{\partial x}\mathbf{i} + \dfrac{\partial\phi}{\partial y}\mathbf{j} + \dfrac{\partial\phi}{\partial z}\mathbf{k} = 2x\mathbf{i} + 2y\mathbf{j} + 2z\mathbf{k}$ and

$$|\nabla\phi| = \sqrt{4x^2 + 4y^2 + 4z^2} = 2\sqrt{x^2 + y^2 + z^2} = 2\sqrt{9} = 6$$

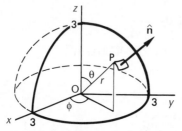

$$\therefore \; \hat{\mathbf{n}} = \frac{1}{6}(2x\mathbf{i} + 2y\mathbf{j} + 2z\mathbf{k})$$

$$= \frac{1}{3}(x\mathbf{i} + y\mathbf{j} + z\mathbf{k})$$

$$
\begin{aligned}
\int_S \mathbf{F} \cdot \mathrm{d}\mathbf{S} = \int_S \mathbf{F} \cdot \hat{\mathbf{n}} \, \mathrm{d}S &= \int_S (y\mathbf{i} + 2\mathbf{j} + \mathbf{k}) \cdot \frac{1}{3}(x\mathbf{i} + y\mathbf{j} + z\mathbf{k}) \, \mathrm{d}S \\
&= \frac{1}{3} \int_S (xy + 2y + z) \, \mathrm{d}S
\end{aligned}
$$

Before integrating over the surface, we convert to spherical polar coordinates.

$$x = \ldots\ldots\ldots\ldots; \qquad y = \ldots\ldots\ldots\ldots$$
$$z = \ldots\ldots\ldots\ldots; \qquad \mathrm{d}S = \ldots\ldots\ldots\ldots$$

34

$$x = 3\sin\theta\cos\phi; \qquad y = 3\sin\theta\sin\phi$$
$$z = 3\cos\theta; \qquad dS = 9\sin\theta\,d\theta\,d\phi$$

Limits of θ and ϕ are $\quad\theta = 0$ to $\dfrac{\pi}{2}$; $\quad\phi = 0$ to $\dfrac{\pi}{2}$.

$$\therefore \int_S \mathbf{F}\cdot d\mathbf{S} = \frac{1}{3}\int_0^{\pi/2}\int_0^{\pi/2}(9\sin^2\theta\sin\phi\cos\phi + 6\sin\theta\sin\phi$$
$$+ 3\cos\theta)\,9\sin\theta\,d\theta\,d\phi$$

$$= 9\int_0^{\pi/2}\int_0^{\pi/2}(3\sin^3\theta\sin\phi\cos\phi + 2\sin^2\theta\sin\phi$$
$$+ \sin\theta\cos\theta)\,d\theta\,d\phi$$

$$= \ldots\ldots\ldots\ldots$$

Complete the integral

35

$$\int_S \mathbf{F}\cdot d\mathbf{S} = 9\left(1 + \frac{3\pi}{4}\right)$$

Because

$$\int_S \mathbf{F}\cdot d\mathbf{S} = 9\int_0^{\pi/2}\left(2\sin\phi\cos\phi + \frac{\pi}{2}\sin\phi + \frac{1}{2}\right)d\phi$$

$$= 9\left[\sin^2\phi - \frac{\pi}{2}\cos\phi - \frac{\phi}{2}\right]_0^{\pi/2} = 9\left(1 + \frac{3\pi}{4}\right)$$

Example 2

Evaluate $\displaystyle\int_S \mathbf{F}\cdot d\mathbf{S}$ where $\mathbf{F} = 2y\mathbf{j} + z\mathbf{k}$ and S is the surface $x^2 + y^2 = 4$ in the first two octants bounded by the planes $z = 0$, $z = 5$ and $y = 0$.

$\phi: x^2 + y^2 - 4 = 0 \qquad \hat{\mathbf{n}} = \dfrac{\nabla\phi}{|\nabla\phi|}$

$\nabla\phi = \dfrac{\partial\phi}{\partial x}\mathbf{i} + \dfrac{\partial\phi}{\partial y}\mathbf{j} + \dfrac{\partial\phi}{\partial z}\mathbf{k} = 2x\mathbf{i} + 2y\mathbf{j}$

$\therefore |\nabla\phi| = \sqrt{4x^2 + 4y^2} = 2\sqrt{x^2 + y^2}$

$\qquad = 2\sqrt{4} = 4$

$\therefore \hat{\mathbf{n}} = \dfrac{\nabla\phi}{|\nabla\phi|} = \dfrac{2x\mathbf{i} + 2y\mathbf{j}}{4} = \dfrac{1}{2}(x\mathbf{i} + y\mathbf{j})$

$\therefore \displaystyle\int_S \mathbf{F}\cdot d\mathbf{S} = \int_S \mathbf{F}\cdot\hat{\mathbf{n}}\,dS = \ldots\ldots\ldots\ldots$

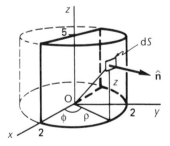

36

$$\int_S y^2 \, dS$$

Because

$$\int_S \mathbf{F} \cdot \hat{\mathbf{n}} \, dS = \int_S (2y\mathbf{j} + z\mathbf{k}) \cdot \frac{1}{2}(x\mathbf{i} + y\mathbf{j}) \, dS$$

$$= \frac{1}{2} \int_S (2y^2) \, dS = \int_S y^2 \, dS$$

This is clearly a case for using cylindrical polar coordinates.

$$x = \ldots\ldots\ldots; \qquad y = \ldots\ldots\ldots$$
$$z = \ldots\ldots\ldots; \qquad dS = \ldots\ldots\ldots$$

37

$$x = 2\cos\phi; \quad y = 2\sin\phi$$
$$z = z; \qquad dS = 2\,d\phi\,dz$$

$$\therefore \int_S \mathbf{F} \cdot d\mathbf{S} = \int_S y^2 \, dS = \int_S \int 4\sin^2\phi\, 2\,d\phi\,dz = 8\int_S \int \sin^2\phi\,d\phi\,dz$$

Limits: $\phi = 0$ to $\phi = \pi$; $z = 0$ to $z = 5$

$$\therefore \int_S \mathbf{F} \cdot d\mathbf{S} = \ldots\ldots\ldots$$

38

$$20\pi$$

Because

$$\int_S \mathbf{F} \cdot d\mathbf{S} = 4 \int_{z=0}^5 \int_{\phi=0}^\pi (1 - \cos 2\phi)\, d\phi\, dz$$

$$= 4 \int_0^5 \left[\phi - \frac{\sin 2\phi}{2}\right]_0^\pi dz$$

$$= 4 \int_0^5 \pi \, dz = 4\pi \left[z\right]_0^5 = 20\pi$$

Example 3

Evaluate $\displaystyle\int_S \mathbf{F} \cdot d\mathbf{S}$ where $\mathbf{F}$ is the field $x^2\mathbf{i} - y\mathbf{j} + 2z\mathbf{k}$ and S is the surface $2x + y + 2z = 2$ bounded by $x = 0$, $y = 0$, $z = 0$ in the first octant.

We can sketch the diagram by putting $x = 0$, $y = 0$, $z = 0$ in turn in the equation for S.

When $x = 0$ $y + 2z = 2$ $z = 1 - \dfrac{y}{2}$

$y = 0$ $x + z = 1$ $z = 1 - x$

$z = 0$ $2x + y = 2$ $y = 2 - 2x$

So the diagram is ....

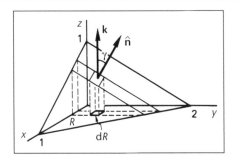

39

$$\mathbf{F} = x^2\mathbf{i} - y\mathbf{j} + 2z\mathbf{k}; \qquad \phi: \quad 2x + y + 2z - 2 = 0$$

$$\nabla\phi = \frac{\partial\phi}{\partial x}\mathbf{i} + \frac{\partial\phi}{\partial y}\mathbf{j} + \frac{\partial\phi}{\partial z}\mathbf{k} = 2\mathbf{i} + \mathbf{j} + 2\mathbf{k} \qquad |\nabla\phi| = 3$$

$$\int_S \mathbf{F} \cdot d\mathbf{S} = \int_S \mathbf{F} \cdot \hat{\mathbf{n}}\, dS$$

$$= \ldots\ldots\ldots\ldots \text{ (next stage)}$$

40

$$\boxed{\frac{1}{3}\int_S (2x^2 - y + 4z)\, dS}$$

Because

$$\int_S \mathbf{F} \cdot \hat{\mathbf{n}}\, dS = \int_S (x^2\mathbf{i} - y\mathbf{j} + 2z\mathbf{k}) \cdot \frac{1}{3}(2\mathbf{i} + \mathbf{j} + 2\mathbf{k})\, dS$$

$$= \frac{1}{3}\int_S (2x^2 - y + 4z)\, dS$$

If we now project the element of surface dS onto the x–y plane

$$dR = dS\cos\gamma \quad \cos\gamma = \hat{\mathbf{n}} \cdot \mathbf{k} \quad \therefore\ dR = \hat{\mathbf{n}} \cdot \mathbf{k}\, dS \quad \therefore\ dS = \frac{dx\, dy}{\hat{\mathbf{n}} \cdot \mathbf{k}}$$

$$\therefore\ \hat{\mathbf{n}} \cdot \mathbf{k} = \frac{1}{3}(2\mathbf{i} + \mathbf{j} + 2\mathbf{k}) \cdot (\mathbf{k}) = \frac{2}{3} \quad \therefore\ dS = \frac{3}{2}\, dx\, dy$$

Using these new relationships, $\displaystyle\int_S \mathbf{F} \cdot d\mathbf{S} = \int_S \mathbf{F} \cdot \hat{\mathbf{n}}\, dS$

$$= \ldots\ldots\ldots\ldots$$

41

$$\boxed{\int_R \int \frac{1}{2}(2x^2 - y + 4z)\,\mathrm{d}x\,\mathrm{d}y}$$

Because

$$\int_S \mathbf{F} \cdot \hat{\mathbf{n}}\,\mathrm{d}S = \frac{1}{3}\int_S (2x^2 - y + 4z)\,\mathrm{d}S$$

$$= \frac{1}{3}\int_R \int (2x^2 - y + 4z)\frac{3}{2}\,\mathrm{d}x\,\mathrm{d}y$$

$$= \frac{1}{2}\int_R \int (2x^2 - y + 4z)\,\mathrm{d}x\,\mathrm{d}y$$

Limits: $y = 0$ to $y = 2 - 2x$; $x = 0$ to $x = 1$

$$\therefore \int_S \mathbf{F} \cdot \hat{\mathbf{n}}\,\mathrm{d}S = \frac{1}{2}\int_0^1 \int_0^{2-2x} (2x^2 - y + 4z)\,\mathrm{d}y\,\mathrm{d}x$$

But $2x + y + 2z = 2$ $\therefore z = \frac{1}{2}(2 - 2x - y)$

$$\therefore \int_S \mathbf{F} \cdot \hat{\mathbf{n}}\,\mathrm{d}S = \dots\dots\dots$$

Complete the integration

42

$$\boxed{\dfrac{1}{2}}$$

Here is the rest of the working.

$$\int_S \mathbf{F} \cdot \mathrm{d}\mathbf{S} = \int_S \mathbf{F} \cdot \hat{\mathbf{n}}\,\mathrm{d}S = \frac{1}{2}\int_0^1 \int_0^{2-2x} (2x^2 - y + 4 - 4x - 2y)\,\mathrm{d}y\,\mathrm{d}x$$

$$= \frac{1}{2}\int_0^1 \int_0^{2-2x} (2x^2 - 4x + 4 - 3y)\,\mathrm{d}y\,\mathrm{d}x$$

$$= \frac{1}{2}\int_0^1 \left[(2x^2 - 4x + 4)y - \frac{3y^2}{2}\right]_0^{2-2x}\,\mathrm{d}x$$

$$= \frac{1}{2}\int_0^1 (4x^2 - 8x + 8 - 4x^3 + 8x^2 - 8x - 6 + 12x - 6x^2)\,\mathrm{d}x$$

$$= \frac{1}{2}\int_0^1 (6x^2 - 4x^3 - 4x + 2)\,\mathrm{d}x = \int_0^1 (3x^2 - 2x^3 - 2x + 1)\,\mathrm{d}x$$

$$= \left[x^3 - \frac{x^4}{2} - x^2 + x\right]_0^1 = \frac{1}{2}$$

While we are concerned with vector fields, let us move on to a further point of interest.

Conservative vector fields

In general, the value of the line integral $\int_c \mathbf{F} \cdot d\mathbf{r}$ between two stated **43** points A and B depends on the particular path of integration followed.

If, however, the line integral between A and B is independent of the path of integration between the two end points, then the vector field **F** is said to be *conservative*.

It follows that, for a closed path in a conservative field, $\oint_c \mathbf{F} \cdot d\mathbf{r} = 0$.

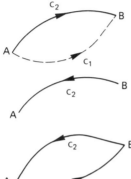

Because, if the field is conservative

$$\int_{c_1(AB)} \mathbf{F} \cdot d\mathbf{r} = \int_{c_2(AB)} \mathbf{F} \cdot d\mathbf{r}$$

But $\int_{c_2(BA)} \mathbf{F} \cdot d\mathbf{r} = -\int_{c_2(AB)} \mathbf{F} \cdot d\mathbf{r}$

Hence, for the closed path $\mathbf{AB}_{c_1} + \mathbf{BA}_{c_2}$

$$\oint \mathbf{F} \cdot d\mathbf{r} = \int_{c_1(AB)} \mathbf{F} \cdot d\mathbf{r} + \int_{c_2(BA)} \mathbf{F} \cdot d\mathbf{r}$$

$$= \int_{c_1(AB)} \mathbf{F} \cdot d\mathbf{r} - \int_{c_2(AB)} \mathbf{F} \cdot d\mathbf{r}$$

$$= \int_{c_1(AB)} \mathbf{F} \cdot d\mathbf{r} - \int_{c_1(AB)} \mathbf{F} \cdot d\mathbf{r} = 0$$

$$\therefore \quad \oint \mathbf{F} \cdot d\mathbf{r} = 0$$

Note that this result holds good only for a closed curve and when the vector field is a conservative field.

Now for an example.

Example

If $\mathbf{F} = 2xyz\mathbf{i} + x^2 z\mathbf{j} + x^2 y\mathbf{k}$, evaluate the line integral $\int \mathbf{F} \cdot d\mathbf{r}$ between A $(0, 0, 0)$ and B $(2, 4, 6)$

(a) along the curve c whose parametric equations are $x = u$, $y = u^2$, $z = 3u$

(b) along the three straight lines c_1: $(0, 0, 0)$ to $(2, 0, 0)$; c_2: $(2, 0, 0)$ to $(2, 4, 0)$; c_3: $(2, 4, 0)$ to $(2, 4, 6)$.

Hence determine whether or not **F** is a conservative field.

First draw the diagram

. . . .

44

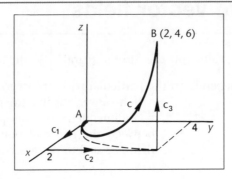

(a) $\mathbf{F} = 2xyz\mathbf{i} + x^2z\mathbf{j} + x^2y\mathbf{k}$

$$x = u; \qquad y = u^2; \qquad z = 3u$$

$$\therefore \ dx = du; \qquad dy = 2u\,du; \qquad dz = 3\,du.$$

$$\mathbf{F} \cdot d\mathbf{r} = (2xyz\mathbf{i} + x^2z\mathbf{j} + x^2y\mathbf{k}) \cdot (\mathbf{i}\,dx + \mathbf{j}\,dy + \mathbf{k}\,dz)$$

$$= 2xyz\,dx + x^2z\,dy + x^2y\,dz$$

Using the transformations shown above, we can now express $\mathbf{F} \cdot d\mathbf{r}$ in terms of u.

$$\mathbf{F} \cdot d\mathbf{r} = \ldots\ldots\ldots\ldots$$

45

$$\boxed{15u^4\,du}$$

Because

$$2xyz\,dx = (2u)(u^2)(3u)\,du = 6u^4\,du$$

$$x^2z\,dy = (u^2)(3u)(2u)\,du = 6u^4\,du$$

$$x^2y\,dz = (u^2)(u^2)3\,du \quad = 3u^4\,du$$

$$\therefore \ \mathbf{F} \cdot d\mathbf{r} = 6u^4\,du + 6u^4\,du + 3u^4\,du = 15u^4\,du$$

The limits of integration in u are

$$\ldots\ldots\ldots\ldots$$

46

$$\boxed{u = 0 \ \text{to} \ u = 2}$$

$$\therefore \ \int_c \mathbf{F} \cdot d\mathbf{r} = \int_0^2 15u^4\,du = \left[3u^5\right]_0^2 = 96 \qquad \int_c \mathbf{F} \cdot d\mathbf{r} = 96$$

(b) The diagram for (b) is as shown. We consider each straight line section in turn.

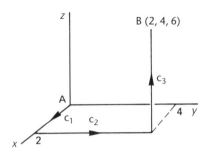

$$\int \mathbf{F} \cdot d\mathbf{r} = \int (2xyz\, dx + x^2 z\, dy + x^2 y\, dz)$$

c_1: $(0,0,0)$ to $(2,0,0)$; $\quad y = 0$, $z = 0$, $dy = 0$, $dz = 0$

$$\therefore \quad \int_{c_1} \mathbf{F} \cdot d\mathbf{r} = 0 + 0 + 0 = 0$$

In the same way, we evaluate the line integral along c_2 and c_3.

$$\int_{c_2} \mathbf{F} \cdot d\mathbf{r} = \ldots\ldots\ldots ; \qquad \int_{c_3} \mathbf{F} \cdot d\mathbf{r} = \ldots\ldots\ldots$$

47

$$\boxed{\int_{c_2} \mathbf{F} \cdot d\mathbf{r} = 0; \qquad \int_{c_3} \mathbf{F} \cdot d\mathbf{r} = 96}$$

Because we have $\displaystyle\int \mathbf{F} \cdot d\mathbf{r} = \int (2xyz\, dx + x^2 z\, dy + x^2 y\, dz)$

c_2: $(2,0,0)$ to $(2,4,0)$; $\quad x = 2$, $\quad z = 0$, $\quad dx = 0$, $\quad dz = 0$

$$\therefore \quad \int_{c_2} \mathbf{F} \cdot d\mathbf{r} = 0 + 0 + 0 = 0$$

$$\int_{c_2} \mathbf{F} \cdot d\mathbf{r} = 0$$

c_3: $(2,4,0)$ to $(2,4,6)$; $\quad x = 2$, $\quad y = 4$, $\quad dx = 0$, $\quad dy = 0$

$$\therefore \quad \int_{c_3} \mathbf{F} \cdot d\mathbf{r} = 0 + 0 + \int_0^6 16\, dz = \Big[16z\Big]_0^6 = 96$$

$$\int_{c_3} \mathbf{F} \cdot d\mathbf{r} = 96$$

Collecting the three results together

$$\int_{c_1 + c_2 + c_3} \mathbf{F} \cdot d\mathbf{r} = 0 + 0 + 96 \qquad \therefore \quad \int_{c_1 + c_2 + c_3} \mathbf{F} \cdot d\mathbf{r} = 96$$

▶

In this particular example, the value of the line integral is independent of the two paths we have used joining the same two end points and indicates that **F** may be a conservative field. It follows that

$$\int_c \mathbf{F} \cdot d\mathbf{r} - \int_{c_1 + c_2 + c_3} \mathbf{F} \cdot d\mathbf{r} = 0 \quad \text{i.e.} \quad \oint \mathbf{F} \cdot d\mathbf{r} = 0$$

So, if **F** is a conservative field, $\oint \mathbf{F} \cdot d\mathbf{r} = 0$

Make a note of this for future use

48

Two tests can be applied to establish that a given vector field is conservative.

If **F** is a conservative field

(a) curl **F** = 0

(b) **F** can be expressed as grad V where V is a scalar field to be determined.

For example, in the work we have just completed, we showed that $\mathbf{F} = 2xyz\mathbf{i} + x^2z\mathbf{j} + x^2y\mathbf{k}$ is a conservative field.

(a) If we determine curl **F** in this case, we have

$$\text{curl } \mathbf{F} = \dots\dots\dots\dots$$

49

$$\boxed{\text{curl } \mathbf{F} = 0}$$

Because

$$\text{curl } \mathbf{F} = \begin{vmatrix} \mathbf{i} & \mathbf{j} & \mathbf{k} \\ \dfrac{\partial}{\partial x} & \dfrac{\partial}{\partial y} & \dfrac{\partial}{\partial z} \\ 2xyz & x^2z & x^2y \end{vmatrix}$$

$$= (x^2 - x^2)\mathbf{i} - (2xy - 2xy)\mathbf{j} + (2xz - 2xz)\mathbf{k} = 0$$

$$\therefore \quad \text{curl } \mathbf{F} = 0$$

(b) We can attempt to express **F** as grad V where V is a scalar in x, y, z. If $V = f(x, y, z)$

$$\text{grad } V = \frac{\partial V}{\partial x}\mathbf{i} + \frac{\partial V}{\partial y}\mathbf{j} + \frac{\partial V}{\partial z}\mathbf{k}$$

and we have $\mathbf{F} = 2xyz\mathbf{i} + x^2z\mathbf{j} + x^2y\mathbf{k}$

$$\therefore \quad \frac{\partial V}{\partial x} = 2xyz \qquad \therefore \quad V = x^2yz + f(y, z)$$

$$\frac{\partial V}{\partial y} = x^2z \qquad \therefore \quad V = \dots\dots\dots\dots$$

$$\frac{\partial V}{\partial z} = x^2y \qquad \therefore \quad V = \dots\dots\dots\dots$$

We therefore have to find a scalar function V that satisfies the three requirements.

$$V = \dots\dots\dots\dots$$

$$\boxed{V = x^2yz}$$

50

Because

$$\frac{\partial V}{\partial x} = 2xyz \qquad \therefore \; V = x^2yz + f(y, z)$$

$$\frac{\partial V}{\partial y} = x^2z \qquad \therefore \; V = x^2yz + g(x, z)$$

$$\frac{\partial V}{\partial z} = x^2y \qquad \therefore \; V = x^2yz + h(x, y)$$

These three are satisfied if $f(y, z) = g(z, x) = h(x, y) = 0$

$$\therefore \; \mathbf{F} = \text{grad } V \; \text{ where } V = x^2yz$$

So two tests can be applied to determine whether or not a vector field is conservative. They are

(a) ....
(b) ....

$$\boxed{\begin{aligned} &\text{(a) } \text{curl } \mathbf{F} = 0 \\ &\text{(b) } \mathbf{F} = \text{grad } V \end{aligned}}$$

51

Any one of these conditions can be applied as is convenient.

Now what about these?

Exercise

Determine which of the following vector fields are conservative.

(a) $\mathbf{F} = (x + y)\mathbf{i} + (y - z)\mathbf{j} + (x + y + z)\mathbf{k}$
(b) $\mathbf{F} = (2xz + y)\mathbf{i} + (z + x)\mathbf{j} + (x^2 + y)\mathbf{k}$
(c) $\mathbf{F} = y\sin z\,\mathbf{i} + x\sin z\,\mathbf{j} + (xy\cos z + 2z)\mathbf{k}$
(d) $\mathbf{F} = 2xy\mathbf{i} + (x^2 + 4yz)\mathbf{j} + 2y^2z\mathbf{k}$
(e) $\mathbf{F} = y\cos x\cos z\,\mathbf{i} + \sin x\cos z\,\mathbf{j} - y\sin x\sin z\,\mathbf{k}$.

Complete all five and check your findings with the next frame.

$$\boxed{\text{(a) No } \quad \text{(b) Yes } \quad \text{(c) Yes } \quad \text{(d) No } \quad \text{(e) Yes}}$$

52

▶

Divergence theorem
(Gauss' theorem)

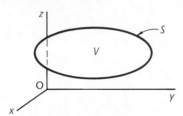

For a closed surface S, enclosing a region V in a vector field $\mathbf{F}$,

$$\int_V \operatorname{div} \mathbf{F} \, dV = \int_S \mathbf{F} \cdot d\mathbf{S}$$

In general, this means that the volume integral (triple integral) on the left-hand side can be expressed as a surface integral (double integral) on the right-hand side. Let us work through one or two examples.

Example 1

Verify the divergence theorem for the vector field $\mathbf{F} = x^2\mathbf{i} + z\mathbf{j} + y\mathbf{k}$ taken over the region bounded by the planes $z = 0$, $z = 2$, $x = 0$, $x = 1$, $y = 0$, $y = 3$.

Start off, as always, by sketching the relevant diagram, which is

. . . .

53

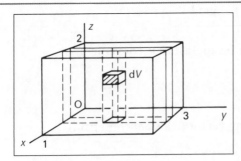

$dV = dx \, dy \, dz$

We have to show that

$$\int_V \operatorname{div} \mathbf{F} \, dV = \int_S \mathbf{F} \cdot d\mathbf{S}$$

(a) To find $\displaystyle\int_V \operatorname{div} \mathbf{F} \, dV$

$$\operatorname{div} \mathbf{F} = \nabla \cdot \mathbf{F} = \left(\frac{\partial}{\partial x}\mathbf{i} + \frac{\partial}{\partial y}\mathbf{j} \, \frac{\partial}{\partial z}\mathbf{k} \right) \cdot (x^2\mathbf{i} + z\mathbf{j} + y\mathbf{k})$$

$$= \frac{\partial}{\partial x}(x^2) + \frac{\partial}{\partial y}(z) + \frac{\partial}{\partial z}(y) = 2x + 0 + 0 = 2x$$

$$\therefore \int_V \operatorname{div} \mathbf{F} \, dV = \int_V 2x \, dV = \iiint_V 2x \, dz \, dy \, dx$$

Inserting the limits and completing the integration

$$\int_V \operatorname{div} \mathbf{F} \, dV = \dots\dots\dots\dots$$

54

$$\int_V \operatorname{div} \mathbf{F} \, dV = 6$$

Because

$$\int_V \operatorname{div} \mathbf{F} \, dV = \int_0^1 \int_0^3 \int_0^2 2x \, dz \, dy \, dx = \int_0^1 \int_0^3 \left[2xz \right]_0^2 \, dy \, dx$$

$$= \int_0^1 \left[4xy \right]_0^3 \, dx = \int_0^1 12x \, dx = \left[6x^2 \right]_0^1 = 6$$

Now we have to find $\displaystyle\int_S \mathbf{F} \cdot d\mathbf{S}$

(b) To find $\displaystyle\int_S \mathbf{F} \cdot d\mathbf{S}$ i.e. $\displaystyle\int_S \mathbf{F} \cdot \hat{\mathbf{n}} \, dS$

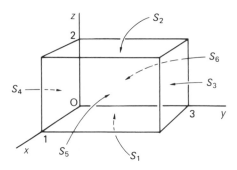

The enclosing surface S consists of six separate plane faces denoted as $S_1, S_2, \ldots, S_6$ as shown. We consider each face in turn.

$\mathbf{F} = x^2 \mathbf{i} + z \mathbf{j} + y \mathbf{k}$

(1) S_1 (base): $z = 0$; $\hat{\mathbf{n}} = -\mathbf{k}$ (outwards and downwards)

$$\therefore \ \mathbf{F} = x^2 \mathbf{i} + y \mathbf{k} \qquad dS_1 = dx \, dy$$

$$\therefore \ \int_{S_1} \mathbf{F} \cdot \hat{\mathbf{n}} \, dS = \iint_{S_1} (x^2 \mathbf{i} + y \mathbf{k}) \cdot (-\mathbf{k}) \, dy \, dx$$

$$= \int_0^1 \int_0^3 (-y) \, dy \, dx$$

$$= \int_0^1 \left[-\frac{y^2}{2} \right]_0^3 \, dx$$

$$= -\frac{9}{2}$$

(2) S_2 (top): $z = 2$; $\hat{\mathbf{n}} = \mathbf{k}$ $\qquad dS_2 = dx \, dy$

$$\therefore \ \int_{S_2} \mathbf{F} \cdot \hat{\mathbf{n}} \, dS = \ldots \ldots \ldots$$

55

$$\boxed{\dfrac{9}{2}}$$

Because

$$\int_{S_2} \mathbf{F} \cdot \hat{\mathbf{n}}\, dS = \iint_{S_2} (x^2\mathbf{i} + 2\mathbf{j} + y\mathbf{k}) \cdot (\mathbf{k})\, dy\, dx$$

$$= \int_0^1 \int_0^3 y\, dy\, dx = \frac{9}{2}$$

So we go on.

(3) S_3 (right-hand end):　$y = 3$;　$\hat{\mathbf{n}} = \mathbf{j}$　$dS_3 = dx\, dz$

$$\mathbf{F} = x^2\mathbf{i} + z\mathbf{j} + y\mathbf{k}$$

$$\therefore \int_{S_3} \mathbf{F} \cdot \hat{\mathbf{n}}\, dS = \iint_{S_3} (x^2\mathbf{i} + z\mathbf{j} + 3\mathbf{k}) \cdot (\mathbf{j})\, dz\, dx$$

$$= \int_0^1 \int_0^2 z\, dz\, dx$$

$$= \int_0^1 \left[\frac{z^2}{2}\right]_0^2 dx = \int_0^1 2\, dx = 2$$

(4) S_4 (left-hand end):　$y = 0$;　$\hat{\mathbf{n}} = -\mathbf{j}$　$dS_4 = dx\, dz$

$$\therefore \int_{S_4} \mathbf{F} \cdot \hat{\mathbf{n}}\, dS = \ldots\ldots\ldots\ldots$$

56

$$\boxed{-2}$$

Because

$$\int_{S_4} \mathbf{F} \cdot \hat{\mathbf{n}}\, dS = \iint_{S_4} (x^2\mathbf{i} + z\mathbf{j} + y\mathbf{k}) \cdot (-\mathbf{j})\, dz\, dx = \int_0^1 \int_0^2 (-z)\, dz\, dx$$

$$= \int_0^1 \left[-\frac{z^2}{2}\right]_0^2 dx = \int_0^1 (-2)\, dx = -2$$

Now for the remaining two sides S_5 and S_6.

Evaluate these in the same manner, obtaining

$$\int_{S_5} \mathbf{F} \cdot \hat{\mathbf{n}}\, dS = \ldots\ldots\ldots\ldots$$

$$\int_{S_6} \mathbf{F} \cdot \hat{\mathbf{n}}\, dS = \ldots\ldots\ldots\ldots$$

$$\int_{S_5} \mathbf{F} \cdot \hat{\mathbf{n}} \, dS = 6; \qquad \int_{S_6} \mathbf{F} \cdot \hat{\mathbf{n}} \, dS = 0$$

Check:

(5) S_5 (front): $\quad x = 1; \qquad \hat{\mathbf{n}} = \mathbf{i} \qquad dS_5 = dy \, dz$

$$\therefore \int_{S_5} \mathbf{F} \cdot \hat{\mathbf{n}} \, dS = \iint_{S_5} (\mathbf{i} + z\mathbf{j} + y\mathbf{k}) \cdot (\mathbf{i}) \, dy \, dz = \iint_{S_5} 1 \, dy \, dz = 6$$

(6) S_6 (back): $\quad x = 0; \qquad \hat{\mathbf{n}} = -\mathbf{i} \qquad dS_6 = dy \, dz$

$$\therefore \int_{S_6} \mathbf{F} \cdot \hat{\mathbf{n}} \, dS = \iint_{S_6} (z\mathbf{j} + y\mathbf{k}) \cdot (-\mathbf{i}) \, dy \, dz = \iint_{S_6} 0 \, dy \, dz = 0$$

Now on to the next frame where we will collect our results together

58

For the whole surface S we therefore have

$$\int_S \mathbf{F} \cdot dS = -\frac{9}{2} + \frac{9}{2} + 2 - 2 + 6 + 0 = 6$$

and from our previous work in section (a) $\displaystyle\int_V \operatorname{div} \mathbf{F} \, dV = 6$

We have therefore verified as required that, in this example

$$\int_V \operatorname{div} \mathbf{F} \, dV = \int_S \mathbf{F} \cdot dS$$

We have made rather a meal of this since we have set out the working in detail. In practice, the actual writing can often be considerably simplified. Let us move on to another example.

Example 2

Verify the Gauss divergence theorem for the vector field $\mathbf{F} = x\mathbf{i} + 2\mathbf{j} + z^2\mathbf{k}$ taken over the region bounded by the planes $z = 0$, $z = 4$, $x = 0$, $y = 0$ and the surface $x^2 + y^2 = 4$ in the first octant.

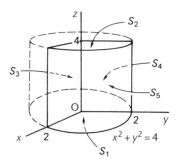

Divergence theorem

$$\int_V \operatorname{div} \mathbf{F} \, dV = \int_S \mathbf{F} \cdot dS$$

S consists of five surfaces $S_1, S_2, \ldots, S_5$ as shown.

(a) $\operatorname{div} \mathbf{F} = \nabla \cdot \mathbf{F} = \left(\dfrac{\partial}{\partial x}\mathbf{i} + \dfrac{\partial}{\partial y}\mathbf{j} + \dfrac{\partial}{\partial z}\mathbf{k} \right) \cdot (x\mathbf{i} + 2\mathbf{j} + z^2\mathbf{k})$

$\qquad = \ldots\ldots\ldots$

59

$$1 + 2z$$

$$\therefore \int_V \text{div } \mathbf{F} \, dV = \int_V \nabla \cdot \mathbf{F} \, dV = \iiint_V (1 + 2z) \, dx \, dy \, dz$$

Changing to cylindrical polar coordinates (ρ, ϕ, z)

$$x = \rho \cos \phi \qquad y = \rho \sin \phi \qquad z = z \qquad dV = \rho \, d\rho \, d\phi \, dz$$

Transforming the variables and inserting the appropriate limits, we then have

$$\int_V \text{div } \mathbf{F} \, dV = \ldots\ldots\ldots$$

Finish it

60

$$20\pi$$

Because

$$\int_V \text{div } \mathbf{F} \, dV = \int_0^{\pi/2} \int_0^2 \int_0^4 (1 + 2z) \, dz \, \rho \, d\rho \, d\phi$$

$$= \int_0^{\pi/2} \int_0^2 \left[z + z^2 \right]_0^4 \rho \, d\rho \, d\phi = \int_0^{\pi/2} \int_0^2 20\rho \, d\rho \, d\phi$$

$$= \int_0^{\pi/2} \left[10\rho^2 \right]_0^2 d\phi = \int_0^{\pi/2} 40 \, d\phi = 20\pi \qquad\qquad (1)$$

(b) Now we evaluate $\displaystyle\int_S \mathbf{F} \cdot dS$ over the closed surface.

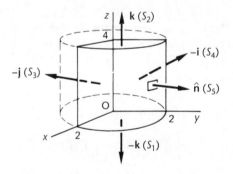

The unit normal vector for each surface is shown.

$$\mathbf{F} = x\mathbf{i} + 2\mathbf{j} + z^2\mathbf{k}$$

(1) S_1: $z = 0$; $\hat{\mathbf{n}} = -\mathbf{k}$ $\mathbf{F} = x\mathbf{i} + 2\mathbf{j}$

$$\therefore \int_{S_1} \mathbf{F} \cdot \hat{\mathbf{n}} \, dS = \int_{S_1} (x\mathbf{i} + 2\mathbf{j}) \cdot (-\mathbf{k}) \, dS = 0$$

▶

(2) S_2: $z = 4$; $\quad \hat{\mathbf{n}} = \mathbf{k}$ $\quad \mathbf{F} = x\mathbf{i} + 2\mathbf{j} + 16\mathbf{k}$

$$\therefore \int_{S_2} \mathbf{F} \cdot \hat{\mathbf{n}} \, dS = \int_{S_2} (x\mathbf{i} + 2\mathbf{j} + 16\mathbf{k}) \cdot (\mathbf{k}) \, dS = \int_{S_2} 16 \, dS$$

$$= 16\left(\frac{\pi 4}{4}\right) = 16\pi$$

In the same way for S_3: $\quad \displaystyle\int_{S_3} \mathbf{F} \cdot \hat{\mathbf{n}} \, dS = \ldots\ldots\ldots$

and for S_4: $\quad \displaystyle\int_{S_4} \mathbf{F} \cdot \hat{\mathbf{n}} \, dS = \ldots\ldots\ldots$

$$\int_{S_3} \mathbf{F} \cdot \hat{\mathbf{n}} \, dS = -16; \qquad \int_{S_4} \mathbf{F} \cdot \hat{\mathbf{n}} \, dS = 0$$	**61**

Because we have

(3) S_3: $y = 0$; $\quad \hat{\mathbf{n}} = -\mathbf{j}$ $\quad \mathbf{F} = x\mathbf{i} + 2\mathbf{j} + z^2\mathbf{k}$

$$\therefore \int_{S_3} \mathbf{F} \cdot \hat{\mathbf{n}} \, dS = \int_{S_3} (x\mathbf{i} + 2\mathbf{j} + z^2\mathbf{k}) \cdot (-\mathbf{j}) \, dS$$

$$= \int_{S_3} (-2) \, dS = -2(8) = -16$$

(4) S_4: $x = 0$; $\quad \hat{\mathbf{n}} = -\mathbf{i}$ $\quad \mathbf{F} = 2\mathbf{j} + z^2\mathbf{k}$

$$\therefore \int_{S_4} \mathbf{F} \cdot \hat{\mathbf{n}} \, dS = \int_{S_4} (2\mathbf{j} + z^2\mathbf{k}) \cdot (-\mathbf{i}) \, dS = 0$$

Finally we have

(5) S_5: $x^2 + y^2 - 4 = 0$ $\quad \hat{\mathbf{n}} = \ldots\ldots\ldots$

$$\hat{\mathbf{n}} = \frac{1}{2}(x\mathbf{i} + y\mathbf{j})$$	**62**

Because

$$x^2 + y^2 - 4 = 0 \qquad \hat{\mathbf{n}} = \frac{\nabla S}{|\nabla S|} = \frac{2x\mathbf{i} + 2y\mathbf{j}}{\sqrt{4x^2 + 4y^2}} = \frac{x\mathbf{i} + y\mathbf{j}}{2}$$

$$\therefore \int_{S_5} \mathbf{F} \cdot \hat{\mathbf{n}} \, dS = \int_{S_5} (x\mathbf{i} + 2\mathbf{j} + z^2\mathbf{k}) \cdot \left(\frac{x\mathbf{i} + y\mathbf{j}}{2}\right) \, dS = \frac{1}{2}\int_{S_5} (x^2 + 2y) \, dS$$

Converting to cylindrical polar coordinates, this gives

$$\int_{S_5} \mathbf{F} \cdot \hat{\mathbf{n}} \, dS = \ldots\ldots\ldots$$

63

$$4\pi + 16$$

Because we have

$$\int_{S_5} \mathbf{F} \cdot \hat{\mathbf{n}} \, dS = \frac{1}{2} \int_{S_5} (x^2 + 2y) \, dS$$

also $x = 2\cos\phi;$ $y = 2\sin\phi$

$z = z;$ $dS = 2 \, d\phi \, dz$

$$\therefore \int_{S_5} \mathbf{F} \cdot \hat{\mathbf{n}} \, dS = \frac{1}{2} \int_0^4 \int_0^{\pi/2} (4\cos^2\phi + 4\sin\phi) \, 2 \, d\phi \, dz$$

$$= 2 \int_0^4 \int_0^{\pi/2} \{(1 + \cos 2\phi) + 2\sin\phi\} \, d\phi \, dz$$

$$= 2 \int_0^4 \left[\left(\phi - \frac{\sin 2\phi}{2}\right) - 2\cos\phi \right]_0^{\pi/2} \, dz$$

$$= 2 \int_0^4 \left(\frac{\pi}{2} + 2\right) \, dz = 4\pi + 16$$

Therefore, for the total surface S

$$\int_S \mathbf{F} \cdot \hat{\mathbf{n}} \, dS = 0 + 16\pi - 16 + 0 + 4\pi + 16 = 20\pi \tag{2}$$

$$\therefore \int_V \operatorname{div} \mathbf{F} \, dV = \int_S \mathbf{F} \cdot d\mathbf{S} = 20\pi$$

Other examples are worked in much the same way. You will remember that, for a closed surface, the normal vectors at all points are drawn in an *outward* direction.

Now we move on to a further important theorem.

Stokes' theorem

64

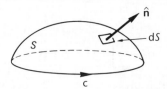

If $\mathbf{F}$ is a vector field existing over an open surface S and around its boundary, closed curve c, then

$$\int_S \operatorname{curl} \mathbf{F} \cdot d\mathbf{S} = \oint_c \mathbf{F} \cdot d\mathbf{r}$$

This means that we can express a surface integral in terms of a line integral round the boundary curve.

The proof of this theorem is rather lengthy and is to be found in the Appendix. Let us demonstrate its application in the following examples.

Example 1

A hemisphere S is defined by $x^2 + y^2 + z^2 = 4 \ (z \geq 0)$. A vector field $\mathbf{F} = 2y\mathbf{i} - x\mathbf{j} + xz\mathbf{k}$ exists over the surface and around its boundary c.

Verify Stokes' theorem, that $\displaystyle\int_S \text{curl } \mathbf{F} \cdot d\mathbf{S} = \oint_c \mathbf{F} \cdot d\mathbf{r}$.

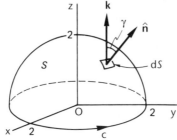

$S: \quad x^2 + y^2 + z^2 - 4 = 0$

$\mathbf{F} = 2y\mathbf{i} - x\mathbf{j} + xz\mathbf{k}$

c is the circle $x^2 + y^2 = 4$.

(a) $\displaystyle\oint_c \mathbf{F} \cdot d\mathbf{r} = \int_c (2y\mathbf{i} - x\mathbf{j} + xz\mathbf{k}) \cdot (\mathbf{i} \, dx + \mathbf{j} \, dy + \mathbf{k} \, dz)$

$\displaystyle \qquad = \int_c (2y \, dx - x \, dy + xz \, dz)$

Converting to polar coordinates

$x = 2\cos\theta;$ $\qquad y = 2\sin\theta;$ $\qquad z = 0$

$dx = -2\sin\theta \, d\theta;$ $\qquad dy = 2\cos\theta \, d\theta;$ $\qquad$ Limits $\theta = 0$ to 2π

Making the substitutions and completing the integral

$$\oint_c \mathbf{F} \cdot d\mathbf{r} = \ldots\ldots\ldots\ldots$$

65

$$\boxed{\oint_c \mathbf{F} \cdot d\mathbf{r} = -12\pi}$$

Because

$\displaystyle\oint_c \mathbf{F} \cdot d\mathbf{r} = \int_0^{2\pi} (4\sin\theta[-2\sin\theta \, d\theta] - 2\cos\theta \, 2\cos\theta \, d\theta)$

$\displaystyle \qquad = -4\int_0^{2\pi} (2\sin^2\theta + \cos^2\theta) \, d\theta$

$\displaystyle \qquad = -4\int_0^{2\pi} (1 + \sin^2\theta) \, d\theta = -2\int_0^{2\pi} (3 - \cos 2\theta) \, d\theta$

$\displaystyle \qquad = -2\left[3\theta - \frac{\sin 2\theta}{2}\right]_0^{2\pi} = -12\pi \qquad\qquad (1)$

On to the next frame

66

(b) Now we determine $\int_S \text{curl } \mathbf{F} \cdot d\mathbf{S}$

$$\int \text{curl } \mathbf{F} \cdot d\mathbf{S} = \int \text{curl } \mathbf{F} \cdot \hat{\mathbf{n}} \, dS \qquad \mathbf{F} = 2y\mathbf{i} - x\mathbf{j} + xz\mathbf{k}$$

$$\therefore \text{ curl } \mathbf{F} = \dots\dots\dots$$

67

$$\boxed{\text{curl } \mathbf{F} = -z\mathbf{j} - 3\mathbf{k}}$$

Because

$$\text{curl } \mathbf{F} = \begin{vmatrix} \mathbf{i} & \mathbf{j} & \mathbf{k} \\ \dfrac{\partial}{\partial x} & \dfrac{\partial}{\partial y} & \dfrac{\partial}{\partial z} \\ 2y & -x & xz \end{vmatrix} = \mathbf{i}(0 - 0) - \mathbf{j}(z - 0) + \mathbf{k}(-1 - 2) = -z\mathbf{j} - 3\mathbf{k}$$

Now $\hat{\mathbf{n}} = \dfrac{\nabla S}{|\nabla S|} = \dfrac{2x\mathbf{i} + 2y\mathbf{j} + 2z\mathbf{k}}{\sqrt{4x^2 + 4y^2 + 4z^2}} = \dfrac{x\mathbf{i} + y\mathbf{j} + z\mathbf{k}}{2}$

Then $\displaystyle\int_S \text{curl } \mathbf{F} \cdot \hat{\mathbf{n}} \, dS = \int_S (-z\mathbf{j} - 3\mathbf{k}) \cdot \left(\dfrac{x\mathbf{i} + y\mathbf{j} + z\mathbf{k}}{2}\right) dS$

$$= \dfrac{1}{2} \int_S (-yz - 3z) \, dS$$

Expressing this in spherical polar coordinates and integrating, we get

$$\int_S \text{curl } \mathbf{F} \cdot \hat{\mathbf{n}} \, dS = \dots\dots\dots$$

68

$$\boxed{-12\pi}$$

Because

$$x = 2\sin\theta \cos\phi; \quad y = 2\sin\theta \sin\phi; \quad z = 2\cos\theta; \quad dS = 4\sin\theta \, d\theta \, d\phi$$

$$\therefore \int_S \text{curl } \mathbf{F} \cdot \hat{\mathbf{n}} \, dS = \dfrac{1}{2} \int_S (-2\sin\theta \sin\phi \, 2\cos\theta - 6\cos\theta)4\sin\theta \, d\theta \, d\phi$$

$$= -4 \int_0^{2\pi} \int_0^{\pi/2} (2\sin^2\theta \cos\theta \sin\phi + 3\sin\theta \cos\theta) \, d\theta \, d\phi$$

$$= -4 \int_0^{2\pi} \left[\dfrac{2\sin^3\theta \sin\phi}{3} + \dfrac{3\sin^2\theta}{2}\right]_0^{\pi/2} d\phi$$

$$= -4 \int_0^{2\pi} \left(\dfrac{2}{3}\sin\phi + \dfrac{3}{2}\right) d\phi = -12\pi \qquad\qquad (2)$$

So we have from our two results (1) and (2)

$$\int_S \text{curl } \mathbf{F} \cdot d\mathbf{S} = \oint_c \mathbf{F} \cdot d\mathbf{r}$$

Before we proceed with another example, let us clarify a point relating to the direction of unit normal vectors now that we are dealing with surfaces.

So on to the next frame

Direction of unit normal vectors to a surface *S*

When we were dealing with the divergence theorem, the normal vectors were drawn in a direction outward from the enclosed region.

With an open surface as we now have, there is in fact no inward or outward direction. With any general surface, a normal vector can be drawn in either of two opposite directions. To avoid confusion, a convention must therefore be agreed upon and the established rule is as follows.

A unit normal $\hat{n}$ is drawn perpendicular to the surface *S* at any point in the direction indicated by applying a right-handed screw sense to the direction of integration round the boundary c.

Having noted that point, we can now deal with the next example.

Example 2

A surface consists of five sections formed by the planes $x = 0$, $x = 1$, $y = 0$, $y = 3$, $z = 2$ in the first octant. If the vector field $\mathbf{F} = y\mathbf{i} + z^2\mathbf{j} + xy\mathbf{k}$ exists over the surface and around its boundary, verify Stokes' theorem.

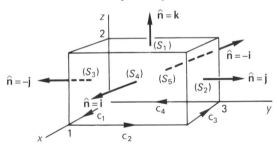

If we progress round the boundary along c_1, c_2, c_3, c_4 in an anti-clockwise manner, the normals to the surfaces will be as shown.

We have to verify that $\displaystyle\int_S \operatorname{curl} \mathbf{F} \cdot d\mathbf{S} = \oint_c \mathbf{F} \cdot d\mathbf{r}$

(a) We will start off by finding $\displaystyle\oint_c \mathbf{F} \cdot d\mathbf{r}$

$$\int \mathbf{F} \cdot d\mathbf{r} = \ldots\ldots\ldots\ldots$$

70

$$\int \mathbf{F} \cdot d\mathbf{r} = \int (y\,dx + z^2\,dy + xy\,dz)$$

(1) Along c_1: $y = 0$; $z = 0$; $dy = 0$; $dz = 0$

$$\therefore \int_{c_1} \mathbf{F} \cdot d\mathbf{r} = \int (0 + 0 + 0) = 0$$

(2) Along c_2: $x = 1$; $z = 0$; $dx = 0$; $dz = 0$

$$\therefore \int_{c_2} \mathbf{F} \cdot d\mathbf{r} = \int (0 + 0 + 0) = 0$$

In the same way

$$\int_{c_3} \mathbf{F} \cdot d\mathbf{r} = \ldots\ldots\ldots \quad \text{and} \quad \int_{c_4} \mathbf{F} \cdot d\mathbf{r} = \ldots\ldots\ldots$$

71

$$\int_{c_3} \mathbf{F} \cdot d\mathbf{r} = -3; \quad \int_{c_4} \mathbf{F} \cdot d\mathbf{r} = 0$$

Because

(3) Along c_3: $y = 3$; $z = 0$; $dy = 0$; $dz = 0$

$$\therefore \int_{c_3} \mathbf{F} \cdot d\mathbf{r} = \int_1^0 (3\,dx + 0 + 0) = \left[3x\right]_1^0 = -3$$

(4) Along c_4: $x = 0$; $z = 0$; $dx = 0$; $dz = 0$

$$\therefore \int_{c_4} \mathbf{F} \cdot d\mathbf{r} = \int (0 + 0 + 0) = 0$$

$$\therefore \oint_c \mathbf{F} \cdot d\mathbf{r} = 0 + 0 - 3 + 0 = -3$$

$$\oint_c \mathbf{F} \cdot d\mathbf{r} = -3 \tag{1}$$

(b) Now we have to find $\displaystyle\int_S \text{curl } \mathbf{F} \cdot d\mathbf{S}$.

First we need an expression for curl $\mathbf{F}$.

$$\mathbf{F} = y\mathbf{i} + z^2\mathbf{j} + xy\mathbf{k}$$

$$\therefore \text{ curl } \mathbf{F} = \ldots\ldots\ldots$$

$$\boxed{\text{curl } \mathbf{F} = (x - 2z)\mathbf{i} - y\mathbf{j} - \mathbf{k}}$$ **72**

Because

$$\text{curl } \mathbf{F} = \nabla \times \mathbf{F} = \begin{vmatrix} \mathbf{i} & \mathbf{j} & \mathbf{k} \\ \dfrac{\partial}{\partial x} & \dfrac{\partial}{\partial y} & \dfrac{\partial}{\partial z} \\ y & z^2 & xy \end{vmatrix}$$

$$= \mathbf{i}(x - 2z) - \mathbf{j}(y - 0) + \mathbf{k}(0 - 1) = (x - 2z)\mathbf{i} - y\mathbf{j} - \mathbf{k}$$

Then, for each section, we obtain $\displaystyle\int \text{curl } \mathbf{F} \cdot \mathrm{d}\mathbf{S} = \int \text{curl } \mathbf{F} \cdot \hat{\mathbf{n}}\, \mathrm{d}S$

(1) S_1 (top): $\hat{\mathbf{n}} = \mathbf{k}$

$$\therefore \int_{S_1} \text{curl } \mathbf{F} \cdot \hat{\mathbf{n}}\, \mathrm{d}S = \ldots\ldots\ldots$$

$$\boxed{-3}$$ **73**

Because

$$\int_{S_1} \text{curl } \mathbf{F} \cdot \hat{\mathbf{n}}\, \mathrm{d}S = \int_{S_1} \{(x - 2z)\mathbf{i} - y\mathbf{j} - \mathbf{k}\} \cdot (\mathbf{k})\, \mathrm{d}S$$

$$= \int_{S_1} (-1)\, \mathrm{d}S = -(\text{area of } S_1) = -3$$

Then, likewise

(2) S_2 (right-hand end): $\hat{\mathbf{n}} = \mathbf{j}$

$$\therefore \int_{S_2} \text{curl } \mathbf{F} \cdot \hat{\mathbf{n}}\, \mathrm{d}S = \int_{S_2} \{(x - 2z)\mathbf{i} - y\mathbf{j} - \mathbf{k}\} \cdot (\mathbf{j})\, \mathrm{d}S$$

$$= \int_{S_2} (-y)\, \mathrm{d}S$$

But $y = 3$ for this section

$$\therefore \int_{S_2} \text{curl } \mathbf{F} \cdot \hat{\mathbf{n}}\, \mathrm{d}S = \int_{S_2} (-3)\, \mathrm{d}S = (-3)(2) = -6$$

(3) S_3 (left-hand end): $\hat{\mathbf{n}} = -\mathbf{j}$

$$\therefore \int_{S_3} \text{curl } \mathbf{F} \cdot \hat{\mathbf{n}}\, \mathrm{d}S = \ldots\ldots\ldots$$

74

$$\boxed{0}$$

Because

$$\int_{S_3} \text{curl } \mathbf{F} \cdot \hat{\mathbf{n}} \, dS = \int_{S_3} \{(x - 2z)\mathbf{i} - y\mathbf{j} - \mathbf{k}\} \cdot (-\mathbf{j}) \, dS$$

$$= \int_{S_3} y \, dS$$

But $y = 0$ over S_3

$$\therefore \int_{S_3} \text{curl } \mathbf{F} \cdot \hat{\mathbf{n}} \, dS = 0$$

Working in the same way

$$\int_{S_4} \text{curl } \mathbf{F} \cdot \hat{\mathbf{n}} \, dS = \ldots\ldots\ldots; \quad \int_{S_5} \text{curl } \mathbf{F} \cdot \hat{\mathbf{n}} \, dS = \ldots\ldots\ldots$$

75

$$\boxed{\int_{S_4} \text{curl } \mathbf{F} \cdot \hat{\mathbf{n}} \, dS = -6; \quad \int_{S_5} \text{curl } \mathbf{F} \cdot \hat{\mathbf{n}} \, dS = 12}$$

Because

(4) S_4 (front): $\hat{\mathbf{n}} = \mathbf{i}$

$$\therefore \int_{S_4} \text{curl } \mathbf{F} \cdot \hat{\mathbf{n}} \, dS = \int_{S_4} \{(x - 2z)\mathbf{i} - y\mathbf{j} - \mathbf{k}\} \cdot (\mathbf{i}) \, dS$$

$$= \int_{S_4} (x - 2z) \, dS$$

But $x = 1$ over S_4

$$\therefore \int_{S_4} \text{curl } \mathbf{F} \cdot \hat{\mathbf{n}} \, dS = \int_0^3 \int_0^2 (1 - 2z) \, dz \, dy = \int_0^3 \left[z - z^2 \right]_0^2 dy$$

$$= \int_0^3 (-2) \, dy = \left[-2y \right]_0^3 = -6$$

(5) S_5 (back): $\hat{\mathbf{n}} = -\mathbf{i}$ with $x = 0$ over S_5

Similar working to that above gives $\int_{S_5} \text{curl } \mathbf{F} \cdot \hat{\mathbf{n}} \, dS = 12$

Finally, collecting the five results together gives

$$\int_S \text{curl } \mathbf{F} \cdot \hat{\mathbf{n}} \, dS = \ldots\ldots\ldots$$

$$\int_S \text{curl } \mathbf{F} \cdot \hat{\mathbf{n}} \, dS = -3 - 6 + 0 - 6 + 12 = -3 \qquad (2)$$

76

So, referring back to our result for section (a) we see that

$$\int_S \text{curl } \mathbf{F} \cdot d\mathbf{S} = \oint_c \mathbf{F} \cdot d\mathbf{r}$$

Of course we can, on occasions, make use of Stokes' theorem to lighten the working – as in the next example.

Example 3

A surface S consists of that part of the cylinder $x^2 + y^2 = 9$ between $z = 0$ and $z = 4$ for $y \geq 0$ and the two semicircles of radius 3 in the planes $z = 0$ and $z = 4$. If $\mathbf{F} = z\mathbf{i} + xy\mathbf{j} + xz\mathbf{k}$, evaluate $\displaystyle\int_S \text{curl } \mathbf{F} \cdot d\mathbf{S}$ over the surface.

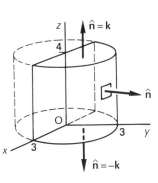

The surface S consists of three sections

(a) the curved surface of the cylinder
(b) the top and bottom semicircles.

We could therefore evaluate

$$\int_S \text{curl } \mathbf{F} \cdot d\mathbf{S}$$

over each of these separately.

However, we know by Stokes' theorem that

$$\int_S \text{curl } \mathbf{F} \cdot d\mathbf{S} = \ldots\ldots\ldots\ldots$$

$$\oint_c \mathbf{F} \cdot d\mathbf{r} \text{ where c is the boundary of } S$$

77

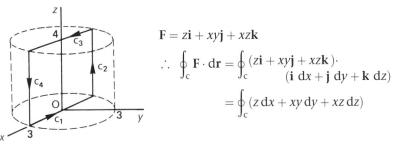

$$\mathbf{F} = z\mathbf{i} + xy\mathbf{j} + xz\mathbf{k}$$

$$\therefore \oint_c \mathbf{F} \cdot d\mathbf{r} = \oint_c (z\mathbf{i} + xy\mathbf{j} + xz\mathbf{k}) \cdot$$
$$(\mathbf{i} \, dx + \mathbf{j} \, dy + \mathbf{k} \, dz)$$

$$= \oint_c (z \, dx + xy \, dy + xz \, dz)$$

Now we can work through this easily enough, taking c_1, c_2, c_3, c_4 in turn, and summing the results, which gives

$$\int_S \text{curl } \mathbf{F} \cdot d\mathbf{S} = \oint_c \mathbf{F} \cdot d\mathbf{r} = \ldots\ldots\ldots\ldots$$

78 $\boxed{-24}$

Here is the working in detail. $\oint_c \mathbf{F} \cdot \mathbf{dr} = \oint_c (z\,dx + xy\,dy + xz\,dz)$

(1) c_1: $y = 0$; $z = 0$; $dy = 0$; $dz = 0$

$$\int_{c_1} \mathbf{F} \cdot \mathbf{dr} = \int_{c_1} (0 + 0 + 0) = 0$$

(2) c_2: $x = -3$; $y = 0$; $dx = 0$; $dy = 0$

$$\int_{c_2} \mathbf{F} \cdot \mathbf{dr} = \int_{c_2} (0 + 0 - 3z\,dz) = \left[\frac{-3z^2}{2}\right]_0^4 = -24$$

(3) c_3: $y = 0$; $z = 4$; $dy = 0$; $dz = 0$

$$\int_{c_3} \mathbf{F} \cdot \mathbf{dr} = \int_{c_3} (4\,dx + 0 + 0) = \int_{-3}^{3} 4\,dx = 24$$

(4) c_4: $x = 3$; $y = 0$; $dx = 0$; $dy = 0$

$$\int_{c_4} \mathbf{F} \cdot \mathbf{dr} = \int_{c_4} (0 + 0 + 3z\,dz) = \left[\frac{3z^2}{2}\right]_4^0 = -24$$

Totalling up these four results, we have

$$\oint_c \mathbf{F} \cdot \mathbf{dr} = 0 - 24 + 24 - 24 = -24$$

But $\int_S \text{curl } \mathbf{F} \cdot \mathbf{dS} = \oint_c \mathbf{F} \cdot \mathbf{dr}$ $\therefore \int_S \text{curl } \mathbf{F} \cdot \mathbf{dS} = -24$

This working is a good deal easier than calculating $\int_S \text{curl } \mathbf{F} \cdot \mathbf{dS}$ over the three separate surfaces direct.

So, if you have not already done so, make a note of Stokes' theorem:

$$\int_S \text{curl } \mathbf{F} \cdot \mathbf{dS} = \oint_c \mathbf{F} \cdot \mathbf{dr}$$

Then on to the next section of the work

Green's theorem

79

Green's theorem enables an integral over a plane area to be expressed in terms of a line integral round its boundary curve.

We showed in Programme 19 that, if P and Q are two single-valued functions of x and y, continuous over a plane surface S, and c is its boundary curve, then

$$\oint_c (P\,dx + Q\,dy) = \iint_S \left(\frac{\partial Q}{\partial x} - \frac{\partial P}{\partial y}\right) dx\,dy$$

where the line integral is taken round c in an anticlockwise manner.

In vector terms, this becomes:

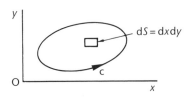

S is a two-dimensional space enclosed by a simple closed curve c.

$$dS = dx\,dy$$
$$dS = \hat{n}\,dS = \mathbf{k}\,dx\,dy$$

If $\mathbf{F} = P\mathbf{i} + Q\mathbf{j}$ where $P = P(x,\ y)$ and $Q = Q(x,\ y)$ then

$$\text{curl } \mathbf{F} = \ldots\ldots\ldots\ldots$$

$$\boxed{\mathbf{k}\left(\frac{\partial Q}{\partial x} - \frac{\partial P}{\partial y}\right)}$$

Because

$$\text{curl } \mathbf{F} = \begin{vmatrix} \mathbf{i} & \mathbf{j} & \mathbf{k} \\ \dfrac{\partial}{\partial x} & \dfrac{\partial}{\partial y} & \dfrac{\partial}{\partial z} \\ P & Q & 0 \end{vmatrix}$$

$$= \mathbf{i}\left(0 - \frac{\partial Q}{\partial z}\right) - \mathbf{j}\left(0 - \frac{\partial P}{\partial z}\right) + \mathbf{k}\left(\frac{\partial Q}{\partial x} - \frac{\partial P}{\partial y}\right)$$

But in the x–y plane, $\dfrac{\partial Q}{\partial z} = \dfrac{\partial P}{\partial z} = 0$. $\therefore$ curl $\mathbf{F} = \mathbf{k}\left(\dfrac{\partial Q}{\partial x} - \dfrac{\partial P}{\partial y}\right)$

So $\displaystyle\int_S \text{curl } \mathbf{F} \cdot d\mathbf{S} = \int \text{curl } \mathbf{F} \cdot \hat{n}\,dS$ and in the x–y plane, $\hat{n} = \mathbf{k}$

$$\therefore \int_S \text{curl } \mathbf{F} \cdot d\mathbf{S} = \int_S \mathbf{k}\left(\frac{\partial Q}{\partial x} - \frac{\partial P}{\partial y}\right) \cdot (\mathbf{k})\,dS = \iint_S \left(\frac{\partial Q}{\partial x} - \frac{\partial P}{\partial y}\right) dx\,dy$$

$$\therefore \int_S \text{curl } \mathbf{F} \cdot d\mathbf{S} = \iint_S \left(\frac{\partial Q}{\partial x} - \frac{\partial P}{\partial y}\right) dx\,dy \tag{1}$$

Now by Stokes' theorem $\ldots\ldots\ldots\ldots$

81

$$\int_S \text{curl } \mathbf{F} \cdot d\mathbf{S} = \oint_c \mathbf{F} \cdot d\mathbf{r}$$

and, in this case, $\oint_c \mathbf{F} \cdot d\mathbf{r} = \oint_c (P\mathbf{i} + Q\mathbf{j}) \cdot (\mathbf{i}\, dx + \mathbf{j}\, dy + \mathbf{k}\, dz)$

$$= \oint_c (P\, dx + Q\, dy)$$

$$\therefore \quad \oint_c \mathbf{F} \cdot d\mathbf{r} = \oint_c (P\, dx + Q\, dy) \tag{2}$$

Therefore from (1) and (2)

Stokes' theorem $\int_S \text{curl } \mathbf{F} \cdot d\mathbf{S} = \oint_c \mathbf{F} \cdot d\mathbf{r}$ in two dimensions becomes

Green's theorem $\iint_S \left(\dfrac{\partial Q}{\partial x} - \dfrac{\partial P}{\partial y} \right) dx\, dy = \oint_c (P\, dx + Q\, dy)$

Example

Verify Green's theorem for the integral $\oint_c \{(x^2 + y^2)\, dx + (x + 2y)\, dy\}$ taken round the boundary curve c defined by

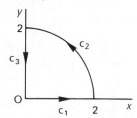

$$\begin{array}{ll} y = 0 & 0 \le x \le 2 \\ x^2 + y^2 = 4 & 0 \le x \le 2 \\ x = 0 & 0 \le y \le 2. \end{array}$$

Green's theorem: $\iint_S \left(\dfrac{\partial Q}{\partial x} - \dfrac{\partial P}{\partial y} \right) dx\, dy = \oint_c (P\, dx + Q\, dy)$

In this case $(x^2 + y^2)\, dx + (x + 2y)\, dy = P\, dx + Q\, dy$

$\therefore \ P = x^2 + y^2 \quad$ and $\quad Q = x + 2y$

We now take c_1, c_2, c_3 in turn.

(1) c_1: $y = 0;$ $\quad dy = 0$

$$\therefore \quad \int_{c_1} (P\, dx + Q\, dy) = \int_0^2 x^2\, dx = \left[\frac{x^3}{3} \right]_0^2 = \frac{8}{3}$$

(2) c_2: $x^2 + y^2 = 4 \quad \therefore \ y^2 = 4 - x^2 \quad \therefore \ y = (4 - x^2)^{1/2}$

$x + 2y = x + 2(4 - x^2)^{1/2}$

$dy = \dfrac{1}{2}(4 - x^2)^{-1/2}(-2x)\, dx = \dfrac{-x}{\sqrt{4 - x^2}}\, dx$

$$\therefore \quad \int_{c_2} (P\, dx + Q\, dy) = \ldots\ldots\ldots$$

Make any necessary substitutions and evaluate the line integral for c_2.

$$\boxed{\pi - 4} \qquad \textbf{82}$$

Because we have

$$\int_{c_2} (P\,dx + Q\,dy) = \int_{c_2} \left\{ 4 + (x + 2\sqrt{4 - x^2}) \left(\frac{-x}{\sqrt{4 - x^2}} \right) \right\} dx$$

$$= \int_{c_2} \left\{ 4 - 2x - \frac{x^2}{\sqrt{4 - x^2}} \right\} dx$$

Putting $x = 2\sin\theta$, $\quad \sqrt{4 - x^2} = 2\cos\theta \quad dx = 2\cos\theta\,d\theta$

Limits: $x = 2$, $\theta = \dfrac{\pi}{2}$; $\quad x = 0$, $\theta = 0$.

$$\therefore \int_{c_2} (P\,dx + Q\,dy) = \int_{\pi/2}^{0} \left\{ 4 - 4\sin\theta - \frac{4\sin^2\theta}{2\cos\theta} \right\} 2\cos\theta\,d\theta$$

$$= 4 \left[2\sin\theta - \sin^2\theta - \frac{1}{2} \left(\theta - \frac{\sin 2\theta}{2} \right) \right]_{\pi/2}^{0}$$

$$= 4 \left[-\left(2 - 1 - \frac{\pi}{4} \right) \right] = \pi - 4$$

Finally

(3) c_3: $\quad x = 0$; $\quad dx = 0$

$$\therefore \int_{c_3} (P\,dx + Q\,dy) = \int_{2}^{0} 2y\,dy = \left[y^2 \right]_{2}^{0} = -4$$

$\therefore$ Collecting our three partial results

$$\oint_{c} (P\,dx + Q\,dy) = \frac{8}{3} + \pi - 4 - 4 = \pi - \frac{16}{3} \tag{1}$$

That is one part done. Now we have to evaluate $\displaystyle\iint_{S} \left(\frac{\partial Q}{\partial x} - \frac{\partial P}{\partial y} \right) dx\,dy$

$$P = x^2 + y^2 \qquad \therefore \frac{\partial P}{\partial y} = 2y$$

$$Q = x + 2y \qquad \therefore \frac{\partial Q}{\partial x} = 1$$

$$\therefore \iint_{S} \left(\frac{\partial Q}{\partial x} - \frac{\partial P}{\partial y} \right) dx\,dy = \iint_{S} (1 - 2y)\,dy\,dx$$

It will be more convenient to work in polar coordinates, so we make the substitutions

$$x = r\cos\theta; \quad y = r\sin\theta; \quad dS = dx\,dy = r\,dr\,d\theta$$

$$\therefore \iint_{S} \left(\frac{\partial Q}{\partial x} - \frac{\partial P}{\partial y} \right) dx\,dy = \int_{0}^{\pi/2} \int_{0}^{2} (1 - 2r\sin\theta)r\,dr\,d\theta$$

$$= \ldots\ldots\ldots\ldots$$

Complete it

83

$$\pi - \frac{16}{3}$$

Here it is:

$$\iint_S \left(\frac{\partial Q}{\partial x} - \frac{\partial P}{\partial y}\right) dx\, dy = \int_0^{\pi/2} \int_0^2 (r - 2r^2 \sin\theta)\, dr\, d\theta$$

$$= \int_0^{\pi/2} \left[\frac{r^2}{2} - \frac{2r^3}{3} \sin\theta\right]_0^2 d\theta$$

$$= \int_0^{\pi/2} \left\{2 - \frac{16}{3} \sin\theta\right\} d\theta$$

$$= \left[2\theta + \frac{16}{3} \cos\theta\right]_0^{\pi/2} = \pi - \frac{16}{3} \qquad (2)$$

So we have established once again that

$$\oint_c (P\, dx + Q\, dy) = \iint_S \left(\frac{\partial Q}{\partial x} - \frac{\partial P}{\partial y}\right) dx\, dy$$

And that brings us to the end of this particular Programme. We have covered a number of important sections, so check carefully down the **Revision summary** and the **Can you?** checklist, and then work through the **Test exercise** that follows. The **Further problems** provide valuable additional practice.

 Revision summary 23

84

1 *Line integrals*

(a) Scalar field V: $\displaystyle\int_c V\, d\mathbf{r}$

The curve c is expressed in parametric form.

$$d\mathbf{r} = \mathbf{i}\, dx + \mathbf{j}\, dy + \mathbf{k}\, dz$$

(b) Vector field **F**: $\displaystyle\int_c \mathbf{F} \cdot d\mathbf{r}$

$$\mathbf{F} = F_x \mathbf{i} + F_y \mathbf{j} + F_z \mathbf{k}$$
$$d\mathbf{r} = \mathbf{i}\, dx + \mathbf{j}\, dy + \mathbf{k}\, dz$$
$$\mathbf{F} \cdot d\mathbf{r} = F_x\, dx + F_y\, dy + F_z\, dz$$

2 *Volume integrals*

F is a vector field; V a closed region with boundary surface S.

$$\int_V \mathbf{F}\, dV = \int_{x_1}^{x_2} \int_{y_1}^{y_2} \int_{z_1}^{z_2} \mathbf{F}\, dz\, dy\, dx$$

3 *Surface integrals* (surface defined by $\phi(x, y, z) = \text{constant}$)

(a) Scalar field $V(x, y, z)$:

$$\int_S V\, d\mathbf{S} = \int_S V\hat{\mathbf{n}}\, dS; \qquad \hat{\mathbf{n}} = \frac{\nabla\phi}{|\nabla\phi|} = \frac{\text{grad}\,\phi}{|\text{grad}\,\phi|}$$

(b) Vector field $\mathbf{F} = F_x\mathbf{i} + F_y\mathbf{j} + F_z\mathbf{k}$:

$$\int_S \mathbf{F}\cdot d\mathbf{S} = \int_S \mathbf{F}\cdot\hat{\mathbf{n}}\, dS; \qquad \hat{\mathbf{n}} = \frac{\nabla\phi}{|\nabla\phi|}$$

4 *Polar coordinates*

(a) Plane polar coordinates (r, θ)

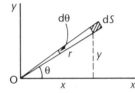

$$x = r\cos\theta; \quad y = r\sin\theta$$
$$dS = r\, dr\, d\theta$$

(b) Cylindrical polar coordinates (ρ, ϕ, z)

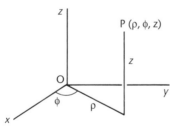

$$x = \rho\cos\phi$$
$$y = \rho\sin\phi$$
$$z = z$$
$$dS = \rho\, d\phi\, dz$$
$$dV = \rho\, d\rho\, d\phi\, dz$$

(c) Spherical polar coordinates (r, θ, ϕ)

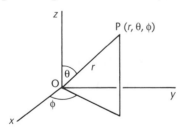

$$x = r\sin\theta\cos\phi$$
$$y = r\sin\theta\sin\phi$$
$$z = r\cos\theta$$
$$dS = r^2\sin\theta\, d\theta\, d\phi$$
$$dV = r^2\sin\theta\, dr\, d\theta\, d\phi$$

5 *Conservative vector fields*

A vector field **F** is conservative if

(a) $\oint_c \mathbf{F}\cdot d\mathbf{r} = 0$ for all closed curves

(b) $\text{curl}\,\mathbf{F} = 0$

(c) $\mathbf{F} = \text{grad}\,V$ where V is a scalar.

6 *Divergence theorem* (Gauss' theorem)

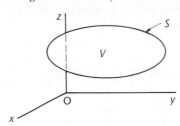

Closed surface S enclosing a region V in a vector field **F**.

$$\int_V \text{div } \mathbf{F}\, dV = \int_S \mathbf{F} \cdot d\mathbf{S}$$

7 *Stokes' theorem*

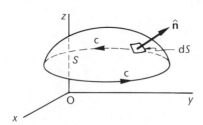

An open surface S bounded by a simple closed curve c, then

$$\int_S \text{curl } \mathbf{F} \cdot d\mathbf{S} = \oint_c \mathbf{F} \cdot d\mathbf{r}$$

8 *Green's theorem*

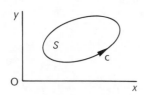

The curve c is a simple closed curve enclosing a plane space S in the x–y plane. P and Q are functions of both x and y.

$$\text{Then } \iint_S \left(\frac{\partial Q}{\partial x} - \frac{\partial P}{\partial y} \right) dx\, dy = \oint_c (P\, dx + Q\, dy).$$

 Can you?

---

85 **Checklist 23**

Check this list before and after you try the end of Programme test.

On a scale of 1 to 5 how confident are you that you can: **Frames**

- Evaluate the line integral of a scalar and a vector field in Cartesian coordinates? | 1 | to | 20 |
  Yes ☐ ☐ ☐ ☐ ☐ *No*

- Evaluate the volume integral of a vector field? | 21 | to | 27 |
  Yes ☐ ☐ ☐ ☐ ☐ *No*

- Evaluate the surface integral of a scalar and a vector field? | 28 | to | 42 |
  Yes ☐ ☐ ☐ ☐ ☐ *No*

▶

- Determine whether or not a vector field is a conservative
  vector field?   43 to 52
  Yes ☐   ☐   ☐   ☐   ☐   No

- Apply Gauss' divergence theorem?   52 to 63
  Yes ☐   ☐   ☐   ☐   ☐   No

- Apply Stokes' theorem?   64 to 68
  Yes ☐   ☐   ☐   ☐   ☐   No

- Determine the direction of unit normal vectors to a surface?   69 to 78
  Yes ☐   ☐   ☐   ☐   ☐   No

- Apply Green's theorem in the plane?   79 to 83
  Yes ☐   ☐   ☐   ☐   ☐   No

 # Test exercise 23

86

1   If $V = x^3y + 2xy^2 + yz$, evaluate $\int_c V\,d\mathbf{r}$ between A $(0, 0, 0)$ and B $(2, 1, -3)$
    along the curve with parametric equations $x = 2t$, $y = t^2$, $z = -3t^3$.

2   If $\mathbf{F} = x^2y^3\,\mathbf{i} + yz^2\,\mathbf{j} + zx^2\,\mathbf{k}$, evaluate $\int_c \mathbf{F}\cdot d\mathbf{r}$ along the curve $x = 3u^2$, $y = u$,
    $z = 2u^3$ between A $(3, -1, -2)$ and B $(3, 1, 2)$.

3   Evaluate $\int_V \mathbf{F}\,dV$ where $\mathbf{F} = 3\mathbf{i} + z\mathbf{j} + 2y\mathbf{k}$ and V is the region bounded by the
    planes $z = 0$, $z = 3$ and the surface $x^2 + y^2 = 4$.

4   If V is the scalar field $V = xyz^2$, evaluate $\int_S V\,d\mathbf{S}$ over the surface S defined by
    $x^2 + y^2 = 9$ between $z = 0$ and $z = 2$ in the first octant.

5   Evaluate $\int_S \mathbf{F}\cdot d\mathbf{S}$ over the surface S defined by $x^2 + y^2 + z^2 = 4$ for $z \geq 0$ and
    bounded by $x = 0$, $y = 0$, $z = 0$ in the first octant where $\mathbf{F} = x\mathbf{i} + 2z\mathbf{j} + y\mathbf{k}$.

6   Determine which of the following vector fields are conservative.
    (a) $\mathbf{F} = (2xy + z)\mathbf{i} + (x^2 + 2yz)\mathbf{j} + (x + y^2)\mathbf{k}$
    (b) $\mathbf{F} = (yz + 2y)\mathbf{i} + (xz + 2x)\mathbf{j} + (xy + 3)\mathbf{k}$
    (c) $\mathbf{F} = (yz^2 + 3)\mathbf{i} + (xz^2 + 2)\mathbf{j} + (2xyz + 4)\mathbf{k}$.

7   By the use of the divergence theorem, determine $\int_S \mathbf{F}\cdot d\mathbf{S}$ where
    $\mathbf{F} = x\mathbf{i} + xy\mathbf{j} + 2\mathbf{k}$,
    taken over the region bounded by the planes $z = 0$, $z = 4$, $x = 0$, $y = 0$ and the
    surface $x^2 + y^2 = 9$ in the first octant.   ▶

8   A surface consists of parts of the planes $x = 0$, $x = 2$, $y = 0$, $y = 2$ and $z = 3 - y$
    in the region $z \geq 0$. Apply Stokes' theorem to evaluate $\int_S$ curl $\mathbf{F} \cdot d\mathbf{S}$ over the
    surface where $\mathbf{F} = 2x\mathbf{i} + xz\mathbf{j} + yz\mathbf{k}$ where S lies in the $z = 0$ plane.

9   Verify Green's theorem in the plane for the integral

$$\oint_c \{(xy^2 - 2x)\,dx + (x + 2xy^2)\,dy\}$$

where c is the square with vertices at $(1, 1)$, $(-1, 1)$, $(-1, -1)$ and $(1, -1)$.

 # Further problems 23

---

87   1   If $V = x^2yz$, evaluate $\int_c V \, d\mathbf{r}$ between A $(0, 0, 0)$ and B $(6, 2, 4)$

(a)   along the straight lines c_1: $(0, 0, 0)$ to $(6, 0, 0)$
                                     c_2: $(6, 0, 0)$ to $(6, 2, 0)$
                                     c_3: $(6, 2, 0)$ to $(6, 2, 4)$

(b)   along the path c_4 having parametric equations $x = 3t$, $y = t$, $z = 2t$.

2   If $V = xy^2 + yz$, evaluate to one decimal place $\int_c V \, d\mathbf{r}$ along the curve c having
    parametric equations $x = 2t^2$, $y = 4t$, $z = 3t + 5$ between A $(0, 0, 5)$ and
    B $(8, 8, 11)$.

3   Evaluate to one decimal place the integral $\int_c (xyz + 4x^2y) \, d\mathbf{r}$ along the curve c
    with parametric equations $x = 2u$, $y = u^2$, $z = 3u^3$ between A $(2, 1, 3)$ and
    B $(4, 4, 24)$.

4   If $\mathbf{F} = xy\mathbf{i} + yz\mathbf{j} + 3xyz\mathbf{k}$, evaluate $\int_c \mathbf{F} \cdot d\mathbf{r}$ between A $(0, 2, 0)$ and B $(3, 6, 1)$
    where c has the parametric equations $x = 3u$, $y = 4u + 2$, $z = u^2$.

5   $\mathbf{F} = x^2\mathbf{i} - 2xy\mathbf{j} + yz\mathbf{k}$. Evaluate $\int_c \mathbf{F} \cdot d\mathbf{r}$ between A $(2, 1, 2)$ and B $(4, 4, 5)$
    where c is the path with parametric equations $x = 2u$, $y = u^2$, $z = 3u - 1$.

6   A unit particle is moved in an anticlockwise manner round a circle with centre
    $(0, 0, 4)$ and radius 2 in the plane $z = 4$ in a force field defined as
    $\mathbf{F} = (xy + z)\mathbf{i} + (2x + y)\mathbf{j} + (x + y + z)\mathbf{k}$. Find the work done.

7   Evaluate $\int_V \mathbf{F}\,dV$ where $\mathbf{F} = \mathbf{i} - y\mathbf{j} + \mathbf{k}$ and V is the region bounded by the plane
    $z = 0$ and the hemisphere $x^2 + y^2 + z^2 = 4$, for $z \geq 0$.

8   V is the region bounded by the planes $x = 0$, $y = 0$, $z = 0$ and the surfaces
    $y = 4 - x^2$ $(z \geq 0)$ and $y = 4 - z^2$ $(y \geq 0)$.

    If $\mathbf{F} = 2\mathbf{i} + y^2\mathbf{j} - \mathbf{k}$, evaluate $\displaystyle\int_V \mathbf{F}\,dV$ throughout the region.

9   If $\mathbf{F} = 3\mathbf{i} + 2\mathbf{j} - 2x\mathbf{k}$, evaluate $\displaystyle\int_V \mathbf{F}\,dV$ where V is the region bounded by the

    planes $y = 0$, $z = 0$, $z = 4 - y$ $(z \geq 0)$ and the surface $x^2 + y^2 = 16$.

10 A scalar field $V = x + y$ exists over a surface S defined by $x^2 + y^2 + z^2 = 9$,
    bounded by the planes $x = 0$, $y = 0$, $z = 0$ in the first octant. Evaluate

    $\displaystyle\int_S V\,d\mathbf{S}$ over the curved surface.

11 A surface S is defined by $y^2 + z = 4$ and is bounded by the planes $x = 0$, $x = 3$,

    $y = 0$, $z = 0$ in the first octant. Evaluate $\displaystyle\int_S V\,d\mathbf{S}$ over this curved surface where

    V denotes the scalar field $V = x^2 y z$.

12 Evaluate $\displaystyle\int_S \operatorname{curl} \mathbf{F} \cdot d\mathbf{S}$ over the surface S defined by $2x + 2y + z = 2$ and

    bounded by $x = 0$, $y = 0$, $z = 0$ in the first octant and where
    $\quad \mathbf{F} = y^2\mathbf{i} + 2yz\mathbf{j} + xy\mathbf{k}$.

13 Evaluate $\displaystyle\int_S \mathbf{F} \cdot d\mathbf{S}$ over the hemisphere defined by $x^2 + y^2 + z^2 = 25$ with $z \geq 0$,

    where $\mathbf{F} = (x + y)\mathbf{i} - 2z\mathbf{j} + y\mathbf{k}$.

14 A vector field $\mathbf{F} = 2x\mathbf{i} + z\mathbf{j} + y\mathbf{k}$ exists over a surface S defined by
    $\quad x^2 + y^2 + z^2 = 16$, bounded by the planes $z = 0$, $z = 3$, $x = 0$, $y = 0$.

    Evaluate $\displaystyle\int_S \mathbf{F} \cdot d\mathbf{S}$ over the stated curved surface.

15 Evaluate $\displaystyle\int_S \mathbf{F} \cdot d\mathbf{S}$, where $\mathbf{F}$ is the vector field $x^2\mathbf{i} + 2z\mathbf{j} - y\mathbf{k}$, over the curved

    surface S defined by $x^2 + y^2 = 25$ and bounded by $z = 0$, $z = 6$, $y \geq 3$.

16 A region V is defined by the quartersphere $x^2 + y^2 + z^2 = 16$, $z \geq 0$, $y \geq 0$ and
    the planes $z = 0$, $y = 0$. A vector field $\mathbf{F} = xy\mathbf{i} + y^2\mathbf{j} + \mathbf{k}$ exists throughout and
    on the boundary of the region. Verify the Gauss divergence theorem for the
    region stated.

17 A surface consists of parts of the planes $x = 0$, $x = 1$, $y = 0$, $y = 2$, $z = 1$ in the
    first octant. If $\mathbf{F} = y\mathbf{i} + x^2 z\mathbf{j} + xy\mathbf{k}$, verify Stokes' theorem.

18 S is the surface $z = x^2 + y^2$ bounded by the planes $z = 0$ and $z = 4$. Verify Stokes'
    theorem for a vector field $\mathbf{F} = xy\mathbf{i} + x^3\mathbf{j} + xz\mathbf{k}$.

19 A vector field $\mathbf{F} = xy\mathbf{i} + z^2\mathbf{j} + xyz\mathbf{k}$ exists over the surfaces $x^2 + y^2 + z^2 = a^2$, $x = 0$ and $y = 0$ in the first octant. Verify Stokes' theorem that

$$\int_S \text{curl } \mathbf{F} \cdot d\mathbf{S} = \oint_c \mathbf{F} \cdot d\mathbf{r}.$$

20 A surface is defined by $z^2 = 4(x^2 + y^2)$ where $0 \le z \le 6$. If a vector field $\mathbf{F} = z\mathbf{i} + xy^2\mathbf{j} + x^2z\mathbf{k}$ exists over the surface and on the boundary circle c, show that $\oint_c \mathbf{F} \cdot d\mathbf{r} = \int_S \text{curl } \mathbf{F} \cdot d\mathbf{S}$.

21 Verify Green's theorem in the plane for the integral

$$\oint_c \left\{ (x - y) \, dx - (y^2 + xy) \, dy \right\}$$

where c is the circle with unit radius, centred on the origin.

Vector analysis 3

Learning outcomes

When you have completed this Programme you will be able to:

- Derive the family of curves of constant coordinates for curvilinear coordinates
- Derive unit base vectors and scale factors in orthogonal curvilinear coordinates
- Obtain the element of arc ds and the element of volume dV in orthogonal curvilinear coordinates
- Obtain expressions for the operators grad, div and curl in orthogonal curvilinear coordinates

1

This short Programme is an extension of the two previous ones and may not be required for all courses. It can well be bypassed without adversely affecting the rest of the work.

Curvilinear coordinates

Let us consider two variables u and v, each of which is a function of x and y

i.e. $u = f(x, y)$
$\ v = g(x, y)$

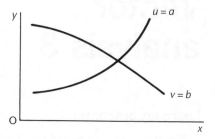

If u and v are each assigned a constant value a and b, the equations will, in general,
define two intersecting curves.

If u and v are each given several such values, the equations define a network of curves covering the x–y plane.

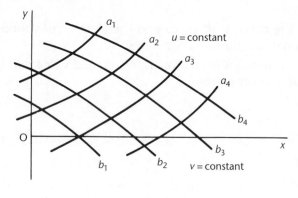

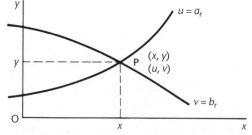

A pair of curves $u = a_r$ and $v = b_r$ pass through each point in the plane. Hence, any point in the plane can be expressed in *rectangular coordinates* (x, y) or in *curvilinear coordinates* (u, v).

Let us see how this works out in an example, so move on

Example 1   **2**

Let us consider the case where $u = xy$ and $v = x^2 - y$.

(a) With $u = xy$, if we put $u = 4$, then $y = \dfrac{4}{x}$ and we can plot y against x to obtain the relevant curve.

Similarly, putting $u = 8, 16, 32, \ldots$ we can build up a family of curves, all of the pattern $u = xy$.

x		0·5	1·0	2·0	3·0	4·0
y	$u = 4$	8	4	2	1·33	1·0
	$u = 8$	16	8	4	2·67	2
	$u = 16$	32	16	8	5·33	4
	$u = 32$	64	32	16	10·67	8

If we plot these on graph paper between $x = 0$ and $x = 4$ with a range of y from $y = 0$ to $y = 20$, we obtain

. . . .

3

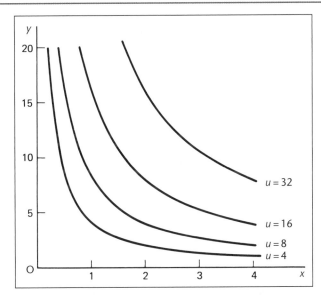

Note that each graph is labelled with its individual u-value.

(b) With $v = x^2 - y$, we proceed in just the same way. We rewrite the equation as $y = x^2 - v$; assign values such as $8, 4, 0, -4, -8, -12, -16, \ldots$ to v; and draw the relevant curve in each case. If we do that for $x = 0$ to $x = 4$ and limit the y-values to the range $y = 0$ to $y = 20$, we obtain the family of curves

. . . .

4

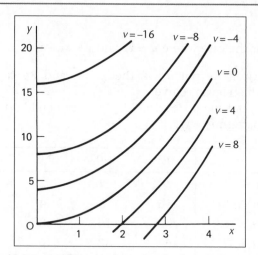

The table of function values is as follows.

x			0	1	2	3	4
y	$v =$	8	−8	−7	−4	1	8
	$v =$	4	−4	−3	0	5	12
	$v =$	0	0	1	4	9	16
	$v = -$	4	4	5	8	13	20
	$v = -$	8	8	9	12	17	24
	$v = -12$		12	13	16	21	28
	$v = -16$		16	17	20	25	32

Note again that we label each graph with its own v-value.

This again is a family of curves with the common pattern $v = x^2 - y$, the members being distinguished from each other by the value assigned to v in each case.

Now we draw both sets of curves on a common set of x–y axes, taking

the range of x from $x = 0$ to $x = 4$

and   the range of y from $y = 0$ to $y = 20$.

It is worthwhile taking a little time over it – and good practice!

When you have the complete picture, move on to the next frame

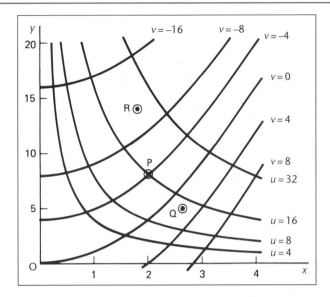

5

The position of any point in the plane can now be stated in two ways. For example, the point P has Cartesian rectangular coordinates $x = 2$, $y = 8$. It can also be stated in curvilinear coordinates $u = 16$, $v = -4$, for it is at the point of intersection of the two curves corresponding to $u = 16$ and $v = -4$.

Likewise, for the point Q, the position in rectangular coordinates is $x = 2{\cdot}65$, $y = 5{\cdot}0$ and for its position in curvilinear coordinates we must estimate it within the network. Approximate values are $u = 13$, $v = 2$.

Similarly, the curvilinear coordinates of R $(x = 1{\cdot}8, y = 14)$ are approximately

$$u = \ldots\ldots\ldots; \quad v = \ldots\ldots\ldots$$

$$\boxed{u = 26; \quad v = -11}$$

Their actual values are in fact $u = 25{\cdot}2$ and $v = -10{\cdot}76$.

Now let us deal with another example.

Example 2

If $u = x^2 + 2y$ and $v = y - (x + 1)^2$, these can be rewritten as $y = \frac{1}{2}(u - x^2)$ and $y = v + (x + 1)^2$. We can now plot the family of curves, say between $x = 0$ and $x = 4$, with $u = 5(5)30$ and $v = -20(5)5$, i.e. values of u from 5 to 30 at intervals of 5 units and values of v from -20 to 5 at intervals of 5 units.

The resulting network is easily obtained and appears as

$$\ldots\ldots\ldots$$

8

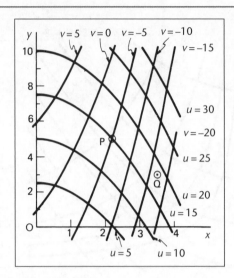

For P, the rectangular coordinates are $(x = 2\cdot18, y = 5\cdot1)$
and the curvilinear coordinates are $(u = 15, v = -5)$.
For Q, the rectangular coordinates are ....
and the curvilinear coordinates are ....

9

$$\boxed{\text{Q: } (x = 3\cdot5, y = 3\cdot0); \quad (u = 18\cdot5, v = -17)}$$

Orthogonal curvilinear coordinates

If the coordinate curves for u and v forming the network cross at right angles, the system of coordinates is said to be *orthogonal*. The test for orthogonality is given by the dot product of the vectors formed from the partial derivatives. This is, if

$$\frac{\partial u}{\partial x}\frac{\partial v}{\partial x} + \frac{\partial u}{\partial y}\frac{\partial v}{\partial y} = 0 \text{ then } u \text{ and } v \text{ are orthogonal.}$$

Example 3

Given the curvilinear coordinates u and v where $u = xy$ and $v = x^2 - y^2$ then

u and v form a coordinate system that is ....

$$\boxed{\text{orthogonal}}$$ **10**

Because

$$u = xy \;\; \text{so} \;\; \frac{\partial u}{\partial x} = y \;\; \text{and} \;\; \frac{\partial u}{\partial y} = x, \;\; v = x^2 - y^2 \;\; \text{so} \;\; \frac{\partial v}{\partial x} = 2x \;\; \text{and} \;\; \frac{\partial v}{\partial y} = -2y.$$

Then $\dfrac{\partial u}{\partial x}\dfrac{\partial v}{\partial x} + \dfrac{\partial u}{\partial y}\dfrac{\partial v}{\partial y} = 2xy - 2xy = 0$ and so u and v form a coordinate

system that is orthogonal.

Example 4

Given the curvilinear coordinates u and v where $u = x^2 + 2y$ and $v = y - (x + 1)^2$ then

u and v form a coordinate system that is ....

$$\boxed{\text{not orthogonal}}$$ **11**

Because

$$u = x^2 + 2y \;\; \text{so} \;\; \frac{\partial u}{\partial x} = 2x \;\; \text{and} \;\; \frac{\partial u}{\partial y} = 2, \;\; v = y - (x + 1)^2 \;\; \text{so} \;\; \frac{\partial v}{\partial x} = -2(x + 1)$$

and $\dfrac{\partial v}{\partial y} = 1.$

Then

$$\frac{\partial u}{\partial x}\frac{\partial v}{\partial x} + \frac{\partial u}{\partial y}\frac{\partial v}{\partial y} = -4x(x + 1) + 2 \neq 0 \;\; \text{and so} \;\; u \;\; \text{and} \;\; v \;\; \text{form a coordinate}$$

system that is not orthogonal.

Let us extend these ideas to three dimensions. Move on

Orthogonal coordinate systems in space

Any vector **F** can be expressed in terms of its components in three mutually perpendicular directions, which have normally been the directions of the coordinate axes, i.e. **12**

$$\mathbf{F} = F_x\mathbf{i} + F_y\mathbf{j} + F_z\mathbf{k}$$

where **i**, **j**, **k** are the unit vectors parallel to the x, y, z axes respectively.

Situations can arise, however, where the directions of the unit vectors do not remain fixed, but vary from point to point in space according to prescribed conditions. Examples of this occur in cylindrical and spherical polar coordinates, with which we are already familiar.

1 *Cylindrical polar coordinates* (ρ, ϕ, z)
Let P be a point with cylindrical co-ordinates (ρ, ϕ, z) as shown. The position of P is a function of the three variables ρ, ϕ, z

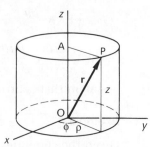

(a) If ϕ and z remain constant and ρ varies, then P will move out along AP by an amount $\dfrac{\partial \mathbf{r}}{\partial \rho}$ and the unit vector $\mathbf{I}$ in this direction will be given by

$$\mathbf{I} = \frac{\partial \mathbf{r}}{\partial \rho} \bigg/ \left| \frac{\partial \mathbf{r}}{\partial \rho} \right|$$

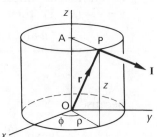

(b) If, instead, ρ and z remain constant and ϕ varies, P will move

. . . .

13

round the circle with AP as radius

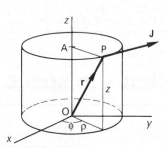

$\dfrac{\partial \mathbf{r}}{\partial \phi}$ is therefore a vector along the tangent to the circle at P and the unit vector $\mathbf{J}$ at P will be given by

$$\mathbf{J} = \frac{\partial \mathbf{r}}{\partial \phi} \bigg/ \left| \frac{\partial \mathbf{r}}{\partial \phi} \right|$$

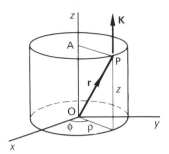

(c) Finally, if ρ and ϕ remain constant and z increases, the vector $\dfrac{\partial \mathbf{r}}{\partial z}$ will be to the z-axis and the unit vector $\mathbf{K}$ in this direction will be given by

$$\mathbf{K} = \frac{\partial \mathbf{r}}{\partial z} \bigg/ \left|\frac{\partial \mathbf{r}}{\partial z}\right|$$

Putting our three unit vectors on to one diagram, we have

. . . .

14

Note that $\mathbf{I}$, $\mathbf{J}$, $\mathbf{K}$ are mutually perpendicular and form a right-handed set. But note also that, unlike the unit vectors $\mathbf{i}$, $\mathbf{j}$, $\mathbf{k}$ in the Cartesian system, the unit vectors $\mathbf{I}$, $\mathbf{J}$, $\mathbf{K}$, or *base vectors* as they are called, are not fixed in directions, but change as the position of P changes.

So we have, for cylindrical polar coordinates

$$\mathbf{I} = \frac{\partial \mathbf{r}}{\partial \rho} \bigg/ \left|\frac{\partial \mathbf{r}}{\partial \rho}\right|$$

$$\mathbf{J} = \frac{\partial \mathbf{r}}{\partial \phi} \bigg/ \left|\frac{\partial \mathbf{r}}{\partial \phi}\right|$$

$$\mathbf{K} = \frac{\partial \mathbf{r}}{\partial z} \bigg/ \left|\frac{\partial \mathbf{r}}{\partial z}\right|$$

If $\mathbf{F}$ is a vector associated with P, then $\mathbf{F}(\mathbf{r}) = F_\rho \mathbf{I} + F_\phi \mathbf{J} + F_z \mathbf{K}$ where F_ρ, F_ϕ, F_z are the components of $\mathbf{F}$ in the directions of the unit base vectors $\mathbf{I}$, $\mathbf{J}$, $\mathbf{K}$.

Now let us attend to spherical coordinates in the same way.

15   2 *Spherical polar coordinates* (r, θ, ϕ)

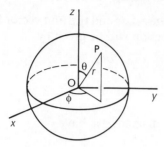

P is a function of the three variables r, θ, ϕ.

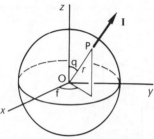

(a) If θ and ϕ remain constant and r increases, P moves outwards in the direction OP. $\dfrac{\partial \mathbf{r}}{\partial r}$ is thus a vector normal to the surface of the sphere at P and the unit vector **I** in that direction is therefore

$$\mathbf{I} = \frac{\partial \mathbf{r}}{\partial r} \Big/ \left|\frac{\partial \mathbf{r}}{\partial r}\right|$$

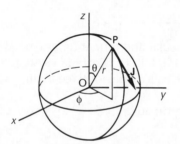

(b) If r and ϕ remain constant and θ increases, P will move along the 'meridian' through P, i.e. $\dfrac{\partial \mathbf{r}}{\partial \theta}$ is a tangent vector to this circle at P and the unit vector **J** is given by

$$\mathbf{J} = \frac{\partial \mathbf{r}}{\partial \theta} \Big/ \left|\frac{\partial \mathbf{r}}{\partial \theta}\right|$$

(c) If r and θ remain constant and ϕ increases, P will move

....

16 | along the circle through P perpendicular to the *z*-axis |

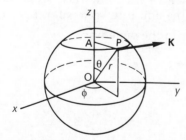

$\dfrac{\partial \mathbf{r}}{\partial \phi}$ is therefore a tangent vector at P and the unit vector **K** in this direction is given by

$$\mathbf{K} = \frac{\partial \mathbf{r}}{\partial \phi} \Big/ \left|\frac{\partial \mathbf{r}}{\partial \phi}\right|$$

So, putting the three results on one diagram, we have ....

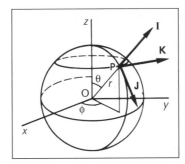

17

Once again, the three unit vectors at P (base vectors) are mutually perpendicular and form a right-handed set. Their directions in space, however, change as the position of P changes.

A vector **F** associated with P can therefore be expressed as $\mathbf{F} = F_r\mathbf{I} + F_\theta\mathbf{J} + F_\phi\mathbf{K}$ where F_r, F_θ, F_ϕ are the components of **F** in the directions of the base vectors **I**, **J**, **K**.

Both cylindrical and spherical polar coordinate systems are

. . . .

| orthogonal |

18

Scale factors

Collecting the recent results together, we have:

1 For cylindrical polar coordinates, the unit base vectors are

$$\mathbf{I} = \frac{\partial \mathbf{r}}{\partial \rho} \bigg/ \left|\frac{\partial \mathbf{r}}{\partial \rho}\right| = \frac{1}{h_\rho}\frac{\partial \mathbf{r}}{\partial \rho} \qquad \text{where } h_\rho = \left|\frac{\partial \mathbf{r}}{\partial \rho}\right|$$

$$\mathbf{J} = \frac{\partial \mathbf{r}}{\partial \phi} \bigg/ \left|\frac{\partial \mathbf{r}}{\partial \phi}\right| = \frac{1}{h_\phi}\frac{\partial \mathbf{r}}{\partial \phi} \qquad \text{where } h_\phi = \left|\frac{\partial \mathbf{r}}{\partial \phi}\right|$$

$$\mathbf{K} = \frac{\partial \mathbf{r}}{\partial z} \bigg/ \left|\frac{\partial \mathbf{r}}{\partial z}\right| = \frac{1}{h_z}\frac{\partial \mathbf{r}}{\partial z} \qquad \text{where } h_z = \left|\frac{\partial \mathbf{r}}{\partial z}\right|$$

2 For spherical polar coordinates, the unit base vectors are

$$\mathbf{I} = \frac{\partial \mathbf{r}}{\partial r} \bigg/ \left|\frac{\partial \mathbf{r}}{\partial r}\right| = \frac{1}{h_r}\frac{\partial \mathbf{r}}{\partial r} \qquad \text{where } h_r = \left|\frac{\partial \mathbf{r}}{\partial r}\right|$$

$$\mathbf{J} = \frac{\partial \mathbf{r}}{\partial \theta} \bigg/ \left|\frac{\partial \mathbf{r}}{\partial \theta}\right| = \frac{1}{h_\theta}\frac{\partial \mathbf{r}}{\partial \theta} \qquad \text{where } h_\theta = \left|\frac{\partial \mathbf{r}}{\partial \theta}\right|$$

$$\mathbf{K} = \frac{\partial \mathbf{r}}{\partial \phi} \bigg/ \left|\frac{\partial \mathbf{r}}{\partial \phi}\right| = \frac{1}{h_\phi}\frac{\partial \mathbf{r}}{\partial \phi} \qquad \text{where } h_\phi = \left|\frac{\partial \mathbf{r}}{\partial \phi}\right|$$

In each case, h is called the *scale factor*.

Move on

19 Scale factors for coordinate systems

1 *Rectangular coordinates* (x, y, z)
With rectangular coordinates, $h_x = h_y = h_z = 1$.

2 *Cylindrical coordinates* (ρ, ϕ, z)

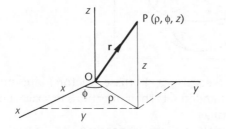

$$\mathbf{r} = x\mathbf{i} + y\mathbf{j} + z\mathbf{k}$$
$$x = \rho \cos \phi$$
$$y = \rho \sin \phi$$
$$z = z$$

$$\therefore \quad \mathbf{r} = \rho \cos \phi \, \mathbf{i} + \rho \sin \phi \, \mathbf{j} + z \, \mathbf{k}$$

$$\mathbf{I} = \frac{\partial \mathbf{r}}{\partial \rho} \Big/ \left|\frac{\partial \mathbf{r}}{\partial \rho}\right| = \frac{1}{h_\rho} \frac{\partial \mathbf{r}}{\partial \rho} \qquad h_\rho = \left|\frac{\partial \mathbf{r}}{\partial \rho}\right| = |\cos \phi \, \mathbf{i} + \sin \phi \, \mathbf{j}|$$
$$= (\cos^2 \phi + \sin^2 \phi)^{1/2} = 1$$
$$\therefore \; h_\rho = 1$$

$$\mathbf{J} = \frac{\partial \mathbf{r}}{\partial \phi} \Big/ \left|\frac{\partial \mathbf{r}}{\partial \phi}\right| = \frac{1}{h_\phi} \frac{\partial \mathbf{r}}{\partial \phi} \qquad h_\phi = \left|\frac{\partial \mathbf{r}}{\partial \phi}\right| = |-\rho \sin \phi \, \mathbf{i} + \rho \cos \phi \, \mathbf{j}|$$
$$= (\rho^2 \sin^2 \phi + \rho^2 \cos^2 \phi)^{1/2} = \rho$$
$$\therefore \; h_\phi = \rho$$

$$\mathbf{K} = \frac{\partial \mathbf{r}}{\partial z} \Big/ \left|\frac{\partial \mathbf{r}}{\partial z}\right| = \frac{1}{h_z} \frac{\partial \mathbf{r}}{\partial z} \qquad h_z = \left|\frac{\partial \mathbf{r}}{\partial z}\right| = |\mathbf{k}| = 1$$
$$\therefore \; h_z = 1$$

$$\therefore \; h_\rho = 1; \; h_\phi = \rho; \; h_z = 1$$

3 *Spherical coordinates* (r, θ, ϕ)

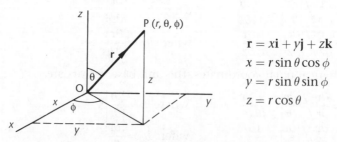

$$\mathbf{r} = x\mathbf{i} + y\mathbf{j} + z\mathbf{k}$$
$$x = r \sin \theta \cos \phi$$
$$y = r \sin \theta \sin \phi$$
$$z = r \cos \theta$$

$$\therefore \quad \mathbf{r} = r \sin \theta \cos \phi \, \mathbf{i} + r \sin \theta \sin \phi \, \mathbf{j} + r \cos \theta \, \mathbf{k}$$

Then working as before

$$h_r = \ldots\ldots\ldots\ldots; \quad h_\theta = \ldots\ldots\ldots\ldots; \quad h_\phi = \ldots\ldots\ldots\ldots$$

$$\boxed{h_r = 1; \quad h_\theta = r; \quad h_\phi = r\sin\theta}$$

Because

$$\mathbf{r} = r\sin\theta\cos\phi\,\mathbf{i} + r\sin\theta\sin\phi\,\mathbf{j} + r\cos\theta\,\mathbf{k}$$

$$\mathbf{I} = \frac{\partial\mathbf{r}}{\partial r}\Big/\left|\frac{\partial\mathbf{r}}{\partial r}\right| = \frac{1}{h_r}\frac{\partial\mathbf{r}}{\partial r}$$

$$h_r = \left|\frac{\partial\mathbf{r}}{\partial r}\right| = |\sin\theta\cos\phi\,\mathbf{i} + \sin\theta\sin\phi\,\mathbf{j} + \cos\theta\,\mathbf{k}|$$
$$= (\sin^2\theta\cos^2\phi + \sin^2\theta\,\sin^2\phi + \cos^2\theta)^{1/2}$$
$$= (\sin^2\theta + \cos^2\theta)^{1/2} = 1$$
$$\therefore\ h_r = 1$$

$$\mathbf{J} = \frac{\partial\mathbf{r}}{\partial\theta}\Big/\left|\frac{\partial\mathbf{r}}{\partial\theta}\right| = \frac{1}{h_\theta}\frac{\partial\mathbf{r}}{\partial\theta}$$

$$h_\theta = \left|\frac{\partial\mathbf{r}}{\partial\theta}\right| = |r\cos\theta\,\cos\phi\,\mathbf{i} + r\,\cos\theta\,\sin\phi\,\mathbf{j} - r\,\sin\theta\,\mathbf{k}|$$
$$= (r^2\cos^2\theta\cos^2\phi + r^2\cos^2\theta\sin^2\phi + r^2\sin^2\theta)^{1/2}$$
$$= (r^2\cos^2\theta + r^2\sin^2\theta)^{1/2} = r$$
$$\therefore\ h_\theta = r$$

$$\mathbf{K} = \frac{\partial\mathbf{r}}{\partial\phi}\Big/\left|\frac{\partial\mathbf{r}}{\partial\phi}\right| = \frac{1}{h_\phi}\frac{\partial\mathbf{r}}{\partial\phi}$$

$$h_\phi = \left|\frac{\partial\mathbf{r}}{\partial\phi}\right| = |-r\sin\theta\sin\phi\,\mathbf{i} + r\sin\theta\cos\phi\,\mathbf{j}|$$
$$= (r^2\sin^2\theta\sin^2\phi + r^2\sin^2\theta\cos^2\phi)^{1/2}$$
$$= (r^2\sin^2\theta)^{1/2} = r\sin\theta$$
$$\therefore\ h_\phi = r\sin\theta$$

$$\therefore\ h_r = 1; \quad h_\theta = r; \quad h_\phi = r\sin\theta$$

So: (a) for cylindrical coordinates

$$\mathbf{I} = \frac{\partial\mathbf{r}}{\partial\rho}; \quad \mathbf{J} = \frac{1}{\rho}\frac{\partial\mathbf{r}}{\partial\phi}; \quad \mathbf{K} = \frac{\partial\mathbf{r}}{\partial z}$$

(b) for spherical coordinates

$$\mathbf{I} = \frac{\partial\mathbf{r}}{\partial r}; \quad \mathbf{J} = \frac{1}{r}\frac{\partial\mathbf{r}}{\partial\theta}; \quad \mathbf{K} = \frac{1}{r\sin\theta}\frac{\partial\mathbf{r}}{\partial\phi}$$

General curvilinear coordinate system (*u*, *v*, *w*)

21

Any system of coordinates can be treated in like manner to obtain expressions for the appropriate unit vectors **I**, **J**, **K**.

$$\mathbf{I} = \frac{\partial \mathbf{r}}{\partial u} \bigg/ \left|\frac{\partial \mathbf{r}}{\partial u}\right|; \quad \mathbf{J} = \frac{\partial \mathbf{r}}{\partial v} \bigg/ \left|\frac{\partial \mathbf{r}}{\partial v}\right|; \quad \mathbf{K} = \frac{\partial \mathbf{r}}{\partial w} \bigg/ \left|\frac{\partial \mathbf{r}}{\partial w}\right|$$

These unit vectors are not always at right angles to each other.

If they are mutually perpendicular, the coordinate system is

. . . .

22

orthogonal

Unit vectors **I**, **J**, **K** are orthogonal if

$$\mathbf{I} \cdot \mathbf{J} = \mathbf{J} \cdot \mathbf{K} = \mathbf{K} \cdot \mathbf{I} = 0$$

Exercise

Determine the unit base vectors in the directions of the following vectors and determine whether the vectors are orthogonal.

1	$\mathbf{i} - 2\mathbf{j} + 4\mathbf{k}$	2	$2\mathbf{i} - 3\mathbf{j} + 2\mathbf{k}$
	$2\mathbf{i} + 3\mathbf{j} + \mathbf{k}$		$\mathbf{i} + 2\mathbf{j} + 2\mathbf{k}$
	$-2\mathbf{i} + \mathbf{j} + \mathbf{k}$		$-10\mathbf{i} - 2\mathbf{j} + 7\mathbf{k}$

3	$4\mathbf{i} + 2\mathbf{j} - \mathbf{k}$	4	$3\mathbf{i} + 2\mathbf{j} + \mathbf{k}$
	$3\mathbf{i} - 5\mathbf{j} + 2\mathbf{k}$		$\mathbf{i} - 3\mathbf{j} + 3\mathbf{k}$
	$\mathbf{i} + 2\mathbf{j} + 6\mathbf{k}$		$6\mathbf{i} + \mathbf{j} - \mathbf{k}$

23

The results are as follows:

1 $\mathbf{I} = \dfrac{1}{\sqrt{21}}(\mathbf{i} - 2\mathbf{j} + 4\mathbf{k}); \quad \mathbf{J} = \dfrac{1}{\sqrt{14}}(2\mathbf{i} + 3\mathbf{j} + \mathbf{k});$

   $\mathbf{K} = \dfrac{1}{\sqrt{6}}(-2\mathbf{i} + \mathbf{j} + \mathbf{k})$

   $\mathbf{I} \cdot \mathbf{J} = 0; \quad \mathbf{J} \cdot \mathbf{K} = 0; \quad \mathbf{K} \cdot \mathbf{I} = 0 \quad \therefore$ orthogonal

2 $\mathbf{I} = \dfrac{1}{\sqrt{17}}(2\mathbf{i} - 3\mathbf{j} + 2\mathbf{k}); \quad \mathbf{J} = \dfrac{1}{3}(\mathbf{i} + 2\mathbf{j} + 2\mathbf{k});$

   $\mathbf{K} = \dfrac{1}{\sqrt{153}}(-10\mathbf{i} + 2\mathbf{j} + 7\mathbf{k})$

   $\mathbf{I} \cdot \mathbf{J} = 0; \quad \mathbf{J} \cdot \mathbf{K} = 0; \quad \mathbf{K} \cdot \mathbf{I} = 0 \quad \therefore$ orthogonal

▶

3 $I = \dfrac{1}{\sqrt{21}}(4\mathbf{i} + 2\mathbf{j} - \mathbf{k}); \quad J = \dfrac{1}{\sqrt{38}}(3\mathbf{i} - 5\mathbf{j} + 2\mathbf{k});$

$K = \dfrac{1}{\sqrt{41}}(\mathbf{i} + 2\mathbf{j} + 6\mathbf{k})$

$I \cdot J = 0; \quad J \cdot K \neq 0 \qquad \therefore \text{ not orthogonal}$

4 $I = \dfrac{1}{\sqrt{14}}(3\mathbf{i} + 2\mathbf{j} + \mathbf{k}); \quad J = \dfrac{1}{\sqrt{19}}(\mathbf{i} - 3\mathbf{j} + 3\mathbf{k});$

$K = \dfrac{1}{\sqrt{38}}(6\mathbf{i} + \mathbf{j} - \mathbf{k})$

$I \cdot J = 0; \quad J \cdot K = 0; \quad K \cdot I \neq 0 \quad \therefore \text{ not orthogonal}$

Transformation equations

In general coordinates, the transformation equations are of the form

24

$$x = f(u, v, w); \quad y = g(u, v, w); \quad z = h(u, v, w)$$

where the functions f, g, h are continuous and single-valued, and whose partial derivatives are continuous.

Then $\mathbf{r} = x\mathbf{i} + y\mathbf{j} + z\mathbf{k} = f(u, v, w)\mathbf{i} + g(u, v, w)\mathbf{j} + h(u, v, w)\mathbf{k}$ and coordinate curves can be formed by keeping two of the three variables constant.

Now $\mathbf{r} = x\mathbf{i} + y\mathbf{j} + z\mathbf{k} \qquad \therefore \quad \mathrm{d}\mathbf{r} = \dfrac{\partial \mathbf{r}}{\partial u}\mathrm{d}u + \dfrac{\partial \mathbf{r}}{\partial v}\mathrm{d}v + \dfrac{\partial \mathbf{r}}{\partial w}\mathrm{d}w$ (1)

$\dfrac{\partial \mathbf{r}}{\partial u}$ is a tangent vector to the u-coordinate curve at P

$\dfrac{\partial \mathbf{r}}{\partial v}$ is a tangent vector to the v-coordinate curve at P

$\dfrac{\partial \mathbf{r}}{\partial w}$ is a tangent vector to the w-coordinate curve at P

$I = \dfrac{\partial \mathbf{r}}{\partial u} \Big/ \left|\dfrac{\partial \mathbf{r}}{\partial u}\right| \qquad \therefore \quad \dfrac{\partial \mathbf{r}}{\partial u} = h_u I \text{ where } h_u = \left|\dfrac{\partial \mathbf{r}}{\partial u}\right|$

$J = \dfrac{\partial \mathbf{r}}{\partial v} \Big/ \left|\dfrac{\partial \mathbf{r}}{\partial v}\right| \qquad \therefore \quad \dfrac{\partial \mathbf{r}}{\partial v} = h_v J \text{ where } h_v = \left|\dfrac{\partial \mathbf{r}}{\partial v}\right|$

$K = \dfrac{\partial \mathbf{r}}{\partial w} \Big/ \left|\dfrac{\partial \mathbf{r}}{\partial w}\right| \qquad \therefore \quad \dfrac{\partial \mathbf{r}}{\partial w} = h_w K \text{ where } h_w = \left|\dfrac{\partial \mathbf{r}}{\partial w}\right|$

Then (1) above becomes

$$\mathrm{d}\mathbf{r} = h_u \mathrm{d}u\, I + h_v \mathrm{d}v\, J + h_w \mathrm{d}w\, K$$

where, as before, h_u, h_v, h_w are the scale factors.

Element of arc d*s* and element of volume d*V* in orthogonal curvilinear coordinates

25

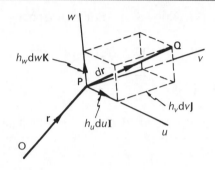

(a) *Element of arc* d*s*

Element of arc d*s* from P to Q is given by

$$\mathbf{dr} = h_u\, du\, \mathbf{I} + h_v\, dv\, \mathbf{J} + h_w dw\, \mathbf{K}$$

$$\therefore\ \mathbf{dr} \cdot \mathbf{dr} = (h_u du\, \mathbf{I} + h_v dv\, \mathbf{J} + h_w dw\, \mathbf{K}) \cdot (h_u du\, \mathbf{I}$$
$$+ h_v dv\, \mathbf{J} + h_w dw\, \mathbf{K})$$

$$\therefore\ ds^2 = h_u^2 du^2 + h_v^2 dv^2 + h_w^2 dw^2$$

$$\therefore\ ds = (h_u^2 du^2 + h_v^2 dv^2 + h_w^2 dw^2)^{1/2}$$

(b) *Element of volume* d*V*

$$dV = (h_u du\, \mathbf{I}) \cdot (h_v dv\, \mathbf{J} \times h_w dw\, \mathbf{K})$$
$$= (h_u du\, \mathbf{I}) \cdot (h_v dv\, h_w dw\, \mathbf{I}) = h_u du\, h_v dv\, h_w dw$$

$$\therefore\ dV = h_u\, h_v\, h_w\, du dv dw$$

Note also that

$$dV = \left| \frac{\partial \mathbf{r}}{\partial u} \cdot \left(\frac{\partial \mathbf{r}}{\partial v} \times \frac{\partial \mathbf{r}}{\partial w} \right) \right| du\, dv\, dw$$

$$= \frac{\partial(x,\, y,\, z)}{\partial(u,\, v,\, w)}\, du\, dv\, dw$$

where $\dfrac{\partial(x,\, y,\, z)}{\partial(u,\, v,\, w)}$ is the Jacobian of the transformation.

Grad, div and curl in orthogonal curvilinear coordinates

(a) *Grad V* (∇V)

26

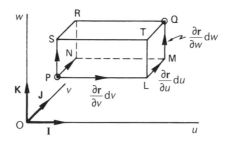

Let a scalar field V exist in space and let dV be the change in V from P to Q. If the position vector of P is **r** then that of Q is $\mathbf{r} + d\mathbf{r}$.

Then $dV = \dfrac{\partial V}{\partial u}\,du + \dfrac{\partial V}{\partial v}\,dv + \dfrac{\partial V}{\partial w}\,dw$

Let grad $V = \nabla V = (\nabla V)_u\mathbf{I} + (\nabla V)_v\mathbf{J} + (\nabla V)_w\mathbf{K}$

where $(\nabla V)_{u,v,w}$ are the components of grad V in the u, v, w directions.

Also $\quad d\mathbf{r} = \dfrac{\partial \mathbf{r}}{\partial u}\,du + \dfrac{\partial \mathbf{r}}{\partial v}\,dv + \dfrac{\partial \mathbf{r}}{\partial w}\,dw$

But $\quad \dfrac{\partial \mathbf{r}}{\partial u} = \left|\dfrac{\partial \mathbf{r}}{\partial u}\right|\mathbf{I} = h_u\,\mathbf{I}; \qquad \dfrac{\partial \mathbf{r}}{\partial v} = \left|\dfrac{\partial \mathbf{r}}{\partial v}\right|\mathbf{J} = h_v\,\mathbf{J};$

and $\quad \dfrac{\partial \mathbf{r}}{\partial w} = \left|\dfrac{\partial \mathbf{r}}{\partial w}\right|\mathbf{K} = h_w\,\mathbf{K}.$

$\therefore \quad d\mathbf{r} = h_u\,du\,\mathbf{I} + h_v\,dv\,\mathbf{J} + h_w\,dw\,\mathbf{K}$

We have previously established that $dV = \text{grad } V \cdot d\mathbf{r}$

$\therefore \quad dV = \{(\nabla V)_u\mathbf{I} + (\nabla V)_v\mathbf{J} + (\nabla V)_w\mathbf{K}\} \cdot$

$$\{h_u du\mathbf{I} + h_v dv\mathbf{J} + h_w dw\mathbf{K}\}$$

$$= (\nabla V)_u h_u\,du + (\nabla V)_v h_v\,dv + (\nabla V)_w h_w\,dw$$

But $\quad dV = \dfrac{\partial V}{\partial u}\,du + \dfrac{\partial V}{\partial v}\,dv + \dfrac{\partial V}{\partial w}\,dw$

▶

∴ Equating coefficients, we then have

$$\frac{\partial V}{\partial u} = (\nabla V)_u h_u \qquad \therefore \ (\nabla V)_u = \frac{1}{h_u} \frac{\partial V}{\partial u}$$

$$\frac{\partial V}{\partial v} = (\nabla V)_v h_v \qquad \therefore \ (\nabla V)_v = \frac{1}{h_v} \frac{\partial V}{\partial v}$$

$$\frac{\partial V}{\partial w} = (\nabla V)_w h_w \qquad \therefore \ (\nabla V)_w = \frac{1}{h_w} \frac{\partial V}{\partial w}$$

$$\therefore \ \text{grad } V = \nabla V = \frac{1}{h_u} \frac{\partial V}{\partial u} \mathbf{I} + \frac{1}{h_v} \frac{\partial V}{\partial v} \mathbf{J} + \frac{1}{h_w} \frac{\partial V}{\partial w} \mathbf{K}$$

i.e. grad operator $\nabla = \dfrac{\mathbf{I}}{h_u} \dfrac{\partial}{\partial u} + \dfrac{\mathbf{J}}{h_v} \dfrac{\partial}{\partial v} + \dfrac{\mathbf{K}}{h_w} \dfrac{\partial}{\partial w}$

27

Other results we state without proof.

(b) *Div* **F** $(\nabla \cdot \mathbf{F})$

$$\text{div } \mathbf{F} = \nabla \cdot \mathbf{F}$$

$$= \frac{1}{h_u h_v h_w} \left\{ \frac{\partial}{\partial u} (h_v h_w F_u) + \frac{\partial}{\partial v} (h_u h_w F_v) + \frac{\partial}{\partial w} (h_u h_v F_w) \right\}$$

Example 1

Show that the curvilinear expression for div **F** agrees with the earlier definition in Cartesian coordinates.

In Cartesian coordinates x, y, z we have $h_x = h_y = h_z = \ldots\ldots\ldots\ldots$ so that

$$\text{div } \mathbf{F} = \ldots\ldots\ldots\ldots$$

28

$$\boxed{\begin{array}{c} h_x = h_y = h_z = 1 \text{ so that} \\[4pt] \text{div } \mathbf{F} = \dfrac{\partial F_x}{\partial x} + \dfrac{\partial F_y}{\partial y} + \dfrac{\partial F_z}{\partial z} \end{array}}$$

(c) *Curl* **F** $(\nabla \times \mathbf{F})$

$$\text{curl } \mathbf{F} = \nabla \times \mathbf{F} = \frac{1}{h_u h_v h_w} \begin{vmatrix} h_u \mathbf{I} & h_v \mathbf{J} & h_w \mathbf{K} \\ \dfrac{\partial}{\partial u} & \dfrac{\partial}{\partial v} & \dfrac{\partial}{\partial w} \\ h_u F_u & h_v F_v & h_w F_w \end{vmatrix}$$

Example 2

Show that the curvilinear expression for curl **F** agrees with the earlier definition in Cartesian coordinates.

In Cartesian coordinates x, y, z we have $h_x = h_y = h_z = \ldots\ldots\ldots\ldots$ and **I**, **J**, **K** $= \ldots, \ \ldots, \ \ldots$ so that

$$\text{curl } \mathbf{F} = \ldots\ldots\ldots\ldots$$

$$h_x = h_y = h_z = 1 \text{ and } \mathbf{I}, \mathbf{J}, \mathbf{K} = \mathbf{i}, \mathbf{j}, \mathbf{k} \text{ so that}$$

$$\text{curl } \mathbf{F} = \mathbf{i}\left(\frac{\partial F_z}{\partial y} - \frac{\partial F_y}{\partial z}\right) + \mathbf{j}\left(\frac{\partial F_x}{\partial z} - \frac{\partial F_z}{\partial x}\right) + \mathbf{k}\left(\frac{\partial F_y}{\partial x} - \frac{\partial F_x}{\partial y}\right)$$

29

Because in Cartesians

$$h_x = h_y = h_z = 1 \quad \text{and} \quad \mathbf{I}, \mathbf{j}, \mathbf{K} = \mathbf{i}, \mathbf{j}, \mathbf{k} \quad \text{so that}$$

$$\nabla \times \mathbf{F} = \frac{1}{h_u h_v h_w} \begin{vmatrix} h_u \mathbf{I} & h_v \mathbf{J} & h_w \mathbf{K} \\ \dfrac{\partial}{\partial u} & \dfrac{\partial}{\partial v} & \dfrac{\partial}{\partial w} \\ h_u F_u & h_v F_v & h_w F_w \end{vmatrix}$$

$$= \begin{vmatrix} \mathbf{i} & \mathbf{j} & \mathbf{k} \\ \dfrac{\partial}{\partial x} & \dfrac{\partial}{\partial y} & \dfrac{\partial}{\partial z} \\ F_x & F_y & F_z \end{vmatrix}$$

$$= \mathbf{i}\left(\frac{\partial F_z}{\partial y} - \frac{\partial F_y}{\partial z}\right) + \mathbf{j}\left(\frac{\partial F_x}{\partial z} - \frac{\partial F_z}{\partial x}\right) + \mathbf{k}\left(\frac{\partial F_y}{\partial x} - \frac{\partial F_x}{\partial y}\right)$$

(d) *Div grad V* $(\nabla^2 V)$

$$\text{div grad } V = \nabla \cdot (\nabla V) = \nabla^2 V$$

$$= \frac{1}{h_u h_v h_w}\left\{\frac{\partial}{\partial u}\left(\frac{h_v h_w}{h_u} \cdot \frac{\partial V}{\partial u}\right) + \frac{\partial}{\partial v}\left(\frac{h_u h_w}{h_v} \cdot \frac{\partial V}{\partial v}\right) + \frac{\partial}{\partial w}\left(\frac{h_u h_v}{h_w} \cdot \frac{\partial V}{\partial w}\right)\right\}$$

Example 3

Show that the curvilinear expression for $\nabla^2 V$ agrees with the earlier definition in Cartesian coordinates.

In Cartesian coordinates x, y, z we have $h_x = h_y = h_z = \ldots\ldots\ldots$ so that

$$\nabla^2 V = \ldots\ldots\ldots\ldots$$

$$h_x = h_y = h_z = 1 \text{ so that}$$

$$\nabla^2 V = \frac{\partial^2 V}{\partial x^2} + \frac{\partial^2 V}{\partial y^2} + \frac{\partial^2 V}{\partial z^2}$$

30

Let's try another example, this time in coordinates other than Cartesians.

Example 4

If $V(u, v, w) = u + v^2 + w^3$ with scale factors $h_u = 2$, $h_v = 1$, $h_w = 1$, find $\nabla^2 V$ at the point (5, 3, 4).

There is very little to it. All we have to do is to determine the various partial derivatives and substitute in the expression above with relevant values.

$$\text{div grad } V = \ldots\ldots\ldots\ldots$$

31

$$\boxed{26}$$

Because

$$\nabla^2 V = \frac{1}{h_u\, h_v\, h_w} \left\{ \frac{\partial}{\partial u}\left(\frac{h_v\, h_w}{h_u} \cdot \frac{\partial V}{\partial u}\right) + \frac{\partial}{\partial v}\left(\frac{h_u\, h_w}{h_v} \cdot \frac{\partial V}{\partial v}\right) + \frac{\partial}{\partial w}\left(\frac{h_u\, h_v}{h_w} \cdot \frac{\partial V}{\partial w}\right) \right\}$$

In this case, $V = u + v^2 + w^3$ $\therefore \dfrac{\partial V}{\partial u} = 1;\quad \dfrac{\partial V}{\partial v} = 2v;\quad \dfrac{\partial V}{\partial w} = 3w^2$

Also $h_u = 2,\ h_v = 1,\ h_w = 1$

$$\therefore\ \nabla^2 V = \frac{1}{2}\left\{ \frac{\partial}{\partial u}\left(\frac{1}{2}\right) + \frac{\partial}{\partial v}(4v) + \frac{\partial}{\partial w}(6w^2) \right\}$$
$$= \tfrac{1}{2}\{0 + 4 + 12w\}$$

$\therefore$ At $w = 4,\quad \nabla^2 V = 26$

That is all there is to it. Here is another.

Example 5

If $V = (u^2 + v^2)w^3$ with $h_u = 3,\ h_v = 1,\ h_w = 2$, find div grad V at the point $(2,\ -2,\ 1)$.

$$\nabla^2 V = \ldots\ldots\ldots$$

32

$$\boxed{14\tfrac{2}{9}}$$

Because

$$V = (u^2 + v^2)w^3 \quad \therefore\ \frac{\partial V}{\partial u} = 2uw^3;\quad \frac{\partial V}{\partial v} = 2vw^3;\quad \frac{\partial V}{\partial w} = 3(u^2 + v^2)w^2$$

also $h_u = 3,\ h_v = 1,\ h_w = 2$

$$\therefore\ \nabla^2 V = \frac{1}{6}\left\{ \frac{\partial}{\partial u}\left(\frac{2}{3}\frac{\partial V}{\partial u}\right) + \frac{\partial}{\partial v}\left(6\frac{\partial V}{\partial v}\right) + \frac{\partial}{\partial w}\left(\frac{3}{2}\frac{\partial V}{\partial w}\right) \right\}$$
$$= \frac{1}{6}\left\{ \frac{\partial}{\partial u}\left(\frac{4}{3}uw^3\right) + \frac{\partial}{\partial v}(12vw^3) + \frac{\partial}{\partial w}\left(\frac{9}{2}(u^2 + v^2)w^2\right) \right\}$$

$\therefore$ at $(2,\ -2,\ 1)$

$$\nabla^2 V = \tfrac{1}{6}\{(\tfrac{4}{3}w^3) + (12w^3) + 9(u^2 + v^2)w\}$$
$$= \tfrac{1}{6}\{\tfrac{4}{3} + 12 + 72\} = \frac{256}{18} = 14\tfrac{2}{9}$$

Particular orthogonal systems

We can apply the general results for div, grad and curl to special coordinate systems by inserting the appropriate scale factors – as we shall now see.

(a) *Cartesian rectangular coordinate system* **33**

If we replace u, v, w by x, y, z and insert values of $h_x = h_y = h_z = 1$, we obtain expressions for grad, div and curl in rectangular coordinates, so that

$$\text{grad } V = \ldots\ldots\ldots ; \quad \text{div } \mathbf{F} = \ldots\ldots\ldots ; \quad \text{curl } \mathbf{F} = \ldots\ldots\ldots$$

34

$$\text{grad } V = \frac{\partial V}{\partial x}\mathbf{i} + \frac{\partial V}{\partial y}\mathbf{j} + \frac{\partial V}{\partial z}\mathbf{k}$$

$$\text{div } \mathbf{F} = \frac{\partial F_x}{\partial x} + \frac{\partial F_y}{\partial y} + \frac{\partial F_z}{\partial z}$$

$$\text{curl } \mathbf{F} = \begin{vmatrix} \mathbf{i} & \mathbf{j} & \mathbf{k} \\ \dfrac{\partial}{\partial x} & \dfrac{\partial}{\partial y} & \dfrac{\partial}{\partial z} \\ F_x & F_y & F_z \end{vmatrix}$$

$$\nabla^2 V = \frac{\partial^2 V}{\partial x^2} + \frac{\partial^2 V}{\partial y^2} + \frac{\partial^2 V}{\partial z^2}$$

all of which you will surely recognise.

(b) *Cylindrical polar coordinate system*

Here we simply replace u, v, w with ρ, ϕ, z and insert $h_u = h_\rho = 1$, $h_v = h_\phi = \rho$, $h_w = h_z = 1$ giving

$$\text{grad } V = \ldots\ldots\ldots ; \quad \text{div } \mathbf{F} = \ldots\ldots\ldots ;$$
$$\text{curl } \mathbf{F} = \ldots\ldots\ldots$$

35

$$\text{grad } V = \frac{\partial V}{\partial \rho}\mathbf{I} + \frac{1}{\rho}\frac{\partial V}{\partial \phi}\mathbf{J} + \frac{\partial V}{\partial z}\mathbf{K}$$

$$\text{div } \mathbf{F} = \frac{1}{\rho}\left\{ \frac{\partial}{\partial \rho}(\rho F_\rho) + \frac{\partial}{\partial \phi}(F_\phi) + \frac{\partial}{\partial z}(\rho F_z) \right\}$$

$$\text{curl } \mathbf{F} = \frac{1}{\rho}\begin{vmatrix} \mathbf{I} & \rho\mathbf{J} & \mathbf{K} \\ \dfrac{\partial}{\partial \rho} & \dfrac{\partial}{\partial \phi} & \dfrac{\partial}{\partial z} \\ F_\rho & \rho F_\phi & F_z \end{vmatrix}$$

$$\nabla^2 V = \frac{\partial^2 V}{\partial \rho^2} + \frac{1}{\rho}\frac{\partial V}{\partial \rho} + \frac{1}{\rho^2}\frac{\partial^2 V}{\partial \phi^2} + \frac{\partial^2 V}{\partial z^2}$$

(c) *Spherical polar coordinate system*

Replacing u, v, w with r, θ, ϕ with $h_r = 1$, $h_\theta = r$, $h_\phi = r\sin\theta$,

$$\text{grad } V = \ldots\ldots\ldots ; \quad \text{div } \mathbf{F} = \ldots\ldots\ldots ;$$
$$\text{curl } \mathbf{F} = \ldots\ldots\ldots$$

36

$$\operatorname{grad} V = \frac{\partial V}{\partial r}\mathbf{I} + \frac{1}{r}\frac{\partial V}{\partial \theta}\mathbf{J} + \frac{1}{r\sin\theta}\frac{\partial V}{\partial \phi}\mathbf{K}$$

$$\operatorname{div}\mathbf{F} = \frac{1}{r^2\sin\theta}\left\{\frac{\partial}{\partial r}(r^2\sin\theta\,F_r) + \frac{\partial}{\partial \theta}(r\sin\theta\,F_\theta) + \frac{\partial}{\partial \phi}(rF_\phi)\right\}$$

$$\operatorname{curl}\mathbf{F} = \frac{1}{r^2\sin\theta}\begin{vmatrix} \mathbf{I} & r\mathbf{J} & r\sin\theta\,\mathbf{K} \\ \dfrac{\partial}{\partial r} & \dfrac{\partial}{\partial \theta} & \dfrac{\partial}{\partial \phi} \\ F_r & rF_\theta & r\sin\theta\,F_\phi \end{vmatrix}$$

$$\nabla^2 V = \frac{\partial^2 V}{\partial r^2} + \frac{2}{r}\frac{\partial V}{\partial r} + \frac{1}{r^2}\frac{\partial^2 V}{\partial \theta^2} + \frac{\cot\theta}{r^2}\frac{\partial V}{\partial \theta} + \frac{1}{r^2\sin\theta}\frac{\partial^2 V}{\partial \phi^2}$$

The results we have compiled are sometimes written in slightly different forms, but they are, of course, equivalent.

That brings us to the end of this Programme which is designed as an introduction to the topic of curvilinear coordinates. It has considerable applications, but these are beyond the scope of this present course of study.

The **Revision summary** follows as usual. Make any further notes as necessary: then you can work through the **Can you?** checklist and the **Test exercise** without difficulty. The Programme ends with the usual **Further problems**.

Revision summary 24

37

1 *Curvilinear coordinates in two dimensions*

$$u = f(x,y); \quad v = g(x,y)$$

2 *Orthogonal coordinate system in space*

(a) *Cartesian rectangular coordinates* (x,y,z)

$\mathbf{F} = F_x\mathbf{i} + F_y\mathbf{j} + F_z\mathbf{k}$ Scale factors $h_x = h_y = h_z = 1$

(b) *Cylindrical polar coordinates* (ρ, ϕ, z)

$\mathbf{r} = \rho\cos\phi\,\mathbf{i} + \rho\,\sin\phi\,\mathbf{j} + z\,\mathbf{k}$

Base unit vectors: Scale factors:

$$\mathbf{I} = \frac{\partial \mathbf{r}}{\partial \rho}\bigg/\left|\frac{\partial \mathbf{r}}{\partial \rho}\right| \qquad\qquad h_\rho = \left|\frac{\partial \mathbf{r}}{\partial \rho}\right| = 1$$

$$\mathbf{J} = \frac{\partial \mathbf{r}}{\partial \phi}\bigg/\left|\frac{\partial \mathbf{r}}{\partial \phi}\right| \qquad\qquad h_\phi = \left|\frac{\partial \mathbf{r}}{\partial \phi}\right| = \rho$$

$$\mathbf{K} = \frac{\partial \mathbf{r}}{\partial z}\bigg/\left|\frac{\partial \mathbf{r}}{\partial z}\right| \qquad\qquad h_z = \left|\frac{\partial \mathbf{r}}{\partial z}\right| = 1$$

$\mathbf{F} = F_\rho\,\mathbf{I} + F_\phi\,\mathbf{J} + F_z\,\mathbf{K}$

▶

(c) *Spherical polar coordinates* (r, θ, ϕ)

$$\mathbf{r} = r\sin\theta\,\cos\phi\,\mathbf{i} + r\sin\theta\,\sin\phi\,\mathbf{j} + r\cos\theta\,\mathbf{k}$$

Base unit vectors: Scale factors:

$$\mathbf{I} = \frac{\partial\mathbf{r}}{\partial r}\Big/\left|\frac{\partial\mathbf{r}}{\partial r}\right| \qquad\qquad h_r = \left|\frac{\partial\mathbf{r}}{\partial r}\right| = 1$$

$$\mathbf{J} = \frac{\partial\mathbf{r}}{\partial\theta}\Big/\left|\frac{\partial\mathbf{r}}{\partial\theta}\right| \qquad\qquad h_\theta = \left|\frac{\partial\mathbf{r}}{\partial\theta}\right| = r$$

$$\mathbf{K} = \frac{\partial\mathbf{r}}{\partial\phi}\Big/\left|\frac{\partial\mathbf{r}}{\partial\phi}\right| \qquad\qquad h_\phi = \left|\frac{\partial\mathbf{r}}{\partial\phi}\right| = r\sin\theta$$

$$\mathbf{F} = F_r\,\mathbf{I} + F_\theta\,\mathbf{J} + F_\phi\,\mathbf{K}$$

3 *General orthogonal curvilinear coordinates* (u, v, w)

$$x = f(u, v, w); \quad y = g(u, v, w); \quad w = h(u, v, w)$$

$$\mathbf{r} = x\mathbf{i} + y\mathbf{j} + z\mathbf{k}$$

$$\frac{\partial\mathbf{r}}{\partial u} = h_u\mathbf{I} \quad \text{where} \quad h_u = \left|\frac{\partial\mathbf{r}}{\partial u}\right|$$

$$\frac{\partial\mathbf{r}}{\partial v} = h_v\mathbf{J} \quad \text{where} \quad h_v = \left|\frac{\partial\mathbf{r}}{\partial v}\right|$$

$$\frac{\partial\mathbf{r}}{\partial w} = h_w\mathbf{K} \quad \text{where} \quad h_w = \left|\frac{\partial\mathbf{r}}{\partial w}\right|$$

Element of arc: $\mathrm{d}s = (h_u^2\,\mathrm{d}u^2 + h_v^2\,\mathrm{d}v^2 + h_w^2\,\mathrm{d}w^2)^{1/2}$

Element of volume: $\mathrm{d}V = h_u h_v h_w\,\mathrm{d}u\,\mathrm{d}v\,\mathrm{d}w$

$$= \frac{\partial(x, y, z)}{\partial(u, v, w)}\,\mathrm{d}u\,\mathrm{d}v\,\mathrm{d}w$$

4 *Grad, div and curl in orthogonal curvilinear coordinates*

(a) Grad $V = \nabla V = \dfrac{1}{h_u}\dfrac{\partial V}{\partial u}\mathbf{I} + \dfrac{1}{h_v}\dfrac{\partial V}{\partial v}\mathbf{J} + \dfrac{1}{h_w}\dfrac{\partial V}{\partial w}\mathbf{K}$

grad operator $= \nabla = \dfrac{\mathbf{I}}{h_u}\dfrac{\partial}{\partial u} + \dfrac{\mathbf{J}}{h_v}\dfrac{\partial}{\partial v} + \dfrac{\mathbf{K}}{h_w}\dfrac{\partial}{\partial w}$

(b) Div $\mathbf{F} = \dfrac{1}{h_u h_v h_w}\left\{\dfrac{\partial}{\partial u}(h_v h_w F_u) + \dfrac{\partial}{\partial v}(h_w h_u F_v) + \dfrac{\partial}{\partial w}(h_u h_v F_w)\right\}$

(c) Curl $\mathbf{F} = \dfrac{1}{h_u h_v h_w}\begin{vmatrix} h_u\mathbf{I} & h_v\mathbf{J} & h_w\mathbf{K} \\ \dfrac{\partial}{\partial u} & \dfrac{\partial}{\partial v} & \dfrac{\partial}{\partial w} \\ h_u F_u & h_v F_v & h_w F_w \end{vmatrix}$

(d) Div grad $V = \nabla \cdot \nabla V = \nabla^2 V$

$$= \frac{1}{h_u h_v h_w}\left\{\frac{\partial}{\partial u}\left(\frac{h_v h_w}{h_u}\cdot\frac{\partial V}{\partial u}\right) + \frac{\partial}{\partial v}\left(\frac{h_u h_w}{h_v}\cdot\frac{\partial V}{\partial v}\right) + \frac{\partial}{\partial w}\left(\frac{h_u h_v}{h_w}\cdot\frac{\partial V}{\partial w}\right)\right\}$$

▶

5   *Grad, div and curl in cylindrical and spherical coordinates*
    (a) *Cylindrical coordinates* (ρ, ϕ, z)

$$\text{grad } V = \frac{\partial V}{\partial \rho}\mathbf{I} + \frac{1}{\rho}\frac{\partial V}{\partial \phi}\mathbf{J} + \frac{\partial V}{\partial z}\mathbf{K}$$

$$\text{div } \mathbf{F} = \frac{1}{\rho}\left\{\frac{\partial(\rho F_\rho)}{\partial \rho}\right\} + \frac{1}{\rho}\left\{\frac{\partial F_\phi}{\partial \phi}\right\} + \frac{\partial F_z}{\partial z}$$

$$\text{curl } \mathbf{F} = \frac{1}{\rho}\begin{vmatrix} \mathbf{I} & \rho\mathbf{J} & \mathbf{K} \\ \dfrac{\partial}{\partial \rho} & \dfrac{\partial}{\partial \phi} & \dfrac{\partial}{\partial z} \\ F_\rho & \rho F_\phi & F_z \end{vmatrix}$$

$$\nabla^2 V = \frac{\partial^2 V}{\partial \rho^2} + \frac{1}{\rho}\frac{\partial V}{\partial \rho} + \frac{1}{\rho^2}\frac{\partial^2 V}{\partial \phi^2} + \frac{\partial^2 V}{\partial z^2}$$

(b) *Spherical coordinates* (r, θ, ϕ)

$$\text{grad } V = \frac{\partial V}{\partial r}\mathbf{I} + \frac{1}{r}\frac{\partial V}{\partial \theta}\mathbf{J} + \frac{1}{r\sin\theta}\frac{\partial V}{\partial \phi}\mathbf{K}$$

$$\text{div } \mathbf{F} = \frac{1}{r^2}\frac{\partial}{\partial r}(r^2 F_r) + \frac{1}{r\sin\theta}\frac{\partial}{\partial \theta}(\sin\theta\, F_\theta) + \frac{1}{r\sin\theta}\frac{\partial}{\partial \phi}(F_\phi)$$

$$\text{curl } \mathbf{F} = \frac{1}{r^2\sin\theta}\begin{vmatrix} \mathbf{I} & r\mathbf{J} & r\sin\theta\,\mathbf{K} \\ \dfrac{\partial}{\partial r} & \dfrac{\partial}{\partial \theta} & \dfrac{\partial}{\partial \phi} \\ F_r & rF_\theta & r\sin\theta\, F_\phi \end{vmatrix}$$

$$\nabla^2 V = \frac{\partial^2 V}{\partial r^2} + \frac{2}{r}\frac{\partial V}{\partial r} + \frac{1}{r^2}\frac{\partial^2 V}{\partial \theta^2} + \frac{\cot\theta}{r^2}\frac{\partial V}{\partial \theta} + \frac{1}{r^2\sin^2\theta}\frac{\partial^2 V}{\partial \phi^2}$$

 # Can you?

---

38   Checklist 24

Check this list before and after you try the end of Programme test

On a scale of 1 to 5 how confident are you that you can: **Frames**

• Derive the family of curves of constant coordinates for
  curvilinear coordinates?   1 to 11
  Yes ☐   ☐   ☐   ☐   ☐   *No*

• Derive unit base vectors and scale factors in orthogonal
  curvilinear coordinates?   12 to 24
  Yes ☐   ☐   ☐   ☐   ☐   *No*

▶

- Obtain the element of arc d*s* and the element of volume d*V* in orthogonal curvilinear coordinates?   [25]
  Yes ☐ ☐ ☐ ☐ ☐   *No*

- Obtain expressions for the operators grad, div and curl in orthogonal curvilinear coordinates?   [26] to [36]
  Yes ☐ ☐ ☐ ☐ ☐   *No*

 Test exercise 24

1 Determine the unit vectors in the directions of the following three vectors and **39** test whether they form an orthogonal set.

   $3\mathbf{i} - 2\mathbf{j} + \mathbf{k}$

   $\mathbf{i} + 2\mathbf{j} + \mathbf{k}$

   $-2\mathbf{i} - \mathbf{j} + 4\mathbf{k}$.

2 If $\mathbf{r} = u\sin 2\theta\,\mathbf{i} + u\cos 2\theta\,\mathbf{j} + v^2\,\mathbf{k}$, determine the scale factors h_u, h_v, h_θ.

3 If P is a point $\mathbf{r} = \rho\cos\phi\,\mathbf{i} + \rho\sin\phi\,\mathbf{j} + z\,\mathbf{k}$ and a scalar field $V = \rho^2 z\sin 2\phi$ exists in space, using cylindrical polar coordinates (ρ, ϕ, z) determine grad V at the point at which $\rho = 1$, $\phi = \pi/4$, $z = 2$.

4 A vector field $\mathbf{F}$ is given in cylindrical coordinates by

   $\mathbf{F} = \rho\cos\phi\,\mathbf{I} + \rho\,\sin 2\phi\,\mathbf{J} + z\,\mathbf{K}$

   Determine   (a) div $\mathbf{F}$;   (b) curl $\mathbf{F}$.

5 Using spherical coordinates (r, θ, ϕ) determine expressions for

   (a) an element of arc d*s*;

   (b) an element of volume d*V*.

6 If V is a scalar field such that $V = u^2 v w^3$ and scale factors are $h_u = 1$, $h_v = 2$, $h_w = 4$, determine $\nabla^2 V$ at the point $(2, 3, -1)$.

Further problems 24

40 **1** Determine whether the following sets of three vectors are orthogonal.

(a) $4\mathbf{i} - 2\mathbf{j} - \mathbf{k}$ (b) $2\mathbf{i} + 3\mathbf{j} - \mathbf{k}$

  $3\mathbf{i} + 5\mathbf{j} + 2\mathbf{k}$ $4\mathbf{i} - 2\mathbf{j} + 2\mathbf{k}$

  $\mathbf{i} - 11\mathbf{j} + 26\mathbf{k}$ $\mathbf{i} + 4\mathbf{j} + 2\mathbf{k}$

2 If $V(u, v, w) = v^3 w^2 \sin 2u$ with scale factors $h_u = 3$, $h_v = 1$, $h_w = 2$, determine div grad V at the point $(\pi/4, -1, 3)$.

3 A scalar field $V = \dfrac{u^2 e^{2w}}{v}$ exists in space. If the relevant scale factors are $h_u = 2$, $h_v = 3$, $h_w = 1$, determine the value of $\nabla^2 V$ at the point $(1, 2, 0)$.

4 If $\mathbf{r} = x\mathbf{i} + y\mathbf{j} + z\mathbf{k}$ and $x = r\sin\theta\cos\phi$, $y = r\sin\theta\sin\phi$, $z = r\cos\theta$ in spherical polar coordinates (r, θ, ϕ), prove that, for any vector field $\mathbf{F}$ where

$$\mathbf{F} = F_x\mathbf{i} + F_y\mathbf{j} + F_z\mathbf{k} = F_r\mathbf{I} + F_\theta\mathbf{J} + F_\phi\mathbf{K}$$

then $F_x = F_r\sin\theta\cos\phi + F_\theta\cos\theta\cos\phi - F_\phi\sin\phi$

  $F_y = F_r\sin\theta\sin\phi + F_\theta\cos\theta\sin\phi + F_\phi\cos\phi$

  $F_z = F_r\cos\theta - F_\theta\sin\theta.$

5 If V is a scalar field, determine an expression for $\nabla^2 V$

(a) in cylindrical polar coordinates

(b) in spherical polar coordinates.

6 Transformation equations from rectangular coordinates (x, y, z) to parabolic cylindrical coordinates (u, v, w) are

$$x = \frac{u^2 - v^2}{2}; \quad y = uv; \quad z = w$$

V is a scalar field and $\mathbf{F}$ a vector field.

(a) Prove that the (u, v, w) system is orthogonal

(b) Determine the scale factors

(c) Find div $\mathbf{F}$

(d) Obtain an expression for $\nabla^2 V$.

Complex analysis 1

Learning outcomes

When you have completed this Programme you will be able to:

- Recognise the transformation equation in the form
  $w = f(z) = u(x, y) + jv(x, y)$
- Illustrate the image of a point in the complex z-plane under a complex mapping onto the w-plane
- Map a straight line in the z-plane onto the w-plane under the transformation $w = f(z)$
- Identify complex mappings that form translations, magnifications, rotations and their combinations
- Deal with the non-linear transformations $w = z^2$, $w = 1/z$, $w = 1/(z - a)$ and $w = (az + b)/(cz + d)$

Prerequisite: Engineering Mathematics (Sixth Edition)
Programmes 1 Complex numbers 1, 2 Complex numbers 2 and 3 Hyperbolic functions

1

The foundations of complex numbers and their application to hyperbolic functions were treated fully in Programmes 1, 2 and 3 of *Engineering Mathematics (Sixth Edition)* and these provide valuable revision should you feel it to be necessary before embarking on the new work.

It will be assumed that you are already familiar with the material covered in those previous Programmes and it would be a wise move to work through the relevant Test exercises to refresh your memory on this all-important part of the course.

Functions of a complex variable

For a function of a single real variable $f(x)$ we can construct the graph of the function by plotting points against two mutually perpendicular Cartesian axes, the x-axis and the $f(x)$-axis. For a function of a single complex variable $w = f(z) = u(x, y) + jv(x, y)$ we have four real variables, x, y, u and v. For example if $z = x + jy$ and $f(z) = z^2$ then

$$f(z) = (x + jy)^2$$
$$= x^2 + 2jxy + (jy)^2$$
$$= x^2 - y^2 + 2jxy$$

and so

$$u(x, y) = x^2 - y^2$$
$$\text{and} \quad v(x, y) = 2xy$$

We cannot plot the graph of the function $f(z)$ against a single set of axes because to do so we would be required to draw four mutually perpendicular axes which is not possible. Instead, we resort to plotting z-values against x- and y-axes in the complex z-plane and to plotting the corresponding values of $w = f(z)$ against u- and v-axes in the complex w-plane. Accordingly, values of z are plotted on one plane and the corresponding values of $f(z)$ are plotted on another plane. So in our example above for a particular value of z, for example, $z = 4 + j3$

$$u = \ldots\ldots\ldots$$
$$v = \ldots\ldots\ldots$$

$$\boxed{u = 7 \quad v = 24}$$ **2**

Because with $z = 4 + j3$, $x = 4$ and $y = 3$. Then $u = 16 - 9 = 7$ and $v = 24$.

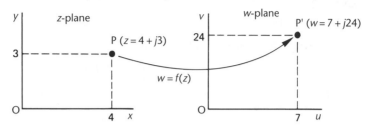

Therefore, z (where $z = x + jy$) and w (where $w = u + jv$) are two complex variables related by the equation $w = f(z)$.

Any other point in the z-plane will similarly be transformed into a corresponding point in the w-plane, the resulting position P′ depending on

(a) the initial position of P

(b) the relationship $w = f(z)$, called the *transformation equation* or *transformation function*.

Complex mapping

The transformation of P in the z-plane onto P′ in the w-plane is said to be a *mapping* of P onto P′ under the transformation $w = f(z)$ and P′ is sometimes referred to as the *image* of P.

Example 1

Determine the image of the point P, $z = 3 + j2$, on the w-plane under the transformation $w = 3z + 2 - j$.

$$w = u + jv = f(z) = 3z + 2 - j$$
$$= 3(x + jy) + 2 - j$$

so that, for this example,

$$u = \ldots\ldots\ldots\ldots; \quad v = \ldots\ldots\ldots\ldots$$

$$\boxed{u = 3x + 2; \quad v = 3y - 1}$$ **3**

Then the point P $(z = 3 + j2)$ transforms onto $\ldots\ldots\ldots\ldots$

4

$$w = 11 + j5$$

Because

$$z = 3 + j2 \quad \therefore \ x = 3, \ y = 2$$
$$u = 3x + 2 = 11; \quad v = 3y - 1 = 5; \quad \therefore \ w = 11 + j5$$

We can illustrate the transformation thus:

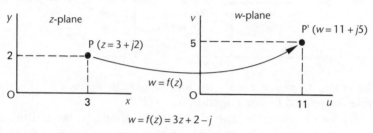

$$w = f(z) = 3z + 2 - j$$

Here is another.

Example 2

Map the points A $(z = -2 + j)$ and B $(z = 3 + j4)$ onto the *w*-plane under the transformation $w = j2z + 3$ and illustrate the transformation on a diagram.

This is no different from the previous example. Complete the job and check with the next frame.

5

$$A' \ (w = 1 - j4); \quad B' \ (w = -5 + j6)$$

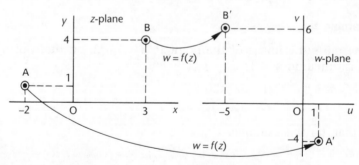

Because

$$w = f(z) = j2z + 3 = j2(x + jy) + 3 = (3 - 2y) + j2x$$
$$w = u + jv \quad \therefore \ u = 3 - 2y; \quad v = 2x$$
$$\text{A: } x = -2, \ y = 1 \quad \therefore \ A': u = 3 - 2 = 1; \ v = -4 \quad \therefore \ A': w = 1 - j4$$
$$\text{B: } x = 3, \ y = 4 \quad \therefore \ B': u = 3 - 8 = -5; \ v = 6 \quad \therefore \ B': w = -5 + j6$$

There now follows a short practice exercise. Work all four of the items before you check the results. There is no need to illustrate the transformation in each case.

So move on

Exercise **6**

Map the following points in the z-plane onto the w-plane under the transformation $w = f(z)$ stated in each case.

1 $z = 4 - j2$ under $w = j3z + j2$

2 $z = -2 - j$ under $w = jz + 3$

3 $z = 3 + j2$ under $w = (1 + j)z - 2$

4 $z = 2 + j$ under $w = z^2$.

---

1 $w = 6 + j14$	**2** $w = 4 - j2$	**7**
3 $w = -1 + j5$	**4** $w = 3 + j4$	

That was easy enough. Now let us extend the ideas.

Mapping of a straight line in the z-plane onto the w-plane under the transformation $w = f(z)$

A typical example will show the method.

Example 1

To map the straight line joining A $(-2 + j)$ and B $(3 + j6)$ in the z-plane onto the w-plane when $w = 3 + j2z$.

We first of all map the end points A and B onto the w-plane to obtain A$'$ and B$'$ as in the previous cases.

$$\text{A}': w = \dots\dots\dots$$
$$\text{B}': w = \dots\dots\dots$$

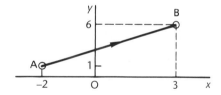

A$'$: $w = 1 - j4$; B$'$: $w = -9 + j6$	**8**

Because

    (1) A: $z = -2 + j$ $w = 3 + j2z$

        $\therefore$ A$'$: $w = 3 + j2(-2 + j) = 3 - j4 - 2 = 1 - j4$

    (2) B: $z = 3 + j6$

        $\therefore$ B$'$: $w = 3 + j2(3 + j6) = 3 + j6 - 12 = -9 + j6$

Then, if we illustrate the transformations on a diagram, as before, we get

$$\dots\dots\dots$$

9

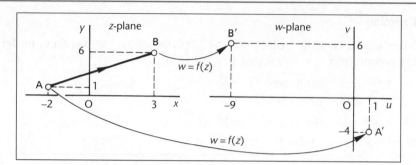

As z moves along the line A to B in the z-plane, we cannot assume that its image in the w-plane travels along a straight line from A′ to B′. As yet, we have no evidence of what the path is. We therefore have to find a general point $w = u + jv$ in the w-plane corresponding to a general point $z = x + jy$ in the z-plane.

$$w = u + jv = f(z) = 3 + j2z$$
$$= \ldots\ldots\ldots\ldots\ldots$$

10

$$\boxed{w = u + jv = (3 - 2y) + j2x}$$

Because

$$w = 3 + j2(x + jy) = 3 + j2x - 2y = (3 - 2y) + j2x$$
$$\therefore\ u = 3 - 2y \ \text{ and } \ v = 2x$$

Rearranging these results, we also have $y = \dfrac{3 - u}{2};\quad x = \dfrac{v}{2}.$

Now the Cartesian equation of AB is $y = x + 3$ and substituting from the previous line, we have $\dfrac{3 - u}{2} = \dfrac{v}{2} + 3$ which simplifies to $\ldots\ldots\ldots\ldots$

11

$$\boxed{v = -u - 3}$$

which is the equation of a straight line, so, in this case, the path joining A′ and B′ *is* in fact a straight line.

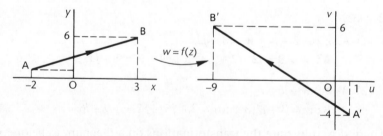

Note that it is useful to attach arrow heads to show the corresponding direction of progression in the transformation.

On to the next

Example 2   **12**

If $w = z^2$, find the path traced out by w as z moves along the straight line joining A $(2 + j0)$ and B $(0 + j2)$.

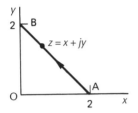

Cartesian equation of AB is

$$y = 2 - x$$

First we transform the two end points A and B onto A′ and B′ in the w-plane.

A′: ....;   B′: ....

$$\boxed{\text{A′: } w = 4 + j0; \quad \text{B′: } w = -4 + j0}$$ **13**

Because

$$w = z^2 \qquad \text{A: } z = 2 \qquad \therefore \text{ A′: } w = 2^2 = 4$$
$$\qquad\qquad \text{B: } z = j2 \qquad \therefore \text{ B′: } w = (j2)^2 = -4$$

So we have

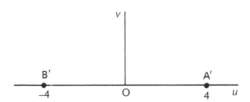

Now we have to find the path from A′ to B′.

The Cartesian equation of AB in the z-plane is $y = 2 - x$.

Also $w = z^2 = (x + jy)^2 = (x^2 - y^2) + j2xy$

$$\therefore \; u = x^2 - y^2 \quad \text{and} \quad v = 2xy$$

Substituting $y = 2 - x$ in these results we can express u and v in terms of x.

$$u = ...; \quad v = ...$$

14

$$u = 4x - 4; \quad v = 4x - 2x^2$$

So, from the first of these $\quad x = \dfrac{u+4}{4}$

Substituting in the second $\quad v = 4\left(\dfrac{u+4}{4}\right) - 2\left(\dfrac{u+4}{4}\right)^2$

$$= u + 4 - \frac{1}{8}(u^2 + 8u + 16)$$

$$= -\frac{1}{8}(u^2 - 16)$$

Therefore the path is $v = -\dfrac{1}{8}(u^2 - 16)$ which is a parabola for which at $u = 0$, $v = 2$.

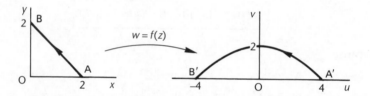

Note that a straight line in the z-plane does not always map onto a straight line in the w-plane. It depends on the particular transformation equation $w = f(z)$.

If the transformation is a *linear equation, $w = f(z) = az + b$,* where a and b may themselves be real or complex, then a straight line in the z-plane maps onto a corresponding straight line in the w-plane.

Example 3

A triangle in the z-plane has vertices at A $(z = 0)$, B $(z = 3)$ and C $(z = 3 + j2)$. Determine the image of this triangle in the w-plane under the transformation equation $w = (2 + j)z$.

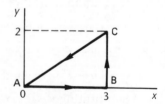

$$w = u + jv = f(z) = (2 + j)z = (2 + j)(x + jy) = (2x - y) + j(2y + x)$$

$$\therefore \ u = 2x - y; \qquad v = 2y + x$$

We now transform each vertex in turn onto the w-plane to determine A', B' and C'.

These are A': ....; B': ....; C': ....

$$\boxed{A':\ w = 0; \quad B':\ w = 6 + j3; \quad C':\ w = 4 + j7}$$

15

The transformation is linear (of the form $w = az$) so A'B', B'C' and C'A' are straight lines and the transformation can be illustrated in the diagram

. . . .

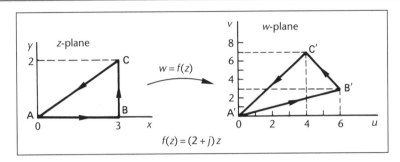

16

$$f(z) = (2 + j)z$$

All very straightforward. Let us now take a more detailed look at linear transformations.

Types of transformation of the form *w* = *az* + *b*

where the constants a and b may be real or complex.

1 *Translation*

Let $a = 1$ and $b = 2 - j$ i.e. $w = z + (2 - j)$.

If we apply this to the straight line joining A $(0 + j)$ and B $(2 + j3)$ in the z-plane, then

$$w = x + jy + 2 - j$$
$$= (x + 2) + j(y - 1)$$

so the corresponding end points A' and B' in the w-plane are

A': . . . . ;   B': . . . .

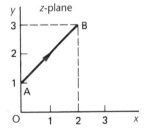

17

$$\boxed{A': \ w = 2; \quad B': \ w = 4 + j2}$$

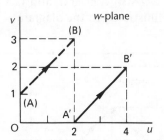

The transformed line $A'B'$ is then as shown. The broken line $(A)(B)$ indicates the position of the original line AB in the z-plane.

Note that the whole line AB has moved two units to the right and one unit downwards, while retaining its original magnitude (length) and direction.

Such a transformation is called a *translation* and occurs whenever the transformation equation is of the form $w = z + b$. The degree of translation is given by the value of b – in this case $(2 - j)$, i.e. 2 units along the positive real axis and 1 unit in the direction of the negative imaginary axis.

On to the next frame

18   **2   *Magnification***

Consider now $w = az + b$ where $b = 0$ and a is real, e.g. $w = 2z$.

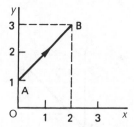

Applying the transformation to the same line AB as before, we have

$$w = u + jv = 2z = 2(x + jy)$$
$$\therefore \ u = 2x \quad \text{and} \quad v = 2y$$

Transforming the end points A $(0 + j)$ and B $(2 + j3)$ onto A' and B' in the w-plane, we have

$$A': \ \ldots\ldots\ldots\ldots; \quad B': \ \ldots\ldots\ldots\ldots$$

and the w-plane diagram becomes

$$\ldots\ldots\ldots\ldots$$

19

$$\boxed{\text{A}': \ w = j2; \quad \text{B}': \ w = 4 + j6}$$

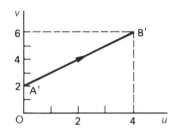

Note that (a) all distances in the *z*-plane are magnified by a factor 2, and (b) the direction of A'B' is that of AB unchanged. Any such transformation $w = az$ where a is real, is said to be a *magnification* by the factor a.

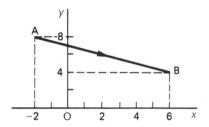

So, if we apply the transformation $w = z/2$ to AB shown here, we can map AB onto A'B' in the *w*-plane and obtain

. . . .

Sketch the result

20

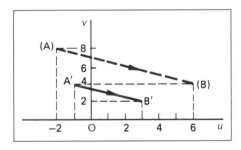

3 *Rotation*

Consider next $w = az + b$ with $b = 0$ and a complex,

e.g. $w = jz$.

$$w = u + jv = jz$$
$$= j(x + jy)$$
$$= -y + jx$$

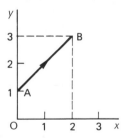

Transforming the end points as usual, we can sketch the original line AB and the mapping A'B', which gives . . . .

21

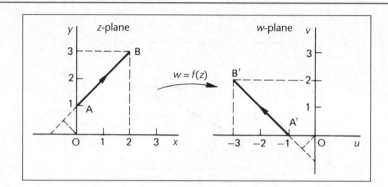

A′ is the point $w = -1 + j0$; B′ is the point $w = -3 + j2$

Note AB $= 2\sqrt{2}$ A′B′ $= 2\sqrt{2}$

Slope of AB $= m = 1$ Slope of A′B′ $= m_1 = -1$

$mm_1 = 1(-1) = -1$

Therefore in transformation by $w = jz$, AB retains its original length but is rotated about the origin, in this case through $90°$ in a positive (anticlockwise) direction.

Some degree of rotation always occurs when the transformation equation is of the form $w = az + b$ with a complex.

Move on to the next frame

22

4 Combined magnification and rotation

If $w = (a + jb)z$, the effect of transformation is

(a) magnification $\mid a + jb \mid = \sqrt{a^2 + b^2}$

(b) rotation anticlockwise through $\arg(a + jb)$, i.e. $\arctan \dfrac{b}{a}$.

Let us see this with an example.

Example

Map the straight line joining A $(0 + j2)$ and B $(4 + j6)$ in the z-plane onto the w-plane under the transformation $w = (3 + j2)z$.

The working is just as before. Draw the z-plane and w-plane diagrams, which give

. . . .

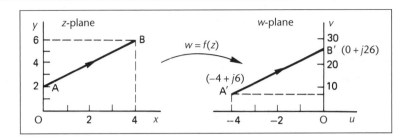

23

$w = (3 + j2)z$

$\therefore u + jv = (3 + j2)(x + jy) = (3x - 2y) + j(2x + 3y)$

$\therefore u = 3x - 2y \quad$ and $\quad v = 2x + 3y$

$\quad$ A: $\ z = 0 + j2$, i.e. $x = 0$, $\ y = 2$

$\therefore$ A': $u = -4$, $v = 6 \quad \therefore$ A': $(-4 + j6)$

$\quad$ B: $\ z = 4 + j6$, i.e. $\ x = 4$, $\ y = 6$

$\therefore$ B': $u = 0$, $v = 26 \quad \therefore$ B': $(0 + j26)$

By a simple application of Pythagoras, we can now calculate the lengths of AB and A'B', and then determine the magnification factor $(A'B')/(AB)$.

$\quad$ AB $\ = \ldots\ldots\ldots\ldots$; A'B' $= \ldots\ldots\ldots\ldots$; magnification $= \ldots\ldots\ldots\ldots$

24

$$\boxed{\text{AB} = 4\sqrt{2};\ \ \text{A'B'} = 4\sqrt{26};\quad \text{mag} = \sqrt{13}}$$

Because

$\quad \text{AB} = \sqrt{16 + 16} = \sqrt{32} = 4\sqrt{2}$

$\quad \text{A'B'} = \sqrt{16 + 400} = \sqrt{416} = 4\sqrt{26}$

$\quad \therefore$ magnification $= \dfrac{4\sqrt{26}}{4\sqrt{2}} = \sqrt{13}$

Also $| a + jb | = | 3 + j2 | = \sqrt{9 + 4} = \sqrt{13} \quad \therefore$ mag $= | a + jb |$

Now let us check the rotation.

$\quad$ For AB $\quad \tan \theta_1 = 1 \quad \therefore \ \theta_1 = 45° = 0.7854$ radians

$\quad$ For A'B' $\quad \tan \theta_2 = 5 \quad \therefore \ \theta_2 = 78° \ 41' = 1.3733$ radians

$\quad \therefore$ rotation $= \theta_2 - \theta_1 = 1.3733 - 0.7854 = 0.5879$

$\quad$ i.e. rotation $= 0.5879$ radians

Also $\arg (a + jb) = \arg (3 + j2) = \ldots\ldots\ldots\ldots$

25

$$\boxed{0.5879 \text{ radians}}$$

Because $\arg(3 + j2) = \arctan\frac{2}{3} = 33°\ 41' = 0.5879$ radians.

So, in transformation $w = (a + jb)z = (3 + j2)z$

(a) AB is magnified by $|a + jb|$, i.e. $\sqrt{13}$

(b) AB is rotated anticlockwise through $\arg(a + jb)$, i.e. $\arg(3 + j2)$ i.e. 0.5879 radians.

5 Combined magnification, rotation and translation

The work we have just done can be extended to include all three effects of transformation.

In general, a transformation equation $w = az + b$, where a and b are each real or complex, results in

magnification $|a|$;   rotation $\arg a$;   translation b

Therefore, if $w = (3 + j)z + 2 - j$

$$\text{magnification} = \ldots\ldots\ldots\ldots; \quad \text{rotation} = \ldots\ldots\ldots\ldots;$$

$$\text{translation} = \ldots\ldots\ldots\ldots$$

26

$$\boxed{\begin{array}{l} \text{mag} = \sqrt{10} = 3.162; \text{rotation} = 18°\ 26' = 0.3218 \text{ radians};\\ \text{translation} = 2 \text{ units to right, } 1 \text{ unit downwards} \end{array}}$$

Because

(a) magnification $= |3 + j| = \sqrt{9 + 1} = \sqrt{10} = 3.162$

(b) rotation $= \arg(3 + j) = \arctan\dfrac{1}{3} = 18°\ 26' = 0.3218$ radians

(c) translation $= 2 - j$, i.e. 2 to the right, 1 downwards.

Let us work through an example in detail.

Example 1

The straight line joining A $(-2 - j3)$ and B $(3 + j)$ in the z-plane is subjected to the linear transformation equation

$$w = (1 + j2)z + 3 - j4$$

Illustrate the mapping onto the w-plane and state the resulting magnification, rotation and translation involved.

The first part is just like examples we have already done. So,

(a) transform the end points A and B onto A$'$ and B$'$ in the w-plane

(b) join A$'$ and B$'$ with a straight line, since AB is a straight line and the transformation equation is linear.

That can be done without trouble, the final diagram being $\ldots\ldots\ldots\ldots$

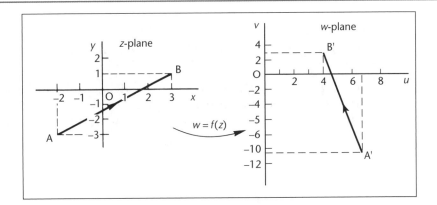

Check the working.　$w = (1 + j2)z + 3 - j4$

A:　$z = x + jy$

　　$= -2 - j3$

A′:　$w = u + jv = (1 + j2)(-2 - j3) + 3 - j4$

　　$= -2 - j7 + 6 + 3 - j4$

　　$= 7 - j11$

B:　$z = x + jy$

　　$= 3 + j$

B′:　$w = u + jv = (1 + j2)(3 + j) + 3 - j4$

　　$= 3 + j7 - 2 + 3 - j4$

　　$= 4 + j3$

Now for the second part of the problem, we have to state the magnification, rotation and translation when $w = (1 + j2)z + 3 - j4$. We remember that the 'tailpiece', i.e. $3 - j4$, independent of z, represents the

. . . .

$\boxed{\text{translation}}$ 　　　**28**

So, for the moment, we concentrate on $w = (1 + j2)z$, which determines the magnification and rotation. This tells us that

$$\text{magnification} = \dots \dots \dots$$
$$\text{rotation} = \dots \dots \dots$$

29

$$\text{mag} = \mid a \mid = \mid 1 + j2 \mid = \sqrt{1 + 4} = \sqrt{5} = 2 \cdot 236$$
$$\text{rotation} = \arg a = \arctan\left(\tfrac{2}{1}\right) = 63° \ 26' = 1 \cdot 107 \text{ radians}$$

The translation is given by $(3 - j4)$, i.e. 3 units to the right, 4 units downwards.

We can in fact see the intermediate steps if we deal first with the transformation $w = (1 + j2)z$ and subsequently with the translation $w = 3 - j4$.

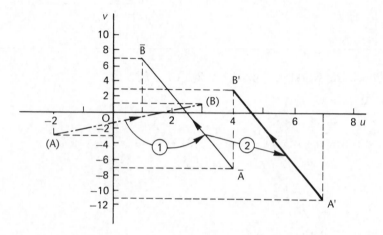

Under $w = (1 + j2)z$, A and B map onto $\overline{A}$ and $\overline{B}$ where $\overline{A}$ is $w = 4 - j7$ and $\overline{B}$ is $w = 1 + j7$.

Then the translation $w = 3 - j4$ moves all points 3 units to the right and 4 units downwards, so that $\overline{A}$ and $\overline{B}$ now map onto A' and B' where A' is $w = 7 - j11$ and B' is $w = 4 + j3$.

Normally, there is no need to analyse the transformation into intermediate steps.

Now for –

Example 2

Map the straight line joining A $(1 + j2)$ and B $(4 + j)$ in the *z*-plane onto the *w*-plane using the transformation equation

$$w = (2 - j3)z - 4 + j5$$

and state the magnification, rotation and translation involved.

There are no snags. Complete the working and check with the next frame.

Here is the complete working.

30

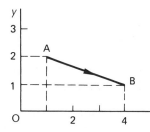

$w = (2 - j3)z - 4 + j5$
A: $z = 1 + j2$
B: $z = 4 + j$

A: $z = 1 + j2$
A′: $w = (2 - j3)(1 + j2) - 4 + j5 = 2 + j + 6 - 4 + j5 = 4 + j6$
B: $z = 4 + j$
B′: $w = (2 - j3)(4 + j) - 4 + j5 = 8 - j10 + 3 - 4 + j5 = 7 - j5$

So we have

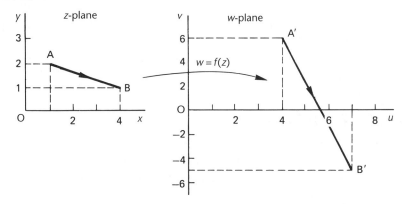

Also we have

(a) magnification $= | 2 - j3 | = \sqrt{4 + 9} = \sqrt{13} = 3·606$

(b) rotation $= \arg(2 - j3) = \arctan(\frac{-3}{2}) = -56° \; 19'$

$= 0·9828$ radians clockwise

(c) translation $= -4 + j5$ i.e. 4 units to left, 5 units upwards

All very straightforward. Before we move on, here is a short revision exercise.

Exercise

Calculate (a) the magnification, (b) the rotation, (c) the translation involved in each of the following transformations.

1   $w = (1 - j2)z + 2 - j3$ 4 $w = (j - 4)z + j2 - 3$

2   $w = (4 + j3)z - 2 + j5$ 5 $w = j2z + 4 - j$

3   $w = (2 - j3)z - 1 - j$ 6 $w = (5 + j2)z + j(j3 - 4)$.

Complete all six and then check the results with the next frame.

31   Results:

1 $w = (1 - j2)z + 2 - j3$

(a) magnitude $= | 1 - j2 | = \sqrt{1 + 4} = \sqrt{5} = 2\cdot236$

(b) rotation $= \arg(1 - j2) = \arctan(-2) = -63° \, 26'$

$= 1\cdot107$ radians clockwise

(c) translation $= 2 - j3$, i.e. 2 units to right, 3 units downwards.

The others are done in the same way and give the following results.

No.	Magnitude	Rotation (rad)	Translation
2	5	0·6435 ac	2L, 5U
3	3·606	0·9828 c	1L, 1D
4	4·123	0·2450 c	3L, 2U
5	2	1·5708 ac	4R, 1D
6	5·385	0·3805 ac	3L, 4D

Now let us start a new section, so on to the next frame

Non-linear transformations

32   So far, we have concentrated on linear transformations of the form $w = az + b$. We can now proceed to something rather more interesting.

1   *Transformation $w = z^2$ (refer to Frame 12)*

The general principles are those we have used before. An example will show the development.

Example 1

The straight line AB in the z-plane as shown is mapped onto the w-plane by $w = z^2$.

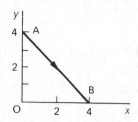

As before, we start by transforming the end points onto A$'$ and B$'$ in the w-plane.

A$'$: $w = \ldots\ldots\ldots$

B$'$: $w = \ldots\ldots\ldots$

33

$$A': \ w = -16; \quad B': \ w = 16$$

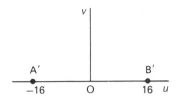

We cannot however assume that AB maps onto the straight line A'B', since the transformation is not linear. We therefore have to deal with a general point.

$$w = u + jv = z^2 = (x + jy)^2 = x^2 + j2xy - y^2 = (x^2 - y^2) + j2xy$$
$$\therefore \ u = x^2 - y^2 \quad \text{and} \quad v = 2xy$$

The Cartesian equation of AB in the z-plane is $y = 4 - x$. So, substituting in the results of the previous line, we can express u and v in terms of x.

$$u = \ldots\ldots\ldots\ldots; \quad v = \ldots\ldots\ldots\ldots$$

34

$$u = 8x - 16; \quad v = 8x - 2x^2$$

The first gives $x = \dfrac{u + 16}{8}$ and substituting this in the expression for v

gives ....

35

$$v = -\tfrac{1}{32}u^2 + 8$$

Because

$$v = 8\left(\frac{u + 16}{8}\right) - 2\left(\frac{u + 16}{8}\right)^2 = u + 16 - \frac{u^2}{32} - u - 8$$
$$\therefore \ v = -\frac{u^2}{32} + 8$$

which is an 'inverted' parabola, symmetrical about the v-axis, with $v = 8$ at $u = 0$. The mapping is therefore

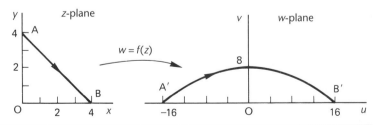

36

Example 2

AB is a straight line in the z-plane joining the origin A to the point B $(a + jb)$. Obtain the mapping of AB onto the w-plane under the transformation $w = z^2$.

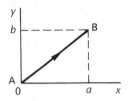

As always, first map the end points.

A′: $w = 0$

B′: $w = (a + jb)^2 = (a^2 - b^2) + j2ab$

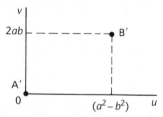

Now to find the path joining A′ and B′, we consider a general point $z = x + jy$.

$$w = u + jv = z^2$$
$$= (x + jy)^2$$
$$= (x^2 - y^2) + j2xy$$
$$\therefore \ u = x^2 - y^2 \quad \text{and} \quad v = 2xy$$

The equation of AB is $y = \dfrac{b}{a}x$. We can therefore find u and v in terms of x and hence v in terms of u.

$$u = \ldots\ldots\ldots\ldots$$
$$v = \ldots\ldots\ldots\ldots$$
$$v = f(u) = \ldots\ldots\ldots\ldots$$

37

$$u = \left(\frac{a^2 - b^2}{a^2}\right)x^2; \quad v = \left(\frac{2b}{a}\right)x^2; \quad v = \left(\frac{2ab}{a^2 - b^2}\right)u$$

Because

$$u = x^2 - y^2 = x^2 - \left(\frac{b^2}{a^2}\right)x^2 = \left(\frac{a^2 - b^2}{a^2}\right)x^2$$

$$v = 2xy = 2x\left(\frac{b}{a}\right)x = \left(\frac{2b}{a}\right)x^2$$

From the expression for u, $x^2 = \left(\frac{a^2}{a^2 - b^2}\right)u \quad \therefore \quad v = \frac{2b}{a}\left(\frac{a^2}{a^2 - b^2}\right)u$

$$\therefore \quad v = \left(\frac{2ab}{a^2 - b^2}\right)u \quad \text{which is of the form} \quad v = ku.$$

$A'B'$ is therefore a straight line through the origin.

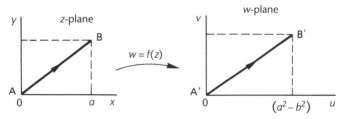

Therefore, under the transformation $w = z^2$, a straight line through the origin in the z-plane maps onto a straight line through the origin in the w-plane, whereas a straight line not passing through the origin maps onto a parabola.

This is worth remembering, so make a note of it

Example 3

38

A triangle consisting of AB, BC, CA in the z-plane is mapped onto the w-plane by the transformation $w = z^2$.

The transformation is $w = z^2$.

$$\therefore \quad w = (x + jy)^2 = (x^2 - y^2) + j2xy$$

$$= u + jv$$

$$\therefore \quad u = x^2 - y^2 \quad \text{and} \quad v = 2xy$$

First we can map the end points A, B, C onto A′, B′, C′ in the w-plane.

A′: ....
B′: ....
C′: ....

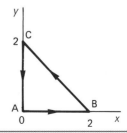

39

$$\boxed{\text{A}': \ w = 0; \quad \text{B}': \ w = 4; \quad \text{C}': \ w = -4}$$

So we establish

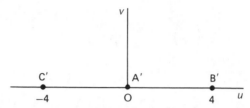

To find the paths joining these three transformed end points, we consider each of the sides of the triangle in turn.

(a) AB: Equation of AB is $y = 0$ $\therefore \ u = x^2; \quad v = 0$
     $\therefore$ Each point in AB maps onto a point between A' and B' for which $v = 0$, i.e. part of the u-axis.

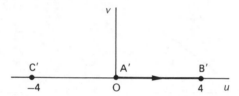

(b) BC: Equation of BC is $y = 2 - x$
     Substitute in $u = x^2 - y^2$ and $v = 2xy$ and determine v as a function of u.

$$u = \ldots\ldots\ldots\ldots$$
$$v = \ldots\ldots\ldots\ldots$$
$$v = f(u) = \ldots\ldots\ldots\ldots$$

40

$$\boxed{u = 4x - 4; \quad v = 4x - 2x^2; \quad v = 2 - \frac{u^2}{8}}$$

Because

$$u = x^2 - y^2 = x^2 - (2 - x)^2 = 4x - 4 \quad \therefore \ x = \frac{u + 4}{4}$$

$$v = 2xy = 2x(2 - x) = 4x - 2x^2$$

$$\therefore \ v = 4\left(\frac{u + 4}{4}\right) - 2\left(\frac{u + 4}{4}\right)^2 = 2 - \frac{u^2}{8}$$

Therefore, the path joining B' to C' is an

$$\ldots\ldots\ldots\ldots$$

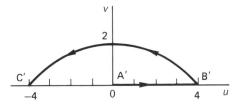

41

$\boxed{\text{inverted parabola}}$

$v = 2 - \dfrac{u^2}{8}$ ∴ at $u = 0$, $v = 2$ and the w-plane diagram now becomes

To complete the mapping, we have still to deal with CA. This transforms onto

. . . .

42

$\boxed{\text{the } u\text{-axis between C}' \text{ and A}'}$

(c) CA: Equation of CA is $x = 0$ ∴ $u = -y^2$, $v = 0$
∴ Each point between C and A maps onto the negative part of the u-axis between C$'$ and A$'$.

So finally we have

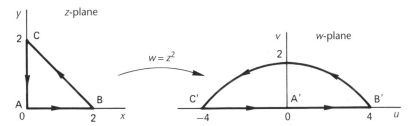

Mapping of regions

In this last example, the three lines AB, BC and CA form the boundary of a triangular region and we have seen how this boundary maps onto the boundary A$'$B$'$C$'$A$'$ in the w-plane. What we do not know yet is whether the points internal to the triangle map to points internal to the figure in the w-plane or to points external to it.

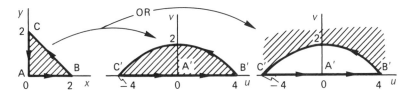

▶

In the *z*-plane, the region is on the left-hand side as we proceed round the figure in the direction of the arrows ABCA. The region on the left-hand side as we proceed round the figure A′B′C′A′ in the *w*-plane determines that the transformed region in this case is, in fact, the internal region.

So

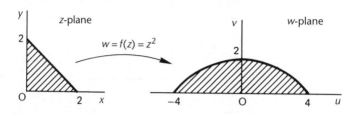

Therefore, every point in the region shaded in the *z*-plane maps onto a corresponding point in the region shaded in the *w*-plane.

43

2 *Transformation* $w = \dfrac{1}{z}$ *(inversion)*

Example 1

A straight line joining A $(-j)$ and B $(2+j)$ in the *z*-plane is mapped onto the *w*-plane by the transformation equation $w = \dfrac{1}{z}$.

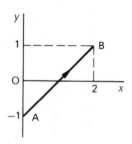

Proceeding as before

$$w = \frac{1}{z}$$

$$\therefore\ u + jv = \frac{1}{x + jy}$$

$$= \frac{x - jy}{x^2 + y^2}$$

$$\therefore\ u = \frac{x}{x^2 + y^2}; \quad v = \frac{-y}{x^2 + y^2}$$

First we map the end points A and B onto the *w*-plane.

$$A':\ w = \ldots\ldots\ldots\ldots$$
$$B':\ w = \ldots\ldots\ldots\ldots$$

$$\boxed{A'\!: \; w = j; \quad B'\!: \; w = \frac{2}{5} - j\frac{1}{5}}$$

Because

A: $x = 0, \; y = -1$　　∴ A′: $u = 0, \; v = 1$　　∴ A′ is $w = j$

B: $x = 2, \; y = 1$　　∴ B′: $u = \frac{2}{5}, \; v = -\frac{1}{5}$　　∴ B′ is $w = \frac{2}{5} - j\frac{1}{5}$

So far then we have

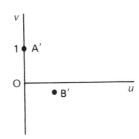

To determine the path A′B′, we can proceed as follows

$$w = \frac{1}{z} \quad \therefore \; z = \frac{1}{w} \quad \text{i.e.} \quad x + jy = \frac{1}{u + jv} = \frac{u - jv}{u^2 + v^2}$$

$$\therefore \; x = \frac{u}{u^2 + v^2} \quad \text{and} \quad y = \frac{-v}{u^2 + v^2}$$

The equation of AB is $y = x - 1$

$$\therefore \quad \frac{-v}{u^2 + v^2} = \frac{u}{u^2 + v^2} - 1$$

which simplifies into . . . .

$$\boxed{u^2 + v^2 - u - v = 0}$$

Because

$$\frac{-v}{u^2 + v^2} = \frac{u}{u^2 + v^2} - 1 \quad \therefore \quad -v = u - u^2 - v^2$$

$$\therefore \; u^2 + v^2 - u - v = 0$$

We can write this as $(u^2 - u) + (v^2 - v) = 0$ and completing the square in each bracket this becomes

$$\left(u - \frac{1}{2}\right)^2 + \left(v - \frac{1}{2}\right)^2 = \frac{1}{2}$$

which we recognise as the equation of a . . . .

46

> circle with centre $\left(\dfrac{1}{2},\dfrac{1}{2}\right)$ and radius $\dfrac{1}{\sqrt{2}}$

The path joining A′ and B′ is therefore an arc of this circle.

But we still have problems, for it could be the minor arc or the major arc.

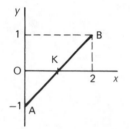

To decide which is correct, we take a further convenient point on the original line AB and determine its image on the *w*-plane.

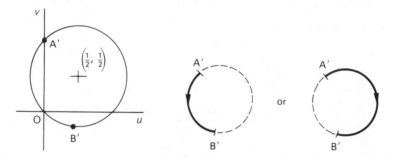

For instance, for K, $x = 1$, $y = 0$

$\therefore$ For K′, $u = \dfrac{x}{x^2 + y^2} = 1$

$$v = \dfrac{-y}{x^2 + y^2} = 0$$

$\therefore$ K′ is the point $w = 1$

The path is, therefore, the major arc A′K′B′ developed in the direction indicated.

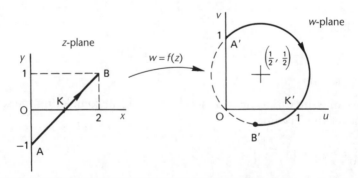

If we consider the line AB of the previous example extended to infinity in each direction, its image in the *w*-plane would then be the complete circle. **47**

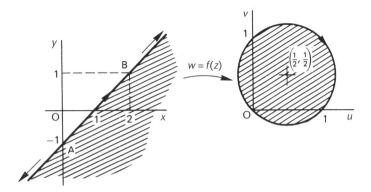

Furthermore, the line AB cuts the entire *z*-plane into two regions and

(a) the region on the right-hand side of the line relative to the arrowed direction maps onto the region inside the circle in the *w*-plane

(b) the region on the left-hand side of the line maps onto

. . . .

$$\boxed{\text{the region outside the circle in the } w\text{-plane}}$$ **48**

Let us now consider a general case.

Example 2

Determine the image in the *w*-plane of a circle in the *z*-plane under the inversion transformation $w = \dfrac{1}{z}$.

The general equation of a circle is

$$x^2 + y^2 + 2gx + 2fy + c = 0$$

with centre $(-g, -f)$

and radius $\sqrt{g^2 + f^2 - c}$.

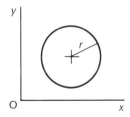

It is convenient at times to write this as

$$A(x^2 + y^2) + Dx + Ey + F = 0$$

in which case

centre is . . . . and radius is . . . .

49

$$\boxed{\text{centre}\left(-\frac{D}{2A},\ -\frac{E}{2A}\right); \quad \text{radius} = \frac{1}{2A}\sqrt{D^2 + E^2 - 4AF}}$$

Because

$$g = \frac{D}{2A}, \quad f = \frac{E}{2A}, \quad c = \frac{F}{A}.$$

As before we have $w = \dfrac{1}{z}$ $\quad \therefore\ z = \dfrac{1}{w}$

$$\therefore\ x + jy = \frac{1}{u + jv} = \frac{u - jv}{u^2 + v^2} \quad \therefore\ x = \frac{u}{u^2 + v^2}; \quad y = \frac{-v}{u^2 + v^2}$$

Then $A(x^2 + y^2) + Dx + Ey + F = 0$

becomes $\dots\dots\dots$

Simplify it as far as possible

50

$$\boxed{A + Du - Ev + F(u^2 + v^2) = 0}$$

Because we have

$$\frac{A(u^2 + v^2)}{(u^2 + v^2)^2} + \frac{Du}{u^2 + v^2} - \frac{Ev}{u^2 + v^2} + F = 0$$

$$\therefore\ A + Du - Ev + F(u^2 + v^2) = 0$$

Changing the order of terms, this can be written

$$F(u^2 + v^2) + Du - Ev + A = 0$$

which is the equation of a circle with

centre $\dots\dots\dots$; radius $\dots\dots\dots$

51

$$\boxed{\text{centre}\left(-\frac{D}{2F},\ \frac{E}{2F}\right); \quad \text{radius}\ \frac{1}{2F}\sqrt{D^2 + E^2 - 4FA}}$$

Thus any circle in the z-plane transforms, with $w = \dfrac{1}{z}$, onto another circle in the w-plane.

We have already seen previously that, under inversion, a straight line also maps onto a circle. This may be regarded as a special case of the general result, if we accept a straight line as the circumference of a circle of $\dots\dots\dots$ radius.

$$\boxed{\text{infinite}}$$

Because

$$A(x^2 + y^2) + Dx + Ey + F = 0$$

If $A = 0$, this becomes $Dx + Ey + F = 0$ i.e. a straight line

and also the centre $\left(-\dfrac{D}{2A}, -\dfrac{E}{2A} \right)$ becomes infinite

and the radius $\dfrac{1}{2A} \sqrt{D^2 + E^2 - 4AF}$ becomes infinite.

Therefore, combining the results we have obtained, we have this conclusion:

> Under inversion $w = \dfrac{1}{z}$, a circle or a straight line in the z-plane transforms
> onto a circle or a straight line in the w-plane.

Now for one more example.

Example 3

A circle in the z-plane has its centre at $z = 3$ and a radius of 2 units.
Determine its image in the w-plane when transformed by $w = \dfrac{1}{z}$.

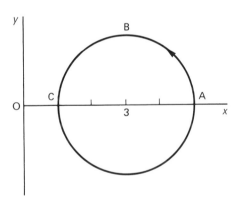

Equation of the circle is

$$(x - 3)^2 + y^2 = 4$$
$$x^2 - 6x + 9 + y^2 = 4$$
$$x^2 + y^2 - 6x + 5 = 0.$$

Using $w = \dfrac{1}{z}$, we can obtain x and y in terms of u and v.

$$x = \ldots\ldots\ldots\ldots; \quad y = \ldots\ldots\ldots\ldots$$

53

$$x = \frac{u}{u^2 + v^2}; \quad y = \frac{-v}{u^2 + v^2}$$

Because $w = \dfrac{1}{z}$,

$$\therefore \ z = \frac{1}{w}$$

$$\therefore \ x + jy = \frac{1}{u + jv}$$

$$= \frac{u - jv}{u^2 + v^2}$$

$$\therefore \ x = \frac{u}{u^2 + v^2}; \quad y = \frac{-v}{u^2 + v^2}$$

Substituting these in the equation of the circle, we get a relationship between u and v, which is

. . . .

54

$$5(u^2 + v^2) - 6u + 1 = 0$$

Because the circle is $x^2 + y^2 - 6x + 5 = 0$

$$\therefore \ \frac{u^2}{(u^2 + v^2)^2} + \frac{v^2}{(u^2 + v^2)^2} - \frac{6u}{u^2 + v^2} + 5 = 0$$

$$\frac{1}{u^2 + v^2} - \frac{6u}{u^2 + v^2} + 5 = 0$$

$$5(u^2 + v^2) - 6u + 1 = 0$$

This is of the form $A(u^2 + v^2) + Du + Ev + F = 0$

where $A = 5,\ D = -6,\ E = 0,\ F = 1$.

Therefore, the centre is . . . .

and the radius is . . . .

55

$$\boxed{\text{centre} = \left(\frac{3}{5}, 0\right); \quad \text{radius} = \frac{2}{5}}$$

Because the centre is $\left(-\dfrac{D}{2A}, -\dfrac{E}{2A}\right) = \left(\dfrac{6}{10}, 0\right)$ i.e. $\left(\dfrac{3}{5}, 0\right)$

and the radius $= \dfrac{1}{2A}\sqrt{D^2 + E^2 - 4AF} = \dfrac{1}{10}\sqrt{36 + 0 - 20} = \dfrac{2}{5}$.

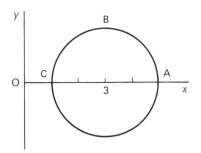 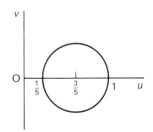

Taking three sample points A, B, C as shown, we can map these onto the
w-plane using $u = \dfrac{x}{x^2 + y^2}$ and $v = \dfrac{-y}{x^2 + y^2}$.

       A′: ....;   B′ : ....;   C′: ....

56

$$\boxed{A' : \left(\frac{1}{5}, 0\right); \quad B' : \left(\frac{3}{13}, -\frac{2}{13}\right); \quad C' : (1, 0)}$$

So we finally have

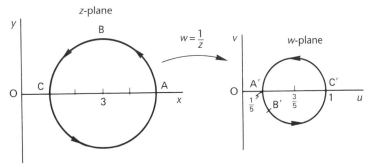

3   *Transformation* $w = \dfrac{1}{z - a}$

An extension of the method we have just applied occurs with transformations
of the form $w = \dfrac{1}{z - a}$ where a is real or complex.

Example

A circle $|z| = 1$ in the z-plane is mapped onto the w-plane by $w = \dfrac{1}{z-2}$.

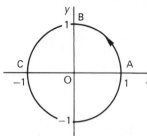

$$w = \frac{1}{z-2} \quad \therefore \quad z - 2 = \frac{1}{w}$$

$$x + jy - 2 = \frac{1}{u+jv}$$

$$(x-2) + jy = \frac{u-jv}{u^2 + v^2}$$

$$\therefore \quad x = \frac{u}{u^2+v^2} + 2; \quad y = \frac{-v}{u^2+v^2}$$

Cartesian equation of the circle is $x^2 + y^2 = 1$.

We then substitute the expressions for x and y in terms of u and v and obtain the relationship between u and v, which is ....

57

$$\boxed{3(u^2 + v^2) + 4u + 1 = 0}$$

Because we have $\left\{ \dfrac{u + 2(u^2+v^2)}{u^2+v^2} \right\}^2 + \left\{ \dfrac{-v}{u^2+v^2} \right\}^2 = 1$

$$\{u + 2(u^2+v^2)\}^2 + v^2 = (u^2+v^2)^2$$

$$u^2 + 4u(u^2+v^2) + 4(u^2+v^2)^2 + v^2 = (u^2+v^2)^2$$

$$1 + 4u + 4(u^2+v^2) = u^2 + v^2$$

$$3(u^2+v^2) + 4u + 1 = 0$$

This can be expressed as

$$u^2 + \frac{4}{3}u + v^2 + \frac{1}{3} = 0$$

$$\left(u + \frac{2}{3}\right)^2 + v^2 = \left(\frac{1}{3}\right)^2$$

which is a circle with centre $\left(-\dfrac{2}{3},\, 0\right)$ and radius $\dfrac{1}{3}$.

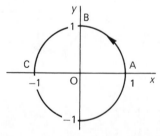
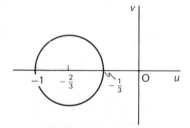

To determine the direction of development relative to the arrowed direction in the z-plane, we consider the mapping of three sample points A, B, C as shown onto the w-plane, giving A′, B′, C′.

A′: ....;   B′: ....;   C′: ....

58

$$A': \; w = (-1, 0); \quad B': \; w = \left(-\frac{2}{5}, -\frac{1}{5}\right); \quad C': \; w = \left(\frac{1}{3}, 0\right)$$

Because

A: $z = 1$ $\qquad \therefore \; w = \dfrac{1}{z-2} = -1 \qquad \therefore \; A' = (-1, 0)$

B: $z = j$ $\qquad \therefore \; w = \dfrac{1}{j-2} = \dfrac{j+2}{-5} \qquad \therefore \; B' = \left(-\dfrac{2}{5}, -\dfrac{1}{5}\right)$

C: $z = -1$ $\qquad \therefore \; w = -\dfrac{1}{3} \qquad\qquad\quad \therefore \; C' = \left(-\dfrac{1}{3}, 0\right)$

Whereupon we have

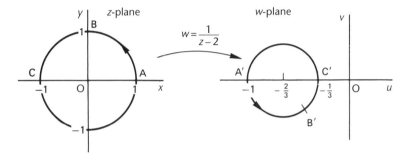

We now have one further transformation which is important, so move on to the next frame for a fresh start

59

4 Bilinear transformation $\quad w = \dfrac{az+b}{cz+d}$

Transformation of the form $w = \dfrac{az+b}{cz+d}$ where a, b, c, d are, in general, complex.

Note that

(a) if $cz + d = 1$, $w = az + b$, i.e. the general linear transformation

(b) if $az + b = 1$, $w = \dfrac{1}{cz+d}$, i.e. the form of inversion just considered.

Example

Determine the image in the w-plane of the circle $|z| = 2$ in the z-plane under the transformation $w = \dfrac{z+j}{z-j}$ and show the region in the w-plane onto which the region within the circle is mapped.

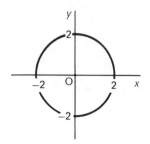

We begin in very much the same way as before by expressing u and v in terms of x and y.

$$u \ldots \ldots \ldots ; \quad v = \ldots \ldots \ldots$$

60

$$\boxed{u = \frac{x^2 + y^2 - 1}{x^2 + y^2 - 2y + 1}; \quad v = \frac{2x}{x^2 + y^2 - 2y + 1}}$$

Because

$$w = u + jv = \frac{z+j}{z-j} = \frac{x+j(y+1)}{x+j(y-1)}$$

$$= \frac{\{x + j(y+1)\}\{x - j(y-1)\}}{\{x + j(y-1)\}\{x - j(y-1)\}}$$

$$= \frac{x^2 + jx(y + 1 - y + 1) + y^2 - 1}{x^2 + (y-1)^2}$$

$$= \frac{x^2 + y^2 - 1 + j2x}{x^2 + y^2 - 2y + 1}$$

$$\therefore \ u = \frac{x^2 + y^2 - 1}{x^2 + y^2 - 2y + 1} \quad \text{and} \quad v = \frac{2x}{x^2 + y^2 - 2y + 1}$$

But the equation of the circle is $x^2 + y^2 = 4$, so these expressions simplify to

$$u = \ldots \ldots \ldots \quad \text{and} \quad v = \ldots \ldots \ldots$$

61

$$\boxed{u = \frac{3}{5 - 2y}; \quad v = \frac{2x}{5 - 2y}}$$

From these, we can form expressions for x and y in terms of u and v.

$$x = \ldots \ldots \ldots ; \quad y = \ldots \ldots \ldots$$

$$x = \frac{3v}{2u}; \quad y = \frac{5u - 3}{2u}$$

62

Because, from the first, $y = \dfrac{5u - 3}{2u}$ and substituting in the second gives

$x = \dfrac{3v}{2u}$.

But $x^2 + y^2 = 4$ ∴ $\dfrac{9v^2}{4u^2} + \dfrac{(5u - 3)^2}{4u^2} = 4$

which can be simplified to . . . .

$$9(u^2 + v^2) - 30u + 9 = 0$$

63

Because

$\quad 9v^2 + 25u^2 - 30u + 9 = 16u^2$ ∴ $9(u^2 + v^2) - 30u + 9 = 0.$

Dividing through by 9, we can now rearrange this to

$$\left(u^2 - \frac{30}{9}u\right) + v^2 + 1 = 0$$

i.e. $\quad \left(u - \dfrac{5}{3}\right)^2 + v^2 + 1 - \dfrac{25}{9} = 0$

$$\left(u - \frac{5}{3}\right)^2 + v^2 = \left(\frac{4}{3}\right)^2$$

which, you will recognise, is a circle in the *w*-plane with

centre . . . . and radius . . . .

$$\text{centre} = \left(\frac{5}{3}, 0\right); \quad \text{radius} = \frac{4}{3}$$

64

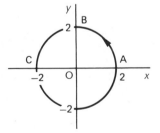
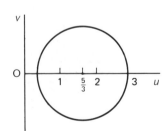

To find the direction of development, we map three sample points A, B, C onto A′, B′, C′ as usual.

A′: . . . . ; B′: . . . . ; C′: . . . .

65

$$A': w = \frac{3}{5} + j\frac{4}{5}; \quad B': w = 3; \quad C': w = \frac{3}{5} - j\frac{4}{5}$$

Because

A: $z = 2$ $\therefore w = \dfrac{2+j}{2-j} = \dfrac{(2+j)^2}{5} = \dfrac{4+j4-1}{5} = \frac{3}{5} + j\frac{4}{5}$ i.e. A'

B: $z = j2$ $\therefore w = \dfrac{j2+j}{j2-j} = \dfrac{j3}{j} = 3$ $\therefore w = 3$ i.e. B'

C: $z = -2$ $\therefore w = \dfrac{-2+j}{-2-j} = \dfrac{2-j}{2+j} = \dfrac{(2-j)^2}{5} = \frac{3}{5} - j\frac{4}{5}$ i.e. C'

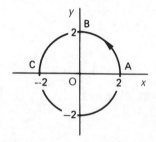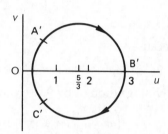

So an anticlockwise progression in the z-plane becomes a clockwise progression in the w-plane with this particular example.

Now we can complete the problem, for the region inside the circle in the z-plane maps onto .... in the w-plane.

66

the region outside the circle

Because the enclosed region in the z-plane is on the left-hand side of the direction of progression. The region on the left-hand side of the direction of progression in the w-plane is thus the region outside the transformed circle.

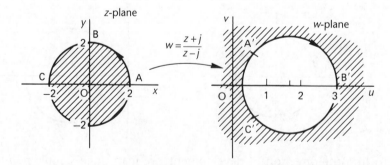

▶

And that brings us successfully to the end of this Programme. We shall pursue the topic further in the following Programme. Meanwhile, all that remains is to check down the **Revision summary** and the **Can you?** checklist before working through the **Test exercise**. All very straightforward. The **Further problems** will give you valuable additional practice.

 # Revision summary 25

1 *Transformation equation*   **67**

   $$z = x + jy \qquad w = u + jv$$

   The transformation equation is the relationship between z and w, i.e. $w = f(z)$.

2 *Linear transformation* $w = az + b$ where a and b are real or complex. A straight line in the z-plane maps onto a corresponding straight line in the w-plane.

3 *Types of transformation* $w = az + b$
   (a) *magnification* – given by $|a|$
   (b) *rotation* – given by arg a
   (c) *translation* – given by b.

4 *Non-linear transformation*
   (a) $w = z^2$
      A straight line through the origin maps onto a corresponding straight line through the origin in the w-plane. A straight line not passing through the origin maps onto a parabola.

   (b) $w = \dfrac{1}{z}$ (inversion)
      A straight line or a circle maps onto a straight line or a circle in the w-plane.
      A straight line may be regarded as a circle of infinite radius.

   (c) $w = \dfrac{az + b}{cz + d}$ (bilinear transformation) – with a, b, c, d real or complex.

5 *Mapping of a region* depends on the direction of development. Right-hand regions map onto right-hand regions: left-hand regions onto left-hand regions.

Can you?

68 **Checklist 25**

Check this list before and after you try the end of Programme test.

On a scale of 1 to 5 how confident are you that you can: **Frames**

- Recognise the transformation equation in the form
  $w = f(z) = u(x, y) + jv(x, y)$?   [1] and [2]
  Yes ☐   ☐   ☐   ☐   ☐   *No*

- Illustrate the image of a point in the complex z-plane under a
  complex mapping onto the w-plane?   [2] to [7]
  Yes ☐   ☐   ☐   ☐   ☐   *No*

- Map a straight line in the z-plane onto the w-plane under the
  transformation $w = f(z)$?   [7] to [16]
  Yes ☐   ☐   ☐   ☐   ☐   *No*

- Identify complex mappings that form translations,
  magnifications, rotations and their combinations?   [16] to [31]
  Yes ☐   ☐   ☐   ☐   ☐   *No*

- Deal with the non-linear transformations $w = z^2$, $w = 1/z$,
  $w = 1/(z - a)$ and $w = (az + b)/(cz + d)$?   [32] to [66]
  Yes ☐   ☐   ☐   ☐   ☐   *No*

Test exercise 25

69 **1** Map the following points in the z-plane onto the w-plane under the
       transformation $w = f(z)$.
       (a) $z = 3 + j2$; $w = 2z - j6$ (c) $z = j(1 - j)$; $w = (2 + j)z - 1$
       (b) $z = -2 + j$; $w = 4 + jz$ (d) $z = j - 2$; $w = (1 - j)(z + 3)$.

2 Map the straight line joining A $(2 - j)$ and B $(4 - j3)$ in the z-plane onto the
       w-plane using the transformation $w = (1 + j2)z + 1 - j3$. State the
       magnification, rotation and translation involved.

3 A triangle ABC in the z-plane as shown is mapped onto the w-plane under the
       transformation $w = z^2$.

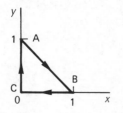

Determine the image in the w-plane and indicate the
mapping of the interior triangular region ABC.

4 Map the straight line joining A ($z = j$) and B ($z = 3 + j4$) in the z-plane onto the w-plane under the inversion transformation $w = \dfrac{1}{z}$.

Sketch the image of AB in the w-plane.

5 The unit circle $|z| = 1$ in the z-plane is mapped onto the w-plane by $w = \dfrac{1}{z - j2}$.

Determine (a) the position of the centre and (b) the radius of the circle obtained.

6 The circle $|z| = 2$ is mapped onto the w-plane by the transformation $w = \dfrac{z + j2}{z + j}$.

Determine the centre and radius of the resulting circle in the w-plane.

Further problems 25

1 A triangle ABC in the z-plane with vertices A ($-1 - j$), B ($2 + j2$), C ($-1 + j2$) is mapped onto the w-plane under the transformation $w = (1 - j)z + (1 + j2)$. Determine the image A'B'C' of ABC in the w-plane.

70

2 The straight line joining A ($1 + j2$) and B ($4 - j3$) in the z-plane is mapped onto the w-plane by the transformation equation $w = (2 + j5)z$. Determine (a) the images of A and B, (b) the magnification, rotation and translation involved.

3 Map the straight line joining A ($-2 + j3$) and B ($1 + j2$) in the z-plane onto the w-plane using the transformation equation
$$w = (-3 + j)z + 2 + j4.$$
State the magnification, rotation and translation occurring in the process.

4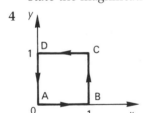

Transform the square ABCD in the z-plane onto the w-plane under the transformation $w = z^2$.

5 Map the square ABCD in the z-plane onto the w-plane using the transformation $w = 2z^2 + 2$.

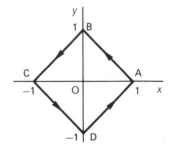

6 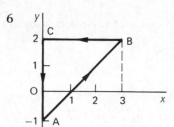 The triangle ABC in the z-plane is mapped onto the w-plane by the transformation $w = j2z^2 + 1$. Determine the image of ABC in the w-plane.

7 A circle in the z-plane has its centre at the point $(-\frac{3}{4} - j)$ and radius $\frac{7}{4}$. Show that its Cartesian equation can be expressed as

$$2(x^2 + y^2) + 3x + 4y - 3 = 0$$

Determine the image of the circle in the w-plane under the inversion transformation $w = \dfrac{1}{z}$.

8 The transformation $w = \dfrac{1}{z - 1}$ is applied to the circle $|z| = 2$ in the z-plane. Determine

(a) the image of the circle in the w-plane

(b) the region in the w-plane onto which the region enclosed within the circle in the z-plane is mapped.

9 The circle $|z| = 4$ is described in the z-plane in an anticlockwise manner. Obtain its image in the w-plane under the transformation $w = \dfrac{z + 1}{z - 2}$ and state the direction of development.

10 The bilinear transformation $w = \dfrac{z - j}{z + j2}$ is applied to the circle $|z| = 3$ in the z-plane. Determine the equation of the image in the w-plane and state its centre and radius.

11 The unit circle $|z| = 1$ in the z-plane is mapped onto the w-plane under the transformation $w = \dfrac{z - 1}{z - 3}$. Determine the equation of its image and the region onto which the region within the circle is mapped.

12 Obtain the image of the unit circle $|z| = 1$ in the z-plane under the transformation $w = \dfrac{z + j3}{z - j2}$.

13 The circle $|z| = 2$ is mapped onto the w-plane by the transformation $w = \dfrac{z + j}{2z - j}$. Determine

(a) the image of the circle in the w-plane

(b) the mapping of the region enclosed by $|z| = 2$.

14 Show that the transformation equation $w = \dfrac{z - a}{z - b}$ where $z = x + jy$, $a = 1 + j4$ and $b = 2 + j3$, transforms the circle $(x - 3)^2 + (y - 5)^2 = 5$ into a straight line through the origin in the w-plane.

Complex analysis 2

Learning outcomes

When you have completed this Programme you will be able to:

- Appreciate when the derivative of a function of a complex variable exists

- Understand the notions of regular functions and singularities and be able to obtain the derivative of a regular function from first principles

- Derive the Cauchy–Riemann equations and apply them to find the derivative of a regular function

- Understand the notion of an harmonic function and derive a conjugate function

- Evaluate line and contour integrals in the complex plane

- Derive and apply Cauchy's theorem

- Apply Cauchy's theorem to contours around regions that contain singularities

- Define the essential characteristics of and conditions for a conformal mapping

- Locate critical points of a function of a complex variable

- Determine the image in the w-plane of a figure in the z-plane under a conformal transformation $w = f(z)$

- Describe and apply the Schwarz–Christoffel transformation

1   In the previous Programme we introduced the ideas of mapping from one complex plane to another and considered some of the more common transformation functions. Now we pursue our consideration of the complex variable a little further.

Differentiation of a complex function

In differentiation of a function of a single real variable, $y = f(x)$, the derivative of y with respect to x can be defined as the limiting value of $\dfrac{(y + \delta y) - y}{\delta x}$ as δx tends to zero.

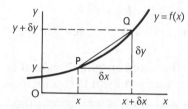

$$y = f(x) \quad \delta y = f(x + \delta x) - f(x)$$

$$\text{i.e.} \quad \frac{dy}{dx} = \lim_{\delta x \to 0} \left\{ \frac{f(x + \delta x) - f(x)}{\delta x} \right\}$$

In considering the differentiation of a function of a complex variable, $w = f(z)$, the derivative of w with respect to z can similarly be defined as the limiting value of .... as δz tends to zero.

2

$$\boxed{\frac{(w + \delta w) - w}{\delta z} \quad \text{i.e.} \quad \frac{f(z + \delta z) - f(z)}{\delta z}}$$

Now, of course, we are dealing in vectors.

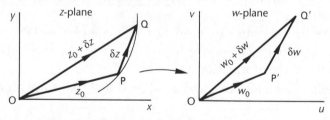

If P and Q in the z-plane map onto P' and Q' in the w-plane, then

$$P'Q' = \delta w = (w_0 + \delta w) - w_0 = f(z_0 + \delta z) - f(z_0)$$

Therefore, the derivative of w at P' $(z = z_0)$ is the limiting value of $\dfrac{\delta w}{\delta z}$ as

$\delta z \to 0$, i.e. $\left[\dfrac{dw}{dz} \right]_{z_0} = \lim_{\delta z \to 0} \left\{ \dfrac{f(z_0 + \delta z) - f(z_0)}{\delta z} \right\} = \lim_{Q \to P} \left(\dfrac{P'Q'}{PQ} \right)$

If this limiting value exists – which is not always the case as we shall see – the function $f(z)$ is said to be *differentiable at P*.

▶

Also, if $w = f(z)$ and $f'(z)$ has a limit for all points z_0 within a given region for which $w = f(z)$ is defined, then $f(z)$ is said to be differentiable in that region. From this, it follows that the limit exists whatever the path of approach from Q $(z = z_0 + \delta z)$ to P $(z = z_0)$.

Regular function

A function $w = f(z)$ is said to be *regular* (or analytic) at a point $z = z_0$, if it is defined and single-valued, and has a derivative at every point at and around z_0. Points in a region where $f(z)$ ceases to be regular are called *singular points*, or *singularities*.

A function of a complex variable that is analytic over the entire finite complex plane is called an *entire* function. Examples of entire functions are polynomials, e^z, $\sin z$ and $\cos z$.

We have introduced quite a few new definitions, so let us pause here while you make a note of them. We shall be meeting the various terms quite often.

In those cases where a derivative exists, the usual rules of differentiation apply. For example, the derivative of $w = z^2$ can be found from first principles in the normal way.

3

$$w = z^2 \quad \therefore \quad w + \delta w = (z + \delta z)^2 = z^2 + 2z\delta z + \delta z^2$$

$$\therefore \quad \delta w = 2z\delta z + \delta z^2 \quad \therefore \quad \frac{\delta w}{\delta z} = 2z + \delta z$$

$$\therefore \quad \frac{dw}{dz} = \underset{\delta z \to 0}{Lim} (2z + \delta z) = 2z \quad \text{and does not depend on the path along}$$

which δz tends to zero.

That was elementary. Here is a rather different one.

Example

To find the derivative of $w = z\bar{z}$ where $z = x + jy$ and $\bar{z} = x - jy$.

We have $\quad w = z\bar{z} \quad \therefore \quad w + \delta w = (z + \delta z)(\bar{z} + \delta\bar{z})$ from which

$$\frac{\delta w}{\delta z} = \ldots\ldots\ldots\ldots$$

$$\boxed{\frac{\delta w}{\delta z} = \bar{z} + z\frac{\delta\bar{z}}{\delta z} + \delta\bar{z}}$$

4

Because

$$w + \delta w = (z + \delta z)(\bar{z} + \delta\bar{z}) = z\bar{z} + \bar{z}\delta z + z\delta\bar{z} + \delta z\delta\bar{z}$$

$$\therefore \quad \delta w = \bar{z}\delta z + z\delta\bar{z} + \delta z\delta\bar{z} \quad \therefore \quad \frac{\delta w}{\delta z} = \bar{z} + z\frac{\delta\bar{z}}{\delta z} + \delta\bar{z}$$

Now since $z = x + jy$ and $\bar{z} = x - jy$, we can express $\frac{\delta w}{\delta z}$ in terms of x and y. $\quad \frac{\delta w}{\delta z} = \ldots\ldots\ldots\ldots$

5

$$\frac{\delta w}{\delta z} = (x - jy) + (x + jy)\left\{\frac{\delta x - j\delta y}{\delta x + j\delta y}\right\} + \delta x - j\delta y$$

Because

$$\left.\begin{array}{l} z = x + jy \quad \therefore \ \delta z = \delta x + j\delta y \\ \bar{z} = x - jy \quad \therefore \ \delta\bar{z} = \delta x - j\delta y \end{array}\right\} \quad \therefore \ \frac{\delta\bar{z}}{\delta z} = \frac{\delta x - j\delta y}{\delta x + j\delta y}$$

Then $\dfrac{\delta w}{\delta z} = \bar{z} + z\dfrac{\delta\bar{z}}{\delta z} + \delta\bar{z}$ gives the expression quoted above.

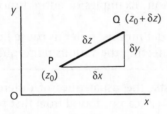

The next step is to reduce δz to zero. But δz consists of $\delta x + j\delta y$ and so reducing δz to zero can be done in one of two ways.

(1) First let $\delta y \to 0$ and afterwards let $\delta x \to 0$.

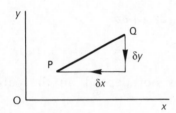

If $\delta y \to 0$, $\quad \dfrac{\delta w}{\delta z} = x - jy + (x + jy)\dfrac{\delta x}{\delta x} + \delta\bar{x}$

Then $\dfrac{dw}{dz} = \underset{\delta x \to 0}{Lim} \{x - jy + x + jy + \delta x\}$

$= \ldots\ldots\ldots$

$$\boxed{\frac{dw}{dz} = 2x}$$

6

On the other hand, we could have reduced δz to zero in the second way.

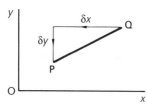

(2) First let $\delta x \to 0$ and afterwards let $\delta y \to 0$.

We have

$$\frac{\delta w}{\delta z} = x - jy + (x + jy)\left\{\frac{\delta x - j\delta y}{\delta x + j\delta y}\right\} + \delta x - j\delta y$$

If $\delta x \to 0$

$$\frac{\delta w}{\delta z} = x - jy + (x + jy)(-1) - j\delta y = -j2y - j\delta y$$

Then

$$\frac{dw}{dz} = \underset{\delta y \to 0}{Lim} \{-j2j - j\delta y\} = -j2y$$

So, in the first case, $\dfrac{dw}{dz} = 2x$ and in the second case $\dfrac{dw}{dz} = -j2y$.

These two results are clearly not the same for all values of x and y – with one exception, i.e.

. . . .

$$\boxed{\text{when } x = y = 0}$$

7

Therefore $w = z\bar{z}$ is a function that has no specific derivative, except at $z = 0$ – and there are others. It would be convenient, therefore, to have some form of test to see whether a particular function $w = f(z)$ has a derivative $f'(z)$ at $z = z_0$. This useful tool is provided by the Cauchy–Riemann equations.

Cauchy–Riemann equations

The development is very much along the same lines as in the previous example. If $w = f(z) = u + jv$, we have to establish conditions for $w = f(z)$ to have a derivative at a given point $z = z_0$.

$$w = u + jv \quad \therefore \quad \delta w = \delta u + j\delta v; \qquad z = x + jy \quad \therefore \quad \delta z = \delta x + j\delta y$$

Then $\quad f'(z) = \dfrac{dw}{dz} = \underset{\delta z \to 0}{Lim}\left\{\dfrac{\delta u + j\delta v}{\delta z}\right\} = \underset{\substack{\delta x \to 0 \\ \delta y \to 0}}{Lim}\left\{\dfrac{\delta u + j\delta v}{\delta x + j\delta y}\right\}$ \hfill (1)

(a) Let $\delta x \to 0$, followed by $\delta y \to 0$

Then from (1) above, $\quad f'(z) = \dfrac{dw}{dz} = \ldots\ldots\ldots$

8

$$\frac{\mathrm{d}w}{\mathrm{d}z} = \frac{\partial v}{\partial y} - j\frac{\partial u}{\partial y}$$

Because

$$f'(z) = \underset{\delta y \to 0}{Lim} \left\{ \frac{\delta u + j\delta v}{j\delta y} \right\} = \underset{\delta y \to 0}{Lim} \left\{ \frac{\delta v}{\delta y} - j\frac{\delta u}{\delta y} \right\} = \frac{\partial v}{\partial y} - j\frac{\partial u}{\partial y} \qquad (2)$$

We use the 'partial' notation since u and v are functions of both x and y.

Or (b) Let $\delta y \to 0$, followed by $\delta x \to 0$.

This gives . . . .

9

$$\frac{\mathrm{d}w}{\mathrm{d}z} = \frac{\partial u}{\partial x} + j\frac{\partial v}{\partial x}$$

Because

$$f'(z) = \underset{\delta x \to 0}{Lim} \left\{ \frac{\delta u + j\delta v}{\delta x} \right\} = \underset{\delta x \to 0}{Lim} \left\{ \frac{\delta u}{\delta x} + j\frac{\delta v}{\delta x} \right\} = \frac{\partial u}{\partial x} + j\frac{\partial v}{\partial x} \qquad (3)$$

If the results (2) and (3) are to have the same value for $f'(z)$ irrespective of the path chosen for δz to tend to zero, then

. . . .

10

$$\frac{\partial u}{\partial x} + j\frac{\partial v}{\partial x} = \frac{\partial v}{\partial y} - j\frac{\partial u}{\partial y}$$

Equating real and imaginary parts, this gives

$$\frac{\partial u}{\partial x} = \frac{\partial v}{\partial y} \quad \text{and} \quad \frac{\partial v}{\partial x} = -\frac{\partial u}{\partial y}$$

These are the *Cauchy–Riemann equations*.

So, to sum up:

A necessary condition for $w = f(z) = u + jv$ to be regular at $z = z_0$ is that u, v and their partial derivatives are continuous and that in the neighbourhood of $z = z_0$

$$\frac{\partial u}{\partial x} = \frac{\partial v}{\partial y} \quad \text{and} \quad \frac{\partial v}{\partial x} = -\frac{\partial u}{\partial y}$$

Make a note of this important result – then move on to the next frame

We said earlier that where a function fails to be regular, a *singular point*, or *singularity* occurs, for example where $w = f(z)$ is not continuous or where the Cauchy–Riemann test fails.

11

Exercise

Determine where each of the following functions fails to be regular, i.e. where singularities occur.

1 $w = z^2 - 4$

4 $w = \dfrac{1}{(z-2)(z-3)}$

2 $w = \dfrac{z}{z-2}$

5 $w = z\bar{z}$

3 $w = \dfrac{z+5}{z+1}$

6 $w = \dfrac{x + jy}{x^2 + y^2}$

Finish all six: then check with the next frame

Conclusions:

12

1 Putting $z = x + jy$, the Cauchy–Riemann conditions are satisfied everywhere. Therefore, no singularity in $w = z^2 - 4$.

2 The function becomes discontinuous at $z = 2$. Singularity at $z = 2$.

3 The function is discontinuous at $z = -1$. Singularity at $z = -1$.

4 Singularities at $z = 2$ and $z = 3$.

5 We have already seen that $w = z\bar{z}$ has no derivative for all values of z apart from $z = 0$. All points on $w = z\bar{z}$ are singularities.

6 Singularity occurs where $x^2 + y^2 = 0$, i.e. $x = 0$ and $y = 0$ $\therefore$ $z = 0$. At all other points the Cauchy–Riemann equations do not hold.

Harmonic functions

If a function of two real variables $f(x, y)$ satisfies Laplace's equation

13

$$\frac{\partial^2 f(x, y)}{\partial x^2} + \frac{\partial^2 f(x, y)}{\partial y^2} = 0$$

then we say that $f(x, y)$ is an *harmonic* function. It is relatively straightforward to demonstrate that the real and imaginary parts of an analytic function are both harmonic.

▶

Let $f(z) = u(x, y) + jv(x, y)$ be an analytic function in some region of the z-plane. Because $f(z)$ is analytic the Cauchy–Riemann equations hold true. That is

$$\frac{\partial u}{\partial x} = \frac{\partial v}{\partial y} \quad \text{and} \quad \frac{\partial u}{\partial y} = -\frac{\partial v}{\partial x}$$

Differentiating the first with respect to x and the second with respect to y shows us that

$$\frac{\partial^2 u}{\partial x^2} = \frac{\partial^2 v}{\partial x \partial y} \quad \text{and} \quad \frac{\partial^2 u}{\partial y^2} = -\frac{\partial^2 v}{\partial y \partial x} = -\frac{\partial^2 v}{\partial x \partial y} = -\frac{\partial^2 u}{\partial x^2}$$

and so $\dfrac{\partial^2 u}{\partial x^2} + \dfrac{\partial^2 u}{\partial y^2} = 0$

By a similar reasoning

$$\frac{\partial^2 \cdots}{\partial x^2} + \frac{\partial^2 \cdots}{\partial y^2} = 0$$

14

$$\boxed{\dfrac{\partial^2 v}{\partial x^2} + \dfrac{\partial^2 v}{\partial y^2} = 0}$$

Because

$$-\frac{\partial^2 v}{\partial x^2} = \frac{\partial^2 u}{\partial x \partial y} \quad \text{and} \quad \frac{\partial^2 v}{\partial y^2} = \frac{\partial^2 u}{\partial y \partial x} = \frac{\partial^2 u}{\partial x \partial y} = -\frac{\partial^2 v}{\partial x^2}$$

and so $\dfrac{\partial^2 v}{\partial x^2} + \dfrac{\partial^2 v}{\partial y^2} = 0$

The functions $u(x, y)$ and $v(x, y)$ are called *conjugate* functions. In addition, the curves $u = $ constant, $v = $ constant are orthogonal.

Example 1

Show that the real and imaginary parts of the function defined by $f(z) = z^2$ are harmonic.

$$f(z) = z^2$$
$$= (x + jy)^2$$
$$= (x^2 - y^2) + 2jxy$$

and so $u = x^2 - y^2$ and $v = 2xy$ and therefore

$$\frac{\partial^2 u}{\partial x^2} + \frac{\partial^2 u}{\partial y^2} = \ldots\ldots\ldots \quad \text{and} \quad \frac{\partial^2 v}{\partial x^2} + \frac{\partial^2 v}{\partial y^2} = \ldots\ldots\ldots$$

15

$$\frac{\partial^2 u}{\partial x^2} + \frac{\partial^2 u}{\partial y^2} = 0 \text{ and } \frac{\partial^2 v}{\partial x^2} + \frac{\partial^2 v}{\partial y^2} = 0$$

Because

$$\frac{\partial u}{\partial x} = 2x \text{ so } \frac{\partial^2 u}{\partial x^2} = 2 \text{ and } \frac{\partial u}{\partial y} = -2y \text{ so } \frac{\partial^2 u}{\partial y^2} = -2$$

therefore $\dfrac{\partial^2 u}{\partial x^2} + \dfrac{\partial^2 u}{\partial y^2} = 0$

and

$$\frac{\partial v}{\partial x} = 2y \text{ so } \frac{\partial^2 v}{\partial x^2} = 0 \text{ and } \frac{\partial v}{\partial y} = 2x \text{ so } \frac{\partial^2 v}{\partial y^2} = 0$$

therefore $\dfrac{\partial^2 v}{\partial x^2} + \dfrac{\partial^2 v}{\partial y^2} = 0$

Example 2

Show that $u(x, y) = x^3 y - y^3 x$ is an harmonic function and find the function $v(x, y)$ that ensures that $f(z) = u(x, u) + jv(x, y)$ is analytic. That is, find the function $v(x, y)$ that is conjugate to $u(x, y)$.

$$\frac{\partial^2 u}{\partial x^2} + \frac{\partial^2 u}{\partial y^2} = \ldots\ldots\ldots\ldots$$

16

$$\frac{\partial^2 u}{\partial x^2} + \frac{\partial^2 u}{\partial y^2} = 0$$

Because

$$\frac{\partial u}{\partial x} = 3x^2 y - y^3 \text{ so } \frac{\partial^2 u}{\partial x^2} = 6xy \text{ and } \frac{\partial u}{\partial y} = x^3 - 3y^2 x \text{ so } \frac{\partial^2 u}{\partial y^2} = -6xy$$

therefore $\dfrac{\partial^2 u}{\partial x^2} + \dfrac{\partial^2 u}{\partial y^2} = 0$

This means that $u(x, y) = x^3 y - y^3 x$ is harmonic.

Now, if $f(z) = u(x, u) + jv(x, y)$ is analytic then $u(x, y)$ and $v(x, y)$ satisfy the .... equations.

17

| Cauchy–Riemann |

That is

$$\frac{\partial u}{\partial x} = 3x^2y - y^3 = \frac{\partial v}{\partial y}$$

and

$$\frac{\partial u}{\partial y} = x^3 - 3y^2x = -\frac{\partial v}{\partial x}$$

Integrating $\frac{\partial v}{\partial y} = 3x^2y - y^3$ with respect to y gives

$$v(x, y) = \ldots\ldots\ldots$$

18

$$v(x, y) = \frac{3}{2}x^2y^2 - \frac{1}{4}y^4 + a(x)$$

Because

$\frac{\partial v}{\partial y} = 3x^2y - y^3$ and so x is treated as a constant and the integral of y^n is $y^{n+1}/(n+1)$.

Did you miss the constant term in the form of $a(x)$? *Because x is treated as a constant, the integration determines v up to an expression involving x.* Differentiate the result with respect to y and you will reclaim the original form for $\frac{\partial v}{\partial y}$.

Now, differentiating this expression with respect to x gives

$$\frac{\partial v}{\partial x} = \ldots\ldots\ldots$$

19

$$\frac{\partial v}{\partial x} = 3xy^2 + a'(x)$$

Because

$v(x, y) = \frac{3}{2}x^2y^2 - \frac{1}{4}y^4 + a(x)$ and so $\frac{\partial v}{\partial x} = 3xy^2 + a'(x)$ and this is

equal to $-\frac{\partial u}{\partial y}$. Now $-\frac{\partial u}{\partial y} = -x^3 + 3y^2x$ and so

$$a'(x) = \ldots\ldots\ldots \text{ giving } a(x) = \ldots\ldots\ldots$$
$$\text{Therefore } v(x, y) = \ldots\ldots\ldots$$

$$a'(x) = -x^3 \text{ giving } a(x) = -\frac{x^4}{4} + C.$$

$$\text{Therefore } v(x, y) = \frac{3x^2y^2}{2} - \frac{y^4}{4} - \frac{x^4}{4} + C$$

20

Because

Comparing $\dfrac{\partial v}{\partial x} = 3xy^2 + a'(x)$ and $-\dfrac{\partial u}{\partial y} = -x^3 + 3y^2x$

where $\dfrac{\partial v}{\partial x} = -\dfrac{\partial u}{\partial y}$ then it is seen that $a'(x) = -x^3$.

Therefore $a(x) = -\dfrac{x^4}{4} + C$ giving $v(x, y) = \dfrac{3x^2y^2}{2} - \dfrac{y^4}{4} - \dfrac{x^4}{4} + C$

Try one for yourself.

Example 3

Given $u(x, y) = e^{-x}\cos y$, show that $u(x, y)$ is an harmonic function and find the function $v(x, y)$ that ensures that $f(z) = u(x, y) + jv(x, y)$ is analytic. That is, find the function $v(x, y)$ that is conjugate to $u(x, y)$.

$$\frac{\partial^2 \ldots}{\partial x^2} + \frac{\partial^2 \ldots}{\partial y^2} = \ldots\ldots\ldots\ldots$$

$$\frac{\partial^2 u}{\partial x^2} + \frac{\partial^2 u}{\partial y^2} = 0$$

21

Because

$u = e^{-x}\cos y$ so $\dfrac{\partial u}{\partial x} = -e^{-x}\cos y$ and $\dfrac{\partial^2 u}{\partial x^2} = e^{-x}\cos y$.

Also $\dfrac{\partial u}{\partial y} = -e^{-x}\sin y$ so $\dfrac{\partial^2 u}{\partial y^2} = -e^{-x}\cos y$. Therefore $\dfrac{\partial^2 u}{\partial x^2} + \dfrac{\partial^2 u}{\partial y^2} = 0$, that is $u(x, y)$ is harmonic. The conjugate function $v(x, y)$ is then

$$v(x, y) = \ldots\ldots\ldots\ldots$$

$$v = -e^{-x}\sin y + C$$

22

Because

By the Cauchy–Riemann equation $\dfrac{\partial u}{\partial x} = \dfrac{\partial v}{\partial y} = -e^{-x}\cos y$. Integrating with respect to y gives $v = -e^{-x}\sin y + a(x)$. Differentiating this with respect to x gives $\dfrac{\partial v}{\partial x} = e^{-x}\sin y + a'(x)$.

Now, by the other Cauchy–Riemann equation $\dfrac{\partial v}{\partial x} = -\dfrac{\partial u}{\partial y} = e^{-x}\sin y$, so that $a'(x) = 0$ giving $a(x) = C$. Therefore, $v = -e^{-x}\sin y + C$.

Now we shall look at complex integration. Move to the next frame

Complex integration

23

At the beginning of this Programme, we defined differentiation with respect to z in the case of a complex function, since z is a function of two independent variables x and y, i.e. $z = x + jy$. Complex integration is approached in the same way.

$$z = x + jy \quad \text{and} \quad w = f(z) = u + jv \text{ where } u \text{ and } v \text{ are also functions of } x \text{ and } y.$$

Also $\quad dz = dx + j\,dy \quad \text{and} \quad dw = du + j\,dv$

$$\therefore \quad \int w\,dz = \int f(z)\,dz = \int (u + jv)(dx + j\,dy)$$

$$= \int \{(u\,dx - v\,dy) + j(v\,dx + u\,dy)\}$$

$$\therefore \quad \int f(z)\,dz = \int (u\,dx - v\,dy) + j\int (v\,dx + u\,dy)$$

That is, the integral reduces to two real-variable integrals

$$\int (u\,dx - v\,dy) \quad \text{and} \quad \int (v\,dx + u\,dy)$$

Note that each of these two integrals is of the general form $\int (P\,dx + Q\,dy)$ which we met before during our work on *line integrals* and, in the complex plane, this rather neatly leads us into *contour integration*.

Let us make a fresh start

---

24 ## Contour integration – line integrals in the *z*-plane

If z moves along the curve c in the z-plane and at each position z has associated with it a function of z, i.e. $f(z)$, then summing up $f(z)$ for all such points between A and B means that we are evaluating a line integral in the z-plane between A $(z = z_1)$ and B $(z = z_2)$ along the curve c, i.e. we are evaluating

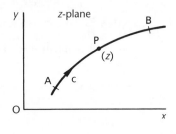

$\displaystyle\int_c f(z)\,dz$ where c is the particular path joining A to B.

The evaluation of line integrals in the complex plane is known as *contour integration*. Let us see how it works in practice.

Example 25

Evaluate the integral $\displaystyle\int_c f(z)\,dz$ where $f(z) = (z-j)^2$ and c is the straight line joining A $(z=0)$ to B $(z=1+j2)$.

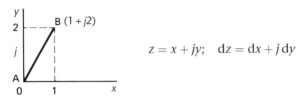

$z = x + jy; \quad dz = dx + j\,dy$

$f(z) = (z-j)^2 = \{x + j(y-1)\}^2 = x^2 - (y-1)^2 + j\,2x(y-1)$

$\therefore \; I = \displaystyle\int \{(x^2 - y^2 + 2y - 1) + j(2xy - 2x)\}\{dx + j\,dy\}$

$\qquad = \displaystyle\int \{(x^2 - y^2 + 2y - 1)\,dx - (2xy - 2x)\,dy\}$

$\qquad\qquad + j\displaystyle\int \{(2xy - 2x)\,dx + (x^2 - y^2 + 2y - 1)\,dy\}$

Now the equation of AB is $y = 2x$. $\therefore \; dy = 2\,dx$ and substituting these in the expression for I, between the limits $x = 0$ and $x = 1$, gives

$$I = \ldots\ldots\ldots\ldots \qquad \text{Finish it.}$$

26

$$\boxed{I = \frac{1}{3}(-2 + j)}$$

Because

$I = \displaystyle\int_0^1 \{(x^2 - 4x^2 + 4x - 1)\,dx - (4x^2 - 2x)2\,dx\}$

$\qquad + j\displaystyle\int_0^1 \{(4x^2 - 2x)\,dx + (2x^2 - 8x^2 + 8x - 2)\,dx\}$

$\qquad = \displaystyle\int_0^1 (-11x^2 + 8x - 1)\,dx + j\displaystyle\int_0^1 (-2x^2 + 6x - 2)\,dx$

and this, by elementary integration, gives $I = \frac{1}{3}(-2 + j)$.

Now you will remember that, in general, the value of a line integral depends on the path of integration between the end points, but that the line integral $\displaystyle\int (P\,dx + Q\,dy)$ is independent of the path of integration in a simply connected region if $\dfrac{\partial P}{\partial y} = \dfrac{\partial Q}{\partial x}$ throughout the region.

▶

In our example

$$I = \int \{(x^2 - y^2 + 2y - 1)\,dx - (2xy - 2x)\,dy\}$$

$$+ j \int \{(2xy - 2x)\,dx + (x^2 - y^2 + 2y - 1)\,dy\} \equiv I_1 + j I_2$$

If we apply the test to I_1, we get ....

27

$$\boxed{\frac{\partial P}{\partial y} = \frac{\partial Q}{\partial x}}$$

Because

for $I_1 = \int \{(x^2 - y^2 + 2y - 1)\,dx - (2xy - 2x)\,dx\} \equiv \int (P\,dx + Q\,dy)$

$$\left. \begin{array}{l} \therefore\ P = x^2 - y^2 + 2y - 1 \quad \therefore\ \dfrac{\partial P}{\partial y} = -2y + 2 \\[4mm] Q = -2xy + 2x \qquad\qquad \therefore\ \dfrac{\partial Q}{\partial x} = -2y + 2 \end{array} \right\} \quad \therefore\ \frac{\partial P}{\partial y} = \frac{\partial Q}{\partial x}$$

Similarly

for $I_2 = \int \{(2xy - 2x)\,dx + (x^2 - y^2 + 2y - 1)dy\} \equiv \int (P\,dx + Q\,dy)$

$$\left. \begin{array}{l} \therefore\ P = 2xy - 2x \qquad\qquad \therefore\ \dfrac{\partial P}{\partial y} = 2x \\[4mm] Q = x^2 - y^2 + 2y - 1 \ \therefore\ \dfrac{\partial Q}{\partial x} = 2x \end{array} \right\} \quad \therefore\ \frac{\partial P}{\partial y} = \frac{\partial Q}{\partial x}$$

Therefore, in this example, the value of the line integral is independent of the path of integration.

Just to satisfy our conscience, determine the value of the line integral between the same two end points, but along the parabola $y = 2x^2$.

$$f(z) = (z - j)^2$$
$$y = 2x^2 \quad \therefore\ dy = 4x\,dx$$

As before we have

$$I = \int \{(x^2 - y^2 + 2y - 1)\,dx - (2xy - 2x)\,dy\}$$

$$+ j \int (2xy - 2x)\,dx + (x^2 - y^2 + 2y - 1)\,dy\}$$

Substituting $y = 2x^2$ and $dy = 4x\,dx$, the evaluation gives

$$I = \ldots\ldots\ldots\ldots$$

28

$$\boxed{I = \frac{1}{3}(-2+j)}$$

We have

$$I = \int_0^1 \{(x^2 - 4x^4 + 4x^2 - 1)\,dx - (4x^3 - 2x)4x\,dx\}$$

$$+ j\int_0^1 \{(4x^3 - 2x)\,dx + (x^2 - 4x^4 + 4x^2 - 1)4x\,dx\}$$

$$= \int_0^1 (-20x^4 + 13x^2 - 1)\,dx + j\int_0^1 (-16x^5 + 24x^3 - 6x)\,dx$$

The rest is easy enough, giving $I = \frac{1}{3}(-2+j)$ which is, of course, the same result as before. Note that all results in Frames 25–28 can be obtained very easily by integrating the function of z with respect to z.

For example, the integral $\int_c f(z)\,dz$ where $f(z) = (z-j)^2$ and c is the straight line joining A $(z=0)$ to B $(z=1+j2)$ can be evaluated as

$$\int_c f(z)\,dz = \int_{z=0}^{1+j2} (z-j)^2\,dz$$

$$= \left[\frac{(z-j)^3}{3}\right]_0^{1+j2}$$

$$= \left(\frac{(1+j2-j)^3}{3} - \frac{(-j)^3}{3}\right)$$

$$= \frac{1}{3}(-2+j)$$

Now on to the next frame

Cauchy's theorem

29

We have already seen that if $w = f(z)$ where, as usual, $w = u + jv$ and $z = x + jy$, then $dz = dx + jdy$ and

$$\int f(z)\,dz = \int (u + jv)(dx + j\,dy)$$

$$= \int (u\,dx - v\,dy) + j\int (v\,dx + u\,dy)$$

If c is a closed curve as the path of integration, then

$$\oint_c f(z)\,dz = \oint_c (u\,dx - u\,dy) + j\oint_c (v\,dx + u\,dy)$$

Applying Green's theorem to each of the two integrals on the right-hand side in turn, we have

(a) $\oint_c (u\,dx - v\,dy) = \iint_S \left(-\dfrac{\partial v}{\partial x} - \dfrac{\partial u}{\partial y} \right) dx\,dy$

where S is the region enclosed by the curve c.

Also, if $f(z)$ is regular at every point within and on c, then the Cauchy–Riemann equations give

$$\dfrac{\partial u}{\partial y} = -\dfrac{\partial v}{\partial x} \quad \text{and therefore} \quad -\dfrac{\partial v}{\partial x} - \dfrac{\partial u}{\partial y} = 0$$

$$\therefore \quad \oint_c (u\,dx - v\,dy) = 0 \tag{1}$$

(b) Similarly, with the second integral, we have

$$\cdots\cdots\cdots$$

30

$$\boxed{\oint_c (v\,dx + u\,dy) = 0}$$

Because

$$\oint_c (v\,dx + u\,dy) = \iint_S \left(\dfrac{\partial u}{\partial x} - \dfrac{\partial v}{\partial y} \right) dx\,dy$$

Again, if $f(z)$ is regular at every point within and on c, then the Cauchy–Riemann equations give

$$\dfrac{\partial u}{\partial x} = \dfrac{\partial v}{\partial y} \quad \text{and therefore} \quad \dfrac{\partial u}{\partial x} - \dfrac{\partial v}{\partial y} = 0$$

$$\therefore \quad \oint_c (v\,dx + u\,dy) = 0 \tag{2}$$

Combining the two results (1) and (2) we have the following result.

If $f(z)$ is regular at every point within and on a closed curve c, then

$$\oint_c f(z)\,dz = 0$$

This is Cauchy's theorem. Make a note of the result;
then we can see an example

Example 1   **31**

Verify Cauchy's theorem by evaluating the integral $\oint_C f(z)\,dz$ where $f(z) = z^2$

around the square formed by joining the points $z = 1$, $z = 2$, $z = 2 + j$, $z = 1 + j$.

$$z = x + jy$$
$$z^2 = x^2 - y^2 + j\,2xy$$
$$dz = dx + j\,dy$$

$$\oint_C f(z)\,dz = \oint_C z^2\,dz = \oint_C \{x^2 - y^2 + j\,2xy\}\{dx + j\,dy\}$$
$$= \oint_C \{(x^2 - y^2)\,dx - 2xy\,dy\} + j\oint_C \{2xy\,dx + (x^2 - y^2)\,dy\}$$

We now take each of the sides in turn.

(a) AB: $y = 0$ $\therefore$ $dy = 0$

$$\therefore \int_{AB} f(z)\,dz = \int_1^2 x^2 dx = \left[\frac{x^3}{3}\right]_1^2 = \frac{8}{3} - \frac{1}{3} = \frac{7}{3}$$

(b) BC: $x = 2$ $\therefore$ $dx = 0$

$$\therefore \int_{BC} f(z)\,dz = \int_0^1 (-4y\,dy) + j\int_0^1 (4 - y^2)\,dy$$
$$= \left[-2y^2\right]_0^1 + j\left[4y - \frac{y^3}{3}\right]_0^1$$
$$= -2 + j\left(4 - \frac{1}{3}\right) = -2 + j\frac{11}{3}$$

Continuing in the same way, the results for the remaining two sides are

$$\dots\dots\dots \text{ and } \dots\dots\dots$$

32

$$\boxed{\text{CD: } -\frac{4}{3} - j3; \quad \text{DA: } 1 - j\frac{2}{3}}$$

Because

(c) CD: $y = 1$ $\therefore$ $dy = 0$

$$\therefore \int_{CD} f(z)\,dz = \int_2^1 (x^2 - 1)\,dx + j\int_2^1 2x\,dx$$
$$= \left[\frac{x^3}{3} - x\right]_2^1 + j\left[x^2\right]_2^1 = -\frac{4}{3} - j3$$

▶

(d) DA: $x = 1$ $\therefore$ $dx = 0$

$$\therefore \int_{DA} f(z)\,dz = \int_1^0 (-2y\,dy) + j\int_1^0 (1 - y^2)\,dy$$

$$= \left[-y^2\right]_1^0 + j\left[y - \frac{y^3}{3}\right]_1^0 = 1 - j\tfrac{2}{3}$$

So, collecting the four results, $\oint_c f(z)\,dz = \ldots\ldots\ldots$

33

$$\boxed{\oint_c f(z)\,dz = 0}$$

Because

$$\oint_c f(z)\,dz = \frac{7}{3} + \left(-2 + j\frac{11}{3}\right) + \left(-\frac{4}{3} - j3\right) + \left(1 - j\frac{2}{3}\right) = 0$$

Example 2

A region in the z-plane has a boundary c consisting of

(a) OA joining $z = 0$ to $z = 2$

(b) AB a quadrant of the circle $|z| = 2$ from $z = 2$ to $z = j2$

(c) BO joining $z = j2$ to $z = 0$.

Verify Cauchy's theorem by evaluating the integral $\int_c (z^2 + 1)\,dz$

(1) along the arc from A to B

(2) along BO and OA.

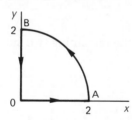

$$f(z) = z^2 + 1 = (x + jy)^2 + 1$$

$$= (x^2 - y^2 + 1) + j\,2xy$$

$$z = x + jy \quad \therefore \quad dz = dx + j\,dy$$

So the general expression for $\int f(z)\,dz = \ldots\ldots\ldots$

34

$$\int \{(x^2 - y^2 + 1) + j\,2xy\}\{dx + j\,dy\}$$

$$= \int \{(x^2 - y^2 + 1)\,dx - 2xy\,dy\} + j \int \{2xy\,dx + (x^2 - y^2 + 1)\,dy\}$$

(1) Arc AB: $x^2 + y^2 = 4$ $\therefore y^2 = 4 - x^2$ $\therefore y = \sqrt{4 - x^2}$

$$dy = \frac{1}{2}(4 - x^2)^{-1/2}(-2x)\,dx \qquad \therefore \;\; dy = \frac{-x}{\sqrt{4 - x^2}}\,dx$$

$$\therefore \;\; \int_{AB} f(z)\,dz$$

$$= \int_2^0 \left\{ (x^2 - 4 + x^2 + 1)\,dx - 2x\sqrt{4 - x^2}\left(\frac{(-x)}{\sqrt{4 - x^2}} \right) dx \right\}$$

$$\qquad + j \int_2^0 \left\{ 2x\sqrt{4 - x^2}\,dx + (x^2 - 4 + x^2 + 1)\left(\frac{(1 - x)}{\sqrt{4 - x^2}} \right) dx \right\}$$

$$= \int_2^0 (4x^2 - 3)\,dx + j \int_2^0 \frac{11x - 4x^3}{\sqrt{4 - x^2}}\,dx = -\frac{14}{3} + j\,I_1$$

Now we must attend to $I_1 = \displaystyle\int_2^0 \frac{11x - 4x^3}{\sqrt{4 - x^2}}\,dx$.

Substituting $x = 2\sin\theta$ and $dx = 2\cos\theta\,d\theta$ with appropriate limits we have

. . . .

35

$$I_1 = -\frac{2}{3}$$

Because

$$I_1 = \int_{\pi/2}^0 \left(\frac{22\sin\theta - 32\sin^3\theta}{2\cos\theta} \right) 2\cos\theta\,d\theta$$

$$= \int_0^{\pi/2} (32\sin^3\theta - 22\sin\theta)\,d\theta$$

$$= 32\,\frac{2}{(3)(1)} + \Big[22\cos\theta \Big]_0^{\pi/2} = \frac{64}{3} - 22 = -\frac{2}{3}$$

$$\therefore \;\; \int_{AB} f(z)\,dz = -4\frac{2}{3} - j\frac{2}{3} = -\frac{2}{3}(7 + j)$$

(2) Along BO and OA. Complete this section on your own in the same way.

$$\int_{BO} f(z)\,dz = \ldots\ldots\ldots\ldots; \quad \int_{OA} f(z)\,dz = \ldots\ldots\ldots\ldots$$

36

$$\int_{BO} f(z)\,dz = j\frac{2}{3}; \quad \int_{OA} f(z)\,dz = 4\frac{2}{3}$$

Because we have

BO: $x = 0$ $\therefore$ $dx = 0$

$$\therefore \int_{BO} f(z)\,dz = j\int_2^1 (1 - y^2)\,dy = j\left[y - \frac{y^3}{3} \right]_2^0 = j\frac{2}{3}$$

OA: $y = 0$ $\therefore$ $dy = 0$

$$\therefore \int_{OA} f(z)\,dz = \int_0^2 (x^2 + 1)\,dx = \left[\frac{x^3}{3} + x \right]_0^2 = 4\frac{2}{3}$$

Collecting the results together, therefore

$$\int_{AB} f(z)\,dz = -\frac{14}{3} - j\frac{2}{3}$$

$$\int_{BO+OA} f(z)\,dz = j\frac{2}{3} + 4\frac{2}{3} = \frac{14}{3} + j\frac{2}{3}$$

$$\therefore \oint_c f(z)\,dz = \int_{AB} f(z)\,dz + \int_{BO+OA} f(z)\,dz = 0$$

which, once again, verifies Cauchy's theorem.

Just by way of revision, Cauchy's theorem actually states that

. . . .

37

If $f(z)$ is *regular* at every point within and on
a closed curve c, then $\oint_c f(z)\,dz = 0$

In our examples so far, $f(z)$ has been regular and no problems have arisen. Let us now consider a case where one or more singularities occur within the region enclosed by the curve c.

Deformation of contours at singularities

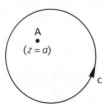

If c is the boundary curve (or *contour*) of a region and $f(z)$ is regular for all points within and on the contour, then the evaluation of $\oint_c f(z)\,dz$ around the contour is straightforward.

However, if $f(z) = \dfrac{1}{z - a}$, where a is a complex constant, and point A corresponds to $z = a$, then at A, $f(z)$ ceases to be regular and a singularity occurs at that point.

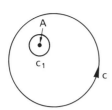

We can isolate A in a very small region within a contour c_1 and then $f(z)$ will be regular at all points within the region c and outside c_1. But the original region is now no longer simply connected (it now has a 'hole' in it) and this was one of our initial conditions.

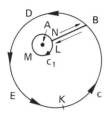

However, all is not lost! We select a suitable point B on the contour c and join it to the inner contour c_1. If we now consider the integration $\int f(z)\,dz$ starting from a point K and proceeding anticlockwise, the path of integration can be taken as K B L M N B D E K.

Therefore

$$\int f(z)\,dz = I = I_{KB} + I_{BL} + I_{LMN} + I_{NB} + I_{BDEK} = \ldots\ldots\ldots\ldots$$

$$\boxed{0}$$

38

The function $f(z)$ is now regular at all points within and on the deformed contour. Remember that the inner contour c_1 can be made as small as we wish.

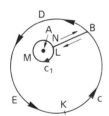

Note that $I_{NB} = -I_{BL}$, being in opposite directions, and these therefore cancel out.

The previous result then becomes

$$I_{KB} + I_{LMN} + I_{BDEK} = 0 \qquad \text{i.e.} \qquad I_{KB} + I_{BDEK} + I_{LMN} = 0$$

But $I_{KB} + I_{BDEK} = \oint_c f(z)\,dz$ and $I_{LMN} = \oint_{c_1} f(z)\,dz$

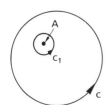

$$\therefore \oint_c f(z)\,dz + \oint_{c_1} f(z)\,dz = 0$$

$$\therefore \oint_c f(z)\,dz - \oint_{c_1} f(z)\,dz = 0$$

$$\therefore \oint_c f(z)\,dz = \oint_{c_1} f(z)\,dz$$

The process can, of course, be extended to cases with more than one such singularity.

 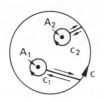 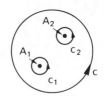

The corresponding result then becomes

$$\oint_c f(z)\,dz = \oint_{c_1} f(z)\,dz + \oint_{c_2} f(z)\,dz \dots \text{etc.}$$

Now let us apply these ideas to an example.

Example 1

Consider the integral $\oint_c f(z)\,dz$ where $f(z) = \dfrac{1}{z}$, evaluated round a closed contour in the z-plane.

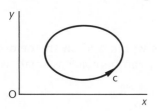

We first check the function $f(z) = \dfrac{1}{z}$ for singularities and find at once that

....

39

> At $z = 0$, $f(z) = \dfrac{1}{z}$ ceases to be regular and a singularity occurs at that point

The actual position of the closed contour is not specified in the problem, so there are two possibilities: either the contour does enclose the origin, or it does not.

Let us consider them in turn.

(a) The contour does not enclose the origin.

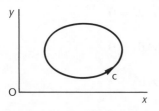

No difficulty arises here and by Cauchy's theorem

....

40

$$\oint_c f(z)\,dz = 0$$

(b) If the contour *does* enclose the origin, the singularity must be taken into account. Then

$$\oint_c f(z)\,dz = \oint_{c_1} f(z)\,dz = \oint_{c_1} \frac{1}{z}\,dz$$

and we attend to evaluating $\oint_{c_1} \dfrac{1}{z}\,dz$

where c_1 is a small circle of radius r entirely within the region bounded by c.

If we take an enlarged view of the small circle c_1, we have $z = x + jy$ which can be expressed in polar form .... and in exponential form ....

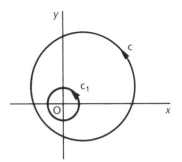

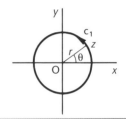

41

$$z = r\,(\cos\theta + j\sin\theta)$$
$$z = re^{j\theta}$$

Using $z = re^{j\theta}$ then $dz = jre^{j\theta}d\theta$ and $\oint_{c_1} \dfrac{1}{z}\,dz = \ldots\ldots\ldots$

Complete it

42

$$j2\pi$$

Because

$$\oint_{c_1} \frac{1}{z}\,dz = \int_0^{2\pi} \frac{1}{re^{j\theta}}\{jre^{j\theta}\}d\theta = \int_0^{2\pi} j\,d\theta = j2\pi$$

$$\therefore \oint_c \frac{1}{z}\,dz = \oint_{c_1} \frac{1}{z}\,dz = j2\pi$$

So we have:

(a) $\oint_c \dfrac{1}{z}\,dz = 0$ if the contour c does not enclose the origin

(b) $\oint_c \dfrac{1}{z}\,dz = j2\pi$ if the contour c does enclose the origin.

These two constitute an important result, so note them well

43

Example 2

Consider the integral $\oint_c f(z)\,dz$ where $f(z) = \dfrac{1}{z^n}$ $(n = 2, 3, 4, \ldots)$.

Again, a singularity clearly occurs at $z = 0$ and again also we have two possible cases.

(a) If the contour c does not enclose the origin, then by Cauchy's theorem

$$\oint_c f(z)\,dz = 0.$$

(b) If the contour c does enclose the origin, then we proceed very much as before.

Using $z = re^{j\theta}$, $dz = jre^{j\theta}d\theta$ and $z^n = r^n e^{jn\theta}$

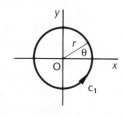

Then $\oint_c f(z)\,dz = \oint_{c_1} f(z)\,dz$

$$= \int_0^{2\pi} \frac{1}{r^n e^{jn\theta}} \{jre^{j\theta}\}\,d\theta$$

$$= \frac{j}{r^{n-1}} \int_0^{2\pi} e^{-j(n-1)\theta}\,d\theta$$

$$= \frac{-1}{(n-1)r^{n-1}} \left[e^{-j(n-1)\theta}\right]_0^{2\pi}$$

$$= \ldots\ldots\ldots\ldots$$

Finish it off

44

$$\boxed{0}$$

Because

$$\oint_c \frac{1}{z^n}\,dz = \frac{-1}{(n-1)r^{n-1}} \{e^{-j(n-1)2\pi} - 1\}$$

$$= \frac{-1}{(n-1)r^{n-1}} \{\cos(n-1)2\pi - j\sin(n-1)2\pi - 1\}$$

$$= 0 \quad \text{since } \left.\begin{array}{l} \cos(n-1)2\pi = 1 \\[4pt] \sin(n-1)2\pi = 0 \end{array}\right\} \; n = 2, 3, 4, \ldots$$

So $\oint_c \dfrac{1}{z^n}\,dz = 0$ for all positive integer values of n other than $n = 1$, where c is any closed contour.

The particular case when $n = 1$ we have seen in Example 1.

Now we can easily cope with this next example.

Example 3

Consider $\oint_c f(z)\,dz$ where $f(z) = \dfrac{1}{(z-a)^n}$ for $n = 1, 2, 3, \ldots$

This is a simple extension of the previous piece of work. Here we see that a singularity occurs at $z = a$ and yet again we have two cases to consider.

(a)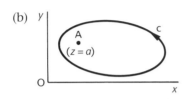

If the contour c does not enclose $z = a$, then by Cauchy's theorem

$$\oint_c f(z)\,dz = 0$$

(b)

If c encloses A $(z = a)$ we consider separately the cases when

(1) $n = 1$ and (2) $n > 1$.

(1) If $n = 1$, $\quad \oint_c f(z)\,dz = \oint_c \dfrac{1}{z-a}\,dz$

Putting $z - a = w$ $\quad \therefore$ $dz = dw$ $\quad \therefore$ $\oint_c \dfrac{1}{z-a}\,dz = \oint_c \dfrac{1}{w}\,dw$

and this we have already established has a value ....

$$\boxed{j2\pi}$$

45

(2) If $n > 1$, $\quad \oint_c f(z)\,dz = \oint_c \dfrac{1}{(z-a)^n}\,dz = \oint_c \dfrac{1}{w^n}\,dw = 0$ for $n \neq 1$.

So collecting our results together, we have the following.

For $\oint_c f(z)\,dz$, where $f(z) = \dfrac{1}{(z-a)^n}$, $n = 1, 2, 3, \ldots$ and c is a closed contour

$$\oint_c \frac{1}{(z-a)^n}\,dz \quad \begin{aligned} &= 0 &\quad n \neq 1 \\ &= 0 &\quad n = 1 \text{ and c does not enclose } z = a \\ &= j2\pi &\quad n = 1 \text{ and c does enclose } z = a. \end{aligned}$$

You will notice that this is a more general result and includes the results obtained from Examples 1 and 2. *Make a note of it, therefore: it is quite important.*

Then on to Example 4

46

Example 4

Finally, we can go one stage further and consider the contour integral of functions such as $f(z) = \dfrac{z - j - 4}{(z + j)(z - 2)}$.

First we express $f(z)$ in partial fractions

$$\frac{z - j - 4}{(z + j)(z - 2)} = \frac{A}{z + j} + \frac{B}{z - 2}$$

One quick way of finding A and B is by the 'cover up' method.

(a) *To find A*, temporarily cover up the denominator $(z + j)$ in the partial fraction $\dfrac{A}{[z + j]}$ and in the function $\dfrac{z - j - 4}{[z + j](z - 2)}$ and substitute $z + j = 0$, i.e. $z = -j$ in the remainder of the function.

$$A = \frac{-j - j - 4}{-j - 2} = \frac{4 + j2}{2 + j} = 2 \quad \therefore \quad A = 2$$

(b) *To find B*, cover up the denominator $(z - 2)$ in the partial fraction $\dfrac{B}{[z - 2]}$ and in the function $\dfrac{z - j - 4}{(z + j)[z - 2]}$ and substitute $z - 2 = 0$, i.e. $z = 2$ in the remainder of the function.

$$B = \dots\dots\dots\dots$$

47

$$\boxed{B = -1}$$

Because

$$B = \frac{2 - j - 4}{2 + j}$$

$$= \frac{-2 - j}{2 + j}$$

$$= -1$$

Therefore the function $f(z)$ becomes

$$f(z) = \frac{z - j - 4}{(z + j)(z - 2)} \equiv \frac{2}{z + j} - \frac{1}{z - 2}$$

Now we can see that there are singularities at $\dots\dots\dots\dots$

$$\boxed{z = -j \quad \text{and} \quad z = 2}$$

48

Denote the singularities by L and M.

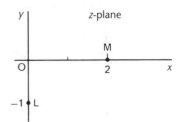

$$\therefore \oint_c \frac{z - j - 4}{(z + j)(z - 2)} \, dz = \oint_c \left\{ \frac{2}{z + j} - \frac{1}{z - 2} \right\} dz$$

$$= \oint_c \left\{ 2\left(\frac{1}{z + j} \right) - \frac{1}{z - 2} \right\} dz$$

So we now have *four* cases to consider, depending on whether L, M, neither, or both, are enclosed within the contour c.

(a) *Neither L nor M enclosed*

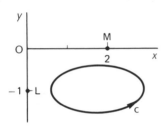

Then, once again, by Cauchy's theorem

$$\oint_c f(z) \, dz = 0$$

(b) *L enclosed but not M*

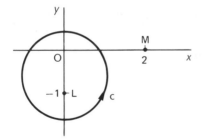

Then, in this case

$$\oint_c f(z) \, dz = 2(j\,2\pi) - 0 = j\,4\pi$$

(c) *M enclosed but not L*

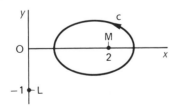

Here

$$\oint_c f(z) \, dz = 0 - (j\,2\pi) = -j\,2\pi$$

▶

(d) *Both L and M enclosed*

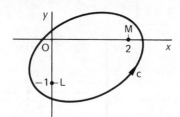

In this case

$$\oint_c f(z)\,dz = \ldots\ldots\ldots\ldots$$

49

$$\boxed{j\,2\pi}$$

Because, when both L and M are enclosed

$$\oint_c f(z)\,dz = \oint_c \left\{ 2\left(\frac{1}{z+j}\right) - \frac{1}{z-2} \right\} dz$$

$$= 2(j\,2\pi) - j\,2\pi$$

$$= j\,2\pi$$

The key is provided by the results we established earlier.

$$\oint_c \frac{1}{(z-a)^n}\,dz \quad = \ldots\ldots\ldots \text{ if } \ldots\ldots\ldots$$
$$= \ldots\ldots\ldots \text{ if } \ldots\ldots\ldots$$
$$= \ldots\ldots\ldots \text{ if } \ldots\ldots\ldots$$

50

$$\oint_c \frac{1}{(z-a)^n}\,dz \quad \begin{aligned} &= 0 &&\text{if } n \neq 1 \\ &= 0 &&\text{if } n = 1 \text{ and c does not enclose } z = a \\ &= j\,2\pi &&\text{if } n = 1 \text{ and c does enclose } z = a. \end{aligned}$$

Now for something somewhat different.

Conformal transformation (conformal mapping)

A mapping from the z-plane onto the w-plane is said to be *conformal* if the angles between lines in the z-plane are preserved both in magnitude and in sense of rotation when transformed onto the corresponding lines in the w-plane.

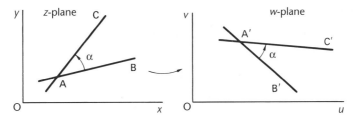

The angle between two intersecting curves in the z-plane is defined by the angle α $(0 \le \alpha \le \pi)$ between their two tangents at the point of intersection, and this is preserved.

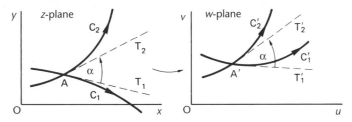

The essential characteristic of a conformal mapping is that

. . . .

| angles are preserved both in magnitude and in sense of rotation | **51** |

Conditions for conformal transformation

The conditions necessary in order that a transformation shall be conformal are as follows.

1   The transformation function $w = f(z)$ must be a regular function of z. That is, it must be defined and single-valued, have a continuous derivative at every point in the region and satisfy the Cauchy–Riemann equations.

2   The derivative $\dfrac{dw}{dz}$ must not be zero, i.e. $f'(z) \ne 0$ at a point of intersection.

▶

Critical points

A point at which $f'(z) = 0$ is called a *critical point* and, at such a point, the transformation is not conformal.

So, if $w = f(z)$ is a regular function, then, except for points at which $f'(z) = 0$, the transformation function will preserve both the magnitude of the angle and its sense of rotation.

Now for a short exercise by way of practice.

Exercise

Determine critical points (if any) which occur in the following transformations $w = f(z)$.

1 $f(z) = (z - 1)^2$ **5** $f(z) = (2z + 3)^3$

2 $f(z) = e^z$ **6** $f(z) = z^3 + 6z + 9$

3 $f(z) = \dfrac{1}{z^2}$ **7** $f(z) = \dfrac{z - j}{z + j}$

4 $f(z) = z + \dfrac{1}{z}$ **8** $f(z) = (z + 3)(z - j)$.

Finish the whole set before checking with the results in the next frame.

52

1	$z = 1$	**5**	$z = -\frac{3}{2}$
2	none	**6**	$z = \pm j\sqrt{2}$
3	none	**7**	none
4	$z = \pm 1$	**8**	$z = \frac{1}{2}(j - 3)$

All that is required is to differentiate each function and to find for which values of z, $f'(z) = 0$.

Now one or two simple examples on conformal mapping.

Example 1

Linear transformation $w = az + b$, $a \neq 0$, a and b complex.

(1) Cauchy–Riemann conditions satisfied.

(2) $f'(z) = a$ i.e. not zero $\therefore$ no critical points.

Therefore, the transformation $w = az + b$ provides conformal mapping throughout the entire z-plane.

Example 2

Non-linear transformation $w = z^2$.

First check for singularities and critical points. These, if any, occur at

....

no singularities; critical point at $z = 0$

Because

$$f'(z) = 2z \quad \therefore \ f'(z) = 0 \text{ at } z = 0.$$

Therefore, the transformation is not conformal at the origin.

If we choose to express z in exponential form $z = x + jy = re^{j\theta}$, then $w = z^2 = r^2 e^{j2\theta}$, i.e. r is squared and the angle doubled.

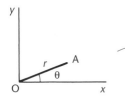

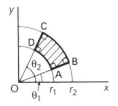

So ABCD, a section of an annulus of inner and outer radii r_1 and r_2 respectively, will be mapped onto

. . . .

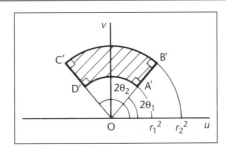

54

The angles at the origin are doubled, but notice that the right angles at A, B, C, D are preserved at A′, B′, C′, D′, i.e. the transformation there is conformal.

Example 3

Consider the mapping of the circle $|z| = 1$ under the transformation $w = z + \dfrac{4}{z}$ onto the w-plane.

First, as always, check for singularities and critical points. We find

. . . .

55

> singularity at $z = 0$; critical points at $z = \pm 2$

A singularity occurs at $z = 0$, i.e. $f'(z)$ does not exist at $z = 0$. Also

$$f(z) = z + \frac{4}{z} \qquad \therefore f'(z) = 1 - \frac{4}{z^2} \qquad \therefore f'(z) = 0 \text{ at } z = \pm 2.$$

Therefore the transformation is not conformal at $z = 0$ and at $z = \pm 2$.

In fact, if we carry out the transformation $w = z + \frac{4}{z}$ on the unit circle $|z| = 1$, we get . . . .

Complete it: it is good revision

56

> the ellipse $\dfrac{u^2}{5^2} + \dfrac{v^2}{3^2} = 1$

Because

$$w = u + jv = z + \frac{4}{z}$$

$$= x + jy + \frac{4}{x + jy}$$

$$= x + jy + \frac{4(x - jy)}{x^2 + y^2}$$

$$\therefore u = x + \frac{4x}{x^2 + y^2}; \qquad v = y - \frac{4y}{x^2 + y^2}$$

$$|z| = 1 \quad \therefore x^2 + y^2 = 1 \quad \therefore u = x(1 + 4) = 5x; \quad v = y(1 - 4) = -3y$$

$$\therefore x = \frac{u}{5} \quad \text{and} \quad y = -\frac{v}{3}$$

Then $x^2 + y^2 = 1$ gives $\dfrac{u^2}{5^2} + \dfrac{v^2}{3^2} = 1$

The image of the unit circle is therefore an ellipse with centre at the origin; semi major axis 5; semi minor axis 3.

Now let us move on to a new section

Schwarz–Christoffel transformation

57

Example 1

Consider a semi-infinite strip on BC as base, the arrows at A and D indicating that the ordinate boundaries extend to infinity in the positive y-direction and that progression round the boundary is to be taken in the direction indicated.

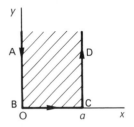

Let us apply the transformation $w = -\cos\dfrac{\pi z}{a}$ to the shaded region.

Then
$$w = u + jv = -\cos\frac{\pi z}{a}$$
$$= -\cos\frac{\pi(x + jy)}{a}$$
$$= -\left\{\cos\frac{\pi x}{a}\cos\frac{j\pi y}{a} - \sin\frac{\pi x}{a}\sin\frac{j\pi y}{a}\right\}$$

Now $\cos j\theta = \cosh\theta$ and $\sin j\theta = j\sinh\theta$.

$$\therefore \ w = u + jv$$
$$= -\cos\frac{\pi x}{a}\cosh\frac{\pi y}{a} + j\sin\frac{\pi x}{a}\sinh\frac{\pi y}{a}$$
$$\therefore \ u = -\cos\frac{\pi x}{a}\cosh\frac{\pi y}{a}; \qquad v = \sin\frac{\pi x}{a}\sinh\frac{\pi y}{a}$$

So B and C map onto B′ and C′ where

$$B' = \ldots\ldots\ldots\ldots; \ C' = \ldots\ldots\ldots\ldots$$

$$\boxed{B': \ u = -1, \ v = 0; \quad C': \ u = 1, \ v = 0}$$

58

Because

(1) at B, $x = 0, \ y = 0 \ \therefore \ u = -(1)(1) = -1; \ v = (0)(0) = 0$

and (2) at C, $x = a, \ y = 0 \ \therefore \ u = -(-1)(1) = 1; \ v = (0)(0) = 0$

So we have

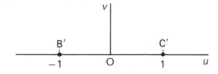

Now we map AB, BC, CD onto the *w*-plane giving A′B′, B′C′, C′D′.

(a) AB: $x = 0$ ∴ A′B′: $u = -\cosh\dfrac{\pi y}{a}$; $v = 0$

  ∴ As *y* decreases from ∞ to 0, *u* increases from −∞ to −1.

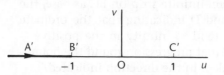

(b) BC: $y = 0$ ∴ B′C′: $u = -\cos\dfrac{\pi x}{a}$; $v = 0$

  ∴ As *x* increases from 0 to *a*, *u* increases from −1 to 1.

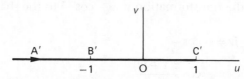

(c) CD: In the same way you can map CD and C′D′ in the *w*-plane and the mapping then becomes . . . .

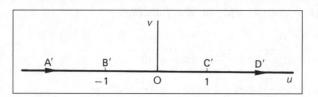

Because

  CD: $x = a$ ∴ C′D′: $u = \cosh\dfrac{\pi y}{a}$; $v = 0.$

Therefore, as *y* increases from 0 to ∞, *u* increases from 1 to ∞.

  Notice the direction of the arrows. These correspond to the directed travel round the boundary shown in the *z*-plane.

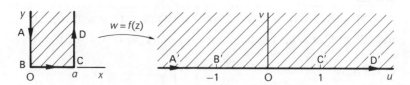

The shaded region in the *z*-plane is on the left-hand side of the boundary as traversed. This maps onto the left-hand side of the image on the *w*-plane, i.e. the entire upper half of the plane.

  Note that $\dfrac{dw}{dz} = \dfrac{\pi}{a}\sin\dfrac{\pi z}{a}$ ∴ at B ($z = 0$) and C ($z = a$), $\dfrac{dw}{dz} = 0.$

Therefore, the conformal property does not hold at these points. The internal angle at B and at C is $\dfrac{\pi}{2}$, while at B′ and C′ it is π .

▶

Example 2

Consider an infinite strip in the z-plane bounded by the real axis and $z = ja$

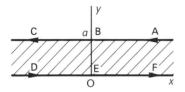

Note the arrows. The boundary comes from $+\infty$ (A) and continues to $-\infty$ (C); then returns from $-\infty$ (D) to $+\infty$ (F).

The strip can be considered as a closed figure with the left- and right- hand vertices at infinity.

We now map the infinite strip onto the w-plane by the transformation $w = e^{\pi z/a}$.

$$\therefore \ w = u + jv = e^{\pi z/a}, \text{ from which}$$

$$u = \ldots\ldots\ldots; \ v = \ldots\ldots\ldots$$

$$\boxed{u = e^{\pi x/a} \cos\frac{\pi y}{a}; \quad v = e^{\pi x/a} \sin\frac{\pi y}{a}}$$

60

Because

$$u + jv = e^{\pi z/a}$$
$$= e^{\pi(x+jy)/a}$$
$$= e^{\pi x/a} e^{j\pi y/a}$$
$$= e^{\pi x/a}\left(\cos\frac{\pi y}{a} + j\sin\frac{\pi y}{a}\right)$$
$$\therefore \ u = e^{\pi x/a}\cos\frac{\pi y}{a}; \quad v = e^{\pi x/a}\sin\frac{\pi y}{a}$$

Now we map points B and E onto B$'$ and E$'$.

(1) B: $x = 0$, $y = a$　$\therefore$　B$'$: $u = -1$, $v = 0$

(2) E: $x = 0$, $y = 0$　$\therefore$　E$'$: $u = 1$, $v = 0$

i.e.

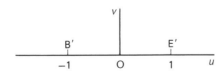

Now we map the lines AB, BC, DE, EF onto the *w*-plane.

(a) AB:　$y = a$　$\therefore$　$u = -e^{\pi x/a}$,　$v = 0$

　　　$\therefore$ As x decreases from $+\infty$ to 0, u increases from $-\infty$ to -1.

(b) BC:　$y = a$　$\therefore$　$u = -e^{\pi x/a}$,　$v = 0$ (as for AB)

　　　$\therefore$ As x decreases from 0 to $-\infty$, u increases from -1 to 0.

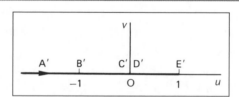

(c) Now there is DE which maps onto

　　　. . . . .

61

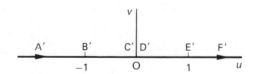

Because

(c) DE:　$y = 0$　$\therefore$　$u = e^{\pi x/a}$,　$v = 0$

　　　$\therefore$ As x increases from $-\infty$ to 0, u increases from 0 to 1.

(d) EF:　$y = 0$　$\therefore$　$u = e^{\pi x/a}$,　$v = 0$ (as for DE)

　　　$\therefore$ As x increases from 0 to $+\infty$, u increases from 1 to $+\infty$.

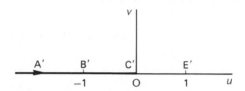

Notice that C and D map to the same point, namely $u = v = 0$.

Finally, what about the shaded region in the *z*-plane? This maps onto

　　　. . . . .

the upper half of the *w*-plane

because it is on the left-hand side of the directed boundary in the *z*-plane.

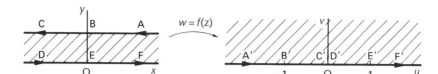

The previous two examples have been simple cases of the application of the Schwarz–Christoffel transformation under which any polygon in the *z*-plane can be made to map onto the entire *upper half* of the *w*-plane and the boundary of the polygon onto the *real axis* of the *w*-plane.

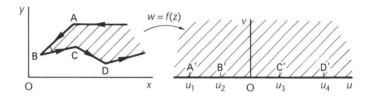

The process depends, of course, on the right choice of transformation function for any particular polygon, which can be defined by its vertices and the internal angle at each vertex.

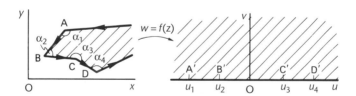

The Schwarz–Christoffel transformation function is given by

$$\frac{dz}{dw} = A(w - u_1)^{\alpha_1/\pi - 1}(w - u_2)^{\alpha_2/\pi - 1}(w - u_3)^{\alpha_3/\pi - 1} \dots$$

$$\therefore \; z = A \int (w - u_1)^{\alpha_1/\pi - 1}(w - u_2)^{\alpha_2/\pi - 1} \dots (w - u_n)^{\alpha_n/\pi - 1} dw + B$$

where *A* and *B* are complex constants, determined by the physical properties of the polygon.

This is not as bad as it looks!

Make a careful note of it: then we will apply it

63   Here it is again.

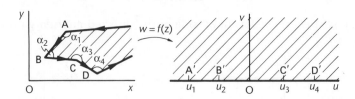

$$\frac{dz}{dw} = A(w - u_1)^{\alpha_1/\pi - 1}(w - u_2)^{\alpha_2/\pi - 1}(w - u_3)^{\alpha_3/\pi - 1} \ldots$$

$$\therefore \ z = A \int (w - u_1)^{\alpha_1/\pi - 1}(w - u_2)^{\alpha_2/\pi - 1} \ldots (w - u_n)^{\alpha_n/\pi - 1} dw + B$$

where A and B are complex constants.

Three other points also have to be noted.

1   Any three points u_1, u_2, u_3 on the u-axis can be selected as required.
2   It is convenient to choose one such point, u_n, at infinity, in which case the relevant factor in the integral above does not occur.
3   Infinite open polygons are regarded as limiting cases of closed polygons where one (or more) vertex is taken to infinity.

Open polygons

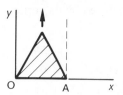

We have already introduced these in Examples 1 and 2 of this section.

In Example 1, the semi-infinite strip is a case of a triangle with one vertex that is

. . . .

64   taken to infinity in the y-direction

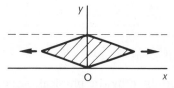

In Example 2, the infinite strip is a case of a double triangle, or quadrilateral, with two vertices taken to infinity.

An open polygon with n sides with one vertex at infinity will have $(n-1)$ internal angles.

An open polygon with n sides with two vertices at infinity will have $(n-2)$ internal angles.

Now for an example to see how all this works.

Example 3 **65**

To determine the transformation that will map the semi-infinite strip ABCD onto the w-plane so that the images of B and C occur at $u = -1$ and $u = 1$, respectively, and the shaded region maps onto the upper half of the w-plane.

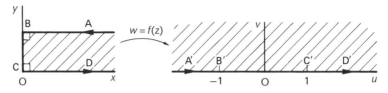

In this case, B' is $u_1 = -1$ and C' is $u_2 = 1$.

The corresponding internal angles are:

at B $(z = ja)$, $\alpha_1 = \dfrac{\pi}{2}$ and at C $(z = 0)$, $\alpha_2 = \dfrac{\pi}{2}$.

So we have

$$\frac{dz}{dw} = A(w + 1)^{(\pi/2)/\pi - 1}(w - 1)^{(\pi/2)/\pi - 1} \quad \text{where } A \text{ is a complex constant}$$

$$= A(w + 1)^{-1/2}(w - 1)^{-1/2}$$

$$= A(w^2 - 1)^{-1/2}$$

$$= K(1 - w^2)^{-1/2} = \frac{K}{\sqrt{1 - w^2}}$$

$$\therefore z = \int \frac{K}{\sqrt{1 - w^2}} dw = \dots\dots\dots$$

$$\boxed{z = K \arcsin w + \bar{B}}$$ **66**

$$\therefore \arcsin w = \frac{z - \bar{B}}{K} \qquad \therefore w = \sin \frac{z - \bar{B}}{K}$$

Now we have to find $\bar{B}$ and K.

(a) We require B $(z = ja)$ to map onto B' $(w = -1)$

$$\therefore -1 = \sin \frac{ja - \bar{B}}{K}$$

$$\therefore \frac{ja - \bar{B}}{K} = -\frac{\pi}{2} \qquad \therefore 2ja - 2\bar{B} = -K\pi \tag{1}$$

(b) We also require C $(z = 0)$ to map onto C' $(w = 1)$ $\therefore 1 = \sin \dfrac{0 - \bar{B}}{K}$

$$\therefore -\frac{\bar{B}}{K} = \frac{\pi}{2} \qquad \therefore -2\bar{B} = K\pi \tag{2}$$

Then, from (1) and (2), $\bar{B} = \dots\dots\dots; K = \dots\dots\dots$

67

$$\boxed{\overline{B} = \frac{ja}{2}; \quad K = -\frac{ja}{\pi}}$$

$$\therefore \ w = \sin\left\{\frac{z - (ja)/2}{-ja/\pi}\right\} = \sin\left\{jz\frac{\pi}{a} + \frac{\pi}{2}\right\} = \cos\frac{jz\pi}{a}$$

But $\cos j\theta = \cosh\theta$ $\therefore$ $w = \cosh\frac{\pi z}{a}$

To verify that this is the required transformation, let us apply it to the figure given in the z-plane.

We will do that in the next frame

68   We have

$$w = u + jv = \cosh\frac{\pi z}{a} = \cosh\frac{(x + jy)\pi}{a}$$

$$\therefore \ u + jv = \cosh\frac{x\pi}{a}\cosh\frac{jy\pi}{a} + \sinh\frac{x\pi}{a}\sinh\frac{jy\pi}{a}$$

But $\cosh j\theta = \cosh\theta$ and $\sinh j\theta = j\sin\theta$

$$\therefore \ u + jv = \cosh\frac{x\pi}{a}\cos\frac{y\pi}{a} + j\sinh\frac{x\pi}{a}\sin\frac{y\pi}{a}$$

$$\therefore \ u = \cosh\frac{x\pi}{a}\cos\frac{y\pi}{a}; \quad v = \sinh\frac{x\pi}{a}\sin\frac{y\pi}{a}$$

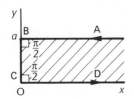

First map the points B and C onto B′ and C′ in the w-plane.

$$\text{B}': \ldots\ldots\ldots\ldots; \ \text{C}': \ldots\ldots\ldots\ldots$$

69

$$\boxed{\text{B}': \ u = -1, v = 0; \quad \text{C}': \ u = 1, v = 0}$$

Because

B: $x = 0, y = a$ $\therefore$ B′: $u = \cos\pi = -1, v = 0$ $\therefore$ B′: $u = -1, v = 0$
C: $x = 0, y = 0$ $\therefore$ C′: $u = 1, v = 0$ $\qquad\qquad \therefore$ C′: $u = 1, v = 0$.

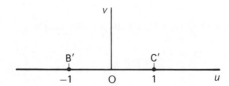

Now we map AB, BC, CD in turn.

(a) AB: $y = a$ ∴ $u = -\cosh\dfrac{x\pi}{a}$, $v = 0$

  ∴ As x decreases from ∞ to 0, u increases from $-\infty$ to -1.

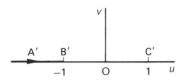

(b) BC: ⎱
(c) CD: ⎰ Complete the working and show the mapped region

  which is ....

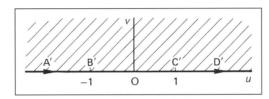

70

Because we have

(b) BC: $x = 0$ ∴ $u = \cos\dfrac{y\pi}{a}$, $v = 0$

  ∴ As y decreases from a to 0, u increases from -1 to 1.

  CD: $y = 0$ ∴ $u = \cosh\dfrac{x\pi}{a}$, $v = 0$

  ∴ As x increases from 0 to ∞, u increases from 1 to ∞.

In each plane, the shaded region is on the left-hand side of the boundary.
We will now finish with one further example.

So move on

Example 4

71

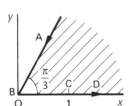
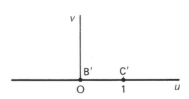

Determine the transformation function $w = f(z)$ that maps the infinite sector
in the z-plane onto the upper half of the w-plane with points B and C mapping
onto B′ and C′ as shown.

The transformation function $w = f(z)$ is given by ....

72

$$\frac{dz}{dw} = A(w - u_1)^{\alpha_1/\pi - 1}(w - u_2)^{\alpha_2/\pi - 1} \dots (w - u_n)^{\alpha_n/\pi - 1}$$

At B, $\alpha_1 = \dfrac{\pi}{3}$. At C, $\alpha_2 = \pi$.

With that reminder, you can now work through on your own, just as we did before, finally obtaining

$$w = \dots\dots\dots$$

73

$$w = z^3$$

Check with the working.

$$\frac{dz}{dw} = A(w - 0)^{(\pi/3)/\pi - 1}(w - 1)^{\pi/\pi - 1}$$
$$= Aw^{-2/3}(w - 1)^0$$
$$= Aw^{-2/3}$$
$$\therefore \quad z = 3Aw^{1/3} + \overline{B}$$
$$= Kw^{1/3} + \overline{B}$$
$$\therefore \quad w = \left(\frac{z - \overline{B}}{K}\right)^3$$

To find $\overline{B}$ and K

(a) At B: $z = 0$ At B': $w = 0$ $\therefore$ $0 = \left(\dfrac{-\overline{B}}{K}\right)^3$ $\therefore$ $\overline{B} = 0$ $\therefore$ $w = \left(\dfrac{z}{K}\right)^3$

(b) At C: $z = 1$ At C': $w = 1$ $\therefore$ $1 = \left(\dfrac{1}{K}\right)^3$ $\therefore$ $K = 1$ $\therefore$ $w = z^3$

$\therefore$ the transformation function is $w = z^3$

Finally, as a check – and a little more valuable practice – apply the function $w = z^3$ to the region shaded in the z-plane.

$$w = u + jv = (x + jy)^3 = x^3 + 3x^2(jy) + 3x(jy)^2 + (jy)^3$$

$$\therefore \quad u = \dots\dots\dots; \quad v = \dots\dots\dots$$

$$u = x^3 - 3xy^2; \quad v = 3x^2y - y^3$$

74

At B: $x = 0$, $y = 0$ $\therefore$ $u = 0$, $v = 0$ $\therefore$ B': $u = 0$, $v = 0$

At C: $x = 1$, $y = 0$ $\therefore$ $u = 1$, $v = 0$ $\therefore$ C': $u = 1$, $v = 0$

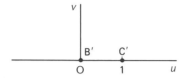

Now we map AB, BC, CD onto A'B', B'C', C'D'.

AB: $y = \sqrt{3}x$ $\therefore$ $u = x^3 - 9x^3 = -8x^3$, $v = 0$

$\therefore$ As x decreases from ∞ to 0, u increases from $-\infty$ to 0.

You can now deal with BC and CD in the same way and finally show the transformed region.

So we get ....

Here is the remaining working.

75

BC: $y = 0$ $\therefore$ $u = x^3$, $v = 0$

$\therefore$ As x increases from 0 to 1, u increases from 0 to 1.

CD: $y = 0$ $\therefore$ $u = x^3$, $v = 0$

$\therefore$ As x increases from 1 to ∞, u increases from 1 to ∞.

So we have

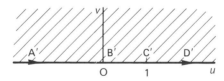

The shaded region is to the left of the directed boundary in the *z*-plane. This therefore maps onto the region to the left of the directed real axis in the *w*-plane, i.e. the upper half of the plane.

We have just touched on the fringe of the work on Schwarz– Christoffel transformation. The whole topic of mapping between planes has applications in fluid mechanics, heat conduction, electromagnetic theory, etc. and it is at times convenient to solve a problem relating to the *z*-plane by transforming to the upper half of the *w*-plane and later to transform back to the *z*-plane. The transformation function can be operated in either direction.

And that is it. The **Revision summary** follows and the **Can you?** checklist. Then on to the **Test exercise** and the **Further problems** for additional practice.

Revision summary 26

76

1 *Differentiation of a complex function*

$$w = f(z) \qquad \frac{dw}{dz} = f'(z) = \underset{\delta z \to 0}{Lim} \left\{ \frac{f(z_0 + \delta z) - f(z_0)}{\delta z} \right\}$$

2 *Regular (or analytic) function*
   $w = f(z)$ is *regular* at z_0 if it is defined, single-valued and has a derivative at every point at and around $z = z_0$.

3 *Singularities* or singular points – points at which $f(z)$ ceases to be regular.

4 *Cauchy–Riemann equations* test whether $w = f(z)$ has a derivative $f'(z)$ at $z = z_0$. $w = u + jv = f(z)$ where $z = x + jy$.

   Then $\dfrac{\partial u}{\partial x} = \dfrac{\partial v}{\partial y}$ and $\dfrac{\partial v}{\partial x} = -\dfrac{\partial u}{\partial y}$

5 If a function of two real variables $f(x, y)$ satisfies Laplace's equation $\dfrac{\partial^2 f(x, y)}{\partial x^2} + \dfrac{\partial^2 f(x, y)}{\partial y^2} = 0$ then $f(x, y)$ is an harmonic function. The real and imaginary parts of an analytic function are both harmonic and form a conjugate pair of functions.

6 *Complex integration*

$$\int w \, dz = \int f(z) dz = \int (u \, dx - v \, dy) + j \int (v \, dx + u \, dy)$$

7 *Contour integration* – evaluation of line integrals in the z-plane.

8 *Cauchy's theorem* If $f(z)$ is regular at every point within and on closed curve c, then $\displaystyle\oint_c f(z) \, dz = 0$.

9 *Deformation of contours*

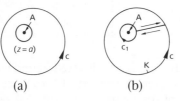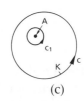

(a)   (b)   (c)

(a) Singularity at A
(b) Restored to a closed curve

(c) $\displaystyle\oint_c f(z) \, dz = \oint_{c_1} f(z) \, dz.$

For $\displaystyle\oint_c f(z) \, dz$ where $f(z) = \dfrac{1}{(z - a)^n}$ $n = 1, 2, 3, \ldots$

$\displaystyle\oint_c \dfrac{1}{(z-a)^n} \, dz$ $= 0$ if $n \neq 1$
$= 0$ if $n = 1$ and c does not enclose $z = a$
$= j2\pi$ if $n = 1$ and c does enclose $z = a$.

10 *Conformal transformation* – mapping in which angles are preserved in size and sense of rotation.

Conditions

1 $w = f(z)$ must be a regular function of z.

2 $f'(z)$, i.e. $\dfrac{\mathrm{d}w}{\mathrm{d}z}$, $\neq 0$ at the point of intersection.

If $f'(z) = 0$ at $z = z_0$, then z_0 is a *critical point*.

11 *Schwarz–Christoffel transformation* maps any polygon in the z-plane onto the entire *upper half* of the w-plane and the boundary of the polygon onto the *real axis* of the w-plane.

$$\frac{\mathrm{d}z}{\mathrm{d}w} = A(w - u_1)^{\alpha_1/\pi - 1}(w - u_2)^{\alpha_2/\pi - 1} \ldots (w - u_n)^{\alpha_n/\pi - 1}$$

1 Any three points u_1, u_2, u_3 can be selected on the u-axis.
2 One such point can be chosen at infinity.
3 Infinite open polygons are regarded as limiting cases of closed polygons.

 Can you?

Checklist 26   77

Check this list before and after you try the end of Programme test.

On a scale of 1 to 5 how confident are you that you can: **Frames**

- Appreciate when the derivative of a function of a complex variable exists?   1 to 3
  Yes ☐ ☐ ☐ ☐ ☐ No

- Understand the notions of regular functions and singularities and be able to obtain the derivative of a regular function from first principles?   3 to 6
  Yes ☐ ☐ ☐ ☐ ☐ No

- Derive the Cauchy–Riemann equations and apply them to find the derivative of a regular function?   7 to 12
  Yes ☐ ☐ ☐ ☐ ☐ No

- Understand the notion of an harmonic function and derive a conjugate function?   13 to 22
  Yes ☐ ☐ ☐ ☐ ☐ No

- Evaluate line and contour integrals in the complex plane?   23 to 28
  Yes ☐ ☐ ☐ ☐ ☐ No

- Derive and apply Cauchy's theorem?   29 to 36
  Yes ☐ ☐ ☐ ☐ ☐ No ▶

- Apply Cauchy's theorem to contours around regions that contain singularities?

  Yes ☐ ☐ ☐ ☐ ☐ No   37 to 49

- Define the essential characteristics of and conditions for a conformal mapping?

  Yes ☐ ☐ ☐ ☐ ☐ No   50 and 51

- Locate critical points of a function of a complex variable?

  Yes ☐ ☐ ☐ ☐ ☐ No   51 and 52

- Determine the image in the w-plane of a figure in the z-plane under a conformal transformation $w = f(z)$?

  Yes ☐ ☐ ☐ ☐ ☐ No   52 to 56

- Describe and apply the Schwarz–Christoffel transformation?

  Yes ☐ ☐ ☐ ☐ ☐ No   57 to 75

 Text exercise 26

78

1 Determine where each of the following functions fails to be regular.

(a) $w = z^3 + 4$

(b) $w = \dfrac{z}{z + 5}$

(c) $w = e^{2z+4}$

(d) $w = \dfrac{z - 2}{(z - 4)(z + 1)}$

(e) $w = \dfrac{x - jy}{x^2 + y^2}$.

2 Demonstrate that each of the following is harmonic and obtain the conjugate function.

(a) $u(x, y) = \sinh x \cos y$

(b) $u(x, y) = 4y(1 + 3x)$.

3 Verify Cauchy's theorem by evaluating $\oint_c f(z)\, dz$ where $f(z) = z^2$ round the rectangle formed by joining the points $z = 2 + j$, $z = 2 + j4$, $z = j4$, $z = j$.

4 Evaluate the integral $\oint_c f(z)dz$ where $f(z) = \dfrac{3z - 6 - j}{(z - j)(z - 3)}$ round the contour $|z| = 2$.

5 Determine critical points, if any, at which the following transformation functions $w = f(z)$ fail to be conformal.

(a) $w = z^4$

(b) $w = z^3 - 3z$

(c) $w = e^{1-z}$

(d) $w = z + \dfrac{2}{z}$

(e) $w = e^{(z^2)}$

(f) $w = \dfrac{z + j}{z - j}$.

6 Determine the Schwarz–Christoffel transformation function $w = f(z)$ that will map the semi-infinite strip shaded in the z-plane onto the upper half of the w-plane, so that the image of B is B' ($w = -1$) and that of C is C' ($w = 0$). Obtain the image of the point D.

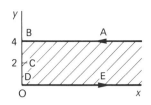

 Further problems 26

1 Verify Cauchy's theorem for the closed path c consisting of three straight lines joining A $(1 + j)$, B $(3 + j3)$, C $(-1 + j3)$ where $f(z) = z - 1 + j$. **79**

2 If $z = 2 + jy$ is mapped onto the w-plane under the transformation $w = f(z) = \dfrac{1}{z}$, show that the locus of w is a circle with centre $w = 0.25$ and radius 0.25.

3 Determine the image in the w-plane of the circle $|z - 2| = 1$ in the z-plane under the transformation $w = (1 - j)z + 3$.

4 The unit circle $|z| = 1$ in the z-plane is generated in an anticlockwise manner from the point A $(z = 1)$ and is transformed onto the w-plane by $w = \dfrac{z}{z - 2}$. Determine the locus of w and the direction in which it is generated.

5 Find the conjugate function of each of the following.
 (a) $u(x, y) = x^2 - 2x - y^2$
 (b) $u(x, y) = x^3 - 3xy^2 - x^2 + y^2 + x$
 (c) $u(x, y) = 2y(x - 1)$
 (d) $u(x, y) = e^{x^2 - y^2} \cos 2xy$.

6 Evaluate $\oint_c f(z)\,dz$ where $f(z) = \dfrac{5z - 2 - j3}{(z - j)(z - 1)}$ around the closed contour c for the two cases when
 (a) c is the path $|z| = 2$
 (b) c is the path $|z - 1| = 1$.

7 If $f(z) = \dfrac{5z + j}{(z - j)(z + j2)}$, evaluate $\oint_c f(z)\,dz$ along the contours
 (a) $|z - 1| = 1$; (b) $|z| = \dfrac{3}{2}$; (c) $|z| = 3$.

8   If $z = x + jy$ and $w = f(z)$, show that, if $\dfrac{j(w + z)}{w - z}$ is entirely real, then $|w| = |z|$.

9   Evaluate $\displaystyle\oint_c f(z)\,dz$, where $f(z) = \dfrac{3z - j5}{(z + 1 - j2)(z - 2 - j)}$, around the perimeter of the rectangle formed by the lines $z = 1$, $z = j3$, $z = -2$, $z = -j$.

10   If $f(z) = \dfrac{8z^2 - 2}{z(z - 1)(z + 1)}$, evaluate $\displaystyle\oint_c f(z)\,dz$ along the contour c where c is the triangle joining the points $z = 2$, $z = j$, $z = -1 - j$.

11   (a)   For the transformation $w = z + \dfrac{1}{z}$, state (1) singularities, (2) critical points.

       (b)   Apply $w = z + \dfrac{1}{z}$ to map the circle $|z| = 2$ onto the w-plane.

12   Find the images in the w-plane of (a) the line $y = 0$ and (b) the line $y = x$ that result from the mapping $w = \dfrac{z - j}{z + j}$. Show that the curves intersect at the points $(\pm 1,\ 0)$ in the w-plane and determine the angle at which they intersect.

13   Use the transformation $w = \dfrac{j(1 + z)}{1 - z}$ to map the unit circle $|z| = 1$ in the z-plane onto the w-plane. Determine also the image in the w-plane of the region bounded by $|z| = 1$ and inside the circle.

14   Determine the transformation that will map the semi-infinite strip shown, onto the upper half of the w-plane, where the image of B is B' $(w = -1)$ and that of C is C' $(w = 1)$.

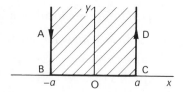

Complex analysis 3

Learning outcomes

When you have completed this Programme you will be able to:

- Expand a function of a complex variable about the origin in a Maclaurin series

- Determine the circle and radius of convergence of a Maclaurin series expansion

- Recognise singular points in the form of poles of order n, removable and essential singularities

- Expand a function of a complex variable about a point in the complex plane in a Taylor series, transforming the coordinates with a shift of origin

- Expand a function of a complex variable about a singular point in a Laurent series

- Recognise the principal and analytic parts of the Laurent series and link the form of the principal part to the type of singularity

- Recognise the residue of a Laurent series and state the Residue theorem

- Calculate the residues at the poles of an expression without resort to deriving the Laurent series

- Evaluate certain types of real integrals using the Residue theorem

Maclaurin series

1

You will recall that the Maclaurin series expansion of the function of a real variable x with output $f(x)$ is given as

$$f(x) = f(0) + xf'(0) + x^2 \frac{f''(0)}{2!} + x^3 \frac{f'''(0)}{3!} + \cdots + x^n \frac{f^{(n)}(0)}{n!} + \cdots$$

This is an infinite series expansion of $f(x)$ about the point $x = 0$. Because the series on the right-hand side of this equation contains an infinite number of terms, the right-hand side may only converge for a restricted set of values of x. Consequently, this expansion is only valid for that restricted set of values. For example, the expression $f(x) = (1 - x)^{-1}$ has the Maclaurin series expansion

. . . .

2

$$\boxed{f(x) = 1 + x + x^2 + x^3 + \cdots + x^n + \cdots}$$

Because

$$f(x) = (1 - x)^{-1} \text{ and so } f(0) = (1 - 0)^{-1} = 1$$
$$f'(x) = (1 - x)^{-2} \text{ and so } f'(0) = (1 - 0)^{-2} = 1$$
$$f''(x) = 2(1 - x)^{-3} \text{ and so } f''(0) = 2(1 - 0)^{-3} = 2$$
$$f'''(x) = 3!(1 - x)^{-4} \text{ and so } f'''(0) = 3!(1 - 0)^{-4} = 3!$$
$$\vdots \qquad\qquad\qquad \vdots$$
$$f^{(n)}(x) = n!(1 - x)^{-(n+1)} \text{ and so } f^{(n)}(0) = n!(1 - 0)^{-(n+1)} = n!$$

Therefore, substituting into the Maclaurin series expansion, we find

$$f(x) = f(0) + xf'(0) + x^2 \frac{f''(0)}{2!} + x^3 \frac{f'''(0)}{3!} + \cdots + x^n \frac{f^{(n)}(0)}{n!} + \cdots$$
$$= 1 + x \times 1 + x^2 \times \frac{2!}{2!} + x^3 \times \frac{3!}{3!} + \cdots + x^n \times \frac{n!}{n!} + \cdots$$
$$= 1 + x + x^2 + x^3 + \cdots + x^n + \cdots$$

This same result could also be derived by using the binomial theorem or even by performing the long division of 1 by $1 - x$. However, performing the algorithmic procedure is one thing, but knowing that the result of the procedure is valid is another. To determine the validity of the expansion we resort to convergence tests, and in this case we use the ratio test. To refresh your memory, the ratio test for the infinite series

$$f(x) = a_0(x) + a_1(x) + a_2(x) + a_3(x) + \cdots + a_n(x) + \cdots$$

is that given

$$\underset{n \to \infty}{Lim} \left| \frac{a_{n+1}(x)}{a_n(x)} \right| = L \text{ then if } \quad L < 1 \text{ the series converges}$$
$$L > 1 \text{ the series diverges}$$
$$L = 1 \text{ the test fails and an alternative}$$
$$\text{convergence test is required.}$$

▶

Applying the ratio test to the Maclaurin series expansion

$$f(x) = 1 + x + x^2 + x^3 + \cdots + x^n + \cdots$$

tells us that

The series converges for ....

The series diverges for ....

The test fails for ....

> The series converges for $-1 < x < 1$
>
> The series diverges for $x < -1$ or $x > 1$
>
> The test fails for $x = \pm 1$

3

Because

$$\underset{n \to \infty}{Lim} \left| \frac{a_{n+1}(x)}{a_n(x)} \right| = \underset{n \to \infty}{Lim} \left| \frac{x^{n+1}}{x^n} \right| = \underset{n \to \infty}{Lim} |x| = |x|, \text{ so}$$

if $|x| < 1$, that is $-1 < x < 1$, the series converges and so the expansion is valid
$|x| > 1$, that is $x < -1$ or $x > 1$, the series diverges and so the expansion is invalid
$|x| = 1$, that is $x = \pm 1$, the ratio test fails to give a conclusion.

By inspection, when $x = 1$ the series clearly diverges and when $x = -1$ the sum of terms alternates between 1 and 0 as each successive term is added. Clearly the series does not converge and so, therefore, it must diverge when $x = -1$.

Everything that has been said about the Maclaurin series expansion of an expression involving a real variable x can equally be said about an expression involving a complex variable z. That is, if $f(z)$ is a function in the complex variable z, analytic at $z = 0$, then the Maclaurin series expansion is

$$f(z) = f(0) + zf'(0) + z^2 \frac{f''(0)}{2!} + z^3 \frac{f'''(0)}{3!} + \cdots$$

So, the Maclaurin series expansion of $f(z) = \sin z$ is ....

4

$$f(z) = z - \frac{z^3}{3!} + \frac{z^5}{5!} - \cdots + \frac{(-1)^n z^{2n+1}}{(2n+1)!} + \cdots$$

Because

$f(z) = \sin z$ and so $f(0) = \sin 0 = 0$

$f'(z) = \cos z$ and so $f'(0) = \cos 0 = 1$

$f''(z) = -\sin z$ and so $f''(0) = -\sin 0 = 0$

$f'''(z) = -\cos z$ and so $f'''(0) = -\cos 0 = -1$

$$\vdots \qquad\qquad \vdots$$

Therefore

$$f(z) = f(0) + zf'(0) + z^2 \frac{f''(0)}{2!} + z^3 \frac{f'''(0)}{3!} + \cdots$$

$$= 0 + z \times 1 + z^2 \times \frac{0}{2!} + z^3 \times \frac{(-1)}{3!} + \cdots$$

$$= z - \frac{z^3}{3!} + \frac{z^5}{5!} - \cdots + \frac{(-1)^n z^{2n+1}}{(2n+1)!} + \cdots$$

Furthermore, applying the ratio test tells us that this series expansion is valid for ....

5

all finite values of z

Because

$$\underset{n\to\infty}{Lim} \left| \frac{a_{n+1}(z)}{a_n(z)} \right| = \underset{n\to\infty}{Lim} \left| \frac{(-1)^{n+1} z^{2(n+1)+1}/[2(n+1)+1]!}{(-1)^n z^{2n+1}/[2n+1]!} \right|$$

$$= \underset{n\to\infty}{Lim} \left| \frac{z^2}{(2n+3)(2n+2)} \right| = 0 < 1$$

So the expansion is valid for all finite values of z.

Try this one. The Maclaurin series expansion of $f(z) = \ln(1+z)$ is

$$\ln(1+z) = \ldots\ldots\ldots\ldots$$

6

$$\ln(1+z) = z - \frac{z^2}{2} + \frac{z^3}{3} - \cdots + \frac{(-1)^{n+1}z^n}{n} + \cdots \quad n = 1, 2, \ldots$$

Because

$f(z) = \ln(1+z)$ and so $f(0) = (1+0) = 0$

$f'(z) = (1+z)^{-1}$ and so $f'(0) = (1+0)^{-1} = 1$

$f''(z) = -(1+z)^{-2}$ and so $f''(0) = -(1+0)^{-2} = -1$

$f'''(z) = 2(1+z)^{-3}$ and so $f'''(0) = 2(1+0)^{-3} = 2$

$f^{(iv)}(z) = -3!(1+z)^{-4}$ and so $f^{(iv)}(0) = -3!(1+0)^{-4} = -3!$

$$\vdots \qquad\qquad\qquad \vdots$$

$f^{(n)}(z) = (-1)^{n+1}n!(1+z)^{-n}$ and so $f^{(n)}(0) = (-1)^{n+1}n!(1+0)^{-n}$

$$= (-1)^{n+1}n!$$

Therefore

$$\ln(1+z) = z - \frac{z^2}{2} + \frac{z^3}{3} - \cdots + \frac{(-1)^{n+1}z^n}{n} + \cdots$$

This series is valid for

. . . .

7

$$\boxed{|z| < 1}$$

Because

$$\operatorname*{Lim}_{n\to\infty} \left| \frac{a_{n+1}(z)}{a_n(z)} \right| = \operatorname*{Lim}_{n\to\infty} \left| \frac{(-1)^{n+2}z^{n+1}/[n+1]}{(-1)^{n+1}z^n/[n]} \right| = \operatorname*{Lim}_{n\to\infty} \left| \frac{nz}{n+1} \right| = |z|$$

So if $|z| < 1$ the series converges and so the expansion is valid

$|z| > 1$ the series diverges and so the expansion is invalid

$|z| = 1$ the ratio test fails

We shall look at the case $|z| = 1$ a little later.

Move to the next frame

Radius of convergence

8

We have just seen that the Maclaurin expansion of $\ln(1 + z)$ is valid for $|z| < 1$. This inequality defines the interior of a circle of radius 1 centred on the origin, namely $z = 1e^{j\theta}$.

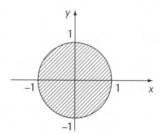

This means that the expansion is valid for all z-values lying within this circle. The radius of the circle within which a series expansion is valid is called the *radius of convergence* of the series and the circle is called the *circle of convergence*.

Example

To find the infinite series expansion and radius of convergence of the expression $f(z) = \dfrac{z}{(1 - 3z)^2}$, we progress in stages, noting that $\dfrac{z}{(1 - 3z)^2} = z(1 - 3z)^{-2}$. We expand $(1 - 3z)^{-2}$ first.

By the binomial theorem, the expansion of $(1 - 3z)^{-2}$ is

$$(1 - 3z)^{-2} = \ldots\ldots\ldots\ldots$$

9

$$\boxed{(1 - 3z)^{-2} = 1 + 6z + 27z^2 + 108z^3 + 405z^4 + \cdots}$$

Because

$$(1 - 3z)^{-2} = \left(1 + (-2) \times (-3z) + \frac{(-2)(-3) \times (-3z)^2}{2!}\right.$$

$$\left. + \frac{(-2)(-3)(-4) \times (-3z)^3}{3!} + \ldots\right)$$

$$= \left(1 + 6z + 3(-3z)^2 - 4(-3z)^3 + 5(-3z)^4 + \ldots\right.$$

$$\left. + (-1)^n(n + 1)(-3)^n z^n + \ldots\right)$$

$$= 1 + 6z + 27z^2 + 108z^3 + 405z^4 + \ldots + (n + 1)3^n z^n + \ldots$$

and so

$$z(1 - 3z)^{-2} = z + 6z^2 + 27z^3 + 108z^4 + 405z^5 + \ldots + (n + 1)z^n z^{n+1} + \ldots$$

The radius of convergence is then $\ldots\ldots\ldots\ldots$

Because

The general term of the expansion is $a_n(z) = (n+1)3^n z^{n+1}$ and so the ratio test tells us that

$$\underset{n\to\infty}{Lim} \left| \frac{a_{n+1}(z)}{a_n(z)} \right| = \underset{n\to\infty}{Lim} \left| \frac{(n+2)3^{n+1}z^{n+2}}{(n+1)3^n z^{n+1}} \right| = \underset{n\to\infty}{Lim} \left| \frac{3(n+2)z}{(n+1)} \right| = |3z|$$

So, if $|3z| < 1$, that is $|z| < 1/3$, then the series converges and the expansion is valid. The radius of convergence is therefore $1/3$.

Move to the next frame

Singular points

Any point at which $f(z)$ fails to be analytic, that is where the derivative does not exist, is called a *singular point* (also called a singularity). For example

$$f(z) = \frac{1}{z-1}$$

is analytic everywhere in the finite complex plane except at the point $z = 1$ where not only is the derivative $f'(z)$ not defined but neither is $f(z)$. Accordingly, the point $z = 1$ is a singular point. There are different types of singular points, for now we shall look at just two of them.

11

Poles

If $f(z)$ has a singular point at z_0 and for some natural number n, $\underset{z\to z_0}{Lim} \{(z - z_0)^n f(z)\} = L \neq 0$ then the singular point is called a *pole of order n*. For example

$$f(z) = \frac{2z}{(z+4)^2}$$

has a singular point at $z = -4$ and because

$$\underset{z\to -4}{Lim} \left\{ (z+4)^2 f(z) \right\} = \underset{z\to -4}{Lim} \{2z\} = -8 \neq 0$$

the singularity is a *pole of order* 2 (also called a *double pole*).

▶

Removable singularities

If $f(z)$ has a singular point at z_0 and $\underset{z \to z_0}{Lim} \{f(z)\}$ exists then the singular point is called a *removable singularity*. For example

$$f(z) = \frac{\sin z}{z}$$

has a singular point at $z = 0$. However, $\underset{z \to 0}{Lim} \left\{ \frac{\sin z}{z} \right\} = 1$ and so the singularity at $z = 0$ is a removable singularity. We can see this from the Maclaurin series expansion of $f(z)$ where

$$f(z) = \frac{\sin z}{z} = \frac{1}{z} \left(z - \frac{z^3}{3!} + \frac{z^5}{5!} - \cdots \right) = 1 - \frac{z^2}{3!} + \frac{z^4}{5!} - \cdots$$

While we cannot substitute $z = 0$ into $f(z) = \frac{\sin z}{z}$, we can define $f(0) = 1$ in complete consistency with the series expansion. In this sense the singularity at $z = 0$ is removable by virtue of the fact that we can assign a value to $f(z)$ at the singularity which is consistent with the series expansion.

Move to the next frame

Circle of convergence

12   When an expression is expanded in a Maclaurin series, the circle of convergence is always centred on the origin and the radius of convergence is determined by the location of the first singular point met as $|z|$ increases from $|z| = 0$. For example, the Maclaurin series expansion of $f(z) = \ln(1 + z)$ is

$$\ln(1 + z) = z - \frac{z^2}{2} + \frac{z^3}{3} - \cdots + \frac{(-1)^{n+1} z^n}{n} + \cdots$$

which is valid inside the circle of convergence $|z| = 1$. The first singular point met by this function as $|z|$ increases from zero is at $z = -1$, for at that point $\ln(1 + z)$ is not defined and the series

$$-1 - \frac{1}{2} - \frac{1}{3} - \cdots - \frac{1}{n} - \cdots$$

diverges – it is the negative of the harmonic series. Hence the radius of convergence is 1. When $z = 1$, substitution into the series expansion gives

$$\ln 2 = 1 - \frac{1}{2} + \frac{1}{3} - \cdots + \frac{(-1)^{n+1}}{n} + \cdots$$

▶

The right-hand side is the alternating harmonic series which we know converges by the *alternating sign test* which states that if the magnitude of the terms decreases and the signs alternate then the series converges. Now we know that it converges to $\ln 2$. Notice that the circle of convergence is identified by the location of the *first* singularity as $|z|$ increases from $|z| = 0$. This does not mean that the function is singular at all points on the circle of convergence.

There are times when it is desirable to have a series expansion of an expression that is singular at the origin. Because the Maclaurin expansion requires the function to be analytic everywhere within the circle of convergence which is centred on the origin, we cannot use that method. Fortunately, we do have a method of expanding a function about *any point* in the complex plane – this is Taylor's expansion.

Move to the next frame

Taylor's series

Provided $f(z)$ is analytic inside and on a simple closed curve c, the Taylor series expansion of $f(z)$ about the point z_0 which is interior to c is given as

$$f(z) = f(z_0) + (z - z_0)f'(z_0) + \frac{(z - z_0)^2 f''(z_0)}{2!} + \cdots$$
$$+ \frac{(z - z_0)^n f^{(n)}(z_0)}{n!} + \cdots$$

13

where here, the point z_0 is the centre of the circle of convergence. The circle of convergence is given as $|z - z_0| = R$. That is $z - z_0 = Re^{j\theta}$ or $z = z_0 + Re^{j\theta}$ where R is the radius of convergence.

Notice that Maclaurin's series is a special case of Taylor's series where $z_0 = 0$.

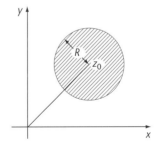

Example

Expand $f(z) = \dfrac{1}{z + 1}$ in a Taylor series about the point $z = 1$ and find the values of z for which the expansion is valid.

The simplest way of doing this is to perform a coordinate transformation that moves the origin of the new coordinate to the point $z = 1$ and then derive the series about the new origin. To do this we define a new complex variable $u = z - 1$ so that $z = u + 1$ and so

$$\frac{1}{z + 1} \text{ becomes } \frac{1}{u + 2} = (2 + u)^{-1} = \frac{1}{2}\left(1 + \frac{u}{2}\right)^{-1}.$$

The expansion of this expression can now be derived using either Maclaurin or, as here, the binomial theorem to obtain

$$\frac{1}{u+2} = \frac{1}{2}\left(1 + (-1)\frac{u}{2} + \frac{(-1)(-2)}{2!}\left(\frac{u}{2}\right)^2 + \cdots\right)$$

$$= \frac{1}{2} - \frac{u}{4} + \frac{u^2}{8} - \frac{u^3}{16} + \cdots$$

Transforming back to the original variable z gives

$$\frac{1}{z+1} = \frac{1}{2} - \frac{z-1}{4} + \frac{(z-1)^2}{8} - \frac{(z-1)^3}{16} + \cdots$$

The circle of convergence is given by $\left|\frac{u}{2}\right| = 1$, that is $\left|\frac{z-1}{2}\right| = 1$ or $|z-1| = 2$.

Consequently, this series expansion is valid provided z is inside the circle defined by

$$z - 1 = 2e^{j\theta} \text{ that is } z = 1 + 2e^{j\theta}$$

By the same reasoning, the Taylor series expansion of $f(z) = \cos z$ about the point $z = \pi/3$ is

....

14

$$\boxed{\frac{1}{2}\left(1 - \sqrt{3}(z - \pi/3) - \frac{(z-\pi/3)^2}{2!} + \sqrt{3}\frac{(z-\pi/3)^3}{3!} + \frac{(z-\pi/3)^4}{4!} - \cdots\right)}$$

Because

If $u = z - \pi/3$ then

$\cos z = \cos(u + \pi/3)$

$= \cos u \cos \pi/3 - \sin u \sin \pi/3$

$= \frac{1}{2}\left(\cos u - \sqrt{3}\sin u\right)$

$= \frac{1}{2}\left(\left[1 - \frac{u^2}{2!} + \frac{u^4}{4!} - \cdots\right] - \sqrt{3}\left[u - \frac{u^3}{3!} + \frac{u^5}{5!} - \cdots\right]\right)$

$= \frac{1}{2}\left(1 - \frac{u^2}{2!} + \frac{u^4}{4!} - \cdots - \sqrt{3}u + \sqrt{3}\frac{u^3}{3!} - \sqrt{3}\frac{u^5}{5!} - \cdots\right)$

$= \frac{1}{2}\left(1 - \sqrt{3}u - \frac{u^2}{2!} + \sqrt{3}\frac{u^3}{3!} + \frac{u^4}{4!} - \sqrt{3}\frac{u^5}{5!} - \cdots\right)$

$= \frac{1}{2}\left(1 - \sqrt{3}(z - \pi/3) - \frac{(z-\pi/3)^2}{2!} + \sqrt{3}\frac{(z-\pi/3)^3}{3!}\right.$

$\left. + \frac{(z-\pi/3)^4}{4!} - \cdots\right)$ for $z < \infty$

Laurent's series

15

Sometimes a valid series expansion of a function is required within a specific region of the complex plane that contains a singular point. In this case we cannot avoid the singular point as we did with Taylor's series by expanding about an alternative non-singular point, because then we move away from part of the specified region. To accommodate this case we can use the *Laurent series expansion* which provides a series expansion valid within an annular region *centred on the singular point*.

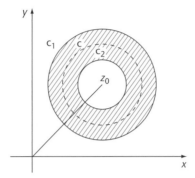

Let $f(z)$ be singular at $z = z_0$ and let c_1 and c_2 be two concentric circles centred on z_0. Then if $f(z)$ is analytic in the annular region between c_1 and c_2 and if c is any concentric circle lying within the annular region between c_1 and c_2 we can expand $f(z)$ as a Laurent series in the form

$$f(z) = \cdots + \frac{a_{-2}}{(z-z_0)^2} + \frac{a_{-1}}{(z-z_0)} + a_0 + a_1(z-z_0) + a_2(z-z_0)^2 + \cdots$$

$$= \sum_{n=-\infty}^{\infty} a_n(z-z_0)^n$$

where $a_n = \dfrac{1}{2\pi j}\displaystyle\oint_c \frac{f(z)}{(z-z_0)^{n+1}}\,\mathrm{d}z$

Example

Expand $\dfrac{e^{3z}}{(z-2)^4}$ in a Laurent series about the point $z = 2$ and determine the nature of the singularity at $z = 2$.

$f(z) = \dfrac{e^{3z}}{(z-2)^4}$ and $f'(z) = \dfrac{e^{3z}(3z-10)}{(z-2)^5}$ so $f(z)$ is analytic everywhere except at $z = 2$. The first thing we must do is to transform the coordinate system by shifting the origin to the point $z = 2$ by defining $u = z - 2$ so that $z = u + 2$. Then

$$\frac{e^{3z}}{(z-2)^4} = \frac{e^{3(u+2)}}{u^4} = e^6 \frac{e^{3u}}{u^4}.$$

▶

Now we can expand using the Maclaurin series expansion

$$= \frac{e^6}{u^4}\left\{1 + 3u + \frac{(3u)^2}{2!} + \frac{(3u)^3}{3!} + \frac{(3u)^4}{4!} + \frac{(3u)^5}{5!} + \cdots\right\}$$

$$= e^6\left\{\frac{1}{u^4} + \frac{3u}{u^4} + \frac{(3u)^2}{2!u^4} + \frac{(3u)^3}{3!u^4} + \frac{(3u)^4}{4!u^4} + \frac{(3u)^5}{5!u^4} + \cdots\right\}$$

$$= e^6\left\{\frac{1}{u^4} + \frac{3}{u^3} + \frac{9}{2u^2} + \frac{27}{6u} + \frac{81}{24} + \frac{243u}{120} + \cdots\right\}$$

$$= e^6\left\{\frac{1}{(z-2)^4} + \frac{3}{(z-2)^3} + \frac{9}{2(z-2)^2} + \frac{9}{2(z-2)} + \frac{27}{8} + \frac{81(z-2)}{40} + \cdots\right\}$$

This series converges for all finite z except $z = 2$ at which point there is a pole of order 4.

The part of the Laurent series that contains negative powers of the variable is called the *principal part* of the series and the remaining terms constitute what is called the *analytic part* of the series. *If, in the principal part the highest power of $1/z$ is n, then the function possesses a pole of order n; and if the principal part contains an infinite number of terms, the function possesses an* **essential singularity**.

Now you try one

16

The Laurent series expansion of $z^2 \cos\dfrac{1}{z}$ about the point $z = 0$

is .... valid for ....

at which point there is ....

17

$$\boxed{\begin{array}{c} z^2 - \dfrac{1}{2!} + \dfrac{1}{4!z^2} - \dfrac{1}{6!z^4} + \cdots \text{ valid for all } z \neq 0 \\ \text{at which point there is an essential singularity} \end{array}}$$

Because

$f(z) = z^2 \cos\dfrac{1}{z}$ and $f'(z) = 2z \cos\dfrac{1}{z} + \sin\dfrac{1}{z}$ and so $f(z)$ is analytic everywhere except at $z = 0$. Expanding about $z = 0$ gives

$$z^2 \cos\frac{1}{z} = z^2\left(1 - \frac{(1/z)^2}{2!} + \frac{(1/z)^4}{4!} - \frac{(1/z)^6}{6!} + \cdots\right)$$

$$= z^2 - \frac{1}{2!} + \frac{1}{4!z^2} - \frac{1}{6!z^4} + \cdots$$

valid for all $z \neq 0$, at which point there is an essential singularity because there is an infinity of terms in the principal part of the series.

Try another. The Laurent series expansion of $\dfrac{z}{(z+2)(z+4)}$ valid for

$2 < |z| < 4$ is ....

$$\cdots + \frac{8}{z^4} - \frac{4}{z^3} + \frac{2}{z^2} - \frac{1}{z} + \frac{1}{2} - \frac{z}{8} + \frac{z^2}{32} - \frac{z^3}{128} + \cdots$$

Because

$$\frac{z}{(z+2)(z+4)} = \frac{2}{z+4} - \frac{1}{z+2} \qquad \text{(separating into partial fractions)}$$

If $|z| > 2$ then we can write $\dfrac{1}{z+2} = \dfrac{1}{z(1+2/z)} = \dfrac{(1+2/z)^{-1}}{z}$

and because $|z| > 2$, that is, $|2/z| < 1$, we can now use the binomial theorem

$$\frac{1}{z+2} = \frac{1}{z(1+2/z)} = \frac{1}{z}\left\{1 - \frac{2}{z} + \frac{4}{z^2} - \frac{8}{z^3} + \cdots\right\} = \frac{1}{z} - \frac{2}{z^2} + \frac{4}{z^3} - \frac{8}{z^4} + \cdots$$

and if $|z| < 4$ then

$$\frac{2}{z+4} = \frac{1}{2(1+z/4)} = \frac{1}{2}\left\{1 - \frac{z}{4} + \frac{z^2}{16} - \frac{z^3}{64} + \cdots\right\}$$

$$= \frac{1}{2} - \frac{z}{8} + \frac{z^2}{32} - \frac{z^3}{128} + \cdots$$

Note the expansion of $(1 + z/4)^{-1}$ which is valid for $|z/4| < 1$, that is $|z| < 4$.

The first expansion for $|z| > 2$ is still valid for $|z| < 4$ since $4 > 2$ and the second expansion for $|z| < 4$ is still valid for $|z| > 2$ since $2 < 4$. Consequently, if $2 < |z| < 4$, then, by subtracting the first series from the second

$$\frac{z}{(z+2)(z+4)} = \frac{2}{z+4} - \frac{1}{z+2}$$

$$= \left\{\frac{1}{2} - \frac{z}{8} + \frac{z^2}{32} - \frac{z^3}{128} + \cdots\right\} - \left\{\frac{1}{z} - \frac{2}{z^2} + \frac{4}{z^3} - \frac{8}{z^4} + \cdots\right\}$$

$$= \cdots + \frac{8}{z^4} - \frac{4}{z^3} + \frac{2}{z^2} - \frac{1}{z} + \frac{1}{2} - \frac{z}{8} + \frac{z^2}{32} - \frac{z^3}{128} + \cdots$$

Take care here! You may be tempted to think that this displays an essential singularity at $z = 0$. This is not the case because the expansion is only valid inside the annular region $2 < |z| < 4$ centred on the origin. Consequently, the point $z = 0$ is outside this region and the series expansion is invalid at that point.

The series expansion of the same function valid for $|z| < 2$ is

$$\cdots\cdots\cdots\cdots$$

19

$$\boxed{\dfrac{z}{8} - \dfrac{3z^2}{32} + \dfrac{7z^3}{128} + \cdots}$$

Because

If $|z| < 2$ then $\dfrac{1}{z+2} = \dfrac{1}{2(1+z/2)} = \dfrac{1}{2}\left\{1 - \dfrac{z}{2} + \dfrac{z^2}{4} - \dfrac{z^3}{8} + \cdots\right\}$

$$= \dfrac{1}{2} - \dfrac{z}{4} + \dfrac{z^2}{8} - \dfrac{z^3}{16} + \cdots$$

We have already seen that if $|z| < 4$ then

$$\dfrac{2}{z+4} = \dfrac{1}{2} - \dfrac{z}{8} + \dfrac{z^2}{32} - \dfrac{z^3}{128} + \cdots$$

This is still valid for $|z| < 2$ since $2 < 4$. Conseqently, if $|z| < 2$, then, by subtracting the first series from the second

$$\dfrac{z}{(z+2)(z+4)} = \dfrac{2}{z+4} - \dfrac{1}{z+2}$$

$$= \left\{\dfrac{1}{2} - \dfrac{z}{8} + \dfrac{z^2}{32} - \dfrac{z^3}{128} + \cdots\right\} - \left\{\dfrac{1}{2} - \dfrac{z}{4} + \dfrac{z^2}{8} - \dfrac{z^3}{16} + \cdots\right\}$$

$$= \dfrac{z}{8} - \dfrac{3z^2}{32} + \dfrac{7z^3}{128} - \cdots$$

Notice that for different regions of convergence we obtain different series expansions. Furthermore, each series expansion is unique within its own particular radius of convergence.

Try one more just to make sure that you can derive these expansions.

The Laurent series of $\dfrac{1 - \cos(z-6)}{(z-6)^2}$ about the point $z = 6$ is

.... valid for .... at which point there is ....

20

$$\boxed{\begin{array}{l}\dfrac{1}{2!} - \dfrac{(z-6)^2}{4!} + \dfrac{(z-6)^4}{6!} - \cdots \text{ valid for all } z \neq 6 \\[2mm] \text{at which point there is a removable singularity}\end{array}}$$

Because

If we let $u = z - 6$ then

$$\dfrac{1 - \cos(z-6)}{(z-6)^2} = \dfrac{1 - \cos u}{u^2}$$

$$= \dfrac{1}{u^2}\left\{1 - \left(1 - \dfrac{u^2}{2!} + \dfrac{u^4}{4!} - \dfrac{u^6}{6!} + \cdots\right)\right\}$$

$$= \dfrac{1}{2!} - \dfrac{u^2}{4!} + \dfrac{u^4}{6!} - \cdots$$

$$= \dfrac{1}{2!} - \dfrac{(z-6)^2}{4!} + \dfrac{(z-6)^4}{6!} - \cdots$$

▶

This is valid for all finite values of $z \neq 6$ at which point there is a removable singularity which can be removed by defining

$\dfrac{1 - \cos(z - 6)}{(z - 6)^2}$ at $z = 6$ as $\dfrac{1}{2!}$. Notice that here the principal part has

no terms, so that the Laurent series is identical to the Taylor series.

Residues

In the Laurent series **21**

$$f(z) = \cdots + \frac{a_{-2}}{(z - z_0)^2} + \frac{a_{-1}}{(z - z_0)} + a_0 + a_1(z - z_0) + a_2(z - z_0)^2 + \cdots$$

the coefficient a_{-1} is referred to as the *residue* of $f(z)$ for reasons that will soon become apparent. Recall the integral in Frame 45 of Programme 21 which states that if the simple closed contour c has z_0 as an interior point, then

$$\oint_c \frac{dz}{(z - z_0)^n} = 2\pi j \delta_{n1}$$

where the Kronecker delta $\delta_{n1} = \begin{cases} 1 & \text{if } n = 1 \\ 0 & \text{if } n \neq 1 \end{cases}$. Applying this fact to the

Laurent series of $f(z)$ yields

$$\oint_c f(z)\,dz = \oint_c \left[\cdots + \frac{a_{-2}}{(z - z_0)^2} + \frac{a_{-1}}{(z - z_0)} + a_0 + a_1(z - z_0) \right.$$

$$\left. + a_2(z - z_0)^2 + \cdots \right] dz$$

$$= \cdots + \oint_c \frac{a_{-2}\,dz}{(z - z_0)^2} + \oint_c \frac{a_{-1}\,dz}{(z - z_0)} + \oint_c a_0\,dz$$

$$+ \oint_c a_1(z - z_0)\,dz + \oint_c a_2(z - z_0)^2 dz + \cdots$$

$$= \cdots + 0 + 2\pi j a_{-1} + 0 + 0 + 0 + \cdots$$

$$= 2\pi j a_{-1}$$

That is, provided $f(z)$ is analytic at all points inside and on the simple closed contour c, apart from the single isolated singularity at z_0 which is interior to c, then

$$\oint_c f(z)\,dz = 2\pi j a_{-1}$$

Hence the name *residue* for a_{-1} because it is all that remains when the Laurent series is integrated term by term. This statement is called the **Residue theorem** and it has many far reaching consequences – we shall see some of these later. For now, just try an example.

▶

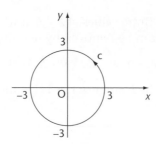

If c is a circle, centred on the origin and of radius 3, then

$$\oint_c \frac{z\,dz}{(z+2)(z+4)} = \ldots\ldots\ldots\ldots$$

22

$$\boxed{\oint_c \frac{z\,dz}{(z+2)(z+4)} = -2\pi j}$$

Because

The circle $|z| = 3$ lies within the annular region $2 < |z| < 4$ and we have already found the Laurent series for the integrand valid for $2 < |z| < 4$ in Frame 18, namely

$$\frac{z}{(z+2)(z+4)} = \frac{2}{z+4} - \frac{1}{z+2}$$

$$= \cdots + \frac{8}{z^4} - \frac{4}{z^3} + \frac{2}{z^2} - \frac{1}{z} + \frac{1}{2} - \frac{z}{8} + \frac{z^2}{32} - \frac{z^3}{128} + \cdots$$

Here the residue is $a_{-1} = -1$ and so $\oint_c \dfrac{z\,dz}{(z+2)(z+4)} = 2\pi j(-1) = -2\pi j$ where

c lies entirely within the region of convergence.

The Residue theorem extends to the case where the contour contains a finite number of singularities. If $f(z)$ is analytic inside and on the simple closed contour c except at the finite number of points $z_0, z_1, z_2, \ldots$, each with a Laurent series expansion and each with corresponding residues $\overset{(0)}{a}_{-1}$, $\overset{(1)}{a}_{-1}$, $\overset{(2)}{a}_{-1}, \ldots$ then

$$\oint_c f(z)\,dz = 2\pi j\left\{\overset{(0)}{a}_{-1} + \overset{(1)}{a}_{-1} + \overset{(2)}{a}_{-1}\right\} = 2\pi j\{\text{sum of residues inside c}\}$$

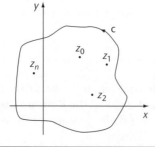

What could be more straightforward? Next frame

Calculating residues

23

When evaluating these integrals the major part of the exercise is to find the residues, and it would be very tedious if we had to find a Laurent series for each and every singularity. Fortunately there is a simpler method for poles. If $f(z)$ is analytic inside and on the simple closed contour c except at the interior point z_0 at which there is a pole of order n, then

$$a_{-1} = \lim_{z \to z_0} \left[\frac{1}{(n-1)!} \frac{d^{n-1}}{dz^{n-1}} \left((z - z_0)^n f(z) \right) \right]$$

Example

Find the residues at all the poles of $f(z) = \dfrac{3z}{(z+2)^2(z^2-1)}$.

$f(z)$ has a pole of order 2 (a double pole) at $z = -2$ and two poles of order 1 (simple poles) at $z = \pm 1$.

At $z = -2$ the residue is $a_{-1} = \lim_{z \to -2} \left[\dfrac{1}{(2-1)!} \dfrac{d^{2-1}}{dz^{2-1}} \left((z+2)^2 f(z) \right) \right]$

$$= \lim_{z \to -2} \left[\frac{d}{dz} \left(\frac{3z}{z^2 - 1} \right) \right]$$

$$= \lim_{z \to -2} \left[\frac{3(z^2 - 1) - 6z^2}{(z^2 - 1)^2} \right]$$

$$= \frac{3(4 - 1) - 24}{(4 - 1)^2} = -\frac{5}{3}$$

At $z = 1$ the residue is ....

$$\boxed{\frac{1}{6}}$$

Because

At $z = 1$ the residue is $a_{-1} = \lim_{z \to 1} \left[\dfrac{1}{(1-1)!} \dfrac{d^{1-1}}{dz^{1-1}} \left((z-1)f(z) \right) \right]$

$$= \lim_{z \to 1} \left[\frac{d^0}{dz^0} \left(\frac{3z}{(z+2)^2(z+1)} \right) \right]$$

The zeroth derivative of an expression is the expression itself

$$= \lim_{z \to 1} \left[\frac{3z}{(z+2)^2(z+1)} \right]$$

$$= \frac{3}{(3)^2(2)} = \frac{1}{6}$$

At $z = -1$ the residue is ....

25

$$\boxed{\dfrac{3}{2}}$$

Because

$$\text{At } z = -1 \text{ the residue is } a_{-1} = \underset{z \to -1}{Lim}\left[\frac{1}{(1-1)!}\frac{d^{1-1}}{dz^{1-1}}((z+1)f(z))\right]$$

$$= \underset{z \to -1}{Lim}\left[\frac{d^0}{dz^0}\left(\frac{3z}{(z+2)^2(z-1)}\right)\right]$$

$$= \underset{z \to -1}{Lim}\left[\frac{3z}{(z+2)^2(z-1)}\right]$$

$$= \frac{-3}{(1)^2(-2)}$$

$$= \frac{3}{2}$$

Move to the next frame

Integrals of real functions

26 The Residue theorem can be very usefully employed to evaluate integrals of real functions that cannot be evaluated using the real calculus. Even when an integral is susceptible to evaluation by the real calculus, the use of the residue calculus can often save a great amount of effort. We shall look at three types of real integral and in each case we shall proceed by example.

Integrals of the form $\displaystyle\int_0^{2\pi} F(\cos\theta, \sin\theta)\, d\theta$

Example

Evaluate $\displaystyle\int_0^{2\pi} \frac{1}{4\cos\theta - 5}\, d\theta.$

To evaluate this integral we make use of the exponential representation of a complex number of unit length, namely $z = e^{j\theta}$, and the exponential form of the trigonometric functions

$$\cos\theta = \frac{e^{j\theta} + e^{-j\theta}}{2} = \frac{z + z^{-1}}{2} \text{ and } \sin\theta = \frac{e^{j\theta} - e^{-j\theta}}{2j} = \frac{z - z^{-1}}{2j},$$

and finally $dz = je^{j\theta}\, d\theta = jz\, d\theta$ so that $d\theta = dz/jz$

Using these relations we can transform the real integral from 0 to 2π into a contour integral in the complex plane where the contour c is the *unit circle centred on the origin*. That is

$$\int_0^{2\pi} \frac{1}{4\cos\theta - 5} \, d\theta = \oint_c \frac{1}{4\frac{z + z^{-1}}{2} - 5} \times \frac{dz}{jz}$$

$$= -j \oint_c \frac{1}{2z^2 - 5z + 2} \, dz$$

$$= -j \oint_c \frac{1}{(2z - 1)(z - 2)} \, dz$$

The complex integrand has two simple poles, one at $z = \dfrac{1}{2}$ which is inside the contour c and another at $z = 2$ which is outside the contour c. Using the Residue theorem

$$-j \oint_c \frac{1}{(2z - 1)(z - 2)} \, dz = -j \times 2\pi j \times \{\text{residue at } z = 1/2\}$$

The residue at $z = 1/2$ is

$$\underset{z \to 1/2}{Lim} \left\{ (z - 1/2) \frac{1}{(2z - 1)(z - 2)} \right\} = \underset{z \to 1/2}{Lim} \left\{ \frac{1}{2(z - 2)} \right\}$$

$$= -\frac{1}{3}$$

so that

$$\int_0^{2\pi} \frac{1}{4\cos\theta - 5} \, d\theta = -j \oint_c \frac{1}{(2z - 1)(z - 2)} \, dz$$

$$= -j \times 2\pi j \times \{\text{residue at } z = 1/2\}$$

$$= -2\pi/3$$

Now you try one

$$\int_0^{2\pi} \frac{d\theta}{2 + \cos\theta} = \dots\dots\dots$$

27

$$\boxed{\dfrac{2\pi}{\sqrt{3}}}$$

Because

$$\int_0^{2\pi} \frac{d\theta}{2 + \cos\theta} = \oint_c \frac{dz/jz}{2 + \dfrac{z + z^{-1}}{2}} \qquad \text{where c is the unit circle centred on the}$$
$$\text{origin.}$$

$$= -j \oint_c \frac{2\,dz}{z^2 + 4z + 1}$$

$$= -j \oint_c \frac{2\,dz}{\left(z + 2 - \sqrt{3}\right)\left(z + 2 + \sqrt{3}\right)}$$

The integrand has two simple poles, one at $z = -2 + \sqrt{3}$ which is inside c and another at $z = -2 - \sqrt{3}$ which is outside c. Therefore

$$-j \oint_c \frac{2\,dz}{\left(z + 2 - \sqrt{3}\right)\left(z + 2 + \sqrt{3}\right)} = -j \times 2\pi j \times \left\{ \text{residue at } z = -2 + \sqrt{3} \right\}$$

The residue is

$$\operatorname*{Lim}_{z \to -2 + \sqrt{3}} \left\{ \left(z + 2 - \sqrt{3}\right) \frac{2}{\left(z + 2 - \sqrt{3}\right)\left(z + 2 + \sqrt{3}\right)} \right\}$$

$$= \operatorname*{Lim}_{z \to -2 + \sqrt{3}} \left\{ \frac{2}{\left(z + 2 + \sqrt{3}\right)} \right\} = \frac{1}{\sqrt{3}} \quad \text{and so}$$

$$\int_0^{2\pi} \frac{d\theta}{2 + \cos\theta} = -j \oint_c \frac{2\,dz}{\left(z + 2 - \sqrt{3}\right)\left(z + 2 + \sqrt{3}\right)} = -j \times 2\pi j \times \frac{1}{\sqrt{3}}$$

$$= 2\pi \times \frac{1}{\sqrt{3}} = \frac{2\pi}{\sqrt{3}}$$

---

28   **Integrals of the form** $\displaystyle\int_{-\infty}^{\infty} F(x)\,dx$

Example

Evaluate $\displaystyle\int_{-\infty}^{\infty} \frac{1}{1 + x^4}\,dx$.

To evaluate this integral we must consider the integral $\displaystyle\oint_c \frac{1}{1 + z^4}\,dz$ where c is the contour shown in the figure, so that

$$\oint_c \frac{1}{1 + z^4}\,dz = \int_s \frac{dz}{1 + z^4} + \int_{-R}^{R} \frac{dx}{1 + x^4} = 2\pi j \{\text{sum of residues inside c}\}$$

Notice that along the real axis between $-R$ and R, $z = x$. Provided $R > 1$ we can evaluate this integral using the Residue theorem. That is

$$\oint_c \frac{1}{1 + z^4}\,dz = 2\pi j \times \{\text{sum of residues inside c}\}$$

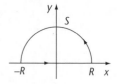

29

$$\boxed{\dfrac{\pi}{\sqrt{2}}}$$

Because

The integrand $\dfrac{1}{1+z^4}$ possesses four simple poles at $z = e^{\pi j/4}$, $e^{3\pi j/4}$, $e^{5\pi j/4}$, $e^{7\pi j/4}$ of which only the first two are inside c.

The residue at $z = e^{\pi j/4}$ is $\underset{z \to e^{\pi j/4}}{Lim} \left\{ \left(z - e^{\pi j/4} \right) \times \dfrac{1}{1+z^4} \right\}$

$$= \underset{z \to e^{\pi j/4}}{Lim} \left\{ \dfrac{1}{4z^3} \right\} \text{ by L'Hôpital's rule}$$

$$= \dfrac{e^{-3\pi j/4}}{4}$$

The residue at $z = e^{3\pi j/4}$ is $\underset{z \to e^{3\pi j/4}}{Lim} \left\{ \left(z - e^{3\pi j/4} \right) \times \dfrac{1}{1+z^4} \right\}$

$$= \underset{z \to e^{3\pi j/4}}{Lim} \left\{ \dfrac{1}{4z^3} \right\} \text{ by L'Hôpital's rule}$$

$$= \dfrac{e^{-9\pi j/4}}{4} = \dfrac{e^{-\pi j/4}}{4}$$

Therefore

$$\oint_c \dfrac{1}{1+z^4} \, dz = 2\pi j \times \left\{ \dfrac{1}{4} \left(e^{-3\pi j/4} + e^{-\pi j/4} \right) \right\}$$

Now $e^{-3\pi j/4} = \cos \dfrac{3\pi}{4} - j \sin \dfrac{3\pi}{4} = -\dfrac{1}{\sqrt{2}} - \dfrac{j}{\sqrt{2}}$ and

$$e^{-\pi j/4} = \cos \dfrac{\pi}{4} - j \sin \dfrac{\pi}{4} = \dfrac{1}{\sqrt{2}} - \dfrac{j}{\sqrt{2}} \text{ and so}$$

$$\oint_c \dfrac{1}{1+z^4} \, dz = 2\pi j \times \left\{ \dfrac{1}{4} \left(\dfrac{-2j}{\sqrt{2}} \right) \right\} = \dfrac{\pi}{\sqrt{2}}$$

We now look at the components of this integral in the next frame

30

We now recognise that

$$\oint_c \dfrac{1}{1+z^4} \, dz = \int_{-R}^{R} \dfrac{1}{1+x^4} \, dx + \int_S \dfrac{1}{1+z^4} \, dz$$

because $z = x$ along the real line.

Now we let R increase indefinitely and take limits, so that

$$\underset{R \to \infty}{Lim} \oint_c \dfrac{1}{1+z^4} \, dz = \int_{-\infty}^{\infty} \dfrac{1}{1+x^4} \, dx + \underset{R \to \infty}{Lim} \int_S \dfrac{1}{1+z^4} \, dz = \dfrac{\pi}{\sqrt{2}}$$

because the value of the contour integral is independent of the value of R. We shall now proceed to show that $\underset{R \to \infty}{Lim} \int_S \dfrac{1}{1+z^4} \, dz = 0$.

▶

Writing $z = Re^{j\theta}$ so that, on S, $dz = Re^{j\theta} d\theta$, the limit of the integral becomes

$$\underset{R\to\infty}{Lim} \int_S \frac{Re^{j\theta}}{1 + R^4 e^{j4\theta}} d\theta = 0$$

Notice that the requirement that ensures that the integral along the semicircle vanishes in the limit is equivalent to the requirement that the degree of the denominator be at least two degrees higher than the numerator.

Now you try one.

$$\int_{-\infty}^{\infty} \frac{x^2 \, dx}{(x^2 + 1)^2} = \dots\dots\dots$$

31

$$\boxed{\dfrac{\pi}{2}}$$

Because

Consider the integral $\displaystyle\oint_c \frac{z^2 \, dz}{(z^2 + 1)^2}$ where the contour c is the same semi-circular contour as in the previous example. Here the integrand has two double poles at $z = j$ and $z = -j$ but only the pole at $z = j$ is inside the contour. The residue at $z = j$ is

$$\underset{z\to j}{Lim} \left\{ \frac{d}{dz} (z - j)^2 \frac{z^2}{(z - j)^2 (z + j)^2} \right\} = \underset{z\to j}{Lim} \left\{ \frac{2z(z + j)^2 - z^2 2(z + j)}{(z + j)^4} \right\}$$

$$= -\frac{j}{4}$$

Therefore

$$\oint_c \frac{z^2 \, dz}{(z^2 + 1)^2} = 2\pi j \left(-\frac{j}{4} \right) = \frac{\pi}{2}$$

Taking limits

$$\underset{R\to\infty}{Lim} \oint_c \frac{z^2 \, dz}{(z^2 + 1)^2} = \int_{-\infty}^{\infty} \frac{x^2 \, dx}{(x^2 + 1)^2} + \underset{R\to\infty}{Lim} \int_S \frac{z^2 \, dz}{(z^2 + 1)^2} = \frac{\pi}{2}$$

Where, in the second integral on the right-hand side, the degree of the denominator is two higher than the degree of the numerator, and so

$$\underset{R\to\infty}{Lim} \int_S \frac{z^2 \, dz}{(z^2 + 1)^2} = 0, \text{ therefore } \int_{-\infty}^{\infty} \frac{x^2 \, dx}{(x^2 + 1)^2} = \frac{\pi}{2}$$

Integrals of the form $\displaystyle\int_{-\infty}^{\infty} F(x) \begin{Bmatrix} \sin x \\ \cos x \end{Bmatrix} dx$ **32**

These integrals are often referred to as Fourier integrals because of their appearances within Fourier analysis.

Example

Evaluate $\displaystyle\int_{-\infty}^{\infty} \frac{\cos kx}{a^2 + x^2}\, dx$ where $a > 0$ and $k > 0$.

To evaluate this integral we consider the contour integral $\displaystyle\oint_c \frac{e^{jkz}}{a^2 + z^2}\, dz$ where

c is the semicircular contour of the previous problems and whose integrand possesses two simple poles at $z = aj$ and $z = -aj$ of which only the first is inside the contour. Consequently

$$\oint_c \frac{e^{jkz}}{a^2 + z^2}\, dz = 2\pi j\{\text{residue at } z = aj\} = \ldots\ldots\ldots\ldots$$

$$\boxed{\dfrac{\pi e^{-ka}}{a}}$$ **33**

Because

The residue at $z = aj$ is

$$\underset{z \to aj}{Lim}\left\{ (z - aj)\frac{e^{jkz}}{a^2 + z^2} \right\} = \underset{z \to aj}{Lim}\left\{ \frac{e^{jkz}}{z + aj} \right\} = \frac{e^{jk(aj)}}{2aj} = -\frac{je^{-ka}}{2a} \text{ and so}$$

$$\oint_c \frac{e^{jkz}}{a^2 + z^2}\, dz = 2\pi j\left\{ -\frac{je^{-ka}}{2a} \right\} = \frac{\pi e^{-ka}}{a}$$

Taking limits as $R \to \infty$

$$\underset{R \to \infty}{Lim}\oint_c \frac{e^{jkz}}{a^2 + z^2}\, dz = \int_{-\infty}^{\infty} \frac{e^{jkz}}{a^2 + z^2}\, dz + \underset{R \to \infty}{Lim}\int_S \frac{e^{jkz}}{a^2 + z^2}\, dz = \frac{\pi e^{-ka}}{a}$$

In the second integral on the right-hand side, the degree of the denominator is two higher than the degree of the numerator, and so

$$\underset{R \to \infty}{Lim}\int_S \frac{e^{jkz}}{a^2 + z^2}\, dz = 0, \text{ therefore } \int_{-\infty}^{\infty} \frac{e^{jkx}}{a^2 + x^2}\, dx = \frac{\pi e^{-ka}}{a}. \text{ That is}$$

$$\int_{-\infty}^{\infty} \frac{\cos kx + j\sin kx}{a^2 + x^2}\, dx = \frac{\pi e^{-ka}}{a} = 2\pi j\{\text{residue at } z = aj\}.$$

Consequently

$$\int_{-\infty}^{\infty} \frac{\cos kx}{a^2 + x^2}\, dx = \frac{\pi e^{-ka}}{a} = -2\pi \operatorname{Im}\{\text{residue at } z = aj\} \text{ and}$$

$$\int_{-\infty}^{\infty} \frac{\sin kx}{a^2 + x^2}\, dx = 0 = 2\pi \operatorname{Re}\{\text{residue at } z = aj\}$$

▶

Notice that e^{jkz} is easier to use than $\cos kx = (e^{jkx} + e^{-jkx})/2$, and it also gives the solution to the related integral with $\cos kx$ replaced with $\sin kx$.

Finally, to finish off the Programme, here is one for you to try.

$$\int_{-\infty}^{\infty} \frac{\cos \pi x}{x^2 + x + 1} \, dx = \ldots\ldots\ldots$$

34

$$\boxed{0}$$

Because

Consider $\oint_c \dfrac{e^{j\pi z}}{z^2 + z + 1} \, dz$ where c is the semicircular contour of the previous problem. The integrand is singular at the simple poles $z = (-1 \pm j\sqrt{3})/2$ where only $z = (-1 + j\sqrt{3})/2$ is inside the contour. The residue at $z = (-1 + j\sqrt{3})/2$ is then

$$\operatorname*{Lim}_{z \to (-1+j\sqrt{3})/2} \left\{ \left(z - \left[-1 + j\sqrt{3} \right]/2 \right) \frac{e^{j\pi z}}{z^2 + z + 1} \right\}$$

$$= \operatorname*{Lim}_{z \to (-1+j\sqrt{3})/2} \left\{ \frac{e^{j\pi z}}{z - \left[-1 - j\sqrt{3} \right]/2} \right\}$$

$$= \frac{e^{j\pi(-1+j\sqrt{3})/2}}{j\sqrt{3}}$$

$$= \frac{e^{-j\pi/2} e^{-\sqrt{3}\pi/2}}{j\sqrt{3}}$$

$$= -\frac{e^{-\sqrt{3}\pi/2}}{\sqrt{3}} \qquad \text{since } e^{-j\pi/2} = -j$$

Therefore

$$\oint_c \frac{e^{j\pi z}}{z^2 + z + 1} \, dz = 2\pi j \left\{ \frac{e^{-\sqrt{3}\pi/2}}{\sqrt{3}} \right\} = -j \frac{2\pi e^{-\sqrt{3}\pi/2}}{\sqrt{3}}$$

that is

$$\oint_c \frac{e^{j\pi z}}{z^2 + z + 1} \, dz = \oint_c \frac{\cos \pi z + j \sin \pi z}{z^2 + z + 1} \, dz = -j \frac{2\pi e^{-\sqrt{3}\pi/2}}{\sqrt{3}}$$

and so

$$\oint_c \frac{\cos \pi z}{z^2 + z + 1} \, dz = 0 \text{ and } \oint_c \frac{\sin \pi z}{z^2 + z + 1} \, dz = -\frac{2\pi e^{-\sqrt{3}\pi/2}}{\sqrt{3}}$$

Note that, again, the contribution from the contour integral along the semicircle is zero.

The **Revision summary** now follows. Check it through in conjunction with the **Can you?** checklist before going on to the **Test exercise**. The **Further problems** provide additional practice.

Revision summary 27

35

1   *Maclaurin series*

The Maclaurin series expansion of a function of a complex variable z is

$$f(z) = f(0) + zf'(0) + z^2 \frac{f''(0)}{2!} + z^3 \frac{f'''(0)}{3!} + \cdots$$

2   *Ratio test for convergence*

The ratio test for convergence of a series of terms of a complex variable

$$f(z) = a_0(z) + a_1(z) + a_2(z) + a_3(z) + \cdots + a_n(z) + \cdots$$

is that given

$$\underset{n \to \infty}{Lim} \left| \frac{a_{n+1}(z)}{a_n(z)} \right| = L$$

then if $L < 1$ the series converges and so the expansion is valid

$\qquad\qquad$ $L > 1$ the series diverges and so the expansion is invalid

$\qquad\qquad$ $L = 1$ the ratio test fails to give a conclusion.

3   *Radius and circle of convergence*

The radius of the circle within which a series expansion is valid is called the *radius of convergence* of the series and the circle is called the *circle of convergence*. The radius of convergence can be found using the ratio test for convergence.

4   *Singular points*

Any point at which $f(z)$ fails to be analytic, that is where the derivative does not exist, is called a *singular point*.

> *Poles*
>
> If $f(z)$ has a singular point at z_0 and for some natural number n
>
> $$\underset{z \to z_0}{Lim} \left\{ (z - z_0)^n f(z) \right\} = L \neq 0$$
>
> then the singular point (also called a singularity) is called *a pole of order n*.
>
> *Removable singularity*
>
> If $f(z)$ has a singular point at z_0 but $\underset{z \to z_0}{Lim} \{ f(z) \}$ exists then the singular point is called a *removable singularity*.

5   *Circle of convergence*

When an expression is expanded in a Maclaurin series, the *circle of convergence* is always centred on the origin and the *radius of convergence* is determined by the location of the first singular point met as z moves out from the origin.

▶

6 *Taylor's series*

Provided $f(z)$ is analytic inside and on a simple closed curve c, the Taylor series expansion of $f(z)$ about a point z_0 which is interior to c is given as

$$f(z) = f(z_0) + (z - z_0)f'(z_0) + \frac{(z - z_0)^2 f''(z_0)}{2!} + \cdots$$
$$+ \frac{(z - z_0)^n f^{(n)}(z_0)}{n!} + \cdots$$

where, here, the expansion is about the point z_0 which is the centre of the circle of convergence. The circle of convergence is given as $|z - z_0| = R$ where R is the radius of convergence. Maclaurin's series is a special case of Taylor's series where $z_0 = 0$.

7 *Laurent's series*

The *Laurent series expansion* provides a series expansion valid within an annular region *centred on the singular point*.

Let $f(z)$ be singular at $z = z_0$ and let c_1 and c_2 be two concentric circles centred on z_0. Then if $f(z)$ is analytic in the annular region between c_1 and c_2 and c is any concentric circle lying within the annular region between c_1 and c_2 we can expand $f(z)$ as a Laurent series in the form

$$f(z) = \cdots + \frac{a_{-2}}{(z - z_0)^2} + \frac{a_{-1}}{(z - z_0)} + a_0 + a_1(z - z_0) + a_2(z - z_0)^2 + \cdots$$
$$= \sum_{n \to -\infty}^{\infty} a_n(z - z_0)^n \text{ where } a_n = \frac{1}{2\pi j} \oint_c \frac{f(z)}{(z - z_0)^{n+1}} \, dz$$

8 *Residues*

In the Laurent series

$$f(z) = \cdots + \frac{a_{-2}}{(z - z_0)^2} + \frac{a_{-1}}{(z - z_0)} + a_0 + a_1(z - z_0) + a_2(z - z_0)^2 + \cdots$$

the coefficient a_{-1} is referred to as the *residue* of $f(z)$.

Residue theorem

Provided $f(z)$ is analytic at all points inside and on the simple closed contour c, apart from the single isolated singularity at z_0 which is interior to c, then

$$\oint_c f(z) \, dz = 2\pi j a_{-1}$$

9 The Residue theorem extends to the case where the contour contains a finite number of singularities. If $f(z)$ is analytic inside and on the simple closed contour c except at the finite number of points $z_0, z_1, z_2, \cdots$ each with a Laurent series expansion and each with corresponding residues $a_{-1}^{(0)}, a_{-1}^{(1)}, a_{-1}^{(2)}, \cdots$ then

$$\oint_c f(z) \, dz = 2\pi j \left\{ a_{-1}^{(0)} + a_{-1}^{(1)} + a_{-1}^{(2)} + \cdots \right\}$$

▶

10 *Calculating residues*

$$a_{-1} = \underset{z \to z_0}{Lim} \left[\frac{1}{(n-1)!} \frac{d^{n-1}}{dz^{n-1}} \left((z - z_0)^n f(z) \right) \right]$$

11 *Real integrals*

The Residue theorem can be very usefully employed to evaluate integrals of real functions.

Integrals of the form $\displaystyle\int_0^{2\pi} F(\cos\theta, \ \sin\theta)\, d\theta$

Use $z = e^{j\theta}$ and the exponential form of the trigonometric functions

$$\cos\theta = \frac{e^{j\theta} + e^{-j\theta}}{2} = \frac{z + z^{-1}}{2}, \qquad \sin\theta = \frac{e^{j\theta} - e^{-j\theta}}{2j} = \frac{z - z^{-1}}{2j} \qquad \text{and}$$

$$dz = je^{j\theta}\, d\theta = jz\, d\theta$$

so that $d\theta = dz/jz$. Convert the integral into a contour integral around the unit circle centred on the origin and use the Residue theorem.

Integrals of the form $\displaystyle\int_{-\infty}^{\infty} F(x)\, dx$ and $\displaystyle\int_{-\infty}^{\infty} F(x) \begin{cases} \sin x \\ \cos x \end{cases} dx$

Consider integrals of the form $\displaystyle\oint_c F(z)\, dz$ and $\displaystyle\oint_c F(z)e^{jz}\, dz$ respectively, where the contour c is a semicircle with the diameter lying along the real axis. The principle is that the integral can be evaluated by the Residue theorem and then the contour can be expanded to cover the required extent of the real axis, the integration along the semicircle giving a zero contribution.

 # Can you?

Checklist 27

36

Check this list before and after you try the end of Programme test

On a scale of 1 to 5 how confident are you that you can: **Frames**

- Expand a function of a complex variable about the origin in a Maclaurin series?   1 to 7
  Yes ☐ ☐ ☐ ☐ ☐ No

- Determine the circle and radius of convergence of a Maclaurin series expansion?   8 to 10
  Yes ☐ ☐ ☐ ☐ ☐ No

- Recognise singular points in the form of poles of order n, removable and essential singularities?   11
  Yes ☐ ☐ ☐ ☐ ☐ No

▶

- Expand a function of a complex variable about a point in the complex plane in a Taylor series, transforming the coordinates with a shift of origin?   12 to 14
  Yes ☐   ☐   ☐   ☐   ☐   No

- Expand a function of a complex variable about a singular point in a Laurent series?   15
  Yes ☐   ☐   ☐   ☐   ☐   No

- Recognise the principal and analytic parts of the Laurent series and link the form of the principal part to the type of singularity?   16 to 20
  Yes ☐   ☐   ☐   ☐   ☐   No

- Recognise the residue of a Laurent series and state the Residue theorem?   21 and 22
  Yes ☐   ☐   ☐   ☐   ☐   No

- Calculate the residues at the poles of an expression without resort to deriving the Laurent series?   23 to 25
  Yes ☐   ☐   ☐   ☐   ☐   No

- Evaluate certain types of real integrals using the Residue theorem?   26 to 34
  Yes ☐   ☐   ☐   ☐   ☐   No

 Test exercise 27

37   1   Expand each of the following in a Maclaurin series and determine the radius and the circle of convergence in each case.

(a) $f(z) = e^z$

(b) $f(z) = \ln(1 + 4z)$.

2   Determine the location and nature of the singular points in each of the following.

(a) $f(z) = \dfrac{3z}{(z + 1)^5}$

(b) $f(z) = z^{10}e^{1/z}$

(c) $f(z) = z\sin(1/z)$

(d) $f(z) = \dfrac{1 - \cos z}{z^2}$

3   Expand $f(z) = \sin z$ in a Taylor series about the point $z = \pi/4$ and determine the radius of convergence.

▶

4 Expand each of the following in a Laurent series. In (a) and (c) determine the nature of the singularity from the principal part of the series.

(a) $f(z) = (5 - z) \cos \dfrac{1}{z+3}$ about the point $z = -3$

(b) $f(z) = \dfrac{2z}{(z+1)(z+3)}$ valid for $1 < |z| < 3$

(c) $f(z) = \dfrac{1}{z^3(z-2)^2}$ about the point $z = 2$.

5 Calculate the residues at each of the singularities of

$$f(z) = \frac{3z - 1}{z^2(z+1)^2(z-1)}.$$

6 Evaluate each of the following integrals.

(a) $\displaystyle\int_0^{2\pi} \frac{d\theta}{5\cos\theta - 13}$

(b) $\displaystyle\int_{-\infty}^{\infty} \frac{dx}{x^2 + x + 1}$

(c) $\displaystyle\int_{-\infty}^{\infty} \frac{\cos 3x}{x^4 + 2x^2 + 1}\, dx$

 Further problems 27

1 For each of the following find the Maclaurin series expansion and determine the radius of convergence.
38

(a) $\sinh z$

(b) $\tan z$

(c) $\ln\left(\dfrac{1+z}{1-z}\right)$

(d) a^z, where $a > 0$

(e) $\dfrac{15z^2}{(5 - 3z)^3}$.

2 By using the appropriate Maclaurin series expansions, show that

(a) $(\cos z)' = -\sin z$

(b) $\cos z = \dfrac{e^z + e^{-z}}{2}$

(c) $(e^z)' = e^z$.

3 Given the series expansion for $(1 + z)^{-1}$

    (a) show by integration that this is compatible with the series expansion for $\ln(1 + z)$

    (b) by differentiation find $\displaystyle\sum_{n=1}^{\infty}(-1)^n n z^n$ and $\displaystyle\sum_{n=1}^{\infty}(-1)^n n^2 z^n$.

4 Use the ratio test to test each of the following for convergence.

    (a) $\displaystyle\sum_{n=0}^{\infty}\frac{(2n)!}{(n!)^2}z^n$ (d) $\displaystyle\sum_{n=0}^{\infty}\frac{(\cos n\pi)z^n}{2n - 1}$

    (b) $\displaystyle\sum_{n=0}^{\infty}\frac{z^n}{1 - 3n}$ (e) $\displaystyle\sum_{n=1}^{\infty}\frac{(-1)^n z^n}{(n + 1)!}$

    (c) $\displaystyle\sum_{n=0}^{\infty}\frac{n^2 z^n}{1 - 3n}$

5 Find the Taylor series about the point indicated of each of the following.

    (a) e^z about the point $z = 2$

    (b) $\cos z$ about the point $z = \pi/6$

    (c) $(z - 3)\sin(z + 3)$ about the point $z = 3$

    (d) $(2z - 5)^{-1}$ about the point $z = 1/3$

    (e) $(2z - 5)^{-1}$ about the point $z = 3$.

6 Find the series expansion of $z \ln z$ valid for $|z - 1| < 1$.

7 Find the circle of convergence of each of the following when expanded in a Taylor series about the point indicated.

    (a) $e^{-z}\cos(z - 2)$ about the point $z = 1$

    (b) $\dfrac{z^3}{(z^2 + 6)}$ about the point $z = 0$

    (c) $\dfrac{z - 2}{(z - 6)(z - 4)}$ about the point $z = 5$

    (d) $\dfrac{z^2}{(e^z + 1)}$ about the point $z = 0$.

8 Locate and classify all of the singularities of each of the following.

    (a) $\dfrac{(z - 1)^3}{z^2(z^2 - 1)^2}$

    (b) $z^{-2}e^{-1/z}$.

9 Find the Laurent series about the point indicated of each of the following.

(a) $\dfrac{1}{z}\sin\left(\dfrac{1}{z}\right)$ about the point $z = 0$

(b) $\dfrac{1}{2z - 3}$ about the point $z = 3/2$

(c) $\dfrac{z}{(z - 2)(z - 3)}$ about the point $z = 3$.

10 Find the Laurent series of $\dfrac{z - 1}{(z + 2)(z + 5)}$ that is valid for

(a) $2 < |z| < 5$

(b) $|z| > 5$

(c) $|z| < 2$.

11 Evaluate each of the following integrals.

(a) $\displaystyle\int_0^{2\pi} \dfrac{d\theta}{2 + \sin\theta}$

(b) $\displaystyle\int_0^{2\pi} \dfrac{d\theta}{\alpha + \beta\sin\theta}$ for $\alpha > |\beta|$

(c) $\displaystyle\int_0^{2\pi} \dfrac{d\theta}{1 + \alpha^2 - 2\alpha\cos\theta}$ where $0 < \alpha < 1$

(d) $\displaystyle\int_0^{2\pi} \dfrac{\sin^2\theta d\theta}{5 - 4\cos\theta}$

(e) $\displaystyle\int_0^{2\pi} \dfrac{d\theta}{5 - 3\cos\theta}$

(f) $\displaystyle\int_{-\infty}^{\infty} \dfrac{dx}{x^2 + 6x + 13}$

(g) $\displaystyle\int_{-\infty}^{\infty} \dfrac{x^2 dx}{x^4 + 6x^2 + 13}$

(h) $\displaystyle\int_{-\infty}^{\infty} \dfrac{x^2 dx}{(x^2 + 4)^2}$

(i) $\displaystyle\int_{-\infty}^{\infty} \dfrac{x^2 + x + 1}{x^4 + x^2 + 1}\, dx$

(j) $\displaystyle\int_{-\infty}^{\infty} \dfrac{dx}{x^6 + 1}$

(k) $\displaystyle\int_{-\infty}^{\infty} \dfrac{x^2 \sin\pi x\, dx}{x^4 + 6x^2 + 13}$

(l) $\displaystyle\int_{-\infty}^{\infty} \dfrac{\sin\pi x}{x^4 + x^2 + 1}\, dx$.

Optimization and linear programming

Learning outcomes

When you have completed this Programme you will be able to:

- Describe an optimization problem in terms of the objective function and a set of constraints
- Algebraically manipulate and graphically describe inequalities
- Solve a linear programming problem in two real variables
- Use the simplex method to describe a linear programming problem in two real variables as a problem in two real variables with two slack variables
- Set up the simplex tableau and compute the simplex
- Use the simplex method to solve a linear programming problem in three real variables with three slack variables
- Introduce artificial variables into the solution method as and when the need arises
- Solve minimisation problems using the simplex method
- Construct the algebraic form of the objective function and the constraints for a problem stated in words

Optimization

An *optimization problem* is one requiring the determination of the *optimal (maximum or minimum) value* of a given function, called the *objective function*, subject to a set of stated restrictions, or *constraints*, placed on the variables concerned.

In practice, for example, we may need to maximise an objective function representing units of output in a manufacturing situation, subject to constraints reflecting the availability of labour, machine time, stocks of raw materials, transport conditions, etc.

1

Linear programming (or linear optimization)

Linear programming is a method of solving an optimization problem when the objective function is a *linear function* and the constraints are *linear equations* or *linear inequalities*.

In this Programme, we shall restrict our considerations to problems of this type that form an important introduction to the much wider study of operational research.

Let us consider a simple example, so move on to the next frame

A simple linear programming problem may look like this:

2

$$\text{Maximise} \quad P = x + 2y \qquad \text{(objective function)}$$

$$\text{subject to} \quad \left. \begin{array}{l} y \le 3 \\ x + y \le 5 \\ x - 2y \le 2 \\ x \ge 0; \, y \ge 0 \end{array} \right\} \quad \text{(constraints)}$$

The last two constraints, i.e. $x \ge 0$ and $y \ge 0$, apply to all linear programming (LP) problems and indicate that the problem variables, x and y, are restricted to non-negative values: they may have zero or positive values, but NOT negative values. These two constraints are often combined and written $x, y \ge 0$ – or omitted altogether since they are taken for granted in all LP problems.

Before we proceed, we will take a brief look at linear inequalities in general.

On, then, to Frame 3

3 Linear inequalities

In most respects, *linear inequalities* can be manipulated in the same manner as can equations.

(a) Both sides may be increased or decreased by a common term, e.g.

$$2x \leq y + 4 \quad \therefore \quad 2x - y \leq 4$$

(b) Both sides may be multiplied or divided by a positive factor, e.g.

$$4x + 6y \geq 12 \quad \therefore \quad 2x + 3y \geq 6$$

But NOTE this:

(c) If both sides are multiplied or divided by a negative factor, e.g. (-1), then the inequality sign must be reversed, i.e. $\geq$ becomes $\leq$ and vice versa.

Here, then, is a short exercise.

Exercise

Simplify the following inequalities so that each right-hand side consists of a positive constant term only.

(a) $3x - 5 \leq 4y$ (b) $2(x + 2y) \leq -8$

(c) $4x - 6y \leq -10$ (d) $2x + 3 \geq -(y + 4)$

(e) $-(x - 3y + 5) \geq 2x + 4y - 6$

Check the results in the next frame

4

(a) $3x - 4y \leq 5$	(b) $-x - 2y \geq 4$
(c) $-2x + 3y \geq 5$	(d) $-2x - y \leq 7$
(e) $3x + y \leq 1$	

Graphical representation of linear inequalities

Consider the inequality $y - 2x \leq 3$. We can add $2x$ to each side, so that $y \leq 2x + 3$.

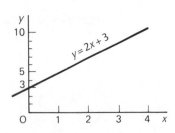

The equation $y = 2x + 3$ can be represented by a straight line dividing the x–y plane into two parts.

For all points on the line,

$$y = 2x + 3.$$

For all points below the line,

. . . .

$$\boxed{y < 2x + 3}$$ **5**

$\therefore\ y \le 2x + 3$ indicates all points on or below the straight line, but excludes all points above it. We can indicate this exclusion zone by shading the upper side of the line.

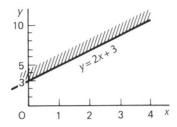

Arguing in much the same way, $x - 2y \le 2$ can be rewritten as $y \ge \dfrac{x}{2} - 1$ and we can draw the line $y = \dfrac{x}{2} - 1$ and shade in the exclusion zone

. . . .

$$\boxed{\text{below the line}}$$ **6**

i.e.

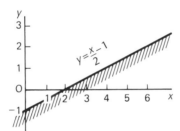

Example 1

The problem we quoted earlier in Frame 2 was

Maximise $P = x + 2y$ (*objective function*)

subject to $\left.\begin{array}{l} y \le 3 \\ x + y \le 5 \\ x - 2y \le 2 \\ x \ge 0;\ y \ge 0 \end{array}\right\}$ (*constraints*)

Now, on a common pair of x and y axes, we can represent the five constraints and shade in the exclusion zone for each. We then have the composite diagram

. . . .

7

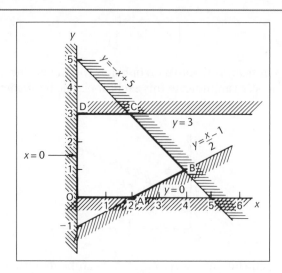

The coordinates of all points on the boundary of the polygon OABCD, or within the figure so formed, satisfy the system of constraints. The set of variables for each such point is called a *feasible point* or *feasible solution* and the figure OABCD is the *feasible domain* or *feasible polygon*.

Note these definitions.

8

Our problem now is to find the particular point within this domain that makes the objective function $P = x + 2y$ a maximum. The equation can be rewritten as $y = -\dfrac{x}{2} + \dfrac{P}{2}$ and this represents a set of parallel lines with different values of the intercept $\dfrac{P}{2}$.

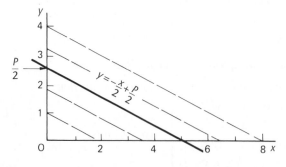

If we draw one sample line of this set to cross the feasible polygon we have just obtained, we get, using $P = 3$

. . . .

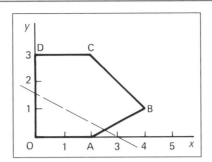

9

We can increase the value of P and hence raise the objective line up the page until it passes through the extreme point C. Any further increase in the value of P would take the line outside the feasible polygon and hence fail to conform to the given set of constraints.

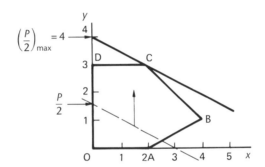

In this example, then, point C gives the optimal solution.

From the graph it can be seen that the line with maximal P passes through the point of intersection of the two lines $y = 3$ and $y = -x + 5$. This means that $y = 3$, $x = 2$ and so $P_{max} = x + 2y = 8$.

A graphical method of solution is clearly limited to linear programming problems involving two variables only. However, it is a useful introduction to other techniques, so let us deal with another example.

Example 2

Maximise $\quad P = x + 4y$

subject to $\quad -x + 2y \leq 6$

$\qquad\qquad 5x + 4y \leq 40$

$\qquad\qquad x, y \geq 0$

First of all, plot the appropriate straight line graphs to obtain the feasible polygon. This gives

. . . .

10

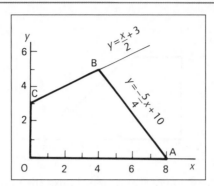

The objective function $P = x + 4y$ can be expressed in the form $y = -\dfrac{x}{4} + \dfrac{P}{4}$ and its graph added to the feasible polygon, as before. We then obtain

....

11

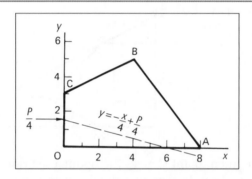

The line $y = -\dfrac{x}{4} + \dfrac{P}{4}$ is then raised to give the optimal solution, which is

....

12

$$\boxed{P_{\max} = 24 \text{ with } x = 4, y = 5}$$

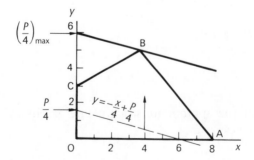

▶

From the graph it can be seen that the line with maximal P passes through the point of intersection of the two lines $y = \frac{x}{2} + 3$ and $y = -\frac{5x}{4} + 10$. That is $x = 4$, $y = 5$ and so $P_{\max} = x + 4y = 24$.

As easy as that.

Now this one.

Example 3

Minimise $P = -4x + 6y$

subject to $-x + 6y \geq 24$

                $2x - y \leq 7$

                $x + 8y \leq 80$

                $x, y \geq 0$

It is very much as before. Complete it on your own.

$$\boxed{P_{\min} = 6 \quad \text{with} \quad x = 6,\ y = 5}$$

13

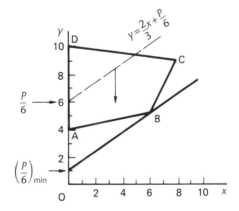

To obtain the minimum optimal value of P, the graph of the objective function is, of course, lowered to the appropriate extreme point.

    In practice, linear programming problems usually contain many more variables than the two we have so far considered and a computational method is then required. One such technique is the *simplex method* and the remainder of this Programme will be devoted to the steps necessary to put it into practice.

So move on to Frame 14

The simplex method

14 The first step in the *simplex method* is to ensure that each constraint is written with a *positive* right-hand side constant term. Then we express all inequalities as equations by the introduction of *slack variables*.

For example, $-x + 2y \leq 6$ can be written $-x + 2y + w_1 \quad = 6$

and $5x + 4y \leq 40$ can be written $5x + 4y \quad + w_2 = 40$

where w_1 and w_2 are positive (or zero) variables with unit coefficients, required to make up the left-hand side to the value of the right-hand side constant term. The new variables, w_1 and w_2, are called *slack variables*.

Let us look again at the problem we solved earlier.

Example 1

Maximise $P = x + 4y$

subject to $-x + 2y \leq 6$

$\qquad\qquad 5x + 4y \leq 40$ (as always, $x, y \geq 0$)

The constraints now become $-x + 2y + w_1 \quad = 6$

$\qquad\qquad\qquad\qquad\qquad 5x + 4y \quad + w_2 = 40$

and the objective function $P - x - 4y \qquad = 0$

From this, we can now begin to form the simplex tableau (or table).

So make a note of the above information – and then move on

---

15 Setting up the simplex tableau

(a) *Framework* First construct a framework with the headings shown.

x	y	w_1	w_2	b	*check*

Next, we enter, in the framework, the coefficients of the problem variables and of the slack variables in the constraints, together with the right-hand side constants in the column headed b. (Ignore the *check* column for the time being.)

So we have

. . . .

16

x	y	w_1	w_2	b	check
−1	2	1	0	6	
5	4	0	1	40	

Problem variables   Slack variables   Const.

body   unity matrix

(b) *Check column* The right-hand side column is included to provide a check on the numerical calculations as we develop the simplex, so, for each row, total up the entries in that row, including the constant column, and enter the sum in the check column.

Do that

17

Basis	x	y	w_1	w_2	b	check
w_1	− 1	2	1	0	6	8
w_2	5	4	0	1	40	50

(c) *Starting basic solution* The two constraints now contain four variables, but if we start by letting x and y each be zero, then we have the temporary solutions, $w_1 = 6$ and $w_2 = 40$, and we indicate these variables in the extra left-hand side column, as shown.

Note (1) The columns with the slack variables form a unity matrix.
(2) There are now four variables, x, y, w_1, w_2 ($n = 4$).
(3) There are two constraints ($m = 2$).
(4) We put ($n − m$) variables, i.e. two variables (x and y), equal to zero as a start.

Finally, we have to deal with the objective function,
so move on to the next frame

18

(d) *The objective function* The objective function, $P = x + 4y$, is written $P − x − 4y = 0$ and the coefficients of this form the bottom row, or *index row*, of the tableau, thus

Basis	x	y	w_1	w_2	b	check
w_1	− 1	2	1	0	6	8
w_2	5	4	0	1	40	50
P	− 1	−4	0	0	0	− 5

Complete your tableau, if you have not already done so, and then we will see how the computation is carried out.

19   Computation of the simplex

1   First we select the column containing the most negative entry in the index row: in this case -4. This is called the *key column* and we enclose it as shown.

Basis	x	y	w_1	w_2	b	check
w_1	$-\frac{1}{2}$	1	$\frac{1}{2}$	0	3	4
w_2	5	4	0	1	40	50
P	-1	-4	0	0	0	-5

                                                 └─── key column

2   In each row, we now divide the value in the b column by the positive entry in the key column: the smaller ratio determines the *key row*.

$$\text{For row 1 } (w_1), \quad r = \frac{6}{2} = 3$$
$$\text{row 2 } (w_2), \quad r = \frac{40}{4} = 10$$

$\left.\right\}$ $\therefore$ row 1 is the key row. Enclose it as shown below.

		pivot				
Basis	x	y	w_1	w_2	b	check
w_1	-1	2	1	0	6	8 ← key row
w_2	5	4	0	1	40	50
P	-1	-4	0	0	0	-5

The number at the intersection of the key column and key row is the *key number* or *pivot*: in this case the number 2.

3   We now divide all entries in the key row by the pivot to reduce the pivot to *unity* – which we then circle. The new version of the key row is sometimes called the *main row*. The rest of the tableau remains unchanged, so we then get . . . .

20

                                  unit pivot

Basis	x	y	w_1	w_2	b	check
w_1	$-\frac{1}{2}$	$①$	$\frac{1}{2}$	0	3	4 ← main row
w_2	5	4	0	1	40	50
P	-1	-4	0	0	0	-5 ← index

                                     └─── key column

So far, so good. Now we deal with the actual calculations.

Next frame

4 Using the main row, we now operate on the remaining rows of the tableau, including the index row, to reduce the other entries in the key column to zero. Note that the main row remains unaltered. The new value in any position in the other rows, including the b column and the check column, can be calculated as follows:

21

New number = old number – the product of the corresponding entries in the main row and key column

K is replaced by $K - (A \times B)$

For example, in the second row (w_2):

$$5 \text{ is replaced by } \quad 5 - \left(-\tfrac{1}{2}\right)(4) = 5 + 2 = 7$$
$$4 \text{ is replaced by } \quad 4 - (1)(4) \quad = 4 - 4 = 0$$
$$0 \text{ is replaced by } \quad 0 - \left(\tfrac{1}{2}\right)(4) \quad = 0 - 2 = -2$$
$$1 \text{ is replaced by } \quad 1 - (0)(4) \quad = 1 - 0 = 1$$
$$40 \text{ is replaced by } 40 - (3)(4) \quad = 40 - 12 = 28$$
$$50 \text{ is replaced by } 50 - (4)(4) \quad = 50 - 16 = 34$$

and, in the third row (P):

$$-1 \text{ is replaced by } \; -1 - \left(-\tfrac{1}{2}\right)(-4) = -1 - 2 = -3.$$

Completing the operations for rows (w_2) and (P), we have

. . . .

Basis	x	y	w_1	w_2	b	check
w_1	$-\tfrac{1}{2}$	1	$\tfrac{1}{2}$	0	3	4
w_2	7	0	-2	1	28	34
P	-3	0	2	0	12	11

22

Now confirm that the new values in the check column are, indeed, the sums of the entries in the corresponding rows. If not, there is a mistake somewhere in the working to be corrected before we proceed.

If all is well, move on to the next frame

23

5 *Change of basic variables*   In its final form, the key column consists of a single 1 and the remaining entries zero. This is in the column headed y which indicates that the basic variable w_1 in the main row can be replaced by y.

Basis	x	y	w_1	w_2	b	check
y ~~w_1~~	$-\frac{1}{2}$	1	$\frac{1}{2}$	0	3	4
w_2	7	0	-2	1	28	34
P	-3	0	2	0	12	11

Note that there are two columns containing a single 1 and the rest 0. These are headed y and w_2 which are now also the basic variables in the left-hand side column. Reading the values in the b column therefore gives a basic solution $y = 3$ and $w_2 = 28$, and at this stage $P = 12$. Any variable not listed in the basis column is zero. One basic solution at this stage is therefore $x = 0$, $y = 3$, $w_2 = 28$. However, we are not finished.

The index row (P) still contains another negative entry, so we have to repeat the simplex process using the same steps as before.

Basis	x	y	w_1	w_2	b	check	
y	$-\frac{1}{2}$	1	$\frac{1}{2}$	0	3	4	
w_2	7	0	-2	1	28	34	← key row
P	-3	0	2	0	12	11	

              └── key column

Now divide the key row by the key number (7) to reduce the pivot to a unit pivot. This gives

. . . .

24

Basis	x	y	w_1	w_2	b	check	
y	$-\frac{1}{2}$	1	$\frac{1}{2}$	0	3	4	
w_2	①	0	$-\frac{2}{7}$	$\frac{1}{7}$	4	$\frac{34}{7}$	← main row
P	-3	0	2	0	12	11	

Using the main row, operate on the remaining rows (including the index row) to reduce the other entries in the key column to zero. Complete that stage and we have

. . . .

25

Basis	x	y	w_1	w_2	b	check
y	0	1	$\frac{5}{14}$	$\frac{1}{14}$	5	$\frac{45}{7}$
w_2	1	0	$-\frac{2}{7}$	$\frac{1}{7}$	4	$\frac{34}{7}$
P	0	0	$\frac{8}{7}$	$\frac{3}{7}$	24	$\frac{179}{7}$

Again, at this stage, check your working by totalling up the entries in each row and satisfy yourself that the sum agrees with the value in the check column.

Note that w_2 in the basis column can now be replaced by x which was the heading of the column containing the last unit pivot.

So finally, we have

Basis	x	y	w_1	w_2	b	check
y	0	1	$\frac{5}{14}$	$\frac{1}{14}$	5	$\frac{45}{7}$
x	1	0	$-\frac{2}{7}$	$\frac{1}{7}$	4	$\frac{34}{7}$
P	0	0	$\frac{8}{7}$	$\frac{3}{7}$	24	$\frac{179}{7}$

A new basic solution now emerges as $x = 4$, $y = 5$.

Furthermore, since there is no further negative entry in the index row, this is also the optimal solution and the optimal value of P is given in the b column, i.e. $P_{max} = 24$.

For interest, you may wish to compare this result with that obtained in Frame 12.

We have been through the simplex operation in some detail by way of **26** explanation. Many problems involve more than just two variables, but the method of computation is basically the same, being an iterative process which is repeated until the index row contains no negative entry, at which point the optimal value of the objective function has been attained.

The problem we have just solved would normally look like this:

$$\text{Maximise} \quad P = x + 4y$$
$$\text{subject to} \quad -x + 2y \leq 6$$
$$5x + 4y \leq 40 \quad (x, y \geq 0)$$

Entering slack variables, etc., this is written

$$-x + 2y + w_1 \qquad = 6$$
$$5x + 4y \qquad + w_2 = 40$$
$$P - x - 4y \qquad = 0$$

The complete tableau is given in the next frame

27

Basis	x	y	w_1	w_2	b	check
w_1	-1	2	1	0	6	8
w_2	5	4	0	1	40	50
P	-1	-4	0	0	0	-5
y ~~w_1~~	$-\frac{1}{2}$	$①$	$\frac{1}{2}$	0	3	4
w_2	5	4	0	1	40	50
P	-1	-4	0	0	0	-5
y ~~w_1~~	$-\frac{1}{2}$	1	$\frac{1}{2}$	0	3	4
w_2	7	0	-2	1	28	34
P	-3	0	2	0	12	11
y	$-\frac{1}{2}$	1	$\frac{1}{2}$	0	3	4
w_2	$①$	0	$-\frac{2}{7}$	$\frac{1}{7}$	4	$\frac{34}{7}$
P	-3	0	2	0	12	11
y	0	1	$\frac{5}{14}$	$\frac{1}{14}$	5	$\frac{45}{7}$
x ~~w_2~~	1	0	$-\frac{2}{7}$	$\frac{1}{7}$	4	$\frac{34}{7}$
P	0	0	$\frac{8}{7}$	$\frac{3}{7}$	24	$\frac{179}{7}$

$$P_{\max} = 24 \text{ with } x = 4, \ y = 5$$

Now for another example – so move on to Frame 28

28

Here is one for you to do on your own. The method is just the same as before so you will have no difficulty.

Example 2

Maximise $P = 4x + 3y$

subject to $-x + \ y \le 4$

$\qquad\qquad x + 2y \le 14$

$\qquad\qquad 2x + \ y \le 16 \quad (x, y \ge 0)$

We have three inequalities this time, so we shall need to introduce three slack variables. Converting the inequalities into equations, we obtain

. . . .

29

$$
\begin{aligned}
-x + \ y + w_1 \qquad\qquad &= 4 \\
x + 2y \qquad + w_2 \qquad &= 14 \\
2x + \ y \qquad\qquad + w_3 &= 16
\end{aligned}
$$

Then we set out the simplex framework with appropriate headings, i.e.

. . . .

30

Basis	x	y	w_1	w_2	w_3	b	check

Remembering that the index row uses $P - 4x - 3y = 0$, we can set out the first tableau. Choosing x and y, as usual, to be zero for a start, we have

. . . .

31

Basis	x	y	w_1	w_2	w_3	b	check
w_1	−1	1	1	0	0	4	5
w_2	1	2	0	1	0	14	18
w_3	2	1	0	0	1	16	20
P	−4	−3	0	0	0	0	−7

Carry on now and complete the working on this first tableau.

Check with the next frame

32   Here is the working so far.

Basis	x	y	w_1	w_2	w_3	b	*check*
w_1	-1	1	1	0	0	4	5
w_2	1	2	0	1	0	14	18
w_3	2	1	0	0	1	16	20
P	-4	-3	0	0	0	0	-7
w_1	-1	1	1	0	0	4	5
w_2	1	2	0	1	0	14	18
w_3	①	$\frac{1}{2}$	0	0	$\frac{1}{2}$	8	10
P	-4	-3	0	0	0	0	-7
w_1	0	$\frac{3}{2}$	1	0	$\frac{1}{2}$	12	15
w_2	0	$\frac{3}{2}$	0	1	$-\frac{1}{2}$	6	8
x w_3	1	$\frac{1}{2}$	0	0	$\frac{1}{2}$	8	10
P	0	-1	0	0	2	32	33

(a) The basic variable (w_3) of the unit pivot can now be replaced by the variable at the heading of the unit pivot (x).

(b) We see there is still a negative value in the index row, so we repeat the process for this second tableau.

Now you can finish it off

Check to see if you agree.

Basis	x	y	w_1	w_2	w_3	b	check
w_1	0	$\frac{3}{2}$	1	0	$\frac{1}{2}$	12	15
w_2	0	$\frac{3}{2}$	0	1	$-\frac{1}{2}$	6	8
x	1	$\frac{1}{2}$	0	0	$\frac{1}{2}$	8	10
P	0	-1	0	0	2	32	33
w_1	0	$\frac{3}{2}$	1	0	$\frac{1}{2}$	12	15
w_2	0	①	0	$\frac{2}{3}$	$-\frac{1}{3}$	4	$\frac{16}{3}$
x	1	$\frac{1}{2}$	0	0	$\frac{1}{2}$	8	10
P	0	-1	0	0	2	32	33
w_1	0	0	1	-1	1	6	7
y ~~w_2~~	0	1	0	$\frac{2}{3}$	$-\frac{1}{3}$	4	$\frac{16}{3}$
x	1	0	0	$-\frac{1}{3}$	$\frac{2}{3}$	6	$\frac{22}{3}$
P	0	0	0	$\frac{2}{3}$	$\frac{5}{3}$	36	$\frac{115}{3}$

The basic variable (w_2) can now be replaced by y, being the heading of the unit pivot column.

There is no further negative entry in the index row: therefore, the optimal value of P has been attained.

$$\therefore \ P_{\max} = 36 \quad \text{with} \quad x = 6, \ y = 4.$$

Note: We also see that $w_1 = 6$, since the unity matrix has headings x, y, w_1. The full result, therefore, is

$$P_{\max} = 36 \quad \text{with} \quad x = 6, \ y = 4, \ w_1 = 6, \ w_2 = 0, \ w_3 = 0$$

though we do not normally require this extra information.

The meaning of $w_1 = 0$ and $w_2 = 0$ is that the second and third constraints become

$$x + 2y = 14 \ \text{and} \ 2x + y = 16 \ \text{respectively rather than}$$
$$x + 2y \le 14 \ \text{and} \ 2x + y \le 16.$$

The meaning of $w_1 = 6$ gives the first constraint as $-x + y < 14$ rather than $-x + y \le 14$.

Now we will extend the simplex method to an example involving three problem variables.

Next frame

34 Simplex with three problem variables

Maximise $P = p_1 x + p_2 y + p_3 z$

subject to $a_{11}x + a_{12}y + a_{13}z \le b_1$

$$a_{21}x + a_{22}y + a_{23}z \le b_2$$

$$a_{31}x + a_{32}y + a_{33}z \le b_3$$

$$x, y, z \ge 0$$

Introducing slack variables we have

$$
\begin{aligned}
a_{11}x + a_{12}y + a_{13}z + w_1 &= b_1 \\
a_{21}x + a_{22}y + a_{23}z \quad\; + w_2 \quad\;\; &= b_2 \\
a_{31}x + a_{32}y + a_{33}z \quad\quad\quad\; + w_3 &= b_3 \\
P - p_1 x - p_2 y - p_3 z \quad\quad\quad\quad &= 0
\end{aligned}
$$

If there is now a total of n variables and m constraints, then at least $(n - m)$ variables are equated to zero. The remainder form the basic variable column entries. Equating x, y, z to zero, then, the basic variables are w_1, w_2, w_3.

Basis	x	y	z	w_1	w_2	w_3	b	check
w_1	a_{11}	a_{12}	a_{13}	1	0	0	b_1	
w_2	a_{21}	a_{22}	a_{23}	0	1	0	b_2	
w_3	a_{31}	a_{32}	a_{33}	0	0	1	b_3	
P	$-p_1$	$-p_2$	$-p_3$	0	0	0	0	

The variables in the basis column are the variables heading the unity matrix. The method is exactly as before.

(a) Select the most negative entry in the index row to determine the *key column*.

(b) Divide the entries in the constant column (b) by the corresponding positive entries in the key column. The smallest positive ratio determines the *key row*.

(c) The entry at the intersection of the key column and the key row is the *key number* or *pivot*.

(d) Divide each entry in the key row by the pivot to reduce the key number to a *unit pivot*. The revised key row is now called the *main row*.

(e) Use the main row to operate on the remaining rows to reduce all other entries in the key column to zero.

(f) Repeat steps (a) to (e) until no negative entry remains in the index row.

Now for an example

Example 1   **35**

Maximise $P = 2x + 6y + 4z$

subject to $2x + 5y + 2z \leq 38$

$4x + 2y + 3z \leq 57$

$x + 3y + 5z \leq 57$

$x, y, z \geq 0$

Rewriting the inequalities as equations gives

       ....

36

$$2x + 5y + 2z + w_1 \qquad\qquad = 38$$
$$4x + 2y + 3z \qquad + w_2 \qquad = 57$$
$$x + 3y + 5z \qquad\qquad + w_3 = 57$$

We also have $P - 2x - 6y - 4z = 0$, so we can now set up the simplex tableau ready for solution. That is ....

37

Basis	x	y	z	w_1	w_2	w_3	b	*check*
w_1	2	5	2	1	0	0	38	48
w_2	4	2	3	0	1	0	57	67
w_3	1	3	5	0	0	1	57	67
P	−2	−6	−4	0	0	0	0	− 12

Now we just apply the normal simplex routine until there is no negative entry in the index row.

Remember:

(1) to replace the basic variables as the problem variables become available at each stage, and

(2) any variable not appearing in the basis column has zero value.

Now you can work the solution right through and then check the result with the next frame. Take your time: there are no snags.

38

Basis	x	y	z	w_1	w_2	w_3	b	check
w_1	2	5	2	1	0	0	38	48
w_2	4	2	3	0	1	0	57	67
w_3	1	3	5	0	0	1	57	67
P	-2	-6	-4	0	0	0	0	-12
w_1	$\frac{2}{5}$	$①$	$\frac{2}{5}$	$\frac{1}{5}$	0	0	$\frac{38}{5}$	$\frac{48}{5}$
w_2	4	2	3	0	1	0	57	67
w_3	1	3	5	0	0	1	57	67
P	-2	-6	-4	0	0	0	0	-12
$y\ \cancel{w_1}$	$\frac{2}{5}$	1	$\frac{2}{5}$	$\frac{1}{5}$	0	0	$\frac{38}{5}$	$\frac{48}{5}$
w_2	$\frac{16}{5}$	0	$\frac{11}{5}$	$-\frac{2}{5}$	1	0	$\frac{209}{5}$	$\frac{239}{5}$
w_3	$-\frac{1}{5}$	0	$\frac{19}{5}$	$-\frac{3}{5}$	0	1	$\frac{171}{5}$	$\frac{191}{5}$
P	$\frac{2}{5}$	0	$-\frac{8}{5}$	$\frac{6}{5}$	0	0	$\frac{228}{5}$	$\frac{228}{5}$
y	$\frac{2}{5}$	1	$\frac{2}{5}$	$\frac{1}{5}$	0	0	$\frac{38}{5}$	$\frac{48}{5}$
w_2	$\frac{16}{5}$	0	$\frac{11}{5}$	$-\frac{2}{5}$	1	0	$\frac{209}{5}$	$\frac{239}{5}$
w_3	$-\frac{1}{19}$	0	$①$	$-\frac{3}{19}$	0	$\frac{5}{19}$	9	$\frac{191}{19}$
P	$\frac{2}{5}$	0	$-\frac{8}{5}$	$\frac{6}{5}$	0	0	$\frac{228}{5}$	$\frac{228}{5}$
y	$\frac{8}{19}$	1	0	$\frac{5}{19}$	0	$-\frac{2}{19}$	4	$\frac{106}{19}$
w_2	$\frac{63}{19}$	0	0	$-\frac{1}{19}$	1	$-\frac{11}{19}$	22	$\frac{448}{19}$
$z\ \cancel{w_3}$	$-\frac{1}{19}$	0	1	$-\frac{3}{19}$	0	$\frac{5}{19}$	9	$\frac{191}{19}$
P	$\frac{16}{19}$	0	0	$\frac{18}{19}$	0	$\frac{8}{19}$	60	$\frac{1172}{19}$

$$\therefore\ P_{\max} = 60 \ \text{ with } \ x = 0,\ y = 4,\ z = 9.$$

Do you agree?

If so, on to the next frame

39   **Example 2**

Maximise $P = 3x + 4y + 5z$

subject to $2x + 4y + 3z \le 80$

$\qquad\qquad 4x + 2y + \ z \le 48$

$\qquad\qquad\ x + \ y + 2z \le 40$

$\qquad\qquad\quad x,\ y,\ z \ge 0$

It is much the same as before. Work through it carefully and then check with the next frame.

$$2x + 4y + 3z + w_1 \qquad\qquad = 80$$
$$4x + 2y + \ z \qquad + w_2 \qquad = 48$$
$$x + \ y + 2z \qquad\qquad + w_3 = 40$$
$$P - 3x - 4y - 5z \qquad\qquad\qquad = 0$$

40

Basis	x	y	z	w_1	w_2	w_3	b	check
w_1	2	4	3	1	0	0	80	90
w_2	4	2	1	0	1	0	48	56
w_3	1	1	2	0	0	1	40	45
P	-3	-4	-5	0	0	0	0	-12
w_1	2	4	3	1	0	0	80	90
w_2	4	2	1	0	1	0	48	56
w_3	$\frac{1}{2}$	$\frac{1}{2}$	1	0	0	$\frac{1}{2}$	20	$\frac{45}{2}$
P	-3	-4	-5	0	0	0	0	-12
w_1	$\frac{1}{2}$	$\frac{5}{2}$	0	1	0	$-\frac{3}{2}$	20	$\frac{45}{2}$
w_2	$\frac{7}{2}$	$\frac{3}{2}$	0	0	1	$-\frac{1}{2}$	28	$\frac{67}{2}$
$z \ \cancel{w_3}$	$\frac{1}{2}$	$\frac{1}{2}$	1	0	0	$\frac{1}{2}$	20	$\frac{45}{2}$
P	$-\frac{1}{2}$	$-\frac{3}{2}$	0	0	0	$\frac{5}{2}$	100	$\frac{201}{2}$
w_1	$\frac{1}{5}$	1	0	$\frac{2}{5}$	0	$-\frac{3}{5}$	8	9
w_2	$\frac{7}{2}$	$\frac{3}{2}$	0	0	1	$-\frac{1}{2}$	28	$\frac{67}{2}$
z	$\frac{1}{2}$	$\frac{1}{2}$	1	0	0	$\frac{1}{2}$	20	$\frac{45}{2}$
P	$-\frac{1}{2}$	$-\frac{3}{2}$	0	0	0	$\frac{5}{2}$	100	$\frac{201}{2}$
$y \ \cancel{w_1}$	$\frac{1}{5}$	1	0	$\frac{2}{5}$	0	$-\frac{3}{5}$	8	9
w_2	$\frac{16}{5}$	0	0	$-\frac{3}{5}$	1	$\frac{2}{5}$	16	20
z	$\frac{2}{5}$	0	1	$-\frac{1}{5}$	0	$\frac{4}{5}$	16	18
P	$-\frac{1}{5}$	0	0	$\frac{3}{5}$	0	$\frac{8}{5}$	112	114
y	$\frac{1}{5}$	1	0	$\frac{2}{5}$	0	$-\frac{3}{5}$	8	9
w_2	1	0	0	$-\frac{3}{16}$	$\frac{5}{16}$	$\frac{1}{8}$	5	$\frac{25}{4}$
z	$\frac{2}{5}$	0	1	$-\frac{1}{5}$	0	$\frac{4}{5}$	16	18
P	$-\frac{1}{5}$	0	0	$\frac{3}{5}$	0	$\frac{8}{5}$	112	114
y	0	1	0	$\frac{7}{16}$	$-\frac{1}{16}$	$-\frac{5}{8}$	7	$\frac{31}{4}$
$x \ \cancel{w_2}$	1	0	0	$-\frac{3}{16}$	$\frac{5}{16}$	$\frac{1}{8}$	5	$\frac{25}{4}$
z	0	0	1	$-\frac{1}{8}$	$-\frac{1}{8}$	$\frac{3}{4}$	14	$\frac{31}{2}$
P	0	0	0	$\frac{9}{16}$	$\frac{1}{16}$	$\frac{13}{8}$	113	$\frac{461}{4}$

$$\therefore \ P_{max} = 113 \quad \text{with } x = 5, \ y = 7, \ z = 14.$$

Now let us meet a further complication. *Next frame*

41   Artificial variables

So far, our approach to each problem has been the same.

(a) We first of all convert the 'less than' inequalities into equations by the inclusion of slack variables.

(b) If there are now n variables and m constraints, then at least $(n - m)$ variables are equated to zero – usually x, y, z, etc. – so that the initial basic solution is given by the slack variables, the coefficients of which form the unity matrix in the tableau.

(c) We then proceed by the simplex method to convert the basic solution to one containing the problem variables, the tableau entries for which now form a new unity matrix.

(d) The method is repeated as necessary. When no negative entry remains in the index row, the value of P denoted in the constant column is the optimal value of the objective function.

Now let us look at this example.

Example 1

Maximise $\quad P = 7x + 4y$

subject to $\quad 2x + \ y \le 150$

$$4x + 3y \le 350$$

$$x + \ y \ge \ \ 80 \quad (x, y \ge 0)$$

Converting the inequalities to equations, we have

. . . .

42

$$
\boxed{
\begin{aligned}
2x + \ y + w_1 \qquad\qquad &= 150 \\
4x + 3y \qquad + w_2 \qquad &= 350 \\
x + \ y \qquad\qquad - w_3 &= \ \ 80
\end{aligned}
}
$$

Also, of course, $P - 7x - 4y = 0$.

NOTE that since the third constraint is a 'greater than' statement, we must subtract the slack positive variable (w_3) to form the equation.

Alternatively, we could have written the inequality as $-x - y \le -80$ so that $-x - y + w_3 = -80$ and so $x + y - w_3 = 80$.

Forming the first tableau, in the usual manner, we obtain

. . . .

Basis	x	y	w_1	w_2	w_3	b	*check*
w_1	2	1	1	0	0	150	154
w_2	4	3	0	1	0	350	358
w_3	1	1	0	0	-1	80	81
P	-7	-4	0	0	0	0	-11

43

Now we are stuck, for we do not have a unity matrix to start off with. The entry in the w_3 column is -1 and not the necessary $+1$, and no amount of manipulation will help since the entries in the constant column (b) are, by definition, positive.

So how can we find a starting technique?

Let us restate the problem.

Maximise $P = 7x + 4y$

subject to
$$2x + y + w_1 \qquad\qquad = 150$$
$$4x + 3y \qquad + w_2 \qquad = 350$$
$$x + y \qquad\qquad - w_3 = 80$$

44

The trouble comes in the third constraint by virtue of the negative sign of the slack variable. To save the situation, we introduce a new small positive variable (w_4) so that w_1, w_2 and w_4 will give rise to a unity matrix and the simplex computation can then proceed. Of course, w_4 is ficticious, is extremely small and cannot appear in the final basic solution. To establish this, we include in the objective function a new term $-Mw_4$, where M is an extremely large positive value which will ensure that w_4 will ultimately vanish. So we now write

$$P = 7x + 4y - Mw_4$$

The new variable, w_4, is called an *artificial variable*: it is introduced solely so that the simplex procedure can be carried out; and it must not appear in the final basic solution listed in the basis column.

The third constraint above now becomes

$$x + y \qquad\qquad - w_3 + w_4 = 80$$

Next frame

45   We now have

$$2x + y + w_1 \qquad\qquad\qquad = 150$$
$$4x + 3y \qquad + w_2 \qquad\qquad = 350$$
$$x + y \qquad\quad - w_3 + \quad w_4 = \quad 80$$
$$P - 7x - 4y \qquad\qquad + Mw_4 = \quad 0$$

Forming the tableau in the usual way:

Basis	x	y	w_1	w_2	w_3	w_4	b	check
w_1	2	1	1	0	0	0	150	154
w_2	4	3	0	1	0	0	350	358
w_4	1	1	0	0	−1	1	80	82
P	−7	−4	0	0	0	M	0	$M - 11$

Note that:

(1) The columns headed w_1, w_2, w_4 now form the unity matrix.

(2) There are now 6 variables and 3 constraints, i.e. $n = 6$ and $m = 3$. At least $(n - m)$, i.e. $6 - 3 = 3$, variables are put equal to zero. We start off with x, y, w_3 as zero and w_1, w_2, w_4 form the first basic solution with the values given in the b column.

We now proceed in the normal way. Solve the first tableau and check the results so far in the following frame.

Basis	x	y	w_1	w_2	w_3	w_4	b	check
w_1	2	1	1	0	0	0	150	154
w_2	4	3	0	1	0	0	350	358
w_4	1	1	0	0	-1	1	80	82
P	-7	-4	0	0	0	M	0	$M - 11$
w_1	①	$\frac{1}{2}$	$\frac{1}{2}$	0	0	0	75	77
w_2	4	3	0	1	0	0	350	358
w_4	1	1	0	0	-1	1	80	82
P	-7	-4	0	0	0	M	0	$M - 11$
x ~~w_1~~	1	$\frac{1}{2}$	$\frac{1}{2}$	0	0	0	75	77
w_2	0	1	-2	1	0	0	50	50
w_4	0	$\frac{1}{2}$	$-\frac{1}{2}$	0	-1	1	5	5
P	0	$-\frac{1}{2}$	$\frac{7}{2}$	0	0	M	525	$M + 528$

The basic variable w_1 can be replaced by x and we continue as before to remove the further negative entry in the index row.

Do that

Basis	x	y	w_1	w_2	w_3	w_4	b	check
x	1	$\frac{1}{2}$	$\frac{1}{2}$	0	0	0	75	77
w_2	0	1	-2	1	0	0	50	50
w_4	0	$\frac{1}{2}$	$-\frac{1}{2}$	0	-1	1	5	5
P	0	$-\frac{1}{2}$	$\frac{7}{2}$	0	0	M	525	$M + 528$
x	1	$\frac{1}{2}$	$\frac{1}{2}$	0	0	0	75	77
w_2	0	1	-2	1	0	0	50	50
w_4	0	①	-1	0	-2	2	10	10
P	0	$-\frac{1}{2}$	$\frac{7}{2}$	0	0	M	525	$M + 528$
x	1	0	1	0	1	-1	70	72
w_2	0	0	-1	1	2	-2	40	40
y ~~w_4~~	0	1	-1	0	-2	2	10	10
P	0	0	3	0	-1	$M + 1$	530	$M + 533$

The basic variable w_4 is now replaced by y (the column of the last unit pivot). We now have another negative entry in the index row, so we have to perform the simplex calculation yet again.

For the next round, we get . . . .

48

Basis	x	y	w_1	w_2	w_3	w_4	b	check
x	1	0	1	0	1	−1	70	72
w_2	0	0	−1	1	2	−2	40	40
y	0	1	−1	0	−2	2	10	10
P	0	0	3	0	−1	$M+1$	530	$M+533$
x	1	0	1	0	1	−1	70	72
w_2	0	0	$-\frac{1}{2}$	$\frac{1}{2}$	①	−1	20	20
y	0	1	−1	0	−2	2	10	10
P	0	0	3	0	−1	$M+1$	530	$M+533$
x	1	0	$\frac{3}{2}$	$-\frac{1}{2}$	0	0	50	52
w_3 ~~w_2~~	0	0	$-\frac{1}{2}$	$\frac{1}{2}$	1	−1	20	20
y	0	1	−2	1	0	0	50	50
P	0	0	$\frac{5}{2}$	$\frac{1}{2}$	0	M	550	$M+553$

No further negative entry remains in the index row. The optimal solution has been found, i.e.

$P_{\text{max}} = 550$ with $x = 50$, $y = 50$.

In addition, we see that $w_3 = 20$ while $w_1 = w_2 = w_4 = 0$ since they do not occur in the basic variable column.

Notice, also, that w_4, the artificial variable, does not figure in the optimal solution – as indeed it must not.

Next frame

49

Here is one for you to do in the same way.

Example 2

Maximise $P = 2x + 5y$
subject to $x + 4y \leq 60$
            $3x + 2y \leq 40$
            $x + \ y \geq 12 \quad (x, y \geq 0)$

Work right through it, just as before. The result is

. . . .

$$\boxed{P_{\max} = 78 \quad \text{with} \quad x = 4, \, y = 14}$$

Because

$$
\begin{aligned}
x + 4y + w_1 \qquad\qquad\qquad &= 60 \\
3x + 2y \quad + w_2 \qquad\qquad &= 40 \\
x + \, y \qquad\quad - w_3 + \, w_4 &= 12 \\
P - 2x - 5y \qquad\qquad + Mw_4 &= \, 0
\end{aligned}
$$

Basis	x	y	w_1	w_2	w_3	w_4	b	check
w_1	1	4	1	0	0	0	60	66
w_2	3	2	0	1	0	0	40	46
w_4	1	(1)	0	0	−1	1	12	14
P	−2	−5	0	0	0	M	0	$M - 7$
w_1	−3	0	1	0	4	−4	12	10
w_2	1	0	0	1	2	−2	16	18
y w_4	1	1	0	0	−1	1	12	14
P	3	0	0	0	−5	$M + 5$	60	$M + 63$
w_1	$-\frac{3}{4}$	0	$\frac{1}{4}$	0	(1)	−1	3	$\frac{5}{2}$
w_2	1	0	0	1	2	−2	16	18
y	1	1	0	0	−1	1	12	14
P	3	0	0	0	−5	$M + 5$	60	$M + 63$
w_3 w_1	$-\frac{3}{4}$	0	$\frac{1}{4}$	0	1	−1	3	$\frac{5}{2}$
w_2	$\frac{5}{2}$	0	$-\frac{1}{2}$	1	0	0	10	13
y	$\frac{1}{4}$	1	$\frac{1}{4}$	0	0	0	15	$\frac{33}{2}$
P	$-\frac{3}{4}$	0	$\frac{5}{4}$	0	0	M	75	$M + \frac{151}{2}$
w_3	$-\frac{3}{4}$	0	$\frac{1}{4}$	0	1	−1	3	$\frac{5}{2}$
w_2	(1)	0	$-\frac{1}{5}$	$\frac{2}{5}$	0	0	4	$\frac{26}{2}$
y	$\frac{1}{4}$	1	$\frac{1}{4}$	0	0	0	15	$\frac{33}{2}$
P	$-\frac{3}{4}$	0	$\frac{5}{4}$	0	0	M	75	$M + \frac{151}{2}$
w_3	0	0	$\frac{1}{10}$	$\frac{3}{10}$	1	−1	6	$\frac{32}{5}$
x w_2	1	0	$-\frac{1}{5}$	$\frac{2}{5}$	0	0	4	$\frac{26}{5}$
y	0	1	$\frac{3}{10}$	$-\frac{1}{10}$	0	0	14	$\frac{76}{5}$
P	0	0	$\frac{11}{10}$	$\frac{3}{10}$	0	M	78	$M + \frac{397}{5}$

$$\therefore \; P_{\max} = 78 \quad \text{with} \quad x = 4, \, y = 14.$$

On to the next frame

51   We are not always as lucky as we were in the previous two examples and other steps sometimes have to be taken to remove the artificial variable. Consider the following case.

Example 3

Maximise $\quad\quad P = 8x + 4y$

subject to $\quad 2x + 3y \leq 120$

$$x + y \leq 45$$

$$-3x + 5y \geq 25 \quad (x, y \geq 0)$$

Inserting the slack variables and artificial variable as required, we have

$$............$$

52

$$
\begin{array}{l}
2x + 3y + w_1 \quad\quad\quad\quad\quad\quad = 120 \\
x + y \quad\quad + w_2 \quad\quad\quad\quad = 45 \\
-3x + 5y \quad\quad\quad - w_3 + \quad w_4 = 25 \\
P - 8x - 4y \quad\quad\quad\quad\quad + Mw_4 = \quad 0
\end{array}
$$

That is very much as before, so work through it and check with the next frame.

---

53   Here it is.

Basis	x	y	w_1	w_2	w_3	w_4	b	*check*
w_1	2	3	1	0	0	0	120	126
w_2	①	1	0	1	0	0	45	48
w_4	−3	5	0	0	−1	1	25	27
P	−8	−4	0	0	0	M	0	$M - 12$
w_1	0	1	1	−2	0	0	30	30
$x\ \cancel{w_2}$	1	1	0	1	0	0	45	48
* ⟶ w_4	0	8	0	3	−1	1	160	171
P	0	4	0	8	0	M	360	$M + 372$

There is no further negative entry in the index row, so it looks as though the optimal solution has been attained. However, the artificial variable w_4 still remains in the basic variable column at * and thus must be removed. Therefore, we take the entry at the junction of the y column and the w_4 row as the pivot and proceed to eliminate w_4 by simplifying the tableau a stage further.

If we do that, we get ....

54

Basis	x	y	w_1	w_2	w_3	w_4	b	check
w_1	0	1	1	-2	0	0	30	30
x	1	1	0	1	0	0	45	48
$*\to w_4$	0	[8]	0	3	-1	1	160	171
P	0	4	0	8	0	M	360	$M+372$
w_1	0	1	1	-2	0	0	30	30
x	1	1	0	1	0	0	45	48
$*\to w_4$	0	(1)	0	$\frac{3}{8}$	$-\frac{1}{8}$	$\frac{1}{8}$	20	$\frac{171}{8}$
P	0	4	0	8	0	M	360	$M+372$
w_1	0	0	1	$-\frac{19}{8}$	$\frac{1}{8}$	$-\frac{1}{8}$	10	$\frac{69}{8}$
x	1	0	0	$\frac{5}{8}$	$\frac{1}{8}$	$-\frac{1}{8}$	25	$\frac{213}{8}$
$y \;\cancel{w_4}$	0	1	0	$\frac{3}{8}$	$-\frac{1}{8}$	$\frac{1}{8}$	50	$\frac{171}{8}$
P	0	0	0	$\frac{13}{2}$	$\frac{1}{2}$	$M-\frac{1}{2}$	280	$M+\frac{573}{2}$

The artificial variable, w_4, is now replaced by y in the basic variable column and the optimal solution has been reached.

$\therefore P_{\max} = 280$ with $x = 25$, $y = 20$.

Next frame

Now here is one for you to deal with.

55

Example 4

Maximise $P = 10x + 2y$

subject to $\quad -x + 2y \le \ \ 60$

$\qquad\qquad\quad 5x + 4y \le 260$

$\qquad\qquad\ -x + 8y \ge \ \ 80 \quad (x, y \ge 0)$

Work through it as before and see if you agree with the solution in the next frame.

56

$$-x + 2y + w_1 \qquad\qquad = 60$$
$$5x + 4y \qquad + w_2 \qquad\qquad = 260$$
$$-x + 8y \qquad\qquad - w_3 + \; w_4 = 80$$
$$P - 10x - 2y \qquad\qquad + Mw_4 = \; 0$$

Basis	x	y	w_1	w_2	w_3	w_4	b	check
w_1	-1	2	1	0	0	0	60	62
w_2	5	4	0	1	0	0	260	270
w_4	-1	8	0	0	-1	1	80	87
P	-10	-2	0	0	0	M	0	$M - 12$
w_1	-1	2	1	0	0	0	60	62
w_2	①	$\frac{4}{5}$	0	$\frac{1}{5}$	0	0	52	54
w_4	-1	8	0	0	-1	1	80	87
P	-10	-2	0	0	0	M	0	$M - 12$
w_1	0	$\frac{14}{5}$	1	$\frac{1}{5}$	0	0	112	116
x $\overline{w_2}$	1	$\frac{4}{5}$	0	$\frac{1}{5}$	0	0	52	54
⋆→ w_4	0	$\frac{44}{5}$	0	$\frac{1}{5}$	-1	1	132	141
P	0	6	0	2	0	M	520	$M + 528$
w_1	0	$\frac{14}{5}$	1	$\frac{1}{5}$	0	0	112	116
x	1	$\frac{4}{5}$	0	$\frac{1}{5}$	0	0	52	54
⋆→ w_4	0	①	0	$\frac{1}{44}$	$-\frac{5}{44}$	$\frac{5}{44}$	15	$\frac{705}{44}$
P	0	6	0	2	0	M	520	$M + 528$
w_1	0	0	1	$\frac{3}{22}$	$\frac{7}{22}$	$-\frac{7}{22}$	70	$\frac{1565}{22}$
x	1	0	0	$\frac{2}{11}$	$\frac{1}{11}$	$-\frac{1}{11}$	40	$\frac{453}{11}$
y $\overline{w_4}$	0	1	0	$\frac{1}{44}$	$-\frac{5}{44}$	$\frac{5}{44}$	15	$\frac{705}{44}$
P	0	0	0	$\frac{41}{22}$	$\frac{15}{22}$	$M - \frac{15}{22}$	430	$M + \frac{9501}{22}$

$$\therefore \; P_{\max} = 430 \quad \text{with} \quad x = 40, \; y = 15.$$

Now for another example

Example 5

57

This one is slightly different, so take note.

Maximise $P = 11x + 15y$

subject to $\quad 3x + 5y \leq 130$

$$-4x + 5y \geq 25$$

$$x + 5y \geq 75 \quad (x, y \geq 0)$$

In this problem, notice that there are two 'greater than' inequalities so that there will be two slack variables to be subtracted and two artificial variables to be incorporated. In the objective function, we can use the same factor, M, for both artificial variables, since neither of those two variables will appear in the final optimal solution. So, we have:

$$\begin{aligned}
3x + 5y + w_1 &= 130 \\
-4x + 5y - w_2 + w_4 &= 25 \\
x + 5y - w_3 + w_5 &= 75 \\
P - 11x - 15y + Mw_4 + Mw_5 &= 0
\end{aligned}$$

w_1, w_4, w_5 now form the unity matrix from which to start. The method is just the same as in previous examples, so finish it off.

58

Basis	x	y	w_1	w_2	w_3	w_4	w_5	b	check
w_1	3	5	1	0	0	0	0	130	139
w_4	-4	5	0	-1	0	1	0	25	26
w_5	1	5	0	0	-1	0	1	75	81
P	-11	-15	0	0	0	M	M	0	$2M-26$
w_1	3	5	1	0	0	0	0	130	139
w_4	$-\frac{4}{5}$	①	0	$-\frac{1}{5}$	0	$\frac{1}{5}$	0	5	$\frac{26}{5}$
w_5	1	5	0	0	-1	0	1	75	81
P	-11	-15	0	0	0	M	M	0	$2M-26$
w_1	7	0	1	1	0	-1	0	105	113
y $\cancel{w_4}$	$-\frac{4}{5}$	1	0	$-\frac{1}{5}$	0	$\frac{1}{5}$	0	5	$\frac{26}{5}$
w_5	5	0	0	1	-1	-1	1	50	55
P	-23	0	0	-3	0	$M+3$	M	75	$2M+52$
w_1	7	0	1	1	0	-1	0	105	113
y	$-\frac{4}{5}$	1	0	$-\frac{1}{5}$	0	$\frac{1}{5}$	0	5	$\frac{26}{5}$
w_5	①	0	0	$\frac{1}{5}$	$-\frac{1}{5}$	$-\frac{1}{5}$	$\frac{1}{5}$	10	11
P	-23	0	0	-3	0	$M+3$	M	75	$2M+52$
w_1	0	0	1	$-\frac{2}{5}$	$\frac{7}{5}$	$\frac{2}{5}$	$-\frac{7}{5}$	35	36
y	0	1	0	$-\frac{1}{25}$	$-\frac{4}{25}$	$\frac{1}{25}$	$\frac{4}{25}$	13	14
x $\cancel{w_5}$	1	0	0	$\frac{1}{5}$	$-\frac{1}{5}$	$-\frac{1}{5}$	$\frac{1}{5}$	10	11
P	0	0	0	$\frac{8}{5}$	$-\frac{23}{5}$	$M-\frac{8}{5}$	$M+\frac{23}{5}$	305	$2M+305$
w_1	0	0	$\frac{5}{7}$	$-\frac{2}{7}$	①	$\frac{2}{7}$	-1	25	$\frac{180}{7}$
y	0	1	0	$-\frac{1}{25}$	$-\frac{4}{25}$	$\frac{1}{25}$	$\frac{4}{25}$	13	14
x	1	0	0	$\frac{1}{5}$	$-\frac{1}{5}$	$-\frac{1}{5}$	$\frac{1}{5}$	10	11
P	0	0	0	$\frac{8}{5}$	$-\frac{23}{5}$	$M-\frac{8}{5}$	$M+\frac{23}{5}$	305	$2M+305$
w_3 $\cancel{w_1}$	0	0	$\frac{5}{7}$	$-\frac{2}{7}$	1	$\frac{2}{7}$	-1	25	$\frac{180}{7}$
y	0	1	$\frac{4}{35}$	$-\frac{3}{35}$	0	$\frac{3}{35}$	0	17	$\frac{634}{35}$
x	1	0	$\frac{1}{7}$	$\frac{1}{7}$	0	$-\frac{1}{7}$	0	15	$\frac{113}{7}$
P	0	0	$\frac{23}{7}$	$\frac{2}{7}$	0	$M-\frac{2}{7}$	M	420	$2M+\frac{2963}{7}$

So there it is. $P_{\max} = 420$ with $x = 15$, $y = 17$.

Incidentally, also, $w_3 = 25$ and $w_1 = w_2 = w_4 = w_5 = 0$.

Our examples on the use of artificial variables have so far concerned only two problem variables, x and y. The method, however, is exactly the same when more problem variables are involved, though, naturally, the solution then becomes somewhat longer.

Here is one for you to work through: it brings in most of what we have covered and provides excellent revision. The result is given in the next frame.

59

Example 6

Maximise $\quad P = 24x + 21y + 30z$

subject to $\quad 12x + 4y + 8z \leq 240$

$\qquad\qquad 8x + 3y + 3z \leq 140$

$\qquad\qquad 6x + 2y + 3z \geq 110 \quad (x,\ y,\ z \geq 0)$

$$\boxed{P_{\max} = 750 \ \text{ with } \ x = 10,\ y = 10,\ z = 10}$$

60

The simplex technique is designed to maximise a given objective function in the light of stated constraints. However, a problem requiring the minimisation of an objective function (denoting costs, machine idling time, etc.) can easily be converted for solution by the same method.

For this, move on to the next frame

Minimisation

61

If P denotes the objective function to be minimised, we write Q as the negative of this function. $Q_{\max}$ is then determined by the usual simplex method and, finally, the negative value of $Q_{\max}$ is the value of the required $P_{\min}$.

i.e. Write $Q = -P$. Determine $Q_{\max}$ in the normal way.

Then $\quad P_{\min} = -(Q_{\max})$.

Example 1

Minimise $\quad P = -3x + 4y$

subject to $\quad x + 3y \leq 54$

$\qquad\qquad 3x + \ y \leq 34$

$\qquad\qquad -x + 2y \geq 12 \quad (x,\ y \geq 0)$

First write $Q = -P$, i.e. $Q = 3x - 4y$, and maximise Q.

Inserting the usual slack variables and artificial variable as needed, we have

$\cdots\cdots\cdots\cdots$

62

$$
\begin{aligned}
x + 3y + w_1 &= 54 \\
3x + y + w_2 &= 34 \\
-x + 2y - w_3 + w_4 &= 12 \\
Q - 3x + 4y + Mw_4 &= 0
\end{aligned}
$$

Now we just carry out the usual simplex routine to evaluate Q_{max} and hence P_{min}, since $P_{min} = -(Q_{max})$. $P_{min} = \dots\dots\dots$

63

$$\boxed{P_{min} = 16 \quad \text{with} \quad x = 8, \, y = 10}$$

Because $Q_{max} = -16$ and hence $P_{min} = -(Q_{max}) = 16$.

The full working is available in the next frame, should you need to refer to it.

If not, move straight on to Frame 65

64

Basis	x	y	w_1	w_2	w_3	w_4	b	check
w_1	1	3	1	0	0	0	54	59
w_2	3	1	0	1	0	0	34	39
w_4	−1	2	0	0	−1	1	12	13
Q	−3	4	0	0	0	M	0	M + 1
w_1	1	3	1	0	0	0	54	59
w_2	①	$\frac{1}{3}$	0	$\frac{1}{3}$	0	0	$\frac{34}{3}$	13
w_4	−1	2	0	0	−1	1	12	13
Q	−3	4	0	0	0	M	0	M + 1
w_1	0	$\frac{8}{3}$	1	$-\frac{1}{3}$	0	0	$\frac{128}{3}$	46
x w_2	1	$\frac{1}{3}$	0	$\frac{1}{3}$	0	0	$\frac{34}{2}$	13
w_4	0	$\boxed{\frac{7}{3}}$	0	$\frac{1}{3}$	−1	1	$\frac{70}{3}$	26
Q	0	5	0	1	0	M	34	M + 40
w_1	0	$\frac{8}{3}$	1	$-\frac{1}{3}$	0	0	$\frac{128}{3}$	46
x	1	$\frac{1}{3}$	0	$\frac{1}{3}$	0	0	$\frac{34}{3}$	13
w_4	0	①	0	$\frac{1}{7}$	$-\frac{3}{7}$	$\frac{3}{7}$	10	$\frac{78}{7}$
Q	0	5	0	1	0	M	34	M + 40
w_1	0	0	1	$-\frac{5}{7}$	$\frac{8}{7}$	$-\frac{8}{7}$	16	$\frac{114}{7}$
x	1	0	0	$\frac{2}{7}$	$\frac{1}{7}$	$-\frac{1}{7}$	8	$\frac{65}{7}$
y w_4	0	1	0	$\frac{1}{7}$	$-\frac{3}{7}$	$\frac{3}{7}$	10	$\frac{78}{7}$
Q	0	0	0	$\frac{2}{7}$	$\frac{15}{7}$	$M - \frac{15}{7}$	−16	$M - \frac{110}{7}$

$Q_{max} = -16$ $\therefore$ $P_{min} = 16$ with $x = 8, \, y = 10$.

Example 2   **65**

Minimise $P = -2x + 8y$

subject to $3x + 4y \leq 80$

             $-3x + 4y \geq 8$

               $x + 4y \geq 40$ $(x, y \geq 0)$

Note that we have two constraints that are 'greater than' inequalities, so, beside the slack variables, we shall need two artificial variables.

The three constraints in their new form therefore become

$\dotsc\dotsc\dotsc\dotsc$

$$
\begin{aligned}
3x + 4y + w_1 &= 80 \\
-3x + 4y - w_2 + w_4 &= 8 \\
x + 4y - w_3 + w_5 &= 40
\end{aligned}
$$

66

and, in the subsequent manipulation, we must see that w_4 and w_5 disappear from the basic solution before the optimal solution is obtained.

The objective function P is now replaced by $Q\ (= -P)$ and the new form of Q is written as

$\dotsc\dotsc\dotsc\dotsc$

$$Q - 2x + 8y + Mw_4 + Mw_5 = 0$$

67

because

$$P = -2x + 8y \quad \therefore \quad Q = -P = 2x - 8y$$

and with the artificial variables $Q = 2x - 8y - Mw_4 - Mw_5$.

$$\therefore \quad Q - 2x + 8y + Mw_4 + Mw_5 = 0$$

In this example, w_1, w_4, w_5 will form the unity matrix, so work through the solution in the usual way. Simplify the initial tableau and then refer to the next frame.

68

Basis	x	y	w_1	w_2	w_3	w_4	w_5	b	check
w_1	3	4	1	0	0	0	0	80	88
w_4	-3	4	0	-1	0	1	0	8	9
w_5	1	4	0	0	-1	0	1	40	45
Q	-2	8	0	0	0	M	M	0	$2M+6$
w_1	①	$\frac{4}{3}$	$\frac{1}{3}$	0	0	0	0	$\frac{80}{3}$	$\frac{88}{3}$
w_4	-3	4	0	-1	0	1	0	8	9
w_5	1	4	0	0	-1	0	1	40	45
Q	-2	8	0	0	0	M	M	0	$2M+6$
$x\ \cancel{w_1}$	1	$\frac{4}{3}$	$\frac{1}{3}$	0	0	0	0	$\frac{80}{3}$	$\frac{88}{3}$
w_4	0	8	1	-1	0	1	0	88	97
w_5	0	$\frac{8}{3}$	$-\frac{1}{3}$	0	-1	0	1	$\frac{40}{3}$	$\frac{47}{3}$
Q	0	$\frac{32}{3}$	$\frac{2}{3}$	0	0	M	M	$\frac{160}{3}$	$2M+\frac{194}{3}$

At this stage, there is no further negative entry in the index row, but we still must get rid of w_4 and w_5 from the basic variable column. Let us start by dealing with w_5.

We will take the entry $\frac{8}{3}$ at the intersection of the w_5 row and the y column as the next pivot and launch forth on the next stage. Complete the second stage and then again refer to the next frame.

69

Here is the working of stage 2.

Basis	x	y	w_1	w_2	w_3	w_4	w_5	b	check
x	1	$\frac{4}{3}$	$\frac{1}{3}$	0	0	0	0	$\frac{80}{3}$	$\frac{88}{3}$
w_4	0	8	1	-1	0	1	0	88	97
w_5	0	$\frac{8}{3}$	$-\frac{1}{3}$	0	-1	0	1	$\frac{40}{3}$	$\frac{47}{3}$
Q	0	$\frac{32}{3}$	$\frac{2}{3}$	0	0	M	M	$\frac{160}{3}$	$2M+\frac{194}{3}$
x	1	$\frac{4}{3}$	$\frac{1}{3}$	0	0	0	0	$\frac{80}{3}$	$\frac{88}{3}$
w_4	0	8	1	-1	0	1	0	88	97
w_5	0	①	$-\frac{1}{8}$	0	$-\frac{3}{8}$	0	$\frac{3}{8}$	5	$\frac{47}{8}$
Q	0	$\frac{32}{3}$	$\frac{2}{3}$	0	0	M	M	$\frac{160}{3}$	$2M+\frac{194}{3}$
x	1	0	$\frac{1}{2}$	0	$\frac{1}{2}$	0	$-\frac{1}{2}$	20	$\frac{43}{2}$
w_4	0	0	2	-1	3	1	-3	48	50
$y\ \cancel{w_5}$	0	1	$-\frac{1}{8}$	0	$-\frac{3}{8}$	0	$\frac{3}{8}$	5	$\frac{47}{8}$
Q	0	0	2	0	4	M	$M-4$	0	$2M+2$

▶

At this point, w_5 is replaced by y in the basic variable column.

Now we deal with w_4 by taking the entry 2 at the junction of the w_4 row and the w_1 column as the next pivot. That should do the trick, so finish off the solution and check with the next frame.

$$\boxed{P_{\min} = 48 \quad \text{with} \quad x = 8, y = 8}$$

70

Basis	x	y	w_1	w_2	w_3	w_4	w_5	b	check
x	1	0	$\frac{1}{2}$	0	$\frac{1}{2}$	0	$-\frac{1}{2}$	20	$\frac{43}{2}$
$\star \to w_4$	0	0	$\boxed{2}$	-1	3	1	-3	48	50
y	0	1	$-\frac{1}{8}$	0	$-\frac{3}{8}$	0	$\frac{3}{8}$	5	$\frac{47}{8}$
Q	0	0	2	0	4	M	$M-4$	0	$2M+2$
x	1	0	$\frac{1}{2}$	0	$\frac{1}{2}$	0	$-\frac{1}{2}$	20	$\frac{43}{2}$
w_4	0	0	$①$	$-\frac{1}{2}$	$\frac{3}{2}$	$\frac{1}{2}$	$-\frac{3}{2}$	24	25
y	0	1	$-\frac{1}{8}$	0	$-\frac{3}{8}$	0	$\frac{3}{8}$	5	$\frac{47}{8}$
Q	0	0	2	0	4	M	$M-4$	0	$2M+2$
x	1	0	0	$\frac{1}{4}$	$-\frac{1}{4}$	$-\frac{1}{4}$	$\frac{1}{4}$	8	9
$w_1 \;\cancel{w_4}$	0	0	1	$-\frac{1}{2}$	$\frac{3}{2}$	$\frac{1}{2}$	$-\frac{3}{2}$	24	25
y	0	1	0	$-\frac{1}{16}$	$-\frac{3}{16}$	$\frac{1}{16}$	$\frac{3}{16}$	8	9
Q	0	0	0	1	1	$M-1$	$M-1$	-48	$2M-48$

w_4 in the basic variable column is now replaced by w_1, so the conditions are satisfied at last. From the final tableau, we have

$Q_{\max} = -48$ But $P_{\min} = -(Q_{\max}) = 48$

$\therefore \; P_{\min} = 48$ with $x = 8, y = 8$.

By this means, then, we can solve minimisation problems by the simplex method and so widen the scope of this valuable technique.

Applications

So far we have seen how to solve a typical linear programming problem by the simplex method, when the data are presented as a linear objective function and a number of linear constraints in the form of equations or inequalities. A practical problem, however, must first be interpreted into algebraic form and we conclude the Programme with a brief reference to this initial requirement. Let us consider the following example.

71

▶

Example 1

A firm manufactures two types of couplings, A and B, each of which requires processing time on lathes, grinders and polishers. The machine times needed for each type of coupling are given in the table.

Coupling type	Time required (hours)		
	Lathe	Grinder	Polisher
A	2	8	5
B	5	5	2

The total machine time available is 250 hours on lathes, 310 hours on grinders and 160 hours on polishers. The net profit per coupling of type A is £9 and of type B £10.

Determine

(a) the number of each type to be produced to maximise profit

(b) the maximum profit.

If we let x = the number of type A units to be produced

$\qquad y$ = the number of type B units to be produced

the objective function to be maximised can be expressed as ....

72

$$\boxed{P = 9x + 10y}$$

Now we have to sort out the constraints from the given data.

Total time available on lathes = 250 hours

$\qquad \therefore \ 2x + 5y \leq 250 \quad \text{(lathes)}$

Similar statements for the grinders and polishers are ....

73

$$\boxed{\begin{array}{l} 8x + 5y \leq 310 \ \text{(grinders)} \\ 5x + 2y \leq 160 \ \text{(polishers)} \end{array}}$$

The problem now can be expressed as

$\qquad$ Maximise $\qquad P = 9x + 10y$

$\qquad$ subject to $\quad 2x + 5y \leq 250$

$\qquad\qquad\qquad\quad 8x + 5y \leq 310$

$\qquad\qquad\qquad\quad 5x + 2y \leq 160 \quad (x, y \geq 0)$

Then we go through the usual process. Inserting slack variables to convert the inequalities into equations, we have ....

74

$$2x + 5y + w_1 \qquad\qquad = 250$$
$$8x + 5y \quad + w_2 \qquad = 310$$
$$5x + 2y \qquad\quad + w_3 = 160$$
$$P - 9x - 10y \qquad\qquad = \quad 0$$

and the solution then develops in the usual way. Work through it carefully – it is all good practice – and see if you agree with the result given in the next frame.

The result is . . . .

75

$$P_{\max} = 550 \quad \text{with} \quad x = 10, y = 46$$

The maximum profit of £550 occurs with a manufacturing schedule of

      10 couplings of type A

and   46 couplings of type B.

Now for another, so move on.

Example 2

76

A firm produces three types of pumps, A, B, C, each of which requires the four processes of turning, drilling, assembling and testing.

Pump type	Process time (hours) per pump				Profit per pump £
	Turning	Drilling	Assembling	Testing	
A	2	1	3	4	84
B	1	1	4	3	72
C	2	1	2	2	52
Total available time (h/week)	98	60	145	160	

From the information given in the table, determine

(a) the weekly output of each type of pump to maximise profit

(b) the maximum profit.

So, if we let $x =$ the number of pumps, type A

           $y =$ the number of pumps, type B

           $z =$ the number of pumps, type C

we can interpret the problem into its algebraic form, which is

         . . . .

77

> Maximise $P = 84x + 72y + 52z$
>
> subject to $\quad 2x + y + 2z \leq 98$
>
> $\qquad\qquad x + y + z \leq 60$
>
> $\qquad\quad 3x + 4y + 2z \leq 145$
>
> $\qquad\quad 4x + 3y + 2z \leq 160 \qquad (x,\ y,\ z \geq 0)$

Inserting the slack variables and expressing the problem as equations, we have

. . . .

78

> $2x + y + 2z + w_1 \qquad\qquad\qquad\quad = 98$
>
> $x + y + z \qquad + w_2 \qquad\qquad\quad = 60$
>
> $3x + 4y + 2z \qquad\qquad + w_3 \qquad = 145$
>
> $4x + 3y + 2z \qquad\qquad\qquad + w_4 = 160$
>
> $P - 84x - 72y - 52z \qquad\qquad\qquad\qquad = 0$

Now you can proceed to set up the simplex tableau and solve the problem on your own in the usual manner. It is very similar to the other examples you have worked earlier in the Programme.

The result you no doubt get is

. . . .

79

> $P_{max} = 3652 \quad \text{with} \quad x = 23,\ y = 8,\ z = 22$

i.e. by producing 23 pumps, type A

$\qquad\qquad$ 8 pumps, type B

$\qquad\qquad$ 22 pumps, type C

the maximum profit of £ 3652 is attained.

Care with the calculations and constant use of the check column provide the key to avoiding errors in the working.

That completes the Programme. Check down the **Revision summary** that comes next, in conjunction with the **Can you?** checklist, before working through the **Test exercise** that follows thereafter. As usual, a set of **Further problems** provides further necessary practice in these useful techniques.

Revision summary 28

80

1 *Optimization* – determination of an optimal value (maximum or minimum) of an objective function subject to a set of constraints.

2 *Linear programming (linear optimization)* – optimization where the objective function is a linear function and the constraints are linear equations or linear inequalities.

3 *Inequalities* – multiplying or dividing both sides by a negative factor $(-k)$ reverses the inequality, i.e. $\geq$ becomes $\leq$ and $\leq$ becomes $\geq$.

4 *Problem variables* $(x, y, z,$ etc.$)$ are always non-negative.

5 *Feasible solution* – a set of variables that satisfies all the given constraints.

6 *Optimal solution* – a feasible solution for which the objective function becomes a maximum (or minimum) within the constraints.

7 *Basic feasible solution* – a feasible solution for which at least $(n - m)$ of the total variables are zero, where
n = total number of variables in the constraints
m = number of constraints.

8 *Basis* – collection of the m variables which are not put equal to zero.

9 *Basic solution* – solution obtained by equating $(n - m)$ variables to zero and solving for the remaining m variables.

10 *Graphical solution*
(a) Constraints – graphs of constraints form the feasible polygon or feasible domain.

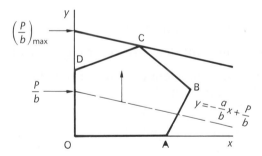

Feasible point or feasible solution – coordinates of all points within the feasible polygon or on its boundary (OABCD).

(b) Objective function $P = ax + by$ $\therefore$ $y = -\dfrac{a}{b}x + \dfrac{P}{b}$ represented by a set of parallel lines, slope $-\dfrac{a}{b}$, intercept $\dfrac{P}{b}$. Line through the extreme point C gives $P_{\max}$, the optimal value of P.

▶

11 *Slack variable* – non-negative variable added to, or subtracted from, a linear inequality to form a linear equation.

12 *Simplex method of solution* – computation.

Refer back to Frame 34.

Where necessary, we multiply an inequality by (–1), with consequent reversal of inequality sign, to ensure that the right-hand side constant term $b_i \geq 0$.

13 *Artificial variable* – to convert a 'greater than' inequality to an equation, the slack variable required must be subtracted. To complete the unity matrix in the tableau, a further artificial variable w_i is included to allow the simplex procedure to continue. Such artificial variables must be eliminated before the optimal solution is finally attained.

The objective function $P = ax + by$ becomes $P = ax + by - Mw_i$.

14 *Minimisation* – If P is the objective function to be minimised
(a) write $Q = -P$
(b) maximise Q by the usual simplex method
(c) then $Q_{max} = (-P)_{max} = -(P_{min})$

i.e. $P_{min} = -(Q_{max})$.

 # Can you?

81 ## Checklist 28

Check this list before and after you try the end of Programme test.

On a scale of 1 to 5 how confident are you that you can: **Frames**

• Describe an optimization problem in terms of the objective function and a set of constraints? $\boxed{1}$ and $\boxed{2}$
   Yes ☐ ☐ ☐ ☐ ☐   *No*

• Algebraically manipulate and graphically describe inequalities? $\boxed{3}$ to $\boxed{6}$
   Yes ☐ ☐ ☐ ☐ ☐   *No*

• Solve a linear programming problem in two real variables? $\boxed{6}$ to $\boxed{13}$
   Yes ☐ ☐ ☐ ☐ ☐   *No*

• Use the simplex method to describe a linear programming problem in two real variables as a problem in two real variables with two slack variables? $\boxed{14}$
   Yes ☐ ☐ ☐ ☐ ☐   *No*

• Set up the simplex tableau and compute the simplex? $\boxed{15}$ to $\boxed{32}$
   Yes ☐ ☐ ☐ ☐ ☐   *No*

▶

- Use the simplex method to solve a linear programming
  problem in three real variables with three slack variables?
  $\boxed{33}$ to $\boxed{40}$
  Yes ☐ ☐ ☐ ☐ ☐ *No*

- Introduce artificial variables into the solution method as and
  when the need arises?
  $\boxed{41}$ to $\boxed{60}$
  Yes ☐ ☐ ☐ ☐ ☐ *No*

- Solve minimisation problems using the simplex method?
  $\boxed{61}$ to $\boxed{70}$
  Yes ☐ ☐ ☐ ☐ ☐ *No*

- Construct the algebraic form of the objective function and the
  constraints for a problem stated in words?
  $\boxed{71}$ to $\boxed{79}$
  Yes ☐ ☐ ☐ ☐ ☐ *No*

 # Test exercise 28

---

1 Using a *graphical method*, maximise $P = x + 2y$ subject to the constraints

$$-3x + 4y \leq \ 8$$
$$x + 4y \leq 16$$
$$3x + 2y \leq 18$$
$$x, y \geq \ 0.$$

82

Note: Use the *simplex method* to solve Exercises **2** to **6**. In each case, all variables are
non-negative.

2 Maximise $\quad P = -3x + 4y$
  subject to $\quad 3x - 2y \leq 15$
  $\qquad\qquad\quad x + \ y \leq 10$
  $\qquad\qquad -x + 4y \leq 15$
  $\qquad\qquad -2x + \ y \leq \ 2.$

3 Maximise $\quad P = 8x + 12y + 10z$
  subject to $\quad 4x + 3y + 2z \leq 64$
  $\qquad\qquad\ 2x + \ y + 4z \leq 48$
  $\qquad\qquad\ x + 2y + \ z \leq 24.$

4 Maximise $\quad P = 44x + 20y$
  subject to $\quad 12x + 6y \leq 84$
  $\qquad\qquad\ 3x + 2y \geq 24.$

5 Minimise $\quad P = 3y - 4x$
  subject to $\quad x + 4y \leq 60$
  $\qquad\qquad\ 2x + \ y \leq 22$
  $\qquad\qquad -x + \ y \geq \ 7.$

▶

6 A firm makes two types of containers, A and B, each of which requires cutting, assembly and finishing. The maximum available machine capacity in hours per week for each process is: cutting 50, assembly 84, finishing 72.

The process times for one unit of each type are as follows:

| | Time in hours | |
Process	A	B
Cutting	2	5
Assembly	4	8
Finishing	4	5

If the profit margin is £600 per unit A and £1000 per unit B, determine

(a) the optimum weekly output of containers

(b) the maximum profit.

 ## Further problems 28

83 *All variables in the following problems are non-negative.*

Graphical Solution

1 Maximise $P = -x + 8y$

subject to $-3x + 4y \leq 10$

$-x + 4y \leq 14$

$3x + 2y \leq 21$

$3x + y \leq 18.$

2 Maximise $P = -4x + 8y$

subject to $x + 3y \leq 57$

$7x + 4y \leq 110$

$-x + 5y \leq 40.$

3 Maximise $P = 5x + 4y$

subject to $x - 2y \leq 2$

$3x - 4y \leq 8$

$5x + 6y \leq 45$

$x + 3y \leq 18.$

Simplex Solution

4 Maximise $P = 2x + y$
  subject to $x + 4y \leq 24$
  $x + y \leq 9$
  $x - y \leq 3$
  $x - 2y \leq 2.$

5 Maximise $P = -3x + 4y$
  subject to $3x - 4y \leq 12$
  $5x + 4y \leq 36$
  $-x + 3y \leq 8$
  $-3x + y \leq 0.$

6 Maximise $P = x + 2y$
  subject to $-2x + y \leq 1$
  $-x + y \leq 2$
  $x + y \leq 6$
  $2x - 3y \leq 2.$

7 Maximise $P = 4y - 3x$
  subject to $x - 2y \leq 0$
  $x - y \leq 2$
  $x + 2y \leq 14$
  $-x + 2y \leq 6$
  $-3x + 2y \leq 2.$

8 Maximise $P = 3x + 4y + 5z$
  subject to $5x + 4y + 8z \leq 40$
  $3x + 2y + 12z \leq 30$
  $y \leq 8.$

9 Maximise $P = 3x + 4y + 3z$
  subject to $2x + 3y + 4z \leq 58$
  $4x + 2y + 3z \leq 51$
  $3x + 4y + 2z \leq 62.$

10 Maximise $P = 4x + 3y + 3z$
  subject to $4x + y + 2z \leq 40$
  $x + 4y + z \leq 50$
  $2x + 3y + 4z \leq 60.$

11 Maximise $P = 5 \cdot 3x + 3 \cdot 6y + 2 \cdot 0z$
  subject to $2 \cdot 1x + 4 \cdot 3y + 1 \cdot 5z \leq 70$
  $3 \cdot 2x + 1 \cdot 4y + 2 \cdot 2z \leq 60$
  $1 \cdot 6x + 6 \cdot 2y + 3 \cdot 1z \leq 100.$

Artificial Variables

12 Maximise $P = 8x + 5y$
  subject to $2x + y \leq 80$
  $x + 3y \leq 90$
  $x + y \geq 30.$

13 Maximise $P = 12x + 8y$
  subject to $x + 2y \leq 20$
  $4x - y \leq 8$
  $-x + y \geq 1.$

14 Maximise $P = 3x + 4y$
  subject to $x + 4y \leq 76$
  $-5x + 8y \geq 40$
  $-x + 4y \geq 32.$

15 Minimise $P = 4x + 5y$
  subject to $x + 2y \leq 63$
  $3x + y \leq 70$
  $2x + y \geq 42$
  $x + 4y \geq 84.$

16 Maximise $P = 65x - 23y$
  subject to $5x - y \leq 30$
  $10x + 4y \geq 150.$

17 Maximise $P = 24x - 8y$
  subject to $x + 3y \leq 360$
  $2x + y \leq 850$
  $-5x + 25y \geq 320.$

▶

18   Maximise $P = 4x + 2y$

     subject to $x + 2y \le 60$

                 $3x + 2y \le 80$

               $-3x + 10y \ge 40.$

19   Maximise $P = 18x + 40y + 24z$

     subject to $5x + 2y + 4z \le 63$

                 $2x + 4y + 2z \le 42$

                 $2x + 3y + z \ge 35.$

20   Maximise $P = 60x + 45y + 25z$

     subject to $4x + 8y + 2z \le 160$

                $6x + 3y + 4z \le 168$

                $4x + 3y + 3z \ge 128.$

21   Maximise $P = 12x + 8y - 10z$

     subject to $4x + 2y - 3z \le 210$

               $6x + 8y + z \le 630$

               $2x - y + 4z \ge 210$

                $x + y + z \le 180.$

Minimisation

22   Minimise $P = -4x + 3y$

     subject to $x + 4y \le 20$

               $2x + y \le 12$

                $x - y \le 3.$

23   Minimise $P = -5x + 8y$

     subject to $x + 2y \le 40$

                 $3x + 2y \le 48$

               $-x + 4y \ge 40.$

24   Minimise $P = -4x + 8y$

     subject to $-5x + 4y \le 32$

                $7x + 4y \le 80$

              $-x + 8y \ge 40.$

25   Minimise $P = 2x + 8y$

     subject to $-x + 2y \le 24$

               $7x + 6y \le 132$

             $-x + 2y \ge 4$

               $x + 2y \ge 12.$

26   Minimise $P = 4x - 8y + 5z$

     subject to $2x + 3y + z \le 70$

              $x + 2y + 2z \le 60$

             $3x + 4y + z \le 84$

               $x + y + z \ge 33.$

27   Minimise $P = 6x - 5y - 3z$

     subject to $5x + 8y + 4z \le 220$

              $2x + y + 6z \le 154$

             $4x + 2y + z \ge 77$

               $x + y + 2z \ge 55.$

▶

Applications

28 A firm manufacturing two types of switching module, A and B, is under contract to produce a daily output of at least 35 modules in all. Assembly and testing times for each type of module are as follows:

Module type	Processing time (hours)	
	Assembly	*Testing*
A	1·0	2·0
B	2·0	1·0

Available staff resources provide a daily maximum of 80 hours for assembly and 55 hours for testing.

The profit on the sale of each A-module is £40 and of each B-module £50. Determine

(a) the daily production schedule for maximum profit.

(b) the maximum daily profit.

29 Three different types of coupling units are produced by a firm. The times required for machining, polishing and assembling a unit of each type are included in the information given in the following table.

Type of unit	Process time (hours) per unit			Profit (£) per unit
	Machining	*Polishing*	*Assembling*	
A	4	1	2	110
B	2	3	1	100
C	3	2	4	120
Available time (h/week)	320	250	280	

The firm is required to supply a total of at least 100 units of mixed types each week. Determine

(a) the weekly output of each type to maximise profit

(b) the maximum weekly profit.

30 A firm makes three types of wooden cabinets, A, B, C, with profit margins of
£35, £30, £24 per unit respectively.

| | Time in hours per cabinet | | |
Process	A	B	C
Preparation	2	5	4
Assembly	2	3	2
Finishing	5	4	3

The manufacturer has 25 men available for preparation, 20 men for assembly
and 30 men for polishing, and all staff work a 40 hour week. To remain
competitive, at least 300 cabinets in all must be produced each week.
Determine

(a) the number of each model to be manufactured each week in order to
maximise the profit

(b) the maximum weekly profit.

Appendix

1 Green's theorem

If P and Q are two functions in x and y, finite and continuous inside a region R and on its boundary c in the x–y plane, with continuous first partial derivatives, then Green's theorem states that

$$\iint\limits_{R} \left(\frac{\partial P}{\partial y} - \frac{\partial Q}{\partial x} \right) dx\, dy = -\oint_{c} \{Pdx + Qdy\}$$

Proof of Green's theorem

Let the lower boundary of the region be the curve $y_1 = f(x)$ and the upper boundary the curve $y_2 = F(x)$.

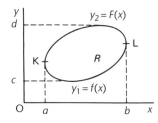
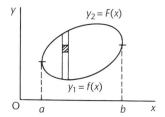

Using vertical strips, we then have

$$\iint\limits_{R} \frac{\partial P}{\partial y} dx\, dy = \int_{a}^{b} \int_{y_1}^{y_2} \frac{\partial P}{\partial y} dy\, dx = \int_{a}^{b} \left[P \right]_{y_1=f(x)}^{y_2=F(x)} dx$$

$$= \int_{a}^{b} \{P(x, y_2) - P(x, y_1)\}\, dx$$

$$= -\int_{a}^{b} P(x, y_1)\, dx - \int_{b}^{a} P(x, y_2)\, dx$$

$$= -\left\{ \int_{a}^{b} P(x, y_1)\, dx + \int_{b}^{a} P(x, y_2)\, dx \right\}$$

$$= -\oint P(x, y)\, dx \tag{1}$$

Similarly, using horizontal strips, we have

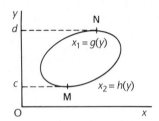
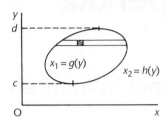

$$\iint\limits_R \frac{\partial Q}{\partial x}\, dx\, dy = \int_c^d \int_{x_1}^{x_2} \frac{\partial Q}{\partial y}\, dx\, dy$$

$$= \int_c^d \left[Q \right]_{x_1=g(y)}^{x_2=h(y)} dy$$

where $x_1 = g(y)$ and $x_2 = h(y)$ are the left-hand and right-hand portions of the boundary curve c.

$$\therefore \quad \iint\limits_R \frac{\partial Q}{\partial x}\, dx\, dy = \int_c^d Q(x_2, y)\, dy - \int_c^d Q(x_1, y)\, dy$$

$$= \int_c^d Q(x_2, y)\, dy + \int_d^c Q(x_1, y)\, dy$$

$$= \oint_c Q(x, y)\, dy \tag{2}$$

$$\therefore \quad \iint\limits_R \left(\frac{\partial P}{\partial y} - \frac{\partial Q}{\partial x} \right) dx\, dy = -\oint_c P(x, y)\, dx - \oint_c Q(x, y)\, dy$$

$$= -\oint_c \{ P\, dx - Q\, dy \}$$

2 Proof that $\sec \gamma = \sqrt{1+\left(\dfrac{\partial z}{\partial x}\right)^2+\left(\dfrac{\partial z}{\partial y}\right)^2}$

Let α, β, γ be the angles that OP makes with the x, y and z axes respectively.

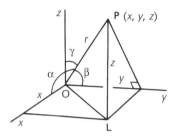

Then $x = r\cos\alpha$; $y = r\cos\beta$; $z = r\cos\gamma$

Also $x^2 + y^2 + z^2 = r^2$

If $r = 1$ unit, then $x^2 + y^2 + z^2 = 1$ $\quad\therefore\ z^2 = 1 - x^2 - y^2$

$$\therefore\ z = (1 - x^2 - y^2)^{1/2}$$

$$\frac{\partial z}{\partial x} = \frac{1}{2}(1 - x^2 - y^2)^{-1/2}(-2x)$$

$$= \frac{-x}{\sqrt{1 - x^2 - y^2}}$$

$$\frac{\partial z}{\partial y} = \frac{1}{2}(1 - x^2 - y^2)^{-1/2}(-2y)$$

$$= \frac{-y}{\sqrt{1 - x^2 - y^2}}$$

$$\therefore\ 1 + \left(\frac{\partial z}{\partial x}\right)^2 + \left(\frac{\partial z}{\partial y}\right)^2 = 1 + \frac{x^2}{1 - x^2 - y^2} + \frac{y^2}{1 - x^2 - y^2}$$

$$= \frac{1 - x^2 - y^2 + x^2 + y^2}{1 - x^2 - y^2}$$

$$= \frac{1}{1 - x^2 - y^2} = \frac{1}{z^2}$$

But, with $r = 1$, $z = \cos\gamma$ $\quad\therefore\ \dfrac{1}{z^2} = \sec^2\gamma$

$$\therefore\ \sec\gamma = \sqrt{1 + \left(\frac{\partial z}{\partial x}\right)^2 + \left(\frac{\partial z}{\partial y}\right)^2}$$

3 Vector triple products

(a) $\mathbf{A} \times (\mathbf{B} \times \mathbf{C}) = (\mathbf{A} \cdot \mathbf{C})\,\mathbf{B} - (\mathbf{A} \cdot \mathbf{B})\,\mathbf{C}$

(b) $(\mathbf{A} \times \mathbf{B}) \times \mathbf{C} = (\mathbf{C} \cdot \mathbf{A})\,\mathbf{B} - (\mathbf{C} \cdot \mathbf{B})\,\mathbf{A}$

Let $\quad \mathbf{A} = a_x\mathbf{i} + a_y\mathbf{j} + a_z\mathbf{k}; \quad \mathbf{B} = b_x\mathbf{i} + b_y\mathbf{j} + b_z\mathbf{k};$

$\quad \mathbf{C} = c_x\mathbf{i} + c_y\mathbf{j} + c_z\mathbf{k}$

Then $\quad \mathbf{B} \times \mathbf{C} = (b_x\mathbf{i} + b_y\mathbf{j} + b_z\mathbf{k}) \times (c_x\mathbf{i} + c_y\mathbf{j} + c_z\mathbf{k})$

$$= \begin{vmatrix} \mathbf{i} & \mathbf{j} & \mathbf{k} \\ b_x & b_y & b_z \\ c_x & c_y & c_z \end{vmatrix}$$

$$= \mathbf{i} \begin{vmatrix} b_y & b_z \\ c_y & c_z \end{vmatrix} - \mathbf{j} \begin{vmatrix} b_x & b_z \\ c_x & c_z \end{vmatrix} + \mathbf{k} \begin{vmatrix} b_x & b_y \\ c_x & c_y \end{vmatrix}$$

Then $\quad \mathbf{A} \times (\mathbf{B} \times \mathbf{C}) = \begin{vmatrix} \mathbf{i} & \mathbf{j} & \mathbf{k} \\ a_x & a_y & a_z \\ \begin{vmatrix} b_y & b_z \\ c_y & c_z \end{vmatrix} & \begin{vmatrix} b_z & b_x \\ c_z & c_x \end{vmatrix} & \begin{vmatrix} b_x & b_y \\ c_x & c_y \end{vmatrix} \end{vmatrix}$

$$= \mathbf{i} \left\{ a_y \begin{vmatrix} b_x & b_y \\ c_x & c_y \end{vmatrix} - a_z \begin{vmatrix} b_z & b_x \\ c_z & c_x \end{vmatrix} \right\} - \mathbf{j} \left\{ a_x \begin{vmatrix} b_x & b_y \\ c_x & c_y \end{vmatrix} - a_z \begin{vmatrix} b_y & b_z \\ c_y & c_z \end{vmatrix} \right\}$$

$$+ \mathbf{k} \left\{ a_x \begin{vmatrix} b_z & b_x \\ c_z & c_x \end{vmatrix} - a_y \begin{vmatrix} b_y & b_z \\ c_y & c_z \end{vmatrix} \right\}$$

$$= \mathbf{i}\,\{a_y(b_xc_y - b_yc_x) - a_z(b_zc_x - b_xc_z)\}$$

$$+ \mathbf{j}\,\{a_z(b_yc_z - c_yb_z) - a_x(b_xc_y - b_yc_x)\}$$

$$+ \mathbf{k}\,\{a_x(b_zc_x - b_xc_z) - a_y(b_yc_z - b_zc_y)\}$$

$$= \mathbf{i}\,\{b_xa_xc_x + b_xa_yc_y + b_xa_zc_z - c_xa_xb_x - c_xa_yb_y - c_xa_zb_z\}$$

$$+ \mathbf{j}\,\{b_ya_xc_x + b_ya_yc_y + b_ya_zc_z - c_ya_xb_x - c_ya_yb_y - c_ya_zb_z\}$$

$$+ \mathbf{k}\,\{b_za_xc_x + b_za_yc_y + b_za_zc_z - c_za_xb_x - c_za_yb_y - c_za_zb_z\}$$

$$= \mathbf{i}\,\{b_x(a_xc_x + a_yc_y + a_zc_z) - c_x(a_xb_x + a_yb_y + a_zb_z)\}$$

$$+ \mathbf{j}\,\{b_y(a_xc_x + a_yc_y + a_zc_z) - c_y(a_xb_x + a_yb_y + a_zb_z)\}$$

$$+ \mathbf{k}\,\{b_z(a_xc_x + a_yc_y + a_zc_z) - c_z(a_xb_x + a_yb_y + a_zb_z)\}$$

Now $\mathbf{A} \cdot \mathbf{C} = (a_x\mathbf{i} + a_y\mathbf{j} + a_z\mathbf{k}) \cdot (c_x\mathbf{i} + c_y\mathbf{j} + c_z\mathbf{k})$

$$= a_x c_x + a_y c_y + a_z c_z$$

and similarly $\mathbf{A} \cdot \mathbf{B} = a_x b_x + a_y b_y + a_z b_z$

$\therefore \; \mathbf{A} \times (\mathbf{B} \times \mathbf{C}) = \mathbf{i}\{b_x(\mathbf{A} \cdot \mathbf{C}) - c_x(\mathbf{A} \cdot \mathbf{B})\}$

$$+ \mathbf{j}\{b_y(\mathbf{A} \cdot \mathbf{C}) - c_y(\mathbf{A} \cdot \mathbf{B})\}$$

$$+ \mathbf{k}\{b_z(\mathbf{A} \cdot \mathbf{C}) - c_z(\mathbf{A} \cdot \mathbf{B})\}.$$

$\therefore \; \mathbf{A} \times (\mathbf{B} \times \mathbf{C}) = (\mathbf{A} \cdot \mathbf{C})\{\mathbf{i}b_x + \mathbf{j}b_y + \mathbf{k}b_z\} - (\mathbf{A} \cdot \mathbf{B})\{\mathbf{i}c_x + \mathbf{j}c_y + \mathbf{k}c_z\}$

$\therefore \; \mathbf{A} \times (\mathbf{B} \times \mathbf{C}) = (\mathbf{A} \cdot \mathbf{C})\,\mathbf{B} - (\mathbf{A} \cdot \mathbf{B})\,\mathbf{C}$

In the same way, it can be established that

$(\mathbf{A} \times \mathbf{B}) \times \mathbf{C} = (\mathbf{C} \cdot \mathbf{A})\,\mathbf{B} - (\mathbf{C} \cdot \mathbf{B})\,\mathbf{A}$

4 Divergence theorem (Gauss' theorem)

To prove that $\displaystyle\int_V \operatorname{div}\mathbf{F}\,dV = \int_S \mathbf{F} \cdot d\mathbf{S}$ for the region V bounded by the surface S.

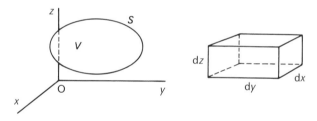

Consider an element of volume $dV = dx\,dy\,dz$ and let the components of $\mathbf{F}$ in the x, y and z directions be denoted by $F_x\mathbf{i}$, $F_y\mathbf{j}$ and $F_z\mathbf{k}$ respectively at any point P. We then determine $\displaystyle\int \mathbf{F} \cdot d\mathbf{S}$ over the element dV and finally sum the results for all such elements throughout the region.

(a) S_1: $dS_1 = dy\,dz$; $\mathbf{n} = \mathbf{i}$

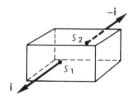

$(\mathbf{F} \cdot d\mathbf{S})_1 = (F_x\mathbf{i} + F_y\mathbf{j} + F_z\mathbf{k}) \cdot (\mathbf{n})\,dS_1$

$= (F_x\mathbf{i} + F_y\mathbf{j} + F_z\mathbf{k}) \cdot (\mathbf{i})\,dS_1$

$= F_x\,dS_1$

(b) S_2 : $dS_2 = dy\,dz;$ $\mathbf{n} = -\mathbf{i}$

$$\therefore\ (\mathbf{F} \cdot d\mathbf{S})_2 = (F_x\mathbf{i} + F_y\mathbf{j} + F_z\mathbf{k}) \cdot (-\mathbf{i})\,dS_2$$
$$= -F_x\,dS_2$$

Combining these two results, we have

$$(\mathbf{F} \cdot d\mathbf{S})_1 + (\mathbf{F} \cdot d\mathbf{S})_2 = (F_x\,dS)_1 - (F_x\,dS)_2$$

$$= \frac{\partial}{\partial x}(F_x\,dS)\,dx$$

$$\therefore\ \int_{S_1+S_2} \mathbf{F} \cdot d\mathbf{S} = \frac{\partial F_x}{\partial x}\,dS\,dx = \left(\frac{\partial F_x}{\partial x}\right)dx\,dy\,dz \tag{1}$$

Similarly, for S_3 and S_4 we have

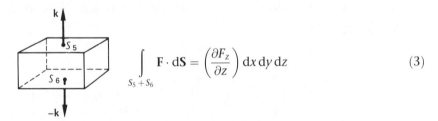

$$\int_{S_3+S_4} \mathbf{F} \cdot d\mathbf{S} = \left(\frac{\partial F_y}{\partial y}\right)dx\,dy\,dz \tag{2}$$

and for S_5 and S_6

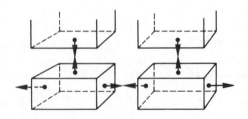

$$\int_{S_5+S_6} \mathbf{F} \cdot d\mathbf{S} = \left(\frac{\partial F_z}{\partial z}\right)dx\,dy\,dz \tag{3}$$

These three results together cover the total surface of the element dV.

$$\int_{S_1 \dots S_6} \mathbf{F} \cdot d\mathbf{S} = \left(\frac{\partial F_x}{\partial x} + \frac{\partial F_y}{\partial y} + \frac{\partial F_z}{\partial z}\right)dx\,dy\,dz = \text{div}\,\mathbf{F}\,dV$$

Finally, summing the results for all such elements throughout the region with $dV \to 0$ and $d\mathbf{S} \to 0$, we obtain

$$\int_V \text{div}\,\mathbf{F}\,dV = \sum \int \mathbf{F} \cdot d\mathbf{S} \quad \text{with } d\mathbf{S} \to 0$$

On the common boundaries between adjacent elements, the values of $\int \mathbf{F} \cdot d\mathbf{S}$ cancel out. On the boundary surface, however, there are no such adjacent faces and the integral $\oint_S \mathbf{F} \cdot d\mathbf{S}$ remains.

$$\therefore \int_V \operatorname{div} \mathbf{F} \, dV = \int_S \mathbf{F} \cdot d\mathbf{S}$$

5 Stokes' theorem

If $\mathbf{F}$ is a single-valued vector field, continuous and differentiable over an open surface S and on the boundary c of the surface, then

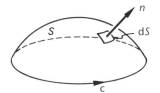

$$\int_S \operatorname{curl} \mathbf{F} \cdot d\mathbf{S} = \oint_c \mathbf{F} \cdot d\mathbf{r}$$

Proof of Stokes' theorem

Consider the surface S divided into small rectangular elements and let ABCD be one such element. If axes of reference x and y be arranged to coincide with AB and AD respectively as shown, a third axis z will then be normal to the surface at A.

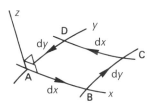

If   AB = dx, then d$\mathbf{x}$ = $\mathbf{i}$ dx and

if   AD = dy, then d$\mathbf{y}$ = $\mathbf{j}$ dy.

Let $\mathbf{F}_a$ denote the vector field at A; $\mathbf{F}_b$ that at B; $\mathbf{F}_c$ that at C; and $\mathbf{F}_d$ that at D. Now consider each side in turn.

$$\text{AB:} \quad \mathbf{F} \cdot d\mathbf{r} = \mathbf{F}_a \cdot d\mathbf{x} = \{F_{ax}\mathbf{i} + F_{ay}\mathbf{j} + F_{az}\mathbf{k}\} \cdot \{\mathbf{i} \, dx\} = F_{ax} \, dx$$
$$\text{BC:} \quad \mathbf{F} \cdot d\mathbf{r} = \mathbf{F}_b \cdot d\mathbf{y} = \{F_{bx}\mathbf{i} + F_{by}\mathbf{j} + F_{bz}\mathbf{k}\} \cdot \{\mathbf{j} \, dy\} = F_{by} \, dy$$
$$\text{CD:} \quad \mathbf{F} \cdot d\mathbf{r} = \mathbf{F}_c \cdot d\mathbf{x} = \{F_{cx}\mathbf{i} + F_{cy}\mathbf{j} + F_{cz}\mathbf{k}\} \cdot \{-\mathbf{i} \, dx\} = -F_{cx} \, dx$$
$$\text{DA:} \quad \mathbf{F} \cdot d\mathbf{r} = \mathbf{F}_d \cdot d\mathbf{y} = \{F_{dx}\mathbf{i} + F_{dy}\mathbf{j} + F_{dz}\mathbf{k}\} \cdot \{-\mathbf{j} \, dy\} = -F_{dy} \, dy$$

(a) AB + CD:

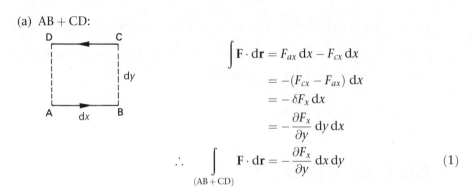

$$\int \mathbf{F} \cdot d\mathbf{r} = F_{ax}\, dx - F_{cx}\, dx$$

$$= -(F_{cx} - F_{ax})\, dx$$

$$= -\delta F_x\, dx$$

$$= -\frac{\partial F_x}{\partial y}\, dy\, dx$$

$$\therefore \int_{(AB+CD)} \mathbf{F} \cdot d\mathbf{r} = -\frac{\partial F_x}{\partial y}\, dx\, dy \qquad (1)$$

(b) BC + DA:

$$\int \mathbf{F} \cdot d\mathbf{r} = F_{by}\, dy - F_{dy}\, dy$$

$$= (F_{by} - F_{dy})\, dy$$

$$= \delta F_y\, dy$$

$$= \frac{\partial F_y}{\partial x}\, dx\, dy$$

$$\therefore \int_{(BC+DA)} \mathbf{F} \cdot d\mathbf{r} = \frac{\partial F_y}{\partial x}\, dx\, dy \qquad (2)$$

Adding these two results together for the complete rectangle, we have

$$\int_{(ABCD)} \mathbf{F} \cdot d\mathbf{r} = \left\{ \frac{\partial F_y}{\partial x} - \frac{\partial F_x}{\partial y} \right\} dx\, dy \qquad (3)$$

Now curl $\mathbf{F} = \begin{vmatrix} \mathbf{i} & \mathbf{j} & \mathbf{k} \\ \dfrac{\partial}{\partial x} & \dfrac{\partial}{\partial y} & \dfrac{\partial}{\partial z} \\ F_x & F_y & F_z \end{vmatrix}$

$$= \mathbf{i}\left(\frac{\partial F_z}{\partial y} - \frac{\partial F_y}{\partial z} \right) - \mathbf{j}\left(\frac{\partial F_z}{\partial x} - \frac{\partial F_x}{\partial z} \right) + \mathbf{k}\left(\frac{\partial F_y}{\partial x} - \frac{\partial F_x}{\partial y} \right)$$

$$\therefore \left\{ \frac{\partial F_y}{\partial x} - \frac{\partial F_x}{\partial y} \right\} = (\text{curl } \mathbf{F}) \cdot (\mathbf{k}) \qquad (4)$$

From (3) $\qquad \int_{ABCD} \mathbf{F} \cdot d\mathbf{r} = \text{curl } \mathbf{F} \cdot \mathbf{k}\, dx\, dy = \text{curl } \mathbf{F} \cdot d\mathbf{S}$

Summing for all such elements over the surface

$$\int_S \text{curl } \mathbf{F} \cdot d\mathbf{S} = \underset{dr \to 0}{Lim} \sum \left\{ \int_{ABCD} \mathbf{F} \cdot d\mathbf{r} \right\} \qquad (5)$$

$\displaystyle\int \mathbf{F}\cdot d\mathbf{r}$ on boundary lines between adjacent rectangular elements will cancel out, except on the boundary curve c of the surface S. The integral then becomes $\displaystyle\oint_{c} \mathbf{F}\cdot d\mathbf{r}$.

$$\therefore \int_{S} \operatorname{curl} \mathbf{F}\cdot d\mathbf{S} = \oint_{c} \mathbf{F}\cdot d\mathbf{r}$$

Answers

Test exercise 1 (page 42)

1 $x = -1 - j\sqrt{3}$; $x^2 + 2x + 4 = 0$ **2** $x = -4, 6, 3/2$ **3** $x = -1\cdot\dot{6} = -5/3$
4 $x \approx 1\cdot710$ **5** $x \approx 0\cdot454304$ **6** $x \approx 1\cdot317672$ **7** (a) $39\cdot375$ (b) $103\cdot392$
  (c) $481\cdot528$ **8** $-12\cdot8$

Further problems 1 (page 43)

1 $\dfrac{-1+j\sqrt{3}}{2}$, $\dfrac{-1-j}{\sqrt{2}}$, $x^4 + (1+\sqrt{2})x^3 + (2+\sqrt{2})x^2 + (1+\sqrt{2})x + 1 = 0$

2 $x = 1, 6, -2$ **3** $p = -5, q = -1$ **4** $p = 4, q = 9$ **5** $x = 2, 3, -3$
6 $x = 1, -3, 9$ **7** $y^3 - 5y^2 + 17y - 13 = 0$ **8** $y^3 - 13y^2 + 52y - 60 = 0$
9 $x = \frac{1}{2}, \frac{3}{2}, -1$ **10** $x = -2, 4, 8$ **11** $2y^3 - 15y^2 + 25y = 0$ **13** $0\cdot8934$
14 $x = 2\cdot732, -0\cdot732, -2\cdot000$ **15** $y^3 - 3y + 2 = 0$; $x = -4, -1, -1$
16 $x = 1\cdot646$ **17** (a) $-0\cdot6736$ (b) $0\cdot3717$
18 (a) $-2\cdot3301, 0\cdot2016, 2\cdot1284$ (b) $1, -0\cdot50 \pm j1\cdot66$
  (c) $-2\cdot115, 0\cdot254, 1\cdot861$ **19** (a) $-4\cdot104, -0\cdot9481 \pm j0\cdot5652$
  (b) $0\cdot5, -1\cdot5, -1\cdot5$ (c) $0\cdot25, 1 \pm j3$ **20** (a) $-2\cdot456$ (b) $1\cdot765$
  (c) $0\cdot739$ (d) $1\cdot812$ (e) $1\cdot8175$ (f) $0\cdot5170$ (g) $0\cdot8449$ (h) $0\cdot8806$
21 (a) $32\cdot872$ (b) $204\cdot328$ (c) $381\cdot429$ **22** (a) $-1\cdot375$ and $81\cdot104$
  (b) $136\cdot971$ and $-363\cdot429$ **23** (a) $-6\cdot048$ (b) $461\cdot496$
24 (a) 133 and $-9\cdot048$ (b) $0\cdot136$ and $-65\cdot433$ (c) $-199\cdot112$ and -867
25 $0\cdot02768$ **26** $-1\cdot0670$ **27** (a) $-2\cdot54846$ (b) $-2\cdot41734$ (c) $-1\cdot87134$

Test exercise 2 (page 90)

1 (a) $\dfrac{32-2s}{s^2-16}$ (b) $\dfrac{s+4}{s^2+16}$ (c) $\dfrac{1}{s^4}\{4s^3 - s^2 + 4s + 6\}$ (d) $\dfrac{s+2}{s^2+4s+29}$

  (e) $\dfrac{6s}{(s^2+9)^2}$ (f) $\ln\left\{\dfrac{s+2}{s+1}\right\}$ **2** (a) $2e^{3t} - e^{4t}$ (b) $2\cos\sqrt{2}t + \dfrac{5}{\sqrt{2}}\sin\sqrt{2}t - e^t$

  (c) $e^t(3t+2) - e^{3t}$ (d) $\frac{1}{8}\{e^t(17\cos 2t + 9\sin 2t) - e^{3t}\}$ **3** (a) $x = e^{-2t} + e^{-3t}$

  (b) $x = \frac{1}{12}\{13e^{2t} - \cos 2t - \sin 2t\}$ (c) $x = \frac{1}{6} - \frac{5}{3}e^{3t} + \frac{5}{2}e^{4t}$ (d) $x = e^t\left(1 - t + \dfrac{t^3}{6}\right)$

4 $x = \frac{1}{2}\{9\cos t - 7\sin t - e^{-3t}\}$ $y = 3\sin t - 2\cos t + e^{-2t}$

Further problems 2 (page 90)

1 (a) $\dfrac{s-4}{s^2-8s+20}$ (b) $\dfrac{4s}{(s^2+4)^2}$ (c) $\dfrac{6}{s^4} + \dfrac{8}{s^3} + \dfrac{5}{s}$ (d) $\dfrac{4s^2 - 24s + 38}{(s-3)^3}$

  (e) $\dfrac{2s^3 - 6s}{(s^2+1)^3}$ (f) $\ln\sqrt{\dfrac{s+2}{s-2}}$ **2** (a) $e^{2t} + e^{4t}$ (b) $3e^{4t} + 2$ (c) $e^{2t}\left\{\dfrac{3t^2}{2} + 2t + 1\right\}$

  (d) $e^{-t}\{2\cos t - 5\sin t\} - 2e^{2t}$ (e) $\frac{1}{3}(\cos t - \cos 2t)$ (f) $e^{-2t}\{\cos 4t - \frac{7}{4}\sin 4t\}$
3 $x = 4e^{4t} - 2$ **4** $x = \frac{35}{78}e^{4t/3} - \frac{3}{26}\{\cos 2t + \frac{2}{3}\sin 2t\}$ **5** $x = e^t(2t+1) + 2t + 4 + \cos t$

6   $x = \frac{3}{2}e^{4t} - e^{3t} - \frac{1}{2}e^{2t}$ 7 $x = \frac{4}{5}\cos 3t + \sin 3t + \frac{1}{5}\cos 2t$

8   $x = \frac{1}{5}\{e^{2t} - e^t(\cos 2t - 2\sin 2t)\}$ 9 $x = \frac{1}{8}\{2t^2 - 4t + 3 + e^{-2t}(4t^2 + 6t + 1)\}$

10   $x = \frac{2}{5}\{2(e^{-4t} - 1)\cos 4t + (e^{-4t} + 1)\sin 4t\}$ 11 $x = (2t + 1)\cos 5t + t\sin 5t$

12   $x = \frac{1}{13}\{2e^{2t} + 3e^{-2t} - 5(\cos 3t - \sin 3t)\}$ $y = \frac{1}{13}\{5(\cos 3t + \sin 3t) - 3e^{2t} - 2e^{-2t}\}$

13   $x = \frac{1}{6}\{7e^{-6t} + 5\}$ $y = \frac{1}{3}\{7e^{-6t} + 5\}$ 14 $x = 10e^{-4t} + 2$ $y = 5e^{-4t} + 3$

15   $x = e^{-2t} - e^t + 2t$ $y = 3e^t + \frac{1}{2}e^{-2t} + t - \frac{7}{2}$ 16 $x = 5e^t + 3e^{-t}$ $y = 4e^t - e^{-t}$

17   $x = 4\cos t - 2\sin t - \frac{1}{3}\{8e^{-t} + e^{2t}\}$ $y = 6\cos t + 2\sin t - \frac{4}{3}\{2e^{-t} + e^{2t}\}$

18   $x = \frac{5}{3}\{\cos 2t + \sin 2t - \cosh\sqrt{2}t - \sqrt{2}\sinh\sqrt{2}t\}$

19   $y = \frac{1}{5}\{3\sin 2t - 4\cos 2t + \frac{4}{3}\sin 3t + \frac{48}{7}\cos 3t\} - \frac{4}{7}\cos 4t$

20   $x = \cos\left(t\sqrt{\frac{3}{10}}\right) + \frac{3}{4}\cos(t\sqrt{6})$ $y = \frac{5}{4}\cos\left(t\sqrt{\frac{3}{10}}\right) - \frac{1}{4}\cos(t\sqrt{6})$

Test exercise 3 (page 120)

1   (a)

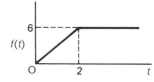

$$F(s) = \frac{3}{s^2}\{1 - e^{-2s}\}$$

(b)

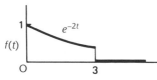

$$F(s) = \frac{1}{s+2}\{1 - e^{-6}e^{-3s}\}$$

(c)

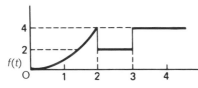

$$F(s) = \frac{2}{s^3} - 2e^{-2s}\left\{\frac{1}{s^3} + \frac{2}{s^2} - \frac{1}{s}\right\}$$
$$+ \frac{2}{s}e^{-3s}$$

(d)

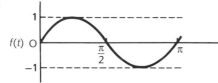

$$F(s) = \frac{2}{s^2 + 4}\{1 - e^{-\pi s}\}$$

2   $f(t) = 2 \cdot u(t) - 5 \cdot u(t - 1) + 8 \cdot u(t - 3)$

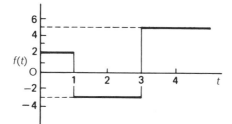

3 $f(t) = t \cdot u(t) + 3(t-2) \cdot u(t-2) - (t-3) \cdot u(t-3) - 3(t-5) \cdot u(t-5)$

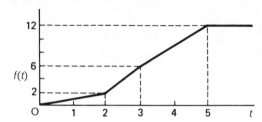

4 $f(t) = 2 \cdot u(t) - 2 \cdot u(t-1) + 2 \cdot u(t-3) - 2 \cdot u(t-4)$
$\qquad + 2 \cdot u(t-6) - 2 \cdot u(t-7) + \dots$

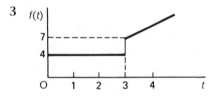

$\left. \begin{array}{ll} f(t) = 2 & 0 < t < 1 \\ \quad = 0 & 1 < t < 3 \end{array} \right\} \qquad f(t) = f(t+3)$

5 $f(t) = u(t)\sinh t - u(t-1)\sinh(t-1)$ **6** $\dfrac{t\sin 4t}{4}$

Further problems 3 (page 121)

1 $f(t) = 3 \cdot u(t) + 2(t-2) \cdot u(t-2) - 2(t-5) \cdot u(t-5)$

2 $f(t) = t \cdot u(t) - (t-1) \cdot u(t-1) + (t-2) \cdot u(t-2) - (t-3) \cdot u(t-3)$

3

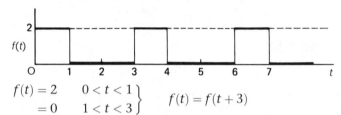

$F(s) = \dfrac{4}{s} + \dfrac{3e^{-3s}}{s} + \dfrac{2e^{-3s}}{s^2}$

4 (a) $f(t) = t^2 \cdot u(t) - (t^2 - 5t) \cdot u(t-3)$

   (b) $f(t) = \cos t \cdot u(t) + (\cos 2t - \cos t) \cdot u(t-\pi) + (\cos 3t - \cos 2t) \cdot u(t-2\pi)$

5 $F(s) = e^{-2s}\left\{\dfrac{1}{s^2} + \dfrac{3}{s}\right\} - e^{-3s}\left\{\dfrac{1}{s^2} + \dfrac{4}{s}\right\}$

6 (a) $f(t) = t^2 \cdot u(t) - t^2 \cdot u(t-2) + 4 \cdot u(t-2) - 4 \cdot u(t-5)$

   (b) $F(s) = \dfrac{2}{s^3} - \dfrac{2e^{-2s}}{s^3} - \dfrac{4e^{-2s}}{s^2} - \dfrac{4e^{-5s}}{s}$

7 (a) $\left(\dfrac{2-t}{t} + e^{-2t}\right)u(t) - \left(\dfrac{3-t}{t-1} + e^{-2(t-1)}\right)u(t-1)$

   (b) $(1 - e^{2t} + 2te^{2t})\dfrac{u(t)}{4} - \left(1 - e^{2(t-1)} + 2(t-1)e^{2(t-1)}\right)\dfrac{u(t-1)}{4}$

   (c) $(3e^t - 3e^{-t} - \sin 3t)\dfrac{u(t)}{20} - (3e^{t-1} - 3e^{-(t-1)} - \sin 3(t-1))\dfrac{u(t-1)}{20}$

   (d) $(t-2)u(t-1) + \dfrac{u(t-1)e^{-(t-1)/2}}{\sqrt{3}}\left\{\sqrt{3}\cos\dfrac{\sqrt{3}t}{2} - \sin\dfrac{\sqrt{3}t}{2}\right\}$

8 (a) $\dfrac{1}{2}\{1 - e^{-t}(\cos t + \sin t)\}$ (b) $\dfrac{1}{2}\{\cosh\sqrt{2}t - 1\}$ (c) $-\dfrac{\sqrt{2}}{5}\left\{\sinh\dfrac{2t}{\sqrt{3}} + \sinh\dfrac{t}{\sqrt{2}}\right\}$

Test exercise 4 (page 153)

1 $F(s) = \dfrac{2(1 - e^{-2s} - 2se^{-2s})}{s^2(1 - e^{-4s})}$ **2** (a) e^{-6} (b) 0   (c) 11

3 (a) $F(s) = 4e^{-3s}$ (b) $F(s) = e^{-2(3+s)}$

4

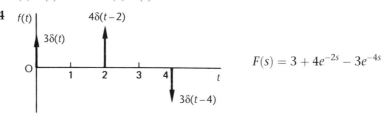

$F(s) = 3 + 4e^{-2s} - 3e^{-4s}$

5 $x = e^{-3t}\{4\sin t - \cos t\}$ **6** $x = 3e^4 e^{-t} \cdot u(t - 4) + e^{-2t}\{2 \cdot u(t) - 3e^8 \cdot u(t - 4)\}$

7 (a) $f(t) = \sin t$, frequency 1 radian per unit of time, period 2π units of time

   (b) $f(t) = \dfrac{18}{\sqrt{53}} e^{-t/6} \sin\left(\dfrac{\sqrt{53}}{6}\right) t$, frequency $\dfrac{\sqrt{53}}{6}$ radian per unit of time,

   period $\dfrac{12\pi}{\sqrt{53}}$ units of time

8 Transient solution $\dfrac{e^{-t}}{19}\left(32\sqrt{2}\sin\sqrt{2}t - 40\cos\sqrt{2}t\right)$, steady-state solution $\dfrac{2}{19}e^{5t}$

Further problems 4 (page 154)

2 $L\{f(t)\} = \dfrac{a(1 + e^{-\pi s})}{(s^2 + 1)(1 - e^{-\pi s})}$ **3** (a) $F(s) = \dfrac{1}{s^2} - \dfrac{w}{s}\left\{\dfrac{e^{-ws}}{1 - e^{-ws}}\right\}$

   (b) $F(s) = \dfrac{1 - e^{2(1-s)\pi}}{(s - 1)(1 - e^{-2\pi s})}$ (c) $F(s) = \dfrac{1 - e^{-s}(s + 1)}{s^2(1 - e^{-2s})}$

   (d) $F(s) = \dfrac{1}{1 - e^{-3s}}\left\{\dfrac{2}{s^3} - \dfrac{2e^{-2s}}{s^3} - \dfrac{4e^{-2s}}{s^2} - \dfrac{4e^{-3s}}{s}\right\}$

4 $x = \dfrac{P}{M\omega}\sin\omega t$ **5** $i = \dfrac{E}{L}\cos\left(\dfrac{t}{\sqrt{LC}}\right)$

6 $x = 2e^{-2t}\{1 + 10e^8 \cdot u(t - 4)\} - 2e^{-3t}\{1 + 10e^{12} \cdot u(t - 4)\}$

7 (a) $f(t) = 4\sqrt{3}\sin\dfrac{t}{2\sqrt{3}} - \cos\dfrac{t}{2\sqrt{3}}$, frequency $\dfrac{1}{2\sqrt{3}}$ radian per unit of time,

   period $4\pi\sqrt{3}$ units of time   (b) $f(t) = 2\cos 2\sqrt{3}t - \dfrac{1}{2\sqrt{3}}\sin 2\sqrt{3}t$,

   frequency $2\sqrt{3}$ radian per unit of time, period $\pi\sqrt{3}$ units of time

8 (a) $f(t) = -4 \cdot 48\sin 0 \cdot 69t + 1 \cdot 06\cos 0 \cdot 69t$ (b) $f(t) = \dfrac{\pi}{(3/2)^{\frac{1}{4}}}\sin[(1 \cdot 5)^{\frac{1}{4}}t]$

9 Transient solution $e^{-3t/8}\left(\dfrac{421}{9\sqrt{23}}\sin\dfrac{\sqrt{23}}{8}t - \dfrac{1}{9}\cos\dfrac{\sqrt{23}}{8}t\right)$, steady-state solution $\dfrac{1}{9}e^t$

Test exercise 5 (page 190)

1 (a) $f(n+1) = f(n) + 5$ (b) $f(n+1) = f(n) - 4$ (c) $f(n+1) = 3^{-1}f(n)$

2 (a) order 3: 1, −1, 3, 1, 4, 13 (b) order 2: 0, 1, 5, 21, 79, 275

   (c) order 2: −2, 5, 32, −25, −406, −103

3 $f(n) = \dfrac{1}{2}\left(-3 \times 2^{n+2} + 3^{n+3} + 2n + 5\right)$ 4 (a) $\dfrac{z}{z+1}$ (b) $\dfrac{4z(z-a) - 2z(z-1)^2}{(z-a)(z-1)^2}$

   (c) $\dfrac{z(4-3z)}{(z-1)^2}$ (d) $\dfrac{25z}{z-5}$ 5 $\{2k + 3 - 2^{k+1}\}$ 6 $f(n) = (2^{n-1}n - 2^{n+1} + 3)u(n)$

7 $\dfrac{z \sin T}{z^2 - 2z \cos T + 1}$

Further problems 5 (page 190)

1 $\dfrac{z}{z+a}$ provided $|z| > |a|$ 2 (a) $\left\{\dfrac{1}{12}u_k - \dfrac{3}{4}(-3)^k + \dfrac{2}{3}(-2)^k\right\}$

   (b) $\left\{\dfrac{1}{4}u_k - \dfrac{k}{2} + \dfrac{3}{4}(1/3)^k\right\}$ (c) $\left\{\dfrac{2}{3}(3^k) + \dfrac{1}{3}(-3)^k - 2k\right\}$

3 $\left\{\dfrac{1}{2}(1+j)(-j)^{k-1} + \dfrac{1}{2}(1-j)(j)^{k-1}\right\}$ 4 (a) $\left\{u_k + \dfrac{3}{2}k(-2)^k\right\}$

   (b) $\left\{\dfrac{1}{9}u_k - \dfrac{5}{6}k(-2)^k + \dfrac{8}{9}(-2)^k\right\}$ 5 (a) $\dfrac{z^2}{z^2 - 1}$ (b) $\dfrac{z}{z^2 - 1}$ (c) $\dfrac{z^7 + z^5 + z^4 + 1}{z^7}$

   (d) $\dfrac{z^7 + z^6 + z^5 + z + 1}{z^7}$ (e) $\dfrac{z^7 + z^6 + z^5 + z + 1}{z^{10}}$ (f) $\dfrac{z^6 + z^5 + z + 1}{z^6}$

6 (a) $\{x_k\} = \left\{\dfrac{1}{2}\left((-3)^k - 2(-2)^k + (-1)^k\right)\right\}$ for $k \geq 1$

   (b) $\{x_k\} = \left\{\dfrac{1}{2}\left((-3^{k+1} - (-2)^{k+2} + (-1)^{k+1}\right)\right\}$ (c) $\{x_k\} = \{10(3^k) - 7(2^k)\}$

   (d) $\{x_k\} = \{6(2^k) - 3u_k\}$ 9 3 10 $-\dfrac{2}{7}$ 13 (a) $\dfrac{z \sinh T}{z^2 - 2z \cosh T + 1}$

   (b) $\dfrac{z(z - \cosh aT)}{z^2 - 2z \cosh aT + 1}$ (c) $\dfrac{ze^{-aT}(ze^{aT} - \cosh bT)}{z^2 - 2ze^{-aT} \cosh bT + e^{-2aT}}$

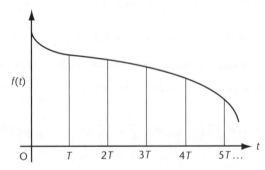

14 (a) $f(n) = \dfrac{1}{12}\left(1 + 8(-2)^n - 9(-3)^n\right)u(n)$ (b) $f(n) = \dfrac{1}{4}\left(3^{1-n} - 2n + 1\right)u(n)$

   (c) $f(n) = \dfrac{1}{12}\left(10(3^n) + 5(-3)^n - 3\right)u(n)$ (d) $f(n) = \dfrac{1}{9}\left(3n + 1 - (-5)^n\right)u(n)$

16 $g(n) = \dfrac{1}{5}\left(3^n - (-2)^n\right)u(n)$ **18** $\dfrac{81(z-2)}{z^2(z-3)^2}$

19 $f(n) = \dfrac{1}{\sqrt{5}}\left\{\left(\dfrac{1+\sqrt{5}}{2}\right)^n - \left(\dfrac{1-\sqrt{5}}{2}\right)^n\right\}$

Test exercise 6 (page 233)

1 (a) linear and time-invariant   (b) non-linear, shift-invariant
(c) non-linear, not time-variant   (d) non-linear, shift-invariant
(e) linear, not time-invariant   (f) linear, not shift-invariant
(g) linear, time-invariant   (h) linear, not shift-invariant

2 (a) $y_{zi}(t) = 2e^{3t}$, $y_{zs}(t) = \dfrac{1}{27}\left\{2e^{3t} - (9t^2 + 6t + 2)\right\}$, not time-invariant

(b) $y_{zi}(t) = \dfrac{4}{3}\left(e^t - e^{4t}\right)$, $y_{zs}(t) = \dfrac{1}{102}\left\{2e^{4t} - 17e^t + 3(3\sin t + 5\cos t)\right\}$,

not time-invariant   (c) $y_{zi}(t) = 0$, $y_{zs}(t) = e^{-4t/5} - e^{-t}$, time-invariant

(d) $y_{zi}(t) = 0$, $y_{zs}(t) = 1 - (1+t)e^{-t}$, time-invariant   **3** $\dfrac{e^{-3t}}{3}(e^9 - 1)$

4 $H(s) = \dfrac{e^{-s}}{s^2}(s+1)$; $y(t) = \dfrac{t(t+2)}{2}\left\{u(t-1) - 2u(t-2) + u(t-3)\right\}$

5 $H(s) = \dfrac{1}{(s+4)(s-1)}$; $y(t) = 2e^t - 5e^{-2t} + 3e^{-4t}$

6 $H[z] = \displaystyle\sum_{k=5}^{\infty} kz^{-k}$; $y[n] = \dfrac{1}{9}\left(12n - 4 + 4^{1-n}\right)u[n]$ **7** $h[n] = 2^n u[n]$

8 $y[n] - 2y[n-1] + y[n-2] = x[n-1]$

Further problems 6 (page 234)

1 All non-zero values   **2** All values of a but only $b = 0$
3 Linear but not time-invariant   **4** $\dfrac{\sinh 3t}{3}$ **5** Yes   **6** Yes   **13** $\dfrac{4e^{jn\omega_0}}{4 - e^{jn\omega_0}}$

14 $y(t) = \dfrac{t^2}{2}e^{-t}$ **15** $\dfrac{1}{a}\exp\left(-\dfrac{t}{a}\right)u(t)$ **16** $y(t) = G\left(1 - e^{-t/T}\right)$

17 $y[n] = (1 - \alpha^{n+1})u(n)$ **18** $y[n] = (20 + 140(0.93)^n)u[n]$ **19** $y[n] = nu[n]$
20 $h(t) = u(t)$: $y(t) = (t-1)u(t-1)$
21 $h[n] = nu[n] - 2(n-1)u[n-1]$: $y[n] = \dfrac{u[n]}{4}(2n - 1 + 3^n)$

Test exercise 7 (page 263)

1 Amplitude $\sqrt{2}$, period $\dfrac{8\pi}{3}$

2 $f(x) = \begin{cases} 1 & : \ 0 \le x < 2 \\ 3 - x & : \ 2 \le x < 3 \\ x - 3 & : \ 3 \le x < 4 \end{cases}$

$f(x + 4) = f(x)$

3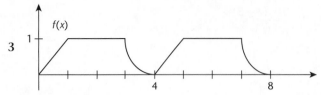

4 (a) yes (b) yes (c) no (d) yes (e) no (f) no

5 $f(x) = 2\pi - 4\{\sin x + \frac{1}{2}\sin 2x + \frac{1}{3}\sin 3x + \ldots\}$ **6** 1

Further problems 7 (page 264)

1 (a) $f(x) = \begin{cases} x & : & 0 \le x < 2 \\ 2 & : & 2 \le x < 4 \end{cases}$

$f(x+4) = f(x)$

(b) $f(x) = \begin{cases} -3 & : & 0 \le x < 3 \\ x-6 & : & 3 \le x < 6 \\ 9-x & : & 6 \le x < 9 \end{cases}$

$f(x+9) = f(x)$

(c) $f(x) = \begin{cases} x+5 & : & -4 \le x < -3 \\ 2 & : & -3 \le x < -1 \\ -2x & : & -1 \le x < 0 \\ 1 & : & 0 \le x < 1 \end{cases}$

$f(x+5) = f(x)$

2 (a)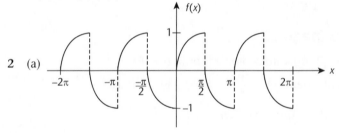

(b)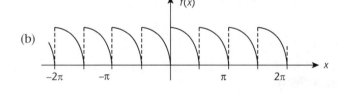

(c)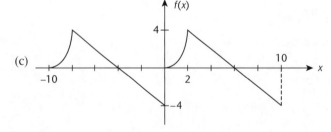

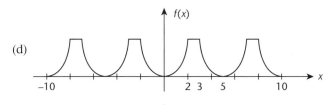

(d)

(e)

3 $f(x) = \dfrac{2}{\pi}\left\{\sin x + \dfrac{1}{2}\sin 2x + \dfrac{1}{3}\sin 3x + \ldots\right\}$

4 $f(x) = \dfrac{\pi}{2} + \dfrac{4}{\pi}\left\{\cos x + \dfrac{1}{9}\cos 3x + \dfrac{1}{25}\cos 5x + \ldots\right\}$

5 $f(x) = \dfrac{2A}{\pi}\left\{1 - 2\left(\dfrac{1}{1\times 3}\cos 2x + \dfrac{1}{3\times 5}\cos 4x + \ldots\right)\right\}$

6 $f(x) = \dfrac{\pi}{4} - \dfrac{2}{\pi}\left\{\cos x + \dfrac{1}{9}\cos 3x + \dfrac{1}{25}\cos 5x + \ldots\right\}$

$\qquad\qquad + \left\{\sin x - \dfrac{1}{2}\sin 2x + \dfrac{1}{3}\sin 3x - \ldots\right\}$

7 $f(x) = \dfrac{2}{\pi}\left\{\dfrac{1}{2} + \dfrac{\pi}{4}\cos x + \dfrac{1}{1\times 3}\cos 2x - \dfrac{1}{3\times 5}\cos 4x + \ldots\right\}$

8 $f(x) = \dfrac{\pi^2}{3} - 4\left\{\cos x - \dfrac{1}{4}\cos 2x + \dfrac{1}{9}\cos 3x - \dfrac{1}{16}\cos 4x + \ldots\right\}$

9 $f(x) = 7 - \dfrac{6}{\pi}\left\{\sin x - \dfrac{1}{2}\sin 2x + \dfrac{1}{3}\sin 3x - \dfrac{1}{4}\sin 4x + \ldots\right\}$

10 $f(x) = -\left\{\sin x + \dfrac{1}{2}\sin 2x + \dfrac{1}{3}\sin 3x + \dfrac{1}{4}\sin 4x + \ldots\right\}$

11 $f(x) = \dfrac{4\pi^2}{3} + 4\left\{\cos x + \dfrac{1}{4}\cos 2x + \dfrac{1}{9}\cos 3x + \ldots\right\}$

$\qquad\qquad\qquad\qquad - 4\pi\left\{\sin x + \dfrac{1}{2}\sin 2x + \dfrac{1}{3}\sin 3x + \ldots\right\}$

Test exercise 8 (page 294)

1 $ft = \dfrac{4}{3} + \displaystyle\sum_{n=1}^{\infty}\dfrac{4}{\pi^2 n^2}\{\cos n\pi t + n\sin n\pi t\}$ 2 (a) odd (b) odd (c) even (d) neither

(e) neither (f) even 3 (a) $f(x) = \dfrac{\pi}{2} + \dfrac{4}{\pi}\left\{\cos x + \dfrac{1}{9}\cos 3x + \dfrac{1}{25}\cos 5x + \ldots\right\}$

(b) $f(x) = -2\left\{\sin x - \dfrac{1}{2}\sin 2x + \dfrac{1}{3}\sin 3x - \ldots\right\}$ 4 (a) cosine terms only

(b) sine terms only; odd harmonics only (c) odd harmonics only
(d) odd harmonics only

5 $f(t) = \dfrac{1}{2} - \dfrac{1}{\omega^2}\left\{\cos\omega t + \dfrac{1}{9}\cos 3\omega t + \dfrac{1}{25}\cos 5\omega t + \dots\right\}$

$\qquad\qquad + \dfrac{1}{\omega}\left\{\sin\omega t - \dfrac{1}{2}\sin 2\omega t + \dfrac{1}{3}\sin 3\omega t - \dots\right\}$ where $\omega = \pi/2$

Further problems 8 (page 295)

1 $f(t) = -1 - \dfrac{16}{\pi}\left\{\sin\omega t + \dfrac{1}{3}\sin 3\omega t + \dfrac{1}{5}\sin 5\omega t + \ \dots\right\}$ where $\omega = \pi/2$

2 $f(x) = \dfrac{4}{\pi}\left\{\dfrac{1}{2} - \dfrac{1}{1\times 3}\cos 2x - \dfrac{1}{3\times 5}\cos 4x - \dfrac{1}{5\times 7}\cos 6x - \ \dots\right\}$

3 $i = f(t) = \dfrac{A}{\pi}\left\{1 + \dfrac{\pi}{2}\sin\omega t - 2\left(\dfrac{1}{1\times 3}\cos 2\omega t + \dfrac{1}{3\times 5}\cos 4\omega t + \dfrac{1}{5\times 7}\cos 6\omega t + \ \dots\right)\right\}$

   where $\omega = \dfrac{2\pi}{T}$

4 $f(x) = \dfrac{3a}{\pi}\left\{\sin 2x + \dfrac{1}{2}\sin 4x + \dfrac{1}{4}\sin 8x + \dfrac{1}{5}\sin 10x + \ \dots\right\}$

5 (a) $f(x) = \dfrac{\pi^2}{6} - \left\{\cos 2x + \dfrac{1}{4}\cos 4x + \dfrac{1}{9}\cos 6x + \ \dots\right\}$

   (b) $f(x) = \dfrac{8}{\pi}\left\{\sin x + \dfrac{1}{3^3}\sin 3x + \dfrac{1}{5^3}\sin 5x + \ \dots\right\}$

6 $f(x) = \dfrac{2}{\pi}\left\{\dfrac{1}{2} + \dfrac{\pi}{4}\cos x + \dfrac{1}{1\times 3}\cos 2x - \dfrac{1}{3\times 5}\cos 4x + \dots\right\}$

7 $f(x) = -\dfrac{1}{\pi} + \dfrac{1}{2}\cos x - \dfrac{2}{3\pi}\cos 2x + \dfrac{2}{15\pi}\cos 4x$

8 $f(x) = \dfrac{4}{\pi}\left\{\sin x - \dfrac{1}{9}\sin 3x + \dfrac{1}{25}\sin 5x - \ \dots\right\}$

9 $f(t) = -\dfrac{4}{\pi^2}\left\{\cos\pi t + \dfrac{1}{9}\cos 3\pi t + \dots\right\} + \dfrac{2}{\pi}\left\{2\sin\pi t - \dfrac{1}{2}\sin 2\pi t + \dots\right\}$

10 $f(t) = \dfrac{2}{3} + \dfrac{4}{\pi^2}\left\{\cos\pi t - \dfrac{1}{4}\cos 2\pi t + \dfrac{1}{9}\cos 3\pi t - \ \dots\right\}$

11 $f(t) = -\dfrac{2}{\pi}\left\{\sin\omega t - \sin 2\omega t + \dfrac{1}{3}\sin 3\omega t + \dfrac{1}{5}\sin 5\omega t \ \dots\right\}$ where $\omega = \pi/2$

12 $f(t) = 1 - 1{\cdot}17\cos\omega t + 0{\cdot}328\cos 2\omega t + \ \dots$
$\qquad\qquad + 0{\cdot}282\sin\omega t + 0{\cdot}288\sin 2\omega t - 0{\cdot}318\sin 3\omega t + \ \dots$ where $\omega = \pi/3$

Test exercise 9 (page 331)

1 $f(t) = \dfrac{1}{2} + \dfrac{j}{2\pi}\displaystyle\sum_{\substack{n=-\infty \\ n\neq 0}}^{\infty}\dfrac{e^{j2n\pi t}}{n}$ **2** $F(\omega) = \sqrt{\dfrac{2}{\pi}}\dfrac{(a-j\omega)\sinh(a+j\omega)}{a^2+\omega^2}$

3 $\sqrt{\dfrac{2}{\pi}}\left(\dfrac{\sinh a\cos\omega + j\sin\omega\cosh a}{a+j\omega}\right)$ **4** $-\dfrac{j}{2}(F(\omega+\omega_0) - F(\omega-\omega_0))$

5 $2\sqrt{2\pi}(e^t - e^{4t})u(t)$ **6** $F_c(\omega) = \sqrt{\dfrac{2}{\pi}}\dfrac{k}{k^2\omega^2}$, $F_s(\omega) = \sqrt{\dfrac{2}{\pi}}\dfrac{\omega}{k^2+\omega^2}$

Further problems 9 (page 332)

3

$$f(t) = -\frac{2}{\pi}\sum_{n=-\infty}^{\infty}\frac{1}{4n^2-1}e^{2\pi nt}$$

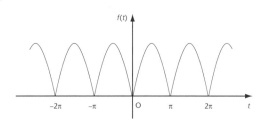

4

$$f(t) = -\frac{4j}{\pi}\sum_{n=-\infty}^{\infty}\frac{n}{4n^2-1}e^{j2\pi nt}$$

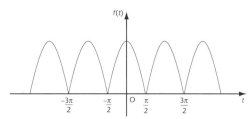

5

$$f(t) = -\frac{e^{2\pi}-1}{2\pi}\sum_{n=-\infty}^{\infty}\frac{1+jn}{1+n^2}e^{j\pi nt}$$

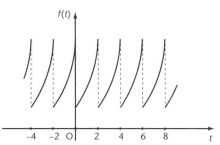

6

$$f(t) = \frac{1}{2} + \frac{1}{2\pi}\sum_{\substack{n=-\infty \\ n\neq 0}}^{\infty}\frac{1}{n}e^{j(n\omega_0 t+\pi/2)}$$

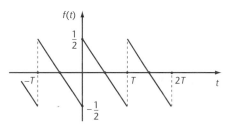

8 $\dfrac{(e^2-1)\cos\omega + \omega(e^2+1)\sin\omega}{\sqrt{2\pi}e(\omega^2+1)}$ **9** $\dfrac{j\left(\omega(e^2-1)\cos\omega - (e^2+1)\sin\omega\right)}{\sqrt{2\pi}e(\omega^2+1)}$

10 $\sqrt{\dfrac{\pi}{2}}\left(\dfrac{1+e^{-j\omega}}{\pi^2-\omega^2}\right)$ **11** $\dfrac{\sqrt{2\pi}\cos(\omega/2)}{\pi^2-\omega^2}$

12

$$F(\omega) = \frac{2a}{a^2+\omega^2}$$

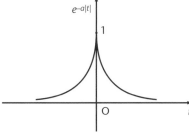

13 (a)

(b) $f(t) = u(t-1) - 2u(t) + u(t+1)$ (c) $F(\omega) = \dfrac{4j}{\omega}\sin^2(\omega/2)$

14 $\dfrac{j}{\sqrt{2\pi}(k^2 - \omega^2)}\left(\omega[1 - \cos\pi(k+\omega)] - jk\sin\pi(k-\omega)\right)$

20 $F_s(\omega) = \sqrt{\dfrac{2}{\pi}}\dfrac{1}{a^2 + \omega^2}\left(e^a(a\sin\omega t - \omega\cos\omega t) + \omega\right)$

$F_c(\omega) = \sqrt{\dfrac{2}{\pi}}\dfrac{1}{a^2 + \omega^2}\left(e^a(a\cos\omega t + \omega\sin\omega t) - a\right)$

21 $F_s(\omega) = 0$ $F_c(\omega) = 2\cos\omega\operatorname{sinc}t$

Test exercise 10 (page 355)

2 $y = a_0\left\{1 + \dfrac{5x^2}{2} + \dfrac{15x^4}{8} + \dfrac{5x^6}{16} + \dots\right\} + a_1\left\{x + \dfrac{4x^3}{3} + \dfrac{8x^5}{15} + \dots\right\}$

3 (a) $y(x) = Ax + \dfrac{B}{x^2}$ (b) $y(x) = Ax^{3/2} + \dfrac{B}{x^3} + \dfrac{x^2}{5}$ (c) $y(x) = \left(\dfrac{3}{2} - \dfrac{7\ln x}{2\ln 2}\right)x + 3x^2 - \dfrac{x^3}{2}$

Further problems 10 (page 355)

1 $y_5 = 64e^{4x}\{16x^3 + 60x^2 + 60x + 15\}$

2 $y_n = (-1)^n e^{-x}\{x^3 - 3nx^2 + n(n-1)3x - n(n-1)(n-2)\}$, $n > 3$

3 $y_4 = 480x + 96$ **4** $y_6 = -\{(x^4 - 180x^2 + 360)\cos x + (24x^3 - 480x)\sin x\}$

5 $y_4 = -4e^{-x}\sin x$ **6** $y_3 = 2x(13 + 12\ln x)$ **8** $y_6 = -1018$

10 (a) $y_{2n} = \{x^2 + 2n(2n-1)\}\sinh x + 4nx\cosh x$

(b) $y_{2n} = \{x^3 + 6n(2n-1)x\}\cosh x + \{6nx^2 + 2n(2n-1)(2n-2)\}\sinh x$

11 $y_6 = 2^5 e^{2x}\{2x^3 + 24x^2 + 81x + 75\}$ **12** $y_3 = 2\sqrt{2}a^3 e^{-ax}\{\cos(ax + \pi/4)\}$

14 $y = y_0\left\{1 + \dfrac{9x^2}{2} + \dfrac{15x^4}{8} - \dfrac{7x^6}{16} + \dfrac{27x^8}{128} + \dots\right\} + y_1\left\{x + \dfrac{4x^3}{3}\right\}$

15 $y = A(1 + x^2) + Be^{-x}$

16 $y = y_0\left\{1 + \dfrac{3^2 \times x^2}{2!} + \dfrac{3^2 \times 5^2 \times x^4}{4!} + \dfrac{3^2 \times 5^2 \times 7^2 \times x^6}{6!} + \dots\right\}$

$+ y_1\left\{x + \dfrac{4^2 \times x^3}{3!} + \dfrac{4^2 \times 6^2 \times x^5}{5!} + \dots\right\}$

17 $y = y_1 x + y_0\left\{1 - x^2 - \dfrac{x^4}{3} - \dfrac{x^6}{5} - \dfrac{x^8}{7} - \dots\right\}$

18 $y = y_0 \left\{ 1 - \dfrac{2x}{2^2} + \dfrac{2^2 \times x^4}{2^2 \times 4^2} - \dfrac{2^3 \times x^6}{2^2 \times 4^2 \times 6^2} + \cdots \right\}$

$\qquad + y_1 \left\{ x - \dfrac{2x^3}{3^2} + \dfrac{2^2 \times x^5}{3^2 \times 5^2} - \dfrac{2^3 \times x^7}{3^2 \times 5^2 \times 7^2} + \cdots \right\}$

19 $y(x) = Ax + Bx^4 - x^3$ **20** $y(x) = Ax^{-2/3} + Bx^{-3/2} + \dfrac{2x^3}{99} - \dfrac{3x^2}{56}$

21 $y(x) = Ax^{(1/2)+j\sqrt{11}/2} + Bx^{(1/2)-j\sqrt{11}/2} + \dfrac{4x^3}{9}$

22 $y(x) = Ax^{-(1/2)+j\sqrt{3}/2} + Bx^{-(1/2)-j\sqrt{3}/2} + \dfrac{x}{3}$

Test exercise 11 (page 377)

1 (a) $y = A \left\{ 1 - \dfrac{x}{1 \times 2} + \dfrac{x^2}{(1 \times 2)(2 \times 5)} - \dfrac{x^3}{(1 \times 2)(2 \times 5)(3 \times 8)} + \cdots \right\}$

$\qquad + Bx^{\frac{1}{3}} \left\{ 1 - \dfrac{x}{1 \times 4} + \dfrac{x^2}{(1 \times 4)(2 \times 7)} - \dfrac{x^3}{(1 \times 4)(2 \times 7)(3 \times 10)} + \cdots \right\}$

$\quad$ (b) $y = a_0 \left\{ 1 - \dfrac{x^4}{3 \times 4} + \dfrac{x^8}{(3 \times 4)(7 \times 8)} + \cdots \right\}$

$\qquad\qquad\qquad + a_1 \left\{ x - \dfrac{x^5}{4 \times 5} + \dfrac{x^9}{(4 \times 5)(8 \times 9)} + \cdots \right\}$

$\quad$ (c) $y_A = A \left\{ -\dfrac{1}{2} - \dfrac{x}{6} - \cdots \right\}$ $y_B = B \left\{ \ln x \left(-\dfrac{1}{2} - \dfrac{x}{6} - \cdots \right) + x^{-2} \left(1 - x + \dfrac{x^2}{4} + \cdots \right) \right\}$

Further problems 11 (page 377)

1 (a) $y = A \left\{ 1 + x + \dfrac{x^2}{2 \times 4} + \dfrac{x^3}{(2 \times 3)(4 \times 7)} + \dfrac{x^4}{(2 \times 3 \times 4)(4 \times 7 \times 10)} + \cdots \right\}$

$\qquad + Bx^{\frac{2}{3}} \left\{ 1 + \dfrac{x}{1 \times 5} + \dfrac{x^2}{(1 \times 2)(5 \times 8)} + \dfrac{x^3}{(1 \times 2 \times 3)(5 \times 8 \times 11)} + \cdots \right\}$

$\quad$ (b) $y = a_0 \left\{ 1 - \dfrac{x^2}{2!} + \dfrac{x^4}{4!} + \cdots \right\} + a_1 \left\{ x - \dfrac{x^3}{3!} + \cdots \right\}$

$\quad$ (c) $y = a_0 \left\{ 1 + \dfrac{x^3}{2 \times 3} + \dfrac{x^6}{(2 \times 3)(5 \times 6)} + \cdots \right\}$

$\qquad\qquad + a_1 \left\{ x + \dfrac{x^4}{3 \times 4} + \dfrac{x^7}{(3 \times 4)(6 \times 7)} + \cdots \right\}$

$\quad$ (d) $y = A \left\{ 1 - \dfrac{x}{1 \times 4} + \dfrac{x^2}{(1 \times 2)(4 \times 7)} - \dfrac{x^3}{(1 \times 2 \times 3)(4 \times 7 \times 10)} + \cdots \right\}$

$\qquad + Bx^{-\frac{1}{3}} \left\{ 1 - \dfrac{x}{1 \times 2} + \dfrac{x^2}{(1 \times 2)(2 \times 5)} - \dfrac{x^3}{(1 \times 2 \times 3)(2 \times 5 \times 8)} + \cdots \right\}$

$\quad$ (e) $y = a_1 x + a_0 \left\{ 1 - \dfrac{x^2}{2!} - \dfrac{x^4}{4!} - \dfrac{3x^6}{6!} - \dfrac{(3)(5)x^8}{8!} + \cdots \right\}$

(f) $y = u + v$ where $u = A\left\{\dfrac{-x^4}{4!\,3!} + \dfrac{x^5}{5!\,3!} - \cdots\right\}$

$$v = B\left\{\ln x\left(\dfrac{-x^4}{4!\,3!} + \dfrac{x^5}{5!\,3!} - \cdots\right) + \left(1 + \dfrac{x}{1 \times 3} + \dfrac{x^2}{(1 \times 2)(2 \times 3)} + \cdots\right)\right\}$$

(g) $y = u + v$ where $u = A\left\{1 + \dfrac{3x}{1^2} + \dfrac{3^2 \times x^2}{1^2 \times 2^2} + \dfrac{3^3 \times x^3}{1^2 \times 2^2 \times 3^2} + \cdots\right\}$

$$v = B\left\{\ln x\left(1 + \dfrac{3x}{1^2} + \dfrac{3^2 \times x^2}{1^2 \times 2^2} + \dfrac{3^3 \times x^3}{1^2 \times 2^2 \times 3^2} + \cdots\right)\right.$$
$$\left. - \left(\dfrac{2 \times 3x}{1^2} + \dfrac{3 \times 3^2 \times x^2}{1^2 \times 2^2} + \dfrac{11 \times 3^3 \times x^3}{1^2 \times 2^2 \times 3^3} + \cdots\right)\right\}$$

Test exercise 12 (page 396)

1   $P_2(x) = \dfrac{3x^2 - 1}{2},\ P_3(x) = \dfrac{5x^3 - 3x}{2}$ 2 $\dfrac{1}{3}P_0(x) - \dfrac{4}{3}P_2(x)$

Further problems 12 (page 396)

1   eigenfunctions: $y_n(x) = A_n \cos\sqrt{\lambda_n}\,x$;   eigenvalues: $\lambda_n = \dfrac{(2n+1)^2\pi^2}{4}$

2   $H_0 = 1,\ H_1 = 2x,\ H_2 = 4x^2 - 2,\ H_3 = 8x^3 - 12x$

3   $L_0 = 1,\ L_1 = 1 - x,\ L_2 = 2 - 4x + x^2,\ L_3 = 6 - 18x + 9x^2 - x^3$

Text exercise 13 (page 436)

1

x	y
0	1·0
0·1	1·1
0·2	1·211
0·3	1·3352
0·4	1·4753
0·5	1·6343

2

x	y
1	0
1·2	0·204
1·4	0·4211
1·6	0·6600
1·8	0·9264
2·0	1·2243

3

x	y
0	1·0
0·1	1·2052
0·2	1·4214
0·3	1·6499
0·4	1·8918
0·5	2·1487

4

x	y
2·0	3·0
2·1	3·005
2·2	3·0195
2·3	3·0427
2·4	3·0736
2·5	3·1117

5

x	y
1·0	0
1·1	0·1052
1·2	0·2215
1·3	0·3401
1·4	0·4717
1·5	0·6180

6

x	y
0·0	1·0000
0·1	1·0101
0·2	1·0202
0·3	1·0305
0·4	1·0408
0·5	1·0513
0·6	1·0619
0·7	1·0726
0·8	1·0834
0·9	1·0943
1·0	1·1053

Further problems 13 (page 437)

1

x	y
0	1·0
0·2	0·8
0·4	0·72
0·6	0·736
0·8	0·8288
1·0	0·9830

2

x	y
0	1·4
0·1	1·596
0·2	0·8707
0·3	2·2607
0·4	2·8318
0·5	3·7136

3

x	y
1·0	2·0
1·2	2·0333
1·4	2·1143
1·6	2·2250
1·8	2·3556
2·0	2·5000

4

x	y
0	0·5
0·1	0·543
0·2	0·5716
0·3	0·5863
0·4	0·5878
0·5	0·5768

5

x	y
0	1·0
0·1	1·1022
0·2	1·2085
0·3	1·3179
0·4	1·4296
0·5	1·5428

6

x	y
1·0	1·0
1·1	1·1871
1·2	1·3531
1·3	1·5033
1·4	1·6411
1·5	1·7688

7

x	y
0	0
0·1	0·1002
0·2	0·2015
0·3	0·3048
0·4	0·4110
0·5	0·5214

8

x	y
0	1·0
0·2	0·8562
0·4	0·8110
0·6	0·8465
0·8	0·9480
1·0	1·1037

9

x	y
0	1·0
0·1	0·9138
0·2	0·8512
0·3	0·8076
0·4	0·7798
0·5	0·7653

10

x	y
0	0·4
0·2	0·4259
0·4	0·4374
0·6	0·4319
0·8	0·4085
1·0	0·3689

11

x	y
1·0	2·0
1·2	2·4197
1·4	2·8776
1·6	3·3724
1·8	3·9027
2·0	4·4677

12

x	y
0	1·0
0·2	1·1997
0·4	1·3951
0·6	1·5778
0·8	1·7358
1·0	1·8540

13

x	y
0	1·0
0·2	1·1679
0·4	1·2902
0·6	1·3817
0·8	1·4497
1·0	1·4983

14

x	y
0	1·0
0·1	1·11
0·2	1·2422
0·3	1·4013
0·4	1·5937
0·5	1·8271

15

x	y
0	3·0
0·1	2·88
0·2	2·5224
0·3	1·9368
0·4	1·1424
0·5	0·1683

16

x	y
0	0
0·2	0·1987
0·4	0·3897
0·6	0·5665
0·8	0·7246
1·0	0·8624

17

x	y
0	1·0
0·2	1·1972
0·4	1·3771
0·6	1·5220
0·8	1·6161
1·0	1·6487

18

x	y
0	2·0
0·1	2·0845
0·2	2·1367
0·3	2·1554
0·4	2·1407
0·5	2·0943

19

x	y
0	1·0
0·2	1·0367
0·4	1·1373
0·6	1·2958
0·8	1·5145
1·0	1·8029

20

x	y
1·0	0
1·2	0·1833
1·4	0·3428
1·6	0·4875
1·8	0·6222
2·0	0·7500

21

x	y
1·0	2·0000
1·2	2·0333
1·4	2·1121
1·6	2·2219
1·8	2·3522
2·0	2·4965

22

x	y
0·0	1·0000
0·2	0·8600
0·4	0·8118
0·6	0·8452
0·8	0·9454
1·0	1·1002

23

x	y
1·0	2·0000
1·2	2·4191
1·4	2·8769
1·6	3·3715
1·8	3·9018
2·0	4·4666

Test exercise 14 (page 479)

2 $145·7 \pm 2·6$ mm   **3** $5·8$ m/s   **4** $\dfrac{-2(x+y)}{2x+3y}$; $\dfrac{-2}{(2x+3y)^3}$

5 $\dfrac{x}{2(x^2-y^2)}$; $\dfrac{-y}{4(x^2-y^2)}$; $\dfrac{-y}{2(x^2-y^2)}$; $\dfrac{x}{4(x^2-y^2)}$

6 (a) $(-1, 1)$, saddle; $(-1, -\frac{4}{3})$, min   (b) an infinity of maxima along the line
$y = 5x/2$ when $z = 4$ **7** $1·10$ m $\times$ $1·10$ m $\times$ $0·825$ m high

8 $u = \dfrac{8}{7}$, $x = \dfrac{6}{7}$, $y = -\dfrac{4}{7}$, $z = \dfrac{2}{7}$

Further problems 14 (page 480)

1 $(8x \cos x - 6y \sin x)/J$; $-(4x^3 \cos y + 6x \sin y)/J$;
$J = 4x \cos x \sin y + 2x^2 y \sin x \cos y$ **2** $e^{3y}/2(xe^{3y} + e^{-3y})$; $e^{-3y}/2(xe^{3y} + e^{-3y})$;
$-1/3(xe^{3y} + e^{-3y})$; $x/3(xe^{3y} + e^{-3y})$

5 $(2e^{-x} \sinh 2x \sin 3y + 3ye^{-x} \cosh 2x \cos 3y)/(1 + 3y^2)$;
$\{-4ye^x \sinh 2x \sin 3y + 3e^x(1 + y^2) \cosh 2x \cos 3y\}/2(1 + 3y^2)$

7 (a) $(4, -4, -11)$, min   (b) $(1, -2, 4)$, saddle   (c) $(\frac{10}{7}, \frac{6}{7}, \frac{97}{7})$, max

8 $(0, 0)$, saddle; $(2, 0)$, min; $(-2, 0)$, min   **9** $(2, 1)$, max; $(-\frac{2}{3}, -\frac{1}{3})$, min

10 $(0, 0)$; $(3, 3)$; $(-3, -3)$, all saddle points

11 (a) $(1, 0)$, saddle; $(1, 1)$, min; $(-2, \frac{1}{2})$, saddle; $(-\frac{7}{5}, \frac{1}{5})$, max

(b) $(0, 0)$, max; $(1, 1)$; $(1, -1)$; $(-1, 1)$; $(-1, -1)$, all four saddle points

12 (a) A point of inflexion at the origin   (b) An infinity of maxima along the line $y = x/4$ when $z = 6$ (c) The value of z ranges from -1 to 1 and has an infinity of stationary points lying on the circles $x^2 + y^2 = n\pi$. When n is even the stationary points are maxima and when n is odd the stationary points are minima. There is also a single maximum at $(0, 0, 1)$

13 $x = 66.7$ mm; $\theta = \dfrac{\pi}{3}$ **14** $l = h = \dfrac{1}{5\pi}\sqrt[3]{60\pi^2 V}$; $d = l\sqrt{5}$ **15** $l = 1.00$ cm;

$d = 4.48$ cm; $\theta = 48° 11'$ **16**   cube of side $\dfrac{2r}{\sqrt{3}}$; $V_{\max} = \dfrac{8r^3}{3\sqrt{3}}$

17 (a) $u = \dfrac{64}{27}$; $x = y = z = \pm\dfrac{2}{\sqrt{3}}$; $u = \dfrac{9}{7}$, $x = y = \pm\dfrac{3}{\sqrt{14}}$ (b) $u = 9$;

$x = \pm\dfrac{3}{\sqrt{2}}$, $y = \mp\dfrac{3}{\sqrt{2}}$

Test exercise 15 (page 516)

1 (a) $u = 2x^4(t - 2) + 4xt + e^{2t}$ (b) $u = 2\sin 2x \cdot (e^y - 1) + \sin x + y^2$

2 $u(x, t) = \dfrac{16}{\pi^2}\displaystyle\sum_{r=1}^{\infty}\dfrac{1}{r^2}\cdot\sin\dfrac{r\pi}{2}\cdot\sin\dfrac{r\pi x}{10}\cdot\cos\dfrac{r\pi t}{10}$

3 $u(x, t) = \dfrac{100}{\pi}\displaystyle\sum_{r=1}^{\infty}(-1)^{r+1}\cdot\dfrac{1}{r}\sin\dfrac{\lambda x}{c}\cdot e^{-\lambda^2 t}$ where $\lambda = \dfrac{r\pi c}{2}$

4 $u(x, y) = \displaystyle\sum_{r=1}^{\infty}\dfrac{20}{r\pi}\cdot\operatorname{cosech} r\pi\cdot\sin\dfrac{r\pi x}{2}\cdot\sinh\dfrac{r\pi y}{2}$ with $r = 1, 3, 5, \ldots$

5 $v(r, \theta) = 5r^3\cos 3\theta$

Further problems 15 (page 517)

2 $u(x, t) = \dfrac{32}{\pi^3}\displaystyle\sum_{r=1}^{\infty}\dfrac{1}{r^3}\cdot\sin\dfrac{r\pi x}{2}\cdot\cos\dfrac{3r\pi t}{2}$ (r odd)

3 $u(x, t) = \dfrac{2}{25\pi^2}\displaystyle\sum_{r=1}^{\infty}\dfrac{1}{r^2}\cdot\sin\dfrac{r\pi}{2}\cdot\sin\dfrac{r\pi x}{4}\cdot\cos\dfrac{5r\pi t}{2}$

4 $u(x, t) = \dfrac{25}{2\pi^2}\displaystyle\sum_{r=1}^{\infty}\dfrac{1}{r^2}\cdot\sin\dfrac{r\pi}{5}\cdot\sin\dfrac{r\pi x}{10}\cdot\cos\dfrac{cr\pi t}{10}$

5 $u(x, t) = \dfrac{800}{\pi^3}\displaystyle\sum_{r=1}^{\infty}\dfrac{1}{r^3}\cdot\sin\dfrac{r\pi x}{10}\cdot e^{-4\lambda^2 t}$ with $r = 1, 3, 5, \ldots$ where $\lambda = \dfrac{r\pi}{10}$

6 $u(x, t) = \dfrac{16}{\pi^2}\displaystyle\sum_{r=1}^{\infty}\dfrac{1}{r^2}\cdot\sin\dfrac{r\pi}{2}\cdot\sin\dfrac{r\pi x}{10}\cdot e^{-r^2 c^2\pi^2 t/100}$ with $r = 1, 3, 5, \ldots$

7 $u(x, y) = \dfrac{128}{\pi^3}\displaystyle\sum_{r=1}^{\infty}\dfrac{1}{r^3}\cdot\operatorname{cosech}\dfrac{r\pi}{2}\cdot\sinh\dfrac{r\pi}{4}(2 - y)\cdot\sin\dfrac{r\pi x}{4}$ with $r = 1, 3, 5, \ldots$

8 $u(x, y) = \dfrac{200}{\pi^3}\displaystyle\sum_{r=1}^{\infty}\dfrac{1}{r^3}\cdot\operatorname{cosech}\dfrac{2r\pi}{5}\cdot\sin\dfrac{r\pi x}{5}\cdot\sinh\dfrac{r\pi}{5}(y - 2)$ with $r = 1, 3, 5, \ldots$

9 $v(r, \theta) = -4r\cos\theta + r^2\sin 2\theta$ **10** $v(r, \theta) = 3(1 - r^2\cos 2\theta)$

Test exercise 16 (page 560)

1 (a) solutions unique (b) infinite number of solutions **2** $x_1 = -4$, $x_2 = 2$, $x_3 = -3$ **3** $x_1 = -2$, $x_2 = 2$, $x_3 = 3$ **4** $x_1 = -3$, $x_2 = 4$, $x_3 = -2$

5 $x_1 = 1$, $x_2 = -2$, $x_3 = 2$

6 $a = 2$, $b = -1$, $c = 5$, $d = 0$, $e = 4$

7 (a) $\begin{bmatrix} -8 \\ 1 \end{bmatrix}$ (b) $\begin{bmatrix} 7.196 \\ -0.464 \end{bmatrix}$

Further problems 16 (page 561)

1 $x_1 = 1$, $x_2 = -4$, $x_3 = 3$ **2** (a) $x_1 = 3$, $x_2 = 1$, $x_3 = -4$ (b) $x_1 = 4$, $x_2 = -2$, $x_3 = -1$ **3** (a) $x_1 = 4$, $x_2 = 2$, $x_3 = 5$, $x_4 = 3$ (b) $x_1 = 5$, $x_2 = -4$, $x_3 = 1$, $x_4 = 3$ (c) $x_1 = 3$, $x_2 = -2$, $x_3 = 0$, $x_4 = 5$
4 (a) $x_1 = -3$, $x_2 = 1$, $x_3 = 3$ (b) $x_1 = 5$, $x_2 = 2$, $x_3 = -1$ (c) $x_1 = 4$, $x_2 = 3$, $x_3 = -1$, $x_4 = -2$

Test exercise 17 (page 591)

1 $\lambda_1 = 1$, $\lambda_2 = -2$, $\lambda_3 = 3$. $x_1 = \begin{bmatrix} 1 \\ 0 \\ -1 \end{bmatrix}$; $x_2 = \begin{bmatrix} 1 \\ -1 \\ 3 \end{bmatrix}$; $x_3 = \begin{bmatrix} 3 \\ 2 \\ -1 \end{bmatrix}$

2 $\mathbf{M} = \begin{bmatrix} 1 & 1 \\ -2 & 1 \end{bmatrix}$; $\mathbf{M}^{-1} = \begin{bmatrix} 1/3 & -1/3 \\ 2/3 & 1/3 \end{bmatrix}$; $\mathbf{M}^{-1}\mathbf{AM} = \begin{bmatrix} -1 & 0 \\ 0 & 5 \end{bmatrix}$

3 $f_1(x) = -\dfrac{10}{3}e^{6x} + \dfrac{1}{3}e^{3x}$; $f_2(x) = \dfrac{5}{3}e^{6x} + \dfrac{1}{3}e^{3x}$

4 $f_1(x) = \dfrac{1}{3}\cos\sqrt{5}x + \dfrac{4}{3\sqrt{5}}\sin\sqrt{5}x + \dfrac{2}{3}\cosh 2x + \dfrac{1}{3}\sinh 2x$

$f_2(x) = -\dfrac{1}{3}\cos\sqrt{5}x - \dfrac{4}{3\sqrt{5}}\sin\sqrt{5}x + \dfrac{1}{3}\cosh 2x + \dfrac{1}{6}\sinh 2x$

Further problems 17 (page 591)

1 (a) $f_1(x) = \dfrac{1}{4}(5e^x - e^{-3x})$; $f_2(x) = \dfrac{1}{4}(e^x - e^{-3x})$

(b) $f_1(x) = \dfrac{9}{5}(e^{-6x} - e^{4x})$; $f_2(x) = -\dfrac{1}{5}(e^{-6x} + 9e^{4x})$

(c) $f_1(x) = \dfrac{1}{2}(5e^{4x} - 3e^{2x})$; $f_2(x) = \dfrac{2}{3}e^x - \dfrac{3}{2}e^{2x} + \dfrac{5}{6}e^{4x}$;

$f_3(x) = 4e^x - \dfrac{9}{2}e^{2x} + \dfrac{5}{2}e^{4x}$ (d) $f_1(x) = 3e^{-2x} - e^{4x} + 2e^{7x}$;

$f_2(x) = -e^{-2x} - \dfrac{4}{3}e^{4x} + \dfrac{1}{3}e^{7x}$; $f_3(x) = e^{-2x} - \dfrac{5}{3}e^{4x} - \dfrac{1}{3}e^{7x}$ **2** $\lambda = 0, 7, 13$

3 (a) $f_1(x) = \dfrac{3}{5}\cosh\sqrt{2}x + \dfrac{9}{5\sqrt{2}}\sinh\sqrt{2}x + \dfrac{2}{5}\cosh\sqrt{7}x + \dfrac{11}{5\sqrt{7}}\sinh\sqrt{7}x$;

$f_2(x) = -\dfrac{2}{5}\cosh\sqrt{2}x - \dfrac{6}{5\sqrt{2}}\sinh\sqrt{2}x + \dfrac{2}{5}\cosh\sqrt{7}x + \dfrac{11}{5\sqrt{7}}\sinh\sqrt{7}x$

(b) $f_1(x) = -\dfrac{5}{12}\cos 2\sqrt{2}x + \dfrac{5}{12\sqrt{2}}\sin 2\sqrt{2}x + \dfrac{5}{12}\cosh 2x + \dfrac{1}{12}\sinh 2x;$

$f_2(x) = \dfrac{1}{6}\cos 2\sqrt{2}x - \dfrac{1}{6\sqrt{2}}\sin 2\sqrt{2}x + \dfrac{5}{6}\cosh 2x + \dfrac{1}{6}\sinh 2x$

(c) $f_1(x) = -\dfrac{35}{16}\cosh x + \dfrac{7}{16}\sinh x + \dfrac{35}{12}\cosh\sqrt{3}x - \dfrac{7}{12\sqrt{3}}\sinh\sqrt{3}x$

$+ \dfrac{13}{48}\cosh 3x + \dfrac{7}{144}\sinh 3x;\quad f_2(x) = \dfrac{5}{16}\cosh x - \dfrac{1}{16}\sinh x + \dfrac{5}{12}\cosh\sqrt{3}x$

$-\dfrac{1}{12\sqrt{3}}\sinh\sqrt{3}x + \dfrac{13}{48}\cosh 3x + \dfrac{7}{144}\sinh 3x;\quad f_3(x) = \dfrac{25}{16}\cosh x - \dfrac{5}{16}\sinh x$

$-\dfrac{35}{12}\cosh\sqrt{3}x + \dfrac{7}{12\sqrt{3}}\sinh\sqrt{3}x + \dfrac{65}{48}\cosh 3x + \dfrac{35}{144}\sinh 3x$

(d) $f_1(x) = -\dfrac{9}{8}\cos x + \dfrac{9}{4}\sin x + \dfrac{19}{10}\cos\sqrt{3}x - \dfrac{12}{5\sqrt{3}}\sin\sqrt{3}x + \dfrac{9}{40}\cosh\sqrt{7}x$

$+\dfrac{3}{20\sqrt{7}}\sinh\sqrt{7}x;\quad f_2(x) = \dfrac{15}{16}\cos x - \dfrac{15}{8}\sin x - \dfrac{19}{20}\cos\sqrt{3}x + \dfrac{6}{5\sqrt{3}}\sin\sqrt{3}x$

$+\dfrac{81}{80}\cosh\sqrt{7}x + \dfrac{27}{40\sqrt{7}}\sinh\sqrt{7}x;\quad f_3(x) = -\dfrac{3}{8}\cos x + \dfrac{3}{4}\sin x + \dfrac{3}{8}\cosh\sqrt{7}x$

$+\dfrac{1}{4\sqrt{7}}\sinh\sqrt{7}x$

Test exercise 18 (page 636)

1 $f(1/4, 1/3) = -19/12,\ f(1/2, 1/3) = -5/6,\ f(3/4, 1/3) = -1/12,$
$f(1/4, 2/3) = 1/12,\ f(1/2, 2/3) = 5/6,\ f(3/4, 2/3) = 19/12$

2 $f(1/3, 1/3) = 4,\ f(2/3, 1/3) = 17/3,\ f(1, 1/3) = 26/3,\ f(1/3, 2/3) = 2/3,$
$f(2/3, 2/3) = 3,\ f(1, 2/3) = 16/3$ **3** (a) parabolic (b) hyperbolic
(c) parabolic (d) hyperbolic (e) elliptic **4** $f(1/3, 1/3) = -1\cdot61728,$
$f(2/3, 1/3) = -1\cdot18519,\ f(1, 1/3) = -0\cdot82716,\ f(1/3, 2/3) = -1\cdot61728,$
$f(2/3, 2/3) = -1\cdot18519,\ f(1, 2/3) = -0\cdot82716$

5

t\x	0·0	0·2	0·4	0·6	0·8	1·0	1·2
0·00	0·00000	0·04000	0·16000	0·36000	0·64000	1·00000	0·89000
0·02	0·00000	0·08000	0·20000	0·40000	0·68000	0·76500	0·93000
0·04	0·00000	0·10000	0·24000	0·44000	0·58250	0·80500	0·83250
0·06	0·00000	0·12000	0·27000	0·41125	0·62250	0·70750	0·87250
0·08	0·00000	0·13500	0·26563	0·44625	0·55938	0·74750	0·80938
0·10	0·00000	0·13281	0·29063	0·41250	0·59688	0·68438	0·84688
0·12	0·00000	0·14531	0·27266	0·44375	0·54844	0·72188	0·79844
0·14	0·00000	0·13633	0·29453	0·41055	0·58281	0·67344	0·83281
0·16	0·00000	0·14727	0·27344	0·43867	0·54199	0·70781	0·79199

6

t\x	0·00	0·20	0·40	0·60	0·80	1·00
0·000	1·000000	0·840000	0·760000	0·760000	0·840000	1·000000
0·040	1·000000	0·898182	0·832727	0·832727	0·898182	1·000000
0·080	1·000000	0·929917	0·886942	0·886942	0·929917	1·000000
0·120	1·000000	0·952517	0·923125	0·923125	0·952517	1·000000
0·160	1·000000	0·967729	0·94779	0·94779	0·967729	1·000000
0·200	1·000000	0·978081	0·964533	0·964533	0·978081	1·000000

Further problems 18 (page 637)

1

x\y	0·00	0·33	0·67	1·00
0·00	−3·0000	−2·3333	−1·6667	−1·0000
0·25	−2·7500	−2·0833	−1·4167	−0·7500
0·50	−2·5000	−1·8333	−1·1667	−0·5000
0·75	−2·2500	−1·5833	−0·9167	−0·2500
1·00	−2·0000	−1·3333	−0·6667	0·0000

2

x\y	0·00	0·33	0·67	1·00
0·00	4·0000	7·3333	10·6667	14·0000
0·33	6·3333	9·6667	13·0000	16·3333
0·67	8·6667	12·0000	15·3333	18·6667
1·00	11·0000	14·3333	17·6667	21·0000

3

x\y	0·00	0·33	0·67	1·00
0·00	−1·0000	−1·0000	−1·0000	−1·0000
0·33	−0·6667	−0·7500	−0·8000	−0·8333
0·67	−0·3333	−0·5000	−0·6000	−0·6667
1·00	0·0000	−0·2500	−0·4000	−0·5000

4

x\y	0·00	0·33	0·67	1·00
0·00	0·0000	0·0000	0·0000	0·0000
0·25	0·0000	−0·0069	−0·0694	−0·1875
0·50	0·0000	0·0278	−0·0556	−0·2500
0·75	0·0000	0·1042	0·0417	−0·1875
1·00	0·0000	0·2222	0·2222	0·0000

5

x\y	0·00	0·33	0·67	1·00
0·00	15·0000	16·6667	18·3333	20·0000
0·33	17·3333	19·0000	20·6667	22·3333
0·67	19·6667	21·3333	23·0000	24·6667
1·00	22·0000	23·6667	25·3333	27·0000

6

x\y	0·00	0·33	0·67	1·00
0·00	21·0000	20·0000	19·0000	18·0000
0·33	22·6667	21·6667	20·6667	19·6667
0·67	24·3333	23·3333	22·3333	21·3333
1·00	26·0000	25·0000	24·0000	23·0000

7

x\y	0·00	0·33	0·67	1·00
0·00	4·0000	4·0000	4·0000	4·0000
0·33	4·2222	4·1111	3·7778	3·2222
0·67	4·8889	4·6667	4·0000	2·8889
1·00	6·0000	5·6667	4·6667	3·0000

8

x\y	0·00	0·33	0·67	1·00
0·00	0·0000	0·0000	0·0000	0·0000
0·33	0·0000	0·0000	−0·0741	−0·2963
0·67	0·0000	0·0741	0·0000	−0·3704
1·00	0·0000	0·2963	0·3704	0·0000

9

x\y	0·00	0·33	0·67	1·00
0·00	0·0000	−0·5556	−2·2222	−5·0000
0·33	0·3333	−0·2222	−1·8889	−4·6667
0·67	1·3333	0·7778	−0·8889	−3·6667
1·00	3·0000	2·4444	0·7778	−2·0000

10

x\y	0·00	0·33	0·67	1·00
0·00	−1·0000	−1·0000	−1·0000	−1·0000
0·33	−1·0000	−0·7037	−0·3333	0·1111
0·67	−1·0000	−0·3333	0·4815	1·4444
1·00	−1·0000	0·1111	1·4444	3·0000

11

x\y	0·00	0·33	0·67	1·00
0·00	0·0000	0·0000	0·0000	0·0000
0·33	0·1111	0·1050	0·0873	0·0600
0·67	0·4444	0·4200	0·3493	0·2401
1·00	1·0000	0·9450	0·7859	0·5403

12

x\y	0·00	0·33	0·67	1·00
0·00	0·0000	0·0370	0·2963	1·0000
0·33	0·0370	0·1481	0·5556	1·4815
0·67	0·2963	0·5556	1·1852	2·4074
1·00	1·0000	1·4815	2·4074	4·0000

13

x\y	0·00	0·33	0·67	1·00
0·00	0·0000	0·0000	0·0000	0·0000
0·33	0·0000	0·1111	0·2222	0·3333
0·67	0·0000	0·2222	0·4444	0·6667
1·00	0·0000	0·3333	0·6667	1·0000

14

x\y	0·00	0·33	0·67	1·00
0·00	0·0000	0·0000	0·0000	0·0000
0·33	0·0000	0·0000	−0·0741	−0·2222
0·67	0·0000	0·0741	0·0000	−0·2222
1·00	0·0000	0·2222	0·2222	0·0000

15

t\x	0·00	0·20	0·40	0·60	0·80	1·00
0·00	0·0000	−0·1600	−0·2400	−0·2400	−0·1600	0·0000
0·02	0·0400	−0·1200	−0·2000	−0·2000	−0·1200	0·0400
0·04	0·0800	−0·0800	−0·1600	−0·1600	−0·0800	0·0800
0·06	0·1200	−0·0400	−0·1200	−0·1200	−0·0400	0·1200
0·08	0·1600	0·0000	−0·0800	−0·0800	0·0000	0·1600
0·10	0·2000	0·0400	−0·0400	−0·0400	0·0400	0·2000
0·12	0·2400	0·0800	0·0000	0·0000	0·0800	0·2400
0·14	0·2800	0·1200	0·0400	0·0400	0·1200	0·2800
0·16	0·3200	0·1600	0·0800	0·0800	0·1600	0·3200
0·18	0·3600	0·2000	0·1200	0·1200	0·2000	0·3600
0·20	0·4000	0·2400	0·1600	0·1600	0·2400	0·4000

16

t\x	0·00	0·20	0·40	0·60	0·80	1·00
0·00	0·0000	0·1987	0·3894	0·5646	0·7174	0·8415
0·02	0·0000	0·1983	0·3886	0·5635	0·7159	0·8398
0·04	0·0000	0·1979	0·3879	0·5624	0·7145	0·8381
0·06	0·0000	0·1975	0·3871	0·5613	0·7131	0·8364
0·08	0·0000	0·1971	0·3863	0·5601	0·7116	0·8348
0·10	0·0000	0·1967	0·3855	0·5590	0·7102	0·8331
0·12	0·0000	0·1963	0·3848	0·5579	0·7088	0·8314
0·14	0·0000	0·1959	0·3840	0·5568	0·7074	0·8298
0·16	0·0000	0·1955	0·3832	0·5557	0·7060	0·8281
0·18	0·0000	0·1951	0·3825	0·5546	0·7046	0·8265
0·20	0·0000	0·1947	0·3817	0·5535	0·7032	0·8248

17

t\x	0·00	0·20	0·40	0·60	0·80	1·00
0·00	0·0000	0·3830	0·7596	1·1239	1·4698	1·7916
0·02	0·0000	0·3798	0·7534	1·1147	1·4578	1·7770
0·04	0·0000	0·3767	0·7473	1·1056	1·4459	1·7624
0·06	0·0000	0·3736	0·7412	1·0966	1·4341	1·7481
0·08	0·0000	0·3706	0·7351	1·0876	1·4223	1·7338
0·10	0·0000	0·3676	0·7291	1·0787	1·4107	1·7196
0·12	0·0000	0·3646	0·7232	1·0699	1·3992	1·7056
0·14	0·0000	0·3616	0·7173	1·0612	1·3878	1·6916
0·16	0·0000	0·3586	0·7114	1·0525	1·3764	1·6778
0·18	0·0000	0·3557	0·7056	1·0439	1·3652	1·6641
0·20	0·0000	0·3528	0·6998	1·0354	1·3541	1·6505

18

t\x	0·00	0·20	0·40	0·60	0·80	1·00
0·00	−1·0000	−0·7600	−0·4400	−0·0400	0·4400	1·0000
0·04	−0·9200	−0·6800	−0·3600	0·0400	0·5200	1·0800
0·08	−0·8400	−0·6000	−0·2800	0·1200	0·6000	1·1600
0·12	−0·7600	−0·5200	−0·2000	0·2000	0·6800	1·2400
0·16	−0·6800	−0·4400	−0·1200	0·2800	0·7600	1·3200
0·20	−0·6000	−0·3600	−0·0400	0·3600	0·8400	1·4000
0·24	−0·5200	−0·2800	0·0400	0·4400	0·9200	1·4800
0·28	−0·4400	−0·2000	0·1200	0·5200	1·0000	1·5600
0·32	−0·3600	−0·1200	0·2000	0·6000	1·0800	1·6400
0·36	−0·2800	−0·0400	0·2800	0·6800	1·1600	1·7200
0·40	−0·2000	0·0400	0·3600	0·7600	1·2400	1·8000
0·44	−0·1200	0·1200	0·4400	0·8400	1·3200	1·8800
0·48	−0·0400	0·2000	0·5200	0·9200	1·4000	1·9600
0·52	0·0400	0·2800	0·6000	1·0000	1·4800	2·0400
0·56	0·1200	0·3600	0·6800	1·0800	1·5600	2·1200
0·60	0·2000	0·4400	0·7600	1·1600	1·6400	2·2000

19

t\x	0·00	0·10	0·20	0·30	0·40	0·50	0·60	0·70	0·80	0·90	1·00
0·00	0·0000	−0·9000	−1·6000	−2·1000	−2·4000	−2·5000	−2·4000	−2·1000	−1·6000	−0·9000	0·0000
0·02	0·4000	−0·5000	−1·2000	−1·7000	−2·0000	−2·1000	−2·0000	−1·7000	−1·2000	−0·5000	0·4000
0·04	0·8000	−0·1000	−0·8000	−1·3000	−1·6000	−1·7000	−1·6000	−1·3000	−0·8000	−0·1000	0·8000
0·06	1·2000	0·3000	−0·4000	−0·9000	−1·2000	−1·3000	−1·2000	−0·9000	−0·4000	0·3000	1·2000
0·08	1·6000	0·7000	0·0000	−0·5000	−0·8000	−0·9000	−0·8000	−0·5000	0·0000	0·7000	1·6000
0·10	2·0000	1·1000	0·4000	−0·1000	−0·4000	−0·5000	−0·4000	−0·1000	0·4000	1·1000	2·0000
0·12	2·4000	1·5000	0·8000	0·3000	0·0000	−0·1000	0·0000	0·3000	0·8000	1·5000	2·4000
0·14	2·8000	1·9000	1·2000	0·7000	0·4000	0·3000	0·4000	0·7000	1·2000	1·9000	2·8000

20

t\x	0·00	0·10	0·20	0·30	0·40	0·50	0·60	0·70	0·80	0·90	1·00
0·00	0·0000	30·9017	58·7785	80·9017	95·1057	100·0000	95·1057	80·9017	58·7785	30·9017	0·0000
0·04	0·0000	20·8224	39·6065	54·5136	64·0846	67·3825	64·0846	54·5136	39·6065	20·8224	0·0000
0·08	0·0000	14·0306	26·6878	36·7327	43·1818	45·4041	43·1818	36·7327	26·6878	14·0306	0·0000
0·12	0·0000	9·4542	17·9829	24·7514	29·0970	30·5944	29·0970	24·7514	17·9829	9·4542	0·0000
0·16	0·0000	6·3705	12·1174	16·6781	19·6063	20·6153	19·6063	16·6781	12·1174	6·3705	0·0000
0·20	0·0000	4·2926	8·1650	11·2381	13·2112	13·8911	13·2112	11·2381	8·1650	4·2926	0·0000
0·24	0·0000	2·8925	5·5018	7·5725	8·9021	9·3602	8·9021	7·5725	5·5018	2·8925	0·0000
0·28	0·0000	1·9490	3·7072	5·1026	5·9984	6·3071	5·9984	5·1026	3·7072	1·9490	0·0000
0·32	0·0000	1·3133	2·4980	3·4382	4·0419	4·2499	4·0419	3·4382	2·4980	1·3133	0·0000
0·36	0·0000	0·8849	1·6832	2·3168	2·7235	2·8637	2·7235	2·3168	1·6832	0·8849	0·0000
0·40	0·0000	0·5963	1·1342	1·5611	1·8352	1·9296	1·8352	1·5611	1·1342	0·5963	0·0000
0·44	0·0000	0·4018	0·7643	1·0519	1·2366	1·3002	1·2366	1·0519	0·7643	0·4018	0·0000
0·48	0·0000	0·2707	0·5150	0·7088	0·8332	0·8761	0·8332	0·7088	0·5150	0·2707	0·0000
0·52	0·0000	0·1824	0·3470	0·4776	0·5615	0·5904	0·5615	0·4776	0·3470	0·1824	0·0000
0·56	0·0000	0·1229	0·2338	0·3218	0·3783	0·3978	0·3783	0·3218	0·2338	0·1229	0·0000
0·60	0·0000	0·0828	0·1576	0·2169	0·2549	0·2680	0·2549	0·2169	0·1576	0·0828	0·0000

Test exercise 19 (page 689)

1 (a) $dz = 4x^3 \cos 3y\, dx - 3x^4 \sin 3y\, dy$ (b) $dz = 2e^{2y}\{2\cos 4x\, dx + \sin 4x\, dy\}$
(c) $dz = xw^2\{2yw\, dx + xw\, dy + 3xy\, dw\}$ 2 (a) $z = x^3y^4 + 4x^2 - 5y^3$
(b) $z = x^2 \cos 4y + 2\cos 3x + 4y^2$ (c) not exact differential
3 9 square units 4 (a) 278·6 (b) $\pi/2$ (c) 22·5 (d) 48 (e) -21
(f) -54π 5 Area $= \frac{5}{12}$ square units 6 (a) 2 (b) 0

Further problems 19 (page 690)

1 14 2 1·6 3 $\frac{\pi}{36}\{9 - 4\sqrt{3}\}$ 4 $\frac{1}{2}\{\pi^4 + 4\}$ 5 $\frac{9\pi}{256}$ 6 $\frac{1}{2} \cdot \ln 2$

7 $2 - \pi/2$ 8 $\frac{1}{8}$ 9 14 10 (a) 39·24 (b) 0 11 $\frac{2}{3}$

Test exercise 20 (page 732)

1 $4\sqrt{2}\pi$ 2 $a(\pi/2)^2$ 3 (a) (1) $(4·47, 0·464, 3)$ (2) $(5·92, 0·564, 0·322)$
(b) (1) $(3·54, 3·54, 3)$ (2) $(-0·832, 1·82, 3·46)$ 4 12π

5 $a^3(8 - 3a)\pi/12$ 6 (a) $I = \iint v(1 + u)(1 + u + v)\, dv\, dv$

(b) $I = \iiint \frac{(2u + v)(v - 4w)}{vw}\, du\, dv\, dw$

Further problems 20 (page 732)

1 $4\sqrt{5}\pi$ 2 $\left(\frac{a}{2}, \frac{a}{2}, \frac{a}{2}\right)$ 3 $10\sqrt{61}$ 4 $\frac{4\sqrt{22}\pi}{3}$ 5 $\frac{\pi}{24}(5\sqrt{5} - 1)$ 6 $\pi\sqrt{5}$

7 $16a^2$ **8** $2a^2(\pi - 2)$ **9** $4\pi(a + b)\sqrt{a^2 - b^2}$ **10** 45π **11** $\dfrac{11}{30}$ **12** $\dfrac{\pi a^4}{2}$

13 $2\left(\pi - \dfrac{4}{3}\right)$ **14** $\bar{x} = \bar{y} = \bar{z} = \dfrac{3a}{8}$ **15** $\dfrac{\pi a^3}{3}\{4\sqrt{2} - 3\}$ **16** $\dfrac{4\pi abc}{3}$ **17** $\dfrac{2a^3}{3}$

18 $\dfrac{1}{4}\displaystyle\iint (u^2 + v^2)\,du\,dv$ **19** $u^2 v\,du\,dv\,dw$ **20** $\bar{z} = -\dfrac{a}{5}$ **21** $\dfrac{7}{18}$ **22** $2 - \dfrac{\pi}{2}$

23 $\dfrac{1}{4}(\sqrt{2} - 1)$

Test exercise 21 (page 769)

1 (a) $\dfrac{20}{3}$ (b) $\dfrac{2}{3}$ (c) -2 (d) 120 (e) $\dfrac{15\sqrt{\pi}}{2048}$ **2** (a) $\dfrac{256}{315}$ (b) $\dfrac{1}{40}$ (c) $\dfrac{2}{105}$

3 (a) $\dfrac{1}{\sqrt{2}}\cdot K\left(\dfrac{1}{\sqrt{2}}\right)$ (b) $\dfrac{1}{2}\cdot F\left(\dfrac{1}{2}, \dfrac{\sqrt{3}}{2}\right)$ **4** (a) 0 (b) 1 **5** (a) $F\left(\sqrt{2}, \dfrac{\pi}{4}\right)$

  (b) $\dfrac{1}{2}F\left(\dfrac{1}{2}, \dfrac{\sqrt{3}}{2}\right)$

Further problems 21 (page 769)

1 (a) 6 (b) $-\dfrac{1}{2}$ (c) $0\cdot4$ (d) 24 (e) $\dfrac{315}{4}$ **2** (a) 6 (b) $\dfrac{8}{81}$ (c) $\dfrac{\sqrt{2\pi}}{16}$ (d) 4

4 (a) $\dfrac{1}{8960}$ (b) $\dfrac{\sqrt{2\pi}}{64}$ (c) $\dfrac{8}{315}$ (d) $\dfrac{2}{7}$ (e) $\dfrac{1}{63}$ (f) $\dfrac{\pi}{432} = 0\cdot00727$

8 (a) $\sqrt{5}\cdot E\left(\dfrac{2}{\sqrt{5}}\right)$ (b) $\sqrt{2}\cdot K\left(\dfrac{1}{\sqrt{2}}\right) = 2\cdot622$ (c) $2\cdot E\left(\dfrac{1}{2}, 1\right) = 2\cdot935$

  (d) $\dfrac{1}{4}\cdot F\left(\dfrac{3}{4}, \dfrac{2}{3}\right) = 0\cdot193$ (e) $\dfrac{1}{\sqrt{5}}\cdot F\left(\dfrac{2}{\sqrt{5}}, 1\right)$ (f) $\dfrac{1}{\sqrt{2}}\cdot F\left(\dfrac{1}{\sqrt{2}}, \dfrac{\pi}{6}\right)$

  (g) $\dfrac{1}{\sqrt{2}}\cdot\left\{F\left(\dfrac{1}{\sqrt{2}}, \dfrac{\pi}{3}\right) - F\left(\dfrac{1}{\sqrt{2}}, \dfrac{\pi}{4}\right)\right\}$ **9** $\dfrac{1}{2}\cdot\left\{F\left(\dfrac{\sqrt{3}}{2}, \dfrac{\pi}{2}\right) - F\left(\dfrac{\sqrt{3}}{2}, \dfrac{\pi}{4}\right)\right\}$

10 (a) $\dfrac{1}{\sqrt{3}}\cdot F\left(\dfrac{1}{\sqrt{3}}, \dfrac{1}{2}\right) = 0\cdot307$ (b) $\dfrac{1}{\sqrt{3}}\cdot\left\{F\left(\dfrac{1}{\sqrt{3}}, 1\right) - F\left(\dfrac{1}{\sqrt{3}}, \dfrac{1}{2}\right)\right\}$

  (c) $\dfrac{1}{\sqrt{34}}\cdot K\left(\dfrac{3}{\sqrt{34}}\right) = 0\cdot2905$ (d) $\dfrac{1}{\sqrt{7}}\left\{F\left(\sqrt{\dfrac{3}{7}}, \dfrac{\pi}{2}\right) - F\left(\sqrt{\dfrac{3}{7}}, \dfrac{\pi}{6}\right)\right\}$

Text exercise 22 (page 815)

1 (a) -15 (b) $-16\mathbf{i} + 10\mathbf{j} + 17\mathbf{k}$ **2** (a) 9 (b) $-(47\mathbf{i} + 17\mathbf{j} + 29\mathbf{k})$

3 $\mathbf{A}\cdot(\mathbf{B}\times\mathbf{C}) = 0$ $\therefore$ vectors coplanar **4** (a) $4\mathbf{i} - 4\mathbf{j} + 24\mathbf{k}$

  (b) $2\mathbf{i} - 2\mathbf{j} + 24\mathbf{k}$ (c) $24\cdot66$ **5** $\mathbf{T} = \dfrac{1}{\sqrt{66}}(4\mathbf{i} + \mathbf{j} + 7\mathbf{k})$ **6** $\dfrac{8}{5}(25\mathbf{i} - 6\mathbf{j} - 15\mathbf{k})$

7 $5\cdot08$ **8** $\dfrac{1}{\sqrt{101}}(2\mathbf{i} + 4\mathbf{j} + 9\mathbf{k})$ **9** (a) $14\mathbf{i} - 12\mathbf{j} - 30\mathbf{k}$ (b) 8

  (c) $5\mathbf{i} - 2\mathbf{j} - 4\mathbf{k}$ (d) $7\mathbf{i} + 2\mathbf{j} + 3\mathbf{k}$ (e) $3\mathbf{i} + 2\mathbf{j} + \mathbf{k}$

Further problems 22 (page 815)

1 61 2 $29\mathbf{i} - 10\mathbf{j} + 16\mathbf{k}$ 3 (a) $22\mathbf{i} + 14\mathbf{j} + 2\mathbf{k}$ (b) $-2\mathbf{i} + 14\mathbf{j} - 22\mathbf{k}$

4 (a) $2x\mathbf{i} + 3\mathbf{j} + \cos x\,\mathbf{k}$ (b) $2\mathbf{i} - \sin x\,\mathbf{k}$ (c) $(4x^2 + 9 + \cos^2 x)^{1/2}$
   (d) $34 + \sin 2$ 5 (a) $2 - 2u - 9u^2$
   (b) $(3u^2 + 4u + 3)\mathbf{i} + (3u^2 + 6)\mathbf{j} + (1 - 2u)\mathbf{k}$ (c) $\mathbf{i} - 2\mathbf{j} + (3 - 2u)\mathbf{k}$

6 $\dfrac{1}{5\sqrt{21}}(2\mathbf{i} - 20\mathbf{j} + 11\mathbf{k})$ 7 $\dfrac{-1}{\sqrt{129}}(10\mathbf{i} + 2\mathbf{j} - 5\mathbf{k})$ 8 $\dfrac{-1}{\sqrt{126}}(5\mathbf{i} - \mathbf{j} + 10\mathbf{k})$

9 $\dfrac{-1}{\sqrt{601}}(12\mathbf{i} + 4\mathbf{j} - 21\mathbf{k})$ 10 $-8\cdot285$ 11 $-9\cdot165$ 12 (a) 15 (b) -33 (c) 7

13 (a) $-6\mathbf{i} + 4\mathbf{j} - 7\mathbf{k}$ (b) $62\mathbf{i} + 10\mathbf{j} - 38\mathbf{k}$ (c) $18\mathbf{i} - 21\mathbf{j} + 10\mathbf{k}$

14 (a) $12\mathbf{i} - 4\mathbf{j} + 4\mathbf{k}$ (b) $24\mathbf{i} - 4\mathbf{j}$ (c) 144

15 (a) $(2\sin 2)\mathbf{i} + 2e^3\mathbf{j} + (\cos 2 + e^3)\mathbf{k}$ (b) $(4\sin^2 2 + \cos^2 2 + 2e^3\cos 2 + 5e^6)^{1/2}$

16 $-5\cdot014$ 17 $p = \dfrac{1}{\sqrt{29}}(3\mathbf{i} + 2\mathbf{j} - 4\mathbf{k}); \ q = \dfrac{1}{\sqrt{38}}(6\mathbf{i} - \mathbf{j} + \mathbf{k}); \ \theta = 68° \, 48'$

18 (a) $(2t + 3)\mathbf{i} - (6\cos 3t)\mathbf{j} + 6e^{2t}\mathbf{k}$ (b) $2\mathbf{i} + (18\sin 3t)\mathbf{j} + 12e^{2t}\mathbf{k}$ (c) $12\cdot17$

20 $-4x\mathbf{i} + 4z\mathbf{k}$ 21 $(2\cos 5\cdot5)\mathbf{i} - (6\sin 5\cdot5)\mathbf{j} - (6\sin 5\cdot5)\mathbf{k}$ 22 $p = 6$

23 (a) (1) $p = 15/4$ (2) $p = -33$ (b) $\dfrac{1}{7}(3\mathbf{i} - 2\mathbf{j} + 6\mathbf{k})$

Test exercise 23 (page 865)

1 $3\mathbf{i} + \dfrac{18}{7}\mathbf{j} - \dfrac{81}{8}\mathbf{k}$ 2 12 3 $18\pi(2\mathbf{i} + \mathbf{j})$ 4 $24(\mathbf{i} + \mathbf{j})$ 5 $8 + \dfrac{4\pi}{3}$

6 all conservative 7 $36\left(\dfrac{\pi}{4} + 1\right)$ 8 0

Further problems 23 (page 866)

1 (a) $576\mathbf{k}$ (b) $\dfrac{576}{5}(3\mathbf{i} + \mathbf{j} + 2\mathbf{k})$ 2 $1771\mathbf{i} + 1107\mathbf{j} + 830\cdot4\mathbf{k}$

3 $416\cdot1\mathbf{i} + 718\cdot5\mathbf{j} + 5679\mathbf{k}$ 4 $46\cdot9$ 5 $-4\cdot18$ 6 8π 7 $\dfrac{16\pi}{3}(\mathbf{i} + \mathbf{k})$

8 $\dfrac{1}{3}(48\mathbf{i} + 64\mathbf{j} - 24\mathbf{k})$ 9 $64\left(\dfrac{\pi}{4} - \dfrac{1}{3}\right)(6\mathbf{i} + 4\mathbf{j})$

10 $\dfrac{9}{2}\{(\pi + 2)\mathbf{i} + (\pi + 2)\mathbf{j} + 4\mathbf{k}\}$ 11 $\dfrac{12}{5}(32\mathbf{j} + 15\mathbf{k})$ 12 -1 13 $\dfrac{250}{3}\pi$

14 $\dfrac{1}{6}(117\pi + 256 - 28\sqrt{7}) = 91\cdot58$ 15 -80 16 96π 17 -2 18 12π

19 $-\dfrac{a^3}{3}$ 20 $\dfrac{81\pi}{4}$

Test exercise 24 (page 893)

1 yes, an orthogonal set 2 $h_u = 1, \ h_v = 2v, \ h_\theta = 2u$ 3 $4\mathbf{I} + \mathbf{K}$

4 (a) $(2\cos\phi + 2\cos 2\phi + 1)$ (b) $(2\sin 2\phi + \sin\phi)\mathbf{K}$

5 (a) $(\mathrm{d}s)^2 = (\mathrm{d}r)^2 + r^2(\mathrm{d}\theta)^2 + r^2\sin^2\theta\,(\mathrm{d}\phi)^2$ (b) $\mathrm{d}V = r^2\sin\theta\,\mathrm{d}r\,\mathrm{d}\theta\,\mathrm{d}\phi$

6 $-10\cdot5$

Further problems 24 (page 894)

1 (a) yes (b) no   2 $-50 \cdot 5$ 3 $2\dfrac{5}{18}$

5 (a) $\nabla^2 V = \dfrac{\partial^2 V}{\partial \rho^2} + \dfrac{1}{\rho} \cdot \dfrac{\partial V}{\partial \rho} + \dfrac{1}{\rho^2} \cdot \dfrac{\partial^2 V}{\partial \phi^2} + \dfrac{\partial^2 V}{\partial z^2}$

   (b) $\nabla^2 V = \dfrac{1}{r^2} \cdot \dfrac{\partial}{\partial r} \left(r^2 \dfrac{\partial V}{\partial r} \right) + \dfrac{1}{r^2 \sin \theta} \cdot \dfrac{\partial}{\partial \theta} \left(\sin \theta \dfrac{\partial V}{\partial \theta} \right) + \dfrac{1}{r^2 \sin^2 \theta} \cdot \dfrac{\partial^2 V}{\partial \phi^2}$

6 (b) $h_u = h_v = \sqrt{u^2 + v^2};\ h_w = 1$

   (c) $\operatorname{div} F = \dfrac{1}{u^2 + v^2} \left\{ \dfrac{\partial}{\partial u} \left(\sqrt{u^2 + v^2} \cdot \dfrac{\partial F_u}{\partial u} \right) + \dfrac{\partial}{\partial v} \left(\sqrt{u^2 + v^2} \cdot \dfrac{\partial F_v}{\partial v} \right) \right\} + \dfrac{\partial F_w}{\partial w}$

   (d) $\nabla^2 V = \dfrac{1}{u^2 + v^2} \left\{ \dfrac{\partial^2 V}{\partial u^2} + \dfrac{\partial^2 V}{\partial v^2} \right\} + \dfrac{\partial^2 V}{\partial w^2}$

Test exercise 25 (page 932)

1 (a) $w = 6 - j2$ (b) $w = 3 - j2$ (c) $w = j3$ (d) $w = 2$
2 Magnification $= 2 \cdot 236$; rotation $= 63° \, 26'$; translation $= 1$ unit to right,

   3 units downwards

3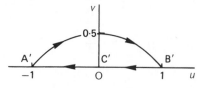

   $v = \dfrac{1}{2}(1 - u^2)$

4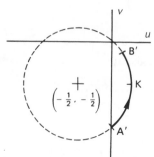

   Minor arc of circle, centre

   $\left(-\dfrac{1}{2}, -\dfrac{1}{2} \right)$, radius $\dfrac{1}{\sqrt{2}} = 0 \cdot 7071$,

5 (a) centre $\left(u = 0,\ v = \dfrac{2}{3} \right)$ (b) radius $\dfrac{1}{3}$ 6 centre $\left(u = \dfrac{2}{3},\ v = 0 \right)$; radius $\dfrac{2}{3}$

Further problems 25 (page 933)

1 Triangle A'B'C' with A' $(-1 + j2)$, B' $(5 + j2)$, C' $(2 + j5)$
2 (a) A' $(-8 + j9)$; B' $(23 + j14)$
   (b) Magnification $= \sqrt{29} = 5 \cdot 385$; rotation $= 68° 12'$; translation $=$ nil
3 Straight line joining A' $(5 - j7)$ to B' $(-3 - j)$; magnification $= 3 \cdot 162$; rotation
   $= 161° 34'$ anticlockwise; translation $= 2$ to right, 4 upwards

4

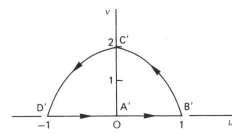

A′ B′: $v = 0$

B′C′: $u = 1 - \dfrac{v^2}{4}$

C′D′: $u = \dfrac{v^2}{4} - 1$

D′A′: $v = 0$

5

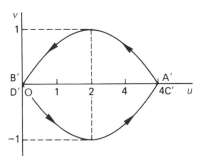

A′B′ and C′D′: $v = \dfrac{1}{4}(4u - u^2)$

B′C′ and D′A′: $v = \dfrac{1}{4}(u^2 - 4u)$

6 A′ $(1 - j2)$; B′ $(-23 + j10)$; C′ $(1 - j8)$ A′B′: $u = 2 - \dfrac{v^2}{4}$;

B′C′: $v = \dfrac{(u-1)^2}{32} - 8$; C′A′: $u = 1$ **7** circle, centre $\left(\dfrac{1}{2} - j\dfrac{2}{3}\right)$, radius $\dfrac{7}{6}$

8 (a) circle, centre $\left(\dfrac{1}{3} - j0\right)$, radius $\dfrac{2}{3}$ (b) region outside the circle in (a)

9 circle, centre $\left(\dfrac{3}{2} + j0\right)$, radius 1; clockwise development

10 circle, $u^2 + v^2 - \dfrac{22u}{5} + \dfrac{8}{5} = 0$, centre $\left(\dfrac{11}{5} + j0\right)$, radius $\dfrac{9}{5}$

11 circle, $u^2 + v^2 - \dfrac{u}{2} = 0$,   centre $\left(\dfrac{1}{4} + j0\right)$,   radius $\dfrac{1}{4}$; region inside this circle

12 circle, centre $\left(-\dfrac{7}{3} + j0\right)$, radius $\dfrac{5}{3}$

13 (a) circle, centre $\left(\dfrac{3}{5},\ 0\right)$, radius $\dfrac{2}{5}$, developed clockwise

(b) region outside the circle in (a)

14 $v = -\dfrac{u}{3}$

Test exercise 26 (page 980)

1 (a) regular at all points   (b) $z = -5$ (c) regular at all points
(d) $z = -1$ and $z = 4$ (e) $z = 0$, where $z = x + jy$
2 (a) $v(x, y) = \cosh x \cos y + C$ (b) $v(x, y) = 6(y^2 - x^2) - 4x + C$ **4** $j4\pi$
5 (a) $z = 0$ (b) $z = \pm1$ (c) no critical point   (d) $z = \pm\sqrt{2}$ (e) $z = 0$

(f) no critical point   **6** $w = \cosh\dfrac{\pi z}{4}$; D′: $w = 1$

Further problems 26 (page 981)

3   circle, centre $(5, -2)$, radius $\sqrt{2}$ 4   circle, centre $\left(-\dfrac{1}{3}, 0\right)$,

radius $\dfrac{2}{3}$, anticlockwise   5   (a) $v(x, y) = 2y(x - 1) + C$

(b) $v(x, y) = 3x^2 y - y^3 - 2xy + y + C$ (c) $v(x, y) = x^2 - 2x - y^2 + C$

(d) $v(x, y) = e^{x^2 - y^2} \sin 2xy + C$ 6   (a) $j10\pi$ (b) $j6\pi$ 7   (a) 0 (b) $j4\pi$

(c) $j10\pi$ 9 $j2\pi$ 10 $j10\pi$ 11   (a) (1) $z = 0$ (2) $z = \pm 1$

(b) ellipse, centre $(0, 0)$, semi major axis $\frac{5}{2}$, semi minor axis $\frac{3}{2}$

12   (a) $u^2 + v^2 = 1$ (b) $u^2 + (v - 1)^2 = 2$; $\theta = 45°$.   13   Unit circle becomes the
real axis on the w-plane. Region within the circle maps onto the upper half

plane   14 $w = \sin \dfrac{z\pi}{2a}$

Test exercise 27 (page 1010)

1   (a) $f(z) = 1 + z + \dfrac{z^2}{2!} + \dfrac{z^3}{3!} + \ldots + \dfrac{z^n}{n!} + \ldots$ valid for $|z| < \infty$

(b) $f(z) = 4z - \dfrac{(4z)^2}{2} + \dfrac{(4z)^3}{3} - \ldots + \dfrac{(-1)^{n+1}(4z)^n}{n} + \ldots$ valid for $|z| < 1/4$

2   (a) pole of order 5 at $z = -1$ (b) essential singularity at $z = 0$
(c) essential singularity at $z = 0$ (d) removable singularity at $z = 0$

3   $f(z) = \dfrac{1}{\sqrt{2}} \left\{ 1 + (z - \pi/4) - \dfrac{(z - \pi/4)^2}{2!} - \dfrac{(z - \pi/4)^3}{3!} \right.$

$\left. + \dfrac{(z - \pi/4)^4}{4!} + \dfrac{(z - \pi/4)^5}{5!} - \ldots \right\}$; valid for $|z| < \infty$

4   (a) $f(z) = -(z + 3) + 8 + \dfrac{1}{2(z + 3)} - \dfrac{4}{(z + 3)^2} - \dfrac{1}{24(z + 3)^3} + \dfrac{1}{3(z + 3)^4} + \ldots$;
essential singularity

(b) $f(z) = \dfrac{3}{z + 3} - \dfrac{1}{z + 1} = \ldots - \dfrac{1}{z^3} + \dfrac{1}{z^2} - \dfrac{1}{z} + 1 - \dfrac{z}{3} + \dfrac{z^2}{9} - \dfrac{z^3}{27} + \ldots$

(c) $f(z) = \dfrac{1}{8(z - 2)^2} - \dfrac{3}{16(z - 2)} + \dfrac{3}{16} - \dfrac{5(z - 2)}{32} + \dfrac{15(z - 2)^2}{64} + \ldots$; pole of order 2

5   double pole at $z = 0$; residue -4, double pole at $z = -1$, residue $7/2$, single pole
at $z = 1$, residue $1/2$ 6   (a) $-\pi/6$ (b) $2\pi/\sqrt{3}$
(c) $2\pi e^{-3}$

Further problems 27 (page 1011)

1   (a) $z + \dfrac{z^3}{3!} + \dfrac{z^5}{5!} + \ldots + \dfrac{z^{2n+1}}{(2n + 1)!} + \ldots$, $|z| < \infty$

(b) $z + \dfrac{z^3}{3} + \dfrac{2z^5}{15} + \dfrac{17z^7}{315} + \ldots$, $|z| < \pi/2$

(c) $2\left\{ z + \dfrac{z^3}{3} + \dfrac{z^5}{5} + \ldots + \dfrac{z^{2n+1}}{2n + 1} + \ldots \right\}$, $|z| < 1$

(d) $1 + z \ln a + \dfrac{z^2 (\ln a)^2}{2!} + \dfrac{z^3 (\ln a)^3}{3!} + \ldots + \dfrac{z^n (\ln a)^n}{n!} + \ldots, \; |z| < \infty$

(e) $\dfrac{3z^2}{25} + \dfrac{27z^3}{125} + \dfrac{162z^4}{625} + \dfrac{810z^5}{3125} + \ldots, \; |z| < 5/3;$

$-\dfrac{5}{9z} - \dfrac{25}{9z^2} - \dfrac{250}{27z^3} - \dfrac{6250}{243z^4} - \ldots, \; |z| > 5/3$ **3** (b) $-\dfrac{z}{(z+1)^2}, \dfrac{z(z-1)}{(z+1)^3}$

4 (a) convergent for $|z| < \infty$ (b) convergent for $|z| < 1$ (c) convergent for $|z| < 1$ (d) convergent for $|z| < 1$ (e) convergent for $|z| < \infty$

5 (a) $e^2 \left\{ 1 + (z - 2) + \dfrac{(z-2)^2}{2!} + \dfrac{(z-2)^3}{3!} + \ldots + \dfrac{(z-2)^n}{n!} + \ldots \right\}$

(b) $\dfrac{\sqrt{3}}{2} - \dfrac{(z - \pi/6)}{2} - \dfrac{\sqrt{3}(z - \pi/6)^2}{2 \times 2!} + \dfrac{(z - \pi/6)^3}{2 \times 3!} + \dfrac{\sqrt{3}(z - \pi/6)^4}{2 \times 4!} + \ldots$

(c) $(z - 3) \sin 6 + (z - 3)^2 \cos 6 - \dfrac{(z - 3)^3 \sin 6}{2!} - \dfrac{(z - 3)^4 \cos 6}{3!}$

$+ \dfrac{(z - 3)^5 \sin 6}{4!} + \ldots$ (d) $-\left\{ \dfrac{3}{13} + 2 \left(\dfrac{3}{13} \right)^2 (z - 1/3) \right.$

$\left. + 4 \left(\dfrac{3}{13} \right)^3 (z - 1/3)^2 + \ldots + 2^n \left(\dfrac{3}{13} \right)^{n+1} (z - 1/3)^n + \ldots \right\}$

(e) $1 - 2(z - 3) + 4(z - 3)^2 + \ldots + (-2)^n (z - 3)^n + \ldots$

6 $(z - 1) + \dfrac{(z-1)^2}{1 \times 2} - \dfrac{(z-1)^3}{2 \times 3} + \dfrac{(z-1)^4}{3 \times 4} - \dfrac{(z-1)^5}{4 \times 5} + \ldots$ **7** (a) $z = \infty$

(b) $|z| = \sqrt{6}$ (c) $|z - 5| = 1$ (d) $z = \infty$ **8** (a) poles of order 2 at $z = 0$ and $z = -1$, removable singularity at $z = \pm 1$ (b) essential singularity at $z = 0$ **9** (a) $\dfrac{1}{z^2} - \dfrac{1}{z^4 3!} + \dfrac{1}{z^6 5!} - \dfrac{1}{z^8 7!} + \ldots, \; |z| > 0$

(b) $\dfrac{1}{2} \left(z - \dfrac{3}{2} \right)^{-1}, \; |2z - 3| > 0$

(c) $\dfrac{3}{z - 3} - 2\{ 1 - (z - 3) + (z - 3)^2 - (z - 3)^3 + \ldots \}, \; 0 < |z - 3| < 1$

10 (a) $\ldots + \dfrac{8}{z^4} - \dfrac{4}{z^3} + \dfrac{2}{z^2} - \dfrac{1}{z} + \dfrac{2}{5} - \dfrac{2z}{25} + \dfrac{2z^2}{125} - \dfrac{2z^3}{625} + \ldots$

(b) $\dfrac{1}{z} - \dfrac{8}{z^2} + \dfrac{46}{z^3} - \dfrac{242}{z^4} + \ldots$ (c) $-\dfrac{1}{10} + \dfrac{17z}{100} - \dfrac{109z^2}{1000} + \dfrac{593z^3}{10000} - \ldots$

11 (a) $2\pi/\sqrt{3}$ (b) $\dfrac{2\pi}{\sqrt{\alpha^2 - \beta^2}}$ (c) $\dfrac{2\pi}{|\alpha^2 - 1|}$ (d) $\pi/4$ (e) $\pi/2$ (f) $\pi/2$

(g) $\pi \sqrt{\sqrt{13}/8 - 3/8}$ (h) $\pi/4$ (i) $2\pi/\sqrt{3}$ (j) $2\pi/3$ (k) 0 (l) 0

Test exercise 28 (page 1057)

1 $P_{\max} = 10$ $(x = 4, \; y = 3)$ **2** $P_{\max} = 13$ $(x = 1, \; y = 4)$
3 $P_{\max} = 188$ $(x = 10, \; y = 4, \; z = 6)$ **4** $P_{\max} = 296$ $(x = 4, \; y = 6)$
5 $P_{\min} = 16$ $(x = 5, \; y = 12)$ **6** (a) 13 type A + 4 type B (b) £11,800

Further problems 28 (page 1058)

1 $P_{max} = 32$ $(x = 4,\ y = 9/2)$ 2 $P_{max} = 64$ $(x = 0,\ y = 8)$
3 $P_{max} = 40$ $(x = 6,\ y = 5/2)$ 4 $P_{max} = 15$ $(x = 6,\ y = 3)$
5 $P_{max} = 9$ $(x = 1,\ y = 3)$ 6 $P_{max} = 10$ $(x = 2,\ y = 4)$
7 $P_{max} = 10$ $(x = 2,\ y = 4)$ 8 $P_{max} = 37$ $(x = 0,\ y = 8,\ z = 1)$
9 $P_{max} = 67$ $(x = 4,\ y = 10,\ z = 5)$ 10 $P_{max} = 65$ $(x = 5,\ y = 10,\ z = 5)$
11 $P_{max} = 11 \cdot 568$ $(x = 29/22,\ y = 14/11,\ z = 0)$ to 3 $s.f.$
12 $P_{max} = 340$ $(x = 30,\ y = 20)$ 13 $P_{max} = 112$ $(x = 4,\ y = 8)$
14 $P_{max} = 108$ $(x = 16,\ y = 15)$ 15 $P_{min} = 138$ $(x = 12,\ y = 18)$
16 $P_{max} = 240$ $(x = 9,\ y = 15)$ 17 $P_{max} = 4400$ $(x = 201,\ y = 53)$
18 $P_{max} = 100$ $(x = 20,\ y = 10)$ 19 $P_{max} = 410$ $(x = 9,\ y = 5,\ z = 2)$
20 $P_{max} = 1560$ $(x = 11,\ y = 10,\ z = 18)$
21 $P_{max} = 660$ $(x = 60,\ y = 30,\ z = 30)$ 22 $P_{min} = -14$ $(x = 5,\ y = 2)$
23 $P_{min} = 56$ $(x = 8,\ y = 12)$ 24 $P_{min} = 16$ $(x = 8,\ y = 6)$
25 $P_{min} = 40$ $(x = 4,\ y = 4)$ 26 $P_{min} = -10$ $(x = 6,\ y = 13,\ z = 14)$
27 $P_{min} = -75$ $(x = 8,\ y = 12,\ z = 21)$
28 (a) 10 type A + 35 type B   (b) £2150
29 (a) 22 type A + 44 type B + 48 type C   (b) £12,580
30 (a) 129 type A + 0 type B + 185 type C;   (b) £8955

Index